Practice Problems
for the Civil Engineering PE Exam

A Companion to the *Civil Engineering Reference Manual*

Thirteenth Edition

Michael R. Lindeburg, PE

The Power to Pass®
www.ppi2pass.com

Professional Publications, Inc. • Belmont, California

Benefit by Registering This Book with PPI

- Get book updates and corrections.
- Hear the latest exam news.
- Obtain exclusive exam tips and strategies.
- Receive special discounts.

Register your book at **www.ppi2pass.com/register**.

Report Errors and View Corrections for This Book

PPI is grateful to every reader who notifies us of a possible error. Your feedback allows us to improve the quality and accuracy of our products. You can report errata and view corrections at **www.ppi2pass.com/errata**.

**PRACTICE PROBLEMS FOR THE CIVIL ENGINEERING PE EXAM:
A COMPANION TO THE CIVIL ENGINEERING REFERENCE MANUAL**

Thirteenth Edition

Current printing of this edition: 2

Printing History

edition number	printing number	update
12	1	New edition. Major updates. Copyright update.
13	1	New edition. Code updates. Copyright update.
13	2	Minor corrections.

PPI
1250 Fifth Avenue, Belmont, CA 94002
(650) 593-9119
www.ppi2pass.com

ISBN: 978-1-59126-382-1
Library of Congress Control Number: 2012937451

Topics

Topic I: Background and Support

Topic II: Water Resources

Topic III: Environmental

Topic IV: Geotechnical

Topic V: Structural

Topic VI: Transportation

Topic VII: Construction

Topic VIII: Systems, Management, and Professional

Background and Support

Water Resources

Environmental

Geotechnical

Structural

Transportation

Construction

Systems, Mgmt, and Professional

Where do I find help solving these Practice Problems?

Practice Problems for the Civil Engineering PE Exam presents complete, step-by-step solutions for more than 750 problems to help you prepare for the Civil PE exam. You can find all the background information, including charts and tables of data, that you need to solve these problems in the *Civil Engineering Reference Manual*.

The *Civil Engineering Reference Manual* may be purchased from PPI at **www.ppi2pass.com** or from your favorite bookstore.

Table of Contents

Preface to the Thirteenth Edition vii

Acknowledgments ix

Recommended References for the Exam xi

Codes Used to Prepare This Book xiii

How to Use This Book xv

Topic I: Background and Support

Systems of Units 1-1
Engineering Drawing Practice 2-1
Algebra 3-1
Linear Algebra 4-1
Vectors 5-1
Trigonometry 6-1
Analytic Geometry 7-1
Differential Calculus 8-1
Integral Calculus 9-1
Differential Equations 10-1
Probability and Statistical Analysis of Data 11-1
Numerical Analysis 12-1
Energy, Work, and Power 13-1

Topic II: Water Resources

Fluid Properties 14-1
Fluid Statics 15-1
Fluid Flow Parameters 16-1
Fluid Dynamics 17-1
Hydraulic Machines 18-1
Open Channel Flow 19-1
Meteorology, Climatology, and Hydrology 20-1
Groundwater 21-1
Inorganic Chemistry 22-1
Organic Chemistry *(no problems)* 23-1
Combustion and Incineration 24-1
Water Supply Quality and Testing 25-1
Water Supply Treatment and Distribution 26-1

Topic III: Environmental

Biochemistry, Biology, and Bacteriology 27-1
Wastewater Quantity and Quality 28-1
Wastewater Treatment: Equipment and Processes . . . 29-1
Activated Sludge and Sludge Processing 30-1
Municipal Solid Waste 31-1
Pollutants in the Environment 32-1
Storage and Disposition of
    Hazardous Materials *(no problems)* 33-1
Environmental Remediation *(no problems)* 34-1

Topic IV: Geotechnical

Soil Properties and Testing 35-1
Shallow Foundations 36-1
Rigid Retaining Walls 37-1
Piles and Deep Foundations 38-1
Excavations 39-1
Special Soil Topics 40-1

Topic V: Structural

Determinate Statics 41-1
Properties of Areas 42-1
Material Testing 43-1
Strength of Materials 44-1
Basic Elements of Design 45-1
Structural Analysis I 46-1
Structural Analysis II 47-1
Properties of Concrete and Reinforcing Steel 48-1
Concrete Proportioning, Mixing, and Placing 49-1
Reinforced Concrete: Beams 50-1
Reinforced Concrete: Slabs 51-1
Reinforced Concrete: Short Columns 52-1
Reinforced Concrete: Long Columns 53-1
Reinforced Concrete: Walls and Retaining Walls 54-1
Reinforced Concrete: Footings 55-1
Prestressed Concrete 56-1
Composite Concrete and Steel Bridge Girders 57-1
Structural Steel: Introduction *(no problems)* 58-1
Structural Steel: Beams 59-1
Structural Steel: Tension Members 60-1
Structural Steel: Compression Members 61-1
Structural Steel: Beam-Columns 62-1
Structural Steel: Built-Up Sections 63-1
Structural Steel: Composite Beams 64-1
Structural Steel: Connectors 65-1
Structural Steel: Welding 66-1
Properties of Masonry *(no problems)* 67-1
Masonry Walls 68-1
Masonry Columns 69-1

Topic VI: Transportation

Properties of Solid Bodies 70-1
Kinematics 71-1
Kinetics 72-1
Roads and Highways: Capacity Analysis 73-1
Bridges: Condition and Rating *(no problems)* 74-1
Vehicle Dynamics and Accident Analysis 75-1
Flexible Pavement Design 76-1
Rigid Pavement Design 77-1

Plane Surveying 78-1
Horizontal, Compound, Vertical, and Spiral Curves . . 79-1

Topic VII: Construction

Construction Earthwork 80-1
Construction Staking and Layout 81-1
Building Codes and Materials Testing 82-1
Construction and Jobsite Safety 83-1

Topic VIII: Systems, Management, and Professional

Electrical Systems and Equipment 84-1
Instrumentation and Measurements 85-1
Project Management, Budgeting, and Scheduling 86-1
Engineering Economic Analysis 87-1
Professional Services, Contracts, and
    Engineering Law 88-1
Engineering Ethics 89-1
Engineering Licensing in the
    United States   *(no problems)* 90-1

Preface to the Thirteenth Edition

When I wrote the first edition of my first book more than 35 years ago, I naively considered my job to be done. Thirteen editions of this book, *Practice Problems for the Civil Engineering PE Exam*, have taught me otherwise: changes in scope and body of knowledge, improvements, and corrections are inevitable.

This edition encompasses revisions and updates made to the twelfth edition. Problems have been revised so that this book is consistent with revisions made to the thirteenth edition of the *Civil Engineering Reference Manual* and the NCEES-adopted codes and standards. (The actual codes and standards used to develop this book are listed in "Codes Used to Prepare This Book.")

The revisions to the structural chapters reflect NCEES' reliance on specific editions and releases of ACI 318, ACI 530 and 530.1, AISC *Steel Construction Manual*, ASCE7, IBC, NDS, *PCI Design Handbook*, and *AASHTO LRFD Bridge Design Specifications*.

For concrete, solutions based on ACI 318 App. C may no longer be used on the exam. Therefore, those methods and their accompanying examples have been removed from this edition. This book provides solutions using only ACI 318 Chap. 9 (the "unified") methods.

For steel, you may still use either LRFD or ASD on the exam. This book solves problems using both methods so that regardless of which method you choose to study—strength design or allowable stress design—you will be supported.

For masonry, only ASD may be used on the exam, with the exception that strength design (SD) may be used for walls with out-of-plane loads. This book provides ASD solutions (followed by SD solutions for additional reference).

The transportation chapters reflect NCEES' reliance on specific editions of AASHTO *A Policy on Geometric Design of Highways and Streets* (Green Book), AASHTO *Roadside Design Guide*, FHWA *Manual on Uniform Traffic Control Devices* (MUTCD), ITE *Traffic Engineering Handbook*, NRC *Highway Capacity Manual*, and PCA *Design and Control of Concrete Mixtures*. Revised editions of AASHTO *Guide for Design of Pavement Structures* and AI *The Asphalt Handbook* have been reinstated by NCEES as exam design standards, and the *Flexible Pavement Design* and *Rigid Pavement Design* chapters in this book are based on these reinstated editions.

The construction chapters reflect NCEES' reliance on specific editions of ACI 318, ACI 347, ACI SP-4, AISC *Steel Construction Manual*, ASCE 37, NDS, CMWB *Standard Practice for Bracing Masonry Walls During Construction*, and OSHA *Safety and Health Regulations for Construction*. These chapters were revised to include additional material and to make them more consistent with NCEES' revised construction design standards. Still, the majority of what NCEES considers to be "construction" was already present in this book and continues to be covered in other chapters. Earthwork, foundations, slope stability, compaction, temporary structures, and other geotechnical subjects are in their own chapters, as are formwork, engineering economics, construction law, and many other construction subjects. Although there is no single chapter titled "Construction" that contains every construction problem NCEES thinks you should know how to solve, this book presents the Construction topic to the same degree of detail as other topics—regardless of where you read it.

Problems in multiple-choice and essay formats are both included. Multiple-choice problems mirror the actual exam and require knowledge of only one or two narrow concepts, equations, or facts. Longer, more complex essay problems are used to integrate individual concepts, illustrate procedures, introduce exceptions and limitations, and show how data are obtained and chosen. If you are in the middle of a long essay solution and happen to wonder, "How could this be an exam question?", just consider the contents of any two inches of the column you're working in. That two inches would make a great six-minute multiple-choice question.

As with each new edition, some of the problem statements have been changed to include additional information in order to more closely simulate the exam and to eliminate the need to make assumptions that might get you off track. Suggestions for improvements from readers of previous editions have been incorporated, as have corrections of typos and, of course, errata.

I've tried to anticipate your questions by adding strategic commentary throughout many solutions. My daily dose of fan mail (as I like to view the correspondence) brings questions that show me how to improve logic and flow. In these solutions, you won't find any professor chalkboard talk such as "the interested reader may prove the expression in his/her own time," or "it is patently obvious what the next seven steps should be, so we'll skip them." Although I might leave out some interim steps (e.g., omitting the steps to solve for the roots of a quadratic or higher-order equation, which is something your calculator can do), for the most part, the solutions are complete down to the algebraic manipulations.

The chapters in this edition parallel those in the *Civil Engineering Reference Manual*. This book contains numerous references to the equations, tables, and figures in the *Civil Engineering Reference Manual*, and this book uses the same nomenclature, data, and methods. Though independent, these two books intentionally complement each other. Because this book is essentially an independent collection of solved practice problems that you can take with you anywhere, even if you don't have the *Civil Engineering Reference Manual* with you while you review this book's solutions, you should be able to follow the solution steps in most of the chapters. Except where the intended equation is obvious (e.g., $A = \pi r^2$ and $v = Q/A$), the actual equation is given in variable format before the numbers are plugged in.

As usual, I've already begun thinking about the next edition of this book. Even before it arrives, however, you can keep up with changes to this book by logging onto PPI's website at **www.ppi2pass.com/errata**.

I have done my best to make *Practice Problems for the Civil Engineering PE Exam* and the *Civil Engineering Reference Manual* the most useful books in your library. Now, it's up to you.

Michael R. Lindeburg, PE

Acknowledgments

This is a book that is singularly designed to help people. It is not inconsistent, then, that I would be helped immeasurably by other people in bringing it out. So, who helped me with this new edition? Wow. Where do I start?

Like many of the previous editions, I depended upon three subject matter experts from industry and academia to review existing material: James W. Giancaspro, PhD, PE; Andy Richardson, PE; and Megan Synnestvedt. Dr. Giancaspro, assistant professor of civil engineering at the University of Miami, who recently had his own exam preparation book, *Structural Depth Practice Exams for the Civil PE Exam*, published by PPI, revised the concrete chapters in accordance with ACI 318-08. Mr. Richardson, a practicing structural engineer in private practice (Structural Innovations, LLC, Beaufort, SC), who has just finished conducting an online civil engineering review course for PPI, updated the masonry chapters to be consistent with ACI 530-08 and the IBC. PPI's own Megan Synnestvedt, a civil engineer and project manager, updated the concrete chapters to be consistent with the fifth edition of the *LRFD Bridge Design Specifications* and reviewed and revised my new chapter on bridge rating. James, Andy, and Megan helped PPI get this new edition into print before the next examination, and their expertise helped me ensure that the updates were complete and accurate.

The editorial and production departments of many large publishers are giant clearing houses where the managing editors are one-star generals coordinating armies of outside independent editors, illustrators, typesetters, and other sundry contractors. Not so at PPI. This edition, as all previous editions, was produced entirely in-house. Julia White was the four-star general—this book's editorial project manager. She coordinated the work of other internal PPI editors, engineers, illustrators, designers, typesetters, proofreaders, and paginators. She was the buck-stops-here editor and proofreader. Accordingly, there isn't a symbol, comma, or line that she didn't review and measure against her internal scales of accuracy and precision. Considering that she started as a copy editor only a year and a half ago, Julia's advancement and assignment to this important project indicate what she has already shown herself capable of and the confidence that PPI has in her.

As I mentioned, sometimes the lion's share of the work falls to the publisher due to a reorganization of existing material. Two of the changes to this book (those of adding a new chapter and of removing all structural concrete content that used ACI 318 App. C methodology) required extensive reformatting and repagination of existing content. Julia had her own army of publishing professionals—they were just all in the same building. Each followed the battle plan by exhibiting subject comprehension and militarily precise adherence to style, attention to detail, and consistency. I'll list them by function, in alphabetical order, and I hope that each knows that I understand and appreciate his/her contribution.

Paginating: Tom Bergstrom, Lisa Devoto Farrell, Kate Hayes, Tyler Hayes, Chelsea Logan, Scott Marley, and Magnolia Molcan

Proofreading: William Bergstrom, Lisa Devoto Farrell, Tyler Hayes, Chelsea Logan, Scott Marley, and Magnolia Molcan

Illustrating: Tom Bergstrom, Kate Hayes, and Amy Schwertman La Russa

Cover design: Amy Schwertman La Russa

Management: Sarah Hubbard, Director of Product Development and Implementation, and Cathy Schrott, Production Services Manager

Continuing with the military analogy, this edition incorporates the comments, questions, suggestions, and errata submitted by many people on the battlefront who have used the previous edition for their own preparations. As an author, I am humbled to know that these individuals have read the previous edition in such detail as to notice typos, illogic, and other errata, and that they subsequently took the time to share their observations with me. Their suggestions have been incorporated into this edition, and their attention to detail will benefit you and all future readers. The following is a partial list (again in alphabetical order), of some of the soldiers who have improved this book through their comments: John McFarland, PE; Waldon Wong.

This edition shares a common developmental heritage with all of its previous editions. There are hundreds and hundreds of additional people that I mentioned by name in the acknowledgments of those old books. They're not forgotten, and their names will live on in the tens of thousands of old editions that remain in widespread circulation. For this edition, though, there isn't a single contributor that I intentionally excluded. Still, I could have slipped up and forgotten to mention you. I hope you'll let me know if you should have been credited, but were inadvertently left out. I'd appreciate the opportunity to list your name in the next printing of this edition.

Near the end of the acknowledgments, after mentioning a lot of people who contributed to the book and, therefore, could be blamed for a variety of types of errors, it is common for an author to say something like, "I take responsibility for all of the errors you find in this book." Or, "All of the mistakes are mine." This is certainly true, given the process of publishing, since the author sees and approves the final version before his/her book goes to the printer. You would think that after 35 years of writing, I would have figured out by now how to write something without making a mistake and how to proofread without missing those blunders that are so obvious to readers. However, such perfection continues to elude me. So, yes. The finger points straight at me.

All I can say instead is that I'll do my best to respond to any suggestions and errata that you report through PPI's website, **www.ppi2pass.com/errata**. I'd love to see your name in the acknowledgments for the next edition.

Thank you, everyone!

Michael R. Lindeburg, PE

Recommended References for the Exam

STRUCTURAL

AASHTO LRFD Bridge Design Specifications. American Association of State Highway and Transportation Officials (AASHTO)

Building Code Requirements for Masonry Structures (ACI 530) and *Specifications for Masonry Structures* (ACI 530.1). American Concrete Institute

Building Code Requirements for Structural Concrete (ACI 318) *with Commentary* (ACI 318R). American Concrete Institute

ACI Design Handbook (ACI SP-17). American Concrete Institute

International Building Code. International Code Council

Minimum Design Loads for Buildings and Other Structures (ASCE7). American Society of Civil Engineers

National Design Specification for Wood Construction ASD/LRFD. American Forest & Paper Association/American Wood Council

Notes on ACI 318: Building Code Requirements for Structural Concrete with Design Applications. Portland Cement Association

PCI Design Handbook. Precast/Prestressed Concrete Institute

Steel Construction Manual. American Institute of Steel Construction, Inc.

TRANSPORTATION

A Policy on Geometric Design of Highways and Streets ("AASHTO Green Book"). AASHTO

Design and Control of Concrete Mixtures. Portland Cement Association

Guide for Design of Pavement Structures. AASHTO

Highway Capacity Manual. Transportation Research Board/National Research Council

Manual on Uniform Traffic Control Devices (MUTCD). U.S. Department of Transportation, Federal Highway Administration (FHWA)

Roadside Design Guide. AASHTO

The Asphalt Handbook, Manual MS-4. The Asphalt Institute

Thickness Design for Concrete Highway and Street Pavements. Portland Cement Association

Traffic Engineering Handbook. Institute of Transportation Engineers

Trip Generation. Institute of Transportation Engineers

WATER RESOURCES

Handbook of Hydraulics. Ernest F. Brater, Horace W. King, James E. Lindell, and C. Y. Wei

Urban Hydrology for Small Watersheds (Technical Release TR-55). U.S. Department of Agriculture, Natural Resources Conservation Service (previously, Soil Conservation Service)

ENVIRONMENTAL

Recommended Standards for Wastewater Facilities ("10 States' Standards"). Health Education Services Division, Health Resources, Inc.

Standard Methods for the Examination of Water and Wastewater. A joint publication of the American Public Health Association (APHA), the American Water Works Association (AWWA), and the Water Environment Federation (WEF)

Wastewater Engineering: Treatment and Reuse (Metcalf and Eddy). George Tchobanoglous, Franklin L. Burton, and H. David Stensel

GEOTECHNICAL

Geotechnical Engineering Procedures for Foundation Design of Buildings and Structures (UFC 3-220-01N, previously NAVFAC Design Manuals DM 7.01/DM 7.1 and DM 7.02/DM 7.2). Department of the Navy, Naval Facilities Engineering Command

Soil Mechanics (UFC 3-220-10N, previously NAVFAC Design Manual DM 7.3). Department of the Navy, Naval Facilities Engineering Command

CONSTRUCTION

Building Code Requirements for Structural Concrete. American Concrete Institute.

Design Loads on Structures During Construction (ASCE 37). American Society of Civil Engineers

Formwork for Concrete (ACI SP-4). American Concrete Institute

Guide to Formwork for Concrete (ACI 347). American Concrete Institute (in ACI SP-4, 7th edition appendix)

Manual on Uniform Traffic Control Devices—Part 6: Temporary Traffic Control (MUTCD-Pt 6). U.S. Department of Transportation, Federal Highway Administration (FHWA)

National Design Specification for Wood Construction ASD/LRFD (NDS). American Forest & Paper Association/American Wood Council

Safety and Health Regulations for Construction (OSHA), 29 CFR Part 1926 (U.S. Federal version). U.S. Department of Labor

Standard Practice for Bracing Masonry Walls Under Construction (CMWB). Council for Masonry Wall Bracing, Mason Contractors Association of America

Steel Construction Manual (AISC). American Institute of Steel Construction, Inc.

Codes Used to Prepare This Book[1]

The information that was used to write and update this book was based on the exam specifications at the time of publication. However, as with engineering practice itself, the PE examination is not always based on the most current codes or cutting-edge technology. Similarly, codes, standards, and regulations adopted by state and local agencies often lag issuance by several years. It is likely that the codes that are most current, the codes that you use in practice, and the codes that are the basis of your exam will all be different.

PPI lists on its website the dates and editions of the codes, standards, and regulations on which NCEES has announced the PE exams are based. It is your responsibility to find out which codes are relevant to your exam. In the meantime, here are the codes that have been incorporated into this edition.

STRUCTURAL DESIGN STANDARDS

AASHTO: *AASHTO LRFD Bridge Design Specifications*, Fifth ed., 2010, American Association of State Highway and Transportation Officials, Washington, DC

ACI 318:[2] *Building Code Requirements for Structural Concrete*, 2008, American Concrete Institute, Farmington Hills, MI

ACI 530:[3] *Building Code Requirements for Masonry Structures*, 2008, and ACI 530.1: *Specifications for Masonry Structures*, 2008, American Concrete Institute, Detroit, MI

AISC:[4] *Steel Construction Manual*, Thirteenth ed., 2005, American Institute of Steel Construction, Inc., Chicago, IL

ASCE7: *Minimum Design Loads for Buildings and Other Structures*, 2005, American Society of Civil Engineers, Reston, VA

IBC: *International Building Code*, 2009 ed. (without supplements), International Code Council, Inc., Falls Church, VA

NDS:[5] *National Design Specification for Wood Construction ASD/LRFD*, 2005 ed., and *National Design Specification Supplement, Design Values for Wood Construction*, 2005 ed., American Forest & Paper Association/American Wood Council, Washington, DC

PCI: *PCI Design Handbook: Precast and Prestressed Concrete*, Sixth ed., 2004, Precast/Prestressed Concrete Institute, Chicago, IL

TRANSPORTATION DESIGN STANDARDS

AASHTO: *AASHTO Guide for Design of Pavement Structures* (GDPS-4-M), 1993, and 1998 supplement, American Association of State Highway and Transportation Officials, Washington, DC

AASHTO: *A Policy on Geometric Design of Highways and Streets*, Fifth ed., 2004, American Association of State Highway and Transportation Officials, Washington, DC

AASHTO: *Roadside Design Guide*, Third ed., 2002, with 2006 Chapter 6 update, American Association of State Highway and Transportation Officials, Washington, DC

AI: *The Asphalt Handbook* (MS-4), Seventh ed., 2007, Asphalt Institute, Lexington, KY

HCM: *Highway Capacity Manual* (HCM 2000), 2000 ed. (with changes adopted through November 14, 2007), Transportation Research Board, National Research Council, Washington, DC

ITE: *Traffic Engineering Handbook*, Sixth ed., 2009, Institute of Transportation Engineers, Washington, DC

MUTCD: *Manual on Uniform Traffic Control Devices*, 2009, U.S. Dept. of Transportation, Federal Highway Administration, Washington, DC

PCA: *Design and Control of Concrete Mixtures*, Fourteenth ed., 2002 (rev. 2008), Portland Cement Association, Skokie, IL

CONSTRUCTION DESIGN STANDARDS

ACI 318: *Building Code Requirements for Structural Concrete*, 2008, American Concrete Institute, Farmington Hills, MI

---

[1]This is a list of the codes and standards used to prepare this book. The *Recommended References for the Exam* section contains a complete list of the references that you should consider bringing to the exam.

[2]ACI 318 App. C solving methods may not be used on the exam.

[3]Allowable stress design (ASD) methods must be used on the exam, except that strength design (SD) Sec. 3.3.5 may be used for walls with out-of-plane loads.

[4]Either ASD or LRFD may be used on the exam.

[5]ASD methods for wood design must be used on the exam.

ACI 347: *Guide to Formwork for Concrete*, 2004, American Concrete Institute, Farmington Hills, MI (in ACI SP-4, Seventh ed. appendix)

ACI SP-4: *Formwork for Concrete*, Seventh ed., 2005, American Concrete Institute, Farmington Hills, MI

AISC: *Steel Construction Manual*, Thirteenth ed., 2005, American Institute of Steel Construction, Inc., Chicago, IL

ASCE 37: *Design Loads on Structures During Construction*, 2002, American Society of Civil Engineers, Reston, VA

CMWB: *Standard Practice for Bracing Masonry Walls Under Construction*, 2001, Council for Masonry Wall Bracing, Mason Contractors Association of America, Lombard, IL

MUTCD-Pt 6: *Manual on Uniform Traffic Control Devices*—Part 6, "Temporary Traffic Control," 2009, U.S. Dept. of Transportation, Federal Highway Administration, Washington, DC

NDS: *National Design Specification for Wood Construction ASD/LRFD*, 2005 ed., American Forest & Paper Association/American Wood Council, Washington, DC

OSHA: *Safety and Health Regulations for Construction*, 29 CFR Part 1926 (U.S. Federal version), U.S. Department of Labor, Washington, DC

How to Use This Book

This book is primarily a companion to the *Civil Engineering Reference Manual*. As a tool for preparing for an engineering licensing exam, there are a few, but not very many, ways to use it.

Before it grew to behemoth size, I envisioned this book being taken to work, on the bus or train, on business trips, and on weekend pleasure getaways to the beach. I figured that it would "carry" a lot easier than the big *Civil Engineering Reference Manual*. Though I don't think you'll be taking this book on any backpacking trips, you might still find yourself working problems in the cafeteria during your lunch break.

The big issue is whether you really work the practice problems or not. Some people think they can read a problem statement, think about it for about ten seconds, read the solution, and then say "Yes, that's what I was thinking of, and that's what I would have done." Sadly, these people find out too late that the human brain doesn't learn very efficiently in that manner. Under pressure, they find they know and remember too little. For real learning, you have to spend some time with a stubby pencil.

There are so many places where you can get messed up solving a problem. Maybe it's in the use of your calculator, like pushing log instead of ln, or forgetting to set the angle to radians instead of degrees, and so on. Maybe it's rusty math. What is $\ln e(x)$, anyway? How do you factor a polynomial? Maybe it's in finding the data needed or the proper unit conversion. Maybe it's just trying to find out if that proprietary building code

equation expects L to be in feet or inches. Conquering these things takes time—more time than you may want to spend when time is at a premium. And unfortunately, most people have to make a mistake once so that they don't make it again.

Even if you do decide to get your hands dirty and actually work the problems (as opposed to skimming through them), you'll have to decide how much reliance you place on the published solutions. It's tempting to turn to a solution when you get slowed down by details or stumped by the subject material. You'll probably want to maximize the number of problems you solve by spending as little time as you can on each problem. After all, optimization is the engineering way. However, I want you to struggle a little bit more than that—not because I want to see you suffer, but because the optimization is in getting you through your licensing exam. We're not trying to find the optimally minimum study time.

Studying a new subject is analogous to using a machete to cut a path through a dense jungle. By doing the work, you develop pathways that weren't there before. It's a lot different than just looking at the route on a map. You actually get nowhere by looking at a map. But cut that path once, and you're in business until the jungle overgrowth closes in again.

So, do the problems. All of them. Do them twice, once in U.S. customary units, and then again in SI units. Don't look at the answers until you've sweated a little. And, let's not have any whining. Please.

1 Systems of Units

PRACTICE PROBLEMS

1. Convert 250°F to degrees Celsius.

(A) 115°C

(B) 121°C

(C) 124°C

(D) 420°C

2. Convert the Stefan-Boltzmann constant (0.1713×10^{-8} Btu/ft^2-hr-°R^4) from English to SI units.

(A) 5.14×10^{-10} W/m^2·K^4

(B) 0.95×10^{-8} W/m^2·K^4

(C) 5.67×10^{-8} W/m^2·K^4

(D) 7.33×10^{-6} W/m^2·K^4

3. How many U.S. tons (2000 lbm per ton) of coal with a heating value of 13,000 Btu/lbm must be burned to provide as much energy as a complete nuclear conversion of 1 g of its mass? (Hint: Use Einstein's equation: $E = mc^2$.)

(A) 1.7 tons

(B) 14 tons

(C) 779 tons

(D) 3300 tons

SOLUTIONS

1. The conversion to degrees Celsius is given by

$$°C = \tfrac{5}{9}(°F - 32°F)$$
$$= \left(\tfrac{5}{9}\right)(250°F - 32°F)$$
$$= \boxed{121.1°C \quad (121°C)}$$

The answer is (B).

2. The Stefan-Boltzmann constant represents a certain amount of energy (in Btu). Since it has hr-ft^2-°R^4 in the denominator, the energy is reported on a per hour basis and represents power. The energy is an areal value because it is reported per square foot. Since a square meter is larger than a square foot, the energy must be multiplied by a number larger than 1.0 in order to put it into a per square meter basis. (For example, on any given day, the energy that can be derived from a square meter of sunlight is greater than can be derived from a square foot of sunlight.) $1/0.3048$ is the conversion between feet and meters, and it is larger than 1.0.

The Stefan-Boltzmann constant is reported on a per degree basis, and the same logic applies. One Kelvin (K) is a larger (longer) interval on the temperature scale than one degree Rankine. The amount of energy per Kelvin is larger than the amount of energy per degree Rankine. Therefore, the energy must be multiplied by a number greater than 1.0 to report it on a "per Kelvin" basis. This requires multiplying by $^9/_5$. (An incorrect conversion will result if the unit in the denominator is thought of as an actual temperature. Any given numerical value of temperature in degrees Rankine is larger than the same temperature in Kelvins. However, this problem does not require a temperature conversion. It requires a unit conversion.)

Use the following conversion factors.

$$1 \text{ Btu/hr} = 0.2931 \text{ W}$$
$$1 \text{ ft} = 0.3048 \text{ m}$$
$$\Delta T_K = \tfrac{5}{9}\Delta T_{°R}$$

Background and Support

Performing the conversion gives

$$\sigma = \left(0.1713 \times 10^{-8} \; \frac{\text{Btu}}{\text{hr-ft}^2\text{-}^\circ\text{R}^4}\right)\left(0.2931 \; \frac{\text{W}}{\frac{\text{Btu}}{\text{hr}}}\right)$$

$$\times \left(\frac{1 \; \text{ft}}{0.3048 \; \text{m}}\right)^2 \left(\frac{1}{\frac{5}{9} \frac{\Delta T_K}{\Delta T_{^\circ R}}}\right)^4$$

$$= \boxed{5.67 \times 10^{-8} \; \text{W/m}^2 \cdot \text{K}^4}$$

The answer is (C).

3. The energy produced from the nuclear conversion of any quantity of mass is given as

$$E = mc^2$$

The speed of light, c, is 3×10^8 m/s.

For a mass of 1 g (0.001 kg),

$$E = mc^2$$

$$= (0.001 \; \text{kg})\left(3 \times 10^8 \; \frac{\text{m}}{\text{s}}\right)^2$$

$$= 9 \times 10^{13} \; \text{J}$$

Convert to U.S. customary units with the conversion 1 Btu = 1055 J.

$$E = \frac{9 \times 10^{13} \; \text{J}}{1055.1 \; \frac{\text{J}}{\text{Btu}}}$$

$$= 8.53 \times 10^{10} \; \text{Btu}$$

The number of tons of 13,000 Btu/lbm coal is

$$\frac{8.53 \times 10^{10} \; \text{Btu}}{\left(13,000 \; \frac{\text{Btu}}{\text{lbm}}\right)\left(2000 \; \frac{\text{lbm}}{\text{ton}}\right)} = \boxed{3281 \; \text{tons} \quad (3300 \; \text{tons})}$$

The answer is (D).

2 Engineering Drawing Practice

PRACTICE PROBLEMS

1. Two views of an object are shown. Prepare a free-hand drawing of the missing third view.

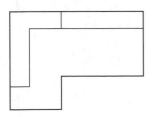

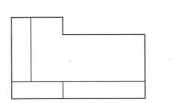

2. Two views of an object are shown. Prepare a free-hand drawing of the missing third view.

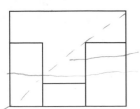

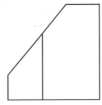

3. Two views of an object are shown. Prepare a free-hand drawing of the missing third view.

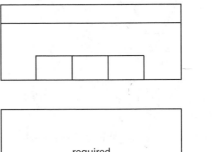

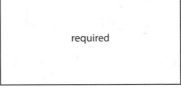

4. A pictorial sketch of an object is shown. Prepare a freehand three-view orthographic drawing set.

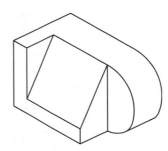

5. Plan and elevation views of an object are shown.

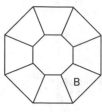

plan view

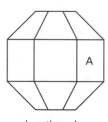

elevation view

(a) Prepare a freehand drawing of the view that shows surface A in true shape.

(b) Referring to part (a), what is this view of surface A called?

(c) Prepare a freehand drawing of the view that shows surface B in true shape.

(d) Referring to part (c), what is the view of surface B called?

6. Prepare a freehand isometric drawing of the object shown. Hint: Lay off horizontal and vertical (i.e., isometric) lines along the isometric axes shown.

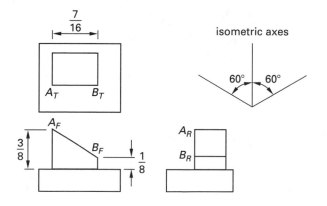

7. Two lines in an oblique position are shown. Prepare a freehand right auxiliary normal view.

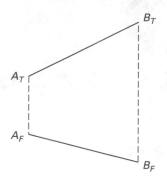

8. Two lines in an oblique position are shown. Prepare a freehand rear auxiliary normal view.

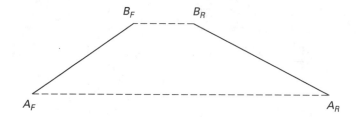

9. A parallel-scale nomograph has been prepared to solve the equation $D = 1.075\sqrt{WH}$. Use the nomograph to estimate the value of D when $H = 10$ and $W = 40$.

$$D = 1.075 \sqrt{WH}$$

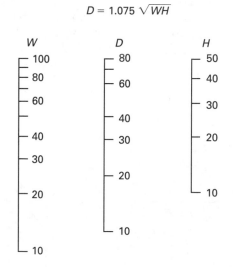

10. How would the following surface finish designation be interpreted?

(A) The waviness height limit is 0.0013 μm.

(B) The lay is perpendicular.

(C) The surface roughness height limit is 0.003 μin.

(D) 32% of the surface may exceed the specification.

11. How would the following surface finish designation be interpreted?

(A) The surface is rolled.

(B) The maximum roughness width is 0.6 μm.

(C) The piece is to be used exactly as it comes directly from the manufacturing process.

(D) The sample length is 0.6 m.

12. How would the following surface finish designation be interpreted?

(A) The maximum waviness is 0.4 μm.

(B) The surface is to be stress relieved by bead blasting.

(C) No part of the item may exceed the specification.

(D) The maximum length that must be tested is 40% of the item's length.

SOLUTIONS

1. The complete set of views is

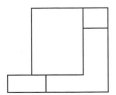

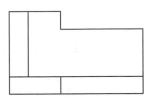

2. The complete set of views is

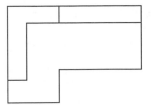

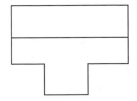

3. The complete set of views is

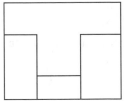

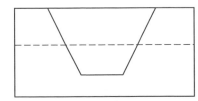

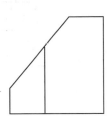

4. The complete set of views is

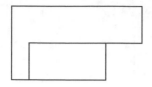
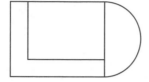

5. (a) Draw the view showing surface A's true shape.

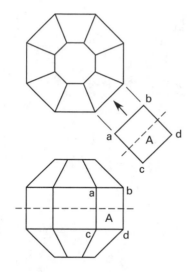

(b) An [elevation auxiliary] view shows surface A in true shape.

(c) Draw the view showing surface B's true shape.

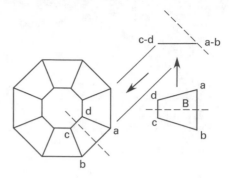

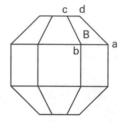

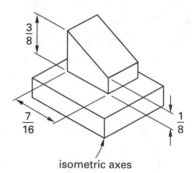

(d) An [oblique] view shows surface B in true shape.

6. The isometric drawing of the object is as shown.

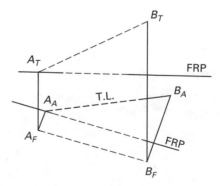

isometric axes

7. The right auxiliary normal view is as shown.

8. The auxiliary normal view is as shown.

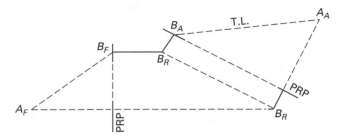

9. A straight line through the points $H = 10$ and $W = 40$ intersects the D-scale at approximately $\boxed{21.5.}$

10. The values 0.0013 and 0.003 are too "fine" to be in micrometers. These values must be in microinches. The roughness height limit is 32 μin. The waviness height limit is 0.0013 μin. The roughness width limit is 0.003 μin. The lay is perpendicular.

The answer is (B).

11. The circle in the crook of the surface finish arrow means that no resurfacing is permitted. The maximum height of the roughness is 0.6 μm. In addition, since no length of sampling is specified, the default limit of five sample lengths (i.e., between five highest peaks and five lowest vallies) applies. Also, since no other value is specified, a maximum of 16% (the default value) of the item may exceed the maximum surface roughness.

The answer is (C).

12. The triangle in the crook of the surface finish arrow means that the surface is to be machined. The maximum roughness height is 0.4 μm. The "max" means that the entire piece must be satisfy the maximum specification (the 16% rule does not apply).

The answer is (C).

3 Algebra

PRACTICE PROBLEMS

Roots of Quadratic Equations

1. What are the roots of the quadratic equation $x^2 - 7x - 44 = 0$?

Logarithm Identities

2. What is the value of x that satisfies the expression $17.3 = e^{1.1x}$?

Series

3. Calculate the following sum.

$$\sum_{j=1}^{5}\left((j+1)^2 - 1\right)$$

(A) 15

(B) 24

(C) 35

(D) 85

Logarithms

4. If every 0.1 sec a quantity increases by 0.1% of its current value, calculate the doubling time.

(A) 14 sec

(B) 69 sec

(C) 690 sec

(D) 69,000 sec

SOLUTIONS

1. *method 1:* Use inspection.

Assuming that the roots are integers, there are only a few ways to arrive at 44 by multiplication: 1×44, 2×22, and 4×11. Of these, the difference of 4 and 11 is -7. Therefore, the quadratic equation can be factored into $(x-11)(x+4) = 0$. The roots are $+11$ and -4.

method 2: Use the quadratic formula, where $a = 1$, $b = -7$, and $c = -44$.

$$x_1, x_2 = \frac{-b \pm \sqrt{b^2 - 4ac}}{2a}$$

$$= \frac{-(-7) \pm \sqrt{(-7)^2 - (4)(1)(-44)}}{(2)(1)}$$

$$= \frac{7 \pm 15}{2}$$

$$= -4, 11$$

method 3: Use completing the square.

$$x^2 - 7x - 44 = 0$$

$$x^2 - 7x = 44$$

$$\left(x - \frac{7}{2}\right)^2 = 44 + \left(\frac{7}{2}\right)^2$$

$$\left(x - \frac{7}{2}\right)^2 = 56.25$$

$$x - 3.5 = \sqrt{56.25}$$

$$x = \pm 7.5 + 3.5$$

$$= 11, -4$$

2. Take the natural logarithm of both sides, using the identity $\log_b b^n = n$.

$$17.3 = e^{1.2x}$$

$$\ln 17.3 = \ln e^{1.1x}$$

$$= 1.1x$$

$$x = \frac{\ln 17.3}{1.1}$$

$$= \ln e^{1.1x} \quad (\approx 2.592)$$

3. Let $S_n = (j+1)^2 - 1$.

For $j = 1$,

$$S_1 = (1+1)^2 - 1 = 3$$

For $j = 2$,

$$S_2 = (2+1)^2 - 1 = 8$$

For $j = 3$,

$$S_3 = (3+1)^2 - 1 = 15$$

For $j = 4$,

$$S_4 = (4+1)^2 - 1 = 24$$

For $j = 5$,

$$S_5 = (5+1)^2 - 1 = 35$$

Substituting the above expressions gives

$$\sum_{j=1}^{5} \left((j+1)^2 - 1 \right) = \sum_{j=1}^{5} S_j$$
$$= S_1 + S_2 + S_3 + S_4 + S_5$$
$$= 3 + 8 + 15 + 24 + 35$$
$$= \boxed{85}$$

The answer is (D).

4. Let n represent the number of elapsed periods of 0.1 sec, and let y_n represent the amount present after n periods.

y_0 represents the initial quantity.

$$y_1 = 1.001 y_0$$
$$y_2 = 1.001 y_1 = (1.001)(1.001 y_0) = (1.001)^2 y_0$$

Therefore, by induction,

$$y_n = (1.001)^n y_0$$

The expression for a doubling of the original quantity is

$$2y_0 = y_n$$

Substitute for y_n.

$$2y_0 = (1.001)^n y_0$$
$$2 = (1.001)^n$$

Take the logarithm of both sides.

$$\log 2 = \log(1.001)^n$$
$$= n \log 1.001$$

Solve for n.

$$n = \frac{\log 2}{\log 1.001} = 693.5$$

Since each period is 0.1 sec, the time is given by

$$t = n(0.1 \text{ sec}) = (693.5)(0.1 \text{ sec})$$
$$= 69.35 \text{ sec}$$

The answer is (B).

 **Linear Algebra**

PRACTICE PROBLEMS

Determinants

1. What is the determinant of matrix $\mathbf{A}$?

$$\mathbf{A} = \begin{bmatrix} 8 & 2 & 0 & 0 \\ 2 & 8 & 2 & 0 \\ 0 & 2 & 8 & 2 \\ 0 & 0 & 2 & 4 \end{bmatrix}$$

    (A) 459

    (B) 832

    (C) 1552

    (D) 1776

Simultaneous Linear Equations

2. Use Cramer's rule to solve for the values of x, y, and z that simultaneously satisfy the following equations.

$$x + y = 4$$
$$x + z - 1 = 0$$
$$2z - y + 3x = 4$$

    (A) $(x, y, z) = (3, 2, 1)$

    (B) $(x, y, z) = (-3, -1, 2)$

    (C) $(x, y, z) = (3, -1, -3)$

    (D) $(x, y, z) = (-1, -3, 2)$

SOLUTIONS

1. Expand by cofactors of the first row since there are two zeros in that row.

$$|\mathbf{A}| = 8 \begin{vmatrix} 8 & 2 & 0 \\ 2 & 8 & 2 \\ 0 & 2 & 4 \end{vmatrix} - 2 \begin{vmatrix} 2 & 0 & 0 \\ 2 & 8 & 2 \\ 0 & 2 & 4 \end{vmatrix} + 0 - 0$$

By first row:

$$\begin{vmatrix} 8 & 2 & 0 \\ 2 & 8 & 2 \\ 0 & 2 & 4 \end{vmatrix} = (8)\big((8)(4) - (2)(2)\big) - (2)\big((2)(4) - (2)(0)\big)$$

$$= (8)(28) - (2)(8)$$
$$= 208$$

By first row:

$$\begin{vmatrix} 2 & 0 & 0 \\ 2 & 8 & 2 \\ 0 & 2 & 4 \end{vmatrix} = (2)\big((8)(4) - (2)(2)\big)$$

$$= 56$$

$$|\mathbf{A}| = (8)(208) - (2)(56) = \boxed{1552}$$

The answer is (C).

2. Rearrange the equations.

$$x + y = -4$$
$$x + z = 1$$
$$3x - y + 2z = 4$$

Write the set of equations in matrix form: $\mathbf{AX} = \mathbf{B}$.

$$\begin{bmatrix} 1 & 1 & 0 \\ 1 & 0 & 1 \\ 3 & -1 & 2 \end{bmatrix} \begin{bmatrix} x \\ y \\ z \end{bmatrix} = \begin{bmatrix} -4 \\ 1 \\ 4 \end{bmatrix}$$

Find the determinant of the matrix $\mathbf{A}$.

$$|\mathbf{A}| = \begin{vmatrix} 1 & 1 & 0 \\ 1 & 0 & 1 \\ 3 & -1 & 2 \end{vmatrix}$$

$$= 1 \begin{vmatrix} 0 & 1 \\ -1 & 2 \end{vmatrix} - 1 \begin{vmatrix} 1 & 0 \\ -1 & 2 \end{vmatrix} + 3 \begin{vmatrix} 1 & 0 \\ 0 & 1 \end{vmatrix}$$

$$= (1)\big((0)(2) - (1)(-1)\big)$$
$$\quad - (1)\big((1)(2) - (-1)(0)\big)$$
$$\quad + (3)\big((1)(1) - (0)(0)\big)$$
$$= (1)(1) - (1)(2) + (3)(1)$$
$$= 1 - 2 + 3$$
$$= 2$$

Find the determinant of the substitutional matrix $\mathbf{A}_1$.

$$|\mathbf{A}_1| = \begin{vmatrix} -4 & 1 & 0 \\ 1 & 0 & 1 \\ 4 & -1 & 2 \end{vmatrix}$$

$$= -4 \begin{vmatrix} 0 & 1 \\ -1 & 2 \end{vmatrix} - 1 \begin{vmatrix} 1 & 0 \\ -1 & 2 \end{vmatrix} + 4 \begin{vmatrix} 1 & 0 \\ 0 & 1 \end{vmatrix}$$

$$= (-4)\big((0)(2) - (1)(-1)\big)$$
$$\quad - (1)\big((1)(2) - (-1)(0)\big)$$
$$\quad + (4)\big((1)(1) - (0)(0)\big)$$
$$= (-4)(1) - (1)(2) + (4)(1)$$
$$= -4 - 2 + 4$$
$$= -2$$

Find the determinant of the substitutional matrix $\mathbf{A}_2$.

$$|\mathbf{A}_2| = \begin{vmatrix} 1 & -4 & 0 \\ 1 & 1 & 1 \\ 3 & 4 & 2 \end{vmatrix}$$

$$= 1 \begin{vmatrix} 1 & 1 \\ 4 & 2 \end{vmatrix} - 1 \begin{vmatrix} -4 & 0 \\ 4 & 2 \end{vmatrix} + 3 \begin{vmatrix} -4 & 0 \\ 1 & 1 \end{vmatrix}$$

$$= (1)\big((1)(2) - (4)(1)\big)$$
$$\quad - (1)\big((-4)(2) - (4)(0)\big)$$
$$\quad + (3)\big((-4)(1) - (1)(0)\big)$$
$$= (1)(-2) - (1)(-8) + (3)(-4)$$
$$= -2 + 8 - 12$$
$$= -6$$

Find the determinant of the substitutional matrix $\mathbf{A}_3$.

$$|\mathbf{A}_3| = \begin{vmatrix} 1 & 1 & -4 \\ 1 & 0 & 1 \\ 3 & -1 & 4 \end{vmatrix}$$

$$= 1 \begin{vmatrix} 0 & 1 \\ -1 & 4 \end{vmatrix} - 1 \begin{vmatrix} 1 & -4 \\ -1 & 4 \end{vmatrix} + 3 \begin{vmatrix} 1 & -4 \\ 0 & 1 \end{vmatrix}$$

$$= (1)\big((0)(4) - (-1)(1)\big)$$
$$\quad - (1)\big((1)(4) - (-1)(-4)\big)$$
$$\quad + (3)\big((1)(1) - (0)(-4)\big)$$
$$= (1)(1) - (1)(0) + (3)(1)$$
$$= 1 - 0 + 3$$
$$= 4$$

Use Cramer's rule.

$$x = \frac{|\mathbf{A}_1|}{|\mathbf{A}|} = \frac{-2}{2} = \boxed{-1}$$

$$y = \frac{|\mathbf{A}_2|}{|\mathbf{A}|} = \frac{-6}{2} = \boxed{-3}$$

$$z = \frac{|\mathbf{A}_3|}{|\mathbf{A}|} = \frac{4}{2} = \boxed{2}$$

The answer is (D).

5 Vectors

PRACTICE PROBLEMS

1. Calculate the dot products for the following vector pairs.

(a) $\mathbf{V}_1 = 2\mathbf{i} + 3\mathbf{j}$; $\mathbf{V}_2 = 5\mathbf{i} - 2\mathbf{j}$

(b) $\mathbf{V}_1 = 1\mathbf{i} + 4\mathbf{j}$; $\mathbf{V}_2 = 9\mathbf{i} - 3\mathbf{j}$

(c) $\mathbf{V}_1 = 7\mathbf{i} - 3\mathbf{j}$; $\mathbf{V}_2 = 3\mathbf{i} + 4\mathbf{j}$

(d) $\mathbf{V}_1 = 2\mathbf{i} - 3\mathbf{j} + 6\mathbf{k}$; $\mathbf{V}_2 = 8\mathbf{i} + 2\mathbf{j} - 3\mathbf{k}$

(e) $\mathbf{V}_1 = 6\mathbf{i} + 2\mathbf{j} + 3\mathbf{k}$; $\mathbf{V}_2 = \mathbf{i} + \mathbf{k}$

2. What is the angle between the vectors in Prob. 1(a), Prob. 1(b), and Prob. 1(c)?

3. Calculate the cross products for each of the vector pairs in Prob. 1.

SOLUTIONS

1. (a) The dot product is

$$\mathbf{V}_1 \cdot \mathbf{V}_2 = \mathbf{V}_{1x}\mathbf{V}_{2x} + \mathbf{V}_{1y}\mathbf{V}_{2y}$$
$$= (2)(5) + (3)(-2)$$
$$= \boxed{4}$$

(b) The dot product is

$$\mathbf{V}_1 \cdot \mathbf{V}_2 = (1)(9) + (4)(-3)$$
$$= \boxed{-3}$$

(c) The dot product is

$$\mathbf{V}_1 \cdot \mathbf{V}_2 = (7)(3) + (-3)(4)$$
$$= \boxed{9}$$

(d) The dot product is

$$\mathbf{V}_1 \cdot \mathbf{V}_2 = \mathbf{V}_{1x}\mathbf{V}_{2x} + \mathbf{V}_{1y}\mathbf{V}_{2y} + \mathbf{V}_{1z}\mathbf{V}_{2z}$$
$$= (2)(8) + (-3)(2) + (6)(-3)$$
$$= \boxed{-8}$$

(e) The dot product is

$$\mathbf{V}_1 \cdot \mathbf{V}_2 = (6)(1) + (2)(0) + (3)(1)$$
$$= \boxed{9}$$

2. (a) The angle between the vectors is

$$\cos\phi = \frac{\mathbf{V}_1 \cdot \mathbf{V}_2}{|\mathbf{V}_1|\,|\mathbf{V}_2|} = \frac{4}{\sqrt{2^2 + 3^2}\sqrt{5^2 + (-2)^2}}$$
$$= 0.206$$
$$\phi = \arccos 0.206 = \boxed{78.1°}$$

(b) The angle between the vectors is

$$\cos \phi = \frac{\mathbf{V}_1 \cdot \mathbf{V}_2}{|\mathbf{V}_1||\mathbf{V}_2|} = \frac{-3}{\sqrt{1^2 + 4^2}\sqrt{9^2 + (-3)^2}}$$

$$= -0.0767$$

$$\phi = \arccos(-0.0767) = \boxed{94.4°}$$

(c) The angle between the vectors is

$$\cos \phi = \frac{\mathbf{V}_1 \cdot \mathbf{V}_2}{|\mathbf{V}_1||\mathbf{V}_2|} = \frac{9}{\sqrt{7^2 + (-3)^2}\sqrt{3^2 + 4^2}}$$

$$= 0.236$$

$$\phi = \arccos 0.236 = \boxed{76.3°}$$

3. (a) The cross product is

$$\mathbf{V}_1 \times \mathbf{V}_2 = \begin{vmatrix} \mathbf{i} & \mathbf{V}_{1x} & \mathbf{V}_{2x} \\ \mathbf{j} & \mathbf{V}_{1y} & \mathbf{V}_{2y} \\ \mathbf{k} & \mathbf{V}_{1z} & \mathbf{V}_{2z} \end{vmatrix}$$

$$= \begin{vmatrix} \mathbf{i} & 2 & 5 \\ \mathbf{j} & 3 & -2 \\ \mathbf{k} & 0 & 0 \end{vmatrix}$$

Expand by the third row.

$$\mathbf{V}_1 \times \mathbf{V}_2 = \mathbf{k} \begin{vmatrix} 2 & 5 \\ 3 & -2 \end{vmatrix} = \boxed{-19\mathbf{k}}$$

(b) The cross product is

$$\mathbf{V}_1 \times \mathbf{V}_2 = \begin{vmatrix} \mathbf{i} & 1 & 9 \\ \mathbf{j} & 4 & -3 \\ \mathbf{k} & 0 & 0 \end{vmatrix}$$

Expand by the third row.

$$\mathbf{V}_1 \times \mathbf{V}_2 = \mathbf{k} \begin{vmatrix} 1 & 9 \\ 4 & -3 \end{vmatrix} = \boxed{-39\mathbf{k}}$$

(c) The cross product is

$$\mathbf{V}_1 \times \mathbf{V}_2 = \begin{vmatrix} \mathbf{i} & \mathbf{V}_{1x} & \mathbf{V}_{2x} \\ \mathbf{j} & \mathbf{V}_{1y} & \mathbf{V}_{2y} \\ \mathbf{k} & \mathbf{V}_{1z} & \mathbf{V}_{2z} \end{vmatrix} = \begin{vmatrix} \mathbf{i} & 7 & 3 \\ \mathbf{j} & -3 & 4 \\ \mathbf{k} & 0 & 0 \end{vmatrix}$$

Expand by the third row.

$$\mathbf{V}_1 \times \mathbf{V}_2 = \mathbf{k} \begin{vmatrix} 7 & 3 \\ -3 & 4 \end{vmatrix} = \boxed{37\mathbf{k}}$$

(d) The cross product is

$$\mathbf{V}_1 \times \mathbf{V}_2 = \begin{vmatrix} \mathbf{i} & \mathbf{V}_{1x} & \mathbf{V}_{2x} \\ \mathbf{j} & \mathbf{V}_{1y} & \mathbf{V}_{2y} \\ \mathbf{k} & \mathbf{V}_{1z} & \mathbf{V}_{2z} \end{vmatrix} = \begin{vmatrix} \mathbf{i} & 2 & 8 \\ \mathbf{j} & -3 & 2 \\ \mathbf{k} & 6 & -3 \end{vmatrix}$$

Expand by the first column.

$$\mathbf{V}_1 \times \mathbf{V}_2 = \mathbf{i} \begin{vmatrix} -3 & 2 \\ 6 & -3 \end{vmatrix} - \mathbf{j} \begin{vmatrix} 2 & 8 \\ 6 & -3 \end{vmatrix} + \mathbf{k} \begin{vmatrix} 2 & 8 \\ -3 & 2 \end{vmatrix}$$

$$= \boxed{-3\mathbf{i} + 54\mathbf{j} + 28\mathbf{k}}$$

(e) The cross product is

$$\mathbf{V}_1 \times \mathbf{V}_2 = \begin{vmatrix} \mathbf{i} & \mathbf{V}_{1x} & \mathbf{V}_{2x} \\ \mathbf{j} & \mathbf{V}_{1y} & \mathbf{V}_{2y} \\ \mathbf{k} & \mathbf{V}_{1z} & \mathbf{V}_{2z} \end{vmatrix} = \begin{vmatrix} \mathbf{i} & 6 & 1 \\ \mathbf{j} & 2 & 0 \\ \mathbf{k} & 3 & 1 \end{vmatrix}$$

Expand by the second row.

$$\mathbf{V}_1 \times \mathbf{V}_2 = -\mathbf{j} \begin{vmatrix} 6 & 1 \\ 3 & 1 \end{vmatrix} + (2) \begin{vmatrix} \mathbf{i} & 1 \\ \mathbf{k} & 1 \end{vmatrix}$$

$$= \boxed{2\mathbf{i} - 3\mathbf{j} - 2\mathbf{k}}$$

6 Trigonometry

PRACTICE PROBLEMS

1. A 5 lbm (5 kg) block sits on a 20° incline without slipping. (a) Draw the free-body diagram with respect to axes parallel and perpendicular to the surface of the incline. (b) Determine the magnitude of the frictional force (holding the block stationary) on the block.

    (A) 1.71 lbf (16.8 N)

    (B) 3.35 lbf (32.9 N)

    (C) 4.70 lbf (46.1 N)

    (D) 5.00 lbf (49.1 N)

2. Complete the following calculation related to a catenary cable.

$$S = c\left(\cosh \frac{a}{h} - 1\right) = (245 \text{ ft})\left(\cosh \frac{50 \text{ ft}}{245 \text{ ft}} - 1\right)$$

3. Part of a turn-around area in a parking lot is shaped like a circular segment. The segment has a central angle of 120° and a radius of 75 ft. If the circular segment is to receive a special surface treatment, what area will be treated?

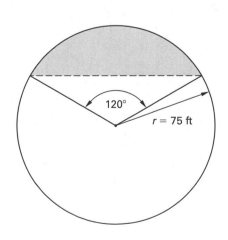

SOLUTIONS

1. (a) The free-body diagram is

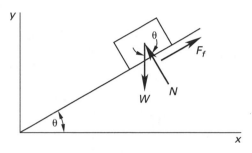

Customary U.S. Solution

(b) The mass of the block is $m = 5$ lbm. The angle of inclination is $\theta = 20°$. The weight is

$$W = \frac{mg}{g_c} = \frac{(5 \text{ lbm})\left(32.2 \frac{\text{ft}}{\text{sec}^2}\right)}{32.2 \frac{\text{lbm-ft}}{\text{lbf-sec}^2}}$$

$$= 5 \text{ lbf}$$

The frictional force is

$$F_f = W\sin\theta = (5 \text{ lbf})(\sin 20°)$$

$$= \boxed{1.71 \text{ lbf}}$$

The answer is (A).

SI Solution

(b) The mass of the block is $m = 5$ kg. The angle of inclination is $\theta = 20°$. The gravitational force is

$$W = mg = (5 \text{ kg})\left(9.81 \frac{\text{m}}{\text{s}^2}\right)$$

$$= 49.1 \text{ N}$$

The frictional force is

$$F_f = W\sin\theta = (49.1 \text{ N})(\sin 20°)$$

$$= \boxed{16.8 \text{ N}}$$

The answer is (A).

2. This calculation contains a hyperbolic cosine function.

$$S = c\left(\cosh\frac{a}{h} - 1\right) = (245 \text{ ft})\left(\cosh\frac{50 \text{ ft}}{245 \text{ ft}} - 1\right)$$
$$= (245 \text{ ft})(1.0209 - 1) = 5.12 \text{ ft}$$

3. From App. 7.A, the area of a circular segment is

$$A = \tfrac{1}{2}r^2(\phi - \sin\phi)$$

Since ϕ appears in the expression by itself, it must be expressed in radians.

$$\phi = \frac{(120°)(2\pi)}{360°} = 2.094 \text{ rad}$$

$$A = \tfrac{1}{2}r^2(\phi - \sin\phi)$$
$$= \left(\tfrac{1}{2}\right)(75 \text{ ft})^2(2.094 \text{ rad} - \sin 120°)$$
$$= 3453.7 \text{ ft}^2$$

7 Analytic Geometry

PRACTICE PROBLEMS

1. The diameter of a sphere and the base of a cone are equal. What percentage of that diameter must the cone's height be so that both volumes are equal?

  (A) 133%

  (B) 150%

  (C) 166%

  (D) 200%

2. The distance between the entrance and exit points on a horizontal roadway curve is 747 ft. The radius of the curve is 400 ft. What is the central angle between the entrance and exit points?

3. A vertical roadway crest curve starts deviating from a constant grade at station 103 (i.e., 10,300 ft from an initial benchmark). At sta 103+62, the curve is 2.11 ft lower than the tangent (i.e., from the straight line extension of the constant grade). How far will the curve be from the tangent at sta 103+87?

4. A pile driving hammer emits 143 W of sound power when operating. What is the maximum areal sound power density at the ground when 10.7 m of pile remains to be driven? Assume isotropic reflection and disregard reflected power.

SOLUTIONS

1. Let d be the diameter of the sphere and the base of the cone. Use App. 7.B.

The volume of the sphere is given by

$$V_{\text{sphere}} = \tfrac{4}{3}\pi r^3 = \tfrac{4}{3}\pi\left(\frac{d}{2}\right)^3$$
$$= \tfrac{1}{6}\pi d^3$$

The volume of the circular cone is given by

$$V_{\text{cone}} = \tfrac{1}{3}\pi r^2 h = \tfrac{1}{3}\pi\left(\frac{d}{2}\right)^2 h$$
$$= \tfrac{1}{12}\pi d^2 h$$

Since the volume of the sphere and cone are equal,

$$V_{\text{cone}} = V_{\text{sphere}}$$
$$\tfrac{1}{12}\pi d^2 h = \tfrac{1}{6}\pi d^3$$
$$h = 2d$$

The height of the cone must be $\boxed{200\%}$ of the diameter of the sphere.

The answer is (D).

2. Horizontal roadway curves are circular arcs. The circumference (perimeter) of an entire circle with a radius of 400 ft is

$$p = 2\pi r = (2\pi)(400 \text{ ft}) = 2513.3 \text{ ft}$$

From a ratio of curve length to angles,

$$\phi = \left(\frac{747 \text{ ft}}{2513.3 \text{ ft}}\right)(360°) = 107°$$

3. Vertical roadway curves are parabolic arcs. Parabolas are second-degree polynomials. Deviations, y, from a baseline are proportional to the square of the separation distance. That is, $y \propto x^2$.

$$\frac{y_1}{y_2} = \left(\frac{x_1}{x_2}\right)^2$$

$$y_2 = y_1 \left(\frac{x_2}{x_1}\right)^2 = (2.11 \text{ ft})\left(\frac{87 \text{ ft}}{62 \text{ ft}}\right)^2 = 4.15 \text{ ft}$$

4. The power is emitted isotropically, spherically, in all directions. The maximum sound power will occur at the surface, adjacent to the pile. The surface area of a sphere with a radius of 10.7 m is

$$A = 4\pi r^2 = (4\pi)(10.7 \text{ m})^2 = 1438.7 \text{ m}^2$$

The areal power density is

$$\rho_P = \frac{P}{A} = \frac{143 \text{ W}}{1438.7 \text{ m}^2} = 0.0994 \text{ W/m}^2$$

8 Differential Calculus

PRACTICE PROBLEMS

1. Find all minima, maxima, and inflection points for

$$y = x^3 - 9x^2 - 3$$

(A) maximum at $x = 0$
    inflection at $x = -3$
    minimum at $x = -6$

(B) maximum at $x = 0$
    inflection at $x = 3$
    minimum at $x = 6$

(C) minimum at $x = 0$
    inflection at $x = 3$
    maximum at $x = 6$

(D) minimum at $x = 3$
    inflection at $x = 0$
    maximum at $x = -3$

2. The equation for the elevation above mean sea level of a sag vertical roadway curve is

$$y(x) = 0.56x^2 - 3.2x + 708.28$$

y is measured in feet, and x is measured in 100 ft stations past the beginning of the curve. What is the elevation of the turning point (i.e., the lowest point on the curve)?

(A) 702 ft

(B) 704 ft

(C) 705 ft

(D) 706 ft

3. A car drives on a highway with a legal speed limit of 100 km/h. The fuel usage, Q, (in liters per 100 kilometers driven) of a car driven at speed v (in km/h) is

$$Q(v) = \frac{1750v}{v^2 + 6700}$$

At what legal speed should the car travel in order to maximize the fuel efficiency?

(A) 82 km/h

(B) 87 km/h

(C) 93 km/h

(D) 100 km/h

4. A chemical feed storage tank is needed with a volume of 3000 ft³ (gross of fittings). The tank will be formed as a circular cylinder with barrel length, L, capped by two hemispherical ends of radius, r. The manufacturing cost per unit area of hemispherical ends is double that of the cylinder. What dimensions will minimize the manufacturing cost?

(A) radius = $4^1/_2$ ft; cylinder barrel length = 42 ft

(B) radius = 5 ft; cylinder barrel length = $31^1/_2$ ft

(C) radius = $5^1/_2$ ft; cylinder barrel length = $22^1/_2$ ft

(D) radius = 6 ft; cylinder barrel length = $18^1/_2$ ft

SOLUTIONS

1. Determine the critical points by taking the first derivative of the function and setting it equal to zero.

$$\frac{dy}{dx} = 3x^2 - 18x = 3x(x - 6)$$

$$3x(x - 6) = 0$$

$$x(x - 6) = 0$$

The critical points are located at $x = 0$ and $x = 6$.

Determine the inflection points by setting the second derivative equal to zero. Take the second derivative.

$$\frac{d^2y}{dx^2} = \left(\frac{d}{dx}\right)\left(\frac{dy}{dx}\right) = \frac{d}{dx}(3x^2 - 18x)$$

$$= 6x - 18$$

Set the second derivative equal to zero.

$$\frac{d^2y}{dx^2} = 0 = 6x - 18 = (6)(x - 3)$$

$$(6)(x - 3) = 0$$

$$x - 3 = 0$$

$$x = 3$$

The inflection point is at $\boxed{x = 3.}$

Determine the local maximum and minimum by substituting the critical points into the expression for the second derivative.

At the critical point $x = 0$,

$$\left.\frac{d^2y}{dx^2}\right|_{x=0} = (6)(x - 3) = (6)(0 - 3) = -18$$

Since $-18 < 0$, $\boxed{x = 0}$ is a local maximum.

At the critical point $x = 6$,

$$\left.\frac{d^2y}{dx^2}\right|_{x=6} = (6)(x - 3) = (6)(6 - 3) = 18$$

Since $18 > 0$, $\boxed{x = 6}$ is a local minimum.

The answer is (B).

2. Set the derivative of the curve's equation to zero.

$$\frac{dy(x)}{dx} = \frac{d}{dx}(0.56x^2 - 3.2x + 708.28) = 1.12x - 3.2$$

$$x_c = \frac{3.2}{1.12} = 2.857 \text{ sta}$$

Insert x_c into the elevation equation.

$$y_{min} = y(x_c) = (0.56)(2.857)^2 - (3.2)(2.857) + 708.28$$

$$= 703.71 \text{ ft} \quad (704 \text{ ft})$$

The answer is (B).

3. Use the quotient rule to calculate the derivative.

$$\mathbf{D}\left(\frac{f(x)}{g(x)}\right) = \frac{g(x)\mathbf{D}f(x) - f(x)\mathbf{D}g(x)}{(g(x))^2}$$

$$\frac{dQ(v)}{dv} = \frac{d}{dv}\left(\frac{1750v}{v^2 + 6700}\right)$$

$$= \frac{(v^2 + 6700)(1750) - (1750v)(2v)}{(v^2 + 6700)^2}$$

Combining terms and simplifying,

$$\frac{dQ(v)}{dv} = \frac{-1750v^2 + 11,725,000}{v^4 + 13,400v^2 + 44,890,000}$$

Set the derivative of the fuel consumption equation to zero. In order for the derivative to be zero, the numerator must be zero.

$$-1750v^2 + 11,725,000 = 0$$

$$v = \sqrt{\frac{11,725,000}{1750}} = 81.85 \quad (82 \text{ km/h})$$

Maximizing the fuel efficiency is the same as minimizing the fuel usage. It is not known if setting $dQ(v)/dt = 0$ results in a minimum or maximum. While using $d^2Q(v)/dt^2$ is possible, it is easier to plot the points.

v	Q(v)
82 km/h	10.689 L
87 km/h	10.669 L
93 km/h	10.603 L
100 km/h	10.479 L

Clearly, $Q(82 \text{ km/h})$ is a maximum, and the minimum fuel usage occurs at the endpoint of the range, at $\boxed{100 \text{ km/h.}}$

The answer is (D).

4. The volume of the tank will be the combined volume of a cylinder and a sphere. The cylinder and sphere have the same radius.

$$V = \pi r^2 L + \tfrac{4}{3}\pi r^3 = 3000 \text{ ft}^3$$

For any given cost per unit area (arbitrarily selected as $1/\text{ft}^2$), the cost function is

$$C(r, L) = \left(1 \ \frac{\$}{\text{ft}^2}\right)(A_{\text{cylinder}} + 2A_{\text{sphere}}) = 2\pi r L + 2(4\pi r^2)$$

$$= 2\pi r L + 8\pi r^2$$

Write the volume equation in terms of L.

$$L = \frac{3000 - \tfrac{4}{3}\pi r^3}{\pi r^2}$$

Substitute L into the cost equation to get a cost function of a single variable.

$$C(r) = 2\pi r L + 8\pi r^2 = 2\pi r \left(\frac{3000 - \tfrac{4}{3}\pi r^3}{\pi r^2}\right) + 8\pi r^2$$

$$= \frac{6000}{r} - \frac{8\pi r^2}{3} + 8\pi r^2$$

$$= \frac{6000}{r} + \frac{16\pi r^2}{3}$$

Find the optimal value of r by setting the first derivative of the cost function equal to zero.

$$\frac{dC(r)}{dr} = \frac{d}{dr}\left(\frac{6000}{r} + \frac{16\pi r^2}{3}\right)$$

$$= \frac{-6000}{r^2} - \frac{32\pi r}{3} = 0$$

$$\frac{6000}{r^2} = \frac{32\pi r}{3}$$

Cross-multiply and solve for the optimal value of r.

$$32\pi r^3 = 18,000$$

$$r = \sqrt[3]{\frac{18,000}{32\pi}}$$

$$= 5.636 \text{ ft} \quad \boxed{\left(5\tfrac{1}{2} \text{ ft}\right)}$$

Calculate the optimal value of the barrel length, L.

$$L = \frac{3000 - \tfrac{4}{3}\pi r^3}{\pi r^2}$$

$$= \frac{3000 - \tfrac{4}{3}\pi(5.636 \text{ ft})^3}{\pi(5.636 \text{ ft})^2}$$

$$= 22.55 \text{ ft} \quad \boxed{\left(22\tfrac{1}{2} \text{ ft}\right)}$$

The answer is (C).

Background and Support

Integral Calculus

PRACTICE PROBLEMS

1. Find the integrals.

(a) $\int \sqrt{1-x}\, dx$

(b) $\int \dfrac{x}{x^2+1}\, dx$

(c) $\int \dfrac{x^2}{x^2+x-6}\, dx$

2. Calculate the definite integrals.

(a) $\int_1^3 (x^2+4x)\, dx$

(b) $\int_{-2}^2 (x^3+1)\, dx$

(c) $\int_1^2 (4x^3-3x^2)\, dx$

3. Find the area bounded by $x=1$, $x=3$, $y+x+1=0$, and $y=6x-x^2$.

4. Find a_0 (the first term of a Fourier series approximation, corresponding to the waveform's average value) for the two waveforms shown.

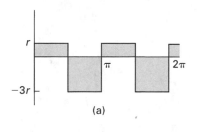

(a)

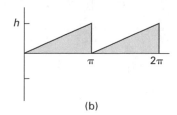

(b)

5. For each of the two waveforms shown, determine if its Fourier series is of type A, B, or C.

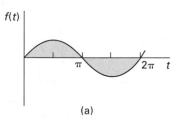

(a)

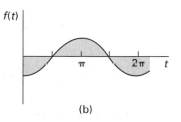

(b)

type A: $f(t) = a_0 + a_2\cos 2t + b_2\sin 2t$
$\qquad + a_4\cos 4t + b_4\sin 4t + \cdots$

type B: $f(t) = a_0 + b_1\sin t + b_2\sin 2t + b_3\sin 3t + \cdots$

type C: $f(t) = a_0 + a_1\cos t + a_2\cos 2t + a_3\cos 3t + \cdots$

SOLUTIONS

1. (a) The integral is

$$\int \sqrt{1-x}\,dx = \int (1-x)^{1/2}\,dx$$
$$= \boxed{-\tfrac{2}{3}(1-x)^{3/2} + C}$$

(b) The integral is

$$\int \frac{x}{x^2+1}\,dx = \tfrac{1}{2}\int \frac{2x}{x^2+1}\,dx$$
$$= \boxed{\tfrac{1}{2}\ln|(x^2+1)| + C}$$

(c) The integral is

$$\frac{x^2}{x^2+x-6} = 1 - \frac{x-6}{x^2+x-6}$$
$$= 1 - \frac{x-6}{(x+3)(x-2)}$$
$$= 1 - \frac{\tfrac{9}{5}}{x+3} + \frac{\tfrac{4}{5}}{x-2}$$

$$\int \frac{x^2}{x^2+x-6}\,dx = \int \left(1 - \frac{\tfrac{9}{5}}{x+3} + \frac{\tfrac{4}{5}}{x-2}\right)dx$$
$$= \int dx - \int \frac{\tfrac{9}{5}}{x+3}\,dx + \int \frac{\tfrac{4}{5}}{x-2}\,dx$$
$$= \boxed{x - \tfrac{9}{5}\ln|(x+3)| + \tfrac{4}{5}\ln|(x-2)| + C}$$

2. (a) The definite integral is

$$\int_1^3 (x^2+4x)\,dx = \left(\frac{x^3}{3} + 2x^2\right)\Bigg|_1^3 = \boxed{24\tfrac{2}{3}}$$

(b) The definite integral is

$$\int_{-2}^2 (x^3+1)\,dx = \left(\frac{x^4}{4} + x\right)\Bigg|_{-2}^2 = \boxed{4}$$

(c) The definite integral is

$$\int_1^2 (4x^3 - 3x^2)\,dx = \left(x^4 - x^3\right)\Bigg|_1^2 = \boxed{8}$$

3. The bounded area is as follows.

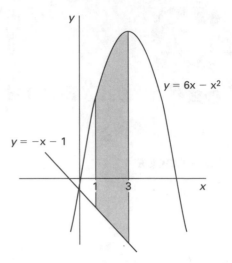

$$\text{area} = \int_1^3 \left((6x - x^2) - (-x-1)\right)dx$$
$$= \int_1^3 (-x^2 + 7x + 1)\,dx$$
$$= \left(-\frac{x^3}{3} + \tfrac{7}{2}x^2 + x\right)\Bigg|_1^3 = \boxed{21\tfrac{1}{3}}$$

4. For waveform (a):

$$a_0 = \frac{1}{2\pi}\int_0^{2\pi} f(t)\,dt$$
$$= \frac{1}{\pi}\int_0^{\pi} f(t)\,dt$$
$$= \frac{1}{\pi}\left((r)\left(\frac{\pi}{2}\right) + (-3r)\left(\frac{\pi}{2}\right)\right)$$
$$= \boxed{-r}$$

For waveform (b):

$$a_0 = \frac{1}{2\pi}\int_0^{2\pi} f(t)\,dt$$
$$= \frac{1}{\pi}\int_0^{\pi} f(t)\,dt$$
$$= \left(\frac{1}{\pi}\right)\left(\tfrac{1}{2}\pi h\right)$$
$$= \boxed{h/2}$$

5. For waveform (a):
Since $f(t) = -f(-t)$, it is $\boxed{\text{type B.}}$

For waveform (b):
Since $f(t) = f(-t)$, it is $\boxed{\text{type C.}}$

10 Differential Equations

PRACTICE PROBLEMS

1. Solve the following differential equation for y.

$$y'' - 4y' - 12y = 0$$

(A) $A_1 e^{6x} + A_2 e^{-2x}$

(B) $A_1 e^{-6x} + A_2 e^{2x}$

(C) $A_1 e^{6x} + A_2 e^{2x}$

(D) $A_1 e^{-6x} + A_2 e^{-2x}$

2. Solve the following differential equation for y.

$$y' - y = 2xe^{2x} \quad y(0) = 1$$

(A) $y = 2e^{-2x}(x-1) + 3e^{-x}$

(B) $y = 2e^{2x}(x-1) + 3e^{x}$

(C) $y = -2e^{-2x}(x-1) + 3e^{-x}$

(D) $y = 2e^{2x}(x-1) + 3e^{-x}$

3. The oscillation exhibited by the top story of a certain building in free motion is given by the following differential equation.

$$x'' + 2x' + 2x = 0 \quad x(0) = 0 \quad x'(0) = 1$$

(a) What is x as a function of time?

(A) $e^{-2t} \sin t$

(B) $e^{t} \sin t$

(C) $e^{-t} \sin t$

(D) $e^{-t} \sin t + e^{-t} \cos t$

(b) What is the building's fundamental natural frequency of vibration?

(A) $^1/_2$

(B) 1

(C) $\sqrt{2}$

(D) 2

(c) What is the amplitude of oscillation?

(A) 0.32

(B) 0.54

(C) 1.7

(D) 6.6

(d) What is x as a function of time if a lateral wind load is applied with a form of $\sin t$?

(A) $\frac{6}{5}e^{-t}\sin t + \frac{2}{5}e^{-t}\cos t$

(B) $\frac{6}{5}e^{t}\sin t + \frac{2}{5}e^{t}\cos t$

(C) $\frac{2}{5}e^{t}\sin t + \frac{6}{5}e^{-t}\cos t + \frac{2}{5}\sin t - \frac{1}{5}\cos t$

(D) $\frac{6}{5}e^{-t}\sin t + \frac{2}{5}e^{-t}\cos t + \frac{1}{5}\sin t - \frac{2}{5}\cos t$

4. (*Time limit: one hour*) A 90 lbm (40 kg) bag of a chemical is accidentally dropped in an aerating lagoon. The chemical is water soluble and nonreacting. The lagoon is 120 ft (35 m) in diameter and filled to a depth of 10 ft (3 m). The aerators circulate and distribute the chemical evenly throughout the lagoon.

Water enters the lagoon at a rate of 30 gal/min (115 L/min). Fully mixed water is pumped into a reservoir at a rate of 30 gal/min (115 L/min).

The established safe concentration of this chemical is 1 ppb (part per billion). Most nearly, how many days will it take for the concentration of the discharge water to reach this level?

(A) 25 days

(B) 50 days

(C) 100 days

(D) 200 days

5. A tank contains 100 gal (100 L) of brine made by dissolving 60 lbm (60 kg) of salt in pure water. Salt water with a concentration of 1 lbm/gal (1 kg/L) enters the tank at a rate of 2 gal/min (2 L/min). A well-stirred mixture is drawn from the tank at a rate of 3 gal/min (3 L/min). Find the mass of salt in the tank after 1 hr.

(A) 13 lbm (13 kg)

(B) 37 lbm (37 kg)

(C) 43 lbm (43 kg)

(D) 51 lbm (51 kg)

SOLUTIONS

1. Obtain the characteristic equation by replacing each derivative with a polynomial term of equal degree.

$$r^2 - 4r - 12 = 0$$

Factor the characteristic equation.

$$(r - 6)(r + 2) = 0$$

The roots are $r_1 = 6$ and $r_2 = -2$.

Since the roots are real and distinct, the solution is

$$y = A_1 e^{r_1 x} + A_2 e^{r_2 x}$$
$$= \boxed{A_1 e^{6x} + A_2 e^{-2x}}$$

The answer is (A).

2. The equation is a first-order linear differential equation of the form

$$y' + p(x)y = g(x)$$
$$p(x) = -1$$
$$g(x) = 2xe^{2x}$$

The integration factor $u(x)$ is given by

$$u(x) = \exp\left(\int p(x)\,dx\right)$$
$$= \exp\left(\int (-1)\,dx\right)$$
$$= e^{-x}$$

The closed form of the solution is given by

$$y = \frac{1}{u(x)}\left(\int u(x)g(x)\,dx + C\right)$$
$$= \frac{1}{e^{-x}}\left(\int (e^{-x})(2xe^{2x})\,dx + C\right)$$
$$= e^x\left(2(xe^x - e^x) + C\right)$$
$$= e^x\left(2e^x(x - 1) + C\right)$$
$$= 2e^{2x}(x - 1) + Ce^x$$

Apply the initial condition $y(0) = 1$ to obtain the integration constant C.

$$y(0) = 2e^{(2)(0)}(0 - 1) + Ce^0 = 1$$
$$(2)(1)(-1) + C(1) = 1$$
$$-2 + C = 1$$
$$C = 3$$

Substituting in the value for the integration constant C, the solution is

$$\boxed{y = 2e^{2x}(x - 1) + 3e^x}$$

The answer is (B).

3. (a) The differential equation is a homogeneous second-order linear differential equation with constant coefficients. Write the characteristic equation.

$$r^2 + 2r + 2 = 0$$

This is a quadratic equation of the form $ar^2 + br + c = 0$ where $a = 1$, $b = 2$, and $c = 2$.

Solve for r.

$$r = \frac{-b \pm \sqrt{b^2 - 4ac}}{2a}$$
$$= \frac{-2 \pm \sqrt{(2)^2 - (4)(1)(2)}}{(2)(1)}$$
$$= \frac{-2 \pm \sqrt{4 - 8}}{2}$$
$$= \frac{-2 \pm \sqrt{-4}}{2}$$
$$= \frac{-2 \pm 2\sqrt{-1}}{2}$$
$$= -1 \pm \sqrt{-1}$$
$$= -1 \pm i$$
$$r_1 = -1 + i, \text{ and } r_2 = -1 - i$$

Since the roots are imaginary and of the form $\alpha + i\omega$ and $\alpha - i\omega$ where $\alpha = -1$ and $\omega = 1$, the general form of the solution is given by

$$x(t) = A_1 e^{\alpha t}\cos \omega t + A_2 e^{\alpha t}\sin \omega t$$
$$= A_1 e^{-1t}\cos(t) + A_2 e^{-1t}\sin(t)$$
$$= A_1 e^{-t}\cos t + A_2 e^{-t}\sin t$$

Apply the initial conditions $x(0) = 0$ and $x'(0) = 1$ to solve for A_1 and A_2.

First, apply the initial condition $x(0) = 0$.

$$x(t) = A_1 e^0 \cos 0 + A_2 e^0 \sin 0 = 0$$
$$A_1(1)(1) + A_2(1)(0) = 0$$
$$A_1 = 0$$

Substituting, the solution of the differential equation becomes

$$x(t) = A_2 e^{-t} \sin t$$

To apply the second initial condition, take the first derivative.

$$x'(t) = \frac{d}{dt}(A_2 e^{-t} \sin t)$$
$$= A_2 \frac{d}{dt}(e^{-t} \sin t)$$
$$= A_2 \left(\sin t \frac{d}{dt}(e^{-t}) + e^{-t} \frac{d}{dt} \sin t \right)$$
$$= A_2 \left(\sin t(-e^{-t}) + e^{-t}(\cos t) \right)$$
$$= A_2(e^{-t})(-\sin t + \cos t)$$

Apply the initial condition, $x'(0) = 1$.

$$x(0) = A_2 e^0(-\sin 0 + \cos 0) = 1$$
$$A_2(1)(0 + 1) = 1$$
$$A_2 = 1$$

The solution is

$$x(t) = A_2 e^{-t} \sin t$$
$$= (1)e^{-t} \sin t$$
$$= \boxed{e^{-t} \sin t}$$

The answer is (C).

(b) To determine the natural frequency, set the damping term to zero. The equation has the form

$$x'' + 2x = 0$$

This equation has a general solution of the form

$$x(t) = x_0 \cos \omega t + \left(\frac{v_0}{\omega}\right)\sin \omega t$$

ω is the natural frequency. Given the equation $x'' + 2x = 0$, the characteristic equation is

$$r^2 + 2 = 0$$
$$r = \sqrt{-2}$$
$$= \pm\sqrt{2}i$$

Since the roots are imaginary and of the form $\alpha + i\omega$ and $\alpha - i\omega$ where $\alpha = 0$ and $\omega = \sqrt{2}$, the general form of the solution is given by

$$x(t) = A_1 e^{\alpha t}\cos\omega t + A_2 e^{\alpha t}\sin\omega t$$
$$= A_1 e^{0t}\cos\sqrt{2}t + A_2 e^{0t}\sin\sqrt{2}t$$
$$= A_1(1)\cos\sqrt{2}t + A_2(1)\sin\sqrt{2}t$$
$$= A_1\cos\sqrt{2}t + A_2\sin\sqrt{2}t$$

Apply the initial conditions, $x(0) = 0$ and $x'(0) = 1$, to solve for A_1 and A_2. Applying the initial condition $x(0) = 0$ gives

$$x(0) = A_1\cos\big((\sqrt{2})(0)\big) + A_2\sin\big((\sqrt{2})(0)\big) = 0$$
$$A_1\cos 0 + A_2\sin 0 = 0$$
$$A_1(1) + A_2(0) = 0$$
$$A_1 = 0$$

Substituting, the solution of the differential equation becomes

$$x(t) = A_2\sin\sqrt{2}t$$

To apply the second initial condition, take the first derivative.

$$x'(t) = \frac{d}{dt}\big(A_2\sin\sqrt{2}t\big)$$
$$= A_2\sqrt{2}\cos\sqrt{2}t$$

Apply the second initial condition, $x'(0) = 1$.

$$x'(0) = A_2\sqrt{2}\cos\big((\sqrt{2})(0)\big) = 1$$
$$A_2\sqrt{2}\cos(0) = 1$$
$$A_2(\sqrt{2})(1) = 1$$
$$A_2\sqrt{2} = 1$$
$$A_2 = \frac{1}{\sqrt{2}} = \frac{\sqrt{2}}{2}$$

Substituting, the undamped solution becomes

$$x(t) = \frac{\sqrt{2}}{2}\sin\sqrt{2}t$$

Therefore, the undamped natural frequency is $\omega = \boxed{\sqrt{2}.}$

The answer is (C).

(c) The amplitude of the oscillation is the maximum displacement.

Take the derivative of the solution, $x(t) = e^{-t}\sin t$.

$$x'(t) = \frac{d}{dt}(e^{-t}\sin t)$$
$$= \sin t\frac{d}{dt}(e^{-t}) + e^{-t}\frac{d}{dt}\sin t$$
$$= \sin t(-e^{-t}) + e^{-t}\cos t$$
$$= e^{-t}(\cos t - \sin t)$$

The maximum displacement occurs at $x'(t) = 0$.

Since $e^{-t} \neq 0$ except as t approaches infinity,

$$\cos t - \sin t = 0$$
$$\tan t = 1$$
$$t = \tan^{-1}(1)$$
$$= 0.785 \text{ rad}$$

At $t = 0.785$ rad, the displacement is maximum. Substitute into the orginal solution to obtain a value for the maximum displacement.

$$x(0.785) = e^{-0.785}\sin 0.785$$
$$= 0.322$$

The amplitude is $\boxed{0.322\ (0.32).}$

The answer is (A).

(d) (An alternative solution using Laplace transforms follows this solution.) The application of a lateral wind load with the form $\sin t$ revises the differential equation to the form

$$x'' + 2x' + 2x = \sin t$$

Express the solution as the sum of the complementary x_c and particular x_p solutions.

$$x(t) = x_c(t) + x_p(t)$$

From part (a),

$$x_c(t) = A_1 e^{-t}\cos t + A_2 e^{-t}\sin t$$

The general form of the particular solution is given by

$$x_p(t) = x^s(A_3 \cos t + A_4 \sin t)$$

Determine the value of s; check to see if the terms of the particular solution solve the homogeneous equation.

Examine the term $A_3 \cos(t)$.

Take the first derivative.

$$\frac{d}{dx}(A_3 \cos t) = -A_3 \sin t$$

Take the second derivative.

$$\frac{d}{dx}\left(\frac{d}{dx}(A_3 \cos t)\right) = \frac{d}{dx}(-A_3 \sin t)$$
$$= -A_3 \cos t$$

Substitute the terms into the homogeneous equation.

$$x'' + 2x' + 2x = -A_3 \cos t + (2)(-A_3 \sin t)$$
$$+ (2)(-A_3 \cos t)$$
$$= A_3 \cos t - 2A_3 \sin t$$
$$\neq 0$$

Except for the trival solution $A_3 = 0$, the term $A_3 \cos t$ does not solve the homogeneous equation.

Examine the second term, $A_4 \sin t$.

Take the first derivative.

$$\frac{d}{dx}(A_4 \sin t) = A_4 \cos t$$

Take the second derivative.

$$\frac{d}{dx}\left(\frac{d}{dx}(A_4 \sin t)\right) = \frac{d}{dx}(A_4 \cos t)$$
$$= -A_4 \sin t$$

Substitute the terms into the homogeneous equation.

$$x'' + 2x' + 2x = -A_4 \sin t + (2)(A_4 \cos t)$$
$$+ (2)(A_4 \sin t)$$
$$= A_4 \sin t + 2A_4 \cos t$$
$$\neq 0$$

Except for the trivial solution $A_4 = 0$, the term $A_4 \sin t$ does not solve the homogeneous equation.

Neither of the terms satisfies the homogeneous equation $s = 0$; therefore, the particular solution is of the form

$$x_p(t) = A_3 \cos t + A_4 \sin t$$

Use the method of undetermined coefficients to solve for A_3 and A_4. Take the first derivative.

$$x'_p(t) = \frac{d}{dx}(A_3 \cos t + A_4 \sin t)$$
$$= -A_3 \sin t + A_4 \cos t$$

Take the second derivative.

$$x''_p(t) = \frac{d}{dx}\left(\frac{d}{dx}(A_3 \cos t + A_4 \sin t)\right)$$
$$= \frac{d}{dx}(-A_3 \sin t + A_4 \cos t)$$
$$= -A_3 \cos t - A_4 \sin t$$

Substitute the expressions for the derivatives into the differential equation.

$$x'' + 2x' + 2x = (-A_3 \cos t - A_4 \sin t)$$
$$+ (2)(-A_3 \sin t + A_4 \cos t)$$
$$+ (2)(A_3 \cos t + A_4 \sin t)$$
$$= \sin t$$

Rearranging terms gives

$$(-A_3 + 2A_4 + 2A_3)\cos t$$
$$+ (-A_4 - 2A_3 + 2A_4)\sin t = \sin t$$
$$(A_3 + 2A_4)\cos t + (-2A_3 + A_4)\sin t = \sin t$$

Equating coefficients gives

$$A_3 + 2A_4 = 0$$
$$-2A_3 + A_4 = 1$$

Multiplying the first equation by 2 and adding equations gives

$$2A_3 + 4A_4 = 0$$
$$\underline{+(-2A_3 + A_4) = 1}$$
$$5A_4 = 1 \text{ or } A_4 = \tfrac{1}{5}$$

From the first equation for $A_4 = \frac{1}{5}$, $A_3 + (2)(\frac{1}{5}) = 0$, and $A_3 = -\frac{2}{5}$.

Substituting for the coefficients, the particular solution becomes

$$x_p(t) = -\tfrac{2}{5}\cos t + \tfrac{1}{5}\sin t \quad /$$

Combining the complementary and particular solutions gives

$$x(t) = x_c(t) + x_p(t)$$
$$= A_1 e^{-t}\cos t + A_2 e^{-t}\sin t - \tfrac{2}{5}\cos t + \tfrac{1}{5}\sin t$$

Apply the initial conditions to solve for the coefficients A_1 and A_2; then apply the first initial condition, $x(0) = 0$.

$$x(t) = A_1 e^0 \cos 0 + A_2 e^0 \sin 0$$
$$-\tfrac{2}{5}\cos 0 + \tfrac{1}{5}\sin 0 = 0$$
$$A_1(1)(1) + A_2(1)(0) + \left(-\tfrac{2}{5}\right)(1) + \left(\tfrac{1}{5}\right)(0) = 0$$
$$A_1 - \tfrac{2}{5} = 0$$
$$A_1 = \tfrac{2}{5}$$

Substituting for A_1, the solution becomes

$$x(t) = \tfrac{2}{5}e^{-t}\cos t + A_2 e^{-t}\sin t - \tfrac{2}{5}\cos t + \tfrac{1}{5}\sin t$$

Take the first derivative.

$$x'(t) = \frac{d}{dx}\left(\tfrac{2}{5}e^{-t}\cos t + A_2 e^{-t}\sin t\right)$$
$$+ \left(-\tfrac{2}{5}\cos t + \tfrac{1}{5}\sin t\right)$$
$$= \left(\tfrac{2}{5}\right)(-e^{-t}\cos t - e^{-t}\sin t)$$
$$+ A_2(-e^{-t}\sin t + e^{-t}\cos t)$$
$$+ \left(-\tfrac{2}{5}\right)(-\sin t) + \tfrac{1}{5}\cos t$$

Apply the second initial condition, $x'(0) = 1$.

$$x'(0) = \left(\tfrac{2}{5}\right)(-e^0\cos 0 - e^0\sin 0)$$
$$+ A_2(-e^0\sin 0 + e^0\cos 0)$$
$$+ \left(-\tfrac{2}{5}\right)(-\sin 0) + \tfrac{1}{5}\cos 0$$
$$= 1$$

$$\left(\tfrac{2}{5}\right)\left(-(1)(1) - (1)(0)\right)$$
$$+ A_2\left(-(1)(0) + (1)(1)\right) + \left(-\tfrac{2}{5}\right)(0) + \left(\tfrac{1}{5}\right)(1) = 1$$
$$\left(\tfrac{2}{5}\right)(-1) + A_2(1) + \left(\tfrac{1}{5}\right) = 1$$
$$A_2 = \tfrac{6}{5}$$

Substituting for A_2, the solution becomes

$$\boxed{x(t) = \tfrac{2}{5}e^{-t}\cos t + \tfrac{6}{5}e^{-t}\sin t - \tfrac{2}{5}\cos t + \tfrac{1}{5}\sin t}$$

The answer is (D).

(d) *Alternate solution:*

Use the Laplace transform method.

$$x'' + 2x' + 2x = \sin t$$
$$\mathcal{L}(x'') + 2\mathcal{L}(x') + 2\mathcal{L}(x) = \mathcal{L}(\sin t)$$
$$s^2\mathcal{L}(x) - 1 + 2s\mathcal{L}(x) + 2\mathcal{L}(x) = \frac{1}{s^2+1}$$
$$\mathcal{L}(x)(s^2 + 2s + 2) - 1 = \frac{1}{s^2+1}$$
$$\mathcal{L}(x) = \frac{1}{s^2+2s+2} + \frac{1}{(s^2+1)(s^2+2s+2)}$$
$$= \frac{1}{(s+1)^2+1} + \frac{1}{(s^2+1)(s^2+2s+2)}$$

Use partial fractions to expand the second term.

$$\frac{1}{(s^2+1)(s^2+2s+2)} = \frac{A_1+B_1 s}{s^2+1} + \frac{A_2+B_2 s}{s^2+2s+2}$$

Cross multiply.

$$= \frac{\begin{array}{c} A_1 s^2 + 2A_1 s + 2A_1 + B_1 s^3 + 2B_1 s^2 \\ + A_2 s^2 + A_2 + B_2 s^3 + B_2 s \end{array}}{(s^2+1)(s^2+2s+2)}$$

$$= \frac{\begin{array}{c} s^3(B_1+B_2) + s^2(A_1+A_2+2B_1) \\ + s(2A_1+2B_1+B_2) + 2A_1+A_2 \end{array}}{(s^2+1)(s^2+2s+2)}$$

Compare numerators to obtain the following four simultaneous equations.

$$B_1 + B_2 = 0$$
$$A_1 + A_2 + 2B_1 = 0$$
$$2A_1 + 2B_1 + B_2 = 0$$
$$2A_1 + A_2 = 1$$

Use Cramer's rule to find A_1.

$$A_1 = \frac{\begin{vmatrix} 0 & 0 & 1 & 1 \\ 0 & 1 & 2 & 0 \\ 0 & 0 & 2 & 1 \\ 1 & 1 & 0 & 0 \end{vmatrix}}{\begin{vmatrix} 0 & 0 & 1 & 1 \\ 1 & 1 & 2 & 0 \\ 2 & 0 & 2 & 1 \\ 2 & 1 & 0 & 0 \end{vmatrix}} = \frac{-1}{-5} = \frac{1}{5}$$

The rest of the coefficients are found similarly.

$$A_1 = \tfrac{1}{5}$$
$$A_2 = \tfrac{3}{5}$$
$$B_1 = -\tfrac{2}{5}$$
$$B_2 = \tfrac{2}{5}$$

Then,

$$\mathcal{L}(x) = \frac{1}{(s+1)^2 + 1} + \frac{\tfrac{1}{5}}{s^2 + 1} + \frac{-\tfrac{2}{5}s}{s^2 + 1}$$
$$+ \frac{\tfrac{3}{5}}{s^2 + 2s + 2} + \frac{\tfrac{2}{5}s}{s^2 + 2s + 2}$$

Take the inverse transform.

$$x(t) = \mathcal{L}^{-1}\{\mathcal{L}(x)\}$$
$$= e^{-t}\sin t + \tfrac{1}{5}\sin t - \tfrac{2}{5}\cos t + \tfrac{3}{5}e^{-t}\sin t$$
$$+ \tfrac{2}{5}(e^{-t}\cos t - e^{-t}\sin t)$$
$$= \boxed{\tfrac{6}{5}e^{-t}\sin t + \tfrac{2}{5}e^{-t}\cos t + \tfrac{1}{5}\sin t - \tfrac{2}{5}\cos t}$$

The answer is (D).

4. *Customary U.S. Solution*

The differential equation is

$$m'(t) = a(t) - \frac{m(t)o(t)}{V(t)}$$

$a(t) =$ rate of addition of chemical

$m(t) =$ mass of chemical at time t

$o(t) =$ volumetric flow out of the lagoon, 30 gpm

$V(t) =$ volume in the lagoon at time t

Water flows into the lagoon at a rate of 30 gpm, and a water-chemical mix flows out of the lagoon at a rate of 30 gpm. Therefore, the volume of the lagoon at time t is equal to the initial volume.

$$V(t) = \left(\frac{\pi}{4}\right)(\text{diameter of lagoon})^2(\text{depth of lagoon})$$
$$= \left(\frac{\pi}{4}\right)(120 \text{ ft})^2(10 \text{ ft})$$
$$= 113{,}097 \text{ ft}^3$$

Using a conversion factor of 7.48 gal/ft^3 gives

$$o(t) = \frac{30 \; \frac{\text{gal}}{\text{min}}}{7.48 \; \frac{\text{gal}}{\text{ft}^3}}$$
$$= 4.01 \text{ ft}^3/\text{min}$$

Substituting into the general form of the differential equation gives

$$m'(t) = a(t) - \frac{m(t)o(t)}{V(t)}$$
$$= (0) - m(t)\left(\frac{4.01 \; \frac{\text{ft}^3}{\text{min}}}{113{,}097 \text{ ft}^3}\right)$$
$$= -\left(\frac{3.55 \times 10^{-5}}{\text{min}}\right)m(t)$$

$$m'(t) + \left(\frac{3.55 \times 10^{-5}}{\text{min}}\right)m(t) = 0$$

The differential equation of the problem has the following characteristic equation.

$$r + \frac{3.55 \times 10^{-5}}{\text{min}} = 0$$
$$r = -3.55 \times 10^{-5}/\text{min}$$

The general form of the solution is given by

$$m(t) = Ae^{rt}$$

Substituting the root, r, gives

$$m(t) = Ae^{(-3.55 \times 10^{-5}/\text{min})t}$$

Apply the initial condition $m(0) = 90$ lbm at time $t = 0$.

$$m(0) = Ae^{(-3.55 \times 10^{-5}/\text{min})(0)} = 90 \text{ lbm}$$
$$Ae^0 = 90 \text{ lbm}$$
$$A = 90 \text{ lbm}$$

Therefore,

$$m(t) = (90 \text{ lbm}) e^{(-3.55 \times 10^{-5}/\text{min})t}$$

Solve for t.

$$\frac{m(t)}{90 \text{ lbm}} = e^{(-3.55 \times 10^{-5}/\text{min})t}$$

$$\ln\left(\frac{m(t)}{90 \text{ lbm}}\right) = \ln\left(e^{(-3.55 \times 10^{-5}/\text{min})t}\right)$$

$$= \left(\frac{-3.55 \times 10^{-5}}{\text{min}}\right)t$$

$$t = \frac{\ln\left(\dfrac{m(t)}{90 \text{ lbm}}\right)}{\dfrac{-3.55 \times 10^{-5}}{\text{min}}}$$

The initial mass of the water in the lagoon is given by

$$m_i = V\rho = (113{,}097 \text{ ft}^3)\left(62.4 \frac{\text{lbm}}{\text{ft}^3}\right)$$

$$= 7.05 \times 10^6 \text{ lbm}$$

The final mass of chemicals at a concentration of 1 ppb is

$$m_f = \frac{7.06 \times 10^6 \text{ lbm}}{1 \times 10^9}$$

$$= 7.06 \times 10^{-3} \text{ lbm}$$

Find the time required to achieve a mass of 7.06×10^{-3} lbm.

$$t = \left(\frac{\ln\left(\dfrac{m(t)}{90 \text{ lbm}}\right)}{\dfrac{-3.55 \times 10^{-5}}{\text{min}}}\right)\left(\frac{1 \text{ hr}}{60 \text{ min}}\right)\left(\frac{1 \text{ day}}{24 \text{ hr}}\right)$$

$$= \left(\frac{\ln\left(\dfrac{7.06 \times 10^{-3} \text{ lbm}}{90 \text{ lbm}}\right)}{\dfrac{-3.55 \times 10^{-5}}{\text{min}}}\right)\left(\frac{1 \text{ hr}}{60 \text{ min}}\right)\left(\frac{1 \text{ day}}{24 \text{ hr}}\right)$$

$$= \boxed{185 \text{ days} \quad (200 \text{ days})}$$

The answer is (D).

SI Solution

The differential equation is

$$m'(t) = a(t) - \frac{m(t)o(t)}{V(t)}$$

$a(t) =$ rate of addition of chemical

$m(t) =$ mass of chemical at time t

$o(t) =$ volumetric flow out of the lagoon, 115 L/min

$V(t) =$ volume in the lagoon at time t

Water flows into the lagoon at a rate of 115 L/min, and a water-chemical mix flows out of the lagoon at a rate of 115 L/min. Therefore, the volume of the lagoon at time t is equal to the initial volume.

$$V(t) = \left(\frac{\pi}{4}\right)(\text{diameter of lagoon})^2(\text{depth of lagoon})$$

$$= \left(\frac{\pi}{4}\right)(35 \text{ m})^2(3 \text{ m})$$

$$= 2886 \text{ m}^3$$

Using a conversion factor of 1000 L/m³ gives

$$o(t) = \frac{115 \dfrac{\text{L}}{\text{min}}}{1000 \dfrac{\text{L}}{\text{m}^3}}$$

$$= 0.115 \text{ m}^3/\text{min}$$

Substitute into the general form of the differential equation.

$$m'(t) = a(t) - \frac{m(t)o(t)}{V(t)}$$

$$= 0 - m(t)\left(\frac{0.115 \dfrac{\text{m}^3}{\text{min}}}{2886 \text{ m}^3}\right)$$

$$= -\left(\frac{3.985 \times 10^{-5}}{\text{min}}\right)m(t)$$

$$m'(t) + \left(\frac{3.985 \times 10^{-5}}{\text{min}}\right)m(t) = 0$$

The differential equation of the problem has the following characteristic equation.

$$r + \frac{3.985 \times 10^{-5}}{\text{min}} = 0$$

$$r = -3.985 \times 10^{-5}/\text{min}$$

The general form of the solution is given by

$$m(t) = A e^{rt}$$

Substituting in for the root, r, gives

$$m(t) = A e^{(-3.985 \times 10^{-5}/\text{min})t}$$

Apply the initial condition $m(0) = 40$ kg at time $t = 0$.

$$m(0) = Ae^{(-3.985 \times 10^{-5}/\text{min})(0)} = 40 \text{ kg}$$

$$Ae^0 = 40 \text{ kg}$$

$$A = 40 \text{ kg}$$

Therefore,

$$m(t) = (40 \text{ kg})e^{(-3.985 \times 10^{-5}/\text{min})t}$$

Solve for t.

$$\frac{m(t)}{40 \text{ kg}} = e^{(-3.985 \times 10^{-5}/\text{min})t}$$

$$\ln\left(\frac{m(t)}{40 \text{ kg}}\right) = \ln\left(e^{(-3.985 \times 10^{-5}/\text{min})t}\right)$$

$$= \left(\frac{-3.985 \times 10^{-5}}{\text{min}}\right)t$$

$$t = \frac{\ln\left(\dfrac{m(t)}{40 \text{ kg}}\right)}{\dfrac{-3.985 \times 10^{-5}}{\text{min}}}$$

The initial mass of water in the lagoon is given by

$$m_i = V\rho = (2886 \text{ m}^3)\left(1000 \ \frac{\text{kg}}{\text{m}^3}\right)$$

$$= 2.886 \times 10^6 \text{ kg}$$

The final mass of chemicals at a concentration of 1 ppb is

$$m_f = \frac{2.886 \times 10^6 \text{ kg}}{1 \times 10^9}$$

$$= 2.886 \times 10^{-3} \text{ kg}$$

Find the time required to achieve a mass of 2.886×10^{-3} kg.

$$t = \left(\frac{\ln\left(\dfrac{m(t)}{40 \text{ kg}}\right)}{\dfrac{-3.985 \times 10^{-5}}{\text{min}}}\right)\left(\frac{1 \text{ h}}{60 \text{ min}}\right)\left(\frac{1 \text{ d}}{24 \text{ h}}\right)$$

$$= \left(\frac{\ln\left(\dfrac{2.886 \times 10^{-3} \text{ kg}}{40 \text{ kg}}\right)}{\dfrac{-3.985 \times 10^{-5}}{\text{min}}}\right)\left(\frac{1 \text{ h}}{60 \text{ min}}\right)\left(\frac{1 \text{ d}}{24 \text{ h}}\right)$$

$$= \boxed{166 \text{ days} \quad (200 \text{ days})}$$

The answer is (D).

5. Let

$$m(t) = \text{mass of salt in tank at time } t$$

$$m_0 = 60 \text{ mass units}$$

$$m'(t) = \text{rate at which salt content is changing}$$

Two mass units of salt enter each minute, and three volumes leave each minute. The amount of salt leaving each minute is

$$\left(3 \ \frac{\text{vol}}{\text{min}}\right)\left(\text{concentration in } \frac{\text{mass}}{\text{vol}}\right)$$

$$= \left(3 \ \frac{\text{vol}}{\text{min}}\right)\left(\frac{\text{salt content}}{\text{volume}}\right)$$

$$= \left(3 \ \frac{\text{vol}}{\text{min}}\right)\left(\frac{m(t)}{100 - t}\right)$$

$$m'(t) = 2 - (3)\left(\frac{m(t)}{100 - t}\right) \text{ or } m'(t) + \frac{3m(t)}{100 - t}$$

$$= 2 \text{ mass/min}$$

This is a first-order linear differential equation. The integrating factor is

$$m = \exp\left(3\int \frac{dt}{100 - t}\right)$$

$$= \exp\left((3)\left(-\ln(100 - t)\right)\right)$$

$$= (100 - t)^{-3}$$

$$m(t) = (100 - t)^3\left(2\int \frac{dt}{(100 - t)^3} + k\right)$$

$$= 100 - t + (k)(100 - t)^3$$

But $m = 60$ mass units at $t = 0$, so $k = -0.00004$.

$$m(t) = 100 - t - (0.00004)(100 - t)^3$$

At $t = 60$ min,

$$m = 100 - 60 \text{ min} - (0.00004)(100 - 60 \text{ min})^3$$

$$= \boxed{37.44 \ (37) \text{ mass units}}$$

The answer is (B).

11 Probability and Statistical Analysis of Data

PRACTICE PROBLEMS

Probability

1. Four military recruits whose respective shoe sizes are 7, 8, 9, and 10 report to the supply clerk to be issued boots. The supply clerk selects one pair of boots in each of the four required sizes and hands them at random to the recruits.

(a) What is the probability that all recruits will receive boots of an incorrect size?

(A) 0.25

(B) 0.38

(C) 0.45

(D) 0.61

(b) What is the probability that exactly three recruits will receive boots of the correct size?

(A) 0

(B) 0.063

(C) 0.17

(D) 0.25

Probability Distributions

2. The time taken by a toll taker to collect the toll from vehicles crossing a bridge is an exponential distribution with a mean of 23 sec. What is the probability that a random vehicle will be processed in 25 sec or more (i.e., will take longer than 25 sec)?

(A) 0.17

(B) 0.25

(C) 0.34

(D) 0.52

3. The number of cars entering a toll plaza on a bridge during the hour after midnight follows a Poisson distribution with a mean of 20.

(a) What is the probability that exactly 17 cars will pass through the toll plaza during that hour on any given night?

(A) 0.076

(B) 0.12

(C) 0.16

(D) 0.23

(b) What is the percent probability that three or fewer cars will pass through the toll plaza at that hour on any given night?

(A) 0.00032%

(B) 0.0019%

(C) 0.079%

(D) 0.11%

4. A mechanical component exhibits a negative exponential failure distribution with a mean time to failure of 1000 hr. What is the maximum operating time such that the reliability remains above 99%?

(A) 3.3 hr

(B) 5.6 hr

(C) 8.1 hr

(D) 10 hr

5. (*Time limit: one hour*) A survey field crew measures one leg of a traverse four times. The following results are obtained.

repetition	measurement	direction
1	1249.529	forward
2	1249.494	backward
3	1249.384	forward
4	1249.348	backward

The crew chief is under orders to obtain readings with confidence limits of 90%.

(a) Which readings are acceptable?

(A) No readings are acceptable.

(B) Two readings are acceptable.

(C) Three readings are acceptable.

(D) All four readings are acceptable.

(b) Which readings are unacceptable?

(A) No readings are unacceptable.

(B) One reading is unacceptable.

(C) Two readings are unacceptable.

(D) All four readings are unacceptable.

(c) Explain how to determine which readings are unacceptable.

(A) Readings inside the 90% confidence limits are unacceptable.

(B) Readings outside the 90% confidence limits are unacceptable.

(C) Readings outside the upper 90% confidence limit are unacceptable.

(D) Readings outside the lower 90% confidence limit are unacceptable.

(d) What is the most probable value of the distance?

(A) 1249.399

(B) 1249.410

(C) 1249.439

(D) 1249.452

(e) What is the error in the most probable value (at 90% confidence)?

(A) 0.08

(B) 0.11

(C) 0.14

(D) 0.19

(f) If the distance is one side of a square traverse whose sides are all equal, what is the most probable closure error?

(A) 0.14

(B) 0.20

(C) 0.28

(D) 0.35

(g) What is the most probable error of part (f) expressed as a fraction?

(A) 1:17,600

(B) 1:14,200

(C) 1:12,500

(D) 1:10,900

(h) What is the order of accuracy of the closure?

(A) first order

(B) second order

(C) third order

(D) fourth order

(i) Define accuracy and distinguish it from precision.

(A) If an experiment can be repeated with identical results, the results are considered accurate.

(B) If an experiment has a small bias, the results are considered precise.

(C) If an experiment is precise, it cannot also be accurate.

(D) If an experiment is unaffected by experimental error, the results are accurate.

(j) Which of the following is an example of systematic error?

(A) measuring river depth as a motorized ski boat passes by

(B) using a steel tape that is too short to measure consecutive distances

(C) locating magnetic north near a large iron ore deposit along an overland route

(D) determining local wastewater BOD after a toxic spill

Statistical Analysis

6. (*Time limit: one hour*) California law requires a statistical analysis of the average speed driven by motorists on a road prior to the use of radar speed control. The following speeds (all in mph) were observed in a random sample of 40 cars.

44, 48, 26, 25, 20, 43, 40, 42, 29, 39, 23, 26, 24, 47, 45, 28, 29, 41, 38, 36, 27, 44, 42, 43, 29, 37, 34, 31, 33, 30, 42, 43, 28, 41, 29, 36, 35, 30, 32, 31

(a) Tabulate the frequency distribution and the cumulative frequency distribution of the data.

(b) Draw the frequency histogram.

(c) Draw the frequency polygon.

(d) Draw the cumulative frequency graph.

(e) What is the upper quartile speed?

   (A) 30 mph

   (B) 35 mph

   (C) 40 mph

   (D) 45 mph

(f) What is the mean speed?

   (A) 31 mph

   (B) 33 mph

   (C) 35 mph

   (D) 37 mph

(g) What is the standard deviation of the sample data?

   (A) 2.1 mph

   (B) 6.1 mph

   (C) 6.8 mph

   (D) 7.4 mph

(h) What is the sample standard deviation?

   (A) 7.5 mph

   (B) 18 mph

   (C) 35 mph

   (D) 56 mph

(i) What is the sample variance?

   (A) 56 mi^2/hr^2

   (B) 324 mi^2/hr^2

   (C) 1225 mi^2/hr^2

   (D) 3136 mi^2/hr^2

7. A spot speed study is conducted for a stretch of roadway. During a normal day, the speeds were found to be normally distributed with a mean of 46 and a standard deviation of 3.

(a) What is the 50th percentile speed?

   (A) 39

   (B) 43

   (C) 46

   (D) 49

(b) What is the 85th percentile speed?

   (A) 47.1

   (B) 48.3

   (C) 49.1

   (D) 52.7

(c) What is the upper two standard deviation speed?

   (A) 47.2

   (B) 49.3

   (C) 51.1

   (D) 52.0

(d) The daily average speeds for the same stretch of roadway on consecutive normal days were determined by sampling 25 vehicles each day. What is the upper two-standard deviation average speed?

   (A) 46.6

   (B) 47.2

   (C) 52.0

   (D) 54.7

8. The diameters of bolt holes drilled in structural steel members are normally distributed with a mean of 0.502 in and a standard deviation of 0.005 in. Holes are out of specification if their diameters are less than 0.497 in or more than 0.507 in.

(a) What is the probability that a hole chosen at random will be out of specification?

   (A) 0.16

   (B) 0.22

   (C) 0.32

   (D) 0.68

(b) What is the probability that two holes out of a sample of 15 will be out of specification?

   (A) 0.074

   (B) 0.12

   (C) 0.15

   (D) 0.32

Hypothesis Testing

9. 100 bearings were tested to failure. The average life was 1520 hours, and the standard deviation was 120 hours. The manufacturer claims a 1600 hour life. Evaluate using confidence limits of 95% and 99%.

(A) The claim is accurate at both 95% and 99% confidence.

(B) The claim is inaccurate only at 95%.

(C) The claim is inaccurate only at 99%.

(D) The claim is inaccurate at both 95% and 99% confidence.

Curve Fitting

10. (a) Find the best equation for a line passing through the points given. (b) Find the correlation coefficient.

x	y
400	370
800	780
1250	1210
1600	1560
2000	1980
2500	2450
4000	3950

11. Find the best equation for a line passing through the points given.

s	t
20	43
18	141
16	385
14	1099

12. The number of vehicles lining up behind a flashing railroad crossing has been observed for five trains of different lengths, as given. What is the mathematical formula that relates the two variables?

no. of cars in train	no. of vehicles
2	14.8
5	18.0
8	20.4
12	23.0
27	29.9

13. The following yield data are obtained from five identical treatment plants. (a) Develop a mathematical equation to correlate the yield and average temperature. (b) What is the correlation coefficient?

treatment plant	average temperature, T	average yield, Y
1	207.1	92.30
2	210.3	92.58
3	200.4	91.56
4	201.1	91.63
5	203.4	91.83

14. The following data are obtained from a soil compaction test. What is the mathematical formula that relates the two variables?

x	y
-1	0
0	1
1	1.4
2	1.7
3	2
4	2.2
5	2.4
6	2.6
7	2.8
8	3

15. Two resistances, the meter resistance and a shunt resistor, are connected in parallel in an ammeter. Most of the current passing through the meter goes through the shunt resistor. In order to determine the accuracy of the resistance of shunt resistors being manufactured for a line of ammeters, a manufacturer tests a sample of 100 shunt resistors. The numbers of shunt resistors with the resistance indicated (to the nearest hundredth of an ohm) are as follows.

0.200 Ω, 1; 0.210 Ω, 3; 0.220 Ω, 5; 0.230 Ω, 10; 0.240 Ω, 17; 0.250 Ω, 40; 0.260 Ω, 13; 0.270 Ω, 6; 0.280 Ω, 3; 0.290 Ω, 2.

(a) What is the mean resistance?

(A) 0.235 Ω

(B) 0.247 Ω

(C) 0.251 Ω

(D) 0.259 Ω

(b) What is the sample standard deviation?

(A) 0.0003 Ω

(B) 0.010 Ω

(C) 0.016 Ω

(D) 0.24 Ω

(c) What is the median resistance?

(A) 0.22 Ω

(B) 0.24 Ω

(C) 0.25 Ω

(D) 0.26 Ω

(d) What is the sample variance?

(A) 0.00027 Ω^2

(B) 0.0083 Ω^2

(C) 0.0114 Ω^2

(D) 0.0163 Ω^2

SOLUTIONS

1. (a) There are 4! = 24 different possible outcomes. By enumeration, there are 9 completely wrong combinations.

$$p\{\text{all wrong}\} = \frac{9}{24} = \boxed{0.375 \quad (38)}$$

sizes

correct →	7	8	9	10	all wrong
sizes issued	7	8	9	10	
	7	8	10	9	
	7	9	8	10	
	7	9	10	8	
	7	10	8	9	
	7	10	9	8	
	8	9	10	7	X
	8	9	7	10	
	8	10	9	7	
	8	10	7	9	X
	8	7	9	10	
	8	7	10	9	X
	9	10	7	8	X
	9	10	8	7	X
	9	7	10	8	X
	9	7	8	10	
	9	8	7	10	
	9	8	10	7	
	10	7	8	9	X
	10	7	9	8	
	10	8	7	9	
	10	8	9	7	
	10	9	8	7	X
	10	9	7	8	X

The answer is (B).

(b) If three recruits get the correct size, the fourth recruit will also since there will be only one pair remaining.

$$p\{\text{exactly 3}\} = \boxed{0}$$

The answer is (A).

2. For an exponential distribution function, the mean is given as

$$\mu = \frac{1}{\lambda}$$

Using Eq. 11.39, for a mean of 23,

$$\mu = 23 = \frac{1}{\lambda}$$

$$\lambda = \frac{1}{23} = 0.0435$$

Using Eq. 11.38, for an exponential distribution function,

$$p = F(x) = 1 - e^{-\lambda x}$$
$$p\{X < x\} = 1 - p$$
$$p\{X > x\} = 1 - F(x)$$
$$= 1 - (1 - e^{-\lambda x})$$
$$= e^{-\lambda x}$$

The probability of a random vehicle being processed in 25 sec or more is given by

$$p\{x > 25\} = e^{-(0.0435)(25)}$$
$$= \boxed{0.337 \quad (0.34)}$$

The answer is (C).

3. (a) The distribution is a Poisson distribution with an average of $\lambda = 20$.

The probability for a Poisson distribution is given by Eq. 11.51.

$$p\{x\} = f(x) = \frac{e^{-\lambda}\lambda^x}{x!}$$

Therefore, the probability of 17 cars is

$$p\{x = 17\} = f(17) = \frac{e^{-20}20^{17}}{17!}$$
$$= \boxed{0.076}$$

The answer is (A).

(b) The probability of three or fewer cars is given by

$$p\{x \le 3\} = p\{x = 0\} + p\{x = 1\} + p\{x = 2\}$$
$$+ p\{x = 3\}$$
$$= f(0) + f(1) + f(2) + f(3)$$
$$= \frac{e^{-20}20^0}{0!} + \frac{e^{-20}20^1}{1!}$$
$$+ \frac{e^{-20}20^2}{2!} + \frac{e^{-20}20^3}{3!}$$
$$= 2 \times 10^{-9} + 4.1 \times 10^{-8}$$
$$+ 4.12 \times 10^{-7} + 2.75 \times 10^{-6}$$
$$= 3.2 \times 10^{-6}$$
$$= \boxed{0.0000032 \quad (0.00032\%)}$$

The answer is (A).

4. Using Eq. 11.48,

$$\lambda = \frac{1}{\text{MTTF}} = \frac{1}{1000} = 0.001$$

Using Eq. 11.49, the reliability function is

$$R\{t\} = e^{-\lambda t} = e^{-0.001t}$$

Since the reliability is greater than 99%,

$$e^{-0.001t} > 0.99$$
$$\ln e^{-0.001t} > \ln 0.99$$
$$-0.001t > \ln 0.99$$
$$t < -1000 \ln 0.99$$
$$t < 10.05 \quad (10)$$

The maximum operating time such that the reliability remains above 99% is $\boxed{10 \text{ hr.}}$

The answer is (D).

5. (a) Find the average using Eq. 11.56.

$$\bar{x} = \frac{\sum x_i}{n}$$
$$= \frac{1249.529 + 1249.494 + 1249.384 + 1249.348}{4}$$
$$= 1249.439$$

Since the sample population is small, use Eq. 11.61 to find the sample standard deviation.

$$s = \sqrt{\frac{\sum(x_i - \bar{x})^2}{n-1}}$$

$$= \sqrt{\frac{\begin{matrix}(1249.529 - 1249.439)^2 \\ + (1249.494 - 1249.439)^2 \\ + (1249.384 - 1249.439)^2 \\ + (1249.348 - 1249.439)^2\end{matrix}}{4-1}}$$

$$= 0.08647$$

From Table 11.1 or App. 11.A, a two-tail 90% confidence limit falls within $1.645s$ of $\bar{x}$.

$$1249.439 \pm (1.645)(0.08647) = 1249.439 \pm 0.142$$

Therefore, (1249.297, 1249.581) is the 90% confidence range.

By observation, all the readings fall within the 90% confidence range.

The answer is (D).

(b) No readings are unacceptable.

The answer is (A).

(c) Readings outside the 90% confidence limits are unacceptable.

The answer is (B).

(d) The unbiased estimate of the most probable distance is $\boxed{1249.439.}$

The answer is (C).

(e) The error in the most probable value for the 90% confidence range is $\boxed{0.142 \ (0.14).}$

The answer is (C).

(f) If the surveying crew places a marker, measures a distance x, places a second marker, and then measures the same distance x back to the original marker, the ending point should coincide with the original marker. If, due to measurement errors, the ending and starting points do not coincide, the difference is the closure error.

In this example, the survey crew moves around the four sides of a square, so there are two measurements in the x-direction and two measurements in the y-direction. If the errors E_1 and E_2 are known for two measurements,

x_1 and x_2, the error associated with the sum or difference $x_1 \pm x_2$ is

$$E\{x_1 \pm x_2\} = \sqrt{E_1^2 + E_2^2}$$

In this case, the error in the x-direction is

$$E_x = \sqrt{(0.1422)^2 + (0.1422)^2}$$
$$= 0.2011$$

The error in the y-direction is calculated the same way and is also 0.2011. E_x and E_y are combined by the Pythagorean theorem to yield

$$E_{\text{closure}} = \sqrt{(0.2011)^2 + (0.2011)^2}$$
$$= \boxed{0.2844 \quad (0.28)}$$

The answer is (C).

(g) In surveying, error may be expressed as a fraction of one or more legs of the traverse. Assume that the total of all four legs is to be used as the basis.

$$\frac{0.2844}{(4)(1249)} = \boxed{\frac{1}{17,567} \quad \left(\frac{1}{17,600}\right)}$$

The answer is (A).

(h) In surveying, a class 1 third-order error is smaller than 1/10,000. The error of 1/17,567 is smaller than the third-order error; therefore, the error is within the third-order accuracy.

The answer is (C).

(i) An experiment is accurate if it is unchanged by experimental error. Precision is concerned with the repeatability of the experimental results. If an experiment is repeated with identical results, the experiment is said to be precise. However, it is possible to have a highly precise experiment with a large bias.

The answer is (D).

(j) A systematic error is one that is always present and is unchanged from sample to sample. For example, a steel tape that is 0.02 ft short introduces a systematic error.

The answer is (B).

6. (a) Tabulate the frequency distribution and the cumulative frequency distribution of the data.

The lowest speed is 20 mph and the highest speed is 48 mph; therefore, the range is 28 mph. Choose 10 cells with a width of 3 mph.

midpoint	interval (mph)	frequency	cumulative frequency	cumulative percent
21	20–22	1	1	3
24	23–25	3	4	10
27	26–28	5	9	23
30	29–31	8	17	43
33	32–34	3	20	50
36	35–37	4	24	60
39	38–40	3	27	68
42	41–43	8	35	88
45	44–46	3	38	95
48	47–49	2	40	100

(b) Draw the frequency histogram.

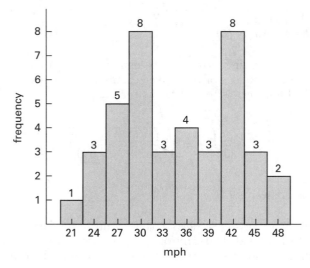

(c) Draw the frequency polygon.

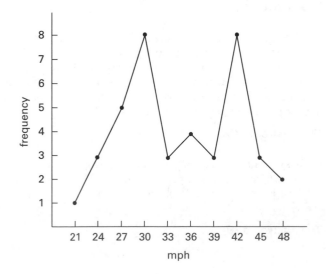

(d) Use the table in part (a).

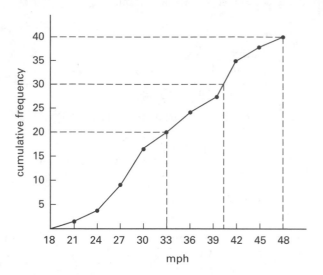

(e) From the cumulative frequency graph in part (d), the upper quartile speed occurs at 30 cars or 75%, which corresponds to approximately 40 mph.

The answer is (C).

(f) Calculate the following quantities.

$$\sum x_i = 1390 \text{ mi/hr}$$
$$n = 40$$

The mean is computed using Eq. 11.56.

$$\overline{x} = \frac{\sum x_i}{n} = \frac{1390 \; \frac{\text{mi}}{\text{hr}}}{40}$$
$$= \boxed{34.75 \text{ mph} \quad (35 \text{ mph})}$$

The answer is (C).

(g) The standard deviation of the sample data is given by Eq. 11.60.

$$\sigma = \sqrt{\frac{\sum x^2}{n} - \mu^2}$$
$$\sum x^2 = 50{,}496 \text{ mi}^2/\text{hr}^2$$

Use the sample mean as an unbiased estimator of the population mean, μ.

$$\sigma = \sqrt{\frac{\sum x^2}{n} - \mu^2}$$

$$= \sqrt{\frac{50{,}496 \dfrac{\text{mi}^2}{\text{hr}^2}}{40} - \left(34.75 \dfrac{\text{mi}}{\text{hr}}\right)^2}$$

$$= \boxed{7.405 \text{ mph} \quad (7.4 \text{ mph})}$$

The answer is (D).

(h) The sample standard deviation is given by

$$s = \sqrt{\frac{\sum x^2 - \dfrac{\left(\sum x\right)^2}{n}}{n-1}}$$

$$= \sqrt{\frac{50{,}496 \dfrac{\text{mi}^2}{\text{hr}^2} - \dfrac{\left(1390 \dfrac{\text{mi}}{\text{hr}}\right)^2}{40}}{40-1}}$$

$$= \boxed{7.5 \text{ mph}}$$

The answer is (A).

(i) The sample variance is given by the square of the sample standard deviation.

$$s^2 = \left(7.5 \dfrac{\text{mi}}{\text{hr}}\right)^2$$

$$= \boxed{56.25 \text{ mi}^2/\text{hr}^2 \quad (56 \text{ mi}^2/\text{hr}^2)}$$

The answer is (A).

7. (a) The 50th percentile speed is the median speed, $\boxed{46,}$ which for a symmetrical normal distribution is the mean speed.

The answer is (C).

(b) The 85th percentile speed is the speed that is exceeded by only 15% of the measurements. Since this is a normal distribution, App. 11.A can be used. 15% in the upper tail corresponds to 35% between the mean and the 85th percentile. From App. 11.A, this occurs at approximately 1.04σ. The 85th percentile speed is

$$x_{85\%} = \mu + 1.04\sigma = 46 + (1.04)(3)$$

$$= \boxed{49.12 \quad (49.1)}$$

The answer is (C).

(c) The upper 2σ speed is

$$x_{2\sigma} = \mu + 2\sigma = 46 + (2)(3)$$

$$= \boxed{52.0 \quad (52)}$$

The answer is (D).

(d) According to the central limit theorem, the mean of the average speeds is the same as the distribution mean, and the standard deviation of sample means (from Eq. 11.66) is

$$s_{\bar{x}} = \frac{\sigma_x}{\sqrt{n}} = \frac{3}{\sqrt{25}}$$

$$= 0.6$$

The upper two-standard deviation average speed is

$$\bar{x}_{2\sigma} = \mu + 2\sigma_{\bar{x}} = 46 + (2)(0.6)$$

$$= \boxed{47.2}$$

The answer is (B).

8. (a) From Eq. 11.43,

$$z = \frac{x_o - \mu}{6}$$

$$z_{\text{upper}} = \frac{0.507 \text{ in} - 0.502 \text{ in}}{0.005 \text{ in}} = +1$$

From App. 11.A, the area outside $z = +1$ is

$$0.5 - 0.3413 = 0.1587$$

Since these are symmetrical limits, $z_{\text{lower}} = -1$.

$$\text{total fraction defective} = (2)(0.1587)$$

$$= \boxed{0.3174 \quad (0.32)}$$

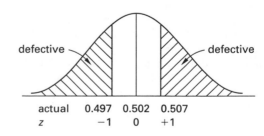

The answer is (C).

(b) This is a binomial problem.

$$p = p\{\text{defective}\} = 0.3174$$

$$q = 1 - p = 0.6826$$

From Eq. 11.28,

$$p\{x\} = f(x) = \binom{n}{x}\hat{p}^x\hat{q}^{n-x}$$

$$f(2) = \binom{15}{2}(0.3174)^2(0.6826)^{13}$$

$$= \left(\frac{15!}{13!2!}\right)(0.3174)^2(0.6826)^{13}$$

$$= \boxed{0.0739 \quad (0.074)}$$

The answer is (A).

9. This is a typical hypothesis test of two sample population means. The two populations are the original population the manufacturer used to determine the 1600 hr average life value and the new population the sample was taken from. The mean ($\overline{x} = 1520$ hr) of the sample and its standard deviation ($s = 120$ hr) are known, but the mean and standard deviation of a population of average lifetimes are unknown.

Assume that the average lifetime population mean and the sample mean are identical.

$$\overline{x} = \mu = 1520 \text{ hr}$$

The standard deviation of the average lifetime population is

$$\sigma_{\overline{x}} = \frac{s}{\sqrt{n}} = \frac{120 \text{ hr}}{\sqrt{100}} = 12 \text{ hr}$$

The manufacturer can be reasonably sure that the claim of a 1600 hr average life is justified if the average test life is near 1600 hr. "Reasonably sure" must be evaluated based on acceptable probability of being incorrect. If the manufacturer is willing to be wrong with a 5% probability, then a 95% confidence level is required.

Since the direction of bias is known, a one-tailed test is required. To determine if the mean has shifted downward, test the hypothesis that 1600 hr is within the 95% limit of a distribution with a mean of 1520 hr and a standard deviation of 12 hr. From a standard normal table, 5% of a standard normal distribution is outside of $z = 1.645$. Therefore, the 95% confidence limit is

$$1520 \text{ hr} + (1.645)(12 \text{ hr}) = 1540 \text{ hr}$$

The manufacturer can be 95% certain that the average lifetime of the bearings is less than 1600 hr.

If the manufacturer is willing to be wrong with a probability of only 1%, then a 99% confidence limit is

required. From the normal table, $z = 2.33$, and the 99% confidence limit is

$$1520 \text{ hr} + (2.33)(12 \text{ hr}) = 1548 \text{ hr}$$

The manufacturer can be 99% certain that the average bearing life is less than 1600 hr.

The answer is (D).

10. (a) Plot the data points to determine if the relationship is linear.

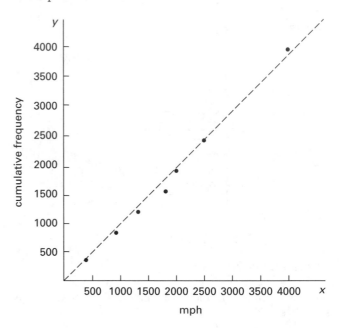

The data appear to be essentially linear. The slope, m, and the y-intercept, b, can be determined using linear regression.

The individual terms are

$$n = 7$$

$$\sum x_i = 400 + 800 + 1250 + 1600 + 2000$$
$$+ 2500 + 4000$$
$$= 12{,}550$$

$$\left(\sum x_i\right)^2 = (12{,}550)^2$$
$$= 1.575 \times 10^8$$

From Eq. 11.56,

$$\overline{x} = \frac{\sum x_i}{n} = \frac{12{,}550}{7}$$
$$= 1792.9$$

$$\sum x_i^2 = (400)^2 + (800)^2 + (1250)^2 + (1600)^2$$
$$+ (2000)^2 + (2500)^2 + (4000)^2$$
$$= 3.117 \times 10^7$$

Similarly,

$$\sum y_i = 370 + 780 + 1210 + 1560 + 1980$$
$$+ 2450 + 3950$$
$$= 12{,}300$$

$$\left(\sum y_i\right)^2 = (12{,}300)^2 = 1.513 \times 10^8$$

$$\overline{y} = \frac{\sum y_i}{n} = \frac{12{,}300}{7} = 1757.1$$

$$\sum y_i^2 = (370)^2 + (780)^2 + (1210)^2 + (1560)^2$$
$$+ (1980)^2 + (2450)^2 + (3950)^2$$
$$= 3.017 \times 10^7$$

Also,

$$\sum x_i y_i = (400)(370) + (800)(780) + (1250)(1210)$$
$$+ (1600)(1560) + (2000)(1980)$$
$$+ (2500)(2450) + (4000)(3950)$$
$$= 3.067 \times 10^7$$

Using Eq. 11.75, the slope is

$$m = \frac{n\sum x_i y_i - \sum x_i \sum y_i}{n\sum x_i^2 - \left(\sum x_i\right)^2}$$
$$= \frac{(7)(3.067 \times 10^7) - (12{,}550)(12{,}300)}{(7)(3.117 \times 10^7) - (12{,}550)^2}$$
$$= 0.994$$

Using Eq. 11.76, the y-intercept is

$$b = \overline{y} - m\overline{x}$$
$$= 1757.1 - (0.994)(1792.9)$$
$$= -25.0$$

The least squares equation of the line is

$$y = mx + b$$
$$= \boxed{0.994x - 25.0}$$

(b) Using Eq. 11.77, the correlation coefficient is

$$r = \frac{n\sum x_i y_i - \sum x_i \sum y_i}{\sqrt{\left(n\sum x_i^2 - \left(\sum x_i\right)^2\right)\left(n\sum y_i^2 - \left(\sum y_i\right)^2\right)}}$$
$$= \frac{(7)(3.067 \times 10^7) - (12{,}500)(12{,}300)}{\sqrt{\begin{array}{c}\left((7)(3.117 \times 10^7) - (12{,}500)^2\right)\\ \times \left((7)(3.017 \times 10^7) - (12{,}300)^2\right)\end{array}}}$$
$$\approx \boxed{1.00}$$

11. Plotting the data shows that the relationship is nonlinear.

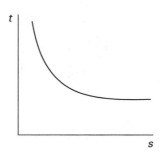

This appears to be an exponential with the form

$$t = ae^{bs}$$

Take the natural log of both sides.

$$\ln t = \ln ae^{bs}$$
$$= \ln a + \ln e^{bs}$$
$$= \ln a + bs$$

But, $\ln a$ is just a constant, c.

$$\ln t = c + bs$$

Make the transformation $R = \ln t$.

$$R = c + bs$$

s	R
20	3.76
18	4.95
16	5.95
14	7.00

This is linear.

$$n = 4$$

$$\sum s_i = 20 + 18 + 16 + 14 = 68$$

$$\bar{s} = \frac{\sum s}{n} = \frac{68}{4} = 17$$

$$\sum s_i^2 = (20)^2 + (18)^2 + (16)^2 + (14)^2 = 1176$$

$$\left(\sum s_i\right)^2 = (68)^2 = 4624$$

$$\sum R_i = 3.76 + 4.95 + 5.95 + 7.00 = 21.66$$

$$\bar{R} = \frac{\sum R_i}{n} = \frac{21.66}{4} = 5.415$$

$$\sum R_i^2 = (3.76)^2 + (4.95)^2 + (5.95)^2 + (7.00)^2$$
$$= 123.04$$

$$\left(\sum R_i\right)^2 = (21.66)^2 = 469.16$$

$$\sum s_i R_i = (20)(3.76) + (18)(4.95) + (16)(5.95)$$
$$+ (14)(7.00)$$
$$= 357.5$$

The slope, b, of the transformed line is

$$b = \frac{n\sum s_i R_i - \sum s_i \sum R_i}{n\sum s_i^2 - \left(\sum s_i\right)^2}$$

$$= \frac{(4)(357.5) - (68)(21.66)}{(4)(1176) - (68)^2}$$

$$= -0.536$$

The intercept is

$$c = \bar{R} - b\bar{s} = 5.415 - (-0.536)(17)$$
$$= 14.527$$

The transformed equation is

$$R = c + bs$$
$$= 14.527 - 0.536s$$
$$\boxed{\ln t = 14.527 - 0.536s}$$

12. The first step is to graph the data.

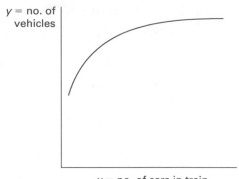

It is assumed that the relationship between the variables has the form $y = a + b\log x$. Therefore, the variable change $z = \log x$ is made, resulting in the following set of data.

z	y
0.301	14.8
0.699	18.0
0.903	20.4
1.079	23.0
1.431	29.9

$$\sum z_i = 4.413$$

$$\sum y_i = 106.1$$

$$\sum z_i^2 = 4.6082$$

$$\sum y_i^2 = 2382.2$$

$$\left(\sum z_i\right)^2 = 19.475$$

$$\left(\sum y_i\right)^2 = 11{,}257.2$$

$$\bar{z} = 0.8826$$

$$\bar{y} = 21.22$$

$$\sum z_i y_i = 103.06$$

$$n = 5$$

Using Eq. 11.75, the slope is

$$m = \frac{n\sum z_i y_i - \sum z_i \sum y_i}{n\sum z_i^2 - \left(\sum z_i\right)^2}$$

$$= \frac{(5)(103.06) - (4.413)(106.1)}{(5)(4.6082) - 19.475}$$

$$= 13.20$$

The y-intercept is

$$b = \bar{y} - m\bar{z}$$
$$= 21.22 - (13.20)(0.8826)$$
$$= 9.570$$

The resulting equation is

$$y = 9.570 + 13.20z$$

The relationship between x and y is approximately

$$\boxed{y = 9.570 + 13.20\log x}$$

This is not an optimal correlation, as better correlation coefficients can be obtained if other assumptions about the form of the equation are made. For example, $y = 9.1 + 4\sqrt{x}$ has a better correlation coefficient.

13. (a) Plot the data to verify that they are linear.

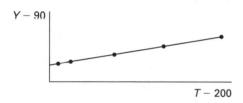

x	y
$T - 200$	$Y - 90$
7.1	2.30
10.3	2.58
0.4	1.56
1.1	1.63
3.4	1.83

step 1: Calculate the following quantities.

$$\sum x_i = 22.3 \qquad \sum y_i = 9.9$$
$$\sum x_i^2 = 169.43 \qquad \sum y_i^2 = 20.39$$
$$\left(\sum x_i\right)^2 = 497.29 \qquad \left(\sum y_i\right)^2 = 98.01$$
$$\bar{x} = \frac{22.3}{5} = 4.46 \qquad \bar{y} = 1.98$$
$$\sum x_i y_i = 51.54$$

step 2: From Eq. 11.75, the slope is

$$m = \frac{n\sum x_i y_i - \sum x_i \sum y_i}{n\sum x_i^2 - \left(\sum x_i\right)^2}$$
$$= \frac{(5)(51.54) - (22.3)(9.9)}{(5)(169.43) - 497.29}$$
$$= 0.1055$$

step 3: From Eq. 11.76, the y-intercept is

$$b = \bar{y} - m\bar{x}$$
$$= 1.98 - (0.1055)(4.46)$$
$$= 1.509$$

The equation of the line is

$$y = mx + b$$
$$= 0.1055x + 1.509$$
$$Y - 90 = (0.1055)(T - 200) + 1.509$$
$$Y = \boxed{0.1055\,T + 70.409}$$

(b) *step 4:* Use Eq. 11.77 to get the correlation coefficient.

$$r = \frac{n\sum x_i y_i - \sum x_i \sum y_i}{\sqrt{\left(n\sum x_i^2 - \left(\sum x_i\right)^2\right)\left(n\sum y_i^2 - \left(\sum y_i\right)^2\right)}}$$
$$= \frac{(5)(51.54) - (22.3)(9.9)}{\sqrt{\left((5)(169.43) - 497.29\right)\left((5)(20.39) - 98.01\right)}}$$
$$= \boxed{0.995}$$

14. Plot the data to see if they are linear.

This looks like it could be of the form

$$y = a + b\sqrt{x}$$

However, when x is negative (as in the first point), the function is imaginary. Try shifting the curve to the right, replacing x with $x + 1$.

$$y = a + bz$$

$$z = \sqrt{x + 1}$$

z	y
0	0
1	1
1.414	1.4
1.732	1.7
2	2
2.236	2.2
2.45	2.4
2.65	2.6
2.83	2.8
3	3

Since $y \approx z$, the relationship is

$$\boxed{y = \sqrt{x + 1}}$$

In this problem, the answer was found accidentally. Usually, regression would be necessary.

15. (a) For convenience, tabulate the frequency-weighted values of R and R^2.

R	f	fR	fR^2
0.200	1	0.200	0.0400
0.210	3	0.360	0.1323
0.220	5	1.100	0.2420
0.230	10	2.300	0.5290
0.240	17	4.080	0.9792
0.250	40	10.000	2.5000
0.260	13	3.380	0.8788
0.270	6	1.620	0.4374
0.280	3	0.840	0.2352
0.290	2	0.580	0.1682
	100	24.730	6.1421

The mean resistance is

$$\overline{R} = \frac{\sum fR}{\sum f} = \frac{24.730 \ \Omega}{100}$$

$$= \boxed{0.2473 \ \Omega \quad (0.247 \ \Omega)}$$

The answer is (B).

(b) The sample standard deviation is given by Eq. 11.61.

$$s = \sqrt{\frac{\sum fR^2 - \dfrac{\left(\sum fR\right)^2}{n}}{n - 1}}$$

$$= \sqrt{\frac{6.1421 \ \Omega^2 - \dfrac{(24.73 \ \Omega)^2}{100}}{99}}$$

$$= \boxed{0.0163 \ \Omega \quad (0.016 \ \Omega)}$$

The answer is (C).

(c) The 50th and 51st values are both 0.25 Ω. The median is $\boxed{0.25 \ \Omega.}$

The answer is (C).

(d) The sample variance is

$$s^2 = (0.0163 \ \Omega)^2 = \boxed{0.0002656 \ \Omega^2 \quad (0.00027 \ \Omega^2)}$$

The answer is (A).

12 Numerical Analysis

PRACTICE PROBLEMS

1. A function is given as $y = 3x^{0.93} + 4.2$. What is the percent relative error if the value of y at $x = 2.7$ is found by using straight-line interpolation between $x = 2$ and $x = 3$?

(A) 0.06%

(B) 0.18%

(C) 2.5%

(D) 5.4%

2. Given the following data points, find y by straight-line interpolation for $x = 2.75$.

x	y
1	4
2	6
3	2
4	−14

(A) 2.1

(B) 2.4

(C) 2.7

(D) 3.0

3. Using the bisection method, find all of the roots of $f(x) = 0$ to the nearest 0.000005.

$$f(x) = x^3 + 2x^2 + 8x - 2$$

4. The increase in concentration of mixed-liquor suspended solids (MLSS) in an activated sludge aeration tank as a function of time is given in the table. Use a second-order Lagrangian interpolation to estimate the MLSS after 16 min of aeration.

t (min)	MLSS (mg/L)
0	0
10	227
15	362
20	517
22.5	602
30	901

SOLUTIONS

1. The actual value at $x = 2.7$ is given by

$$y(x) = 3x^{0.93} + 4.2$$
$$y(2.7) = (3)(2.7)^{0.93} + 4.2$$
$$= 11.756$$

At $x = 3$,

$$y(3) = (3)(3)^{0.93} + 4.2$$
$$= 12.534$$

At $x = 2$,

$$y(2) = (3)(2)^{0.93} + 4.2$$
$$= 9.916$$

Use straight-line interpolation.

$$\frac{x_2 - x}{x_2 - x_1} = \frac{y_2 - y}{y_2 - y_1}$$
$$\frac{3 - 2.7}{3 - 2} = \frac{12.534 - y}{12.534 - 9.916}$$
$$y = 11.749$$

The relative error is given by

$$\frac{\text{actual value} - \text{predicted value}}{\text{actual value}} = \frac{11.756 - 11.749}{11.756}$$
$$= \boxed{0.0006 \quad (0.06\%)}$$

The answer is (A).

2. Let $x_1 = 2$; therefore, from the table of data points, $y_1 = 6$. Let $x_2 = 3$; therefore, from the table of data points, $y_2 = 2$.

Let $x = 2.75$. By straight-line interpolation,

$$\frac{x_2 - x}{x_2 - x_1} = \frac{y_2 - y}{y_2 - y_1}$$

$$\frac{3 - 2.75}{3 - 2} = \frac{2 - y}{2 - 6}$$

$$\boxed{y = 3}$$

The answer is (D).

3. Use the equation $f(x) = x^3 + 2x^2 + 8x - 2$ to try to find an interval in which there is a root.

x	$f(x)$
0	−2
1	9

A root exists in the interval $[0, 1]$.

Try $x = \frac{1}{2}(0 + 1) = 0.5$.

$$f(0.5) = (0.5)^3 + (2)(0.5)^2 + (8)(0.5) - 2 = 2.625$$

A root exists in $[0, 0.5]$.

Try $x = 0.25$.

$$f(0.25) = (0.25)^3 + (2)(0.25)^2 + (8)(0.25) - 2 = 0.1406$$

A root exists in $[0, 0.25]$.

Try $x = 0.125$.

$$f(0.125) = (0.125)^3 + (2)(0.125)^2 + (8)(0.125) - 2$$
$$= -0.967$$

A root exists in $[0.125, 0.25]$.

Try $x = \frac{1}{2}(0.125 + 0.25) = 0.1875$.

Continuing,

$$f(0.1875) = -0.42 \quad [0.1875, 0.25]$$
$$f(0.21875) = -0.144 \quad [0.21875, 0.25]$$
$$f(0.234375) = -0.002 \quad [\text{This is close enough.}]$$

One root is $x_1 \approx \boxed{0.234375}$.

Try to find the other two roots. Use long division to factor the polynomial.

$$
\begin{array}{r}
x^2 + 2.234375x + 8.52368 \\
x - 0.234375 \overline{\smash{)}\ x^3 + \quad 2x^2 + \quad\quad 8x - 2} \\
\underline{-(x^3 - 0.234375x^2)} \\
2.234375x^2 + \quad\quad 8x \\
\underline{-(2.234375x^2 - 0.52368x)} \\
8.52368x - 2 \\
\underline{-(8.52368x - 1.9977)} \\
\approx 0
\end{array}
$$

Use the quadratic equation to find the roots of $x^2 + 2.234375x + 8.52368$.

$$x_2, x_3 = \frac{-2.234375 \pm \sqrt{(2.234375)^2 - (4)(1)(8.52368)}}{(2)(1)}$$

$$= \boxed{-1.117189 \pm i2.697327} \quad [\text{both imaginary}]$$

4. Choose the three data points that bracket $t = 16$ as closely as possible. These three points are $t_0 = 10$, $t_1 = 15$, and $t_2 = 20$.

	$i = 0$	$i = 1$	$i = 2$
$k = 0$: $S_0(16) = -0.08$	$\left(\frac{16 - 10}{10 - 10}\right)$	$\left(\frac{16 - 15}{10 - 15}\right)$	$\left(\frac{16 - 20}{10 - 20}\right)$
$k = 1$: $S_1(16) = 0.96$	$\left(\frac{16 - 10}{15 - 10}\right)$	$\left(\frac{16 - 15}{15 - 15}\right)$	$\left(\frac{16 - 20}{15 - 20}\right)$
$k = 2$: $S_2(16) = 0.12$	$\left(\frac{16 - 10}{20 - 10}\right)$	$\left(\frac{16 - 15}{20 - 15}\right)$	$\left(\frac{16 - 20}{20 - 20}\right)$

Use Eq. 12.3.

$$S = S(10)S_0(16) + S(15)S_1(16) + S(20)S_2(16)$$
$$= \left(227\ \frac{\text{mg}}{\text{L}}\right)(-0.08) + \left(362\ \frac{\text{mg}}{\text{L}}\right)(0.96)$$
$$+ \left(517\ \frac{\text{mg}}{\text{L}}\right)(0.12)$$
$$= 391\ \text{mg/L}$$

13 Energy, Work, and Power

PRACTICE PROBLEMS

Energy

1. A solid cast-iron sphere ($\rho = 0.256$ lbm/in^3 (7090 kg/m^3)) of 10 in (25 cm) diameter travels without friction at 30 ft/sec (9 m/s) horizontally. What is its kinetic energy?

(A) 900 ft-lbf (1.2 kJ)

(B) 1200 ft-lbf (1.6 kJ)

(C) 1600 ft-lbf (2.0 kJ)

(D) 1900 ft-lbf (2.3 kJ)

Work

2. What work is done when a balloon carries a 12 lbm (5.2 kg) load to 40,000 ft (12 000 m) height?

(A) 2.4×10^5 ft-lbf (300 kJ)

(B) 4.8×10^5 ft-lbf (610 kJ)

(C) 7.7×10^5 ft-lbf (980 kJ)

(D) 9.9×10^5 ft-lbf (1.3 MJ)

3. Find the compression of a spring if a 100 lbm (50 kg) weight is dropped from 8 ft (2 m) onto a spring with a constant of 33.33 lbf/in (5.837×10^3 N/m).

(A) 27 in (0.67 m)

(B) 34 in (0.85 m)

(C) 39 in (0.90 m)

(D) 45 in (1.1 m)

4. A punch press flywheel operates at 300 rpm with a moment of inertia of 15 slug-ft^2 (20 kg·m^2). Find the speed in rpm to which the wheel will be reduced after a sudden punching requiring 4500 ft-lbf (6100 J) of work.

(A) 160 rpm

(B) 190 rpm

(C) 220 rpm

(D) 310 rpm

5. A force of 550 lbf (2500 N) making a 40° angle (upward) from the horizontal pushes a box 20 ft (6 m) across the floor. What work is done?

(A) 2200 ft-lbf (3.0 kJ)

(B) 3700 ft-lbf (5.2 kJ)

(C) 4200 ft-lbf (6.0 kJ)

(D) 8400 ft-lbf (12 kJ)

6. A 1000 ft long (300 m long) cable has a mass of 2 lbm per foot (3 kg/m) and is suspended from a winding drum down into a vertical shaft. What work must be done to rewind the cable?

(A) 0.50×10^6 ft-lbf (0.6 MJ)

(B) 0.75×10^6 ft-lbf (0.9 MJ)

(C) 1×10^6 ft-lbf (1.3 MJ)

(D) 2×10^6 ft-lbf (2.6 MJ)

Power

7. What volume in ft^3 (m^3) of water can be pumped to a 130 ft (40 m) height in 1 hr by a 7 hp (5 kW) pump? Assume 85% efficiency.

(A) 1500 ft^3 (40 m^3)

(B) 1800 ft^3 (49 m^3)

(C) 2000 ft^3 (54 m^3)

(D) 2400 ft^3 (65 m^3)

8. What power in horsepower (kW) is required to lift a 3300 lbm (1500 kg) mass 250 ft (80 m) in 14 sec?

(A) 40 hp (30 kW)

(B) 70 hp (53 kW)

(C) 90 hp (68 kW)

(D) 110 hp (84 kW)

SOLUTIONS

1. *Customary U.S. Solution*

Since there is no friction, there is no rotation. The sphere slides.

$$E_{\text{kinetic}} = \tfrac{1}{2}\left(\frac{m}{g_c}\right)v^2 = \tfrac{1}{2}\left(\frac{V\rho}{g_c}\right)v^2$$

$$= \left(\tfrac{1}{2}\right)\left(\tfrac{4}{3}\pi r^3\right)\left(\frac{\rho}{g_c}\right)v^2 = \left(\tfrac{2}{3}\pi\right)\left(\frac{10 \text{ in}}{2}\right)^3\left(\frac{\rho}{g_c}\right)v^2$$

$$= \left(\tfrac{2}{3}\pi\right)\left(\frac{10 \text{ in}}{2}\right)^3\left(\frac{0.256 \frac{\text{lbm}}{\text{in}^3}}{32.2 \frac{\text{lbm-ft}}{\text{lbf-sec}^2}}\right)\left(30 \frac{\text{ft}}{\text{sec}}\right)^2$$

$$= \boxed{1873 \text{ ft-lbf} \quad (1900 \text{ ft-lbf})}$$

The answer is (D).

SI Solution

Since there is no friction, there is no rotation. The sphere slides.

$$E_{\text{kinetic}} = \tfrac{1}{2}mv^2 = \tfrac{1}{2}(\rho V)v^2$$

$$= \left(\tfrac{1}{2}\right)\rho\left(\tfrac{4}{3}\pi r^3\right)v^2$$

$$= \left(\tfrac{2}{3}\pi\right)\left(\frac{0.25 \text{ m}}{2}\right)^3\left(7090 \frac{\text{kg}}{\text{m}^3}\right)\left(9 \frac{\text{m}}{\text{s}}\right)^2$$

$$= \boxed{2349 \text{ J} \quad (2.3 \text{ kJ})}$$

The answer is (D).

2. *Customary U.S. Solution*

From Eq. 13.12, the work done by the balloon is

$$W = \Delta E_{\text{potential}} = \frac{mg\Delta h}{g_c}$$

$$= \frac{(12 \text{ lbm})\left(32.2 \frac{\text{ft}}{\text{sec}^2}\right)(40{,}000 \text{ ft})}{32.2 \frac{\text{lbm-ft}}{\text{lbf-sec}^2}}$$

$$= \boxed{4.8 \times 10^5 \text{ ft-lbf}}$$

The answer is (B).

SI Solution

From Eq. 13.12, the work done by the balloon is

$$W = \Delta E_{\text{potential}} = mg\Delta h$$

$$= \frac{(5.2 \text{ kg})\left(9.81 \frac{\text{m}}{\text{s}^2}\right)(12\,000 \text{ m})}{1000 \frac{\text{J}}{\text{kJ}}}$$

$$= \boxed{612.1 \text{ kJ} \quad (610 \text{ kJ})}$$

The answer is (B).

3. *Customary U.S. Solution*

Equating the potential energy to the energy of the spring gives

$$\Delta E_{\text{potential}} = \Delta E_{\text{spring}}$$

$$W(\Delta h + \Delta x) = \tfrac{1}{2}k(\Delta x)^2$$

Then, rearranging and using $W = m(g/g_c)$,

$$\tfrac{1}{2}k(\Delta x)^2 - W\Delta x - W\Delta h = 0$$

$$\left(\tfrac{1}{2}\right)\left(33.33 \frac{\text{lbf}}{\text{in}}\right)(\Delta x)^2$$

$$- (100 \text{ lbm})\left(\frac{32.2 \frac{\text{ft}}{\text{sec}^2}}{32.2 \frac{\text{lbm-ft}}{\text{lbf-sec}^2}}\right)\Delta x$$

$$- (100 \text{ lbm})\left(\frac{32.2 \frac{\text{ft}}{\text{sec}^2}}{32.2 \frac{\text{lbm-ft}}{\text{lbf-sec}^2}}\right)(8 \text{ ft})\left(12 \frac{\text{in}}{\text{ft}}\right) = 0$$

$$16.665\Delta x^2 - 100\Delta x = 9600$$

Complete the square.

$$\Delta x^2 - 6\Delta x = 576$$

$$(\Delta x - 3)^2 = 576 + 9$$

$$\Delta x - 3 = \sqrt{585} = \pm 24.2$$

$$\Delta x = \boxed{27.2 \text{ in} \quad (27 \text{ in})}$$

The answer is (A).

SI Solution

Equating the potential energy to the energy of the spring gives

$$\Delta E_{\text{potential}} = \Delta E_{\text{spring}}$$

$$mg(\Delta h + \Delta x) = \tfrac{1}{2}k(\Delta x)^2$$

Then, rearranging,

$$\tfrac{1}{2}k(\Delta x)^2 - mg\Delta x - mg\Delta h = 0$$

$$\left(\tfrac{1}{2}\right)\left(5.837 \times 10^3 \ \frac{\text{N}}{\text{m}}\right)(\Delta x)^2 - (50 \ \text{kg})\left(9.81 \ \frac{\text{m}}{\text{s}^2}\right)\Delta x$$

$$- (50 \ \text{kg})\left(9.81 \ \frac{\text{m}}{\text{s}^2}\right)(2 \ \text{m}) = 0$$

$$2918.5\Delta x^2 - 490.5\Delta x - 981.0 = 0$$

$$\Delta x^2 - 0.1681\Delta x = 0.3361$$

$$(\Delta x - 0.08403)^2 = 0.3361 + (0.08403)^2$$

$$= 0.3432$$

$$\Delta x - 0.08403 = \sqrt{0.3432} = \pm 0.5858$$

$$\Delta x = \boxed{0.6699 \ \text{m} \quad (0.67 \ \text{m})}$$

The answer is (A).

4. *Customary U.S. Solution*

The work done by the wheel is equal to the change in the rotational energy.

$$W_{\text{done by wheel}} = \Delta E_{\text{rotational}}$$

$$= \tfrac{1}{2}I\omega_{\text{initial}}^2 - \tfrac{1}{2}I\omega_{\text{final}}^2$$

The final angular velocity is found from

$$\omega_{\text{final}} = \sqrt{\omega_{\text{initial}}^2 - \frac{2W}{I}} = 2\pi f$$

The final speed is

$$f_{\text{final}} = \left(\frac{1}{2\pi}\right)\left(60 \ \frac{\text{sec}}{\text{min}}\right)$$

$$\times \sqrt{\left(\left(2\pi \ \frac{\text{rad}}{\text{rev}}\right)\left(\frac{300 \ \frac{\text{rev}}{\text{min}}}{60 \ \frac{\text{sec}}{\text{min}}}\right)\right)^2 - \frac{(2)(4500 \ \text{ft-lbf})}{15 \ \text{slug-ft}^2}}$$

$$= \boxed{187.8 \ \text{rpm} \quad (190 \ \text{rpm})}$$

The answer is (B).

SI Solution

The work done by the wheel is equal to the change in the rotational energy.

$$W_{\text{done by wheel}} = \Delta E_{\text{rotational}}$$

$$= \tfrac{1}{2}I\omega_{\text{initial}}^2 - \tfrac{1}{2}I\omega_{\text{final}}^2$$

The final angular velocity is found from

$$\omega_{\text{final}} = \sqrt{\omega_{\text{initial}}^2 - \frac{2W}{I}} = 2\pi f$$

The final speed is

$$f_{\text{final}} = \left(\frac{1}{2\pi}\right)\left(60 \ \frac{\text{sec}}{\text{min}}\right)$$

$$\times \sqrt{\left((2\pi)\left(\frac{300 \ \frac{\text{rev}}{\text{min}}}{60 \ \frac{\text{s}}{\text{min}}}\right)\right)^2 - \frac{(2)(6100 \ \text{J})}{20 \ \text{kg·m}^2}}$$

$$= \boxed{185.4 \ \text{rpm} \quad (190 \ \text{rpm})}$$

The answer is (B).

5. *Customary U.S. Solution*

The work done is

$$W_{\text{done on box}} = F_x\Delta x = F(\cos\theta)\Delta x$$

$$= (550 \ \text{lbf})(\cos 40°)(20 \ \text{ft})$$

$$= \boxed{8426 \ \text{ft-lbf} \quad (8400 \ \text{ft-lbf})}$$

The answer is (D).

SI Solution

The work done is

$$W_{\text{done on box}} = F_x\Delta x = F(\cos\theta)\Delta x$$

$$= \frac{(2500 \ \text{N})(\cos 40°)(6 \ \text{m})}{1000 \ \frac{\text{J}}{\text{kJ}}}$$

$$= \boxed{11.49 \ \text{kJ}}$$

The answer is (D).

6. *Customary U.S. Solution*

The work done to rewind the cable is

$$W_{\text{to rewind cable}} = \int_0^l F \, dh$$
$$= \int_0^l ((l-h)w) \, dh$$
$$= \tfrac{1}{2}wl^2$$
$$= \left(\tfrac{1}{2}\right)\left(2 \frac{\text{lbf}}{\text{ft}}\right)(1000 \text{ ft})^2$$
$$= \boxed{10^6 \text{ ft-lbf}}$$

The answer is (C).

SI Solution

The work done to rewind the cable is

$$W_{\text{to rewind cable}} = \int_0^l F \, dh$$
$$= \int_0^l ((l-h)m_l g) \, dh$$
$$= \tfrac{1}{2}m_l g l^2$$
$$= \left(\tfrac{1}{2}\right)\left(3 \frac{\text{kg}}{\text{m}}\right)\left(9.81 \frac{\text{m}}{\text{s}^2}\right)(300 \text{ m})^2$$
$$= \boxed{1.32 \times 10^6 \text{ J} \quad (1.32 \text{ MJ})}$$

The answer is (C).

7. *Customary U.S. Solution*

The volume of water is found from the following equations.

$$P_{\text{actual}} \Delta t = W_{\text{done by pump}}$$
$$\eta P_{\text{ideal}} \Delta t = \Delta E_{\text{potential}}$$
$$= \frac{mg\Delta h}{g_c}$$
$$= \frac{(\rho V)g\Delta h}{g_c}$$

$$V = \frac{\eta P_{\text{ideal}} \Delta t}{\dfrac{\rho g \Delta h}{g_c}}$$

$$= \frac{(0.85)(7 \text{ hp})\left(550 \frac{\text{ft-lbf}}{\text{hp-sec}}\right)(1 \text{ hr})\left(3600 \frac{\text{sec}}{\text{hr}}\right)}{\dfrac{\left(62.4 \frac{\text{lbm}}{\text{ft}^3}\right)\left(32.2 \frac{\text{ft}}{\text{sec}^2}\right)(130 \text{ ft})}{32.2 \frac{\text{lbm-ft}}{\text{lbf-sec}^2}}}$$

$$= \boxed{1452 \text{ ft}^3 \quad (1500 \text{ ft}^3)}$$

The answer is (A).

SI Solution

The volume of water is found from the following equations.

$$P_{\text{actual}} \Delta t = W_{\text{done by pump}}$$
$$\eta P_{\text{ideal}} \Delta t = \Delta E_{\text{potential}}$$
$$= mg\Delta h$$
$$= (\rho V)g\Delta h$$

$$V = \frac{\eta P_{\text{ideal}} \Delta t}{\rho g \Delta h}$$

$$= \frac{(0.85)(5 \text{ kW})\left(1000 \frac{\text{W}}{\text{kW}}\right)(1 \text{ h})\left(3600 \frac{\text{s}}{\text{h}}\right)}{\left(1000 \frac{\text{kg}}{\text{m}^3}\right)\left(9.81 \frac{\text{m}}{\text{s}^2}\right)(40 \text{ m})}$$

$$= \boxed{39.0 \text{ m}^3 \quad (40 \text{ m}^3)}$$

The answer is (A).

8. *Customary U.S. Solution*

The power required to lift the mass is

$$P\Delta t = W = \frac{mg\Delta h}{g_c}$$
$$P = \frac{mg\Delta h}{g_c \Delta t}$$

$$= \frac{(3300 \text{ lbm})\left(32.2 \frac{\text{ft}}{\text{sec}^2}\right)(250 \text{ ft})}{\left(32.2 \frac{\text{lbm-ft}}{\text{sec}^2\text{-lbf}}\right)(14 \text{ sec})\left(550 \frac{\text{ft-lbf}}{\text{hp-sec}}\right)}$$

$$= \boxed{107 \text{ hp} \quad (110 \text{ hp})}$$

The answer is (D).

SI Solution

The power required to lift the mass is

$$P\Delta t = W = mg\Delta h$$

$$P = \frac{mg\Delta h}{\Delta t}$$

$$= \frac{(1500 \text{ kg})\left(9.81 \ \frac{\text{m}}{\text{s}^2}\right)(80 \text{ m})}{(14 \text{ s})\left(1000 \ \frac{\text{W}}{\text{kW}}\right)}$$

$$= \boxed{84.1 \text{ kW} \quad (84 \text{ kW})}$$

The answer is (D).

14 Fluid Properties

PRACTICE PROBLEMS

(Use $g = 32.2$ ft/sec^2 or 9.81 m/s^2 unless told to do otherwise in the problem.)

Pressure

1. Atmospheric pressure is 14.7 lbf/in^2 (101.3 kPa). What is the absolute pressure if a gauge reads 8.7 psi (60 kPa) vacuum?

(A) 4 psia (27 kPa)

(B) 6 psia (41 kPa) ✓

(C) 8 psia (55 kPa)

(D) 10 psia (68 kPa)

Viscosity

2. Calculate the kinematic viscosity of air at $80°$F ($27°$C) and 70 psia (480 kPa). The specific gas constant for air is 53.3 ft-lbf/lbm-$°$R (287 J/kg·K).

(A) 3.54×10^{-5} ft^2/sec (3.30×10^{-6} m^2/s)

(B) 4.25×10^{-5} ft^2/sec (3.96×10^{-6} m^2/s)

(C) 4.96×10^{-5} ft^2/sec (4.62×10^{-6} m^2/s)

(D) 6.37×10^{-5} ft^2/sec (5.94×10^{-6} m^2/s)

Solutions

3. Volumes of an 8% solution, a 10% solution, and a 20% solution of nitric acid are to be combined in order to get 100 mL of a 12% solution. Concentrations are volumetric. The 8% solution contributes half of the total volume of nitric acid contributed by the 10% and 20% solutions. What volume of 10% acid solution is required?

(A) 20 mL

(B) 30 mL

(C) 50 mL

(D) 80 mL

SOLUTIONS

1. *Customary U.S. Solution*

$$p_{gage} = -8.7 \text{ lbf/in}^2$$
$$p_{atmospheric} = 14.7 \text{ lbf/in}^2$$

The relationship between absolute, gage, and atmospheric pressure is

$$p_{absolute} = p_{gage} + p_{atmospheric}$$
$$= -8.7 \frac{\text{lbf}}{\text{in}^2} + 14.7 \frac{\text{lbf}}{\text{in}^2}$$
$$= \boxed{6 \text{ lbf/in}^2 \quad (6 \text{ psia})}$$

The answer is (B).

SI Solution

$$p_{gage} = -60 \text{ kPa}$$
$$p_{atmospheric} = 101.3 \text{ kPa}$$

The relationship between absolute, gage, and atmospheric pressure is

$$p_{absolute} = p_{gage} + p_{atmospheric}$$
$$= -60 \text{ kPa} + 101.3 \text{ kPa}$$
$$= \boxed{41.3 \text{ kPa} \quad (41 \text{ kPa})}$$

The answer is (B).

2. *Customary U.S. Solution*

From App. 14.D, for air at 14.7 psia and $80°$F, the absolute viscosity (independent of pressure) is $\mu = 3.85 \times 10^{-7}$ lbf-sec/ft^2.

Determine the density of air at 70 psia and $80°$F. (Assume an ideal gas.)

$$\rho = \frac{p}{RT}$$

Substituting gives

$$\rho = \frac{\left(70 \; \frac{\text{lbf}}{\text{in}^2}\right)\left(12 \; \frac{\text{in}}{\text{ft}}\right)^2}{\left(53.3 \; \frac{\text{ft-lbf}}{\text{lbm-}^\circ\text{R}}\right)(80^\circ\text{F} + 460^\circ)}$$

$$= 0.350 \; \text{lbm/ft}^3$$

The kinematic viscosity, ν, is related to the absolute viscosity by

$$\nu = \frac{\mu g_c}{\rho}$$

$$= \frac{\left(3.85 \times 10^{-7} \; \frac{\text{lbf-sec}}{\text{ft}^2}\right)\left(32.2 \; \frac{\text{lbm-ft}}{\text{lbf-sec}^2}\right)}{0.350 \; \frac{\text{lbm}}{\text{ft}^3}}$$

$$= \boxed{3.54 \times 10^{-5} \; \text{ft}^2/\text{sec}}$$

The answer is (A).

SI Solution

From App. 14.E, for air at 480 kPa and 27°C, the absolute viscosity (independent of pressure) is $\mu = 1.84 \times 10^{-5}$ Pa·s.

Determine the density of air at 480 kPa and 27°C. (Assume an ideal gas.)

$$\rho = \frac{p}{RT}$$

Substituting gives

$$\rho = \frac{(480 \; \text{kPa})\left(1000 \; \frac{\text{Pa}}{\text{kPa}}\right)}{\left(287 \; \frac{\text{J}}{\text{kg·K}}\right)(27^\circ\text{C} + 273^\circ)}$$

$$= 5.575 \; \text{kg/m}^3$$

The kinematic viscosity, ν, is related to the absolute viscosity by

$$\nu = \frac{\mu}{\rho}$$

$$= \frac{1.84 \times 10^{-5} \; \text{Pa·s}}{5.575 \; \frac{\text{kg}}{\text{m}^3}}$$

$$= \boxed{3.30 \times 10^{-6} \; \text{m}^2/\text{s}}$$

The answer is (A).

3. Let

$$x = \text{volume of 8\% solution}$$
$$0.08x = \text{volume of nitric acid contributed by 8\%}$$
$$\text{solution}$$
$$y = \text{volume of 10\% solution}$$
$$0.10y = \text{volume of nitric acid contributed by 10\%}$$
$$\text{solution}$$
$$z = \text{volume of 20\% solution}$$
$$0.20z = \text{volume of nitric acid contributed by 20\%}$$
$$\text{solution}$$

The three conditions that must be satisfied are

$$x + y + z = 100 \; \text{mL}$$
$$0.08x + 0.10y + 0.20z = (0.12)(100 \; \text{mL}) = 12 \; \text{mL}$$
$$0.08x = \left(\tfrac{1}{2}\right)(0.10y + 0.20z)$$

Simplifying these equations,

$$x + y + z = 100$$
$$4x + 5y + 10z = 600$$
$$8x - 5y - 10z = 0$$

Adding the second and third equations gives

$$12x = 600$$
$$x = 50 \; \text{mL}$$

Work with the first two equations to get

$$y + z = 100 - 50 = 50$$
$$5y + 10z = 600 - (4)(50) = 400$$

Multiplying the top equation by -5 and adding to the bottom equation,

$$5z = 150$$
$$z = 30 \; \text{mL}$$

From the first equation,

$$y = \boxed{20 \; \text{mL}}$$

The answer is (A).

15 **Fluid Statics**

PRACTICE PROBLEMS

(Use $g = 32.2$ ft/sec^2 or 9.81 m/s^2 unless told to do otherwise in the problem.)

1. A 4000 lbm (1800 kg) blimp contains 10,000 lbm (4500 kg) of hydrogen (specific gas constant = 766.5 ft-lbf/lbm-°R (4124 J/kg·K)) at 56°F (13°C) and 30.2 in Hg (770 mm Hg). What is its lift if the hydrogen and air are in thermal and pressure equilibrium?

(A) 7.6×10^3 lbf (3.4×10^4 N)

(B) 1.2×10^4 lbf (5.3×10^4 N)

(C) 1.3×10^5 lbf (5.7×10^5 N)

(D) 1.7×10^5 lbf (7.7×10^5 N)

2. A hollow 6 ft (1.8 m) diameter sphere floats half-submerged in seawater. What mass of concrete is required as an external anchor to just submerge the sphere completely?

(A) 2700 lbm (1200 kg)

(B) 4200 lbm (1900 kg)

(C) 5500 lbm (2500 kg)

(D) 6300 lbm (2700 kg)

SOLUTIONS

1. *Customary U.S. Solution*

The lift of the hydrogen-filled blimp, F_{lift}, is equal to the difference between the buoyant force, F_b, and the weight of the hydrogen contained in the blimp, W_H.

$$F_{\text{lift}} = F_b - W_H - W_{\text{blimp}}$$

The weight of the hydrogen is calculated from the mass of hydrogen by

$$W_H = \frac{m_H g}{g_c} = \frac{(10{,}000 \text{ lbm})\left(32.2 \ \dfrac{\text{ft}}{\text{sec}^2}\right)}{32.2 \ \dfrac{\text{lbm-ft}}{\text{lbf-sec}^2}}$$

$$= 10{,}000 \text{ lbm}$$

The buoyant force is equal to the weight of the displaced air. The volume of the air displaced is equal to the volume of hydrogen enclosed in the blimp.

The temperature of the hydrogen is given as 56°F. Convert to absolute temperature (°R).

$$T = 56°\text{F} + 460° = 516°\text{R}$$

The pressure of the hydrogen is given as 30.2 in Hg. Convert the pressure to units of pounds per square foot.

$$p = \frac{(30.2 \text{ in Hg})\left(12 \ \dfrac{\text{in}}{\text{ft}}\right)^2}{2.036 \ \dfrac{\text{in Hg}}{\dfrac{\text{lbf}}{\text{in}^2}}}$$

$$= 2136 \text{ lbf/ft}^2$$

Compute the volume of hydrogen from the ideal gas law.

$$V_H = \frac{m_H R T}{p}$$

$$= \frac{(10{,}000 \text{ lbm})\left(766.5 \ \dfrac{\text{ft-lbf}}{\text{lbm-°R}}\right)(516°\text{R})}{2136 \ \dfrac{\text{lbf}}{\text{ft}^2}}$$

$$= 1.85 \times 10^6 \text{ ft}^3$$

Since the volume of the hydrogen contained in the blimp is equal to the air displaced, the air displaced can be computed from the ideal gas equation. Since the air and hydrogen are in thermal and pressure equilibrium, the temperature and pressure are equal to the values given for the hydrogen.

For air, $R = 53.35$ ft-lbf/lbm-°R.

$$m_{\text{air}} = \frac{pV_{\text{H}}}{RT} = \frac{\left(2136 \ \frac{\text{lbf}}{\text{ft}^2}\right)\left(1.85 \times 10^6 \ \text{ft}^3\right)}{\left(53.35 \ \frac{\text{ft-lbf}}{\text{lbm-°R}}\right)(516°\text{R})}$$

$$= 1.435 \times 10^5 \ \text{lbm}$$

The buoyant force is equal to the weight of the air.

$$F_b = W_{\text{air}} = \frac{m_{\text{air}}g}{g_c}$$

$$= \frac{\left(1.435 \times 10^5 \ \text{lbm}\right)\left(32.2 \ \frac{\text{ft}}{\text{sec}^2}\right)}{32.2 \ \frac{\text{lbm-ft}}{\text{lbf-sec}^2}}$$

$$= 1.435 \times 10^5 \ \text{lbf}$$

The lift is

$$F_{\text{lift}} = F_b - W_{\text{H}} - W_{\text{blimp}}$$

$$= 1.435 \times 10^5 \ \text{lbf} - 10{,}000 \ \text{lbf} - 4000 \ \text{lbf}$$

$$= \boxed{1.295 \times 10^5 \ \text{lbf} \quad (1.3 \times 10^5 \ \text{lbf})}$$

The answer is (C).

SI Solution

The lift of the hydrogen-filled blimp, F_{lift}, is equal to the difference between the buoyant force, F_b, and the weight of the hydrogen contained in the blimp, W_{H}.

$$F_{\text{lift}} = F_b - W_{\text{H}} - W_{\text{blimp}}$$

The weight of the hydrogen is calculated from the mass of hydrogen by

$$W_{\text{H}} = m_{\text{H}}g = (4500 \ \text{kg})\left(9.81 \ \frac{\text{m}}{\text{s}^2}\right)$$

$$= 44\,145 \ \text{N}$$

The buoyant force is equal to the weight of the displaced air. The volume of the air displaced is equal to the volume of hydrogen enclosed in the blimp.

The temperature of the hydrogen is given as 13°C. Convert to absolute temperature (K).

$$T = 13°\text{C} + 273° = 286\text{K}$$

The pressure of the hydrogen is given as 770 mm Hg. Convert the pressure to units of pascals.

$$p = \frac{(770 \ \text{mm Hg})\left(133.4 \ \frac{\text{kPa}}{\text{m}}\right)}{1000 \ \frac{\text{mm}}{\text{m}}}$$

$$= 102.7 \ \text{kPa}$$

Compute the volume of hydrogen from the ideal gas law.

$$V_{\text{H}} = \frac{m_{\text{H}}RT}{p}$$

$$= \frac{(4500 \ \text{kg})\left(4124 \ \frac{\text{J}}{\text{kg·K}}\right)(286\text{K})}{(102.7 \ \text{kPa})\left(1000 \ \frac{\text{Pa}}{\text{kPa}}\right)}$$

$$= 5.168 \times 10^4 \ \text{m}^3$$

Since the volume of the hydrogen contained in the blimp is equal to the air displaced, the air displaced can be computed from the ideal gas equation. Since the air and hydrogen are assumed to be in thermal and pressure equilibrium, the temperature and pressure are equal to the values given for the hydrogen.

For air, $R = 287.03$ J/kg·K.

$$m_{\text{air}} = \frac{pV_{\text{H}}}{RT}$$

$$= \frac{(102.7 \ \text{kPa})\left(1000 \ \frac{\text{Pa}}{\text{kPa}}\right)\left(5.168 \times 10^4 \ \text{m}^3\right)}{\left(287.03 \ \frac{\text{J}}{\text{kg·K}}\right)(286\text{K})}$$

$$= 6.465 \times 10^4 \ \text{kg}$$

The buoyant force is equal to the weight of the air.

$$F_b = W_{\text{air}} = m_{\text{air}}g$$

$$= \left(6.465 \times 10^4 \ \text{kg}\right)\left(9.81 \ \frac{\text{m}}{\text{s}^2}\right)$$

$$= 6.34 \times 10^5 \ \text{N}$$

The lift is

$$F_{\text{lift}} = F_b - W_{\text{H}} - W_{\text{blimp}}$$

$$= 6.34 \times 10^5 \ \text{N} - 44\,145 \ \text{N} - (1800 \ \text{kg})\left(9.81 \ \frac{\text{m}}{\text{s}^2}\right)$$

$$= \boxed{5.7 \times 10^5 \ \text{N}}$$

The answer is (C).

2. *Customary U.S. Solution*

The weight of the sphere is equal to the weight of the displaced volume of water when floating.

The buoyant force is given by

$$F_b = \frac{\rho g V_{\text{displaced}}}{g_c}$$

Since the sphere is half submerged,

$$W_{\text{sphere}} = \tfrac{1}{2}\left(\frac{\rho g V_{\text{sphere}}}{g_c}\right)$$

For seawater, $\rho = 64.0 \text{ lbm/ft}^3$.

The volume of the sphere is

$$V_{\text{sphere}} = \frac{\pi}{6}d^3 = \left(\frac{\pi}{6}\right)(6 \text{ ft})^3$$
$$= 113.1 \text{ ft}^3$$

The weight of the sphere is

$$W_{\text{sphere}} = \tfrac{1}{2}\left(\frac{\rho g V_{\text{sphere}}}{g_c}\right)$$
$$= \left(\tfrac{1}{2}\right)\left(\frac{\left(64.0 \,\frac{\text{lbm}}{\text{ft}^3}\right)\left(32.2 \,\frac{\text{ft}}{\text{sec}^2}\right)(113.1 \text{ ft}^3)}{32.2 \,\frac{\text{lbm-ft}}{\text{lbf-sec}^2}}\right)$$
$$= 3619 \text{ lbf}$$

The buoyant force equation for a fully submerged sphere and anchor can be solved for the concrete volume.

$$W_{\text{sphere}} + W_{\text{concrete}} = (V_{\text{sphere}} + V_{\text{concrete}})\rho_{\text{water}}$$

$$W_{\text{sphere}} + \rho_{\text{concrete}} V_{\text{concrete}}\left(\frac{g}{g_c}\right)$$
$$= (V_{\text{sphere}} + V_{\text{concrete}})\rho_{\text{water}}\left(\frac{g}{g_c}\right)$$

$$3619 \text{ lbf} + \left(150 \,\frac{\text{lbm}}{\text{ft}^3}\right)V_{\text{concrete}}\left(\frac{32.2 \,\frac{\text{ft}}{\text{sec}^2}}{32.2 \,\frac{\text{ft-lbm}}{\text{lbf-sec}^2}}\right)$$

$$= (113.1 \text{ ft}^3 + V_{\text{concrete}})\left(64.0 \,\frac{\text{lbm}}{\text{ft}^3}\right)\left(\frac{32.2 \,\frac{\text{ft}}{\text{sec}^2}}{32.2 \,\frac{\text{ft-lbm}}{\text{lbf-sec}^2}}\right)$$

$$V_{\text{concrete}} = 42.09 \text{ ft}^3$$
$$m_{\text{concrete}} = \rho_{\text{concrete}} V_{\text{concrete}}$$
$$= \left(150 \,\frac{\text{lbm}}{\text{ft}^3}\right)(42.09 \text{ ft}^3)$$
$$= \boxed{6314 \text{ lbm} \quad (6300 \text{ lbm})}$$

The answer is (D).

SI Solution

The weight of the sphere is equal to the weight of the displaced volume of water when floating.

The buoyant force is given by

$$F_b = \rho g V_{\text{displaced}}$$

Since the sphere is half submerged,

$$W_{\text{sphere}} = \tfrac{1}{2}\rho g V_{\text{sphere}}$$

For seawater, $\rho = 1025 \text{ kg/m}^3$.

The volume of the sphere is

$$V_{\text{sphere}} = \frac{\pi}{6}d^3 = \left(\frac{\pi}{6}\right)(1.8 \text{ m})^3$$
$$= 3.054 \text{ m}^3$$

The weight of the sphere required is

$$W_{\text{sphere}} = \tfrac{1}{2}\rho g V_{\text{sphere}}$$
$$= \left(\tfrac{1}{2}\right)\left(1025 \,\frac{\text{kg}}{\text{m}^3}\right)\left(9.81 \,\frac{\text{m}}{\text{s}^2}\right)(3.054 \text{ m}^3)$$
$$= 15\,354 \text{ N}$$

The buoyant force equation for a fully submerged sphere and anchor can be solved for the concrete volume.

$$W_{\text{sphere}} + W_{\text{concrete}} = (V_{\text{sphere}} + V_{\text{concrete}})\rho_{\text{water}}$$
$$W_{\text{sphere}} + \rho_{\text{concrete}} g V_{\text{concrete}} = g(V_{\text{sphere}} + V_{\text{concrete}})\rho_{\text{water}}$$

$$15\,354 \text{ N} + \left(2400 \,\frac{\text{kg}}{\text{m}^3}\right)\left(9.81 \,\frac{\text{m}}{\text{s}^2}\right)V_{\text{concrete}}$$
$$= (3.054 \text{ m}^3 + V_{\text{concrete}})\left(1025 \,\frac{\text{kg}}{\text{m}^3}\right)\left(9.81 \,\frac{\text{m}}{\text{s}^2}\right)$$

$$V_{\text{concrete}} = 1.138 \text{ m}^3$$
$$m_{\text{concrete}} = \rho_{\text{concrete}} V_{\text{concrete}}$$
$$= \left(2400 \,\frac{\text{kg}}{\text{m}^3}\right)(1.138 \text{ m}^3)$$
$$= \boxed{2731 \text{ kg} \quad (2700 \text{ kg})}$$

The answer is (D).

16 Fluid Flow Parameters

PRACTICE PROBLEMS

Use the following values unless told to do otherwise in the problem:

$$g = 32.2 \text{ ft/sec}^2 \ (9.81 \text{ m/s}^2)$$

$$\rho_{\text{water}} = 62.4 \text{ lbm/ft}^3 \ (1000 \text{ kg/m}^3)$$

$$p_{\text{atmospheric}} = 14.7 \text{ psia} \ (101.3 \text{ kPa})$$

Hydraulic Radius

1. A 10 in (25 cm) composition pipe is compressed by a tree root into an elliptical cross section until its inside height is only 7.2 in (18 cm). What is its approximate hydraulic radius when flowing half full?

(A) 2.2 in (5.5 cm)

(B) 2.7 in (6.9 cm)

(C) 3.2 in (8.1 cm)

(D) 4.5 in (11.4 cm)

2. A pipe with an inside diameter of 18.812 in contains water to a depth of 15.7 in. What is the hydraulic radius? (Work in customary U.S. units only.)

(A) 4.4 in

(B) 5.1 in

(C) 5.7 in

(D) 6.5 in

SOLUTIONS

1. *Customary U.S. Solution*

The perimeter of the pipe is

$$p = \pi d = \pi(10 \text{ in})$$
$$= 31.42 \text{ in}$$

If the pipe is flowing half-full, the wetted perimeter becomes

$$\text{wetted perimeter} = \tfrac{1}{2}p = \left(\tfrac{1}{2}\right)(31.42 \text{ in})$$
$$= 15.71 \text{ in}$$

The ellipse will have a minor axis, b, equal to one-half the height of the compressed pipe or

$$b = \frac{7.2 \text{ in}}{2} = 3.6 \text{ in}$$

When the pipe is compressed, the perimeter of the pipe will remain constant. The perimeter of an ellipse is given by

$$p \approx 2\pi\sqrt{\tfrac{1}{2}(a^2 + b^2)}$$

Solve for the major axis.

$$a = \sqrt{2\left(\frac{p}{2\pi}\right)^2 - b^2}$$
$$= \sqrt{(2)\left(\frac{31.42 \text{ in}}{2\pi}\right)^2 - (3.6 \text{ in})^2}$$
$$= 6.09 \text{ in}$$

The flow area or area of the ellipse is given by

$$\text{flow area} = \tfrac{1}{2}\pi ab$$
$$= \tfrac{1}{2}\pi(6.09 \text{ in})(3.6 \text{ in})$$
$$= 34.4 \text{ in}^2$$

The hydraulic radius is

$$r_h = \frac{\text{area in flow}}{\text{wetted perimeter}} = \frac{34.4 \text{ in}^2}{15.7 \text{ in}}$$

$$= \boxed{2.19 \text{ in} \quad (2.2 \text{ in})}$$

The answer is (A).

SI Solution

The perimeter of the pipe is

$$p = \pi d = \pi(25 \text{ cm})$$

$$= 78.54 \text{ cm}$$

If the pipe is flowing half-full, the wetted perimeter becomes

$$\text{wetted perimeter} = \tfrac{1}{2}p = \left(\tfrac{1}{2}\right)(78.54 \text{ cm})$$

$$= 39.27 \text{ cm}$$

Assume the compressed pipe is an elliptical cross section. The ellipse will have a minor axis, b, equal to one-half the height of the compressed pipe or

$$b = \frac{18 \text{ cm}}{2} = 9 \text{ cm}$$

When the pipe is compressed, the perimeter of the pipe will remain constant. The perimeter of an ellipse is given by

$$p \approx 2\pi \sqrt{\tfrac{1}{2}(a^2 + b^2)}$$

Solve for the major axis.

$$a = \sqrt{2\left(\frac{p}{2\pi}\right)^2 - b^2}$$

$$= \sqrt{(2)\left(\frac{78.54 \text{ cm}}{2\pi}\right)^2 - (9 \text{ cm})^2}$$

$$= 15.2 \text{ cm}$$

The flow area or area of the ellipse is given by

$$\text{flow area} = \tfrac{1}{2}\pi ab = \tfrac{1}{2}\pi(15.2 \text{ cm})(9 \text{ cm})$$

$$= 214.9 \text{ cm}^2$$

The hydraulic radius is

$$r_h = \frac{\text{area in flow}}{\text{wetted perimeter}}$$

$$= \frac{214.9 \text{ cm}^2}{39.27 \text{ cm}}$$

$$= \boxed{5.47 \text{ cm} \quad (5.5 \text{ cm})}$$

The answer is (A).

2. *method 1:* Use App. 7.A for a circular segment.

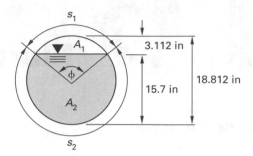

$$r = \frac{d}{2} = \frac{18.812 \text{ in}}{2} = 9.406 \text{ in}$$

$$\phi = 2 \arccos \frac{r - d}{r}$$

$$= 2 \arccos \frac{9.406 \text{ in} - 3.112 \text{ in}}{9.406 \text{ in}}$$

$$= 1.675 \text{ rad}$$

$$\sin \phi = 0.9946$$

$$A_1 = \tfrac{1}{2}r^2(\phi - \sin \phi)$$

$$= \left(\tfrac{1}{2}\right)(9.406 \text{ in})^2(1.675 - 0.9946)$$

$$= 30.1 \text{ in}^2$$

$$A_{\text{total}} = A_1 + A_2 = \frac{\pi}{4}D^2 = \left(\frac{\pi}{4}\right)(18.812 \text{ in})^2$$

$$= 277.95 \text{ in}^2$$

$$A_2 = A_{\text{total}} - A_1 = 277.95 \text{ in}^2 - 30.1 \text{ in}^2$$

$$= 247.85 \text{ in}^2$$

$$s_1 = r\phi = (9.406 \text{ in})(1.675) = 15.76 \text{ in}$$

$$s_{\text{total}} = s_1 + s_2 = \pi D = \pi(18.812 \text{ in}) = 59.1 \text{ in}$$

$$s_2 = s_{\text{total}} - s_1 = 59.1 \text{ in} - 15.76 \text{ in} = 43.34 \text{ in}$$

$$r_h = \frac{A_2}{s_2} = \frac{247.85 \text{ in}^2}{43.34 \text{ in}} = \boxed{5.719 \text{ in}}$$

method 2: Use App. 16.A.

$$\frac{d}{D} = \frac{15.7 \text{ in}}{18.812 \text{ in}} = 0.83$$

From App. 16.A, $r_h/D = 0.3041$.

$$r_h = (0.3041)(18.812 \text{ in}) = \boxed{5.72 \text{ in} \quad (5.7 \text{ in})}$$

The answer is (C).

17 Fluid Dynamics

Water Resources

PRACTICE PROBLEMS

(Use $g = 32.2$ ft/sec^2 (9.81 m/s^2) and 60°F (16°C) water unless told to do otherwise in the problem.)

Conservation of Energy

1. 5 ft^3/sec (130 L/s) of water flows through a schedule-40 steel pipe that changes gradually in diameter from 6 in at point A to 18 in at point B. Point B is 15 ft (4.6 m) higher than point A. The respective pressures at points A and B are 10 psia (70 kPa) and 7 psia (48.3 kPa). All minor losses are insignificant. What are the direction of flow and velocity at point A?

(A) 3.2 ft/sec (1 m/s); from A to B

(B) 25 ft/sec (7 m/s); from A to B

(C) 3.2 ft/sec (1 m/s); from B to A

(D) 25 ft/sec (7 m/s); from B to A

2. Points A and B are separated by 3000 ft of new 6 in schedule-40 steel pipe. 750 gal/min of 60°F water flows from point A to point B. Point B is 60 ft above point A. What must be the pressure at point A if the pressure at B must be 50 psig?

(A) 90 psig

(B) 100 psig

(C) 120 psig

(D) 170 psig

3. (*Time limit: one hour*) A pipe network connects junctions A, B, C, and D as shown. All pipe sections have a *C*-value of 150. Water can be added and removed at any of the junctions to achieve the flows listed. Water flows from point A to point D. No flows are backward. The minimum allowable pressure anywhere in the system is 20 psig. All minor losses are insignificant. For simplification, use the nominal pipe sizes.

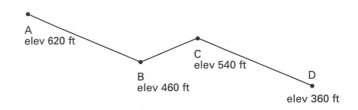

pipe section	length	diameter (in)	flow
A to B	20,000 ft	6	120 gal/min
B to C	10,000 ft	6	160 gal/min
C to D	30,000 ft	4	120 gal/min

(a) What is the pressure at point A?

(A) 14 psig

(B) 23 psig

(C) 30 psig

(D) 47 psig

(b) What is the elevation of the hydraulic grade line at point A referenced to point D?

(A) 280 ft

(B) 330 ft

(C) 470 ft

(D) 610 ft

Friction Loss

4. 1.5 ft^3/sec (40 L/s) of 70°F (20°C) water flows through 1200 ft (355 m) of 6 in (nominal) diameter new schedule-40 steel pipe. What is the friction loss?

(A) 4 ft (1.2 m)

(B) 18 ft (5.2 m)

(C) 36 ft (9.5 m)

(D) 70 ft (21 m)

5. 500 gal/min (30 L/s) of 100°F (40°C) water flows through 300 ft (90 m) of 6 in schedule-40 pipe. The pipe contains two 6 in flanged steel elbows, two full-open gate valves, a full-open 90° angle valve, and a swing check valve. The discharge is located 20 ft (6 m) higher than the entrance. What is the pressure difference between the two ends of the pipe?

(A) 12 psi (78 kPa)

(B) 21 psi (140 kPa)

(C) 45 psi (310 kPa)

(D) 87 psi (600 kPa)

6. 70°F (20°C) air is flowing at 60 ft/sec (18 m/s) through 300 ft (90 m) of 6 in schedule-40 pipe. The pipe contains two 6 in (0.15 m) flanged steel elbows, two full-open gate valves, a full-open 90° angle valve, and a swing check valve. The discharge is located 20 ft (6 m) higher than the entrance and the average air density is 0.075 lbm/ft^3. The air experiences only small increases in pressure above atmospheric. What is the pressure difference between the two ends of the pipe?

(A) 0.26 psi (1.8 kPa)

(B) 0.49 psi (3.2 kPa)

(C) 1.5 psi (10 kPa)

(D) 13 psi (90 kPa)

Reservoirs

7. Three reservoirs (A, B, and C) are interconnected with a common junction (point D) at elevation 25 ft above an arbitrary reference point. The water levels for reservoirs A, B, and C are at elevations of 50 ft, 40 ft, and 22 ft, respectively. The pipe from reservoir A to the junction is 800 ft of 3 in (nominal) steel pipe. The pipe from reservoir B to the junction is 500 ft of 10 in (nominal) steel pipe. The pipe from reservoir C to the junction is 1000 ft of 4 in (nominal) steel pipe. All pipes are schedule-40 with a friction factor of 0.02. All minor losses and velocity heads can be neglected. What are the direction of flow and the pressure at point D?

(A) out of reservoir B; 500 psf

(B) out of reservoir B; 930 psf

(C) into reservoir B; 1100 psf

(D) into reservoir B; 1260 psf

Water Hammer

8. (*Time limit: one hour*) A cast-iron pipe has an inside diameter of 24 in (600 mm) and a wall thickness of 0.75 in (20 mm). The pipe's modulus of elasticity is 20×10^6 psi (140 GPa). The pipeline is 500 ft (150 m) long. 70°F (20°C) water is flowing at 6 ft/sec (2 m/s).

(a) If a valve is closed instantaneously, what will be the pressure increase experienced in the pipe?

(A) 48 psi (330 kPa)

(B) 140 psi (970 kPa)

(C) 320 psi (2.5 MPa)

(D) 470 psi (3.2 MPa)

(b) If the pipe is 500 ft (150 m) long, over what length of time must the valve be closed to create a pressure equivalent to instantaneous closure?

(A) 0.25 sec

(B) 0.68 sec

(C) 1.6 sec

(D) 2.1 sec

Parallel Pipe Systems

9. 8 MGD (millions of gallons per day) (350 L/s) of 70°F (20°C) water flows into the new schedule-40 steel pipe network shown. Minor losses are insignificant.

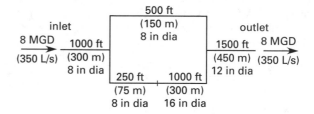

(a) What quantity of water is flowing in the upper branch?

(A) 1.2 ft^3/sec (0.034 m^3/s)

(B) 2.9 ft^3/sec (0.081 m^3/s)

(C) 4.1 ft^3/sec (0.11 m^3/s)

(D) 5.3 ft^3/sec (0.15 m^3/s)

(b) What is the energy loss per unit mass between the inlet and the outlet?

(A) 120 ft-lbf/lbm (0.37 kJ/kg)

(B) 300 ft-lbf/lbm (0.90 kJ/kg)

(C) 480 ft-lbf/lbm (1.4 kJ/kg)

(D) 570 ft-lbf/lbm (1.7 kJ/kg)

Pipe Networks

10. A single-loop pipe network is shown. The distance between each junction is 1000 ft. All junctions are on the same elevation. All pipes have a *C*-value of 100. The volumetric flow rates are to be determined to within 2 gal/min.

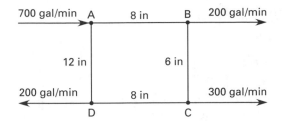

What is the flow rate between junctions B and C?

(A) 36 gal/min from B to C

(B) 58 gal/min from B to C

(C) 84 gal/min from C to B

(D) 112 gal/min from C to B

11. (*Time limit: one hour*) A double-loop pipe network is shown. The distance between each junction is 1000 ft. The water temperature is 60°F. Elevations and some pressure are known for the junctions. All pipes have a *C*-value of 100.

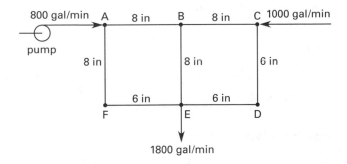

point	pressure	elevation
A		200 ft
B		150 ft
C	40 psig	300 ft
D		150 ft
E		200 ft
F		150 ft

(a) What is the flow rate between junctions B and E?

(A) 540 gal/min

(B) 620 gal/min

(C) 810 gal/min

(D) 980 gal/min

(b) What is the pressure at point D?

(A) 51 psig

(B) 73 psig

(C) 96 psig

(D) 120 psig

(c) If the pump receives water at 20 psig (140 kPa), what hydraulic power is required?

(A) 14 hp

(B) 28 hp

(C) 35 hp

(D) 67 hp

12. (*Time limit: one hour*) The water distribution network shown consists of class A cast-iron pipe installed 1.5 years ago. 1.5 MGD of 50°F water enters at junction A and leaves at junction B. The minimum acceptable pressure at point D is 40 psig (280 kPa).

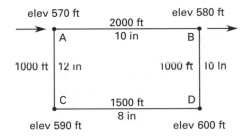

(a) What is the flow rate between junctions A and B?

(A) 320 gal/min

(B) 480 gal/min

(C) 590 gal/min

(D) 660 gal/min

(b) What is the pressure at point B?

(A) 32 psig

(B) 48 psig

(C) 57 psig

(D) 88 psig

(c) What is the pressure at point A?

(A) 39 psig

(B) 55 psig

(C) 69 psig

(D) 110 psig

Tank Discharge

13. The velocity of discharge from a fire hose is 50 ft/sec (15 m/s). The hose is oriented 45° from the horizontal. Disregarding air friction, what is the maximum range of the discharge?

(A) 45 ft (14 m)

(B) 78 ft (23 m)

(C) 91 ft (27 m)

(D) 110 ft (33 m)

14. A full cylindrical tank that is 40 ft (12 m) high has a constant diameter of 20 ft. The tank has a 4 in (100 mm) diameter hole in its bottom. The coefficient of discharge for the hole is 0.98. How long will it take for the water level to drop from 40 ft to 20 ft (12 m to 6 m)?

(A) 950 sec

(B) 1200 sec

(C) 1450 sec

(D) 1700 sec

Venturi Meters

15. A venturi meter with an 8 in diameter throat is installed in a 12 in diameter water line. The venturi is perfectly smooth, so that the discharge coefficient is 1.00. An attached mercury manometer registers a 4 in differential. What is the volumetric flow rate?

(A) 1.7 ft^3/sec

(B) 5.2 ft^3/sec

(C) 6.4 ft^3/sec

(D) 18 ft^3/sec

16. 60°F (15°C) benzene (specific gravity at 60°F (15°C) of 0.885) flows through an 8 in/3^1/$_2$ in (200 mm/90 mm) venturi meter whose coefficient of discharge is 0.99. A mercury manometer indicates a 4 in difference in the heights of the mercury columns. What is the volumetric flow rate of the benzene?

(A) 1.2 ft^3/sec (34 L/s)

(B) 9.1 ft^3/sec (250 L/s)

(C) 13 ft^3/sec (360 L/s)

(D) 27 ft^3/sec (760 L/s)

Orifice Meters

17. A sharp-edged orifice meter with a 0.2 ft diameter opening is installed in a 1 ft diameter pipe. 70°F water approaches the orifice at 2 ft/sec. What is the indicated pressure drop across the orifice meter?

(A) 5.9 psi

(B) 13 psi

(C) 22 psi

(D) 47 psi

18. A mercury manometer is used to measure a pressure difference across an orifice meter in a water line. The difference in mercury levels is 7 in (17.8 cm). What is the pressure differential?

(A) 1.7 psi (12 kPa)

(B) 3.2 psi (22 kPa)

(C) 7.9 psi (55 kPa)

(D) 23 psi (160 kPa)

19. A sharp-edged ISA orifice is used in a schedule-40 steel 12 in (300 mm inside diameter) water line. (Figure 17.28 is applicable.) The water temperature is 70°F (20°C), and the flow rate is 10 ft^3/sec (250 L/s). The differential pressure change across the orifice (to the vena contracta) must not exceed 25 ft (7.5 m). What is the smallest orifice that can be used?

(A) 5.5 in (14 cm)

(B) 7.3 in (19 cm)

(C) 8.1 in (20 cm)

(D) 8.9 in (23 cm)

Impulse-Momentum

20. A pipe necks down from 24 in at point A to 12 in at point B. 8 ft^3/sec of 60°F water flows from point A to point B. The pressure head at point A is 20 ft. Friction is insignificant over the distance between points A and B. What are the magnitude and direction of the resultant force on the water?

(A) 2900 lbf, toward A

(B) 3500 lbf, toward A

(C) 2900 lbf, toward B

(D) 3500 lbf, toward B

21. A 2 in (50 mm) diameter horizontal water jet has an absolute velocity (with respect to a stationary point) of 40 ft/sec (12 m/s) as it strikes a curved blade. The blade is moving horizontally away with an absolute velocity of 15 ft/sec (4.5 m/s). Water is deflected 60° from the horizontal. What is the force on the blade?

(A) 18 lbf (80 N)

(B) 26 lbf (110 N)

(C) 35 lbf (160 N)

(D) 47 lbf (210 N)

Pumps

22. 2000 gal/min (125 L/s) of brine with a specific gravity of 1.2 passes through an 85% efficient pump. The centerlines of the pump's 12 in inlet and 8 in outlet are at the same elevation. The inlet suction gauge indicates 6 in (150 mm) of mercury below atmospheric. The discharge pressure gauge is located 4 ft (1.2 m) above the centerline of the pump's outlet and indicates 20 psig (138 kPa). All pipes are schedule-40. What is the input power to the pump?

(A) 12 hp (8.9 kW)

(B) 36 hp (26 kW)

(C) 52 hp (39 kW)

(D) 87 hp (65 kW)

Turbines

23. 100 ft³/sec (2.6 L/s) of water passes through a horizontal turbine. The water's pressure is reduced from 30 psig (210 kPa) to 5 psig (35 kPa) vacuum. Disregarding friction, velocity, and other factors, what power is generated?

(A) 110 hp (82 kW)

(B) 380 hp (280 kW)

(C) 730 hp (540 kW)

(D) 920 hp (640 kW)

Drag

24. (*Time limit: one hour*) A car traveling through 70°F (20°C) air has the following characteristics.

frontal area	28 ft² (2.6 m²)
mass	3300 lbm (1500 kg)
drag coefficient	0.42
rolling resistance	1% of weight
engine thermal efficiency	28%
fuel heating value	115,000 Btu/gal (460 MJ/L)

(a) Considering only the drag and rolling resistance, what is the fuel consumption when the car is traveling at 55 mi/hr (90 km/h)?

(A) 0.026 gal/mi (0.0043 L/km)

(B) 0.038 gal/mi (0.0065 L/km)

(C) 0.051 gal/mi (0.0097 L/km)

(D) 0.13 gal/mi (0.022 L/km)

(b) What is the percentage increase in fuel consumption at 65 mi/hr (105 km/h) compared to at 55 mi/hr (90 km/h)?

(A) 10%

(B) 20%

(C) 30%

(D) 40%

Water Resources

SOLUTIONS

1. *Customary U.S. Solution*

For schedule-40 pipe,

$$D_A = 0.5054 \text{ ft}$$

$$D_B = 1.4063 \text{ ft}$$

Let point A be at zero elevation.

The total energy at point A from Bernoulli's equation is

$$E_{t,A} = E_p + E_v + E_z = \frac{p_A}{\rho} + \frac{v_A^2}{2g_c} + \frac{z_A g}{g_c}$$

At point A, the diameter is 6 in. The velocity at point A is

$$\dot{V} = v_A A_A = v_A \left(\frac{\pi}{4}\right) D_A^2$$

$$v_A = \left(\frac{4}{\pi}\right)\left(\frac{\dot{V}}{D_A^2}\right) = \left(\frac{4}{\pi}\right)\left(\frac{5 \frac{\text{ft}^3}{\text{sec}}}{(0.5054 \text{ ft})^2}\right)$$

$$= \boxed{24.9 \text{ ft/sec} \quad (25 \text{ ft/sec})}$$

$$p_A = \left(10 \frac{\text{lbf}}{\text{in}^2}\right)\left(12 \frac{\text{in}}{\text{ft}}\right)^2 = 1440 \text{ lbf/ft}^2$$

$$z_A = 0$$

For water, $\rho \approx 62.4 \text{ lbm/ft}^3$.

$$E_{t,A} = \frac{p_A}{\rho} + \frac{v_A^2}{2g_c} + \frac{z_A g}{g_c}$$

$$= \frac{1440 \frac{\text{lbf}}{\text{ft}^2}}{62.4 \frac{\text{lbm}}{\text{ft}^3}} + \frac{\left(24.9 \frac{\text{ft}}{\text{sec}}\right)^2}{(2)\left(32.2 \frac{\text{lbm-ft}}{\text{lbf-sec}^2}\right)} + 0$$

$$= 32.7 \text{ ft-lbf/lbm}$$

Similarly, the total energy at point B is

$$v_B = \left(\frac{4}{\pi}\right)\left(\frac{\dot{V}}{D_B^2}\right) = \left(\frac{4}{\pi}\right)\left(\frac{5 \frac{\text{ft}^3}{\text{sec}}}{(1.4063 \text{ ft})^2}\right)$$

$$= 3.22 \text{ ft/sec}$$

$$p_B = \left(7 \frac{\text{lbf}}{\text{in}^2}\right)\left(12 \frac{\text{in}}{\text{ft}}\right)^2 = 1008 \text{ lbf/ft}^2$$

$$z_B = 15 \text{ ft}$$

$$E_{t,B} = \frac{p_B}{\rho} + \frac{v_B^2}{2g_c} + \frac{z_B g}{g_c}$$

$$= \frac{1008 \frac{\text{lbf}}{\text{ft}^2}}{62.4 \frac{\text{lbm}}{\text{ft}^3}} + \frac{\left(3.22 \frac{\text{ft}}{\text{sec}}\right)^2}{(2)\left(32.2 \frac{\text{lbm-ft}}{\text{lbf-sec}^2}\right)}$$

$$+ \frac{(15 \text{ ft})\left(32.2 \frac{\text{ft}}{\text{sec}^2}\right)}{32.2 \frac{\text{lbm-ft}}{\text{lbf-sec}^2}}$$

$$= 31.3 \text{ ft-lbf/lbm}$$

Since $E_{t,A} > E_{t,B}$, $\boxed{\text{the flow is from point A to point B.}}$

The answer is (B).

SI Solution

Let point A be at zero elevation.

The total energy at point A from Bernoulli's equation is

$$E_{t,A} = E_p + E_v + E_z = \frac{p_A}{\rho} + \frac{v_A^2}{2} + z_A g$$

At point A, from App. 16.C, the diameter is 154 mm (0.154 m). The velocity at point A is

$$\dot{V} = v_A A_A = v_A \left(\frac{\pi}{4}\right) D_A^2$$

$$v_A = \left(\frac{4}{\pi}\right)\left(\frac{\dot{V}}{D_A^2}\right) = \left(\frac{4}{\pi}\right)\left(\frac{130 \frac{\text{L}}{\text{s}}}{(0.154 \text{ m})^2 \left(1000 \frac{\text{L}}{\text{m}^3}\right)}\right)$$

$$= \boxed{6.98 \text{ m/s} \quad (7 \text{ m/s})}$$

$$p_A = 70 \text{ kPa} \quad (70\,000 \text{ Pa})$$

$$z_A = 0$$

For water, $\rho = 1000 \text{ kg/m}^3$.

$$E_{t,A} = \frac{p_A}{\rho} + \frac{v_A^2}{2} + z_A g$$

$$= \frac{70\,000 \text{ Pa}}{1000 \frac{\text{kg}}{\text{m}^3}} + \frac{\left(6.98 \frac{\text{m}}{\text{s}}\right)^2}{2} + 0$$

$$= 94.36 \text{ J/kg}$$

Similarly, at B, the diameter is 0.429 m. The total energy at point B is

$$v_B = \left(\frac{4}{\pi}\right)\left(\frac{\dot{V}}{D_B^2}\right)$$

$$= \left(\frac{4}{\pi}\right)\left(\frac{130 \ \frac{L}{s}}{(0.429 \ \text{m})^2\left(1000 \ \frac{L}{m^3}\right)}\right)$$

$$= 0.90 \ \text{m/s}$$

$$p_B = 48.3 \ \text{kPa} \quad (48\,300 \ \text{Pa})$$

$$z_B = 4.6 \ \text{m}$$

$$E_{t,B} = \frac{p_B}{\rho} + \frac{v_B^2}{2} + z_B g$$

$$= \frac{48\,300 \ \text{Pa}}{1000 \ \frac{\text{kg}}{\text{m}^3}} + \frac{\left(0.90 \ \frac{\text{m}}{\text{s}}\right)^2}{2}$$

$$\qquad + (4.6 \ \text{m})\left(9.81 \ \frac{\text{m}}{\text{s}^2}\right)$$

$$= 93.8 \ \text{J/kg}$$

Since $E_{t,A} > E_{t,B}$, the flow is from point A to point B.

The answer is (B).

2.
$$\dot{V} = \left(750 \ \frac{\text{gal}}{\text{min}}\right)\left(0.002228 \ \frac{\text{ft}^3\text{-min}}{\text{sec-gal}}\right)$$

$$= 1.671 \ \text{ft}^3/\text{sec}$$

$D = 0.5054$ ft and $A = 0.2006 \ \text{ft}^2$.

$$v = \frac{\dot{V}}{A} = \frac{1.671 \ \frac{\text{ft}^3}{\text{sec}}}{0.2006 \ \text{ft}^2} = 8.33 \ \text{ft/sec}$$

For 60°F water,

$$\nu = 1.217 \times 10^{-5} \ \text{ft}^2/\text{sec}$$

$$\text{Re} = \frac{vD}{\nu} = \frac{\left(8.33 \ \frac{\text{ft}}{\text{sec}}\right)(0.5054 \ \text{ft})}{1.217 \times 10^{-5} \ \frac{\text{ft}^2}{\text{sec}}} = 3.46 \times 10^5$$

For steel,

$$\epsilon = 0.0002$$

$$\frac{\epsilon}{D} = \frac{0.0002 \ \text{ft}}{0.5054 \ \text{ft}} \approx 0.0004$$

$$f = 0.0175$$

From Eq. 17.22,

$$h_f = \frac{fLv^2}{2Dg} = \frac{(0.0175)(3000 \ \text{ft})\left(8.33 \ \frac{\text{ft}}{\text{sec}}\right)^2}{(2)(0.5054 \ \text{ft})\left(32.2 \ \frac{\text{ft}}{\text{sec}^2}\right)}$$

$$= 109.8 \ \text{ft}$$

Use the Bernoulli equation. Since velocity is the same at points A and B, it may be omitted.

$$\frac{\left(12 \ \frac{\text{in}}{\text{ft}}\right)^2 p_1}{62.37 \ \frac{\text{lbf}}{\text{ft}^3}} = \frac{\left(50 \ \frac{\text{lbf}}{\text{in}^2}\right)\left(12 \ \frac{\text{in}}{\text{ft}}\right)^2}{62.37 \ \frac{\text{lbf}}{\text{ft}^3}}$$

$$\qquad\qquad\qquad + 60 \ \text{ft} + 109.8 \ \text{ft}$$

$$p_1 = \boxed{123.5 \ \text{lbf/in}^2 \quad (120 \ \text{psig})}$$

The answer is (C).

3. (a) From Eq. 17.31, the friction loss from A to B is

$$h_{f,\text{ft}} = \frac{10.44 L_{\text{ft}} Q_{\text{gpm}}^{1.85}}{C^{1.85} d_{\text{in}}^{4.87}}$$

$$h_{f,\text{A-B}} = \frac{(10.44)(20,000 \ \text{ft})\left(120 \ \frac{\text{gal}}{\text{min}}\right)^{1.85}}{(150)^{1.85}(6 \ \text{in})^{4.87}}$$

$$= 22.4 \ \text{ft}$$

Calculate the velocity head.

$$v = \frac{\dot{V}}{A} = \frac{\left(120 \ \frac{\text{gal}}{\text{min}}\right)\left(0.002228 \ \frac{\text{ft}^3\text{-min}}{\text{sec-gal}}\right)}{\left(\frac{\pi}{4}\right)\left(\frac{6 \ \text{in}}{12 \ \frac{\text{in}}{\text{ft}}}\right)^2}$$

$$= 1.36 \ \text{ft/sec}$$

$$h_v = \frac{v^2}{2g} = \frac{\left(1.36 \ \frac{\text{ft}}{\text{sec}}\right)^2}{(2)\left(32.2 \ \frac{\text{ft}}{\text{sec}^2}\right)}$$

$$= 0.029 \ \text{ft}$$

Velocity heads are low and can be disregarded. The friction loss from B to C is

$$h_{f,\text{B-C}} = \frac{(10.44)(10{,}000\ \text{ft})\left(160\ \dfrac{\text{gal}}{\text{min}}\right)^{1.85}}{(150)^{1.85}(6\ \text{in})^{4.87}}$$
$$= 19.10\ \text{ft}$$

For C to D,

$$h_{f,\text{C-D}} = \frac{(10.44)(30{,}000\ \text{ft})\left(120\ \dfrac{\text{gal}}{\text{min}}\right)^{1.85}}{(150)^{1.85}(4\ \text{in})^{4.87}}$$
$$= 242.4\ \text{ft}$$

Assume a pressure of 20 psig at point A.

$$h_{p,\text{A}} = \frac{\left(20\ \dfrac{\text{lbf}}{\text{in}^2}\right)\left(12\ \dfrac{\text{in}}{\text{ft}}\right)^2}{62.4\ \dfrac{\text{lbf}}{\text{ft}^3}} = 46.2\ \text{ft}$$

From the Bernoulli equation, ignoring velocity head,

$$h_{p,\text{A}} + z_\text{A} = h_{p,\text{B}} + z_\text{B} + h_{f,\text{A-B}}$$
$$46.2\ \text{ft} + 620\ \text{ft} = h_{p,\text{B}} + 460\ \text{ft} + 22.4\ \text{ft}$$
$$h_{p,\text{B}} = 183.8\ \text{ft}$$

$$p_\text{B} = \gamma h = \frac{\left(62.4\ \dfrac{\text{lbf}}{\text{ft}^3}\right)(183.8\ \text{ft})}{\left(12\ \dfrac{\text{in}}{\text{ft}}\right)^2}$$
$$= 79.6\ \text{lbf/in}^2 \quad (\text{psig})$$

For B to C,

$$183.8\ \text{ft} + 460\ \text{ft} = h_{p,\text{C}} + 540\ \text{ft} + 19.10\ \text{ft}$$
$$h_{p,\text{C}} = 84.7\ \text{ft}$$

$$p_\text{C} = \frac{(84.7\ \text{ft})\left(62.4\ \dfrac{\text{lbf}}{\text{ft}^3}\right)}{\left(12\ \dfrac{\text{in}}{\text{ft}}\right)^2}$$
$$= 36.7\ \text{lbf/in}^2 \quad (\text{psig})$$

For C to D,

$$84.7\ \text{ft} + 540\ \text{ft} = h_{p,\text{D}} + 360\ \text{ft} + 242.4\ \text{ft}$$
$$h_{p,\text{D}} = 22.3\ \text{ft}$$

$$p_\text{D} = \frac{(22.3\ \text{ft})\left(62.4\ \dfrac{\text{lbf}}{\text{ft}^3}\right)}{\left(12\ \dfrac{\text{in}}{\text{ft}}\right)^2}$$
$$= 9.7\ \text{lbf/in}^2 \quad (\text{psig}) \quad [\text{too low}]$$

p_D is too low; therefore, add $20\ \text{lbf/in}^2 - 9.7\ \text{lbf/in}^2 = 10.3\ \text{lbf/in}^2$ (psig) to each point.

$$p_\text{A} = 20.0\ \frac{\text{lbf}}{\text{in}^2} + 10.3\ \frac{\text{lbf}}{\text{in}^2} = \boxed{30.3\ \text{lbf/in}^2 \quad (30\ \text{psig})}$$

$$p_\text{B} = 79.6\ \frac{\text{lbf}}{\text{in}^2} + 10.3\ \frac{\text{lbf}}{\text{in}^2} = 89.9\ \text{lbf/in}^2 \quad (\text{psig})$$

$$p_\text{C} = 36.7\ \frac{\text{lbf}}{\text{in}^2} + 10.3\ \frac{\text{lbf}}{\text{in}^2} = 47.0\ \text{lbf/in}^2 \quad (\text{psig})$$

$$p_\text{D} = 9.7\ \frac{\text{lbf}}{\text{in}^2} + 10.3\ \frac{\text{lbf}}{\text{in}^2} = 20.0\ \text{lbf/in}^2 \quad (\text{psig})$$

The answer is (C).

(b) The elevation of the hydraulic grade line above point D is the sum of the potential and static heads.

$$\Delta h_{\text{A-D}} = z_\text{A} - z_\text{D} + \frac{p_\text{A} - p_\text{D}}{\gamma}$$
$$= 620\ \text{ft} - 360\ \text{ft}$$
$$+ \frac{\left(30.3\ \dfrac{\text{lbf}}{\text{in}^2} - 20\ \dfrac{\text{lbf}}{\text{in}^2}\right)\left(12\ \dfrac{\text{in}}{\text{ft}}\right)^2}{62.4\ \dfrac{\text{lbf}}{\text{ft}^3}}$$
$$= \boxed{283.8\ \text{ft} \quad (280\ \text{ft})}$$

The answer is (A).

4. *Customary U.S. Solution*

For 6 in schedule-40 pipe, the internal diameter, D_i, is 0.5054 ft. The internal area is 0.2006 ft^2.

The velocity, v, is calculated from the volumetric flow, $\dot{V}$, and the flow area, A_i, by

$$\text{v} = \frac{\dot{V}}{A_i} = \frac{1.5\ \dfrac{\text{ft}^3}{\text{sec}}}{0.2006\ \text{ft}^2}$$
$$= 7.48\ \text{ft/sec}$$

Use App. 14.A. For water at 70°F, the kinematic viscosity, ν, is 1.059×10^{-5} ft^2/sec.

Calculate the Reynolds number.

$$\text{Re} = \frac{D_i \text{v}}{\nu} = \frac{(0.5054 \text{ ft})\left(7.48 \frac{\text{ft}}{\text{sec}}\right)}{1.059 \times 10^{-5} \frac{\text{ft}^2}{\text{sec}}}$$

$$= 3.57 \times 10^5$$

Since $\text{Re} > 2100$, the flow is turbulent. The friction loss coefficient can be determined from the Moody diagram.

For new steel pipe, the specific roughness, ϵ, is 0.0002 ft.

The relative roughness is

$$\frac{\epsilon}{D_i} = \frac{0.0002 \text{ ft}}{0.5054 \text{ ft}} = 0.0004$$

From the Moody diagram with $\text{Re} = 3.57 \times 10^5$ and $\epsilon/D_i = 0.0004$, the friction factor, f, is 0.0174.

Use Darcy's equation to compute the frictional loss.

$$h_f = \frac{fL\text{v}^2}{2D_i g} = \frac{(0.0174)(1200 \text{ ft})\left(7.48 \frac{\text{ft}}{\text{sec}}\right)^2}{(2)(0.5054 \text{ ft})\left(32.2 \frac{\text{ft}}{\text{sec}^2}\right)}$$

$$= \boxed{35.9 \text{ ft} \quad (36 \text{ ft})}$$

The answer is (C).

SI Solution

For 6 in pipe, the internal diameter is 154.1 mm, and the internal area is $186.5 \times 10^{-4} \text{ m}^2$.

The velocity, v, is calculated from the volumetric flow, $\dot{V}$, and the flow area, A_i, by

$$\text{v} = \frac{\dot{V}}{A_i} = \frac{40 \frac{\text{L}}{\text{s}}}{(186.5 \times 10^{-4} \text{ m}^2)\left(1000 \frac{\text{L}}{\text{m}^3}\right)}$$

$$= 2.145 \text{ m/s}$$

For water at 20°C, the kinematic viscosity is

$$\nu = \frac{\mu}{\rho} = \frac{1.0050 \times 10^{-3} \text{ Pa·s}}{998.23 \frac{\text{kg}}{\text{m}^3}}$$

$$= 1.007 \times 10^{-6} \text{ m}^2/\text{s}$$

Calculate the Reynolds number.

$$\text{Re} = \frac{D_i \text{v}}{\nu} = \frac{(154.1 \text{ mm})\left(2.145 \frac{\text{m}}{\text{s}}\right)}{\left(1.007 \times 10^{-6} \frac{\text{m}^2}{\text{s}}\right)\left(1000 \frac{\text{mm}}{\text{m}}\right)}$$

$$= 3.282 \times 10^5$$

Since $\text{Re} > 2100$, the flow is turbulent. The friction loss coefficient can be determined from the Moody diagram.

For new steel pipe, the specific roughness, ϵ, is 6.0×10^{-5} m.

The relative roughness is

$$\frac{\epsilon}{D_i} = \frac{6.0 \times 10^{-5} \text{ m}}{0.1541 \text{ m}} = 0.0004$$

From the Moody diagram with $\text{Re} = 3.28 \times 10^5$ and $\epsilon/D_i = 0.0004$, the friction factor, f, is 0.0175.

Use Darcy's equation to compute the frictional loss.

$$h_f = \frac{fL\text{v}^2}{2D_i g} = \frac{(0.0175)(355 \text{ m})\left(2.145 \frac{\text{m}}{\text{s}}\right)^2}{(2)(0.1541 \text{ m})\left(9.81 \frac{\text{m}}{\text{s}^2}\right)}$$

$$= \boxed{9.45 \text{ m} \quad (9.5 \text{ m})}$$

The answer is (C).

5. *Customary U.S. Solution*

For 6 in schedule-40 pipe, the internal diameter, D_i, is 0.5054 ft. The internal area is 0.2006 ft^2.

The velocity, v, is calculated from the volumetric flow, $\dot{V}$, and the flow area, A_i.

Convert the volumetric flow rate from gal/min to ft^3/sec.

$$\dot{V} = \frac{500 \frac{\text{gal}}{\text{min}}}{\left(60 \frac{\text{sec}}{\text{min}}\right)\left(7.48 \frac{\text{gal}}{\text{ft}^3}\right)} = 1.114 \text{ ft}^3/\text{sec}$$

The velocity is

$$\text{v} = \frac{\dot{V}}{A_i} = \frac{1.114 \frac{\text{ft}^3}{\text{sec}}}{0.2006 \text{ ft}^2}$$

$$= 5.55 \text{ ft/sec}$$

Use App. 14.A. For water at 100°F, the kinematic viscosity, ν, is 0.739×10^{-5} ft^2/sec, and the density is 62.00 lbm/ft^2.

Calculate the Reynolds number.

$$\text{Re} = \frac{D_i \text{v}}{\nu} = \frac{(0.5054 \text{ ft})\left(5.55 \frac{\text{ft}}{\text{sec}}\right)}{0.739 \times 10^{-5} \frac{\text{ft}^2}{\text{sec}}}$$

$$= 3.80 \times 10^5$$

Since Re > 2100, the flow is turbulent. The friction loss coefficient can be determined from the Moody diagram.

For new steel pipe, the specific roughness, ϵ, is 0.0002 ft.

The relative roughness is

$$\frac{\epsilon}{D_i} = \frac{0.0002 \text{ ft}}{0.5054 \text{ ft}}$$

$$= 0.0004$$

From the Moody diagram with Re = 3.80×10^5 and $\epsilon/D_i = 0.0004$, the friction factor, f, is 0.0173.

Use App. 17.D. The equivalent lengths of the valves and fittings are

standard radius elbow	2×8.9 ft =	17.8 ft
gate valve (fully open)	2×3.2 ft =	6.4 ft
90° angle valve (fully open)	1×63.0 ft =	63.0 ft
swing check valve	1×63.0 ft =	63.0 ft
		150.2 ft

The equivalent pipe length is the sum of the straight run of pipe and the equivalent length of pipe for the valves and fittings.

$$L_e = L + L_{\text{fittings}} = 300 \text{ ft} + 150.2 \text{ ft}$$

$$= 450.2 \text{ ft}$$

Use Darcy's equation to compute the frictional loss.

$$h_f = \frac{f L_e \text{v}^2}{2 D_i g} = \frac{(0.0173)(450.2 \text{ ft})\left(5.55 \frac{\text{ft}}{\text{sec}}\right)^2}{(2)(0.5054 \text{ ft})\left(32.2 \frac{\text{ft}}{\text{sec}^2}\right)}$$

$$= 7.37 \text{ ft}$$

The total difference in head is the sum of the head loss through the pipe, valves, and fittings and the change in elevation.

$$\Delta h = h_f + \Delta z = 7.37 \text{ ft} + 20 \text{ ft}$$

$$= 27.37 \text{ ft}$$

The pressure difference between the entrance and discharge is

$$\Delta p = \gamma \Delta h = \rho h \times \frac{g}{g_c}$$

$$= \frac{\left(62.0 \frac{\text{lbm}}{\text{ft}^3}\right)(27.37 \text{ ft})}{\left(12 \frac{\text{in}}{\text{ft}}\right)^2} \times \frac{32.2 \frac{\text{ft}}{\text{sec}^2}}{32.2 \frac{\text{lbm-ft}}{\text{lbm-sec}^2}}$$

$$= \boxed{11.8 \text{ lbf/in}^2 \quad (12 \text{ psi})}$$

The answer is (A).

SI Solution

For 6 in pipe, the internal diameter is 154.1 mm (0.1541 m).

The internal area is $186.5 \times 10^{-4} \text{ m}^2$.

The velocity, v, is calculated from the volumetric flow, $\dot{V}$, and the flow area, A_i.

$$\text{v} = \frac{\dot{V}}{A_i} = \frac{30 \frac{\text{L}}{\text{s}}}{(186.5 \times 10^{-4} \text{ m}^2)\left(1000 \frac{\text{L}}{\text{m}^3}\right)}$$

$$= 1.61 \text{ m/s}$$

Use App. 14.B. For water at 40°C, the kinematic viscosity is $6.611 \times 10^{-7} \text{ m}^2/\text{s}$, and the density is 992.25 kg/m^3.

Calculate the Reynolds number.

$$\text{Re} = \frac{D_i \text{v}}{\nu} = \frac{(0.1541 \text{ m})\left(1.61 \frac{\text{m}}{\text{s}}\right)}{6.611 \times 10^{-7} \frac{\text{m}^2}{\text{s}}}$$

$$= 3.75 \times 10^5$$

Since Re > 2100, the flow is turbulent. The friction loss coefficient can be determined from the Moody diagram.

For new steel pipe, the specific roughness is 6.0×10^{-5} m.

The relative roughness is

$$\frac{\epsilon}{D_i} = \frac{6.0 \times 10^{-5} \text{ m}}{0.1541 \text{ m}} = 0.0004$$

From the Moody diagram with Re = 3.75×10^5 and $\epsilon/D_i = 0.0004$, the friction factor, f, is 0.0173.

Use App. 17.D. The equivalent lengths of the valves and fittings are

standard radius elbow	2×2.7 m =	5.4 m
gate valve (fully open)	2×1.0 m =	2.0 m
90° angle valve (fully open)	1×18.9 m =	18.9 m
swing check valve	1×18.9 m =	18.9 m
		45.2 m

The equivalent pipe length is the sum of the straight run of pipe and the equivalent length of pipe for the valves and fittings.

$$L_e = L + L_{\text{fittings}} = 90 \text{ m} + 45.2 \text{ m}$$
$$= 135.2 \text{ m}$$

Use Darcy's equation to compute the frictional loss.

$$h_f = \frac{fL_e\text{v}^2}{2D_ig} = \frac{(0.0173)(135.2 \text{ m})\left(1.61 \ \frac{\text{m}}{\text{s}}\right)^2}{(2)(0.1541 \text{ m})\left(9.81 \ \frac{\text{m}}{\text{s}^2}\right)}$$
$$= 2.01 \text{ m}$$

The total difference in head is the sum of the head loss through the pipe, valves, and fittings and the change in elevation.

$$\Delta h = h_f + \Delta z = 2.01 \text{ m} + 6 \text{ m}$$
$$= 8.01 \text{ m}$$

The pressure difference between the entrance and discharge is

$$\Delta p = \rho hg = \left(992.25 \ \frac{\text{kg}}{\text{m}^3}\right)(8.01 \text{ m})\left(9.81 \ \frac{\text{m}}{\text{s}^2}\right)$$
$$= \boxed{77\,969 \text{ Pa} \quad (78 \text{ kPa})}$$

The answer is (A).

6. *Customary U.S. Solution*

For 6 in schedule-40 pipe, the internal diameter, D_i, is 0.5054 ft. The internal area is 0.2006 ft².

Use App. 14.D. For air at 70°F and atmospheric pressure, the kinematic viscosity is 16.15×10^{-5} ft²/sec.

Calculate the Reynolds number.

$$\text{Re} = \frac{D_i\text{v}}{\nu} = \frac{(0.5054 \text{ ft})\left(60 \ \frac{\text{ft}}{\text{sec}}\right)}{16.15 \times 10^{-5} \ \frac{\text{ft}^2}{\text{sec}}}$$
$$= 1.88 \times 10^5$$

Since Re > 2100, the flow is turbulent. The friction loss coefficient can be determined from the Moody diagram.

For new steel pipe, the specific roughness, ϵ, is 0.0002 ft.

The relative roughness is

$$\frac{\epsilon}{D_i} = \frac{0.0002 \text{ ft}}{0.5054 \text{ ft}} = 0.0004$$

From the Moody diagram with Re = 1.88×10^5 and $\epsilon/D_i = 0.0004$, the friction factor, f, is 0.0184.

Use App. 17.D. The equivalent lengths of the valves and fittings are

standard radius elbow	2×8.9 ft =	17.8 ft
gate valve (fully open)	2×3.2 ft =	6.4 ft
90° angle valve (fully open)	1×63.0 ft =	63.0 ft
swing check valve	1×63.0 ft =	63.0 ft
		150.2 ft

The equivalent pipe length is the sum of the straight run of pipe and the equivalent length of pipe for the valves and fittings.

$$L_e = L + L_{\text{fittings}} = 300 \text{ ft} + 150.2 \text{ ft}$$
$$= 450.2 \text{ ft}$$

Use Darcy's equation to compute the frictional loss.

$$h_f = \frac{fL_e\text{v}^2}{2D_ig} = \frac{(0.0184)(450.2 \text{ ft})\left(60 \ \frac{\text{ft}}{\text{sec}}\right)^2}{(2)(0.5054 \text{ ft})\left(32.2 \ \frac{\text{ft}}{\text{sec}^2}\right)}$$
$$= 916.2 \text{ ft}$$

The difference in head is the sum of the head loss through the pipe, valves, and fittings and the change in elevation.

$$h = h_f + \Delta z = 916.2 \text{ ft} + 20 \text{ ft}$$
$$= 936.2 \text{ ft}$$

The pressure difference between the entrance and discharge is

$$\Delta p = \gamma h = \rho h \times \frac{g}{g_c}$$
$$= \frac{\left(0.075 \ \frac{\text{lbm}}{\text{ft}^3}\right)(936.2 \text{ ft})}{\left(12 \ \frac{\text{in}}{\text{ft}}\right)^2} \times \frac{32.2 \ \frac{\text{ft}}{\text{sec}^2}}{32.2 \ \frac{\text{lbm-ft}}{\text{lbf-sec}^2}}$$
$$= \boxed{0.49 \text{ lbf/in}^2 \quad (0.49 \text{ psi})}$$

The answer is (B).

SI Solution

Use App. 16.C. For 6 in pipe, the internal diameter, D_i, is 154.1 mm (0.1541 m), and the internal area is 186.5×10^{-4} m^2.

Use App. 14.E. For air at 20°C, the kinematic viscosity, ν, is 1.51×10^{-5} m^2/s.

Calculate the Reynolds number.

$$\text{Re} = \frac{D_i \text{v}}{\nu} = \frac{(0.1541 \text{ m})\left(18 \; \dfrac{\text{m}}{\text{s}}\right)}{1.51 \times 10^{-5} \; \dfrac{\text{m}^2}{\text{s}}}$$

$$= 1.84 \times 10^5$$

Since Re > 2100, the flow is turbulent. The friction loss coefficient can be determined from the Moody diagram.

For new steel pipe, the specific roughness, ϵ, is 6.0×10^{-5} m.

The relative roughness is

$$\frac{\epsilon}{D_i} = \frac{6.0 \times 10^{-5}}{0.1541 \text{ m}} = 0.0004$$

From the Moody diagram with Re $= 1.84 \times 10^5$ and $\epsilon/D_i = 0.0004$, the friction factor, f, is 0.0185.

Compute the equivalent lengths of the valves and fittings. (Convert from App. 17.D.)

standard radius elbow	2×2.7 m $=$	5.4 m
gate valve (fully open)	2×1.0 m $=$	2.0 m
90° angle valve (fully open)	1×18.9 m $=$	18.9 m
swing check valve	1×18.9 m $=$	18.9 m
		45.2 m

The equivalent pipe length is the sum of the straight run of pipe and the equivalent length of pipe for the valves and fittings.

$$L_e = L + L_{\text{fittings}} = 90 \text{ m} + 45.2 \text{ m}$$

$$= 135.2 \text{ m}$$

Use Darcy's equation to compute the frictional loss.

$$h_f = \frac{f L_e \text{v}^2}{2 D_i g} = \frac{(0.0185)(135.2 \text{ m})\left(18 \; \dfrac{\text{m}}{\text{s}}\right)^2}{(2)(0.1541 \text{ m})\left(9.81 \; \dfrac{\text{m}}{\text{s}^2}\right)}$$

$$= 268.0 \text{ m}$$

The difference in head is the sum of the head loss through the pipe, valves, and fittings and the change in elevation.

$$\Delta h = h_f + \Delta z = 268.0 \text{ m} + 6 \text{ m}$$

$$= 274.0 \text{ m}$$

Assume the density of the air, ρ, is approximately 1.20 kg/m^3.

The pressure difference between the entrance and discharge is

$$\Delta p = \rho h g = \left(1.20 \; \frac{\text{kg}}{\text{m}^3}\right)(274.0 \text{ m})\left(9.81 \; \frac{\text{m}}{\text{s}^2}\right)$$

$$= \boxed{3226 \text{ Pa} \quad (3.2 \text{ kPa})}$$

The answer is (B).

7. Assume that flows from reservoirs A and B are toward D and then toward C. Then,

$$\dot{V}_{\text{A-D}} + \dot{V}_{\text{B-D}} = \dot{V}_{\text{D-C}}$$

Or,

$$A_{\text{A}}\text{v}_{\text{A-D}} + A_{\text{B}}\text{v}_{\text{B-D}} - A_{\text{C}}\text{v}_{\text{D-C}} = 0$$

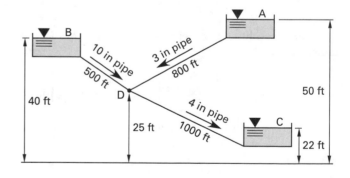

From App. 16.B, for schedule-40 pipe,

$$A_{\text{A}} = 0.05134 \text{ ft}^2 \qquad D_{\text{A}} = 0.2557 \text{ ft}$$

$$A_{\text{B}} = 0.5476 \text{ ft}^2 \qquad D_{\text{B}} = 0.8350 \text{ ft}$$

$$A_{\text{C}} = 0.08841 \text{ ft}^2 \qquad D_{\text{C}} = 0.3355 \text{ ft}$$

$$0.05134\text{v}_{\text{A-D}} + 0.5476\text{v}_{\text{B-D}} - 0.08841\text{v}_{\text{D-C}} = 0 \quad \text{[Eq. I]}$$

Ignoring the velocity heads, the conservation of energy equation between A and D is

$$z_A = \frac{p_D}{\gamma} + z_D + h_{f,\text{A-D}}$$

$$50 \text{ ft} = \frac{p_D}{62.4 \frac{\text{lbf}}{\text{ft}^3}} + 25 \text{ ft} + \frac{(0.02)(800 \text{ ft})v_{\text{A-D}}^2}{(2)(0.2557 \text{ ft})\left(32.2 \frac{\text{ft}}{\text{sec}^2}\right)}$$

Or,

$$v_{\text{A-D}} = \sqrt{25.73 - 0.0165 p_D} \quad [\text{Eq. II}]$$

Similarly, for B–D,

$$40 \text{ ft} = \frac{p_D}{62.4 \frac{\text{lbf}}{\text{ft}^3}} + 25 \text{ ft} + \frac{(0.02)(500 \text{ ft})v_{\text{B-D}}^2}{(2)(0.8350 \text{ ft})\left(32.2 \frac{\text{ft}}{\text{sec}^2}\right)}$$

Or,

$$v_{\text{B-D}} = \sqrt{80.66 - 0.0862 p_D} \quad [\text{Eq. III}]$$

For D–C,

$$22 \text{ ft} = \frac{p_D}{62.4 \frac{\text{lbf}}{\text{ft}^3}} + 25 \text{ ft} - \frac{(0.02)(1000 \text{ ft})v_{\text{D-C}}^2}{(2)(0.3355 \text{ ft})\left(32.2 \frac{\text{ft}}{\text{sec}^2}\right)}$$

Or,

$$v_{\text{D-C}} = \sqrt{3.24 + 0.0173 p_D} \quad [\text{Eq. IV}]$$

Equations I, II, III, and IV must be solved simultaneously. To do this, assume a value for p_D. This value then determines all three velocities in Eqs. II, III, and IV. These velocities are substituted into Eq. I. A trial and error solution yields

$$v_{\text{A-D}} = 3.21 \text{ ft/sec}$$
$$v_{\text{B-D}} = 0.408 \text{ ft/sec}$$
$$v_{\text{D-C}} = 4.40 \text{ ft/sec}$$
$$\boxed{p_D = 933.8 \text{ lbf/ft}^2 \quad (930 \text{ psf})}$$
$$\boxed{\text{Flow is from B to D.}}$$

The answer is (B).

8. *Customary U.S. Solution*

(a) The composite modulus of elasticity of the pipe and water is given by Eq. 17.208.

For water at 70°F, $E_{\text{water}} = 320 \times 10^3 \text{ lbf/in}^2$.

For cast-iron pipe, $E_{\text{pipe}} = 20 \times 10^6 \text{ lbf/in}^2$.

$$E = \frac{E_{\text{water}} t_{\text{pipe}} E_{\text{pipe}}}{t_{\text{pipe}} E_{\text{pipe}} + D_{\text{pipe}} E_{\text{water}}}$$

$$= \frac{\left(320 \times 10^3 \frac{\text{lbf}}{\text{in}^2}\right)(0.75 \text{ in})\left(20 \times 10^6 \frac{\text{lbf}}{\text{in}^2}\right)}{(0.75 \text{ in})\left(20 \times 10^6 \frac{\text{lbf}}{\text{in}^2}\right)}$$
$$+ (24 \text{ in})\left(320 \times 10^3 \frac{\text{lbf}}{\text{in}^2}\right)$$

$$= 2.12 \times 10^5 \text{ lbf/in}^2$$

The speed of sound in the pipe is

$$a = \sqrt{\frac{E g_c}{\rho}}$$

$$= \sqrt{\frac{\left(2.12 \times 10^5 \frac{\text{lbf}}{\text{in}^2}\right)\left(12 \frac{\text{in}}{\text{ft}}\right)^2\left(32.2 \frac{\text{lbm-ft}}{\text{lbf-sec}^2}\right)}{62.3 \frac{\text{lbm}}{\text{ft}^3}}}$$

$$= 3972 \text{ ft/sec}$$

The maximum pressure is given by Eq. 17.207.

$$\Delta p = \frac{\rho a \Delta v}{g_c}$$

$$= \frac{\left(62.3 \frac{\text{lbm}}{\text{ft}^3}\right)\left(3972 \frac{\text{ft}}{\text{sec}}\right)\left(6 \frac{\text{ft}}{\text{sec}}\right)}{\left(32.2 \frac{\text{lbm-ft}}{\text{lbf-sec}^2}\right)\left(12 \frac{\text{in}}{\text{ft}}\right)^2}$$

$$= \boxed{320.2 \text{ lbf/in}^2 \quad (320 \text{ psi})}$$

The answer is (C).

(b) The length of time the pressure is constant at the valve is given by

$$t = \frac{2L}{a} = \frac{(2)(500 \text{ ft})}{3972 \frac{\text{ft}}{\text{sec}}}$$

$$= \boxed{0.25 \text{ sec}}$$

The answer is (A).

SI Solution

(a) The composite modulus of elasticity of the pipe and water is given by Eq. 17.208.

For water at 20°C, $E_{\text{water}} = 2.2 \times 10^9 \text{ Pa}$.

For cast-iron pipe, $E_{\text{pipe}} = 1.4 \times 10^{11}$ Pa.

$$E = \frac{E_{\text{water}} t_{\text{pipe}} E_{\text{pipe}}}{t_{\text{pipe}} E_{\text{pipe}} + D_{\text{pipe}} E_{\text{water}}}$$

$$= \frac{(2.2 \times 10^9 \text{ Pa})(0.02 \text{ m})(1.4 \times 10^{11} \text{ Pa})}{(0.02 \text{ m})(1.4 \times 10^{11} \text{ Pa}) + (0.6 \text{ m})(2.2 \times 10^9 \text{ Pa})}$$

$$= 1.50 \times 10^9 \text{ Pa}$$

The speed of sound in the pipe is

$$a = \sqrt{\frac{E}{\rho}} = \sqrt{\frac{1.50 \times 10^9 \text{ Pa}}{1000 \dfrac{\text{kg}}{\text{m}^3}}}$$

$$= 1225 \text{ m/s}$$

The maximum pressure is given by

$$\Delta p = \rho a \Delta v$$

$$= \left(1000 \ \frac{\text{kg}}{\text{m}^3}\right)\left(1225 \ \frac{\text{m}}{\text{s}}\right)\left(2 \ \frac{\text{m}}{\text{s}}\right)$$

$$= \boxed{2.45 \times 10^6 \text{ Pa} \quad (2.5 \text{ MPa})}$$

The answer is (C).

(b) The length of time the pressure is constant at the valve is given by

$$t = \frac{2L}{a} = \frac{(2)(150 \text{ m})}{1225 \ \dfrac{\text{m}}{\text{s}}}$$

$$= \boxed{0.25 \text{ s}}$$

The answer is (A).

9. *Customary U.S. Solution*

(a) First, it is necessary to collect data on schedule-40 pipe and water. The fluid viscosity, pipe dimensions, and other parameters can be found in various appendices in Chap. 14 and Chap. 16. At 70°F water, $\nu = 1.059 \times 10^{-5}$ ft^2/sec.

From Table 17.2, $\epsilon = 0.0002$ ft.

8 in pipe	$D = 0.6651$ ft	
	$A = 0.3474$ ft^2	
12 in pipe	$D = 0.9948$ ft	
	$A = 0.7773$ ft^2	
16 in pipe	$D = 1.25$ ft	
	$A = 1.2272$ ft^2	

The flow quantity is converted from gallons per minute to cubic feet per second.

$$\dot{V} = \frac{(8 \text{ MGD})\left(10^6 \ \dfrac{\frac{\text{gal}}{\text{day}}}{\text{MGD}}\right)\left(0.002228 \ \dfrac{\frac{\text{ft}^3}{\text{sec}}}{\frac{\text{gal}}{\text{min}}}\right)}{\left(24 \ \dfrac{\text{hr}}{\text{day}}\right)\left(60 \ \dfrac{\text{min}}{\text{hr}}\right)}$$

$$= 12.378 \text{ ft}^3/\text{sec}$$

For the inlet pipe, the velocity is

$$v = \frac{\dot{V}}{A} = \frac{12.378 \ \dfrac{\text{ft}^3}{\text{sec}}}{0.3474 \text{ ft}^2} = 35.63 \text{ ft/sec}$$

The Reynolds number is

$$Re = \frac{Dv}{\nu} = \frac{(0.6651 \text{ ft})\left(35.63 \ \dfrac{\text{ft}}{\text{sec}}\right)}{1.059 \times 10^{-5} \ \dfrac{\text{ft}^2}{\text{sec}}}$$

$$= 2.24 \times 10^6$$

The relative roughness is

$$\frac{\epsilon}{D} = \frac{0.0002 \text{ ft}}{0.6651 \text{ ft}} = 0.0003$$

From the Moody diagram, $f = 0.015$.

Equation 17.23(b) is used to calculate the frictional energy loss.

$$E_{f,1} = h_f \times \frac{g}{g_c} = \frac{f L v^2}{2 D g_c}$$

$$= \frac{(0.015)(1000 \text{ ft})\left(35.63 \ \dfrac{\text{ft}}{\text{sec}}\right)^2}{(2)(0.6651 \text{ ft})\left(32.2 \ \dfrac{\text{lbm-ft}}{\text{lbf-sec}^2}\right)}$$

$$= 444.6 \text{ ft-lbf/lbm}$$

For the outlet pipe, the velocity is

$$v = \frac{\dot{V}}{A} = \frac{12.378 \ \dfrac{\text{ft}^3}{\text{sec}}}{0.7773 \text{ ft}^2} = 15.92 \text{ ft/sec}$$

The Reynolds number is

$$\mathrm{Re} = \frac{Dv}{\nu} = \frac{(0.9948 \text{ ft})\left(15.92 \dfrac{\text{ft}}{\text{sec}}\right)}{1.059 \times 10^{-5} \dfrac{\text{ft}^2}{\text{sec}}}$$

$$= 1.5 \times 10^6$$

The relative roughness is

$$\frac{\epsilon}{D} = \frac{0.0002 \text{ ft}}{0.9948 \text{ ft}} = 0.0002$$

From the Moody diagram, $f = 0.014$.

Equation 17.23(b) is used to calculate the frictional energy loss.

$$E_{f,2} = h_f \times \frac{g}{g_c} = \frac{fLv^2}{2Dg_c}$$

$$= \frac{(0.014)(1500 \text{ ft})\left(15.92 \dfrac{\text{ft}}{\text{sec}}\right)^2}{(2)(0.9948 \text{ ft})\left(32.2 \dfrac{\text{lbm-ft}}{\text{lbf-sec}^2}\right)}$$

$$= 83.1 \text{ ft-lbf/lbm}$$

Assume a 50% split through the two branches. In the upper branch, the velocity is

$$v = \frac{\dot{V}}{A} = \frac{\left(\frac{1}{2}\right)\left(12.378 \dfrac{\text{ft}^3}{\text{sec}}\right)}{0.3474 \text{ ft}^2} = 17.81 \text{ ft/sec}$$

The Reynolds number is

$$\mathrm{Re} = \frac{Dv}{\nu} = \frac{(0.6651 \text{ ft})\left(17.81 \dfrac{\text{ft}}{\text{sec}}\right)}{1.059 \times 10^{-5} \dfrac{\text{ft}^2}{\text{sec}}}$$

$$= 1.1 \times 10^6$$

The relative roughness is

$$\frac{\epsilon}{D} = \frac{0.0002 \text{ ft}}{0.6651 \text{ ft}} = 0.0003$$

From the Moody diagram, $f = 0.015$.

For the 16 in pipe in the lower branch, the velocity is

$$v = \frac{\dot{V}}{A} = \frac{\left(\frac{1}{2}\right)\left(12.378 \dfrac{\text{ft}^3}{\text{sec}}\right)}{1.2272 \text{ ft}^2} = 5.04 \text{ ft/sec}$$

The Reynolds number is

$$\mathrm{Re} = \frac{Dv}{\nu} = \frac{(1.25 \text{ ft})\left(5.04 \dfrac{\text{ft}}{\text{sec}}\right)}{1.059 \times 10^{-5} \dfrac{\text{ft}^2}{\text{sec}}}$$

$$= 5.95 \times 10^5$$

The relative roughness is

$$\frac{\epsilon}{D} = \frac{0.0002 \text{ ft}}{1.25 \text{ ft}} = 0.00016$$

From the Moody diagram, $f = 0.015$.

These values of f for the two branches are fairly insensitive to changes in $\dot{V}$, so they will be used for the rest of the problem in the upper branch.

Equation 17.23(b) is used to calculate the frictional energy loss in the upper branch.

$$E_{f,\text{upper}} = h_f \times \frac{g}{g_c} = \frac{fLv^2}{2Dg_c}$$

$$= \frac{(0.015)(500 \text{ ft})\left(17.81 \dfrac{\text{ft}}{\text{sec}}\right)^2}{(2)(0.6651 \text{ ft})\left(32.2 \dfrac{\text{lbm-ft}}{\text{lbf-sec}^2}\right)}$$

$$= 55.5 \text{ ft-lbf/lbm}$$

To calculate a loss for any other flow in the upper branch,

$$E_{f,\text{upper 2}} = E_{f,\text{upper}} \left(\frac{\dot{V}}{\left(\frac{1}{2}\right)\left(12.378 \dfrac{\text{ft}^3}{\text{sec}}\right)}\right)^2$$

$$= \left(55.5 \dfrac{\text{ft-lbf}}{\text{lbm}}\right)\left(\frac{\dot{V}}{6.189 \dfrac{\text{ft}^3}{\text{sec}}}\right)^2$$

$$= 1.45 \dot{V}^2$$

Similarly, for the lower branch, in the 8 in section,

$$E_{f,\text{lower,8 in}} = h_f \times \frac{g}{g_c} = \frac{fLv^2}{2Dg_c}$$

$$= \frac{(0.015)(250 \text{ ft})\left(17.81 \dfrac{\text{ft}}{\text{sec}}\right)^2}{(2)(0.6651 \text{ ft})\left(32.2 \dfrac{\text{lbm-ft}}{\text{lbf-sec}^2}\right)}$$

$$= 27.8 \text{ ft-lbf/lbm}$$

For the lower branch, in the 16 in section,

$$E_{f,\text{lower},16\,\text{in}} = h_f \times \frac{g}{g_c} = \frac{fLv^2}{2Dg_c}$$

$$= \frac{(0.015)(1000\ \text{ft})\left(5.04\ \dfrac{\text{ft}}{\text{sec}}\right)^2}{(2)(1.25\ \text{ft})\left(32.2\ \dfrac{\text{lbm-ft}}{\text{lbf-sec}^2}\right)}$$

$$= 4.7\ \text{ft-lbf/lbm}$$

The total loss in the lower branch is

$$E_{f,\text{lower}} = E_{f,\text{lower},8\,\text{in}} + E_{f,\text{lower},16\,\text{in}}$$

$$= 27.8\ \frac{\text{ft-lbf}}{\text{lbm}} + 4.7\ \frac{\text{ft-lbf}}{\text{lbm}}$$

$$= 32.5\ \text{ft-lbf/lbm}$$

To calculate a loss for any other flow in the lower branch,

$$E_{f,\text{lower 2}} = E_{f,\text{lower}} \left(\frac{\dot{V}}{\left(\frac{1}{2}\right)\left(12.378\ \dfrac{\text{ft}^3}{\text{sec}}\right)} \right)^2$$

$$= \left(32.5\ \frac{\text{ft-lbf}}{\text{lbm}}\right) \left(\frac{\dot{V}}{6.189\ \dfrac{\text{ft}^3}{\text{sec}}} \right)^2$$

$$= 0.85\,\dot{V}^2$$

Let x be the fraction flowing in the upper branch. Then, because the friction losses are equal,

$$E_{f,\text{upper 2}} = E_{f,\text{lower 2}}$$

$$1.45x^2 = (0.85)(1-x)^2$$

$$x = 0.432$$

$$\dot{V}_{\text{upper}} = (0.432)\left(12.378\ \frac{\text{ft}^3}{\text{sec}}\right)$$

$$= \boxed{5.347\ \text{ft}^3/\text{sec} \quad (5.3\ \text{ft}^3/\text{sec})}$$

The answer is (D).

(b) $\qquad \dot{V}_{\text{lower}} = (1 - 0.432)\left(12.378\ \dfrac{\text{ft}^3}{\text{sec}}\right)$

$$= 7.03\ \text{ft}^3/\text{sec}$$

$$E_{f,\text{total}} = E_{f,1} + E_{f,\text{lower 2}} + E_{f,2}$$

$$E_{f,\text{lower 2}} = 0.85\,\dot{V}_{\text{lower}}^2$$

$$= (0.85)\left(7.03\ \frac{\text{ft}^3}{\text{sec}}\right)^2$$

$$= 42.0\ \text{ft}$$

$$E_{f,\text{total}} = 444.6\ \frac{\text{ft-lbf}}{\text{lbm}} + 42.0\ \frac{\text{ft-lbf}}{\text{lbm}} + 83.1\ \frac{\text{ft-lbf}}{\text{lbm}}$$

$$= \boxed{569.7\ \text{ft-lbf/lbm} \quad (570\ \text{ft-lbf/lbm})}$$

The answer is (D).

SI Solution

(a) First, it is necessary to collect data on schedule-40 pipe and water. The fluid viscosity, pipe dimensions, and other parameters can be found in various appendices in Chap. 14 and Chap. 16. At 20°C water, $\nu = 1.007 \times 10^{-6}\ \text{m}^2/\text{s}$.

From Table 17.2, $\epsilon = 6 \times 10^{-5}\ \text{m}$.

$$
\begin{array}{ll}
\text{8 in pipe} & D = 202.7\ \text{mm} \\
& A = 322.7 \times 10^{-4}\ \text{m}^2 \\
\text{12 in pipe} & D = 303.2\ \text{mm} \\
& A = 721.9 \times 10^{-4}\ \text{m}^2 \\
\text{16 in pipe} & D = 381\ \text{mm} \\
& A = 1140 \times 10^{-4}\ \text{m}^2
\end{array}
$$

For the inlet pipe, the velocity is

$$v = \frac{\dot{V}}{A} = \frac{350\ \dfrac{\text{L}}{\text{s}}}{(322.7 \times 10^{-4}\ \text{m}^2)\left(1000\ \dfrac{\text{L}}{\text{m}^3}\right)} = 10.85\ \text{m/s}$$

The Reynolds number is

$$Re = \frac{Dv}{\nu} = \frac{(0.2027\ \text{m})\left(10.85\ \dfrac{\text{m}}{\text{s}}\right)}{1.007 \times 10^{-6}\ \dfrac{\text{m}^2}{\text{s}}}$$

$$= 2.18 \times 10^6$$

The relative roughness is

$$\frac{\epsilon}{D} = \frac{6 \times 10^{-5}\ \text{m}}{0.2027\ \text{m}} = 0.0003$$

From the Moody diagram, $f = 0.015$.

Equation 17.23(a) is used to calculate the frictional energy loss.

$$E_{f,1} = h_f g = \frac{fLv^2}{2D}$$

$$= \frac{(0.015)(300\ \text{m})\left(10.85\ \dfrac{\text{m}}{\text{s}}\right)^2}{(2)(0.2027\ \text{m})}$$

$$= 1307\ \text{J/kg}$$

For the outlet pipe, the velocity is

$$v = \frac{\dot{V}}{A} = \frac{350 \frac{L}{s}}{\left(721.9 \times 10^{-4} \text{ m}^2\right)\left(1000 \frac{L}{m^3}\right)} = 4.848 \text{ m/s}$$

The Reynolds number is

$$\text{Re} = \frac{Dv}{\nu} = \frac{(0.3032 \text{ m})\left(4.848 \frac{m}{s}\right)}{1.007 \times 10^{-6} \frac{m^2}{s}}$$
$$= 1.46 \times 10^6$$

The relative roughness is

$$\frac{\epsilon}{D} = \frac{6 \times 10^{-5} \text{ m}}{0.3032 \text{ m}} = 0.0002$$

From the Moody diagram, $f = 0.014$.

Equation 17.23(a) is used to calculate the frictional energy loss.

$$E_{f,2} = h_f g = \frac{fLv^2}{2D}$$
$$= \frac{(0.014)(450 \text{ m})\left(4.848 \frac{m}{s}\right)^2}{(2)(0.3032 \text{ m})}$$
$$= 244.2 \text{ J/kg}$$

Assume a 50% split through the two branches. In the upper branch, the velocity is

$$v = \frac{\dot{V}}{A} = \frac{\left(\frac{1}{2}\right)\left(350 \frac{L}{s}\right)}{\left(322.7 \times 10^{-4} \text{ m}^2\right)\left(1000 \frac{L}{m^3}\right)} = 5.423 \text{ m/s}$$

The Reynolds number is

$$\text{Re} = \frac{Dv}{\nu} = \frac{(0.2027 \text{ m})\left(5.423 \frac{m}{s}\right)}{1.007 \times 10^{-6} \frac{m^2}{s}}$$
$$= 1.1 \times 10^6$$

The relative roughness is

$$\frac{\epsilon}{D} = \frac{6 \times 10^{-5} \text{ m}}{0.2027 \text{ m}} = 0.0003$$

From the Moody diagram, $f = 0.015$.

For the 16 in pipe in the lower branch, the velocity is

$$v = \frac{\dot{V}}{A} = \frac{\left(\frac{1}{2}\right)\left(350 \frac{L}{s}\right)}{\left(1140 \times 10^{-4} \text{ m}^2\right)\left(1000 \frac{L}{m^3}\right)} = 1.535 \text{ m/s}$$

The Reynolds number is

$$\text{Re} = \frac{Dv}{\nu} = \frac{(0.381 \text{ m})\left(1.535 \frac{m}{s}\right)}{1.007 \times 10^{-6} \frac{m^2}{s}}$$
$$= 5.81 \times 10^5$$

The relative roughness is

$$\frac{\epsilon}{D} = \frac{6 \times 10^{-5} \text{ m}}{0.381 \text{ m}} = 0.00016$$

From the Moody diagram, $f = 0.015$.

These values of f for the two branches are fairly insensitive to changes in $\dot{V}$, so they will be used for the rest of the problem in the upper branch.

Equation 17.23(a) is used to calculate the frictional energy loss in the upper branch.

$$E_{f,\text{upper}} = h_f g = \frac{fLv^2}{2D}$$
$$= \frac{(0.015)(150 \text{ m})\left(5.423 \frac{m}{s}\right)^2}{(2)(0.2027 \text{ m})}$$
$$= 163.2 \text{ J/kg}$$

To calculate a loss for any other flow in the upper branch,

$$E_{f,\text{upper }2} = E_{f,\text{upper}}\left(\frac{\dot{V}}{\left(\frac{1}{2}\right)\left(0.350 \frac{m^3}{s}\right)}\right)^2$$
$$= \left(163.2 \frac{J}{kg}\right)\left(\frac{\dot{V}}{0.175 \frac{m^3}{s}}\right)^2$$
$$= 5329\dot{V}^2$$

Similarly, for the lower branch, in the 8 in section,

$$E_{f,\text{lower,8 in}} = h_f g = \frac{fLv^2}{2D} = \frac{(0.015)(75 \text{ m})\left(5.423 \frac{m}{s}\right)^2}{(2)(0.2027 \text{ m})}$$
$$= 81.61 \text{ J/kg}$$

For the lower branch, in the 16 in section,

$$E_{f,\text{lower},16\,\text{in}} = h_f g = \frac{fLv^2}{2D} = \frac{(0.015)(300 \text{ m})\left(1.585 \ \frac{\text{m}}{\text{s}}\right)^2}{(2)(0.381 \text{ m})}$$

$$= 14.84 \text{ J/kg}$$

The total loss in the lower branch is

$$E_{f,\text{lower}} = E_{f,\text{lower},8\,\text{in}} + E_{f,\text{lower},16\,\text{in}}$$

$$= 81.61 \ \frac{\text{J}}{\text{kg}} + 14.84 \ \frac{\text{J}}{\text{kg}}$$

$$= 96.45 \text{ J/kg}$$

To calculate a loss for any other flow in the lower branch,

$$E_{f,\text{lower }2} = E_{f,\text{lower}} \left(\frac{\dot{V}}{\left(\frac{1}{2}\right)\left(0.350 \ \frac{\text{m}^3}{\text{s}}\right)} \right)^2$$

$$= \left(96.45 \ \frac{\text{J}}{\text{kg}} \right)\left(\frac{\dot{V}}{0.175 \ \frac{\text{m}^3}{\text{s}}} \right)^2$$

$$= 3149 \dot{V}^2$$

Let x be the fraction flowing in the upper branch. Then, because the friction losses are equal,

$$E_{f,\text{upper }2} = E_{f,\text{lower }2}$$

$$5329 x^2 = (3149)(1-x)^2$$

$$x = 0.435$$

$$\dot{V}_{\text{upper}} = (0.435)\left(0.350 \ \frac{\text{m}^3}{\text{s}}\right)$$

$$= \boxed{0.15 \text{ m}^3/\text{s}}$$

The answer is (D).

(b) $\qquad \dot{V}_{\text{lower}} = (1-0.435)\left(0.350 \ \frac{\text{m}^3}{\text{s}}\right)$

$$= 0.198 \text{ m}^3/\text{s}$$

$$E_{f,\text{total}} = E_{f,1} + E_{f,\text{lower }2} + E_{f,2}$$

$$E_{f,\text{lower }2} = 3149 \dot{V}^2_{\text{lower}}$$

$$= (3149)\left(0.198 \ \frac{\text{m}^3}{\text{s}}\right)^2$$

$$= 123.5 \text{ J/kg}$$

$$E_{f,\text{total}} = 1307 \ \frac{\text{J}}{\text{kg}} + 123.5 \ \frac{\text{J}}{\text{kg}} + 244.2 \ \frac{\text{J}}{\text{kg}}$$

$$= \boxed{1675 \text{ J/kg} \ (1.7 \text{ kJ/kg})}$$

The answer is (D).

10. *steps 1, 2, and 3:*

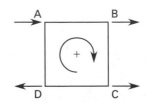

step 4: There is only one loop: ABCD.

step 5: Use Eq. 17.134.

$$K' = \frac{10.44L}{d^{4.87} C^{1.85}}$$

pipe AB: $\quad K' = \dfrac{(10.44)(1000 \text{ ft})}{(8 \text{ in})^{4.87}(100)^{1.85}}$

$$= 8.33 \times 10^{-5}$$

pipe BC: $\quad K' = \dfrac{(10.44)(1000 \text{ ft})}{(6 \text{ in})^{4.87}(100)^{1.85}}$

$$= 3.38 \times 10^{-4}$$

pipe CD: $\quad K' = 8.33 \times 10^{-5} \quad$ [same as AB]

pipe DA: $\quad K' = \dfrac{(10.44)(1000 \text{ ft})}{(12 \text{ in})^{4.87}(100)^{1.85}}$

$$= 1.16 \times 10^{-5}$$

step 6: Assume the flows are as shown in the following illustration.

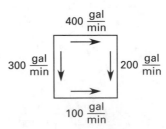

step 7: Use Eq. 17.142.

$$\delta = \frac{-\sum K' \dot{V}_a^n}{n\sum \left| K' \dot{V}_a^{n-1} \right|}$$

$$= \frac{\begin{array}{l} -\Big((8.33 \times 10^{-5})(400)^{1.85} + (3.38 \times 10^{-4})(200)^{1.85} \\ \quad - (8.33 \times 10^{-5})(100)^{1.85} \\ \quad - (1.16 \times 10^{-5})(300)^{1.85}\Big) \end{array}}{\begin{array}{l} (1.85)\Big((8.33 \times 10^{-5})(400)^{0.85} \\ \quad + (3.38 \times 10^{-4})(200)^{0.85} \\ \quad + (8.33 \times 10^{-5})(100)^{0.85} \\ \quad + (1.16 \times 10^{-5})(300)^{0.85}\Big) \end{array}}$$

$$= \frac{-10.67}{(1.85)\left(4.98 \times 10^{-2}\right)}$$

$$= -116 \text{ gal/min}$$

step 8: The adjusted flows are shown.

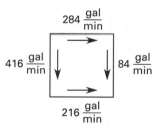

step 7: $\delta = -24$ gal/min

step 8: The adjusted flows are shown.

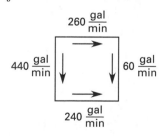

step 7: $\delta = -2$ gal/min   [small enough]

step 8: The final adjusted flows are shown.

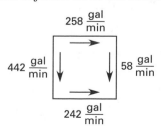

The answer is (B).

11. (a) This is a Hardy Cross problem. The pressure at point C does not change the solution procedure.

step 1: The Hazen-Williams roughness coefficient is given.

step 2: Choose clockwise as positive.

step 3: Nodes are already numbered.

step 4: Choose the loops as shown.

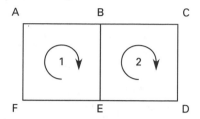

step 5: $\dot{V}$ is in gal/min, so use Eq. 17.134. d is in inches. L is in feet.

Each pipe has the same length.

$$K' = \frac{10.44L}{d^{4.87} C^{1.85}}$$

$$K'_{8\text{in}} = \frac{(10.44)(1000 \text{ ft})}{(8 \text{ in})^{4.87}(100)^{1.85}} = 8.33 \times 10^{-5}$$

$$K'_{6\text{in}} = \frac{(10.44)(1000 \text{ ft})}{(6 \text{ in})^{4.87}(100)^{1.85}} = 3.38 \times 10^{-4}$$

$$K'_{CE} = 2K'_{6\text{in}} = 6.76 \times 10^{-4}$$

$$K'_{EA} = K'_{6\text{in}} + K'_{8\text{in}} = 4.21 \times 10^{-4}$$

step 6: Assume the flows shown.

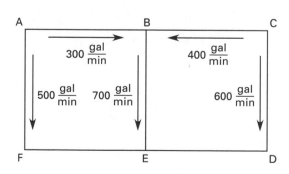

$$AB = 300 \text{ gal/min}$$
$$BE = 700 \text{ gal/min}$$
$$AF = AE = 500 \text{ gal/min}$$
$$CF = CE = 600 \text{ gal/min}$$
$$CB = 400 \text{ gal/min}$$

step 7: If the elevations are included as part of the head loss,

$$\sum h = \sum K' \dot{V}_a^n + \delta \sum n K' \dot{V}_a^{n-1} + z_2 - z_1 = 0$$

However, since the loop closes on itself, $z_2 = z_1$, and the elevations can be omitted.

First iteration:

Loop 1:

$$\delta_1 = \frac{-\left((8.33 \times 10^{-5})(300)^{1.85} + (8.33 \times 10^{-5})(700)^{1.85} - (4.21 \times 10^{-4})(500)^{1.85}\right)}{(1.85)\left((8.33 \times 10^{-5})(300)^{0.85} + (8.33 \times 10^{-5})(700)^{0.85} + (4.21 \times 10^{-4})(500)^{0.85}\right)}$$

$$= \frac{-(-22.97)}{0.213}$$

$$= +108 \text{ gal/min}$$

Loop 2:

$$\delta_2 = \frac{-\left((6.76 \times 10^{-4})(600)^{1.85} - (8.33 \times 10^{-5})(700)^{1.85} - (8.33 \times 10^{-5})(400)^{1.85}\right)}{(1.85)\left((6.76 \times 10^{-4})(600)^{0.85} + (8.33 \times 10^{-5})(700)^{0.85} + 8.33 \times 10^{-5})(400)^{0.85}\right)}$$

$$= \frac{-72.52}{0.353}$$

$$= -205 \text{ gal/min}$$

Second iteration:

AB:	300 +	108	= 408 gal/min
BE:	700 +	108 −(−205)	= 1013 gal/min
AE:	500 −	108	= 392 gal/min
CE:	600 +	(−205)	= 395 gal/min
CB:	400 −	(−205)	= 605 gal/min

Loop 1: $\quad \delta_1 = \dfrac{-9.48}{0.205} = -46 \text{ gal/min}$

Loop 2:

$$\delta_2 = \frac{-1.08}{0.292} = -3.7 \text{ gal/min} \quad [\text{round to } -4]$$

Third iteration:

AB:	408 + (−46)		= 362 gal/min
BE:	1013 + (−46)	−(−4)	= 971 gal/min
AE:	392 − (−46)		= 438 gal/min
CE:	395 +	(−4)	= 391 gal/min
CB:	605 −	(−4)	= 609 gal/min

Loop 1:

$$\delta_1 = \frac{-0.066}{0.213} = -0.31 \text{ gal/min} \quad [\text{round to } 0]$$

Loop 2:

$$\delta_2 = \frac{-2.4}{0.29} = -8.3 \text{ gal/min} \quad [\text{round to } -8]$$

Use the following flows.

AB:	362 +	0		= 362 gal/min
BE:	971 +	0	−(−8) =	979 gal/min (980 gal/min)
AE:	438 −	0		= 438 gal/min
CE:	391 +	(−8)		= 383 gal/min
CB:	609 −	(−8)		= 617 gal/min

The answer is (D).

(b) The friction loss in each section is

$$h_{f,\text{AB}} = (8.33 \times 10^{-5})(362)^{1.85} = 4.5 \text{ ft}$$

$$h_{f,\text{BE}} = (8.33 \times 10^{-5})(979)^{1.85} = 28.4 \text{ ft}$$

$$h_{f,\text{AF}} = (8.33 \times 10^{-5})(438)^{1.85} = 6.4 \text{ ft}$$

$$h_{f,\text{FE}} = (3.38 \times 10^{-4})(438)^{1.85} = 26.0 \text{ ft}$$

$$h_{f,\text{CD}} = h_{f,\text{DE}} = (3.38 \times 10^{-4})(383)^{1.85} = 20.3 \text{ ft}$$

$$h_{f,\text{CB}} = (8.33 \times 10^{-5})(617)^{1.85} = 12.1 \text{ ft}$$

$$\gamma = 62.37 \text{ lbf/ft}^3 \quad [\text{at } 60°\text{F}]$$

The pressure at C is 40 psig.

$$h_C = \frac{\left(40 \; \frac{\text{lbf}}{\text{in}^2}\right)\left(12 \; \frac{\text{in}}{\text{ft}}\right)^2}{62.37 \; \frac{\text{lbf}}{\text{ft}^3}} = 92.4 \text{ ft}$$

Next, use the energy continuity equation between adjacent points. The sign of the friction head depends on the loop and flow directions.

$$h = h_C + z_C - z - h_f$$

$h_D = 92.4 \text{ ft} + 300 \text{ ft} - 150 \text{ ft} - 20.3 \text{ ft} = 222.1 \text{ ft}$

$h_E = 222.1 \text{ ft} + 150 \text{ ft} - 200 \text{ ft} - 20.3 \text{ ft} = 151.8 \text{ ft}$

$h_F = 151.8 \text{ ft} + 200 \text{ ft} - 150 \text{ ft} + 26 \text{ ft} = 227.8 \text{ ft}$

$h_B = 92.4 \text{ ft} + 300 \text{ ft} - 150 \text{ ft} - 12.1 \text{ ft} = 230.3 \text{ ft}$

$h_A = 230.3 \text{ ft} + 150 \text{ ft} - 200 \text{ ft} + 4.5 \text{ ft} = 184.8 \text{ ft}$

Using $p = \gamma h$,

$$p_A = \frac{\left(62.37 \; \frac{\text{lbf}}{\text{ft}^3}\right)(184.8 \text{ ft})}{\left(12 \; \frac{\text{in}}{\text{ft}}\right)^2} = 80.0 \text{ lbf/in}^2 \quad \text{(psig)}$$

$$p_B = \frac{\left(62.37 \; \frac{\text{lbf}}{\text{ft}^3}\right)(230.3 \text{ ft})}{\left(12 \; \frac{\text{in}}{\text{ft}}\right)^2} = 99.7 \text{ psig}$$

$p_C = 40 \text{ psig} \quad \text{[given]}$

$$p_D = \frac{\left(62.37 \; \frac{\text{lbf}}{\text{ft}^3}\right)(222.1 \text{ ft})}{\left(12 \; \frac{\text{in}}{\text{ft}}\right)^2} = \boxed{96.2 \text{ psig (96 psig)}}$$

$$p_E = \frac{\left(62.37 \; \frac{\text{lbf}}{\text{ft}^3}\right)(151.8 \text{ ft})}{\left(12 \; \frac{\text{in}}{\text{ft}}\right)^2} = 65.7 \text{ psig}$$

$$p_F = \frac{\left(62.37 \; \frac{\text{lbf}}{\text{ft}^3}\right)(227.8 \text{ ft})}{\left(12 \; \frac{\text{in}}{\text{ft}}\right)^2} = 98.7 \text{ psig}$$

The answer is (C).

(c) The pressure increase across the pump is

$$\left(80 \; \frac{\text{lbf}}{\text{in}^2} - 20 \; \frac{\text{lbf}}{\text{in}^2}\right)\left(12 \; \frac{\text{in}}{\text{ft}}\right)^2 = 8640 \text{ lbf/ft}^2$$

Use Table 18.5 to find the hydraulic horsepower.

$$P = \frac{\Delta p Q}{2.468 \times 10^5}$$

$$= \frac{\left(8640 \; \frac{\text{lbf}}{\text{ft}^2}\right)\left(800 \; \frac{\text{gal}}{\text{min}}\right)}{2.468 \times 10^5 \; \frac{\text{lbf-gal}}{\text{min-ft}^2\text{-hp}}}$$

$$= \boxed{28 \text{ hp}}$$

The answer is (B).

12. (a) This could be solved as a parallel pipe problem or as a pipe network problem.

Pipe network solution

step 1: Use the Darcy equation since Hazen-Williams coefficients are not given (or assume *C*-values based on the age of the pipe). (See Eq. 17.135.)

step 2: Clockwise is positive.

step 3: All nodes are lettered.

step 4: There is only one loop.

step 5: $\epsilon \approx 0.0008$ ft. Assume full turbulence. For class-A cast-iron pipe, $D_i = 10.1$ in.

$$\frac{\epsilon}{D_i} = \frac{0.0008 \text{ ft}}{\dfrac{10.1 \text{ in}}{12 \; \frac{\text{in}}{\text{ft}}}} \approx 0.001$$

$f \approx 0.020$ for full turbulence. (See Fig. 17.4.)

$$K' = \frac{1.251 \times 10^{-7} f L}{D^5}$$

$$K'_{AB} = (1.251 \times 10^{-7})\left(\frac{(0.02)(2000 \text{ ft})}{\left(\dfrac{10.1 \text{ in}}{12 \; \frac{\text{in}}{\text{ft}}}\right)^5}\right) = 11.8 \times 10^{-6}$$

$$K'_{BD} = (1.251 \times 10^{-7})\left(\frac{(0.02)(1000 \text{ ft})}{\left(\dfrac{10.1 \text{ in}}{12 \; \frac{\text{in}}{\text{ft}}}\right)^5}\right) = 5.92 \times 10^{-6}$$

$$K'_{DC} = (1.251 \times 10^{-7}) \left(\frac{(0.02)(1500 \text{ ft})}{\left(\dfrac{8.13 \text{ in}}{12 \, \frac{\text{in}}{\text{ft}}} \right)^5} \right) = 26.3 \times 10^{-6}$$

$$K'_{CA} = (1.251 \times 10^{-7}) \left(\frac{(0.02)(1000 \text{ ft})}{\left(\dfrac{12.12 \text{ in}}{12 \, \frac{\text{in}}{\text{ft}}} \right)^5} \right) = 2.38 \times 10^{-6}$$

step 6: Assume $\dot{V}_{AB} = 1$ MGD.

$$\dot{V}_{ACDB} = 0.5 \text{ MGD}$$

$$\frac{\dot{V}_{AB}}{\dot{V}_{ACDB}} = \frac{1 \text{ MGD}}{0.5 \text{ MGD}} = 2$$

Convert $\dot{V}$ to gal/min.

$$\dot{V}_{AB} = \frac{(1 \text{ MGD}) \left(1 \times 10^6 \, \frac{\text{gal}}{\text{MG}} \right)}{\left(24 \, \frac{\text{hr}}{\text{day}} \right) \left(60 \, \frac{\text{min}}{\text{hr}} \right)} = 694 \text{ gal/min}$$

$$\dot{V}_{ACDB} = \frac{(0.5 \text{ MGD}) \left(1 \times 10^6 \, \frac{\text{gal}}{\text{MG}} \right)}{\left(24 \, \frac{\text{hr}}{\text{day}} \right) \left(60 \, \frac{\text{min}}{\text{hr}} \right)} = 347 \text{ gal/min}$$

step 7: There is only one loop.

$$694 \, \frac{\text{gal}}{\text{min}} = (2) \left(347 \, \frac{\text{gal}}{\text{min}} \right)$$

$$\left(694 \, \frac{\text{gal}}{\text{min}} \right)^2 = (4) \left(347 \, \frac{\text{gal}}{\text{min}} \right)^2$$

$$\delta = \frac{\begin{array}{c} (-1)(1 \times 10^{-6})(347)^2 \\ \times \big((11.8)(4) - 5.92 - 26.3 - 2.38 \big) \end{array}}{\begin{array}{c} (2)(1 \times 10^{-6})(347) \\ \times \big((11.8)(2) + 5.92 + 26.3 + 2.38 \big) \end{array}}$$

$$= -37.6 \text{ gal/min} \quad [\text{use } -38 \text{ gal/min}]$$

step 8:

$$\dot{V}_{AB} = 694 \, \frac{\text{gal}}{\text{min}} + \left(-38 \, \frac{\text{gal}}{\text{min}} \right) = 656 \text{ gal/min}$$

$$\dot{V}_{ACDB} = 347 \, \frac{\text{gal}}{\text{min}} - \left(-38 \, \frac{\text{gal}}{\text{min}} \right) = 385 \text{ gal/min}$$

Repeat step 7.

$$\frac{\dot{V}_{AB}}{\dot{V}_{ACDB}} = \frac{656 \, \frac{\text{gal}}{\text{min}}}{385 \, \frac{\text{gal}}{\text{min}}} = 1.7$$

$$(1.7)^2 = 2.90$$

$$656 \, \frac{\text{gal}}{\text{min}} = (1.70) \left(385 \, \frac{\text{gal}}{\text{min}} \right)$$

$$\left(656 \, \frac{\text{gal}}{\text{min}} \right)^2 = (2.90) \left(385 \, \frac{\text{gal}}{\text{min}} \right)^2$$

$$\delta = \frac{\begin{array}{c} (-1)(1 \times 10^{-6})(385)^2 \\ \times \big((11.8)(2.90) - 5.92 - 26.3 - 2.38 \big) \end{array}}{\begin{array}{c} (2)(1 \times 10^{-6})(385) \\ \times \big((11.8)(1.70) + 5.92 + 26.3 + 2.38 \big) \end{array}}$$

$$= -1.3 \quad (\approx 0)$$

$$\dot{V}_{AB} = \boxed{656 \text{ gal/min} \quad (660 \text{ gal/min})}$$

$$\dot{V}_{ACDB} = 385 \text{ gal/min}$$

Check the Reynolds number in leg AB to verify that $f = 0.02$ was a good choice.

$$A_{10 \text{ in pipe}} = \left(\frac{\pi}{4} \right) \left(\frac{10.1 \text{ in}}{12 \, \frac{\text{in}}{\text{ft}}} \right)^2 = 0.5564 \quad [\text{cast-iron pipe}]$$

The flow rate is

$$\left(656 \, \frac{\text{gal}}{\text{min}} \right) \left(0.002228 \, \frac{\text{ft}^3\text{-min}}{\text{sec-gal}} \right) = 1.46 \text{ ft}^3/\text{sec}$$

$$\text{v} = \frac{\dot{V}}{A} = \frac{1.46 \, \frac{\text{ft}^3}{\text{sec}}}{0.5564 \text{ ft}^2} = 2.62 \text{ ft/sec} \quad [\text{reasonable}]$$

For 50°F water,

$$\nu = 1.410 \times 10^{-5} \text{ ft}^2/\text{sec}$$

$$\text{Re} = \frac{\text{v}D}{\nu} = \frac{\left(2.62 \, \frac{\text{ft}}{\text{sec}} \right) \left(\dfrac{10.1 \text{ in}}{12 \, \frac{\text{in}}{\text{ft}}} \right)}{1.410 \times 10^{-5} \, \frac{\text{ft}^2}{\text{sec}}}$$

$$= 1.56 \times 10^5 \quad [\text{turbulent}]$$

The assumption of full turbulence made in step 5 is justified.

The answer is (D).

Alternative closed-form solution

Use the Darcy equation. Assume $\epsilon = 0.0008$ ft. The relative roughness is

$$\frac{\epsilon}{D} = \frac{0.0008 \text{ ft}}{\dfrac{10.1 \text{ in}}{12 \, \dfrac{\text{in}}{\text{ft}}}} = 0.00095 \quad [\text{use } 0.001]$$

Assume $v_{max} = 5$ ft/sec and 50°F temperature. The Reynolds number is

$$\text{Re} = \frac{vD}{\nu} = \frac{\left(5 \, \dfrac{\text{ft}}{\text{sec}}\right)\left(\dfrac{10.1 \text{ in}}{12 \, \dfrac{\text{in}}{\text{ft}}}\right)}{1.410 \times 10^{-5} \, \dfrac{\text{ft}^2}{\text{sec}}} = 2.98 \times 10^5$$

From the Moody diagram, $f \approx 0.0205$.

$$
\begin{aligned}
h_{f,\text{AB}} &= \frac{fLv^2}{2Dg} \\
&= \frac{(0.0205)(2000 \text{ ft}) \dot{V}^2_{\text{AB}}}{(2)\left(\dfrac{10.1 \text{ in}}{12 \, \dfrac{\text{in}}{\text{ft}}}\right)(0.556 \text{ ft}^2)^2\left(32.2 \, \dfrac{\text{ft}}{\text{sec}^2}\right)} \\
&= 2.446 \, \dot{V}^2_{\text{AB}}
\end{aligned}
$$

$$
\begin{aligned}
h_{f,\text{ACDB}} &= \frac{fLv^2}{2Dg} \\
&= \frac{(0.0205)(1000 \text{ ft}) \dot{V}^2_{\text{ACDB}}}{(2)\left(\dfrac{12.12 \text{ in}}{12 \, \dfrac{\text{in}}{\text{ft}}}\right)(0.801 \text{ ft}^2)^2\left(32.2 \, \dfrac{\text{ft}}{\text{sec}^2}\right)} \\
&\quad + \frac{(0.0205)(1500 \text{ ft}) \dot{V}^2_{\text{ACDB}}}{(2)\left(\dfrac{8.13 \text{ in}}{12 \, \dfrac{\text{in}}{\text{ft}}}\right)(0.360 \text{ ft}^2)^2\left(32.2 \, \dfrac{\text{ft}}{\text{sec}^2}\right)} \\
&\quad + \frac{(0.0205)(1000 \text{ ft}) \dot{V}^2_{\text{ACDB}}}{(2)\left(\dfrac{10.1 \text{ in}}{12 \, \dfrac{\text{in}}{\text{ft}}}\right)(0.556 \text{ ft}^2)^2\left(32.2 \, \dfrac{\text{ft}}{\text{sec}^2}\right)} \\
&= 0.4912 \, \dot{V}^2_{\text{ACDB}} + 5.438 \, \dot{V}^2_{\text{ACDB}} + 1.223 \, \dot{V}^2_{\text{ACDB}} \\
&= 7.152 \, \dot{V}^2_{\text{ACDB}}
\end{aligned}
$$

$$h_{f,\text{AB}} = h_{f,\text{ACDB}}$$

$$2.446 \, \dot{V}^2_{\text{AB}} = 7.152 \, \dot{V}^2_{\text{ACDB}}$$

$$\dot{V}_{\text{AB}} = \sqrt{\frac{7.152}{2.446}} \dot{V}_{\text{ACDB}} = 1.71 \dot{V}_{\text{ACDB}} \quad [\text{Eq. I}]$$

The total flow rate is

$$\frac{1.5 \text{ MGD}}{\left(24 \, \dfrac{\text{hr}}{\text{day}}\right)\left(60 \, \dfrac{\text{min}}{\text{hr}}\right)} = 1041.7 \text{ gal/min}$$

$$\dot{V}_{\text{AB}} + \dot{V}_{\text{ACDB}} = 1041.7 \text{ gal/min} \quad [\text{Eq. II}]$$

Solving Eqs. I and II simultaneously,

$$\dot{V}_{\text{AB}} = \boxed{657.3 \text{ gal/min} \quad (660 \text{ gal/min})}$$

$$\dot{V}_{\text{ACDB}} = 384.4 \text{ gal/min}$$

This answer is insensitive to the $v_{max} = 5$ ft/sec assumption. A second iteration using actual velocities from these flow rates does not change the answer.

The same technique can be used with the Hazen-Williams equation and an assumed value of C. If $C = 100$ is used, then

$$\dot{V}_{\text{AB}} = \boxed{725.4 \text{ gal/min}}$$

$$\dot{V}_{\text{ACDB}} = 316.3 \text{ gal/min}$$

The answer is (D).

(b) $f \approx 0.021$.

$$h_{f,\text{AB}} = \frac{fLv^2}{2Dg} = \frac{(0.021)(2000 \text{ ft})\left(2.62 \, \dfrac{\text{ft}}{\text{sec}}\right)^2}{(2)\left(\dfrac{10.1 \text{ in}}{12 \, \dfrac{\text{in}}{\text{ft}}}\right)\left(32.2 \, \dfrac{\text{ft}}{\text{sec}^2}\right)}$$

$$= 5.32 \text{ ft}$$

For leg BD, use $f = 0.021$ (assumed).

$$
\begin{aligned}
v &= \frac{\dot{V}}{A} = \frac{\left(385 \, \dfrac{\text{gal}}{\text{min}}\right)\left(0.002228 \, \dfrac{\text{ft}^3\text{-min}}{\text{sec-gal}}\right)}{0.5564 \text{ ft}^2} \\
&= 1.54 \text{ ft/sec}
\end{aligned}
$$

$$h_{f,\text{DB}} = \frac{fLv^2}{2Dg} = \frac{(0.021)(1000 \text{ ft})\left(1.54 \, \dfrac{\text{ft}}{\text{sec}}\right)^2}{(2)\left(\dfrac{10.1 \text{ in}}{12 \, \dfrac{\text{in}}{\text{ft}}}\right)\left(32.2 \, \dfrac{\text{ft}}{\text{sec}^2}\right)}$$

$$= 0.9 \text{ ft}$$

Water Resources

At 50°F, $\gamma = 62.4$ lbf/ft^3. From the Bernoulli equation (omitting the velocity term),

$$\frac{p_B}{\gamma} + z_B + h_{f,DB} = \frac{p_D}{\gamma} + z_D$$

$$p_B = \left(\frac{62.4 \frac{\text{lbf}}{\text{ft}^3}}{\left(12 \frac{\text{in}}{\text{ft}}\right)^2}\right)$$

$$\times \left(\frac{\left(40 \frac{\text{lbf}}{\text{in}^2}\right)\left(12 \frac{\text{in}}{\text{ft}}\right)^2}{62.4 \frac{\text{lbf}}{\text{ft}^3}} + 600 \text{ ft} - 0.9 \text{ ft} - 580 \text{ ft}\right)$$

$$= \boxed{48.3 \text{ lbf/in}^2 \quad (48 \text{ psig})}$$

The answer is (B).

(c) $\quad \dfrac{p_A}{\gamma} + z_A = \dfrac{p_B}{\gamma} + z_B + h_{f,AB}$

$$p_A = \left(\frac{62.4 \frac{\text{lbf}}{\text{ft}^3}}{\left(12 \frac{\text{in}}{\text{ft}}\right)^2}\right)$$

$$\times \left(\frac{\left(48.3 \frac{\text{lbf}}{\text{in}^2}\right)\left(12 \frac{\text{in}}{\text{ft}}\right)^2}{62.4 \frac{\text{lbf}}{\text{ft}^3}} + 580 \text{ ft} + 5.32 \text{ ft} - 570 \text{ ft}\right)$$

$$= 54.9 \text{ lbf/in}^2 \quad (55 \text{ psi})$$

The answer is (B).

13. *Customary U.S. Solution*

Use projectile equations.

From Table 71.2, the maximum range of the discharge is given by

$$R = v_o^2\left(\frac{\sin 2\phi}{g}\right) = \left(50 \frac{\text{ft}}{\text{sec}}\right)^2\left(\frac{\sin(2)(45°)}{32.2 \frac{\text{ft}}{\text{sec}^2}}\right)$$

$$= \boxed{77.64 \text{ ft} \quad (78 \text{ ft})}$$

The answer is (B).

SI Solution

Use projectile equations.

From Table 71.2, the maximum range of the discharge is given by

$$R = v_o^2\left(\frac{\sin 2\phi}{g}\right)$$

$$= \left(15 \frac{\text{m}}{\text{s}}\right)^2\left(\frac{\sin(2)(45°)}{9.81 \frac{\text{m}}{\text{s}^2}}\right)$$

$$= \boxed{22.94 \text{ m} \quad (23 \text{ m})}$$

The answer is (B).

14. $\quad A_o = \left(\dfrac{\pi}{4}\right)\left(\dfrac{4 \text{ in}}{12 \frac{\text{in}}{\text{ft}}}\right)^2 = 0.08727 \text{ ft}^2$

$$A_t = \left(\frac{\pi}{4}\right)(20 \text{ ft})^2 = 314.16 \text{ ft}^2$$

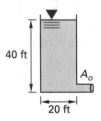

The time it takes to drop from 40 ft to 20 ft is given by Eq. 17.85.

$$t = \frac{2A_t(\sqrt{z_1} - \sqrt{z_2})}{C_d A_o \sqrt{2g}}$$

$$= \frac{(2)(314.16 \text{ ft}^2)(\sqrt{40 \text{ ft}} - \sqrt{20 \text{ ft}})}{(0.98)(0.08727 \text{ ft}^2)\sqrt{(2)\left(32.2 \frac{\text{ft}}{\text{sec}^2}\right)}}$$

$$= \boxed{1696 \text{ sec} \quad (1700 \text{ sec})}$$

The answer is (D).

15. $\quad C_d = 1.00 \quad$ [given]

$$F_{va} = \frac{1}{\sqrt{1 - \left(\dfrac{D_2}{D_1}\right)^4}} = \frac{1}{\sqrt{1 - \left(\dfrac{8 \text{ in}}{12 \text{ in}}\right)^4}}$$

$$= 1.116$$

$$A_2 = \left(\frac{\pi}{4}\right)\left(\frac{8 \text{ in}}{12 \frac{\text{in}}{\text{ft}}}\right)^2 = 0.3491 \text{ ft}^2$$

The specific weight of mercury is 0.491 lbf/in^3; the specific weight of water is 0.0361 lbf/in^3.

$$p_1 - p_2 = \Delta(\gamma h)$$

$$= \left(\begin{array}{c} \left(0.491 \dfrac{\text{lbf}}{\text{in}^3}\right)(4 \text{ in}) \\ - \left(0.0361 \dfrac{\text{lbf}}{\text{in}^3}\right)(4 \text{ in}) \end{array} \right) \left(12 \dfrac{\text{in}}{\text{ft}}\right)^2$$

$$= 262.0 \text{ lbf/ft}^2$$

From Eq. 17.155,

$$\dot{V} = F_{\text{va}} C_d A_2 \sqrt{\frac{2g(p_1 - p_2)}{\gamma}}$$

$$= (1.116)(1)(0.3491 \text{ ft}^2)$$

$$\times \sqrt{\frac{(2)\left(32.2 \dfrac{\text{ft}}{\text{sec}^2}\right)\left(262 \dfrac{\text{lbf}}{\text{ft}^2}\right)}{62.4 \dfrac{\text{lbf}}{\text{ft}^3}}}$$

$$= \boxed{6.406 \text{ ft}^3/\text{sec} \quad (6.4 \text{ ft}^3/\text{sec})}$$

The answer is (C).

16. *Customary U.S. Solution*

The volumetric flow rate of benzene through the venturi meter is given by

$$\dot{V} = C_f A_2 \sqrt{\frac{2g(\rho_m - \rho)h}{\rho}}$$

The density of mercury, ρ_m, at 60°F is approximately 848 lbm/ft^3.

The density of the benzene at 60°F is

$$\rho = (\text{SG})\rho_{\text{water}} = (0.885)\left(62.4 \dfrac{\text{lbm}}{\text{ft}^3}\right)$$

$$= 55.22 \text{ lbm/ft}^3$$

The throat area is

$$A_2 = \frac{\pi D_2^2}{4} = \frac{\pi \left(\dfrac{3.5 \text{ in}}{12 \dfrac{\text{in}}{\text{ft}}}\right)^2}{4}$$

$$= 0.0668 \text{ ft}^2$$

The flow coefficient is defined as

$$C_f = \frac{C_d}{\sqrt{1 - \beta^4}}$$

β is the ratio of the throat to inlet diameters.

$$\beta = \frac{3.5 \text{ in}}{8 \text{ in}} = 0.4375$$

$$C_f = \frac{C_d}{\sqrt{1 - \beta^4}} = \frac{0.99}{\sqrt{1 - (0.4375)^4}}$$

$$= 1.00865$$

Find the volumetric flow of benzene.

$$\dot{V} = C_f A_2 \sqrt{\frac{2g(\rho_m - \rho)h}{\rho}}$$

$$= (1.00865)(0.0668 \text{ ft}^2)$$

$$\times \sqrt{\frac{(2)\left(32.2 \dfrac{\text{ft}}{\text{sec}^2}\right)\left(848 \dfrac{\text{lbm}}{\text{ft}^3} - 55.22 \dfrac{\text{lbm}}{\text{ft}^3}\right)}{\left(55.22 \dfrac{\text{lbm}}{\text{ft}^3}\right)\left(12 \dfrac{\text{in}}{\text{ft}}\right)}}$$

$$= \boxed{1.183 \text{ ft}^3/\text{sec} \quad (1.2 \text{ ft}^3/\text{sec})}$$

The answer is (A).

SI Solution

The volumetric flow rate of benzene through the venturi meter is given by

$$\dot{V} = C_f A_2 \sqrt{\frac{2g(\rho_m - \rho)h}{\rho}}$$

ρ_m is the density of mercury at 15°C; ρ_m is approximately 13 600 kg/m^3.

The density of the benzene at 15°C is

$$\rho = (\text{SG})\rho_{\text{water}} = (0.885)\left(1000 \dfrac{\text{kg}}{\text{m}^3}\right)$$

$$= 885 \text{ kg/m}^3$$

The throat area is

$$A_2 = \frac{\pi D_2^2}{4} = \frac{\pi (0.09 \text{ m})^2}{4}$$

$$= 0.0064 \text{ m}^2$$

The flow coefficient is defined as

$$C_f = \frac{C_d}{\sqrt{1 - \beta^4}}$$

Water Resources

β is the ratio of the throat to inlet diameters.

$$\beta = \frac{9 \text{ cm}}{20 \text{ cm}} = 0.45$$

$$C_f = \frac{C_d}{\sqrt{1 - \beta^4}} = \frac{0.99}{\sqrt{1 - (0.45)^4}}$$

$$= 1.01094$$

Find the volumetric flow of benzene.

$$\dot{V} = C_f A_2 \sqrt{\frac{2g(\rho_m - \rho)h}{\rho}}$$

$$= (1.01094)(0.0064 \text{ m}^2)$$

$$\times \sqrt{\frac{(2)\left(9.81 \frac{\text{m}}{\text{s}^2}\right)}{\times \left(13\,600 \frac{\text{kg}}{\text{m}^3} - 885 \frac{\text{kg}}{\text{m}^3}\right)(0.1 \text{ m})}{885 \frac{\text{kg}}{\text{m}^3}}}$$

$$= \boxed{0.0344 \text{ m}^3/\text{s} \quad (34 \text{ L/s})}$$

The answer is (A).

17. For 70°F water,

$$\nu = 1.059 \times 10^{-5} \text{ ft}^2/\text{sec}$$

$$\gamma = 62.3 \text{ lbf/ft}^3$$

$$D_o = 0.2 \text{ ft}$$

$$v_o = v\left(\frac{D}{D_o}\right)^2 = \left(2 \frac{\text{ft}}{\text{sec}}\right)\left(\frac{1 \text{ ft}}{0.2 \text{ ft}}\right)^2$$

$$= 50 \text{ ft/sec}$$

$$\text{Re} = \frac{D_o v_o}{\nu} = \frac{(0.2 \text{ ft})\left(50 \frac{\text{ft}}{\text{sec}}\right)}{1.059 \times 10^{-5} \frac{\text{ft}^2}{\text{sec}}} = 9.44 \times 10^5$$

$$A_o = \left(\frac{\pi}{4}\right)(0.2 \text{ ft})^2 = 0.0314 \text{ ft}^2$$

$$A_p = \left(\frac{\pi}{4}\right)(1 \text{ ft})^2 = 0.7854 \text{ ft}^2$$

$$\frac{A_o}{A_p} = \frac{0.0314 \text{ ft}^2}{0.7854 \text{ ft}^2} = 0.040$$

From Fig. 17.28,

$$C_f \approx 0.60$$

$$\dot{V} = Av = (0.7854 \text{ ft}^2)\left(2 \frac{\text{ft}}{\text{sec}}\right) = 1.571 \text{ ft}^3/\text{sec}$$

From Eq. 17.163, substituting $\gamma = pg/g_c$,

$$p_p - p_o = \left(\frac{\gamma}{2g}\right)\left(\frac{\dot{V}}{C_f A_o}\right)^2$$

$$= \frac{\left(\frac{62.3 \frac{\text{lbf}}{\text{ft}^3}}{(2)\left(32.2 \frac{\text{ft}}{\text{sec}^2}\right)}\right)\left(\frac{1.571 \frac{\text{ft}^3}{\text{sec}}}{(0.60)(0.0314 \text{ ft}^2)}\right)^2}{\left(12 \frac{\text{in}}{\text{ft}}\right)^2}$$

$$= \boxed{46.7 \text{ lbf/in}^2 \quad (47 \text{ psi})}$$

The answer is (D).

18. *Customary U.S. Solution*

The densities of mercury and water are

$$\rho_{\text{mercury}} = 848 \text{ lbm/ft}^3$$

$$\rho_{\text{water}} = 62.4 \text{ lbm/ft}^3$$

The manometer tube is filled with water above the mercury column. The pressure differential across the orifice meter is

$$\Delta p = p_1 - p_2 = (\rho_{\text{mercury}} - \rho_{\text{water}})h \times \frac{g}{g_c}$$

$$= \frac{\left(848 \frac{\text{lbm}}{\text{ft}^3} - 62.4 \frac{\text{lbm}}{\text{ft}^3}\right)(7 \text{ in})}{12 \frac{\text{in}}{\text{ft}}} \times \frac{32.2 \frac{\text{ft}}{\text{sec}^2}}{32.2 \frac{\text{lbm-ft}}{\text{lbf-sec}^2}}$$

$$= \boxed{458.3 \text{ lbf/ft}^2 \quad (3.2 \text{ psi})}$$

The answer is (B).

SI Solution

The densities of mercury and water are

$$\rho_{\text{mercury}} = 13\,600 \text{ kg/m}^3$$

$$\rho_{\text{water}} = 1000 \text{ kg/m}^3$$

The manometer tube is filled with water above the mercury column. The pressure differential across the orifice meter is given by

$$\Delta p = p_1 - p_2 = (\rho_{\text{mercury}} - \rho_{\text{water}})hg$$

$$= \left(13\,600\ \frac{\text{kg}}{\text{m}^3} - 1000\ \frac{\text{kg}}{\text{m}^3}\right)(0.178\ \text{m})\left(9.81\ \frac{\text{m}}{\text{s}^2}\right)$$

$$= \boxed{22\,002\ \text{Pa}\quad(22\ \text{kPa})}$$

The answer is (B).

19. *Customary U.S. Solution*

Use App. 16.B. For 12 in pipe,

$$D_i = 0.99483\ \text{ft}$$

$$A_i = 0.7773\ \text{ft}^2$$

The velocity is

$$\text{v} = \frac{\dot{V}}{A} = \frac{10\ \frac{\text{ft}^3}{\text{sec}}}{0.7773\ \text{ft}^2}$$

$$= 12.87\ \text{ft/sec}$$

For water at 70°F, $\nu = 1.059 \times 10^{-5}\ \text{ft}^2/\text{sec}$.

The Reynolds number in the pipe is

$$\text{Re} = \frac{\text{v}D_i}{\nu} = \frac{\left(12.87\ \dfrac{\text{ft}}{\text{sec}}\right)(0.99483\ \text{ft})}{1.059 \times 10^{-5}\ \dfrac{\text{ft}^2}{\text{sec}}}$$

$$= 1.21 \times 10^6\quad\text{[fully turbulent]}$$

Flow through the orifice will have a higher Reynolds number and will also be turbulent.

The volumetric flow rate through a sharp-edged orifice is

$$\dot{V} = C_f A_o \sqrt{\frac{2g(\rho_m - \rho)h}{\rho}}$$

$$= C_f A_o \sqrt{\frac{2g_c(p_1 - p_2)}{\rho}}$$

Rearranging,

$$C_f A_o = \frac{\dot{V}}{\sqrt{\dfrac{2g_c(p_1 - p_2)}{\rho}}}$$

p_1 is the upstream pressure, and p_2 is the downstream pressure.

The maximum head loss must not exceed 25 ft; therefore,

$$\frac{\dfrac{g_c}{g} \times (p_1 - p_2)}{\rho} = 25\ \text{ft}$$

$$\frac{g_c(p_1 - p_2)}{\rho} = (25\ \text{ft})g$$

Substituting,

$$C_f A_o = \frac{10\ \dfrac{\text{ft}^3}{\text{sec}}}{\sqrt{(2)\left(32.2\ \dfrac{\text{ft}}{\text{sec}^2}\right)(25\ \text{ft})}}$$

$$= 0.249\ \text{ft}^2$$

Both C_f and A_o depend on the orifice diameter.

For a 7 in diameter orifice,

$$A_o = \frac{\pi D_o^2}{4} = \frac{\pi\left(\dfrac{7\ \text{in}}{12\ \frac{\text{in}}{\text{ft}}}\right)^2}{4}$$

$$= 0.267\ \text{ft}^2$$

$$\frac{A_o}{A_1} = \frac{0.267\ \text{ft}^2}{0.7773\ \text{ft}^2} = 0.343$$

From a chart of flow coefficients (such as Fig. 17.28), for $A_o/A_1 = 0.343$ and fully turbulent flow,

$$C_f = 0.645$$

$$C_f A_o = (0.645)(0.267\ \text{ft}^2) = 0.172\ \text{ft}^2 < 0.249\ \text{ft}^2$$

Therefore, a 7 in diameter orifice is too small.

Try a 9 in diameter orifice.

$$A_o = \frac{\pi D_o^2}{4} = \frac{\pi\left(\dfrac{9\ \text{in}}{12\ \frac{\text{in}}{\text{ft}}}\right)^2}{4}$$

$$= 0.442\ \text{ft}^2$$

$$\frac{A_o}{A_1} = \frac{0.442\ \text{ft}^2}{0.7773\ \text{ft}^2} = 0.569$$

From Fig. 17.28, for $A_o/A_1 = 0.569$ and fully turbulent flow,

$$C_f = 0.73$$

$$C_f A_o = (0.73)(0.442 \text{ ft}^2) = 0.323 \text{ ft}^2 > 0.249 \text{ ft}^2$$

Therefore, a 9 in orifice is too large.

Interpolating gives

$$D_o = 7 \text{ in} + \frac{(9 \text{ in} - 7 \text{ in})(0.249 \text{ ft}^2 - 0.172 \text{ ft}^2)}{0.323 \text{ ft}^2 - 0.172 \text{ ft}^2}$$

$$= 8.0 \text{ in}$$

Further iterations yield

$$D_o \approx \boxed{8.1 \text{ in}}$$

$$C_f A_o = 0.243 \text{ ft}^2$$

The answer is (C).

SI Solution

For 300 mm inside diameter pipe, $D_i = 0.3$ m.

The velocity is

$$\text{v} = \frac{\dot{V}}{A} = \frac{\dot{V}}{\dfrac{\pi D_i^2}{4}}$$

$$= \frac{250 \, \dfrac{\text{L}}{\text{s}}}{\left(\dfrac{\pi(0.3 \text{ m})^2}{4}\right)\left(1000 \, \dfrac{\text{L}}{\text{m}^3}\right)}$$

$$= 3.54 \text{ m/s}$$

From App. 14.B, for water at 20°C,

$$\nu = 1.007 \times 10^{-6} \text{ m}^2/\text{s}$$

The Reynolds number in the pipe is

$$\text{Re} = \frac{\text{v} D_i}{\nu} = \frac{\left(3.54 \, \dfrac{\text{m}}{\text{s}}\right)(0.3 \text{ m})}{1.007 \times 10^{-6} \, \dfrac{\text{m}^2}{\text{s}}}$$

$$= 1.05 \times 10^6 \quad \text{[fully turbulent]}$$

Flow through the orifice will have a higher Reynolds number and also be turbulent.

The volumetric flow rate through a sharp-edged orifice is

$$\dot{V} = C_f A_o \sqrt{\frac{2g(\rho_m - \rho)h}{\rho}}$$

$$= C_f A_o \sqrt{\frac{2(p_1 - p_2)}{\rho}}$$

Rearranging,

$$C_f A_o = \frac{\dot{V}}{\sqrt{\dfrac{2(p_1 - p_2)}{\rho}}}$$

p_1 is the upstream pressure, and p_2 is the downstream pressure.

The maximum head loss must not exceed 7.5 m; therefore,

$$\frac{p_1 - p_2}{g\rho} = 7.5 \text{ m}$$

$$\frac{p_1 - p_2}{\rho} = (7.5 \text{ m})g$$

Substituting,

$$C_f A_o = \frac{0.25 \, \dfrac{\text{m}^3}{\text{s}}}{\sqrt{(2)\left(9.81 \, \dfrac{\text{m}}{\text{s}^2}\right)(7.5 \text{ m})}}$$

$$= 0.021 \text{ m}^2$$

Both C_f and A_o depend on the orifice diameter.

For an 18 cm diameter orifice,

$$A_o = \frac{\pi D_o^2}{4} = \frac{\pi(0.18 \text{ m})^2}{4} = 0.0254 \text{ m}^2$$

$$\frac{A_o}{A_1} = \frac{0.0254 \text{ m}^2}{0.0707 \text{ m}^2} = 0.359$$

From a chart of flow coefficients (such as Fig. 17.28), for $A_o/A_1 = 0.359$ and fully turbulent flow,

$$C_f = 0.65$$

$$C_f A_o = (0.65)(0.0254 \text{ m}^2) = 0.0165 \text{ m}^2 < 0.021 \text{ m}^2$$

Therefore, an 18 cm diameter orifice is too small.

Try a 23 cm diameter orifice.

$$A_o = \frac{\pi D_o^2}{4} = \frac{\pi (0.23 \text{ m})^2}{4} = 0.0415 \text{ m}^2$$

$$\frac{A_o}{A_1} = \frac{0.0415 \text{ m}^2}{0.0707 \text{ m}^2} = 0.587$$

From Fig. 17.28, for $A_o/A_1 = 0.587$ and fully turbulent flow,

$$C_f = 0.73$$

$$C_f A_o = (0.73)(0.0415 \text{ m}^2) = 0.0303 \text{ m}^2 > 0.021 \text{ m}^2$$

Therefore, a 23 cm orifice is too large.

Interpolating gives

$$D_o = 18 \text{ cm}$$

$$+ (23 \text{ cm} - 18 \text{ cm}) \left(\frac{0.021 \text{ m}^2 - 0.0165 \text{ m}^2}{0.0303 \text{ m}^2 - 0.0165 \text{ m}^2} \right)$$

$$= 19.6 \text{ cm}$$

Further iteration yields

$$D_o = \boxed{20 \text{ cm}}$$
$$C_f = 0.675$$
$$C_f A_o = 0.021 \text{ m}^2$$

The answer is (C).

20. $\quad A_A = \left(\frac{\pi}{4} \right) \left(\frac{24 \text{ in}}{12 \frac{\text{in}}{\text{ft}}} \right)^2 = 3.142 \text{ ft}^2$

$$A_B = \left(\frac{\pi}{4} \right) \left(\frac{12 \text{ in}}{12 \frac{\text{in}}{\text{ft}}} \right)^2 = 0.7854 \text{ ft}^2$$

$$v_A = \frac{\dot{V}}{A} = \frac{8 \frac{\text{ft}^3}{\text{sec}}}{3.142 \text{ ft}^2} = 2.546 \text{ ft/sec}$$

$$p_A = \gamma h = \left(62.4 \frac{\text{lbf}}{\text{ft}^3} \right)(20 \text{ ft}) = 1248 \text{ lbf/ft}^2$$

$$v_B = \frac{\dot{V}}{A} = \frac{8 \frac{\text{ft}^3}{\text{sec}}}{0.7854 \text{ ft}^2} = 10.19 \text{ ft/sec}$$

Using the Bernoulli equation to solve for p_B,

$$p_B = p_A + \frac{v_A^2}{2g} - \frac{v_B^2}{2g}$$

$$= 1248 \frac{\text{lbf}}{\text{ft}^2} - \left(\frac{\left(10.19 \frac{\text{ft}}{\text{sec}} \right)^2 - \left(2.546 \frac{\text{ft}}{\text{sec}} \right)^2}{(2) \left(32.2 \frac{\text{ft}}{\text{sec}^2} \right)} \right)$$

$$\times \left(62.4 \frac{\text{lbf}}{\text{ft}^3} \right)$$

$$= 1153.7 \text{ lbf/ft}^2$$

With $\theta = 0°$, from Eq. 17.203,

$$F_x = p_B A_B - p_A A_A + \frac{\dot{m}(v_B - v_A)}{g_c}$$

$$= \left(1153.7 \frac{\text{lbf}}{\text{ft}^2} \right)(0.7854 \text{ ft}^2) - \left(1248 \frac{\text{lbf}}{\text{ft}^2} \right)(3.142 \text{ ft}^2)$$

$$+ \frac{\left(\left(8 \frac{\text{ft}^3}{\text{sec}} \right) \left(62.4 \frac{\text{lbf}}{\text{ft}^3} \right) \right) \left(10.19 \frac{\text{ft}}{\text{sec}} - 2.546 \frac{\text{ft}}{\text{sec}} \right)}{32.2 \frac{\text{ft}}{\text{sec}^2}}$$

$$= \boxed{-2897 \text{ lbf } (2900 \text{ lbf}) \text{ on the fluid (toward A)}}$$

$$F_y = 0$$

The answer is (A).

21. *Customary U.S. Solution*

The mass flow rate of the water is

$$\dot{m} = \rho \dot{V} = \rho v A = \frac{\rho v \pi D^2}{4}$$

$$= \frac{\left(62.4 \frac{\text{lbm}}{\text{ft}^3} \right) \left(40 \frac{\text{ft}}{\text{sec}} \right) \left(\pi \left(\frac{2 \text{ in}}{12 \frac{\text{in}}{\text{ft}}} \right)^2 \right)}{4}$$

$$= 54.45 \text{ lbm/sec}$$

The effective mass flow rate of the water is

$$\dot{m}_{\text{eff}} = \left(\frac{v - v_b}{v} \right) \dot{m}$$

$$= \left(\frac{40 \frac{\text{ft}}{\text{sec}} - 15 \frac{\text{ft}}{\text{sec}}}{40 \frac{\text{ft}}{\text{sec}}} \right) \left(54.45 \frac{\text{lbm}}{\text{sec}} \right)$$

$$= 34.0 \text{ lbm/sec}$$

Water Resources

The force in the (horizontal) x-direction is given by

$$F_x = \frac{\dot{m}_{\text{eff}}(\text{v} - \text{v}_b)(\cos\theta - 1)}{g_c}$$

$$= \frac{\left(34.0\ \dfrac{\text{lbm}}{\text{sec}}\right)\left(40\ \dfrac{\text{ft}}{\text{sec}} - 15\ \dfrac{\text{ft}}{\text{sec}}\right)(\cos 60° - 1)}{32.2\ \dfrac{\text{lbm-ft}}{\text{lbf-sec}^2}}$$

$$= -13.2\ \text{lbf}\quad \text{[the force is acting to the left]}$$

The force in the (vertical) y-direction is given by

$$F_y = \frac{\dot{m}_{\text{eff}}(\text{v} - \text{v}_b)\sin\theta}{g_c}$$

$$= \frac{\left(34.0\ \dfrac{\text{lbm}}{\text{sec}}\right)\left(40\ \dfrac{\text{ft}}{\text{sec}} - 15\ \dfrac{\text{ft}}{\text{sec}}\right)(\sin 60°)}{32.2\ \dfrac{\text{lbm-ft}}{\text{lbf-sec}^2}}$$

$$= 22.9\ \text{lbf}\quad \text{[the force is acting upward]}$$

The net resultant force is

$$F = \sqrt{F_x^2 + F_y^2}$$

$$= \sqrt{(-13.2\ \text{lbf})^2 + (22.9\ \text{lbf})^2}$$

$$= \boxed{26.4\ \text{lbf}\quad (26\ \text{lbf})}$$

The answer is (B).

SI Solution

The mass flow rate of the water is

$$\dot{m} = \rho\dot{V} = \rho\text{v}A = \frac{\rho\text{v}\pi D^2}{4}$$

$$= \frac{\left(1000\ \dfrac{\text{kg}}{\text{m}^3}\right)\left(12\ \dfrac{\text{m}}{\text{s}}\right)\left(\pi(0.05\ \text{m})^2\right)}{4}$$

$$= 23.56\ \text{kg/s}$$

The effective mass flow rate of the water is

$$\dot{m}_{\text{eff}} = \left(\frac{\text{v} - \text{v}_b}{\text{v}}\right)\dot{m}$$

$$= \left(\frac{12\ \dfrac{\text{m}}{\text{s}} - 4.5\ \dfrac{\text{m}}{\text{s}}}{12\ \dfrac{\text{m}}{\text{s}}}\right)\left(23.56\ \dfrac{\text{kg}}{\text{s}}\right)$$

$$= 14.73\ \text{kg/s}$$

The force in the (horizontal) x-direction is given by

$$F_x = \dot{m}_{\text{eff}}(\text{v} - \text{v}_b)(\cos\theta - 1)$$

$$= \left(14.73\ \frac{\text{kg}}{\text{s}}\right)\left(12\ \frac{\text{m}}{\text{s}} - 4.5\ \frac{\text{m}}{\text{s}}\right)(\cos 60° - 1)$$

$$= -55.2\ \text{N}\quad \text{[the force is acting to the left]}$$

The force in the (vertical) y-direction is given by

$$F_y = \dot{m}_{\text{eff}}(\text{v} - \text{v}_b)\sin\theta$$

$$= \left(14.73\ \frac{\text{kg}}{\text{s}}\right)\left(12\ \frac{\text{m}}{\text{s}} - 4.5\ \frac{\text{m}}{\text{s}}\right)(\sin 60°)$$

$$= 95.7\ \text{N}\quad \text{[the force is acting upward]}$$

The net resultant force is

$$F = \sqrt{F_x^2 + F_y^2}$$

$$= \sqrt{(-55.2\ \text{N})^2 + (95.7\ \text{N})^2}$$

$$= \boxed{110.5\ \text{N}\quad (110\ \text{N})}$$

The answer is (B).

22. *Customary U.S. Solution*

For schedule-40 pipe,

$$D_i = 0.9948\ \text{ft}$$

$$A_i = 0.7773\ \text{ft}^2$$

$$\text{v} = \frac{\dot{V}}{A_i} = \frac{\left(2000\ \dfrac{\text{gal}}{\text{min}}\right)\left(0.002228\ \dfrac{\text{ft}^3\text{-min}}{\text{sec-gal}}\right)}{0.7773\ \text{ft}^2}$$

$$= 5.73\ \text{ft/sec}$$

(The pressures are in terms of gage pressure, and the density of mercury is 0.491 lbm/in³.)

$$p_i = \left(14.7\ \frac{\text{lbf}}{\text{in}^2} - \frac{(6\ \text{in})\left(0.491\ \dfrac{\text{lbm}}{\text{in}^3}\right)\left(32.2\ \dfrac{\text{ft}}{\text{sec}^2}\right)}{32.2\ \dfrac{\text{lbm-ft}}{\text{lbf-sec}^2}}\right)$$

$$\times \left(12\ \frac{\text{in}}{\text{ft}}\right)^2$$

$$= 1692.6\ \text{lbf/ft}^2$$

$$E_{ti} = \frac{p_i}{\rho} + \frac{\text{v}_i^2}{2g_c} + \frac{z_i g}{g_c}$$

Since the pump inlet and outlet are at the same elevation, use $z = 0$ and $\rho = (SG)\rho_{\text{water}}$.

$$E_{ti} = \frac{p_i}{(SG)\rho_{\text{water}}} + \frac{v_i^2}{2g_c} + 0$$

$$= \frac{1692.6 \ \frac{\text{lbf}}{\text{ft}^2}}{(1.2)\left(62.4 \ \frac{\text{lbm}}{\text{ft}^3}\right)} + \frac{\left(5.73 \ \frac{\text{ft}}{\text{sec}}\right)^2}{(2)\left(32.2 \ \frac{\text{lbm-ft}}{\text{lbf-sec}^2}\right)}$$

$$= 23.11 \ \text{ft-lbf/lbm}$$

Calculate the total head at the inlet.

$$h_{ti} = E_{ti} \times \frac{g_c}{g}$$

$$= 23.11 \ \frac{\text{ft-lbf}}{\text{lbm}} \times \frac{32.2 \ \frac{\text{lbm-ft}}{\text{lbf-sec}^2}}{32.2 \ \frac{\text{ft}}{\text{sec}^2}}$$

$$= 23.11 \ \text{ft}$$

At the outlet side of the pump,

$$D_o = 0.6651 \ \text{ft}$$
$$A_o = 0.3474 \ \text{ft}^2$$
$$Q = v_o A_o$$

$$v_o = \frac{Q}{A_o} = \frac{\left(2000 \ \frac{\text{gal}}{\text{min}}\right)\left(0.002228 \ \frac{\text{ft}^3\text{-min}}{\text{sec-gal}}\right)}{0.3474 \ \text{ft}^2}$$

$$= 12.83 \ \text{ft/sec}$$

(The pressures are in terms of gage pressure and the gauge is located 4 ft above the pump outlet, which adds 4 ft of pressure head at the pump outlet.)

$$p_o = \left(14.7 \ \frac{\text{lbf}}{\text{in}^2} + 20 \ \frac{\text{lbf}}{\text{in}^2}\right)\left(12 \ \frac{\text{in}}{\text{ft}}\right)^2$$

$$+ 4 \ \text{ft} \left(\frac{(1.2)\left(62.4 \ \frac{\text{lbm}}{\text{ft}^3}\right)\left(32.2 \ \frac{\text{ft}}{\text{sec}^2}\right)}{32.2 \ \frac{\text{lbm-ft}}{\text{lbf-sec}^2}}\right)$$

$$= 5296 \ \text{lbf/ft}^2$$

$$E_{to} = \frac{p_o}{\rho} + \frac{v_o^2}{2g_c} + \frac{z_o g}{g_c}$$

Since the pump inlet and outlet are at the same elevation, use $z = 0$ and $\rho = (SG)\rho_{\text{water}}$.

$$E_{to} = \frac{p_o}{(SG)\rho_{\text{water}}} + \frac{v_o^2}{2g_c} + 0$$

$$= \frac{5296 \ \frac{\text{lbf}}{\text{ft}^2}}{(1.2)\left(62.4 \ \frac{\text{lbm}}{\text{ft}^3}\right)} + \frac{\left(12.83 \ \frac{\text{ft}}{\text{sec}}\right)^2}{(2)\left(32.2 \ \frac{\text{lbm-ft}}{\text{lbf-sec}^2}\right)}$$

$$= 73.28 \ \text{ft-lbf/lbm}$$

Calculate the total head at the outlet.

$$h_{to} = E_{to} \times \frac{g_c}{g}$$

$$= 73.28 \ \frac{\text{ft-lbf}}{\text{lbm}} \times \frac{32.2 \ \frac{\text{lbm-ft}}{\text{lbf-sec}^2}}{32.2 \ \frac{\text{ft}}{\text{sec}^2}}$$

$$= 73.28 \ \text{ft}$$

Compute the total head across the pump.

$$\Delta h = h_{to} - h_{ti}$$
$$= 73.28 \ \text{ft} - 23.11 \ \text{ft}$$
$$= 50.17 \ \text{ft}$$

The mass flow rate is

$$\dot{m} = \rho \dot{V}$$
$$= (SG)\rho_{\text{water}} \dot{V}$$
$$= (1.2)\left(62.4 \ \frac{\text{lbm}}{\text{ft}^3}\right)\left(2000 \ \frac{\text{gal}}{\text{min}}\right)\left(0.002228 \ \frac{\text{ft}^3\text{-min}}{\text{sec-gal}}\right)$$
$$= 333.7 \ \text{lbm/sec}$$

The power input to the pump is

$$P = \frac{\Delta h \dot{m} \times \frac{g}{g_c}}{\eta}$$

$$= \frac{(50.17 \ \text{ft})\left(333.7 \ \frac{\text{lbm}}{\text{sec}}\right) \times \frac{32.2 \ \frac{\text{ft}}{\text{sec}^2}}{32.2 \ \frac{\text{lbm-ft}}{\text{lbf-sec}^2}}}{(0.85)\left(550 \ \frac{\text{ft-lbf}}{\text{hp-sec}}\right)}$$

$$= \boxed{35.8 \ \text{hp} \quad (36 \ \text{hp})}$$

(It is not necessary to use absolute pressures as has been done in this solution.)

The answer is (B).

SI Solution

For schedule-40 pipe,

$$D_i = 303.2 \text{ mm}$$
$$A_i = 0.0722 \text{ m}^2$$
$$v = \frac{\dot{V}}{A_i} = \frac{0.125 \, \frac{\text{m}^3}{\text{s}}}{0.0722 \text{ m}^2}$$
$$= 1.73 \text{ m/s}$$

(The pressures are in terms of gage pressure, and the density of mercury is $13\,600 \text{ kg/m}^3$.)

$$p_i = 1.013 \times 10^5 \text{ Pa} - (0.15 \text{ m})\left(13\,600 \, \frac{\text{kg}}{\text{m}^3}\right)\left(9.81 \, \frac{\text{m}}{\text{s}^2}\right)$$
$$= 8.13 \times 10^4 \text{ Pa}$$
$$E_{ti} = \frac{p}{\rho} + \frac{v_i^2}{2} + z_i g$$

Since the pump inlet and outlet are at the same elevation, use $z = 0$ and $\rho = (\text{SG})\rho_{\text{water}}$.

$$E_{ti} = \frac{p}{(\text{SG})\rho_{\text{water}}} + \frac{v_i^2}{2} + 0$$
$$= \frac{8.13 \times 10^4 \text{ Pa}}{(1.2)\left(1000 \, \frac{\text{kg}}{\text{m}^3}\right)} + \frac{\left(1.73 \, \frac{\text{m}}{\text{s}}\right)^2}{2}$$
$$= 69.2 \text{ J/kg}$$

The total head at the inlet is

$$h_{ti} = \frac{E_{ti}}{g} = \frac{69.2 \, \frac{\text{J}}{\text{kg}}}{9.81 \, \frac{\text{m}}{\text{s}^2}}$$
$$= 7.05 \text{ m}$$

Assume the pipe nominal diameter is equal to the internal diameter. On the outlet side of the pump,

$$D_i = 202.7 \text{ mm}$$
$$A_o = 0.0323 \text{ m}^2$$
$$v_o = \frac{\dot{V}}{A_o} = \frac{0.125 \, \frac{\text{m}^3}{\text{s}}}{0.0323 \text{ m}^2}$$
$$= 3.87 \text{ m/s}$$

(The pressures are in terms of gage pressure and the gauge is located 1.2 m above the pump outlet, which adds 1.2 m of pressure head at the pump outlet.)

$$p_o = 1.013 \times 10^5 \text{ Pa} + 138 \times 10^3 \text{ Pa}$$
$$+ (1.2 \text{ m})\left((1.2)\left(1000 \, \frac{\text{kg}}{\text{m}^3}\right)\left(9.81 \, \frac{\text{m}}{\text{s}^2}\right)\right)$$
$$= 2.53 \times 10^5 \text{ Pa}$$
$$E_{to} = \frac{p_o}{\rho} + \frac{v_o^2}{2} + z_o g$$

Since the pump inlet and outlet are at the same elevation, use $z = 0$ and $\rho = (\text{SG})\rho_{\text{water}}$.

$$E_{to} = \frac{p_o}{(\text{SG})\rho_{\text{water}}} + \frac{v_o^2}{2} + 0$$
$$= \frac{2.53 \times 10^5 \text{ Pa}}{(1.2)\left(1000 \, \frac{\text{kg}}{\text{m}^3}\right)} + \frac{\left(3.87 \, \frac{\text{m}}{\text{s}}\right)^2}{2}$$
$$= 218.3 \text{ J/kg}$$

The total head at the outlet is

$$h_{to} = \frac{E_{to}}{g} = \frac{218.3 \, \frac{\text{J}}{\text{kg}}}{9.81 \, \frac{\text{m}}{\text{s}^2}}$$
$$= 22.25 \text{ m}$$

The total head required across the pump is

$$\Delta h = h_{to} - h_{ti}$$
$$= 22.25 \text{ m} - 7.05 \text{ m}$$
$$= 15.2 \text{ m}$$

The mass flow rate is

$$\dot{m} = \rho \dot{V}$$

In terms of the specific gravity, the mass flow rate is

$$\dot{m} = (\text{SG})\rho_{\text{water}} Q$$
$$= (1.2)\left(1000 \, \frac{\text{kg}}{\text{m}^3}\right)\left(0.125 \, \frac{\text{m}^3}{\text{s}}\right)$$
$$= 150 \text{ kg/s}$$

The power input to the pump is

$$P = \frac{\Delta h \dot{m} g}{\eta} = \frac{(15.2 \text{ m})\left(150 \frac{\text{kg}}{\text{s}}\right)\left(9.81 \frac{\text{m}}{\text{s}^2}\right)}{0.85}$$

$$= \boxed{26\,313 \text{ W} \quad (26 \text{ kW})}$$

(It is not necessary to use absolute pressures as has been done in this solution.)

The answer is (B).

23. *Customary U.S. Solution*

The mass flow rate is

$$\dot{m} = \dot{V}\rho = \left(100 \frac{\text{ft}^3}{\text{sec}}\right)\left(62.4 \frac{\text{lbm}}{\text{ft}^3}\right)$$

$$= 6240 \text{ lbm/sec}$$

The head loss across the horizontal turbine is

$$h_{\text{loss}} = \frac{\Delta p}{\rho} \times \frac{g_c}{g}$$

$$= \frac{\left(30 \frac{\text{lbf}}{\text{in}^2} - \left(-5 \frac{\text{lbf}}{\text{in}^2}\right)\right)\left(12 \frac{\text{in}}{\text{ft}}\right)^2}{62.4 \frac{\text{lbm}}{\text{ft}^3}}$$

$$\times \frac{32.2 \frac{\text{lbm-ft}}{\text{lbf-sec}^2}}{32.2 \frac{\text{ft}}{\text{sec}^2}}$$

$$= 80.77 \text{ ft}$$

From Table 18.5, the power developed by the turbine is

$$P = \dot{m} h_{\text{loss}} \times \frac{g}{g_c}$$

$$= \frac{\left(6240 \frac{\text{lbm}}{\text{sec}}\right)(80.77 \text{ ft})}{550 \frac{\text{ft-lbf}}{\text{hp-sec}}} \times \frac{32.2 \frac{\text{ft}}{\text{sec}^2}}{32.2 \frac{\text{lbm-ft}}{\text{lbf-sec}^2}}$$

$$= \boxed{916 \text{ hp} \quad (920 \text{ hp})}$$

The answer is (D).

SI Solution

The mass flow rate is

$$\dot{m} = \dot{V}\rho = \left(2.6 \frac{\text{m}^3}{\text{s}}\right)\left(1000 \frac{\text{kg}}{\text{m}^3}\right)$$

$$= 2600 \text{ kg/s}$$

The head loss across the horizontal turbine is

$$h_{\text{loss}} = \frac{\Delta p}{\rho g} = \frac{\left(210 \text{ kPa} - (-35 \text{ kPa})\right)\left(1000 \frac{\text{Pa}}{\text{kPa}}\right)}{\left(1000 \frac{\text{kg}}{\text{m}^3}\right)\left(9.81 \frac{\text{m}}{\text{s}^2}\right)}$$

$$= 25.0 \text{ m}$$

From Table 18.5, the power developed by the turbine is

$$P = \dot{m} h_{\text{loss}} g = \left(2600 \frac{\text{kg}}{\text{s}}\right)(25.0 \text{ m})\left(9.81 \frac{\text{m}}{\text{s}^2}\right)$$

$$= \boxed{637\,650 \text{ W} \quad (640 \text{ kW})}$$

The answer is (D).

24. *Customary U.S. Solution*

(a) For air at 70°F,

$$\rho = \frac{p}{RT} = \frac{\left(14.7 \frac{\text{lbf}}{\text{in}^2}\right)\left(12 \frac{\text{in}}{\text{ft}}\right)^2}{\left(53.35 \frac{\text{ft-lbf}}{\text{lbm-°R}}\right)(70°\text{F} + 460°)}$$

$$= 0.0749 \text{ lbm/ft}^3$$

$$v = \frac{\left(55 \frac{\text{mi}}{\text{hr}}\right)\left(5280 \frac{\text{ft}}{\text{mi}}\right)}{3600 \frac{\text{sec}}{\text{hr}}}$$

$$= 80.67 \text{ ft/sec}$$

The drag is

$$F_D = \frac{C_D A \rho v^2}{2g_c}$$

$$= \frac{(0.42)(28 \text{ ft}^2)\left(0.0749 \frac{\text{lbm}}{\text{ft}^3}\right)\left(80.67 \frac{\text{ft}}{\text{sec}}\right)^2}{(2)\left(32.2 \frac{\text{lbm-ft}}{\text{lbf-sec}^2}\right)}$$

$$= 89.0 \text{ lbf}$$

Water Resources

The total resisting force is

$$F = F_D + \text{rolling resistance}$$

$$= 89.0 \text{ lbf} + (0.01)(3300 \text{ lbm}) \times \frac{g}{g_c}$$

$$= 89.0 \text{ lbf} + (0.01)(3300 \text{ lbm}) \times \frac{32.2 \dfrac{\text{ft}}{\text{sec}^2}}{32.2 \dfrac{\text{lbm-ft}}{\text{lbf-sec}^2}}$$

$$= 122.0 \text{ lbf}$$

The power required is

$$P = Fv = \frac{(122.0 \text{ lbf})\left(80.67 \dfrac{\text{ft}}{\text{sec}}\right)}{778.26 \dfrac{\text{ft-lbf}}{\text{Btu}}}$$

$$= 12.65 \text{ Btu/sec}$$

The energy available from the fuel is

$$E_A = (\text{engine thermal efficiency})(\text{fuel heating value})$$

$$= (0.28)\left(115{,}000 \dfrac{\text{Btu}}{\text{gal}}\right)$$

$$= 32{,}200 \text{ Btu/gal}$$

The fuel consumption at 55 mi/hr is

$$\frac{P}{E_A} = \frac{12.65 \dfrac{\text{Btu}}{\text{sec}}}{32{,}200 \dfrac{\text{Btu}}{\text{gal}}} = 3.93 \times 10^{-4} \text{ gal/sec}$$

The fuel consumption is

$$\frac{\left(3.93 \times 10^{-4} \dfrac{\text{gal}}{\text{sec}}\right)\left(3600 \dfrac{\text{sec}}{\text{hr}}\right)}{55 \dfrac{\text{mi}}{\text{hr}}}$$

$$= \boxed{0.0257 \text{ gal/mi} \quad (0.026 \text{ gal/mi})}$$

The answer is (A).

(b)

$$v = \frac{\left(65 \dfrac{\text{mi}}{\text{hr}}\right)\left(5280 \dfrac{\text{ft}}{\text{mi}}\right)}{3600 \dfrac{\text{sec}}{\text{hr}}}$$

$$= 95.33 \text{ ft/sec}$$

$$F_D = \frac{C_D A \rho v^2}{2g_c}$$

$$= \frac{(0.42)(28 \text{ ft}^2)\left(0.0749 \dfrac{\text{lbm}}{\text{ft}^3}\right)\left(95.33 \dfrac{\text{ft}}{\text{sec}}\right)^2}{(2)\left(32.2 \dfrac{\text{lbm-ft}}{\text{lbf-sec}^2}\right)}$$

$$= 124.3 \text{ lbf}$$

The total resisting force is

$$F = F_D + \text{rolling resistance}$$

$$= 124.3 \text{ lbf} + (0.01)(3300 \text{ lbm}) \times \frac{g}{g_c}$$

$$= 124.3 \text{ lbf} + (0.01)(3300 \text{ lbm}) \times \frac{32.2 \dfrac{\text{ft}}{\text{sec}^2}}{32.2 \dfrac{\text{lbm-ft}}{\text{lbf-sec}^2}}$$

$$= 157.3 \text{ lbf}$$

The power required is

$$P = Fv = \frac{(157.3 \text{ lbf})\left(95.33 \dfrac{\text{ft}}{\text{sec}}\right)}{778.26 \dfrac{\text{ft-lbf}}{\text{Btu}}}$$

$$= 19.27 \text{ Btu/sec}$$

The fuel consumption at 65 mi/hr is

$$\frac{P}{E_A} = \frac{19.27 \dfrac{\text{Btu}}{\text{sec}}}{32{,}200 \dfrac{\text{Btu}}{\text{gal}}} = 5.98 \times 10^{-4} \text{ gal/sec}$$

The fuel consumption is

$$\frac{\left(5.98 \times 10^{-4} \dfrac{\text{gal}}{\text{sec}}\right)\left(3600 \dfrac{\text{sec}}{\text{hr}}\right)}{65 \dfrac{\text{mi}}{\text{hr}}} = 0.0331 \text{ gal/mi}$$

The relative difference between the fuel consumption at 55 mi/hr and 65 mi/hr is

$$\frac{0.0331 \dfrac{\text{gal}}{\text{mi}} - 0.0257 \dfrac{\text{gal}}{\text{mi}}}{0.0257 \dfrac{\text{gal}}{\text{mi}}} = \boxed{0.288 \quad (30\%)}$$

The answer is (C).

SI Solution

(a) For air at 20°C,

$$\rho = \frac{p}{RT} = \frac{1.013 \times 10^5 \text{ Pa}}{\left(287.03 \dfrac{\text{J}}{\text{kg·K}}\right)(20°\text{C} + 273°)}$$

$$= 1.205 \text{ kg/m}^3$$

$$v = \frac{\left(90 \dfrac{\text{km}}{\text{h}}\right)\left(1000 \dfrac{\text{m}}{\text{km}}\right)}{3600 \dfrac{\text{s}}{\text{h}}}$$

$$= 25.0 \text{ m/s}$$

The drag is

$$F_D = \frac{C_D A \rho v^2}{2} = \frac{(0.42)(2.6 \text{ m}^2)\left(1.205 \dfrac{\text{kg}}{\text{m}^3}\right)\left(25.0 \dfrac{\text{m}}{\text{s}}\right)^2}{2}$$

$$= 411.2 \text{ N}$$

The total resisting force is

$$F = F_D + \text{rolling resistance} \times g$$

$$= 411.2 \text{ N} + (0.01)(1500 \text{ kg})g$$

$$= 411.2 \text{ N} + (0.01)(1500 \text{ kg})\left(9.81 \dfrac{\text{m}}{\text{s}^2}\right)$$

$$= 558.4 \text{ N}$$

The power required is

$$P = Fv = (558.4 \text{ N})\left(25 \dfrac{\text{m}}{\text{s}}\right)$$

$$= 13\,960 \text{ W}$$

The energy available from the fuel is

$$E_A = (\text{engine thermal efficiency})(\text{fuel heating value})$$

$$= (0.28)\left(4.6 \times 10^8 \dfrac{\text{J}}{\text{L}}\right)$$

$$= 1.288 \times 10^8 \text{ J/L}$$

The fuel consumption at 90 km/h is

$$\frac{P}{E_A} = \frac{13\,960 \text{ W}}{1.288 \times 10^8 \dfrac{\text{J}}{\text{L}}} = 1.08 \times 10^{-4} \text{ L/s}$$

The fuel consumption is

$$\frac{\left(1.08 \times 10^{-4} \dfrac{\text{L}}{\text{s}}\right)\left(3600 \dfrac{\text{s}}{\text{h}}\right)}{90 \dfrac{\text{km}}{\text{h}}} = \boxed{0.0043 \text{ L/km}}$$

The answer is (A).

(b) Similarly, the fuel consumption at 105 km/h is

$$v = \frac{\left(105 \dfrac{\text{km}}{\text{h}}\right)\left(1000 \dfrac{\text{m}}{\text{km}}\right)}{3600 \dfrac{\text{s}}{\text{h}}}$$

$$= 29.2 \text{ m/s}$$

$$D = \frac{C_D A \rho v^2}{2} = \frac{(0.42)(2.6 \text{ m}^2)\left(1.205 \dfrac{\text{kg}}{\text{m}^3}\right)\left(29.2 \dfrac{\text{m}}{\text{s}}\right)^2}{2}$$

$$= 561.0 \text{ N}$$

The total resisting force is

$$F = F_D + \text{rolling resistance} \times g$$

$$= 561.0 \text{ N} + (0.01)(1500 \text{ kg})g$$

$$= 561.0 \text{ N} + (0.01)(1500 \text{ kg})\left(9.81 \dfrac{\text{m}}{\text{s}^2}\right)$$

$$= 708.2 \text{ N}$$

The power required is

$$P = Fv = (708.2 \text{ N})\left(29.2 \dfrac{\text{m}}{\text{s}}\right)$$

$$= 20\,679 \text{ W}$$

The fuel consumption at 105 km/h is

$$\frac{P}{E_A} = \frac{20\,679 \text{ W}}{1.288 \times 10^8 \dfrac{\text{J}}{\text{L}}} = 1.61 \times 10^{-4} \text{ L/s}$$

The fuel consumption is

$$\frac{\left(1.61 \times 10^{-4} \dfrac{\text{L}}{\text{s}}\right)\left(3600 \dfrac{\text{s}}{\text{h}}\right)}{105 \dfrac{\text{km}}{\text{h}}} = 0.00552 \text{ L/km}$$

The relative difference between the fuel consumption at 90 km/h and 105 km/h is

$$\frac{0.00552 \dfrac{\text{L}}{\text{km}} - 0.00434 \dfrac{\text{L}}{\text{km}}}{0.00434 \dfrac{\text{L}}{\text{km}}} = \boxed{0.272 \quad (30\%)}$$

The answer is (C).

18 Hydraulic Machines

PRACTICE PROBLEMS

Pumping Power

1. 2000 gal/min of 60°F thickened sludge with a specific gravity of 1.2 flows through a pump with an inlet diameter of 12 in and an outlet of 8 in. The centerlines of the inlet and outlet are at the same elevation. The inlet pressure is 8 in of mercury (vacuum). A discharge pressure gauge located 4 ft above the pump discharge centerline reads 20 psig. The pump efficiency is 85%. All pipes are schedule-40. What is the input power of the pump?

(A) 26 hp

(B) 31 hp

(C) 37 hp

(D) 53 hp

2. 1.25 ft³/sec (35 L/s) of 70°F (21°C) water is pumped from the bottom of a tank through 700 ft (230 m) of 4 in (102.3 mm) schedule-40 steel pipe. The line includes a 50 ft (15 m) rise in elevation, two right-angle elbows, a wide-open gate valve, and a swing check valve. All fittings and valves are regular screwed. The inlet pressure is 50 psig (345 kPa), and a working pressure of 20 psig (140 kPa) is needed at the end of the pipe. What is the hydraulic power for this pumping application?

(A) 16 hp (13 kW)

(B) 23 hp (17 kW)

(C) 49 hp (37 kW)

(D) 66 hp (50 kW)

3. 80 gal/min (5 L/s) of 80°F (27°C) water is lifted 12 ft (4 m) vertically by a pump through a total length of 50 ft (15 m) of a 2 in (5.1 cm) diameter smooth rubber hose. The discharge end of the hose is submerged in 8 ft (2.5 m) of water as shown. What head is added by the pump?

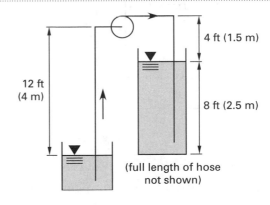

4 ft (1.5 m)

12 ft (4 m)

8 ft (2.5 m)

(full length of hose not shown)

(A) 10 ft (3.0 m)

(B) 13 ft (4.3 m)

(C) 22 ft (6.6 m)

(D) 31 ft (9.3 m)

4. A 20 hp motor drives a centrifugal pump. The pump discharges 60°F (16°C) water at 12 ft/sec (4 m/s) into a 6 in (15.2 cm) steel schedule-40 line. The inlet is 8 in (20.3 cm) schedule-40 steel pipe. The pump suction is 5 psi (35 kPa) below standard atmospheric pressure. The friction and fitting head loss in the system is 10 ft (3.3 m). The pump efficiency is 70%. The suction and discharge lines are at the same elevation. What is the maximum height above the pump inlet that water is available with that velocity at standard atmospheric pressure?

(A) 28 ft (6.9 m)

(B) 37 ft (11 m)

(C) 49 ft (15 m)

(D) 81 ft (25 m)

5. (*Time limit: one hour*) A pump station is used to fill a tank on a hill from a lake below. The flow rate is 10,000 gal/hr (10.5 L/s) of 60°F (16°C) water. The atmospheric pressure is 14.7 psia (101 kPa). The pump is 12 ft (4 m) above the lake, and the tank surface level is 350 ft (115 m) above the pump. The suction and discharge lines are 4 in (10.2 cm) diameter schedule-40 steel pipe. The equivalent length of the inlet line between the lake and the pump is 300 ft (100 m). The total equivalent length between the lake and the tank is 7000 ft (2300 m), including all fittings, bends, screens, and valves. The cost of electricity is $0.04 per kW-hr. The overall efficiency of the pump and motor set is 70%.

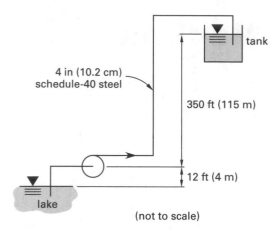

4 in (10.2 cm) schedule-40 steel

350 ft (115 m)

12 ft (4 m)

tank

lake

(not to scale)

(a) What does it cost to operate the pump for one hour?

- (A) $0.1
- (B) $1
- (C) $3
- (D) $6

(b) What motor power is required?

- (A) 10 hp (7.5 kW)
- (B) 30 hp (25 kW)
- (C) 50 hp (40 kW)
- (D) 75 hp (60 kW)

(c) What is the NPSHA for this application?

- (A) 4 ft (1.2 m)
- (B) 8 ft (2.4 m)
- (C) 12 ft (3.6 m)
- (D) 16 ft (4.5 m)

6. (*Time limit: one hour*) A town with a stable, constant population of 10,000 produces sewage at the average rate of 100 gallons per capita day (gpcd), with peak flows of 250 gpcd. The pipe to the pumping station is 5000 ft in length and has a *C*-value of 130. The elevation drop along the length is 48 ft. Minor losses in infiltration are insignificant. The pump's maximum suction lift is 10 ft.

(a) If all diameters are available and the pipe flows 100% full under gravity flow, what minimum pipe diameter is required?

- (A) 8 in
- (B) 12 in
- (C) 14 in
- (D) 18 in

(b) If constant-speed pumps are used, what is the minimum number of pumps (disregarding spares and backups) that should be used?

- (A) 2
- (B) 3
- (C) 4
- (D) 5

(c) If variable-speed pumps are used, what is the minimum number of pumps (disregarding spares and backups) that should be used?

- (A) 1
- (B) 2
- (C) 3
- (D) 4

(d) If three constant-speed pumps are used, with a fourth as backup, and the pump-motor set efficiency is 60%, what motor power is required?

- (A) 2 hp
- (B) 3 hp
- (C) 5 hp
- (D) 8 hp

(e) If two variable-speed pumps are used, with a third as backup, and the pump-motor set efficiency is 80%, what motor power is required?

- (A) 3 hp
- (B) 8 hp
- (C) 12 hp
- (D) 18 hp

(f) Which of the following are ways of controlling sump pump on-off cycles?

I.   detecting sump levels

II.   detecting pressure in the sump

III. detecting incoming flow rates

IV.   using fixed run times

V.   detecting outgoing flow rates

VI.   operating manually

   (A) I, II, and III

   (B) I, II, IV, and V

   (C) I, III, and V

   (D) I, II, IV, V, and VI

(g) With intermittent fan operation, approximately how many air changes should the wet well receive per hour?

   (A) 6

   (B) 12

   (C) 20

   (D) 30

(h) With continuous fan operation, approximately how many air changes should the dry well receive per hour?

   (A) 6

   (B) 12

   (C) 20

   (D) 30

7. (*Time limit: one hour*) A pump transfers 3.5 MGD of filtered water from the clear well of a 10 ft × 20 ft (plan) rapid sand filter to a higher elevation. The pump efficiency is 85%, and the motor driving pump has an efficiency of 90%. Minor losses are insignificant. Refer to the following illustration for additional information.

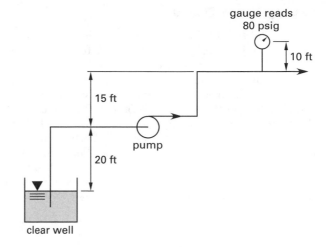

(a) What is the static suction lift?

   (A) 15 ft

   (B) 20 ft

   (C) 35 ft

   (D) 40 ft

(b) What is the static discharge head?

   (A) 15 ft

   (B) 20 ft

   (C) 35 ft

   (D) 40 ft

(c) Based on the information given, what is the approximate total dynamic head?

   (A) 45 ft

   (B) 185 ft

   (C) 210 ft

   (D) 230 ft

(d) What motor power is required?

   (A) 50 hp

   (B) 100 hp

   (C) 150 hp

   (D) 200 hp

Pumping Other Fluids

8. (*Time limit: one hour*) Gasoline with a specific gravity of 0.7 and viscosity of 6×10^{-6} ft^2/sec (5.6×10^{-7} m^2/s) is transferred from a tanker to a storage tank. The interior of the storage tank is maintained at atmospheric pressure by a vapor-recovery system. The free surface in the storage tank is 60 ft (20 m) above the tanker's free surface. The pipe consists of 500 ft (170 m) of 3 in (7.62 cm) schedule-40 steel pipe with six flanged elbows and two wide-open gate valves. The pump and motor both have individual efficiencies of 88%. Electricity costs \$0.045 per kW-hr. The pump's performance data (based on cold, clear water) are known.

flow rate	head
gpm (L/s)	ft (m)
0 (0)	127 (42)
100 (6.3)	124 (41)
200 (12)	117 (39)
300 (18)	108 (36)
400 (24)	96 (32)
500 (30)	80 (27)
600 (36)	55 (18)

(a) What is the transfer rate?

(A) 150 gal/min (9.2 L/s)

(B) 180 gal/min (11 L/s)

(C) 200 gal/min (12 L/s)

(D) 230 gal/min (14 L/s)

(b) What is the total cost of operating the pump for one hour?

(A) \$0.20

(B) \$0.80

(C) \$1.30

(D) \$2.70

Specific Speed

9. A double-suction water pump moving 300 gal/sec (1.1 kL/s) turns at 900 rpm. The pump adds 20 ft (7 m) of head to the water. What is the specific speed?

(A) 3000 rpm (52 rpm)

(B) 6000 rpm (100 rpm)

(C) 9000 rpm (160 rpm)

(D) 12,000 rpm (210 rpm)

Cavitation

10. 100 gal/min (6.3 L/s) of pressurized hot water at 281°F and 80 psia (138°C and 550 kPa) is drawn through 30 ft (10 m) of 1.5 in (3.81 cm) schedule-40 steel pipe into a 2 psig (14 kPa) tank. The inlet and outlet are both 20 ft (6 m) below the surface of the water when the tank is full. The inlet line contains a square mouth inlet, two wide-open gate valves, and two long-radius elbows. All components are regular screwed. The pump's NPSHR is 10 ft (3 m) for this application. The kinematic viscosity of 281°F (138°C) water is 0.239×10^{-5} ft^2/sec (0.222×10^{-6} m^2/s) and the vapor pressure is 50.02 psia (3.431 bar). Will the pump cavitate?

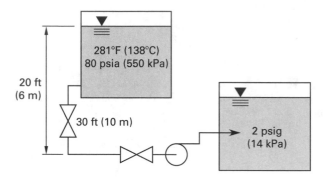

(A) yes; NPSHA = 4 ft (1.2 m)

(B) yes; NPSHA = 9 ft (2.7 m)

(C) no; NPSHA = 24 ft (7.2 m)

(D) no; NPSHA = 68 ft (21 m)

11. The velocity of the tip of a marine propeller is 4.2 times the boat velocity. The propeller is located 8 ft (3 m) below the surface. The temperature of the seawater is 68°F (20°C). The density is approximately 64.0 lbm/ft^3 (1024 kg/m^3), and the salt content is 2.5% by weight. What is the practical maximum boat velocity, as limited strictly by cavitation?

(A) 9.1 ft/sec (2.7 m/s)

(B) 12 ft/sec (3.8 m/s)

(C) 15 ft/sec (4.5 m/s)

(D) 22 ft/sec (6.6 m/s)

Pump and System Curves

12. The inlet of a centrifugal water pump is 7 ft (2.3 m) above the free surface from which it draws. The suction point is a submerged pipe. The supply line consists of 12 ft (4 m) of 2 in (5.08 cm) schedule-40 steel pipe and contains one long-radius elbow and one check valve. The discharge line is 2 in (5.08 cm) schedule-40 steel pipe and includes two long-radius elbows and an 80 ft (27 m) run. The discharge is 20 ft (6.3 m) above the free surface and is to the open atmosphere. All components are regular screwed. The water temperature is 70°F (21°C). The following pump curve data are applicable.

flow rate gpm (L/s)	head ft (m)
0 (0)	110 (37)
10 (0.6)	108 (36)
20 (1.2)	105 (35)
30 (1.8)	102 (34)
40 (2.4)	98 (33)
50 (3.2)	93 (31)
60 (3.6)	87 (29)
70 (4.4)	79 (26)
80 (4.8)	66 (22)
90 (5.7)	50 (17)

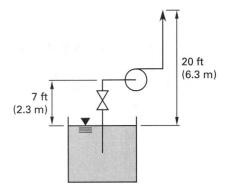

(a) What is the flow rate?

(A) 44 gal/min (2.9 L/s)

(B) 69 gal/min (4.5 L/s)

(C) 82 gal/min (5.5 L/s)

(D) 95 gal/min (6.2 L/s)

(b) What can be said about the use of this pump in this installation?

(A) A different pump should be used.

(B) The pump is operating near its most efficient point.

(C) Pressure fluctuations could result from surging.

(D) Overloading will not be a problem.

Affinity Laws

13. A pump was intended to run at 1750 rpm when driven by a 0.5 hp (0.37 kW) motor. What is the required power rating of a motor that will turn the pump at 2000 rpm?

(A) 0.25 hp (0.19 kW)

(B) 0.45 hp (0.34 kW)

(C) 0.65 hp (0.49 kW)

(D) 0.75 hp (0.55 kW)

14. (*Time limit: one hour*) A centrifugal pump running at 1400 rpm has the curve shown. The pump will be installed in an existing pipeline with known head requirements given by the formula $H = 30 + 2Q^2$. H is the system head in feet of water. Q is the flow rate in cubic feet per second.

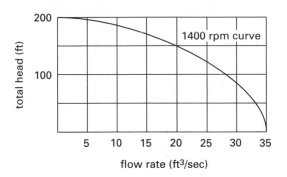

(a) What will be the flow rate if the pump is turned at 1400 rpm?

(A) 2000 gal/min

(B) 3500 gal/min

(C) 4000 gal/min

(D) 4500 gal/min

(b) What power is required to drive the pump?

(A) 190 hp

(B) 210 hp

(C) 230 hp

(D) 260 hp

(c) What will be the flow rate if the pump is turned at 1200 rpm?

(A) 2000 gal/min

(B) 3500 gal/min

(C) 4000 gal/min

(D) 4500 gal/min

Turbines

15. A horizontal turbine reduces 100 ft³/sec of water from 30 psia to 5 psia. Friction is negligible. What power is developed?

(A) 350 hp

(B) 500 hp

(C) 650 hp

(D) 800 hp

16. 1000 ft³/sec of 60°F water flows from a high reservoir through a hydroelectric turbine installation, exiting 625 ft lower. The head loss due to friction is 58 ft. The turbine efficiency is 89%. What power is developed in the turbines?

(A) 40 kW

(B) 18 MW

(C) 43 MW

(D) 71 MW

17. Water at 500 psig and 60°F (3.5 MPa and 16°C) drives a 250 hp (185 kW) turbine at 1750 rpm against a back pressure of 30 psig (210 kPa). The water discharges through a 4 in (100 mm) diameter nozzle at 35 ft/sec (10.5 m/s). The water is deflected 80° by a single blade moving directly away at 10 ft/sec (3 m/s).

(a) What is the specific speed?

(A) 4 (17)

(B) 25 (85)

(C) 75 (260)

(D) 230 (770)

(b) What is the total force acting on a single blade?

(A) 100 lbf (450 N)

(B) 140 lbf (570 N)

(C) 160 lbf (720 N)

(D) 280 lbf (1300 N)

18. (*Time limit: one hour*) A Francis-design hydraulic reaction turbine with 22 in (560 mm) diameter blades runs at 610 rpm. The turbine develops 250 hp (185 kW) when 25 ft³/sec (700 L/s) of water flow through it. The pressure head at the turbine entrance is 92.5 ft (30.8 m). The elevation of the turbine above the tailwater level is 5.26 ft (1.75 m). The inlet and outlet velocity are both 12 ft/sec (3.6 m/s).

(a) What is the effective head?

(A) 90 ft (30 m)

(B) 95 ft (31 m)

(C) 100 ft (33 m)

(D) 105 ft (35 m)

(b) What is the overall turbine efficiency?

(A) 81%

(B) 88%

(C) 93%

(D) 96%

(c) What will be the turbine speed if the effective head is 225 ft (75 m)?

(A) 600 rpm

(B) 920 rpm

(C) 1100 rpm

(D) 1400 rpm

(d) What horsepower is developed if the effective head is 225 ft (75 m)?

(A) 560 hp (420 kW)

(B) 630 hp (470 kW)

(C) 750 hp (560 kW)

(D) 840 hp (630 kW)

(e) What is the flow rate if the effective head is 225 ft (75 m)?

(A) 25 ft³/sec (700 L/s)

(B) 38 ft³/sec (1100 L/s)

(C) 56 ft³/sec (1600 L/s)

(D) 64 ft³/sec (1800 L/s)

SOLUTIONS

1. The flow rate is

$$\dot{V} = \left(2000 \ \frac{\text{gal}}{\text{min}}\right)\left(0.002228 \ \frac{\frac{\text{ft}^3}{\text{sec}}}{\frac{\text{gal}}{\text{min}}}\right)$$

$$= 4.456 \ \text{ft}^3/\text{sec}$$

From App. 16.B,

$$12'': D_1 = 0.9948 \ \text{ft} \qquad A_1 = 0.7773 \ \text{ft}^2$$
$$8'': D_2 = 0.6651 \ \text{ft} \qquad A_2 = 0.3473 \ \text{ft}^2$$

$$p_1 = \left(14.7 \ \frac{\text{lbf}}{\text{in}^2} - (8 \ \text{in})\left(0.491 \ \frac{\text{lbf}}{\text{in}^3}\right)\right)\left(12 \ \frac{\text{in}}{\text{ft}}\right)^2$$
$$= 1551.2 \ \text{lbf}/\text{ft}^2$$

$$p_2 = \left(14.7 \ \frac{\text{lbf}}{\text{in}^2} + 20 \ \frac{\text{lbf}}{\text{in}^2}\right)\left(12 \ \frac{\text{in}}{\text{ft}}\right)^2$$
$$+ (4 \ \text{ft})(1.2)\left(62.4 \ \frac{\text{lbf}}{\text{ft}^3}\right)$$
$$= 5296.3 \ \text{lbf}/\text{ft}^2$$

$$v_1 = \frac{\dot{V}}{A_1} = \frac{4.456 \ \frac{\text{ft}^3}{\text{sec}}}{0.7773 \ \text{ft}^2} = 5.73 \ \text{ft}/\text{sec}$$

$$v_2 = \frac{\dot{V}}{A_2} = \frac{4.456 \ \frac{\text{ft}^3}{\text{sec}}}{0.3473 \ \text{ft}^2} = 12.83 \ \text{ft}/\text{sec}$$

From Eq. 18.9, the total heads (in feet of sludge) at points 1 and 2 are

$$h_{t,1} = h_{t,s} = \frac{p_1}{\gamma} + \frac{v_1^2}{2g}$$

$$= \frac{1551.2 \ \frac{\text{lbf}}{\text{ft}^2}}{\left(62.4 \ \frac{\text{lbf}}{\text{ft}^3}\right)(1.2)} + \frac{\left(5.73 \ \frac{\text{ft}}{\text{sec}}\right)^2}{(2)\left(32.2 \ \frac{\text{ft}}{\text{sec}^2}\right)}$$

$$= 21.23 \ \text{ft}$$

$$h_{t,2} = h_{t,d} = \frac{p_2}{\gamma} + \frac{v_2^2}{2g}$$

$$= \frac{5296.3 \ \frac{\text{lbf}}{\text{ft}^2}}{\left(62.4 \ \frac{\text{lbf}}{\text{ft}^3}\right)(1.2)} + \frac{\left(12.83 \ \frac{\text{ft}}{\text{sec}}\right)^2}{(2)\left(32.2 \ \frac{\text{ft}}{\text{sec}^2}\right)}$$

$$= 73.29 \ \text{ft}$$

The pump must add 73.29 ft − 21.23 ft = 52.06 ft of head (sludge head).

The power required is given in Table 18.5.

$$P_{\text{ideal}} = \frac{\Delta p \dot{V}}{550}$$

$$= \frac{(52.06 \ \text{ft})(1.2)\left(62.4 \ \frac{\text{lbf}}{\text{ft}^3}\right)\left(4.456 \ \frac{\text{ft}^3}{\text{sec}}\right)}{550 \ \frac{\text{ft-lbf}}{\text{hp-sec}}}$$

$$= 31.58 \ \text{hp}$$

The input horsepower is

$$P_{\text{in}} = \frac{P_{\text{ideal}}}{\eta} = \frac{31.58 \ \text{hp}}{0.85} = \boxed{37.15 \ \text{hp} \quad (37 \ \text{hp})}$$

The answer is (C).

2. *Customary U.S. Solution*

From App. 16.B, data for 4 in schedule-40 steel pipe are

$$D_i = 0.3355 \ \text{ft}$$
$$A_i = 0.08841 \ \text{ft}^2$$

The velocity in the pipe is

$$v = \frac{\dot{V}}{A} = \frac{1.25 \ \frac{\text{ft}^3}{\text{sec}}}{0.08841 \ \text{ft}^2} = 14.139 \ \text{ft}/\text{sec}$$

From Table 17.3, typical equivalent lengths for schedule-40, screwed steel fittings for 4 in pipes are

$$90° \ \text{elbow: } 13 \ \text{ft}$$
$$\text{gate valve: } 2.5 \ \text{ft}$$
$$\text{check valve: } 38 \ \text{ft}$$

The total equivalent length is

$$(2)(13 \ \text{ft}) + (1)(2.5 \ \text{ft}) + (1)(38 \ \text{ft}) = 66.5 \ \text{ft}$$

At 70°F, from App. 14.A, the density of water is 62.3 lbm/ft^3, and the kinematic viscosity of water, ν, is 1.059×10^{-5} ft^2/sec. The Reynolds number is

$$\text{Re} = \frac{Dv}{\nu} = \frac{(0.3355 \ \text{ft})\left(14.139 \ \frac{\text{ft}}{\text{sec}}\right)}{1.059 \times 10^{-5} \ \frac{\text{ft}^2}{\text{sec}}}$$

$$= 4.479 \times 10^5$$

From App. 17.A, for steel, $\epsilon = 0.0002$ ft.

Water Resources

So,

$$\frac{\epsilon}{D} = \frac{0.0002 \text{ ft}}{0.3355 \text{ ft}} \approx 0.0006$$

From App. 17.B, the friction factor is $f = 0.01835$. The friction head is given by Eq. 18.6.

$$h_f = \frac{fLv^2}{2Dg}$$

$$= \frac{(0.01835)(700 \text{ ft} + 66.5 \text{ ft})\left(14.139 \frac{\text{ft}}{\text{sec}}\right)^2}{(2)(0.3355 \text{ ft})\left(32.2 \frac{\text{ft}}{\text{sec}^2}\right)}$$

$$= 130.1 \text{ ft}$$

The total dynamic head is given by Eq. 18.8. s is taken as the bottom of the supply tank. d is taken as the end of the discharge pipe.

$$h = \frac{(p_d - p_s)g_c}{\rho g} + \frac{v_d^2 - v_s^2}{2g} + z_d - z_s$$

$$v_s \approx 0$$

$$z_d - z_s = 50 \text{ ft} \quad \text{[given as rise in elevation]}$$

The discharge and suction pressures are

$$p_d = 20 \text{ psig}$$

$$p_s = 50 \text{ psig}$$

The pressure head added by the pump is

$$h = \frac{(p_d - p_s)g_c}{\rho g} + \frac{v_d^2 - v_s^2}{2g} + z_d - z_s$$

$$= \frac{\left(20 \frac{\text{lbf}}{\text{in}^2} - 50 \frac{\text{lbf}}{\text{in}^2}\right)\left(12 \frac{\text{in}}{\text{ft}}\right)^2 \left(32.2 \frac{\text{lbm-ft}}{\text{lbf-sec}^2}\right)}{\left(62.3 \frac{\text{lbm}}{\text{ft}^3}\right)\left(32.2 \frac{\text{ft}}{\text{sec}^2}\right)}$$

$$+ \frac{\left(14.139 \frac{\text{ft}}{\text{sec}}\right)^2}{(2)\left(32.2 \frac{\text{ft}}{\text{sec}^2}\right)} + 50 \text{ ft}$$

$$= -16.2 \text{ ft}$$

The head added is

$$h_A = h + h_f = -16.2 \text{ ft} + 130.1 \text{ ft}$$

$$= 113.9 \text{ ft}$$

The mass flow rate is

$$\dot{m} = \rho \dot{V} = \left(62.3 \frac{\text{lbm}}{\text{ft}^3}\right)\left(1.25 \frac{\text{ft}^3}{\text{sec}}\right)$$

$$= 77.875 \text{ lbm/sec}$$

From Table 18.5, the hydraulic horsepower is

$$\text{WHP} = \frac{h_A \dot{m}}{550} \times \frac{g}{g_c}$$

$$= \frac{(113.9 \text{ ft})\left(77.875 \frac{\text{lbm}}{\text{sec}}\right)}{550 \frac{\text{ft-lbf}}{\text{hp-sec}}}$$

$$\times \frac{32.2 \frac{\text{ft}}{\text{sec}^2}}{32.2 \frac{\text{lbm-ft}}{\text{lbf-sec}^2}}$$

$$= \boxed{16.13 \text{ hp} \quad (16 \text{ hp})}$$

The answer is (A).

SI Solution

From App. 16.C, data for 4 in schedule-40 steel pipe are

$$D_i = 102.3 \text{ mm}$$

$$A_i = 82.19 \times 10^{-4} \text{ m}^2$$

The velocity in the pipe is

$$v = \frac{\dot{V}}{A} = \frac{35 \frac{\text{L}}{\text{s}}}{(82.19 \times 10^{-4} \text{ m}^2)\left(1000 \frac{\text{L}}{\text{m}^3}\right)} = 4.26 \text{ m/s}$$

From Table 17.3, typical equivalent lengths for schedule-40, screwed steel fittings for 4 in pipes are

$$90° \text{ elbow: } 13 \text{ ft}$$

$$\text{gate valve: } 2.5 \text{ ft}$$

$$\text{check valve: } 38 \text{ ft}$$

The total equivalent length is

$$(2)(13 \text{ ft}) + (1)(2.5 \text{ ft}) + (1)(38 \text{ ft}) = 66.5 \text{ ft}$$

$$(66.5 \text{ ft})\left(0.3048 \frac{\text{m}}{\text{ft}}\right) = 20.27 \text{ m}$$

At 21°C, from App. 14.B, the water properties are

$$\rho = 998 \text{ kg/m}^3$$

$$\mu = 0.9827 \times 10^{-3} \text{ Pa·s}$$

$$\nu = \frac{\mu}{\rho} = \frac{0.9827 \times 10^{-3} \text{ Pa·s}}{998 \frac{\text{kg}}{\text{m}^3}}$$

$$= 9.85 \times 10^{-7} \text{ m}^2/\text{s}$$

The Reynolds number is

$$\mathrm{Re} = \frac{D\mathrm{v}}{\nu} = \frac{(102.3 \text{ mm})\left(4.26 \frac{\text{m}}{\text{s}}\right)}{\left(9.85 \times 10^{-7} \frac{\text{m}^2}{\text{s}}\right)\left(1000 \frac{\text{mm}}{\text{m}}\right)}$$

$$= 4.424 \times 10^5$$

From Table 17.2, for steel, $\epsilon = 6 \times 10^{-5}$ m. So,

$$\frac{\epsilon}{D} = \frac{\left(6 \times 10^{-5} \text{ m}\right)\left(1000 \frac{\text{mm}}{\text{m}}\right)}{102.3 \text{ mm}} = 0.0006$$

From App. 17.B, the friction factor is $f = 0.01836$.
From Eq. 18.6, the friction head is

$$h_f = \frac{fL\mathrm{v}^2}{2Dg}$$

$$= \frac{(0.01836)(230 \text{ m} + 20.27 \text{ m})\left(4.26 \frac{\text{m}}{\text{s}}\right)^2\left(1000 \frac{\text{mm}}{\text{m}}\right)}{(2)(102.3 \text{ mm})\left(9.81 \frac{\text{m}}{\text{s}^2}\right)}$$

$$= 41.5 \text{ m}$$

The total dynamic head is given by Eq. 18.8. s is taken as the bottom of the supply tank. d is taken as the end of the discharge pipe.

$$h = \frac{p_d - p_s}{\rho g} + \frac{\mathrm{v}_d^2 - \mathrm{v}_s^2}{2g} + z_d - z_s$$

$$\mathrm{v}_s \approx 0$$

$$z_d - z_s = 15 \text{ m} \quad \text{[given as rise in elevation]}$$

The difference between discharge and suction pressure is

$$p_d - p_s = 140 \text{ kPa} - 345 \text{ kPa} = -205 \text{ kPa}$$

$$h = \frac{(-205 \text{ kPa})\left(1000 \frac{\text{Pa}}{\text{kPa}}\right)}{\left(998 \frac{\text{kg}}{\text{m}^3}\right)\left(9.81 \frac{\text{m}}{\text{s}^2}\right)}$$

$$+ \frac{\left(4.26 \frac{\text{m}}{\text{s}}\right)^2}{(2)\left(9.81 \frac{\text{m}}{\text{s}^2}\right)} + 15 \text{ m}$$

$$= -5.0 \text{ m}$$

The head added by the pump is

$$h_A = h + h_f = -5.0 \text{ m} + 41.5 \text{ m}$$

$$= 36.5 \text{ m}$$

The mass flow rate is

$$\dot{m} = \rho \dot{V} = \frac{\left(998 \frac{\text{kg}}{\text{m}^3}\right)\left(35 \frac{\text{L}}{\text{s}}\right)}{1000 \frac{\text{L}}{\text{m}^3}}$$

$$= 34.93 \text{ kg/s}$$

From Table 18.6, the hydraulic power is

$$\mathrm{WkW} = \frac{9.81 h_A \dot{m}}{1000}$$

$$= \frac{\left(9.81 \frac{\text{m}}{\text{s}^2}\right)(36.5 \text{ m})\left(34.93 \frac{\text{kg}}{\text{s}}\right)}{1000 \frac{\text{W}}{\text{kW}}}$$

$$= \boxed{12.51 \text{ kW} \quad (13 \text{ kW})}$$

The answer is (A).

3. *Customary U.S. Solution*

The area of the rubber hose is

$$A = \frac{\pi D^2}{4} = \frac{\pi\left(\dfrac{2 \text{ in}}{12 \frac{\text{in}}{\text{ft}}}\right)^2}{4}$$

$$= 0.0218 \text{ ft}^2$$

The velocity of water in the hose is

$$\mathrm{v} = \frac{\dot{V}}{A} = \frac{\left(80 \frac{\text{gal}}{\text{min}}\right)\left(0.002228 \dfrac{\frac{\text{ft}^3}{\text{sec}}}{\frac{\text{gal}}{\text{min}}}\right)}{0.0218 \text{ ft}^2}$$

$$= 8.176 \text{ ft/sec}$$

At 80°F from App. 14.A, the kinematic viscosity of water is $\nu = 0.93 \times 10^{-5}$ ft²/sec.

The Reynolds number is

$$\mathrm{Re} = \frac{\mathrm{v}D}{\nu} = \frac{\left(8.176 \frac{\text{ft}}{\text{sec}}\right)(2 \text{ in})}{\left(0.93 \times 10^{-5} \frac{\text{ft}^2}{\text{sec}}\right)\left(12 \frac{\text{in}}{\text{ft}}\right)}$$

$$= 1.47 \times 10^5$$

Since the rubber hose is smooth, from App. 17.B, the friction factor is $f = 0.0166$.

From Eq. 18.6, the friction head is

$$h_f = \frac{fLv^2}{2Dg} = \frac{(0.0166)(50 \text{ ft})\left(8.176 \, \frac{\text{ft}}{\text{sec}}\right)^2 \left(12 \, \frac{\text{in}}{\text{ft}}\right)}{(2)(2 \text{ in})\left(32.2 \, \frac{\text{ft}}{\text{sec}^2}\right)}$$

$$= 5.17 \text{ ft}$$

Neglecting entrance and exit losses, the head added by the pump is

$$h_A = h_f + h_z = 5.17 \text{ ft} + 12 \text{ ft} - 4 \text{ ft}$$
$$= \boxed{13.17 \text{ ft} \quad (13 \text{ ft})}$$

The answer is (B).

SI Solution

The area of the rubber hose is

$$A = \frac{\pi D^2}{4} = \frac{\pi (5.1 \text{ cm})^2}{(4)\left(100 \, \frac{\text{cm}}{\text{in}}\right)^2} = 0.00204 \text{ m}^2$$

The velocity of water in the hose is

$$v = \frac{\dot{V}}{A} = \frac{5 \, \frac{\text{L}}{\text{s}}}{(0.00204 \text{ m}^2)\left(1000 \, \frac{\text{L}}{\text{m}^3}\right)} = 2.45 \text{ m/s}$$

At 27°C from App. 14.B, the kinematic viscosity of water is $\nu = 8.60 \times 10^{-7} \text{ m}^2/\text{s}$.

The Reynolds number is

$$\text{Re} = \frac{vD}{\nu} = \frac{\left(2.45 \, \frac{\text{m}}{\text{s}}\right)(5.1 \text{ cm})}{\left(8.60 \times 10^{-7} \, \frac{\text{m}^2}{\text{s}}\right)\left(100 \, \frac{\text{cm}}{\text{in}}\right)} = 1.45 \times 10^5$$

Since the rubber hose is smooth, from App. 17.B, the friction factor is $f \approx 0.0166$.

From Eq. 18.6, the friction head is

$$h_f = \frac{fLv^2}{2Dg} = \frac{(0.0166)(15 \text{ m})\left(2.45 \, \frac{\text{m}}{\text{s}}\right)^2 \left(100 \, \frac{\text{cm}}{\text{in}}\right)}{(2)(5.1 \text{ cm})\left(9.81 \, \frac{\text{m}}{\text{s}^2}\right)}$$

$$= 1.49 \text{ m}$$

Neglecting entrance and exit losses, the head added by the pump is

$$h_A = h_f + h_z = 1.49 \text{ m} + 4 \text{ m} - 1.5 \text{ m}$$
$$= \boxed{3.99 \text{ m} \quad (4.3 \text{ m})}$$

The answer is (B).

4. *Customary U.S. Solution*

From App. 16.B, the diameters (inside) for 8 in and 6 in schedule-40 steel pipe are

$$D_1 = 7.981 \text{ in}$$
$$D_2 = 6.065 \text{ in}$$

At 60°F from App. 14.A, the density of water is 62.37 lbm/ft^3.

The mass flow rate through 6 in pipe is

$$\dot{m} = A_2 v_2 \rho = \frac{\left(\frac{\pi (6.065 \text{ in})^2}{4}\right)\left(12 \, \frac{\text{ft}}{\text{sec}}\right)\left(62.37 \, \frac{\text{lbm}}{\text{ft}^3}\right)}{\left(12 \, \frac{\text{in}}{\text{ft}}\right)^2}$$

$$= 150.2 \text{ lbm/sec}$$

The inlet (suction) pressure is

$$(14.7 \text{ psia} - 5 \text{ psig})\left(12 \, \frac{\text{in}}{\text{ft}}\right)^2 = 1397 \text{ lbf/ft}^2$$

From Table 18.5, the head added by the pump is

$$h_A = \frac{550(\text{BHP})\eta}{\dot{m}} \times \frac{g_c}{g}$$

$$= \frac{\left(550 \, \frac{\text{ft-lbf}}{\text{hp-sec}}\right)(20 \text{ hp})(0.70)}{150.2 \, \frac{\text{lbm}}{\text{sec}}} \times \frac{32.2 \, \frac{\text{lbm-ft}}{\text{lbf-sec}^2}}{32.2 \, \frac{\text{ft}}{\text{sec}^2}}$$

$$= 51.26 \text{ ft}$$

At 1 (pump inlet):

$$p_1 = 1397 \text{ lbf/ft}^2 \quad [\text{absolute}]$$
$$z_1 = 0$$
$$v_1 = \frac{v_2 A_2}{A_1} = v_2 \left(\frac{D_2}{D_1}\right)^2 = \left(12 \, \frac{\text{ft}}{\text{sec}}\right)\left(\frac{6.065 \text{ in}}{7.981 \text{ in}}\right)^2$$
$$= 6.93 \text{ ft/sec}$$

At 2 (pump outlet):

$$p_2 \quad [\text{unknown}]$$
$$v_2 = 12 \text{ ft/sec} \quad [\text{given}]$$
$$z_2 = z_1 = 0$$
$$h_{f,1-2} = 0$$

Let z_3 be the additional head above atmospheric. From Eq. 18.9(b), the head added by the pump is

$$h_A = \frac{(p_2 - p_1)g_c}{\rho g} + \frac{v_2^2 - v_1^2}{2g} + z_2 - z_1 + h_{f,1-2}$$

$$51.26 \text{ ft} = \frac{\left(p_2 - 1397 \frac{\text{lbf}}{\text{ft}^2}\right)\left(32.2 \frac{\text{lbm-ft}}{\text{lbf-sec}^2}\right)}{\left(62.37 \frac{\text{lbm}}{\text{ft}^3}\right)\left(32.2 \frac{\text{ft}}{\text{sec}^2}\right)}$$

$$+ \frac{\left(12 \frac{\text{ft}}{\text{sec}}\right)^2 - \left(6.93 \frac{\text{ft}}{\text{sec}}\right)^2}{(2)\left(32.2 \frac{\text{ft}}{\text{sec}^2}\right)}$$

$$+ 0 - 0 + 0$$

$$p_2 = 4501.2 \text{ ft}$$

At 3 (discharge):

$$p_3 = \left(14.7 \frac{\text{lbf}}{\text{in}^2}\right)\left(12 \frac{\text{in}}{\text{ft}}\right)^2 = 2117 \text{ lbf/ft}^2$$

$$v_3 = 12 \text{ ft/sec} \quad \text{[given]}$$

$$z_3 \quad \text{[unknown]}$$

$$h_{f,2-3} = 10 \text{ ft}$$

$$h_A = 0 \quad \text{[no pump between points 2 and 3]}$$

$$h_A = \frac{(p_3 - p_2)g_c}{\rho g} + \frac{v_3^2 - v_2^2}{2g} + z_3 - z_2 + h_{f,2-3}$$

$$0 = \frac{\left(2117 \frac{\text{lbf}}{\text{ft}^2} - 4501.2 \frac{\text{lbf}}{\text{ft}^2}\right)\left(32.2 \frac{\text{lbm-ft}}{\text{lbf-sec}^2}\right)}{\left(62.37 \frac{\text{lbm}}{\text{ft}^3}\right)\left(32.2 \frac{\text{ft}}{\text{sec}^2}\right)}$$

$$+ \frac{\left(12 \frac{\text{ft}}{\text{sec}}\right)^2 - \left(12 \frac{\text{ft}}{\text{sec}}\right)^2}{(2)\left(32.2 \frac{\text{ft}}{\text{sec}^2}\right)}$$

$$+ z_3 - 0 + 10 \text{ ft}$$

$$z_3 = 28.2 \text{ ft} \quad (28 \text{ ft})$$

z_3 could have been found directly without determining the intermediate pressure, p_2. This method is illustrated in the SI solution.

The answer is (A).

SI Solution

From App. 16.C, the inside diameters for 8 in and 6 in steel schedule-40 pipe are

$$D_1 = 202.7 \text{ mm}$$

$$D_2 = 154.1 \text{ mm}$$

At 16°C from App. 14.B, the density of water is 998.83 kg/m³.

The mass flow rate through the 6 in pipe is

$$\dot{m} = A_2 v_2 \rho = \frac{\left(\frac{\pi(154.1 \text{ mm})^2}{4}\right)\left(4 \frac{\text{m}}{\text{s}}\right)\left(998.83 \frac{\text{kg}}{\text{m}^3}\right)}{\left(1000 \frac{\text{mm}}{\text{m}}\right)^2}$$

$$= 74.5 \text{ kg/s}$$

The inlet (suction) pressure is

$$101.3 \text{ kPa} - 35 \text{ kPa} = 66.3 \text{ kPa}$$

From Table 18.6, the head added by the pump is

$$h_A = \frac{1000(\text{BkW})\eta}{9.81 \dot{m}}$$

$$= \frac{\left(1000 \frac{\text{W}}{\text{kW}}\right)(20 \text{ hp})\left(0.7457 \frac{\text{kW}}{\text{hp}}\right)(0.70)}{\left(9.81 \frac{\text{m}}{\text{s}^2}\right)\left(74.5 \frac{\text{kg}}{\text{s}}\right)}$$

$$= 14.28 \text{ m}$$

At 1:

$$p_1 = 66.3 \text{ kPa}$$

$$z_1 = 0$$

$$v_1 = v_2\left(\frac{A_2}{A_1}\right) = v_2\left(\frac{D_2}{D_1}\right)^2 = \left(4 \frac{\text{m}}{\text{s}}\right)\left(\frac{154.1 \text{ mm}}{202.7 \text{ mm}}\right)^2$$

$$= 2.31 \text{ m/s}$$

At 2:

$$p_2 = 101.3 \text{ kPa}$$

$$v_2 = 4 \text{ m/s} \quad \text{[given]}$$

From Eq. 18.9(a), the head added by the pump is

$$h_A = \frac{p_2 - p_1}{\rho g} + \frac{v_2^2 - v_1^2}{2g} + z_2 - z_1 + h_f + z_3$$

$$14.28 \text{ m} = \frac{(101.3 \text{ kPa} - 66.3 \text{ kPa})\left(1000 \frac{\text{Pa}}{\text{kPa}}\right)}{\left(998.83 \frac{\text{kg}}{\text{m}^3}\right)\left(9.81 \frac{\text{m}}{\text{s}^2}\right)}$$

$$+ \frac{\left(4 \frac{\text{m}}{\text{s}}\right)^2 - \left(2.31 \frac{\text{m}}{\text{s}}\right)^2}{(2)\left(9.81 \frac{\text{m}}{\text{s}^2}\right)}$$

$$+ 0 - 0 + 3.3 \text{ m} + z_3$$

$$z_3 = \boxed{6.86 \text{ m} \quad (6.9 \text{ m})}$$

The answer is (A).

5. *Customary U.S. Solution*

(a) The flow rate is

$$\dot{V} = \left(10{,}000 \ \frac{\text{gal}}{\text{hr}}\right)\left(0.1337 \ \frac{\text{ft}^3}{\text{gal}}\right) = 1337 \ \text{ft}^3/\text{hr}$$

From App. 16.B, data for 4 in schedule-40 steel pipe are

$$D_i = 0.3355 \ \text{ft}$$
$$A_i = 0.08841 \ \text{ft}^2$$

The velocity in the pipe is

$$\text{v} = \frac{\dot{V}}{A} = \frac{1337 \ \dfrac{\text{ft}^3}{\text{hr}}}{(0.08841 \ \text{ft}^2)\left(3600 \ \dfrac{\text{sec}}{\text{hr}}\right)} = 4.20 \ \text{ft/sec}$$

From App. 14.A, the kinematic viscosity of water at 60°F is

$$\nu = 1.217 \times 10^{-5} \ \text{ft}^2/\text{sec}$$
$$\rho = 62.37 \ \text{lbm/ft}^3$$

The Reynolds number is

$$\text{Re} = \frac{D\text{v}}{\nu} = \frac{(0.3355 \ \text{ft})\left(4.20 \ \dfrac{\text{ft}}{\text{sec}}\right)}{1.217 \times 10^{-5} \ \dfrac{\text{ft}^2}{\text{sec}}} = 1.16 \times 10^5$$

From App. 17.A, for welded and seamless steel, $\epsilon = 0.0002$ ft.

$$\frac{\epsilon}{D} = \frac{0.0002 \ \text{ft}}{0.3355 \ \text{ft}} \approx 0.0006$$

From App. 17.B, the friction factor, f, is 0.0205. The 7000 ft of equivalent length includes the pipe between the lake and the pump. The friction head is

$$h_f = \frac{f L \text{v}^2}{2Dg} = \frac{(0.0205)(7000 \ \text{ft})\left(4.20 \ \dfrac{\text{ft}}{\text{sec}}\right)^2}{(2)(0.3355 \ \text{ft})\left(32.2 \ \dfrac{\text{ft}}{\text{sec}^2}\right)} = 117.2 \ \text{ft}$$

The head added by the pump is

$$h_A = h_f + h_z = 117.2 \ \text{ft} + 12 \ \text{ft} + 350 \ \text{ft} = 479.2 \ \text{ft}$$

From Table 18.5, the hydraulic horsepower is

$$\text{WHP} = \frac{h_A Q(\text{SG})}{3956} = \frac{(479.2 \ \text{ft})\left(10{,}000 \ \dfrac{\text{gal}}{\text{hr}}\right)(1)}{\left(3956 \ \dfrac{\text{ft-gal}}{\text{hp-min}}\right)\left(60 \ \dfrac{\text{min}}{\text{hr}}\right)}$$

$$= 20.2 \ \text{hp}$$

From Eq. 18.16, the overall efficiency of the pump is

$$\eta = \frac{\text{WHP}}{\text{EHP}}$$

$$\text{EHP} = \frac{20.2 \ \text{hp}}{0.7}$$

$$= 28.9 \ \text{hp}$$

At $0.04/kW-hr, power costs for 1 hr are

$$(28.9 \ \text{hp})\left(0.7457 \ \frac{\text{kW}}{\text{hp}}\right)(1 \ \text{hr})\left(0.04 \ \frac{\$}{\text{kW-hr}}\right)$$

$$= \boxed{\$0.86 \ (\$1) \ \text{per hour}}$$

The answer is (B).

(b) The motor horsepower, EHP, is 28.9 hp. Select the next higher standard motor size. Use a $\boxed{\text{30 hp motor.}}$

The answer is (B).

(c) From Eq. 18.5(b),

$$h_{\text{atm}} = \frac{p_{\text{atm}}}{\rho} \times \frac{g_c}{g} = \frac{\left(14.7 \ \dfrac{\text{lbf}}{\text{in}^2}\right)\left(12 \ \dfrac{\text{in}}{\text{ft}}\right)^2}{62.37 \ \dfrac{\text{lbm}}{\text{ft}^3}} \times \frac{32.2 \ \dfrac{\text{lbm-ft}}{\text{lbf-sec}^2}}{32.2 \ \dfrac{\text{ft}}{\text{sec}^2}}$$

$$= 33.94 \ \text{ft}$$

The friction losses due to 300 ft is

$$h_{f(s)} = \left(\frac{300 \ \text{ft}}{7000 \ \text{ft}}\right)h_f = \left(\frac{300 \ \text{ft}}{7000 \ \text{ft}}\right)(117.2 \ \text{ft}) = 5.0 \ \text{ft}$$

From App. 14.A, the vapor pressure head at 60°F is 0.59 ft.

The NPSHA from Eq. 18.31(a) is

$$\text{NPSHA} = h_{\text{atm}} + h_{z(s)} - h_{f(s)} - h_{\text{vp}}$$

$$= 33.94 \ \text{ft} - 12 \ \text{ft} - 5.0 \ \text{ft} - 0.59 \ \text{ft}$$

$$= \boxed{16.35 \ \text{ft} \quad (16 \ \text{ft})}$$

The answer is (D).

SI Solution

(a) From App. 16.C, data for 4 in schedule-40 steel pipe are

$$D_i = 102.3 \ \text{mm}$$
$$A_i = 82.19 \times 10^{-4} \ \text{m}^2$$

The velocity in the pipe is

$$\text{v} = \frac{\dot{V}}{A} = \frac{10.5 \ \dfrac{\text{L}}{\text{s}}}{(82.19 \times 10^{-4} \ \text{m}^2)\left(1000 \ \dfrac{\text{L}}{\text{m}^3}\right)} = 1.28 \ \text{m/s}$$

From App. 14.B, at 16°C the water data are

$$\rho = 998.83 \text{ kg/m}^3$$
$$\mu = 1.1261 \times 10^{-3} \text{ Pa·s}$$

The Reynolds number is

$$\text{Re} = \frac{\rho v D}{\mu}$$

$$= \frac{\left(998.83 \dfrac{\text{kg}}{\text{m}^3}\right)\left(1.28 \dfrac{\text{m}}{\text{s}}\right)(102.3 \text{ mm})}{(1.1261 \times 10^{-3} \text{ Pa·s})\left(1000 \dfrac{\text{mm}}{\text{m}}\right)}$$

$$= 1.16 \times 10^5$$

From Table 17.2, for welded and seamless steel, $\epsilon = 6.0 \times 10^{-5}$ m.

$$\frac{\epsilon}{D} = \frac{\left(6.0 \times 10^{-5} \text{ m}\right)\left(1000 \dfrac{\text{mm}}{\text{m}}\right)}{102.3 \text{ mm}} \approx 0.0006$$

From App. 17.B, the friction factor is $f = 0.0205$.

From Eq. 18.6, the friction head is

$$h_f = \frac{fLv^2}{2Dg} = \frac{(0.0205)(2300 \text{ m})\left(1.28 \dfrac{\text{m}}{\text{s}}\right)^2\left(1000 \dfrac{\text{mm}}{\text{m}}\right)}{(2)(102.3 \text{ mm})\left(9.81 \dfrac{\text{m}}{\text{s}^2}\right)}$$

$$= 38.5 \text{ m}$$

The head added by the pump is

$$h_A = h_f + h_z = 38.5 \text{ m} + 4 \text{ m} + 115 \text{ m}$$
$$= 157.5 \text{ m}$$

From Table 18.6, the hydraulic power is

$$\text{WkW} = \frac{9.81 h_A Q(\text{SG})}{1000}$$

$$= \frac{\left(9.81 \dfrac{\text{m}}{\text{s}^2}\right)(157.5 \text{ m})\left(10.5 \dfrac{\text{L}}{\text{s}}\right)(1)}{1000 \dfrac{\text{W·L}}{\text{kW·kg}}}$$

$$= 16.22 \text{ kW}$$

From Eq. 18.16, the electrical power is

$$\text{EHP} = \frac{\text{WHP}}{\eta_{\text{overall}}} = \frac{16.22 \text{ kW}}{0.7}$$
$$= 23.2 \text{ kW}$$

At \$0.04/kW·h, power costs for 1 h are

$$(23.2 \text{ kW})(1 \text{ h})\left(0.04 \ \frac{\$}{\text{kW·h}}\right) = \boxed{\$0.93 \ (\$1) \text{ per hour}}$$

The answer is (B).

(b) The required motor power is 23.2 kW. Select the next higher standard motor size. $\boxed{\text{Use a 25 kW motor.}}$

The answer is (B).

(c) From Eq. 18.5(a),

$$h_{\text{atm}} = \frac{p}{\rho g} = \frac{(101 \text{ kPa})\left(1000 \dfrac{\text{Pa}}{\text{kPa}}\right)}{\left(998.83 \dfrac{\text{kg}}{\text{m}^3}\right)\left(9.81 \dfrac{\text{m}}{\text{s}^2}\right)}$$

$$= 10.31 \text{ m}$$

The friction loss due to 100 m is

$$h_{f(s)} = \left(\frac{100 \text{ m}}{2300 \text{ m}}\right)h_f = \left(\frac{100 \text{ m}}{2300 \text{ m}}\right)(38.5 \text{ m})$$
$$= 1.67 \text{ m}$$

The vapor pressure at 16°C is 0.01818 bar.

From Eq. 18.5(a),

$$h_{\text{vp}} = \frac{p_{\text{vp}}}{g\rho} = \frac{(0.01818 \text{ bar})\left(1 \times 10^5 \dfrac{\text{Pa}}{\text{bar}}\right)}{\left(9.81 \dfrac{\text{m}}{\text{s}^2}\right)\left(998.83 \dfrac{\text{kg}}{\text{m}^3}\right)}$$

$$= 0.19 \text{ m}$$

The NPSHA from Eq. 18.31(a) is

$$\text{NPSHA} = h_{\text{atm}} + h_{z(s)} - h_{f(s)} - h_{\text{vp}}$$
$$= 10.31 \text{ m} - 4 \text{ m} - 1.67 \text{ m} - 0.19 \text{ m}$$
$$= \boxed{4.45 \text{ m} \quad (4.5 \text{ m})}$$

The answer is (D).

6. (a) Sewers are usually gravity-flow systems. $h_f = \Delta z = 48$ ft since $\Delta p = 0$ and $\Delta v = 0$ for open channel flow.

$$Q = \frac{\left(250 \dfrac{\text{gal}}{\text{person-day}}\right)(10,000 \text{ people})}{\left(24 \dfrac{\text{hr}}{\text{day}}\right)\left(60 \dfrac{\text{min}}{\text{hr}}\right)}$$

$$= 1736 \text{ gal/min}$$

Solving for d from Eq. 17.31,

$$d_{in}^{4.87} = \frac{10.44 L \dot{V}^{1.85}}{C^{1.85} h_f}$$

$$= \frac{(10.44)(5000 \text{ ft})\left(1736 \dfrac{\text{gal}}{\text{min}}\right)^{1.85}}{(130)^{1.85}(48 \text{ ft})}$$

$$= 131{,}462$$

$$d = \boxed{11.24 \text{ in} \quad [\text{round to 12 in minimum}]}$$

The answer is (B).

(b) Without having a specific pump curve, the number of pumps can only be specified based on general rules. Use the *Ten States' Standards*, which states:

- No station will have less than two identical pumps.

- Capacity must be met with one pump out of service.

- Provision must be made in order to alternate pumps automatically.

$$\boxed{\text{Two pumps are required, plus spares.}}$$

The answer is (A).

(c) With a variable speed pump, it will be possible to adjust to the wide variations in flow (100–250 gpcd). It may be possible to operate with one pump. However, TSS still requires two.

The answer is (B).

(d) With three constant speed pumps,

$$Q = \frac{1736 \dfrac{\text{gal}}{\text{min}}}{3}$$

$$= 579 \text{ gal/min at maximum capacity}$$

The pump only has to lift the sewage 10 ft. To get to the pump, the sewage descended 48 ft under the influence of gravity. From Table 18.5, assuming specific gravity ≈ 1.00,

$$\text{rated motor power} = \frac{h_A Q(\text{SG})}{3956\eta} = \frac{(10 \text{ ft})\left(579 \dfrac{\text{gal}}{\text{min}}\right)(1)}{\left(3956 \dfrac{\text{gal-ft}}{\text{min-hp}}\right)(0.60)}$$

$$= \boxed{2.44 \text{ hp} \quad (3 \text{ hp})}$$

The answer is (B).

(e) With two variable-speed pumps,

$$Q = \frac{1736 \dfrac{\text{gal}}{\text{min}}}{2}$$

$$= 868 \text{ gal/min}$$

$$\text{rated motor power} = \frac{h_A Q(\text{SG})}{3956\eta} = \frac{(10 \text{ ft})\left(868 \dfrac{\text{gal}}{\text{min}}\right)(1)}{\left(3956 \dfrac{\text{gal-ft}}{\text{min-hp}}\right)(0.80)}$$

$$= \boxed{2.74 \text{ hp} \quad (3 \text{ hp})}$$

The answer is (A).

(f) Incoming flow rate is independent of sump level.

The answer is (D).

(g) In the wet well, the pump (and perhaps motor) is submerged. Forced ventilation air will prevent a concentration of explosive methane. From the *Ten States' Standards*, if intermittent, the wet well should have $\boxed{30}$ air changes per hour.

The answer is (D).

(h) If continuous, the dry well should have $\boxed{6}$ air changes per hour.

The answer is (A).

7. (a) The static suction lift, $h_{p(s)}$, is $\boxed{20 \text{ ft.}}$

The answer is (B).

(b) The static discharge head, $h_{p(d)}$, is $\boxed{15 \text{ ft.}}$

The answer is (A).

(c) There is no pipe size specified, so h_v cannot be calculated. Even so, v is typically in the 5–10 ft/sec range, and $h_v \approx 0$.

Since pipe lengths are not given, assume $h_f \approx 0$.

$$20 \text{ ft} + 15 \text{ ft} + \frac{\left(80 \dfrac{\text{lbf}}{\text{in}^2}\right)\left(12 \dfrac{\text{in}}{\text{ft}}\right)^2}{62.4 \dfrac{\text{lbf}}{\text{ft}^3}} + 10 \text{ ft}$$

$$= \boxed{229.6 \text{ ft } (230 \text{ ft}) \text{ of water}}$$

The answer is (D).

(d) The flow rate is

$$\frac{(3.5 \text{ MGD})\left(62.4 \frac{\text{lbf}}{\text{ft}^3}\right)}{0.64632 \frac{\text{MGD}}{\frac{\text{ft}^3}{\text{sec}}}} = 337.9 \text{ lbf/sec}$$

The rated motor output power does not depend on the motor efficiency. From Table 18.5,

$$P = \frac{h_A \dot{m}}{550\eta_{\text{pump}}} = \frac{(229.6 \text{ ft})\left(337.9 \frac{\text{lbf}}{\text{sec}}\right)}{\left(550 \frac{\text{ft-lbf}}{\text{hp-sec}}\right)(0.85)}$$

$$= 166.0 \text{ hp}$$

$$\boxed{\text{Use a 200 hp motor.}}$$

The answer is (D).

8. *Customary U.S. Solution*

(a) From App. 16.B, the pipe data for 3 in schedule-40 steel pipe are

$$D_i = 0.2557 \text{ ft}$$

$$A_i = 0.05134 \text{ ft}^2$$

From App. 17.D, the equivalent lengths for various fittings are

$$\text{flanged elbow, } L_e = 4.4 \text{ ft}$$

$$\text{wide-open gate valve, } L_e = 2.8 \text{ ft}$$

The total equivalent length of pipe and fittings is

$$L_e = 500 \text{ ft} + (6)(4.4 \text{ ft}) + (2)(2.8 \text{ ft})$$

$$= 532 \text{ ft}$$

As a first estimate, assume the flow rate is 100 gal/min.

The velocity in the pipe is

$$v = \frac{\dot{V}}{A} = \frac{\left(100 \frac{\text{gal}}{\text{min}}\right)\left(0.002228 \frac{\frac{\text{ft}^3}{\text{sec}}}{\frac{\text{gal}}{\text{min}}}\right)}{0.05134 \text{ ft}^2}$$

$$= 4.34 \text{ ft/sec}$$

The Reynolds number is

$$\text{Re} = \frac{vD}{\nu} = \frac{\left(4.34 \frac{\text{ft}}{\text{sec}}\right)(0.2557 \text{ ft})}{6 \times 10^{-6} \frac{\text{ft}^2}{\text{sec}}}$$

$$= 1.85 \times 10^5$$

From App. 17.A, $\epsilon = 0.0002$ ft.

So,

$$\frac{\epsilon}{D} = \frac{0.0002 \text{ ft}}{0.2557 \text{ ft}} \approx 0.0008$$

From the friction factor table, $f \approx 0.0204$.

For higher flow rates, f approaches 0.0186. Since the chosen flow rate was almost the lowest, $f = 0.0186$ should be used.

From Eq. 18.6, the friction head loss is

$$h_f = \frac{fLv^2}{2Dg} = \frac{(0.0186)(532 \text{ ft})\left(4.34 \frac{\text{ft}}{\text{sec}}\right)^2}{(2)(0.2557 \text{ ft})\left(32.2 \frac{\text{ft}}{\text{sec}^2}\right)}$$

$$= 11.3 \text{ ft of gasoline}$$

This neglects the small velocity head. The other system points can be found using Eq. 18.43.

$$\frac{h_{f_1}}{h_{f_2}} = \left(\frac{Q_1}{Q_2}\right)^2$$

$$h_{f_2} = h_{f_1}\left(\frac{Q_2}{100 \frac{\text{gal}}{\text{min}}}\right)^2$$

$$= (11.3 \text{ ft})\left(\frac{Q_2}{100 \frac{\text{gal}}{\text{min}}}\right)^2$$

$$= 0.00113 Q_2^2$$

Q	h_f	$h_f + 60$
(gal/min)	(ft)	(ft)
100	11.3	71.3
200	45.2	105.2
300	101.7	161.7
400	180.8	240.8
500	282.5	342.5
600	406.8	466.8

The pump's characteristic curve is independent of the liquid's specific gravity. Plot the system and pump curves.

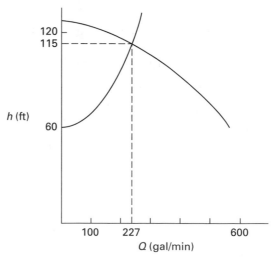

$$h = 115 \text{ ft}$$

$$\boxed{Q = 227 \text{ gal/min} \quad (230 \text{ gal/min})}$$

(This value could be used to determine a new friction factor.)

The answer is (D).

(b) From Table 18.5, the hydraulic horsepower is

$$\text{WHP} = \frac{h_A Q(\text{SG})}{3956} = \frac{(115 \text{ ft})\left(227 \ \frac{\text{gal}}{\text{min}}\right)(0.7)}{3956 \ \frac{\text{ft-gal}}{\text{hp-min}}}$$

$$= 4.62 \text{ hp}$$

The cost per hour is

$$\frac{(4.62 \text{ hp})\left(0.7457 \ \frac{\text{kW}}{\text{hp}}\right)}{(0.88)(0.88)} \times (1 \text{ hr})\left(0.045 \ \frac{\$}{\text{kW-hr}}\right) = \boxed{\$0.20}$$

The answer is (A).

SI Solution

(a) From App. 16.C, the pipe data for 3 in schedule-40 pipe are

$$D_i = 77.92 \text{ mm}$$
$$A_i = 47.69 \times 10^{-4} \text{ m}^2$$

From App. 17.D, the equivalent lengths for various fittings are

$$\text{flanged elbow, } L_e = 4.4 \text{ ft}$$
$$\text{wide-open gate valve, } L_e = 2.8 \text{ ft}$$

The total equivalent length of pipe and fittings is

$$L_e = 170 \text{ m} + (6)(4.4 \text{ ft})\left(0.3048 \ \frac{\text{m}}{\text{ft}}\right)$$
$$+ (2)(2.8 \text{ ft})\left(0.3048 \ \frac{\text{m}}{\text{ft}}\right)$$
$$= 179.8 \text{ m}$$

As a first estimate, assume flow rate is 6.3 L/s. The velocity in the pipe is

$$\text{v} = \frac{\dot{V}}{A}$$

$$= \frac{6.3 \ \frac{\text{L}}{\text{s}}}{(47.69 \times 10^{-4} \text{ m}^2)\left(1000 \ \frac{\text{L}}{\text{m}^3}\right)}$$

$$= 1.32 \text{ m/s}$$

The Reynolds number is

$$\text{Re} = \frac{\text{v}D}{\nu}$$

$$= \frac{\left(1.32 \ \frac{\text{m}}{\text{s}}\right)(77.92 \text{ mm})}{\left(5.6 \times 10^{-7} \ \frac{\text{m}^2}{\text{s}}\right)\left(1000 \ \frac{\text{mm}}{\text{m}}\right)}$$

$$= 1.84 \times 10^5$$

From Table 17.2, $\epsilon = 6.0 \times 10^{-5}$ m.

$$\frac{\epsilon}{D} = \frac{(6.0 \times 10^{-5} \text{ m})\left(1000 \ \frac{\text{mm}}{\text{m}}\right)}{77.92 \text{ mm}}$$

$$\approx 0.0008$$

From the friction factor table (see App. 17.B), $f = 0.0204$.

For higher flow rates, f approaches 0.0186. Since the chosen flow rate was almost the lowest, $f = 0.0186$ should be used.

From Eq. 18.6, the friction head loss is

$$h_f = \frac{fL\text{v}^2}{2Dg}$$

$$= \frac{(0.0186)(179.8 \text{ m})\left(1.32 \ \frac{\text{m}}{\text{s}}\right)^2\left(1000 \ \frac{\text{mm}}{\text{m}}\right)}{(2)(77.92 \text{ mm})\left(9.81 \ \frac{\text{m}}{\text{s}^2}\right)}$$

$$= 3.80 \text{ m of gasoline}$$

This neglects the small velocity head. The other system points can be found using Eq. 18.43.

$$\frac{h_{f1}}{h_{f2}} = \left(\frac{Q_1}{Q_2}\right)^2$$

$$h_{f2} = h_{f1}\left(\frac{Q_2}{Q_1}\right)^2$$

$$= (3.80 \text{ m})\left(\frac{Q_2}{6.3 \ \frac{\text{L}}{\text{s}}}\right)^2$$

$$= 0.0957 Q_2^2$$

Q (L/s)	h_f (m)	$h_f + 20$ (m)
6.3	3.80	23.80
12	13.78	33.78
18	31.0	51.0
24	55.1	75.1
30	86.1	106.1
36	124.0	144.0

The pump's characteristic curve is independent of the liquid's specific gravity. Plot the system and pump curves.

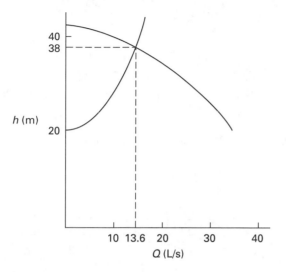

$$h = 38.0 \text{ m}$$

$$\boxed{Q = 13.6 \text{ L/s} \quad (14 \text{ L/s})}$$

(This value could be used to determine a new friction factor.)

The answer is (D).

(b) From Table 18.6, the hydraulic power is

$$\text{WkW} = \frac{9.81 h_A Q (\text{SG})}{1000}$$

$$= \frac{\left(9.81 \ \frac{\text{m}}{\text{s}^2}\right)(38.0 \text{ m})\left(13.6 \ \frac{\text{L}}{\text{s}}\right)(0.7)}{1000 \ \frac{\text{W}}{\text{kW}}}$$

$$= 3.55 \text{ kW}$$

The cost per hour is

$$= \left(\frac{3.55 \text{ kW}}{(0.88)(0.88)}\right)(1 \text{ h})\left(0.045 \ \frac{\$}{\text{kW·h}}\right)$$

$$= \boxed{\$0.21 \quad (\$0.20)}$$

The answer is (A).

9. *Customary U.S. Solution*

From Eq. 18.28(b), the specific speed is

$$n_s = \frac{n\sqrt{Q}}{h_A^{0.75}}$$

For a double-suction pump, Q in the preceding equation is half of the full flow rate.

$$n_s = \frac{\left(900 \ \frac{\text{rev}}{\text{min}}\right)\sqrt{\left(\frac{1}{2}\right)\left(300 \ \frac{\text{gal}}{\text{sec}}\right)\left(60 \ \frac{\text{sec}}{\text{min}}\right)}}{(20 \text{ ft})^{0.75}}$$

$$= \boxed{9028 \text{ rpm} \quad (9000 \text{ rpm})}$$

The answer is (C).

SI Solution

From Eq. 18.28(a), the specific speed is

$$n_s = \frac{n\sqrt{\dot{V}}}{h_A^{0.75}}$$

For a double-suction pump, $\dot{V}$ in the preceding equation is half of the full flow rate.

$$n_s = \frac{\left(900 \ \frac{\text{rev}}{\text{min}}\right)\sqrt{\left(\frac{1}{2}\right)\left(1.1 \ \frac{\text{kL}}{\text{s}}\right)\left(1 \ \frac{\text{m}^3}{\text{kL}}\right)}}{(7 \text{ m})^{0.75}}$$

$$= \boxed{155.1 \text{ rpm} \quad (160 \text{ rpm})}$$

The answer is (C).

10.

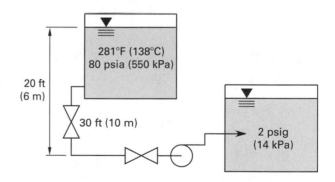

Customary U.S. Solution

From App. 16.B, data for 1.5 in schedule-40 steel pipe are

$$D_i = 0.1342 \text{ ft}$$

$$A_i = 0.01414 \text{ ft}^2$$

The velocity in the pipe is

$$v = \frac{\dot{V}}{A} = \frac{\left(100 \ \frac{\text{gal}}{\text{min}}\right)\left(0.002228 \ \frac{\frac{\text{ft}^3}{\text{sec}}}{\frac{\text{gal}}{\text{min}}}\right)}{0.01414 \ \text{ft}^2}$$

$$= 15.76 \ \text{ft/sec}$$

From App. 17.D, for screwed steel fittings, the approximate equivalent lengths for fittings are

$$\text{inlet (square mouth):} \ L_e = 3.1 \ \text{ft}$$
$$\text{long radius } 90° \text{ elbow:} \ L_e = 3.4 \ \text{ft}$$
$$\text{wide-open gate valves:} \ L_e = 1.2 \ \text{ft}$$

The total equivalent length is

$$30 \ \text{ft} + 3.1 \ \text{ft} + (2)(3.4 \ \text{ft}) + (2)(1.2 \ \text{ft}) = 42.3 \ \text{ft}$$

From App. 17.A, for steel, $\epsilon = 0.0002$ ft, so

$$\frac{\epsilon}{D} = \frac{0.0002 \ \text{ft}}{0.1342 \ \text{ft}} = 0.0015$$

At 281°F, $\nu = 0.239 \times 10^{-5}$ ft²/sec. The Reynolds number is

$$\text{Re} = \frac{D v}{\nu} = \frac{(0.1342 \ \text{ft})\left(15.76 \ \frac{\text{ft}}{\text{sec}}\right)}{0.239 \times 10^{-5} \ \frac{\text{ft}^2}{\text{sec}}}$$

$$= 8.85 \times 10^5$$

From App. 17.B, the friction factor is $f = 0.022$.

From Eq. 18.6, the friction head is

$$h_f = \frac{f L v^2}{2 D g} = \frac{(0.022)(42.3 \ \text{ft})\left(15.76 \ \frac{\text{ft}}{\text{sec}}\right)^2}{(2)(0.1342 \ \text{ft})\left(32.2 \ \frac{\text{ft}}{\text{sec}^2}\right)}$$

$$= 26.74 \ \text{ft}$$

Use App. 18.B. At 281°F, the saturated vapor pressure of pressurized water is

$$p_{\text{vapor}} = 50.02 \ \text{psia}$$

The density of the liquid is the reciprocal of the specific volume, also taken from App. 18.B.

$$\rho = \frac{1}{v_f} = \frac{1}{0.01727 \ \frac{\text{ft}^3}{\text{lbm}}} = 57.9 \ \text{lbm/ft}^3$$

From Eq. 18.5(b),

$$h_{\text{vp}} = \frac{p_{\text{vapor}}}{\rho} \times \frac{g_c}{g}$$

$$= \frac{\left(50.02 \ \frac{\text{lbf}}{\text{in}^2}\right)\left(12 \ \frac{\text{in}}{\text{ft}}\right)^2}{57.9 \ \frac{\text{lbm}}{\text{ft}^3}} \times \frac{32.2 \ \frac{\text{lbm-ft}}{\text{lbf-sec}^2}}{32.2 \ \frac{\text{ft}}{\text{sec}^2}}$$

$$= 124.4 \ \text{ft}$$

From Eq. 18.5(b), the pressure head is

$$h_p = \frac{\rho}{p} \times \frac{g_c}{g}$$

$$= \frac{\left(80 \ \frac{\text{lbf}}{\text{in}^2}\right)\left(12 \ \frac{\text{in}}{\text{ft}}\right)^2}{57.9 \ \frac{\text{lbm}}{\text{ft}^3}} \times \frac{32.2 \ \frac{\text{lbm-ft}}{\text{lbf-sec}^2}}{32.2 \ \frac{\text{ft}}{\text{sec}^2}}$$

$$= 199.0 \ \text{ft}$$

From Eq. 18.31(a), the NPSHA is

$$\text{NPSHA} = h_p + h_{z(s)} - h_{f(s)} - h_{\text{vp}}$$

$$= 199.0 \ \text{ft} + 20 \ \text{ft} - 26.74 \ \text{ft} - 124.4 \ \text{ft}$$

$$= \boxed{67.9 \ \text{ft} \quad (68 \ \text{ft})}$$

Since NPSHR = 10 ft, $\boxed{\text{the pump will not cavitate.}}$

(A pump may not be needed in this configuration.)

The answer is (D).

SI Solution

From App. 16.C, data for 1.5 in schedule-40 steel pipe are

$$D_i = 40.89 \ \text{mm}$$
$$A_i = 13.13 \times 10^{-4} \ \text{m}^2$$

The velocity in the pipe is

$$v = \frac{\dot{V}}{A} = \frac{6.3 \ \frac{\text{L}}{\text{s}}}{(13.13 \times 10^{-4} \ \text{m}^2)\left(1000 \ \frac{\text{L}}{\text{m}^3}\right)}$$

$$= 4.80 \ \text{m/s}$$

From App. 17.D, for screwed steel fittings, the approximate equivalent lengths for fittings are

$$\text{inlet (square mouth):} \ L_e = 3.1 \ \text{ft}$$
$$\text{long radius } 90° \text{ elbow:} \ L_e = 3.4 \ \text{ft}$$
$$\text{wide-open gate valves:} \ L_e = 1.2 \ \text{ft}$$

The total equivalent length is

$$30 \text{ ft} + 3.1 \text{ ft} + (2)(3.4 \text{ ft}) + (2)(1.2 \text{ ft}) = 42.3 \text{ ft}$$

$$(42.3 \text{ ft})\left(0.3048 \frac{\text{m}}{\text{ft}}\right) = 12.89 \text{ m}$$

From Table 17.2, for steel, $\epsilon = 6.0 \times 10^{-5}$ m.

$$\frac{\epsilon}{D} = \frac{(6.0 \times 10^{-5} \text{ m})\left(1000 \frac{\text{mm}}{\text{m}}\right)}{40.89 \text{ mm}}$$

$$\approx 0.0015$$

At 138°C, $\nu = 0.222 \times 10^{-6}$ m²/s. The Reynolds number is

$$\text{Re} = \frac{D\text{v}}{\nu} = \frac{(40.89 \text{ mm})\left(4.80 \frac{\text{m}}{\text{s}}\right)}{\left(0.222 \times 10^{-6} \frac{\text{m}^2}{\text{s}}\right)\left(1000 \frac{\text{mm}}{\text{m}}\right)}$$

$$= 8.84 \times 10^5$$

From App. 17.B, the friction factor is $f = 0.022$.

From Eq. 18.6, the friction head is

$$h_f = \frac{fL\text{v}^2}{2Dg} = \frac{(0.022)(12.89 \text{ m})\left(4.80 \frac{\text{m}}{\text{s}}\right)^2\left(1000 \frac{\text{mm}}{\text{m}}\right)}{(2)(40.89 \text{ mm})\left(9.81 \frac{\text{m}}{\text{s}^2}\right)}$$

$$= 8.14 \text{ m}$$

At 138°C, the saturated vapor pressure of pressurized water is

$$p_{\text{vapor}} = 3.431 \text{ bar} \quad \text{[from steam tables]}$$

The density of the liquid is the reciprocal of the specific volume, also taken from the steam tables.

$$\rho = \frac{1}{v_f} = \frac{(1)\left(100 \frac{\text{cm}}{\text{m}}\right)^3}{\left(1.0777 \frac{\text{cm}^3}{\text{g}}\right)\left(1000 \frac{\text{g}}{\text{kg}}\right)}$$

$$= 927.9 \text{ kg/m}^3$$

From Eq. 18.5(a),

$$h_{\text{vp}} = \frac{p_{\text{vapor}}}{\rho g}$$

$$h_{\text{vp}} = \frac{(3.431 \text{ bar})\left(1 \times 10^5 \frac{\text{Pa}}{\text{bar}}\right)}{\left(927.9 \frac{\text{kg}}{\text{m}^3}\right)\left(9.81 \frac{\text{m}}{\text{s}^2}\right)}$$

$$= 37.69 \text{ m}$$

From Eq. 18.5(a), the pressure head is

$$h_p = \frac{p}{\rho g} = \frac{(550 \text{ kPa})\left(1000 \frac{\text{Pa}}{\text{kPa}}\right)}{\left(927.9 \frac{\text{kg}}{\text{m}^3}\right)\left(9.81 \frac{\text{m}}{\text{s}^2}\right)}$$

$$= 60.42 \text{ m}$$

From Eq. 18.31(a), the NPSHA is

$$\text{NPSHA} = h_p + h_{z(s)} - h_{f(s)} - h_{\text{vp}}$$

$$= 60.42 \text{ m} + 6 \text{ m} - 8.14 \text{ m} - 37.69 \text{ m}$$

$$= \boxed{20.6 \text{ m} \quad (21 \text{ m})}$$

Since NPSHR is 3 m, $\boxed{\text{the pump will not cavitate.}}$

(A pump may not be needed in this configuration.)

The answer is (D).

11. The solvent is the water (fresh), and the solution is the seawater. Since seawater contains approximately $2\frac{1}{2}\%$ salt (NaCl) by weight, 100 lbm of seawater will yield 2.5 lbm salt and 97.5 lbm water. The molecular weight of salt is $23.0 + 35.5 = 58.5$ lbm/lbmol. The number of moles of salt in 100 lbm of seawater is

$$n_{\text{salt}} = \frac{2.5 \text{ lbm}}{58.5 \frac{\text{lbm}}{\text{lbmol}}}$$

$$= 0.043 \text{ lbmol}$$

Similarly, water's molecular weight is 18.016 lbm/lbmol. The number of moles of water is

$$n_{\text{water}} = \frac{97.5 \text{ lbm}}{18.016 \frac{\text{lbm}}{\text{lbmol}}}$$

$$= 5.412 \text{ lbmol}$$

The mole fraction of water is

$$\frac{5.412 \text{ lbmol}}{5.412 \text{ lbmol} + 0.043 \text{ lbmol}} = 0.992$$

Customary U.S. Solution

Cavitation will occur when

$$h_{\text{atm}} - h_{\text{v}} < h_{\text{vp}}$$

The density of seawater is 64.0 lbm/ft³.

Water Resources

From Eq. 18.4(b), the atmospheric head is

$$h_{atm} = \frac{p}{\rho} \times \frac{g_c}{g}$$

$$= \frac{\left(14.7 \frac{lbf}{in^2}\right)\left(12 \frac{in}{ft}\right)^2}{64.0 \frac{lbm}{ft^3}} \times \frac{32.2 \frac{lbm\text{-}ft}{lbf\text{-}sec^2}}{32.2 \frac{ft}{sec^2}}$$

$$= 33.075 \text{ ft} \quad [\text{ft of seawater}]$$

$$h_{depth} = 8 \text{ ft} \quad [\text{given}]$$

From Eq. 18.7, the velocity head is

$$h_v = \frac{v_{propeller}^2}{2g} = \frac{(4.2v_{boat})^2}{(2)\left(32.2 \frac{ft}{sec^2}\right)}$$

$$= 0.2739v_{boat}^2$$

From App. 18.B, the vapor pressure of freshwater at 68°F is $p_{vp} = 0.3391$ psia.

From App. 14.A, the density of water at 68°F is 62.32 lbm/ft³. Raoult's law predicts the actual vapor pressure of the solution.

$$p_{vapor,solution} = \left(\begin{array}{c}\text{mole fraction}\\\text{of the solvent}\end{array}\right)p_{vapor,solvent}$$

$$p_{vapor,seawater} = (0.992)\left(0.3393 \frac{lbf}{in^2}\right)$$

$$= 0.3366 \text{ lbf/in}^2$$

From Eq. 18.5(b), the vapor pressure head is

$$h_{vapor,seawater} = \frac{p}{\rho} \times \frac{g_c}{g}$$

$$= \frac{\left(0.3366 \frac{lbf}{in^2}\right)\left(12 \frac{in}{ft}\right)^2}{64.0 \frac{lbm}{ft^3}}$$

$$\times \frac{32.2 \frac{lbm\text{-}ft}{lbf\text{-}sec^2}}{32.2 \frac{ft}{sec^2}}$$

$$= 0.7574 \text{ ft}$$

Then,

$$8 \text{ ft} + 33.075 \text{ ft}$$

$$- 0.2739v_{boat}^2 = 0.7574 \text{ ft}$$

$$v_{boat} = \boxed{12.13 \text{ ft/sec} \quad (12 \text{ ft/sec})}$$

The answer is (B).

SI Solution

Cavitation will occur when

$$h_{atm} - h_v < h_{vp}$$

The density of seawater is 1024 kg/m³.

From Eq. 18.4(a), the atmospheric head is

$$h_{atm} = \frac{p}{\rho g} = \frac{(101.3 \text{ kPa})\left(1000 \frac{Pa}{kPa}\right)}{\left(1024 \frac{kg}{m^3}\right)\left(9.81 \frac{m}{s^2}\right)}$$

$$= 10.08 \text{ m}$$

$$h_{depth} = 3 \text{ m} \quad [\text{given}]$$

From Eq. 18.7, the velocity head is

$$h_v = \frac{v_{propeller}^2}{2g} = \frac{(4.2v_{boat})^2}{(2)\left(9.81 \frac{m}{s^2}\right)}$$

$$= 0.899v_{boat}^2$$

The vapor pressure of 20°C freshwater is

$$p_{vp} = (0.02339 \text{ bar})\left(100 \frac{kPa}{bar}\right)$$

$$= 2.339 \text{ kPa}$$

From App. 14.B, the density of water at 20°C is 998.23 kg/m³. Raoult's law predicts the actual vapor pressure of the solution.

$$p_{vapor,solution} = p_{vapor,solvent}\left(\begin{array}{c}\text{mole fraction}\\\text{of the solvent}\end{array}\right)$$

The solvent is the freshwater and the solution is the seawater.

The mole fraction of water is 0.992.

$$p_{vapor,seawater} = (2.339 \text{ kPa})(0.992)$$

$$= 2.320 \text{ kPa}$$

From Eq. 18.5(a), the vapor pressure head is

$$h_{vapor,seawater} = \frac{p}{g\rho} = \frac{(2.320 \text{ kPa})\left(1000 \frac{Pa}{kPa}\right)}{\left(9.81 \frac{m}{s^2}\right)\left(1024 \frac{kg}{m^3}\right)}$$

$$= 0.231 \text{ m}$$

Then,

$$3 \text{ m} + 10.08 \text{ m} - 0.899 v_{\text{boat}}^2 = 0.231 \text{ m}$$

$$\boxed{v_{\text{boat}} = 3.78 \text{ m/s} \quad (3.8 \text{ m/s})}$$

The answer is (B).

12. *Customary U.S. Solution*

(a) From App. 17.D, the approximate equivalent lengths of various screwed steel fittings are

  inlet: $L_e = 8.5$ ft [essentially a reentrant inlet]

  check valve: $L_e = 19$ ft

  long radius elbow: $L_e = 3.6$ ft

The total equivalent length of the 2 in line is

$$L_e = 12 \text{ ft} + 8.5 \text{ ft} + 19 \text{ ft} + (3)(3.6 \text{ ft}) + 80 \text{ ft}$$

$$= 130.3 \text{ ft}$$

From App. 16.B, for 2 in schedule-40 pipe, the pipe data are

$$D_i = 0.1723 \text{ ft}$$

$$A_i = 0.0233 \text{ ft}^2$$

Since the flow rate is unknown, it must be assumed in order to find velocity. Assume 90 gal/min.

$$\dot{V} = \left(90 \ \frac{\text{gal}}{\text{min}}\right)\left(0.002228 \ \frac{\frac{\text{ft}^3}{\text{sec}}}{\frac{\text{gal}}{\text{min}}}\right) = 0.2005 \ \text{ft}^3/\text{sec}$$

The velocity is

$$v = \frac{\dot{V}}{A_i} = \frac{0.2005 \ \frac{\text{ft}^3}{\text{sec}}}{0.0233 \ \text{ft}^2} = 8.605 \ \text{ft/sec}$$

From App. 14.A, the kinematic viscosity of water at 70°F is $\nu = 1.059 \times 10^{-5} \ \text{ft}^2/\text{sec}$.

The Reynolds number is

$$\text{Re} = \frac{Dv}{\nu} = \frac{(0.1723 \text{ ft})\left(8.605 \ \frac{\text{ft}}{\text{sec}}\right)}{1.059 \times 10^{-5} \ \frac{\text{ft}^2}{\text{sec}}}$$

$$= 1.4 \times 10^5$$

From Table 17.2 or App. 17.A, the specific roughness of steel pipe is $\epsilon = 0.0002$ ft

$$\frac{\epsilon}{D} = \frac{0.0002 \text{ ft}}{0.1723 \text{ ft}} \approx 0.0012$$

From App. 17.B, $f = 0.022$. At 90 gal/min, the friction loss in the line from Eq. 18.6 is

$$h_f = \frac{fLv^2}{2Dg}$$

$$= \frac{(0.022)(130.3 \text{ ft})\left(8.605 \ \frac{\text{ft}}{\text{sec}}\right)^2}{(2)(0.1723 \text{ ft})\left(32.2 \ \frac{\text{ft}}{\text{sec}^2}\right)}$$

$$= 19.1 \text{ ft}$$

From Eq. 18.7, the velocity head at 90 gal/min is

$$h_v = \frac{v^2}{2g} = \frac{\left(8.605 \ \frac{\text{ft}}{\text{sec}}\right)^2}{(2)\left(32.2 \ \frac{\text{ft}}{\text{sec}^2}\right)} = 1.1 \text{ ft}$$

In general, the friction head and velocity head are proportional to v^2 and Q^2.

$$h_f = (19.1 \text{ ft})\left(\frac{Q_2}{90 \ \frac{\text{gal}}{\text{min}}}\right)^2$$

$$h_v = (1.1 \text{ ft})\left(\frac{Q_2}{90 \ \frac{\text{gal}}{\text{min}}}\right)^2$$

(The 7 ft dimension is included in the 20 ft dimension.)

The total system head is

$$h = h_z + h_v + h_f$$

$$= 20 \text{ ft} + (1.1 \text{ ft} + 19.1 \text{ ft})\left(\frac{Q_2}{90 \ \frac{\text{gal}}{\text{min}}}\right)^2$$

From this equation, the following table for system head can be generated.

Q_2 (gal/min)	system head, h (ft)
0	20.0
10	20.2
20	21.0
30	22.2
40	24.0
50	26.2
60	29.0
70	32.2
80	36.0
90	40.2
100	44.9
110	50.2

The intersection point of the system curve and the pump curve defines the operating flow rate.

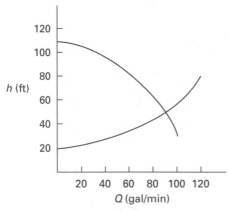

The flow rate is $\boxed{95 \text{ gal/min.}}$

The answer is (D).

(b) The intersection point is not in an efficient range for the pump because it is so far down on the system curve that the pumping efficiency will be low.

$\boxed{\text{A different pump should be used.}}$

The answer is (A).

SI Solution

(a) Use the approximate equivalent lengths of various screwed steel fittings from the customary U.S. solution. The total equivalent length of 5.08 cm schedule-40 pipe is

$$L_e = 4 \text{ m} + \left(8.5 \text{ ft} + 19 \text{ ft} + (3)(3.6 \text{ ft})\right)\left(0.3048 \frac{\text{m}}{\text{ft}}\right)$$

$$+ 27 \text{ m}$$

$$= 42.67 \text{ m}$$

From App. 16.C, for 2 in schedule-40 pipe, the pipe data are

$$D_i = 52.50 \text{ mm}$$

$$A_i = 21.65 \times 10^{-4} \text{ m}^2$$

Since the flow rate is unknown, it must be assumed in order to find velocity. Assume 6 L/s.

$$\dot{V} = \frac{6 \frac{\text{L}}{\text{s}}}{1000 \frac{\text{L}}{\text{m}^3}} = 6 \times 10^{-3} \text{ m}^3/\text{s}$$

The velocity is

$$\text{v} = \frac{\dot{V}}{A_i} = \frac{6 \times 10^{-3} \frac{\text{m}^3}{\text{s}}}{21.65 \times 10^{-4} \text{ m}^2} = 2.77 \text{ m/s}$$

From App. 14.B, the absolute viscosity of water at 21°C is $\mu = 0.9827 \times 10^{-3}$ Pa·s.

The density of water is $\rho = 998$ kg/m^3.

The Reynolds number is

$$\text{Re} = \frac{\rho \text{v} D_i}{\mu} = \frac{\left(998 \frac{\text{kg}}{\text{m}^3}\right)\left(2.77 \frac{\text{m}}{\text{s}}\right)(52.50 \text{ mm})}{\left(0.9827 \times 10^{-3} \text{ Pa·s}\right)\left(1000 \frac{\text{mm}}{\text{m}}\right)}$$

$$= 1.5 \times 10^5$$

From Table 17.2 or App. 17.A, the specific roughness of steel pipe is $\epsilon = 6.0 \times 10^{-5}$ m.

$$\frac{\epsilon}{D} = \frac{\left(6.0 \times 10^{-5} \text{ m}\right)\left(1000 \frac{\text{mm}}{\text{m}}\right)}{52.50 \text{ mm}} \approx 0.0012$$

From App. 17.B, $f = 0.022$. At 6 L/s, the friction loss in the line from Eq. 18.6 is

$$h_f = \frac{fL\text{v}^2}{2Dg}$$

$$= \frac{(0.022)(42.67 \text{ m})\left(2.77 \frac{\text{m}}{\text{s}}\right)^2\left(1000 \frac{\text{mm}}{\text{m}}\right)}{(2)(52.50 \text{ mm})\left(9.81 \frac{\text{m}}{\text{s}^2}\right)}$$

$$= 6.99 \text{ m}$$

At 6 L/s, the velocity head from Eq. 18.7 is

$$h_\text{v} = \frac{\text{v}^2}{2g} = \frac{\left(2.77 \frac{\text{m}}{\text{s}}\right)^2}{(2)\left(9.81 \frac{\text{m}}{\text{s}^2}\right)} = 0.39 \text{ m}$$

In general, the friction head and velocity head are proportional to v^2 and Q^2.

$$h_f = (6.99 \text{ m})\left(\frac{Q_2}{6 \frac{\text{L}}{\text{s}}}\right)^2$$

$$h_\text{v} = (0.39 \text{ m})\left(\frac{Q_2}{6 \frac{\text{L}}{\text{s}}}\right)^2$$

(The 2.3 m dimension is included in the 6.3 m dimension.)

The total system head is

$$h = h_z + h_v + h_f$$

$$= 6.3 \text{ m} + (0.39 \text{ m} + 6.99 \text{ m}) \left(\frac{Q_2}{6 \frac{\text{L}}{\text{s}}} \right)^2$$

From this equation, the following table for the system head can be generated.

Q_2 (L/s)	h (m)
0	6.3
0.6	6.37
1.2	6.60
1.8	6.96
2.4	7.48
3.2	8.40
3.6	8.96
4.4	10.27
4.8	11.02
5.7	12.96
6.0	13.68
6.5	14.96
7.0	16.35
7.5	17.83

The intersection point of the system curve and the pump curve defines the operating flow rate.

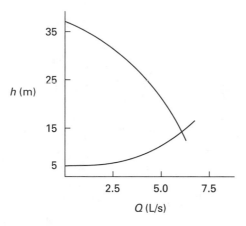

The flow rate is $\boxed{6.2 \text{ L/s.}}$

The answer is (D).

(b) The intersection point is not in an efficient range of the pump because it is so far down on the system curve that the pumping efficiency will be low.

$\boxed{\text{A different pump should be used.}}$

The answer is (A).

13. From Eq. 18.53,

$$P_2 = P_1 \left(\frac{\rho_2 n_2^3 D_2^5}{\rho_1 n_1^3 D_1^5} \right)$$

$$= P_1 \left(\frac{n_2}{n_1} \right)^3 \quad [\rho_2 = \rho_1 \text{ and } D_2 = D_1]$$

Customary U.S. Solution

$$P_2 = (0.5 \text{ hp}) \left(\frac{2000 \frac{\text{rev}}{\text{min}}}{1750 \frac{\text{rev}}{\text{min}}} \right)^3 = \boxed{0.75 \text{ hp}}$$

The answer is (D).

SI Solution

$$P_2 = (0.37 \text{ kW}) \left(\frac{2000 \frac{\text{rev}}{\text{min}}}{1750 \frac{\text{rev}}{\text{min}}} \right)^3$$

$$= \boxed{0.55 \text{ kW}}$$

The answer is (D).

14. (a) Random values of Q are chosen, and the corresponding values of H are determined by the formula $H = 30 + 2Q^2$.

Q (ft³/sec)	H (ft)
0	30
2.5	42.5
5	80
7.5	142.5
10	230
15	480
20	830
25	1280
30	1830

The intersection of the system curve and the 1400 rpm pump curve defines the operating point at that rpm.

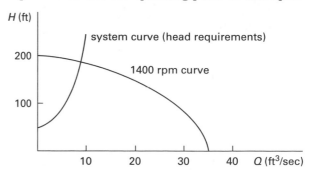

Water Resources

From the intersection of the graphs, at 1400 rpm the flow rate is approximately 9 ft³/sec, and the corresponding head is $30 + (2)(9)^2 \approx 192$ ft.

$$Q = \left(9 \ \frac{\text{ft}^3}{\text{sec}}\right)\left(448.8 \ \frac{\frac{\text{gal}}{\text{min}}}{\frac{\text{ft}^3}{\text{sec}}}\right)$$

$$= \boxed{4039 \ \text{gal/min} \quad (4000 \ \text{gal/min})}$$

The answer is (C).

(b) From Table 18.5, the hydraulic horsepower is

$$\text{WHP} = \frac{h_A \dot{V}(\text{SG})}{8.814}$$

$$= \frac{(192 \ \text{ft})\left(9 \ \frac{\text{ft}^3}{\text{sec}}\right)(1)}{8.814 \ \frac{\text{ft}^4}{\text{hp-sec}}}$$

$$= 196 \ \text{hp}$$

From Eq. 18.28(b), the specific speed is

$$n_s = \frac{n\sqrt{Q}}{h_A^{0.75}}$$

$$= \frac{\left(1400 \ \frac{\text{rev}}{\text{min}}\right)\sqrt{4039 \ \frac{\text{gal}}{\text{min}}}}{(192 \ \text{ft})^{0.75}}$$

$$= 1725$$

From Fig. 18.8 with curve E, $\eta \approx 86\%$.

The minimum pump motor power should be

$$\frac{196 \ \text{hp}}{0.86} = \boxed{228 \ \text{hp} \quad (230 \ \text{hp})}$$

The answer is (C).

(c) From Eq. 18.42,

$$Q_2 = Q_1\left(\frac{n_2}{n_1}\right)$$

$$= \left(4039 \ \frac{\text{gal}}{\text{min}}\right)\left(\frac{1200 \ \frac{\text{rev}}{\text{min}}}{1400 \ \frac{\text{rev}}{\text{min}}}\right)$$

$$= \boxed{3462 \ \text{gal/min} \quad (3500 \ \text{gal/min})}$$

The answer is (B).

15. Since turbines are essentially pumps running backward, use Table 18.5.

$$\Delta p = \left(30 \ \frac{\text{lbf}}{\text{in}^2} - 5 \ \frac{\text{lbf}}{\text{in}^2}\right)\left(12 \ \frac{\text{in}}{\text{ft}}\right)^2$$

$$= 3600 \ \text{lbf/ft}^2$$

$$P = \frac{\Delta p \dot{V}}{550}$$

$$= \frac{\left(3600 \ \frac{\text{lbf}}{\text{ft}^2}\right)\left(100 \ \frac{\text{ft}^3}{\text{sec}}\right)}{550 \ \frac{\text{ft-lbf}}{\text{hp-sec}}}$$

$$= \boxed{654.5 \ \text{hp} \quad (650 \ \text{hp})}$$

The answer is (C).

16. The flow rate is

$$\dot{m} = \rho \dot{V} = \left(62.4 \ \frac{\text{lbm}}{\text{ft}^3}\right)\left(1000 \ \frac{\text{ft}^3}{\text{sec}}\right)$$

$$= 6.24 \times 10^4 \ \text{lbm/sec}$$

The head available for work is

$$\Delta h = 625 \ \text{ft} - 58 \ \text{ft} = 567 \ \text{ft}$$

Use Table 18.5. The power is

$$P = \frac{h_A \dot{m}}{550} \times \frac{g}{g_c}$$

$$= \frac{(0.89)\left(6.24 \times 10^4 \ \frac{\text{lbm}}{\text{sec}}\right)(567 \ \text{ft})}{550 \ \frac{\text{ft-lbf}}{\text{hp-sec}}} \times \frac{32.2 \ \frac{\text{ft}}{\text{sec}^2}}{32.2 \ \frac{\text{lbm-ft}}{\text{lbf-sec}^2}}$$

$$= 57{,}253 \ \text{hp}$$

Convert from hp to kW.

$$P = (57{,}253 \ \text{hp})\left(0.7457 \ \frac{\text{kW}}{\text{hp}}\right)$$

$$= \boxed{4.27 \times 10^4 \ \text{kW} \quad (43 \ \text{MW})}$$

The answer is (C).

17. *Customary U.S. Solution*

(a) From App. 14.A, the density of water at 60°F is 62.37 lbm/ft^3. From Eq. 18.5(b), the head dropped is

$$h = \frac{\Delta p}{\rho} \times \frac{g_c}{g}$$

$$= \frac{\left(500 \ \frac{\text{lbf}}{\text{in}^2} - 30 \ \frac{\text{lbf}}{\text{in}^2}\right)\left(12 \ \frac{\text{in}}{\text{ft}}\right)^2}{62.37 \ \frac{\text{lbm}}{\text{ft}^3}}$$

$$\times \frac{32.2 \ \frac{\text{lbm-ft}}{\text{lbf-sec}^2}}{32.2 \ \frac{\text{ft}}{\text{sec}^2}}$$

$$= 1085 \ \text{ft}$$

From Eq. 18.56(b), the specific speed of a turbine is

$$n_s = \frac{n\sqrt{P}}{h_t^{1.25}} = \frac{\left(1750 \ \frac{\text{rev}}{\text{min}}\right)\sqrt{250 \ \text{hp}}}{(1085 \ \text{ft})^{1.25}}$$

$$= \boxed{4.443 \quad (4)}$$

The answer is (A).

(b) The flow rate, $\dot{V}$, is

$$\dot{V} = A\text{v} = \left(\frac{\pi}{4}\right)\left(\frac{4 \ \text{in}}{12 \ \frac{\text{in}}{\text{ft}}}\right)^2\left(35 \ \frac{\text{ft}}{\text{sec}}\right)$$

$$= 3.054 \ \text{ft}^3/\text{sec}$$

Since the analysis is for a single blade, not the entire turbine, only a portion of the water will catch up with the blade. The flow rate, considering the blade's movement away at 10 ft/sec, is

$$\dot{V}' = \frac{\left(35 \ \frac{\text{ft}}{\text{sec}} - 10 \ \frac{\text{ft}}{\text{sec}}\right)\left(3.054 \ \frac{\text{ft}^3}{\text{sec}}\right)}{35 \ \frac{\text{ft}}{\text{sec}}}$$

$$= 2.181 \ \text{ft}^3/\text{sec}$$

The forces in the x-direction and the y-direction are

$$F_x = \left(\frac{\dot{V}'\rho}{g_c}\right)(\text{v}_j - \text{v}_b)(\cos\theta - 1)$$

$$= \left(\frac{\left(2.181 \ \frac{\text{ft}^3}{\text{sec}}\right)\left(62.37 \ \frac{\text{lbm}}{\text{ft}^3}\right)}{32.2 \ \frac{\text{lbm-ft}}{\text{lbf-sec}^2}}\right)$$

$$\times \left(35 \ \frac{\text{ft}}{\text{sec}} - 10 \ \frac{\text{ft}}{\text{sec}}\right)(\cos 80° - 1)$$

$$= -87.27 \ \text{lbf}$$

$$F_y = \left(\frac{\dot{V}'\rho}{g_c}\right)(\text{v}_j - \text{v}_b)\sin\theta$$

$$= \left(\frac{\left(2.181 \ \frac{\text{ft}^3}{\text{sec}}\right)\left(62.37 \ \frac{\text{lbm}}{\text{ft}^3}\right)}{32.2 \ \frac{\text{lbm-ft}}{\text{lbf-sec}^2}}\right)$$

$$\times \left(35 \ \frac{\text{ft}}{\text{sec}} - 10 \ \frac{\text{ft}}{\text{sec}}\right)\sin 80°$$

$$= 104.0 \ \text{lbf}$$

$$R = \sqrt{F_x^2 + F_y^2}$$

$$= \sqrt{(-87.27 \ \text{lbf})^2 + (104.0 \ \text{lbf})^2}$$

$$= \boxed{135.8 \ \text{lbf} \quad (140 \ \text{lbf})}$$

The answer is (B).

SI Solution

(a) From App. 14.B, the density of water at 16°C is 998.83 kg/m^3. From Eq. 18.5(a), the head dropped is

$$h = \frac{\Delta p}{\rho g}$$

$$= \frac{(3.5 \ \text{MPa})\left(1 \times 10^6 \ \frac{\text{Pa}}{\text{MPa}}\right) - (210 \ \text{kPa})\left(1000 \ \frac{\text{Pa}}{\text{kPa}}\right)}{\left(998.83 \ \frac{\text{kg}}{\text{m}^3}\right)\left(9.81 \ \frac{\text{m}}{\text{s}^2}\right)}$$

$$= 335.8 \ \text{m}$$

From Eq. 18.63(a), the specific speed of a turbine is

$$n_s = \frac{n\sqrt{P}}{h_t^{1.25}}$$

$$= \frac{\left(1750 \ \frac{\text{rev}}{\text{min}}\right)\sqrt{185 \ \text{kW}}}{(335.8 \ \text{m})^{1.25}}$$

$$= \boxed{16.56 \quad (17)}$$

The answer is (A).

(b) The flow rate, $\dot{V}$, is

$$\dot{V} = A\text{v} = \left(\frac{\pi\left(\frac{100 \ \text{mm}}{1000 \ \frac{\text{mm}}{\text{m}}}\right)^2}{4}\right)\left(10.5 \ \frac{\text{m}}{\text{s}}\right)$$

$$= 0.08247 \ \text{m}^3/\text{s}$$

Since the analysis is for a single blade, not the entire turbine, only a portion of the water will catch up with the blade. The flow rate, considering the blade's movement away at 3 m/s, is

$$\dot{V}' = \frac{\left(10.5\ \frac{\text{m}}{\text{s}} - 3\ \frac{\text{m}}{\text{s}}\right)\left(0.08247\ \frac{\text{m}^3}{\text{s}}\right)}{10.5\ \frac{\text{m}}{\text{s}}}$$

$$= 0.05891\ \text{m}^3/\text{s}$$

The forces in the x-direction and the y-direction are

$$F_x = \dot{V}'\rho(\text{v}_j - \text{v}_b)(\cos\theta - 1)$$

$$= \left(0.05891\ \frac{\text{m}^3}{\text{s}}\right)\left(998.83\ \frac{\text{kg}}{\text{m}^3}\right)$$

$$\times\left(10.5\ \frac{\text{m}}{\text{s}} - 3\ \frac{\text{m}}{\text{s}}\right)(\cos 80° - 1)$$

$$= -364.7\ \text{N}$$

$$F_y = \dot{V}'\rho(\text{v}_j - \text{v}_b)\sin\theta$$

$$= \left(0.05891\ \frac{\text{m}^3}{\text{s}}\right)\left(998.83\ \frac{\text{kg}}{\text{m}^3}\right)$$

$$\times\left(10.5\ \frac{\text{m}}{\text{s}} - 3\ \frac{\text{m}}{\text{s}}\right)\sin 80°$$

$$= 434.6\ \text{N}$$

$$R = \sqrt{F_x^2 + F_y^2}$$

$$= \sqrt{(-364.7\ \text{N})^2 + (434.6\ \text{N})^2}$$

$$= \boxed{567.3\ \text{N} \quad (570\ \text{N})}$$

The answer is (B).

18. *Customary U.S. Solution*

(a) The total effective head is due to the pressure head, velocity head, and tailwater head.

$$h_{\text{eff}} = h_p + h_\text{v} - h_{z,\text{tailwater}}$$

$$= 92.5\ \text{ft} + \frac{\left(12\ \frac{\text{ft}}{\text{sec}}\right)^2}{(2)\left(32.2\ \frac{\text{ft}}{\text{sec}^2}\right)} - (-5.26\ \text{ft})$$

$$= \boxed{100\ \text{ft}}$$

The answer is (C).

(b) From Table 18.5, the theoretical hydraulic horse-power is

$$P_{\text{th}} = \frac{h_A\dot{V}(\text{SG})}{8.814}$$

$$= \frac{(100\ \text{ft})\left(25\ \frac{\text{ft}^3}{\text{sec}}\right)(1)}{8.814\ \frac{\text{ft}^4}{\text{hp-sec}}}$$

$$= 283.6\ \text{hp}$$

The overall turbine efficiency is

$$\eta = \frac{P_{\text{brake}}}{P_{\text{th}}} = \frac{250\ \text{hp}}{283.6\ \text{hp}} = \boxed{0.882 \quad (88\%)}$$

The answer is (B).

(c) From Eq. 18.43 (the affinity laws),

$$n_2 = n_1\sqrt{\frac{h_2}{h_1}} = \left(610\ \frac{\text{rev}}{\text{min}}\right)\sqrt{\frac{225\ \text{ft}}{100\ \text{ft}}}$$

$$= \boxed{915\ \text{rpm} \quad (920\ \text{rpm})}$$

The answer is (B).

(d) Combine Eq. 18.43 and Eq. 18.44.

$$P_2 = P_1\left(\frac{h_2}{h_1}\right)^{1.5} = (250\ \text{hp})\left(\frac{225\ \text{ft}}{100\ \text{ft}}\right)^{1.5}$$

$$= \boxed{843.8\ \text{hp} \quad (840\ \text{hp})}$$

The answer is (D).

(e) Combine Eq. 18.42 and Eq. 18.43.

$$Q_2 = Q_1\sqrt{\frac{h_2}{h_1}} = \left(25\ \frac{\text{ft}^3}{\text{sec}}\right)\sqrt{\frac{225\ \text{ft}}{100\ \text{ft}}}$$

$$= \boxed{37.5\ \text{ft}^3/\text{sec} \quad (38\ \text{ft}^3/\text{sec})}$$

The answer is (B).

SI Solution

(a) The total effective head is due to the pressure head, velocity head, and tailwater head.

$$h_{\text{eff}} = h_p + h_\text{v} - h_{z,\text{tailwater}}$$

$$= 30.8\ \text{m} + \frac{\left(3.6\ \frac{\text{m}}{\text{s}}\right)^2}{(2)\left(9.81\ \frac{\text{m}}{\text{s}^2}\right)} - (-1.75\ \text{m})$$

$$= \boxed{33.21\ \text{m} \quad (33\ \text{m})}$$

The answer is (C).

(b) From Table 18.6, the theoretical hydraulic kilowatts are

$$P_{\text{th}} = \frac{9.81 h_A Q (SG)}{1000}$$

$$= \frac{\left(9.81 \ \frac{\text{m}}{\text{s}^2}\right)(33.21 \ \text{m})\left(700 \ \frac{\text{L}}{\text{s}}\right)(1)}{1000 \ \frac{\text{W}}{\text{kW}}}$$

$$= 228.1 \ \text{kW}$$

The overall turbine efficiency is

$$\eta = \frac{P_{\text{brake}}}{P_{\text{th}}} = \frac{185 \ \text{kW}}{228.1 \ \text{kW}}$$

$$= \boxed{0.811 \quad (81\%)}$$

The answer is (A).

(c) From Eq. 18.43 (the affinity laws),

$$n_2 = n_1 \sqrt{\frac{h_2}{h_1}} = \left(610 \ \frac{\text{rev}}{\text{min}}\right)\sqrt{\frac{75 \ \text{m}}{33.21 \ \text{m}}}$$

$$= \boxed{917 \ \text{rpm} \quad (920 \ \text{rpm})}$$

The answer is (B).

(d) Combine Eq. 18.43 and Eq. 18.44.

$$P_2 = P_1 \left(\frac{h_2}{h_1}\right)^{1.5} = (185 \ \text{kW})\left(\frac{75 \ \text{m}}{33.21 \ \text{m}}\right)^{1.5}$$

$$= \boxed{627.8 \ \text{kW} \quad (630 \ \text{kW})}$$

The answer is (D).

(e) Combine Eq. 18.42 and Eq. 18.43.

$$Q_2 = Q_1 \sqrt{\frac{h_2}{h_1}} = \left(700 \ \frac{\text{L}}{\text{s}}\right)\sqrt{\frac{75 \ \text{m}}{33.21 \ \text{m}}}$$

$$= \boxed{1051.9 \ \text{L/s} \quad (1100 \ \text{L/s})}$$

The answer is (B).

19 Open Channel Flow

PRACTICE PROBLEMS

Rectangular Channels

1. A wooden flume ($n = 0.012$) with a rectangular cross section is 2 ft wide. The flume carries 3 ft^3/sec of water down a 1% slope. What is the depth of flow?

    (A) 0.3 ft

    (B) 0.4 ft

    (C) 0.5 ft

    (D) 0.6 ft

2. A rectangular open channel is to be constructed with smooth concrete on a slope of 0.08. The design flow rate is 17 m^3/s. The channel design must be optimum. What are the channel dimensions?

    (A) 0.6 m deep; 1.2 m wide

    (B) 0.8 m deep; 1.6 m wide

    (C) 1.0 m deep; 2.0 m wide

    (D) 1.2 m deep; 2.4 m wide

3. (*Time limit: one hour*) A dam's outfall structure has three rectangular inlets sized 1 ft high × 2 ft wide as shown. The orifice coefficient for each inlet is 0.7. The vertical riser joins with a 100 ft long square box culvert ($n = 0.013$) placed on a 0.05 slope. The box culvert drains into a tailwater basin whose water level remains constant at the top of the culvert opening. During steady-state operation, the water level in the vertical barrel of the structure is 4 ft above the top of the box culvert. The orifice coefficient of the box culvert entrance is $^2/_3$. What are the dimensions of the box culvert?

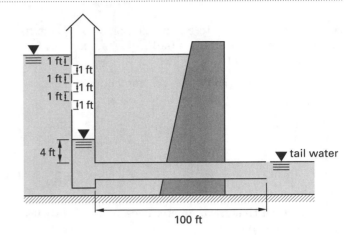

    (A) 1.5 ft × 1.5 ft

    (B) 2.0 ft × 2.0 ft

    (C) 2.5 ft × 2.5 ft

    (D) 4.0 ft × 4.0 ft

Circular Channels

4. A 24 in diameter pipe ($n = 0.013$) was installed 30 years ago on a 0.001 slope. Recent tests indicate that the full-flow capacity of the pipe is 6.0 ft^3/sec.

(a) What was the original full-flow capacity?

    (A) 4.1 ft^3/sec

    (B) 6.3 ft^3/sec

    (C) 6.7 ft^3/sec

    (D) 7.2 ft^3/sec

(b) What was the original full-flow velocity?

    (A) 1.8 ft/sec

    (B) 2.3 ft/sec

    (C) 2.7 ft/sec

    (D) 3.7 ft/sec

(c) What is the current full-flow velocity?

    (A) 1.9 ft/sec

    (B) 2.4 ft/sec

    (C) 3.1 ft/sec

    (D) 5.8 ft/sec

(d) What is the current Manning coefficient?

(A) 0.011

(B) 0.013

(C) 0.016

(D) 0.024

5. A circular sewer is to be installed on a 1% grade. Its Manning coefficient is $n = 0.013$. The maximum full-flow capacity is to be 3.5 ft³/sec.

(a) What size pipe should be recommended?

(A) 8 in

(B) 12 in

(C) 16 in

(D) 18 in

(b) Assuming that the diameter chosen is 12 in, what is the full-flow capacity?

(A) 2.7 ft³/sec

(B) 3.3 ft³/sec

(C) 3.6 ft³/sec

(D) 4.5 ft³/sec

(c) What is the full-flow velocity?

(A) 2.9 ft/sec

(B) 3.3 ft/sec

(C) 4.1 ft/sec

(D) 4.6 ft/sec

(d) What is the depth of flow when the flow is 0.7 ft³/sec?

(A) 4 in

(B) 7 in

(C) 9 in

(D) 12 in

(e) What minimum velocity will prevent solids from settling out in the sewer?

(A) 1.0–1.5 ft/sec

(B) 1.5–2.0 ft/sec

(C) 2.0–2.5 ft/sec

(D) 2.5–3.0 ft/sec

6. A 4 ft diameter concrete storm drain on a 0.02 slope carries water at a depth of 1.5 ft. Manning's roughness coefficient for a full storm drain is $n_{full} = 0.013$, but n varies with depth.

(a) What is the velocity of the water in the storm drain?

(A) 3 ft/sec

(B) 6 ft/sec

(C) 8 ft/sec

(D) 11 ft/sec

(b) What is the maximum velocity that can be achieved in the storm drain?

(A) 8 ft/sec

(B) 13 ft/sec

(C) 17 ft/sec

(D) 21 ft/sec

(c) What is the maximum capacity of the storm drain?

(A) 130 ft³/sec

(B) 160 ft³/sec

(C) 190 ft³/sec

(D) 210 ft³/sec

7. A circular storm sewer ($n = 0.012$) is being designed to carry a peak flow of 5 ft³/sec. To allow for excess capacity, the depth at peak flow is to be 75% of the sewer diameter. The sewer is to be installed in a bed with a slope of 2%. What is the required sewer diameter?

(A) 12 in

(B) 16 in

(C) 20 in

(D) 24 in

8. (*Time limit: one hour*) Flow in a 6 ft diameter, newly formed concrete culvert is 150 ft³/sec at a point where the depth of flow is 3 ft.

(a) What is the hydraulic radius?

(A) 1.1 ft

(B) 1.5 ft

(C) 2.3 ft

(D) 3.0 ft

(b) What is the slope of the pipe?

   (A) 0.004

   (B) 0.008

   (C) 0.010

   (D) 0.030

(c) What type of flow occurs?

   (A) subcritical

   (B) critical

   (C) supercritical

   (D) choked

(d) What is the hydraulic radius if the pipe flows full?

   (A) 1.5 ft

   (B) 3.0 ft

   (C) 4.0 ft

   (D) 6.0 ft

(e) What is the flow quantity if the culvert flows full?

   (A) 100 ft^3/sec

   (B) 120 ft^3/sec

   (C) 200 ft^3/sec

   (D) 300 ft^3/sec

(f) If the flow is 150 ft^3/sec, what is the critical depth?

   (A) 1.1 ft

   (B) 1.5 ft

   (C) 3.3 ft

   (D) 4.3 ft

(g) What is the critical slope?

   (A) 0.002

   (B) 0.003

   (C) 0.004

   (D) 0.05

(h) If the exit depth is less than critical, which culvert flow types are most probable?

   (A) 1, 5

   (B) 1, 2, 3

   (C) 1, 3, 5

   (D) 5 only

(i) What is the Froude number for a depth of flow of 3 ft?

   (A) 0.86

   (B) 0.99

   (C) 1.22

   (D) 2.84

(j) After the water exits the circular culvert, it flows at a velocity of 15 ft/sec and a depth of 2 ft in a rectangular channel. Assuming a hydraulic jump is possible, what will be the depth after the jump?

   (A) 2.5 ft

   (B) 3.0 ft

   (C) 3.3 ft

   (D) 4.4 ft

Trapezoidal Channels

9. The trapezoidal channel shown has a Manning coefficient of $n = 0.013$ and is laid at a slope of 0.002. The depth of flow is 2 ft. What is the flow rate?

   (A) 120 ft^3/sec

   (B) 150 ft^3/sec

   (C) 180 ft^3/sec

   (D) 210 ft^3/sec

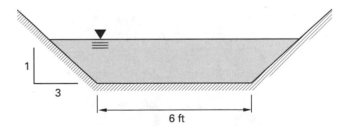

10. A trapezoidal open channel is to be constructed with smooth concrete on a slope of 0.08. The design flow rate is 17 m^3/s. The channel design must be optimum. What are most nearly the channel dimensions?

   (A) depth 0.8 m; base 1.0 m

   (B) depth 1.0 m; base 1.2 m

   (C) depth 1.2 m; base 1.4 m

   (D) depth 1.4 m; base 1.8 m

Weirs

11. A sharp-crested rectangular weir with two end contractions is 5 ft wide. The weir height is 6 ft. The head over the weir is 0.43 ft. What is the flow rate?

(A) 3.9 ft^3/sec

(B) 4.3 ft^3/sec

(C) 4.6 ft^3/sec

(D) 6.4 ft^3/sec

12. A weir is constructed with a trapezoidal opening. The base of the opening is 18 in wide, and the sides have a slope of 4:1 (vertical:horizontal). The depth of flow over the weir is 9 in. What is the rate of discharge?

(A) 1.8 ft^3/sec

(B) 3.3 ft^3/sec

(C) 5.1 ft^3/sec

(D) 8.9 ft^3/sec

13. (*Time limit: one hour*) A 17 ft × 20 ft rectangular tank is fed from the bottom and drains through a double-constricted rectangular weir into a trough. The weir opening is to be designed using the following parameters.

discharge rate:	2 MGD minimum
	4 MGD average
	8 MGD maximum
minimum tank freeboard:	4 in
minimum head over weir:	10 in
maximum weir elevation:	590 ft
tank bottom elevation:	580 ft

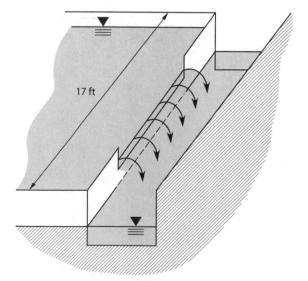

17 ft

(a) What is the required weir opening length?

(A) 1.2 ft

(B) 1.4 ft

(C) 2.1 ft

(D) 2.7 ft

(b) What is the surface elevation in the tank at maximum flow?

(A) 581.9 ft

(B) 589.8 ft

(C) 592.6 ft

(D) 593.1 ft

(c) What is the surface elevation in the tank at minimum flow?

(A) 590.8 ft

(B) 591.2 ft

(C) 591.7 ft

(D) 592.0 ft

Hydraulic Jumps and Drops

14. Water flowing in a rectangular channel 5 ft wide experiences a hydraulic jump. The depth of flow upstream from the jump is 1 ft. The depth of flow downstream of the jump is 2.4 ft. What quantity is flowing?

(A) 39 ft^3/sec

(B) 45 ft^3/sec

(C) 52 ft^3/sec

(D) 57 ft^3/sec

15. A hydraulic jump forms at the toe of a spillway. The water surface levels are 0.2 ft and 6 ft above the apron before and after the jump, respectively. The velocity before the jump is 54.7 ft/sec. What is the energy loss in the jump?

(A) 41 ft-lbf/lbm

(B) 58 ft-lbf/lbm

(C) 65 ft-lbf/lbm

(D) 92 ft-lbf/lbm

16. Water flowing in a rectangular channel 6 ft wide experiences a hydraulic jump. The depth of flow upstream from the jump is 1.2 ft. The depth of flow downstream of the jump is 3.8 ft.

(a) What quantity is flowing?

(A) 75 ft³/sec

(B) 115 ft³/sec

(C) 180 ft³/sec

(D) 210 ft³/sec

(b) What is the energy loss in the jump?

(A) 1 ft-lbf/lbm

(B) 3 ft-lbf/lbm

(C) 5 ft-lbf/lbm

(D) 12 ft-lbf/lbm

Spillways

17. A spillway operates with 2 ft of head. The toe of the spillway is 40 ft below the crest of the spillway. The spillway discharge coefficient is 3.5.

(a) What is the discharge per foot of crest?

(A) 9.9 ft³/sec

(B) 11 ft³/sec

(C) 15 ft³/sec

(D) 27 ft³/sec

(b) What is the depth of flow at the toe?

(A) 0.19 ft

(B) 0.35 ft

(C) 0.78 ft

(D) 1.4 ft

18. 10,000 ft³/sec of water flows over a 100 ft wide spillway and continues down a constant-width discharge chute on a 5% grade. The discharge chute has a Manning coefficient of $n = 0.012$.

(a) What is the normal depth of the water flowing down the chute?

(A) 1.9 ft

(B) 2.2 ft

(C) 2.7 ft

(D) 3.5 ft

(b) What is the critical depth of the water flowing down the chute?

(A) 3.8 ft

(B) 5.1 ft

(C) 6.8 ft

(D) 7.7 ft

(c) What type of flow occurs on the chute?

(A) tranquil

(B) subcritical

(C) rapid

(D) nonsteady

(d) If a hydraulic jump forms at the bottom of the 5% slope where it joins with a horizontal apron, what is the depth after the jump?

(A) 9.8 ft

(B) 12 ft

(C) 16 ft

(D) 21 ft

19. A spillway has a crest length of 60 ft. The stilling basin below the crest has a width of 60 ft and a level bottom. The spillway discharge coefficient is 3.7. The head loss in the chute is 20% of the difference in level between the reservoir surface and the stilling basin bottom.

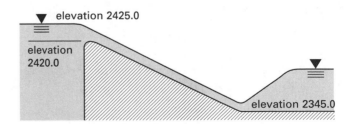

(a) What is the flow rate over the spillway?

(A) 1800 ft³/sec

(B) 2100 ft³/sec

(C) 2500 ft³/sec

(D) 2900 ft³/sec

(b) What is the depth of flow at the toe of the spillway?

(A) 0.65 ft

(B) 1.2 ft

(C) 1.6 ft

(D) 2.0 ft

(c) What tailwater depth is required to cause a hydraulic jump to form at the toe?

(A) 4 ft

(B) 7 ft

(C) 10 ft

(D) 13 ft

(d) What energy is lost in the hydraulic jump?

(A) 30 ft-lbf/lbm

(B) 50 ft-lbf/lbm

(C) 75 ft-lbf/lbm

(D) 100 ft-lbf/lbm

Parshall Flumes

20. A Parshall flume has a throat width of 6 ft. The upstream head measured from the throat floor is 18 in.

(a) What is the flow rate?

(A) 41 ft^3/sec

(B) 46 ft^3/sec

(C) 54 ft^3/sec

(D) 63 ft^3/sec

(b) Which of the following is NOT normally considered to be a feature of Parshall flumes?

(A) They are self-cleansing.

(B) They can be operated by unskilled personnel.

(C) They have low head loss.

(D) They can be placed in temporary installations.

Backwater Curves

21. A rectangular channel 8 ft wide carries a flow of 150 ft^3/sec. The channel slope is 0.0015, and the Manning coefficient is $n = 0.015$. A weir installed across the channel raises the depth at the weir to 6 ft.

(a) What is the normal depth of flow?

(A) 2.9 ft

(B) 3.3 ft

(C) 3.6 ft

(D) 4.5 ft

(b) What is the distance between the weir and the point where the depth is 5 ft?

(A) 610 ft

(B) 840 ft

(C) 1100 ft

(D) 1700 ft

(c) What is the distance between the points where the depth is 4 ft and 5 ft?

(A) 700 ft

(B) 800 ft

(C) 950 ft

(D) 1100 ft

Culverts

22. A 42 in diameter concrete culvert is 250 ft long and laid at a slope of 0.006. The culvert entrance is flush and square-edged. The tailwater level at the outlet is just above the crown of the barrel, and the headwater is 5.0 ft above the crown of the culvert's inlet. What is the capacity?

(A) 60 ft^3/sec

(B) 80 ft^3/sec

(C) 100 ft^3/sec

(D) 120 ft^3/sec

23. (*Time limit: one hour*) A dam spills 2000 m³/s of water over its crest as shown. The crest height is 30 m. The width of the dam is 150 m. The depth at point B is 0.6 m. The stilling basin, constructed of rough concrete, is rectangular in shape, 150 m in width, and has a slope of 0.002.

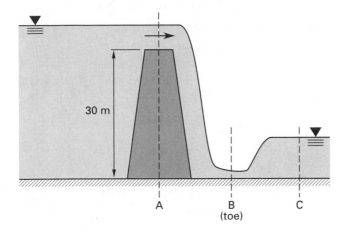

(a) What is the head over the crest?

(A) 0.8 m

(B) 1.2 m

(C) 2.2 m

(D) 3.3 m

(b) What is the total upstream specific energy head?

(A) 32.2 m

(B) 34.1 m

(C) 36.1 m

(D) 38.0 m

(c) What is the total specific energy head at point B?

(A) 25.8 m

(B) 32.2 m

(C) 34.1 m

(D) 43.3 m

(d) What is the friction head loss between point A and point B?

(A) 2.2 m

(B) 4.2 m

(C) 8.3 m

(D) 9.8 m

(e) What is the normal downstream depth?

(A) 2.0 m

(B) 2.6 m

(C) 5.6 m

(D) 10.0 m

(f) What is the critical depth?

(A) 2.0 m

(B) 2.6 m

(C) 7.0 m

(D) 8.3 m

(g) At what point will a hydraulic jump occur?

(A) on the spillway slope

(B) at point B

(C) downstream of point B

(D) nowhere (a hydraulic jump will not occur)

(h) The stilling basin feeds into a natural channel composed of graded material from silt to cobble size. What is the maximum velocity in this channel to prevent erosion?

(A) 1.2 m/s

(B) 6.7 m/s

(C) 15 m/s

(D) 26 m/s

(i) If the natural channel has the same geometry as the stilling basin (i.e., rectangular with a width of 150 m), what slope is required to maintain the maximum velocity and prevent erosion?

(A) 0.00001

(B) 0.00004

(C) 0.0002

(D) 0.002

(j) Assuming that the slope in the natural channel remains at 0.002, what will be the friction loss over a 1000 m reach?

(A) 0.1 m

(B) 0.6 m

(C) 1.5 m

(D) 2.0 m

SOLUTIONS

1.

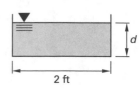

$$R = \frac{2d}{2+2d} = \frac{d}{1+d}$$

From Eq. 19.13,

$$Q = \left(\frac{1.49}{n}\right) A R^{2/3} \sqrt{S}$$

$$3 \; \frac{\text{ft}^3}{\text{sec}} = \left(\frac{1.49}{0.012}\right)(2d)\left(\frac{d}{d+1}\right)^{2/3}\sqrt{0.01}$$

$$0.120805 = \left(\frac{d^{5/2}}{d+1}\right)^{2/3}$$

$$0.042 = \frac{d^{5/2}}{d+1}$$

By trial and error, $d = \boxed{0.314 \text{ ft } (0.3 \text{ ft}).}$

The answer is (A).

2. From App. 19.A, $n = 0.011$. $d = w/2$ for an optimum channel.

$$A = wd = \frac{w^2}{2}$$

$$P = w + 2\left(\frac{w}{2}\right) = 2w$$

$$R = \frac{A}{P} = \frac{\dfrac{w^2}{2}}{2w} = \frac{w}{4}$$

(The units conversion factor of 1.49 is not needed in problems with SI units.)

$$17 \; \frac{\text{m}^3}{\text{s}} = \left(\frac{w^2}{2}\right)\left(\frac{1}{0.011}\right)\left(\frac{w}{4}\right)^{2/3}\sqrt{0.08}$$

$$w = \boxed{1.57 \text{ m } \quad (1.6 \text{ m})}$$

$$d = \boxed{0.79 \text{ m } \quad (0.8 \text{ m})}$$

The answer is (B).

3. Find the flow of the culvert.

$$Q_o = C_d A_o \sqrt{2gh}$$

$$Q_1 = (0.7)(2 \text{ ft}^2)\sqrt{(2)\left(32.2 \; \frac{\text{ft}}{\text{sec}^2}\right)(1.5 \text{ ft})}$$

$$= (11.235)\sqrt{1.5} \quad \text{[top inlet]}$$

$$Q_1 + Q_2 + Q_3 = (11.235)\left(\sqrt{1.5} + \sqrt{3.5} + \sqrt{5.5}\right)$$

$$= 61.13 \text{ ft}^3/\text{sec}$$

Culvert size (and hence R) is unknown. To get a trial value, initially disregard entrance loss and barrel friction.

$$H = 4 \text{ ft} + (100 \text{ ft})(0.05) = 9 \text{ ft}$$

$$v = \sqrt{2gh} = \sqrt{(2)\left(32.2 \; \frac{\text{ft}}{\text{sec}^2}\right)(9 \text{ ft})}$$

$$= 24.07 \text{ ft/sec}$$

$$A = \frac{Q}{C_d v} = \frac{61.13 \; \dfrac{\text{ft}^3}{\text{sec}}}{\left(\dfrac{2}{3}\right)\left(24.07 \; \dfrac{\text{ft}}{\text{sec}}\right)} = 3.81 \text{ ft}^2$$

$$\text{box size} = \sqrt{3.81 \text{ ft}^2} = 1.95 \text{ ft}$$

(Round to $2^{1}/_{4}$ ft^2 to account for friction.)

Due to the hydrostatic pressure, assume the culvert box flows full.

Use Eq. 19.101. $k_e = 0.50$ for a square-edged entrance.

$$R = \frac{(2.25 \text{ ft})^2}{(4)(2.25 \text{ ft})} = 0.5625 \text{ ft}$$

$$v = \sqrt{\frac{H}{\dfrac{1+k_e}{2g} + \dfrac{n^2 L}{2.21 R^{4/3}}}}$$

$$= \sqrt{\frac{9 \text{ ft}}{\dfrac{1+0.5}{(2)\left(32.2 \; \dfrac{\text{ft}}{\text{sec}^2}\right)} + \dfrac{(0.013)^2(100 \text{ ft})}{(2.21)(0.5625 \text{ ft})^{4/3}}}}$$

$$= 15.05 \text{ ft/sec}$$

$$\text{box size} = \sqrt{\frac{61.13 \; \dfrac{\text{ft}^3}{\text{sec}}}{\left(\dfrac{2}{3}\right)\left(15.05 \; \dfrac{\text{ft}}{\text{sec}}\right)}}$$

$$= \boxed{2.47 \text{ ft } \quad (2.5 \text{ ft or more})}$$

The answer is (C).

4. (a) From Eq. 19.13,

$$Q_o = \left(\frac{1.49}{n}\right) A R^{2/3} \sqrt{S}$$

$$= \left(\frac{1.49}{0.013}\right)\left(\frac{\pi}{4}\right)\left(\frac{24 \text{ in}}{12 \frac{\text{in}}{\text{ft}}}\right)^2 \left(\frac{24 \text{ in}}{\left(12 \frac{\text{in}}{\text{ft}}\right)(4)}\right)^{2/3}$$

$$\times \sqrt{0.001}$$

$$= \boxed{7.173 \text{ ft}^3/\text{sec} \quad (7.2 \text{ ft}^3/\text{sec})}$$

The answer is (D).

(b) The original full-flow capacity is

$$v_o = \frac{Q_o}{A} = \frac{7.173 \frac{\text{ft}^3}{\text{sec}}}{\left(\frac{\pi}{4}\right)(2 \text{ ft})^2}$$

$$= \boxed{2.28 \text{ ft/sec} \quad (2.3 \text{ ft/sec})}$$

The answer is (B).

(c) The current full-flow velocity is

$$v_{\text{current}} = \frac{Q_{\text{current}}}{A}$$

$$= \frac{6.0 \frac{\text{ft}^3}{\text{sec}}}{\left(\frac{\pi}{4}\right)(2 \text{ ft})^2}$$

$$= \boxed{1.91 \text{ ft/sec} \quad (1.9 \text{ ft/sec})}$$

The answer is (A).

(d) Since Q is inversely proportional to n,

$$n_{\text{current}} = \left(\frac{Q_o}{Q_{\text{current}}}\right) n_o = \left(\frac{7.173 \frac{\text{ft}^3}{\text{sec}}}{6.0 \frac{\text{ft}^3}{\text{sec}}}\right)(0.013)$$

$$= \boxed{0.01554 \quad (0.016)}$$

The answer is (C).

5. (a) From Eq. 19.16,

$$D = 1.335 \left(\frac{nQ}{\sqrt{S}}\right)^{3/8}$$

$$= (1.335)\left(\frac{\left(3.5 \frac{\text{ft}^3}{\text{sec}}\right)(0.013)}{\sqrt{0.01}}\right)^{3/8}$$

$$= 0.99 \text{ ft}$$

$$\boxed{\text{Use 12 in pipe.}}$$

The answer is (B).

(b) From Eq. 19.13,

$$D = \frac{12 \text{ in}}{12 \frac{\text{in}}{\text{ft}}} = 1 \text{ ft}$$

$$R = \frac{D}{4}$$

$$Q = \left(\frac{1.49}{n}\right) A R^{2/3} \sqrt{S}$$

$$= \left(\frac{1.49}{0.013}\right)\left(\frac{\pi}{4}\right)(1 \text{ ft})^2 \left(\frac{1 \text{ ft}}{4}\right)^{2/3} \sqrt{0.01}$$

$$= \boxed{3.57 \text{ ft}^3/\text{sec} \quad (3.6 \text{ ft}^3/\text{sec})}$$

The answer is (C).

(c) The full-flow velocity is

$$v = \frac{Q}{A} = \frac{3.57 \frac{\text{ft}^3}{\text{sec}}}{\left(\frac{\pi}{4}\right)(1 \text{ ft})^2}$$

$$= \boxed{4.55 \text{ ft/sec} \quad (4.6 \text{ ft/sec})}$$

The answer is (D).

(d) The depth of flow is calculated as follows.

$$\frac{Q}{Q_{\text{full}}} = \frac{0.7 \frac{\text{ft}^3}{\text{sec}}}{3.57 \frac{\text{ft}^3}{\text{sec}}} \approx 0.2$$

From App. 19.C, letting n vary with depth,

$$\frac{d}{D} \approx 0.35$$

$$d = (0.35)(12 \text{ in})$$

$$= \boxed{4.2 \text{ in} \quad (4 \text{ in})}$$

The answer is (A).

(e) To be self-cleansing, $v = \boxed{2.0\text{–}2.5 \text{ ft/sec.}}$

The answer is (C).

6. *Full flowing:*

$$R = \frac{D}{4} = \frac{4 \text{ ft}}{4} = 1 \text{ ft}$$

From Eq. 19.12 and Eq. 19.13,

$$v = \left(\frac{1.49}{n}\right) R^{2/3} \sqrt{S}$$

$$= \left(\frac{1.49}{0.013}\right)(1 \text{ ft})^{2/3} \sqrt{0.02}$$

$$= 16.21 \text{ ft/sec}$$

$$Q = Av = \left(\frac{\pi}{4}\right)(4 \text{ ft})^2 \left(16.21 \frac{\text{ft}}{\text{sec}}\right)$$

$$= 203.7 \text{ ft}^3/\text{sec}$$

Actual:

$$\frac{d}{D} = \frac{1.5 \text{ ft}}{4 \text{ ft}} = 0.375$$

From App. 19.C, assuming that n varies with depth,

$$\frac{\text{v}}{\text{v}_{\text{full}}} = 0.68$$

(a) $\text{v} = (0.68)\left(16.21 \dfrac{\text{ft}}{\text{sec}}\right) = \boxed{11 \text{ ft/sec}}$

The answer is (D).

(b) From App. 19.C, the maximum value of $\text{v}/\text{v}_{\text{full}}$ is 1.04 (at $d/D \approx 0.9$).

$$\text{v}_{\text{max}} = (1.04)\left(16.21 \frac{\text{ft}}{\text{sec}}\right)$$
$$= \boxed{16.86 \text{ ft/sec} \quad (17 \text{ ft/sec})}$$

The answer is (C).

(c) Similarly, $Q_{\text{max}}/Q_{\text{full}} = 1.02$ (at $d/D \approx 0.96$).

$$Q_{\text{max}} = (1.02)\left(203.7 \frac{\text{ft}^3}{\text{sec}}\right)$$
$$= \boxed{207.8 \text{ ft}^3/\text{sec} \quad (210 \text{ ft}^3/\text{sec})}$$

The answer is (D).

7. From App. 16.A, for $Q/Q_{\text{full}} = 0.75$,

$$A = 0.6318D^2$$
$$R = 0.3017D$$

From Eq. 19.13,

$$Q = \left(\frac{1.49}{n}\right)AR^{2/3}\sqrt{S}$$
$$5 \frac{\text{ft}^3}{\text{sec}} = \left(\frac{1.49}{0.012}\right)(0.6318D^2)(0.3017D)^{2/3}\sqrt{0.02}$$
$$1.0 = D^{8/3}$$
$$D = \boxed{1 \text{ ft} \quad (12 \text{ in})}$$

The answer is (A).

(Appendix 19.C can also be used to solve this problem.)

8. (a) For a circular channel flowing half-full, the hydraulic radius is

$$R = \frac{D}{4} = \frac{6 \text{ ft}}{4}$$
$$= \boxed{1.5 \text{ ft}}$$

The answer is (B).

(b) From App. 19.A, $n = 0.012$. From Eq. 19.13,

$$Q = \left(\frac{1.49}{n}\right)AR^{2/3}\sqrt{S}$$

Rearranging and recognizing that the culvert flows half full,

$$S = \left(\frac{Qn}{1.49AR^{2/3}}\right)^2$$
$$= \left(\frac{\left(150 \frac{\text{ft}^3}{\text{sec}}\right)(0.012)}{(1.49)(0.5)\left(\frac{\pi(6 \text{ ft})^2}{4}\right)(1.5 \text{ ft})^{2/3}}\right)^2$$
$$= \boxed{0.0043 \quad (0.004)}$$

The answer is (A).

(c) Equation 19.78 could be used, but App. 19.D is more convenient for use with circular channels. Entering with $Q = 150 \text{ ft}^3/\text{sec}$ and $D = 6 \text{ ft}$, $d_c = 3.3 \text{ ft}$. The actual flow depth is less than the critical depth; therefore, the flow is $\boxed{\text{supercritical.}}$

The answer is (C).

(d) For a circular channel flowing full, the hydraulic radius is the same as for half-full flow.

$$R = \frac{D}{4} = \frac{6 \text{ ft}}{4}$$
$$= \boxed{1.5 \text{ ft}}$$

The answer is (A).

(e) From Eq. 19.13,

$$Q = \left(\frac{1.49}{n}\right)AR^{2/3}\sqrt{S}$$
$$= \left(\frac{1.49}{0.012}\right)\left(\frac{\pi(6 \text{ ft})^2}{4}\right)(1.5 \text{ ft})^{2/3}\sqrt{0.0043}$$
$$= \boxed{301.7 \text{ ft}^3/\text{sec} \quad (300 \text{ ft}^3/\text{sec})}$$

The answer is (D).

(f) This part was solved in part (c): $d_c = \boxed{3.3 \text{ ft.}}$

The answer is (C).

(g) From Eq. 19.13, use $d = d_c$.

$$Q = \left(\frac{1.49}{n}\right)AR^{2/3}\sqrt{S}$$

R is calculated using the formula in Table 19.2.

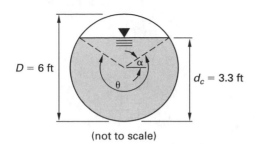

(not to scale)

$$\sin \alpha = \frac{0.3}{3}$$

$$\alpha = 5.74°$$

$$\theta = 180° + 2\alpha = 180° + (2)(5.74°)$$

$$= 191.48°$$

Convert to radians.

$$\theta = (191.48°)\left(\frac{2\pi \text{ rad}}{360°}\right) = 3.34 \text{ rad}$$

$$R = \frac{1}{4}\left(1 - \frac{\sin \theta}{\theta}\right)D = \left(\frac{1}{4}\right)\left(1 - \frac{\sin(3.34 \text{ rad})}{3.34 \text{ rad}}\right)(6 \text{ ft})$$

$$= 1.59 \text{ ft}$$

$$A = \frac{1}{8}(\theta - \sin \theta)D^2 = \left(\frac{1}{8}\right)(3.34 \text{ rad} - \sin(3.34 \text{ rad}))(6 \text{ ft})^2$$

$$= 15.9 \text{ ft}^2$$

$$S = \left(\frac{Qn}{(1.49AR^{2/3})}\right)^2$$

$$= \left(\frac{\left(150 \ \frac{\text{ft}^3}{\text{sec}}\right)(0.012)}{(1.49)(15.9 \text{ ft}^2)(1.59 \text{ ft})^{2/3}}\right)^2$$

$$= \boxed{0.00311 \quad (0.003)}$$

The answer is (B).

(h) The pipe flows partially full, and the exit is not submerged. This eliminates types 4 and 6 and leaves types 1, 2, 3, and 5. Type 2 is eliminated because the flow is not at critical depth at the outlet. Type 3 is tranquil flow throughout, so it is also eliminated. Distinguishing between types $\boxed{1 \text{ and } 5}$ cannot be done without further information about the entrance conditions.

The answer is (A).

(i)
$$\text{v} = \left(\frac{1.49}{n}\right)R^{2/3}\sqrt{S}$$

$$= \left(\frac{1.49}{0.012}\right)(1.5 \text{ ft})^{2/3}\sqrt{0.0043}$$

$$= 10.67 \text{ ft/sec}$$

The characteristic length is the hydraulic depth. For a circular channel flowing half full,

$$L = D_h = \frac{\pi D}{8} = \frac{\pi(6 \text{ ft})}{8}$$

$$= 2.36 \text{ ft}$$

From Eq. 19.79,

$$\text{Fr} = \frac{\text{v}}{\sqrt{gD_h}}$$

$$= \frac{10.67 \ \frac{\text{ft}}{\text{sec}}}{\sqrt{\left(32.2 \ \frac{\text{ft}}{\text{sec}^2}\right)(2.36 \text{ ft})}}$$

$$= \boxed{1.22}$$

Since Fr > 1, the flow is supercritical.

The answer is (C).

(j) From Eq. 19.92,

$$d_2 = -\frac{1}{2}d_1 + \sqrt{\frac{2\text{v}_1^2 d_1}{g} + \frac{d_1^2}{4}}$$

$$= \left(\frac{1}{2}\right)(2 \text{ ft}) + \sqrt{\frac{(2)\left(15 \ \frac{\text{ft}}{\text{sec}}\right)^2(2 \text{ ft})}{32.2 \ \frac{\text{ft}}{\text{sec}^2}} + \frac{(2 \text{ ft})^2}{4}}$$

$$= \boxed{4.38 \text{ ft} \quad (4.4 \text{ ft})}$$

The answer is (D).

9.

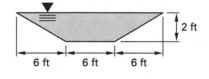

$$A = (6 \text{ ft})(2 \text{ ft}) + \frac{(2)(6 \text{ ft})(2 \text{ ft})}{2} = 24 \text{ ft}^2$$

$$P = 6 \text{ ft} + 2\sqrt{(6 \text{ ft})^2 + (2 \text{ ft})^2} = 18.65 \text{ ft}$$

$$R = \frac{A}{P} = \frac{24 \text{ ft}^2}{18.65 \text{ ft}} = 1.287 \text{ ft}$$

$$Q = (24 \text{ ft}^2)\left(\frac{1.49}{0.013}\right)(1.287 \text{ ft})^{2/3}\sqrt{0.002}$$

$$= \boxed{145.6 \text{ ft}^3/\text{sec} \quad (150 \text{ ft}^3/\text{sec})}$$

The answer is (B).

Alternate solution:

Solve using App. 19.E.

$$m = 3 \ (18.4°) \ \text{column}$$

$$\frac{D}{b} = \frac{2 \ \text{ft}}{6 \ \text{ft}} = 0.33$$

Interpolating, $K \approx 6.69$.

$$Q = K\left(\frac{1}{n}\right) D^{8/3}\sqrt{S}$$

$$= (6.69)\left(\frac{1}{0.013}\right)(2 \ \text{ft})^{8/3}\sqrt{0.002}$$

$$= 146.1 \ \text{ft}^3/\text{sec}$$

The answer is (B).

10. The optimum trapezoidal channel has side slopes of 60° and a depth that is twice the hydraulic radius.

From App. 19.A, $n = 0.011$ for smooth concrete.

Rearranging Eq. 19.13,

$$AR^{2/3} = \frac{Qn}{\sqrt{S}}$$

$$= \frac{\left(17 \ \dfrac{\text{m}^3}{\text{s}}\right)(0.011)}{\sqrt{0.08}}$$

$$= 0.661$$

From Table 19.2,

$$A = \left(b + \frac{d}{\tan\theta}\right)d$$

$$R = \frac{bd\sin\theta + d^2\cos\theta}{b\sin\theta + 2d}$$

Since $R = d/2$ and $b = 2d/\sqrt{3}$,

$$AR^{2/3} = \left(\frac{2d}{\sqrt{3}} + \frac{d}{\tan\theta}\right)d\left(\frac{d}{2}\right)^{2/3} = 0.661$$

Simplifying, $d^{8/3} = 0.606$ with $\theta = 60°$, the optimal channel has a depth, d, of 0.828 m and a base, b, of 0.957 m. Use 0.8 m and 1.0 m.

The answer is (A).

Alternate solution:

Use App. 19.F.

"Optimum" means $\alpha = 60°$ and $D/b = \sqrt{3}/2 = 0.866$.

Double interpolate $K' \approx 1.12$.

This is metric, so from App. 19.F footnote (c),

$$K' = \frac{1.12}{1.486} \approx 0.754$$

$$Q = K'\left(\frac{1}{n}\right) b^{8/3}\sqrt{S}$$

$$17 \ \frac{\text{m}^2}{\text{s}} = (0.754)\left(\frac{1}{0.011}\right)b^{8/3}\sqrt{0.08}$$

$$b = 0.952 \ \text{m}$$

$$d = \frac{\sqrt{3}}{2}b = \left(\frac{\sqrt{3}}{2}\right)(0.952 \ \text{m}) = 0.824 \ \text{m}$$

The answer is (A).

11. From Eq. 19.53,

$$b = b_{\text{actual}} - 0.1NH$$

$$= 5 \ \text{ft} - (0.1)(2)(0.43 \ \text{ft})$$

$$= 4.914 \ \text{ft}$$

From Eq. 19.50,

$$C_1 = \left(0.6035 + 0.0813\left(\frac{H}{Y}\right) + \frac{0.000295}{Y}\right)\left(1 + \frac{0.00361}{H}\right)^{3/2}$$

$$= \left(0.6035 + (0.0813)\left(\frac{0.43 \ \text{ft}}{6 \ \text{ft}}\right) + \frac{0.000295}{6 \ \text{ft}}\right)$$

$$\times \left(1 + \frac{0.00361}{0.43 \ \text{ft}}\right)^{3/2}$$

$$= 0.617$$

From Eq. 19.49,

$$Q = \tfrac{2}{3}C_1 b\sqrt{2g}H^{3/2}$$

$$= \left(\tfrac{2}{3}\right)(0.617)(4.914 \ \text{ft})\sqrt{(2)\left(32.2 \ \frac{\text{ft}}{\text{sec}^2}\right)}(0.43 \ \text{ft})^{3/2}$$

$$= \boxed{4.574 \ \text{ft}^3/\text{sec} \quad (4.6 \ \text{ft}^3/\text{sec})}$$

The answer is (C).

12. This is a Cipoletti weir. From Eq. 19.58,

$$Q = 3.367 bH^{3/2}$$

$$= (3.367)\left(\frac{18 \ \text{in}}{12 \ \frac{\text{in}}{\text{ft}}}\right)\left(\frac{9 \ \text{in}}{12 \ \frac{\text{in}}{\text{ft}}}\right)^{3/2}$$

$$= \boxed{3.28 \ \text{ft}^3/\text{sec} \quad (3.3 \ \text{ft}^3/\text{sec})}$$

The answer is (B).

13.

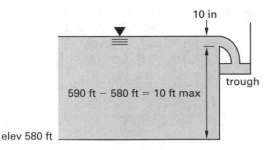

(a) Use Eq. 19.51.

$$C_1 \approx 0.602 + 0.083\left(\frac{H}{Y}\right)$$

$$= 0.602 + (0.083)\left(\frac{10\text{ in}}{(590\text{ ft} - 580\text{ ft})\left(12\ \frac{\text{in}}{\text{ft}}\right)}\right)$$

$$= 0.609$$

Use Eq. 19.49 with $C_1 = 0.609$.

$$Q_{\text{MGD}} = \left(0.6463\ \frac{\text{MGD-sec}}{\text{ft}^3}\right) Q_{\text{ft}^3/\text{sec}}$$

$$= \left(0.6463\ \frac{\text{MGD-sec}}{\text{ft}^3}\right)(\tfrac{2}{3})(0.609)bH^{3/2}$$

$$\times \sqrt{(2)\left(32.2\ \frac{\text{ft}}{\text{sec}^2}\right)}$$

$$= 2.11bH^{3/2}$$

To satisfy the minimum head requirement, the minimum flow must be used. The minimum head requirement will then be automatically satisfied at maximum flow.

At minimum flow, $Q = 2$ MGD.

$$H = \frac{10\text{ in}}{12\ \frac{\text{in}}{\text{ft}}} = 0.833\text{ ft}$$

$$b_{\text{effective}} = \frac{2\text{ MGD}}{(2.11)(0.833\text{ ft})^{3/2}} = 1.25\text{ ft}$$

From Eq. 19.53,

$$b_{\text{actual}} = 1.25\text{ ft} + (0.1)(2)(0.833\text{ ft})$$

$$= \boxed{1.42\text{ ft}\quad(1.4\text{ ft})}$$

The answer is (B).

(b) At maximum flow,

$$8\text{ MGD} = (2.11)(1.25)H^{3/2}$$

$$H \approx 2.1\text{ ft}$$

Recalculate C_1 from Eq. 19.51.

$$C_1 \approx 0.602 + 0.083\left(\frac{H}{Y}\right)$$

$$= 0.602 + (0.083)\left(\frac{2.1\text{ ft}}{590\text{ ft} - 580\text{ ft}}\right)$$

$$= 0.619$$

From Eq. 19.49,

$$Q = \left(0.6463\ \frac{\text{MGD-sec}}{\text{ft}^3}\right)(\tfrac{2}{3})(0.619)bH^{3/2}$$

$$\times \sqrt{(2)\left(32.2\ \frac{\text{ft}}{\text{sec}^2}\right)}$$

$$= 2.14bH^{3/2}$$

Use Eq. 19.53 to calculate $b_{\text{effective}}$. Continue iterating until H stabilizes.

$$b_{\text{effective}} = 1.42\text{ ft} - (0.1)(2)(2.1\text{ ft}) = 1\text{ ft}$$

$$8\text{ MGD} = (2.14)(1)H^{3/2}$$

$$H = 2.41\text{ ft}$$

$$b_{\text{effective}} = 1.42\text{ ft} - (0.1)(2)(2.41\text{ ft}) = 0.94\text{ ft}$$

$$8\text{ MGD} = (2.14)(0.94)H^{3/2}$$

$$H = 2.51\text{ ft}$$

$$b_{\text{effective}} = 1.42\text{ ft} - (0.1)(2)(2.51\text{ ft}) = 0.918\text{ ft}$$

$$8\text{ MGD} = (2.14)(0.918)H^{3/2}$$

$$H = 2.55\text{ ft}$$

$$b_{\text{effective}} = 1.42\text{ ft} - (0.1)(2)(2.55\text{ ft}) = 0.910\text{ ft}$$

$$8\text{ MGD} = (2.14)(0.910)H^{3/2}$$

$$H = 2.56\text{ ft}\quad[\text{acceptable}]$$

Use a freeboard elevation of

$$580\text{ ft} + 10\text{ ft} + 2.56\text{ ft} + \frac{4\text{ in}}{12\ \frac{\text{in}}{\text{ft}}} = 592.89\text{ ft}$$

The elevation at maximum flow is

$$592.89\text{ ft} - \text{freeboard height} = 592.89\text{ ft} - \frac{4\text{ in}}{12\ \frac{\text{in}}{\text{ft}}}$$

$$= \boxed{592.56\text{ ft}\quad(592.6\text{ ft})}$$

The answer is (C).

(c) The channel elevation is

$$580\text{ ft} + 10\text{ ft} = 590\text{ ft}$$

The elevation at minimum flow is

$$590\text{ ft} + 0.833\text{ ft} = \boxed{590.833\text{ ft}\quad(590.8\text{ ft})}$$

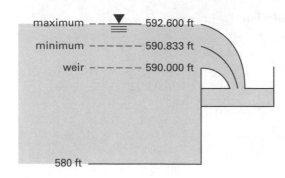

The answer is (A).

14. From Eq. 19.94,

$$v_1 = \sqrt{\left(\frac{gd_2}{2d_1}\right)(d_1 + d_2)}$$

$$= \sqrt{\left(\frac{\left(32.2 \ \frac{\text{ft}}{\text{sec}^2}\right)}{2}\right)\left(\frac{2.4 \ \text{ft}}{1 \ \text{ft}}\right)(1 \ \text{ft} + 2.4 \ \text{ft})}$$

$$= 11.46 \ \text{ft/sec}$$

$$Q = Av = (5 \ \text{ft})(1 \ \text{ft})\left(11.46 \ \frac{\text{ft}}{\text{sec}}\right)$$

$$= \boxed{57.30 \ \text{ft}^3/\text{sec} \quad (57 \ \text{ft}^3/\text{sec})}$$

The answer is (D).

15. From Eq. 19.8,

$$A_1v_1 = A_2v_2$$

$$(0.2 \ \text{ft})(\text{width})\left(54.7 \ \frac{\text{ft}}{\text{sec}}\right) = (6 \ \text{ft})(\text{width})v_2$$

$$v_2 = 1.82 \ \text{ft/sec}$$

From Eq. 19.95,

$$\Delta E = \left(d_1 + \frac{v_1^2}{2g}\right) - \left(d_2 + \frac{v_2^2}{2g}\right)$$

$$= 0.2 \ \text{ft} + \frac{\left(54.7 \ \frac{\text{ft}}{\text{sec}}\right)^2}{(2)\left(32.2 \ \frac{\text{ft}}{\text{sec}^2}\right)}$$

$$- \left(6 \ \text{ft} + \frac{\left(1.82 \ \frac{\text{ft}}{\text{sec}}\right)^2}{(2)\left(32.2 \ \frac{\text{ft}}{\text{sec}^2}\right)}\right)$$

$$= \boxed{40.61 \ \text{ft} \quad (41 \ \text{ft-lbf/lbm})}$$

The answer is (A).

16. (a) From Eq. 19.94,

$$v_1 = \sqrt{\left(\frac{gd_2}{2d_1}\right)(d_1 + d_2)}$$

$$= \sqrt{\left(\frac{\left(32.2 \ \frac{\text{ft}}{\text{sec}^2}\right)(3.8 \ \text{ft})}{(2)(1.2 \ \text{ft})}\right)(1.2 \ \text{ft} + 3.8 \ \text{ft})}$$

$$= 15.97 \ \text{ft/sec}$$

$$Q = Av = (1.2 \ \text{ft})(6 \ \text{ft})\left(15.97 \ \frac{\text{ft}}{\text{sec}}\right)$$

$$= \boxed{115 \ \text{ft}^3/\text{sec}}$$

The answer is (B).

(b)
$$v_2 = \frac{Q}{A} = \frac{115 \ \frac{\text{ft}^3}{\text{sec}}}{(3.8 \ \text{ft})(6 \ \text{ft})} = 5.04 \ \text{ft/sec}$$

From Eq. 19.95,

$$\Delta E = \left(d_1 + \frac{v_1^2}{2g}\right) - \left(d_2 + \frac{v_2^2}{2g}\right)$$

$$= \left(1.2 \ \text{ft} + \frac{\left(15.97 \ \frac{\text{ft}}{\text{sec}}\right)^2}{(2)\left(32.2 \ \frac{\text{ft}}{\text{sec}^2}\right)}\right)$$

$$- \left(3.8 \ \text{ft} + \frac{\left(5.04 \ \frac{\text{ft}}{\text{sec}}\right)^2}{(2)\left(32.2 \ \frac{\text{ft}}{\text{sec}^2}\right)}\right)$$

$$= \boxed{0.97 \ \text{ft} \quad (1 \ \text{ft-lbf/lbm})}$$

The answer is (A).

17. (a) Disregard the velocity of approach. Use $C_s = 3.5$ in Eq. 19.61.

$$Q = C_sbH^{3/2}$$

$$= \left(3.5 \ \frac{\text{ft}^{1/2}}{\text{sec}}\right)(1 \ \text{ft})(2 \ \text{ft})^{3/2}$$

$$= \boxed{9.9 \ \text{ft}^3/\text{sec per ft of width}}$$

The answer is (A).

(b) The upstream energy is

$$E_1 = d = 2 \ \text{ft} \quad [v_1 \approx 0]$$

At the toe, from Eq. 19.73,

$$E_2 = E_1 + z_1 - z_2$$
$$= 2 \text{ ft} + 40 \text{ ft} - 0$$
$$= 42 \text{ ft}$$

$$E_2 = d_2 + \frac{v_2^2}{2g} = d_2 + \frac{Q^2}{2gA_2^2}$$

$$42 \text{ ft} = d_2 + \frac{\left(9.9 \frac{\text{ft}^3}{\text{sec}}\right)^2}{(2)\left(32.2 \frac{\text{ft}}{\text{sec}^2}\right)(1 \text{ ft}^2)d_2^2}$$

By trial and error, $d_2 = \boxed{0.19 \text{ ft.}}$

The answer is (A).

18.

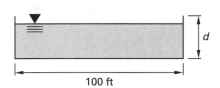

100 ft

(a) $$R = \frac{(100 \text{ ft})d}{100 \text{ ft} + 2d} = \frac{50d}{50 + d}$$

From Eq. 19.13,

$$Q = \left(\frac{1.49}{n}\right)AR^{2/3}\sqrt{S}$$

$$10,000 \frac{\text{ft}^3}{\text{sec}} = \left(\frac{1.49}{0.012}\right)(100d)\left(\frac{50d}{50 + d}\right)^{2/3}\sqrt{0.05}$$

$$0.26538 = \left(\frac{d^{5/2}}{50 + d}\right)^{2/3}$$

$$0.1367 = \frac{d^{5/2}}{50 + d}$$

By trial and error, $d = \boxed{2.2 \text{ ft.}}$

The answer is (B).

(b) From Eq. 19.75,

$$d_c = \sqrt[3]{\frac{Q^2}{gw^2}}$$

$$= \sqrt[3]{\frac{\left(10,000 \frac{\text{ft}^3}{\text{sec}}\right)^2}{\left(32.2 \frac{\text{ft}}{\text{sec}^2}\right)(100 \text{ ft})^2}}$$

$$= \boxed{6.77 \text{ ft} \quad (6.8 \text{ ft})}$$

The answer is (C).

(c) Since $d < d_c$, $\boxed{\text{flow is rapid (shooting).}}$

The answer is (C).

(d) $$v = \frac{Q}{A} = \frac{10,000 \frac{\text{ft}^3}{\text{sec}}}{(100 \text{ ft})(2.2 \text{ ft})} = 45.45 \text{ ft/sec}$$

From Eq. 19.92,

$$d_2 = -\tfrac{1}{2}d_1 + \sqrt{\frac{2v_1^2 d_1}{g} + \frac{d_1^2}{4}}$$

$$= -\left(\tfrac{1}{2}\right)(2.2 \text{ ft})$$

$$+ \sqrt{\frac{(2)\left(45.45 \frac{\text{ft}}{\text{sec}}\right)^2(2.2 \text{ ft})}{32.2 \frac{\text{ft}}{\text{sec}^2}} + \frac{(2.2 \text{ ft})^2}{4}}$$

$$= \boxed{15.74 \text{ ft} \quad (16 \text{ ft})}$$

The answer is (C).

19. (a) From Eq. 19.61 using $C_s = 3.7$,

$$H = 2425 \text{ ft} - 2420 \text{ ft} = 5 \text{ ft}$$
$$Q = C_3 bH^{3/2} = (3.7)(60 \text{ ft})(5 \text{ ft})^{3/2}$$
$$= \boxed{2482 \text{ ft}^3/\text{sec} \quad (2500 \text{ ft}^3/\text{sec})}$$

The answer is (C).

(b) Disregarding the velocity of approach, the initial energy is

$$E_1 = (\text{elev})_1 = 2425 \text{ ft}$$

At the toe (point 2),

$$E_1 = E_2 + h_f$$

$$2425 = (\text{elev})_2 + d_2 + \frac{Q^2}{2gw^2 d_2^2}$$
$$+ (0.20)(2425 \text{ ft} - 2345 \text{ ft})$$

$$(1 - 0.20)$$
$$\times (2425 \text{ ft} - 2345 \text{ ft}) = d_2 + \frac{\left(2482 \frac{\text{ft}^3}{\text{sec}}\right)^2}{(2)\left(32.2 \frac{\text{ft}}{\text{sec}^2}\right)(60 \text{ ft})^2 d_2^2}$$

$$64 \text{ ft} = d_2 + \frac{26.57}{d_2^2}$$

By trial and error, $d_2 = \boxed{0.647 \text{ ft} \ (0.65 \text{ ft}).}$

The answer is (A).

Water Resources

(c) $d_{toe} = d_1$ is implicitly a conjugate depth.

If there was a hydraulic jump, it would be between the conjugate depths, d_1 to d_2.

$$d_1 = 0.647 \text{ ft}$$

$$v_1 = \frac{Q}{A_1} = \frac{Q}{d_1 w} = \frac{2482 \ \frac{\text{ft}^3}{\text{sec}}}{(0.647 \text{ ft})(60 \text{ ft})}$$

$$= 63.94 \text{ ft/sec}$$

From Eq. 19.92,

$$d_2 = -\tfrac{1}{2}d_1 + \sqrt{\frac{2v_1^2 d_1}{g} + \frac{d_1^2}{4}}$$

$$= -\left(\tfrac{1}{2}\right)(0.647 \text{ ft})$$

$$+ \sqrt{\frac{(2)\left(63.9 \ \frac{\text{ft}}{\text{sec}}\right)^2 (0.647 \text{ ft})}{32.2 \ \frac{\text{ft}}{\text{sec}^2}} + \frac{(0.647 \text{ ft})^2}{4}}$$

$$= 12.49 \text{ ft}$$

The answer is (D).

(d) $v_2 = \dfrac{Q}{A_2} = \dfrac{Q}{d_2 w} = \dfrac{2482 \ \frac{\text{ft}^3}{\text{sec}}}{(12.49 \text{ ft})(60 \text{ ft})} = 3.31 \text{ ft/sec}$

Use d_2 and v_2 in Eq. 19.95.

$$\Delta E = \left(d_1 + \frac{v_1^2}{2g}\right) - \left(d_2 + \frac{v_2^2}{2g}\right)$$

$$= \left(0.647 \text{ ft} + \frac{\left(63.9 \ \frac{\text{ft}}{\text{sec}}\right)^2}{(2)\left(32.2 \ \frac{\text{ft}}{\text{sec}^2}\right)}\right)$$

$$- \left(12.49 \text{ ft} + \frac{\left(3.31 \ \frac{\text{ft}}{\text{sec}}\right)^2}{(2)\left(32.2 \ \frac{\text{ft}}{\text{sec}^2}\right)}\right)$$

$$= \boxed{51.4 \text{ ft} \quad (50 \text{ ft-lbf/lbm})}$$

The answer is (B).

20. (a) From Eq. 19.65,

$$n = 1.522b^{0.026}$$

$$= (1.522)(6 \text{ ft})^{0.026}$$

$$= 1.595$$

From Eq. 19.64,

$$Q = KbH_a^n$$

$$= (4)(6 \text{ ft})\left(\frac{18 \text{ in}}{12 \ \frac{\text{in}}{\text{ft}}}\right)^{1.595}$$

$$= \boxed{45.8 \text{ ft}^3/\text{sec} \quad (46 \text{ ft}^3/\text{sec})}$$

The answer is (B).

(b) The flumes are self-cleansing, have low head loss, and can be operated by unskilled personnel. Parshall flumes are usually permanent structures.

The answer is (D).

21. (a) The normal depth is d.

$$R = \frac{8d}{8 + 2d}$$

From Eq. 19.13,

$$Q = \left(\frac{1.49}{n}\right)AR^{2/3}\sqrt{S}$$

$$150 \ \frac{\text{ft}^3}{\text{sec}} = \left(\frac{1.49}{0.015}\right)(8d)\left(\frac{8d}{8 + 2d}\right)^{2/3}\sqrt{0.0015}$$

$$4.875 = d\left(\frac{8d}{8 + 2d}\right)^{2/3}$$

By trial and error, $d = \boxed{3.28 \text{ ft} \ (3.3 \text{ ft}).}$

The answer is (B).

(b) To get the backwater curve, choose depths and compute distances to those depths.

Preliminary parameters:

d	$A = 8d$	$v = \dfrac{Q}{A}$	$\dfrac{v^2}{2g}$	$E = d + \dfrac{v^2}{2g}$	$P = 2d + 8$	R
6	48	3.12	0.151	6.151	20	2.40
5	40	3.75	0.218	5.218	18	2.22
4	32	4.69	0.342	4.342	16	2.00
3.28	26.24	5.72	0.502	3.782	14.56	1.80

The average velocities and average hydraulic radii are

d	v	v_{ave}	R	R_{ave}
6.00	3.12		2.40	
		3.435		2.31
5.00	3.75		2.22	
		4.22		2.11
4.00	4.69		2.00	
		5.21		1.90
3.28	5.72		1.80	

From Eq. 19.88, the average energy gradient is

$$S_{\text{ave}} = \left(\frac{n\text{v}_{\text{ave}}}{1.49 R_{\text{ave}}^{2/3}}\right)^2$$

d	S_{ave}	ΔE	$S_0 - S_{\text{ave}}$
6.00			
	0.000392	0.933	0.001108
5.00			
	0.000667	0.876	0.000833
4.00			
	0.001169	0.552	0.000331
3.28			

From Eq. 19.90,

$$L_{6\text{-}5} = \frac{\Delta E}{S_0 - S_{\text{ave}}} = \frac{0.933 \text{ ft}}{0.001108}$$
$$= \boxed{842 \text{ ft} \quad (840 \text{ ft})}$$

The answer is (B).

(c)
$$L_{5\text{-}4} = \frac{\Delta E}{S_0 - S_{\text{ave}}} = \frac{0.876 \text{ ft}}{0.000833}$$
$$= \boxed{1052 \text{ ft} \quad (1100 \text{ ft})}$$

The answer is (D).

22. For concrete, $n = 0.012$.

The pipe diameter is

$$D = \frac{42 \text{ in}}{12 \dfrac{\text{in}}{\text{ft}}} = 3.5 \text{ ft}$$

The hydraulic radius is

$$R = \frac{D}{4} = \frac{3.5 \text{ ft}}{4}$$
$$= 0.875 \text{ ft}$$

From a table of orifice coefficients, Table 17.5, $C_d = 0.82$.

$$h_1 = 5 \text{ ft} + 3.5 \text{ ft} + (250 \text{ ft})(0.006)$$
$$= 10 \text{ ft}$$

Determine the culvert flow type.

$$\frac{h_1 - z}{D} = \frac{10 \text{ ft} - (250 \text{ ft})(0.006)}{3.5 \text{ ft}}$$
$$= 2.4$$

The tailwater submerges the culvert outlet, so this is type-4 flow (outlet control as per Table 19.9).

$$h_1 = 10 \text{ ft}$$
$$h_4 = D = 3.5 \text{ ft}$$

$$Q = C_d A_0 \sqrt{2g\left(\frac{h_1 - h_4}{1 + \dfrac{29 C_d^2 n^2 L}{R^{4/3}}}\right)}$$

$$= \left(\frac{(0.82)\pi(3.5 \text{ ft})^2}{4}\right)$$

$$\times \sqrt{(2)\left(32.2 \dfrac{\text{ft}}{\text{sec}^2}\right) \times \left(\dfrac{10 \text{ ft} - 3.5 \text{ ft}}{1 + \dfrac{(29)(0.82)^2(0.012)^2(250 \text{ ft})}{(0.875 \text{ ft})^{4/3}}}\right)}$$

$$= 119 \text{ ft}^3/\text{sec}$$

Check to see if the flow rate is limited by inlet geometry. Evaluate type-6 flow. From Eq. 19.108 with $h_3 = d$ and $h_f = 0$,

$$Q = C_d A_o \sqrt{2g(h_1 - h_3 - h_f)}$$
$$= (0.82)\left(\frac{\pi(3.5 \text{ ft})^2}{4}\right)$$
$$\times \sqrt{(2)\left(32.2 \dfrac{\text{ft}}{\text{sec}^2}\right)(10 \text{ ft} - 3.5 \text{ ft} - 0)}$$
$$= 161.4 \text{ ft}^3/\text{sec}$$

The culvert has the capacity.

$$Q = \boxed{119 \text{ ft}^3/\text{sec} \quad (120 \text{ ft}^3/\text{sec})}$$

The answer is (D).

23. (a) Rearrange Eq. 19.61.

$$H = \left(\frac{Q}{C_s b}\right)^{2/3} = \left(\frac{2000 \dfrac{\text{m}^3}{\text{s}}}{(2.2)(150 \text{ m})}\right)^{2/3}$$
$$= \boxed{3.32 \text{ m} \quad (3.3 \text{ m})}$$

The answer is (D).

(b) At the crest,

$$v_1 = \frac{Q}{A_1} = \frac{2000 \ \frac{m^3}{s}}{(150 \ m)(3.32 \ m)}$$

$$= 4.02 \ m/s$$

From Eq. 19.97,

$$E_1 = y_{crest} + H + \frac{v_1^2}{2g}$$

$$= 30 \ m + 3.32 \ m + \frac{\left(4.02 \ \frac{m}{s}\right)^2}{(2)\left(9.81 \ \frac{m}{s^2}\right)}$$

$$= \boxed{34.1 \ m}$$

The answer is (B).

(c) At the toe,

$$v_2 = \frac{Q}{A_2} = \frac{2000 \ \frac{m^3}{s}}{(150 \ m)(0.6 \ m)} = 22.22 \ m/s$$

$$E_2 = d_2 + \frac{v_2^2}{2g}$$

$$= 0.6 \ m + \frac{\left(22.22 \ \frac{m}{s}\right)^2}{(2)\left(9.81 \ \frac{m}{s^2}\right)}$$

$$= \boxed{25.8 \ m}$$

The answer is (A).

(d) The friction loss is

$$E_1 - E_2 = 34.1 \ m - 25.8 \ m$$

$$= \boxed{8.3 \ m}$$

The answer is (C).

(e) From Eq. 19.13,

$$Q = \left(\frac{1}{n}\right) A R^{2/3} \sqrt{S}$$

$$2000 \ \frac{m^3}{s} = \left(\frac{1}{0.016}\right)(150 \ m) d_n \left(\frac{(150 \ m) d_n}{2 d_n + 150 \ m}\right)^{2/3}$$

$$\times \sqrt{0.002}$$

By trial and error, $d_n = \boxed{2.6 \ m.}$

The answer is (B).

(f) From Eq. 19.75,

$$d_c = \left(\frac{Q^2}{gw^2}\right)^{1/3}$$

$$= \left(\frac{\left(2000 \ \frac{m^3}{s}\right)^2}{\left(9.81 \ \frac{m}{s^2}\right)(150 \ m)^2}\right)^{1/3}$$

$$= \boxed{2.6 \ m}$$

The answer is (B).

(g) The normal depth of flow coincides with the critical depth. A hydraulic jump is an abrupt rise from below the critical depth to above the critical depth. In this case the water will rise gradually and no jump occurs.

The answer is (D).

(h) From Table 19.7, the maximum velocity (assuming the water is clear) is

$$v_{max} = \left(4.0 \ \frac{ft}{sec}\right)\left(0.3 \ \frac{m}{ft}\right) = \boxed{1.2 \ m/s}$$

The answer is (A).

(i) From App. 19.A, for natural channels in good condition, $n = 0.025$.

The depth of flow is

$$d = \frac{Q}{vb} = \frac{2000 \ \frac{m^3}{sec}}{\left(1.2 \ \frac{m}{s}\right)(150 \ m)}$$

$$= 11.11 \ m$$

$$R = \frac{bd}{2d + b}$$

$$= \frac{(150 \ m)(11.1 \ m)}{(2)(11.1 \ m) + 150 \ m}$$

$$= 9.67 \ m$$

Rearranging Eq. 19.12,

$$S = \left(\frac{vn}{R^{2/3}}\right)^2$$

$$= \left(\frac{\left(1.2 \ \frac{m}{s}\right)(0.025)}{(9.67 \ m)^{2/3}}\right)^2$$

$$= \boxed{0.000044}$$

The answer is (B).

(j) From Eq. 19.29,

$$h_f = LS = (1000 \ m)(0.002) = \boxed{2.0 \ m}$$

The answer is (D).

Water Resources

PRACTICE PROBLEMS

Hydrographs

1. A 2 h storm over a 111 km^2 area produces a total runoff volume of 4×10^6 m^3 with a peak discharge of 260 m^3/s.

(a) What is the total excess precipitation?

(A) 1.4 cm

(B) 2.6 cm

(C) 3.6 cm

(D) 4.0 cm

(b) What is the unit hydrograph peak discharge?

(A) 72 m^3/s·cm

(B) 120 m^3/s·cm

(C) 210 m^3/s·cm

(D) 260 m^3/s·cm

(c) If a 2 h storm producing 6.5 cm of runoff is to be used to design a culvert, what is the design flood hydrograph volume?

(A) 4.0×10^6 m^3

(B) 7.2×10^6 m^3

(C) 2.6×10^7 m^3

(D) 3.6×10^7 m^3

(d) What is the design discharge?

(A) 89 m^3/s

(B) 130 m^3/s

(C) 260 m^3/s

(D) 470 m^3/s

(e) The recurrence interval of the 6.5 cm storm is 50 yr, and the culvert is to be designed for a 30 yr life. What is the probability that the capacity will be exceeded during the design life?

(A) 0%

(B) 33%

(C) 45%

(D) 92%

(f) The unit hydrograph represents water flowing into a stream from

I.   base flow

II.   evapotranspiration

III.   overland flow

IV.   surface flow

V.   interflow

(A) I only

(B) I and III

(C) II, III, and IV

(D) III, IV, and V

Use the following information for parts (g) through (j).

Stream discharges recorded during a 3 h storm on a 40 km^2 watershed are as follows.

t (h)	Q (m^3/s)
0	17
1	16
2	48
3	90
4	108
5	85
6	58
7	38
8	26
9	20
10	18
11	17
12	16

(g) What is the unit hydrograph peak discharge from surface runoff for this 3 h storm?

(A) 5.2 m³/s·cm

(B) 29 m³/s·cm

(C) 65 m³/s·cm

(D) 120 m³/s·cm

(h) What is the peak discharge for a 6 h storm producing 5 cm of runoff in this watershed?

(A) 92 m³/s

(B) 140 m³/s

(C) 170 m³/s

(D) 300 m³/s

(i) What is the peak discharge for a 5 h storm producing 5 cm of runoff in this watershed?

(A) 63 m³/s

(B) 90 m³/s

(C) 110 m³/s

(D) 260 m³/s

(j) What is the design flood hydrograph volume for a 5 h storm producing 5 cm of runoff in this watershed?

(A) 1.3×10^4 m³

(B) 4.0×10^5 m³

(C) 8.3×10^5 m³

(D) 2.0×10^6 m³

2. (*Time limit: one hour*) A stream gaging station recorded the following discharges from a 1.2 mi² drainage area after a storm.

t (hr)	Q (ft³/sec)
0	102
1	99
2	101
3	215
4	507
5	625
6	455
7	325
8	205
9	145
10	100
11	70
12	55
13	49
14	43
15	38

(a) Draw the actual hydrograph.

(b) Separate the groundwater and surface water.

(c) Draw the unit hydrograph.

(d) What is the length of the direct runoff recession limb?

(A) 5 hr

(B) 6 hr

(C) 10 hr

(D) 13 hr

Rational Equation

3. A 0.5 mi² drainage area has a runoff coefficient of 0.6 and a time of concentration of 60 min. The drainage area is in Steel region no. 3, and a 10 yr storm is to be used for design purposes. What is the peak runoff?

(A) 310 ft³/sec

(B) 390 ft³/sec

(C) 460 ft³/sec

(D) 730 ft³/sec

4. Four contiguous 5 ac watersheds are served by an adjacent 1200 ft storm drain ($n = 0.013$ and slope $= 0.005$). Inlets to the storm drain are placed every 300 ft along the storm drain. The inlet time for each area served by an inlet is 15 min, and the area's runoff coefficient is 0.55. A storm to be used for design purposes has the following characteristics (I is in in/hr, t is in min).

$$I = \frac{100}{t_c + 10}$$

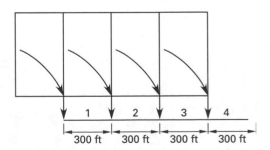

All flows are maximum, and all pipe sizes are available. n is constant. What is the diameter of the last section of storm drain?

(A) 28 in

(B) 32 in

(C) 36 in

(D) 42 in

5. (*Time limit: one hour*) A 75 ac urbanized section of land drains into a rectangular 5 ft × 7 ft channel that directs runoff through a round culvert under a roadway. The culvert is concrete and is 60 in in diameter. It is 36 ft long and placed on a 1% slope.

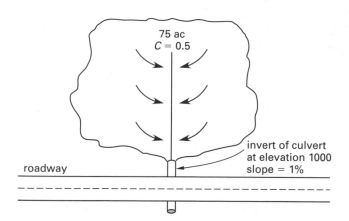

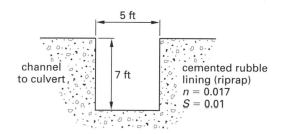

It is desired to evaluate the culvert design based on a 50 yr storm. It is known that the time of concentration to the entrance of the culvert is 30 min. The corresponding rainfall intensity is 5.3 in/hr.

(a) If the minimum road surface elevation is 1010 ft, will the road surface be flooded?

    (A) Yes; the culvert needs 10 ft³/sec additional capacity.

    (B) Yes; the culvert needs 70 ft³/sec additional capacity.

    (C) No; the culvert has 20 ft³/sec excess capacity.

    (D) No; the culvert has 70 ft³/sec excess capacity.

(b) What is the depth of flow elevation upstream at the culvert entrance after 30 min?

    (A) 2.7 ft

    (B) 3.1 ft

    (C) 3.6 ft

    (D) 4.1 ft

(c) If $n_{full} = 0.013$ and n varies with depth of flow, what is the depth of flow at the culvert exit after 30 min?

    (A) 2.7 ft

    (B) 3.1 ft

    (C) 3.6 ft

    (D) 4.1 ft

Hydrograph Synthesis

6. (*Time limit: one hour*) A standard 4 hr storm produces 2 in net of runoff. A stream gaging report is produced from successive sampling every few hours. Two weeks later, the first 4 hr of an 8 hr storm over the same watershed produces 1 in of runoff. The second 4 hr produces 2 in of runoff. Neglecting ground water, draw a hydrograph of the 8 hr storm.

t (hr)	Q (ft³/sec)
0	0
2	100
4	350
6	600
8	420
10	300
12	250
14	150
16	100
18	–
20	50
22	–
24	0

Reservoir Sizing

7. A class A evaporation pan located near a reservoir shows a 1 day evaporation loss of 0.8 in. If the pan coefficient is 0.7, what is the approximate evaporation loss in the reservoir?

    (A) 0.56 in

    (B) 0.63 in

    (C) 0.98 in

    (D) 1.1 in

8. A reservoir has a total capacity of 7 volume units. At the beginning of a study, the reservoir contains 5.5 units. The monthly demand on the reservoir from a nearby city is 0.7 units. The monthly inflow to the reservoir is normally distributed with a mean of 0.9 units and a standard deviation of 0.2 units. Simulate one year of reservoir operation with a 99.97% confidence level.

9. Repeat Prob. 8 assuming that the monthly demand on the reservoir is normally distributed with a mean of 0.7 units and a standard deviation of 0.2 units.

10. (*Time limit: one hour*) A reservoir is needed to provide 20 ac-ft of water each month. The inflow for each of 13 representative months is given. The reservoir starts full. Size the reservoir.

month	inflow (ac-ft)
February	30
March	60
April	20
May	10
June	5
July	10
August	5
September	10
October	20
November	90
December	85
January	75
February	50

(A) 40 ac-ft

(B) 60 ac-ft

(C) 90 ac-ft

(D) 120 ac-ft

11. (*Time limit: one hour*) A reservoir with a constant draft of 240 MG/mi²-yr is being designed. The inflow distribution is known.

month	inflow (MG/mi²)	inflow (m³/m²)
January	20	0.029
February	30	0.044
March	45	0.066
April	30	0.044
May	40	0.058
June	30	0.044
July	15	0.022
August	5	0.007
September	15	0.022
October	60	0.088
November	90	0.132
December	40	0.058

(a) What should be the minimum reservoir size?

(A) 25 MG/mi²

(B) 35 MG/mi²

(C) 45 MG/mi²

(D) 65 MG/mi²

(b) The reservoir will start to spill at the end of

(A) March

(B) June

(C) September

(D) March and October

SOLUTIONS

1. (a) From Eq. 20.21,

$$P_{\text{ave}} = \frac{V}{A_d} = \frac{4 \times 10^6 \text{ m}^3}{(111 \text{ km}^2)\left(1000 \, \dfrac{\text{m}}{\text{km}}\right)^2}$$

$$= \boxed{0.036 \text{ m} \quad (3.6 \text{ cm})}$$

The answer is (C)

(b) The unit hydrograph discharge is the peak discharge divided by the average precipitation.

$$Q_{\text{unit hydrograph}} = \frac{Q_p}{P} = \frac{260 \, \dfrac{\text{m}^3}{\text{s}}}{3.6 \text{ cm}}$$

$$= \boxed{72.2 \text{ m}^3/\text{s·cm} \quad (72 \text{ m}^3/\text{s·cm})}$$

The answer is (A).

(c) The design flood hydrograph volume for a 6.5 cm storm is determined by multiplying the unit hydrograph volume by 6.5 cm. For the unit hydrograph,

$$V_{\text{hydrograph}} = \frac{V}{P} = \frac{4 \times 10^6 \text{ m}^3}{3.6 \text{ cm}}$$

$$= 1.11 \times 10^6 \text{ m}^3/\text{cm}$$

For the 6.5 cm storm,

$$V = \left(1.11 \times 10^6 \, \frac{\text{m}^3}{\text{cm}}\right)(6.5 \text{ cm}) = \boxed{7.2 \times 10^6 \text{ m}^3}$$

The answer is (B).

(d) The design discharge is determined by multiplying the unit hydrograph discharge by 6.5 cm.

$$Q_p = Q_{\text{hydrograph}}(6.5 \text{ cm}) = \left(72.2 \, \frac{\text{m}^3}{\text{s·cm}}\right)(6.5 \text{ cm})$$

$$= \boxed{469 \text{ m}^3/\text{s} \quad (470 \text{ m}^3/\text{s})}$$

The answer is (D).

(e) From Eq. 20.20,

$$p\{F \text{ event in } n \text{ years}\} = 1 - \left(1 - \frac{1}{F}\right)^n$$

$$p\{50 \text{ yr flood in 30 yr}\} = 1 - \left(1 - \frac{1}{50}\right)^{30}$$

$$= \boxed{0.45 \quad (45\%)}$$

The answer is (C).

(f) The unit hydrograph represents all discharge into a stream except for groundwater or base flow, which are separated out. Evapotranspiration refers to water that is returned to the atmosphere and, therefore, is not measured in the stream discharge. The unit hydrograph includes overland flow, surface flow, and interflow.

The answer is (D).

(g) The actual runoff is plotted. To separate base flow from overland flow, use the straight line method. Draw a horizontal line from the start of the rising limb to the falling limb. In the table, subtract the base flow from the overland flow.

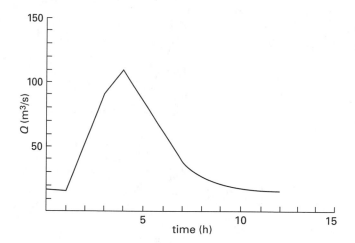

hour	runoff	ground water	surface water	surface water 3.14
0	17	16	1	0.32
1	16	16	0	0
2	48	16	32	10.19
3	90	16	74	23.56
4	108	16	92	29.29
5	85	16	69	21.97
6	58	16	42	13.37
7	38	16	22	7.00
8	26	16	10	3.18
9	20	16	4	1.27
10	18	16	2	0.64
11	17	16	1	0.32
12	16	16	0	0
	total		349 m³/s	

The average precipitation is given by Eq. 20.21.

$$P = \frac{V}{A_d}$$

The volume of runoff is the area under the separated hydrograph curve. Since data are given for each hour, the "width" of each histogram cell is 1 h.

$$V = \left(349 \, \frac{\text{m}^3}{\text{s}}\right)(1 \text{ h})\left(3600 \, \frac{\text{s}}{\text{h}}\right)$$

$$= 1\,256\,400 \text{ m}^3$$

$$P = \frac{1\,256\,400 \text{ m}^3}{(40 \text{ km}^2)\left(1000 \, \dfrac{\text{m}}{\text{km}}\right)^2}$$

$$= 0.0314 \text{ m} \quad (3.14 \text{ cm})$$

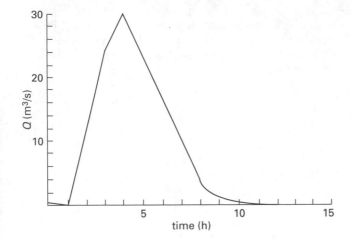

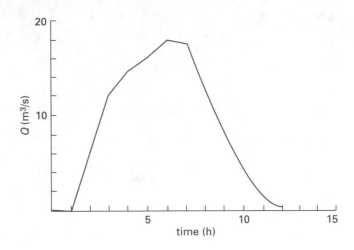

The unit hydrograph peak discharge is

$$\frac{108 \ \dfrac{m^3}{s} - 16 \ \dfrac{m^3}{s}}{3.14 \ cm} = \boxed{29.3 \ m^3/s{\cdot}cm \quad (29 \ m^3/s{\cdot}cm)}$$

The answer is (B).

(h) There are two methods for constructing hydrographs for a longer storm than that of the unit hydrograph. For storms that are whole multiples of the unit hydrograph duration, the lagging storm method is simplest.

In a table, add another unit hydrograph separated by $t_r = 3$ h. Then, add the ordinates for a hydrograph of $Nt_r = (2)(3 \ h) = 6$ h duration. The ordinates must then be divided by $N = 2$ to get the unit hydrograph ordinates.

hour	unit hydrograph 3 h storm	second storm	total	unit hydrograph 6 h storm
0	0.32	0	0.32	0.16
1	0	0	0	0
2	10.19	0	10.19	5.09
3	23.56	0.32	23.88	11.94
4	29.29	0	29.29	14.65
5	21.97	10.19	32.16	16.08
6	13.37	23.56	36.93	18.47
7	7.00	29.29	36.29	18.15
8	3.18	21.97	25.15	12.58
9	1.27	13.37	14.65	7.32
10	0.64	7.00	7.64	3.82
11	0.32	3.18	3.50	1.75
12	0	1.27	1.27	0.64
13	0	0.64	0.64	0.32
14	0	0.32	0.32	0.16
15	0	0	0	0

The peak discharge for a 5 cm storm is calculated from the peak unit hydrograph discharge.

$$Q_{p,\text{unit hydrograph}} = 18.47 \ m^3/s{\cdot}cm$$

$$Q_{p,5 \ cm \ storm} = (5 \ cm)\left(18.47 \ \frac{m^3}{s{\cdot}cm}\right)$$

$$= \boxed{92.35 \ m^3/s \quad (92 \ m^3/s)}$$

The answer is (A).

(i) If a storm is not a multiple of the unit hydrograph duration, the S-curve method must be used to construct a hydrograph for that storm. In the table, add the ordinates of five 3 h unit hydrographs, each offset by 3 h, to produce an S-curve.

hour	3 h unit hydrograph	second storm	third storm	fourth storm	fifth storm	total (S-curve)
0	0.32	0	0	0	0	0.32
1	0	0	0	0	0	0
2	10.19	0	0	0	0	10.19
3	23.56	0.32	0	0	0	23.88
4	29.29	0	0	0	0	29.29
5	21.97	10.19	0	0	0	32.16
6	13.37	23.56	0.32	0	0	37.25
7	7.00	29.29	0	0	0	36.29
8	3.18	21.97	10.19	0	0	35.33
9	1.27	13.37	23.56	0.32	0	38.52
10	0.64	7.00	29.29	0	0	36.93
11	0.32	3.18	21.97	10.19	0	35.66
12	0	1.27	13.37	23.56	0.32	38.52
13	0	0.64	7.00	29.29	0	36.93
14	0	0.32	3.18	21.97	10.19	35.66
15	0	0	1.27	13.37	23.56	38.20

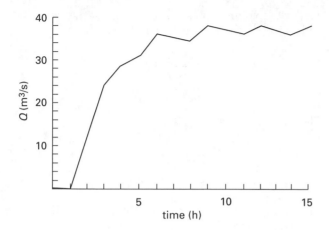

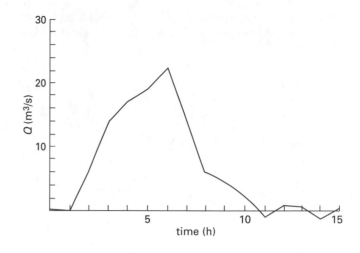

After approximately 5 h, the *S*-curve levels off to a steady value. (It actually alternates slightly between 35 m^3/s and 39 m^3/s.) The 5 h unit hydrograph is calculated by tabulating another *S*-curve lagging the first by 5 h, finding the difference, and scaling to a ratio of 3:5. (The negative values obtained reflect the approximate nature of the method.)

hour	*S*-curve	lagging *S*-curve	5 h unit hydrograph
0	0.32	0	0.19
1	0	0	0
2	10.19	0	6.11
3	23.88	0	14.32
4	29.29	0	17.57
5	32.16	0.32	19.10
6	37.25	0	22.35
7	36.29	10.19	15.66
8	35.33	23.88	6.88
9	38.52	29.29	5.54
10	36.93	32.16	2.87
11	35.66	37.25	−0.96
12	38.52	36.29	1.34
13	36.93	35.33	0.96
14	35.66	38.52	−1.72
15	38.20	36.93	0.96
		total	111.17

The peak discharge for a 5 cm storm is calculated from the peak unit hydrograph discharge.

$$Q_{p,\text{unit hydrograph}} = 22.35 \text{ m}^3/\text{s·cm} \quad \text{[per cm of rainfall]}$$

$$Q_{p,5\,\text{cm storm}} = (5 \text{ cm})\left(22.35 \ \frac{\text{m}^3}{\text{s·cm}}\right)$$

$$= \boxed{111.75 \text{ m}^3/\text{s} \quad (110 \text{ m}^3/\text{s})}$$

The answer is (C).

(j) The design flood volume is derived from the 5 h unit hydrograph.

$$V_{\text{unit hydrograph}} = \left(111.17 \ \frac{\text{m}^3}{\text{s·cm}}\right)(1 \text{ h})\left(3600 \ \frac{\text{s}}{\text{h}}\right)$$

$$= 402\,212 \text{ m}^3/\text{cm}$$

For a 5 cm storm,

$$V_{5\,\text{cm storm}} = \left(402\,212 \ \frac{\text{m}^3}{\text{cm}}\right)(5 \text{ cm})$$

$$= \boxed{2.00 \times 10^6 \text{ m}^3 \quad (2.0 \times 10^6 \text{ m}^3)}$$

The answer is (D).

2. (a) The actual hydrograph is a plot of time versus flow quantity, both of which are given in the problem statement.

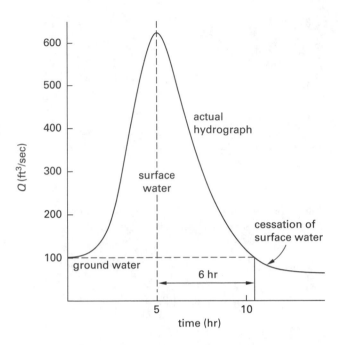

(b) Use the fixed-base method to separate the base flow from the overland flow. The base flow before the storm is projected to a point directly under the peak (in this case a more or less horizontal line), and then a straight line connects the projection to the falling limb. The connection point is between 5 hr and 7 hr later, so use 6 hr.

(c) The ordinates of the unit hydrograph are found by separating the base flow and then dividing the ordinates of the actual hydrograph by the average precipitation. The average precipitation is determined by tabulating the surface water flow and adding to find the total volume.

hour	runoff (ft^3/sec)	ground water (ft^3/sec)	surface water (ft^3/sec)	surface water/2.43 (ft^3/sec)
0	102	100	0	0
1	99	100	0	0
2	101	100	0	0
3	215	100	115	47.3
4	507	100	407	167.5
5	625	100	525	215.0
6	455	93	362	149.0
7	325	87	238	97.9
8	205	80	125	51.4
9	145	74	71	29.2
10	100	67	33	13.6
11	70	60	10	4.1
12	55	55	0	0
13	49	49	0	0
14	43	43	0	0
15	38	38	0	0
		total	$\overline{1886 \ ft^3/sec}$	

The total volume of surface water is the total flow multiplied by the time interval (1 hr).

$$V = \left(1886 \ \frac{ft^3}{sec}\right)(1 \ hr)\left(3600 \ \frac{sec}{hr}\right)$$

$$= 6.79 \times 10^6 \ ft^3$$

The watershed area is

$$A_d = (1.2 \ mi^2)\left(5280 \ \frac{ft}{mi}\right)^2$$

$$= 3.35 \times 10^7 \ ft^2$$

The average precipitation is the total volume divided by the drainage area.

$$P = \frac{(6.79 \times 10^6 \ ft^3)\left(12 \ \frac{in}{ft}\right)}{3.35 \times 10^7 \ ft^2}$$

$$= 2.43 \ in$$

The ordinates for the unit hydrograph are tabulated in the last column of the table.

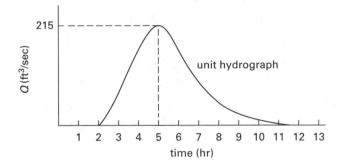

(d) The time base for direct runoff is approximately $11 \ hr - 2 \ hr = 9 \ hr$. The time base includes all the time in which direct runoff is observed, that is, both the rising and falling limbs.

The actual hydrograph is as shown. From the graph, the recession limb starts at $t = 5$ and continues to $t = 11$. Therefore, the length of the direct runoff recession limb is $\boxed{6 \ hr.}$

The answer is (B).

3. Use Table 20.2.

$$K = 170$$

$$b = 23$$

From Eq. 20.14, the intensity is

$$I = \frac{K}{t_c + b} = \frac{170\ \frac{\text{in-min}}{\text{hr}}}{60\ \text{min} + 23\ \text{min}}$$

$$= 2.05\ \text{in/hr}$$

$$A_d = (0.5)(640\ \text{ac}) = 320\ \text{ac}$$

From Eq. 20.36,

$$Q_p = CIA_d = (0.6)\left(2.05\ \frac{\text{in}}{\text{hr}}\right)(320\ \text{ac})$$

$$= \boxed{394\ \text{ft}^3/\text{sec} \quad (390\ \text{ft}^3/\text{sec})}$$

The answer is (B).

4.

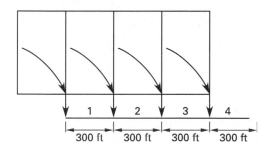

- First inlet:

Peak flow for the first inlet begins at $t_c = 15$ min.

$$I = \frac{K}{t_c + b} = \frac{100\ \frac{\text{in-min}}{\text{hr}}}{15\ \text{min} + 10\ \text{min}}$$

$$= 4\ \text{in/hr}$$

From Eq. 20.36, the runoff is

$$Q_p = CIA_d = (0.55)\left(4\ \frac{\text{in}}{\text{hr}}\right)(5\ \text{ac}) = 11\ \text{ft}^3/\text{sec}$$

Use Eq. 19.16.

$$D = 1.335\left(\frac{nQ}{\sqrt{S}}\right)^{3/8} = (1.335)\left(\frac{(0.013)\left(11\ \frac{\text{ft}^3}{\text{sec}}\right)}{\sqrt{0.005}}\right)^{3/8}$$

$$= 1.74\ \text{ft} \quad (21\ \text{in})$$

$$v_{\text{full}} = \frac{Q}{A} = \frac{11\ \frac{\text{ft}^3}{\text{sec}}}{\left(\frac{\pi}{4}\right)\left(\frac{21\ \text{in}}{12\ \frac{\text{in}}{\text{ft}}}\right)^2}$$

$$= 4.6\ \text{ft/sec}$$

(An integer-inch pipe diameter must be used to calculate velocity, because 1.73 ft diameter pipes are not manufactured.)

- Second inlet:

The flow time from inlet 1 to inlet 2 is

$$t = \frac{L}{v} = \frac{300\ \text{ft}}{\left(4.6\ \frac{\text{ft}}{\text{sec}}\right)\left(60\ \frac{\text{sec}}{\text{min}}\right)} = 1.09\ \text{min}$$

The intensity at $t_c = 15.0$ min + 1.09 min = 16.09 min is

$$I = \frac{K}{t_c + b} = \frac{100\ \frac{\text{in-min}}{\text{hr}}}{16.09\ \text{min} + 10\ \text{min}}$$

$$= 3.83\ \text{in/hr}$$

From Eq. 20.36, the runoff is

$$Q = CIA_d = (0.55)\left(3.83\ \frac{\text{in}}{\text{hr}}\right)(10\ \text{ac})$$

$$= 21.07\ \text{ft}^3/\text{sec}$$

Use Eq. 19.16.

$$D = 1.335\left(\frac{nQ}{\sqrt{S}}\right)^{3/8}$$

$$= (1.335)\left(\frac{(0.013)\left(21.07\ \frac{\text{ft}^3}{\text{sec}}\right)}{\sqrt{0.005}}\right)^{3/8}$$

$$= 2.22\ \text{ft} \quad (27\ \text{in})$$

$$v_{\text{full}} = \frac{Q}{A} = \frac{21.07\ \frac{\text{ft}^3}{\text{sec}}}{\left(\frac{\pi}{4}\right)\left(\frac{27\ \text{in}}{12\ \frac{\text{in}}{\text{ft}}}\right)^2}$$

$$\approx 5.3\ \text{ft/sec}$$

- Third inlet:

The flow time from inlet 2 is

$$t = \frac{L}{v} = \frac{300\ \text{ft}}{\left(5.3\ \frac{\text{ft}}{\text{sec}}\right)\left(60\ \frac{\text{sec}}{\text{min}}\right)} = 0.94\ \text{min}$$

The intensity at $t_c = 16.09 \text{ min} + 0.94 \text{ min} = 17.03 \text{ min}$ is

$$I = \frac{K}{t_c + b} = \frac{100 \ \frac{\text{in-min}}{\text{hr}}}{17.03 \ \text{min} + 10 \ \text{min}}$$
$$= 3.70 \ \text{in/hr}$$

From Eq. 20.36, the runoff is

$$Q = CIA_d = (0.55)\left(3.70 \ \frac{\text{in}}{\text{hr}}\right)(15 \ \text{ac})$$
$$= 30.53 \ \text{ft}^3/\text{sec}$$

Use Eq. 19.16.

$$D = 1.335\left(\frac{nQ}{\sqrt{S}}\right)^{3/8}$$
$$= (1.335)\left(\frac{(0.013)\left(30.53 \ \frac{\text{ft}^3}{\text{sec}}\right)}{\sqrt{0.005}}\right)^{3/8}$$
$$= 2.55 \ \text{ft} \quad (31 \ \text{in})$$

$$v_{\text{full}} = \frac{Q}{A} = \frac{30.53 \ \frac{\text{ft}^3}{\text{sec}}}{\left(\frac{\pi}{4}\right)\left(\frac{31 \ \text{in}}{12 \ \frac{\text{in}}{\text{ft}}}\right)^2}$$
$$\approx 5.82 \ \text{ft/sec}$$

- Fourth inlet:

The flow time from inlet 3 is

$$t = \frac{L}{v} = \frac{300 \ \text{ft}}{\left(5.82 \ \frac{\text{ft}}{\text{sec}}\right)\left(60 \ \frac{\text{sec}}{\text{min}}\right)} = 0.86 \ \text{min}$$

The intensity at $t_c = 17.03 \text{ min} + 0.86 \text{ min} = 17.89 \text{ min}$ is

$$I = \frac{K}{t_c + b} = \frac{100 \ \frac{\text{in-min}}{\text{hr}}}{17.89 \ \text{min} + 10 \ \text{min}}$$
$$= 3.59 \ \text{in/hr}$$

From Eq. 20.36, the runoff is

$$Q = CIA_d = (0.55)\left(3.59 \ \frac{\text{in}}{\text{hr}}\right)(20 \ \text{ac})$$
$$= 39.49 \ \text{ft}^3/\text{sec}$$

Use Eq. 19.16.

$$D = 1.335\left(\frac{nQ}{\sqrt{S}}\right)^{3/8}$$
$$= (1.335)\left(\frac{(0.013)\left(39.49 \ \frac{\text{ft}^3}{\text{sec}}\right)}{\sqrt{0.005}}\right)^{3/8}$$
$$= 2.81 \ \text{ft} \quad (33.7 \ \text{in})$$

Use $\boxed{36 \ \text{in}}$ (standard size) pipe.

The answer is (C).

5.
$$I = 5.3 \ \text{in/hr} \quad \text{[given]}$$
$$C = 0.5 \quad \text{[given]}$$
$$Q = CIA = (0.5)\left(5.3 \ \frac{\text{in}}{\text{hr}}\right)(75 \ \text{ac})$$
$$= 198.8 \ \text{ac-in/hr} \quad (200 \ \text{ft}^3/\text{sec})$$

(a) Take a worst-case approach. See if the culvert has a capacity equal to or greater than 200 ft³/sec when water is at the elevation of the roadway.

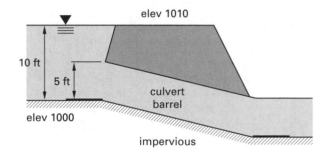

Disregard barrel friction. $K_e = 0.90$ for the projecting, square end. Assume type 6 flow.

$$H = 10 \ \text{ft} + (36 \ \text{ft})(0.01) - 5 \ \text{ft} = 5.36 \ \text{ft}$$

From Eq. 19.101,

$$v = \sqrt{\frac{H}{\frac{1 + k_e}{2g}}} = \sqrt{\frac{5.36 \ \text{ft}}{\frac{1 + 0.9}{(2)\left(32.2 \ \frac{\text{ft}}{\text{sec}^2}\right)}}}$$
$$= 13.48 \ \text{ft/sec}$$

$$Q = Av = \left(\frac{\pi}{4}\right)(5 \ \text{ft})^2\left(13.48 \ \frac{\text{ft}}{\text{sec}}\right) = 265 \ \text{ft}^3/\text{sec}$$

The excess capacity is

$$265 \ \frac{ft^3}{sec} - 200 \ \frac{ft^3}{sec} = 65 \ ft^3/sec \quad [\text{say } 70 \ ft^3/sec]$$

There will be no flooding, as the culvert has excess capacity. The water cannot be maintained at elevation 1010.

The answer is (D).

- Alternate method:

  From Table 17.5,

  $$C_d = 0.72$$

  $$H = 5.36 \ ft$$

  $$Q = C_d A \sqrt{2gh} = (0.72)\left(\frac{\pi}{4}\right)(5 \ ft)^2$$

  $$\times \sqrt{(2)\left(32.2 \ \frac{ft}{sec^2}\right)(5.36 \ ft)}$$

  $$= 263 \ ft^3/sec$$

(b) For the rubble channel,

$$n = 0.017$$

$$S = 0.01$$

$$R = \frac{5d}{2d+5}$$

$$C = \left(\frac{1.49}{0.017}\right)\left(\frac{5d}{2d+5}\right)^{1/6}$$

$$Q = Av = (5d)\left(\frac{1.49}{0.017}\right)\left(\frac{5d}{2d+5}\right)^{2/3}\sqrt{0.01}$$

$$200 \ \frac{ft^3}{sec} = (43.8d)\left(\frac{5d}{2d+5}\right)^{2/3}$$

By trial and error,

$$d = \text{depth in channel} = \boxed{3.55 \ ft \quad (3.6 \ ft)}$$

The answer is (C).

(c) The full capacity of the culvert is calculated from the Chezy-Manning equation.

$$R = \frac{D}{4} = \frac{5 \ ft}{4} = 1.25 \ ft$$

$$A = \frac{\pi}{4}D^2 = \left(\frac{\pi}{4}\right)(5 \ ft)^2 = 19.63 \ ft^2$$

$$n = 0.013$$

From Eq. 19.13,

$$Q_{full} = \left(\frac{1.49}{n}\right)AR^{2/3}\sqrt{S}$$

$$= \left(\frac{1.49}{0.013}\right)(19.63 \ ft^2)(1.25 \ ft)^{2/3}\sqrt{0.01}$$

$$= 261 \ ft^3/sec$$

This is close to (but not the same as) the 265 ft^3/sec calculated for the pressure flow.

$$\frac{Q}{Q_{full}} = \frac{200 \ \dfrac{ft^3}{sec}}{261 \ \dfrac{ft^3}{sec}} = 0.77$$

Use the table of circular channel ratios. From App. 19.C, the depth is

$$\frac{d}{D} \approx 0.72$$

$$d_{barrel} = (0.72)(5 \ ft) = \boxed{3.6 \ ft}$$

The answer is (C).

6. Since the first storm produces 1 in net, and the second storm produces 2 in, divide all runoffs by two. Offset the second storm by 4 hr.

hour	first storm	second storm	total
0	0		0
2	50		50
4	175	0	175
6	300	100	400
8	210	350	560
10	150	600	750
12	125	420	545
14	75	300	375
16	50	250	300
18	≈ 38	150	188
20	25	100	125
22	≈ 12	≈ 75	87
24	0	50	50
26		≈ 25	25
28		0	0

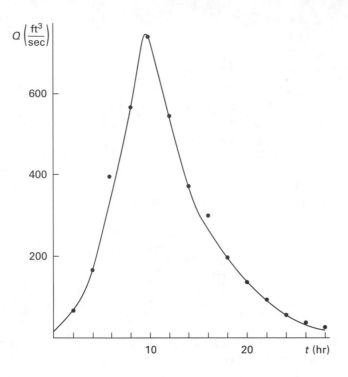

Choose 12 random numbers that are less than 100 from App. 20.B. Use the third row, reading to the right.

- Inflow distribution:

month	random number	corresponding midpoint
1	06	0.55
2	40	0.85
3	18	0.75
4	73	1.05
5	97	1.25
6	72	1.05
7	89	1.15
8	83	1.05
9	24	0.75
10	41	0.85
11	88	1.15
12	86	1.15

Simulate the reservoir operation.

month	starting volume	+ inflow	− constant use	= ending volume	+ spill
1	5.5	0.55	0.7	5.35	
2	5.35	0.85	0.7	5.5	
3	5.5	0.75	0.7	5.55	
4	5.55	1.05	0.7	5.9	
5	5.9	1.25	0.7	6.45	
6	6.45	1.05	0.7	6.8	
7	6.8	1.15	0.7	7.0	0.25
8	7.0	1.05	0.7	7.0	0.35
9	7.0	0.75	0.7	7.0	0.05
10	7.0	0.85	0.7	7.0	0.15
11	7.0	1.15	0.7	7.0	0.45
12	7.0	1.15	0.7	7.0	0.45

7. From Eq. 20.48,

$$E_R = K_p E_p = (0.7)(0.8 \text{ in}) = \boxed{0.56 \text{ in}}$$

The answer is (A).

8. For 99.97% confidence, $z = \pm 3$. The inflow distribution is a normal distribution that extends $\pm 3\sigma$, or from 0.3 to 1.5. Take the cell width as $^1/_2 \sigma$ or 0.1. Then, for the first cell,

actual endpoints: 0.3 to 0.4
midpoint: 0.35
z limits: -3.0 to -2.5
area under curve: $0.5 - 0.49 = 0.01$

The following inflow distribution is produced similarly.

endpoints	midpoint	z limits	area under curve	cum. $\times 100$
0.3 to 0.4	0.35	-3.0 to -2.5	0.01	1
0.4 to 0.5	0.45	-2.5 to -2.0	0.02	3
0.5 to 0.6	0.55	-2.0 to -1.5	0.04	7
0.6 to 0.7	0.65	-1.5 to -1.0	0.09	16
0.7 to 0.8	0.75	-1.0 to -0.5	0.15	31
0.8 to 0.9	0.85	-0.5 to 0	0.19	50
0.9 to 1.0	0.95	0 to 0.5	0.19	69
1.0 to 1.1	1.05	0.5 to 1.0	0.15	84
1.1 to 1.2	1.15	1.0 to 1.5	0.09	93
1.2 to 1.3	1.25	1.5 to 2.0	0.04	97
1.3 to 1.4	1.35	2.0 to 2.5	0.02	99
1.4 to 1.5	1.45	2.5 to 3.0	0.01	100

9. Use the same simulation procedure for inflow as in Prob. 8. Proceed similarly.

- Demand distribution:

endpoints	midpoint	z limits	cum. $\times 100$
0.1 to 0.2	0.15	-3.0 to -2.5	1
0.2 to 0.3	0.25	-2.5 to -2.0	3
0.3 to 0.4	0.35	-2.0 to -1.5	7
0.4 to 0.5	0.45	-1.5 to -1.0	16
0.5 to 0.6	0.55	-1.0 to -0.5	31
0.6 to 0.7	0.65	-0.5 to 0	50
0.7 to 0.8	0.75	0 to 0.5	69
0.8 to 0.9	0.85	0.5 to 1.0	84
0.9 to 1.0	0.95	1.0 to 1.5	93
1.0 to 1.1	1.05	1.5 to 2.0	97
1.1 to 1.2	1.15	2.0 to 2.5	99
1.2 to 1.3	1.25	2.5 to 3.0	100

Choose 12 random numbers. Use the fourth row, reading to the right.

- Demand distribution:

month	random number	use
1	04	0.35
2	75	0.85
3	41	0.65
4	44	0.65
5	89	0.95
6	39	0.65
7	42	0.65
8	09	0.45
9	42	0.65
10	11	0.45
11	58	0.75
12	04	0.35

Simulate the reservoir operation.

month	starting volume	+ inflow	− monthly demand	= ending volume	+ spill
1	5.5	0.55	0.35	5.7	
2	5.7	0.85	0.85	5.7	
3	5.7	0.75	0.65	5.8	
4	5.8	1.05	0.65	6.2	
5	6.2	1.25	0.95	6.5	
6	6.5	1.05	0.65	6.9	
7	6.9	1.15	0.65	7.0	0.4
8	7.0	1.05	0.45	7.0	0.6
9	7.0	0.75	0.65	7.0	0.1
10	7.0	0.85	0.45	7.0	0.4
11	7.0	1.15	0.75	7.0	0.4
12	7.0	1.15	0.35	7.0	0.8

10. Note that all dates correspond to the end of the month. Draw the mass diagram.

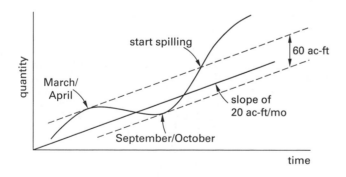

If the reservoir had been full in March or April, the maximum shortfall would have been $\boxed{60 \text{ ac-ft,}}$ and the reservoir would be essentially empty in September and October.

The answer is (B).

11. (a) Assume the reservoir is initially empty. Draw the mass diagram.

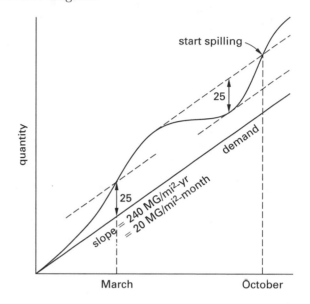

$$\text{capacity} = \boxed{25 \text{ MG/mi}^2}$$

The answer is (A).

(b) The reservoir would start spilling around the end of March. It would also begin spilling around the end of October.

The answer is (D).

PRACTICE PROBLEMS

Wells

1. A well extends from the ground surface at elevation 383 ft through a gravel bed to a layer of bedrock at elevation 289 ft. The screened well is 1500 ft from a river whose surface level is 363 ft. The well is pumped by a 10 in diameter schedule-40 steel pipe which draws 120,000 gal/day. The hydraulic conductivity of the well is 1600 gal/day-ft^2. The pump discharges into a piping network whose friction head is 100 ft. What net power is required for steady flow?

    (A) 1.1 hp

    (B) 2.6 hp

    (C) 7.8 hp

    (D) 15 hp

2. (*Time limit: one hour*) An aquifer consists of a homogeneous material that is 300 ft thick. The surface of the water table in this aquifer is 100 ft below ground surface. An 18 in diameter well extends through the top 100 ft and then 200 ft below the water table, for a total depth of 300 ft. The aquifer transmissivity is 10,000 gal/day-ft. The well's radius of influence is 900 ft with a 20 ft drawdown at the well.

(a) The geologic formation in which the well is installed is called

    (A) an aquifuge

    (B) an artesian well

    (C) a connate aquifer

    (D) an unconfined aquifer

(b) What is the hydraulic conductivity of the aquifer?

    (A) 50 gal/day-ft^2

    (B) 500 gal/day-ft^2

    (C) 1300 gal/day-ft^2

    (D) 10,000 gal/day-ft^2

(c) What steady discharge is possible?

    (A) 0.26 ft^3/sec

    (B) 0.52 ft^3/sec

    (C) 1.2 ft^3/sec

    (D) 5.2 ft^3/sec

(d) What is the drawdown 100 ft from the well?

    (A) 1 ft

    (B) 6 ft

    (C) 13 ft

    (D) 18 ft

(e) Assuming a reasonable pump efficiency, what horse-power motor should be selected to achieve the steady discharge?

    (A) 3.6 hp

    (B) 5.5 hp

    (C) 7.5 hp

    (D) 12.2 hp

The following applies to parts (f) through (j).

The same well conditions are found at an adjacent site where the aquifer extends 100 ft below a layer of low permeability that is 200 ft thick. The piezometric surface is 100 ft below the ground surface.

(f) The geologic formation in which the well is installed is called

    (A) an aquifuge

    (B) a confined aquifer

    (C) a spring

    (D) a vadose well

(g) The 18 in diameter well extends 300 ft below the ground surface to the bottom of the aquifer. The aquifer transmissivity is 10,000 gal/day-ft. What is the hydraulic conductivity of the aquifer?

    (A) 50 gal/day-ft^2

    (B) 100 gal/day-ft^2

    (C) 500 gal/day-ft^2

    (D) 800 gal/day-ft^2

(h) What is the steady discharge in the well if the radius of influence is 900 ft with a 20 ft drawdown at the well?

(A) 0.27 ft³/sec

(B) 0.62 ft³/sec

(C) 1.2 ft³/sec

(D) 3.5 ft³/sec

(i) At what distance from the well is the drawdown equal to 10 ft?

(A) 3 ft

(B) 26 ft

(C) 130 ft

(D) 450 ft

(j) After pumping is stopped, groundwater flows in one direction through the aquifer. If the aquifer is only 100 ft wide and the porosity is 0.4, what is the area in clear flow?

(A) 2000 ft²

(B) 4000 ft²

(C) 8000 ft²

(D) 20,000 ft²

Flow Nets

3. (*Time limit: one hour*) An impervious concrete dam is shown. The dam reduces seepage by using two impervious sheets extending 5 m below the dam bottom.

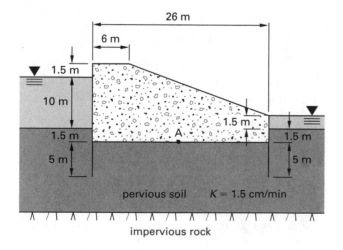

(a) Sketch the flow net.

(b) Determine the seepage per meter of width.

(A) 0.03 m³/min

(B) 0.1 m³/min

(C) 0.3 m³/min

(D) 0.9 m³/min

(c) What is the uplift pressure on the dam at point A, midway between the left and right edges?

(A) 17 kPa

(B) 33 kPa

(C) 71 kPa

(D) 91 kPa

4. (*Time limit: one hour*) A cofferdam in a river is shown. Sheet piles extend below the mud line, and the cofferdam floor is unlined. (The figure is drawn to scale.) Draw the initial flow net.

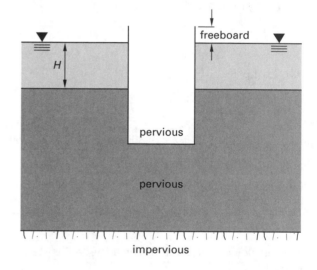

5. (*Time limit: one hour*) The concrete dam shown is drawn to scale. The silt has a hydraulic conductivity of 0.15 ft/hr.

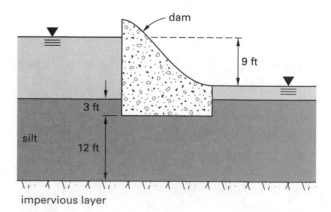

(a) Sketch the flow net for the dam.

(b) What is the approximate seepage rate per foot of width?

    (A) 0.05 ft^3/hr-ft width

    (B) 0.20 ft^3/hr-ft width

    (C) 0.40 ft^3/hr-ft width

    (D) 0.90 ft^3/hr-ft width

SOLUTIONS

1. The well is constructed as shown.

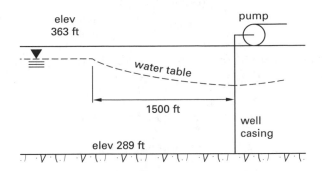

From App. 16.B, for schedule-40 pipe,

$$D_i = 0.835 \text{ ft}$$

$$A_i = 0.5476 \text{ ft}^2$$

$$Q = \left(120{,}000 \; \frac{\text{gal}}{\text{day}}\right)\left(1.547 \times 10^{-6} \; \frac{\text{ft}^3\text{-day}}{\text{sec-gal}}\right)$$

$$= 0.1856 \text{ ft}^3/\text{sec}$$

$$K_p = \left(1600 \; \frac{\text{gal}}{\text{day-ft}^2}\right)\left(0.1337 \; \frac{\text{ft}^3}{\text{gal}}\right)$$

$$= 213.9 \text{ ft}^3/\text{day-ft}^2$$

$$y_1 = 363 \text{ ft} - 289 \text{ ft} = 74 \text{ ft}$$

$$r_1 = 1500 \text{ ft}$$

$$r_2 = \frac{0.835 \text{ ft}}{2} = 0.4175 \text{ ft}$$

From Eq. 21.25,

$$Q = \frac{\pi K(y_1^2 - y_2^2)}{\ln \dfrac{r_1}{r_2}}$$

$$0.1856 \; \frac{\text{ft}^3}{\text{sec}} = \frac{\pi\left(213.9 \; \dfrac{\text{ft}^3}{\text{day-ft}^2}\right)\left((74 \text{ ft})^2 - y_2^2\right)}{\ln\left(\dfrac{1500 \text{ ft}}{0.4175 \text{ ft}}\right)\left(86{,}400 \; \dfrac{\text{sec}}{\text{day}}\right)}$$

$$195.36 \text{ ft}^2 = (74 \text{ ft})^2 - y_2^2$$

$$y_2 = 72.67 \text{ ft}$$

The drawdown is

$$d = 74 \text{ ft} - 72.67 \text{ ft} = 1.33 \text{ ft}$$

d is small compared to y.

The velocity of the water in the pipe is

$$v = \frac{Q}{A} = \frac{0.1856 \ \frac{ft^3}{sec}}{0.5476 \ ft^2} = 0.34 \ ft/sec$$

This velocity is too small to include the velocity head. The suction lift is

$$383 \ ft - 363 \ ft + 1.33 \ ft = 21.33 \ ft$$

$$H = 21.33 \ ft + 100 \ ft = 121.33 \ ft$$

The mass flow of the water is

$$\dot{m} = \left(0.1856 \ \frac{ft^3}{sec} \right) \left(62.4 \ \frac{lbm}{ft^3} \right) = 11.58 \ lbm/sec$$

From Table 18.5, the net water horsepower is

$$P = \frac{h_A \dot{m}}{550} \times \frac{g}{g_c}$$

$$= \frac{(121.33 \ ft) \left(11.58 \ \frac{lbm}{sec} \right)}{550 \ \frac{ft\text{-}lbf}{hp\text{-}sec}} \times \frac{32.2 \ \frac{ft}{sec^2}}{32.2 \ \frac{lbm\text{-}ft}{lbf\text{-}sec^2}}$$

$$= \boxed{2.55 \ hp \quad (2.6 \ hp)}$$

The answer is (B).

2. (a) The geologic formation is an unconfined aquifer. If the aquifer were overlain by an impermeable layer, it would be a confined, or artesian, aquifer.

The answer is (D).

(b) From Eq. 21.13,

$$K = \frac{T}{Y} = \frac{10,000 \ \frac{gal}{day\text{-}ft}}{200 \ ft}$$

$$= \boxed{50 \ gal/day\text{-}ft^2}$$

The answer is (A).

(c)
$$y_1 = 200 \ ft \ at \ r_1 = 900 \ ft$$
$$y_2 = 200 \ ft - 20 \ ft = 180 \ ft \ at \ r_2$$
$$r_2 = \frac{\frac{18 \ in}{2}}{12 \ \frac{in}{ft}} = 0.75 \ ft$$

From Eq. 21.25,

$$Q = \frac{\pi K (y_1^2 - y_2^2)}{\ln \frac{r_1}{r_2}}$$

$$= \frac{\pi \left(50 \ \frac{gal}{day\text{-}ft^2} \right) \left(0.1337 \ \frac{ft^3}{gal} \right)}{\ln \left(\frac{900 \ ft}{0.75 \ ft} \right) \left(86,400 \ \frac{sec}{day} \right)}$$

$$= \boxed{0.261 \ ft^3/sec \quad (0.26 \ ft^3/sec)}$$

The answer is (A).

(d)
$$y_1 = 200 \ ft \ at \ r_1$$
$$r_1 = 900 \ ft$$
$$r_2 = 100 \ ft$$
$$Q = 0.261 \ ft^3/sec$$

Rearranging Eq. 21.25,

$$y_2^2 = y_1^2 - \frac{Q \ln \frac{r_1}{r_2}}{\pi K}$$

$$= (200 \ ft)^2$$

$$- \frac{\left(0.261 \ \frac{ft^3}{sec} \right) \ln \left(\frac{900 \ ft}{100 \ ft} \right) \left(86,400 \ \frac{sec}{day} \right)}{\pi \left(50 \ \frac{gal}{day\text{-}ft^2} \right) \left(0.1337 \ \frac{ft^3}{gal} \right)}$$

$$y_2 = 194 \ ft$$

The drawdown is

$$200 \ ft - 194 \ ft = \boxed{6 \ ft}$$

The answer is (B).

(e) This is a low-capacity pump, so the efficiency will be low. Therefore, assume a pump efficiency of 0.65. From Table 18.5, the hydraulic horsepower is

$$P = \frac{h_A \dot{V} (SG)}{8.814}$$

$$= \frac{(100 \ ft + 20 \ ft) \left(0.261 \ \frac{ft^3}{sec} \right) (1)}{\left(8.814 \ \frac{sec}{ft^4\text{-}hp} \right) (0.65)}$$

$$= 5.47 \ hp$$

Choose a standard motor size of $\boxed{7.5 \ hp.}$

The answer is (C).

(f) The geologic formation is a confined aquifer. The well is an artesian well.

The answer is (B).

(g) The aquifer depth, Y, is 100 ft. Y is the thickness of the aquifer, not the height of the water table or piezometric surface.

From Eq. 21.13,

$$K = \frac{T}{Y}$$

$$= \frac{10{,}000 \ \dfrac{\text{gal}}{\text{day-ft}}}{100 \ \text{ft}}$$

$$= \boxed{100 \ \text{gal/day-ft}^2}$$

The answer is (B).

(h) The well radius of influence is

$$r_1 = 900 \ \text{ft}$$

$$y_1 = 200 \ \text{ft at } r_1$$

Calculate the well casing radius.

$$r_2 = \frac{\dfrac{18 \ \text{in}}{2}}{12 \ \dfrac{\text{in}}{\text{ft}}} = 0.75 \ \text{ft}$$

$$y_2 = 200 \ \text{ft} - 20 \ \text{ft} = 180 \ \text{ft at } r_2$$

From Eq. 21.27,

$$Q = \frac{2\pi K Y(y_1 - y_2)}{\ln \dfrac{r_1}{r_2}}$$

$$= \frac{2\pi (100 \ \text{ft}) \left(100 \ \dfrac{\text{gal}}{\text{day-ft}^2} \right)(200 \ \text{ft} - 180 \ \text{ft})}{\ln \left(\dfrac{900 \ \text{ft}}{0.75 \ \text{ft}} \right)}$$

$$= 177{,}239 \ \text{gal/day}$$

$$Q = \frac{\left(177{,}239 \ \dfrac{\text{gal}}{\text{day}} \right)\left(0.002228 \ \dfrac{\text{ft}^3\text{-min}}{\text{gal-sec}} \right)}{\left(24 \ \dfrac{\text{hr}}{\text{day}} \right)\left(60 \ \dfrac{\text{min}}{\text{hr}} \right)}$$

$$= \boxed{0.274 \ \text{ft}^3/\text{sec} \quad (0.27 \ \text{ft}^3/\text{sec})}$$

The answer is (A).

(i) Rearrange Eq. 21.27 to find the drawdown distance, r_2.

$$y_2 = 200 \ \text{ft} - 10 \ \text{ft} = 190 \ \text{ft}$$

$$\ln \frac{r_1}{r_2} = 2\pi K(y_1 - y_2)\left(\frac{Y}{Q} \right)$$

$$= \frac{2\pi \left(100 \ \dfrac{\text{gal}}{\text{day-ft}^2} \right)(200 \ \text{ft} - 190 \ \text{ft})(100 \ \text{ft})}{177{,}239 \ \dfrac{\text{gal}}{\text{day}}}$$

$$= 3.545$$

$$r_2 = \frac{r_1}{e^{3.545}} = \frac{900 \ \text{ft}}{e^{3.545}}$$

$$= \boxed{26 \ \text{ft}}$$

The answer is (B).

(j) Darcy's law assumes the total cross-sectional area. To obtain the cross-sectional area of the pores (voids), multiply by the porosity.

$$A_{\text{clear flow}} = nA = nbY$$

$$= (0.4)(100 \ \text{ft})(100 \ \text{ft})$$

$$= \boxed{4000 \ \text{ft}^2}$$

The answer is (B).

3. (a) One possible flow net is shown.

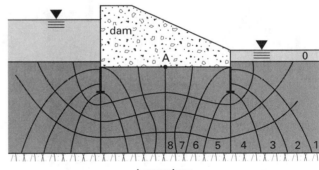

(b) Use Eq. 21.33 to solve for the flow rate.

$$Q = KH\left(\frac{N_f}{N_p} \right)$$

From the flow net as drawn, $N_f = 4$ and $N_p = 16$.

$$H = 10 \ \text{m} - 1.5 \ \text{m} = 8.5 \ \text{m}$$

$$Q = \frac{\left(1.5 \ \dfrac{\text{cm}}{\text{min}} \right)(8.5 \ \text{m})\left(\dfrac{4}{16} \right)\left(1 \ \dfrac{\text{m}}{\text{m}} \right)}{100 \ \dfrac{\text{cm}}{\text{m}}}$$

$$= 0.031875 \ \text{m}^3/\text{min}$$

Use $\boxed{0.03 \ \text{m}^3/\text{min}}$ per meter width of wall.

The answer is (A).

(c) Point A is located at an elevation 3 m lower than the downstream water level. From Eq. 21.36,

$$p_u = \left(\left(\frac{j}{N_p}\right)H + z\right)\rho_w g$$

$$= \left(\left(\frac{8}{16}\right)(8.5 \text{ m}) + 3.0 \text{ m}\right)\left(1000 \text{ } \frac{\text{kg}}{\text{m}^3}\right)\left(9.81 \text{ } \frac{\text{m}}{\text{s}^2}\right)$$

$$= \boxed{71\,122 \text{ Pa} \quad (71 \text{ kPa})}$$

The answer is (C).

4. The initial flow net is drawn as shown.

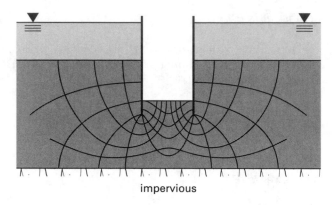

impervious

5. (a) The dam's flow net is drawn as shown.

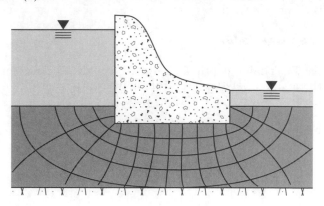

(b) From Eq. 21.33, with 4 flow paths and 13 equipotential drops,

$$Q = KH\left(\frac{N_f}{N_p}\right)$$

$$= \left(0.15 \text{ } \frac{\text{ft}}{\text{hr}}\right)(9.0 \text{ ft})\left(\frac{4}{13}\right)$$

$$= \boxed{0.42 \text{ ft}^3/\text{hr-ft width} \quad (0.40 \text{ ft}^3/\text{hr-ft width})}$$

The answer is (C).

22 Inorganic Chemistry

PRACTICE PROBLEMS

Empirical Formula Development

1. The gravimetric analysis of a compound is 40% carbon, 6.7% hydrogen, and 53.3% oxygen. What is the simplest formula for the compound?

(A) HCO

(B) HCO_2

(C) CH_2O

(D) CHO_2

SOLUTIONS

1. Calculate the mole ratios of the atoms by assuming there are 100 g of sample.

For 100 g of sample,

substance	mass	$\dfrac{m}{AW}$ = no. moles	mole ratio
C	40 g	$\dfrac{40}{12} = 3.33$	1
H	6.7 g	$\dfrac{6.7}{1} = 6.7$	2
O	53.3 g	$\dfrac{53.3}{16} = 3.33$	1

The empirical formula is $\boxed{CH_2O.}$

The answer is (C).

23 Organic Chemistry

PRACTICE PROBLEMS

There are no problems in this book corresponding to Chap. 23 of the *Civil Engineering Reference Manual*.

24 Combustion and Incineration

PRACTICE PROBLEMS

1. Methane (MW = 16.043) with a heating value of 24,000 Btu/lbm (55.8 MJ/kg) is burned with a 50% efficiency. If the heat of vaporization of any water vapor formed is recovered, how much water (specific heat of 1 Btu/lbm-°F (4.1868 kJ/kg·°C)) can be heated from 60°F to 200°F (15°C to 95°C) when 7 ft^3 (200 L) at 60°F and 14.73 psia are burned?

(A) 25 lbm (11 kg)

(B) 35 lbm (16 kg)

(C) 50 lbm (23 kg)

(D) 95 lbm (43 kg)

2. 15 lbm/hr (6.8 kg/h) of propane (C$_3$H$_8$, MW = 44.097) is burned stoichiometrically in air. What volume of dry carbon dioxide (CO$_2$, MW = 44.011) is formed after cooling to 70°F (21°C) and 14.7 psia (101 kPa)?

(A) 180 ft^3/hr (5.0 m^3/h)

(B) 270 ft^3/hr (7.6 m^3/h)

(C) 390 ft^3/hr (11 m^3/h)

(D) 450 ft^3/hr (13 m^3/h)

3. In a particular installation, 30% excess air at 15 psia (103 kPa) and 100°F (40°C) is needed for the combustion of methane. How much nitrogen (MW = 28.016) passes through the furnace if methane is burned at the rate of 4000 ft^3/hr (31 L/s)?

(A) 270 lbm/hr (0.033 kg/s)

(B) 930 lbm/hr (0.11 kg/s)

(C) 1800 lbm/hr (0.22 kg/s)

(D) 2700 lbm/hr (0.34 kg/s)

4. Propane (C$_3$H$_8$) is burned with 20% excess air. What is the gravimetric percentage of carbon dioxide in the flue gas?

(A) 8%

(B) 12%

(C) 15%

(D) 22%

5. (*Time limit: one hour*) A natural gas is 93% methane, 3.73% nitrogen, 1.82% hydrogen, 0.45% carbon monoxide, 0.35% oxygen, 0.25% ethylene, 0.22% carbon dioxide, and 0.18% hydrogen sulfide by volume. The gas is burned with 40% excess air. Atmospheric air is 60°F and at standard atmospheric pressure.

(a) What is the gas density?

(A) 0.017 lbm/ft^3

(B) 0.043 lbm/ft^3

(C) 0.069 lbm/ft^3

(D) 0.110 lbm/ft^3

(b) What are the theoretical air requirements?

(A) 5 ft^3 air/ft^3 fuel

(B) 7 ft^3 air/ft^3 fuel

(C) 9 ft^3 air/ft^3 fuel

(D) 13 ft^3 air/ft^3 fuel

(c) What is the percentage of CO$_2$ in the flue gas (wet basis)?

(A) 6.9%

(B) 7.7%

(C) 8.1%

(D) 11%

(d) What is the percentage of CO$_2$ in the flue gas (dry basis)?

(A) 6.9%

(B) 7.7%

(C) 8.1%

(D) 11%

6. (*Time limit: one hour*) A utility boiler burns coal with an ultimate analysis of 76.56% carbon, 7.7% oxygen, 6.1% silicon, 5.5% hydrogen, 2.44% sulfur, and 1.7% nitrogen. 410 lbm/hr of refuse are removed with a composition of 30% carbon and 0% sulfur. All the sulfur and the remaining carbon is burned. The power plant has the following characteristics.

> coal feed rate: 15,300 lbm/hr
> electric power rating: 17 MW
> generator efficiency: 95%
> steam generator efficiency: 86%
> cooling water rate: 225 ft^3/sec

(a) What is the emission rate of solid particulates in lbm/hr?

- (A) 23 lbm/hr
- (B) 150 lbm/hr
- (C) 810 lbm/hr
- (D) 1700 lbm/hr

(b) How much sulfur dioxide is produced per hour?

- (A) 220 lbm/hr
- (B) 340 lbm/hr
- (C) 750 lbm/hr
- (D) 1100 lbm/hr

(c) What is the temperature rise of the cooling water?

- (A) 2.4°F
- (B) 6.5°F
- (C) 9.8°F
- (D) 13°F

(d) What efficiency must the flue gas particulate collectors have in order to meet a limit of 0.1 lbm of particulates per million Btu per hour (0.155 kg/MW)?

- (A) 93.1%
- (B) 97.4%
- (C) 98.8%
- (D) 99.1%

SOLUTIONS

1. *Customary U.S. Solution*

$$T = 60°F + 460° = 520°R$$

$$p = 14.73 \text{ psia}$$

The specific gas constant, R, is calculated from the universal gas constant, R^*, and the molecular weight.

$$R = \frac{R^*}{\text{MW}} = \frac{1545.35 \ \frac{\text{ft-lbf}}{\text{lbmol-°R}}}{16.043 \ \frac{\text{lbm}}{\text{lbmol}}}$$

$$= 96.33 \text{ ft-lbf/lbm-°R}$$

$$m = \frac{pV}{RT}$$

$$= \frac{\left(14.73 \ \frac{\text{lbf}}{\text{in}^2}\right)\left(12 \ \frac{\text{in}}{\text{ft}}\right)^2 (7 \text{ ft}^3)}{\left(96.33 \ \frac{\text{ft-lbf}}{\text{lbm-°R}}\right)(520°R)}$$

$$= 0.296 \text{ lbm}$$

The energy available from methane is

$$Q = \eta m(\text{HHV})$$

$$= (0.5)(0.296 \text{ lbm})\left(24{,}000 \ \frac{\text{Btu}}{\text{lbm}}\right)$$

$$= 3552 \text{ Btu}$$

This energy is used by water to heat from 60°F to 200°F.

$$Q = m_{\text{water}} c_p (T_2 - T_1)$$

$$m_{\text{water}} = \frac{3552 \text{ Btu}}{\left(1 \ \frac{\text{Btu}}{\text{lbm-°F}}\right)(200°F - 60°F)}$$

$$= \boxed{25.37 \text{ lbm} \quad (25 \text{ lbm})}$$

The answer is (A).

SI Solution

$$T = (60°F + 460°)\left(\frac{1\text{K}}{1.8°R}\right) = 288.9\text{K}$$

$$p = (14.73 \text{ psia})\left(\frac{101.3 \text{ kPa}}{14.7 \text{ psia}}\right) = 101.51 \text{ kPa}$$

The specific gas constant, R, is calculated from the universal gas constant, R^*, and the molecular weight.

$$R = \frac{R^*}{MW} = \frac{8314.5 \; \frac{J}{kmol \cdot K}}{16.043 \; \frac{kg}{kmol}}$$

$$= 518.26 \; J/kg \cdot K$$

$$m = \frac{pV}{RT}$$

$$= \frac{(101.51 \; kPa)\left(1000 \; \frac{Pa}{kPa}\right)(200 \; L)}{\left(518.26 \; \frac{J}{kg \cdot K}\right)(288.9K)\left(1000 \; \frac{L}{m^3}\right)}$$

$$= 0.136 \; kg$$

The energy available from methane is

$$Q = \eta m (HHV)$$

$$= (0.5)(0.136 \; kg)\left(55.8 \; \frac{MJ}{kg}\right)\left(1000 \; \frac{kJ}{MJ}\right)$$

$$= 3794 \; kJ$$

This energy is used by water to heat it from 15°C to 95°C.

$$Q = m_{water} \, c_p (T_2 - T_1)$$

$$m_{water} = \frac{3794 \; kJ}{\left(4.1868 \; \frac{kJ}{kg \cdot C}\right)(95°C - 15°C)}$$

$$= \boxed{11.33 \; kg \quad (11 \; kg)}$$

The answer is (A).

2. *Customary U.S. Solution*

From Table 24.7,

C$_3$H$_8$	+	5O$_2$	→	3CO$_2$	+	4H$_2$O
MW 44.097		(5)(32)		(3)(44.011)		
44.097		160		132.033		

The amount of carbon dioxide produced is 132.033 lbm/44.097 lbm propane. For 15 lbm/hr of propane, the amount of carbon dioxide produced is

$$\left(\frac{132.033 \; lbm}{44.097 \; lbm}\right)\left(15 \; \frac{lbm}{hr}\right) = 44.91 \; lbm/hr$$

$$R = \frac{R^*}{MW} = \frac{1545.35 \; \frac{ft\text{-}lbf}{lbmol\text{-}°R}}{44.011 \; \frac{lbm}{lbmol}}$$

$$= 35.11 \; ft\text{-}lbf/lbm\text{-}°R$$

$$T = 70°F + 460° = 530°R$$

$$\dot{V} = \frac{\dot{m}RT}{p}$$

$$= \frac{\left(44.91 \; \frac{lbm}{hr}\right)\left(35.11 \; \frac{ft\text{-}lbf}{lbm\text{-}°R}\right)(530°R)}{\left(14.7 \; \frac{lbf}{in^2}\right)\left(12 \; \frac{in}{ft}\right)^2}$$

$$= \boxed{394.8 \; ft^3/hr \quad (390 \; ft^3/hr)}$$

The answer is (C).

SI Solution

From Table 24.7,

C$_3$H$_8$	+	5O$_2$	→	3CO$_2$	+	4H$_2$O
MW 44.097		(5)(32)		(3)(44.011)		
44.097		160		132.033		

The amount of carbon dioxide produced is 132.033 kg/44.097 kg propane. For 6.8 kg/h of propane, the amount of carbon dioxide produced is

$$\left(\frac{132.033 \; kg}{44.097 \; kg}\right)\left(6.8 \; \frac{kg}{h}\right) = 20.36 \; kg/h$$

$$R = \frac{R^*}{MW} = \frac{8314.5 \; \frac{J}{kmol \cdot K}}{44.01 \; \frac{kg}{kmol}}$$

$$= 188.92 \; J/kg \cdot K$$

$$T = 21°C + 273° = 294K$$

$$V = \frac{mRT}{p}$$

$$= \frac{\left(20.36 \; \frac{kg}{h}\right)\left(188.92 \; \frac{J}{kg \cdot K}\right)(294K)}{(101 \; kPa)\left(1000 \; \frac{Pa}{kPa}\right)}$$

$$= \boxed{11.20 \; m^3/h \quad (11 \; m^3/h)}$$

The answer is (C).

3. Use the balanced chemical reaction equation from Table 24.7.

$$CH_4 + 2O_2 \rightarrow CO_2 + 2H_2O$$

Use Table 24.6 and Table 24.7. With 30% excess air and considering that there are 3.773 volumes of nitrogen for every volume of oxygen, the reaction equation is

$$CH_4 + (1.3)(2)O_2 + (1.3)(2)(3.773)N_2$$
$$\rightarrow CO_2 + 2H_2O + (1.3)(2)(3.773)N_2 + 0.6O_2$$
$$CH_4 + 2.6O_2 + 9.81N_2$$
$$\rightarrow CO_2 + 2H_2O + 9.81N_2 + 0.6O_2$$

Customary U.S. Solution

The volume of nitrogen that accompanies 4000 ft³/hr of entering methane is

$$V_{N_2} = \left(\frac{9.81 \text{ ft}^3 \text{ N}_2}{1 \text{ ft}^3 \text{ CH}_4}\right)\left(4000 \frac{\text{ft}^3}{\text{hr}} \text{ CH}_4\right)$$
$$= 39,240 \text{ ft}^3 \text{ N}_2/\text{hr}$$

This is the "partial volume" of nitrogen in the input stream.

$$R = \frac{R^*}{MW} = \frac{1545.35 \dfrac{\text{ft-lbf}}{\text{lbmol-}^\circ\text{R}}}{28.016 \dfrac{\text{lbm}}{\text{lbmol}}}$$
$$= 55.16 \text{ ft-lbf/lbm-}^\circ\text{R}$$

The absolute temperature is

$$T = 100^\circ\text{F} + 460^\circ = 560^\circ\text{R}$$

$$\dot{m}_{N_2} = \frac{p_{N_2}V_{N_2}}{RT} = \frac{\left(15 \dfrac{\text{lbf}}{\text{in}^2}\right)\left(12 \dfrac{\text{in}}{\text{ft}}\right)^2\left(39,240 \dfrac{\text{ft}^3}{\text{hr}}\right)}{\left(55.16 \dfrac{\text{ft-lbf}}{\text{lbm-}^\circ\text{R}}\right)(560^\circ\text{R})}$$
$$= \boxed{2744 \text{ lbm/hr} \quad (2700 \text{ lbm/hr})}$$

The answer is (D).

SI Solution

The volume of nitrogen that accompanies 31 L/s of entering methane is

$$\frac{\left(\dfrac{9.81 \text{ m}^3 \text{ N}_2}{1 \text{ m}^3 \text{ CH}_4}\right)\left(31 \dfrac{\text{L}}{\text{s}}\right)}{1000 \dfrac{\text{L}}{\text{m}^3}} = 0.3041 \text{ m}^3/\text{s}$$

This is the "partial volume" of nitrogen in the input stream.

$$R = \frac{R^*}{MW} = \frac{8314.5 \dfrac{\text{J}}{\text{kmol·K}}}{28.016 \dfrac{\text{kg}}{\text{kmol}}}$$
$$= 296.8 \text{ J/kg·K}$$

The absolute temperature is

$$T = 40^\circ\text{C} + 273^\circ = 313\text{K}$$

$$\dot{m}_{N_2} = \frac{p_{N_2}V_{N_2}}{RT} = \frac{(103 \text{ kPa})\left(1000 \dfrac{\text{Pa}}{\text{kPa}}\right)\left(0.3041 \dfrac{\text{m}^3}{\text{s}}\right)}{\left(296.8 \dfrac{\text{J}}{\text{kg·K}}\right)(313\text{K})}$$
$$= \boxed{0.337 \text{ kg/s} \quad (0.34 \text{ kg/s})}$$

The answer is (D).

4. From Table 24.7, the balanced chemical reaction equation is

$$C_3H_8 + 5O_2 \rightarrow 3CO_2 + 4H_2O$$

With 20% excess air, the oxygen volume is $(1.2)(5) = 6$.

$$C_3H_8 + 6O_2 \rightarrow 3CO_2 + 4H_2O + O_2$$

From Table 24.6, there are 3.773 volumes of nitrogen for every volume of oxygen.

$$(6)(3.773) = 22.6$$
$$C_3H_8 + 6O_2 + 22.6\,N_2 \rightarrow 3CO_2 + 4H_2O + O_2 + 22.6\,N_2$$

The percentage of carbon dioxide by weight in flue gas is

$$G_{CO_2} = \frac{m_{CO_2}}{m_{\text{total}}} = \frac{B_{CO_2}MW_{CO_2}}{\sum B_i(MW)_i}$$
$$= \frac{(3)(44.011)}{(3)(44.011) + (4)(18.016) + 32 + (22.6)(28.016)}$$
$$= \boxed{0.152 \quad (15\%)}$$

The answer is (C).

5. (a) For methane, $B = 0.93$.

From ideal gas laws and using App. 24.B for methane, $R = 96.32$ ft-lbf/lbm-°R.

$$\rho = \frac{p}{RT} = \frac{\left(14.7 \dfrac{\text{lbf}}{\text{in}^2}\right)\left(12 \dfrac{\text{in}}{\text{ft}}\right)^2}{\left(96.32 \dfrac{\text{ft-lbf}}{\text{lbm-}^\circ\text{R}}\right)(60^\circ\text{F} + 460^\circ)}$$
$$= 0.0422 \text{ lbm/ft}^3$$

From Table 24.9, $K = 9.556$ ft^3 air/ft^3 fuel.

From Table 24.8,

$$\text{products: } 1 \text{ ft}^3 \; CO_2, \; 2 \text{ ft}^3 \; H_2O$$

From App. 24.A, HHV = 1013 Btu/ft^3.

Similar results for all the other fuel components are tabulated in the following table.

The composite density is

$$\rho = \sum B_i \rho_i$$

$$= (0.93)\left(0.0422 \; \frac{\text{lbm}}{\text{ft}^3}\right) + (0.0373)\left(0.0738 \; \frac{\text{lbm}}{\text{ft}^3}\right)$$

$$+ (0.0045)\left(0.0738 \; \frac{\text{lbm}}{\text{ft}^3}\right) + (0.0182)\left(0.0053 \; \frac{\text{lbm}}{\text{ft}^3}\right)$$

$$+ (0.0025)\left(0.0739 \; \frac{\text{lbm}}{\text{ft}^3}\right) + (0.0018)\left(0.0900 \; \frac{\text{lbm}}{\text{ft}^3}\right)$$

$$+ (0.0035)\left(0.0843 \; \frac{\text{lbm}}{\text{ft}^3}\right) + (0.0022)\left(0.1160 \; \frac{\text{lbm}}{\text{ft}^3}\right)$$

$$= \boxed{0.0433 \text{ lbm/ft}^3 \quad (0.043 \text{ lbm/ft}^3)}$$

The answer is (B).

(b) The air is 20.9% oxygen by volume. The theoretical air requirements are

$$\sum B_i V_{\text{air},i} - \frac{O_2 \text{ in fuel}}{0.209}$$

$$= (0.93)(9.556 \text{ ft}^3) + (0.0373)(0 \text{ ft}^3)$$

$$+ (0.0045)(2.389 \text{ ft}^3) + (0.0182)(2.389 \text{ ft}^3)$$

$$+ (0.0025)(14.33 \text{ ft}^3) + (0.0018)(7.167 \text{ ft}^3)$$

$$+ (0.0035)(0 \text{ ft}^3) + (0.0022)(0 \text{ ft}^3) - \frac{0.0035}{0.209}$$

$$= 8.990 \text{ ft}^3 - 0.01675 \text{ ft}^3$$

$$= \boxed{8.9733 \text{ ft}^3 \text{ air/ft}^3 \text{ fuel} \quad (9 \text{ ft}^3 \text{ air/ft}^3 \text{ fuel})}$$

The answer is (C).

Use the following computations to determine (c) and (d). The theoretical oxygen will be

$$(8.9733 \text{ ft}^3)(0.209) = 1.875 \text{ ft}^3/\text{ft}^3$$

The excess oxygen will be

$$(0.4)(1.875 \text{ ft}^3) = 0.75 \text{ ft}^3/\text{ft}^3$$

Similarly, the total nitrogen in the stack gases is

$$(1.4)(0.791)(8.9733 \text{ ft}^3) + 0.034 = 9.971 \text{ ft}^3/\text{ft}^3 \text{ fuel}$$

The stack gases per cubic foot of fuel are

excess O_2	$= 0.7500$ ft^3
excess N_2	$= 9.971$ ft^3
excess SO_2	$= 0.0018$ ft^3

excess CO_2 $\;(0.93)(1 \text{ ft}^3) + (0.0045)(1 \text{ ft}^3)$
$+ (0.0025)(2 \text{ ft}^3)$
$+ (0.0022)(1 \text{ ft}^3) \qquad = 0.942$ ft^3

excess H_2O $\;(0.93)(2 \text{ ft}^3) + (0.0182)(1 \text{ ft}^3)$
$+ (0.0025)(2 \text{ ft}^3)$
$+ (0.0018)(1 \text{ ft}^3) \qquad = 1.885$ ft^3

$$\text{total} \quad - \overline{13.55 \text{ ft}^3}$$

The total wet volume is 13.55 ft^3/ft^3 fuel.

The total dry volume is 11.66 ft^3/ft^3 fuel.

The volumetric analyses are

	O_2	N_2	SO_2	CO_2	H_2O
wet:	$\dfrac{0.7500 \text{ ft}^3}{13.55 \text{ ft}^3}$	$\dfrac{9.971 \text{ ft}^3}{13.55 \text{ ft}^3}$	$\dfrac{0.0018 \text{ ft}^3}{13.55 \text{ ft}^3}$	$\dfrac{0.942 \text{ ft}^3}{13.55 \text{ ft}^3}$	$\dfrac{1.885 \text{ ft}^3}{13.55 \text{ ft}^3}$
	$= 0.0554$	$= 0.736$	≈ 0	$= 0.070$	$= 0.139$
dry:	$\dfrac{0.7500 \text{ ft}^3}{11.66 \text{ ft}^3}$	$\dfrac{9.971 \text{ ft}^3}{11.66 \text{ ft}^3}$	$\dfrac{0.0018 \text{ ft}^3}{11.66 \text{ ft}^3}$	$\dfrac{0.942 \text{ ft}^3}{11.66 \text{ ft}^3}$	$-$
	$= 0.0643$	$= 0.855$	≈ 0	$= 0.081$	$-$

(c) $B_{CO_2,\text{wet}} = \boxed{6.9\%}$

The answer is (A).

(d) $B_{CO_2,\text{dry}} = \boxed{8.1\%}$

The answer is (C).

Reaction Products for Prob. 5

					volumes of products		
gas	B	ρ (lbm/ft^3)	ft^3 air	HHV (Btu/ft^3)	CO_2	H_2O	other
CH_4	0.93	0.0422	9.556	1013	1	2	–
N_2	0.0373	0.0738	–	–	–	–	1 N_2
CO	0.0045	0.0738	2.389	322	1	–	–
H_2	0.0182	0.0053	2.389	325	–	1	–
C_2H_4	0.0025	0.0739	14.33	1614	2	2	–
H_2S	0.0018	0.0900	7.167	647	–	1	1 SO_2
O_2	0.0035	0.0843	–	–	–	–	–
CO_2	0.0022	0.1160	–	–	1	–	–

6. (a) Silicon in ash is SiO_2 with a molecular weight of

$$28.09 \ \frac{\text{lbm}}{\text{lbmol}} + (2)\left(16 \ \frac{\text{lbm}}{\text{lbmol}}\right) = 60.09 \ \text{lbm/lbmol}$$

The oxygen used with 6.1% by mass silicon is

$$\left(\frac{(2)(16 \ \text{lbm})}{28.09 \ \text{lbm}}\right)\left(0.061 \ \frac{\text{lbm}}{\text{lbm coal}}\right)$$

$$= 0.0695 \ \text{lbm/lbm coal}$$

Silicon ash produced per lbm of coal is

$$0.061 \ \frac{\text{lbm}}{\text{lbm coal}} + 0.0695 \ \frac{\text{lbm}}{\text{lbm coal}}$$

$$= 0.1305 \ \text{lbm/lbm coal}$$

Silicon ash produced per hour is

$$\left(0.1305 \ \frac{\text{lbm}}{\text{lbm coal}}\right)\left(15,300 \ \frac{\text{lbm coal}}{\text{hr}}\right)$$

$$= 1996.7 \ \text{lbm/hr}$$

The silicon in 410 lbm/hr refuse is

$$\left(410 \ \frac{\text{lbm}}{\text{hr}}\right)(1 - 0.3) = 287 \ \text{lbm/hr}$$

The emission rate is

$$1996.7 \ \text{lbm/hr} - 287 \ \text{lbm/hr}$$

$$= \boxed{1709.7 \ \text{lbm/hr} \quad (1700 \ \text{lbm/hr})}$$

The answer is (D).

(b) From Table 24.7, the stoichiometric reaction for sulfur is

$$\begin{array}{ccccc} \text{S} & + & \text{O}_2 & \rightarrow & \text{SO}_2 \\ \text{MW} \quad 32 & & 32 & & 64 \end{array}$$

The sulfur dioxide that is produced from 15,300 lbm/hr of coal feed is

$$\left(15,300 \ \frac{\text{lbm}}{\text{hr}}\right)(0.0244 \ \text{lbm S})\left(\frac{64 \ \text{lbm SO}_2}{32 \ \text{lbm S}}\right)$$

$$= \boxed{746.6 \ \text{lbm/hr} \quad (750 \ \text{lbm/hr})}$$

The answer is (C).

(c) From Eq. 24.16(b), the heating value of the fuel is

$$\text{HHV} = 14,093\,G_C + 60,958\left(G_H - \frac{G_O}{8}\right) + 3983\,G_S$$

$$= \left(14,093 \ \frac{\text{Btu}}{\text{lbm}}\right)(0.7656)$$

$$+ \left(60,958 \ \frac{\text{Btu}}{\text{lbm}}\right)\left(0.055 - \frac{0.077}{8}\right)$$

$$+ \left(3983 \ \frac{\text{Btu}}{\text{lbm}}\right)(0.0244)$$

$$= 13,653 \ \text{Btu/lbm}$$

The gross available combustion power is

$$\dot{m}_f(\text{HV}) = \left(15,300 \ \frac{\text{lbm}}{\text{hr}}\right)\left(13,653 \ \frac{\text{Btu}}{\text{lbm}}\right)$$

$$= 2.089 \times 10^8 \ \text{Btu/hr}$$

The carbon in 410 lbm/hr refuse is

$$\left(410 \ \frac{\text{lbm}}{\text{hr}}\right)(0.3) = 123 \ \text{lbm/hr}$$

Power lost in unburned carbon in refuse is $\dot{m}_C(\text{HV})$.

From App. 24.A, the gross heat of combustion for carbon is 14,093 Btu/lbm.

$$\left(123 \ \frac{\text{lbm}}{\text{hr}}\right)\left(14,093 \ \frac{\text{Btu}}{\text{lbm}}\right) = 1.733 \times 10^6 \ \text{Btu/hr}$$

The remaining combustion power is

$$2.089 \times 10^8 \ \frac{\text{Btu}}{\text{hr}} - 1.733 \times 10^6 \ \frac{\text{Btu}}{\text{hr}}$$

$$= 2.072 \times 10^8 \ \text{Btu/hr}$$

Losses in the steam generator and electrical generator will further reduce this to

$$(0.86)\left(2.072 \times 10^8 \ \frac{\text{Btu}}{\text{hr}}\right) = 1.782 \times 10^8 \ \text{Btu/hr}$$

With an electrical output of 17 MW, thermal power removed by cooling water is

$$Q = 1.782 \times 10^8 \ \frac{\text{Btu}}{\text{hr}} - (17 \ \text{MW})\left(1000 \ \frac{\text{kW}}{\text{MW}}\right)$$

$$\times \left(3413 \ \frac{\frac{\text{Btu}}{\text{hr}}}{\text{kW}}\right)$$

$$= 1.202 \times 10^8 \ \text{Btu/hr}$$

The electrical generator is not cooled by the cooling water. Therefore, it is not correct to include the generator efficiency in the calculation of losses.

At 60°F, the specific heat of water is $c_p = 1$ Btu/lbm-°F. The temperature rise of the cooling water is

$$\Delta T = \frac{Q}{\dot{m} c_p}$$

$$= \frac{1.202 \times 10^8 \ \dfrac{\text{Btu}}{\text{hr}}}{\left(225 \ \dfrac{\text{ft}^3}{\text{sec}}\right)\left(62.4 \ \dfrac{\text{lbm}}{\text{ft}^3}\right)\left(3600 \ \dfrac{\text{sec}}{\text{hr}}\right)\left(1 \ \dfrac{\text{Btu}}{\text{lbm-°F}}\right)}$$

$$= \boxed{2.38°\text{F} \quad (2.4°\text{F})}$$

The answer is (A).

(d) Limiting 0.1 lbm of particulates per million Btu per hour, the allowable emission rate is

$$\frac{\left(0.1 \ \dfrac{\text{lbm}}{\text{MBtu}}\right)\left(15{,}300 \ \dfrac{\text{lbm}}{\text{hr}}\right)\left(13{,}653 \ \dfrac{\text{Btu}}{\text{lbm}}\right)}{10^6 \ \dfrac{\text{Btu}}{\text{MBtu}}} = 20.89 \ \text{lbm/hr}$$

The efficiency of the flue gas particulate collectors is

$$\eta = \frac{\text{actual emission rate} \ - \ \text{allowable emission rate}}{\text{actual emission rate}}$$

$$= \frac{1709.7 \ \dfrac{\text{lbm}}{\text{hr}} - 20.89 \ \dfrac{\text{lbm}}{\text{hr}}}{1709.7 \ \dfrac{\text{lbm}}{\text{hr}}}$$

$$= \boxed{0.988 \quad (98.8\%)}$$

The answer is (C).

25 Water Supply Quality and Testing

PRACTICE PROBLEMS

Phosphorus

1. (*Time limit: one hour*) In a study of a small pond to determine phosphorus impact, the following information was collected.

> pond size: 12 ac-ft (14.8×10^6 L)
> watershed area: 4 ac (grassland)
> average annual rainfall: 5 in
> bioavailable P in runoff: 0.01 mg/L
> runoff coefficient: 0.1

Biological processes in the pond biota convert phosphorus to a nonbioavailable form at the rate of 22% per year. Recycling of sediment phosphorus by rooted plants and by anaerobic conditions in the hypolimnion converts 12% per year of the nonbioavailable phosphorus back to bioavailable forms.

Runoff into the pond evaporates during the year, so no change in pond volume occurs.

(a) Starting from an initial condition of 0.1 mg/L bioavailable P, what is the P concentration after five annual cycles?

(A) 0.03 mg/L

(B) 0.11 mg/L

(C) 0.13 mg/L

(D) 0.18 mg/L

(b) What can be done to reduce the P accumulation?

(A) Add chemicals to precipitate phosphorus in the pond.

(B) Add chemicals to combine with phosphorus in the pond.

(C) Reduce use of phosphorus-based fertilizers in the surrounding fields.

(D) Use natural-based soaps and detergents in the home.

Alkalinity

2. (*Time limit: one hour*) Groundwater is used for a water supply. It is taken from the ground at 25°C. The initial properties of the water are as follows.

> CO_2 60 mg/L (as $CaCO_3$)
> pH   7.1

The water is treated for CO_2 removal by spraying into the atmosphere through a nozzle. The final CO_2 concentration is 5.6 mg/L (as $CaCO_3$) at 25°C. The final alkalinity is 200 mg/L (as $CaCO_3$). The first ionization constant of carbonic acid is 4.45×10^{-7}. What is the final pH of the water after spraying and recovery?

(A) 7.3

(B) 7.9

(C) 8.4

(D) 8.8

Solids

3. (*Time limit: one hour*) The solids concentration of a stream water sample is to be determined. The total solids concentration is determined by placing a portion of the sample into a porcelain evaporating dish, drying the sample at 105°C, and igniting the residue by placing the dried sample in a muffle furnace at 550°C. The following masses are recorded.

> mass of empty dish: 50.326 g
> mass of dish and sample: 118.400 g
> mass of dish and dry solids: 50.437 g
> mass of dish and ignited solids: 50.383 g

(a) The total solids concentration is

(A) 900 mg/L

(B) 1000 mg/L

(C) 1100 mg/L

(D) 1600 mg/L

(b) The total volatile solids concentration is

(A) 630 mg/L

(B) 710 mg/L

(C) 790 mg/L

(D) 830 mg/L

(c) The total fixed solids concentration is

(A) 270 mg/L

(B) 300 mg/L

(C) 420 mg/L

(D) 840 mg/L

The suspended solids concentration is determined by filtering a portion of the sample through a glass-fiber filter disk, drying the disk at 105°C, and igniting the residue by placing the dried sample in a muffle furnace at 550°C. The follow masses are recorded.

volume of sample: 30 mL
mass of filter disk: 0.1170 g
mass of disk and dry solids: 0.1278 g
mass of disk and ignited solids: 0.1248 g

(d) The total suspended solids concentration is

(A) 240 mg/L

(B) 360 mg/L

(C) 370 mg/L

(D) 820 mg/L

(e) The volatile suspended solids concentration is

(A) 100 mg/L

(B) 120 mg/L

(C) 230 mg/L

(D) 640 mg/L

(f) The fixed suspended solids concentration is

(A) 120 mg/L

(B) 180 mg/L

(C) 240 mg/L

(D) 260 mg/L

The dissolved solids concentration is determined by filtering a portion of the sample through a glass-fiber filter disk into a porcelain evaporating dish, drying the sample at 105°C, and igniting the residue by placing the dried sample in a muffle furnace at 550°C. The following masses are recorded.

volume of sample: 25 mL
mass of empty dish: 51.494 g
mass of dish and dry solids: 51.524 g
mass of dish and ignited solids: 51.506 g

(g) The total dissolved solids concentration is

(A) 1000 mg/L

(B) 1100 mg/L

(C) 1200 mg/L

(D) 1400 mg/L

(h) The volatile dissolved solids concentration is

(A) 680 mg/L

(B) 720 mg/L

(C) 810 mg/L

(D) 900 mg/L

(i) The fixed dissolved solids concentration is

(A) 230 mg/L

(B) 300 mg/L

(C) 410 mg/L

(D) 480 mg/L

(j) Assume the total solids determined in part (a) is equal to the sum of the total suspended solids determined in part (d) plus the total dissolved solids determined in part (g). Does the total solids concentration equal the sum of the suspended and dissolved solids?

(A) yes

(B) no; probably due to poor laboratory technique

(C) no; probably due to rounding of measured values

(D) no; probably because the samples were representative but not identical

Hardness

4. (*Time limit: one hour*) The laboratory analysis of a water sample is as follows. All concentrations are "as substance."

Ca^{++}	74.0 mg/L
Mg^{++}	18.3 mg/L
Na^{+}	27.6 mg/L
K^{+}	39.1 mg/L
pH	7.8
HCO_3^{-}	274.5 mg/L
SO_4^{--}	72.0 mg/L
Cl^{-}	49.7 mg/L

(a) The hardness of the water in terms of mg/L of calcium carbonate equivalent is

    (A) 1.3 mg/L as $CaCO_3$

    (B) 4.7 mg/L as $CaCO_3$

    (C) 66 mg/L as $CaCO_3$

    (D) 260 mg/L as $CaCO_3$

(b) Based on the laboratory analysis, which of the following can be said?

    (A) It is surprising that no carbonates were found in the sample.

    (B) The cations in solution, when converted to milliequivalents, will equal the anions, when converted to milliequivalents.

    (C) The large amount of bicarbonate in the solution tends to make the water acidic.

    (D) None of the above can be said.

(c) Assuming that hypothetical compounds are formed proportionally to the relative concentrations of ions, the hypothetical concentration of calcium bicarbonate in the sample is

    (A) 3.7 meq/L

    (B) 4.5 meq/L

    (C) 4.7 meq/L

    (D) 74 meq/L

(d) The hypothetical concentration of magnesium bicarbonate is

    (A) 0.8 meq/L

    (B) 1.5 meq/L

    (C) 1.6 meq/L

    (D) 5.2 meq/L

(e) The hypothetical concentration of magnesium sulfate is

    (A) 0.7 meq/L

    (B) 1.5 meq/L

    (C) 4.5 meq/L

    (D) 5.2 meq/L

(f) The hypothetical concentration of sodium sulfate is

    (A) 0.8 meq/L

    (B) 1.2 meq/L

    (C) 1.5 meq/L

    (D) 6.4 meq/L

(g) The hypothetical concentration of sodium chloride is

    (A) 0.2 meq/L

    (B) 0.4 meq/L

    (C) 1.2 meq/L

    (D) 1.4 meq/L

(h) The amount of lime (CaO) necessary in a lime softening process to remove the hardness caused by calcium bicarbonate is

    (A) 17 mg/L

    (B) 28 mg/L

    (C) 95 mg/L

    (D) 100 mg/L

(i) To remove hardness caused by magnesium bicarbonate, it is necessary to raise the pH by adding 35 mg/L of CaO in excess of the stoichiometric requirements. The amount of lime (CaO) necessary to remove the carbonate hardness caused by magnesium bicarbonate is

    (A) 12 mg/L

    (B) 45 mg/L

    (C) 56 mg/L

    (D) 80 mg/L

(j) The water is subsequently recarbonated to reduce the pH. Assume that by this softening process calcium hardness can be reduced to 30 mg/L and magnesium hardness can be reduced to 10 mg/L, both measured in terms of equivalent calcium carbonate. The amount of hardness that remains in the water is

    (A) 10 mg/L as $CaCO_3$

    (B) 30 mg/L as $CaCO_3$

    (C) 45 mg/L as $CaCO_3$

    (D) 75 mg/L as $CaCO_3$

Water Resources

SOLUTIONS

1. (a) $\text{annual rainfall} = (4\text{ ac})(5\text{ in}) = 20\text{ ac-in}$

$$\text{runoff to pond} = (0.1)(20\text{ ac-in})$$
$$\times \left(102{,}790\ \frac{\text{L}}{\text{ac-in}}\right)$$
$$= 205{,}580\text{ L}$$

$$\begin{aligned}\text{bioavailable P} \atop \text{reaching pond} &= (205{,}580\text{ L})\left(0.01\ \frac{\text{mg}}{\text{L}}\right)\\ &= 2056\text{ mg}\end{aligned}$$

$$\begin{aligned}\text{initial bioavailable} \atop \text{P in pond} &= (14.8 \times 10^6\text{ L})\left(0.1\ \frac{\text{mg}}{\text{L}}\right)\\ &= 14.8 \times 10^5\text{ mg}\end{aligned}$$

The total bioavailable P in the pond is

$$2056\text{ mg} + 14.8 \times 10^5\text{ mg} = 14.82056 \times 10^5\text{ mg}$$

Of this, after biological processes and recycling, the following percentage remains.

$$100\% - 22\% + \frac{(12\%)(22\%)}{100\%} = 80.64\%$$

Therefore,

$$(0.8064)(14.82056 \times 10^5\text{ mg}) = 11.95130 \times 10^5\text{ mg}$$

The following table is prepared in a similar manner.

year	P at start of year	P from rainfall	total P available for activity	P at end of year
1	14.8×10^5	2056	14.82056×10^5	11.95130×10^5
2	11.95130×10^5	2056	11.97186×10^5	9.65411×10^5
3	9.65411×10^5	2056	9.67467×10^5	7.80165×10^5
4	7.80165×10^5	2056	7.82221×10^5	6.30783×10^5
5	6.30783×10^5	2056	6.32839×10^5	5.10322×10^5

The concentration after five years will be

$$\frac{5.10322 \times 10^5\text{ mg}}{14.8 \times 10^6\text{ L}} = \boxed{0.0345\text{ mg/L}\quad (0.03\text{ mg/L})}$$

The answer is (A).

(b) To reduce the phosphorus accumulation in the pond, the arrival of additional bioavailable phosphorous must be reduced. This entails watershed management to reduce the amount of phosphorus applied as fertilizer and released through other sources. As eutrophication is a natural process accelerated by the availability of plant nutrients (nitrogen and phosphorus especially), reducing phosphorus can slow the process.

While there are chemical means to alter the forms of phosphorus to nonbioavailable states, this usually is not practical on a large scale and is potentially harmful in itself.

It is unlikely that the pond receives untreated discharge from local homes. Use of phosphate-rich detergents in the home will not affect the pond.

The answer is (C).

2. Since pH < 8.3 (see Eq. 25.2 and Eq. 25.3), all alkalinity is in the bicarbonate (HCO_3^-) form. Therefore, the equilibrium expression for the first ionization of carbonic acid may be used.

$$CO_2 + H_2O \rightarrow H_2CO_3 \rightleftharpoons H^+ + HCO_3^-$$

$$K_1 = \frac{[H^+][HCO_3^-]}{[H_2CO_3]} = 4.45 \times 10^{-7}$$

The coefficients of CO_2 and H_2CO_3 are both 1. The number of moles of each compound is the same. One mole of CO_2 produces one mole of H_2CO_3.

$$[H_2CO_3] = [CO_2] = \frac{5.6\ \frac{\text{mg}}{\text{L}}}{\left(100\ \frac{\text{g}}{\text{mol}}\right)\left(1000\ \frac{\text{mg}}{\text{g}}\right)}$$
$$= 5.6 \times 10^{-5}$$

$$[HCO_3^-] = \frac{200\ \frac{\text{mg}}{\text{L}}}{\left(100\ \frac{\text{g}}{\text{mol}}\right)\left(1000\ \frac{\text{mg}}{\text{g}}\right)}$$
$$= 2 \times 10^{-3}$$

Solve for the hydrogen ion concentration.

$$\begin{aligned}[H^+] &= \frac{K_1[H_2CO_3]}{[HCO_3^-]} = \frac{K_1[H_2CO_3]}{[CO_2]}\\ &= \frac{(4.45 \times 10^{-7})(5.6 \times 10^{-5})}{2 \times 10^{-3}}\\ &= 1.246 \times 10^{-8}\end{aligned}$$

$$pH = -\log[H^+] = -\log(1.246 \times 10^{-8}) = \boxed{7.9}$$

The answer is (B).

3. (a) The density of water is 1 g/mL. The volume of the tested sample is

$$V = \frac{m}{\rho} = \frac{118.4\text{ g} - 50.326\text{ g}}{1\ \frac{\text{g}}{\text{mL}}} = 68.1\text{ mL}$$

The total solids concentration is

$$TS = \frac{(50.437 \text{ g} - 50.326 \text{ g})\left(1000 \frac{\text{mg}}{\text{g}}\right)\left(1000 \frac{\text{mL}}{\text{L}}\right)}{68.1 \text{ mL}}$$

$$= \boxed{1630 \text{ mg/L} \quad (1600 \text{ mg/L})}$$

The answer is (D).

(b) The total volatile solids concentration is

$$TVS = \frac{(50.437 \text{ g} - 50.383 \text{ g})\left(1000 \frac{\text{mg}}{\text{g}}\right)\left(1000 \frac{\text{mL}}{\text{L}}\right)}{68.1 \text{ mL}}$$

$$= \boxed{793 \text{ mg/L} \quad (790 \text{ mg/L})}$$

The answer is (C).

(c) The total fixed solids concentration is

$$TFS = 1630 \frac{\text{mg}}{\text{L}} - 793 \frac{\text{mg}}{\text{L}}$$

$$= \boxed{837 \text{ mg/L} \quad (840 \text{ mg/L})}$$

The answer is (D).

(d) The total suspended solids concentration is

$$TSS = \frac{(0.1278 \text{ g} - 0.1170 \text{ g}) \times \left(1000 \frac{\text{mg}}{\text{g}}\right)\left(1000 \frac{\text{mL}}{\text{L}}\right)}{30 \text{ mL}}$$

$$= \boxed{360 \text{ mg/L}}$$

The answer is (B).

(e) The volatile suspended solids concentration is

$$VSS = \frac{(0.1278 \text{ g} - 0.1248 \text{ g}) \times \left(1000 \frac{\text{mg}}{\text{g}}\right)\left(1000 \frac{\text{mL}}{\text{L}}\right)}{30 \text{ mL}}$$

$$= \boxed{100 \text{ mg/L}}$$

The answer is (A).

(f) The fixed suspended solids concentration is

$$FFS = 360 \frac{\text{mg}}{\text{L}} - 100 \frac{\text{mg}}{\text{L}} = \boxed{260 \text{ mg/L}}$$

The answer is (D).

(g) The total dissolved solids concentration is

$$TDS = \frac{(51.524 \text{ g} - 51.494 \text{ g}) \times \left(1000 \frac{\text{mg}}{\text{g}}\right)\left(1000 \frac{\text{mL}}{\text{L}}\right)}{25 \text{ mL}}$$

$$= \boxed{1200 \text{ mg/L}}$$

The answer is (C).

(h) The volatile dissolved solids concentration is

$$VDS = \frac{(51.524 \text{ g} - 51.506 \text{ g}) \times \left(1000 \frac{\text{mg}}{\text{g}}\right)\left(1000 \frac{\text{mL}}{\text{L}}\right)}{25 \text{ mL}}$$

$$= \boxed{720 \text{ mg/L}}$$

The answer is (B).

(i) The fixed dissolved solids concentration is

$$FDS - 1200 \frac{\text{mg}}{\text{L}} - 720 \frac{\text{mg}}{\text{L}} = \boxed{480 \text{ mg/L}}$$

The answer is (D).

(j) If only the first two procedures were performed, the dissolved solids would be calculated as

$$TDS = TS - TSS = 1630 \frac{\text{mg}}{\text{L}} - 360 \frac{\text{mg}}{\text{L}}$$

$$= 1270 \text{ mg/L} \quad (\text{vs. } 1200 \text{ mg/L})$$

$$VDS = TVS - VSS = 793 \frac{\text{mg}}{\text{L}} - 100 \frac{\text{mg}}{\text{L}}$$

$$= 693 \text{ mg/L} \quad (\text{vs. } 720 \text{ mg/L})$$

$$FDS = TFS - FSS = 837 \frac{\text{mg}}{\text{L}} - 260 \frac{\text{mg}}{\text{L}}$$

$$= 577 \text{ mg/L} \quad (\text{vs. } 480 \text{ mg/L})$$

The results of the three sets of tests are not the same. These differences are too great to be caused by improper rounding or faulty laboratory technique. This is probably the result of using samples that are not truly identical. For this reason, the suspended solids values are mathematically determined from the total solids and dissolved solids tests.

The answer is (D).

4. (a) Hardness is caused by multivalent cations: Ca^{++} and Mg^{++} in this example. Calculate the milliequivalents by dividing the measured concentration by the milliequivalent weight.

$$Ca^{++}: \dfrac{74.0 \, \frac{mg}{L}}{20 \, \frac{mg}{meq}} = 3.7 \text{ meq/L}$$

$$Mg^{++}: \dfrac{18.3 \, \frac{mg}{L}}{12.2 \, \frac{mg}{meq}} = 1.5 \text{ meq/L}$$

$$\text{total hardness} = 3.7 \, \frac{meq}{L} + 1.5 \, \frac{meq}{L} = 5.2 \text{ meq/L}$$

Determine the calcium carbonate equivalent by multiplying by the equivalent weight of calcium carbonate.

$$\text{hardness} = \left(5.2 \, \frac{meq}{L}\right)\left(50 \, \frac{mg}{meq}\right)$$
$$= \boxed{260 \text{ mg/L as } CaCO_3}$$

The answer is (D).

(b) Alkalinity in the form of carbonate radical does not exist at a pH below 8.3. Bicarbonate is a form of alkalinity, and it neutralizes, not creates, acidity. To determine the cation/anion relationship, convert all of the concentrations to milliequivalents by dividing the measured concentrations by the milliequivalent weights.

$$Ca^{++}: \dfrac{74.0 \, \frac{mg}{L}}{20 \, \frac{mg}{meq}} = 3.7 \text{ meq/L}$$

$$Mg^{++}: \dfrac{18.3 \, \frac{mg}{L}}{12.2 \, \frac{mg}{meq}} = 1.5 \text{ meq/L}$$

$$Na^{+}: \dfrac{27.6 \, \frac{mg}{L}}{23 \, \frac{mg}{meq}} = 1.2 \text{ meq/L}$$

$$K^{+}: \dfrac{39.1 \, \frac{mg}{L}}{39.1 \, \frac{mg}{meq}} = 1.0 \text{ meq/L}$$

$$\text{total cations} = 3.7 \, \frac{meq}{L} + 1.5 \, \frac{meq}{L} + 1.2 \, \frac{meq}{L}$$
$$+ 1.0 \, \frac{meq}{L}$$
$$= 7.4 \text{ meq/L}$$

$$HCO_3^{-}: \dfrac{274.5 \, \frac{mg}{L}}{61 \, \frac{mg}{meq}} = 4.5 \text{ meq/L}$$

$$SO_4^{--}: \dfrac{72.0 \, \frac{mg}{L}}{48 \, \frac{mg}{meq}} = 1.5 \text{ meq/L}$$

$$Cl^{-}: \dfrac{49.7 \, \frac{mg}{L}}{35.5 \, \frac{mg}{meq}} = 1.4 \text{ meq/L}$$

$$\text{total anions} = 4.5 \, \frac{meq}{L} + 1.5 \, \frac{meq}{L} + 1.4 \, \frac{meq}{L}$$
$$= 7.4 \text{ meq/L}$$

$\boxed{\text{The total cations equals the total anions.}}$

The answer is (B).

(c) To determine the hypothetical compounds, construct a milliequivalent per liter bar graph for cations and anions, as shown.

0	3.7	5.2	6.4	7.4
Ca^{++} 3.7	Mg^{++} 1.5	Na^{+} 1.2	K^{+} 1.0
HCO_3^{-} 4.5		SO_4^{--} 1.5	Cl^{-} 1.4
0	4.5	6.0	7.4

The hypothetical compounds are determined by moving from left to right.

hypothetical	compound concentration (meq/L)	remarks
$Ca(HCO_3)_2$	3.7	Ca^{++} exhausted
$Mg(HCO_3)_2$	0.8	HCO_3^{-} exhausted
$MgSO_4$	0.7	Mg^{++} exhausted
Na_2SO_4	0.8	SO_4^{-} exhausted
$NaCl$	0.4	Na^{+} exhausted
KCl	1.0	K^{+} and Cl^{-} exhausted

The hypothetical concentration of calcium bicarbonate is $\boxed{3.7 \text{ meq/L.}}$

The answer is (A).

(d) From part (c), the hypothetical concentration of magnesium bicarbonate is $\boxed{0.8 \text{ meq/L.}}$

The answer is (A).

(e) From part (c), the hypothetical concentration of magnesium sulfate is $\boxed{0.7 \text{ meq/L.}}$

The answer is (A).

(f) From part (c), the hypothetical concentration of sodium sulfate is $\boxed{0.8 \text{ meq/L.}}$

The answer is (A).

(g) From part (c), the hypothetical concentration of sodium chloride is $\boxed{0.4 \text{ meq/L.}}$

The answer is (B).

(h) The reactions are

$$CaO + H_2O \rightarrow Ca(OH)_2$$
$$Ca(HCO_3)_2 + Ca(OH)_2 \rightarrow 2CaCO_3\downarrow + 2H_2O$$

One molecule of CaO forms one molecule of $Ca(OH)_2$, which in turn reacts with one molecule of $Ca(HCO_3)_2$. There are 3.7 meq/L of $Ca(HCO_3)_2$ in solution. The equivalent weight of CaO is $(40 + 16)/2 = 28$. Therefore, 3.7 meq/L of CaO are needed.

$$\left(3.7 \ \frac{\text{meq}}{\text{L}} \ CaO\right)\left(28 \ \frac{\text{mg}}{\text{meq}}\right)$$
$$= \boxed{103.6 \text{ mg/L CaO} \quad (100 \text{ mg/L CaO})}$$

The answer is (D).

(i) The reactions are

$$CaO + H_2O \rightarrow Ca(OH)_2$$
$$Mg(HCO_3)_2 + 2Ca(OH)_2 \rightarrow$$
$$2CaCO_3\downarrow + Mg(OH)_2\downarrow + 2H_2O$$

Two molecules of $Ca(OH)_2$ are required for each molecule of $Mg(HCO_3)_2$.

$$\left(0.8 \ \frac{\text{meq}}{\text{L}} \ Mg(HCO_3)_2\right)(2)\left(28 \ \frac{\text{mg}}{\text{meq}}\right)$$
$$+ 35 \ \frac{\text{mg}}{\text{L}} \text{ excess} = \boxed{79.8 \text{ mg/L} \quad (80 \text{ mg/L})}$$

The answer is (D).

(j) The original hardness was in the hypothetical forms of calcium bicarbonate, magnesium bicarbonate, and magnesium sulfate. The bicarbonates have been removed in parts (h) and (i), leaving the residuals of 30 mg/L calcium hardness and 10 mg/L magnesium hardness. However, no attempt was made to remove the noncarbonate magnesium hardness represented by magnesium sulfate (this would have required the addition of soda ash and additional lime).

The residual hardness is as follows.

$$\text{calcium hardness:} \ \frac{30 \ \dfrac{\text{mg}}{\text{L}}}{50 \ \dfrac{\text{mg}}{\text{meq}}} = 0.6 \text{ meq/L}$$

$$\text{magnesium carbonate hardness:} \ \frac{10 \ \dfrac{\text{mg}}{\text{L}}}{50 \ \dfrac{\text{mg}}{\text{meq}}} = 0.2 \text{ meq/L}$$

From part (c), the magnesium noncarbonate hardness is 0.7 meq/L.

$$\text{residual hardness} = 0.6 \ \frac{\text{meq}}{\text{L}} + 0.2 \ \frac{\text{meq}}{\text{L}} + 0.7 \ \frac{\text{meq}}{\text{L}}$$
$$= 1.5 \text{ meq/L}$$
$$\left(1.5 \ \frac{\text{meq}}{\text{L}}\right)\left(50 \ \frac{\text{mg}}{\text{meq}}\right) = \boxed{75 \text{ mg/L as } CaCO_3}$$

The answer is (D).

Water Resources

26 Water Supply Treatment and Distribution

PRACTICE PROBLEMS

Plain Sedimentation

1. A settling tank's overflow rate is 100,000 gal/ft^2-day. Water carrying sediment of various sizes enters the tank. The sediment has the following distribution of settling velocities.

settling velocity (ft/min)	mass fraction remaining
10.0	0.54
5.0	0.45
2.0	0.35
1.0	0.20
0.75	0.10
0.50	0.03

(a) What is the gravimetric percentage of sediment particles completely removed?

(A) 32%

(B) 39%

(C) 47%

(D) 55%

(b) What is the total gravimetric percentage of all sediment particles removed?

(A) 35%

(B) 40%

(C) 50%

(D) 60%

2. A spherical sand particle has a specific gravity of 2.6 and a diameter of 1 mm. What is the settling velocity?

(A) 0.2 ft/sec

(B) 0.7 ft/sec

(C) 1.1 ft/sec

(D) 1.6 ft/sec

3. A mechanically cleaned circular clarifier is to be designed with the following characteristics.

flow rate	2.8 MGD
detention period	2 hr
surface loading	700 gal/ft^2-day

(a) What is the approximate diameter?

(A) 45 ft

(B) 60 ft

(C) 70 ft

(D) 90 ft

(b) What is the approximate depth?

(A) 6 ft

(B) 8 ft

(C) 12 ft

(D) 15 ft

(c) If the initial flow rate is reduced to 1.1 MGD, what is the surface loading?

(A) 190 gal/day-ft^2

(B) 230 gal/day-ft^2

(C) 250 gal/day-ft^2

(D) 280 gal/day-ft^2

(d) If the initial flow rate is reduced to 1.1 MGD, what is the average detention period?

(A) 4 hr

(B) 5 hr

(C) 6 hr

(D) 8 hr

4. (*Time limit: one hour*) A water treatment plant is designed to handle a total flow rate of 1.5 MGD. The current design includes two identical sedimentation basins that run in parallel and have the following characteristics.

plan area	90 ft × 16 ft
depth	12 ft
total weir length	48 ft per basin
three-month sustained average low	70% of the design average daily flow
three-month sustained average high	200% of the design average daily flow

The basins must meet the following government standards.

minimum retention time	4.0 hr
maximum weir load	20,000 gpd/ft
maximum velocity	0.5 ft/min

Do the basins meet the standards?

(A) Yes, specifications are met at both peak and low flows.

(B) No, specifications are not met at low flow.

(C) No, specifications are not met at high flow.

(D) No, specifications are not met at either high or low flow.

Mixing Physics

5. (*Time limit: one hour*) A flocculator tank with a volume of 200,000 ft^3 uses a paddle wheel to mix the coagulant in 60°F water. The operating characteristics are as follows.

mean velocity gradient	45 sec^{-1}
paddle drag coefficient	1.75
paddle tip velocity	2 ft/sec
relative water/paddle velocity	1.5 ft/sec

(a) What is the theoretical power required to drive the paddle?

(A) 12 hp

(B) 17 hp

(C) 60 hp

(D) 120 hp

(b) What is the drag force on the paddle?

(A) 450 lbf

(B) 910 lbf

(C) 5500 lbf

(D) 6400 lbf

(c) What is the required paddle area?

(A) 900 ft^2

(B) 1200 ft^2

(C) 1500 ft^2

(D) 1700 ft^2

Filtration

6. A water treatment plant has four square rapid sand filters. The flow rate is 4 gal/min-ft^2. Each filter has a treatment capacity of 600,000 gal/day. Each filter is backwashed once a day for 8 min. The rate of rise during washing is 24 in/min.

(a) What are the inside dimensions of each filter?

(A) 6 ft × 6 ft

(B) 8 ft × 8 ft

(C) 10 ft × 10 ft

(D) 12 ft × 12 ft

(b) What percentage of the filtered water is used for backwashing?

(A) 2%

(B) 4%

(C) 8%

(D) 10%

7. A water treatment plant has five identical rapid sand filters. Each is square in plan and has a capacity of 1.0 MGD. The application rate is 4 gal/min-ft^2, and the total wetted depth is 10 ft.

(a) What should be the cross-section dimensions of each filter?

(A) 7 ft × 7 ft

(B) 9 ft × 9 ft

(C) 10 ft × 10 ft

(D) 13 ft × 13 ft

(b) Each filter is backwashed every day for 5 min. The rate of rise of the backwash water is 2 ft/min. What percentage of the plant's filtered water will be used for backwashing?

(A) 1.3%

(B) 1.9%

(C) 2.4%

(D) 3.1%

Precipitation Softening

8. A town's water supply has the following ionic concentrations.

Al^{+++}	0.5 mg/L
Ca^{++}	80.2 mg/L
Cl^-	85.9 mg/L
CO_2	19 mg/L
CO_3^{--}	0
Fe^{++}	1.0 mg/L
Fl^-	0
HCO_3^-	185 mg/L
Mg^{++}	24.3 mg/L
Na^+	46.0 mg/L
NO_3^-	0
SO_4^{--}	125 mg/L

(a) What is the total hardness (as $CaCO_3$)?

(A) 160 mg/L

(B) 200 mg/L

(C) 260 mg/L

(D) 300 mg/L

(b) How much slaked lime (as substance) is required to combine with the carbonate hardness?

(A) 45 mg/L

(B) 90 mg/L

(C) 130 mg/L

(D) 150 mg/L

(c) How much soda ash (as substance) is required to react with the carbonate hardness?

(A) 0 mg/L

(B) 15 mg/L

(C) 35 mg/L

(D) 60 mg/L

9. A city's water supply contains the following ionic concentrations.

$Ca(HCO_3)_2$	137 mg/L as $CaCO_3$
$MgSO_4$	72 mg/L as $CaCO_3$
CO_2	0

(a) How much slaked lime will be required to soften 1,000,000 gal of this water to a hardness of 100 mg/L (as $CaCO_3$) if 30 mg/L (as $CaCO_3$) of excess lime is used?

(A) 860 lbm

(B) 930 lbm

(C) 1170 lbm

(D) 1310 lbm

(b) The soda ash required to soften 1,000,000 gal of this water to a hardness of 100 mg/L (as $CaCO_3$) if there is 30 mg/L (as $CaCO_3$) of excess lime is most nearly

(A) 0 lbm

(B) 300 lbm

(C) 500 lbm

(D) 700 lbm

(c) How much soda ash (as $CaCO_3$) is required to soften 1,000,000 gal of this water to a hardness of 50 mg/L (as $CaCO_3$) assuming all carbonate hardness has been removed?

(A) 10 mg/L

(B) 20 mg/L

(C) 40 mg/L

(D) 50 mg/L

Zeolite Softening

10. The water described in Prob. 9 is to be softened using a zeolite process with the following characteristics: exchange capacity, 10,000 grains/ft^3; and salt requirement, 0.5 lbm per 1000 grains hardness removed. How much salt is required to soften the water to 100 mg/L hardness?

(A) 700 lbm/MG

(B) 1800 lbm/MG

(C) 2500 lbm/MG

(D) 3200 lbm/MG

11. The hardness of water from an underground aquifer is to be reduced from 245 mg/L (as $CaCO_3$) to 80 mg/L (as $CaCO_3$) using a zeolite process. The volumetric flow rate is 20,000 gal/day. The process has the following characteristics: resin exchange capacity, 20,000 grains/ft^3; and zeolite volume, 2 ft^3.

(a) What fraction of the water is bypassed around the process?

(A) 0.15

(B) 0.33

(C) 0.67

(D) 0.85

(b) What is the time between regenerations of the softener?

(A) 5 hr

(B) 16 hr

(C) 24 hr

(D) 30 hr

Demand

12. (*Time limit: one hour*) An area is expected to attain a population of 40,000 in 20 years. The cumulative per capita water demand for a peak day in the area is shown in the following diagram.

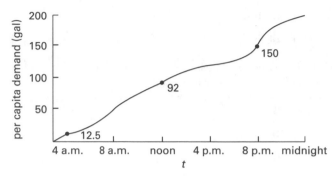

(a) What is the total per capita demand for a peak day?

(A) 160 gal

(B) 200 gal

(C) 240 gal

(D) 290 gal

(b) Assuming uniform operation over 24 hours, what storage volume is required in the treatment plant for all uses, including fire fighting demand?

(A) 1.1 MG

(B) 2.5 MG

(C) 3.2 MG

(D) 5.5 MG

(c) Assume that the pumping station runs uniformly from 4 a.m. until 8 a.m. to fill the storage tanks for the day. What storage is required to meet all uses, including fire fighting? (Use the American Insurance Association equation to calculate fire fighting demand.)

(A) 3.2 MG

(B) 3.8 MG

(C) 5.5 MG

(D) 8.0 MG

13. The water supply for a town of 15,000 people is taken from a river. The average consumption is 110 gpcd. The water has the following characteristics.

turbidity	20–100 NTU (varies)
total hardness	less than 60 mg/L as $CaCO_3$
coliform count	200 to 1000 per 100 mL (varies)

(a) What capacity should the distribution and treatment system have? (Use the American Insurance Association equation to calculate fire fighting demand.)

(A) 6000 gpm

(B) 7500 gpm

(C) 9500 gpm

(D) 12,000 gpm

(b) If the application rate is 4 gal/min-ft^2, what total filter area is required?

(A) 290 ft^2

(B) 580 ft^2

(C) 910 ft^2

(D) 1100 ft^2

(c) Is softening required?

(A) No, 60 mg/L is soft water.

(B) No, softening would interfere with turbidity removal.

(C) Yes, the turbidity and coliform counts would also benefit.

(D) Yes, all municipal water should be softened.

(d) If a chlorine dose of 2 mg/L is required to obtain the desired chlorine residual, how much chlorine is required every 24 hours? (Disregard fire fighting flow.)

(A) 15 lbm

(B) 22 lbm

(C) 28 lbm

(D) 32 lbm

SOLUTIONS

1. (a) Calculate the overflow rate.

$$v^* = \frac{100{,}000 \ \dfrac{\text{gal}}{\text{ft}^2\text{-day}}}{\left(7.48 \ \dfrac{\text{gal}}{\text{ft}^3}\right)\left(24 \ \dfrac{\text{hr}}{\text{day}}\right)\left(60 \ \dfrac{\text{min}}{\text{hr}}\right)}$$

$$= 9.28 \ \text{ft/min}$$

Using interpolation, the mass fraction remaining is

$$x_c = 0.45 + \left(\frac{9.28 \ \dfrac{\text{ft}}{\text{min}} - 5.0 \ \dfrac{\text{ft}}{\text{min}}}{10.0 \ \dfrac{\text{ft}}{\text{min}} - 5.0 \ \dfrac{\text{ft}}{\text{min}}}\right)(0.54 - 0.45)$$

$$= 0.527 \ \text{remains in flow}$$

$$1 - x_c = 1 - 0.527 = 0.473$$

$$\boxed{47\% \ \text{completely removed}}$$

The answer is (C).

(b) Plot the mass fraction remaining versus the settling velocity.

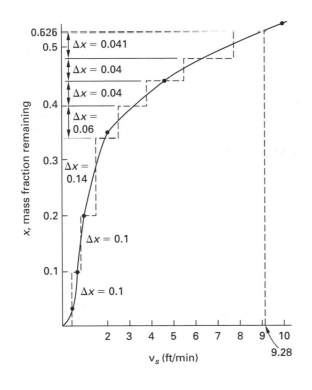

$$x_c \approx 0.525 \quad (52.5\%)$$

This corresponds to the value in part (a).

Determine $\Delta x \, v_t$ by graphical integration.

Δx	v_t	$\Delta x v_t$
0.1	0.5	0.05
0.1	0.8	0.08
0.14	1.4	0.196
0.06	2.5	0.15
0.04	3.75	0.15
0.04	5.5	0.22
0.041	7.75	0.318
	total	1.164

The overall removal efficiency is

$$f = 1 - x_c + \sum \frac{\Delta x v_t}{v^*}$$

$$= 1 - 0.525 + \frac{1.164 \ \dfrac{\text{ft}}{\text{min}}}{9.28 \ \dfrac{\text{ft}}{\text{min}}}$$

$$= 0.60 \quad (60\%)$$

The answer is (D).

2. From Fig. 26.2, $v_s = \boxed{0.7 \ \text{ft/sec.}}$

The answer is (B).

3. (a) The surface area is

$$A_{\text{surface}} = \frac{Q}{v^*} = \frac{2.8 \times 10^6 \ \dfrac{\text{gal}}{\text{day}}}{700 \ \dfrac{\text{gal}}{\text{ft}^2\text{-day}}} = 4000 \ \text{ft}^2$$

Since $A = \dfrac{\pi}{4}D^2$,

$$D = \sqrt{\frac{(4)(4000 \ \text{ft}^2)}{\pi}} = \boxed{71.4 \ \text{ft}}$$

The answer is (C).

(b) From Eq. 26.18, the volume is

$$V = tQ = \frac{\left(2.8 \times 10^6 \ \dfrac{\text{gal}}{\text{day}}\right)(2 \ \text{hr})}{24 \ \dfrac{\text{hr}}{\text{day}}}$$

$$= 2.333 \times 10^5 \ \text{gal}$$

The depth is

$$d = \frac{V}{A_{\text{surface}}} = \frac{2.333 \times 10^5 \text{ gal}}{(4000 \text{ ft}^2)\left(7.48 \frac{\text{gal}}{\text{ft}^3}\right)}$$

$$= \boxed{7.8 \text{ ft} \quad (8 \text{ ft})}$$

The answer is (B).

(c) Use Eq. 26.7.

$$v^* = \frac{Q}{A} = \frac{1.1 \times 10^6 \frac{\text{gal}}{\text{day}}}{4000 \text{ ft}^2}$$

$$= \boxed{275 \text{ gal/day-ft}^2 \quad (280 \text{ gal/day-ft}^2)}$$

The answer is (D).

(d) Use Eq. 26.6.

$$t = \frac{V}{Q} = \frac{(2.333 \times 10^5 \text{ gal})\left(24 \frac{\text{hr}}{\text{day}}\right)}{1.1 \times 10^6 \frac{\text{gal}}{\text{day}}}$$

$$= \boxed{5.09 \text{ hr} \quad (5 \text{ hr})}$$

The answer is (B).

4.   volume per basin $= (90 \text{ ft})(16 \text{ ft})(12 \text{ ft})$

$$= 17{,}280 \text{ ft}^3$$

(The freeboard is not given.)

The area per basin is

$$A = (90 \text{ ft})(16 \text{ ft}) = 1440 \text{ ft}^2$$

The three-month peak flow per basin is

$$(2)\left(\frac{1.5 \text{ MGD}}{2}\right) = 1.5 \text{ MGD}$$

$$(1.5 \text{ MGD})\left(1.547 \frac{\frac{\text{ft}^3}{\text{sec}}}{\text{MGD}}\right) = 2.32 \text{ ft}^3/\text{sec}$$

The detention time at peak flow is given by Eq. 26.6.

$$t = \frac{V}{Q} = \frac{17{,}280 \text{ ft}^3}{\left(2.32 \frac{\text{ft}^3}{\text{sec}}\right)\left(60 \frac{\text{min}}{\text{hr}}\right)\left(60 \frac{\text{sec}}{\text{min}}\right)} = 2.07 \text{ hr}$$

Since 2.07 hr < 4 hr, this is not acceptable.

The weir loading is

$$\frac{1.5 \times 10^6 \frac{\text{gal}}{\text{day}}}{48 \text{ ft}} = 31{,}250 \text{ gal/day-ft}$$

Since 31,250 gal/day-ft > 20,000 gal/day-ft, this is not acceptable either.

The overflow rate is

$$v^* = \frac{Q}{A} = \frac{\left(2.32 \frac{\text{ft}^3}{\text{sec}}\right)\left(60 \frac{\text{sec}}{\text{min}}\right)}{1440 \text{ ft}^2} = 0.0967 \text{ ft/min}$$

Since 0.0967 ft/min < 0.5 ft/min, this is acceptable.

At low flow,

$$\text{flow} = \frac{0.7}{2.0} = 0.35 \quad [35\% \text{ of high flow}]$$

$$t = \frac{2.07 \text{ hr}}{0.35} = 5.91 \text{ hr} \quad [\text{acceptable}]$$

$$\text{weir loading} = (0.35)\left(31{,}250 \frac{\text{gal}}{\text{day-ft}}\right)$$

$$= 10{,}938 \frac{\text{gal}}{\text{day-ft}} \quad [\text{acceptable}]$$

$$v^* = (0.35)\left(0.1 \frac{\text{ft}}{\text{min}}\right) = 0.035 \text{ ft/min}$$

$$[\text{acceptable}]$$

> The basins have been correctly designed for low flow but not for peak flow. One or more basins should be used.

The answer is (C).

5. For 60°F water, $\mu = 2.359 \times 10^{-5}$ lbf-sec/ft^2.

From Eq. 26.25,

$$P = \mu G^2 V_{\text{tank}}$$

$$= \left(2.359 \times 10^{-5} \frac{\text{lbf-sec}}{\text{ft}^2}\right)\left(45 \frac{1}{\text{sec}}\right)^2 (200{,}000 \text{ ft}^3)$$

$$= 9554 \text{ ft-lbf/sec}$$

(a) $$\text{water horsepower} = \frac{9544 \frac{\text{ft-lbf}}{\text{sec}}}{550 \frac{\text{ft-lbf}}{\text{hp-sec}}}$$

$$= \boxed{17.4 \text{ hp} \quad (17 \text{ hp})}$$

The answer is (B).

(b) Since work = force × distance, then power = force × velocity.

$$D = \frac{P}{v} = \frac{9554 \frac{\text{ft-lbf}}{\text{sec}}}{1.5 \frac{\text{ft}}{\text{sec}}} = \boxed{6369 \text{ lbf} \quad (6400 \text{ lbf})}$$

The answer is (D).

(c) Use Eq. 26.20.

$$A = \frac{2gF_D}{C_D \gamma v^2} = \frac{(2)\left(32.2 \frac{\text{ft}}{\text{sec}^2}\right)(6369 \text{ lbf})}{(1.75)\left(62.4 \frac{\text{lbf}}{\text{ft}^3}\right)\left(1.5 \frac{\text{ft}}{\text{sec}}\right)^2}$$
$$= \boxed{1669 \text{ ft}^2 \quad (1700 \text{ ft}^2)}$$

The answer is (D).

6. (a) The required area is

$$A = \frac{Q}{v^*} = \frac{600,000 \frac{\text{gal}}{\text{day}}}{\left(4 \frac{\text{gal}}{\text{min-ft}^2}\right)\left(24 \frac{\text{hr}}{\text{day}}\right)\left(60 \frac{\text{min}}{\text{hr}}\right)}$$
$$= 104.2 \text{ ft}^2$$
$$\boxed{\text{Use 10 ft} \times \text{10 ft.}}$$

The answer is (C).

(b) The required volume is

$$V = tAv = \left(8 \frac{\text{min}}{\text{day}}\right)(100 \text{ ft}^2)\left(2 \frac{\text{ft}}{\text{min}}\right)\left(7.48 \frac{\text{gal}}{\text{ft}^3}\right)$$
$$= 11,968 \text{ gal/day}$$

The backwash fraction is

$$\frac{11,968 \frac{\text{gal}}{\text{day}}}{600,000 \frac{\text{gal}}{\text{day}}} = \boxed{0.01995 \quad (2\%)}$$

The answer is (A).

7. (a) $A = \dfrac{Q}{v^*} = \dfrac{1 \times 10^6 \frac{\text{gal}}{\text{day}}}{\left(24 \frac{\text{hr}}{\text{day}}\right)\left(60 \frac{\text{min}}{\text{hr}}\right)\left(4 \frac{\text{gal}}{\text{min-ft}^2}\right)}$
$$= 173.6 \text{ ft}^2$$
$$\text{width} = \sqrt{173.6 \text{ ft}^2} = 13.2 \text{ ft}$$
$$\boxed{\text{Use 13 ft} \times \text{13 ft.}}$$

The answer is (D).

(b) The water volume is

$$V = tAv = \left(5 \frac{\text{min}}{\text{day}}\right)(173.6 \text{ ft}^2)\left(2 \frac{\text{ft}}{\text{min}}\right)(5 \text{ filters})$$
$$\times \left(7.48 \frac{\text{gal}}{\text{ft}^3}\right)$$
$$= 64,926 \text{ gal/day}$$

The backwash fraction is

$$\frac{64,926 \frac{\text{gal}}{\text{day}}}{5 \times 10^6 \frac{\text{gal}}{\text{day}}} = \boxed{0.013 \quad (1.3\%)}$$

The answer is (A).

8. (a) Using the factors from App. 22.B,

	mg/L as substance		factor		
Ca^{++}:	80.2	×	2.5	=	200.5 mg/L
Mg^{++}:	24.3	×	4.1	=	99.63 mg/L
Fe^{++}:	1	×	1.79	=	1.79 mg/L
Al^{+++}:	0.5	×	5.56	=	2.78 mg/L
			hardness =		304.7 mg/L

The total hardness is approximately $\boxed{300 \text{ mg/L}}$ as $CaCO_3$.

The answer is (D).

(b) To remove the carbonate hardness, the carbon dioxide must first be removed. The CO_2 concentration in $CaCO_3$ equivalents is

$$CO_2: 19 \frac{\text{mg}}{\text{L}} \times 2.27 = 43.13 \text{ mg/L as } CaCO_3$$

Add lime to remove the carbonate hardness. It does not matter whether the HCO_3^- comes from Mg^{++}, Ca^{++}, or Fe^{++}; adding lime will remove it.

There may be Mg^{++}, Ca^{++}, or Fe^{++} ions left over in the form of noncarbonate hardness, but the problem asked for carbonate hardness. Converting from mg/L of substance to mg/L as $CaCO_3$,

$$HCO_3^-: 185 \frac{\text{mg}}{\text{L}} \times 0.82 = 151.7 \text{ mg/L}$$

The total equivalents to be neutralized are

$$43.13 \frac{\text{mg}}{\text{L}} + 151.7 \frac{\text{mg}}{\text{L}} = 194.83 \text{ mg/L}$$

Convert Ca(OH)$_2$ using App. 22.B.

$$\frac{mg}{L} \text{ of Ca(OH)}_2 = \frac{194.83 \frac{mg}{L}}{1.35}$$

$$= \boxed{144.3 \text{ mg/L} \quad (150 \text{ mg/L})}$$

The answer is (D).

(c) | No soda ash is required since it is used to remove noncarbonate hardness. |

The answer is (A).

9. (a) Ca(HCO$_3$)$_2$ and MgSO$_4$ both contribute to hardness. Since 100 mg/L of hardness is the goal, leave all MgSO$_4$ in the water. Take out 137 mg/L + 72 mg/L − 100 mg/L = 109 mg/L of Ca(HCO$_3$)$_2$. From App. 22.B (including the excess even though the reaction is not complete),

$$\text{pure Ca(OH)}_2 = \frac{30 \frac{mg}{L} + 109 \frac{mg}{L}}{1.35}$$

$$= 102.96 \text{ mg/L}$$

$$\left(102.96 \frac{mg}{L}\right)\left(8.345 \frac{lbm\text{-}L}{mg\text{-}MG}\right) = 859 \text{ lbm/MG}$$

The amount of slaked time required is approximately $\boxed{860 \text{ lbm/MG.}}$

The answer is (A).

(b) Since the goal is a residual hardness of 100 mg/L and the magnesium sulfate only contributes 72 mg/L, it does not have to be removed. No soda ash is required.

The answer is (A).

(c) In order to reduce hardness to 50 mg/L, 72 mg/L − 50 mg/L = 22 mg/L of MgSO$_4$ must be removed. (This assumes that all of the carbonate hardness has already been removed.) This will require 22 mg/L (as CaCO$_3$) of soda ash.

The answer is (B).

10. There are 7000 grains in a pound. The hardness removed is

$$137 \frac{mg}{L} + 72 \frac{mg}{L} - 100 \frac{mg}{L} = 109 \text{ mg/L}$$

$$\left(109 \frac{mg}{L}\right)\left(8.345 \frac{lbm\text{-}L}{mg\text{-}MG}\right) = 909.6 \text{ lbm hardness/MG}$$

$$\left(\frac{0.5 \text{ lbm}}{1000 \text{ gr}}\right)\left(909.6 \frac{lbm}{MG}\right)\left(7000 \frac{gr}{lbm}\right)$$

$$= \boxed{3184 \text{ lbm/MG} \quad (3200 \text{ lbm/MG})}$$

The answer is (D).

11. (a) A bypass process is required.

$$\text{fraction bypassed: } \frac{80 \frac{mg}{L}}{245 \frac{mg}{L}} = \boxed{0.327 \quad (0.33)}$$

fraction processed: 1 − 0.327 = 0.673

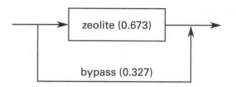

The answer is (B).

(b) The maximum hardness reduction is

$$\frac{(2 \text{ ft}^3)\left(20{,}000 \frac{gr}{ft^3}\right)}{7000 \frac{gr}{lbm}} = 5.71 \text{ lbm}$$

The hardness removal rate is

$$\frac{(0.673)\left(245 \frac{mg}{L}\right)\left(8.345 \times 10^{-6} \frac{lbm}{gal}\right)\left(20{,}000 \frac{gal}{day}\right)}{24 \frac{hr}{day}}$$

$$= 1.15 \text{ lbm/hr}$$

$$t = \frac{5.71 \text{ lbm}}{1.15 \frac{lbm}{hr}} = \boxed{4.97 \text{ hr} \quad (5 \text{ hr})}$$

The answer is (A).

12. (a) The given graph shows that by the end of the day, the cumulative demand has risen to 200 gal per person. The daily demand is $\boxed{200 \text{ gal.}}$

The answer is (B).

(b) The analysis is similar to that used with reservoir sizing.

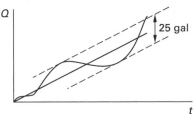

From the cumulative flow quantity, the storage requirement per capita-day is 25 gal.

The population use is

$$\left(25 \ \frac{\text{gal}}{\text{day-person}}\right)(40{,}000 \text{ people}) = 1{,}000{,}000 \text{ gal/day}$$

From Eq. 26.69, the fire fighting requirement is

$$Q = 1020\sqrt{P}(1 - 0.01\sqrt{P})$$
$$= 1020\sqrt{40}(1 - 0.01\sqrt{40})$$
$$= 6043 \text{ gal/min}$$

Therefore, the flow rate is $6043/1000 = 6$ thousands of gallons. Maintain the flow for 4 hr (approximate ISO specifications).

$$\text{capacity} = 1{,}000{,}000 \text{ gal}$$
$$+ \left(6043 \ \frac{\text{gal}}{\text{min}}\right)\left(60 \ \frac{\text{min}}{\text{hr}}\right)(4 \text{ hr})$$
$$= \boxed{2{,}450{,}000 \text{ gal} \quad (2.5 \text{ MG})}$$

The answer is (B).

(c) The pump supplies demand from 4 a.m. until 8 a.m. The storage supplies demand from 8 a.m. until 4 a.m.

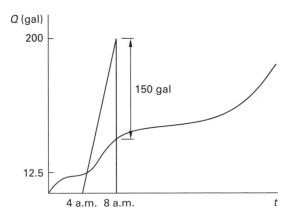

The population use is

$$(150 \text{ gal} + 12.5 \text{ gal})(40{,}000) = 6{,}500{,}000 \text{ gal}$$

Add fire fighting.

$$6{,}500{,}000 \text{ gal} + \left(6043 \ \frac{\text{gal}}{\text{min}}\right)\left(60 \ \frac{\text{min}}{\text{hr}}\right)(4 \text{ hr})$$
$$= \boxed{7{,}950{,}000 \text{ gal} \quad (8.0 \text{ MG})}$$

The answer is (D).

13. (a) From Table 26.7, use 3 as the peak multiplier.

$$\frac{\left(110 \ \frac{\text{gal}}{\text{day}}\right)(15{,}000)(3)}{\left(24 \ \frac{\text{hr}}{\text{day}}\right)\left(60 \ \frac{\text{min}}{\text{hr}}\right)} = 3437 \text{ gal/min}$$

The fire fighting requirements are given by Eq. 26.69.

$$Q = 1020\sqrt{P}(1 - 0.01\sqrt{P})$$
$$= 1020\sqrt{15}(1 - 0.01\sqrt{15})$$
$$= 3797 \text{ gal/min}$$

The total maximum demand for which the distribution system should be designed is

$$3437 \ \frac{\text{gal}}{\text{min}} + 3797 \ \frac{\text{gal}}{\text{min}}$$
$$= \boxed{7234 \text{ gal/min} \quad (7500 \text{ gal/min})}$$

The answer is (B).

(b) The filter area should not be based on the maximum hourly rate since some of the demand during the peak hours can come from storage (clearwell or tanks). Also, fire requirements can bypass the filters if necessary.

$$\left(110 \ \frac{\text{gal}}{\text{day-person}}\right)(15{,}000 \text{ people}) = 1.65 \times 10^6 \text{ gal/day}$$

Using a flow rate of 4 gpm/ft^2, the required filter area is

$$A = \frac{Q}{\text{v}^*} = \frac{1.65 \times 10^6 \ \frac{\text{gal}}{\text{day}}}{\left(4 \ \frac{\text{gal}}{\text{min-ft}^2}\right)\left(24 \ \frac{\text{hr}}{\text{day}}\right)\left(60 \ \frac{\text{min}}{\text{hr}}\right)}$$
$$= \boxed{286.5 \text{ ft}^2 \quad (290 \text{ ft}^2)}$$

The answer is (A).

(c) No, 60 mg/L is soft water.

The answer is (A).

(d) Disregarding the fire flow, the required average (not peak) daily chlorine mass is given by Eq. 26.16.

$$F = \frac{DQ\left(8.345 \ \frac{\text{lbm-L}}{\text{mg-MG}}\right)}{PG}$$

$$= \frac{\left(110 \ \frac{\text{gal}}{\text{day-person}}\right)(15{,}000 \ \text{people})}{\times \left(2 \ \frac{\text{mg}}{\text{L}}\right)\left(8.345 \ \frac{\text{lbm-L}}{\text{mg-MG}}\right)}{10^6 \ \frac{\text{gal}}{\text{MG}}}$$

$$= \boxed{27.54 \ \text{lbm/day}}$$

The answer is (C).

27 Biochemistry, Biology, and Bacteriology

PRACTICE PROBLEMS

1. (*Time limit: one hour*) A fresh wastewater sample containing nitrate ions, sulfate ions, and dissolved oxygen is placed in a sealed jar absent of air.

(a) What is the correct sequence of oxidation of the compounds?

   (A) nitrate, dissolved oxygen, and then sulfate

   (B) sulfate, nitrate, and then dissolved oxygen

   (C) dissolved oxygen, nitrate, and then sulfate

   (D) none of the above

(b) Obnoxious odors will

   (A) appear in the sample when the dissolved oxygen is exhausted

   (B) appear in the sample when the dissolved oxygen and nitrate are exhausted

   (C) appear in the sample when the dissolved oxygen, nitrate, and sulfate are exhausted

   (D) not appear

(c) Bacteria will convert ammonia to nitrate if the bacteria are

   (A) phototropic

   (B) autotrophic

   (C) thermophilic

   (D) obligate anaerobes

(d) Bacteria will generally reduce nitrate to nitrogen gas only if the bacteria are

   (A) photosynthetic

   (B) obligate aerobic

   (C) facultative heterotrophic

   (D) aerobic phototropic

(e) A bacteriophage is a

   (A) bacterial enzyme

   (B) virus that infects bacteria

   (C) mesophilic organism

   (D) virus that stimulates bacterial growth

(f) Algal growth in the wastewater

   (A) will be inhibited when the dissolved oxygen is increased

   (B) will be augmented when nitrifying bacteria are also found in the solution

   (C) will be inhibited when toxins are found in the solution or when the dissolved oxygen is exhausted

   (D) cannot be inhibited by chemical means

(g) In the presence of nitrifying bacteria, nontoxic inorganic compounds, and sunlight, algal growth in the wastewater sample will be

   (A) prevented

   (B) inhibited

   (C) unaffected

   (D) enhanced

(h) The addition of protozoa to the wastewater will

   (A) not change the biochemical composition of the wastewater

   (B) increase the growth of algae in wastewater

   (C) increase the growth of bacteria in wastewater

   (D) decrease the growth of algae and bacteria in the wastewater

(i) Coliform bacteria in wastewater from human, animal, or soil sources

(A) can be distinguished in the multiple-tube fermentation test

(B) can be categorized into only two groups: human/animal and soil

(C) can be distinguished if multiple-tube fermentation and Eschericheiae coli (EC) tests are both used

(D) cannot be distinguished

(j) Under what specific condition would a presence-absence (P-A) coliform test be used on this sample?

(A) if the sample contains known pathogens

(B) when multiple-tube fermentation indicates positive dilution in a presumptive test

(C) when multiple-tube fermentation indicates negative dilution in a presumptive test

(D) if the sample will be used as drinking water

2. A waste stabilization pond will be used in a municipal wastewater treatment system.

(a) Explain the function of bacteria and algae in the stabilization pond's application.

(b) Explain why algae would be a problem if it were present in the discharge from the pond.

(c) Define mechanisms that could be utilized to control algae and that would prove helpful in the development and use of this wastewater treatment system.

(d) Explain what is meant by *facultative pond*.

SOLUTIONS

1. (a) The sequence of oxygen usage reduced by bacteria is dissolved oxygen, nitrate, and then sulfate.

The answer is (C).

(b) Following the sequence of oxidation, obnoxious odors (e.g., hydrogen sulfide) will occur when dissolved oxygen and nitrate are exhausted.

The answer is (B).

(c) Nitrification is performed by autotrophic bacteria to gain energy for growth by synthesis of carbon dioxide in an aerobic environment.

The answer is (B).

(d) Facultative heterotropic bacteria can decompose organic matter to gain energy under anaerobic conditions by removing the oxygen from nitrate, releasing nitrogen gas.

The answer is (C).

(e) A bacteriophage is a virus that infects bacteria.

The answer is (B).

(f) Algae are photosynthetic (gaining energy from light), releasing oxygen during metabolism. They thrive in aerobic environments. The presence of nitrifying bacteria and toxins and the depletion of dissolved oxygen would inhibit the growth process.

The answer is (C).

(g) Nitrifying bacteria are nonphotosynthetic, gaining energy by taking in oxygen to oxidize reduced inorganic nitrogen. The presence of inorganic nutrients would maintain the nitrification process, thereby limiting algal growth.

The answer is (B).

(h) Protozoa consume bacteria and algae in wastewater treatment and in the aquatic food chain.

The answer is (D).

(i) Fecal coliforms from humans and other warm-blooded animals are the same bacterial species. Coliforms originating from the soil can be separated by a confirmatory procedure using EC medium broth incubated at the elevated temperature of 44.5°C (112°F).

The answer is (B).

(j) The presence-absence technique is used for the testing of drinking water.

The answer is (D).

2. (a) A waste stabilization pond's operation is dependent on the reaction of bacteria and algae. Organic matter is metabolized by bacteria to produce the principle products of carbon dioxide, water, and a small amount of ammonia nitrogen. Algae convert sunlight into energy through photosynthesis. They utilize the end products of cell synthesis and other nutrients to synthesize new cells and produce oxygen. The most important role of the algae is in the production of oxygen in the pond for use by aerobic bacteria. In the absence of sunlight, the algae will consume oxygen in the same manner as bacteria. Algae removal is important in producing a high-quality effluent from the pond.

(b) The discharge of algae increases suspended solids in the discharge and may present a problem in meeting water quality criteria. The algae exert an oxygen demand when they settle to the bottom of the stream and undergo respiration.

(c) The following methods have been suggested for control of algae: (1) multiple ponds in series, (2) drawing off of effluent from below the surface by use of a good baffling arrangement to avoid algae concentrations, (3) sand filter or rock filter for algae removal, (4) alum addition and flocculation, (5) microscreening, and (6) chlorination to kill algae. Chlorination may increase BOD loading due to dead algae cells releasing stored organic material.

(d) Facultative ponds have two zones of treatment: an aerobic surface layer in which oxygen is used by aerobic bacteria for waste stabilization and an anaerobic bottom zone in which sludge decomposition occurs. No artificially induced aeration is used.

Environmental

28 Wastewater Quantity and Quality

PRACTICE PROBLEMS

Sewer Velocities and Sizing

1. (*Time limit: one hour*) A town of 10,000 people (125 gpcd) has its own primary treatment plant.

(a) What mass of total solids should the treatment plant expect?

    (A) 57 lbm/day

    (B) 740 lbm/day

    (C) 3800 lbm/day

    (D) 8300 lbm/day

(b) If the town is 4 mi from the treatment plant and 400 ft above it in elevation, what minimum size pipe should be used between the town and the plant, assuming that the pipe flows full? Disregard infiltration.

    (A) 8 in

    (B) 12 in

    (C) 14 in

    (D) 18 in

2. (*Time limit: one hour*) Your client has just completed a subdivision, and his sewage lines hook up to a collector that goes into a trunk. A problem has developed in the first manhole up from the trunk, and the collector pipe overflows periodically. An industrial plant is hooked directly into the trunk, and it is this plant's flow that is making your client's line back up. The subdivision is in a flood plain, and the sewer lines are very flat and cannot be steepened. (a) Describe two possible solutions to this problem. (b) Sketch plan views of your solutions.

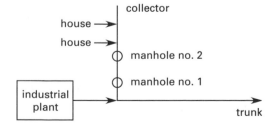

Dilution

3. Town A wants to discharge wastewater into a river 5.79 km upstream of town B. The state standard level for dissolved oxygen is 5 mg/L. Other data pertinent to this problem are given in the following table.

parameter	wastewater	river
flow (m^3/s)	0.280	0.877
ultimate BOD at 28°C (mg/L)	6.44	7.00
DO (mg/L)	1.00	6.00
K_d at 28°C, d^{-1} (base-10)	N/A	0.199
K_r at 28°C, d^{-1} (base-10)	N/A	0.370
velocity (m/s)	N/A	0.650
temperature (°C)	28	28

(a) Will the dissolved oxygen level at town A be reduced below the state standard level?

    (A) yes; $DO_{min} = 4.1$ mg/L

    (B) yes; $DO_{min} = 4.8$ mg/L

    (C) no; $DO_{min} = 5.3$ mg/L

    (D) no; $DO_{min} = 5.9$ mg/L

(b) Will the dissolved oxygen level be below the state standard level at town B?

    (A) yes; $DO_{min} = 4.1$ mg/L

    (B) yes; $DO_{min} = 4.8$ mg/L

    (C) no; $DO_{min} = 5.3$ mg/L

    (D) no; $DO_{min} = 5.9$ mg/L

(c) At any point downstream, will the dissolved oxygen level be below the state standard level?

    (A) yes; $DO_{min} = 4.1$ mg/L

    (B) yes; $DO_{min} = 4.7$ mg/L

    (C) no; $DO_{min} = 5.1$ mg/L

    (D) no; $DO_{min} = 5.5$ mg/L

(d) Determine if the discharge will reduce the dissolved oxygen in the river at town B below the state standard level if the winter river temperature drops to 12°C (before mixing) and all other characteristics remain the same.

(A) yes; $DO_{min} = 4.5$ mg/L

(B) yes; $DO_{min} = 4.9$ mg/L

(C) no; $DO_{min} = 5.0$ mg/L

(D) no; $DO_{min} = 5.9$ mg/L

4. To protect aquatic life, the limit on increase in temperature in a certain stream is 2°C at seasonal low flow. In addition, the stream must not contain more than 0.002 mg/L of un-ionized ammonia.

A manufacturing facility proposes to draw cooling water from the stream and return it to the stream at a higher temperature. They will also discharge some un-ionized ammonia in the cooling water.

The critical low flow in the stream is 1200 m³/s at 20°C. The stream has no measurable un-ionized ammonia in its natural state. The facility intends to withdraw 60 m³/s of cooling water and return it to the stream without loss of mass.

(a) What is the maximum temperature for the discharge water if the stream is not to be damaged?

(A) 40°C

(B) 60°C

(C) 80°C

(D) 90°C

(b) What is the maximum ammonia concentration for the discharge water if the stream is not to be damaged?

(A) 0.04 mg/L

(B) 0.06 mg/L

(C) 0.09 mg/L

(D) 0.13 mg/L

5. A rural town of 10,000 people discharges partially treated wastewater into a stream. The stream has the following characteristics.

minimum flow rate	120 ft³/sec
velocity	3 mi/hr
minimum dissolved oxygen	7.5 mg/L at 15°C
temperature	15°C
BOD	0

The reoxygenation and deoxygenation coefficients (base-10) are

reoxygenation coefficient of stream and effluent mixture	0.2 d⁻¹ at 20°C
deoxygenation coefficient	0.15 d⁻¹ at 20°C

The town's effluent has the following characteristics.

source	volume	BOD at 20°C	temp.
domestic	122 gpcd	0.191 lbm/cd	64°F
infiltration	116,000 gpd		51°F
industrial no. 1	180,000 gpd	800 mg/L	95°F
industrial no. 2	76,000 gpd	1700 mg/L	84°F

(a) What is the domestic waste BOD in mg/L at 20°C?

(A) 140 mg/L

(B) 160 mg/L

(C) 190 mg/L

(D) 220 mg/L

(b) What is the approximate total effluent 20°C BOD in mg/L just before discharge into the stream?

(A) 245 mg/L

(B) 270 mg/L

(C) 295 mg/L

(D) 315 mg/L

(c) What is the average temperature of the effluent just before discharge into the stream?

(A) 20°C

(B) 30°C

(C) 50°C

(D) 70°C

(d) If the wastewater does not contribute to dissolved oxygen at all, how far downstream is the theoretical point of minimum dissolved oxygen concentration?

(A) 40 mi

(B) 70 mi

(C) 80 mi

(D) 120 mi

(e) What is the theoretical minimum dissolved oxygen concentration in the stream?

(A) 4.7 mg/L

(B) 5.5 mg/L

(C) 7.3 mg/L

(D) 8.3 mg/L

Environmental

6. A sewage treatment plant is being designed to handle both domestic and industrial wastewaters. The city population is 20,000, and an excess capacity factor of 15% is to be used for future domestic expansion. The wastewaters have the following characteristics.

source	volume	BOD
domestic	100 gpcd	0.18 lbm/cd
industrial no. 1	1.3 MGD	1100 mg/L
industrial no. 2	1.0 MGD	500 mg/L

(a) What is the design population equivalent for the plant?

(A) 60,000

(B) 75,000

(C) 90,000

(D) 110,000

(b) What is the plant's organic loading?

(A) 2300 lbm/day

(B) 8100 lbm/day

(C) 14,000 lbm/day

(D) 20,000 lbm/day

(c) What is the plant's hydraulic loading?

(A) 4.6 MGD

(B) 5.8 MGD

(C) 6.3 MGD

(D) 7.5 MGD

SOLUTIONS

1. (a)   Assume: 125 gpcd for average flow
800 mg/L total solids

$$\dot{m} = \frac{(10{,}000)\left(125 \ \dfrac{\text{gal}}{\text{day}}\right)\left(800 \ \dfrac{\text{mg}}{\text{L}}\right)\left(8.345 \ \dfrac{\text{lbm-L}}{\text{mg-MG}}\right)}{1 \times 10^6 \ \dfrac{\text{gal}}{\text{MG}}}$$

$$= \boxed{8345 \ \text{lbm/day}}$$

(Minor variations in the assumptions should not affect the answer choice.)

The answer is (D).

(b) $S = \dfrac{400 \ \text{ft}}{(4 \ \text{mi})\left(5280 \ \dfrac{\text{ft}}{\text{mi}}\right)} = 0.01894$

Use Eq. 28.1.

$$\frac{Q_{\text{peak}}}{Q_{\text{ave}}} = \frac{18 + \sqrt{P}}{4 + \sqrt{P}} = \frac{18 + \sqrt{10}}{4 + \sqrt{10}}$$

$$\approx 3.0$$

$$Q_{\text{peak}} = \frac{\left(125 \ \dfrac{\text{gal}}{\text{day}}\right)(3)(10{,}000)\left(0.1337 \ \dfrac{\text{ft}^3}{\text{gal}}\right)}{\left(24 \ \dfrac{\text{hr}}{\text{day}}\right)\left(60 \ \dfrac{\text{min}}{\text{hr}}\right)\left(60 \ \dfrac{\text{sec}}{\text{min}}\right)}$$

$$= 5.80 \ \text{ft}^3/\text{sec}$$

Assume $n = 0.013$. From Eq. 19.16,

$$D = 1.335 \left(\frac{nQ}{\sqrt{S}}\right)^{3/8}$$

$$= (1.335)\left(\frac{(0.013)\left(5.80 \ \dfrac{\text{ft}^3}{\text{sec}}\right)}{\sqrt{0.01894}}\right)^{3/8}\left(12 \ \frac{\text{in}}{\text{ft}}\right)$$

$$= \boxed{12.8 \ \text{in} \quad (14 \ \text{in})}$$

(A 12 in pipe does not have the capacity of a 12.8 in pipe.)

The answer is (C).

2. (a) If the first manhole overflows, the piezometric head at the rim must be less than the head in the trunk. Further up at manhole no. 2, the increase in elevation is sufficient to raise rim no. 2, so the head in the trunk must be lowered.

Alternate solutions:

- relief storage (surge chambers) between trunk and manhole no. 1

- storage at plant and gradual release

- private trunk for plant

- private treatment and discharge for plant

- larger trunk capacity using larger or parallel pipes

- backflow preventors

(b)

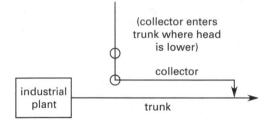

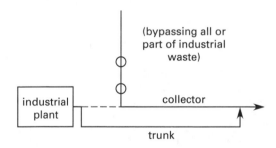

3. (a) Find the dissolved oxygen at town A using Eq. 28.35.

$$\text{DO} = \frac{Q_w \text{DO}_w + Q_r \text{DO}_r}{Q_w + Q_r}$$

$$= \frac{\left(0.280 \ \frac{\text{m}^3}{\text{s}}\right)\left(1 \ \frac{\text{mg}}{\text{L}}\right) + \left(0.877 \ \frac{\text{m}^3}{\text{s}}\right)\left(6 \ \frac{\text{mg}}{\text{L}}\right)}{0.280 \ \frac{\text{m}^3}{\text{s}} + 0.877 \ \frac{\text{m}^3}{\text{s}}}$$

$$= \boxed{4.79 \ \text{mg/L}}$$

So, immediately after mixing, the water will be reduced below the state standard of 5 mg/L.

The answer is (B).

(b) The composite ultimate BOD is

$$\text{BOD}_{u,28°C} = \frac{Q_w \text{BOD}_w + Q_r \text{BOD}_r}{Q_w + Q_r}$$

$$= \frac{\begin{array}{c}\left(0.280 \ \frac{\text{m}^3}{\text{s}}\right)\left(6.44 \ \frac{\text{mg}}{\text{L}}\right) \\ + \left(0.877 \ \frac{\text{m}^3}{\text{s}}\right)\left(7 \ \frac{\text{mg}}{\text{L}}\right)\end{array}}{0.280 \ \frac{\text{m}^3}{\text{s}} + 0.877 \ \frac{\text{m}^3}{\text{s}}}$$

$$= 6.8645 \ \text{mg/L}$$

From App. 22.C, $\text{DO}_{\text{sat}} = 7.92$ mg/L at 28°C. The initial deficit is

$$D_0 = \text{DO}_{\text{sat}} - \text{DO} = 7.92 \ \frac{\text{mg}}{\text{L}} - 4.79 \ \frac{\text{mg}}{\text{L}}$$

$$= 3.13 \ \text{mg/L}$$

Calculate travel time from town A to town B.

$$t = \frac{(5.79 \ \text{km})\left(1000 \ \frac{\text{m}}{\text{km}}\right)}{\left(0.650 \ \frac{\text{m}}{\text{s}}\right)\left(86\,400 \ \frac{\text{s}}{\text{d}}\right)}$$

$$= 0.1031 \ \text{d}$$

Calculate the deficit 5.79 km downstream. Use base-10 exponents because the K values are in base-10. Use Eq. 28.36.

$$D_t = \left(\frac{K_d \text{BOD}_u}{K_r - K_d}\right)\left(10^{-K_d t} - 10^{-K_r t}\right) + D_0\left(10^{-K_r t}\right)$$

$$= \frac{(0.199 \ \text{d}^{-1})\left(6.8645 \ \frac{\text{mg}}{\text{L}}\right)}{0.370 \ \frac{\text{mg}}{\text{L}} - 0.199 \ \frac{\text{mg}}{\text{L}}}$$

$$\times \left(\begin{array}{c}10^{-(0.199 \ \text{d}^{-1})(0.1031 \ \text{d})} \\ - 10^{-(0.370 \ \text{d}^{-1})(0.1031 \ \text{d})}\end{array}\right)$$

$$+ \left(3.13 \ \frac{\text{mg}}{\text{L}}\right)\left(10^{-(0.370 \ \text{d}^{-1})(0.1031 \ \text{d})}\right)$$

$$= 3.17 \ \text{mg/L}$$

Calculate the dissolved oxygen downstream.

$$\text{DO}_{\text{town B}} = \text{DO}_{\text{sat}} - D$$

$$= 7.92 \ \frac{\text{mg}}{\text{L}} - 3.17 \ \frac{\text{mg}}{\text{L}}$$

$$= \boxed{4.75 \ \text{mg/L}}$$

This is still below the state standard of 5 mg/L. More distance is required for reoxygenation to increase the oxygen concentration.

The answer is (B).

(c) Determine the critical time. Use Eq. 28.38.

$$t_c = \left(\frac{1}{K_r - K_d}\right)$$
$$\times \log_{10}\left(\left(\frac{K_d \text{BOD}_u - K_r D_0 + K_d D_0}{K_d \text{BOD}_u}\right)\left(\frac{K_r}{K_d}\right)\right)$$
$$= \left(\frac{1}{0.370\ \text{d}^{-1} - 0.199\ \text{d}^{-1}}\right)$$
$$\times \log_{10}\left(\left(\frac{\begin{array}{c}(0.199\ \text{d}^{-1})\left(6.8645\ \dfrac{\text{mg}}{\text{L}}\right) \\[4pt] -\ (0.370\ \text{d}^{-1})\left(3.13\ \dfrac{\text{mg}}{\text{L}}\right) \\[4pt] +\ (0.199\ \text{d}^{-1})\left(3.13\ \dfrac{\text{mg}}{\text{L}}\right)\end{array}}{(0.199\ \text{d}^{-1})\left(6.8645\ \dfrac{\text{mg}}{\text{L}}\right)}\right)\right.$$
$$\left.\times \left(\frac{0.370\ \text{d}^{-1}}{0.199\ \text{d}^{-1}}\right)\right)$$
$$= 0.3122\ \text{d}$$

Calculate critical deficit and dissolved oxygen. Use Eq. 28.39.

$$D_c = \left(\frac{K_d \text{BOD}_u}{K_r}\right)10^{-K_d t_c}$$
$$= \left(\frac{(0.199\ \text{d}^{-1})\left(6.8645\ \dfrac{\text{mg}}{\text{L}}\right)}{0.370\ \text{d}^{-1}}\right)$$
$$\times \left(10^{-(0.199\ \text{d}^{-1})(0.3122\ \text{d})}\right)$$
$$= 3.20\ \text{mg/L}$$
$$\text{DO} = \text{DO}_{\text{sat}} - D_c$$
$$= 7.92\ \frac{\text{mg}}{\text{L}} - 3.20\ \frac{\text{mg}}{\text{L}}$$
$$= \boxed{4.72\ \text{mg/L}}$$

This is below the state standard of 5 mg/L.

The answer is (B).

(d) From Eq. 28.35, the composite dissolved oxygen is

$$\text{DO} = \frac{Q_w \text{DO}_w + Q_r \text{DO}_r}{Q_w + Q_r}$$
$$= \frac{\left(0.280\ \dfrac{\text{m}^3}{\text{s}}\right)\left(1\ \dfrac{\text{mg}}{\text{L}}\right) + \left(0.877\ \dfrac{\text{m}^3}{\text{s}}\right)\left(6\ \dfrac{\text{mg}}{\text{L}}\right)}{0.280\ \dfrac{\text{m}^3}{\text{s}} + 0.877\ \dfrac{\text{m}^3}{\text{s}}}$$
$$= 4.79\ \text{mg/L} \quad [\text{no change}]$$

Calculate the temperature of the river water and wastewater mixture.

$$T_1 = \frac{Q_w T_w + Q_r T_r}{Q_w + Q_r}$$
$$= \frac{\left(0.280\ \dfrac{\text{m}^3}{\text{s}}\right)(28°\text{C}) + \left(0.877\ \dfrac{\text{m}^3}{\text{s}}\right)(12°\text{C})}{0.280\ \dfrac{\text{m}^3}{\text{s}} + 0.877\ \dfrac{\text{m}^3}{\text{s}}}$$
$$= 15.87°\text{C}$$

Calculate K_d at this temperature. Two different values of θ_d are used for the two temperature ranges. Use Eq. 28.29, written for base-10 constants.

$$K_{d,T_1} = K_{d,T_2}\theta^{T_1 - T_2}$$
$$K_{d,20°\text{C}} = (0.199\ \text{d}^{-1})(1.056)^{20°\text{C} - 28°\text{C}}$$
$$= 0.1287\ \text{d}^{-1}$$
$$K_{d,T_1} = K_{d,T_2}\theta^{T_1 - T_2}$$
$$K_{d,15.87°\text{C}} = (0.1287\ \text{d}^{-1})(1.135)^{15.87°\text{C} - 20°\text{C}}$$
$$= 0.07629\ \text{d}^{-1}$$

Similarly, use Eq. 28.25 to correct K_r.

$$K_{r,T_1} = K_{r,T_2}\theta^{T_1 - T_2}$$
$$K_{r,15.87°\text{C}} = K_{r,28°\text{C}}(1.024)^{15.87°\text{C} - 28°\text{C}}$$
$$= (0.370\ \text{d}^{-1})(1.024)^{15.87°\text{C} - 28°\text{C}}$$
$$= 0.2775\ \text{d}^{-1}$$

Correct BOD_u to the new temperature. Use Eq. 28.33 twice.

$$\text{BOD}_{u,20°\text{C}} = \frac{\text{BOD}_{u,28°\text{C}}}{(0.02)(28°\text{C}) + 0.6}$$
$$\text{BOD}_{u,15.87°\text{C}} = (\text{BOD}_{u,20°\text{C}})((0.02)(15.87°\text{C}) + 0.6)$$
$$= \frac{\left(6.8645\ \dfrac{\text{mg}}{\text{L}}\right)((0.02)(15.87°\text{C}) + 0.6)}{(0.02)(28°\text{C}) + 0.6}$$
$$= 5.43\ \text{mg/L}$$

Environmental

At 15.87°C, the saturation dissolved oxygen is found from App. 22.C. $DO_{sat,15.87°C} = 9.98$ mg/L.

Calculate the initial deficit from Eq. 28.17.

$$D_0 = DO_{sat} - DO = 9.98 \frac{mg}{L} - 4.79 \frac{mg}{L}$$
$$= 5.19 \text{ mg/L}$$

Calculate the deficit at town B. Use Eq. 28.36.

$$D_B = \left(\frac{K_d BOD_u}{K_r - K_d}\right)(10^{-K_d t} - 10^{-K_r t}) + D_0(10^{-K_r t})$$

$$= \left(\frac{(0.07629 \text{ d}^{-1})\left(5.43 \frac{mg}{L}\right)}{0.2775 \text{ d}^{-1} - 0.07629 \text{ d}^{-1}}\right)$$

$$\times \left(\begin{array}{c} 10^{-(0.07629 \text{ d}^{-1})(0.1031 \text{ d})} \\ - 10^{-(0.2775 \text{ d}^{-1})(0.1031 \text{ d})} \end{array}\right)$$

$$+ \left(5.19 \frac{mg}{L}\right)\left(10^{-(0.2775 \text{ d}^{-1})(0.1031 \text{ d})}\right)$$

$$= 4.95 \text{ mg/L}$$

The dissolved oxygen at town B is

$$DO_B = DO_{sat} - D_B$$
$$= 9.98 \frac{mg}{L} - 4.95 \frac{mg}{L}$$
$$= \boxed{5.03 \text{ mg/L} \quad (5.0 \text{ mg/L})}$$

This meets the state standard of 5 mg/L.

The answer is (C).

4. (a)
$$T = 20°C + 2°C = 22°C$$
$$T_1 = 20°C$$
$$Q_1 + Q_2 = 1200 \text{ m}^3/\text{s}$$
$$Q_2 = 60 \text{ m}^3/\text{s}$$
$$Q_1 = 1200 \frac{m^3}{s} - 60 \frac{m^3}{s} = 1140 \text{ m}^3/\text{s}$$
$$T_f = \frac{T_1 Q_1 + T_2 Q_2}{Q_1 + Q_2}$$
$$22°C = \frac{(20°C)\left(1140 \frac{m^3}{s}\right) + C_2\left(60 \frac{m^3}{s}\right)}{1200 \frac{m^3}{s}}$$
$$= \boxed{60°C}$$

The answer is (B).

(b)
$$C = 0 + 0.002 \frac{mg}{L}$$
$$C_1 = 0$$
$$C_f = \frac{C_1 Q_1 + C_2 Q_2}{Q_1 + Q_2}$$

$$0.002 \frac{mg}{L} = \frac{(0)\left(1140 \frac{m^3}{s}\right) + C_2\left(60 \frac{m^3}{s}\right)}{1200 \frac{m^3}{s}}$$

$$= \boxed{0.04 \text{ mg/L}}$$

The answer is (A).

5. (a) The domestic BOD concentration is

$$BOD = \frac{\left(0.191 \frac{lbm}{capita\text{-}day}\right)\left(10^6 \frac{gal}{MG}\right)}{\times \left(0.1198 \frac{mg\text{-}MG}{L\text{-}lbm}\right)}{122 \frac{gal}{capita\text{-}day}}$$

$$= \boxed{187.6 \text{ mg/L}}$$

The answer is (C).

(b) Use Eq. 28.35.

$$BOD = \frac{\sum Q_i BOD_i}{\sum Q_i}$$

$$= \frac{\begin{array}{c} \left(122 \frac{gal}{day}\right)(10{,}000)\left(187.6 \frac{mg}{L}\right) \\ + \left(116{,}000 \frac{gal}{day}\right)(0) \\ + \left(180{,}000 \frac{gal}{day}\right)\left(800 \frac{mg}{L}\right) \\ + \left(76{,}000 \frac{gal}{day}\right)\left(1700 \frac{mg}{L}\right) \end{array}}{\begin{array}{c} \left(122 \frac{gal}{day}\right)(10{,}000) + 116{,}000 \frac{gal}{day} \\ + 0 + 180{,}000 \frac{gal}{day} + 76{,}000 \frac{gal}{day} \end{array}}$$

$$= \boxed{315.4 \text{ mg/L}}$$

The answer is (D).

(c) Use Eq. 28.35.

$$T_{°F} = \frac{\sum Q_i T_i}{\sum Q_i}$$

$$= \frac{\left(122 \ \frac{gal}{day}\right)(10{,}000)(64°F)}{} $$

$$+ \left(116{,}000 \ \frac{gal}{day}\right)(51°F)$$

$$+ \left(180{,}000 \ \frac{gal}{day}\right)(95°F)$$

$$+ \left(76{,}000 \ \frac{gal}{day}\right)(84°F)$$

$$= \frac{}{\left(122 \ \frac{gal}{day}\right)(10{,}000) + 116{,}000 \ \frac{gal}{day}}$$

$$+ 180{,}000 \ \frac{gal}{day} + 76{,}000 \ \frac{gal}{day}$$

$$= 67.5°F$$

$$T_{°C} = \tfrac{5}{9}(T_{°F} - 32°) = \left(\tfrac{5}{9}\right)(67.5°F - 32°F)$$

$$= \boxed{19.7°C}$$

The answer is (A).

(d) The total discharge into the river is

$$\begin{pmatrix} \left(122 \ \frac{gal}{day}\right)(10{,}000) \\[4pt] + 116{,}000 \ \frac{gal}{day} \\[4pt] + 180{,}000 \ \frac{gal}{day} \\[4pt] + 76{,}000 \ \frac{gal}{day} \end{pmatrix} \left(1.547 \times 10^{-6} \ \frac{ft^3\text{-}day}{sec\text{-}day}\right)$$

$$= 2.46 \ ft^3/sec$$

step 1: Find the stream conditions immediately after mixing.

$$BOD_{5,20°C} = \frac{\left(2.46 \ \frac{ft^3}{sec}\right)\left(315.4 \ \frac{mg}{L}\right)}{}$$

$$+ \left(120 \ \frac{ft^3}{sec}\right)(0)$$

$$= \frac{}{2.46 \ \frac{ft^3}{sec} + 120 \ \frac{ft^3}{sec}}$$

$$= 6.34 \ mg/L$$

$$\left(2.46 \ \frac{ft^3}{sec}\right)(0)$$

$$DO = \frac{+ \left(120 \ \frac{ft^3}{sec}\right)\left(7.5 \ \frac{mg}{L}\right)}{2.46 \ \frac{ft^3}{sec} + 120 \ \frac{ft^3}{sec}}$$

$$= 7.35 \ mg/L$$

$$\left(2.46 \ \frac{ft^3}{sec}\right)(19.7°C)$$

$$T = \frac{+ \left(120 \ \frac{ft^3}{sec}\right)(15°C)}{2.46 \ \frac{ft^3}{sec} + 120 \ \frac{ft^3}{sec}}$$

$$= 15.1°C$$

step 2: Calculate the rate constants at 15.1°C. Use Eq. 28.28, written for base-10 constants.

$$K_{d,T} = K_{d,20°C} \ \theta^{T-20°C}$$

$$K_{d,15.1°C} = (0.15 \ day^{-1})(1.135)^{15.1°C-20°C}$$

$$= 0.0807 \ day^{-1}$$

$$K_{r,15.1°C} = (0.2 \ day^{-1})(1.024)^{15.1°C-20°C}$$

$$= 0.178 \ day^{-1}$$

step 3: Estimate BOD_u. Use Eq. 28.31.

$$BOD_{u,20°C} = \frac{BOD_t}{1 - 10^{-K_d t}}$$

$$= \frac{6.34 \ \frac{mg}{L}}{1 - 10^{-(0.15 \ day^{-1})(5 \ days)}}$$

$$= 7.71 \ mg/L$$

Use Eq. 28.33 to convert BOD_u to 15.1°C.

$$BOD_{u,15.1°C} = BOD_{u,20°C}(0.02 T_{°C} + 0.6)$$

$$= \left(7.71 \ \frac{mg}{L}\right)((0.02)(15.1°C) + 0.6)$$

$$= 6.95 \ mg/L$$

step 4: From App. 22.C at 15°C, saturated DO = 10.15 mg/L. Since the actual is 7.35 mg/L, the deficit is

$$D_0 = DO_{sat} - DO = 10.15 \ \frac{mg}{L} = 7.35 \ \frac{mg}{L}$$

$$= 2.8 \ mg/L$$

step 5: Calculate t_c. Use Eq. 28.38.

$$t_c = \left(\frac{1}{K_r - K_d}\right)$$

$$\times \log_{10}\left(\left(\frac{K_d\text{BOD}_u - K_rD_0 + K_dD_0}{K_d\text{BOD}_u}\right)\left(\frac{K_r}{K_d}\right)\right)$$

$$= \left(\frac{1}{0.178 \text{ day}^{-1} - 0.0807 \text{ day}^{-1}}\right)$$

$$\times \log_{10}\left(\begin{pmatrix}\dfrac{\begin{pmatrix}(0.0807 \text{ day}^{-1})\left(6.95 \dfrac{\text{mg}}{\text{L}}\right) \\ -(0.178 \text{ day}^{-1})\left(2.8 \dfrac{\text{mg}}{\text{L}}\right) \\ +(0.0807 \text{ day}^{-1})\left(2.8 \dfrac{\text{mg}}{\text{L}}\right)\end{pmatrix}}{(0.0807 \text{ day}^{-1})\left(6.95 \dfrac{\text{mg}}{\text{L}}\right)} \\ \times \left(\dfrac{0.178 \text{ day}^{-1}}{0.0807 \text{ day}^{-1}}\right)\end{pmatrix}\right)$$

$$= 0.562 \text{ day}$$

step 6: The distance downstream is

$$(0.562 \text{ day})\left(3 \frac{\text{mi}}{\text{hr}}\right)\left(24 \frac{\text{hr}}{\text{day}}\right) = \boxed{40.5 \text{ mi}}$$

The answer is (A).

step 7: Use Eq. 28.39.

$$D_c = \left(\frac{K_d\text{BOD}_u}{K_r}\right)10^{-K_dt_c}$$

$$= \left(\frac{(0.0807 \text{ day}^{-1})\left(6.95 \dfrac{\text{mg}}{\text{L}}\right)}{0.178 \text{ day}^{-1}}\right)$$

$$\times \left(10^{-(0.0807 \text{ day}^{-1})(0.562 \text{ day})}\right)$$

$$= 2.84 \text{ mg/L}$$

step 8:

$$\text{DO}_{\min} = \text{DO}_{\text{sat}} - D_c$$

$$= 10.15 \frac{\text{mg}}{\text{L}} - 2.84 \frac{\text{mg}}{\text{L}}$$

$$= \boxed{7.31 \text{ mg/L}}$$

The answer is (C).

6. (a) Do not apply the population expansion factor to the industrial effluents. Use Eq. 28.3, modified for the given population equivalent (in thousands of people) for the domestic flow contribution.

$$P_{e,1000s} = P_{\text{domestic flow}} + P_{\text{industrial source 1}} + P_{\text{industrial source 2}}$$

$$= (20)(1.15)$$

$$+ \frac{\left(1100 \dfrac{\text{mg}}{\text{L}}\right)\left(1.3 \times 10^6 \dfrac{\text{gal}}{\text{day}}\right) \times \left(8.345 \times 10^{-9} \dfrac{\text{lbm-L}}{\text{MG-mg}}\right)}{0.18 \dfrac{\text{lbm}}{\text{day-person}}}$$

$$+ \frac{(500)\left(1.0 \times 10^6 \dfrac{\text{gal}}{\text{day}}\right) \times \left(8.345 \times 10^{-9} \dfrac{\text{lbm-L}}{\text{MG-mg}}\right)}{0.18 \dfrac{\text{lbm}}{\text{day-person}}}$$

$$= \boxed{112.5 \quad (112,500)}$$

The answer is (D).

(b) Since the plant loading is requested, the organic loading can be given in lbm/day. The population from part (a) is 112,500.

$$L_{\text{BOD}} = (112,500)\left(0.18 \frac{\text{lbm}}{\text{day}}\right) = \boxed{20,250 \text{ lbm/day}}$$

The answer is (D).

(c) $\quad L_H = (20,000)(1.15)\left(100 \dfrac{\text{gal}}{\text{day}}\right)$

$$+ 1.3 \times 10^6 \frac{\text{gal}}{\text{day}} + 1.0 \times 10^6 \frac{\text{gal}}{\text{day}}$$

$$= \boxed{4.6 \times 10^6 \text{ gal/day} \quad (4.6 \text{ MGD})}$$

The answer is (A).

29 Wastewater Treatment: Equipment and Processes

PRACTICE PROBLEMS

Lagoons

1. A cheese factory located in a normally warm state has liquid waste with the following characteristics.

> *waste no. 1*
>
> | volume | 10,000 gal/day |
> |--------|----------------|
> | BOD | 1000 mg/L |
>
> *waste no. 2*
>
> | volume | 25,000 gal/day |
> |--------|----------------|
> | BOD | 250 mg/L |

The factory will use a 4 ft deep, on-site, nonaerated lagoon to stabilize the waste.

(a) What is the total BOD loading?

(A) 60 lbm/day

(B) 85 lbm/day

(C) 110 lbm/day

(D) 140 lbm/day

(b) What lagoon size is required if the BOD loading is 20 lbm BOD/ac-day?

(A) 2.7 ac

(B) 6.8 ac

(C) 8.3 ac

(D) 10.9 ac

(c) What is the detention time?

(A) 5 wk

(B) 14 wk

(C) 36 wk

(D) 52 wk

Trickling Filters

2. It is estimated that the BOD of raw sewage received at a treatment plant serving a population of 20,000 will be 300 mg/L. It is estimated that the per capita BOD loading is 0.17 lbm/day. 30% of the influent BOD is removed by settling. One single-stage high-rate trickling filter is to be used to reduce the plant effluent to 50 mg/L. Recirculation is from the filter effluent to the primary settling influent. *Ten States' Standards* is in effect.

(a) What is the design flow rate?

(A) 0.9 MGD

(B) 1.1 MGD

(C) 1.4 MGD

(D) 2.0 MGD

(b) Assuming a design flow rate of 1.35 MGD, what is the total organic load on the filter?

(A) 1700 lbm/day

(B) 2100 lbm/day

(C) 2400 lbm/day

(D) 3400 lbm/day

(c) Assuming an incoming volume of 1.35 MGD and an organic loading of 60 lbm/day-1000 ft^3, what flow should be recirculated?

(A) 0.6 MGD

(B) 0.9 MGD

(C) 1.2 MGD

(D) 2.2 MGD

(d) What is the overall plant efficiency?

(A) 45%

(B) 76%

(C) 83%

(D) 91%

3. The average wastewater flow from a community of 20,000 is 125 gpcd. The 5-day, 20°C BOD is 250 mg/L. The suspended solids content is 300 mg/L. A final plant effluent of 50 mg/L of BOD is to be achieved through the use of two sets of identical settling tanks and trickling filters operating in parallel. The settling tanks are to be designed to a standard of 1000 gpd/ft^2. The trickling filters are to be 6 ft deep. There is no recirculation.

(a) What settling tank surface area is required?

(A) 2500 ft^2

(B) 3000 ft^2

(C) 3500 ft^2

(D) 4500 ft^2

(b) What settling tank diameter is required?

(A) 40 ft

(B) 50 ft

(C) 65 ft

(D) 80 ft

(c) Estimate the BOD removal in the settling tanks.

(A) 15%

(B) 30%

(C) 45%

(D) 60%

(d) What is the trickling filter diameter?

(A) 65 ft

(B) 75 ft

(C) 85 ft

(D) 95 ft

4. Wastewater from a city with a population of 40,000 has an average daily flow of 4.4 MGD. The sewage has the following characteristics.

BOD$_5$ at 20°C	160 mg/L
COD	800 mg/L
total solids	900 mg/L
suspended solids	180 mg/L
volatile solids	320 mg/L
settleable solids	8 mg/L
pH	7.8

The wastewater is to be treated with primary settling and secondary trickling filtration. The settling basins are to be circular, 8 ft deep, and designed to a standard overflow rate of 1000 gal/day-ft^2.

(a) Assuming two identical basins operating in parallel, what should be the diameter of each sedimentation basin in order to remove 30% of the BOD?

(A) 45 ft

(B) 53 ft

(C) 77 ft

(D) 86 ft

(b) What is the detention time?

(A) 1.4 hr

(B) 1.8 hr

(C) 2.3 hr

(D) 3.1 hr

(c) What is the weir loading?

(A) 8200 gpd/ft

(B) 11,000 gpd/ft

(C) 13,000 gpd/ft

(D) 15,000 gpd/ft

5. (*Time limit: one hour*) A small community has a projected average flow of 1 MGD, with a peaking factor of 2.0. Incoming wastewater has the following properties: BOD, 250 mg/L; grit specific gravity, 2.65; and total suspended solids, 400 mg/L. The community wants to have a wastewater treatment plant consisting of a single aerated grit chamber, a single primary clarifier, two identical circular trickling filters in parallel, and a single secondary clarifier. There will be no equalization basin. Recirculation from the second clarifier to the entrance of the trickling filters will be 100% of the average flow. The final effluent is to have a BOD of 30 mg/L.

(a) What is the peak design flow at the grit chamber?

(A) 1.0 MGD

(B) 1.5 MGD

(C) 2.0 MGD

(D) 2.5 MGD

(b) Determine the aerated grit chamber width assuming the following: a 3 min detention time, a 20 ft length, and a width:depth ratio of 1.25.

(A) 4 ft

(B) 6 ft

(C) 8 ft

(D) 10 ft

(c) Determine the approximate air requirements for the grit chamber in order to capture approximately 95% of the grit.

    (A) 60 cfm

    (B) 140 cfm

    (C) 280 cfm

    (D) 420 cfm

(d) If the clarifier is 12 ft deep, what should be its diameter?

    (A) 50 ft

    (B) 56 ft

    (C) 63 ft

    (D) 81 ft

(e) What is the peak flow entering the trickling filters?

    (A) 1.0 MGD

    (B) 2.0 MGD

    (C) 3.0 MGD

    (D) 4.0 MGD

(f) Assume the primary clarifier removes 30% of the incoming BOD, and the filters see the average flow only. Determine the diameter of the trickling filters assuming a 6 ft deep rock bed.

    (A) 45 ft

    (B) 55 ft

    (C) 65 ft

    (D) 85 ft

(g) Determine the diameter of the final clarifier.

    (A) 35 ft

    (B) 50 ft

    (C) 65 ft

    (D) 80 ft

Recirculating Biological Contactors

6. (*Time limit: one hour*) 1.5 MGD of wastewater with a BOD of 250 mg/L is processed by a high-rate rock trickling filter followed by recirculating biological contactor (RBC) processing. The trickling filter is 75 ft in diameter and 6 ft deep and was designed to NRC standards. The BOD removal efficiency of the RBC process is given by the following equation. (k is 2.45 gal/day-ft^2, and Q (in units of gal/day) does not include recirculation. A is the immersed area of the RBC in ft^2.)

$$\eta_{\text{BOD}} = \frac{1}{\left(1 + \dfrac{kA}{Q}\right)^3}$$

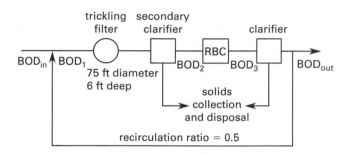

(a) What total RBC surface area is required to achieve an effluent BOD$_{\text{out}}$ of 30 mg/L?

    (A) 4.2×10^4 ft^2

    (B) 8.4×10^4 ft^2

    (C) 1.3×10^5 ft^2

    (D) 2.0×10^5 ft^2

(b) The recirculation pick-up point is relocated from after the final clarifier to after the trickling filter. The efficiency of the RBC process is 65%. Determine the recirculation ratio such that BOD$_{\text{out}}$ is 30 mg/L.

    (A) 25%

    (B) 50%

    (C) 75%

    (D) 100%

(c) If BOD$_2 = 85$ mg/L, BOD$_{\text{out}} = 30$ mg/L, the yield is 0.4 lbm/lbm BOD removed, and the sludge specific gravity is essentially 1.0, what is the approximate sludge volume produced from the clarifiers?

    (A) 4 ft^3/day

    (B) 12 ft^3/day

    (C) 21 ft^3/day

    (D) 35 ft^3/day

SOLUTIONS

1. (a) The total BOD is

$$\left(1000 \ \frac{mg}{L}\right)\left(8.345 \ \frac{lbm\text{-}L}{mg\text{-}MG}\right)\left(\frac{10{,}000 \ \frac{gal}{day}}{1{,}000{,}000 \ \frac{gal}{MG}}\right)$$

$$+ \left(250 \ \frac{mg}{L}\right)\left(8.345 \ \frac{lbm\text{-}L}{mg\text{-}MG}\right)\left(\frac{25{,}000 \ \frac{gal}{day}}{1{,}000{,}000 \ \frac{gal}{MG}}\right)$$

$$= \boxed{135.6 \ lbm/day \quad (140 \ lbm/day)}$$

The answer is (D).

(b) The warm weather and depth contribute to a decrease in pond effectiveness. Assume 20 lbm BOD/ ac-day for a nonaerated stabilization pond. From Eq. 29.4, the required area is

$$A = \frac{Q}{v^*} = \frac{135.6 \ \frac{lbm}{day}}{20 \ \frac{lbm}{ac\text{-}day}} = \boxed{6.78 \ ac \quad (6.8 \ ac)}$$

The answer is (B).

(c) Use Eq. 29.5.

$$t_d = \frac{V}{Q}$$

$$= \frac{(6.78 \ ac)\left(43{,}560 \ \frac{ft^2}{ac}\right)(4 \ ft)\left(7.48 \ \frac{gal}{ft^3}\right)}{\left(35{,}000 \ \frac{gal}{day}\right)\left(7 \ \frac{days}{wk}\right)}$$

$$= \boxed{36.07 \ wk \quad (36 \ wk)}$$

The answer is (C).

2. (a) The design flow rate is

$$Q = \frac{\dot{m}}{C} = \frac{\left(0.17 \ \frac{lbm}{capita\text{-}day}\right)(20{,}000 \ people)}{\left(8.345 \times 10^{-6} \ \frac{lbm\text{-}L}{gal\text{-}mg}\right)\left(300 \ \frac{mg}{L}\right)}$$

$$= \boxed{1.358 \times 10^6 \ gal/day \quad (1.4 \ MGD)}$$

(*Ten States' Standards* specifies 100 gpcd in the absence of other information. In such a case, $Q = (100 \ gal/capita\text{-}day)(20{,}000 \ people) = 2 \times 10^6 \ gal/day$.)

The answer is (C).

(b) The total BOD load leaving the primary clarifier and entering the filter is

$$BOD_i = (1 - 0.30)\left(300 \ \frac{mg}{L}\right) = 210 \ mg/L$$

$$L_{BOD} = (1.35 \ MGD)\left(210 \ \frac{mg}{L}\right)\left(8.345 \ \frac{lbm\text{-}L}{mg\text{-}MG}\right)$$

$$= \boxed{2366 \ lbm/day \quad (2400 \ lbm/day)}$$

The answer is (C).

(c) The efficiency of the filter and secondary clarifier is found from Eq. 29.9.

$$\eta = \frac{S_{ps} - S_o}{S_{ps}} = \frac{210 \ \frac{mg}{L} - 50 \ \frac{mg}{L}}{210 \ \frac{mg}{L}}$$

$$= 0.762 \quad (76.2\%)$$

Use Eq. 29.13.

$$\eta = \frac{1}{1 + 0.0561\sqrt{\dfrac{L_{BOD}}{F}}}$$

$$0.762 = \frac{1}{1 + 0.0561\sqrt{\dfrac{60 \ \dfrac{lbm}{day\text{-}1000 \ ft^3}}{F}}}$$

$$F = 1.936$$

Use Eq. 29.15.

$$F = \frac{1 + R}{\left(1 + wR\right)^2}$$

$$1.936 = \frac{1 + R}{\left(1 + 0.1R\right)^2}$$

$$R = 1.6$$

Use Eq. 29.10.

$$R = \frac{Q_r}{Q_w}$$

$$= (1.6)(1.35 \ MGD)$$

$$= \boxed{2.16 \ MGD \quad (2.2 \ MGD)}$$

The answer is (D).

(d) Use Eq. 29.9.

$$\eta = \frac{S_{ps} - S_o}{S_{ps}}$$

$$= \frac{300 \ \frac{mg}{L} - 50 \ \frac{mg}{L}}{300 \ \frac{mg}{L}}$$

$$= \boxed{0.833 \quad (83\%)}$$

The answer is (C).

3. (a) Use Eq. 29.4. Disregarding variations in peak flow, the average design volume is

$$v^* = \frac{Q}{A} = \frac{\left(125 \ \frac{gal}{capita\text{-}day}\right)(20{,}000 \ \text{people})}{1{,}000{,}000 \ \frac{gal}{MG}}$$

$$= 2.5 \ \text{MGD}$$

The settling tank surface area is

$$\frac{2.5 \times 10^6 \ \frac{gal}{day}}{1000 \ \frac{gal}{day\text{-}ft^2}} = \boxed{2500 \ \text{ft}^2}$$

The answer is (A).

(b) The required diameter when using two tanks in parallel is

$$D = \sqrt{\frac{(4)(2500 \ \text{ft}^2)}{2\pi}} = \boxed{39.9 \ \text{ft (use 40 ft) each}}$$

The answer is (A).

(c) A 30% removal is typical.

The answer is (B).

(d) BOD entering the filter is

$$(1 - 0.30)\left(250 \ \frac{mg}{L}\right) = 175 \ \text{mg/L}$$

The filter efficiency is

$$\eta = \frac{175 \ \frac{mg}{L} - 50 \ \frac{mg}{L}}{175 \ \frac{mg}{L}} = 0.71$$

From Fig. 29.4 with 71% efficiency and $R = 0$, $L_{BOD} = 55 \ \text{lbm/day-1000 ft}^3$.

The total load is found from Eq. 29.12 (rearranged in terms of V_1).

$$V_1 = \frac{QS\left(8.345 \ \frac{lbm\text{-}L}{MG\text{-}mg}\right)(1000)}{L_{BOD}}$$

$$= \frac{(2.5 \ \text{MGD})\left(175 \ \frac{mg}{L}\right)}{\left(55 \ \frac{lbm}{day\text{-}1000 \ ft^3}\right)(2 \ \text{filters})}$$

$$\times \left(8.345 \ \frac{lbm\text{-}L}{MG\text{-}mg}\right)\left(1000 \ \frac{ft^3}{1000 \ ft^3}\right)$$

$$= 33{,}190 \ \text{ft}^3/\text{filter}$$

With a depth of 6 ft, the total required surface area is

$$A = \frac{V}{Z} = \frac{33{,}190 \ \text{ft}^3}{6 \ \text{ft}} = 5532 \ \text{ft}^2$$

The required diameter per filter is

$$D = \sqrt{\frac{4A}{\pi}} = \sqrt{\frac{(4)(5532 \ \text{ft}^2)}{\pi}}$$

$$= \boxed{83.9 \ \text{ft} \quad (85 \ \text{ft})}$$

The answer is (C).

4. (a) There is nothing particularly special about a basin that removes 30% BOD. Choose two basins in parallel, each working with half of the total flow. Choose an overflow rate of 1000 gpd/ft^2. The area per basin is given by Eq. 29.4.

$$A = \frac{Q}{v^*} = \frac{(4.4 \ \text{MGD})\left(10^6 \ \frac{gal}{MG}\right)}{(2 \ \text{basins})\left(1000 \ \frac{gal}{day\text{-}ft^2}\right)}$$

$$= 2200 \ \text{ft}^2$$

$$D = \sqrt{\frac{4}{\pi}A} = \sqrt{\left(\frac{4}{\pi}\right)(2200 \ \text{ft}^2)}$$

$$= \boxed{52.9 \ \text{ft} \quad (53 \ \text{ft})}$$

The answer is (B).

(b) The detention time is given by Eq. 29.5.

$$t_d = \frac{V}{Q}$$

$$= \frac{(2200 \text{ ft}^2)(8 \text{ ft})\left(7.48 \frac{\text{gal}}{\text{ft}^3}\right)\left(24 \frac{\text{hr}}{\text{day}}\right)}{\left(2.2 \frac{\text{MGD}}{\text{tank}}\right)\left(10^6 \frac{\text{gal}}{\text{MG}}\right)}$$

$$= \boxed{1.436 \text{ hr} \quad (1.4 \text{ hr})}$$

The answer is (A).

(c) Find the circumference of the basins.

$$\text{circumference} = \pi D = \pi(52.9 \text{ ft})$$
$$= 166.2 \text{ ft}$$

The weir loading is

$$\text{weir loading} = \frac{2.2 \times 10^6 \frac{\text{gal}}{\text{day}}}{166.2 \text{ ft}}$$

$$= \boxed{13{,}237 \text{ gal/day-ft} \quad (13{,}000 \text{ gpd/ft})}$$

The answer is (C).

5.

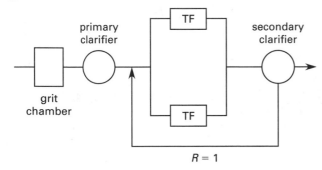

$$R = 1$$

(a) The peak flow is

$$(2)(1 \text{ MGD}) = \boxed{2 \text{ MGD}}$$

The answer is (C).

(b) The peak flow rate per second is

$$Q_{\text{peak}} = \frac{(2)(1 \text{ MGD})\left(10^6 \frac{\text{gal}}{\text{MG}}\right)\left(0.1337 \frac{\text{ft}^3}{\text{gal}}\right)}{\left(24 \frac{\text{hr}}{\text{day}}\right)\left(60 \frac{\text{min}}{\text{hr}}\right)\left(60 \frac{\text{sec}}{\text{min}}\right)}$$

$$= 3.095 \text{ ft}^3/\text{sec}$$

With a detention time of 3 min, the volume of the grit chamber would be

$$V = Qt = \left(3.095 \frac{\text{ft}^3}{\text{sec}}\right)\left(60 \frac{\text{sec}}{\text{min}}\right)(3 \text{ min}) = 557.1 \text{ ft}^3$$

$$557.1 \text{ ft}^3 = (\text{length})(\text{width})(\text{depth})$$
$$= (20 \text{ ft})(1.25)(\text{depth})^2$$
$$\text{water depth} = 4.72 \text{ ft} \quad \left[\text{round to } 4\tfrac{3}{4} \text{ ft}\right]$$
$$\text{chamber width} = (1.25)(4.72 \text{ ft})$$
$$= \boxed{5.9 \text{ ft} \quad (6 \text{ ft})}$$

The answer is (B).

(c) This is a shallow chamber. Use 3 cfm/ft.

$$\left(3 \frac{\text{ft}^3}{\text{min-ft}}\right)(20 \text{ ft}) = \boxed{60 \text{ ft}^3/\text{min (cfm) of air}}$$

The answer is (A).

(d) (Note that the *Ten States' Standards* requires two basins.) Surface loading is the primary design parameter. Choose a surface loading of 1000 gal/day-ft^2. The diameter is

$$A = \frac{Q}{v^*} = \frac{(2 \text{ MGD})\left(10^6 \frac{\text{gal}}{\text{day-MGD}}\right)}{1000 \frac{\text{gal}}{\text{day-ft}^2}}$$

$$= 2000 \text{ ft}^2$$

$$D = \sqrt{\frac{4A}{\pi}} = \sqrt{\frac{(4)(2000 \text{ ft}^2)}{\pi}}$$

$$= \boxed{50.5 \text{ ft} \quad (50 \text{ ft})}$$

The answer is (A).

(e) The volume of the primary clarifier is fixed by the weir height. Therefore, the clarifier does not provide any storage (i.e., no damping of the flow rates). The fluctuations in flow will be passed on to the trickling filters.

$$Q_p + Q_r = 2 \text{ MGD} + 1 \text{ MGD} = 3 \text{ MGD}$$

The answer is (C).

(f) First, assume the primary sedimentation basin removes 30% of the BOD. Find the BOD incoming to the trickle filter to solve for the water flow quantity, Q_w.

$$\text{BOD}_{\text{in}} = (1 - 0.30)\left(250 \ \frac{\text{mg}}{\text{L}}\right) = 175 \ \text{mg/L}$$

$$Q_w = \left(175 \ \frac{\text{mg}}{\text{L}}\right)\left(8.345 \ \frac{\text{lbm-L}}{\text{mg-MG}}\right)(1 \ \text{MGD})$$
$$= 1460 \ \text{lbm/day}$$

Second, find the required trickling filter-clarifier process efficiency.

$$\eta = \frac{\text{BOD}_{\text{in}} - \text{BOD}_{\text{out}}}{\text{BOD}_{\text{in}}} = \frac{175 \ \frac{\text{mg}}{\text{L}} - 30 \ \frac{\text{mg}}{\text{L}}}{175 \ \frac{\text{mg}}{\text{L}}}$$

$$= 0.829 \quad (82.9\%)$$

From Eq. 29.15 with $w = 0.1$ and $R = 1$,

$$F = \frac{1 + R}{(1 + wR)^2} = \frac{1 + 1}{\left(1 + (0.1)(1)\right)^2} = 1.65$$

From Eq. 29.13,

$$\eta = \frac{1}{1 + 0.0561\sqrt{\frac{L_{\text{BOD}}}{F}}}$$

$$0.83 = \frac{1}{1 + 0.0561\sqrt{\frac{L_{\text{BOD}}}{1.65}}}$$

$$L_{\text{BOD}} \approx 22 \ \text{lbm/day-1000 ft}^3$$

Third, find the filter volume.

$$V = \frac{\dot{m}_{\text{BOD}}}{L_{\text{BOD}}}$$
$$= \frac{\left(1460 \ \frac{\text{lbm}}{\text{day}}\right)\left(1000 \ \frac{\text{ft}^3}{1000 \ \text{ft}^3}\right)}{\left(22 \ \frac{\text{lbm}}{\text{day-1000 ft}^3}\right)(2 \ \text{filters})}$$
$$= 33{,}182 \ \text{ft}^3$$

Finally, for 6 ft deep rock bed, the diameter for the trickling filters is

$$D = \sqrt{\frac{\frac{4}{\pi} V}{Z}} = \sqrt{\frac{\left(\frac{4}{\pi}\right)(33{,}182 \ \text{ft}^3)}{6 \ \text{ft}}}$$
$$= \boxed{83.9 \ \text{ft} \quad (85 \ \text{ft})}$$

Use two 85 ft diameter, 6 ft deep filters.

The answer is (D).

(g) For the final clarifier, the maximum overflow rate is 1100 gpd/ft^2, and the minimum depth is 10 ft. Assume volumetric flow fluctuations will be damped out by previous processes.

$$A = \frac{1 \times 10^6 \ \frac{\text{gal}}{\text{day}}}{1100 \ \frac{\text{gal}}{\text{day-ft}^2}} = 909 \ \text{ft}^2$$

$$D = \sqrt{\frac{4}{\pi} A} = \sqrt{\left(\frac{4}{\pi}\right)(909 \ \text{ft}^2)}$$
$$= \boxed{34.0 \ \text{ft} \quad (35 \ \text{ft})}$$

Use a 35 ft diameter, 10 ft deep basin.

The answer is (A).

6. (a) Assume the last clarifier removes only sloughed off biological material and removes no BOD.

$$\text{BOD}_3 = 30 \ \text{mg/L}$$

There is no recirculation that matches the NRC model. The NRC model "recirculation" is from the trickling filter discharge directly back to the entrance to the filter. This problem's recirculation is from several processes beyond the trickling filter. The recirculation increases the BOD loading and dilutes the influent.

The BOD loading, L_{BOD}, to the trickling filter must include recirculation, L_r. Use Eq. 29.12.

$$L_{\text{BOD}} + L_r = \frac{(1.5 \ \text{MGD})\left(250 \ \frac{\text{mg}}{\text{L}}\right)}{\times \left(8.345 \ \frac{\text{lbm-L}}{\text{MG-mg}}\right)\left(1000 \ \frac{\text{ft}^3}{1000 \ \text{ft}^3}\right)}{\pi\left(\frac{75 \ \text{ft}}{2}\right)^2 (6 \ \text{ft})}$$
$$+ \frac{(0.5)(1.5 \ \text{MGD})\left(30 \ \frac{\text{mg}}{\text{L}}\right)}{\times \left(8.345 \ \frac{\text{lbm-L}}{\text{mg-MG}}\right)\left(1000 \ \frac{\text{ft}^3}{1000 \ \text{ft}^3}\right)}{\pi\left(\frac{75 \ \text{ft}}{2}\right)^2 (6 \ \text{ft})}$$

$$= 118.1 \ \frac{\text{lbm}}{\text{day-1000 ft}^3} + 7.1 \ \frac{\text{lbm}}{\text{day-1000 ft}^3}$$
$$= 125.2 \ \text{lbm/day-1000 ft}^3$$

From Eq. 29.15, since $R = 0$, then $F = 1$.

The filter/clarifier efficiency is given by Eq. 29.13.

$$\eta = \frac{1}{1 + 0.0561\sqrt{\dfrac{L_{BOD}}{F}}}$$

$$= \frac{1}{1 + 0.0561\sqrt{\dfrac{125.2\,\dfrac{lbm}{day\text{-}1000\ ft^3}}{1.00}}}$$

$$= 0.61 \quad (61\%)$$

$$BOD_2 = (1 - 0.61)\left(250\,\frac{mg}{L}\right) = 97.5\ mg/L$$

The removal fraction in the RBC must be

$$\eta = \frac{BOD_2 - BOD_3}{BOD_2}$$

$$= \frac{97.5\,\dfrac{mg}{L} - 30\,\dfrac{mg}{L}}{97.5\,\dfrac{mg}{L}}$$

$$= 0.69$$

Solving the given performance equation, the immersed area is

$$\eta_{BOD} = \frac{1}{\left(1 + \dfrac{kA}{Q}\right)^3}$$

$$0.69 = \frac{1}{\left(1 + \dfrac{2.45A}{1.5 \times 10^6}\right)^3}$$

$$A = 80{,}609\ ft^2$$

From Table 29.11, only 40% of the total RBC area is immersed at one time. The total RBC area is

$$A_{total} = \frac{80{,}609\ ft^2}{0.4} = \boxed{201{,}523\ ft^2 \quad (2.0 \times 10^5\ ft^2)}$$

The answer is (D).

(b) This is one of the recirculation modes to which the NRC model applies. Solve the problem backward to get $BOD_{out} = 30$ mg/L.

$$BOD_3 = 30\ mg/L$$

$$BOD_2 = \frac{30\,\dfrac{mg}{L}}{1 - 0.65} = 85.7\ mg/L$$

The efficiency in the NRC model includes the effect of the clarifier, even though the recirculation occurs before the clarifier.

$$\eta_{\text{trickling filter and clarifier}} = \frac{BOD_1 - BOD_2}{BOD_1}$$

$$= \frac{250\,\dfrac{mg}{L} - 85.7\,\dfrac{mg}{L}}{250\,\dfrac{mg}{L}}$$

$$= 0.657$$

In this configuration, the BOD loading does not include the effects of recirculation, as the NRC model places a higher emphasis on organic loading than on hydraulic loading. From part (a), $L_{BOD} = 118.1$ lbm/day-1000 ft^3.

Equation 29.13 could be solved for F, and then that value used in Eq. 29.15 to find R. It is easier to use Fig. 29.4 with $L_{BOD} = 118.1$ and $\eta = 65.7\%$.

$$R \approx \boxed{0.5 \quad (50\%)}$$

The answer is (B).

(c) Approximate the BOD removal of the secondary (first in-line) clarifier.

$$BOD_{entering} = \frac{BOD_2}{1 - \eta} = \frac{85\,\dfrac{mg}{L}}{1 - 0.30} = 121.4\ mg/L$$

$$\text{[assuming 30\% removal]}$$

The sludge production is

$$\left(0.4\,\frac{lbm}{lbm}\right)\left(\begin{array}{c}\left(121.4\,\dfrac{mg}{L} - 85\,\dfrac{mg}{L}\right) \\ + \left(43\,\dfrac{mg}{L} - 30\,\dfrac{mg}{L}\right)\end{array}\right)$$

$$\times \left(8.345\,\frac{lbm\text{-}L}{mg\text{-}MG}\right)(1.5\ MGD)$$

$$= 247.3\ lbm/day$$

The sludge volume is

$$V = \frac{247.3\,\dfrac{lbm}{day}}{62.4\,\dfrac{lbm}{ft^3}}$$

$$= \boxed{3.96\ ft^3/day \quad (4\ ft^3/day)}$$

The answer is (A).

30 Activated Sludge and Sludge Processing

PRACTICE PROBLEMS

Sludge Quantities

1. 33 m³/d of thickened sludge with a suspended solids content of 3.8% and 13 m³/d of anaerobic digester sludge with a suspended solids content of 7.8% are produced in a wastewater treatment plant.

(a) What would be the decrease in sludge volume per year by using a filter press to increase the solids content of the thickened sludge to 24%?

    (A) 4000 m³/yr

    (B) 8000 m³/yr

    (C) 10 000 m³/yr

    (D) 15 000 m³/yr

(b) What volume of digester sludge must be disposed of from sand drying beds that increase the solids concentration to 35%?

    (A) 500 m³/yr

    (B) 1000 m³/yr

    (C) 2000 m³/yr

    (D) 4000 m³/yr

2. An activated sludge plant processes 10 MGD of wastewater with 240 mg/L BOD and 225 mg/L suspended solids. 70% of the suspended solids are inorganic. The discharge from the final clarifier contains 15 mg/L BOD (all organic) and 20 mg/L suspended solids (all inorganic). Primary clarification removes 60% of the suspended solids and 35% of the BOD. The BOD reduction in the primary clarifier does not contribute to sludge production. The sludge produced has a specific gravity of 1.02 and a solids content of 6%. The cell yield (conversion of BOD reduction to biological solids) is 60%. The final clarifier does not reduce BOD.

(a) What is the daily mass of dry sludge solids produced?

    (A) 2500 lbm/day

    (B) 5000 lbm/day

    (C) 10,000 lbm/day

    (D) 20,000 lbm/day

(b) Assuming that the sludge is completely dried and compressed solid, what is the daily sludge volume?

    (A) 200 ft³/day

    (B) 300 ft³/day

    (C) 1000 ft³/day

    (D) 1600 ft³/day

(c) Assuming that the final gravimetric fraction of water in the sludge is 70% and air voids increase the volume by 10%, what is the daily sludge volume?

    (A) 200 ft³/day

    (B) 500 ft³/day

    (C) 1000 ft³/day

    (D) 1600 ft³/day

Environmental

SOLUTIONS

1. (a) Sludge volume is inversely proportional to the solids content. The sludge volume from the filter press is

$$V_2 = V_1\left(\frac{SS_1}{SS_2}\right) = \left(33\ \frac{m^3}{d}\right)\left(\frac{0.038}{0.24}\right)$$
$$= 5.23\ m^3/d$$

The yearly decrease in volume is

$$\left(33\ \frac{m^3}{d} - 5.23\ \frac{m^3}{d}\right)\left(365\ \frac{d}{yr}\right)$$
$$= \boxed{10\,136\ m^3/yr\quad(10\,000\ m^3/yr)}$$

The answer is (C).

(b) The sludge volume from the sand drying bed is

$$V_2 = V_1\left(\frac{SS_1}{SS_2}\right) = \left(13\ \frac{m^3}{d}\right)\left(\frac{0.078}{0.35}\right)$$
$$= 2.90\ m^3/d$$

The yearly disposal volume is

$$\left(2.9\ \frac{m^3}{d}\right)\left(365\ \frac{d}{yr}\right) = \boxed{1059\ m^3/yr\quad(1000\ m^3/yr)}$$

The answer is (B).

2. The BOD and suspended solids are not mutually exclusive. Some of the suspended solids are organic in nature and show up as BOD.

The inorganic suspended solids are

$$(0.70)\left(225\ \frac{mg}{L}\right) = 157.5\ mg/L$$

The inorganic suspended solids that survive primary settling are

$$(1 - 0.60)\left(157.5\ \frac{mg}{L}\right) = 63\ mg/L$$

The BOD that survives primary settling is

$$(1 - 0.35)\left(240\ \frac{mg}{L}\right) = 156\ mg/L$$

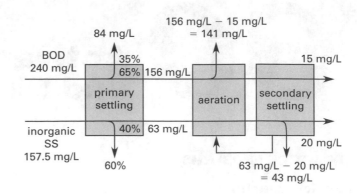

(a) The total dry weight of suspended solids removed in all processes is

$$\left(157.5\ \frac{mg}{L} - 20\ \frac{mg}{L}\right)\left(8.345\ \frac{lbm\text{-}L}{MG\text{-}mg}\right)\left(10\ \frac{mg}{day}\right)$$
$$= 11{,}474\ lbm/day$$

Solids from BOD reduction in the secondary process are

$$Y(\Delta BOD)Q = (0.60)\left(156\ \frac{mg}{L} - 15\ \frac{mg}{L}\right)$$
$$\times \left(10\ \frac{MG}{day}\right)\left(8.345\ \frac{lbm\text{-}L}{mg\text{-}MG}\right)$$
$$= 7060\ lbm/day$$

The total dry sludge mass is

$$11{,}474\ \frac{lbm}{day} + 7060\ \frac{lbm}{day}$$
$$= \boxed{18{,}534\ lbm/day\quad(20{,}000\ lbm/day)}$$

The answer is (D).

(b) The specific gravity of the sludge solids can be found from Eq. 30.47.

$$\frac{1}{SG} = \frac{1-s}{1} + \frac{s}{SG_{solids}}$$
$$\frac{1}{1.02} = \frac{1-0.06}{1} + \frac{0.06}{SG_{solids}}$$
$$SG_{solids} = 1.485$$

The density of the solids is

$$\rho_{solids} = SG_{solids}\rho_{water} = (1.485)\left(62.4\ \frac{lbm}{ft^3}\right)$$
$$= 92.66\ lbm/ft^3$$

The solid dry sludge volume is

$$V_{\text{solid}} = \frac{m}{\rho} = \frac{18{,}534 \, \dfrac{\text{lbm}}{\text{day}}}{92.66 \, \dfrac{\text{lbm}}{\text{ft}^3}}$$

$$= \boxed{200.0 \text{ ft}^3/\text{day} \quad (200 \text{ ft}^3/\text{day})}$$

The answer is (A).

(c) $1 - s$ (the moisture content) is given as 0.70. The disposal volume is

$$V_t = V_{\text{solid}} + V_{\text{water}}$$

$$V_{\text{solid}} = \frac{m_{\text{solid}}}{\rho}$$

$$V_{\text{water}} = \frac{m_{\text{water}}}{62.4 \, \dfrac{\text{lbm}}{\text{ft}^3}} = \frac{0.70 m_t}{62.4 \, \dfrac{\text{lbm}}{\text{ft}^3}}$$

$$= \frac{0.70 m_{\text{solids}}}{\left(62.4 \, \dfrac{\text{lbm}}{\text{ft}^3}\right)(1 - 0.70)}$$

$$V \approx (1.10)\left(\frac{18{,}534 \, \dfrac{\text{lbm}}{\text{day}}}{92.66 \, \dfrac{\text{lbm}}{\text{ft}^3}} + \frac{\left(18{,}534 \, \dfrac{\text{lbm}}{\text{day}}\right)(0.70)}{\left(62.4 \, \dfrac{\text{lbm}}{\text{ft}^3}\right)(1 - 0.70)} \right)$$

$$= \boxed{982.4 \text{ ft}^3/\text{day} \quad (1000 \text{ ft}^3/\text{day})}$$

Notice that the basis of moisture content, s, is the total sludge mass, whereas the traditional "water content" basis used in geotechnical calculations is the solids mass.

The answer is (C).

31 Municipal Solid Waste

PRACTICE PROBLEMS

Landfill Capacity

1. A town has a current population of 10,000, which is expected to double in 15 yr. The town intends to dispose of its municipal solid waste in a 30 ac landfill that will be converted to a park in 20 yr. Solid waste is generated at the rate of 5 lbm/capita-day. The average compacted density in the landfill will be 1000 lbm/yd^3. Disregarding any soil addition for cover and cell construction, how long will it take to fill the landfill to a uniform height of 6 ft?

(A) 2100 days

(B) 4200 days

(C) 6300 days

(D) 9500 days

2. (*Time limit: one hour*) A town of 10,000 people has selected a 50 ac square landfill site to deposit its solid waste. The minimum unused side borders are 50 ft. The landfill currently consists of a square depression with an average depth of 20 ft below the surrounding grade. When the landfill is at final capacity, it will be covered with 10 ft of earth cover. The maximum height of the covered landfill is 20 ft above the surrounding grade. Solid waste is generated at the rate of 5 lbm/capita-day. The average compacted density in the landfill will be 1000 lbm/yd^3.

(a) What is the volumetric capacity of the landfill site?

(A) 1.1×10^6 ft^3

(B) 6.5×10^6 ft^3

(C) 3.1×10^7 ft^3

(D) 5.7×10^7 ft^3

(b) Using a loading factor of 1.25, what is the volume of landfill used each day?

(A) 60 yd^3/day

(B) 120 yd^3/day

(C) 180 yd^3/day

(D) 240 yd^3/day

(c) What is the service life of the landfill site?

(A) 30 yr

(B) 45 yr

(C) 60 yr

(D) 90 yr

Leachate

3. The pressure and temperature within a closed landfill cell are 1 atm and 45°C, respectively. The volumetric fraction of carbon dioxide in the landfill is 40%. The solubility, K_s, of carbon dioxide in leachate at 45°C is 0.8 g/kg. What is the concentration of carbon dioxide in the leachate in units of mg CO_2/L water?

4. Dissolved hydrogen sulfide gas, H_2S, in landfill leachate has a concentration of 270 mg/L. What is this concentration expressed in parts per million (ppm)?

SOLUTIONS

1. Find the rate of increase of waste production.

The mass of waste deposited in the first day will be

$$(10{,}000 \text{ people})\left(5\ \frac{\text{lbm}}{\text{person-day}}\right) = 50{,}000\ \text{lbm/day}$$

The mass of waste deposited on the last day will be

$$(20{,}000 \text{ people})\left(5\ \frac{\text{lbm}}{\text{person-day}}\right) = 100{,}000\ \text{lbm/day}$$

The increase in rate is

$$\frac{\Delta m}{\Delta t} = \frac{100{,}000\ \dfrac{\text{lbm}}{\text{day}} - 50{,}000\ \dfrac{\text{lbm}}{\text{day}}}{(15\ \text{yr})\left(365\ \dfrac{\text{days}}{\text{yr}}\right)} = 9.132\ \text{lbm/day}^2$$

The mass deposited on day t is

$$m_D = 50{,}000 + 9.132(t-1)$$
$$\approx 50{,}000 + 9.132t$$

The cumulative mass deposited is

$$m_t = \int_0^t m_D\, dt = 50{,}000t + \frac{9.132t^2}{2}$$

With a compacted density of 1000 lbm/yd^3 and a loading factor of 1.00 (no soil cover), the capacity of the site with a 6 ft lift is

$$m_{\text{max}} = \frac{(30\ \text{ac})\left(43{,}560\ \dfrac{\text{ft}^2}{\text{ac}}\right)(6\ \text{ft})\left(1000\ \dfrac{\text{lbm}}{\text{yd}^3}\right)}{27\ \dfrac{\text{ft}^3}{\text{yd}^3}}$$
$$= 2.9 \times 10^8\ \text{lbm}$$

The time to fill is found by solving the quadratic equation.

$$2.9 \times 10^8 = 50{,}000t + \frac{9.132t^2}{2}$$
$$t^2 + 10{,}951t = 6.351 \times 10^7$$
$$t = \boxed{4193\ \text{days} \quad (4200\ \text{days})}$$

The answer is (B).

2. (a) The side length of the square disposal site is

$$\text{length} = \sqrt{(50\ \text{ac})\left(43{,}560\ \frac{\text{ft}^2}{\text{ac}}\right)} = 1476\ \text{ft}$$

With 50 ft borders, the usable area is

$$A = \left(1476\ \text{ft} - (2)(50\ \text{ft})\right)^2 = 1.893 \times 10^6\ \text{ft}^2$$

If the site is excavated 20 ft, 10 ft of soil is used as cover, and the maximum above-ground height is 20 ft, the service capacity of compacted waste is

$$(1.893 \times 10^6\ \text{ft}^2)(30\ \text{ft}) = \boxed{5.68 \times 10^7\ \text{ft}^3}$$

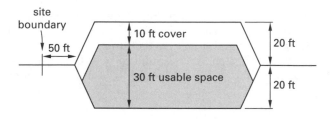

(not to scale or representative of actual construction)

The answer is (D).

(b) The volume of landfill used per day is

$$\frac{(10{,}000 \text{ people})\left(5\ \dfrac{\text{lbm}}{\text{day-person}}\right)(1.25)}{1000\ \dfrac{\text{lbm}}{\text{yd}^3}}$$
$$= \boxed{62.5\ \text{yd}^3/\text{day}}$$

The answer is (A).

(c) The service life is

$$\frac{5.68 \times 10^7\ \text{ft}^3}{\left(27\ \dfrac{\text{ft}^3}{\text{yd}^3}\right)\left(62.5\ \dfrac{\text{yd}^3}{\text{day}}\right)\left(365\ \dfrac{\text{days}}{\text{yr}}\right)} = \boxed{92.2\ \text{yr}}$$

The answer is (D).

3. Use the ideal gas law to find the density of carbon dioxide at the landfill conditions.

$$\rho_{CO_2,g/L} = \frac{p}{RT} = \frac{B_{CO_2} p_t (MW)}{R^* T}$$

$$= \frac{(0.4)(1 \text{ atm})\left(44 \ \frac{g}{mol}\right)}{\left(0.08206 \ \frac{atm \cdot L}{mol \cdot K}\right)(45°C + 273°)}$$

$$= 0.674 \text{ g/L}$$

Use the ideal gas law to determine the volume of 0.8 g of carbon dioxide.

$$V_{CO_2} = \frac{m}{\rho} = \frac{0.8 \text{ g}}{0.674 \ \frac{g}{L}}$$

$$= 1.19 \text{ L}$$

The density of water at 45°C is approximately 990 kg/m³. The volume of 1 kg of water is

$$V_{water} = \frac{m_{water}}{\rho_{water}} = \frac{(1 \text{ kg})\left(1000 \ \frac{L}{m^3}\right)}{990 \ \frac{kg}{m^3}}$$

$$= 1.01 \text{ L}$$

The solubility of carbon dioxide is

$$K_s = \frac{V_{CO_2}}{V_{water}} = \frac{1.19 \text{ L}}{1.01 \text{ L}}$$

$$= 1.18 \text{ L/L}$$

Find the concentration of carbon dioxide from Eq. 31.23.

$$C_{CO_2} = \rho K_s = \left(0.674 \ \frac{g}{L}\right)\left(1.18 \ \frac{L}{L}\right)\left(1000 \ \frac{mg}{L}\right)$$

$$= 795 \text{ mg/L}$$

4. ppm is a concentration based on a mass ratio; that is, milligrams of H_2S per million milligrams of leachate. The quantity of dissolved gas is already expressed in milligrams, so express the leachate quantity in milligrams. For dilute solutions, the leachate can be assumed to be water. Since the density of water is approximately 1000 kg/m³, and there are 1000 liters per cubic meter, a liter of water has a mass of 10^6 mg (i.e., 1 kg). So, a concentration of 270 mg/L is equivalent to 270 ppm.

Environmental

32 Pollutants in the Environment

PRACTICE PROBLEMS

There are no problems in this book corresponding to Chap. 32 of the *Civil Engineering Reference Manual*.

33 Storage and Disposition of Hazardous Materials

PRACTICE PROBLEMS

There are no problems in this book corresponding to Chap. 33 of the *Civil Engineering Reference Manual*.

34 Environmental Remediation

PRACTICE PROBLEMS

There are no problems in this book corresponding to Chap. 34 of the *Civil Engineering Reference Manual*.

35 Soil Properties and Testing

PRACTICE PROBLEMS

Properties and Classification

1. A clay has the following Atterberg limits: liquid limit = 60, plastic limit = 40, and shrinkage limit = 25. The clay shrinks from 15 cm³ to 9.57 cm³ when the moisture content is decreased from the liquid limit to the shrinkage limit in the Atterberg tests. What is the clay's specific gravity (dry)?

(A) 2.0

(B) 2.3

(C) 2.5

(D) 2.7

2. A sample of soil has the following characteristics.

% passing no. 40 screen	95
% passing no. 200 screen	57
liquid limit	37
plastic limit	18

What are the AASHTO classification and group index number?

(A) A-5(6)

(B) A-5(8)

(C) A-6(8) ←

(D) A-7(5)

3. (*Time limit: one hour*) A sample of moist soil was found to have the following characteristics.

volume	0.514 ft³ (as sampled)
mass	56.74 lbm (as sampled)
	48.72 lbm (after oven drying)
specific gravity of solids	2.69

After drying, the soil was run through a set of sieves with the following percentages passing through each sieve.

sieve	% finer by mass
¹/₂ in	52
no. 4	37
no. 10	32
no. 20	23
no. 40	11
no. 60	7
no. 100	4

(a) What is the density of the in situ soil?

(A) 92 lbm/ft³

(B) 98 lbm/ft³

(C) 110 lbm/ft³ ←

(D) 120 lbm/ft³

(b) What is the unit weight of the in situ soil?

(A) 92 lbf/ft³

(B) 98 lbf/ft³

(C) 110 lbf/ft³ ←

(D) 120 lbf/ft³

(c) What is the void ratio of the in situ soil?

(A) 0.77 ←

(B) 0.80

(C) 1.00

(D) 1.60

(d) What is the porosity of the in situ soil?

(A) 0.44 ←

(B) 0.57

(C) 0.77

(D) 0.98

(e) What is the degree of saturation of the in situ soil?

(A) 44%

(B) 57% ←

(C) 78%

(D) 95%

(f) Which of the following illustrations represents the particle size distribution of the soil?

(A)

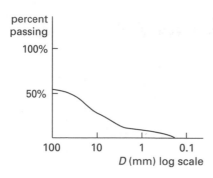

(B)

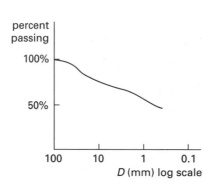

(C)

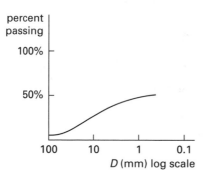

→ (D)

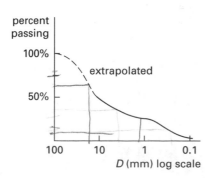

(g) What is the uniformity coefficient?

(A) 35

(B) 40

(C) 43

(D) 58

(h) What is the coefficient of curvature?

(A) 0.28

(B) 0.35

(C) 0.51

(D) 0.68

(i) What is the USCS classification of the soil?

(A) GW

(B) GP ←

(C) SW

(D) SM

(j) What is the AASHTO classification of the soil?

(A) A-1-a ←

(B) A-1-b

(C) A-2-4

(D) A-2-7

Permeability Test

4. A permeability test is conducted with a sample of soil that is 60 mm in diameter and 120 mm long. The head is kept constant at 225 mm. The flow is 1.5 mL in 6.5 min. What is the coefficient of permeability?

(A) 4.4 m/yr

(B) 12 m/yr ←

(C) 17 m/yr

(D) 23 m/yr

Proctor Test

5. A Proctor test was performed on a soil that has a specific gravity of solids of 2.71. The following results were obtained.

water content (%)	total unit weight (lbf/ft^3)
10	98
13	106
16	119
18	125
20	129
22	128
25	123

(a) Plot the moisture versus dry density curve.

(b) What is the maximum dry unit weight?

(A) 97 lbf/ft^3

(B) 103 lbf/ft^3

(C) 108 lbf/ft^3

(D) 112 lbf/ft^3

(c) What is the optimum moisture content?

(A) 12%

(B) 17%

(C) 20%

(D) 24%

(d) What range of moisture is permitted if a contractor must achieve 90% compaction?

(A) 12–23%

(B) 14–26%

(C) 18–28%

(D) 20–24%

6. For the soil in Prob. 5, what volume of water must be added to obtain 1 yd^3 of soil at the maximum density if the soil is originally at 10% water content (dry basis)?

(A) 6 gal

(B) 14 gal

(C) 18 gal

(D) 31 gal

Borrow Soil

7. Specifications on a job require a fill using borrow soil to be compacted to 95% of its standard Proctor maximum dry density. Tests indicate that this maximum is 124.0 lbm/ft^3 when dry. The soil now has 12% moisture. The borrow material has a void ratio of 0.60 and a solid specific gravity of 2.65. What is the minimum volume of borrow soil required per 1.0 ft^3 of fill volume?

(A) 0.90 ft^3

(B) 1.1 ft^3

(C) 1.3 ft^3

(D) 1.5 ft^3

8. (*Time limit: one hour*) Two choices for borrow soil are available.

	borrow A	borrow B
specific weight in place	115 lbf/ft^3	120 lbf/ft^3
specific weight in transport	–	95 lbf/ft^3
void ratio in transport	0.92	–
water content in place	25%	20%
cost to excavate	\$0.20/yd^3	\$0.10/yd^3
cost to haul	\$0.30/yd^3	\$0.40/yd^3
SG of solids	2.7	2.7
maximum Proctor dry density	112 lbf/ft^3	110 lbf/ft^3

It will be necessary to fill a 200,000 yd^3 depression, and the fill material must be compacted to 95% of the standard Proctor (maximum) density. A 10% (dry basis) final moisture content is desired in either case.

(a) What would be the volume of borrow from site A?

(A) 4.8×10^6 ft^3

(B) 5.4×10^6 ft^3

(C) 5.6×10^6 ft^3

(D) 6.3×10^6 ft^3

(b) What would be the volume of borrow from site B?

(A) 4.8×10^6 ft^3

(B) 5.4×10^6 ft^3

(C) 5.6×10^6 ft^3

(D) 6.3×10^6 ft^3

(c) What soil would be cheaper to use?

(A) borrow A by \$6200

(B) borrow A by \$7400

(C) borrow B by \$4500

(D) borrow B by \$19,000

Triaxial Shear Test

9. A triaxial shear test is performed on a well-drained sand sample. At failure, the normal stress on the failure plane was 6260 lbf/ft^2 and the shear stress on the failure plane was 4175 lbf/ft^2.

(a) What is the angle of internal friction?

(A) 16°

(B) 28°

(C) 34°

(D) 42°

Geotechnical

(b) What is the maximum principal stress?

(A) 9000 lbf/ft^2

(B) 14,000 lbf/ft^2

(C) 16,000 lbf/ft^2

(D) 22,000 lbf/ft^2

10. (*Time limit: one hour*) Two series of triaxial shear tests on a soil were performed with the following results. (All values are in lbf/in^2.)

Undrained series:

confining pressure	total axial stress
0	60
50	110
100	160

Drained series:

confining pressure	total axial stress
50	250
100	400
150	550

(a) What is the angle of internal friction for the undrained series?

(A) 0°

(B) 5°

(C) 12°

(D) 23°

(b) What is the angle of internal friction for the drained series?

(A) 10°

(B) 20°

(C) 30°

(D) 40°

(c) What is the cohesion for the undrained series?

(A) 30 psi

(B) 90 psi

(C) 180 psi

(D) 280 psi

(d) What is the cohesion for the drained series?

(A) 17 psi

(B) 23 psi

(C) 29 psi

(D) 45 psi

(e) What is the angle of the failure plane (with respect to the horizontal axis) for the undrained series?

(A) 0°

(B) 15°

(C) 30°

(D) 45°

(f) What is the angle of the failure plane (with respect to the horizontal axis) for the drained series?

(A) 15°

(B) 39°

(C) 45°

(D) 60°

(g) Given a fourth test of the drained sample with a radial confining pressure of 300 lbf/in^2, what is the expected axial load at failure?

(A) 550 lbf/in^2

(B) 1000 lbf/in^2

(C) 1200 lbf/in^2

(D) 1600 lbf/in^2

11. (*Time limit: one hour*) A consolidated, undrained (CU) triaxial test was performed on a normally consolidated, saturated clay. The minor principal stress, σ_3, at the start of the test was 100 kPa. The deviator stress, σ_D, at failure was 60 kPa. The excess pore pressure, ΔU, was measured during the test, and at failure it was 45 kPa.

(a) What is the undrained strength at failure, S_u?

(A) 15 kPa

(B) 30 kPa

(C) 40 kPa

(D) 60 kPa

(b) What is the total minor principal stress at failure, σ_{3f}?

(A) 40 kPa

(B) 55 kPa

(C) 60 kPa

(D) 100 kPa

(c) What is the total major principal stress at failure, σ_{1f}?

(A) 30 kPa

(B) 40 kPa

(C) 130 kPa

(D) 160 kPa

Geotechnical

(d) What is the effective minor principal stress at failure, σ'_{3f}?

(A) 30 kPa

(B) 55 kPa

(C) 60 kPa

(D) 100 kPa

(e) What is the effective major principal stress at failure, σ'_{1f}?

(A) 60 kPa

(B) 100 kPa

(C) 120 kPa

(D) 160 kPa

(f) What is the effective friction angle, ϕ'?

(A) 21°

(B) 22°

(C) 26°

(D) 31°

(g) What is the angle of the effective failure plane, α?

(A) 37°

(B) 45°

(C) 55°

(D) 61°

(h) What is the effective normal stress on the failure plane at failure, $\sigma_{n,f}$?

(A) 66 kPa

(B) 74 kPa

(C) 85 kPa

(D) 130 kPa

(i) What is the effective shear stress on the failure plane at failure, $\tau_{n,f}$?

(A) 28 kPa

(B) 30 kPa

(C) 37 kPa

(D) 60 kPa

(j) What is another name for this type of test?

(A) S-test

(B) R-test

(C) unconfined compressive test

(D) CBR test

Direct Shear Test

12. (*Time limit: one hour*) A sandy clay soil was tested in a direct shear apparatus, with the following results.

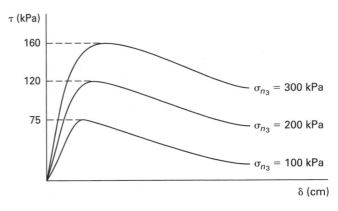

(a) What is the friction angle?

(A) 7°

(B) 14°

(C) 23°

(D) 29°

(b) What is the cohesion intercept?

(A) 24 kPa

(B) 33 kPa

(C) 57 kPa

(D) 86 kPa

Consolidation Test

13. A sample of sand has a relative density of 40% with a solids specific gravity of 2.65. The minimum void ratio is 0.45, and the maximum void ratio is 0.97.

(a) What is the specific weight of the sand in a saturated condition?

(A) 94 lbf/ft^3

(B) 116 lbf/ft^3

(C) 121 lbf/ft^3

(D) 137 lbf/ft^3

(b) If the sand is compacted to a relative density of 65%, what will be the decrease in thickness of a layer 4 ft thick?

(A) 0.1 ft

(B) 0.2 ft

(C) 0.3 ft

(D) 0.4 ft

14. A consolidation test is performed on a soil with the following results.

pressure (lbf/ft^2)	void ratio, e
250	0.755
520	0.754
1040	0.753
2090	0.750
4180	0.740
8350	0.724
16,700	0.704
33,400	0.684
8350	0.691
250	0.710

(a) Graph the curve of stress versus void ratio on log or semilog paper.

(b) What is the preconsolidation pressure?

   (A) 4400 lbf/ft^2

   (B) 6200 lbf/ft^2

   (C) 8900 lbf/ft^2

   (D) 12,000 lbf/ft^2

(c) What is the compression index?

   (A) 0.0053

   (B) 0.090

   (C) 0.17

   (D) 0.25

15. If a nuclear density gauge is used early in the morning immediately after visible ground surface frost has evaporated, the gauge will most likely indicate

   (A) a higher moisture content

   (B) a higher dry density

   (C) a lower voids ratio

   (D) the same as later in the day

16. A nuclear density gauge is UNABLE to measure which of the following without the target density programmed into it?

   (A) wet density

   (B) dry density

   (C) moisture content

   (D) degree of compaction

SOLUTIONS

1. The water reduction is

$$15 \text{ cm}^3 - 9.57 \text{ cm}^3 = 5.43 \text{ cm}^3$$

Since 1 cm^3 of water has a mass of 1 g, the mass loss is 5.43 g. The fractional mass loss (dry basis) is

$$0.60 - 0.25 = 0.35$$

Therefore, from Eq. 35.6, and using a Δw to indicate the change in the water content, the solid mass is

$$m_s = \frac{\Delta m_w}{\Delta w} = \frac{5.43 \text{ g}}{0.35} = 15.5 \text{ g}$$

The water volume at the shrinkage limit is

$$V_w = \frac{m_w}{\rho_w} = \frac{w m_s}{\rho_w} = \frac{(0.25)(15.5 \text{ g})}{1 \, \frac{\text{g}}{\text{cm}^3}}$$

$$= 3.875 \text{ cm}^3$$

Since at and above the shrinkage limit there are no air voids (i.e., all voids are filled with water), the volume of the solid at the shrinkage limit is

$$9.57 \text{ cm}^3 - 3.875 \text{ cm}^3 = 5.695 \text{ cm}^3$$

The density of the solids is

$$\rho = \frac{m}{V} = \frac{15.5 \text{ g}}{5.695 \text{ cm}^3} = 2.72 \text{ g/cm}^3$$

$$\text{SG} = \frac{\rho_s}{\rho_w} = \frac{2.72 \, \frac{\text{g}}{\text{cm}^3}}{1 \, \frac{\text{g}}{\text{cm}^3}} = \boxed{2.72 \quad (2.7)}$$

The answer is (D).

2. The plasticity index is given by Eq. 35.23.

$$\text{PI} = \text{LL} - \text{PL} = 37 - 18 = 19$$

The group index is given by Eq. 35.3.

$$I_g = (F_{200} - 35)(0.2 + 0.005(\text{LL} - 40))$$

$$+ 0.01(F_{200} - 15)(\text{PI} - 10)$$

$$= (57 - 35)(0.2 + (0.005)(37 - 40))$$

$$+ (0.01)(57 - 15)(19 - 10)$$

$$= 7.85 \quad [\text{round to 8}]$$

$$\text{classification} = \boxed{\text{A-6(8)}}$$

The answer is (C).

Geotechnical

3. (a) The density of the in situ soil is

$$\rho = \frac{m_t}{V_t} = \frac{56.74 \text{ lbm}}{0.514 \text{ ft}^3}$$
$$= \boxed{110.4 \text{ lbm/ft}^3 \quad (100 \text{ lbf/ft}^3)}$$

The answer is (C).

(b) The unit weight of the in situ soil is

$$\gamma = \frac{\rho g}{g_c} = \frac{\left(110.4 \dfrac{\text{lbm}}{\text{ft}^3}\right)\left(32.2 \dfrac{\text{ft}}{\text{sec}^2}\right)}{32.2 \dfrac{\text{lbm-ft}}{\text{lbf-sec}^2}}$$
$$= \boxed{110.4 \text{ lbf/ft}^3 \quad (100 \text{ lbf/ft}^3)}$$

The answer is (C).

(c) The void ratio is as follows.

$$m_w = m_t - m_s = 56.74 \text{ lbm} - 48.72 \text{ lbm}$$
$$= 8.02 \text{ lbm}$$
$$V_s = \frac{m_s}{\rho_s} = \frac{48.72 \text{ lbm}}{(2.69)\left(62.4 \dfrac{\text{lbm}}{\text{ft}^3}\right)}$$
$$- 0.29025 \text{ ft}^3$$
$$V_w = \frac{8.02 \text{ lbm}}{62.4 \dfrac{\text{lbm}}{\text{ft}^3}} = 0.12853 \text{ ft}^3$$
$$V_g = 0.514 \text{ ft}^3 - 0.12853 \text{ ft}^3 - 0.29025 \text{ ft}^3$$
$$= 0.09522 \text{ ft}^3$$

From Eq. 35.5,

$$e = \frac{V_g + V_w}{V_s} = \frac{0.09522 \text{ ft}^3 + 0.12853 \text{ ft}^3}{0.29025 \text{ ft}^3}$$
$$= \boxed{0.771 \quad (0.77 \text{ ft})}$$

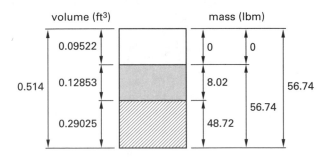

The answer is (A).

(d) Use Eq. 35.4.

$$n = \frac{V_g + V_w}{V_t} = \frac{0.09522 \text{ ft}^3 + 0.12853 \text{ ft}^3}{0.514 \text{ ft}^3}$$
$$= \boxed{0.435 \quad (0.44)}$$

The answer is (A).

(e) Use Eq. 35.7.

$$S = \frac{V_w}{V_g + V_w} \times 100\%$$
$$= \frac{0.12853 \text{ ft}^3}{0.09522 \text{ ft}^3 + 0.12853 \text{ ft}^3} \times 100\%$$
$$= \boxed{57\%}$$

The answer is (B).

(f) Use Table 35.2 to determine the opening size for each sieve. Then, graph the values.

sieve	nominal opening size (mm)	% finer by weight
$\frac{1}{2}$ in	12.5	52
no. 4	4.75	37
no. 10	2.00	32
no. 20	0.85	23
no. 40	0.425	11
no. 60	0.25	7
no. 100	0.15	4

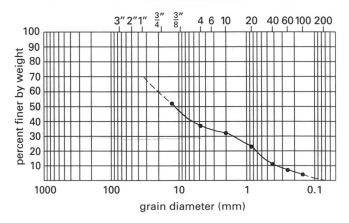

The answer is (D).

(g) Picking off values from the graph,

$$D_{10} = 0.4 \text{ mm} \quad \text{[interpolated]}$$
$$D_{60} \approx 23 \text{ mm} \quad \text{[extrapolated]}$$

From Eq. 35.1,

$$C_u = \frac{D_{60}}{D_{10}} = \frac{23 \text{ mm}}{0.4 \text{ mm}}$$

$$= \boxed{57.5 \quad (58)}$$

The answer is (D).

(h) From the graph, $D_{30} = 1.6$ mm. Use Eq. 35.2.

$$C_z = \frac{D_{30}^2}{D_{10}D_{60}} = \frac{(1.6 \text{ mm})^2}{(0.4 \text{ mm})(23 \text{ mm})}$$

$$= \boxed{0.278 \quad (28)}$$

The answer is (A).

(i) Use Table 35.5 and the particle size indicators from parts (f)–(g).

The soil is coarse because over 50% is coarser than the no. 200 sieve. Over half is larger than the no. 4 sieve, so the soil is gravelly. The amount finer than the no. 200 sieve is between 0% and 5%, so the soil is either in the GW or GP group. From part (g), $C_u = 57.5$ and from part (h), $C_z = 0.278$. Therefore, the first requirement for GW is met but not the second, and the soil is classified as $\boxed{\text{GP.}}$ The uniformity coefficient is too high for the soil to be uniform, so the soil is a gap-graded gravel.

The answer is (B).

(j) Use Table 35.4 and the particle size indicators previously calculated.

The first group from the left consistent with the test data is the correct classification. The soil does not exceed 50% passing the no. 10, 30% passing the no. 40, or 15% passing the no. 200. Therefore the soil is classified as $\boxed{\text{A-1-a.}}$

The answer is (A).

4. From Eq. 35.29,

$$K = \frac{VL}{hAt} = \frac{VL}{h\left(\dfrac{\pi d^2}{4}\right)t}$$

$$= \frac{(1.5 \text{ mL})(120 \text{ mm})}{(225 \text{ mm})\left(\dfrac{\pi(60 \text{ mm})^2}{4}\right)(6.5 \text{ min})}$$

$$\times \frac{\left(1000 \dfrac{\text{mm}^3}{\text{mL}}\right)\left(365 \dfrac{\text{d}}{\text{yr}}\right)\left(24 \dfrac{\text{h}}{\text{d}}\right)\left(60 \dfrac{\text{min}}{\text{h}}\right)}{1000 \dfrac{\text{mm}}{\text{m}}}$$

$$= \boxed{22.88 \text{ m/yr} \quad (23 \text{ m/yr})}$$

The answer is (D).

5. (a) For the first sample, from Eq. 35.11,

$$\gamma_d = \frac{\gamma}{1+w} = \frac{98 \dfrac{\text{lbf}}{\text{ft}^3}}{1+0.10} = 89.1 \text{ lbf/ft}^3$$

Similarly, for the rest of the samples,

w	γ_d
10%	89.1
13%	93.8
16%	102.6
18%	105.9
20%	107.5
22%	104.9
25%	98.4

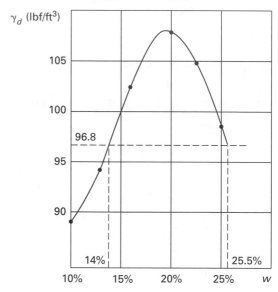

(b) $\gamma_{d,\max} = \boxed{107.5 \text{ lbf/ft}^3 \quad (108 \text{ lbf/ft}^3)}$

The answer is (C).

(c) Depending on how the curve is drawn through the data points, $w = \boxed{19\text{--}20\%}$

The answer is (C).

(d) The range of moisture permitted for 90% compaction is

$$(0.90)\left(107.5 \dfrac{\text{lbf}}{\text{ft}^3}\right) = 96.8 \text{ lbf/ft}^3$$

$$\boxed{14\% \leq w \leq 25.5\%}$$

The answer is (B).

6. At maximum density, the masses of water and solids in 1 ft^3 of soil are

$$m_s = 107.5 \text{ lbm}$$

$$m_w = (0.19)(107.5 \text{ lbm}) = 20.43 \text{ lbm}$$

At 10% water content, $\gamma_d \approx 89.1$ lbf/ft^3. For 1 ft^3,

$$m_s = 89.1 \text{ lbm}$$

$$m_w = (0.10)(89.1 \text{ lbm}) = 8.91 \text{ lbm}$$

To get 1 yd^3 (27 ft^3) of maximum density soil, the volume of 10% moisture soil is

$$\frac{\left(27 \frac{\text{ft}^3}{\text{yd}^3}\right)(107.5 \text{ lbm})}{89.1 \text{ lbm}} = \frac{32.58 \text{ ft}^3/\text{yd}^3 \text{ of}}{10\% \text{ moisture soil}}$$

The required water for 1 yd^3 of soil is

$$(1 \text{ yd}^3) \frac{\left(\begin{array}{c}\left(27 \frac{\text{ft}^3}{\text{yd}^3}\right)\left(20.43 \frac{\text{lbm}}{\text{ft}^3}\right) \\ -\left(32.58 \frac{\text{ft}^3}{\text{yd}}\right)\left(8.91 \frac{\text{lbm}}{\text{ft}^3}\right)\end{array}\right)\left(7.48 \frac{\text{gal}}{\text{ft}^3}\right)}{62.4 \frac{\text{lbm}}{\text{ft}^3}}$$

$$= \boxed{31.33 \text{ gal} \quad (31 \text{ gal})}$$

The answer is (D).

7. Start with 1 ft^3 of fill in final form.

$$\gamma_{d,\text{required}} = (0.95)\left(124.0 \frac{\text{lbf}}{\text{ft}^3}\right) = 117.8 \text{ lbf/ft}^3$$

The weight of solids in 1 ft^3 of fill is

$$W_s = \left(117.8 \frac{\text{lbf}}{\text{ft}^3}\right)(1 \text{ ft}^3) = 117.8 \text{ lbf}$$

Determine the borrow required.

$$\gamma_s = (2.65)\left(62.4 \frac{\text{lbf}}{\text{ft}^3}\right) = 165.36 \text{ lbf/ft}^3$$

$$W_s = 117.8 \text{ lbf} \quad [\text{same as fill}]$$

$$V_s = \frac{W_s}{\gamma_s} = \frac{117.8 \text{ lbf}}{165.36 \frac{\text{lbf}}{\text{ft}^3}} = 0.712 \text{ ft}^3$$

$$V_t = V_s(1 + e) = (0.712 \text{ ft}^3)(1 + 0.6)$$

$$= \boxed{1.139 \text{ ft}^3 \quad (1.1 \text{ ft}^3)}$$

The moisture content is irrelevant.

The answer is (B).

8. (a) Soil A:

Calculate the required fill soil density.

$$\gamma_{F,\text{dry}} = (0.95)\left(112 \frac{\text{lbf}}{\text{ft}^3}\right) = 106.4 \text{ lbf/ft}^3$$

This is also the weight of the solids in the fill.

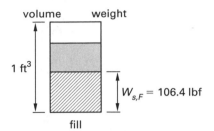

The weight of solids in the fill is the same as the weight of solids in the borrow.

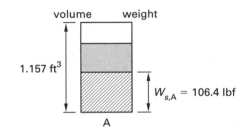

The weight of water in the fill is

$$W_{w,F} = w\,W_{s,F} = (0.10)(106.4 \text{ lbf}) = 10.64 \text{ lbf}$$

Air has essentially no weight, so the total fill weight is

$$W_{t,F} = W_{s,F} + W_{w,F} = 106.4 \text{ lbf} + 10.64 \text{ lbf}$$

$$= 117.04 \text{ lbf}$$

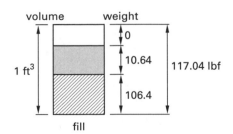

The volume is 200,000 yd^3. The weight of borrow soil solids is

$$(200{,}000 \text{ yd}^3)\left(27 \frac{\text{ft}^3}{\text{yd}^3}\right)\left(106.4 \frac{\text{lbf}}{\text{ft}^3}\right) = 5.75 \times 10^8 \text{ lbf}$$

Determine the borrow soil weight from the requirements.

$$W_{s,A} = W_{s,F} = 106.4 \text{ lbf}$$

$$W_{w,A} = wW_{s,A} = (0.25)(106.4 \text{ lbf}) = 26.6 \text{ lbf}$$

$$W_{a,A} = 0$$

The weight to be excavated to get 1 ft^3 of fill is

$$W_{t,A} = W_{w,A} + W_{s,A} = 26.6 \text{ lbf} + 106.4 \text{ lbf}$$
$$= 133 \text{ lbf}$$

The volume of borrow A per ft^3 of fill is

$$\frac{W_{t,A}}{\gamma_A} = \frac{133 \text{ lbf}}{115 \frac{\text{lbf}}{\text{ft}^3}} = 1.157 \text{ ft}^3$$

The volume of soil A is

$$\left(\frac{1.157 \text{ ft}^3}{1 \text{ ft}^3}\right)(200{,}000 \text{ yd}^3)\left(27 \frac{\text{ft}^3}{\text{yd}^3}\right)$$
$$= \boxed{6.25 \times 10^6 \text{ ft}^3 \quad (6.3 \times 10^6 \text{ ft}^3)}$$

The answer is (D).

(b) Soil B:

$$\gamma_{\text{dry}} = (0.95)\left(110 \frac{\text{lbf}}{\text{ft}^3}\right) = 104.5 \text{ lbf/ft}^3$$

$$W_{s,F} = 104.5 \text{ lbf}$$

$$W_{s,F} = W_{s,B} = 104.5 \text{ lbf}$$

$$W_{w,F} = wW_{s,F} = (0.10)(104.5 \text{ lbf}) = 10.45 \text{ lbf}$$

$$W_{t,F} = 104.5 \text{ lbf} + 10.45 \text{ lbf} = 114.95 \text{ lbf}$$

The weight of borrow soil solids is

$$(200{,}000 \text{ yd}^3)\left(27 \frac{\text{ft}^3}{\text{yd}^3}\right)\left(104.5 \frac{\text{lbf}}{\text{ft}^3}\right) = 5.64 \times 10^8 \text{ lbf}$$

$$W_{s,B} = 104.5 \text{ lbf}$$

$$W_{w,B} = (0.20)(104.5 \text{ lbf}) = 20.9 \text{ lbf}$$

$$W_{t,B} = 20.9 \text{ lbf} + 104.5 \text{ lbf} = 125.4 \text{ lbf}$$

The volume of borrow B per ft^3 of fill is

$$\frac{125.4 \text{ lbf}}{120 \frac{\text{lbf}}{\text{ft}^3}} = 1.045 \text{ ft}^3$$

The volume of soil B is

$$\left(1.045 \frac{\text{ft}^3}{\text{ft}^3}\right)(200{,}000 \text{ yd}^3)\left(27 \frac{\text{ft}^3}{\text{yd}^3}\right)$$
$$= \boxed{5.64 \times 10^6 \text{ ft}^3 \quad (5.6 \times 10^6 \text{ ft}^3)}$$

The answer is (C).

(c) Compute the volume of solids in soil A in transport.

$$V_{s,\text{transport}} = \frac{W}{\gamma_{\text{transport}}} = \frac{5.75 \times 10^8 \text{ lbf}}{(2.70)\left(62.4 \frac{\text{lbf}}{\text{ft}^3}\right)}$$
$$= 3.41 \times 10^6 \text{ ft}^3$$

The volume of voids in transport is

$$V_{\text{void}} = eV_s = (0.92)(3.41 \times 10^6 \text{ ft}^3) = 3.14 \times 10^6 \text{ ft}^3$$

The total transport volume is

$$V_t = V_v + V_s = 3.41 \times 10^6 \text{ ft}^3 + 3.14 \times 10^6 \text{ ft}^3$$
$$= 6.55 \times 10^6 \text{ ft}^3$$

The cost to excavate and transport A is

$$C_A = \left(0.20 \frac{\$}{\text{yd}^3}\right)\left(\frac{6.25 \times 10^6 \text{ ft}^3}{27 \frac{\text{ft}^3}{\text{yd}^3}}\right)$$
$$+ \left(0.30 \frac{\$}{\text{yd}^3}\right)\left(\frac{6.55 \times 10^6 \text{ ft}^3}{27 \frac{\text{ft}^3}{\text{yd}^3}}\right)$$
$$= \$119{,}074$$

The volume of solids in soil B in transport is

$$V_{s,\text{transport}} = \frac{W}{\gamma_{\text{transport}}} = \frac{(5.64 \times 10^6 \text{ ft}^3)\left(120 \frac{\text{lbf}}{\text{ft}^3}\right)}{95 \frac{\text{lbf}}{\text{ft}^3}}$$
$$= 7.13 \times 10^6 \text{ ft}^3$$

Soil A has the lesser transport volume.

Geotechnical

The cost to excavate and transport B is

$$C_{\text{B}} = \left(0.10 \ \frac{\$}{\text{yd}^3} \right) \left(\frac{5.64 \times 10^6 \ \text{ft}^3}{27 \ \frac{\text{ft}^3}{\text{yd}^3}} \right)$$

$$+ \left(0.40 \ \frac{\$}{\text{yd}^3} \right) \left(\frac{7.13 \times 10^6 \ \text{ft}^3}{27 \ \frac{\text{ft}^3}{\text{yd}^3}} \right)$$

$$= \$126{,}518$$

C_{A} is the lesser cost by

$$\$126{,}518 - \$119{,}074 = \boxed{\$7444 \quad (\$7400)}$$

The answer is (B).

9. (a) For a dry or drained sample, $c = 0$. Use Mohr's circle.

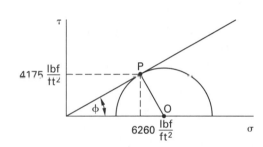

$$\phi = \arctan \frac{4175 \ \frac{\text{lbf}}{\text{ft}^2}}{6260 \ \frac{\text{lbf}}{\text{ft}^2}} = \arctan 0.667 - \boxed{33.7° \quad (34°)}$$

The answer is (C).

(b) The maximum principal stress is calculated as follows.

$$\alpha = 45° + \tfrac{1}{2}\phi$$

$$\text{slope of line PO} = \frac{-1}{0.667} = -1.5$$

$$y = mx + b$$

$$4175 \ \frac{\text{lbf}}{\text{ft}^2} = (-1.5) \left(6260 \ \frac{\text{lbf}}{\text{ft}^2} \right) + b$$

$$b = 13{,}565 \ \frac{\text{lbf}}{\text{ft}^2}$$

$$y = -1.5x + 13{,}565 \ \frac{\text{lbf}}{\text{ft}^2}$$

When $y = 0$,

$$x = \frac{13{,}565 \ \frac{\text{lbf}}{\text{ft}^2}}{1.5} = 9043 \ \text{lbf/ft}^2$$

$$\text{length PO} = \sqrt{ \begin{aligned} &\left(6260 \ \frac{\text{lbf}}{\text{ft}^2} - 9043 \ \frac{\text{lbf}}{\text{ft}^2} \right)^2 \\ &+ \left(4175 \ \frac{\text{lbf}}{\text{ft}^2} \right)^2 \end{aligned} }$$

$$= 5018 \ \frac{\text{lbf}}{\text{ft}^2}$$

$$\begin{aligned} \text{maximum} \\ \text{principal stress} \end{aligned} = 9043 \ \frac{\text{lbf}}{\text{ft}^2} + 5018 \ \frac{\text{lbf}}{\text{ft}^2}$$

$$= \boxed{14{,}061 \ \text{lbf/ft}^2 \quad (14{,}000 \ \text{lbf/ft}^2)}$$

The answer is (B).

10. Graph the results.

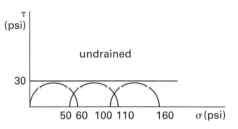

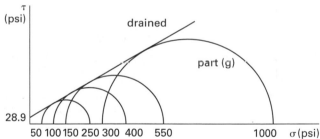

(a) undrained: $\phi = \boxed{0°}$

The answer is (A).

(b) drained: $\phi = \boxed{30°}$ [found using a protractor]

The answer is (C).

(c) undrained: $c = \boxed{30 \ \text{psi}}$

The answer is (A).

(d) drained:

$$c = \boxed{28.9 \ \text{psi} \quad (29 \ \text{psi})} \quad \text{[determined graphically]}$$

The answer is (C).

(e) Use Eq. 35.41.

$$\text{undrained: } \alpha = 45° + \tfrac{1}{2}\phi = 45° + \left(\tfrac{1}{2}\right)(0°)$$
$$= \boxed{45°}$$

The answer is (D).

(f) Use Eq. 35.41.

$$\text{drained: } \alpha = 45° + \tfrac{1}{2}\phi = 45° + \left(\tfrac{1}{2}\right)(30°)$$
$$= \boxed{60°}$$

The answer is (D).

(g) By drawing circles to touch the rupture line and 300 psi by trial and error,

$$\sigma_A = \boxed{1000 \text{ lbf/in}^2}$$

The answer is (B).

11.

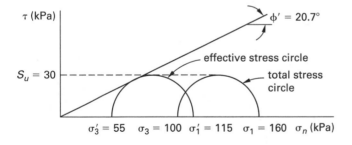

(a) For clays in the undrained condition, $\phi = 0°$ is assumed. Use Eq. 35.42.

$$S_{us} = c = \frac{\sigma_D}{2} = \frac{60 \text{ kPa}}{2}$$
$$= \boxed{30 \text{ kPa}}$$

The answer is (B).

(b) The chamber pressure remains constant during the test, so in terms of total stress, the minor principal stress is the same as at the beginning of the test.

$$\sigma_{3f} = \boxed{100 \text{ kPa}}$$

The answer is (D).

(c) The major principal stress is equal to the minor principal stress plus the deviator stress.

$$\sigma_{1f} = \sigma_{3f} + \sigma_D$$
$$= 100 \text{ kPa} + 60 \text{ kPa}$$
$$= \boxed{160 \text{ kPa}}$$

The total stress conditions can be represented by Mohr's circle. The diameter of the circle is equal to the deviator stress.

The answer is (D).

(d) The sample starts with the minor principal stress equal to 100 kPa, then develops a pore pressure of 45 kPa. Use Eq. 35.17.

$$\sigma_3' = \sigma_3 - u$$
$$= 100 \text{ kPa} - 45 \text{ kPa}$$
$$= \boxed{55 \text{ kPa}}$$

The answer is (B).

(e) The total major principal stress at failure is 160 kPa. Use Eq. 35.17.

$$\sigma_1' = \sigma_1 - u$$
$$= 160 \text{ kPa} - 45 \text{ kPa}$$
$$= \boxed{115 \text{ kPa} \quad (120 \text{ kPa})}$$

The answer is (C).

(f) Mohr's circle for the effective stress conditions in the test can be drawn from the known conditions. The size of the effective and total stress circles is the same; therefore, the diameter of the circle is 60 kPa.

The friction angle can be computed from Eq. 35.40.

$$\frac{\sigma_{1f'}}{\sigma_{3f'}} = \frac{1 + \sin\phi'}{1 - \sin\phi'}$$
$$\frac{115 \text{ kPa}}{55 \text{ kPa}} = \frac{1 + \sin\phi'}{1 - \sin\phi'}$$

By trial and error, $\phi' = \boxed{20.7° \ (20°)}.$

The answer is (A).

(g) The failure angle is the theoretical angle of the plane of failure from the horizontal. Use Eq. 35.41.

$$\alpha = 45° + \tfrac{1}{2}\phi'$$
$$= 45° + \left(\tfrac{1}{2}\right)(20.7°)$$
$$= \boxed{55.3° \quad (55°)}$$

The answer is (C).

Geotechnical

(h) The major and minor principal stresses are known. Use Eq. 35.38.

$$\sigma_\alpha = \tfrac{1}{2}(\sigma_1 + \sigma_3) + \tfrac{1}{2}(\sigma_1 - \sigma_3)\cos 2\alpha$$
$$= \left(\tfrac{1}{2}\right)(115 \text{ kPa} + 55 \text{ kPa})$$
$$\qquad + \left(\tfrac{1}{2}\right)(115 \text{ kPa} - 55 \text{ kPa})\cos\left((2)(55.3°)\right)$$
$$= \boxed{74.4 \text{ kPa} \quad (74 \text{ kPa})}$$

The answer is (B).

(i) Use Eq. 35.39.

$$\tau_\alpha = \tfrac{1}{2}(\sigma_1 - \sigma_3)\sin 2\alpha$$
$$= \left(\tfrac{1}{2}\right)(115 \text{ kPa} - 55 \text{ kPa})\sin\left((2)(55.3°)\right)$$
$$= \boxed{28.1 \text{ kPa} \quad (28 \text{ kPa})}$$

The answer is (A).

(j) Another name for the consolidated, undrained test is the $\boxed{\text{R-test.}}$

The answer is (B).

12. (a) Plot the maximum shear stress from the stress displacement curve versus the normal stress from each individual test.

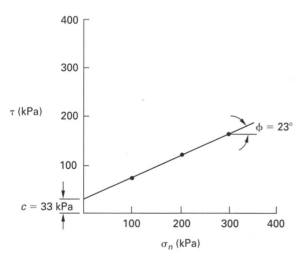

The best fit line through the three data points represents the failure envelope for the soil. Measuring from the plot,

$$\phi = \boxed{23°}$$

The answer is (C).

(b) From the graph in part (a), $c = \boxed{33 \text{ kPa.}}$

The answer is (B).

13. (a) The total (i.e., specific) weight is found as follows.

$$D_r = \frac{e_{max} - e}{e_{max} - e_{min}}$$
$$0.4 = \frac{0.97 - e}{0.97 - 0.45}$$
$$e = 0.76$$

Since $e = V_v/V_s$,

$$V_{\text{total}} = V_s + V_v = V_s + eV_s = (1+e)V_s$$

Consider 1 ft^3 of saturated sand.

$$\text{weight of solids} = \frac{(2.65)\left(62.4 \ \dfrac{\text{lbf}}{\text{ft}^3}\right)}{1 + 0.76} = 93.95 \text{ lbf/ft}^3$$

$$\text{weight of water} = \frac{(0.76)\left(62.4 \ \dfrac{\text{lbf}}{\text{ft}^3}\right)}{1 + 0.76} = 26.95 \text{ lbf/ft}^3$$

The total weight is

$$93.95 \ \frac{\text{lbf}}{\text{ft}^3} + 26.95 \ \frac{\text{lbf}}{\text{ft}^3} = \boxed{120.9 \text{ lbf/ft}^3 \quad (121 \text{ lbf/ft}^3)}$$

Alternate solution:

From Table 35.7,

$$\rho_{\text{sat}} = \frac{(\text{SG} + e)\rho_w}{1 + e}$$

Writing this in terms of specific weight,

$$\gamma_{\text{sat}} = \frac{(2.65 + 0.76)\left(62.4 \ \dfrac{\text{lbf}}{\text{ft}^3}\right)}{1 + 0.76}$$
$$= \boxed{120.9 \text{ lbf/ft}^3 \quad (121 \text{ lbf/ft}^3)}$$

The answer is (C).

(b) The decrease in thickness is found as follows.

$$\frac{\Delta V}{V_{t,40}} = \frac{V_{t,40} - V_{t,65}}{V_{t,40}} = \frac{V_s(1 + e_{40}) - V_s(1 + e_{65})}{V_s(1 + e_{40})}$$
$$= \frac{e_{40} - e_{65}}{1 + e_{40}} = \frac{0.76 - 0.63}{1 + 0.76}$$
$$= 0.0739$$
$$\Delta t = (0.0739)(4 \text{ ft})$$
$$= \boxed{0.295 \text{ ft} \quad (0.3 \text{ ft})}$$

The answer is (C).

Geotechnical

14. (a) Refer to the graph.

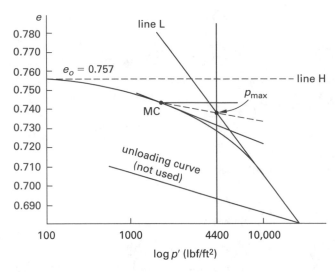

(b) Use the Casagrande procedure to locate p'_{max}.

step 1: Extend the curve to the left, and estimate $e_o = 0.757$.

step 2: Draw a horizontal line from e_o (line H).

step 3: Extend the tail tangent upward (line L).

step 4: Find the point of maximum curvature by inspection (point MC). At point MC, $e = 0.746$ and $p = 3000$ lbf/ft^2.

step 5: Draw horizontal and tangent lines from point MC.

step 6: Bisect the angle (dashed line).

step 7: The intersection of the dashed line and line L gives $p'_{max} = 4400$ lbf/ft^2.

The answer is (A).

(c) Choose two points on line L, which is the virgin compression line. For example, (4400 lbf/ft^2, 0.739) and (10,000 lbf/ft^2, 0.707).

Use Eq. 35.34.

$$C_c = -\frac{e_1 - e_2}{\log \dfrac{p'_1}{p'_2}} = -\frac{0.739 - 0.707}{\log \dfrac{4400 \dfrac{\text{lbf}}{\text{ft}^2}}{10,000 \dfrac{\text{lbf}}{\text{ft}^2}}}$$

$$= \boxed{0.090}$$

The answer is (B).

15. The ground will still be frozen immediately after the surface frost has evaporated. When water freezes, it expands and takes up a larger volume in the material beneath the gauge. Because water is less dense than the measured material, the density will be lower and the nuclear density gauge will indicate a higher moisture content. As the percent moisture result will be elevated, the dry density will be lower, not higher, if the material contains frost.

The answer is (A).

16. The target density is used by the nuclear gauge only to calculate the degree of compaction that has been achieved. The gauge will still give wet density, dry density, moisture, and percent moisture results without the target density entered.

The answer is (D).

36 Shallow Foundations

PRACTICE PROBLEMS

1. A mat foundation is to be used to support a building with dimensions of 80 ft × 40 ft and a 5200 ton total load. The mat is located 8 ft below the ground surface. The soil beneath the mat is a sand with a density of 120 lbf/ft^3 and an average SPT N-value of 18.

(a) What is the allowable bearing capacity of the mat?

(A) 2.2 tons/ft^2

(B) 2.5 tons/ft^2

(C) 2.9 tons/ft^2

(D) 5.4 tons/ft^2

(b) What is the factor of safety against bearing capacity failure?

(A) 1.3

(B) 1.9

(C) 2.2

(D) 4.4

2. A total force of 1600 kN is to be supported by a square footing. The footing is to rest directly on sand that has a density of 1900 kg/m^3 and an angle of internal friction of 38°. Use the Terzaghi bearing capacity factors.

(a) Size the footing.

(A) 1.8 m square

(B) 2.4 m square

(C) 2.7 m square

(D) 3.4 m square

(b) Size the footing if it is to be placed 1.5 m below the surface.

(A) 1.0 m square

(B) 1.1 m square

(C) 1.3 m square

(D) 1.8 m square

3. (*Time limit: one hour*) The three foundations shown are situated in a sandy clay with the following characteristics: saturated specific weight = 108 lbf/ft^3, angle of internal friction = 25°, and cohesion = 400 lbf/ft^2. The water table is 35 ft below the ground surface. Use a factor of safety of 2.5 where required.

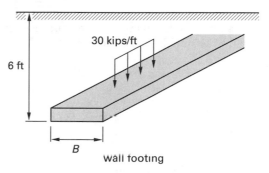

wall footing

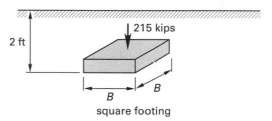

square footing

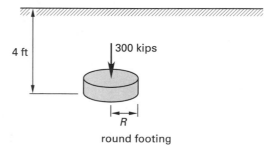

round footing

(a) What is the bearing capacity factor, N_c, according to the Terzaghi and Meyerhof theories, respectively?

(A) 5.7, 5.1

(B) 11, 9.7

(C) 25, 21

(D) 350, 270

(b) What is the bearing capacity factor, N_q, according to the Terzaghi and Meyerhof theories, respectively?

(A) 0, 0

(B) 1.0, 1.0

(C) 3.6, 3.2

(D) 13, 11

(c) What is the bearing capacity factor, N_γ, according to the Terzaghi and Meyerhof theories, respectively?

(A) 0, 0

(B) 1.0, 1.0

(C) 1.9, 1.9

(D) 9.7, 6.8

(d) What should be the width of the wall footing using the Terzaghi factors? Neglect the weight of the footing.

(A) 2.0 ft

(B) 3.5 ft

(C) 3.8 ft

(D) 4.6 ft

(e) What should be the width of the square footing using the Terzaghi factors? Neglect the weight of the footing.

(A) 2.5 ft

(B) 3.2 ft

(C) 4.8 ft

(D) 5.5 ft

(f) What should be the radius of the circular footing, using the Terzaghi factors? Neglect the weight of the footing.

(A) 3.5 ft

(B) 3.8 ft

(C) 4.2 ft

(D) 5.5 ft

(g) What should be the radius of the circular footing using the Meyerhof factors? Neglect the weight of the footing.

(A) 3.5 ft

(B) 3.7 ft

(C) 4.7 ft

(D) 5.8 ft

(h) What is the allowable bearing capacity of the square footing assuming a width of 4 ft and assuming that the water table is at 2.0 ft? Use the Meyerhof factors.

(A) 5300 lbf/ft^2

(B) 6300 lbf/ft^2

(C) 7600 lbf/ft^2

(D) 11,000 lbf/ft^2

(i) What is the allowable bearing capacity of the square footing assuming a width of 4 ft and assuming that the water table is at 1 ft? Use the Meyerhof factors.

(A) 4900 lbf/ft^2

(B) 5100 lbf/ft^2

(C) 6200 lbf/ft^2

(D) 8900 lbf/ft^2

(j) What is the allowable bearing capacity of the circular footing assuming a radius of 4 ft and assuming that the water table is at the ground surface? Use the Meyerhof factors.

(A) 5200 lbf/ft^2

(B) 5700 lbf/ft^2

(C) 7600 lbf/ft^2

(D) 9300 lbf/ft^2

4. (*Time limit: one hour*) A 2 m × 2 m square spread footing is designed to support an axial column load of 600 kN and a moment of 250 kN·m, as shown. The footing is placed 1 m into a sandy soil with a density of 2000 kg/m^3 and an angle of internal friction of 30°. Use bearing capacity factors derived from the Meyerhof theory.

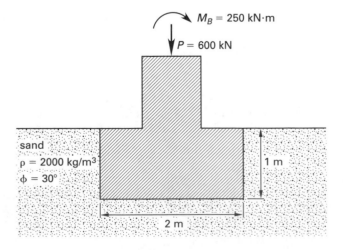

(a) What is the eccentricity of the resultant load?

(A) 0.10 m

(B) 0.25 m

(C) 0.42 m

(D) 0.50 m

(b) What equivalent width should be used for the footing design?

(A) 1.2 m

(B) 1.5 m

(C) 1.8 m

(D) 2.0 m

(c) What is the value of the bearing capacity factor, N_γ, to be used in calculating the bearing capacity?

(A) 16

(B) 20

(C) 22

(D) 35

(d) What is the value of the bearing capacity factor, N_q, to be used in calculating the bearing capacity?

(A) 11

(B) 16

(C) 18

(D) 23

(e) What is the value of the shape factor for N_γ to be used in calculating the bearing capacity?

(A) 0.85

(B) 0.88

(C) 0.90

(D) 1.10

(f) What is the value of the net bearing capacity?

(A) 100 kPa

(B) 300 kPa

(C) 510 kPa

(D) 1300 kPa

(g) Using a factor of safety of 2, what is the value of the allowable bearing capacity?

(A) 50 kPa

(B) 150 kPa

(C) 250 kPa

(D) 600 kPa

(h) Using a factor of safety of 2, what is the total allowable load on the shallow foundation?

(A) 590 kN

(B) 820 kN

(C) 910 kN

(D) 1100 kN

(i) What is the maximum soil pressure on the bottom of the footing?

(A) 90 kPa

(B) 200 kPa

(C) 260 kPa

(D) 340 kPa

(j) What is the factor of safety against bearing capacity failure for this footing?

(A) 1.5

(B) 2.1

(C) 2.7

(D) 3.0

Geotechnical

SOLUTIONS

1. (a) The SPT N-value should be corrected for the overburden pressure. At the base of the mat foundation, the overburden pressure is

$$p_{\text{overburden}} = \gamma D_f = \frac{\left(120 \ \frac{\text{lbf}}{\text{ft}^3}\right)(8 \ \text{ft})}{2000 \ \frac{\text{lbf}}{\text{ton}}}$$

$$= 0.48 \ \text{tons/ft}^2$$

From Table 36.6, the correction factor is $C_n \approx 1.21$.

At a depth of $D_f + B$ below the ground surface, the overburden pressure is

$$p_{\text{overburden}} = \gamma(D_f + B) = \frac{\left(120 \ \frac{\text{lbf}}{\text{ft}^3}\right)(8 \ \text{ft} + 40 \ \text{ft})}{2000 \ \frac{\text{lbf}}{\text{ton}}}$$

$$= 2.88 \ \text{tons/ft}^2$$

From Table 36.6, the correction factor is $C_n \approx 0.63$. This is the factor that should be used for design, since it will result in the lowest N-value.

The net allowable bearing capacity is given by Eq. 36.23.

$$q_{\text{net},a} = 0.22 C_n N = (0.22)(0.63)(18)$$

$$= \boxed{2.49 \ \text{tons/ft}^2}$$

The answer is (B).

(b) By Eq. 36.24,

$$p_{\text{net,actual}} = \frac{P_{\text{total}}}{A_{\text{raft}}} - \gamma D_f$$

$$= \frac{5200 \ \text{tons}}{(80 \ \text{ft})(40 \ \text{ft})} - \frac{\left(120 \ \frac{\text{lbf}}{\text{ft}^3}\right)(8 \ \text{ft})}{2000 \ \frac{\text{lbf}}{\text{ton}}}$$

$$= 1.15 \ \text{tons/ft}^2$$

The factor of safety against bearing capacity failure based on allowable stress is given by Eq. 36.4.

$$\frac{q_{\text{net},a}}{p_{\text{net,actual}}} = \frac{2.49 \ \frac{\text{tons}}{\text{ft}^2}}{1.15 \ \frac{\text{tons}}{\text{ft}^2}}$$

$$= 2.2$$

Since Eq. 36.22 already has a factor of safety of 2 based on the ultimate bearing capacity,

$$F = (2)(2.2) = \boxed{4.4}$$

The answer is (D).

2. (a) If the footing rests on the sand, then $D_f = 0$. From Table 36.2 or Fig. 36.2, for $\phi = 38°$, $N_\gamma \approx 77$.

The shape factor from Table 36.5 is 0.85.

The net bearing capacity equation for sand is given by Eq. 36.10.

$$q_{\text{net}} = \tfrac{1}{2} B \rho g N_\gamma + \rho g D_f (N_q - 1)$$

$$= \frac{\tfrac{1}{2} B \left(1900 \ \frac{\text{kg}}{\text{m}^3}\right)\left(9.81 \ \frac{\text{m}}{\text{s}^2}\right)(77)(0.85) + 0}{1000 \ \frac{\text{Pa}}{\text{kPa}}}$$

$$= 610 B \ \text{kPa}$$

From Eq. 36.4, using $F = 2$,

$$q_a = \frac{q_{\text{net}}}{F}$$

$$q_{\text{net}} = F q_a = (2)\left(\frac{1600 \ \text{kN}}{B^2}\right)$$

$$= 3200/B^2 \quad [\text{in kPa}]$$

Equating these two expressions for q_{net},

$$\frac{3200}{B^2} = 610 B$$

$$B = \boxed{1.74 \ \text{m}}$$

The answer is (A).

(b) From Fig. 36.2 or Table 36.2, $N_q \approx 65$.

The net bearing capacity equation for sand is given by Eq. 36.10.

$$q_{\text{net}} = \tfrac{1}{2} B \rho g N_\gamma + \rho g D_f (N_q - 1)$$

$$= \frac{\tfrac{1}{2} B \left(1900 \ \frac{\text{kg}}{\text{m}^3}\right)\left(9.81 \ \frac{\text{m}}{\text{s}^2}\right)(77)(0.85)}{1000 \ \frac{\text{Pa}}{\text{kPa}}}$$

$$+ \frac{\left(1900 \ \frac{\text{kg}}{\text{m}^3}\right)\left(9.81 \ \frac{\text{m}}{\text{s}^2}\right)(1.5 \ \text{m})(65 - 1)}{1000 \ \frac{\text{Pa}}{\text{kPa}}}$$

$$= 610 B \ \text{kPa} + 1789 \ \text{kPa}$$

Geotechnical

The design capacity is the same as for part (a). Equating the two expressions for q_{net},

$$\frac{3200}{B^2} = 610B \text{ kPa} + 1789 \text{ kPa}$$

$$3200 = 610B^3 + 1789B^2$$

By trial and error, $B = \boxed{1.14 \text{ m.}}$

The answer is (B).

3. (a) From Table 36.2 and Table 36.3, for $\phi = 25°$, $N_c = 25.1$ (Terzaghi) and $N_c = 20.7$ (Meyerhof).

The answer is (C).

(b) From Table 36.2 and Table 36.3, for $\phi = 25°$, $N_q = 12.7$ (Terzaghi) and $N_q = 10.7$ (Meyerhof).

The answer is (D).

(c) From Table 36.2 and Table 36.3, for $\phi = 25°$, $N_\gamma = 9.7$ (Terzaghi) and $N_\gamma = 6.8$ (Meyerhof).

The answer is (D).

(d) The water table is too deep to affect the bearing capacity. The net bearing pressure on the footing is

$$q_{net} = q_{ult} - \gamma D_f$$
$$= \tfrac{1}{2}\gamma B N_\gamma + cN_c + \gamma D_f(N_q - 1)$$

For a wall footing, the shape factor for N_c from Table 36.4 is 1.0; the shape factor for N_γ from Table 36.5 is also 1.0.

$$q_{net} = \left(\tfrac{1}{2}\right)\left(108 \ \frac{\text{lbf}}{\text{ft}^3}\right)B(9.7)(1.0)$$
$$+ \left(400 \ \frac{\text{lbf}}{\text{ft}^2}\right)(25.1)(1.0)$$
$$+ \left(108 \ \frac{\text{lbf}}{\text{ft}^3}\right)(6 \text{ ft})(12.7 - 1)$$
$$= 523.8B + 17,622 \quad [\text{in lbf/ft}^2]$$

By Eq. 36.4, per foot of wall length,

$$q_{net} = Fq_a = \frac{(2.5)(30,000 \text{ lbf})}{B}$$
$$= 75,000/B \quad [\text{in lbf/ft}^2]$$

Therefore,

$$q_{net} = \frac{75,000}{B} = 523.8B + 17,622$$

$$523.8B^2 + 17,622B = 75,000$$
$$B^2 + 33.64B = 143.18$$
$$B = \boxed{3.8 \text{ ft}}$$

The answer is (C).

(e) The water table is too deep to affect the bearing capacity. The net bearing pressure on the footing is

$$q_{net} = q_{ult} - \gamma D_f$$
$$= \tfrac{1}{2}\gamma B N_\gamma + cN_c + \gamma D_f(N_q - 1)$$

For a square footing, the shape factor for N_c from Table 36.4 is 1.25; the shape factor for N_γ from Table 36.5 is 0.85.

$$q_{net} = \left(\tfrac{1}{2}\right)\left(108 \ \frac{\text{lbf}}{\text{ft}^3}\right)B(9.7)(0.85)$$
$$+ \left(400 \ \frac{\text{lbf}}{\text{ft}^2}\right)(25.1)(1.25)$$
$$+ \left(108 \ \frac{\text{lbf}}{\text{ft}^3}\right)(2 \text{ ft})(12.7 - 1)$$
$$= 445.2B + 15,077 \quad [\text{in lbf/ft}^2]$$

By Eq. 36.4,

$$q_{net} = Fq_a = \frac{(2.5)(215,000 \text{ lbf})}{B^2}$$
$$= 537,500/B^2 \quad [\text{in lbf/ft}^2]$$

Therefore,

$$q_{net} = \frac{537,500}{B^2} = 445.2B + 15,077 \ \frac{\text{lbf}}{\text{ft}^2}$$

$$445.2B^3 + 15,077B^2 = 537,500 \quad [\text{in lbf/ft}^2]$$

By trial and error, $B = \boxed{5.5 \text{ ft.}}$

The answer is (D).

(f) The water table is too deep to affect the bearing capacity. The net bearing pressure on the footing is

$$q_{net} = q_{ult} - \gamma D_f$$
$$= \tfrac{1}{2}\gamma B N_\gamma + cN_c + \gamma D_f(N_q - 1)$$

For a circular footing, the shape factor for N_c from Table 36.4 is 1.20; the shape factor for N_γ from Table 36.5 is 0.70.

$$q_{net} = \left(\tfrac{1}{2}\right)\left(108 \ \frac{\text{lbf}}{\text{ft}^3}\right)(2R)(9.7)(0.70)$$
$$+ \left(400 \ \frac{\text{lbf}}{\text{ft}^2}\right)(25.1)(1.20)$$
$$+ \left(108 \ \frac{\text{lbf}}{\text{ft}^3}\right)(4 \text{ ft})(12.7 - 1)$$
$$= 733.3R + 17,102 \quad [\text{in lbf/ft}^2]$$

Geotechnical

By Eq. 36.4,

$$q_{net} = Fq_a$$

$$= \frac{(2.5)(300{,}000 \text{ lbf})}{\pi R^2}$$

$$= 238{,}732/R^2 \quad [\text{in lbf/ft}^2]$$

Therefore,

$$q_{net} = \frac{238{,}732}{R^2} = 733.32R + 17{,}102$$

$$733.3R^3 + 17{,}102R^2 = 238{,}732 \quad [\text{in lbf/ft}^2]$$

By trial and error, $R = \boxed{3.5 \text{ ft.}}$

The answer is (A).

(g) Solve the problem as in part (f), substituting the Meyerhof/Vesic factors for the Terzaghi factors in the bearing capacity equation. The shape factors are the same.

$$q_{net} = \tfrac{1}{2}\gamma B N_\gamma + c N_c + \gamma D_f (N_q - 1)$$

$$= \left(\tfrac{1}{2}\right)\left(108 \; \frac{\text{lbf}}{\text{ft}^3}\right)(2R)(10.8)(0.70)$$

$$+ \left(400 \; \frac{\text{lbf}}{\text{ft}^2}\right)(20.7)(1.20)$$

$$+ \left(108 \; \frac{\text{lbf}}{\text{ft}^3}\right)(4 \text{ ft})(10.7 - 1)$$

$$= 816.5R + 14{,}126 \quad [\text{in lbf/ft}^2]$$

By Eq. 36.4,

$$q_{net} = Fq_a = \frac{(2.5)(300{,}000 \text{ lbf})}{\pi R^2}$$

$$= 238{,}732/R^2 \quad [\text{in lbf/ft}^2]$$

Therefore,

$$q_{net} = \frac{238{,}732}{R^2} = 816.5R + 14{,}126 \; \frac{\text{lbf}}{\text{ft}^2}$$

$$816.5R^3 + 14{,}126R^2 = 238{,}732 \quad [\text{in lbf/ft}^2]$$

By trial and error, $R = \boxed{3.7 \text{ ft.}}$

The answer is (B).

(h) The water table is at the base of the footing. By Eq. 36.1 and Eq. 36.3, the net bearing pressure on the footing is

$$q_{net} = q_{ult} - \gamma D_f$$

$$= \tfrac{1}{2}\gamma_b B N_\gamma + c N_c + \gamma_d D_f (N_q - 1)$$

For a square footing, the shape factor for N_c from Table 36.4 is 1.25; the shape factor for N_γ from Table 36.5 is 0.85.

$$q_{net} = \left(\tfrac{1}{2}\right)\left(108 \; \frac{\text{lbf}}{\text{ft}^3} - 62.4 \; \frac{\text{lbf}}{\text{ft}^3}\right)(4 \text{ ft})(10.8)(0.85)$$

$$+ \left(400 \; \frac{\text{lbf}}{\text{ft}^2}\right)(20.7)(1.25)$$

$$+ \left(108 \; \frac{\text{lbf}}{\text{ft}^3}\right)(2 \text{ ft})(10.7 - 1)$$

$$= 13{,}282 \text{ lbf/ft}^2$$

By Eq. 36.4,

$$q_a = \frac{q_{net}}{F} = \frac{13{,}282 \; \frac{\text{lbf}}{\text{ft}^2}}{2.5}$$

$$= \boxed{5313 \text{ lbf/ft}^2}$$

The answer is (A).

(i) The water table is between the base of the footing and the surface. By Eq. 36.1 and Eq. 36.16, the net bearing pressure on the footing is

$$q_{net} = q_{ult} - \gamma D_f$$

$$= \tfrac{1}{2}\gamma_b B N_\gamma + c N_c$$

$$+ \left(\gamma_d D_f + \left(62.4 \; \frac{\text{lbf}}{\text{ft}^3}\right)(D_w - D_f)\right)(N_q - 1)$$

For a square footing, the shape factor for N_c from Table 36.4 is 1.25; the shape factor for N_γ from Table 36.5 is 0.85.

$$q_{net} = \left(\tfrac{1}{2}\right)\left(108 \; \frac{\text{lbf}}{\text{ft}^3} - 62.4 \; \frac{\text{lbf}}{\text{ft}^3}\right)(4 \text{ ft})(10.8)(0.85)$$

$$+ \left(400 \; \frac{\text{lbf}}{\text{ft}^2}\right)(20.7)(1.25)$$

$$+ \left(\left(108 \; \frac{\text{lbf}}{\text{ft}^3}\right)(2 \text{ ft}) + \left(62.4 \; \frac{\text{lbf}}{\text{ft}^3}\right)(1 \text{ ft} - 2 \text{ ft})\right)$$

$$\times (10.7 - 1)$$

$$= 12{,}677 \text{ lbf/ft}^2$$

By Eq. 36.4,

$$q_a = \frac{q_{net}}{F} = \frac{12{,}677 \; \frac{\text{lbf}}{\text{ft}^2}}{2.5}$$

$$= \boxed{5070 \text{ lbf/ft}^2}$$

The answer is (B).

(j) The water table is at the surface. By Eq. 36.1 and Eq. 36.15, the net bearing pressure on the footing is

$$q_{net} = q_{ult} - \gamma D_f$$

$$= \tfrac{1}{2}\gamma_b B N_\gamma + c N_c + \gamma_b D_f (N_q - 1)$$

The cN_c term is included because the soil has a non-zero cohesion, unlike pure sand.

For a round footing, the shape factor for N_c from Table 36.4 is 1.20, and the shape factor for N_γ from Table 36.5 is 0.70.

$$q_{net} = \left(\tfrac{1}{2}\right)\left(108 \ \frac{lbf}{ft^3} - 62.4 \ \frac{lbf}{ft^3}\right)(2)(4 \ ft)(10.8)(0.70)$$
$$+ \left(400 \ \frac{lbf}{ft^2}\right)(20.7)(1.20)$$
$$+ \left(\left(108 \ \frac{lbf}{ft^3} - 62.4 \ \frac{lbf}{ft^3}\right)(4 \ ft)\right)(10.7 - 1)$$
$$= 13,084 \ lbf/ft^2$$

By Eq. 36.4,

$$q_a = \frac{q_{net}}{F} = \frac{13,084 \ \frac{lbf}{ft^2}}{2.5}$$
$$= \boxed{5234 \ lbf/ft^2}$$

The answer is (A).

4. (a) ($M_L = 0$.) The eccentricity is given by Eq. 36.18.

$$\epsilon_B = \frac{M_B}{P} = \frac{250 \ kN \cdot m}{600 \ kN}$$
$$= \boxed{0.417 \ m \quad (0.42 \ m)}$$

The answer is (C).

(b) $\qquad 0.417 \ m > \dfrac{B}{6} = \dfrac{2 \ m}{6} = 0.333 \ m$

The footing width to be used for analysis of bearing capacity is reduced by twice the eccentricity. Using Eq. 36.19, the effective width is

$$B' = B - 2\epsilon_B = 2 \ m - (2)(0.417 \ m)$$
$$= \boxed{1.17 \ m \quad (1.2 \ m)}$$

The answer is (A).

(c) From Table 36.3, for $\phi = 30°$, $N_\gamma = \boxed{15.7.}$

The answer is (A).

(d) From Table 36.3, for $\phi = 30°$, $N_q = \boxed{18.4.}$

The answer is (C).

(e) The shape factor for N_γ is found by interpolation from Table 36.5 and using the effective width. For

$B'/L' = 1.17 \ m/2.0 \ m = 0.585$. The shape factor decreases as B'/L' increases. Therefore, the shape factor will be

$$0.90 - \left(\frac{0.5 - 0.585}{0.5 - 1.0}\right)(0.90 - 0.85)$$
$$= \boxed{0.8915 \quad (0.90)}$$

($L' = L$ since $M_L = 0$.)

The answer is (C).

(f) The net bearing capacity is given by Eq. 36.10.

(If available, a shape factor for the N_q term can be used.)

$$q_{net} = q_{ult} - \rho g D_f$$
$$= \tfrac{1}{2}B\rho g N_\gamma + \rho g D_f (N_q - 1)$$
$$= \left(\tfrac{1}{2}\right)(1.17 \ m)\left(2000 \ \frac{kg}{m^3}\right)\left(9.81 \ \frac{m}{s^2}\right)(15.7)(0.914)$$
$$+ \left(2000 \ \frac{kg}{m^3}\right)\left(9.81 \ \frac{m}{s^2}\right)(1 \ m)(18.4 - 1)$$
$$= \boxed{506,090 \ Pa \quad (510 \ kPa)}$$

The answer is (C).

(g) The allowable bearing capacity is given by Eq. 36.11.

$$q_a = \frac{q_{net}}{F} = \frac{506,090 \ Pa}{(2)\left(1000 \ \frac{Pa}{kPa}\right)}$$
$$= \boxed{253 \ kPa \quad (250 \ kPa)}$$

The answer is (C).

(h) The allowable load is found using the reduced area.

$$p_a = q_a B' L' = (253 \ kPa)(1.17 \ m)(2.0 \ m)$$
$$= \boxed{592 \ kN \quad (590 \ kN)}$$

The answer is (A).

(i) The maximum pressure beneath the footing is given by Eq. 36.20. The actual footing dimensions are used to calculate the pressure distribution.

$$p_{max} = \left(\frac{P}{BL}\right)\left(1 + \frac{6\epsilon}{B}\right)$$
$$= \left(\frac{600 \ kN}{(2 \ m)(2 \ m)}\right)\left(1 + \frac{(6)(0.417 \ m)}{2 \ m}\right)$$
$$= \boxed{337.7 \ kPa \quad (340 \ kPa)}$$

(This pressure is excessive.)

The answer is (D).

(j) The factor of safety against bearing capacity failure is determined by dividing the net bearing capacity by the maximum contact pressure.

$$F = \frac{q_{\text{net}}}{p_{\text{max}}} = \frac{506 \text{ kPa}}{337.7 \text{ kPa}}$$
$$= \boxed{1.49 \quad (1.5)}$$

The answer is (A).

Geotechnical

37 Rigid Retaining Walls

PRACTICE PROBLEMS

Retaining Wall Analysis

1. A retaining wall is designed for free-draining granular backfill. The internal angle of friction, ϕ, dry unit weight, γ_{dry}, moisture content, w, and degree of saturation, S, are shown in the illustration. After several years of operation, the weepholes become plugged and the water table rises to within 10 ft of the top of the wall. (a) What is the resultant force for the drained case? (b) What is the point of application for the drained case? (c) What is the resultant force for the plugged case? (d) What is the point of application of the resultant force for the plugged case?

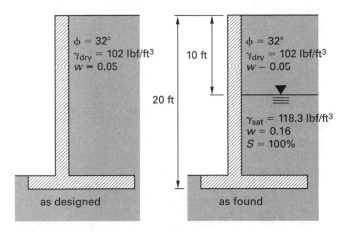

2. (*Time limit: one hour*) A 26 ft high retaining wall holds back sand with a 96 lbf/ft³ drained specific weight. The water table is 10 ft below the top of the wall. The saturated specific weight is 121 lbf/ft³. The angle of internal friction is 36°. (a) What is the active earth resultant? (b) What is the location of the active earth pressure resultant? (c) If the water table elevation could be reduced 16 ft to the bottom of the wall, what would be the reduction in overturning moment?

Retaining Wall Design

3. (*Time limit: one hour*) A reinforced concrete retaining wall is used to support a 14 ft cut in sandy soil. The backfill is level, but a surcharge of 500 lbf/ft² is present for a considerable distance behind the wall. Factors of safety of 1.5 against sliding and overturning are required. Customary and reasonable assumptions regarding the proportions can be made. Passive pressure is to be disregarded. Except for sliding resistance, disregard wall-soil friction. The need for a key must be established.

soil drained specific weight	130 lbf/ft³
angle of internal friction	35°
coefficient of friction against concrete (sliding)	0.5
allowable soil pressure	4500 lbf/ft²
frost line	4 ft below grade

(a) What is the approximate minimum stem height?

(A) 14 ft

(B) 16 ft

(C) 18 ft

(D) 20 ft

For parts (b) through (j), assume a base length, B, of 11.5 ft; a base thickness, d, of 1.75 ft; a stem thickness at the base of 1.75 ft; a stem thickness at the top of 1 ft; a stem height (above base) of 18 ft; and a heel extension (past the back of the stem) of 6.5 ft.

(b) The surcharge is equivalent to what thickness of backfill soil?

(A) 2 ft

(B) 3 ft

(C) 4 ft

(D) 5 ft

(c) What is the horizontal reaction due to the surcharge?

(A) 2400 lbf/ft

(B) 2700 lbf/ft

(C) 3500 lbf/ft

(D) 3900 lbf/ft

(d) What is the active soil resultant?

(A) 5700 lbf/ft

(B) 6800 lbf/ft

(C) 7500 lbf/ft

(D) 8300 lbf/ft

(e) What is the total overturning moment, taken about the toe, per foot of wall?

(A) 26,000 ft-lbf

(B) 45,000 ft-lbf

(C) 70,000 ft-lbf

(D) 190,000 ft-lbf

(f) What is the factor of safety against overturning?

(A) 1.5

(B) 1.8

(C) 2.3

(D) 2.6

(g) What is the maximum vertical pressure at the toe?

(A) 4000 lbf/ft^2

(B) 4500 lbf/ft^2

(C) 5000 lbf/ft^2

(D) 5500 lbf/ft^2

(h) What is the minimum vertical pressure at the heel?

(A) 600 lbf/ft^2

(B) 1300 lbf/ft^2

(C) 2400 lbf/ft^2

(D) 2700 lbf/ft^2

(i) What is the factor of safety against sliding without a key?

(A) 1.1

(B) 1.4

(C) 1.6

(D) 1.8

(j) What is the factor of safety against sliding if a key, 1.75 ft wide and 1.0 ft deep, is used?

(A) 1.2

(B) 1.5

(C) 1.7

(D) 2.1

SOLUTIONS

1. (a) As designed,

Use Eq. 37.7.

$$k_a = \frac{1 - \sin \phi}{1 + \sin \phi} = \frac{1 - \sin 32°}{1 + \sin 32°} = 0.307$$

The soil is drained but not dry.

$$\gamma = (1 + w)\gamma_{\text{dry}} = (1 + 0.05)\left(102 \ \frac{\text{lbf}}{\text{ft}^3}\right)$$
$$= 107.1 \ \text{lbf/ft}^3$$

Use Eq. 37.10.

$$R_a = \tfrac{1}{2}k_a\gamma H^2$$
$$R_a = \left(\tfrac{1}{2}\right)(0.307)\left(107.1 \ \frac{\text{lbf}}{\text{ft}^3}\right)(20 \ \text{ft})^2$$
$$= \boxed{6576 \ \text{lbf/ft}}$$

(b) The point of application for the drained case is

$$\left(\tfrac{1}{3}\right)(20 \ \text{ft}) = \boxed{6.67 \ \text{ft up from the bottom}}$$

(c) As found, the pressure distributions (using Eq. 37.56) are

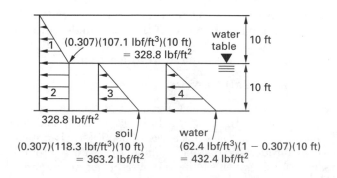

$$R_1 = \left(\tfrac{1}{2}\right)(10 \text{ ft})\left(328.8 \ \frac{\text{lbf}}{\text{ft}^2}\right) = 1644 \text{ lbf/ft}$$

$$[13.33 \text{ ft up}]$$

$$R_2 = (10 \text{ ft})\left(328.8 \ \frac{\text{lbf}}{\text{ft}^2}\right) = 3288 \text{ lbf/ft} \quad [5 \text{ ft up}]$$

$$R_3 + R_4 = \left(\tfrac{1}{2}\right)(10 \text{ ft})\left(363.2 \ \frac{\text{lbf}}{\text{ft}^2} + 432.4 \ \frac{\text{lbf}}{\text{ft}^2}\right)$$

$$= 3978 \text{ lbf/ft} \quad [10/3 \text{ ft up}]$$

$$R_{\text{total}} = 1644 \ \frac{\text{lbf}}{\text{ft}} + 3288 \ \frac{\text{lbf}}{\text{ft}} + 3978 \ \frac{\text{lbf}}{\text{ft}}$$

$$= \boxed{8910 \text{ lbf/ft}}$$

(d) The point of application for the plugged case is

$$\frac{\left(1644 \ \frac{\text{lbf}}{\text{ft}}\right)(13.33 \text{ ft}) + \left(3288 \ \frac{\text{lbf}}{\text{ft}}\right)(5 \text{ ft}) + \left(3978 \ \frac{\text{lbf}}{\text{ft}}\right)\left(\frac{10 \text{ ft}}{3}\right)}{8910 \ \frac{\text{lbf}}{\text{ft}}}$$

$$= \boxed{5.79 \text{ ft up from the bottom}}$$

2. (a) From Eq. 37.7,

$$k_a = \frac{1 - \sin\phi}{1 + \sin\phi} = \frac{1 - \sin 36°}{1 + \sin 36°} = 0.26$$

At $H = 10$ ft,

$$p_h = k_a p_v = k_a \gamma H$$

$$= (0.26)\left(96 \ \frac{\text{lbf}}{\text{ft}^3}\right)(10 \text{ ft})$$

$$= 249.6 \text{ lbf/ft}^2$$

At depth H,

$$p = 249.6 \ \frac{\text{lbf}}{\text{ft}^2}$$

$$+ (0.26)\left(121 \ \frac{\text{lbf}}{\text{ft}^3} - 62.4 \ \frac{\text{lbf}}{\text{ft}^3}\right)(H - 10 \text{ ft})$$

$$+ \left(62.4 \ \frac{\text{lbf}}{\text{ft}^3}\right)(H - 10 \text{ ft})$$

$$= 249.6 \ \frac{\text{lbf}}{\text{ft}^2} + \left(77.6 \ \frac{\text{lbf}}{\text{ft}^3}\right)(H - 10 \text{ ft})$$

At $H = 26$ ft,

$$p = 249.6 \ \frac{\text{lbf}}{\text{ft}^2} + \left(77.6 \ \frac{\text{lbf}}{\text{ft}^3}\right)(26 \text{ ft} - 10 \text{ ft})$$

$$= 249.6 \ \frac{\text{lbf}}{\text{ft}^2} + 1241.6 \ \frac{\text{lbf}}{\text{ft}^2}$$

$$= 1491.2 \text{ lbf/ft}^2$$

The pressure distributions on the wall are as shown.

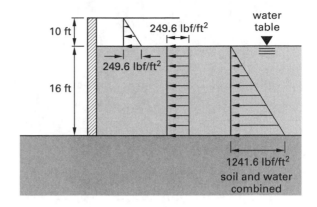

The resultants of each of the preceding three distributions are

$$\left(\tfrac{1}{2}\right)(10 \text{ ft})(249.6 \text{ lbf/ft}^2) = 1248 \text{ lbf/ft,}$$

located $\left(\tfrac{2}{3}\right)(10 \text{ ft}) = 6.67$ ft from the top (19.33 ft from the bottom)

$$(16 \text{ ft})\left(249.6 \ \frac{\text{lbf}}{\text{ft}^2}\right) = 3994 \text{ lbf/ft,}$$

located 8 ft from the bottom

$$\left(\tfrac{1}{2}\right)(16 \text{ ft})\left(1241.6 \ \frac{\text{lbf}}{\text{ft}^2}\right) = 9933 \text{ lbf/ft,}$$

located $\left(\tfrac{1}{3}\right)(16 \text{ ft}) = 5.33$ ft from the bottom

The active resultant is

$$1248 \ \frac{\text{lbf}}{\text{ft}} + 3994 \ \frac{\text{lbf}}{\text{ft}} + 9933 \ \frac{\text{lbf}}{\text{ft}} = \boxed{15{,}175 \text{ lbf/ft}}$$

(b) Taking moments about the base,

$$\text{moment arm} = \frac{\left(1248 \ \frac{\text{lbf}}{\text{ft}}\right)(19.33 \text{ ft}) + \left(3994 \ \frac{\text{lbf}}{\text{ft}}\right)(8 \text{ ft}) + \left(9933 \ \frac{\text{lbf}}{\text{ft}}\right)(5.33 \text{ ft})}{15{,}175 \ \frac{\text{lbf}}{\text{ft}}}$$

$$= \frac{109{,}019 \ \frac{\text{ft-lbf}}{\text{ft}}}{15{,}175 \ \frac{\text{lbf}}{\text{ft}}}$$

$$= \boxed{7.18 \text{ ft up from the bottom}}$$

(c) The original overturning moment is

$$M_{\text{original}} = 109{,}019 \text{ ft-lbf/ft}$$

If the water table were lowered, the active resultant would be

$$R_a = \tfrac{1}{2}k_a\gamma H^2 = \left(\tfrac{1}{2}\right)(0.26)\left(96\ \frac{\text{lbf}}{\text{ft}^3}\right)(26\ \text{ft})^2$$

$$= 8436\ \text{lbf/ft}$$

The resultant would be located $\left(\tfrac{1}{3}\right)(26\ \text{ft}) = 8.67\ \text{ft}$ from the bottom.

$$M_{\text{drained}} = \left(8436\ \frac{\text{lbf}}{\text{ft}}\right)(8.67\ \text{ft})$$

$$= 73{,}140\ \text{ft-lbf/ft}$$

$$M_{\text{reduction}} = 109{,}019\ \frac{\text{ft-lbf}}{\text{ft}} - 73{,}140\ \frac{\text{ft-lbf}}{\text{ft}}$$

$$= \boxed{35{,}879\ \text{ft-lbf/ft}}$$

3. (a) To prevent horizontal frost heave that could move the retaining wall, the base must be at least 4 ft below the surface to be below the frost line. Therefore, the overall height of the wall is at least $\boxed{18\ \text{ft.}}$

The answer is (C).

(b)

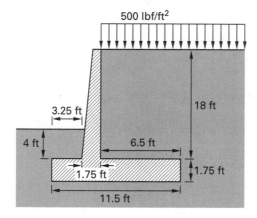

Convert the surcharge to an equivalent weight (depth) of soil.

$$\frac{500\ \dfrac{\text{lbf}}{\text{ft}^2}}{130\ \dfrac{\text{lbf}}{\text{ft}^3}} = \boxed{3.85\ \text{ft} \quad (4\ \text{ft})}$$

The answer is (C).

(c) From Eq. 37.7,

$$k_a = \frac{1 - \sin\phi}{1 + \sin\phi} = \frac{1 - \sin 35°}{1 + \sin 35°} = 0.27$$

The surcharge pressure is

$$p_{\text{surcharge}} = (0.27)\left(500\ \frac{\text{lbf}}{\text{ft}^2}\right) = 135\ \text{lbf/ft}^2$$

The surcharge reaction is

$$R_{\text{surcharge}} = p_{\text{surcharge}} H = \left(135\ \frac{\text{lbf}}{\text{ft}^2}\right)(19.75\ \text{ft})$$

$$= \boxed{2666\ \text{lbf/ft} \quad (2700\ \text{lbf/ft})}$$

The answer is (B).

(d) The active soil resultant is

$$R_a = \tfrac{1}{2}k_a\gamma H^2 = \left(\tfrac{1}{2}\right)(0.27)\left(130\ \frac{\text{lbf}}{\text{ft}^3}\right)(19.75\ \text{ft})^2$$

$$= \boxed{6846\ \text{lbf/ft} \quad (6800\ \text{lbf/ft})}$$

Since the soil above the heel is horizontal, R_a is horizontal and there is no vertical component of R_a.

The answer is (B).

(e) The surcharge resultant acts halfway up from the base, and the active soil resultant acts one-third up from the base.

$$M_{\text{OT}} = R_{\text{surcharge}}\left(\frac{H}{2}\right) + R_{a,h}y_{a,h}$$

$$= \left(2666\ \frac{\text{lbf}}{\text{ft}}\right)\left(\frac{19.75\ \text{ft}}{2}\right) + \left(6846\ \frac{\text{lbf}}{\text{ft}}\right)\left(\frac{19.75\ \text{ft}}{3}\right)$$

$$= \boxed{71{,}396\ \text{ft-lbf} \quad (70{,}000\ \text{ft-lbf})} \quad [\text{per foot of wall}]$$

The answer is (C).

(f) To calculate the factor of safety against overturning, it is necessary to take resisting moments about the pivot point (which, in this case, is the toe).

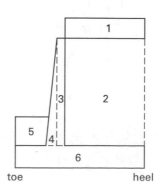

i	area (ft^2)	γ (lbf/ft^3)	W_i (lbf/ft)	x_i from toe (ft)	$M_{\text{resisting}}$ (ft-lbf/ft)
1	$(6.5)(3.85) = 25.03$	130	3254	8.25	26,846
2	$(6.5)(18) = 117$	130	15,210	8.25	125,483
3	$(1)(18) = 18$	150	2700	4.50	12,150
4	$(\frac{1}{2})(18)(0.75) = 6.75$	150	1013	3.75	3799
5	$(3.25)(4) = 13$	130	1690	1.63	2755
6	$(11.5)(1.75) = 20.13$	150	3020	5.75	17,365
			26,887		188,398

(The weight of component 6 disregards the key, which could be included.)

Use Eq. 37.53.

$$F_{\text{OT}} = \frac{M_{\text{resisting}}}{M_{\text{OT}}} = \frac{188,398 \ \dfrac{\text{ft-lbf}}{\text{ft}}}{71,396 \ \dfrac{\text{ft-lbf}}{\text{ft}}}$$

$$= \boxed{2.64 \quad (2.6)} \quad [\text{acceptable}]$$

The answer is (D).

(g) Use Eq. 37.51. There are no vertical components of the active resultant.

$$x_R = \frac{M_{\text{resisting}} - M_{\text{OT}}}{\sum W_i}$$

$$= \frac{188,398 \ \dfrac{\text{ft-lbf}}{\text{ft}} - 71,396 \ \dfrac{\text{ft-lbf}}{\text{ft}}}{26,887 \ \dfrac{\text{lbf}}{\text{ft}}}$$

$$= 4.35 \text{ ft}$$

Use Eq. 37.52.

$$\epsilon = \left| \frac{B}{2} - x_R \right|$$

$$= \left| \frac{11.5 \text{ ft}}{2} - 4.35 \text{ ft} \right|$$

$$= 1.4 \text{ ft}$$

This eccentricity is less than $11.5 \text{ ft}/6 = 1.92 \text{ ft}$, so it is acceptable. Use Eq. 37.54.

$$p_{\text{max,toe}} = \left(\frac{\sum W_i + R_{a,v}}{B} \right) \left(1 + \frac{6\epsilon}{B} \right)$$

$$= \left(\frac{26,887 \ \dfrac{\text{lbf}}{\text{ft}}}{11.5 \text{ ft}} \right) \left(1 + \frac{(6)(1.4 \text{ ft})}{11.5 \text{ ft}} \right)$$

$$= \boxed{4045 \text{ lbf/ft}^2 \quad (4000 \text{ lbf/ft}^2)}$$

The answer is (A).

(h) Use Eq. 37.54.

$$p_{\text{min,heel}} = \left(\frac{\sum W_i + R_{a,v}}{B} \right) \left(1 - \frac{6\epsilon}{B} \right)$$

$$= \left(\frac{26,887 \ \dfrac{\text{lbf}}{\text{ft}}}{11.5 \text{ ft}} \right) \left(1 - \frac{(6)(1.4 \text{ ft})}{11.5 \text{ ft}} \right)$$

$$= \boxed{630 \text{ lbf/ft}^2 \quad (600 \text{ lbf/ft}^2)}$$

The answer is (A).

(i) Since there is no key, the friction between concrete and soil resists sliding. The surcharge force was calculated separately and is not included in $R_{a,h}$.

$$R_{\text{SL}} = f \sum W_i = (0.5)\left(26,887 \ \frac{\text{lbf}}{\text{ft}} \right) = 13,444 \text{ lbf/ft}$$

$$F_{\text{SL}} = \frac{R_{\text{SL}}}{R_{\text{surcharge}} + R_{a,h}} = \frac{13,444 \ \dfrac{\text{lbf}}{\text{ft}}}{2666 \ \dfrac{\text{lbf}}{\text{ft}} + 6846 \ \dfrac{\text{lbf}}{\text{ft}}}$$

$$= \boxed{1.4}$$

(This is less than 1.5, so it is not acceptable.)

The answer is (B).

(j) A key is needed for this design, so extend the stem down about 1 ft.

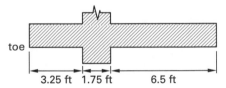

3.25 ft   1.75 ft   6.5 ft

The upward earth pressure distribution on the base is

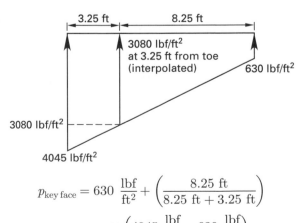

$$p_{\text{key face}} = 630 \ \frac{\text{lbf}}{\text{ft}^2} + \left(\frac{8.25 \text{ ft}}{8.25 \text{ ft} + 3.25 \text{ ft}} \right)$$

$$\times \left(4045 \ \frac{\text{lbf}}{\text{ft}^2} - 630 \ \frac{\text{lbf}}{\text{ft}^2} \right)$$

$$= 3080 \text{ lbf/ft}^2$$

Check the resistance to sliding again. From the toe to the key, the soil must shear. From the toe to the key, the frictional resisting force is found as the product of the total upward normal force and the coefficient of shearing (internal) friction.

$$R_{s_1} = \begin{pmatrix} (3.25 \text{ ft}) \left(3080 \, \dfrac{\text{lbf}}{\text{ft}^2} \right) \\ + (\tfrac{1}{2})(3.25 \text{ ft}) \left(4045 \, \dfrac{\text{lbf}}{\text{ft}^2} - 3080 \, \dfrac{\text{lbf}}{\text{ft}^2} \right) \end{pmatrix}$$
$$\times (\tan 35°)$$
$$= (11{,}578 \text{ lbf})(\tan 35°)$$
$$= 8107 \text{ lbf/ft}$$

From the key to the heel, the base slides on the soil. From the key to the heel, the frictional resisting force is found as the product of the total upward normal force and the coefficient of sliding friction.

$$R_{s_2} = \left(26{,}887 \, \frac{\text{lbf}}{\text{ft}} - 11{,}578 \, \frac{\text{lbf}}{\text{ft}} \right)(0.5) = 7655 \text{ lbf/ft}$$

$$F_{\text{SL}} = \frac{8107 \, \dfrac{\text{lbf}}{\text{ft}} + 7655 \, \dfrac{\text{lbf}}{\text{ft}}}{2666 \, \dfrac{\text{lbf}}{\text{ft}} + 6846 \, \dfrac{\text{lbf}}{\text{ft}}} = \boxed{1.66 \quad (1.7)} \quad [\text{acceptable}]$$

The answer is (C).

38 Piles and Deep Foundations

PRACTICE PROBLEMS

Single Piles

1. A 10.75 in diameter steel pile is driven 65 ft into stiff, insensitive clay. The clay has an undrained shear strength of 1300 lbf/ft². The clay's specific weight is 115 lbf/ft³. The water table is at the ground surface. The entire pile length is effective. (a) What is the end-bearing capacity? (b) What is the friction capacity? (c) What is the allowable bearing capacity?

Pile Groups

2. (*Time limit: one hour*) Tests have shown that a single pile would have, by itself, an uplift capacity of 150 tons and a compressive capacity of 500 tons. 36 piles are installed on a grid with a spacing of 3.5 ft. The pile group is capped and connected at the top by a thick steel-reinforced concrete slab adding its own axial load of 600 tons. Find the maximum moment that the pile group can take in the *x*- and *y*-directions assuming axial load is applied to permit the moment.

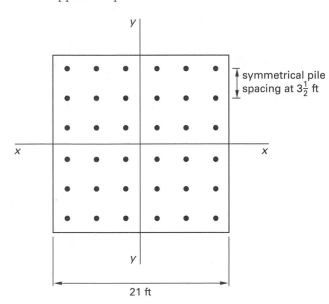

SOLUTIONS

1. (a) $B = \dfrac{10.75 \text{ in}}{12 \, \frac{\text{in}}{\text{ft}}} = 0.896 \text{ ft}$

For saturated clay, $\phi = 0°$.

$$N_c = 9$$

Use Eq. 38.10.

$$Q_p = A_p c N_c$$

$$= \frac{\left(\dfrac{\pi(0.896 \text{ ft})^2}{4}\right)\left(1300 \, \frac{\text{lbf}}{\text{ft}^2}\right)(9)}{1000 \, \frac{\text{lbf}}{\text{kip}}}$$

$$= \boxed{7.38 \text{ kips}}$$

(b) Assume

$$c_A = 0.5c = (0.5)\left(1300 \, \frac{\text{lbf}}{\text{ft}^2}\right)$$

$$= 650 \text{ lbf/ft}^2$$

Use Eq. 38.18.

$$Q_f = p c_A L$$

$$= \frac{\pi\left(\dfrac{10.75 \text{ in}}{12 \, \frac{\text{in}}{\text{ft}}}\right)\left(650 \, \frac{\text{lbf}}{\text{ft}^2}\right)(65 \text{ ft})}{1000 \, \frac{\text{lbf}}{\text{kip}}}$$

$$= \boxed{118.9 \text{ kips}}$$

(c) Using Eq. 38.2, the allowable load is

$$Q_a = \frac{Q_{\text{ult}}}{F} = \frac{7.38 \text{ kips} + 118.9 \text{ kips}}{3}$$

$$= \boxed{42.1 \text{ kips}}$$

2. The moment capacity is limited by the tension capacity of the piles. Assume rotation about the y-y axis. Piles to the left of the y-y axis will be in tension. Work on the basis of one row of six piles.

$$x_A = (0.5)(3.5 \text{ ft}) = 1.75 \text{ ft} = x_D$$

$$x_B = (1.5)(3.5 \text{ ft}) = 5.25 \text{ ft} = x_E$$

$$x_C = (2.5)(3.5 \text{ ft}) = 8.75 \text{ ft} = x_F$$

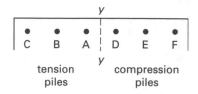

From the parallel axis theorem $(I_o + Ad^2)$ with an insignificant value of I_o, the contribution of each pile is proportional to the square of its distance from the neutral line, y-y.

The force in outermost tension pile C is

$$F_C = \frac{W}{n} + \frac{Mx_C}{X_A^2 + X_B^2 + X_C^2 + X_D^2 + X_E^2 + X_F^2}$$

Use a convention that tension is positive and compression is negative.

$$\frac{150}{\text{tons}} = -\frac{600 \text{ tons}}{36 \text{ piles}}$$

$$+ M \left(\frac{8.75 \text{ ft}}{(2)\left((1.75 \text{ ft})^2 + (5.25 \text{ ft})^2 + (8.75 \text{ ft})^2\right)} \right)$$

$$= -16.67 \text{ tons} + M\left(0.0408 \ \frac{1}{\text{ft}}\right)$$

$$M = 4085 \text{ ft-tons} \quad [\text{per row}]$$

For six rows,

$$M_{\max} = (6)(4085 \text{ ft-tons})$$
$$= \boxed{24{,}510 \text{ ft-tons}}$$

39 Excavations

PRACTICE PROBLEMS

Braced Cuts

1. A 30 ft deep, 40 ft square excavation in sand is being designed. The sand will be dewatered before excavation. The angle of internal friction is 40°, and the specific weight is 121 lbf/ft³. Bracing consists of horizontal lagging supported by 8 in soldier piles separated horizontally by 8 ft. (a) What is the approximate shape of the pressure diagram on the lagging? (b) What is the bending moment on the lagging?

Flexible Bulkheads

2. (*Time limit: one hour*) A 35 ft long sheet pile is driven through 10 ft of clay to bedrock below. The sheet pile supports a 25 ft vertical cut through drained sand. A tie rod is located 8 ft below the surface, terminating at a deadman behind the failure plane. There is no significant water table. Soil parameters are shown in the accompanying illustration. What is the tensile force in the tie rod?

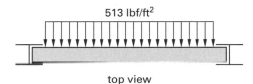

SOLUTIONS

1. (a) Use Eq. 39.1 and Eq. 37.7.

$$p_{max} = 0.65 k_a \gamma H \tan^2\left(45° - \frac{\theta}{2}\right)$$
$$= (0.65)\left(121 \ \frac{lbf}{ft^3}\right)(30 \ ft)\tan^2\left(45° - \frac{40°}{2}\right)$$
$$= 513 \ lbf/ft^2$$

The distribution is rectangular.

(b) Assume simple supports of each horizontal lagging timber.

513 lbf/ft²

top view

From beam equations, the maximum moment per foot of timber width is

$$M = \frac{\left(513 \ \frac{lbf}{ft^2}\right)(8 \ ft)^2}{8} = 4104 \ ft\text{-}lbf$$

2.
$$k_{a,sand} = \frac{1 - \sin 32°}{1 + \sin 32°} = 0.307$$

From Eq. 37.3,

$$p_v = \gamma H = \left(110 \ \frac{lbf}{ft^3}\right)(25 \ ft) = 2750 \ lbf/ft^2$$

From Eq. 37.18,

$$p_{a,25 \ ft} = k_a p_v = (0.307)\left(2750 \ \frac{lbf}{ft^2}\right) = 844 \ lbf/ft^2$$

From Eq. 37.19, the reaction per foot of wall is

$$R = \left(\tfrac{1}{2}\right)(25 \ ft)\left(844 \ \frac{lbf}{ft^2}\right)(1 \ ft) = 10{,}550 \ lbf$$

R acts 10 ft + 25 ft/3 = 18.33 ft up from bedrock.

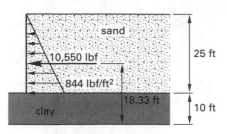

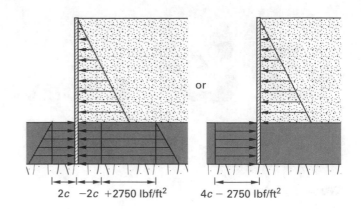

or

$2c$ $-2c$ $+2750$ lbf/ft^2 $4c - 2750$ lbf/ft^2

Determine the horizontal clay pressure.

$$\phi = 0°$$

$$k_a = 1 \quad [\text{since } \phi = 0°]$$

From Eq. 37.8,

$$p_{a,\text{clay}} = p_v - 2c$$

p_v is the surcharge at the top of the clay layer plus the self-load (γH), which is duplicated on the passive side.

$$p_a = 2750 \; \frac{\text{lbf}}{\text{ft}^2} + (\text{clay depth})\left(120 \; \frac{\text{lbf}}{\text{ft}^3}\right) - 2c$$

$$p_p = p_v + 2c$$

$$= (\text{clay depth})\left(120 \; \frac{\text{lbf}}{\text{ft}^3}\right) + 2c$$

$$4c - 2750 \; \frac{\text{lbf}}{\text{ft}^2} = (4)\left(750 \; \frac{\text{lbf}}{\text{ft}^2}\right) - 2750 \; \frac{\text{lbf}}{\text{ft}^2}$$

$$= 250 \; \text{lbf/ft}^2$$

$$R_p = \left(250 \; \frac{\text{lbf}}{\text{ft}^2}\right)(10 \; \text{ft})(1 \; \text{ft}) = 2500 \; \text{lbf}$$

$$F = 10{,}550 \; \text{lbf} - 2500 \; \text{lbf}$$

$$= \boxed{8050 \; \text{lbf (per foot of wall)}}$$

40 Special Soil Topics

PRACTICE PROBLEMS

Pressure at Depths

1. A concentrated vertical load of 6000 lbf is applied at the ground surface. What is the increase in vertical pressure 3.5 ft below the surface and 4 ft from the line of action of the force?

2. A 10 ft × 10 ft footing exerts a pressure of 3000 lbf/ft² on the soil below it. What is the increase in vertical soil stress 10 ft below and 8 ft from the center of the footing?

3. A building weighing 20 tons is placed on the L-shaped slab shown. The soil below the basement is a 100 ft thick layer of dense sand. (a) What is the increase in vertical soil pressure at a depth of 30 ft below point A? (b) What is the increase in vertical soil pressure at a depth of 45 ft below point B?

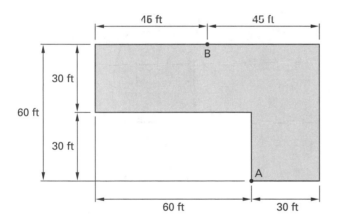

Consolidation

4. A large manufacturing plant with a uniform floor loading of 500 lbf/ft² is supported by a large mat foundation. A 38 ft thick layer of silty soil is underlain by sand and gravel. The plant was constructed on the silty soil 10 ft above the water table. After a number of years and after all settlement of the building had stopped, a series of wells was drilled, which dropped the water table from 10 ft below the surface to 18 ft below the surface. The unit weight of the soil above the water table is 100 lbf/ft³. It is 120 lbf/ft³ below the water table. The silt compression index is 0.02. The void ratio before the water table was lowered was 0.6. How much more will the building settle?

5. The compression index of a normally loaded 16 ft thick clay soil is 0.31. The clay is bounded above and below by layers of sand. When the effective stress on the soil is 2600 lbf/ft², the void ratio is 1.04, and the permeability is 4×10^{-7} mm/s. The stress is increased gradually to 3900 lbf/ft². (a) What is the change in the void ratio? (b) What is the settlement in the clay layer? (c) How much time is required for settlement to reach 75%?

6. At a building site, a rock layer is covered by 8 ft of soft clay. The clay is covered with 15 ft of silty sand. The water table is 18 ft above the rock layer. Properties of the layers are shown on the illustration. A construction firm wants to consolidate the clay layer by surcharging the site with 10 more ft of sandy fill (with a specific weight of 110 lbf/ft³) and by lowering the water table 5 ft. What consolidation will be achieved by this process?

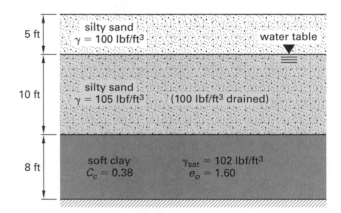

Stresses on Buried Pipes

7. A 12 in concrete sewer pipe is to be installed under a backfill of 11 ft of saturated topsoil with a specific weight of 120 lbf/ft³. The trench width is 2 ft. The pipe strength is 1500 lbf/ft. There are no live loads. A factor of safety of 1.5 is required. Specify the bedding design.

Slope Stability

8. A 2:1 (horizontal:vertical) sloped cut is made in homogeneous saturated clay. The clay's specific weight is 112 lbf/ft³, and its ultimate shear strength is 1100 lbf/ft². The cut is 43 ft deep. The clay extends 15 ft below the toe of the cut to a rock layer. What is the cohesive factor of safety of this cut?

9. (*Time limit: one hour*) A 2 in PVC pipe is to be buried in a trench 18 in wide and $3^{1}/_{2}$ ft deep. The angle of internal friction of the cohesive soil is 18°. The cohesion is 200 lbf/ft². The moisture content is 20%. The dry specific weight of the soil is 115 lbf/ft³. (a) Will the soil stand or slide during trenching if the spoils are placed directly alongside the trench sides? (b) How far away from the trench must the spoils be placed to satisfy OSHA? (c) When the spoils are spread evenly for a distance of 2 ft from the trench, will the soil remain stable? (d) Recommend initial bedding and backfill materials. (e) What additional precautions should be recommended to protect workers entering the trench?

SOLUTIONS

1. Use Eq. 40.1.

$$\Delta p_v = \left(\frac{3P}{2\pi h^2}\right)\left(\frac{1}{1+\left(\frac{r}{h}\right)^2}\right)^{5/2}$$

$$= \left(\frac{(3)(6000 \text{ lbf})}{2\pi(3.5 \text{ ft})^2}\right)\left(\frac{1}{1+\left(\frac{4 \text{ ft}}{3.5 \text{ ft}}\right)^2}\right)^{5/2}$$

$$= \boxed{28.96 \text{ lbf/ft}^2}$$

2. *method 1:* Use Fig. 40.3. About 25 squares are covered. Use Eq. 40.5.

$$p = I(\text{no. of squares})p_{\text{app}}$$

$$= (0.005)(25)\left(3000 \ \frac{\text{lbf}}{\text{ft}^2}\right)$$

$$= \boxed{375 \text{ lbf/ft}^2}$$

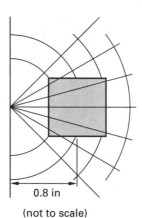

0.8 in

(not to scale)

method 2: Use App. 40.A. At $(-B, +0.8B)$, read $\approx 0.1p$. Use Eq. 40.5.

$$p = (0.1)\left(3000 \ \frac{\text{lbf}}{\text{ft}^2}\right) = \boxed{300 \text{ lbf/ft}^2}$$

3. (a) $A = (90 \text{ ft})(30 \text{ ft}) + (30 \text{ ft})(30 \text{ ft})$

$$= 3600 \text{ ft}^2$$

$$p_{\text{app}} = \frac{P}{A} = \frac{(20 \text{ tons})\left(2000 \ \frac{\text{lbf}}{\text{ton}}\right)}{3600 \text{ ft}^2}$$

$$= 11.11 \text{ lbf/ft}^2$$

Use Fig. 40.3.

Draw the slab such that 1 in = 30 ft. Approximately 46 squares are covered. Use Eq. 40.5.

$$p_{30\,\text{ft}} = I(\text{no. of squares})p_{\text{app}}$$

$$= (0.005)(46)\left(11.11\ \frac{\text{lbf}}{\text{ft}^2}\right)$$

$$= \boxed{2.56\ \text{lbf/ft}^2}$$

(b) Draw the slab such that 1 in = 45 ft. Approximately 63 squares are covered. Use Eq. 40.5.

$$p_{45\,\text{ft}} = I(\text{no. of squares})p_{\text{app}}$$

$$= (0.005)(63)\left(11.11\ \frac{\text{lbf}}{\text{ft}^2}\right)$$

$$= \boxed{3.5\ \text{lbf/ft}^2}$$

4. *steps 1 and 2:* The top 10 ft of soil will have settled under the 500 psf load as much as it is going to. The next 8 ft will settle due to the drop in the water table. The soil under the water table will also experience a reconsolidation. The midpoints of these two settling layers are at depths of 14 ft and 28 ft, respectively. The settlement will be due to the change in effective pressure at these midpoints. Because the raft is so large, its pressure load is assumed not to decrease with depth. Besides, its size was not given.

Before the drop in the water table,

$$p_{o,14\,\text{ft}} = 500\ \frac{\text{lbf}}{\text{ft}^2} + (10\ \text{ft})\left(100\ \frac{\text{lbf}}{\text{ft}^3}\right)$$

$$+ (4\ \text{ft})\left(120\ \frac{\text{lbf}}{\text{ft}^3} - 62.4\ \frac{\text{lbf}}{\text{ft}^3}\right)$$

$$= 1730\ \text{lbf/ft}^2$$

$$p_{o,28\,\text{ft}} = 500\ \frac{\text{lbf}}{\text{ft}^2} + (10\ \text{ft})\left(100\ \frac{\text{lbf}}{\text{ft}^3}\right)$$

$$+ (18\ \text{ft})\left(120\ \frac{\text{lbf}}{\text{ft}^3} - 62.4\ \frac{\text{lbf}}{\text{ft}^3}\right)$$

$$= 2537\ \text{lbf/ft}^2$$

After the drop in the water table,

$$p_{v,14\,\text{ft}} = 500\ \frac{\text{lbf}}{\text{ft}^2} + (14\ \text{ft})\left(100\ \frac{\text{lbf}}{\text{ft}^3}\right)$$

$$= 1900\ \text{lbf/ft}^2$$

$$p_{v,28\,\text{ft}} = 500\ \frac{\text{lbf}}{\text{ft}^2} + (18\ \text{ft})\left(100\ \frac{\text{lbf}}{\text{ft}^3}\right)$$

$$+ (10\ \text{ft})\left(120\ \frac{\text{lbf}}{\text{ft}^3} - 62.4\ \frac{\text{lbf}}{\text{ft}^3}\right)$$

$$= 2876\ \text{lbf/ft}^2$$

step 3: $e_o = 0.6$ [given]

step 4: $C_c = 0.02$ [given]

step 5: Use Eq. 40.16. In the 8 ft layer,

$$S_1 = H(\text{CR})\log_{10}\frac{p_v'}{p_o'}$$

$$= (8\ \text{ft})\left(\frac{0.02}{1 + 0.6}\right)\log_{10}\frac{1900\ \frac{\text{lbf}}{\text{ft}^2}}{1730\ \frac{\text{lbf}}{\text{ft}^2}}$$

$$= 0.00407\ \text{ft}$$

In the submerged layer,

$$S_2 = H(\text{CR})\log_{10}\frac{p_v'}{p_o'}$$

$$= (20\ \text{ft})\left(\frac{0.02}{1 + 0.6}\right)\log_{10}\frac{2876\ \frac{\text{lbf}}{\text{ft}^2}}{2537\ \frac{\text{lbf}}{\text{ft}^2}}$$

$$= 0.01362\ \text{ft}$$

The total settlement is

$$S_1 + S_2 = 0.00407\ \text{ft} + 0.01362\ \text{ft}$$

$$= \boxed{0.0177\ \text{ft}}$$

5. (a) Though reported as a positive number, the compression index represents a negative slope. Use Eq. 40.7.

$$C_c = \frac{e_3 - e_2}{\log_{10}\dfrac{p_2'}{p_3'}}$$

$$-0.31 = \frac{1.04 - e_2}{\log_{10}\dfrac{2600\ \frac{\text{lbf}}{\text{ft}^2}}{3900\ \frac{\text{lbf}}{\text{ft}^2}}}$$

$$e_2 = 0.985$$

$$\Delta e = 0.985 - 1.04 = \boxed{-0.055}$$

(b) Use Eq. 40.15.

$$S = \frac{H\Delta e}{1 + e_o} = \frac{(16\ \text{ft})(-0.055)}{1 + 1.04}$$

$$= \boxed{-0.43\ \text{ft}}$$

(c) Use Eq. 40.22.

$$a_v = \frac{-(e_2 - e_1)}{p_2' - p_1'} = \frac{-(0.985 - 1.04)}{3900\ \frac{\text{lbf}}{\text{ft}^2} - 2600\ \frac{\text{lbf}}{\text{ft}^2}}$$

$$= 4.23 \times 10^{-5}\ \text{ft}^2/\text{lbf}$$

The permeability is

$$K = \frac{4 \times 10^{-7} \frac{\text{mm}}{\text{sec}}}{\left(10 \frac{\text{mm}}{\text{cm}}\right)\left(2.54 \frac{\text{cm}}{\text{in}}\right)\left(12 \frac{\text{in}}{\text{ft}}\right)}$$
$$= 1.31 \times 10^{-9} \text{ ft/sec}$$

$$C_v = \frac{K(1 + e_o)}{a_v \gamma_{\text{water}}} = \frac{\left(1.31 \times 10^{-9} \frac{\text{ft}}{\text{sec}}\right)(1 + 1.04)}{\left(4.23 \times 10^{-5} \frac{\text{ft}^2}{\text{lbf}}\right)\left(62.4 \frac{\text{lbf}}{\text{ft}^3}\right)}$$
$$= 1.01 \times 10^{-6} \text{ ft}^2/\text{sec}$$

For $U_z = 75\%$, $T_v = 0.48$.

$$t = \frac{T_v H_d^2}{C_v}$$
$$= \frac{(0.48)\left(\frac{16 \text{ ft}}{2}\right)^2}{\left(1.01 \times 10^{-6} \frac{\text{ft}^2}{\text{sec}}\right)\left(24 \frac{\text{hr}}{\text{day}}\right)\left(3600 \frac{\text{sec}}{\text{hr}}\right)}$$
$$= \boxed{352 \text{ days}}$$

6. The effective pressure at the midpoint of the clay layer is

$$p_o' = (5 \text{ ft})\left(100 \frac{\text{lbf}}{\text{ft}^3}\right) + (10 \text{ ft})\left(105 \frac{\text{lbf}}{\text{ft}^3} - 62.4 \frac{\text{lbf}}{\text{ft}^3}\right)$$
$$+ \left(\tfrac{1}{2}\right)\left(102 \frac{\text{lbf}}{\text{ft}^3} - 62.4 \frac{\text{lbf}}{\text{ft}^3}\right)(8 \text{ ft})$$
$$= 1084.4 \text{ lbf/ft}^2$$

The final effective pressure is

$$p_{\text{final}}' = (10 \text{ ft})\left(110 \frac{\text{lbf}}{\text{ft}^3}\right) + (10 \text{ ft})\left(100 \frac{\text{lbf}}{\text{ft}^3}\right)$$
$$+ (5 \text{ ft})\left(105 \frac{\text{lbf}}{\text{ft}^3} - 62.4 \frac{\text{lbf}}{\text{ft}^3}\right)$$
$$+ \left(\tfrac{1}{2}\right)\left(102 \frac{\text{lbf}}{\text{ft}^3} - 62.4 \frac{\text{lbf}}{\text{ft}^3}\right)(8 \text{ ft})$$
$$= 2471.4 \text{ lbf/ft}^2$$

Use Eq. 40.16.

$$S = \frac{HC_c \log_{10} \frac{p_{\text{final}}'}{p_o'}}{1 + e_o} = \frac{(8 \text{ ft})(0.38)\log_{10} \frac{2471.4 \frac{\text{lbf}}{\text{ft}^2}}{1084.4 \frac{\text{lbf}}{\text{ft}^2}}}{1 + 1.60}$$
$$= \boxed{0.42 \text{ ft}}$$

C_c can also be estimated from e_o, but the data provided in the problem takes priority.

7. $\dfrac{h}{B} = \dfrac{11 \text{ ft}}{2 \text{ ft}} = 5.5$ [round to 6.0]

Use Table 40.2. For $\gamma = 120 \text{ lbf/ft}^3$, $C = 3.04$. (The load on the pipe is more important than the soil type.)

The actual load is given by Eq. 40.30.

$$w = C\gamma B^2 = (3.04)\left(120 \frac{\text{lbf}}{\text{ft}^3}\right)(2 \text{ ft})^2 = 1459.2 \text{ lbf/ft}$$

Solve for the load factor. Use Eq. 40.36.

$$\text{LF} = \frac{w_{\text{allowable}} F}{\text{known pipe crushing strength}}$$
$$= \frac{\left(1459.2 \frac{\text{lbf}}{\text{ft}}\right)(1.5)}{1500 \frac{\text{lbf}}{\text{ft}}}$$
$$= 1.46$$

$$\boxed{\text{Select Class C or better.}}$$

8. For clay and $\phi = 0°$, the cohesion is the same as the shear strength. ("Saturated clay" is not the same as "submerged clay.")

$$c = 1100 \text{ lbf/ft}^2$$
$$\beta = \arctan\tfrac{1}{2} = 26.6°$$
$$d = \frac{D}{H} = \frac{15 \text{ ft}}{43 \text{ ft}} = 0.35$$

From the Taylor chart, the stability number is $N_o \approx 6.5$. From Eq. 40.28, the cohesion safety factor is

$$F = \frac{N_o c}{\gamma H} = \frac{(6.5)\left(1100 \frac{\text{lbf}}{\text{ft}^2}\right)}{\left(112 \frac{\text{lbf}}{\text{ft}^3}\right)(43 \text{ ft})} = \boxed{1.48}$$

$$\left[\begin{array}{c}\text{within 1.3 to 1.5 range;} \\ \text{acceptable}\end{array}\right]$$

9. $\gamma = (1 + 0.20)\left(115 \frac{\text{lbf}}{\text{ft}^3}\right) = 138 \text{ lbf/ft}^3$

The weight of the soil excavated per foot of trench is

$$\left(\frac{18 \text{ in}}{12 \frac{\text{in}}{\text{ft}}}\right)(3.5 \text{ ft})\left(138 \frac{\text{lbf}}{\text{ft}^3}\right) = 725 \text{ lbf/ft}$$

Half of the spoils will appear on each side of the trench.

$$\dfrac{725\ \dfrac{\text{lbf}}{\text{ft}}}{2} = 362\ \text{lbf/ft per side}$$

(a) Since $\phi > 5°$, failure will be toe slope. Investigate the stability with the trial wedge method. Since the soil is somewhat cohesive ($c > 0$), the failure plane would make an angle of 18° or greater. The horizontal distance, BC, is

$$BC = (3.5\ \text{ft})\tan(90° - 18°) = 10.77\ \text{ft}$$

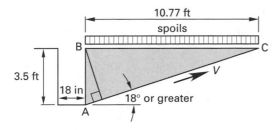

The length of line AC is

$$\sqrt{(3.5\ \text{ft})^2 + (10.77\ \text{ft})^2} = 11.3\ \text{ft}$$

The soil weight in ABC is

$$\left(\tfrac{1}{2}\right)(3.5\ \text{ft})(10.77\ \text{ft})\left(138\ \dfrac{\text{lbf}}{\text{ft}^3}\right) = 2600\ \text{lbf}$$

The average pressure occurs at the average depth.

$$p_v = \left(\dfrac{3.5\ \text{ft}}{2}\right)\left(138\ \dfrac{\text{lbf}}{\text{ft}^3}\right) = 241.5\ \text{lbf/ft}^2$$

The surcharge from the spoils is

$$p_q = \dfrac{362\ \dfrac{\text{lbf}}{\text{ft}}}{10.77\ \text{ft}} = 33.6\ \text{lbf/ft}^2$$

The normal stress on an 18° failure plane would be

$$\sigma_n = (p_v + p_q)\cos 18°$$

The shear strength of the soil is

$$\begin{aligned}
S_{us} &= c + \sigma_n \tan\phi \\
&= 200\ \dfrac{\text{lbf}}{\text{ft}^2} + \left(241.5\ \dfrac{\text{lbf}}{\text{ft}^2} + 33.6\ \dfrac{\text{lbf}}{\text{ft}^2}\right) \\
&\quad \times (\cos 18°)\tan 18° \\
&= 285\ \text{lbf/ft}^2
\end{aligned}$$

The ultimate resisting shear force is

$$V = \left(285\ \dfrac{\text{lbf}}{\text{ft}^2}\right)(11.3\ \text{ft}) = 3221\ \text{lbf/ft}$$

The total vertical weight is

$$W_{\text{wedge}} + W_{\text{spoils}} = 2600\ \dfrac{\text{lbf}}{\text{ft}} + 362\ \dfrac{\text{lbf}}{\text{ft}} = 2962\ \text{lbf/ft}$$

Resolve the soil weight into a force parallel to the assumed shear plane.

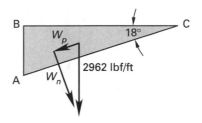

$$W_p = \left(2962\ \dfrac{\text{lbf}}{\text{ft}}\right)\sin 18° = 915\ \text{lbf/ft}$$

Since 915 lbf/ft < 3221 lbf/ft, $\boxed{\text{the soil will not slide.}}$

The factor of safety against sliding is

$$F_{\text{sliding}} = \dfrac{V}{W_p} = \dfrac{3221\ \dfrac{\text{lbf}}{\text{ft}}}{915\ \dfrac{\text{lbf}}{\text{ft}}} = 3.52$$

The failure plane angle is unknown and may not be 18°. This process can be duplicated for other angles.

(b) $\boxed{2\ \text{ft}}$

(c) $\boxed{\text{yes}}$ [see table]

(d) $\boxed{\begin{array}{l}\text{For a flexible pipe of this small size and buried}\\\text{as deep as this, there is no special bedding.}\\\text{Backfill with granular soil to eliminate settling}\\\text{of fill.}\end{array}}$

(e) $\boxed{\begin{array}{l}\bullet\ \text{Place barricades open trench.}\\ \bullet\ \text{Join pipes above ground, not in trench.}\\ \bullet\ \text{Have someone outside of trench spotting}\\ \quad \text{(less than 5 ft deep does not normally get}\\ \quad \text{shoring).}\\ \bullet\ \text{Place exit ladders every 50 ft.}\end{array}}$

Geotechnical

Table for Sol. 9(c)

ϕ	BC (ft)	AC (ft)	soil mass $\left(\frac{\text{lbf}}{\text{ft}^2}\right)$	p_q $\left(\frac{\text{lbf}}{\text{ft}^2}\right)$	S_{us} $\left(\frac{\text{lbf}}{\text{ft}^2}\right)$	V $\left(\frac{\text{lbf}}{\text{ft}}\right)$	W_p $\left(\frac{\text{lbf}}{\text{ft}}\right)$	F	failure?
18°	10.77	11.3	2600	33.6	285	3221	915	3.52	no
30°	6.06	7.0	1463	59.7	285	1995	913	2.19	no
40°	4.17	5.44	1007	86.8	282	1534	880	1.74	no
60°	2.02	4.04	488	179.2	269	1087	736	1.48	no
80°	0.62	3.55	150	584.0	247	877	504	1.74	no

Geotechnical

41 Determinate Statics

PRACTICE PROBLEMS

1. Two towers are located on level ground 100 ft (30 m) apart. They support a transmission line with a mass of 2 lbm/ft (3 kg/m). The midpoint sag is 10 ft (3 m).

(a) What is the midpoint tension?

  (A) 125 lbf (0.55 kN)

  (B) 250 lbf (1.1 kN)

  (C) 375 lbf (1.6 kN)

  (D) 500 lbf (2.2 kN)

(b) What is the maximum tension in the transmission line?

  (A) 170 lbf (0.70 kN)

  (B) 210 lbf (0.86 kN)

  (C) 270 lbf (1.2 kN)

  (D) 330 lbf (1.4 kN)

(c) If the maximum tension is 500 lbf (2200 N), what is the sag in the cable?

  (A) 1.3 ft (0.4 m)

  (B) 3.2 ft (1.0 m)

  (C) 4.0 ft (1.2 m)

  (D) 5.1 ft (1.5 m)

2. Two legs of a tripod are mounted on a vertical wall. Both legs are horizontal. The apex is 12 distance units from the wall. The right leg is 13.4 units long. The wall mounting points are 10 units apart. A third leg is mounted on the wall 6 units to the left of the right upper leg and 9 units below the two top legs. A vertical downward load of 200 is supported at the apex. What is the reaction at the lowest mounting point?

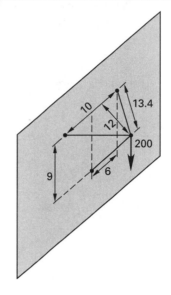

  (A) 120

  (B) 170

  (C) 250

  (D) 330

3. The ideal truss shown is supported by a pinned connection at point D and a roller connection at point C. Loads are applied at points A and F. What is the force in member DE?

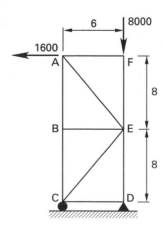

(A) 1200

(B) 2700

(C) 3300

(D) 3700

4. A pin-connected tripod is loaded at the apex by a horizontal force of 1200, as shown. What is the magnitude of the force in member AD?

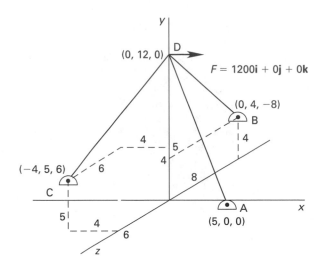

(A) 1100

(B) 1300

(C) 1800

(D) 2500

5. A truss is loaded by forces of 4000 at each upper connection point and forces of 60,000 at each lower connection point, as shown.

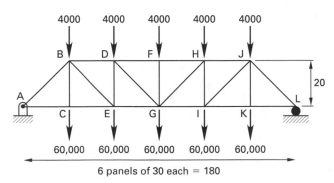

(a) What is the force in member DE?

(A) 36,000

(B) 45,000

(C) 60,000

(D) 160,000

(b) What is the force in member HJ?

(A) 24,000

(B) 60,000

(C) 160,000

(D) 380,000

6. The rigid rod AO is supported by guy wires BO and CO, as shown. (Points A, B, and C are all in the same vertical plane. Points A, O, and C are all in the same horizontal plane. Points A, O, and B are all in the same vertical plane.) Vertical and horizontal forces are 12,000 and 6000, respectively, as carried at the end of the rod. What are the x-, y-, and z-components of the reactions at point C?

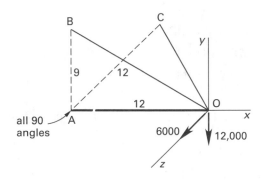

(A) $(C_x, C_y, C_z) = (0, 6000, 0)$

(B) $(C_x, C_y, C_z) = (6000, 0, 6000)$

(C) $(C_x, C_y, C_z) = (6000, 0, 4200)$

(D) $(C_x, C_y, C_z) = (4200, 0, 8500)$

7. When the temperature is 70°F (21.11°C), sections of steel railroad rail are welded end to end to form a continuous, horizontal track that is exactly 1 mi (1.6 km) long. Both ends of the track are constrained by preexisting installed sections of rail. Before the 1 mi section of track can be nailed to the ties, however, the sun warms it to a uniform temperature of 99.14°F (37.30°C). Laborers watch in amazement as the rail pops up in the middle and takes on a parabolic shape. The laborers prop the rail up (so that it does not buckle over) while they take souvenir pictures. How high off the ground is the midpoint of the hot rail?

(A) 0.8 ft (2.4 m)

(B) 2.1 ft (3.6 m)

(C) 17 ft (5.1 m)

(D) 45 ft (14 m)

8. (a) Find the force in member FC.

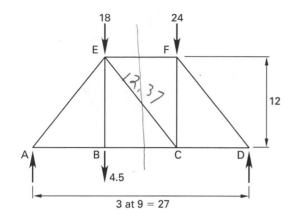

(A) 0.5

(B) 17

(C) 23

(D) 29

(b) What is the force in member CE?

(A) 0.50

(B) 0.63

(C) 11

(D) 17

SOLUTIONS

1. *Customary U.S. Solution*

(a) Use Eq. 41.68 to relate the midpoint sag, S, to the constant, c.

$$S = c\left(\cosh\left(\frac{a}{c}\right) - 1\right)$$

$$10 \text{ ft} = c\left(\cosh\left(\frac{50 \text{ ft}}{c}\right) - 1\right)$$

Solve by trial and error.

$$c = 126.6 \text{ ft}$$

Use Eq. 41.52 and Eq. 41.70 to find the midpoint tension.

$$H = wc = m\left(\frac{g}{g_c}\right)c$$

$$= \left(2 \frac{\text{lbm}}{\text{ft}}\right)\left(\frac{32.2 \frac{\text{ft}}{\text{sec}^2}}{32.2 \frac{\text{lbm-ft}}{\text{lbf-sec}^2}}\right)(126.6 \text{ ft})$$

$$= \boxed{253.2 \text{ lbf} \quad (250 \text{ lbf})}$$

The answer is (B).

(b) Use Eq. 41.72 to find the maximum tension.

$$T = wy = w(c + S) = m\left(\frac{g}{g_c}\right)(c + S)$$

$$= \left(2 \frac{\text{lbm}}{\text{ft}}\right)\left(\frac{32.2 \frac{\text{ft}}{\text{sec}^2}}{32.2 \frac{\text{lbm-ft}}{\text{lbf-sec}^2}}\right)(126.6 \text{ ft} + 10 \text{ ft})$$

$$= \boxed{273.2 \text{ lbf} \quad (270 \text{ lbf})}$$

The answer is (C).

(c) From $T = wy$,

$$y = \frac{T}{w} = \frac{500 \text{ lbf}}{2 \frac{\text{lbf}}{\text{ft}}} = 250 \text{ ft} \quad \text{[at right support]}$$

$$250 \text{ ft} = c\left(\cosh\frac{50 \text{ ft}}{c}\right)$$

By trial and error, $c = 245$ ft.

Substitute into Eq. 41.68.

$$S = c\left(\cosh\left(\frac{a}{c}\right) - 1\right) = (245 \text{ ft})\left(\cosh\left(\frac{50 \text{ ft}}{245 \text{ ft}}\right) - 1\right)$$

$$= \boxed{5.1 \text{ ft}}$$

The answer is (D).

SI Solution

(a) Use Eq. 41.68 to relate the midpoint sag, S, to the constant, c.

$$S = c\left(\cosh\left(\frac{a}{c}\right) - 1\right)$$

$$3 \text{ m} = c\left(\cosh\left(\frac{15 \text{ m}}{c}\right) - 1\right)$$

Solve by trial and error.

$$c = 38.0 \text{ m}$$

Use Eq. 41.52 and Eq. 41.70 to find the midpoint tension.

$$H = wc = mgc = \left(3 \ \frac{\text{kg}}{\text{m}}\right)\left(9.81 \ \frac{\text{m}}{\text{s}^2}\right)(38.0 \text{ m})$$

$$= \boxed{1118.3 \text{ N} \quad (1.1 \text{ kN})}$$

The answer is (B).

(b) Use Eq. 41.72 to find the maximum tension.

$$T = wy = w(c + S) = mg(c + S)$$

$$= \left(3 \ \frac{\text{kg}}{\text{m}}\right)\left(9.81 \ \frac{\text{m}}{\text{s}^2}\right)(38.0 \text{ m} + 3.0 \text{ m})$$

$$= \boxed{1206.6 \text{ N} \quad (1.2 \text{ kN})}$$

The answer is (C).

(c) From $T = wy$,

$$y = \frac{T}{w} = \frac{2200 \text{ N}}{\left(3 \ \frac{\text{kg}}{\text{m}}\right)\left(9.81 \ \frac{\text{m}}{\text{s}^2}\right)}$$

$$= 74.75 \text{ m} \quad [\text{at right support}]$$

$$74.75 \text{ m} = c\left(\cosh\frac{15 \text{ m}}{c}\right)$$

By trial and error, $c = 73.2$ m.

Substitute into Eq. 41.68.

$$S = c\left(\cosh\left(\frac{a}{c}\right) - 1\right)$$

$$= (73.2 \text{ m})\left(\cosh\left(\frac{15 \text{ m}}{73.2 \text{ m}}\right) - 1\right)$$

$$= \boxed{1.54 \text{ m} \quad (1.5 \text{ m})}$$

The answer is (D).

2. *step 1:* Draw the tripod with the origin at the apex.

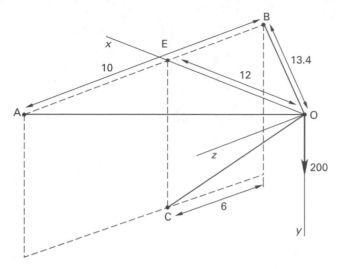

step 2: By inspection, the force components are $F_x = 0$, $F_y = 200$, and $F_z = 0$.

step 3: First, from triangle EBO, length BE is

$$\text{BE} = \sqrt{(13.4 \text{ units})^2 - (12 \text{ units})^2}$$

$$= 5.96 \text{ units} \quad [\text{use 6 units}]$$

The (x, y, z) coordinates of the three support points are

$$\begin{aligned} \text{point A:} & \quad (12, 0, 4) \\ \text{point B:} & \quad (12, 0, -6) \\ \text{point C:} & \quad (12, 9, 0) \end{aligned}$$

step 4: Find the lengths of the legs.

$$\text{AO} = \sqrt{(x_\text{A} - x_\text{O})^2 + (y_\text{A} - y_\text{O})^2 + (z_\text{A} - z_\text{O})^2}$$

$$= \sqrt{(12 \text{ units})^2 + (0 \text{ units})^2 + (4 \text{ units})^2}$$

$$= 12.65 \text{ units}$$

$$\text{BO} = \sqrt{(x_\text{B} - x_\text{O})^2 + (y_\text{B} - y_\text{O})^2 + (z_\text{B} - z_\text{O})^2}$$

$$= \sqrt{(12 \text{ units})^2 + (0 \text{ units})^2 + (-6 \text{ units})^2}$$

$$= 13.4 \text{ units}$$

$$\text{CO} = \sqrt{(x_\text{C} - x_\text{O})^2 + (y_\text{C} - y_\text{O})^2 + (z_\text{C} - z_\text{O})^2}$$

$$= \sqrt{(12 \text{ units})^2 + (9 \text{ units})^2 + (0 \text{ units})^2}$$

$$= 15.0 \text{ units}$$

step 5: Use Eq. 41.76, Eq. 41.77, and Eq. 41.78 to find the direction cosines for each leg.

For leg AO,

$$\cos\theta_{A,x} = \frac{x_A}{AO} = \frac{12 \text{ units}}{12.65 \text{ units}} = 0.949$$

$$\cos\theta_{A,y} = \frac{y_A}{AO} = \frac{0 \text{ units}}{12.65 \text{ units}} = 0$$

$$\cos\theta_{A,z} = \frac{z_A}{AO} = \frac{4 \text{ units}}{12.65 \text{ units}} = 0.316$$

For leg BO,

$$\cos\theta_{B,x} = \frac{x_B}{BO} = \frac{12 \text{ units}}{13.4 \text{ units}} = 0.896$$

$$\cos\theta_{B,y} = \frac{y_B}{BO} = \frac{0 \text{ units}}{13.4 \text{ units}} = 0$$

$$\cos\theta_{B,z} = \frac{z_B}{BO} = \frac{-6 \text{ units}}{13.4 \text{ units}} = -0.448$$

For leg CO,

$$\cos\theta_{C,x} = \frac{x_C}{CO} = \frac{12 \text{ units}}{15.0 \text{ units}} = 0.80$$

$$\cos\theta_{C,y} = \frac{y_C}{CO} = \frac{9 \text{ units}}{15.0 \text{ units}} = 0.60$$

$$\cos\theta_{C,z} = \frac{z_C}{CO} = \frac{0 \text{ units}}{15.0 \text{ units}} = 0$$

steps 6 and 7: Substitute Eq. 41.79, Eq. 41.80, and Eq. 41.81 into equilibrium Eq. 41.82, Eq. 41.83, and Eq. 41.84.

$$F_A\cos\theta_{A,x} + F_B\cos\theta_{B,x} + F_C\cos\theta_{C,x} + F_x = 0$$

$$F_A\cos\theta_{A,y} + F_B\cos\theta_{B,y} + F_C\cos\theta_{C,y} + F_y = 0$$

$$F_A\cos\theta_{A,z} + F_B\cos\theta_{B,z} + F_C\cos\theta_{C,z} + F_z = 0$$

$$0.949F_A + 0.896F_B + 0.80F_C + 0 = 0$$

$$0F_A + 0F_B + 0.60F_C + 200 = 0$$

$$0.316F_A - 0.448F_B + 0F_C + 0 = 0$$

Solve the three equations simultaneously.

$$F_A = 168.6 \quad (T)$$

$$F_B = 118.9 \quad (T)$$

$$F_C = \boxed{-333.3 \quad (C) \quad (330)}$$

The answer is (D).

3. First, find the vertical reaction at point D.

$$\sum M_C = (CD)D_y - (AF)(8000) + (AC)(1600) = 0$$

$$6D_y - (6)(8000) + (16)(1600) = 0$$

Solve for $D_y = 3733.3$.

The free-body diagram of pin D is as follows.

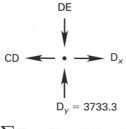

$$\sum F_y = D_y - DE = 0$$

Therefore,

$$DE = D_y = \boxed{3733.3 \quad (C) \quad (3700)}$$

The answer is (D).

4. *step 1:* Move the origin to the apex of the tripod. Call this point O.

step 2: By inspection, the force components are $F_x = 1200$, $F_y = 0$, and $F_z = 0$.

step 3: The $(x,\ y,\ z)$ coordinates of the three support points are

$$\text{point A:} \quad (5,\ -12,\ 0)$$
$$\text{point B:} \quad (0,\ -8,\ -8)$$
$$\text{point C:} \quad (-4,\ -7,\ 6)$$

step 4: Find the lengths of the legs.

$$AO = \sqrt{(x_A - x_O)^2 + (y_A - y_O)^2 + (z_A - z_O)^2}$$

$$= \sqrt{(5)^2 + (-12)^2 + (0)^2}$$

$$= 13.0$$

$$BO = \sqrt{(x_B - x_O)^2 + (y_B - y_O)^2 + (z_B - z_O)^2}$$

$$= \sqrt{(0)^2 + (-8)^2 + (-8)^2}$$

$$= 11.31$$

$$CO = \sqrt{(x_C - x_O)^2 + (y_C - y_O)^2 + (z_C - z_O)^2}$$

$$= \sqrt{(-4)^2 + (-7)^2 + (6)^2}$$

$$= 10.05$$

step 5: Use Eq. 41.76, Eq. 41.77, and Eq. 41.78 to find the direction cosines for each leg.

Structural

For leg AO,

$$\cos\theta_{A,x} = \frac{x_A}{AO} = \frac{5}{13.0} = 0.385$$

$$\cos\theta_{A,y} = \frac{y_A}{AO} = \frac{-12}{13.0} = -0.923$$

$$\cos\theta_{A,z} = \frac{z_A}{AO} = \frac{0}{13.0} = 0$$

For leg BO,

$$\cos\theta_{B,x} = \frac{x_B}{BO} = \frac{0}{11.31} = 0$$

$$\cos\theta_{B,y} = \frac{y_B}{BO} = \frac{-8}{11.31} = -0.707$$

$$\cos\theta_{B,z} = \frac{z_B}{BO} = \frac{-8}{11.31} = -0.707$$

For leg CO,

$$\cos\theta_{C,x} = \frac{x_C}{CO} = \frac{-4}{10.05} = -0.398$$

$$\cos\theta_{C,y} = \frac{y_C}{CO} = \frac{-7}{10.05} = -0.697$$

$$\cos\theta_{C,z} = \frac{z_C}{CO} = \frac{6}{10.05} = 0.597$$

steps 6 and 7: Substitute Eq. 41.79, Eq. 41.80, and Eq. 41.81 into equilibrium Eq. 41.82, Eq. 41.83, and Eq. 41.84.

$$F_A\cos\theta_{A,x} + F_B\cos\theta_{B,x} + F_C\cos\theta_{C,x} + F_x = 0$$
$$F_A\cos\theta_{A,y} + F_B\cos\theta_{B,y} + F_C\cos\theta_{C,y} + F_y = 0$$
$$F_A\cos\theta_{A,z} + F_B\cos\theta_{B,z} + F_C\cos\theta_{C,z} + F_z = 0$$
$$0.385F_A + 0F_B - 0.398F_C + 1200 = 0$$
$$-0.923F_A - 0.707F_B - 0.697F_C + 0 = 0$$
$$0F_A - 0.707F_B + 0.597F_C + 0 = 0$$

Solve the three equations simultaneously.

$$\boxed{F_A = -1793 \quad (C) \quad (1800)}$$
$$F_B = 1080 \quad (T)$$
$$F_C = 1279 \quad (T)$$

The answer is (C).

5. First, find the vertical reactions.

$$\sum F_y = A_y + L_y - (5)(4000) - (5)(60,000) = 0$$

By symmetry, $A_y = L_y$.

$$2A_y = (5)(4000) + (5)(60,000)$$
$$A_y = 160,000$$
$$L_y = 160,000$$

(a) For DE, make a cut in members BD, DE, and EG.

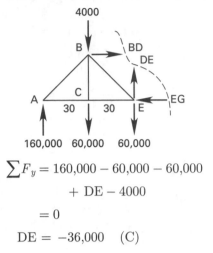

$$\sum F_y = 160,000 - 60,000 - 60,000$$
$$+ DE - 4000$$
$$= 0$$
$$DE = -36,000 \quad (C)$$

The answer is (A).

(b) For HJ, make a cut in members HJ, HI, and GI.

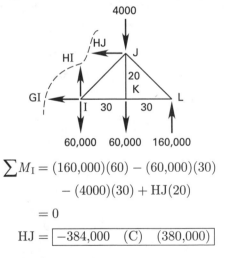

$$\sum M_I = (160,000)(60) - (60,000)(30)$$
$$- (4000)(30) + HJ(20)$$
$$= 0$$
$$HJ = \boxed{-384,000 \quad (C) \quad (380,000)}$$

The answer is (D).

6. First, consider a free-body diagram at point O in the *x-y* plane.

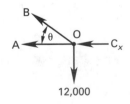

θ is obtained from triangle AOB.

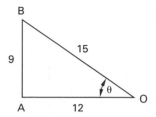

C_x is obtained from triangle AOC.

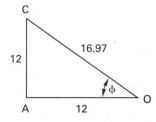

Equilibrium in the x-y plane at point O requires $\sum F_y = 0$.

$$B \sin \theta = 12{,}000$$
$$B = \frac{12{,}000}{\sin \theta} = \frac{12{,}000}{\dfrac{9}{15}}$$
$$= 20{,}000$$
$$B_x = B \cos \theta = (20{,}000)\left(\frac{12}{15}\right) = 16{,}000$$
$$B_y = B \sin \theta = (20{,}000)\left(\frac{9}{15}\right) = 12{,}000$$

Since BO is in the x-y plane, $B_z = 0$.

Next, consider a free-body diagram at point O in the x-z plane.

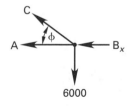

$\sum F_x = 0$:

$$A + B_x + C \cos \phi = 0$$
$$A + C \cos \phi = -B_x = -B \cos \theta$$
$$= (-20{,}000)\left(\frac{12}{15}\right)$$
$$= -16{,}000$$

$\sum F_y = 0$:

$$C \sin \phi = 6000$$
$$C = \frac{6000}{\sin \phi}$$
$$= \frac{6000}{\dfrac{12}{16.97}}$$
$$= 8485$$

Therefore,

$$A = -C \cos \phi - 16{,}000$$
$$= (-8485)\left(\frac{12}{16.97}\right) - 16{,}000$$
$$= -22{,}000$$

Since AO is on the x-axis,

$$A_x = -22{,}000$$
$$A_y = 0$$
$$A_z = 0$$

Solve for C reactions.

$$C_x = C \cos \phi = (8485)\left(\frac{12}{16.97}\right)$$
$$= 6000$$
$$C_z = C \sin \phi = (8485)\left(\frac{12}{16.97}\right)$$
$$= 6000$$

Since CO is in the x-z plane, $C_y = 0$.

The answer is (B).

7. *Customary U.S. Solution*

First, find the amount of thermal expansion. From Table 44.2, the coefficient of thermal expansion for steel is 6.5×10^{-6} $1/°F$. Use Eq. 44.9.

$$\Delta L = \alpha L (T_2 - T_1)$$
$$= \left(6.5 \times 10^{-6} \ \frac{1}{°F}\right)(1)\left(5280 \ \frac{ft}{mi}\right)$$
$$\times (99.14°F - 70°F)$$
$$= 1.000085 \ ft \quad (\approx 1 \ ft)$$

Assume the distributed load is uniform along the length of the rail. This resembles the case of a cable under its own weight. From the parabolic cable figure, as shown in Fig. 41.18, when distance S is small relative to distance a, the problem can be solved by using the parabolic equation, Eq. 41.58.

$$L \approx a\left(1 + \tfrac{2}{3}\left(\tfrac{S}{a}\right)^2 - \tfrac{2}{5}\left(\tfrac{S}{a}\right)^4\right)$$

$$\frac{5280 \text{ ft} + 1 \text{ ft}}{2} \approx (2640 \text{ ft})\left(1 + \tfrac{2}{3}\left(\frac{S}{2640 \text{ ft}}\right)^2\right.$$

$$\left. - \tfrac{2}{5}\left(\frac{S}{2640 \text{ ft}}\right)^4\right)$$

Using trial and error, $S \approx \boxed{44.5 \text{ ft } (45 \text{ ft}).}$

The answer is (D).

SI Solution

First, find the amount of thermal expansion. From Table 44.2, the coefficient of thermal expansion for steel is $11.7 \times 10^{-6} \, 1/°C$. Use Eq. 44.9.

$$\delta = \alpha L_o (T_2 - T_1)$$

$$= \left(11.7 \times 10^{-6} \, \tfrac{1}{°C}\right)(1.6 \text{ km})\left(1000 \, \tfrac{m}{km}\right)$$

$$\times (37.30°C - 21.11°C)$$

$$= 0.30308 \text{ m} \approx 0.30 \text{ m}$$

Assume the distributed load is uniform along the length of the rail. This resembles the case of a cable under its own weight. From the parabolic cable figure, as shown in Fig. 41.18, when distance S is small relative to distance a, the problem can be solved by using the parabolic equation, Eq. 41.58.

$$L \approx a\left(1 + \tfrac{2}{3}\left(\tfrac{S}{a}\right)^2 - \tfrac{2}{5}\left(\tfrac{S}{a}\right)^4\right)$$

$$\frac{1600 \text{ m} + 0.3 \text{ m}}{2} \approx (800 \text{ m})\left(1 + \tfrac{2}{3}\left(\frac{S}{800 \text{ m}}\right)^2\right.$$

$$\left. - \tfrac{2}{5}\left(\frac{S}{800 \text{ m}}\right)^4\right)$$

Using trial and error, $S \approx \boxed{13.6 \text{ m } (14 \text{ m}).}$

The answer is (D).

8. (a) First, find the reactions. Take clockwise moments about A as positive.

$$\sum M_A = (27)(-D_y) + (18)(24) + (9)(22.5) = 0$$

$$D_y = 23.5$$

$$A_y = 18 + 24 + 4.5 - 23.5 = 23$$

Either the method of sections (easiest) or a member-by-member analysis can be used.

The general force triangle is

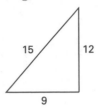

At pin A,

$$AE_y = 23$$

$$AE_x = \left(\tfrac{9}{12}\right)(23) = 17.25$$

$$AE = \left(\tfrac{15}{12}\right)(23) = 28.75 \quad (C)$$

$$AB = AE_x = 17.25 \quad (T)$$

At pin B,

$$BE = 4.5 \quad (T)$$

$$BC = AB = 17.25 \quad (T)$$

At pin D,

$$DF_y = 23.5$$

$$DF_x = \left(\tfrac{9}{12}\right)(23.5) = 17.63$$

$$DF = \left(\tfrac{15}{12}\right)(23.5) = 29.38 \quad (C)$$

$$DC = DF_x = 17.63 \quad (T)$$

At pin F,

$$FE = DF_x = 17.63 \quad (C)$$

$$FC = 24 - DF_y = 24 - 23.5 = \boxed{0.5 \quad (C)}$$

The answer is (A).

(b) At pin C,

$$CE_y = FC = 0.5$$

$$CE = \left(\tfrac{15}{12}\right)(0.5) = \boxed{0.63 \quad (T)}$$

The answer is (B).

42 Properties of Areas

PRACTICE PROBLEMS

1. Locate the x-coordinate of the centroid of the area.

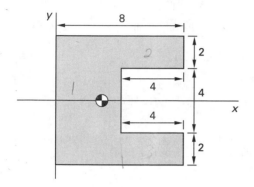

(A) 2.7 units

(B) 2.9 units

(C) 3.1 units

(D) 3.3 units

2. Replace the distributed load with three concentrated loads, and indicate the points of application.

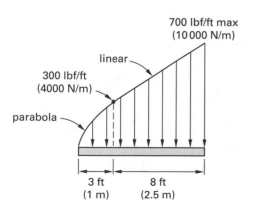

3. Find the centroidal moment of inertia about an axis parallel to the x-axis.

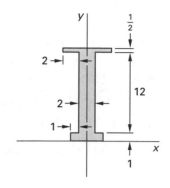

(A) 160 units^4

(B) 290 units^4

(C) 570 units^4

(D) 740 units^4

SOLUTIONS

1. The area is divided into three basic shapes.

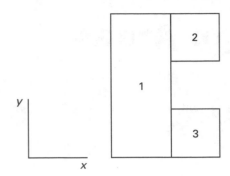

First, calculate the areas of the basic shapes.

$$A_1 = (4)(8) = 32 \text{ units}^2$$
$$A_2 = (4)(2) = 8 \text{ units}^2$$
$$A_3 = (4)(2) = 8 \text{ units}^2$$

Next, find the x-components of the centroids of the basic shapes.

$$x_{c,1} = 2 \text{ units}$$
$$x_{c,2} = 6 \text{ units}$$
$$x_{c,3} = 6 \text{ units}$$

Finally, use Eq. 42.5.

$$x_c = \frac{\sum A_i x_{c,i}}{\sum A_i}$$

$$= \frac{(32 \text{ units}^2)(2 \text{ units}) + (8 \text{ units}^2)(6 \text{ units})}{32 \text{ units}^2 + 8 \text{ units}^2 + 8 \text{ units}^2}$$

$$= \boxed{3.33 \text{ units} \quad (3.3 \text{ units})}$$

The answer is (D).

2. *Customary U.S. Solution*

The parabolic shape is

$$f(x) = \left(300 \ \frac{\text{lbf}}{\text{ft}}\right)\sqrt{\frac{x}{3}}$$

First, use Eq. 42.3 to find the concentrated load given by the area.

$$A = \int f(x)\,dx = \int_{0 \text{ ft}}^{3 \text{ ft}} 300\sqrt{\frac{x}{3}}\,dx$$

$$= \left(\frac{300 \ \frac{\text{lbf}}{\text{ft}}}{\sqrt{3}}\right)\left(\frac{x^{3/2}}{\frac{3}{2}}\bigg|_{0 \text{ ft}}^{3 \text{ ft}}\right)$$

$$= \left(\frac{300 \ \frac{\text{lbf}}{\text{ft}}}{\frac{3}{2}\sqrt{3}}\right)(3^{3/2} - 0^{3/2})$$

$$= \boxed{600 \text{ lbf}} \quad \text{[first concentrated load]}$$

Next, from Eq. 42.4,

$$dA = f(x)\,dx = 300\sqrt{\frac{x}{3}}$$

Finally, use Eq. 42.1 to find the location, x_c, of the concentrated load from the left end.

$$x_c = \frac{\int x\,dA}{A} = \frac{1}{600 \text{ lbf}}\int_{0 \text{ ft}}^{3 \text{ ft}} 300x\sqrt{\frac{x}{3}}\,dx$$

$$= \frac{300}{600\sqrt{3}}\int_{0 \text{ ft}}^{3 \text{ ft}} x^{3/2}\,dx = \frac{300x^{5/2}}{600\sqrt{3}\left(\frac{5}{2}\right)}\bigg|_{0 \text{ ft}}^{3 \text{ ft}}$$

$$= \left(\frac{300 \ \frac{\text{lbf}}{\text{ft}}}{(600 \text{ lbf})\sqrt{3}\left(\frac{5}{2}\right)}\right)(3^{5/2} - 0^{5/2})$$

$$= \boxed{1.8 \text{ ft}} \quad \text{[location]}$$

Alternative solution for the parabola:

Use App. 42.A.

$$A = \frac{2bh}{3} = \frac{(2)\left(300 \ \frac{\text{lbf}}{\text{ft}}\right)(3 \text{ ft})}{3} = \boxed{600 \text{ lbf}}$$

The centroid is located at a distance from the left end of

$$\frac{3h}{5} = \frac{(3)(3 \text{ ft})}{5} = \boxed{1.8 \text{ ft}}$$

The concentrated load for the triangular shape is the area from App. 42.A.

$$A = \frac{bh}{2} = \frac{\left(700 \ \frac{\text{lbf}}{\text{ft}} - 300 \ \frac{\text{lbf}}{\text{ft}}\right)(8 \text{ ft})}{2}$$

$$= \boxed{1600 \text{ lbf}} \quad \text{[second concentrated load]}$$

From App. 42.A, the location of the concentrated load from the right end is

$$\frac{h}{3} = \frac{8 \text{ ft}}{3} = \boxed{2.67 \text{ ft}}$$

The concentrated load for the rectangular shape is the area from App. 42.A.

$$A = bh = \left(300 \ \frac{\text{lbf}}{\text{ft}}\right)(8 \text{ ft})$$

$$= \boxed{2400 \text{ lbf}} \quad \text{[third concentrated load]}$$

From App. 42.A, the location of the concentrated load from the right end is

$$\frac{h}{2} = \frac{8 \text{ ft}}{2} = \boxed{4 \text{ ft}}$$

SI Solution

The parabolic shape is

$$f(x) = \left(4000 \ \frac{\text{N}}{\text{m}}\right)\sqrt{x}$$

First, use Eq. 42.3 to find the concentrated load given by the area.

$$A = \int f(x)\,dx = \int_{0 \text{ m}}^{1 \text{ m}} 4000\sqrt{x}\,dx = \left.\frac{4000 x^{3/2}}{\frac{3}{2}}\right|_{0 \text{ m}}^{1 \text{ m}}$$

$$= \left(\frac{4000 \ \frac{\text{N}}{\text{m}}}{\frac{3}{2}}\right)(1^{3/2} - 0^{3/2})$$

$$= \boxed{2666.7 \text{ N}} \quad \text{[first concentrated load]}$$

Next, from Eq. 42.4,

$$dA = f(x)\,dx = 4000\sqrt{x}\ dx$$

Finally, use Eq. 42.1 to find the location, x_c, of the concentrated load from the left end.

$$x_c = \frac{\int x\,dA}{A} = \frac{1}{2666.7 \text{ N}} \int_{0 \text{ m}}^{1 \text{ m}} 4000 x \sqrt{x}\,dx$$

$$= \frac{4000}{2666.7} \int_{0 \text{ m}}^{1 \text{ m}} x^{3/2}\,dx = \left.\frac{4000 x^{5/2}}{(2666.7)\left(\frac{5}{2}\right)}\right|_{0 \text{ m}}^{1 \text{ m}}$$

$$= \left(\frac{4000 \ \frac{\text{N}}{\text{m}}}{(2666.7 \text{ N})\left(\frac{5}{2}\right)}\right)(1^{5/2} - 0^{5/2})$$

$$= \boxed{0.60 \text{ m}} \quad \text{[location]}$$

Alternative solution for the parabola:

Use App. 42.A.

$$A = \frac{2bh}{3} = \frac{(2)\left(4000 \ \frac{\text{N}}{\text{m}}\right)(1 \text{ m})}{3}$$

$$= \boxed{2666.7 \text{ N}}$$

The centroid is located at a distance from the left end of

$$\frac{3h}{5} = \frac{(3)(1 \text{ m})}{5} = \boxed{0.6 \text{ m}}$$

The concentrated load for the triangular shape is the area from App. 42.A.

$$A = \frac{bh}{2} = \frac{\left(10\,000 \ \frac{\text{N}}{\text{m}} - 4000 \ \frac{\text{N}}{\text{m}}\right)(2.5 \text{ m})}{2}$$

$$= \boxed{7500 \text{ N}} \quad \text{[second concentrated load]}$$

From App. 42.A, the location of the concentrated load from the right end is

$$\frac{h}{3} = \frac{2.5 \text{ m}}{3} = \boxed{0.83 \text{ m}}$$

The concentrated load for the rectangular shape is the area from App. 42.A.

$$A = bh = \left(4000 \ \frac{\text{N}}{\text{m}}\right)(2.5 \text{ m})$$

$$= \boxed{10\,000 \text{ N}} \quad \text{[third concentrated load]}$$

From App. 42.A, the location of the concentrated load from the right end is

$$\frac{h}{2} = \frac{2.5 \text{ m}}{2} = \boxed{1.25 \text{ m}}$$

3. The area is divided into three basic shapes.

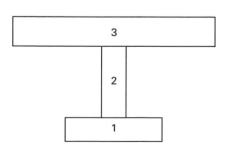

First, calculate the areas of the basic shapes.

$$A_1 = (4)(1) = 4 \text{ units}^2$$
$$A_2 = (2)(12) = 24 \text{ units}^2$$
$$A_3 = (6)(0.5) = 3 \text{ units}^2$$

Next, find the y-components of the centroids of the basic shapes.

$$y_{c,1} = 0.5 \text{ units}$$
$$y_{c,2} = 7 \text{ units}$$
$$y_{c,3} = 13.25 \text{ units}$$

From Eq. 42.6, the centroid of the area is

$$y_c = \frac{\sum A_i y_{c,i}}{\sum A_i} = \frac{(4)(0.5) + (24)(7) + (3)(13.25)}{4 + 24 + 3}$$
$$= 6.77 \text{ units}$$

From App. 42.A, the moment of inertia of basic shape 1 about its own centroid is

$$I_{cx,1} = \frac{bh^3}{12} = \frac{(4)(1)^3}{12} = 0.33 \text{ units}^4$$

The moment of inertia of basic shape 2 about its own centroid is

$$I_{cx,2} = \frac{bh^3}{12} = \frac{(2)(12)^3}{12} = 288 \text{ units}^4$$

The moment of inertia of basic shape 3 about its own centroid is

$$I_{cx,3} = \frac{bh^3}{12} = \frac{(6)(0.5)^3}{12} = 0.063 \text{ units}^4$$

From the parallel axis theorem, Eq. 42.20, the moment of inertia of basic shape 1 about the centroidal axis of the section is

$$I_{x,1} = I_{cx,1} + A_1 d_1^2 = 0.33 + (4)(6.77 - 0.5)^2$$
$$= 157.6 \text{ units}^4$$

The moment of inertia of basic shape 2 about the centroidal axis of the section is

$$I_{x,2} = I_{cx,2} + A_2 d_2^2 = 288 + (24)(7.0 - 6.77)^2$$
$$= 289.3 \text{ units}^4$$

The moment of inertia of basic shape 3 about the centroidal axis of the section is

$$I_{x,3} = I_{cx,3} + A_3 d_3^2 = 0.063 + (3)(13.25 - 6.77)^2$$
$$= 126.0 \text{ units}^4$$

The total moment of inertia about the centroidal axis of the section is

$$I_x = I_{x,1} + I_{x,2} + I_{x,3}$$
$$= 157.6 \text{ units}^4 + 289.3 \text{ units}^4 + 126.0 \text{ units}^4$$
$$= \boxed{572.9 \text{ units}^4 \quad (570 \text{ units}^4)}$$

The answer is (C).

Structural

43 Material Testing

1. The engineering stress and engineering strain for a copper specimen are 20,000 lbf/in² (140 MPa) and 0.0200 in/in (0.0200 mm/mm), respectively. Poisson's ratio for the specimen is 0.3.

(a) What is the true stress?

   (A) 14,000 lbf/in² (98 MPa)

   (B) 18,000 lbf/in² (130 MPa)

   (C) 20,000 lbf/in² (140 MPa)

   (D) 22,000 lbf/in² (160 MPa)

(b) What is the true strain?

   (A) 0.0182 in/in (0.0182 mm/mm)

   (B) 0.0189 in/in (0.0189 mm/mm)

   (C) 0.0194 in/in (0.0194 mm/mm)

   (D) 0.0198 in/in (0.0198 mm/mm)

2. A graph of engineering stress-strain is shown. Poisson's ratio for the material is 0.3.

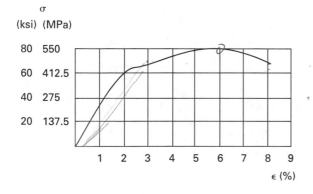

(a) Find the 0.5% yield strength.

   (A) 70,000 lbf/in² (480 MPa)

   (B) 76,000 lbf/in² (530 MPa)

   (C) 84,000 lbf/in² (590 MPa)

   (D) 98,000 lbf/in² (690 MPa)

(b) Find the elastic modulus.

   (A) 2.4 × 10⁶ lbf/in² (17 GPa)

   (B) 2.7 × 10⁶ lbf/in² (19 GPa)

   (C) 2.9 × 10⁶ lbf/in² (20 GPa)

  ➔(D) 3.0 × 10⁶ lbf/in² (21 GPa)

(c) Find the ultimate strength.

   (A) 72,000 lbf/in² (500 MPa)

   (B) 76,000 lbf/in² (530 MPa)

  ➤(C) 80,000 lbf/in² (550 MPa)

   (D) 84,000 lbf/in² (590 MPa)

(d) Find the fracture strength.

   (A) 62,000 lbf/in² (430 MPa)

  ➔(B) 70,000 lbf/in² (480 MPa)

   (C) 76,000 lbf/in² (530 MPa)

   (D) 84,000 lbf/in² (590 MPa)

(e) Find the percentage of elongation at fracture.

   (A) 6%

  ➔(B) 8%

   (C) 10%

   (D) 12%

(f) Find the shear modulus.

   (A) 0.9 × 10⁶ lbf/in² (6.3 GPa)

   (B) 1.1 × 10⁶ lbf/in² (7.7 GPa)

  ➔(C) 1.2 × 10⁶ lbf/in² (7.9 GPa)

   (D) 1.5 × 10⁶ lbf/in² (11 GPa)

(g) Find the toughness.

   (A) 5100 in-lbf/in³ (35 MJ/m³)

  ➔(B) 5700 in-lbf/in³ (38 MJ/m³)

   (C) 6300 in-lbf/in³ (42 MJ/m³)

   (D) 8900 in-lbf/in³ (60 MJ/m³)

Structural

3. A specimen with an unstressed cross-sectional area of 4 in^2 (25 cm^2) necks down to 3.42 in^2 (22 cm^2) before breaking in a standard tensile test. What is the percentage reduction in area of the material?

(A) 9.4% (8.9%)

(B) 10% (9.4%)

(C) 13% (10%)

(D) 15% (12%)

4. A constant 15,000 lbf/in^2 (100 MPa) tensile stress is applied to a specimen. The stress is known to be less than the material's yield strength. The strain is measured at various times. What is the steady-state creep rate for the material?

time (hr)	strain (in/in)
5	0.018
10	0.022
20	0.026
30	0.031
40	0.035
50	0.040
60	0.046
70	0.058

(A) 0.00037 hr^{-1}

(B) 0.00041 hr^{-1}

(C) 0.00046 hr^{-1}

(D) 0.00049 hr^{-1}

SOLUTIONS

1. *Customary U.S. Solution*

(a) The fractional reduction in diameter is

$$\nu e = (0.3)\left(0.020 \ \frac{\text{in}}{\text{in}}\right)$$
$$= 0.006$$

The true stress is given by Eq. 43.6.

$$\sigma = \frac{s}{(1 - \nu e)^2} = \frac{20,000 \ \dfrac{\text{lbf}}{\text{in}^2}}{(1 - 0.006)^2}$$
$$= \boxed{20,242 \ \text{lbf/in}^2 \quad (20,000 \ \text{lbf/ft}^2)}$$

The answer is (C).

(b) The true strain is given by Eq. 43.7.

$$\epsilon = \ln(1 + e) = \ln(1 + 0.020)$$
$$= \boxed{0.0198 \ \text{in/in}}$$

The answer is (D).

SI Solution

(a) The fractional reduction in diameter is

$$\nu e = (0.3)\left(0.020 \ \frac{\text{mm}}{\text{mm}}\right)$$
$$= 0.006$$

The true stress is given by Eq. 43.6.

$$\sigma = \frac{s}{(1 - \nu e)^2} = \frac{140 \ \text{MPa}}{(1 - 0.006)^2}$$
$$= \boxed{141.7 \ \text{MPa} \quad (140 \ \text{MPa})}$$

The answer is (C).

(b) The true strain is given by Eq. 43.7.

$$\epsilon = \ln(1 + e) = \ln(1 + 0.020)$$
$$= \boxed{0.0198 \ \text{mm/mm}}$$

The answer is (D).

Structural

2. *Customary U.S. Solution*

(a) Extend a line from the 0.5% offset strain value parallel to the linear portion of the curve.

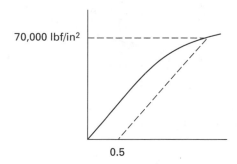

The 0.5% yield strength is $\boxed{70,000 \text{ lbf/in}^2.}$

The answer is (A).

(b) At the elastic limit, the stress is 60,000 lbf/in², and the percent strain is 2. The elastic modulus is

$$E = \frac{\text{stress}}{\text{strain}} = \frac{60,000 \ \frac{\text{lbf}}{\text{in}^2}}{0.02}$$
$$= \boxed{3.0 \times 10^6 \text{ lbf/in}^2}$$

The answer is (D).

(c) The ultimate strength is the highest point of the curve. This value is $\boxed{80,000 \text{ lbf/in}^2.}$

The answer is (C).

(d) The fracture strength is at the end of the curve. This value is $\boxed{70,000 \text{ lbf/in}^2.}$

The answer is (B).

(e) The percent elongation at fracture is determined by extending a straight line parallel to the initial strain line from the fracture point. This gives an approximate value of $\boxed{6\%.}$

The answer is (A).

(f) The shear modulus is given by Eq. 43.22.

$$G = \frac{E}{2(1+\nu)} = \frac{3 \times 10^6 \ \frac{\text{lbf}}{\text{in}^2}}{(2)(1+0.3)}$$
$$= \boxed{1.15 \times 10^6 \text{ lbf/in}^2 \quad (1.2 \times 10^6 \text{ lbf/in}^2)}$$

The answer is (C).

(g) The toughness is the area under the stress-strain curve. Divide the area into squares of 20 ksi × 1%. There are about 25.5 squares covered.

$$(25.5)\left(20,000 \ \frac{\text{lbf}}{\text{in}^2}\right)\left(0.01 \ \frac{\text{in}}{\text{in}}\right) = \boxed{5100 \text{ in-lbf/in}^3}$$

(Using Eq. 43.18 would be inappropriate for a material as elastic as this one.)

The answer is (A).

SI Solution

(a) Extend a line from the 0.5% offset strain value parallel to the linear portion of the curve.

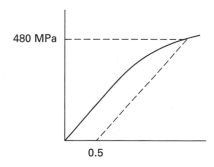

The 0.5% yield strength is $\boxed{480 \text{ MPa.}}$

The answer is (A).

(b) At the elastic limit, the stress is 410 MPa, and the percent strain is 2. The elastic modulus is

$$E = \frac{\text{stress}}{\text{strain}} = \frac{410 \text{ MPa}}{(0.02)\left(1000 \ \frac{\text{MPa}}{\text{GPa}}\right)}$$
$$= \boxed{20.5 \text{ GPa} \quad (21 \text{ GPa})}$$

The answer is (D).

(c) The ultimate strength is the highest point of the curve. This value is $\boxed{550 \text{ MPa.}}$

The answer is (C).

(d) The fracture strength is at the end of the curve. This value is $\boxed{480 \text{ MPa.}}$

The answer is (B).

(e) The percent elongation at fracture is determined by extending a straight line parallel to the initial strain line from the fracture point. This gives an approximate value of $\boxed{6\%.}$

The answer is (A).

(f) The shear modulus is given by Eq. 43.22.

$$G = \frac{E}{2(1+\nu)} = \frac{20.5 \text{ GPa}}{(2)(1+0.3)}$$
$$= \boxed{7.88 \text{ GPa} \quad (7.9 \text{ GPa})}$$

The answer is (C).

(g) The toughness is the area under the stress-strain curve. Divide the area into squares of 137.5 MPa × 1%. There are about 25.5 squares covered.

$$(25.5)(137.5 \text{ MPa})\left(0.01 \frac{\text{m}}{\text{m}}\right) = \boxed{35 \text{ MJ/m}^3}$$

The answer is (A).

3. *Customary U.S. Solution*

Use Eq. 43.13.

$$q_f = \frac{A_o - A_f}{A_o} \times 100\%$$
$$= \frac{4.0 \text{ in}^2 - 3.42 \text{ in}^2}{4.0 \text{ in}^2} \times 100\%$$
$$= \boxed{14.5\% \quad (15\%)}$$

The answer is (D).

SI Solution

Use Eq. 43.13.

$$q_f = \frac{A_o - A_f}{A_o} \times 100\%$$
$$= \frac{25 \text{ cm}^2 - 22 \text{ cm}^2}{25 \text{ cm}^2} \times 100\%$$
$$= \boxed{12\%}$$

The answer is (D).

4. Plot the data and draw a straight line. Disregard the first and last data points, as these represent primary and tertiary creep, respectively.

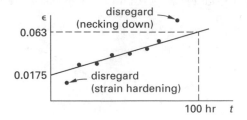

The creep rate is the slope of the line.

$$\text{creep rate} = \frac{\Delta\epsilon}{\Delta t} = \frac{0.063 \frac{\text{in}}{\text{in}} - 0.0175 \frac{\text{in}}{\text{in}}}{100 \text{ hr}}$$
$$= \boxed{0.000455 \text{ 1/hr} \quad (0.00046 \text{ 1/hr})}$$

The answer is (C).

44 Strength of Materials

PRACTICE PROBLEMS

Elastic Deformation

1. A 1 in (25 mm) diameter soft steel rod carries a tensile load of 15,000 lbf (67 kN). The elongation is 0.158 in (4 mm). The modulus of elasticity is 2.9×10^7 lbf/in^2 (200 GPa). What is the total length of the rod?

(A) 239.46 in (5.853 m)

(B) 239.93 in (5.857 m)

(C) 240.03 in (5.861 m)

→(D) 240.07 in (5.865 m)

Thermal Deformation

2. A straight steel beam 200 ft (60 m) long is installed when the temperature is 40°F (4°C). It is supported in such a manner as to allow only 0.5 in (12 mm) longitudinal expansion. Lateral support is provided to prevent buckling. If the temperature increases to 110°F (43°C), what will be the compressive stress in the member?

(A) 3900 lbf/in^2 (28 MPa)

(B) 7200 lbf/in^2 (50 MPa)

(C) 9200 lbf/in^2 (67 MPa)

(D) 12,000 lbf/in^2 (88 MPa)

Shear and Moment Diagrams

3. A beam 25 ft long is simply supported at two points: at its left and 5 ft from its right end. A uniform load of 2 kips/ft extends over a 10 ft length starting from the left end. There is a concentrated 10 kip load at the right end. Draw the shear and moment diagrams.

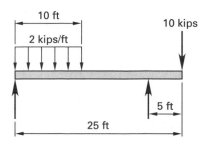

4. A beam 14 ft (3.6 m) long is supported at the left end and 2 ft (0.6 m) from the right end. The beam has a mass of 20 lbm/ft (30 kg/m). A 100 lbf (450 N) load is applied 2 ft (0.6 m) from the left end. An 80 lbf (350 N) load is applied at the right end.

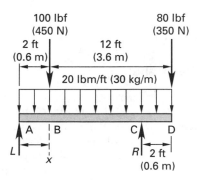

(a) What is the maximum moment?

(A) 150 ft-lbf (200 N·m)

(B) 250 ft-lbf (340 N·m)

(C) 390 ft-lbf (520 N·m)

(D) 830 ft-lbf (1100 N·m)

(b) What is the maximum shear?

(A) 80 lbf (360 N)

(B) 120 lbf (530 N)

(C) 150 lbf (650 N)

(D) 190 lbf (830 N)

Stresses in Beams

5. The allowable stress in a steel beam is 20 ksi, and the maximum moment carried by the beam is 1.5×10^5 ft-lbf. What is the required section modulus?

(A) 90 in^3

(B) 110 in^3

(C) 120 in^3

(D) 350 in^3

6. A simply supported beam 14 ft long carries a uniform load of 200 lbf/ft over its entire length. The beam has a width of 3.625 in and a depth of 7.625 in.

(a) What is the maximum bending stress?

(A) 1700 lbf/in^2

(B) 2500 lbf/in^2

(C) 3700 lbf/in^2

(D) 9200 lbf/in^2

(b) What is the maximum shear stress?

(A) 38 lbf/in^2

(B) 76 lbf/in^2

(C) 150 lbf/in^2

(D) 300 lbf/in^2

Beam Deflections

7. A 40 ft long simply supported steel beam with a moment of inertia of I and modulus of elasticity of E is reinforced along the central 20 ft, leaving the two 10 ft ends unreinforced. The moment of inertia of the reinforced sections is $2I$. A 20,000 lbf load is applied at midspan, 20 ft from each end. Use the conjugate beam method to determine the deflection.

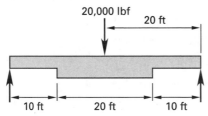

(A) $1.1 \times 10^6/EI$ ft

(B) $2.7 \times 10^6/EI$ ft

(C) $1.5 \times 10^7/EI$ ft

(D) $8.2 \times 10^7/EI$ ft

8. A 17 ft steel beam carries a uniform load of 500 lbf/ft over its entire length and a 2000 lbf concentrated load 5 ft from the right end, as shown. The beam is simply supported at each end. The centroidal moment of inertia of the cross section is 200 in^4. What is the midspan deflection?

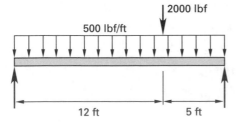

(A) 0.05 in

(B) 0.20 in

(C) 0.40 in

(D) 0.70 in

9. A cantilever beam is 6 ft (1.8 m) in length. The cross section is 6 in wide × 4 in high (150 mm wide × 100 mm high). The modulus of elasticity is 1.5×10^6 psi (10 GPa). The beam is loaded by two concentrated forces: 200 lbf (900 N) located 1 ft (0.3 m) from the free end and 120 lbf (530 N) located 2 ft (0.6 m) from the free end. What is the tip deflection?

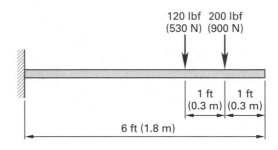

(A) 0.29 in (0.0084 m)

(B) 0.31 in (0.0090 m)

(C) 0.47 in (0.014 m)

(D) 0.55 in (0.015 m)

Truss Deflections

10. A steel truss with pinned joints is constructed as shown. Support R_1 is pinned. Support R_2 is a roller. The length:area ratio for each member is 50 in^{-1}. The modulus of elasticity of the steel is 2.9×10^7 lbf/in^2. Use the virtual work method to determine the vertical deflection at the point where the 10,000 lbf load is applied.

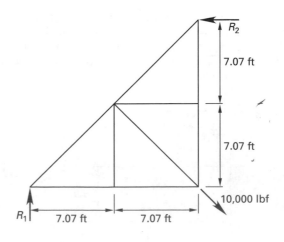

(A) 0.07 in

(B) 0.13 in

(C) 0.24 in

(D) 0.35 in

Composite Structures

11. A bimetallic spring is constructed of an aluminum strip with a $^1/_8$ in $\times$ $1^1/_2$ in (3.2 mm $\times$ 38 mm) cross section bonded on top of a $^1/_{16}$ in $\times$ $1^1/_2$ in (1.6 mm $\times$ 38 mm) hard steel strip. The modulus of elasticity of the aluminum is 10×10^6 lbf/in^2 (70 GPa). The modulus of elasticity of the steel is 30×10^6 lbf/in^2 (200 GPa). What is the equivalent, all-aluminum centroidal area moment of inertia of the cross section?

(A) 2.4×10^{-4} in^4 (190 mm^4)

(B) 5.2×10^{-4} in^4 (390 mm^4)

(C) 1.3×10^{-3} in^4 (550 mm^4)

(D) 2.6×10^{-3} in^4 (1100 mm^4)

12. The reinforced concrete beam illustrated is subjected to a maximum moment of 8125 ft-lbf. The total steel cross sectional area is 1 in^2, assumed to be concentrated 10 in from the top surface. The modulus of elasticity of the concrete is 2×10^6 lbf/in^2, and the modulus of elasticity of the steel is 2.9×10^7 lbf/in^2. The bond between the steel and the concrete is perfect. The beam is balanced such that the concrete and steel fail simultaneously. The load-carrying contribution of concrete in tension is to be disregarded. Analyze the beam using the transformed area method.

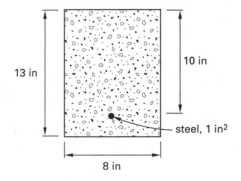

(a) What is the maximum flexural stress in the concrete?

(A) 320 lbf/in^2

(B) 640 lbf/in^2

(C) 910 lbf/in^2

(D) 15,300 lbf/in^2

(b) What is the maximum flexural stress in the steel?

(A) 11,500 lbf/in^2

(B) 16,300 lbf/in^2

(C) 29,500 lbf/in^2

(D) 36,000 lbf/in^2

SOLUTIONS

1. *Customary U.S. Solution*

The cross-sectional area of the rod is

$$A = \frac{\pi}{4} d^2$$

From Eq. 44.4, the unstretched length of the rod is

$$L_o = \frac{\delta E A}{F} = \frac{(0.158 \text{ in})\left(2.9 \times 10^7 \frac{\text{lbf}}{\text{in}^2}\right)\left(\frac{\pi}{4}\right)(1 \text{ in})^2}{15,000 \text{ lbf}}$$
$$= 239.913 \text{ in}$$

The total length of the rod is given by Eq. 44.5

$$L = L_o + \delta = 239.913 \text{ in} + 0.158 \text{ in} = \boxed{240.07 \text{ in}}$$

The answer is (D).

SI Solution

The cross-sectional area of the rod is

$$A = \frac{\pi}{4} d^2$$

From Eq. 44.4, the unstretched length of the rod is

$$L_o = \frac{\delta E A}{F} = \frac{(0.004 \text{ m})\left(20 \times 10^{10} \frac{\text{N}}{\text{m}^2}\right)\left(\frac{\pi}{4}\right)(0.025 \text{ m})^2}{67 \times 10^3 \text{ N}}$$
$$= 5.861 \text{ m}$$

The total length of the rod is given by Eq. 44.5

$$L = L_o + \delta = 5.861 \text{ m} + 0.004 \text{ m} = \boxed{5.865 \text{ m}}$$

The answer is (D).

2. *Customary U.S. Solution*

Use Eq. 44.9 to find the amount of expansion for an unconstrained beam. Use Table 44.2.

$$\Delta L = \alpha L_o(T_2 - T_1)$$
$$= \left(6.5 \times 10^{-6} \frac{1}{°\text{F}}\right)(200 \text{ ft})\left(12 \frac{\text{in}}{\text{ft}}\right)$$
$$\times (110°\text{F} - 40°\text{F})$$
$$= 1.092 \text{ in}$$

The constrained length is

$$\Delta L_c = 1.092 \text{ in} - 0.5 \text{ in} = 0.592 \text{ in}$$

From Eq. 44.14, the thermal strain is

$$\epsilon_{\text{th}} = \frac{\Delta L_c}{L_o + 0.5 \text{ in}} = \frac{0.592 \text{ in}}{(200 \text{ ft})\left(12 \frac{\text{in}}{\text{ft}}\right) + 0.5 \text{ in}}$$
$$= 2.466 \times 10^{-4} \text{ in/in}$$

From Eq. 44.15, the compressive thermal stress is

$$\sigma_{\text{th}} = E\epsilon_{\text{th}} = \left(2.9 \times 10^7 \frac{\text{lbf}}{\text{in}^2}\right)\left(2.466 \times 10^{-4} \frac{\text{in}}{\text{in}}\right)$$
$$= \boxed{7151 \text{ lbf/in}^2 \quad (7400 \text{ lbf/in}^2)}$$

The answer is (B).

SI Solution

Use Eq. 44.9 to find the amount of expansion for an unconstrained beam. Use Table 44.2.

$$\Delta L = \alpha L_o(T_2 - T_1)$$
$$= \left(11.7 \times 10^{-6} \frac{1}{°\text{C}}\right)(60 \text{ m})(43°\text{C} - 4°\text{C})$$
$$= 0.02738 \text{ m}$$

The constrained length is

$$\Delta L_c = 0.02738 \text{ m} - 0.012 \text{ m} = 0.01538 \text{ m}$$

From Eq. 44.14, the thermal strain is

$$\epsilon_{\text{th}} = \frac{\Delta L_c}{L_o + 0.012 \text{ m}} = \frac{0.01538 \text{ m}}{60 \text{ m} + 0.012 \text{ m}}$$
$$= 0.000256 \text{ m/m}$$

From Eq. 44.15, the compressive thermal stress is

$$\sigma_{\text{th}} = E\epsilon_{\text{th}} = (20 \times 10^{10} \text{ Pa})\left(0.000256 \frac{\text{m}}{\text{m}}\right)$$
$$= \boxed{5.12 \times 10^7 \text{ Pa} \quad (51 \text{ MPa})}$$

The answer is (B).

3. The shear diagram is

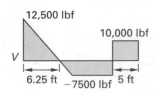

The moment diagram is

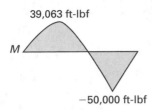

39,063 ft-lbf

M

−50,000 ft-lbf

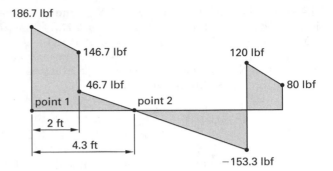

186.7 lbf

146.7 lbf

120 lbf

46.7 lbf

80 lbf

point 1   point 2

2 ft

4.3 ft

−153.3 lbf

4. *Customary U.S. Solution*

(a) First, determine the reactions. The uniform load can be assumed to be concentrated at the center of the beam.

Sum the moments about point A.

$(100 \text{ lbf})(2 \text{ ft}) + (80 \text{ lbf})(14 \text{ ft})$

$$+ \left(20 \; \frac{\text{lbm}}{\text{ft}}\right) \left(\frac{32.2 \; \frac{\text{ft}}{\text{sec}^2}}{32.2 \; \frac{\text{lbm-ft}}{\text{lbf-sec}^2}}\right)(14 \text{ ft})(7 \text{ ft})$$

$$- R(12 \text{ ft}) = 0$$

$$R = 273.3 \text{ lbf}$$

Sum the forces in the vertical direction.

$L + 273.3 \text{ lbf} = 100 \text{ lbf} + 80 \text{ lbf}$

$$+ \left(20 \; \frac{\text{lbm}}{\text{ft}}\right) \left(\frac{32.2 \; \frac{\text{ft}}{\text{sec}^2}}{32.2 \; \frac{\text{lbm-ft}}{\text{lbf-sec}^2}}\right)(14 \text{ ft})$$

$$L = 186.7 \text{ lbf}$$

The shear diagram starts at +186.7 lbf at the left reaction and decreases linearly at a rate of 20 lbf/ft to 146.7 lbf at point B. The concentrated load reduces the shear to 46.7 lbf. The shear then decreases linearly at a rate of 20 lbf/ft to point C. Measuring x from the left, the shear line goes through zero at

$$x = 2 \text{ ft} + \frac{46.7 \text{ lbf}}{20 \; \frac{\text{lbf}}{\text{ft}}} = 4.3 \text{ ft}$$

The shear at the right of the beam at point D is 80 lbf and increases linearly at a rate of 20 lbf/ft to 120 lbf at point C. The reaction, R, at point C decreases the shear to −153.3 lbf. This is sufficient to draw the shear diagram.

From the shear diagram, the maximum moment occurs when the shear is zero. Call this point 2. The moment at the left reaction is zero. Call this point 1. Use Eq. 44.27.

$$M_2 = M_1 + \int_{x_1}^{x_2} V \, dx$$

The integral is the area under the curve from $x_1 = 0$ to $x_2 = 4.3$ ft.

$$M_2 = 0 + (146.7 \text{ lbf})(2 \text{ ft})$$

$$+ \left(\tfrac{1}{2}\right)(186.7 \text{ lbf} - 146.7 \text{ lbf})(2 \text{ ft})$$

$$+ \left(\tfrac{1}{2}\right)(46.7 \text{ lbf})(4.3 \text{ ft} - 2 \text{ ft})$$

$$= \boxed{387.1 \text{ ft-lbf} \quad (390 \text{ ft-lbf})}$$

The answer is (C).

(b) From the shear diagram, the maximum shear is $\boxed{186.7 \text{ lbf } (190 \text{ lbf}).}$

The answer is (D).

SI Solution

(a) First, determine the reactions. The uniform load can be assumed to be concentrated at the center of the beam.

Sum the moments about point A.

$(450 \text{ N})(0.6 \text{ m}) + (350 \text{ N})(4.2 \text{ m})$

$$+ \left(30 \; \frac{\text{kg}}{\text{m}}\right)(4.2 \text{ m})\left(9.81 \; \frac{\text{m}}{\text{s}^2}\right)(2.1 \text{ m})$$

$$- R(3.6 \text{ m}) = 0$$

$$R = 1204.4 \text{ N}$$

Sum the forces in the vertical direction.

$L + 1204.4 \text{ N} = 450 \text{ N} + 350 \text{ N}$

$$+ \left(30 \; \frac{\text{kg}}{\text{m}}\right)(4.2 \text{ m})\left(9.81 \; \frac{\text{m}}{\text{s}^2}\right)$$

$$L = 831.7 \text{ N}$$

Structural

The shear diagram starts at +831.7 N at the left reaction and decreases linearly at a rate of 294.3 N/m to 655.1 N at point B. The concentrated load reduces the shear to 205.1 N. The shear then decreases linearly at a rate of 294.3 N/m to point C. Measuring x from the left, the shear line goes through zero at

$$x = 0.6 \text{ m} + \frac{205.1 \text{ N}}{294.3 \frac{\text{N}}{\text{m}}} = 1.3 \text{ m}$$

The shear at the right of the beam at point D is 350 N and increases linearly at a rate of 294.3 N/m to 526.3 N at point C. The reaction, R, at point C decreases the shear to -677.8 N. This is sufficient to draw the shear diagram.

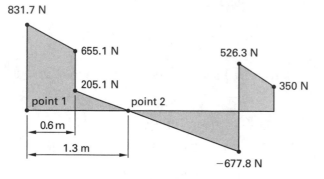

From the shear diagram, the maximum moment occurs when the shear = 0. Call this point 2. The moment at the left reaction = 0. Call this point 1. Use Eq. 44.27.

$$M_2 = M_1 + \int_{x_1}^{x_2} V \, dx$$

The integral is the area under the curve from $x_1 = 0$ to $x_2 = 1.3$ m.

$$\begin{aligned} M_2 = {}& 0 + (655.1 \text{ N})(0.6 \text{ m}) \\ &+ (\tfrac{1}{2})(831.7 \text{ N} - 655.1 \text{ N})(0.6 \text{ m}) \\ &+ (\tfrac{1}{2})(205.1 \text{ N})(1.3 \text{ m} - 0.6 \text{ m}) \\ = {}& \boxed{517.8 \text{ N·m} \quad (520 \text{ N·m})} \end{aligned}$$

The answer is (C).

(b) From the shear diagram, the maximum shear is $\boxed{831.7 \text{ N} (830 \text{ N}).}$

The answer is (D).

5. Use Eq. 44.38.

$$S = \frac{M}{\sigma} = \frac{(1.5 \times 10^5 \text{ ft-lbf})\left(12 \frac{\text{in}}{\text{ft}}\right)}{20,000 \frac{\text{lbf}}{\text{in}^2}}$$

$$= \boxed{90 \text{ in}^3}$$

The answer is (A).

6. (a) The maximum bending stress is

$$I = \frac{bh^3}{12} = \frac{(3.625 \text{ in})(7.625 \text{ in})^3}{12} = 133.92 \text{ in}^4$$

$$\begin{aligned} M_{\max} &= (7 \text{ ft})\left(\frac{7 \text{ ft}}{2}\right)\left(200 \frac{\text{lbf}}{\text{ft}}\right)\left(12 \frac{\text{in}}{\text{ft}}\right) \\ &= 58,800 \text{ in-lbf} \end{aligned}$$

$$\begin{aligned} \sigma_{b,\max} &= \frac{Mc}{I} = \frac{(58,800 \text{ in-lbf})\left(\dfrac{7.625 \text{ in}}{2}\right)}{133.92 \text{ in}^4} \\ &= \boxed{1674 \text{ lbf/in}^2 \quad (1700 \text{ lbf/in}^2)} \end{aligned}$$

The answer is (A).

(b) The maximum shear stress is

$$V_{\max} = \frac{wL}{2} = \frac{(14 \text{ ft})\left(200 \dfrac{\text{lbf}}{\text{ft}}\right)}{2} = 1400 \text{ lbf}$$

$$\tau_{\max} = \frac{3V}{2A} = \frac{(3)(1400 \text{ lbf})}{(2)(3.625 \text{ in})(7.625 \text{ in})}$$

$$= \boxed{76 \text{ lbf/in}^2}$$

The answer is (B).

7. Assume simple supports. Use the conjugate beam method.

step 1:

$$\begin{aligned} M_{\max} &= (20 \text{ ft})\left(\frac{20,000 \text{ lbf}}{2}\right) \\ &= 200,000 \text{ ft-lbf} \end{aligned}$$

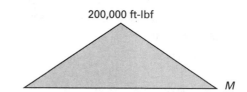

step 2:

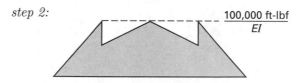

steps 3 and 4:

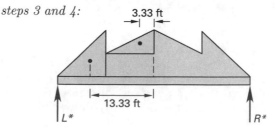

The total load on the conjugate beam is

$$(2)\left(\left(\tfrac{1}{2}\right)(10)\left(\frac{100{,}000}{EI}\right) + \left(\tfrac{1}{2}\right)(10)\left(\frac{50{,}000}{EI}\right)\right.$$
$$\left. + (10)\left(\frac{50{,}000}{EI}\right)\right)$$

$$= \frac{2.5 \times 10^6 \text{ ft}^2\text{-lbf}}{EI}$$

$$L^* = R^* = \frac{2.5 \times 10^6 \text{ ft}^2\text{-lbf}}{2EI} = \frac{1.25 \times 10^6 \text{ ft}^2\text{-lbf}}{EI}$$

step 5: Taking clockwise moments as positive, the conjugate moment at the midpoint is

$$M_{\text{mid}} = (20)\left(\frac{1.25 \times 10^6}{EI}\right)$$
$$- \left(\tfrac{1}{2}\right)(10)\left(\frac{100{,}000}{EI}\right)(13.33)$$
$$- \left(\tfrac{1}{2}\right)(10)\left(\frac{50{,}000}{EI}\right)(3.33)$$
$$- (10)\left(\frac{50{,}000}{EI}\right)(5)$$
$$= \frac{1.5 \times 10^7 \text{ ft}^3\text{-lbf}}{EI}$$

If E is in lbf/ft^2 and I is in ft^4, then

$$\delta = \boxed{\frac{1.5 \times 10^7}{EI} \text{ ft}}$$

The answer is (C).

8. Use superposition.

Uniform load:

$$y_{\text{center}} = y_{\text{max}} = \frac{5wL^4}{384EI}$$

$$= \frac{(5)\left(\dfrac{500 \ \frac{\text{lbf}}{\text{ft}}}{12 \ \frac{\text{in}}{\text{ft}}}\right)\left((17 \text{ ft})\left(12 \ \frac{\text{in}}{\text{ft}}\right)\right)^4}{(384)\left(2.9 \times 10^7 \ \frac{\text{lbf}}{\text{in}^2}\right)(200 \text{ in}^4)}$$

$$= 0.1620 \text{ in}$$

Concentrated load:

Use case 7 from App. 44.A.

$$L = (17 \text{ ft})\left(12 \ \frac{\text{in}}{\text{ft}}\right) = 204 \text{ in}$$

$$b = (5 \text{ ft})\left(12 \ \frac{\text{in}}{\text{ft}}\right) = 60 \text{ in}$$

At midspan,

$$x_a = \left(\frac{17 \text{ ft}}{2}\right)\left(12 \ \frac{\text{in}}{\text{ft}}\right) = 102 \text{ in}$$

$$y_{\text{center}} = y_a$$
$$= \frac{Pbx}{6EIL}\left(L^2 - b^2 - x^2\right)$$
$$= \left(\frac{(2000 \text{ lbf})(60 \text{ in})(102 \text{ in})}{(6)\left(2.9 \times 10^7 \ \frac{\text{lbf}}{\text{in}^2}\right)(200 \text{ in}^4)(204 \text{ in})}\right)$$
$$\times \left((204 \text{ in})^2 - (60 \text{ in})^2 - (102 \text{ in})^2\right)$$
$$= 0.0476 \text{ in}$$

$$y_{\text{total}} = 0.1620 \text{ in} + 0.0476 \text{ in}$$
$$= \boxed{0.2096 \text{ in} \quad (0.20 \text{ in})}$$

The answer is (B).

9. *Customary U.S. Solution*

First, the moment of inertia of the beam cross section is

$$I = \frac{bh^3}{12} = \frac{(6 \text{ in})(4 \text{ in})^3}{12} = 32 \text{ in}^4$$

From case 1 in App. 44.A, the deflection of the 200 lbf load is

$$y_1 = \frac{PL^3}{3EI} = \frac{(200 \text{ lbf})(72 \text{ in} - 12 \text{ in})^3}{(3)\left(1.5 \times 10^6 \ \frac{\text{lbf}}{\text{in}^2}\right)(32 \text{ in}^4)}$$
$$= 0.30 \text{ in}$$

The slope at the 200 lbf load is

$$\theta = \frac{PL^2}{2EI} = \frac{(200 \text{ lbf})(72 \text{ in} - 12 \text{ in})^2}{(2)\left(1.5 \times 10^6 \ \frac{\text{lbf}}{\text{in}^2}\right)(32 \text{ in}^4)}$$
$$= 0.0075 \text{ rad}$$

The additional deflection at the tip of the beam is

$$y_{1'} = (0.0075 \text{ rad})(12 \text{ in}) = 0.09 \text{ in}$$

From case 1 in App. 44.A, the deflection at the 120 lbf load is

$$y_2 = \frac{PL^3}{3EI} = \frac{(120 \text{ lbf})(72 \text{ in} - 24 \text{ in})^3}{(3)\left(1.5 \times 10^6 \ \frac{\text{lbf}}{\text{in}^2}\right)(32 \text{ in}^4)}$$
$$= 0.0922 \text{ in}$$

Structural

The slope at the 120 lbf load is

$$\theta_2 = \frac{PL^2}{2EI} = \frac{(120 \text{ lbf})(72 \text{ in} - 24 \text{ in})^2}{(2)\left(1.5 \times 10^6 \, \frac{\text{lbf}}{\text{in}^2}\right)(32 \text{ in}^4)}$$

$$= 0.00288 \text{ rad}$$

The additional deflection at the tip of the beam is

$$y_{2'} = (0.00288 \text{ rad})(24 \text{ in}) = 0.0691 \text{ in}$$

The total deflection is the sum of the preceding four parts.

$$y_{\text{tip}} = y_1 + y_{1'} + y_2 + y_{2'}$$

$$= 0.30 \text{ in} + 0.09 \text{ in} + 0.0922 \text{ in} + 0.0691 \text{ in}$$

$$= \boxed{0.5513 \text{ in} \quad (0.55 \text{ in})}$$

The answer is (D).

SI Solution

First, the moment of inertia of the beam cross section is

$$I = \frac{bh^3}{12} = \frac{(0.15 \text{ m})(0.10 \text{ m})^3}{12} = 1.25 \times 10^{-5} \text{ m}^4$$

From case 1 in App. 44.A, the deflection at the 900 N load is

$$y_1 = \frac{PL^3}{3EI} = \frac{(900 \text{ N})(1.8 \text{ m} - 0.3 \text{ m})^3}{(3)(10 \times 10^9 \text{ Pa})(1.25 \times 10^{-5} \text{ m}^4)}$$

$$= 0.0081 \text{ m}$$

The slope at the 900 N load is

$$\theta_1 = \frac{PL^2}{2EI} = \frac{(900 \text{ N})(1.8 \text{ m} - 0.3 \text{ m})^2}{(2)(10 \times 10^9 \text{ Pa})(1.25 \times 10^{-5} \text{ m}^4)}$$

$$= 0.0081 \text{ rad}$$

The additional deflection at the tip of the beam is

$$y_{1'} = (0.0081 \text{ rad})(0.3 \text{ m}) = 0.00243 \text{ m}$$

From case 1 in App. 44.A, the deflection at the 530 N load is

$$y_2 = \frac{PL^3}{3EI} = \frac{(530 \text{ N})(1.8 \text{ m} - 0.6 \text{ m})^3}{(3)(10 \times 10^9 \text{ Pa})(1.25 \times 10^{-5} \text{ m}^4)}$$

$$= 0.00244 \text{ m}$$

The slope at the 530 N load is

$$\theta_2 = \frac{PL^2}{2EI} = \frac{(530 \text{ N})(1.8 \text{ m} - 0.6 \text{ m})^2}{(2)(10 \times 10^9 \text{ Pa})(1.25 \times 10^{-5} \text{ m}^4)}$$

$$= 0.00305 \text{ rad}$$

The additional deflection at the tip of the beam is

$$y_{2'} = (0.00305 \text{ rad})(0.6 \text{ m}) = 0.00183 \text{ m}$$

The total deflection is the sum of the preceding four parts.

$$y_{\text{tip}} = y_1 + y_{1'} + y_2 + y_{2'}$$

$$= 0.0081 \text{ m} + 0.00243 \text{ m} + 0.00244 \text{ m} + 0.00183 \text{ m}$$

$$= \boxed{0.0148 \text{ m} \quad (0.015 \text{ m})}$$

The answer is (D).

10. Use the virtual work method. The reactions are

$$R_1 = 7071 \text{ lbf}$$

$$R_2 = 7071 \text{ lbf}$$

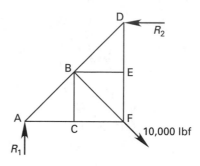

The member forces, S, tabulated below can be found from principles of static equilibrium.

member	force, S (lbf)	u	Su (lbf)
AB	−10,000	−1.414	14,140
AC	7071	0	0
CB	0	0	0
CF	7071	0	0
BF	0	0	0
FE	7071	1	7071
EB	0	0	0
ED	7071	1	7071
BD	−10,000	−1.414	14,140
total			42,422

Apply a vertical load of 1 at point F. Calculate the forces, u, in each member.

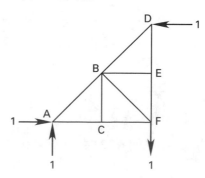

From Eq. 44.53, where $f = 1$,

$$f\delta = \sum \frac{SuL}{AE} = \frac{(42{,}422 \text{ lbf})\left(50 \, \frac{1}{\text{in}}\right)}{2.9 \times 10^7 \, \frac{\text{lbf}}{\text{in}^2}}$$

$$= \boxed{0.0731 \text{ in} \quad (0.07 \text{ in})}$$

($\delta = 0.0693$ in if the strain energy method is used. However, that method is less accurate.)

The answer is (A).

11. *Customary U.S. Solution*

First, determine an equivalent aluminum area for the steel. The ratio of equivalent aluminum area to steel area is the modular ratio.

$$n = \frac{30 \times 10^6 \, \frac{\text{lbf}}{\text{in}^2}}{10 \times 10^6 \, \frac{\text{lbf}}{\text{in}^2}} = 3$$

The equivalent aluminum width to replace the steel is

$$(1.5 \text{ in})(3) = 4.5 \text{ in}$$

The equivalent all-aluminum cross section is

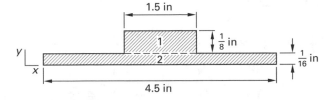

To find the centroid of the section, first calculate the areas of the basic shapes.

$$A_1 = (1.5 \text{ in})\left(\tfrac{1}{8} \text{ in}\right) = 0.1875 \text{ in}^2$$
$$A_2 = (4.5 \text{ in})\left(\tfrac{1}{16} \text{ in}\right) = 0.28125 \text{ in}^2$$

Next, find the y-components of the centroids of the basic shapes.

$$y_{c,1} = \tfrac{1}{16} \text{ in} + \left(\tfrac{1}{2}\right)\left(\tfrac{1}{8} \text{ in}\right) = 0.125 \text{ in}$$

$$y_{c,2} = \frac{\tfrac{1}{16} \text{ in}}{2} = 0.03125 \text{ in}$$

The centroid of the section is

$$y_c = \frac{\displaystyle\sum_i A_i y_{c,i}}{\displaystyle\sum_i A_i}$$

$$= \frac{(0.1875 \text{ in}^2)(0.125 \text{ in}) + (0.28125 \text{ in}^2)(0.03125 \text{ in})}{0.1875 \text{ in}^2 + 0.28125 \text{ in}^2}$$

$$= 0.06875 \text{ in}$$

The moment of inertia of basic shape 1 about its own centroid is

$$I_{cy,1} = \frac{bh^3}{12} = \frac{(1.5 \text{ in})(0.125 \text{ in})^3}{12} = 2.441 \times 10^{-4} \text{ in}^4$$

The moment of inertia of basic shape 2 about its own centroid is

$$I_{cy,2} = \frac{bh^3}{12} = \frac{(4.5 \text{ in})(0.0625 \text{ in})^3}{12} = 9.155 \times 10^{-5} \text{ in}^4$$

The distance from the centroid of basic shape 1 to the section centroid is

$$d_1 = y_{c,1} - y_c = 0.125 \text{ in} - 0.06875 \text{ in} = 0.05625 \text{ in}$$

The distance from the centroid of basic shape 2 to the section centroid is

$$d_2 = y_c - y_{c,2} = 0.06875 \text{ in} - 0.03125 \text{ in} = 0.0375 \text{ in}$$

From the parallel axis theorem, as given in Eq. 42.20, the moment of inertia of basic shape 1 about the centroid of the section is

$$I_{y,1} = I_{yc,1} + A_1 d_1^2$$
$$= 2.441 \times 10^{-4} \text{ in}^4 + (0.1875 \text{ in}^2)(0.05625 \text{ in})^2$$
$$= 8.374 \times 10^{-4} \text{ in}^4$$

The moment of inertia of basic shape 2 about the centroid of the section is

$$I_{y,2} = I_{yc,2} + A_2 d_2^2$$
$$= 9.155 \times 10^{-5} \text{ in}^4 + (0.28125 \text{ in}^2)(0.0375 \text{ in})^2$$
$$= 4.871 \times 10^{-4} \text{ in}^4$$

The total moment of inertia of the section about the centroid of the section is

$$I_y = I_{y,1} + I_{y,2} = 8.374 \times 10^{-4} \text{ in}^4 + 4.871 \times 10^{-4} \text{ in}^4$$

$$= \boxed{1.325 \times 10^{-3} \text{ in}^4 \quad (1.3 \times 10^{-3} \text{ in}^4)}$$

The answer is (C).

SI Solution

First, determine an equivalent aluminum area for the steel. The ratio of equivalent aluminum area to steel area is the modular ratio.

$$n = \frac{20 \times 10^4 \text{ MPa}}{70 \times 10^3 \text{ MPa}} = 2.86$$

Structural

The equivalent aluminum width to replace the steel is

$$(38 \text{ mm})(2.86) = 108.7 \text{ mm}$$

The equivalent all-aluminum cross section is

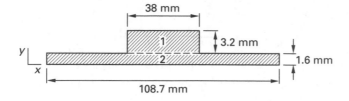

To find the centroid of the section, first calculate the areas of the basic shapes.

$$A_1 = (38 \text{ mm})(3.2 \text{ mm}) = 121.6 \text{ mm}^2$$
$$A_2 = (108.7 \text{ mm})(1.6 \text{ mm}) = 173.9 \text{ mm}^2$$

Next, find the y-components of the centroids of the basic shapes.

$$y_{c,1} = 1.6 \text{ mm} + \left(\tfrac{1}{2}\right)(3.2 \text{ mm}) = 3.2 \text{ mm}$$
$$y_{c,2} = \frac{1.6 \text{ mm}}{2} = 0.8 \text{ mm}$$

The centroid of the section is

$$y_c = \frac{\sum\limits_i A_i y_{c,i}}{\sum\limits_i A_i}$$
$$= \frac{(121.6 \text{ mm}^2)(3.2 \text{ mm}) + (173.9 \text{ mm}^2)(0.8 \text{ mm})}{121.6 \text{ mm}^2 + 173.9 \text{ mm}^2}$$
$$= 1.79 \text{ mm}$$

The moment of inertia of basic shape 1 about its own centroid is

$$I_{cy,1} = \frac{bh^3}{12} = \frac{(38 \text{ mm})(3.2 \text{ mm})^3}{12} = 103.77 \text{ mm}^4$$

The moment of inertia of basic shape 2 about its own centroid is

$$I_{cy,2} = \frac{bh^3}{12} = \frac{(108.7 \text{ mm})(1.6 \text{ mm})^3}{12} = 37.10 \text{ mm}^4$$

The distance from the centroid of basic shape 1 to the section centroid is

$$d_1 = y_{c,1} - y_c = 3.2 \text{ mm} - 1.79 \text{ mm} = 1.41 \text{ mm}$$

The distance from the centroid of basic shape 2 to the section centroid is

$$d_2 = y_c - y_{c,2} = 1.79 \text{ mm} - 0.8 \text{ mm} = 0.99 \text{ mm}$$

From the parallel axis theorem, the moment of inertia of basic shape 1 about the centroid of the section is

$$I_{y,1} = I_{yc,1} + A_1 d_1^2$$
$$= 103.77 \text{ mm}^4 + (121.6 \text{ mm}^2)(1.41 \text{ mm})^2$$
$$= 345.52 \text{ mm}^4$$

The moment of inertia of basic shape 2 about the centroid of the section is

$$I_{y,2} = I_{yc,2} + A_2 d_2^2$$
$$= 37.10 \text{ mm}^4 + (173.9 \text{ mm}^2)(0.99 \text{ mm})^2$$
$$= 207.54 \text{ mm}^4$$

The total moment of inertia of the section about the centroid of the section is

$$I_y = I_{y,1} + I_{y,2} = 345.52 \text{ mm}^4 + 207.54 \text{ mm}^4$$
$$= \boxed{553.06 \text{ mm}^4 \quad (550 \text{ mm}^4)}$$

The answer is (C).

12. (a) Use Eq. 44.55.

$$n = \frac{E_{\text{steel}}}{E_{\text{concrete}}} = \frac{2.9 \times 10^7 \ \frac{\text{lbf}}{\text{in}^2}}{2 \times 10^6 \ \frac{\text{lbf}}{\text{in}^2}} = 14.5$$
$$A_s = 1 \text{ in}^2$$
$$nA_s = (14.5)(1 \text{ in}^2) = 14.5 \text{ in}^2$$

The two relevant sections are

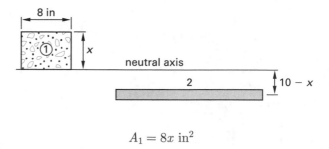

$$A_1 = 8x \text{ in}^2$$
$$A_2 = 14.5 \text{ in}^2$$

Since the neutral axis is the centroid axis of the transformed section, the moment of area above the neutral axis must equal the moment of area below the neutral axis.

$$(8x)\left(\tfrac{1}{2}x\right) = (10 - x)(14.5)$$

$$4x^2 + 14.5x - 145 = 0$$

$$x = 4.4752 \text{ in}$$

$$r = 10 - x = 5.5248 \text{ in}$$

The steel's contribution to I is found from the parallel axis theorem, with the steel treated as a line. Disregarding the concrete below the neutral axis, the centroidal moment of inertia is

$$\begin{aligned}
I_c &= \frac{bh^3}{3} + Ar^2 \\
&= \frac{(8 \text{ in})(4.4752 \text{ in})^3}{3} + (14.5 \text{ in}^2)(5.5248 \text{ in})^2 \\
&= 681.6 \text{ in}^4
\end{aligned}$$

The maximum concrete bending stress is

$$\begin{aligned}
\sigma_c &= \frac{M c_{\text{concrete}}}{I} = \frac{(8125 \text{ ft-lbf})\left(12 \,\frac{\text{in}}{\text{ft}}\right)(4.4752 \text{ in})}{681.6 \text{ in}^4} \\
&= \boxed{640.2 \text{ lbf/in}^2 \quad (640 \text{ lbf/in}^2)}
\end{aligned}$$

The answer is (B).

(b) The maximum steel bending stress is

$$\begin{aligned}
\sigma_s &= \frac{nM c_{\text{steel}}}{I} = \frac{(14.5)(8125 \text{ ft-lbf})\left(12 \,\frac{\text{in}}{\text{ft}}\right)(5.5248 \text{ in})}{681.6 \text{ in}^4} \\
&= \boxed{11{,}459 \text{ lbf/in}^2 \quad (11{,}500 \text{ lbf/in}^2)}
\end{aligned}$$

The answer is (A).

Structural

45 Basic Elements of Design

PRACTICE PROBLEMS

Unless instructed otherwise in a problem, use the following properties:

$$\text{steel: } E = 29 \times 10^6 \text{ lbf/in}^2 \ (20 \times 10^4 \text{ MPa})$$

$$G = 11.5 \times 10^6 \text{ lbf/in}^2 \ (8.0 \times 10^4 \text{ MPa})$$

$$\alpha = 6.5 \times 10^{-6} \ 1/°F \ (1.2 \times 10^{-5} \ 1/°C)$$

$$\nu = 0.3$$

Columns

1. A structural steel member 50 ft (15 m) long is used as a long column to support 75,000 lbf (330 kN). Both ends are built in, and there are no intermediate supports. A factor of safety of 2.5 is used. What is the required moment of inertia?

(A) 67 in⁴ (2.8×10^{-5} m⁴)

(B) 100 in⁴ (4.0×10^{-5} m⁴)

(C) 130 in⁴ (5.6×10^{-5} m⁴)

(D) 190 in⁴ (8.0×10^{-5} m⁴)

2. A long steel column has pinned ends and a yield strength of 36,000 lbf/in² (250 MPa). The column is 25 ft (7.5 m) long and has a cross-sectional area of 25.6 in² (165 cm²), a centroidal moment of inertia of 350 in⁴ (14 600 cm²), and a distance from the neutral axis to the extreme fiber of 7 in (180 mm). It carries an axial concentric load of 100,000 lbf (440 kN) and an eccentric load of 150,000 lbf (660 kN) located 3.33 in (80 mm) from the longitudinal axial axis. Use the secant formula to determine the stress factor of safety.

(A) 1.6

(B) 2.1

(C) 2.4

(D) 3.0

3. A 4 in × 4 in (nominal size) timber post is used to support a sign. The post's modulus of elasticity is 1.5×10^6 lbf/in². One end of the post is embedded in a deep concrete base. The other end supports a sign 9 ft above the ground. Neglect torsion and wind effects. What is the Euler load sign weight that will cause failure by buckling?

(A) 2700 lbf

(B) 3600 lbf

(C) 4000 lbf

(D) 5500 lbf

Bolts

4. A ³/₄-16 UNF steel bolt is used without washers to clamp two rigid steel plates, each 2 in (50 mm) thick. 0.75 in (19 mm) of the threaded section of the bolt remains under the nut. The nut has six threads. Assume half of the bolt in the nut contributes to elongation. The nut is tightened until the stress in the bolt body is 40,000 lbf/in² (280 MPa). The bolt's modulus of elasticity is 2.9×10^7 lbf/in² (200 GPa). How much does the bolt stretch?

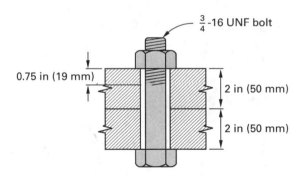

(A) 0.006 in (0.15 mm)

(B) 0.012 in (0.30 mm)

(C) 0.024 in (0.60 mm)

(D) 0.048 in (1.2 mm)

Eccentrically Loaded Bolted Connections

5. The bracket shown is attached to a column with three 0.75 in (19 mm) bolts arranged in an equilateral triangular layout. A force is applied with a moment arm of 20 in (500 mm) measured to the centroid of the bolt group. The maximum shear stress in the bolts is limited to 15,000 lbf/in² (100 MPa). Perform an elastic analysis to determine the maximum force that the connection can support.

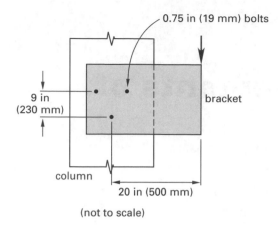

0.75 in (19 mm) bolts

9 in (230 mm)

bracket

column

20 in (500 mm)

(not to scale)

(A) 700 lbf (3.1 kN)

(B) 1200 lbf (5.3 kN)

(C) 2400 lbf (11 kN)

(D) 4700 lbf (20 kN)

Eccentrically Loaded Welded Connections

6. A fillet weld is used to secure a steel bracket to a column. The bracket supports a 10,000 lbf (44 kN) force applied 12 in (300 mm) from the face of the column, as shown. The maximum shear stress in the weld material is 8000 lbf/in^2 (55 MPa). Perform an elastic analysis to determine the size fillet weld required.

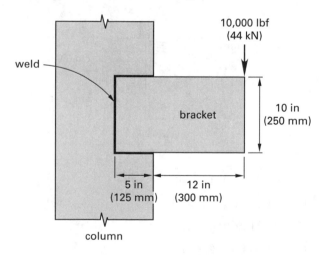

weld

10,000 lbf (44 kN)

bracket

10 in (250 mm)

5 in (125 mm)

12 in (300 mm)

column

(A) $^3/_8$ in (9.5 mm)

(B) $^1/_2$ in (13 mm)

(C) $^5/_8$ in (16 mm)

(D) $^3/_4$ in (19 mm)

Wire Rope

7. (*Time limit: one hour*) Wet concrete (specific weight 150 lbf/ft^3) is mixed at a point that is separated from the pour location by a deep, 200 ft wide ravine. It is decided to use a $^1/_2$ in steel cable and a series of $^1/_2$ in steel cable hangers to support a 6 in schedule-40 steel pipe (specific weight 490 lbf/ft^3). The cables are 6×19 standard hoisting rope with a mass of 0.4 lbm/ft and a breaking strength of 11 tons. The pipe can be assumed to be completely filled with concrete. The cabling geometry is as illustrated. The cable sections act as tension links and have no significant self-weight. Horizontal restraint is infinite, and the impulse effects of periodic pump stroke pulsations are to be neglected. The vertical wire rope hangers remain vertical. The two pipe ends are supported. Use a tensile factor of safety of 5, and consider a 50% loss of strength at cable connections. What is the factor of safety in the cable?

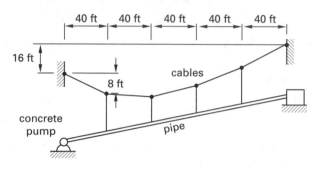

40 ft   40 ft   40 ft   40 ft   40 ft

16 ft

8 ft

cables

concrete pump

pipe

(A) 0.15

(B) 1.5

(C) 1.9

(D) 2.4

SOLUTIONS

1. *Customary U.S. Solution*

The design load for a factor of safety of 2.5 is

$$F = (2.5)(75{,}000 \text{ lbf}) = 187{,}500 \text{ lbf}$$

From Table 45.1, the theoretical end restraint coefficient for built-in ends is $K = 0.5$, and the minimum design value is 0.65.

From Eq. 45.9, the effective length of the column is

$$L' = KL = (0.65)(50 \text{ ft})\left(12 \ \frac{\text{in}}{\text{ft}}\right) = 390 \text{ in}$$

Set the design load equal to the Euler load, F_e, and use Eq. 45.7 to find the required moment of inertia.

$$I = \frac{F_e(L')^2}{\pi^2 E} = \frac{(187{,}500 \text{ lbf})(390 \text{ in})^2}{\pi^2\left(29 \times 10^6 \ \dfrac{\text{lbf}}{\text{in}^2}\right)}$$

$$= \boxed{99.64 \text{ in}^4 \quad (100 \text{ in}^4)}$$

The answer is (B).

SI Solution

The design load for a factor of safety of 2.5 is

$$F = (2.5)(330 \times 10^3 \text{ N}) = 825 \times 10^3 \text{ N}$$

From Table 45.1, the theoretical end restraint coefficient for built-in ends is $K = 0.5$, and the minimum design value is 0.65.

From Eq. 45.9, the effective length of the column is

$$L' = KL = (0.65)(15 \text{ m}) = 9.75 \text{ m}$$

Set the design load equal to the Euler load, F_e, and use Eq. 45.7 to find the required moment of inertia.

$$I = \frac{F_e(L')^2}{\pi^2 E} = \frac{(825 \times 10^3 \text{ N})(9.75 \text{ m})^2}{\pi^2 (20 \times 10^4 \text{ MPa})\left(10^6 \ \dfrac{\text{Pa}}{\text{MPa}}\right)}$$

$$= \boxed{3.973 \times 10^{-5} \text{ m}^4 \quad (4.0 \times 10^{-5} \text{ m}^4)}$$

The answer is (B).

2. *Customary U.S. Solution*

The radius of gyration is

$$r = \sqrt{\frac{I}{A}} = \sqrt{\frac{350 \text{ in}^4}{25.6 \text{ in}^2}} = 3.70 \text{ in}$$

The slenderness ratio is

$$\frac{L}{r} = \frac{(25 \text{ ft})\left(12 \ \dfrac{\text{in}}{\text{ft}}\right)}{3.70 \text{ in}} = 81.08$$

The total buckling load is

$$F = 100{,}000 \text{ lbf} + 150{,}000 \text{ lbf} = 250{,}000 \text{ lbf}$$

From Eq. 45.15,

$$\phi = \tfrac{1}{2}\left(\frac{L}{r}\right)\sqrt{\frac{F}{AE}}$$

$$= \left(\tfrac{1}{2}\right)(81.08)\sqrt{\frac{250{,}000 \text{ lbf}}{(25.6 \text{ in}^2)\left(29 \times 10^6 \ \dfrac{\text{lbf}}{\text{in}^2}\right)}}$$

$$= 0.744 \text{ rad}$$

The eccentricity is

$$e = \frac{M}{F} = \frac{(150{,}000 \text{ lbf})(3.33 \text{ in})}{250{,}000 \text{ lbf}} = 2.0 \text{ in}$$

From Eq. 45.14, the critical column stress is

$$\sigma_{\max} = \left(\frac{F}{A}\right)\left(1 + \left(\frac{ec}{r^2}\right)\sec \phi\right)$$

$$= \left(\frac{250{,}000 \text{ lbf}}{25.6 \text{ in}^2}\right)$$

$$\times \left(1 + \left(\frac{(2.0 \text{ in})(7 \text{ in})}{(3.70 \text{ in})^2}\right)\sec(0.744 \text{ rad})\right)$$

$$= 23{,}339 \text{ lbf/in}^2$$

The stress factor of safety for the column is

$$\text{FS} = \frac{S_y}{\sigma_{\max}} = \frac{36{,}000 \ \dfrac{\text{lbf}}{\text{in}^2}}{23{,}339 \ \dfrac{\text{lbf}}{\text{in}^2}}$$

$$= \boxed{1.54 \quad (1.6)}$$

The answer is (A).

SI Solution

The radius of gyration is

$$r = \sqrt{\frac{I}{A}} = \sqrt{\frac{14\,600 \text{ cm}^4}{165 \text{ cm}^2}} = 9.41 \text{ cm}$$

The slenderness ratio is

$$\frac{L}{r} = \frac{(7.5 \text{ m})\left(100 \ \frac{\text{cm}}{\text{m}}\right)}{9.41 \text{ cm}} = 79.70$$

The total buckling load is

$$F = (440 \text{ kN} + 660 \text{ kN})\left(1000 \ \frac{\text{N}}{\text{kN}}\right)$$
$$= 1.1 \times 10^6 \text{ N}$$

From Eq. 45.15,

$$\phi = \tfrac{1}{2}\left(\frac{L}{r}\right)\sqrt{\frac{F}{AE}}$$

$$= \left(\tfrac{1}{2}\right)(79.70)\sqrt{\frac{1.1 \times 10^6 \text{ N}}{\left(\dfrac{165 \text{ cm}^2}{\left(100 \ \frac{\text{cm}}{\text{m}}\right)^2}\right)(20 \times 10^4 \text{ MPa})} \times \left(10^6 \ \frac{\text{Pa}}{\text{MPa}}\right)}$$

$$= 0.728 \text{ rad}$$

The eccentricity is

$$e = \frac{M}{F} = \frac{(660 \text{ kN})(80 \text{ mm})}{1100 \text{ kN}} = 48 \text{ mm}$$

From Eq. 45.14, the critical column stress is

$$\sigma_{\max} = \left(\frac{F}{A}\right)\left(1 + \left(\frac{ec}{r^2}\right)\sec \phi\right)$$

$$= \left(\frac{1.1 \times 10^6 \text{ N}}{\dfrac{165 \text{ cm}^2}{\left(100 \ \frac{\text{cm}}{\text{m}}\right)^2}}\right)$$

$$\times \left(1 + \left(\frac{(48 \text{ mm})(180 \text{ mm})}{(9.41 \text{ cm})^2\left(10 \ \frac{\text{mm}}{\text{cm}}\right)^2} \times (\sec(0.728 \text{ rad}))\right)\right)$$

$$\times \left(\frac{1 \text{ MPa}}{1 \times 10^6 \text{ Pa}}\right)$$

$$= 153.8 \text{ MPa}$$

The stress factor of safety for the column is

$$\text{FS} = \frac{S_y}{\sigma_{\max}} = \frac{250 \text{ MPa}}{153.8 \text{ MPa}}$$
$$= \boxed{1.63 \quad (1.6)}$$

The answer is (A).

3. A nominal 4 in × 4 in timber post has a 3.5 in × 3.5 in cross section.

$$I = \frac{bh^3}{12} = \frac{(3.5 \text{ in})^4}{12}$$
$$= 12.51 \text{ in}^4 \quad \text{[finished lumber size]}$$

$$r = \sqrt{\frac{I}{A}} = \sqrt{\frac{12.51 \text{ in}^4}{(3.5 \text{ in})^2}} = 1.01 \text{ in}$$

The end-restraint coefficient is $K = 2.1$ (NDS). The slenderness ratio is

$$\frac{L}{r} = \frac{(2.1)(9 \text{ ft})\left(12 \ \frac{\text{in}}{\text{ft}}\right)}{1.01 \text{ in}} = 224.6$$

L/r is well above 100. (Most timber codes limit L/r to 50, so this would not be a permitted application.)

$$F_e = \frac{\pi^2 EI}{L^2} = \frac{\pi^2\left(1.5 \times 10^6 \ \frac{\text{lbf}}{\text{in}^2}\right)(12.51 \text{ in}^4)}{\left((2.1)(9 \text{ ft})\left(12 \ \frac{\text{in}}{\text{ft}}\right)\right)^2}$$

$$= \boxed{3600 \text{ lbf}}$$

The answer is (B).

4. *Customary U.S. Solution*

Since the plates are rigid, they do not deform.

The area of the bolt body is

$$A_1 = \frac{\pi d^2}{4} = \frac{\pi(0.75 \text{ in})^2}{4} = 0.4418 \text{ in}^2$$

The tensile force in the bolt is

$$F = \sigma_2 A_2 = \left(40,000 \ \frac{\text{lbf}}{\text{in}^2}\right)(0.4418 \text{ in}^2)$$
$$= 17,672 \text{ lbf}$$

Structural

From Table 45.5, the stress area (in the threaded section) for a $3/4$-16 UNF bolt is $A_2 = 0.373$ in^2.

The lengths of the unthreaded and threaded sections are

$$L_1 = 2 \text{ in} + 2 \text{ in} - 0.75 \text{ in} = 3.25 \text{ in}$$

$$L_2 = 0.75 \text{ in} + \left(\frac{3 \text{ threads}}{16 \dfrac{\text{threads}}{\text{in}}} \right) = 0.9375 \text{ in}$$

The stiffnesses of the unthreaded and threaded sections are

$$k_1 = \frac{A_1 E}{L_1} = \left(0.4418 \text{ in}^2\right) \left(\frac{2.9 \times 10^7 \dfrac{\text{lbf}}{\text{in}^2}}{3.25 \text{ in}} \right)$$

$$= 3.94 \times 10^6 \text{ lbf/in}$$

$$k_2 = \frac{A_2 E}{L_2} = \left(0.373 \text{ in}^2\right) \left(\frac{2.9 \times 10^7 \dfrac{\text{lbf}}{\text{in}^2}}{0.9375 \text{ in}} \right)$$

$$= 11.5 \times 10^6 \text{ lbf/in}$$

The two parts of the bolt act like two springs in series to clamp the plates together. From Eq. 45.70, the equivalent spring constant is

$$\frac{1}{k_{\text{eq}}} = \frac{1}{k_1} + \frac{1}{k_2}$$

$$k_{\text{eq}} = \frac{k_1 k_2}{k_1 + k_2}$$

$$= \frac{\left(3.94 \times 10^6 \dfrac{\text{lbf}}{\text{in}}\right) \left(11.5 \times 10^6 \dfrac{\text{lbf}}{\text{in}}\right)}{3.94 \times 10^6 \dfrac{\text{lbf}}{\text{in}} + 11.5 \times 10^6 \dfrac{\text{lbf}}{\text{in}}}$$

$$= 2.95 \times 10^6 \text{ lbf/in}$$

The total elongation is

$$\delta = \frac{F}{k_{\text{eq}}} = \frac{17{,}672 \text{ lbf}}{2.95 \times 10^6 \dfrac{\text{lbf}}{\text{in}}}$$

$$= \boxed{0.006 \text{ in}}$$

The answer is (A).

SI Solution

Since the plates are rigid, they do not deform.

The area of the bolt body is

$$A_1 = \frac{\pi d^2}{4} = \frac{\pi (0.75 \text{ in})^2 \left(25.4 \dfrac{\text{mm}}{\text{in}}\right)^2}{4} = 285.0 \text{ mm}^2$$

The tensile force in the bolt is

$$F = \sigma_2 A_2 = \frac{(280 \text{ MPa}) \left(10^6 \dfrac{\text{Pa}}{\text{MPa}}\right) (285.0 \text{ mm}^2)}{\left(1000 \dfrac{\text{mm}}{\text{m}}\right)^2}$$

$$= 79\,800 \text{ N}$$

From Table 45.5, the stress area (in the threaded section) for a $3/4$-16 UNF bolt is $A_2 = 0.373$ in^2.

$$A_2 = \left(0.373 \text{ in}^2\right) \left(25.4 \dfrac{\text{mm}}{\text{in}}\right)^2 = 240.6 \text{ mm}^2$$

The lengths of the unthreaded and threaded sections are

$$L_1 = 50 \text{ mm} + 50 \text{ mm} - 19 \text{ mm} = 81 \text{ mm}$$

$$L_2 = 19 \text{ mm} + \left(\frac{3 \text{ threads}}{16 \dfrac{\text{threads}}{\text{in}}} \right) \left(25.4 \dfrac{\text{mm}}{\text{in}}\right) = 23.76 \text{ mm}$$

The stiffnesses of the unthreaded and threaded sections are

$$k_1 = \frac{A_1 E}{L_1} = \frac{(285.0 \text{ mm}^2)(200 \text{ GPa}) \left(10^9 \dfrac{\text{Pa}}{\text{GPa}}\right)}{(81 \text{ mm}) \left(1000 \dfrac{\text{mm}}{\text{m}}\right)^2}$$

$$= 7.04 \times 10^5 \text{ N/mm}$$

$$k_2 = \frac{A_2 E}{L_2} = \frac{(240.6 \text{ mm}^2)(200 \text{ GPa}) \left(10^9 \dfrac{\text{Pa}}{\text{GPa}}\right)}{(23.76 \text{ mm}) \left(1000 \dfrac{\text{mm}}{\text{m}}\right)^2}$$

$$= 20.25 \times 10^5 \text{ N/mm}$$

Structural

The two parts of the bolt act like two springs in series to clamp the plates together. From Eq. 45.70, the equivalent spring constant is

$$\frac{1}{k_{eq}} = \frac{1}{k_1} + \frac{1}{k_2}$$

$$k_{eq} = \frac{k_1 k_2}{k_1 + k_2}$$

$$= \frac{\left(7.04 \times 10^5 \; \frac{N}{mm}\right)\left(20.25 \times 10^5 \; \frac{N}{mm}\right)}{7.04 \times 10^5 \; \frac{N}{mm} + 20.25 \times 10^5 \; \frac{N}{mm}}$$

$$= 5.22 \times 10^5 \text{ N/mm}$$

The total elongation is

$$\delta = \frac{F}{k_{eq}} = \frac{79\,800 \text{ N}}{5.22 \times 10^5 \; \frac{N}{mm}} = \boxed{0.15 \text{ mm}}$$

The answer is (A).

5. *Customary U.S. Solution*

Find the properties of the bolt area. For an equilateral triangular layout, the distance from the centroid of the bolt group to the center of each bolt is

$$r = \left(\tfrac{2}{3}\right)(9 \text{ in}) = 6 \text{ in}$$

The area of each bolt is

$$A = \frac{\pi d^2}{4} = \frac{\pi (0.75 \text{ in})^2}{4} = 0.442 \text{ in}^2$$

The polar moment of inertia of a bolt about the centroid is

$$J = Ar^2 = (0.442 \text{ in}^2)(6 \text{ in})^2 = 15.91 \text{ in}^4$$

The vertical shear load at each bolt is

$$F_v = \frac{F}{3} = 0.333F \quad \text{[in lbf]}$$

The moment applied to each bolt is

$$M = \frac{(20 \text{ in})F}{3} = 6.67F \quad \text{[in in-lbf]}$$

The vertical shear stress in each bolt is

$$\tau_v = \frac{F_v}{A} = \frac{0.333F}{0.442 \text{ in}^2} = 0.753F \quad \text{[in lbf/in}^2\text{]}$$

The torsional shear stress in each bolt is

$$\tau = \frac{Mr}{J} = \frac{(6.67F)(6 \text{ in})}{15.91 \text{ in}^4} = 2.52F \quad \text{[in lbf/in}^2\text{]}$$

The most highly stressed bolt is the right-most bolt. The shear stress configuration is

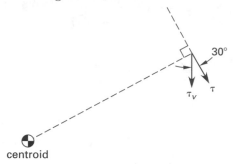

The stresses are combined to find the maximum stress.

$$\tau_{max} = \sqrt{(\tau \sin 30°)^2 + (\tau_v + \tau \cos 30°)^2}$$

$$15{,}000 \; \frac{\text{lbf}}{\text{in}^2} = \sqrt{\begin{array}{l}\left((2.52F)(\sin 30°)\right)^2 \\ + \left(0.753F + (2.52F)(\cos 30°)\right)^2\end{array}}$$

$$= 3.19F$$

$$F = \boxed{4702 \text{ lbf} \quad (4700 \text{ lbf})}$$

The answer is (D).

SI Solution

Find the properties of the bolt area. For an equilateral triangular layout, the distance from the centroid of the bolt group to each bolt is

$$r = \left(\tfrac{2}{3}\right)(230 \text{ mm}) = 153 \text{ mm}$$

The area of each bolt is

$$A = \frac{\pi d^2}{4} = \frac{\pi (19 \text{ mm})^2}{(4)\left(1000 \; \frac{mm}{m}\right)^2}$$

$$= 2.835 \times 10^{-4} \text{ m}^2$$

The polar moment of inertia of a bolt about the centroid is

$$J = Ar^2 = (2.835 \times 10^{-4} \text{ m}^2)(0.153 \text{ m})^2$$

$$= 6.636 \times 10^{-6} \text{ m}^4$$

The vertical shear load at each bolt is

$$F_v = \frac{F}{3} = 0.333F \quad \text{[in N]}$$

The moment applied to each bolt is

$$M = \frac{(0.5 \text{ m})F}{3} = 0.167F \quad \text{[in N·m]}$$

The vertical shear stress in each bolt is

$$\tau_v = \frac{F_v}{A} = \frac{0.333F}{2.835 \times 10^{-4} \text{ m}^2} = 1175F \quad \text{[in Pa]}$$

The torsional shear stress in each bolt is

$$\tau = \frac{Mr}{J} = \frac{(0.167F)(0.153 \text{ m})}{6.636 \times 10^{-6} \text{ m}^4}$$
$$= 3850F \quad \text{[in Pa]}$$

The most highly stressed bolt is the right-most bolt. The shear stress configuration is

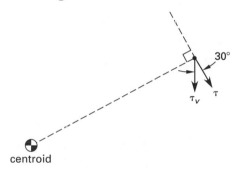

The stresses are combined to find the maximum stress.

$$\tau_{\max} = \sqrt{(\tau \sin 30°)^2 + (\tau_v + \tau \cos 30°)^2}$$

$$\begin{aligned}(100 \text{ MPa}) \\ \times \left(1000 \frac{\text{kPa}}{\text{MPa}}\right)\end{aligned} = \sqrt{\begin{aligned}&\left((3850F)(\sin 30°)\right)^2 \\ &+ \left((1175F) + (3850F)(\cos 30°)\right)^2\end{aligned}}$$

$$= 4903F \quad \text{[in Pa]}$$
$$F = \boxed{20.4 \text{ kN} \quad (20 \text{ kN})}$$

The answer is (D).

---

6. *Customary U.S. Solution*

The fillet weld consists of three basic shapes, each with a throat size of t.

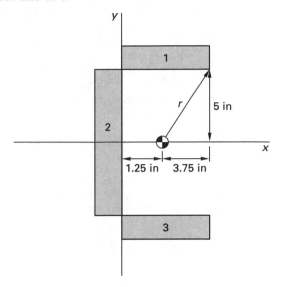

First, find the centroid.

By inspection, $y_c = 0$.

The areas of the basic shapes are

$$A_1 = 5t \quad \text{[in in}^2\text{]}$$
$$A_2 = 10t \quad \text{[in in}^2\text{]}$$
$$A_3 = 5t \quad \text{[in in}^2\text{]}$$

The x-components of the centroids of the basic shapes are

$$x_{c1} = \frac{5 \text{ in}}{2} = 2.5 \text{ in}$$

$$x_{c2} = -\frac{t}{2} = 0 \text{ in} \quad \text{[for small } t\text{]}$$

$$x_{c3} = \frac{5 \text{ in}}{2} = 2.5 \text{ in}$$

$$\begin{aligned} x_c &= \frac{\sum\limits_i A_i x_{ci}}{\sum\limits_i A_i} \\ &= \frac{(5t)(2.5 \text{ in}) + (10t)(0 \text{ in}) + (5t)(2.5 \text{ in})}{5t + 10t + 5t} \\ &= 1.25 \text{ in} \end{aligned}$$

Next, find the centroidal moments of inertia.

The moments of inertia of basic shape 1 about its own centroid are

$$I_{cx1} = \frac{bh^3}{12} = \frac{(5 \text{ in})t^3}{12} = 0.417t^3 \quad \text{[in in}^4\text{]}$$

$$I_{cy1} = \frac{bh^3}{12} = \frac{t(5 \text{ in})^3}{12} = 10.417t \quad \text{[in in}^4\text{]}$$

The moments of inertia of basic shape 2 about its own centroid are

$$I_{cx2} = \frac{bh^3}{12} = \frac{t(10 \text{ in})^3}{12} = 83.333t \quad [\text{in in}^4]$$

$$I_{cy2} = \frac{bh^3}{12} = \frac{(10 \text{ in})t^3}{12} = 0.833t^3 \quad [\text{in in}^4]$$

The moments of inertia of basic shape 3 about its own centroid are the same as for basic shape 1.

$$I_{cx3} = I_{cx1} = 0.417t^3 \quad [\text{in in}^4]$$

$$I_{cy3} = I_{cy1} = 10.417t \quad [\text{in in}^4]$$

From the parallel axis theorem, the moments of inertia of basic shape 1 about the centroidal axis of the section are

$$I_{x1} = I_{cx1} + A_1 d_1^2 = 0.417t^3 + (5t)(5 \text{ in})^2$$
$$= 0.417t^3 + 125t$$

$$I_{y1} = I_{cy1} + A_1 d_1^2$$
$$= 10.417t + 5t\left(\frac{5 \text{ in}}{2} - 1.25 \text{ in}\right)^2$$
$$= 18.23t \quad [\text{in in}^4]$$

The moments of inertia of basic shape 2 about the centroidal axis of the section are

$$I_{x2} = I_{cx2} + A_2 d_2^2 = 83.333t + (10t)(0 \text{ in})$$
$$= 83.333t \quad [\text{in in}^4]$$

$$I_{y2} = I_{cy2} + A_2 d_2^2 = 0.833t^3 + (10t)(1.25 \text{ in})^2$$
$$= 0.833t^3 + 15.625t$$

The moment of inertia of basic shape 3 about the centroidal axis of the section are the same as for basic shape 1.

$$I_{x3} = I_{x1} = 0.417t^3 + 125t$$

$$I_{y3} = I_{y1} = 18.23t \quad [\text{in in}^4]$$

The total moments of inertia about the centroidal axis of the section are

$$I_x = I_{x1} + I_{x2} + I_{x3}$$
$$= 0.417t^3 + 125t + 83.333t + 0.417t^3 + 125t$$
$$= 0.834t^3 + 333.33t$$

$$I_y = I_{y1} + I_{y2} + I_{y3}$$
$$= 18.23t + 0.833t^3 + 15.625t + 18.23t$$
$$= 0.833t^3 + 52.09t$$

Since t will be small and since the coefficient of the t^3 term is smaller than the coefficient of the t term, the higher-order t^3 term may be neglected.

$$I_x = 333.33t \quad [\text{in in}^4]$$
$$I_y = 52.09t \quad [\text{in in}^4]$$

The polar moment of inertia for the section is

$$J = I_x + I_y = 333.33t + 52.09t$$
$$= 385.42t \quad [\text{in in}^4]$$

The maximum shear stress will occur at the right-most point of basic shape 1 since this point is farthest from the centroid of the section.

This distance is

$$r = \sqrt{(5 \text{ in} - 1.25 \text{ in})^2 + \left(\frac{10 \text{ in}}{2}\right)^2}$$
$$= 6.25 \text{ in}$$

The applied moment is

$$M = Fe = (10{,}000 \text{ lbf})(12 \text{ in} + 5 \text{ in} - 1.25 \text{ in})$$
$$= 157{,}500 \text{ in-lbf}$$

The torsional shear stress is

$$\tau = \frac{Mr}{J} = \frac{(157{,}500 \text{ in-lbf})(6.25 \text{ in})}{385.42t}$$
$$= 2554.0/t \quad [\text{in lbf/in}^2]$$

The shear stress is perpendicular to the line from the point to the centroid. The x- and y-components are shown in the following figure.

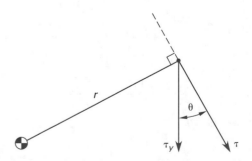

The angle θ is

$$\theta = 90° - \tan^{-1}\left(\frac{5 \text{ in} - 1.25 \text{ in}}{\dfrac{10 \text{ in}}{2}}\right) = 53.1°$$

$$\tau_x = \tau \sin\theta = \left(\frac{2554.0 \ \dfrac{\text{lbf}}{\text{in}^2}}{t}\right)(\sin 53.1°)$$

$$= 2042.4/t \quad [\text{in lbf/in}^2]$$

$$\tau_y = \tau \cos\theta = \left(\frac{2554.0 \ \dfrac{\text{lbf}}{\text{in}^2}}{t}\right)(\cos 53.1°)$$

$$= 1533.5/t \quad [\text{in lbf/in}^2]$$

The vertical shear stress due to the load is

$$\tau_{vy} = \frac{F}{\sum A_i} = \frac{10{,}000 \text{ lbf}}{5t + 10t + 5t}$$

$$= 500/t \quad [\text{in lbf/in}^2]$$

The resultant shear stress is

$$\tau = \sqrt{\tau_x^2 + (\tau_y + \tau_{vy})^2}$$

$$= \sqrt{\left(\frac{2042.4}{t}\right)^2 + \left(\frac{1533.5}{t} + \frac{500}{t}\right)^2}$$

$$= 2882.1/t \quad [\text{in lbf/in}^2]$$

For a maximum shear stress of 8000 lbf/in^2,

$$8000 \ \frac{\text{lbf}}{\text{in}^2} = \frac{2882.1 \ \dfrac{\text{lbf}}{\text{in}^2}}{t}$$

$$t = 0.360 \text{ in}$$

From Eq. 45.48, the required weld size is

$$y = \frac{t}{\dfrac{\sqrt{2}}{2}} = \frac{0.360 \text{ in}}{0.707}$$

$$= \boxed{0.509 \text{ in} \quad (1/2 \text{ in})}$$

The answer is (B).

SI Solution

The fillet weld consists of three basic shapes, each with a throat size of t.

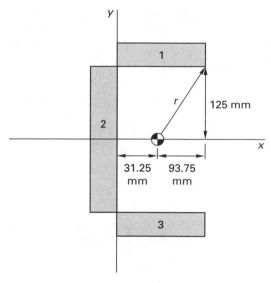

First, find the centroid.

By inspection, $y_c = 0$.

The areas of the basic shapes are

$$A_1 = 125t \quad [\text{in mm}^2]$$

$$A_2 = 250t \quad [\text{in mm}^2]$$

$$A_3 = 125t \quad [\text{in mm}^2]$$

The x-components of the centroids of the basic shapes are

$$x_{c1} = \frac{125 \text{ mm}}{2} = 62.5 \text{ mm}$$

$$x_{c2} = -\frac{t}{2} = 0 \text{ mm} \quad [\text{for small } t]$$

$$x_{c3} = \frac{125 \text{ mm}}{2} = 62.5 \text{ mm}$$

$$x_c = \frac{\displaystyle\sum_i A_i x_{ci}}{\displaystyle\sum A_i}$$

$$= \frac{\begin{array}{c}(125t)(62.5 \text{ mm}) + (250t)(0 \text{ mm}) \\ + (125t)(6.25 \text{ mm})\end{array}}{125t + 250t + 125t}$$

$$= 31.25 \text{ mm}$$

Next, find the centroidal moments of inertia.

The moments of inertia of basic shape 1 about its own centroid are

$$I_{cx1} = \frac{bh^3}{12} = \frac{(125 \text{ mm})t^3}{12} = 10.42t^3 \quad [\text{in mm}^4]$$

$$I_{cy1} = \frac{bh^3}{12} = \frac{t(125 \text{ mm})^3}{12} = 162\,760t \quad [\text{in mm}^4]$$

The moments of inertia of basic shape 2 about its own centroid are

$$I_{cx2} = \frac{bh^3}{12} = \frac{t(250 \text{ mm})^3}{12} = 1\,302\,083t \quad [\text{in mm}^4]$$

$$I_{cy2} = \frac{bh^3}{12} = \frac{(250 \text{ mm})t^3}{12} = 20.83t^3 \quad [\text{in mm}^4]$$

The moments of inertia of basic shape 3 about its own centroid are the same as for basic shape 1.

$$I_{cx3} = I_{cx1} = 10.42t^3 \quad [\text{in mm}^4]$$

$$I_{cy3} = I_{cy1} = 162\,760t \quad [\text{in mm}^4]$$

From the parallel axis theorem, the moments of inertia of basic shape 1 about the centroidal axis of the section are

$$I_{x1} = I_{cx1} + A_1 d_1^2 = 10.42t^3 + (125t)(125 \text{ mm})^2$$

$$= 10.42t^3 + 1\,953\,125t$$

$$I_{y1} = I_{cy1} + A_1 d_1^2$$

$$= 162\,760t + (125t)\left(\frac{125 \text{ mm}}{2} - 31.25 \text{ mm}\right)^2$$

$$= 284\,830t \quad [\text{in mm}^4]$$

The moments of inertia of basic shape 2 about the centroidal axis of the section are

$$I_{x2} = I_{cx2} + A_2 d_2^2 = 1\,302\,083t + (250t)(0 \text{ mm})^2$$

$$= 1\,302\,083t \quad [\text{in mm}^4]$$

$$I_{y2} = I_{cy2} + A_2 d_2^2 = 20.83t^3 + (250t)(31.25 \text{ mm})^2$$

$$= 20.83t^3 + 244\,140t$$

The moments of inertia of basic shape 3 about the centroidal axis of the section are the same as for basic shape 1.

$$I_{x3} = I_{x1} = 10.42t^3 + 1\,953\,125t$$

$$I_{y3} = I_{y1} = 284\,830t \quad [\text{in mm}^4]$$

The total moments of inertia about the centroidal axis of the section are

$$I_x = I_{x1} + I_{x2} + I_{x3}$$

$$= 10.42t^3 + 1\,953\,125t + 1\,302\,083t + 10.42t^3$$

$$\quad + 1\,953\,125t$$

$$= 20.84t^3 + 5\,208\,333t$$

$$I_y = I_{y1} + I_{y2} + I_{y3}$$

$$= 284\,830t + 20.83t^3 + 244\,140t + 284\,830t$$

$$= 20.83t^3 + 813\,800t$$

Since t will be small and since the coefficient of the t^3 term is smaller than the coefficient of the t term, the t^3 term may be neglected.

$$I_x = 5\,208\,333t \quad [\text{in mm}^4]$$

$$I_y = 813\,800t \quad [\text{in mm}^4]$$

The polar moment of inertia for the section is

$$J = I_x + I_y = 5\,208\,333t + 813\,800t$$

$$= 6\,022\,133t \quad [\text{in mm}^4]$$

The maximum shear stress will occur at the right-most point of basic shape 1 since this point is farthest from the centroid of the section.

The distance is

$$r = \sqrt{(125 \text{ mm} - 31.25 \text{ mm})^2 + \left(\frac{250 \text{ mm}}{2}\right)^2}$$

$$= 156.25 \text{ mm}$$

The applied moment is

$$M = Fe = (44 \text{ kN})(300 \text{ mm} + 125 \text{ mm} - 31.25 \text{ mm})$$

$$= 17\,325 \text{ kN·mm}$$

The torsional shear stress is

$$\tau = \frac{Mr}{J}$$

$$= \frac{\left(\frac{(17\,325 \text{ kN·mm})(156.25 \text{ mm})}{6\,022\,133t}\right)\left(1000 \frac{\text{mm}}{\text{m}}\right)^2}{1000 \frac{\text{kPa}}{\text{MPa}}}$$

$$= 449.5/t \quad [\text{in MPa}]$$

The shear stress is perpendicular to the line from the point to the centroid. The x- and y-components are shown in the following figure.

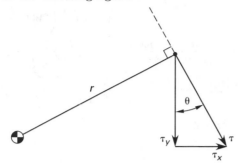

The angle θ is

$$\theta = 90° - \tan^{-1}\left(\dfrac{125\text{ mm} - 31.25\text{ mm}}{\dfrac{250\text{ mm}}{2}}\right) = 53.1°$$

$$\tau_x = \tau\sin\theta = \left(\dfrac{449.5\text{ MPa}}{t}\right)(\sin 53.1°)$$
$$= 359.5/t \quad \text{[in MPa]}$$

$$\tau_y = \tau\cos\theta = \left(\dfrac{449.5\text{ MPa}}{t}\right)(\cos 53.1°)$$
$$= 269.9/t \quad \text{[in MPa]}$$

The vertical shear stress due to the load is

$$\tau_{vy} = \dfrac{F}{\sum A_i}$$

$$= \dfrac{(44\text{ kN})\left(1000\ \dfrac{\text{mm}}{\text{m}}\right)^2}{(125t + 250t + 125t)\left(1000\ \dfrac{\text{kPa}}{\text{MPa}}\right)}$$

$$= 88.0/t \quad \text{[in MPa]}$$

The resultant shear stress is

$$\tau = \sqrt{\tau_x^2 + (\tau_y + \tau_{vy})^2}$$
$$= \sqrt{\left(\dfrac{359.5}{t}\right)^2 + \left(\dfrac{269.9}{t} + \dfrac{88.0}{t}\right)^2}$$
$$= 507.3/t \quad \text{[in MPa]}$$

For a maximum shear stress of 55 MPa,

$$55\text{ MPa} = \dfrac{507.3\text{ MPa}}{t}$$
$$t = 9.22\text{ mm}$$

From Eq. 45.48, the required weld size is

$$y = \dfrac{t}{\dfrac{\sqrt{2}}{2}} = \dfrac{9.22\text{ mm}}{0.707} = \boxed{13.0\text{ mm}}$$

The answer is (B).

7. For the pipe,

$$D_o = 6.625\text{ in}$$
$$D_i = 6.065\text{ in}$$
$$A_i = 0.2006\text{ ft}^2$$

The volume and weight of pipe per 40 ft section is

$$V = \left(\dfrac{\pi}{4}\right)\left(\left(\dfrac{6.625\text{ in}}{12\ \dfrac{\text{in}}{\text{ft}}}\right)^2 - \left(\dfrac{6.065\text{ in}}{12\ \dfrac{\text{in}}{\text{ft}}}\right)^2\right)(40\text{ ft})$$
$$= 1.55\text{ ft}^3$$

$$\text{weight} = \left(490\ \dfrac{\text{lbf}}{\text{ft}^3}\right)(1.55\text{ ft}^3) = 759.5\text{ lbf} \quad \text{[use 760 lbf]}$$

The weight of concrete per 40 ft section is

$$V = (0.2006\text{ ft}^2)(40\text{ ft}) = 8.0\text{ ft}^3$$
$$\text{weight} = \left(150\ \dfrac{\text{lbf}}{\text{ft}^3}\right)(8.0\text{ ft}^3) = 1200\text{ lbf}$$

The load per vertical hanger is

$$760\text{ lbf} + 1200\text{ lbf} = 1960\text{ lbf}$$

Since the horizontal force component is constant, the cable tension will be greatest where the cable is steepest. This appears to be the right-hand support.

At the left cable support,

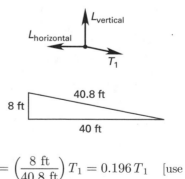

$$L_{\text{vertical}} = \left(\dfrac{8\text{ ft}}{40.8\text{ ft}}\right)T_1 = 0.196\,T_1 \quad \text{[use } 0.20\,T_1\text{]}$$

$$L_{\text{horizontal}} = \left(\dfrac{40\text{ ft}}{40.8\text{ ft}}\right)T_1 = 0.98\,T_1$$

Sum the moments about the right cable support to find T_1. Take counterclockwise moments to be positive.

$$\sum M = (1960 \text{ lbf})(40 \text{ ft} + 80 \text{ ft} + 120 \text{ ft} + 160 \text{ ft})$$
$$\qquad - 16 L_{\text{horizontal}} - 200 L_{\text{vertical}}$$
$$= 784{,}000 - (16)(0.98 T_1) - (200)(0.20 T_1)$$
$$T_1 = 14{,}080 \text{ lbf}$$
$$L_{\text{horizontal}} = 0.98 T_1 = (0.98)(14{,}080 \text{ lbf})$$
$$= 13{,}798 \text{ lbf}$$
$$L_{\text{vertical}} = 0.20 T_1 = (0.20)(14{,}080 \text{ lbf})$$
$$= 2816 \text{ lbf}$$
$$R_{\text{horizontal}} = L_{\text{horizontal}} = 13{,}798 \text{ lbf}$$
$$R_{\text{vertical}} = \sum \text{loads} - L_{\text{vertical}}$$
$$= (4)(1960 \text{ lbf}) - 2816 \text{ lbf}$$
$$= 5024 \text{ lbf}$$
$$T_5 = \sqrt{(13{,}798 \text{ lbf})^2 + (5024 \text{ lbf})^2}$$
$$= 14{,}684 \text{ lbf}$$

The factor of safety includes the stress concentration factor.

$$\text{FS} = \frac{\text{allowable tension}}{\text{actual force}} = \frac{\dfrac{(11 \text{ tons})\left(2000 \ \dfrac{\text{lbf}}{\text{ton}}\right)}{5}}{(2)(14{,}684 \text{ lbf})}$$
$$= \boxed{0.15}$$

The answer is (A).

46 Structural Analysis I

PRACTICE PROBLEMS

Degree of Indeterminacy

1. What is the degree of indeterminacy of the following structure?

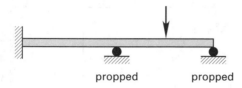

propped   propped

(A) 1

(B) 2

(C) 3

(D) 4

2. What is the degree of indeterminacy of the following truss?

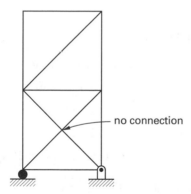

— no connection

(A) 1

(B) 2

(C) 3

(D) 4

Elastic Deformation

3. What is the compressive load if a 1 in × 2 in × 10 in (2 cm × 5 cm × 30 cm) copper tie rod ($E = 18 \times 10^6$ lbf/in^2 (12×10^4 MPa)) experiences a 55°F (30°C) increase from the no-stress temperature?

(A) 8000 lbf (26 kN)

(B) 12,000 lbf (38 kN)

(C) 15,000 lbf (48 kN)

(D) 18,000 lbf (58 kN)

4. A 2 in diameter (5 cm diameter) and 15 in long (40 cm long) steel rod ($E = 29 \times 10^6$ lbf/in^2 (20×10^4 MPa)) supports a 2250 lbf (10 000 N) compressive load.

(a) What is the stress?

(A) 570 lbf/in^2 (4.0 MPa)

(B) 720 lbf/in^2 (5.1 MPa)

(C) 1100 lbf/in^2 (7.8 MPa)

(D) 1400 lbf/in^2 (9.9 MPa)

(b) What is the decrease in length?

(A) 8.7×10^{-5} in (2.3×10^{-6} m)

(B) 1.1×10^{-4} in (3.0×10^{-6} m)

(C) 3.7×10^{-4} in (1.0×10^{-5} m)

(D) 9.3×10^{-4} in (2.6×10^{-5} m)

Consistent Deformation Method

5. (a) Find the force and stress in each member for the structure shown. (b) Find the total deflection for the structure shown.

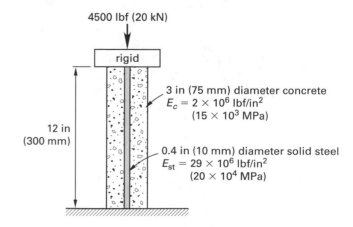

4500 lbf (20 kN)

rigid

12 in
(300 mm)

3 in (75 mm) diameter concrete
$E_c = 2 \times 10^6$ lbf/in^2
(15×10^3 MPa)

0.4 in (10 mm) diameter solid steel
$E_{st} = 29 \times 10^6$ lbf/in^2
(20×10^4 MPa)

6. The concentric pipe assembly shown is rigidly attached at both ends. The rigid ends are unconstrained. What is the stress generated in each pipe if the temperature of the assembly rises 200°F (110°C)?

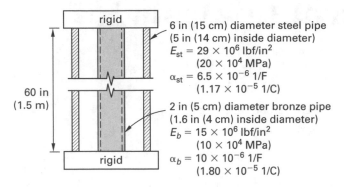

6 in (15 cm) diameter steel pipe
(5 in (14 cm) inside diameter)
$E_{st} = 29 \times 10^6$ lbf/in²
(20 × 10⁴ MPa)
$\alpha_{st} = 6.5 \times 10^{-6}$ 1/F
(1.17 × 10⁻⁵ 1/C)

2 in (5 cm) diameter bronze pipe
(1.6 in (4 cm) inside diameter)
$E_b = 15 \times 10^6$ lbf/in²
(10 × 10⁴ MPa)
$\alpha_b = 10 \times 10^{-6}$ 1/F
(1.80 × 10⁻⁵ 1/C)

7. For the structure shown, what is the load carried by the steel cable? The beam is made of steel, $I = 10$ in⁴ $(4.17 \times 10^6$ mm⁴). The cable cross section is 0.0124 in² (8 mm²). Before the 270 lbf load is applied, the cable is taut but carries no load.

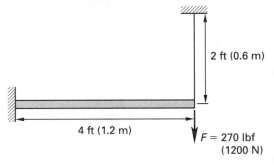

(A) 180 lbf (780 N)

(B) 200 lbf (870 N)

(C) 220 lbf (950 N)

(D) 240 lbf (1.0 kN)

8. A beam is simply supported at its ends and by a column of area A at its center. The beam and column are the same material. If the beam is subject to uniform load, w, what are the reactions and the force in the column?

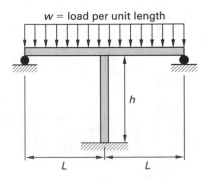

9. A rigid bar is supported by a hinge and two aluminum rods. For the structure shown,

$$a = 6 \text{ ft } (2 \text{ m})$$
$$b = 3 \text{ ft } (1 \text{ m})$$
$$c = 3 \text{ ft } (1 \text{ m})$$
$$d = 12 \text{ ft } (4 \text{ m})$$
$$P = 4500 \text{ lbf } (20 \text{ kN})$$
$$A_{rod} = 0.124 \text{ in}^2 (80 \text{ mm}^2)$$
$$E_{rod} = 10 \times 10^6 \text{ lbf/in}^2 (70 \times 10^3 \text{ MPa})$$

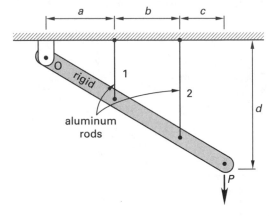

(a) What is the force in each of the aluminum rods?

(A) 3200 lbf (14 kN)

(B) 3600 lbf (16 kN)

(C) 4000 lbf (18 kN)

(D) 4300 lbf (19 kN)

(b) What is the elongation of each of the aluminum rods?

(A) 0.13 in and 0.26 in (3.3 mm and 6.6 mm)

(B) 0.21 in and 0.31 in (5.7 mm and 8.6 mm)

(C) 0.21 in and 0.47 in (5.7 mm and 12 mm)

(D) 0.26 in and 0.62 in (6.6 mm and 16 mm)

10. What are the reactions at each of the supports for the structure? The beams are identical in length, cross-sectional shape and area, and material.

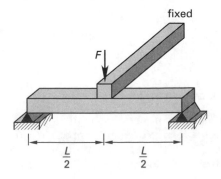

(A) $F/13$

(B) $4F/21$

(C) $8F/17$

(D) $2F/3$

Superposition Method

11. Using the superposition method, find the intermediate support reaction, R_2, for the structure.

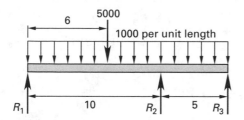

(A) 9000

(B) 11,000

(C) 15,000

(D) 18,000

12. Using the superposition method, determine the reaction, R, at the prop.

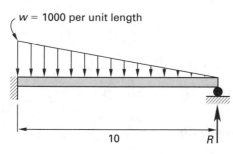

(A) 800

(B) 1000

(C) 1200

(D) 1400

Three-Moment Equation

13. What are the reactions R_1, R_2, and R_3?

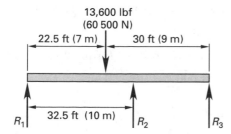

14. What are the reactions R_1, R_2, and R_3?

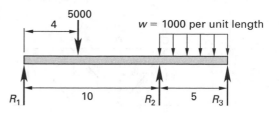

Fixed-End Moments

15. What are the vertical reactions at both ends of the structure?

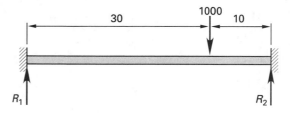

16. The truss shown carries a moving uniform live load of 2 kips/ft and a moving concentrated live load of 15 kips.

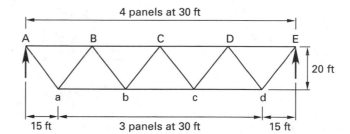

(a) What is the maximum force in member Bb?

(A) 40 kips

(B) 51 kips

(C) 59 kips

(D) 73 kips

(b) What is the maximum force in member BC?

(A) 15 kips

(B) 80 kips

(C) 95 kips

(D) 170 kips

17. The truss shown carries the group of four live loads shown along its bottom chords. What is the maximum force in member CD?

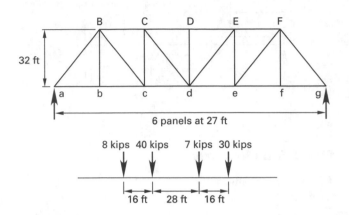

6 panels at 27 ft

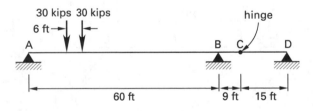

8 kips 40 kips   7 kips 30 kips

16 ft   28 ft   16 ft

(A) 34 kips

(B) 82 kips

(C) 180 kips

(D) 220 kips

18. A moving load consisting of two 30 kip forces separated by a constant 6 ft travels over a two-span bridge. The bridge has an interior expansion joint that can be considered to be an ideal hinge.

(a) What is the maximum moment at point B?

(A) 250 ft-kips

(B) 310 ft-kips

(C) 380 ft-kips

(D) 430 ft-kips

(b) What is the maximum reaction at point B?

(A) 30 kips

(B) 60 kips

(C) 66 kips

(D) 90 kips

SOLUTIONS

1. The degree of indeterminacy is 2. Remove the two props (vertical reactions) in order to make the structure statically determinate.

The answer is (B).

2. From Eq. 46.1,

$$I = r + m - 2j$$
$$= 3 + 10 - (2)(6)$$
$$= \boxed{1}$$

The answer is (A).

3. *Customary U.S. Solution*

The thermal strain is

$$\epsilon_{\text{th}} = \alpha \Delta T = \left(8.9 \times 10^{-6} \, \frac{1}{°F}\right)(55°F) = 0.00049$$

The strain must be counteracted to maintain the rod in its original position.

$$\sigma = E\epsilon_{\text{th}} = \left(18 \times 10^6 \, \frac{\text{lbf}}{\text{in}^2}\right)(0.00049) = 8820 \, \text{lbf/in}^2$$
$$F = \sigma A = \left(8820 \, \frac{\text{lbf}}{\text{in}^2}\right)(1 \text{ in})(2 \text{ in})$$
$$= \boxed{17,640 \text{ lbf} \quad (18,000 \text{ lbf})}$$

The answer is (D).

SI Solution

The thermal strain is

$$\epsilon_{\text{th}} = \alpha \Delta T = \left(16.0 \times 10^{-6} \, \frac{1}{°C}\right)(30°C) = 0.00048$$

The strain must be counteracted to maintain the rod in its original position.

$$\sigma = E\epsilon_{\text{th}} = (12 \times 10^4 \text{ MPa})(0.00048)$$
$$= 57.6 \text{ MPa}$$
$$F = \sigma A = (57.6 \times 10^6 \text{ Pa})(0.02 \text{ m})(0.05 \text{ m})$$
$$= \boxed{57\,600 \text{ N} \quad (58 \text{ kN})}$$

The answer is (D).

4. *Customary U.S. Solution*

(a) The rod's stress is

$$\sigma = \frac{F}{A} = \frac{2250 \text{ lbf}}{\dfrac{\pi(2 \text{ in})^2}{4}}$$

$$= \boxed{716.2 \text{ lbf/in}^2 \quad (720 \text{ lbf/in}^2)}$$

The answer is (B).

(b) The decrease in length can be found by the following equation.

$$\epsilon = \frac{\sigma}{E} = \frac{\delta}{L_o}$$

$$\delta = L_o\left(\frac{\sigma}{E}\right) = (15 \text{ in})\left(\frac{716.2 \ \dfrac{\text{lbf}}{\text{in}^2}}{29 \times 10^6 \ \dfrac{\text{lbf}}{\text{in}^2}}\right)$$

$$= \boxed{3.70 \times 10^{-4} \text{ in}}$$

The answer is (C).

SI Solution

(a) The rod's stress is

$$\sigma = \frac{F}{A} = \frac{10\,000 \text{ N}}{\dfrac{\pi(0.05 \text{ m})^2}{4}}$$

$$= \boxed{5.093 \times 10^6 \text{ Pa} \quad (5.1 \text{ MPa})}$$

The answer is (B).

(b) The decrease in length can be found by the following equation.

$$\epsilon = \frac{\sigma}{E} = \frac{\delta}{L_o}$$

$$\delta = L_o\left(\frac{\sigma}{E}\right) = (0.4 \text{ m})\left(\frac{5.093 \text{ MPa}}{20 \times 10^4 \text{ MPa}}\right)$$

$$= \boxed{1.02 \times 10^{-5} \text{ m}}$$

The answer is (C).

5. *Customary U.S. Solution*

(a) Let F_c and F_{st} be the loads carried by the concrete and steel, respectively.

$$F = F_c + F_{st} = 4500 \text{ lbf} \quad \text{[Eq. I]}$$

The deformation of the steel is

$$\delta_{st} = \frac{F_{st}L}{A_{st}E_{st}}$$

The deformation of the concrete is

$$\delta_c = \frac{F_cL}{A_cE_c}$$

The geometric constraint is $\delta_{st} = \delta_c$, or

$$\frac{F_{st}L}{A_{st}E_{st}} = \frac{F_cL}{A_cE_c} \quad \text{[Eq. II]}$$

Solving Eq. I and Eq. II,

$$F_c = \frac{F}{1 + \dfrac{A_{st}E_{st}}{A_cE_c}}$$

$$= \frac{4500 \text{ lbf}}{1 + \dfrac{\left(\dfrac{\pi(0.4 \text{ in})^2}{4}\right)\left(29 \times 10^6 \ \dfrac{\text{lbf}}{\text{in}^2}\right)}{\left(\dfrac{\pi\left((3.0 \text{ in})^2 - (0.4 \text{ in})^2\right)}{4}\right)\left(2 \times 10^6 \ \dfrac{\text{lbf}}{\text{in}^2}\right)}}$$

$$= \boxed{3564 \text{ lbf}}$$

$$F_{st} = 4500 \text{ lbf} - F_c = 4500 \text{ lbf} - 3564 \text{ lbf} = \boxed{936 \text{ lbf}}$$

$$\sigma_{st} = \frac{F_{st}}{A_{st}} = \frac{936 \text{ lbf}}{\dfrac{\pi(0.4 \text{ in})^2}{4}} = \boxed{7448 \text{ lbf/in}^2}$$

$$\sigma_c = \frac{F_c}{A_c} = \frac{3564 \text{ lbf}}{\dfrac{\pi\left((3 \text{ in})^2 - (0.4 \text{ in})^2\right)}{4}}$$

$$= \boxed{513 \text{ lbf/in}^2}$$

(b) The total deflection is

$$\delta = \frac{F_{st}L}{A_{st}E_{st}} = \frac{(936 \text{ lbf})(12 \text{ in})}{\left(\dfrac{\pi(0.4 \text{ in})^2}{4}\right)\left(29 \times 10^6 \ \dfrac{\text{lbf}}{\text{in}^2}\right)}$$

$$= \boxed{3.08 \times 10^{-3} \text{ in}}$$

SI Solution

(a) Let F_c and F_{st} be the loads carried by the concrete and steel, respectively.

$$F = F_c + F_{st} = 20 \text{ kN} \quad \text{[Eq. I]}$$

The deformation of the steel is

$$\delta_{st} = \frac{F_{st}L}{A_{st}E_{st}}$$

The deformation of the concrete is

$$\delta_c = \frac{F_c L}{A_c E_c}$$

The geometric constraint is $\delta_{st} = \delta_c$, or

$$\frac{F_{st} L}{A_{st} E_{st}} = \frac{F_c L}{A_c E_c} \qquad \text{[Eq. II]}$$

Solving Eq. I and Eq. II,

$$F_c = \frac{F}{1 + \dfrac{A_{st} E_{st}}{A_c E_c}}$$

$$= \frac{20 \text{ kN}}{1 + \dfrac{\left(\dfrac{\pi (0.01 \text{ m})^2}{4}\right)(20 \times 10^4 \text{ MPa})}{\left(\dfrac{\pi\left((0.075 \text{ m})^2 - (0.01 \text{ m})^2\right)}{4}\right)(15 \times 10^3 \text{ MPa})}}$$

$$= \boxed{16.11 \text{ kN}}$$

$$F_{st} = 20 \text{ kN} - F_c = 20 \text{ kN} - 16.11 \text{ kN} = \boxed{3.89 \text{ kN}}$$

$$\sigma_{st} = \frac{F_{st}}{A_{st}} = \frac{3.89 \times 10^{-3} \text{ MN}}{\left(\dfrac{\pi (0.01 \text{ m})^2}{4}\right)} = \boxed{49.5 \text{ MPa}}$$

$$\sigma_c = \frac{F_c}{A_c} = \frac{16.11 \times 10^{-3} \text{ MN}}{\dfrac{\pi\left((0.075 \text{ m})^2 - (0.01 \text{ m})^2\right)}{4}}$$

$$= \boxed{3.71 \text{ MPa}}$$

(b) The total deflection is

$$\delta = \frac{F_{st} L}{A_{st} E_{st}}$$

$$= \frac{(3.89 \times 10^3 \text{ N})(0.3 \text{ m})}{\left(\dfrac{\pi (0.01 \text{ m})^2}{4}\right)(20 \times 10^{10} \text{ Pa})}$$

$$= \boxed{7.43 \times 10^{-5} \text{ m} \quad (74.3 \ \mu\text{m})}$$

6. *Customary U.S. Solution*

$$A_{st} = \frac{\pi\left((6 \text{ in})^2 - (5 \text{ in})^2\right)}{4} = 8.639 \text{ in}^2$$

$$A_b = \frac{\pi\left((2 \text{ in})^2 - (1.6 \text{ in})^2\right)}{4} = 1.131 \text{ in}^2$$

The thermal deformation that would occur if the pipes were free is

$$\delta_b = \alpha_b L \Delta T = \left(10 \times 10^{-6} \ \frac{1}{^\circ\text{F}}\right)(60 \text{ in})(200 ^\circ\text{F})$$

$$= 0.120 \text{ in}$$

$$\delta_{st} = \alpha_{st} L \Delta T = \left(6.5 \times 10^{-6} \ \frac{1}{^\circ\text{F}}\right)(60 \text{ in})(200 ^\circ\text{F})$$

$$= 0.078 \text{ in}$$

The two pipes expand by the same amount, δ, so that $\delta_{st} < \delta < \delta_b$.

$$\delta - \delta_{st} = \frac{F_{st} L}{A_{st} E_{st}} \qquad \text{[Eq. I]}$$

$$\delta_b - \delta = \frac{F_b L}{A_b E_b} \qquad \text{[Eq. II]}$$

F_{st} is a tensile force, and F_b is a compressive force. Since there is no external force,

$$F_{st} = F_b = F$$

Adding Eq. I to Eq. II,

$$\delta_b - \delta_{st} = \frac{FL}{A_{st} E_{st}} + \frac{FL}{A_b E_b}$$

$$= FL\left(\frac{1}{A_{st} E_{st}} + \frac{1}{A_b E_b}\right)$$

$$F = \frac{\dfrac{\delta_b - \delta_{st}}{L}}{\dfrac{1}{A_{st} E_{st}} + \dfrac{1}{A_b E_b}}$$

$$= \frac{\dfrac{0.120 \text{ in} - 0.078 \text{ in}}{60 \text{ in}}}{\dfrac{1}{(8.639 \text{ in}^2)\left(29 \times 10^6 \ \dfrac{\text{lbf}}{\text{in}^2}\right)} + \dfrac{1}{(1.131 \text{ in}^2)\left(15 \times 10^6 \ \dfrac{\text{lbf}}{\text{in}^2}\right)}}$$

$$= 11{,}122 \text{ lbf}$$

$$\sigma_{st} = \frac{F}{A_{st}} = \frac{11{,}122 \text{ lbf}}{8.639 \text{ in}^2} = \boxed{1287 \text{ psi}}$$

$$\sigma_b = \frac{F}{A_b} = \frac{11{,}122 \text{ lbf}}{1.131 \text{ in}^2} = \boxed{9834 \text{ psi}}$$

SI Solution

$$A_{st} = \frac{\pi\left((15 \text{ cm})^2 - (14 \text{ cm})^2\right)}{4} = 22.78 \text{ cm}^2$$

$$= 0.002\,278 \text{ m}^2$$

$$A_b = \frac{\pi\left((5 \text{ cm})^2 - (4 \text{ cm})^2\right)}{4} = 7.069 \text{ cm}^2$$

$$= 0.000\,706\,9 \text{ m}^2$$

The thermal deformation that would occur if the pipes were free is

$$\delta_b = \alpha_b L \Delta T = \left(1.8 \times 10^{-5} \, \frac{1}{°C}\right)(1.5 \text{ m})(110°C)$$

$$= 0.002\,97 \text{ m}$$

$$\delta_{st} = \alpha_{st} L \Delta T = \left(1.17 \times 10^{-5} \, \frac{1}{°C}\right)(1.5 \text{ m})(110°C)$$

$$= 0.001\,93 \text{ m}$$

The two pipes expand by the same amount, δ, so that $\delta_{st} < \delta < \delta_b$.

$$\delta - \delta_{st} = \frac{F_{st} L}{A_{st} E_{st}} \quad \text{[Eq. I]}$$

$$\delta_b - \delta = \frac{F_b L}{A_b E_b} \quad \text{[Eq. II]}$$

F_{st} is a tensile force, and F_b is a compressive force. Since there is no external force,

$$F_{st} = F_b = F$$

Adding Eq. I to Eq. II,

$$\delta_b - \delta_{st} = \frac{FL}{A_{st} E_{st}} + \frac{FL}{A_b E_b}$$

$$= FL \left(\frac{1}{A_{st} E_{st}} + \frac{1}{A_b E_b}\right)$$

$$F = \frac{\dfrac{\delta_b - \delta_{st}}{L}}{\dfrac{1}{A_{st} E_{st}} + \dfrac{1}{A_b E_b}}$$

$$= \frac{\dfrac{0.002\,97 \text{ m} - 0.001\,93 \text{ m}}{1.5 \text{ m}}}{\dfrac{1}{(0.002\,278 \text{ m}^2)(20 \times 10^{10} \text{ Pa})} + \dfrac{1}{(0.000\,706\,9 \text{ m}^2)(10 \times 10^{10} \text{ Pa})}}$$

$$= 4.24 \times 10^4 \text{ N}$$

$$\sigma_{st} = \frac{F}{A_{st}} = \frac{4.24 \times 10^4 \text{ N}}{0.002\,278 \text{ m}^2}$$

$$= \boxed{1.86 \times 10^7 \quad (18.6 \text{ MPa})}$$

$$\sigma_b = \frac{F}{A_b} = \frac{4.24 \times 10^4 \text{ N}}{0.000\,706\,9 \text{ m}^2}$$

$$= \boxed{6 \times 10^7 \text{ Pa} \quad (60.0 \text{ MPa})}$$

7. *Customary U.S. Solution*

The deflection of the beam is $\delta_b = P_b L^3 / 3EI$ where P_b is the net load at the beam tip. If P_c is the tension in the cable,

$$\delta_c = \frac{P_c L_c}{AE}$$

$\delta_c = \delta_b$ is the constraint on the deformation. Therefore,

$$\frac{P_b L_b^3}{3EI} = \frac{P_c L_c}{AE} \quad \text{[Eq. I]}$$

Another equation is the equilibrium equation.

$$F - P_c = P_b \quad \text{[Eq. II]}$$

Solving Eq. 1 and Eq. II simultaneously,

$$P_c = \frac{\dfrac{F L_b^3}{3I}}{\dfrac{L_c}{A} + \dfrac{L_b^3}{3I}}$$

$$= \frac{\dfrac{(270 \text{ lbf})\left((4 \text{ ft})\left(12 \, \frac{\text{in}}{\text{ft}}\right)\right)^3}{(3)(10 \text{ in}^4)}}{\dfrac{(2 \text{ ft})\left(12 \, \frac{\text{in}}{\text{ft}}\right)}{0.0124 \text{ in}^2} + \dfrac{\left((4 \text{ ft})\left(12 \, \frac{\text{in}}{\text{ft}}\right)\right)^3}{(3)(10 \text{ in}^4)}}$$

$$= \boxed{177 \text{ lbf} \quad (180 \text{ lbf})}$$

The answer is (A).

SI Solution

The deflection of the beam is $\delta_b = P_b L^3 / 3EI$ where P_b is the net load at the beam tip. If P_c is the tension in the cable,

$$\delta_c = \frac{P_c L_c}{AE}$$

$\delta_c = \delta_b$ is the constraint on the deformation. Therefore,

$$\frac{P_b L_b^3}{3EI} = \frac{P_c L_c}{AE} \quad \text{[Eq. I]}$$

Another equation is the equilibrium equation.

$$F - P_c = P_b \quad \text{[Eq. II]}$$

Solving Eq. I and Eq. II simultaneously,

$$P_c = \frac{\dfrac{FL_b^3}{3I}}{\dfrac{L_c}{A} + \dfrac{L_b^3}{3I}}$$

$$= \frac{\dfrac{(1200 \text{ N})(1.2 \text{ m})^3 \left(1000 \, \dfrac{\text{mm}}{\text{m}}\right)^4}{(3)(4.17 \times 10^6 \text{ mm}^4)}}{\dfrac{(0.6 \text{ m}) \left(1000 \, \dfrac{\text{mm}}{\text{m}}\right)^2}{8 \text{ mm}^2}}$$

$$+ \frac{(1.2 \text{ m})^3 \left(1000 \, \dfrac{\text{mm}}{\text{m}}\right)^4}{(3)(4.17 \times 10^6 \text{ mm}^4)}$$

$$= \boxed{778 \text{ N} \quad (780 \text{ N})}$$

The answer is (A).

8. Let deflection downward be positive.

The deflection at the center of the beam is

$$\delta_b = \frac{5w(2L)^4}{384EI} - \frac{F(2L)^3}{48EI}$$

F is the force applied by the column at the beam center. The beam deflection is equal to the shortening of the column.

$$\delta_c = \frac{Fh}{EA}$$

Since $\delta_b = \delta_c$,

$$\frac{Fh}{EA} = \frac{5w(2L)^4}{384EI} - \frac{F(2L)^3}{48EI}$$

$$F = \boxed{\frac{5AwL^4}{24hI + 4AL^3}} \quad \text{[Eq. I]}$$

Another equation is the equilibrium equation. Let R_1 and R_2 be the left and right support reactions, respectively, on the beam. By symmetry,

$$R_1 = R_2 = R$$

$$2R + F - 2wL = 0 \quad \text{[Eq. II]}$$

Solving Eq. I and Eq. II simultaneously,

$$R = \frac{2wL - F}{2}$$

$$= \boxed{wL \left(\frac{48hI + 3AL^3}{48hI + 8AL^3}\right)}$$

9. *Customary U.S. Solution*

(a) Let F_1 and F_2 and δ_1 and δ_2 be the tensions and the deformations in the cables, respectively. The moment equilibrium equation is taken at the hinge of the rigid bar and is

$$\sum M_o = aF_1 + (a+b)F_2 - (a+b+c)P = 0 \quad \text{[Eq. I]}$$

The relationship between the elongations is

$$\frac{\delta_1}{a} = \frac{\delta_2}{a+b}$$

This can be rewritten as

$$\frac{F_1 L_1}{AEa} = \frac{F_2 L_2}{AE(a+b)}$$

Since $L_1/a = L_2/(a+b)$,

$$F_1 = F_2 \quad \text{[Eq. II]}$$

Solving Eq. I and Eq. II,

$$aF + (a+b)F - (a+b+c)P = 0$$

$$F = F_1 = F_2 = \left(\frac{a+b+c}{2a+b}\right)P$$

$$= \left(\frac{6 \text{ ft} + 3 \text{ ft} + 3 \text{ ft}}{(2)(6 \text{ ft}) + 3 \text{ ft}}\right)(4500 \text{ lbf})$$

$$= \boxed{3600 \text{ lbf}}$$

The answer is (B).

(b) Take downward as a positive deflection. The slope of the rigid member is

$$\frac{d}{a+b+c}$$

$$\delta_1 = \left(\frac{F_1}{AE}\right)L_1 = \left(\frac{F_1}{AE}\right)\left(\frac{d}{a+b+c}\right)a$$

$$= \left(\frac{3600 \text{ lbf}}{(0.124 \text{ in}^2)\left(10 \times 10^6 \ \frac{\text{lbf}}{\text{in}^2}\right)}\right)$$

$$\times \left(\frac{12 \text{ ft}}{6 \text{ ft} + 3 \text{ ft} + 3 \text{ ft}}\right)(6 \text{ ft})\left(12 \ \frac{\text{in}}{\text{ft}}\right)$$

$$= \boxed{0.21 \text{ in}}$$

$$\delta_2 = \left(\frac{F_2}{AE}\right)L_2 = \left(\frac{F_2}{AE}\right)\left(\frac{d}{a+b+c}\right)(a+b)$$

$$= \left(\frac{3600 \text{ lbf}}{(0.124 \text{ in}^2)\left(10 \times 10^6 \ \frac{\text{lbf}}{\text{in}^2}\right)}\right)$$

$$\times \left(\frac{12 \text{ ft}}{6 \text{ ft} + 3 \text{ ft} + 3 \text{ ft}}\right)(6 \text{ ft} + 3 \text{ ft})\left(12 \ \frac{\text{in}}{\text{ft}}\right)$$

$$= \boxed{0.31 \text{ in}}$$

The answer is (B).

SI Solution

(a) Let F_1 and F_2 and δ_1 and δ_2 be the tensions and the deformations in the cables, respectively. The moment equilibrium equation is taken at the hinge of the rigid bar and is

$$\sum M_o = aF_1 + (a+b)F_2 - (a+b+c)P = 0 \quad \text{[Eq. I]}$$

The relationship between the elongations is

$$\frac{\delta_1}{a} = \frac{\delta_2}{a+b}$$

This can be rewritten as

$$\frac{F_1 L_1}{AEa} = \frac{F_2 L_2}{AE(a+b)}$$

Since

$$\frac{L_1}{a} = \frac{L_2}{a+b}$$

$$F_1 = F_2 \quad \text{[Eq. II]}$$

Solving Eq. I and Eq. II,

$$aF + (a+b)F - (a+b+c)P = 0$$

$$F = F_1 = F_2 = \left(\frac{a+b+c}{2a+b}\right)P$$

$$= \left(\frac{2 \text{ m} + 1 \text{ m} + 1 \text{ m}}{(2)(2 \text{ m}) + 1 \text{ m}}\right)(20 \text{ kN})$$

$$= \boxed{16 \text{ kN}}$$

The answer is (B).

(b) Take downward as a positive deflection. The slope of the rigid member is

$$\frac{d}{a+b+c}$$

$$\delta_1 = \left(\frac{F_1}{AE}\right)L_1 = \left(\frac{F_1}{AE}\right)\left(\frac{d}{a+b+c}\right)a$$

$$= \left(\frac{(16\,000 \text{ N})\left(1000 \ \frac{\text{mm}}{\text{m}}\right)^2}{(80 \text{ mm}^2)(70 \times 10^9 \text{ Pa})}\right)$$

$$\times \left(\frac{4 \text{ m}}{2 \text{ m} + 1 \text{ m} + 1 \text{ m}}\right)(2 \text{ m})\left(1000 \ \frac{\text{mm}}{\text{m}}\right)$$

$$= \boxed{5.7 \text{ mm}}$$

$$\delta_2 = \left(\frac{F_2}{AE}\right)L_2 = \left(\frac{F_2}{AE}\right)\left(\frac{d}{a+b+c}\right)(a+b)$$

$$= \left(\frac{(16\,000 \text{ N})\left(1000 \ \frac{\text{mm}}{\text{m}}\right)^2}{(80 \text{ mm}^2)(70 \times 10^9 \text{ Pa})}\right)$$

$$\times \left(\frac{4 \text{ m}}{2 \text{ m} + 1 \text{ m} + 1 \text{ m}}\right)(2 \text{ m} + 1 \text{ m})\left(1000 \ \frac{\text{mm}}{\text{m}}\right)$$

$$= \boxed{8.6 \text{ mm}}$$

The answer is (B).

10. Let subscript s refer to the supported beam and subscript c refer to the cantilever beam. The deflections are equal.

$$\delta_s = \delta_c$$

$$\frac{P_s L^3}{48EI} = \frac{P_c L^3}{3EI}$$

$$P_s = 16P_c \quad \text{[Eq. I]}$$

P_s and P_c are the net loads exerted on the supported beam center and the cantilever beam tip, respectively. The equilibrium equation for the supported beam is

$$\sum F_{s,y} = 2R - P_s = 0 \quad \text{[Eq. II]}$$

Structural

The equilibrium equation for the cantilever beam is

$$\sum F_{c,y} = P_c = F - P_s \quad \text{[Eq. III]}$$

Solving Eq. I and Eq. III simultaneously,

$$P_c = \frac{F}{17}$$

$$P_s = \frac{16}{17}F$$

From Eq. II,

$$R = \boxed{\frac{8}{17}F}$$

The answer is (C).

11. Assume deflection downward is positive.

step 1: Remove support 2 to make the structure statically determinate.

step 2: The deflection at the location of (removed) support 2 is the sum of the deflections induced by the discrete load and the distributed load.

From a beam deflection table,

$$\delta_{\text{discrete}} = \frac{Pb}{6EIL}\left(\left(\frac{L}{b}\right)(x-a)^3 + (L^2 - b^2)x - x^3\right)$$

$$(P = 5000, L = 15, a = 6, b = 9, x = 10)$$

From a beam deflection table,

$$\delta_{\text{distributed}} = \frac{-w}{24EI}\left(L^3x - 2Lx^3 + x^4\right)$$

step 3: The deflection induced by R_2 alone considered as a load is

$$\delta_{R_2} = \frac{-R_2 a^2 b^2}{3EIL}$$

step 4: The total deflection at the location of support 2 is zero.

$$\delta_{\text{discrete}} + \delta_{\text{distributed}} + \delta_{R_2} = 0$$

$$\left(\frac{(5000)(9)}{(6)(15)}\right)$$

$$\times \left(\left(\frac{15}{9}\right)(10-6)^3 + \left((15)^2 - (9)^2\right)(10) - (10)^3\right)$$

$$+ \left(\frac{1000}{24}\right)\left((15)^3(10) - (2)(15)(10)^3 + (10)^4\right)$$

$$+ \frac{-R_2(10)^2(5)^2}{(3)(15)} = 0$$

$$R_2 = \boxed{15,233 \quad (15,000)}$$

The answer is (C).

12. First, remove the prop to make the structure statically determinate. The deflection induced by the distributed load at the tip is

$$\delta_{\text{distributed}} = \frac{-wL^4}{30EI} \quad \text{[down]}$$

The deflection caused by load R is

$$\delta_R = \frac{RL^3}{3EI} \quad \text{[up]}$$

Since the deflection is actually zero at the tip,

$$\delta_{\text{distributed}} + \delta_R = 0$$

$$\frac{-wL^4}{30EI} + \frac{RL^3}{3EI} = 0$$

$$R = \frac{wL}{10} = \frac{(1000)(10)}{10} = \boxed{1000}$$

The answer is (B).

13. *Customary U.S. Solution*

Use the three-moment method. The first moment of the area is

$$A_1 a = \tfrac{1}{6}Fc(L^2 - c^2)$$

$$= \left(\tfrac{1}{6}\right)(13,600 \text{ lbf})(22.5 \text{ ft})\left((32.5 \text{ ft})^2 - (22.5 \text{ ft})^2\right)$$

$$= 28,050,000 \text{ ft}^3\text{-lbf}$$

Since there is no force between R_2 and R_3, $A_2 b = 0$.

The left and right ends of the beam are simply supported; M_1 and M_3 are zero. Therefore, the three-moment equation becomes

$$2M_2(32.5 \text{ ft} + 20 \text{ ft}) = (-6)\left(\frac{28,050,000 \text{ ft}^3\text{-lbf}}{32.5 \text{ ft}}\right)$$

$$M_2 = -49,318.7 \text{ ft-lbf}$$

M_2 can be written in terms of the load and reactions to the left of support 2.

$$M_2 = (-13,600 \text{ lbf})(10 \text{ ft}) + (32.5 \text{ ft})R_1$$

$$= -49,318.7 \text{ ft-lbf}$$

$$R_1 = \boxed{2667.1 \text{ lbf}}$$

Now that R_1 is known, moments can be taken about support 3 to the left.

$$\sum M_3 = (2667.1 \text{ lbf})(52.5 \text{ ft}) - (13{,}600 \text{ lbf})(30 \text{ ft})$$
$$+ R_2(20 \text{ ft}) = 0$$
$$R_2 = \boxed{13{,}398.9 \text{ lbf}}$$

R_3 can be obtained by taking moments about support 1.

$$\sum M_1 = (22.5 \text{ ft})(13{,}600 \text{ lbf}) - (32.5 \text{ ft})(13{,}398.9 \text{ lbf})$$
$$- (52.5)R_3 = 0$$
$$R_3 = \boxed{-2466.0 \text{ lbf}}$$

SI Solution

Use the three-moment method. The first moment of the area is

$$A_1 a = \tfrac{1}{6} F c (L^2 - c^2)$$
$$= \left(\tfrac{1}{6}\right)(60\,500 \text{ N})(7 \text{ m})\left((10 \text{ m})^2 - (7 \text{ m})^2\right)$$
$$= 3.60 \times 10^6 \text{ N·m}^3$$

Since there is no force between R_2 and R_3, $A_2 b = 0$.

The left and right ends of the beam are simply supported; M_1 and M_3 are zero. Therefore, the three-moment equation becomes

$$2M_2(10 \text{ m} + 6 \text{ m}) = (-6)\left(\frac{3.6 \times 10^6 \text{ N·m}^3}{10 \text{ m}}\right)$$
$$M_2 = -67\,500 \text{ N·m}$$

M_2 can be written in terms of the load and reactions to the left of support 2.

$$M_2 = (-60\,500 \text{ N})(10 \text{ m} - 7 \text{ m}) + (10 \text{ m})R_1$$
$$= -67\,500 \text{ N·m}$$
$$R_1 = \boxed{11\,400 \text{ N}}$$

Now that R_1 is known, moments can be taken about support 3 to the left.

$$\sum M_3 = (11\,400 \text{ N})(16 \text{ m}) - (60\,500 \text{ N})(9 \text{ m})$$
$$+ R_2(6 \text{ m}) = 0$$
$$R_2 = \boxed{60\,350 \text{ N}}$$

R_3 can be obtained by taking moments about support 1.

$$\sum M_1 = (7 \text{ m})(60\,500 \text{ N}) - (10 \text{ m})(60\,350 \text{ N})$$
$$- (16 \text{ m})R_3 = 0$$
$$R_3 = \boxed{-11\,250 \text{ N}}$$

14. Use the three-moment method. The first moments of the areas are

$$A_1 a = \tfrac{1}{6} F c (L_1^2 - c^2) = \left(\tfrac{1}{6}\right)(5000)(4)\left((10)^2 - (4)^2\right)$$
$$= 280{,}000$$
$$A_2 b = \frac{w L_2^4}{24} = \frac{(1000)(5)^4}{24} = 26{,}042$$

The two ends of the beam are simply supported; M_1 and M_3 are zero. Therefore, the three moment equation becomes

$$2M_2(10 + 5) = (-6)\left(\frac{280{,}000}{10} + \frac{26{,}042}{5}\right)$$
$$M_2 = -6642$$

M_2 can also be written in terms of the load and reactions to the left of support 2.

$$M_2 = (-5000)(6) + 10R_1 = -6642$$
$$R_1 = \boxed{2336}$$

Sum the moments around support 3.

$$\sum M_3 = (2336)(15) - (5000)(11) + 5R_2$$
$$- (5000)(2.5) = 0$$
$$R_2 = \boxed{6492}$$

Sum the moments around support 1.

$$\sum M_1 = (5000)(4) - (6492)(10) + (5000)(12.5)$$
$$- 15R_3 = 0$$
$$R_3 = \boxed{1172}$$

15. The equilibrium requirement is

$$R_1 + R_2 = 1000$$

The moment equation at support 1 is

$$\sum M_{R_1} = M_1 + M_2 + (1000)(30) - 40R_2 = 0$$

From a table of fixed-end moments (see App. 47.A),

$$M_1 = \frac{-Fb^2a}{L^2} = \frac{-(1000)(10)^2(30)}{(40)^2} = -1875$$

$$M_2 = \frac{Fa^2b}{L^2} = \frac{(1000)(30)^2(10)}{(40)^2} = 5625$$

$$R_2 = \boxed{843.75}$$

The moment equation at support 2 is

$$\sum M_{R_2} = M_1 + M_2 - (1000)(10) + 40R_1 = 0$$
$$R_1 = \boxed{156.25}$$

The equilibrium requirement is

$$R_1 + R_2 = 1000$$
$$156.25 + 843.75 = 1000 \quad \text{[check]}$$

16. (a) The force in member Bb depends on the shear, V, across the cut shown.

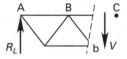

- Influence diagram for shear across panel Bb:

If the unit load is to the right of point C, the reaction R_L will be

$$R_L = \frac{x}{120 \text{ ft}}$$

(x is the distance from the right reaction to the unit load.)

$$V_L = R_L = \frac{x}{120 \text{ ft}}$$

If the unit load is to the left of point B,

$$R_L = \frac{x}{120 \text{ ft}}$$
$$V = R_L - 1 = \frac{x}{120 \text{ ft}} - 1$$

At points B and C,

$$V_B = \frac{90 \text{ ft}}{120 \text{ ft}} - 1 = -0.25$$
$$V_C = \frac{60 \text{ ft}}{120 \text{ ft}} = 0.5$$

The influence diagram for shear in member Bb is

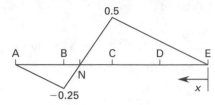

The neutral point, N, is located at

$$x = (2)(30 \text{ ft}) + \left(\frac{0.5}{0.5 + 0.25}\right)(30 \text{ ft}) = 80 \text{ ft}$$

- Maximum shear due to moving uniform load:

The moving load, perhaps representing a stream of cars, is allowed to be over any part or all of the bridge deck. The shear will be maximum in member Bb if the load is distributed from N to E.

The area under the influence line from N to E is

$$\left(\tfrac{1}{2}\right)\left((20 \text{ ft})(0.5) + (2)(30 \text{ ft})(0.5)\right) = 20 \text{ ft}$$

The maximum shear, V, is

$$(20 \text{ ft})\left(2 \ \frac{\text{kips}}{\text{ft}}\right) = 40 \text{ kips}$$

- Maximum shear due to moving concentrated load:

From the influence diagram, maximum shear will occur when the concentrated load is at point C. The shear in panel Bb is

$$(0.5)(15 \text{ kips}) = 7.5 \text{ kips}$$

- Tension in member Bb:

The force triangle is

The total maximum shear across panel Bb is

$$40 \text{ kips} + 7.5 \text{ kips} = 47.5 \text{ kips}$$

The total maximum tension in member Bb is

$$(47.5 \text{ ft})\left(\frac{25 \text{ ft}}{20 \text{ ft}}\right) = \boxed{59.375 \text{ kips}}$$

The answer is (C).

(b) • Influence diagram for moment at point b:

Since the horizontal member BC cannot resist vertical shear, the shear influence diagram previously used will not work.

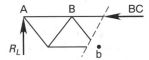

With no loads between A and C, the force in member BC can be found by summing the moments about point b. Taking clockwise moments as positive,

$$\sum M_{\text{b}} = (45 \text{ ft})R_L - (20 \text{ ft})(\text{BC}) = 0$$
$$\text{BC} = \frac{45R_L}{20 \text{ ft}}$$

$45R_L$ is the moment that the moment from force BC opposes. In general,

$$\text{BC} = \frac{M_{\text{b}}}{20 \text{ ft}}$$

If the load is between C and E,

$$R_L = \frac{x}{120 \text{ ft}} \qquad [x \text{ is measured from E}]$$

The moment caused by R_L is

$$M_{\text{b}} = (45 \text{ ft})\left(\frac{x}{120 \text{ ft}}\right) = 0.375x$$

If the load is between A and B, the reaction is

$$R_L = \frac{x}{120 \text{ ft}}$$

The moment at b is also affected by the load between A and B.

$$M_{\text{b}} = 0.375x - (1)(x - 75 \text{ ft})$$
$$= 75 \text{ ft} - 0.625x$$

Plotting these values versus x,

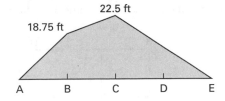

• Maximum moment due to uniform load:

The moment at b is maximum when the entire truss is loaded from A to E. The area under the curve is

$$\left(\tfrac{1}{2}\right)(30 \text{ ft})(18.75 \text{ ft}) + \left(\tfrac{1}{2}\right)(60 \text{ ft})(22.5 \text{ ft})$$
$$+ (30 \text{ ft})(18.75 \text{ ft})$$
$$+ \left(\tfrac{1}{2}\right)(30 \text{ ft})(22.5 \text{ ft} - 18.75 \text{ ft})$$
$$= 1575 \text{ ft}^2$$

The maximum moment is

$$M_{\text{b}} = \left(2 \, \frac{\text{kips}}{\text{ft}}\right)(1575 \text{ ft}^2) = 3150 \text{ ft-kips}$$

• Maximum moment due to concentrated load:

Maximum moment will occur when the load is at C.

$$M_{\text{b}} = (15 \text{ ft})(22.5 \text{ kips}) = 337.5 \text{ ft-kips}$$

• Total maximum moment:

$$M_{\text{b}} = 337.5 \text{ ft-kips} + 3150 \text{ ft-kips} = 3487.5 \text{ ft-kips}$$

• Compression in BC:

$$\text{BC} = \frac{M_{\text{b}}}{20 \text{ ft}} = \frac{3487.5 \text{ ft-kips}}{20 \text{ ft}}$$
$$= \boxed{174.4 \text{ kips} \quad (170 \text{ kips})}$$

The answer is (D).

17. The force in member CD is a function of the moment at point d.

If the unit load is between d and g, the left reaction is

$$R_L = \frac{x}{162 \text{ ft}}$$

The moment at point d is

$$M_{\text{d}} = (3)(27 \text{ ft})\left(\frac{x}{162 \text{ ft}}\right) = 0.5x \quad [\text{d to g}]$$

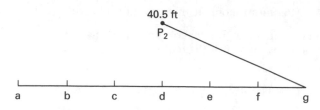

If the unit load is between a and c, the moment at point d is

$$M_{\rm d} = 0.5x - (1)(x - (3)(27 \text{ ft}))$$
$$= 81 \text{ ft} - 0.5x \quad [\text{a to c}]$$

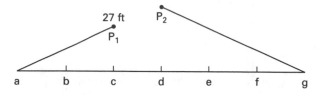

Complete the influence diagram by joining points P_1 and P_2. Observe that the slope of P_1 is the same as that of P_2. This is because point d is at the center of the truss.

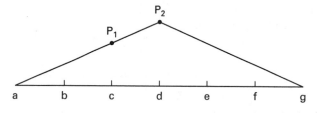

The resultant of the load group and its location are

$$8 \text{ kips} + 40 \text{ kips} + 7 \text{ kips} + 30 \text{ kips} = 85 \text{ kips}$$

$$\frac{\begin{array}{c}(40 \text{ kips})(16 \text{ ft}) + (7 \text{ kips})(16 \text{ ft} + 28 \text{ ft}) \\ + (30 \text{ kips})(16 \text{ ft} + 28 \text{ ft} + 16 \text{ ft})\end{array}}{85 \text{ kips}} = 32.33 \text{ ft}$$

The resultant is located 32.33 ft to the right of the 8 kip load.

Assume the load group moves from right to left.

● Case 1:

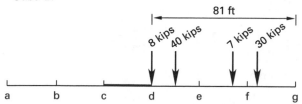

Taking counterclockwise moments as positive,

$$\sum M_{\rm g} = (85 \text{ kips})(81 \text{ ft} - 32.33 \text{ ft}) - (162 \text{ ft})R_L = 0$$
$$R_L = 25.54 \text{ kips}$$

The shear does not change sign under panel cd.

● Case 2:

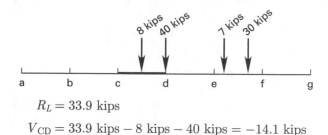

$$R_L = 33.9 \text{ kips}$$

$$V_{\rm CD} = 33.9 \text{ kips} - 8 \text{ kips} - 40 \text{ kips} = -14.1 \text{ kips}$$

Since the shear changes sign (goes through zero), the moment is maximum when the 40 kip load is at point d.

load	x	influence diagram height	moment
8 kips	97 ft	32.5 ft	$(8)(32.5) = 260$ ft-kips
40 kips	81 ft	40.5 ft	$(40)(40.5) = 1620$ ft-kips
7 kips	53 ft	26.5 ft	$(7)(26.5) = 185.5$ ft-kips
30 kips	37 ft	18.5 ft	$(30)(18.5) = 555$ ft-kips
total			2620.5 ft-kips

The compression in CD is

$$CD = \frac{2620.5 \text{ ft-kips}}{32 \text{ ft}} = \boxed{81.9 \text{ kips}}$$

The answer is (B).

Since $7 + 30 < 8 + 40$, if the load moves from left to right it must reach the same position as case 2 for the moment to be maximum. Therefore, the left-to-right analysis is not needed.

● Alternate solution:

Having determined that the 40 kip load should be at point d, find the left reaction.

$$\sum M_{R_R} = (162 \text{ ft})R_L - (8 \text{ kips})(97 \text{ ft})$$
$$- (40 \text{ kips})(81 \text{ ft}) - (7 \text{ kips})(53 \text{ ft})$$
$$- (30 \text{ kips})(37 \text{ ft}) = 0$$
$$R_L = 33.93 \text{ kips}$$

Sum the moments about point d and use the method of sections.

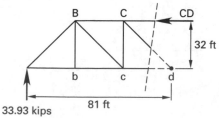

Taking clockwise moments as positive,

$$\sum M_d = (33.93 \text{ kips})(81 \text{ ft}) - (32 \text{ ft})(CD)$$
$$- (8 \text{ kips})(16 \text{ ft}) = 0$$
$$CD = \boxed{81.9 \text{ kips}}$$

The answer is (B).

18. (a) Moment:

Put a hinge at point B and rotate.

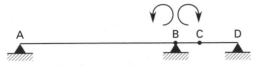

The moment influence diagram is

One of the loads should be at point C.

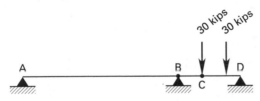

Since the slope of the influence line is less between C and D, the ordinate 6 ft to the right of point C will be larger than the ordinate 6 ft to the left of point C. The reaction at the hinge due to the 30 kip load 6 ft to the right of point C is

$$R = (30 \text{ kips})\left(\frac{15 \text{ ft} - 6 \text{ ft}}{15 \text{ ft}}\right) = 18 \text{ kips}$$

The moment at point B is

$$M_B = (9 \text{ ft})(30 \text{ kips} + 18 \text{ kips})$$
$$= \boxed{432 \text{ ft-kips}}$$

The answer is (D).

(b) Shear:

Use the method of virtual displacement to draw the shear influence diagram. Since the point is a reaction point, lift the point a distance of 1. The shear diagram is

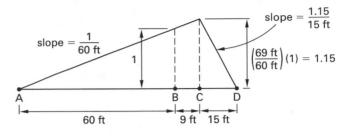

For maximum shear, one load or the other must be at point C.

By inspection, the effect of having both loads to the left of C is greater than having them to the right.

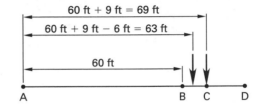

The maximum shear is

$$V_{\max} = (30 \text{ kips})\left(\frac{63 \text{ ft}}{60 \text{ ft}} + \frac{69 \text{ ft}}{60 \text{ ft}}\right) = \boxed{66 \text{ kips}}$$

The answer is (C).

47 Structural Analysis II

PRACTICE PROBLEMS

1. (*Time limit: one hour*) A single-story rigid timber frame is subjected to a uniform load acting over the beam plus a concentrated load at midheight of the left side column. The two columns are solid square, 14 in × 14 in. The beam is 12 in wide and 24 in deep. Determine the moment distribution.

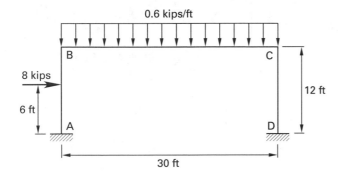

2. A cantilever beam with a cable attached to its free end is loaded uniformly as shown. For the dimensions, properties, and loading indicated, determine (a) the force in the cable and (b) the maximum moment in the beam.

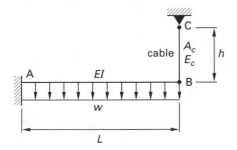

$w = 1$ kip/ft
$L = 10$ ft
$h = 5$ ft
$E = 29{,}000$ kips/in^2
$E_c = 20{,}000$ kips/in^2
$A_c = 0.4$ in^2
$I = 200$ in^4

3. The column is loaded as shown. The product EI has a value of 10^4 ft^2-kips.

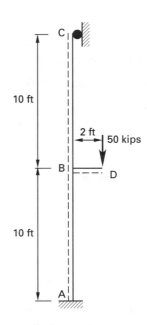

(a) What is the degree of static indeterminacy?

    (A) 0 (determinate)

    (B) 1

    (C) 2

    (D) 3

(b) Using the convention that compression on the side of the dashed line is positive, the moment at point A is most nearly

    (A) −25 ft-kips

    (B) 13 ft-kips

    (C) 25 ft-kips

    (D) 50 ft-kips

(c) The shear at point A is most nearly

    (A) 0 kips

    (B) 5.6 kips

    (C) 10 kips

    (D) 50 kips

(d) The axial force in bar CB is most nearly

(A) 0 kips

(B) 25 kips (tension)

(C) 50 kips (compression)

(D) 50 kips (tension)

(e) The shear in bar CB is most nearly

(A) 0 kips

(B) 5.6 kips

(C) 10 kips

(D) 50 kips

(f) The point of inflection in bar AB is most nearly

(A) at A (there is no inflection point)

(B) 2.2 ft from A

(C) 5 ft from A

(D) 7.5 ft from A

(g) The maximum moment in either AB or BC is most nearly

(A) −56 ft-kips

(B) 50 ft-kips

(C) 56 ft-kips

(D) 100 ft-kips

(h) Using the convention that deflection to the right is positive, the horizontal displacement of point B is most nearly

(A) −0.03 ft

(B) 0 ft

(C) 0.03 ft

(D) 0.06 ft

(i) If the support A is replaced by a pin, the moment at B in member BA will be most nearly

(A) −50 ft-kips

(B) −25 ft-kips

(C) 25 ft-kips

(D) 100 ft-kips

(j) If the support at A is replaced by a pin, the maximum shear in member CB will be most nearly

(A) 0 kips

(B) 5 kips

(C) 7.5 kips

(D) 10 kips

4. The three structures shown are designated as S1, S2, and S3.

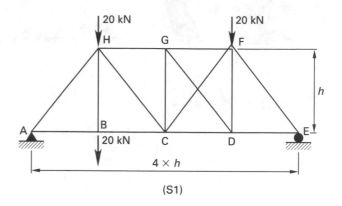

(S1)

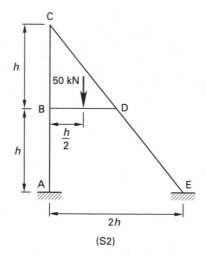

(S2)

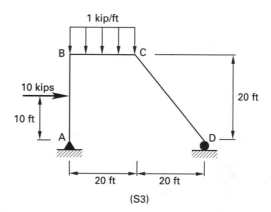

(S3)

(a) The truss S1 is

(A) unstable

(B) statically indeterminate to the first degree

(C) statically indeterminate to the second degree

(D) determinate

(b) The rigid frame S2 is

(A) statically indeterminate to the third degree

(B) statically indeterminate to the sixth degree

(C) statically indeterminate to the ninth degree

(D) determinate

(c) For truss S1, the force in bar HB is most nearly

(A) 0 kN

(B) 20 kN (compression)

(C) 20 kN (tension)

(D) 40 kN (compression)

(d) For the truss S1, the force in bar HC is more nearly

(A) 0 kN

(B) 7.1 kN (compression)

(C) 7.1 kN (tension)

(D) 40 kN (compression)

(e) The degree of kinematic indeterminacy in frame S2 is

(A) 0 (determinate)

(B) 3

(C) 6

(D) 9

(f) The sum of the horizontal reactions in frame S2 is most nearly

(A) 0 kN

(B) 13 kN

(C) 25 kN

(D) 50 kN

(g) For frame S3, the axial force in bar CD is most nearly

(A) 0 kips

(B) 5.3 kips (compression)

(C) 11 kips (compression)

(D) 21 kips (compression)

(h) For frame S3, the maximum shear force on member BC is most nearly

(A) 10 kips

(B) 13 kips

(C) 15 kips

(D) 20 kips

(i) For frame S3, the maximum positive moment on bar BC is most nearly

(A) 0 ft-kips

(B) 50 ft-kips

(C) 75 ft-kips

(D) 180 ft-kips

(j) For the frame S2, if the axial force on bar BC is assumed negligible and the axial force on AB is 5 kN, the maximum shear on beam BD is most nearly

(A) 0 kN

(B) 10 kN

(C) 45 kN

(D) 50 kN

5. All supports on the continuous beam shown are simple. What are the reactions?

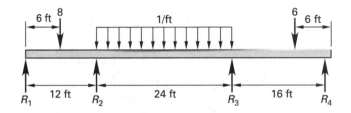

6. In the structure shown, the two support points are pinned and the two joints are rigid. (a) What are the magnitudes of the moments at points B and C? (b) What are the reactions?

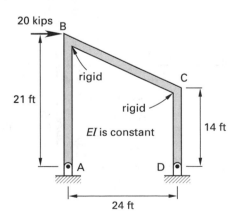

7. A frame with rigid joints is shown. (a) What are the horizontal reactions? (b) What are the joint moments at points B, C, and D?

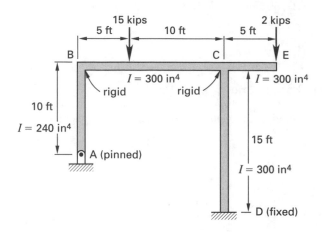

8. What are the magnitudes of the maximum moments on spans AB, BC, and CD?

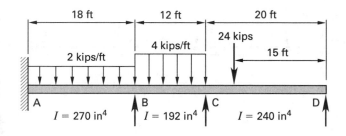

9. Determine the deflection of the frame shown assuming that the two connections are (a) pinned and (b) rigid.

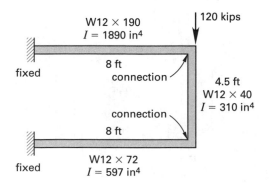

10. (*Time limit: one hour*) A continuous beam (W24 × 76, A36 steel) is loaded with a series of 20 kip point loads as shown. The beam contains two expansion joints that can be assumed to act as frictionless hinges. Determine the (a) maximum shear stress, (b) maximum bending stress, and (c) midpoint deflection.

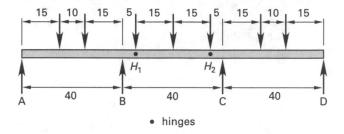

11. (*Time limit: one hour*) All joints are rigid in the structure shown. The beam self-weights are to be disregarded. (a) Determine the horizontal reactions at A, C, and D. (b) Draw the moment diagram on the tension side of the frame.

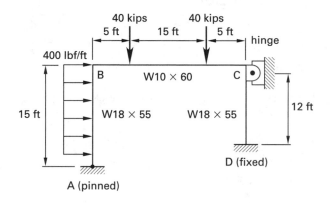

SOLUTIONS

1. Each built-in column base has an unknown horizontal force, vertical force, and moment, for a total of six unknowns. There are three equilibrium equations, so the structure is indeterminate to the third $(6-3=3)$ degree. A flexibility analysis requires the computation of nine independent deflection terms. For a stiffness solution, the degree of kinematic indeterminacy is 6 if axial deformations are considered and is still 3 if axial deformations are neglected. The moment distribution method requires only a single sway restraint, and thus requires the least amount of effort.

The centroidal moments of inertia are

$$I_{\text{column}} = \tfrac{1}{12}bh^3 = \left(\tfrac{1}{12}\right)(14 \text{ in})^4$$
$$= 3201.3 \text{ in}^4$$
$$I_{\text{beam}} = \left(\tfrac{1}{12}\right)(12 \text{ in})(24 \text{ in})^3$$
$$= 13,824 \text{ in}^4$$

Stiffness is $4EI/L$. However, the $4E$ term appears in each stiffness, so that quantity can be omitted. The final units are irrelevant as long as they are consistent.

$$K_{\text{column}} = \frac{I}{L} = \frac{3201.3 \text{ in}^4}{(12 \text{ ft})\left(12 \ \frac{\text{in}}{\text{ft}}\right)}$$
$$= 22.23$$
$$K_{\text{beam}} = \frac{13,824 \text{ in}^4}{(30 \text{ ft})\left(12 \ \frac{\text{in}}{\text{ft}}\right)}$$
$$= 38.4$$

The distribution factors are

$$DF_{\text{AB}} = DF_{\text{DC}} = 0 \quad \text{[fixed ends]}$$
$$DF_{\text{BA}} = DF_{\text{CD}} = \frac{22.23}{22.33 + 38.4} = 0.37$$
$$DF_{\text{BC}} = DF_{\text{CB}} = \frac{38.4}{22.33 + 38.4} = 0.63$$

The fixed-end moments due to loading are found from App. 47.A.

$$\text{FEM}_{\text{AB}} = \text{FEM}_{\text{BA}} = \frac{PL}{8} = \frac{(8 \text{ kips})(12 \text{ ft})}{8}$$
$$= 12 \text{ ft-kips}$$

$$\text{FEM}_{\text{BC}} = \text{FEM}_{\text{CB}} = \frac{wL^2}{12} = \frac{\left(0.6 \ \frac{\text{kips}}{\text{ft}}\right)(30 \text{ ft})^2}{12}$$
$$= 45 \text{ ft-kips}$$

The moment distribution procedure is recorded on the line drawing of the structure.

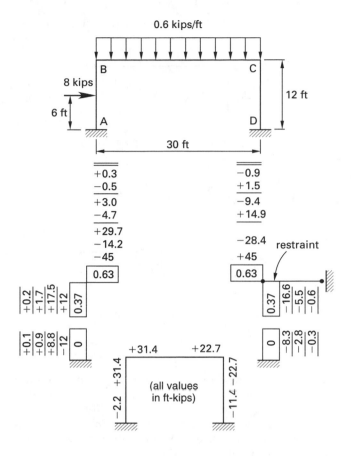

The horizontal reactions at A and D are found from the free-body diagrams of the columns. Summing moments about points B and C, respectively,

$$V_{\text{A}} = \frac{31.4 \text{ ft-kips} - 2.2 \text{ ft-kips} - (8 \text{ kips})(6 \text{ ft})}{12 \text{ ft}}$$
$$= 1.57 \text{ kips}$$
$$V_{\text{D}} = \frac{22.7 \text{ ft-kips} + 11.4 \text{ ft-kips}}{12 \text{ ft}}$$
$$= 2.84 \text{ kips}$$

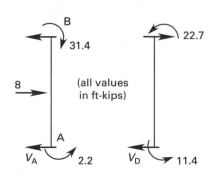

The force in the restraint is found from horizontal equilibrium considerations of the complete structure.

$$F_A = 8 \text{ kips} - 1.57 \text{ kips} - 2.84 \text{ kips} = 3.59 \text{ kips}$$

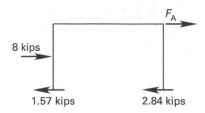

The deformed shape shown results when the restraint is dragged an arbitrary amount.

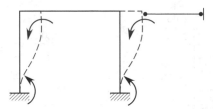

The fixed-end moments that appear on the columns are given by Eq. 47.36. The moments of inertia and lengths are the same for both columns. To simplify the calculations, use an arbitrary value of 100 ft-kips as the fixed-end moment. The fixed-end moments are distributed, and the final results are as shown.

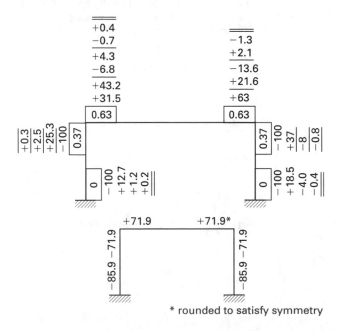

The horizontal reactions in the columns can be found from the free-body diagrams of the column.

$$V_A = V_B = \frac{85.9 \text{ ft-kips} + 71.9 \text{ ft-kips}}{12 \text{ ft}}$$
$$= 13.15 \text{ kips}$$

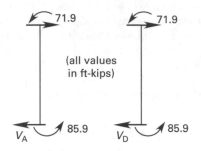

The force in the restraint is found from a free-body diagram of the entire structure.

$$F_B = 13.15 \text{ kips} + 13.15 \text{ kips} = 26.30 \text{ kips}$$

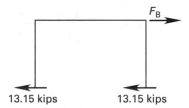

Finally, the effects that derive from force A need to be canceled. The scaling factor is

$$\alpha = \frac{F_A}{F_B} = \frac{3.59 \text{ kips}}{26.30 \text{ kips}} = 0.14$$

At the fixed end on the left side column, for example, the final moment is

$$M_{A,\text{final}} = -2.2 \text{ ft-kips} + (0.14)(-85.9 \text{ ft-kips})$$
$$= -14.2 \text{ ft-kips}$$

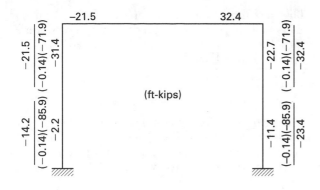

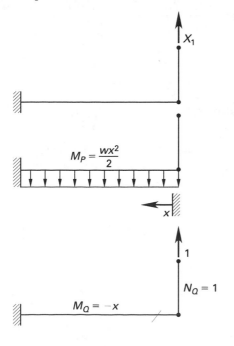

2. (a) Since the cable properties were given, the cable cannot be treated as an unyielding prop. This structure is indeterminate to the first degree, so the flexibility method can be used. First, obtain a general solution for the force in the cable as a function of the parameters involved. Then, the moments in the beam can be calculated.

The illustration shows the selection of the redundant and the internal forces for the $X = 0$ case (the load case) and for the $X_1 = 1$ case.

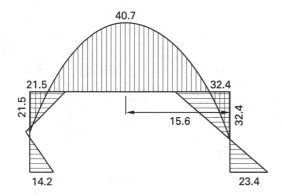

Follow the procedure outlined in the chapter.

$$\delta_1 = \int_0^L \frac{wx^2(-x)}{2EI}\,dx = -\frac{wL^4}{8EI}$$

$$f_{1,1} = \int_0^L \frac{(-x)^2}{EI}\,dx + \int_0^{L_c} \frac{1^2}{A_c E_c}\,dy = \frac{L^3}{3EI} + \frac{L_c}{E_c A_c}$$

Substitute into Eq. 47.24.

$$\left(\frac{L^3}{3EI} + \frac{L_c}{E_c A_c}\right)X_1 = \frac{wL^4}{8EI}$$

Solve for X_1, the force in the cable.

$$X_1 = \frac{wL}{8\left(\dfrac{1}{3} + \dfrac{EIL_c}{E_c A_c L^3}\right)}$$

$$= \frac{\left(1\ \dfrac{\text{kip}}{\text{ft}}\right)(10\ \text{ft})}{(8)\left(\dfrac{1}{3} + \dfrac{\left(29{,}000\ \dfrac{\text{kips}}{\text{in}^2}\right)(200\ \text{in}^4)(5\ \text{ft})\left(12\ \dfrac{\text{in}}{\text{ft}}\right)}{\left(20{,}000\ \dfrac{\text{kips}}{\text{in}^2}\right)(0.4\ \text{in}^2)(120\ \text{in})^3}\right)}$$

$$= \boxed{3.49\ \text{kips}}$$

(b) Once the force in the cable is known, the moment diagram in the beam can be determined. The maximum moment occurs at the fixed end and has a value of $\boxed{15.10\ \text{ft-kips.}}$

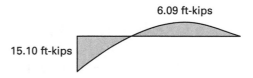

3. (a) When the roller support at C is removed, the structure becomes a statically determinate cantilever.

The degree of indeterminacy is $\boxed{1.}$

The answer is (B).

(b) The 50 kip load introduces a moment of 100 ft-kips at point B. From App. 47.A, case 20, the fixed-end moments due to this applied external moment at each end are

$$a = b = \frac{L}{2}$$

$$\text{FEM}_{AB} = \text{FEM}_{BA} = M\left(\frac{a}{L}\right)3\left(\frac{b}{L} - 1\right)$$

$$= (100 \text{ ft-kips})\left(\frac{1}{2}\right)(3)\left(\frac{1}{2} - 1\right)$$

$$= 25 \text{ ft-kips}$$

Distribute moments as follows: The unbalance at C is 25 ft-kips, so change the sign and multiply by a distribution factor of 1 (since there is a pin at C); carry half (-12.5 ft-kips) of the distributed moment to A; the final moment at A is

$$25 \text{ ft-kips} - 12.5 \text{ ft-kips} =$$

$$\boxed{12.5 \text{ ft-kips} \quad (13 \text{ ft-kips})} \quad \text{[clockwise]}$$

This moment induces compression on the side of the dotted line.

The answer is (B).

(c) Using the fact that the moment at A is 12.5 ft-kips, sum the moments about C.

$$V_A = \frac{12.5 \text{ ft-kips} + 100 \text{ ft-kips}}{20 \text{ ft}}$$

$$= \boxed{5.63 \text{ kips} \quad (5.6 \text{ kips})}$$

The answer is (B).

(d) Since the vertical reaction at C is zero, the axial force in CB is $\boxed{\text{zero.}}$

The answer is (A).

(e) The shear in bar CB is the same as in bar AB: $\boxed{5.63 \text{ kips (5.6 kips).}}$

The answer is (B).

(f) A free-body diagram shows that the moment (in ft-kips) in bar AB is given by $12.5 - 5.625y$, where y is the distance measured from point A. Solving, the point of inflection is located at $y = \boxed{2.22 \text{ ft.}}$

The answer is (B).

(g) The maximum moment occurs at point B on bar BC. This moment equals the product of the reaction at C times the distance to B, or $(5.63 \text{ kips})(10 \text{ ft}) = \boxed{56.3 \text{ ft-kips} \ (56 \text{ ft-kips}).}$

The answer is (C).

(h) Use the dummy load method to compute this deflection. Treating the reaction at C as an applied load, the unit load case consists of a horizontal force acting (to the right) at B on a cantilever fixed at A. For this condition, the moments are

$$m_Q = -10 + y \quad \text{[bar AB]}$$

$$m_Q = 0 \quad \text{[bar BC and bar BD]}$$

Only the moment m_P in bar AB (since the others are multiplied by zero) is needed. From part (f), $m_P = 12.5 - 5.625y$. Therefore, the vertical deflection is given by Eq. 47.9.

$$\Delta_B = \int \frac{m_Q m_P}{EI} \, dy$$

$$= \int_0^{10} \frac{(12.5 - 5.625y)(-10 + y)}{EI} \, dy$$

This is integrated to give the deflection.

$$\Delta_B = \frac{-125y + 34.375y^2 - 1.875y^3}{EI}$$

At $y = 10$,

$$\Delta_B = \frac{(-125)(10) + (34.375)(10)^2 - (1.875)(10)^3}{EI}$$

$$= \frac{312.5 \text{ ft}^3\text{-kips}}{10^4 \text{ ft}^2\text{-kips}}$$

$$= \boxed{0.0325 \text{ ft} \quad (0.03 \text{ ft})}$$

The answer is (C).

(i) If A is replaced by a pin support, the structure becomes statically determinate. The horizontal reactions at A and C are

$$\frac{100 \text{ ft-kips}}{20 \text{ ft}} = 5 \text{ kips} \quad \begin{bmatrix} \text{to the right at A} \\ \text{and to the left at C} \end{bmatrix}$$

The moment at B in member AB is

$$(-5 \text{ kips})(10 \text{ ft}) = \boxed{-50 \text{ ft-kips}}$$

The negative sign indicates tension on the side of the dotted line.

The answer is (A).

(j) As determined in part (i), the horizontal reactions at both ends are $\boxed{5 \text{ kips.}}$

The answer is (B).

4. (a) There are 14 bars, 3 reactions, and 8 joints.

$$\text{bars} + \text{reactions} = 14 + 3 = 17$$
$$2(\text{joints}) = (2)(8) = 16$$

> The truss is statically indeterminate to the first degree.

The answer is (B).

(b)
$$j = \text{no. of joints} = 5$$
$$r - \text{no. of reactions} = 6$$
$$m = \text{no. of members} = 5$$
$$s = \text{no. of special conditions} = 0$$

For rigid frames to be statically determinate, $(3j + s) = (3m + r)$.

In this case,

$$(3m + r) - (3j + s) = \big((3)(5) + 6\big) - \big((3)(5) + 0\big) = 6$$

> The degree of indeterminacy is 6.

The answer is (B).

(c) From an equilibrium of vertical forces at joint B, the force in HB is $\boxed{20 \text{ kN (tension).}}$

The answer is (C).

(d) The reaction at support A is

$$\frac{(20 \text{ kN} + 20 \text{ kN})3h + (20 \text{ kN})h}{4h} = 35 \text{ kN}$$

Taking a vertical section that cuts through HG, HC, and BC and looking to the left side, the vertical component of the force in HC is seen to be 20 kN + 20 kN − 35 kN = 5 kN (upward—which indicates compression). The total force in HC is

$$(5 \text{ kN})\sqrt{2} = \boxed{7.07 \text{ kN} \quad (7.1 \text{ kN})}$$

The answer is (B).

(e) The structure has three free joints, so the degree of kinematic indeterminacy is $3^2 = \boxed{9.}$

The answer is (D).

(f) Since there are no horizontal loads, the horizontal reactions must add to $\boxed{\text{zero.}}$

The answer is (A).

(g) Summing moments about A, the vertical reaction at D is

$$\frac{(10 \text{ ft})(20 \text{ kips}) + (10 \text{ ft})(10 \text{ kips})}{40 \text{ ft}} = 7.5 \text{ kips}$$

Since CD is inclined at 45°, the axial force in CD is

$$\frac{7.5 \text{ kips}}{\sin 45°} = \boxed{10.6 \text{ kips} \quad (11 \text{ kips}) \quad (\text{compression})}$$

The answer is (C).

(h) From part (g), the vertical reaction at A is 20 kips − 7.5 kips = 12.5 kips (upward). The shear in bar BC is 12.5 kips at point B and decreases 1 kip/ft linearly to −7.5 kips at point C. The maximum shear is 12.5 kips, say $\boxed{13 \text{ kips.}}$

The answer is (B).

(i) The maximum positive moment takes place where the shear is zero. From parts (g) and (h), this point is located 7.5 ft to the left of point C. Summing moments about this location to the right (on a free-body diagram),

$$M = (7.5 \text{ kips})(20 \text{ ft} + 7.5 \text{ ft}) - \frac{(7.5 \text{ kips})(7.5 \text{ ft})}{2}$$
$$= \boxed{178.12 \text{ ft-kips} \quad (180 \text{ ft-kips})}$$

The answer is (D)

Structural

(j) Cut a free body through BC and BD. There are two cases to consider, depending on where the cutting plane intersects along BD. If the axial force on BC is assumed to be zero, then the shear on BD at B is 5 kN, changing to $\boxed{45 \text{ kN}}$ to the right of the load.

The answer is (C).

5.

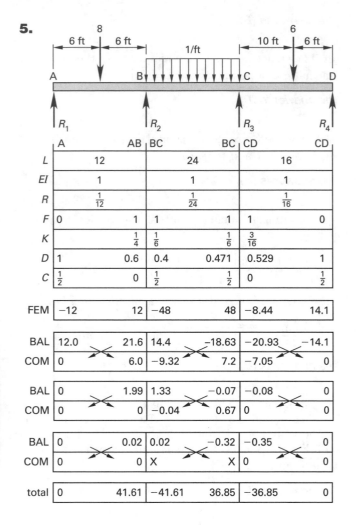

Find reactions.

Member AB:

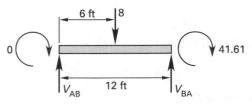

Taking clockwise moments and upward forces as positive,

$$\sum M_A = 0 + 41.61 + (6)(8) - 12 V_{BA} = 0$$
$$V_{BA} = 7.47$$
$$\sum F_y = V_{AB} + 7.47 - 8 = 0$$
$$V_{AB} = 0.53$$

Member BC:

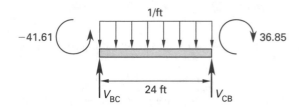

$$\sum M_B = -41.61 + 36.85 + \left(\tfrac{1}{2}\right)(24)^2(1) - 24 V_{CB} = 0$$
$$V_{CB} = 11.8$$
$$\sum F_y = V_{BC} + 11.8 - (24)(1) = 0$$
$$V_{BC} = 12.2$$

Member CD:

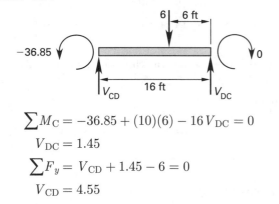

$$\sum M_C = -36.85 + (10)(6) - 16 V_{DC} = 0$$
$$V_{DC} = 1.45$$
$$\sum F_y = V_{CD} + 1.45 - 6 = 0$$
$$V_{CD} = 4.55$$

The reactions are

$R_1 = V_{AB} = 0.53$
$R_2 = V_{BA} + V_{BC} = 7.47 + 12.2 = 19.67$
$R_3 = V_{CB} + V_{CD} = 11.80 + 4.55 = 16.35$
$R_4 = V_{DC} = 1.45$

6. Use the moment distribution worksheet to calculate the end moments produced when a total moment of 100 is distributed to the two vertical members. Draw free bodies.

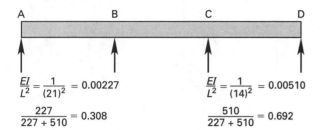

$$\frac{EI}{L^2} = \frac{1}{(21)^2} = 0.00227 \qquad \frac{EI}{L^2} = \frac{1}{(14)^2} = 0.00510$$

$$\frac{227}{227 + 510} = 0.308 \qquad \frac{510}{227 + 510} = 0.692$$

	A	AB	BC		BC	CD	CD
L		21		25		14	
EI		1		1		1	
R		$\frac{1}{21}$		$\frac{1}{25}$		$\frac{1}{14}$	
F	0		1	1		1	1	0
K		$\frac{1}{7}$	$\frac{4}{25}$		$\frac{4}{25}$	$\frac{3}{14}$	
D		0.472	0.528		0.427	0.573	
C	$\frac{1}{2}$	0	$\frac{1}{2}$		$\frac{1}{2}$	0	$\frac{1}{2}$
FEM	30.8		30.8	0		0	69.2	69.2
BAL	−30.8	−14.54	−16.26		−29.55	−39.65	−69.2
COM	0		−15.4	−14.78		−8.13	−34.6	0
BAL	0	14.24	15.94		18.25	24.48	0
COM	0	0	9.13		7.97	0	0
BAL	0	−4.31	−4.82		−3.40	−4.57	0
COM	0	0	−1.70		−2.41	0	0
BAL	0	0.80	0.90		1.03	1.38	0
COM	0	0	0.51		0.45	0	0
BAL	0	−0.24	−0.27		−0.19	−0.26	0
COM	0	0	X		X	0	0
total	0		11.35	−11.35		−15.98	15.98	0

(a) *Member AB:*

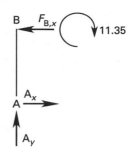

Take clockwise moments as positive.

$$\sum M_A = 11.35 - 21 F_{B,x} = 0$$
$$F_{B,x} = 0.54 \text{ kips}$$

Member DC:

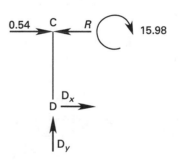

$$\sum M_D = 15.98 + (0.54)(14) - 14R = 0$$
$$R = 1.68 \text{ kips} \quad [\text{to the left}]$$

P_I was 20 to the left to keep the original frame from swaying.

$$\frac{20}{1.68} = 11.90$$

$$M_{AB} = -(11.90)(11.35) = \boxed{-135.07 \text{ ft-kips}}$$
$$M_{CD} = -(11.90)(15.98) = \boxed{-190.16 \text{ ft-kips}}$$

The actual conditions are developed as follows.

(b) *Member AB:*

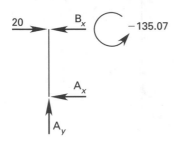

Taking clockwise moments as positive,

$$\sum M_A = -135.07 + (20)(21) - 21B_x = 0$$

$$B_x = 13.57 \text{ kips} \quad [\text{to the left}]$$

$$\sum M_B = 21A_x - 135.07 = 0$$

$$A_x = \boxed{6.43 \text{ kips} \quad [\text{to the left}]}$$

Taking the frame as a whole unit with forces to the right and clockwise moments as positive,

$$\sum F_x = 20 - 6.43 - D_x = 0$$

$$D_x = \boxed{13.57 \text{ kips} \quad [\text{to the left}]}$$

$$\sum M_A = (21)(20) - 24D_y = 0$$

$$D_y = \boxed{17.5 \text{ kips} \quad [\text{up}]}$$

$$A_y = \boxed{-17.5 \text{ kips} \quad [\text{down}]}$$

7. (a) *step 1:* Complete the moment distribution worksheet to find the end moments (assuming sidesway is prevented).

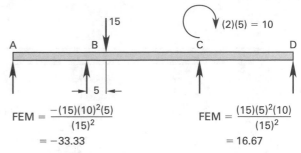

$$\text{FEM} = \frac{-(15)(10)^2(5)}{(15)^2} \qquad \text{FEM} = \frac{(15)(5)^2(10)}{(15)^2}$$

$$= -33.33 \qquad\qquad\qquad = 16.67$$

The applied joint moment is clockwise, so it is positive. The resisting FEM is negative, since the distribution factors are 0.5 for spans BD–DC; 25 goes to each.

	A	AB	BC	BC	CD	CD
L	10		15		15	
EI	240		300		300	
R	24		20		20	
F	0	1	1	1	1	1
K		72	80	80	80	
D		0.474	0.526	0.5	0.5	
C	$\frac{1}{2}$	0	$\frac{1}{2}$	$\frac{1}{2}$	$\frac{1}{2}$	$\frac{1}{2}$
FEM	0	0	−33.33	16.67		0
					−5 −5 (do not add to total)	
BAL	0	15.80	17.53	−3.34	−3.34	0
COM	0	0	−1.67	8.77	0	−1.67
BAL	0	0.79	0.88	−4.39	−4.39	0
COM	0	0	−2.19	0.44	0	−2.19
BAL	0	1.04	1.15	−0.22	−0.22	0
COM	0	0	−0.11	0.58	0	−0.11
BAL	0	0.05	0.06	−0.29	−0.29	0
COM	0	0	X	X	0	X
total	0	17.68	−17.68	18.22	−8.24	−3.97

step 2: Now find the reaction needed to prevent side-sway.

Member AB:

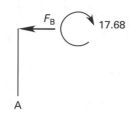

Taking clockwise moments as positive,

$$\sum M_A = 17.68 - 10F_B = 0$$
$$F_B = 1.768 \text{ kips} \quad [\text{to the left}]$$

Member DC:

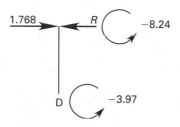

$$\sum M_D = -3.97 - 8.24 + (1.768)(15) - 15R = 0$$
$$R = 0.95 \text{ kips} \quad [\text{to the left}]$$

steps 3 and 4:

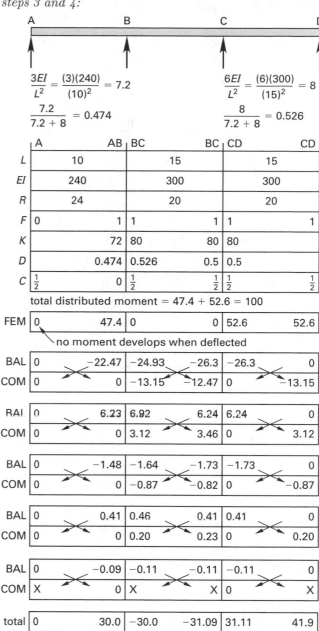

$$\frac{3EI}{L^2} = \frac{(3)(240)}{(10)^2} = 7.2 \qquad \frac{6EI}{L^2} = \frac{(6)(300)}{(15)^2} = 8$$

$$\frac{7.2}{7.2 + 8} = 0.474 \qquad \frac{8}{7.2 + 8} = 0.526$$

	A		AB	BC		BC	CD		CD
L		10			15			15	
EI		240			300			300	
R		24			20			20	
F	0		1	1		1	1		1
K			72	80		80	80		
D			0.474	0.526		0.5	0.5		
C	$\frac{1}{2}$		0	$\frac{1}{2}$		$\frac{1}{2}$	$\frac{1}{2}$		$\frac{1}{2}$

total distributed moment = 47.4 + 52.6 = 100

| FEM | 0 | | 47.4 | 0 | | 0 | 52.6 | | 52.6 |

↳ no moment develops when deflected

| BAL | 0 | | −22.47 | −24.93 | | −26.3 | −26.3 | | 0 |
| COM | 0 | | 0 | −13.15 | | −12.47 | 0 | | −13.15 |

| BAL | 0 | | 6.23 | 6.92 | | 6.24 | 6.24 | | 0 |
| COM | 0 | | 0 | 3.12 | | 3.46 | 0 | | 3.12 |

| BAL | 0 | | −1.48 | −1.64 | | −1.73 | −1.73 | | 0 |
| COM | 0 | | 0 | −0.87 | | −0.82 | 0 | | −0.87 |

| BAL | 0 | | 0.41 | 0.46 | | 0.41 | 0.41 | | 0 |
| COM | 0 | | 0 | 0.20 | | 0.23 | 0 | | 0.20 |

| BAL | 0 | | −0.09 | −0.11 | | −0.11 | −0.11 | | 0 |
| COM | X | | 0 | X | | X | 0 | | X |

| total | 0 | | 30.0 | −30.0 | | −31.09 | 31.11 | | 41.9 |

step 5: Calculate the force required to prevent sidesway.

Member AB:

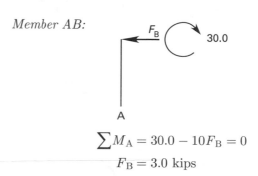

$$\sum M_A = 30.0 - 10F_B = 0$$
$$F_B = 3.0 \text{ kips}$$

Member CD:

$$\sum M_D = 31.09 + 41.9 + (3)(15) - 15R' = 0$$
$$R' = 7.87 \text{ kips} \quad [\text{to the left}]$$

steps 6 and 7:

correction ratio: $\dfrac{0.95}{7.87} = 0.12$

moment at A: $0 + (0.12)(-1)(0) = 0$

moment at B: $17.68 + (0.12)(-1)(30)$

$= \boxed{14.08 \text{ ft-kips}}$

moment at C: $18.22 + (0.12)(-1)(-31.09)$

$= \boxed{21.95 \text{ ft-kips}}$

moment at D: $-3.97 + (0.12)(-1)(41.9)$

$= \boxed{-9.0 \text{ ft-kips}}$

Since R and R' are in the same directions, the derived moments must be reversed in sign.

(b) *Member AB:*

$$\sum M_B = 14.08 \text{ ft-kips} - (10 \text{ ft})A_x = 0$$
$$\boxed{A_x = 1.408 \text{ kips}}$$

Frame as a whole:

$$\sum F_x = 1.408 \text{ kips} + D_x = 0$$
$$\boxed{D_x = -1.408 \text{ kips}}$$

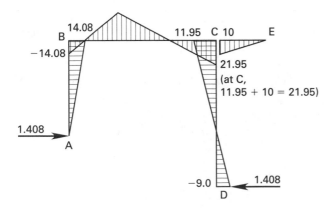

8. The moment distribution worksheet is

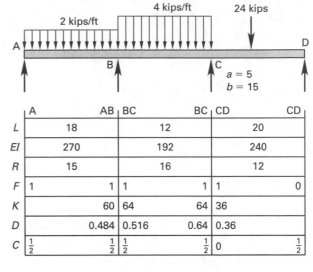

	A		AB	BC		BC	CD		CD
L	18			12			20		
EI	270			192			240		
R	15			16			12		
F	1		1	1		1	1		0
K			60	64		64	36		
D			0.484	0.516		0.64	0.36		
C	$\frac{1}{2}$		$\frac{1}{2}$	$\frac{1}{2}$		$\frac{1}{2}$	0		$\frac{1}{2}$

FEM	−54		54	−48		48	−67.5		22.5

BAL	0		−2.9	−3.1		12.48	7.02		−22.5
COM	−1.45		0	6.24		−1.55	−11.25		0

BAL	0		−3.02	−3.22		8.19	4.61		0
COM	−1.51		0	4.10		−1.61	0		0

BAL	0		−1.98	−2.12		1.03	0.58		0
COM	−0.99		0	0.52		−1.06	0		0

BAL	0		−0.25	−0.27		0.68	0.38		0
COM	−0.13		0	0.34		−0.14	0		0

BAL	0		−0.16	−0.18		0.09	0.05		0
COM	X		0	X		X	X		X

total	−58.08		45.69	−45.69		66.11	−66.11		0

Start at span CD and work to the left.

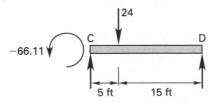

Taking clockwise moments as positive,

$$\text{applied} \sum M_C = -20R_D + (5)(24) - 66.11 = 0$$
$$R_D = 2.69 \text{ kips}$$

$$\text{applied} \sum M_D = 20R_C - (15)(24) - 66.11 = 0$$
$$R_C = 21.31 \text{ kips}$$

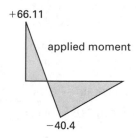

+66.11

applied moment

−40.4

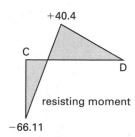

+40.4

C

D

resisting moment

−66.11

For span BC,

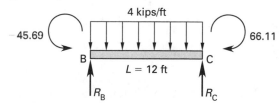

4 kips/ft

−45.69

66.11

B

C

$L = 12$ ft

R_B

R_C

$$\text{applied} \sum M_B = 66.11 - 45.69 + (4)(12)(6) - 12R_C$$
$$= 0$$
$$R_C = 25.7 \text{ kips}$$

$$\text{applied} \sum M_C = 66.11 - 45.69 - (4)(12)(6) + 12R_B$$
$$= 0$$
$$R_B = 22.3 \text{ kips}$$

Taking counterclockwise moments as positive,

$$\text{resisting} \sum M = -66.11 + 25.7x - \left(\tfrac{1}{2}\right)(4)x^2$$

At $x = 12$ ft,

$$M = -66.11 + (25.7)(12) - \left(\tfrac{1}{2}\right)(4)(12)^2$$
$$= -45.71 \text{ ft-kips}$$

The first derivative is

$$\frac{dM}{dx} = 25.7 - 4x = 0$$

M is maximum at

$$x = \frac{25.7}{4} = 6.4 \text{ ft}$$

At $x = 6.4$ ft,

$$M_B = -66.11 + (25.7)(6.4) - \left(\tfrac{1}{2}\right)(4)(6.4)^2$$
$$= 16.45 \text{ ft-kips}$$

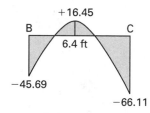

+16.45

B   C

6.4 ft

−45.69

−66.11

resisting moment

For span AB,

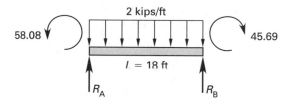

2 kips/ft

58.08

45.69

$l = 18$ ft

R_A

R_B

Taking clockwise moments as positive,

$$\text{applied} \sum M_A = 45.69 - 58.08 + (2)(18)(9) - 18R_B$$
$$= 0$$
$$R_B = 17.31 \text{ kips}$$

$$\text{applied} \sum M_B = 45.69 - 58.08 - (2)(18)(9) + 18R_A$$
$$= 0$$
$$R_A = 18.69 \text{ kips}$$

Taking counterclockwise moments as positive,

$$\text{resisting} \sum M = -45.69 + 17.31x - \left(\tfrac{1}{2}\right)(2)x^2$$

Structural

At $x = 18$ ft,

$$M = -45.69 + (17.31)(18) - \left(\tfrac{1}{2}\right)(2)(18)^2$$
$$= 58.11 \text{ ft-kips}$$

The first derivative is

$$\frac{dM}{dx} = 17.31 - 2x = 0$$

M is maximum at

$$x = \frac{17.31}{2} = 8.66 \text{ ft}$$

At $x = 8.66$ ft,

$$M = -45.69 + (17.31)(8.66) - \left(\tfrac{1}{2}\right)(2)(8.66)^2$$
$$= 29.22 \text{ ft-kips}$$

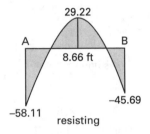

Based on the previous analysis,

$$R_A = 18.69 \text{ kips}$$
$$R_B = 17.31 + 22.3 = 39.61 \text{ kips}$$
$$R_C = 25.70 + 21.31 = 47.01 \text{ kips}$$
$$R_D = 2.69 \text{ kips}$$

The maximum moments are

on span AB:	−58.11 ft-kips
on span BC:	−66.11 ft-kips
on span CD:	−66.11 ft-kips

9. (a) Assume the shapes bend against their largest moment of inertia. Assume the vertical bar remains vertical so that both horizontal members deflect the same amount.

$$W12 \times 190: I = 1890 \text{ in}^4$$
$$W12 \times 72: I = \underline{597 \text{ in}^4}$$
$$\text{total} \quad 2487 \text{ in}^4$$

The two horizontal beams act together like a combined cantilever.

$$L = \left(12 \; \frac{\text{in}}{\text{ft}}\right)(8 \text{ ft}) = 96 \text{ in}$$

$$y_{\max} = \frac{(120{,}000 \text{ lbf})(96 \text{ in})^3}{(3)\left(2.9 \times 10^7 \; \frac{\text{lbf}}{\text{in}^2}\right)(2487 \text{ in}^4)} = \boxed{0.49 \text{ in}}$$

(b) *step 1:* Since there are no fixed-end moments, all total moments are zero.

step 2: In order to prevent all of the sidesway, all of the applied lateral load must be canceled. Therefore, $P_I = 120$ to the left.

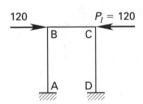

step 3: Apply a total FEM of 100.

For AB,

$$\frac{I}{L^2} = \frac{1890}{(8)^2} = 29.53$$

For DC,

$$\frac{I}{L^2} = \frac{597}{(8)^2} = 9.33$$

Give $29.53/(29.53 + 9.33) = 0.76$ (76%) to AB.

Give $1 - 0.76 = 0.24$ (24%) to DC.

step 4: The moment distribution worksheet for the application of these FEMs is

	A	AB	BC	BC	CD	CD
L		8		4.5		8
EI		1890		310		597
R		236.25		68.89		74.625
F	1	1	1	1	1	1
K	945	945	275.56	275.56	298.5	298.5
D		0.774	0.226		0.48	0.52
C	$\frac{1}{2}$	$\frac{1}{2}$	$\frac{1}{2}$	$\frac{1}{2}$	$\frac{1}{2}$	$\frac{1}{2}$

FEM	−76	−76	0	0	−24	−24

BAL	0	58.82	17.18	11.52	12.48	0
COM	29.41	0	5.76	8.59	0	6.24

BAL	0	−4.46	−1.30	−4.12	−4.47	0
COM	−2.23	0	−2.06	−0.65	0	−2.23

BAL	0	1.59	0.47	0.31	0.34	0
COM	0.79	0	0.15	0.24	0	0.17

BAL	0	−0.12	−0.03	−0.11	−0.13	0
COM	X	0	X	X	0	X

total	−48.03	−20.17	20.17	15.78	−15.78	−19.95

step 5: Take member AB as a free body.

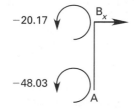

Taking clockwise moments as positive,

$$\sum M_A = 8B_x - 20.17 - 48.03 = 0$$
$$B_x = 8.53 \text{ kips} \quad \text{[to the right]}$$

Take member DC as a free body.

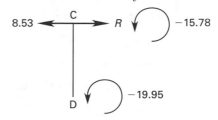

$$\sum M_D = 8R - (8)(8.53) - 15.78 - 19.95 = 0$$
$$R = 13.0 \text{ kips} \quad \text{[to the right]}$$

step 6: The ratio is

$$-\left(\frac{-120}{13.0}\right) = 9.23$$

$$M_A = (9.23)(-48.03) = -443.3 \text{ ft-kips}$$
$$M_B = (9.23)(\pm 20.17) = \pm 186.17 \text{ ft-kips}$$
$$M_C = (9.23)(\pm 15.78) = \pm 145.6 \text{ ft-kips}$$
$$M_D = (9.23)(-19.95) = -184.1 \text{ ft-kips}$$

step 7: Draw the free body of member AB.

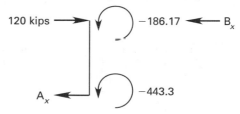

B_x can be found by summing moments about A.

$$\sum M_A = -443.3 \text{ ft-kips} - 186.17 \text{ ft-kips}$$
$$+ (8 \text{ ft})(120 \text{ kips}) - (8 \text{ ft})B_x = 0$$
$$B_x = 41.32 \text{ kips} \quad \text{[to the left]}$$

Break the loads into two parts.

Concentrated loads:

The net load is 120 kips − 41.32 kips = 78.68 kips.

The deflection is

$$y_1 = \frac{FL^3}{3EI} = \frac{(78{,}680 \text{ lbf})(96 \text{ in})^3}{(3)\left(2.9 \times 10^7 \ \frac{\text{lbf}}{\text{in}^2}\right)(1890 \text{ in}^4)}$$
$$= 0.423 \text{ in} \quad \text{[down]}$$

Applied moments:

The moment diagram for the pure moments on the span is

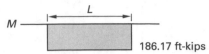

From the area-moment method,

$$EIy = ML\left(\frac{L}{2}\right)$$

$$y_2 = \frac{ML^2}{2EI}$$

The moment at the built-in end does not cause a deflection, so $M = 186.17$ ft-kips.

$$y_2 = \frac{ML^2}{2EI} = \frac{(-186{,}170 \text{ ft-lbf})\left(12 \frac{\text{in}}{\text{ft}}\right)(96 \text{ in})^2}{(2)\left(2.9 \times 10^7 \frac{\text{lbf}}{\text{in}^2}\right)(1890 \text{ in}^4)}$$

$$= -0.188 \text{ in} \quad [\text{up}]$$

$$y_{\text{total}} = 0.423 \text{ in} - 0.188 \text{ in} = \boxed{0.235 \text{ in} \quad [\text{down}]}$$

10. (a) Although the beam has four supports, it is determinate. Since moment $= 0$ at H_1, taking clockwise moments to the left of H_1 as positive,

$$\sum M_{H_1} = (5 \text{ ft})R_B + (45 \text{ ft})R_A$$
$$- (20 \text{ kips})(20 \text{ ft} + 30 \text{ ft}) = 0$$

$$5R_B + 45R_A = 1000 \text{ kips}$$

Since the beam and loading are symmetrical,

$$R_A + R_B = \left(\tfrac{1}{2}\right)(7)(20 \text{ kips}) = 70 \text{ kips}$$

Solving for R_A and R_B simultaneously,

$$R_A = 16.25 \text{ kips} = R_D$$
$$R_B = 53.75 \text{ kips} = R_C$$

For a W24 × 76 beam,

$$A = 22.4 \text{ in}^2$$
$$d = 23.92 \text{ in}$$
$$S = 176 \text{ in}^3$$
$$t_w = 0.440 \text{ in}$$
$$I = 2100 \text{ in}^4$$

The shear diagram is

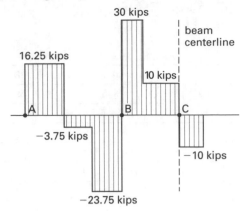

The web takes all the shear. The maximum shear stress is

$$\tau_{\text{max}} = \frac{V}{dt_w} = \frac{30 \times 10^3 \text{ lbf}}{(23.92 \text{ in})(0.440 \text{ in})}$$
$$= \boxed{2850 \text{ lbf/in}^2}$$

(b) The moment is maximum where $V = 0$. Under the first inboard load,

$$M = (15 \text{ ft})(16.25 \text{ kips}) = 243.75 \text{ ft-kips}$$

At R_B,

$$M = (40 \text{ ft})(16.25 \text{ kips}) - (15 \text{ ft} + 25 \text{ ft})(20 \text{ kips})$$
$$= -150 \text{ ft-kips}$$

At the center,

$$M = (15 \text{ ft})\left(\frac{20 \text{ kips}}{2}\right) = 150 \text{ ft-kips}$$

$$M_{\text{max}} = 243.75 \text{ ft-kips}$$

$$\sigma_{\text{max}} = \frac{M}{S} = \frac{(243.75 \times 10^3 \text{ ft-lbf})\left(12 \frac{\text{in}}{\text{ft}}\right)}{176 \text{ in}^3}$$
$$= \boxed{16{,}619 \text{ lbf/in}^2}$$

(c) The deflection between hinges is

$$y_{\text{max}} = \frac{FL^3}{48EI} = \frac{(20 \times 10^3 \text{ lbf})\left((30 \text{ ft})\left(12 \frac{\text{in}}{\text{ft}}\right)\right)^3}{(48)\left(2.9 \times 10^7 \frac{\text{lbf}}{\text{in}^2}\right)(2100 \text{ in}^4)}$$

$$= 0.319 \text{ in}$$

The deflection at the midpoint is composed of four terms.

$$y_1 = \frac{FL^3}{48EI} = \frac{-(20 \times 10^3 \text{ lbf})\left((30 \text{ ft})\left(12 \frac{\text{in}}{\text{ft}}\right)\right)^3}{(48)\left(2.9 \times 10^7 \frac{\text{lbf}}{\text{in}^2}\right)(2100 \text{ in}^4)}$$

$$= -0.319 \text{ in}$$

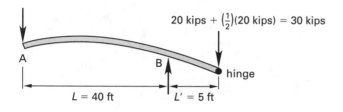

$$y_2 = \left(\frac{FL^2}{3EI}\right)(L + L')$$

$$= \frac{-(30 \times 10^3 \text{ lbf})\left((5 \text{ ft})\left(12 \frac{\text{in}}{\text{ft}}\right)\right)^2}{(3)\left(2.9 \times 10^7 \frac{\text{lbf}}{\text{in}^2}\right)(2100 \text{ in}^4)}$$

$$= -0.319 \text{ in}$$

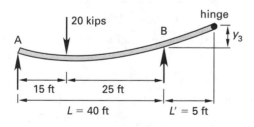

Using App. 44.A, case 7, the slope at B is

$$m_B = \left(\frac{Pab}{6EI}\right)\left(1 + \frac{a}{L}\right)$$

$$= \left(\frac{(20 \times 10^3 \text{ lbf})(15 \text{ ft})\left(12 \frac{\text{in}}{\text{ft}}\right)}{(6)\left(2.9 \times 10^7 \frac{\text{lbf}}{\text{in}^2}\right)(2100 \text{ in}^4)}\right)\left(1 + \frac{15 \text{ ft}}{40 \text{ ft}}\right)$$

$$= 0.00406 \text{ rad}$$

$$y_3 = m_B L' = (0.00406 \text{ rad})(5 \text{ ft})\left(12 \frac{\text{in}}{\text{ft}}\right)$$

$$= 0.244 \text{ in}$$

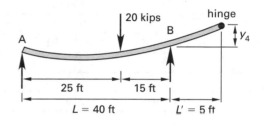

$$m_B = \left(\frac{Pab}{6EI}\right)\left(1 + \frac{a}{L}\right)$$

$$= \left(\frac{(20 \times 10^3 \text{ lbf})(25 \text{ ft})\left(12 \frac{\text{in}}{\text{ft}}\right)}{(6)\left(2.9 \times 10^7 \frac{\text{lbf}}{\text{in}^2}\right)(2100 \text{ in}^4)}\right)\left(1 + \frac{25 \text{ ft}}{40 \text{ ft}}\right)$$

$$= 0.00480 \text{ rad}$$

$$y_4 = M_B L' = (0.00480 \text{ rad})(5 \text{ ft})\left(12 \frac{\text{in}}{\text{ft}}\right)$$

$$= 0.288 \text{ in}$$

The total midpoint deflection is

$$y_t = y_1 + y_2 + y_3 + y_4$$

$$= -0.319 - 0.319 + 0.244 + 0.288$$

$$= \boxed{-0.106 \text{ in}}$$

11. (a) The moment distribution worksheet is

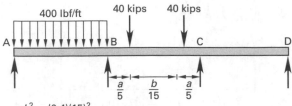

$$\frac{wL^2}{12} = \frac{(0.4)(15)^2}{12} = 7.5 \text{ ft-kips}$$

$$\left(\frac{Pa}{L^2}\right)(2a^2 + 3ab + b^2) = \left(\frac{(40)(5)}{(25)^2}\right)\left((2)(5)^2 + (3)(5)(15) + (15)^2\right)$$
$$= 160 \text{ ft-kips}$$

	AB		BA	BC		CB	CD		DC
L	15			25			12		
EI	890			341			890		
R	59.33			13.64			74.17		
F	0		1	1		1	1		1
K	237.3		178	54.6		54.6	296.7		296.7
D	1		0.765	0.235		0.155	0.845		1
C	$\frac{1}{2}$		0	$\frac{1}{2}$		$\frac{1}{2}$	$\frac{1}{2}$		$\frac{1}{2}$
FEM	−7.5		7.5	−160		160	0		0
BAL	7.5		116.66	35.84		−24.8	−135.2		0
COM	0		3.75	−12.4		17.92	0		−67.6
BAL	0		6.62	2.03		−2.78	−15.14		0
COM	0		0	−1.39		1.01	0		−7.57
BAL	0		1.06	0.33		−0.16	−0.85		0
COM	X		0	X		X	0		X
total	0		136.71	−136.71		151.19	−151.19		−75.17

Taking clockwise moments as positive,

$$\sum M_\mathrm{A} = 136.71 \text{ ft-kips}$$
$$+ \left(\tfrac{1}{2}\right)\left(0.4 \ \frac{\text{kips}}{\text{ft}}\right)(15 \text{ ft})^2$$
$$- (15 \text{ ft})\mathrm{B}_x = 0$$
$$\mathrm{B}_x = 12.11 \text{ kips} \quad [\text{to the left}]$$

$$\left(0.4 \ \frac{\text{kips}}{\text{ft}}\right)(15 \text{ ft}) - 12.11 \text{ kips} + A_x = 0$$
$$A_x = \boxed{6.11 \text{ kips} \quad [\text{to the right}]}$$

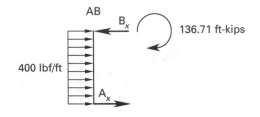

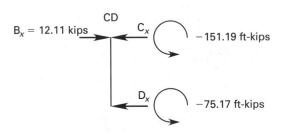

$$\sum M_\mathrm{D} = (12 \text{ ft})(12.11 \text{ kips}) - (12 \text{ ft})R$$
$$- 151.19 \text{ ft-kips} - 75.17 \text{ ft-kips} = 0$$
$$C_x = \boxed{-6.75 \text{ kips} \quad [\text{to the right}]}$$
$$12.11 \text{ kips} + 6.75 \text{ kips} - D_x = 0$$
$$D_x = \boxed{18.86 \text{ kips} \quad [\text{to the left}]}$$

(b)

136.71 ft-kips

151.19 ft-kips

(40)(5) = 200 ft-kips

(40)(5) = 200 ft-kips

63.29 ft-kips

48.81 ft-kips

75.17 ft-kips

48 Properties of Concrete and Reinforcing Steel

PRACTICE PROBLEMS

1. A batch of fine aggregate was sieve graded. The percentages retained on each sieve are shown. What is the sand's fineness modulus?

sieve	percentage retained
4	4
8	11
16	21
30	22
50	24
100	17
dust (pan)	1

(A) 0.99

(B) 1.8

(C) 2.9

(D) 99

2. (*Time limit: one hour*) The 6 in × 6 in concrete beam section shown was tested in a third-point loading apparatus. The failure occurred outside the middle third as shown under a maximum total load of 5000 lbf. The beam was normal-weight concrete with a compressive strength of 5000 psi. Neglect the beam's self-weight.

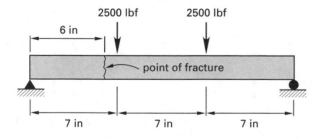

(a) The maximum shear force is most nearly

(A) 625 lbf

(B) 1250 lbf

(C) 2500 lbf

(D) 5000 lbf

(b) The maximum bending moment is most nearly

(A) 5000 in-lbf

(B) 10,000 in-lbf

(C) 15,000 in-lbf

(D) 17,500 in-lbf

(c) The shear force within the middle third of the beam is most nearly

(A) 0 lbf

(B) 625 lbf

(C) 1250 lbf

(D) 2500 lbf

(d) The bending moment within the middle third of the beam

(A) is zero

(B) increases linearly from right to left

(C) is parabolic in shape

(D) is a straight line with zero slope

(e) The modulus of rupture is most nearly

(A) 320 psi

(B) 420 psi

(C) 520 psi

(D) 620 psi

(f) If the fracture occurred within the middle third, the modulus of rupture would most nearly be

(A) 420 psi

(B) 450 psi

(C) 470 psi

(D) 490 psi

(g) The modulus of rupture using the ACI empirical equation is most nearly

(A) 530 psi

(B) 550 psi

(C) 570 psi

(D) 590 psi

(h) If the beam were made of lightweight concrete, the modulus of rupture would most nearly be

(A) 380 psi

(B) 400 psi

(C) 420 psi

(D) 440 psi

(i) The center-point loading test is no longer standard in determining the modulus of rupture for concrete because

(A) it yields higher values of modulus of rupture than the true values

(B) it is difficult to perform accurately in the laboratory

(C) it costs more than the third-point loading test

(D) all of the above

(j) The modulus of rupture test yields a higher value of strength than a direct tensile test or splitting tensile test made on the same specimen because

(A) the assumed stress block shape does not match the real shape

(B) direct tensile tests are sensitive to any accidental eccentricity

(C) the concrete is assumed to be perfectly elastic

(D) all of the above

3. If a concrete has a density of 2300 kg/m^3 and a compressive strength of 15 MPa, what modulus of elasticity is predicted by ACI 318?

(A) 14 MPa

(B) 18 MPa

(C) 28 MPa

(D) 18 GPa

4. (*Time limit: one hour*) A normal-weight concrete specimen tested at 28 days yielded the following results.

(a) A 6 in × 12 in concrete cylinder failed at an axial compressive force of 105,000 lbf. The ultimate compressive strength is most nearly

(A) 1200 psi

(B) 1900 psi

(C) 3700 psi

(D) 4200 psi

(b) A 6 in × 12 in concrete cylinder failed at a transverse force of 45,000 lbf in a splitting tensile cylinder test. The concrete tensile strength is most nearly

(A) 400 psi

(B) 450 psi

(C) 625 psi

(D) 666 psi

(c) A 6 in square unreinforced beam section failed at a force of 5400 lbf on a 21 in span length. A third-point loading test was used, and the fracture occurred within the middle third. The modulus of rupture is most nearly

(A) 150 psi

(B) 400 psi

(C) 525 psi

(D) 900 psi

(d) The ratio of the modulus of rupture to the compressive strength is most nearly

(A) 4%

(B) 5%

(C) 12%

(D) 14%

(e) The ratio of the splitting tensile strength to the compressive strength is most nearly

(A) 8%

(B) 11%

(C) 17%

(D) 20%

(f) What conclusion could be stated based on the answers of parts (d) and (e)?

    (A) Concrete is weak in tension and strong in compression.

    (B) Concrete is weak in tension and strong in shear.

    (C) Concrete is strong in tension and weak in compression.

    (D) None of the above.

(g) Given a compressive strength of 3700 psi, the approximate value of the modulus of rupture is most nearly

    (A) 460 psi

    (B) 500 psi

    (C) 550 psi

    (D) 600 psi

(h) Given a compressive strength of 3700 psi, the approximate value of the splitting tensile strength is most nearly

    (A) 300 psi

    (B) 410 psi

    (C) 500 psi

    (D) 600 psi

(i) From the compression test, the axial strain was 0.0015 in/in and the lateral strain was 0.00027 in/in. Poisson's ratio is most nearly

    (A) 0.00027

    (B) 0.0015

    (C) 0.10

    (D) 0.18

(j) For the beam in part (c), if the fracture occurred 1 in away from the left support, the value of the modulus of rupture obtained from this test

    (A) is less than the correct value

    (B) equals the correct value

    (C) is higher than the correct value

    (D) becomes indeterminate

5. Stress-strain curves are shown for grade 60 steel rebar and 4000 psi normal-weight concrete.

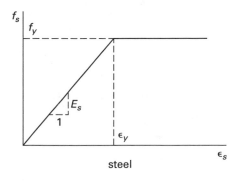

steel

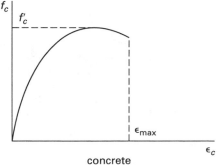

concrete

(a) What is the yield strength of the steel?

    (A) 30 ksi

    (B) 36 ksi

    (C) 50 ksi

    (D) 60 ksi

(b) What is the modulus of elasticity of the steel?

    (A) 10×10^6 psi

    (B) 29×10^6 psi

    (C) 30×10^6 psi

    (D) 36×10^6 psi

(c) What is the steel strain at the yield point?

    (A) 0.002 in/in

    (B) 0.003 in/in

    (C) 0.06 in/in

    (D) 0.09 in/in

(d) What is the concrete's compressive strength?

    (A) 2000 psi

    (B) 4000 psi

    (C) 6000 psi

    (D) 8000 psi

Structural

(e) What is the concrete's modulus of elasticity?

(A) 1.1×10^6 psi

(B) 1.7×10^6 psi

(C) 2.9×10^6 psi

(D) 3.6×10^6 psi

(f) What is the concrete's ultimate compressive strain?

(A) 0.001 in/in

(B) 0.003 in/in

(C) 0.005 in/in

(D) 0.009 in/in

6. The primary difference between components (materials) used to produce portland cement concrete and mortar for masonry is explained best by which of the following statements?

(A) Mortar uses high-alumina cement.

(B) Mortar contains caustic lime.

(C) Mortar contains gypsum.

(D) Mortar contains potassium chloride.

7. The abbreviation "WWF" on a construction document relating to concrete features is most likely to refer to

(A) waterproof walled form

(B) wetted and wired foundation

(C) welded wire fabric

(D) Wisconsin wire frame

8. Two 10 ft wide, untied slabs of concrete are abutting at 80°F. What is their maximum possible separation at 40°F?

(A) 0.03 in

(B) 0.06 in

(C) 0.14 in

(D) 0.35 in

SOLUTIONS

1. Consider the sieves in sequence from finest to coarsest. If the no. 100 sieve had been used first, only the pan (dust) would have passed through, and the retained percentage would have been $100\% - 1\% = 99\%$.

For the no. 50 sieve, the retained amount would have been $100\% - 1\% - 17\% = 82\%$.

The following table is prepared similarly.

sieve	cumulative retained
100	$100\% - 1\% = 99\%$
50	$99\% - 17\% = 82\%$
30	$82\% - 24\% = 58\%$
16	$58\% - 22\% = 36\%$
8	$36\% - 21\% = 15\%$
4	$15\% - 11\% = 4\%$
total	294%

The fineness modulus is $294\%/100\% = \boxed{2.94 \ (2.9).}$

The answer is (C).

2. (a) From the shear diagram, the maximum shear force is $\boxed{2500 \text{ lbf.}}$

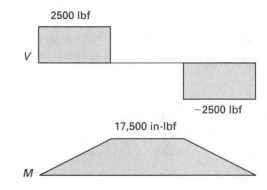

The answer is (C).

(b) From the moment diagram, the maximum moment is $\boxed{17{,}500 \text{ in-lbf.}}$

The answer is (D).

(c) From the shear diagram, the shear force in the middle third is $\boxed{\text{zero.}}$

The answer is (A).

(d) From the moment diagram, the moment is constant in the middle third.

The answer is (D).

(e) Use Eq. 48.6.

$$f_r = \frac{Mc}{I}$$

$$= \frac{(2500 \text{ lbf})(6 \text{ in})\left(\dfrac{6 \text{ in}}{2}\right)}{\dfrac{(6 \text{ in})(6 \text{ in})^3}{12}}$$

$$= \boxed{417 \text{ lbf/in}^2 \quad (420 \text{ psi})}$$

The answer is (B).

(f) Use Eq. 48.6.

$$f_r = \frac{Mc}{I}$$

$$= \frac{(17{,}500 \text{ in-lbf})\left(\dfrac{6 \text{ in}}{2}\right)}{\dfrac{(6 \text{ in})(6 \text{ in})^3}{12}}$$

$$= \boxed{486 \text{ lbf/in}^2 \quad (490 \text{ psi})}$$

The answer is (D).

(g) Use Eq. 48.7.

$$f_r = 7.5\lambda\sqrt{f'_c} = (7.5)(1.0)\sqrt{5000 \; \frac{\text{lbf}}{\text{in}^2}}$$

$$= \boxed{530 \text{ lbf/in}^2 \quad (530 \text{ psi})}$$

The answer is (A).

(h) The modulus of rupture for lightweight concrete is 75% of Eq. 48.7.

$$f_r = (0.75)(7.5)\lambda\sqrt{f'_c} = (0.75)(7.5)(1.0)\sqrt{5000 \; \frac{\text{lbf}}{\text{in}^2}}$$

$$= \boxed{398 \text{ lbf/in}^2 \quad (400 \text{ psi})}$$

The answer is (B).

(i) The center point loading test is difficult to perform accurately in the laboratory and is no longer the standard used to determine the modulus of elasticity.

The answer is (B).

(j) All of the options given are reasons why the modulus of rupture test yields a higher value than a direct tensile or splitting tensile test.

The answer is (D).

3. Use Eq. 48.2. (Equation 48.3 could also be used.)

$$E_c = 0.043w_c^{1.5}\sqrt{f'_c}$$

$$= (0.043)\left(2300 \; \frac{\text{kg}}{\text{m}^3}\right)^{1.5}\sqrt{15 \text{ MPa}}$$

$$= \boxed{18{,}370 \text{ MPa} \quad (18 \text{ GPa})}$$

The answer is (D).

4. (a) Use Eq. 48.1.

$$f'_c = \frac{P}{A} = \frac{105{,}000 \text{ lbf}}{\left(\dfrac{\pi}{4}\right)(6 \text{ in})^2}$$

$$= \boxed{3714 \text{ lbf/in}^2 \quad (3700 \text{ psi})}$$

The answer is (C).

(b) Use Eq. 48.4.

$$f_{ct} = \frac{2P}{\pi DL} = \frac{(2)(45{,}000 \text{ lbf})}{\pi(6 \text{ in})(12 \text{ in})}$$

$$= \boxed{398 \text{ lbf/in}^2 \quad (400 \text{ psi})}$$

The answer is (A).

(c) Use Eq. 48.6.

$$f_r = \frac{Mc}{I} = \frac{\left(\dfrac{5400 \text{ lbf}}{2}\right)\left(\dfrac{21 \text{ in}}{3}\right)\left(\dfrac{6 \text{ in}}{2}\right)}{\dfrac{(6 \text{ in})(6 \text{ in})^3}{12}}$$

$$= \boxed{525 \text{ lbf/in}^2 \quad (525 \text{ psi})}$$

The answer is (C).

(d) The ratio of the modulus of rupture to the compressive strength is

$$\frac{f_r}{f'_c} = \frac{525 \; \dfrac{\text{lbf}}{\text{in}^2}}{3714 \; \dfrac{\text{lbf}}{\text{in}^2}} = \boxed{0.141 \quad (14\%)}$$

The answer is (D).

(e) The ratio of the splitting tensile strength to the compressive strength is

$$\frac{f_{ct}}{f'_c} = \frac{398 \; \dfrac{\text{lbf}}{\text{in}^2}}{3714 \; \dfrac{\text{lbf}}{\text{in}^2}} = \boxed{0.107 \quad (11\%)}$$

The answer is (B).

Structural

(f) Concrete is weak in tension and strong in compression.

The answer is (A).

(g) Use Eq. 48.7.

$$f_r = 7.5\lambda\sqrt{f'_c} = (7.5)(1.0)\sqrt{3700\ \frac{\text{lbf}}{\text{in}^2}}$$
$$= \boxed{456\ \text{lbf/in}^2 \quad (460\ \text{psi})}$$

The answer is (A).

(h) Use Eq. 48.5.

$$f_{ct} = 6.7\lambda\sqrt{f'_c} = (6.7)(1.0)\sqrt{3700\ \frac{\text{lbf}}{\text{in}^2}}$$
$$= \boxed{408\ \text{lbf/in}^2 \quad (410\ \text{psi})}$$

The answer is (B).

(i) Poisson's ratio is

$$\nu = \frac{\text{lateral strain}}{\text{axial strain}} = \frac{0.00027\ \frac{\text{in}}{\text{in}}}{0.0015\ \frac{\text{in}}{\text{in}}}$$
$$= \boxed{0.18}$$

The answer is (D).

(j) If the specimen breaks significantly outside of the middle third of the span, the test results must be discarded. The value of the modulus of rupture becomes indeterminate. Refer to ASTM C78.

The answer is (D).

5. (a) The yield strength for grade 60 steel is $\boxed{60\ \text{ksi.}}$

The answer is (D).

(b) The standard modulus of elasticity for reinforcing steel is $\boxed{29 \times 10^6\ \text{psi.}}$

The answer is (B).

(c) The strain at yield is

$$\epsilon_y = \frac{f_y}{E}$$
$$= \frac{60,000\ \frac{\text{lbf}}{\text{in}^2}}{29 \times 10^6\ \frac{\text{lbf}}{\text{in}^2}}$$
$$= \boxed{0.00207\ \text{in/in} \quad (0.002\ \text{in/in})}$$

The answer is (A).

(d) The compressive strength is the concrete's designation $\boxed{(4000\ \text{psi}).}$

The answer is (B).

(e) Use Eq. 48.3 for normal-weight concrete.

$$E_c = 57,000\sqrt{f'_c} = 57,000\sqrt{4000\ \frac{\text{lbf}}{\text{in}^2}}$$
$$= \boxed{3.6 \times 10^6\ \text{lbf/in}^2 \quad (3.6 \times 10^6\ \text{psi})}$$

The answer is (D).

(f) Concrete is assumed to crack when it reaches a standard strain value of $\boxed{0.003\ \text{in/in.}}$

The answer is (B).

6. A typical mortar mix is one part portland cement, two parts lime hydrate, and eight parts sand by volume. High-alumina cement could be used to produce mortar instead of portland cement, but the mortar set-up time would be greatly reduced just as it is with normal concrete.

The answer is (B).

7. "WWF" is a standard construction abbreviation for "welded wire fabric."

The answer is (C).

8. From Table 44.2, the coefficient of linear thermal expansion of concrete is $6.7 \times 10^{-6}\ °\text{F}^{-1}$. The maximum possible separation would occur if the two slabs were fixed in position at their far edges. Each of the slabs would then recede a distance of

$$\Delta L = \alpha L \Delta T$$
$$= \left(6.7 \times 10^{-6}\ \frac{1}{°\text{F}}\right)(10\ \text{ft})\left(12\ \frac{\text{in}}{\text{ft}}\right)(80°\text{F} - 40°\text{F})$$
$$= 0.032\ \text{in}$$

Since there are two slabs, the total separation is $(2)(0.032\ \text{in}) = \boxed{0.064\ \text{in} \quad (0.06\ \text{in}).}$

The answer is (B).

49 Concrete Proportioning, Mixing, and Placing

PRACTICE PROBLEMS

1. (*Time limit: one hour*) The following information is submitted for a proposed concrete mix.

cement	
specific gravity	3.15
fine aggregate	
fineness modulus	2.65
specific gravity	2.48
absorption	3.0%
coarse aggregate	
specific gravity	2.68
dry bulk density	105 lbf/ft^3
absorption	0.7%
concrete	
slump	5 in
water	5.0 gal/sack
air content	4%
cement content	6.5 sack mix
coarse aggregate	0.57 ft^3/ft^3 bulk

(a) What is the dry weight of the coarse aggregate in 1 yd^3 of concrete?

(A) 1120 lbf/yd^3

(B) 1300 lbf/yd^3

(C) 1500 lbf/yd^3

(D) 1600 lbf/yd^3

(b) What is the absolute volume of the coarse aggregate in 1 yd^3 of concrete?

(A) 8.2 ft^3/yd^3

(B) 8.3 ft^3/yd^3

(C) 8.7 ft^3/yd^3

(D) 9.7 ft^3/yd^3

(c) What is the water-cement ratio?

(A) 0.42

(B) 0.44

(C) 0.46

(D) 0.48

(d) What is the weight of cement in 1 yd^3 of fresh concrete?

(A) 610 lbf/yd^3

(B) 720 lbf/yd^3

(C) 820 lbf/yd^3

(D) 910 lbf/yd^3

(e) What is the absolute volume of fine aggregate in 1 yd^3 of fresh concrete?

(A) 7.8 ft^3

(B) 8.8 ft^3

(C) 9.8 ft^3

(D) 11 ft^3

(f) What is the SSD weight of the sand in 1 yd^3 of fresh concrete?

(A) 1250 lbf

(B) 1270 lbf

(C) 1300 lbf

(D) 1370 lbf

(g) What is the absolute volume of the cement in 1 yd^3 of fresh concrete?

(A) 3.1 ft^3

(B) 3.2 ft^3

(C) 3.3 ft^3

(D) 3.4 ft^3

(h) What is the volume of water designed for use in 1 yd^3 of fresh concrete?

(A) 3.3 ft^3

(B) 3.8 ft^3

(C) 4.3 ft^3

(D) 4.8 ft^3

(i) If the oven-dry weight of the coarse aggregate were 1600 lbf, how much water would it need to absorb to reach SSD conditions?

(A) 5.3 lbf

(B) 11 lbf

(C) 32 lbf

(D) 1100 lbf

(j) The unit weight of this concrete mix is most nearly

(A) 140 lbf/ft^3

(B) 143 lbf/ft^3

(C) 145 lbf/ft^3

(D) 150 lbf/ft^3

2. (*Time limit: one hour*) A trial batch of concrete is needed according to the following requirements and characteristics.

no. of sacks of cement per yd^3 of concrete	7
entrained air	6%
total water-cement ratio	0.54
cement specific gravity	3.15

For economical reasons, two different sizes of coarse aggregate were used. The aggregates have the following characteristics.

	specific gravity	moisture	percentage of aggregate by volume
fine aggregate	2.60	5% excess	36%
$\frac{3}{4}$ in coarse aggregate	2.65	2% excess	18%
1.0 in coarse aggregate	2.62	3% deficit	46%

(a) The weight of cement needed for 2 yd^3 of concrete is most nearly

(A) 660 lbf

(B) 1300 lbf

(C) 1500 lbf

(D) none of the above

(b) The volume of cement required for 2 yd^3 is

(A) 3.4 ft^3

(B) 5.4 ft^3

(C) 6.7 ft^3

(D) 7.0 ft^3

(c) The weight of water needed for 2 yd^3 is

(A) 670 lbf

(B) 710 lbf

(C) 870 lbf

(D) 940 lbf

(d) The volume of water needed for 2 yd^3 is

(A) 8.4 ft^3

(B) 9.7 ft^3

(C) 11 ft^3

(D) 12 ft^3

(e) The absolute volume of fine aggregate in 2 yd^3 is

(A) 11.8 ft^3

(B) 12.5 ft^3

(C) 13.4 ft^3

(D) 14.2 ft^3

(f) The SSD weight of fine aggregate in 2 yd^3 is

(A) 1800 lbf

(B) 1850 lbf

(C) 1910 lbf

(D) 1990 lbf

(g) The absolute volume of the $^3/_4$ in coarse aggregate in 2 yd^3 is

(A) 5.9 ft^3

(B) 6.2 ft^3

(C) 6.8 ft^3

(D) 7.1 ft^3

(h) The absolute volume of the 1 in coarse aggregate in 2 yd^3 is

(A) 15.0 ft^3

(B) 16.8 ft^3

(C) 17.3 ft^3

(D) 18.1 ft^3

Structural

(i) The SSD weight of the $^3/_4$ in coarse aggregate in 2 yd^3 is

(A) 860 lbf

(B) 970 lbf

(C) 1050 lbf

(D) 1120 lbf

(j) The SSD weight of the 1 in coarse aggregate in 2 yd^3 is

(A) 2460 lbf

(B) 2480 lbf

(C) 2510 lbf

(D) 2630 lbf

3. 1.5 yd^3 of portland cement concrete with the following specifications are needed.

cement content	6.5 sacks/yd^3
water-cement ratio	5.75 gal/sack
cement specific gravity	3.10
fine-aggregate specific gravity	2.65
coarse-aggregate specific gravity	2.00
aggregate grading	30% fine; 70% coarse (by volume)
free moisture (SSD basis)	1.5% excess in fine aggregate 3.0% deficient in coarse aggregate
entrained air	5%

(a) What is the as-delivered weight of the cement?

(A) 710 lbf

(B) 850 lbf

(C) 920 lbf

(D) 1100 lbf

(b) What is the as-delivered weight of the fine aggregate?

(A) 1300 lbf

(B) 1500 lbf

(C) 1600 lbf

(D) 1900 lbf

(c) What is the as-delivered weight of the coarse aggregate?

(A) 1850 lbf

(B) 1970 lbf

(C) 2140 lbf

(D) 2220 lbf

(d) What is the as-delivered weight of the water?

(A) 440 lbf

(B) 520 lbf

(C) 690 lbf

(D) 820 lbf

4. Which of the following would a cast-in-place ground slab or floor NOT use to separate steel reinforcing bars or wire mesh from the ground?

I.   precast cementitious blocks

II. metal chairs

III. wood risers

IV. wire bolsters

V.   plastic bar supports

VI. cellulose spacer blocks

(A) III and VI

(B) I, III, and VI

(C) I, II, IV, and V

(D) II, IV, V, and VI

5. Roller compacted concrete is ordered using the following parameters.

water-cement ratio	0.63
weight of water	189 lbf/yd^3
portland cement specific gravity	3.15
class-F fly ash specific gravity	2.26
percent of fly ash in cement	40% (by volume)

The weight of fly ash per cubic yard of roller-compacted concrete is most nearly

(A) 60 lbf/yd^3

(B) 70 lbf/yd^3

(C) 80 lbf/yd^3

(D) 90 lbf/yd^3

Structural

6. While forming a large concrete wall, horizontal sheathing (holding the concrete) is supported by vertical strongbacks (studs) spaced every 22 in, which in turn are supported by horizontal wales spaced every 18 in. Double strands of no. 9 wire (tensile breaking strength is 700 lbf per strand) are used to resist the hydrostatic pressure tending to separate the two halves of the form. The wire is anchored at every third stud. During construction, the maximum concrete pressure on the sheathing is 480 psf. Most nearly, what is the factor of safety in the wire?

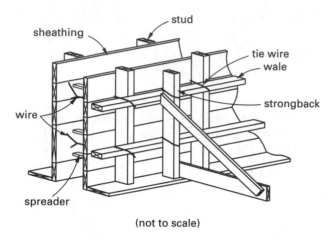

(not to scale)

(A) 0.35

(B) 0.71

(C) 1.4

(D) 2.8

7. Which of the following types of joints typically does NOT use dowels?

I.   construction joints

II.   control joints

III. contraction joints

IV.   expansion joints

V.   isolation joints

   (A) I and V

   (B) II and V

   (C) III and IV

   (D) II, III, IV, and V

8. In addition to location of the pour, tremie concrete poured underwater differs from common concrete poured in the air primarily in its

   (A) placement method and equipment

   (B) components and additives

   (C) specific gravity

   (D) aggregate size

9. In very hot weather, thin floor slabs of newly poured concrete would most normally be

   (A) sprinkled with rock salt or potassium chloride

   (B) sprayed with or submerged in water

   (C) kept insulated by dry burlap insulation

   (D) covered by cube or slush ice

10. In shored construction, which factors do NOT influence the length of time between pouring concrete and stripping the forms?

I.   weather

II.   project specifications

III. developed concrete strength

IV.   ACI 347

V.   project schedule

VI.   local ordinances

   (A) I and III

   (B) I and IV

   (C) II and IV

   (D) V and VI

11. A four-lane concrete highway is paved in two passes, two adjacent lanes at a time. To facilitate connecting with the subsequent pour, the edge of the concrete is fitted with concrete dowels running transverse to the direction of travel, and the edge is keyed. This joint would most likely be referred to as a

   (A) transverse, tied and keyed expansion joint

   (B) transverse, tied and keyed construction joint

   (C) skewed contraction joint

   (D) longitudinal tied construction joint

12. What is the average temperature of fresh, quickly mixed concrete consisting of the following components?

cement	550 lbm added at 80°F
aggregate	3000 lbm added at 65°F
mixing water	145 lbm added at 50°F
free moisture in the aggregates	60 lbm
slush ice	155 lbm
ambient air temperature	95°F
specific heat of mixture and aggregate	0.22 Btu/lbm-°F

    (A) 48°F

    (B) 51°F

    (C) 54°F

    (D) 59°F

13. Normal weight concrete with a 3 in slump is poured at the vertical rate of 4 ft per hour into the form for a 12 ft high retaining wall. The concrete is vibrated to a depth of 4 ft. The temperature is 60°F. Most nearly, what is the maximum hydrostatic pressure experienced by the formwork?

    (A) 500 psf

    (B) 600 psf

    (C) 750 psf

    (D) 1800 psf

SOLUTIONS

1. (a) The weight of the coarse dry aggregate is

$$\left(0.57 \ \frac{ft^3}{ft^3}\right)\left(105 \ \frac{lbf}{ft^3}\right)\left(27 \ \frac{ft^3}{yd^3}\right)$$
$$= \boxed{1616 \ lbf/yd^3 \quad (1600 \ lbf/yd^3)}$$

The answer is (D).

(b) The volume of coarse aggregate is

$$\frac{1616 \ \dfrac{lbf}{yd^3}}{(2.68)\left(62.4 \ \dfrac{lbf}{ft^3}\right)} = \boxed{9.66 \ ft^3/yd^3 \quad (9.7 \ ft^3/yd^3)}$$

The answer is (D).

(c) The water-cement ratio is

$$\frac{\left(5 \ \dfrac{gal}{sack}\right)\left(8.34 \ \dfrac{lbf}{gal}\right)}{94 \ \dfrac{lbf}{sack}} = \boxed{0.44}$$

The answer is (B).

(d) The weight of cement in 1 yd³ is

$$\left(6.5 \ \frac{sacks}{yd^3}\right)\left(94 \ \frac{lbf}{sack}\right) = \boxed{611 \ lbf/yd^3 \quad (610 \ lbf/yd^3)}$$

The answer is (A).

(e) To find the volume of fine aggregate, the volume of all other components must be found.

ingredient	weight (lbf)	volume (ft³)
cement	611	$\dfrac{611 \ lbf}{(3.15)\left(62.4 \ \dfrac{lbf}{ft^3}\right)} = 3.11$
coarse aggregate	1616	9.66 [from part (b)]
water	$(0.44)(611 \ lbf) = 269$	$\dfrac{269 \ lbf}{62.4 \ \dfrac{lbf}{ft^3}} = 4.31$
air	0	$(0.04)(1 \ yd^3)\left(27 \ \dfrac{ft^3}{yd^3}\right)$
		$= \underline{1.08}$
		total $\overline{18.16}$

The volume of fine aggregate in 1 yd^3 of the mix is

$$(1 \text{ yd}^3)\left(27 \frac{\text{ft}^3}{\text{yd}^3}\right) - 18.16 \text{ ft}^3 = \boxed{8.84 \text{ ft}^3 \quad (8.8 \text{ ft}^3)}$$

The answer is (B).

(f) The weight of fine aggregate is

$$(8.84 \text{ ft}^3)\left(62.4 \frac{\text{lbf}}{\text{ft}^3}\right)(2.48) = \boxed{1368 \text{ lbf} \quad (1370 \text{ lbf})}$$

The answer is (D).

(g) Refer to part (e).
The answer is (A).

(h) Refer to part (e).
The answer is (C).

(i) The weight of the water is

$$(0.007)(1600 \text{ lbf}) = \boxed{11.2 \text{ lbf} \quad (11 \text{ lbf})}$$

The answer is (B).

(j) The unit weight of the concrete is

$$\frac{611 \frac{\text{lbf}}{\text{yd}^3} + 1616 \frac{\text{lbf}}{\text{yd}^3} + 269 \frac{\text{lbf}}{\text{yd}^3} + 1368 \frac{\text{lbf}}{\text{yd}^3}}{27 \frac{\text{ft}^3}{\text{yd}^3}}$$

$$= \boxed{143 \text{ lbf/ft}^3}$$

The answer is (B).

2. (a) The weight of cement needed is

$$\left(7 \frac{\text{sacks}}{\text{yd}^3}\right)\left(94 \frac{\text{lbf}}{\text{sack}}\right)(2 \text{ yd}^3) = \boxed{1316 \text{ lbf} \quad (1300 \text{ lbf})}$$

The answer is (B).

(b) The volume of cement required is

$$\frac{1316 \text{ lbf}}{(3.15)\left(62.4 \frac{\text{lbf}}{\text{ft}^3}\right)} = \boxed{6.70 \text{ ft}^3}$$

The answer is (C).

(c) Determine the weights and volumes for 2 yd^3.

ingredient	weight (lbf)	volume (ft^3)
cement	1316	6.70
water	710.6	11.39
air	0	$(0.06)(2 \text{ yd}^3)\left(27 \frac{\text{ft}^3}{\text{yd}^3}\right)$
		$= \underline{3.24}$
		total 21.33

The volume of fine aggregate is

$$\left((2 \text{ yd}^3)\left(27 \frac{\text{ft}^3}{\text{yd}^3}\right) - 21.33 \text{ ft}^3\right)(0.36) = 11.76 \text{ ft}^3$$

The weight of fine aggregate is

$$(11.76 \text{ ft}^3)(2.6)\left(62.4 \frac{\text{lbf}}{\text{yd}^3}\right) = 1907.9 \text{ lbf}$$

The volume of $^3/_4$ in coarse aggregate is

$$\left((2 \text{ yd}^3)\left(27 \frac{\text{ft}^3}{\text{yd}^3}\right) - 21.33 \text{ ft}^3\right)(0.18) = 5.88 \text{ ft}^3$$

The weight of $^3/_4$ in coarse aggregate is

$$(5.88 \text{ ft}^3)(2.65)\left(62.4 \frac{\text{lbf}}{\text{ft}^3}\right) = 972.3 \text{ lbf}$$

The volume of 1 in coarse aggregate is

$$\left((2 \text{ yd}^3)\left(27 \frac{\text{ft}^3}{\text{yd}^3}\right) - 21.33 \text{ ft}^3\right)(0.46) = 15.03 \text{ ft}^3$$

The weight of 1 in coarse aggregate is

$$(15.03 \text{ ft}^3)(2.62)\left(62.4 \frac{\text{lbf}}{\text{ft}^3}\right) = 2457.2 \text{ lbf}$$

The total volume as calculated is

$$21.33 \text{ ft}^3 + 11.76 \text{ ft}^3 + 5.88 \text{ ft}^3 + 15.03 \text{ ft}^3 = 54.00 \text{ ft}^3$$

The total weight is

$$1316 \text{ lbf} + 710.6 \text{ lbf} + 1907.9 \text{ lbf}$$
$$+ 972.3 \text{ lbf} + 2457.2 \text{ lbf}$$
$$= 7364 \text{ lbf}$$

The uncorrected water required is calculated from the water-cement ratio.

$$(0.54)(1316 \text{ lbf}) = 710.6 \text{ lbf} \quad [\text{before correction}]$$

The corrected water depends on the excesses and deficits in the aggregates.

fine aggregate: $(1907.9 \text{ lbf})(1 + 0.05) = 2003.3 \text{ lbf}$

$3/4$ in coarse
aggregate: $(972.3 \text{ lbf})(1 + 0.02) = 991.7 \text{ lbf}$

1 in coarse
aggregate: $(2457.2 \text{ lbf})(1 - 0.03) = 2383.5 \text{ lbf}$

The correction for surface water is

7364 lbf

$- (1316 \text{ lbf} + 2003.2 \text{ lbf} + 991.7 \text{ lbf} + 2383.5 \text{ lbf})$

$= \boxed{669.6 \text{ lbf} \quad (670 \text{ lbf})}$

The answer is (A).

(d) The volume of water is

$$\frac{669.6 \text{ lbf}}{62.4 \dfrac{\text{lbf}}{\text{ft}^3}} = \boxed{10.73 \text{ ft}^3 \quad (11 \text{ ft}^3)}$$

The answer is (C).

(e) From part (c), the absolute volume of fine aggregate is 11.76 ft^3 (11.8 ft^3).

The answer is (A).

(f) From part (c), the SSD weight of fine aggregate is 1907.9 lbf (1910 lbf).

The answer is (C).

(g) From part (c), the absolute volume of the $3/4$ in course aggregate is 5.88 ft^3 (5.9 ft^3).

The answer is (A).

(h) From part (c), the absolute value of the 1 in course aggregate is 15.03 ft^3 (15.0 ft^3).

The answer is (A).

(i) From part (c), the SSD weight of the $3/4$ in course aggregate is 972.3 lbf (970 lbf).

The answer is (B).

(j) From part (c), the SSD weight of the 1 in course aggregate is 2457.2 lbf (2460 lbf).

The answer is (A).

3. On a per cubic yard basis,

$$\begin{array}{l} \text{volume of} \\ \text{cement used} \end{array} = \frac{\left(6.5 \dfrac{\text{sacks}}{\text{yd}^3}\right)\left(94 \dfrac{\text{lbf}}{\text{sack}}\right)}{\left(62.4 \dfrac{\text{lbf}}{\text{ft}^3}\right)(3.10)}$$

$$= 3.16 \text{ ft}^3/\text{yd}^3$$

$$\begin{array}{l} \text{volume of} \\ \text{water used} \end{array} = \frac{\left(6.5 \dfrac{\text{sacks}}{\text{yd}^3}\right)\left(5.75 \dfrac{\text{gal}}{\text{sack}}\right)}{7.48 \dfrac{\text{gal}}{\text{ft}^3}}$$

$$= 5.00 \text{ ft}^3/\text{yd}^3$$

$$\begin{array}{l} \text{volume} \\ \text{of air} \end{array} = (0.05)\left(27 \dfrac{\text{ft}^3}{\text{yd}^3}\right) = 1.35 \text{ ft}^3/\text{yd}^3$$

total: $3.16 \dfrac{\text{ft}^3}{\text{yd}^3} + 5.00 \dfrac{\text{ft}^3}{\text{yd}^3} + 1.35 \dfrac{\text{ft}^3}{\text{yd}^3} = 9.51 \text{ ft}^3/\text{yd}^3$

The remainder of the cubic yard ($27 \text{ ft}^3/\text{yd}^3 - 9.51 \text{ ft}^3/\text{yd}^3 = 17.49 \text{ ft}^3/\text{yd}^3$) must be aggregate. The volumes of fine and coarse aggregates are

fine: $\left(17.49 \dfrac{\text{ft}^3}{\text{yd}^3}\right)(0.30) = 5.25 \text{ ft}^3/\text{yd}^3$

coarse: $\left(17.49 \dfrac{\text{ft}^3}{\text{yd}^3}\right)(0.70) = 12.24 \text{ ft}^3/\text{yd}^3$

The weights are

cement: $\left(3.16 \dfrac{\text{ft}^3}{\text{yd}^3}\right)(3.10)\left(62.4 \dfrac{\text{lbf}}{\text{ft}^3}\right)$

$= 611.3 \text{ lbf/yd}^3$

water: $\left(5.0 \dfrac{\text{ft}^3}{\text{yd}^3}\right)\left(62.4 \dfrac{\text{lbf}}{\text{ft}^3}\right)$

$= 312.0 \text{ lbf/yd}^3$

fine aggregate: $\left(5.25 \dfrac{\text{ft}^3}{\text{yd}^3}\right)(2.65)\left(62.4 \dfrac{\text{lbf}}{\text{ft}^3}\right)$

$= 868.1 \text{ lbf/yd}^3$

coarse aggregate: $\left(12.24 \dfrac{\text{ft}^3}{\text{yd}^3}\right)(2.00)\left(62.4 \dfrac{\text{lbf}}{\text{ft}^3}\right)$

$= 1527.6 \text{ lbf/yd}^3$

Structural

The mix ratio is (611.3/611.3, 868.1/611.3, 1527.6/611.3) or (1:1.42:2.5) by weight.

constituent	ratio	weight per sack cement (lbf)	weight density (lbf/ft^3)	absolute volume (ft^3)
cement	1	94	193.4	0.486
fine	1.42	133.5	165.4	0.807
coarse	2.5	235.0	124.8	1.883
water			$\left(\dfrac{5.75}{7.48}\right)$	$= \underline{0.769}$
				total 3.945

The solid yield is 3.945 ft^3/sack. The yield with 5% air is

$$\frac{3.945 \dfrac{\text{ft}^3}{\text{sack}}}{1 - 0.05} = 4.153 \text{ ft}^3/\text{sack}$$

The number of one-sack batches is

$$\frac{(1.5 \text{ yd}^3)\left(27 \dfrac{\text{ft}^3}{\text{yd}^3}\right)}{4.153 \dfrac{\text{ft}^3}{\text{sack}}} = 9.752 \text{ sacks}$$

(a) The required cement weight is

$$(9.752 \text{ sacks})\left(94 \frac{\text{lbf}}{\text{sack}}\right) = \boxed{916.7 \text{ lbf} \quad (920 \text{ lbf})}$$

The answer is (C).

(b) The required fine aggregate weight, as delivered, is

$$(9.752 \text{ sacks})(1.015)\left(94 \frac{\text{lbf}}{\text{sack}}\right)(1.42)$$
$$= \boxed{1321.2 \text{ lbf} \quad (1300 \text{ lbf})}$$

Dividing this by 1.015 converts to SSD conditions.

The answer is (A).

(c) The required coarse aggregate, as delivered, is

$$(9.752 \text{ sacks})(1 - 0.03)\left(94 \frac{\text{lbf}}{\text{sack}}\right)(2.5)$$
$$= \boxed{2223.0 \quad (2220 \text{ lbf})}$$

Dividing this by $(1 - 0.03)$ converts to SSD conditions.

The answer is (D).

(d) The weight of the excess water in the fine aggregate is

$$(1321.2 \text{ lbf})\left(\frac{0.015}{1 + 0.015}\right) = 19.53 \text{ lbf}$$

The water needed to bring the coarse aggregate to SSD condition is

$$(2223.0 \text{ lbf})\left(\frac{0.03}{1 - 0.03}\right) = 68.75 \text{ lbf}$$

The total weight of water required is

$$\left(\frac{\left(5.75 \dfrac{\text{gal}}{\text{sack}}\right)(9.752 \text{ sacks})}{7.48 \dfrac{\text{gal}}{\text{ft}^3}}\right)\left(62.4 \frac{\text{lbf}}{\text{ft}^3}\right)$$
$$+ 68.75 \text{ lbf} - 19.53 \text{ lbf}$$
$$= \boxed{517.0 \text{ lbf} \quad (520 \text{ lbf})}$$

The answer is (B).

4. After curing, wood and cellulose products would absorb moisture and attract termites. Therefore, they are not used to support bars in slabs. All of the other listed items are used to separate steel reinforcing bars or wire mesh from the ground.

The answer is (A).

5. The total weight of cement per finished cubic yard is

$$\frac{189 \dfrac{\text{lbf}}{\text{yd}^3}}{0.63} = 300 \text{ lbf/yd}^3$$

The volume of cement is

$$V = \frac{W}{\gamma} = \frac{300 \dfrac{\text{lbf}}{\text{yd}^3}}{(3.15)\left(62.4 \dfrac{\text{lbf}}{\text{yd}^3}\right)} = 1.526 \text{ ft}^3/\text{yd}^3$$

The volume of fly ash is

$$V = (0.40)\left(1.526 \frac{\text{ft}^3}{\text{yd}^3}\right) = 0.610 \text{ ft}^3/\text{yd}^3$$

The weight of fly ash is

$$W = V\gamma = \left(0.610 \frac{\text{ft}^3}{\text{yd}^3}\right)\left(62.4 \frac{\text{lbf}}{\text{ft}^3}\right)(2.26)$$
$$= \boxed{86 \text{ lbf/yd}^3 \quad (90 \text{ lbf/yd}^3)}$$

The answer is (D).

6. The full width and length of the wall form is not given, so the analysis is based on a "unit" rectangular area bounded by two adjacent studs and two adjacent wales.

$$A = wh = \frac{(3 \text{ stud spacings})(22 \text{ in})(18 \text{ in})}{\left(12 \frac{\text{in}}{\text{ft}}\right)^2} = 8.25 \text{ ft}^2$$

The force on that area is

$$F = pA = \left(480 \frac{\text{lbf}}{\text{ft}^2}\right)(8.25 \text{ ft}^2) = 3960 \text{ lbf}$$

Each unit area is held by double wires at two corners, for a total of 4 strands. The factor of safety (FS) is

$$FS = \frac{(4 \text{ strands})\left(700 \frac{\text{lbf}}{\text{strand}}\right)}{3960 \text{ lbf}} = 0.707 \quad (0.71)$$

The answer is (B).

7. Control joints, as a method of controlling where cracking will occur, are typically created by saw cutting. Although steel reinforcing may be present at a control joint, dowels will not. Isolation joints separate concrete slabs from columns, footings, and walls. They are used to separate elements that experience differences in loading (soil pressure) and settlement. No connection between the elements should exist. Expansion and contraction joints can be doweled; the slab slides along the dowel bar as the slab expands or contracts. Construction joints are doweled to ensure complete load transfer when subsequent portions of the slab are poured.

The answer is (B).

8. Tremie concrete gets its name from the delivery pipe (the tremie) that reaches from above the water surface to the lowest submerged delivery point. The lower end of the tremie is kept buried in freshly delivered concrete. The concrete is fairly normal in its chemistry and components.

The answer is (A).

9. While ice could be used to reduce temperatures until the cooler nightfall arrives, it is easier and more cost effective to rely on the evaporation of liquid water (i.e., the swamp cooler effect) to lower the temperature.

The answer is (B).

10. The length of time that formwork is left on before being stripped is an engineering decision, under the jurisdiction of the project engineer. Guidance to the

contractor will be placed in the project specifications. The engineer will consider various factors when writing the specifications, including developed concrete strength and loading. ACI 347 recommendations may be considered in some situations when the actual requirements are difficult to determine. The specifications may have options dependent on the local temperature, the need to remove insulating formwork, or to cool the concrete during curing. The project schedule and local ordinances should not be permitted to dictate the time between pouring concrete and stripping the forms.

The answer is (D).

11. The joint runs longitudinally along the highway and is tied by the dowels to the next phase of construction (i.e., the next pour). The key is probably an unnecessary detail, as the dowels will prevent both vertical and lateral separation of the separate pours.

The answer is (D).

12. The mixing process is adiabatic (i.e., there is no loss of energy) if the mixing is performed quickly. The total energy (enthalpy) immediately after mixing is conserved. The adiabatic assumption is the basis of the solution.

The initial energy content of each component is measured with respect to an arbitrary reference point. In this case, $T = 0°F$ is used as a reference point. (This is not the freezing point of water, which is 32°F. It is only a convenient choice of reference.)

Using h for energy (enthalpy), c_p for sensible specific heat, and T for temperature, the thermal energy for each component is

$$h = mc_p T$$

Without knowing the actual sensible specific heats of the cement, aggregate, and mixture, common construction assumptions are made. The specific heats of cement, sand, aggregate, and concrete mixture are in the range of 0.20–0.22 Btu/lbm-°F. 0.22 Btu/lbm-°F is typically used for these types of calculations. (See Table 13.1.) The specific heat of water is 1 Btu/lbm-°F. It is unlikely that the ice would be delivered at a temperature below freezing (i.e., would be subcooled). If slush ice is used, its temperature is automatically known to be 32°F. Ice's heat of fusion is 144 Btu/lbm. The air temperature is not relevant because air is not a significant contributor to the mass of the mixture. Each of the components experiences a temperature change of $\Delta T = T_i - T_f$. The final temperature, T_f, is the same for all components after mixing.

A small fraction of the slush ice will have already melted, but the fraction is unknown and is taken as zero. When the ice melts, it absorbs thermal energy at

Structural

the rate of 144 Btu per pound of ice. (This solution initially assumes that all of the ice melts, which would be an improper assumption only if the components were all initially at near-freezing temperatures.)

The total mass of the mixture is

$$m_t = 550 \text{ lbm} + 3000 \text{ lbm} + 145 \text{ lbm}$$
$$+ 60 \text{ lbm} + 145 \text{ lbm}$$
$$= 3900 \text{ lbm}$$

Following the assumption that the mixture specific heat is 0.22 Btu/ lbm-°F, the final (mixed) energy content of the entire mixture is

$$h_{\text{mixture}} = mc_p T_f = (3900 \text{ lbm})\left(0.22 \ \frac{\text{Btu}}{\text{lbm-°F}}\right) T_f$$
$$= 858 T_f \text{ Btu}$$

The initial energy content of the cement (referenced to 0°F) is

$$h_{\text{cement}} = (550 \text{ lbm})\left(0.22 \ \frac{\text{Btu}}{\text{lbm-°F}}\right)(80°F)$$
$$= 9680 \text{ Btu}$$

Similarly,

$$h_{\text{aggregate}} = (3000 \text{ lbm})\left(0.22 \ \frac{\text{Btu}}{\text{lbm-°F}}\right)(65°F)$$
$$= 42,900 \text{ Btu}$$
$$h_{\text{free moisture}} = (60 \text{ lbm})\left(1 \ \frac{\text{Btu}}{\text{lbm-°F}}\right)(65°F)$$
$$= 3900 \text{ Btu}$$
$$h_{\text{water}} = (145 \text{ lbm})\left(1 \ \frac{\text{Btu}}{\text{lbm-°F}}\right)(50°F)$$
$$= 7250 \text{ Btu}$$
$$h_{\text{slush ice}} = (155 \text{ lbm})\left(1 \ \frac{\text{Btu}}{\text{lbm-°F}}\right)(32°F)$$
$$= 4960 \text{ Btu}$$

When the ice melts, it will absorb

$$\text{melting energy} = (155 \text{ lbm})\left(144 \ \frac{\text{Btu}}{\text{lbm}}\right) = 22,320 \text{ Btu}$$

Equate the initial energy of all of the components to the final energy of the mixture.

$$h_{\text{mixture}} = h_{\text{cement}} + h_{\text{aggregate}} + h_{\text{free moisture}}$$
$$+ h_{\text{water}} + h_{\text{slush ice}} - \text{melting energy}$$
$$858 T_f \text{ Btu} = 9680 \text{ Btu} + 42,900 \text{ Btu} + 3900 \text{ Btu}$$
$$+ 7250 \text{ Btu} + 4960 \text{ Btu} - 22,320 \text{ Btu}$$

The final mixture temperature is

$$\boxed{T_f = 54°F}$$

All of the slush ice will have melted at this temperature.

The answer is (C).

13. Use Eq. 49.2. The empirical equation is not dimensionally consistent.

$$p = 150 + 9000\left(\frac{R}{T}\right) = 150 + \frac{(9000)\left(4 \ \frac{\text{ft}}{\text{hr}}\right)}{60°F}$$
$$= \boxed{750 \text{ lbf/ft}^2 \quad (750 \text{ psf})}$$

The answer is (C).

50 Reinforced Concrete: Beams

PRACTICE PROBLEMS

(Assume all concrete is normal weight unless specified.)

1. The simply supported beam shown spans 20 ft and, in addition to its own weight, carries a uniformly distributed service dead load of 1.75 kips/ft and a uniformly distributed live load of 3.0 kips/ft. Five no. 9 bars running the full length of the beam are used as flexural reinforcement. $f'_c = 4000$ lbf/in^2, and $f_y = 60,000$ lbf/in^2. Number 3 stirrups are used. (a) Determine the theoretical spacing of stirrups at the critical section using the refined expression for the shear strength of the concrete. (b) Determine the theoretical spacing of the stirrups at the critical section using the simplified expression for the concrete shear strength. (c) Identify the portion of the beam where stirrups are not required.

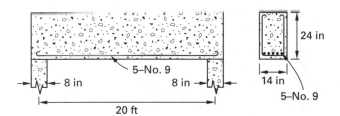

2. (*Time limit: one hour*) A simply supported reinforced concrete beam spans 20 ft as shown. The beam is subjected to a uniform service dead load equal to 2.0 kips/ft (exclusive of beam weight) and to a uniform service live load of 2.4 kips/ft. $f'_c = 3000$ lbf/in^2, and $f_y = 40,000$ lbf/in^2. The depth to the center of the reinforcing layer is 25.5 in.

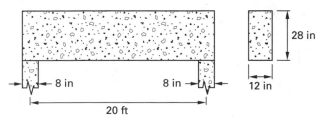

(a) The factored uniform load is most nearly

  (A) 6.7 kips/ft
  (B) 7.4 kips/ft
  (C) 8.0 kips/ft
  (D) 9.2 kips/ft

(b) The tension steel required at the section of maximum moment is most nearly

  (A) 4.5 in^2
  (B) 5.0 in^2
  (C) 5.6 in^2
  (D) 6.1 in^2

(c) The maximum area of tension steel permitted (based on flexure requirements) is most nearly

  (A) 5.6 in^2
  (B) 7.7 in^2
  (C) 8.5 in^2
  (D) 9.2 in^2

(d) The minimum area of tension steel permitted is most nearly

  (A) 1.0 in^2
  (B) 1.3 in^2
  (C) 1.5 in^2
  (D) 1.8 in^2

(e) The maximum uniform factored load that the beam can sustain if no compression steel is used is most nearly

  (A) 8.1 kips/ft
  (B) 9.0 kips/ft
  (C) 10 kips/ft
  (D) 13 kips/ft

(f) The spacing of no. 3 U-shaped stirrups at the location where the shear is maximum is most nearly

  (A) 5.0 in
  (B) 6.0 in
  (C) 6.4 in
  (D) 8.3 in

(g) The total length of the beam where no shear reinforcement is required is most nearly

(A) 10 in

(B) 12 in

(C) 15 in

(D) 20 in

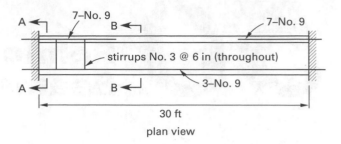

plan view

(h) Based on the maximum total shear permitted in the beam, the maximum uniform factored load that the beam can be designed for, with the given dimensions, is most nearly

(A) 16 kips/ft

(B) 17 kips/ft

(C) 18 kips/ft

(D) 20 kips/ft

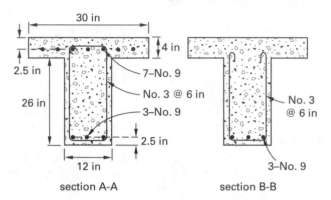

section A-A   section B-B

(i) If the live load is subsequently reduced to 50% of its maximum value, the service moment that should be used to compute the effective moment of inertia is most nearly

(A) 200 ft-kips

(B) 220 ft-kips

(C) 240 ft-kips

(D) 250 ft-kips

(a) The design moment capacity in the positive moment region is most nearly

(A) 300 ft-kips

(B) 360 ft-kips

(C) 400 ft-kips

(D) 410 ft-kips

(j) The cracking moment is most nearly

(A) 40 ft-kips

(B) 45 ft-kips

(C) 51 ft-kips

(D) 54 ft-kips

(b) At capacity, the stress in the compression steel in the negative moment region is most nearly

(A) 35 kips/in^2

(B) 47 kips/in^2

(C) 52 kips/in^2

(D) 60 kips/in^2

3. (*Time limit: one hour*) A reinforced concrete T-beam with a 30 ft span and a 30 in effective width in a floor slab system is fixed at both ends and is reinforced as shown. $f'_c = 3000$ lbf/in^2, and $f_y = 60,000$ lbf/in^2.

(c) The gross moment of inertia is most nearly

(A) 27,000 in^4

(B) 37,000 in^4

(C) 41,000 in^4

(D) 68,000 in^4

(d) The cracked moment of inertia in the positive moment region is most nearly

(A) 10,000 in^4

(B) 13,000 in^4

(C) 15,000 in^4

(D) 20,000 in^4

(e) The cracking moment in the positive moment region is most nearly

(A) 22 ft-kips

(B) 47 ft-kips

(C) 60 ft-kips

(D) 74 ft-kips

(f) The cracking moment in the negative moment region is most nearly

(A) 60 ft-kips

(B) 74 ft-kips

(C) 99 ft-kips

(D) 120 ft-kips

(g) If a uniformly distributed service load of 4 kips/ft acts on the beam, using the refined equation for steel stress, the stress in the tension steel in the positive moment region is most nearly

(A) 18 kips/in^2

(B) 23 kips/in^2

(C) 30 kips/in^2

(D) 35 kips/in^2

(h) The maximum spacing of the tension steel bars based on cracking is most nearly

(A) 8 in

(B) 13 in

(C) 18 in

(D) 21 in

(i) The maximum uniform factored load that the beam can sustain as controlled by the shear reinforcement is most nearly

(A) 4.0 kips/ft

(B) 5.9 kips/ft

(C) 6.5 kips/ft

(D) 7.5 kips/ft

(j) The maximum uniform factored load that the beam can sustain as governed by the flexural capacity is most nearly

(A) 7.1 kips/ft

(B) 8.2 kips/ft

(C) 9.5 kips/ft

(D) 14 kips/ft

4. A singly reinforced rectangular beam carries a moment from dead loads of 50,000 ft-lbf and a moment from live loads of 200,000 ft-lbf. $f'_c = 3000$ lbf/in^2, and $f_y = 50,000$ lbf/in^2. Number 4 stirrups are to be used. Design and detail the cross section.

5. A beam must withstand an ultimate factored moment of 400,000 ft-lbf. $f'_c = 4000$ lbf/in^2, and $f_y = 40,000$ lbf/in^2. Number 3 stirrups will be used. (Do not design the shear reinforcement, check for cracking, or check deflection.) (a) Determine the beam width and depth. (b) Determine the required steel area. (c) How many layers of steel are needed? (d) What is the overall beam depth?

6. The slab shown supports a moment of 20,000 ft-lbf per foot of width. The concrete's unit weight is 145 lbf/ft^3, and $f'_c = 3000$ lbf/in^2. The response is elastic. Use the transformed area method of evaluating the stresses in the (a) concrete and (b) steel.

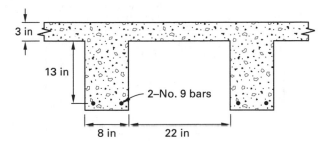

7. (*Time limit: one hour*) A rectangular, singly reinforced beam is shown. The beam supports two concentrated dead loads and a self-weight dead load of 2000 lbf/ft. $f'_c = 4000$ lbf/in^2, and $f_y = 60,000$ lbf/in^2. Number 11 steel bars must be used. The modular ratio is 8. (Do not design shear reinforcement. Do not consider live loading.) (a) Determine the beam width and depth. (b) Determine the required steel area using half of the maximum steel permitted. (c) How many layers of steel are needed?

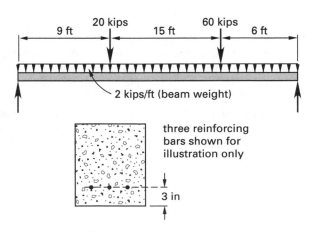

8. (*Time limit: one hour*) A simply supported, singly reinforced rectangular concrete beam supporting a floor is needed to span 27 ft. It carries a dead load of 1 kip/ft (which includes the beam weight) and a live load of 2 kips/ft. $f'_c = 4000$ lbf/in^2, $f_y = 60,000$ lbf/in^2, and

$E_c = 3.6 \times 10^6$ lbf/in^2. The maximum permitted steel reinforcing is to be used. (a) Use no. 11 bars and a beam width of 14 in to design the beam. (b) If the beam has an interior exposure, will cracking be within limits? (c) If 30% of the live load is sustained, calculate the instantaneous and long-term centerline deflections.

9. (*Time limit: one hour*) The beam shown carries a dead load, D, of 1300 lbf/ft (including its own weight) and a live load, L, of 1900 lbf/ft. $f'_c = 3000$ lbf/in^2, and $f_y = 60,000$ lbf/in^2. The clear distance between the two beam supports is 28 ft. The beam's depth is 21 in. (a) What additional shear reinforcement strength is required to be provided by the steel? (b) Over what length of the beam is shear reinforcement not required? (c) Assuming no. 3 bars are used, what stirrup spacing should be used between the face of the support out to a distance of approximately 19 in from the face of the support?

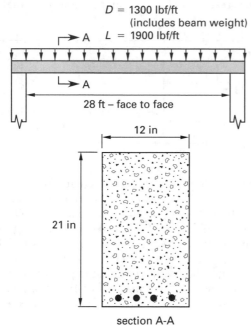

10. (*Time limit: one hour*) The simply supported tension-reinforced beam shown carries two identical sets of dead, D, point loads and a set of live, L, uniform loadings. The beam has a width of 16 in. $f'_c = 3000$ lbf/in^2, and $f_y = 60,000$ lbf/in^2. (a) What is the required flexural strength? (b) Specify the minimum reinforcement depth. (c) Detail the reinforcing steel. (d) Use no. 3 bars to design the shear reinforcing.

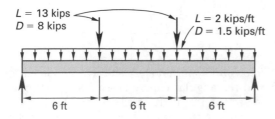

11. (*Time limit: one hour*) The reinforced concrete beam shown in the first illustration supports two 322 kip concentrated loads. (The loads have already been factored and include beam self-weights.) The size of the bearing plates is 18 in × 20 in. Compressive strength of concrete $f'_c = 4$ ksi, and yield strength of reinforcement $f_y = 60$ ksi. A strut-and-tie model for this beam is shown in the second illustration. The strut-and-tie model consists of three struts (AB, BC, and CD), one tie (AD), and four nodes (A, B, C, and D). The effective width of strut BC, w_c, is equal to 7.5 in and the effective width of tie AD, w_t, is equal to 9 in. The vertical position of node A is 9 in/2 = 4.5 in from the bottom of the beam, and for node B it is 7.5 in/2 = 3.75 in from the top of the beam.

(a) Determine if this beam is a deep beam according to code provisions.

(b) Check compression strength at all nodal zones, assuming that no confining reinforcement is provided in the nodal zones.

(c) Check design strength of strut BC.

(d) Determine reinforcement required in tie AD.

(e) Check design strength of strut AB.

(f) Design the anchorage of the tie bars from step (d).

(g) Determine the horizontal and vertical reinforcement required to resist splitting of diagonal struts.

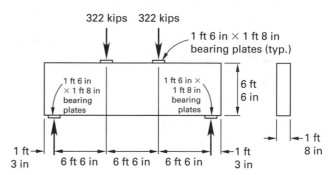

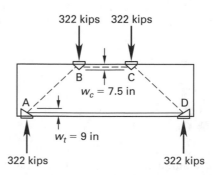

SOLUTIONS

1. (a) The weight of the beam is

$$W = bhw = \frac{(14 \text{ in})(24 \text{ in})\left(0.15 \ \frac{\text{kip}}{\text{ft}^3}\right)}{\left(12 \ \frac{\text{in}}{\text{ft}}\right)^2} = 0.35 \text{ kip/ft}$$

From Eq. 50.2, the factored uniform load is

$$\begin{aligned}
w_u &= 1.2D + 1.6L \\
&= (1.2)\left(1.75 \ \frac{\text{kips}}{\text{ft}} + 0.35 \ \frac{\text{kip}}{\text{ft}}\right) + (1.6)\left(3 \ \frac{\text{kips}}{\text{ft}}\right) \\
&= 7.32 \text{ kips/ft}
\end{aligned}$$

$$\begin{aligned}
1.4D &= (1.4)\left(1.75 \ \frac{\text{kips}}{\text{ft}} + 0.35 \ \frac{\text{kip}}{\text{ft}}\right) \\
&= 2.94 \text{ kips/ft} \quad [\text{does not control}]
\end{aligned}$$

Obtain the approximate shear envelope using the procedure described in Sec. 50.20.

$$\begin{aligned}
\text{maximum reaction} &= \frac{w_u L}{2} \\
&= \frac{\left(7.32 \ \frac{\text{kips}}{\text{ft}}\right)(20 \text{ ft})}{2} \\
&= 73.2 \text{ kips}
\end{aligned}$$

The centerline shear when the factored live load is positioned over half of the span is

$$\begin{aligned}
V &= 1.6\left(\frac{w_u\left(\frac{L}{2}\right)\left(\frac{L}{4}\right)}{L}\right) = \frac{1.6 w_u L}{8} \\
&= \frac{(1.6)\left(3.0 \ \frac{\text{kips}}{\text{ft}}\right)(20 \text{ ft})}{8} \\
&= 12.0 \text{ kips}
\end{aligned}$$

Since the reaction induces compression, the critical section is located at a distance d from the face of the support. The shear envelope, with the shear at the critical section identified, is illustrated. (The effective depth has been taken as $d = h - 2.5 \text{ in} = 21.5 \text{ in}$. The critical shear of 60.2 kips is found by interpolation.)

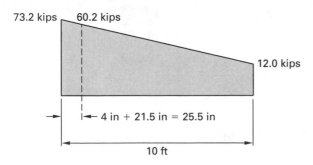

73.2 kips   60.2 kips

12.0 kips

4 in + 21.5 in = 25.5 in

10 ft

The critical section is located (measured from the center of the column) at

$$\frac{4 \text{ in} + 21.5 \text{ in}}{12 \ \frac{\text{in}}{\text{ft}}} = 2.125 \text{ ft}$$

The factored moment at the critical section is

$$\begin{aligned}
M_u &= R_c x - \frac{w_u x^2}{2} \\
&= (73.2 \text{ kips})(2.125 \text{ ft}) - \frac{\left(7.32 \ \frac{\text{kips}}{\text{ft}}\right)(2.125 \text{ ft})^2}{2} \\
&= 139.02 \text{ ft-kips}
\end{aligned}$$

The flexural reinforcement steel ratio at the critical section is given by Eq. 50.24.

$$\begin{aligned}
\rho &= \frac{A_s}{bd} = \frac{5 \text{ in}^2}{(14 \text{ in})(21.5 \text{ in})} \\
&= 0.0166
\end{aligned}$$

From the shear envelope, $V_u = 60.20$ kips. Use Eq. 50.52 to compute the shear strength of the concrete.

$$V_c = \left(1.9\lambda\sqrt{f'_c} + 2500\rho_w\left(\frac{V_u d}{M_u}\right)\right)b_w d \le 3.5\lambda\sqrt{f'_c}b_w d$$

$$= \frac{\begin{pmatrix}(1.9)(1.0)\sqrt{4000 \ \frac{\text{lbf}}{\text{in}^2}} \\ + (2500)(0.0166) \\ \times \left(\frac{(60.2 \text{ kips})(21.5 \text{ in})}{(139.02 \text{ ft-kips})\left(12 \ \frac{\text{in}}{\text{ft}}\right)}\right)\end{pmatrix}\begin{pmatrix}(14 \text{ in}) \\ \times (21.5 \text{ in})\end{pmatrix}}{1000 \ \frac{\text{lbf}}{\text{kip}}}$$

$$= 45.86 \text{ kips}$$

$$3.5\lambda\sqrt{f'_c}b_w d = \frac{(3.5)(1.0)\sqrt{4000 \ \frac{\text{lbf}}{\text{in}^2}}(14 \text{ in})(21.5 \text{ in})}{1000 \ \frac{\text{lbf}}{\text{kip}}}$$

$$= 66.63 \text{ kips} \quad [\text{does not control}]$$

The shear that must be resisted by the shear reinforcement is obtained from Eq. 50.63.

$$\begin{aligned}
V_{s,\text{req}} &= \frac{V_u}{\phi} - V_c = \frac{60.2 \text{ kips}}{0.75} - 45.86 \text{ kips} \\
&= 34.41 \text{ kips}
\end{aligned}$$

The spacing of the no. 3 stirrups is found from Eq. 50.64.

$$s = \frac{A_v f_{yt} d}{V_{s,req}} = \frac{(0.22 \text{ in}^2)\left(60 \frac{\text{kips}}{\text{in}^2}\right)(21.5 \text{ in})}{34.41 \text{ kips}}$$

$$= \boxed{8.25 \text{ in}}$$

Check the maximum spacing, using Eq. 50.59 or Eq. 50.60.

$$4\sqrt{f'_c} b_w d = \frac{4\sqrt{4000 \frac{\text{lbf}}{\text{in}^2}}(14 \text{ in})(21.5 \text{ in})}{1000 \frac{\text{lbf}}{\text{kip}}}$$

$$= 76.15 \text{ kips}$$

Since $V_{s,req} \le 4\sqrt{f'_c} b_w d$, Eq. 50.59 applies, and the maximum spacing permitted is

$$\frac{d}{2} = \frac{21.5 \text{ in}}{2} = 10.75 \text{ in} \quad [\text{does not control}]$$

Check the minimum A_v from Eq. 50.61.

$$A_v = 0.75\sqrt{f'_c}\left(\frac{b_w s}{f_{yt}}\right)$$

$$= 0.75\sqrt{4000 \frac{\text{lbf}}{\text{in}^2}}\left(\frac{(14 \text{ in})(8.25 \text{ in})}{60,000 \frac{\text{lbf}}{\text{in}^2}}\right)$$

$$= 0.0913 \text{ in}^2$$

$$\frac{50 b_w s}{f_{yt}} = \frac{(50)(14 \text{ in})(8.25 \text{ in})}{60,000 \frac{\text{lbf}}{\text{in}^2}}$$

$$= 0.09625 \text{ in}^2 \quad [\text{controls}]$$

0.09625 in^2 is less than the $A_v = 0.22$ in^2 provided by the U-shaped no. 3 stirrup, so the minimum requirement is satisfied.

(b) From Eq. 50.53,

$$V_c = 2\lambda\sqrt{f'_c} b_w d = \frac{(2)(1.0)\sqrt{4000 \frac{\text{lbf}}{\text{in}^2}}(14 \text{ in})(21.5 \text{ in})}{1000 \frac{\text{lbf}}{\text{kip}}}$$

$$= 38.07 \text{ kips}$$

The same procedure as in part (a) is used. From Eq. 50.63 and Eq. 50.64,

$$V_{s,req} = \frac{V_u}{\phi} - V_c = \frac{60.20 \text{ kips}}{0.75} - 38.07 \text{ kips}$$

$$= 42.20 \text{ kips}$$

$$s = \frac{A_v f_{yt} d}{V_{s,req}} = \frac{(0.22 \text{ in}^2)\left(60 \frac{\text{kips}}{\text{in}^2}\right)(21.5 \text{ in})}{42.20 \text{ kips}}$$

$$= \boxed{6.73 \text{ in}}$$

(c) Stirrups are unnecessary when $V_u \le \phi V_c/2$. Use the simplified expression for V_c. From Eq. 50.62,

$$\frac{\phi V_c}{2} = \frac{(0.75)(38.07 \text{ kips})}{2}$$

$$= 14.28 \text{ kips}$$

Using simple geometric calculations, this value of the shear is found in the shear envelope at 4.47 in from the centerline of the beam. The section where stirrups can be omitted is, therefore, $(2)(4.47 \text{ in}) = \boxed{8.94 \text{ in}}$ long in the center of the beam. This is negligible for practical purposes and would not be considered.

2. (a) The weight of the beam is

$$W = bhw = \frac{(12 \text{ in})(28 \text{ in})\left(0.15 \frac{\text{kip}}{\text{ft}^3}\right)}{\left(12 \frac{\text{in}}{\text{ft}}\right)^2} = 0.35 \text{ kip/ft}$$

The factored uniform load is

$$w_u = 1.2D + 1.6L$$

$$= (1.2)\left(2.0 \frac{\text{kips}}{\text{ft}} + 0.35 \frac{\text{kip}}{\text{ft}}\right) + (1.6)\left(2.4 \frac{\text{kips}}{\text{ft}}\right)$$

$$= \boxed{6.66 \text{ kips/ft} \quad (6.7 \text{ kips/ft})}$$

$$w_u = 1.4D = (1.4)\left(2 \frac{\text{kips}}{\text{ft}} + 0.35 \frac{\text{kip}}{\text{ft}}\right)$$

$$= 3.29 \text{ kips/ft} \quad [\text{does not control}]$$

The answer is (A).

(b) The maximum moment is

$$M_u = \frac{w_u L^2}{8} = \frac{\left(6.66 \ \frac{\text{kips}}{\text{ft}}\right)(20 \ \text{ft})^2}{8}$$

$$= 333.0 \ \text{ft-kips}$$

Estimate that $\lambda = 0.1 \, d = (0.1)(25.5 \ \text{in}) = 2.55 \ \text{in}$. From Eq. 50.28,

$$A_s = \frac{M_u}{\phi f_y (d - \lambda)}$$

$$= \frac{(333.0 \ \text{ft-kips})\left(12 \ \frac{\text{in}}{\text{ft}}\right)}{(0.90)\left(40 \ \frac{\text{kips}}{\text{in}^2}\right)(25.5 \ \text{in} - 2.55 \ \text{in})}$$

$$= 4.84 \ \text{in}^2$$

Compute M_n using the standard approach. From Eq. 50.26,

$$A_c = \frac{f_y A_s}{0.85 f'_c} = \frac{(4.84 \ \text{in}^2)\left(40 \ \frac{\text{kips}}{\text{in}^2}\right)}{(0.85)\left(3 \ \frac{\text{kips}}{\text{in}^2}\right)}$$

$$= 75.92 \ \text{in}^2$$

Since the compressed zone is rectangular,

$$a = \frac{A_c}{b} = \frac{75.92 \ \text{in}^2}{12 \ \text{in}}$$

$$= 6.33 \ \text{in}$$

$$\lambda = \frac{a}{2} = \frac{6.33 \ \text{in}}{2} = 3.17 \ \text{in}$$

Verify that the section is tension controlled.

$$c = \frac{a}{\beta_1} = \frac{6.33 \ \text{in}}{0.85} = 7.45 \ \text{in}$$

$$\epsilon_t = 0.003 \left(\frac{d - c}{c}\right) = (0.003)\left(\frac{25.5 \ \text{in} - 7.45 \ \text{in}}{7.45 \ \text{in}}\right)$$

$$= 0.007 > 0.005$$

The section is tension controlled.

The nominal moment capacity may be obtained from Eq. 50.27.

$$M_n = A_s f_y (d - \lambda)$$

$$= \frac{(4.84 \ \text{in}^2)\left(40 \ \frac{\text{kips}}{\text{in}^2}\right)(25.5 \ \text{in} - 3.17 \ \text{in})}{12 \ \frac{\text{in}}{\text{ft}}}$$

$$= 360.26 \ \text{ft-kips}$$

The design moment capacity is

$$\phi M_n = (0.9)(360.26 \ \text{ft-kips})$$

$$= 324.23 \ \text{ft-kips}$$

The adjusted steel area is

$$A_s = \left(\frac{333.0 \ \text{ft-kips}}{324.23 \ \text{ft-kips}}\right)(4.84 \ \text{in}^2)$$

$$= \boxed{4.97 \ \text{in}^2 \quad (5.0 \ \text{in}^2)}$$

The answer is (B).

(c) ACI 318 Sec. 10.3.5 requires that, for nonprestressed flexural members with axial load less than $0.10 f'_c A_g$, the net tensile strain, ϵ_t, at nominal strength must not be less than 0.004.

$$\frac{0.003}{0.004} = \frac{c}{d - c}$$

$$a = \beta_1 c$$

$$A_{s,\text{max}} = \frac{0.85 f'_c b \beta_1 c}{f_y} = \frac{0.85 f'_c b \beta_1 \left(\frac{0.003}{0.007}\right) d}{f_y}$$

$$= \frac{(0.85)\left(3 \ \frac{\text{kips}}{\text{in}^2}\right)(12 \ \text{in})(0.85)}{40 \ \frac{\text{kips}}{\text{in}^2}}$$

$$ \frac{\times \left(\frac{0.003}{0.007}\right)(25.5 \ \text{in})}{40 \ \frac{\text{kips}}{\text{in}^2}}$$

$$= \boxed{7.1 \ \text{in}^2 \quad (7.7 \ \text{in}^2)}$$

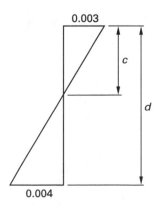

The answer is (B).

(d) Use Eq. 50.13.

$$\frac{3\sqrt{f_c'}b_w d}{f_y} = \frac{3\sqrt{3000\ \frac{\text{lbf}}{\text{in}^2}}(12\ \text{in})(25.5\ \text{in})}{40{,}000\ \frac{\text{lbf}}{\text{in}^2}}$$

$$= 1.26\ \text{in}^2$$

$$\frac{200 b_w d}{f_y} = \frac{(200)(12\ \text{in})(25.5\ \text{in})}{40{,}000\ \frac{\text{lbf}}{\text{in}^2}}$$

$$= \boxed{1.53\ \text{in}^2 \quad (1.5\ \text{in}^2) \quad [\text{controls}]}$$

The answer is (C).

(e) Since $A_{s,\text{max}} = 7.10\ \text{in}^2$, from Eq. 50.26,

$$A_c = \frac{f_y A_s}{0.85 f_c'} = \frac{\left(40\ \frac{\text{kips}}{\text{in}^2}\right)(7.10\ \text{in}^2)}{(0.85)\left(3\ \frac{\text{kips}}{\text{in}^2}\right)} = 111.37\ \text{in}^2$$

Since the compressed zone is rectangular,

$$a = \frac{A_c}{b} = \frac{111.37\ \text{in}^2}{12\ \text{in}} = 9.28\ \text{in}$$

$$\lambda = \frac{a}{2} = \frac{9.28\ \text{in}}{2} = 4.64\ \text{in}$$

The nominal moment capacity is obtained from Eq. 50.27.

$$M_n = A_s f_y (d - \lambda)$$

$$= \frac{(7.10\ \text{in}^2)\left(40\ \frac{\text{kips}}{\text{in}^2}\right)(25.5\ \text{in} - 4.64\ \text{in})}{12\ \frac{\text{in}}{\text{ft}}}$$

$$= 493.69\ \text{ft-kips}$$

From Eq. 50.11, the strength reduction factor corresponding to $\epsilon_t = 0.004$ is

$$\phi = 0.65 + \frac{250}{3}(\epsilon_t - 0.002)$$

$$= 0.65 + \left(\frac{250}{3}\right)(0.004 - 0.002)$$

$$= 0.817$$

The design moment capacity is

$$\phi M_n = (0.817)(493.69\ \text{ft-kips}) = 403.18\ \text{ft-kips}$$

The uniform factored load is

$$w_u = \frac{8\phi M_n}{L^2} = \frac{(8)(403.18\ \text{ft-kips})}{(20\ \text{ft})^2}$$

$$= \boxed{8.06\ \text{kips/ft} \quad (8.1\ \text{kips/ft})}$$

The answer is (A).

(f) Obtain the approximate shear envelope.

$$\text{maximum reaction} = \frac{w_u L}{2}$$

$$= \frac{\left(6.66\ \frac{\text{kips}}{\text{ft}}\right)(20\ \text{ft})}{2}$$

$$= 66.60\ \text{kips}$$

The shear at centerline when the factored live load is positioned over half of the span is

$$V = 1.6\left(\frac{w\left(\frac{L}{2}\right)\left(\frac{L}{4}\right)}{L}\right) = \frac{1.6 w L}{8}$$

$$= \frac{(1.6)\left(2.4\ \frac{\text{kips}}{\text{ft}}\right)(20\ \text{ft})}{8}$$

$$= 9.60\ \text{kips}$$

Since the reaction induces compression, the critical section is located at a distance d from the face of the support. The shear envelope, with the shear at the critical section identified, is illustrated. The effective depth has been taken as $d = h - 2.5\ \text{in}$.

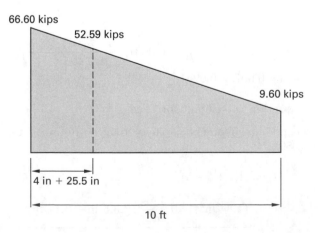

From Eq. 50.53,

$$V_c = 2\lambda\sqrt{f'_c}b_w d = \frac{(2)(1.0)\sqrt{3000\,\frac{\text{lbf}}{\text{in}^2}}(12\text{ in})(25.5\text{ in})}{1000\,\frac{\text{lbf}}{\text{kip}}}$$

$$= 33.52\text{ kips}$$

Using Eq. 50.63 and Eq. 50.64,

$$V_{s,\text{req}} = \frac{V_u}{\phi} - V_c = \frac{52.59\text{ kips}}{0.75} - 33.52\text{ kips}$$

$$= 36.60\text{ kips}$$

$$s = \frac{A_v f_y d}{V_{s,\text{req}}} = \frac{(0.22\text{ in}^2)\left(40\,\frac{\text{kips}}{\text{in}^2}\right)(25.5\text{ in})}{36.60\text{ kips}}$$

$$= \boxed{6.13\text{ in}\quad(6.0\text{ in})}$$

The answer is (B).

(g) Stirrups can be omitted when $V_u \leq \phi V_c/2$. Use the simplified expression for V_c.

$$\frac{\phi V_c}{2} = \frac{(0.75)(33.52\text{ kips})}{2} = 12.57\text{ kips}$$

Using simple geometric calculations, this value of the shear is reached in the shear envelope at 6.25 in from the centerline of the beam. The section where stirrups can be omitted is, therefore, $(2)(6.25\text{ in}) = \boxed{12.50\text{ in}\ (12\text{ in})}$ long in the center of the beam.

The answer is (B).

(h) Use Eq. 50.58.

$$V_{u,\text{max}} = (8 + 2\lambda)\phi\sqrt{f'_c}b_w d$$

$$= \frac{(8 + (2)(1.0))(0.75)\sqrt{3000\,\frac{\text{lbf}}{\text{in}^2}}(12\text{ in})(25.5\text{ in})}{1000\,\frac{\text{lbf}}{\text{kip}}}$$

$$= 125.70\text{ kips}$$

Assuming a uniform scaling of the shear envelope,

$$w_u = \left(\frac{125.70\text{ kips}}{52.59\text{ kips}}\right)\left(6.66\,\frac{\text{kips}}{\text{ft}}\right)$$

$$= \boxed{15.92\text{ kips/ft}\quad(16\text{ kips/ft})}$$

The answer is (A).

(i) Since cracking is not reversible, the moment due to the full service load should be used.

$$\text{service load} = 2\,\frac{\text{kips}}{\text{ft}} + 0.35\,\frac{\text{kip}}{\text{ft}} + 2.4\,\frac{\text{kips}}{\text{ft}}$$

$$= 4.75\text{ kips/ft}$$

$$M_s = \frac{wL^2}{8} = \frac{\left(4.75\,\frac{\text{kips}}{\text{ft}}\right)(20\text{ ft})^2}{8}$$

$$= \boxed{237.5\text{ ft-kips}\quad(240\text{ ft-kips})}$$

The answer is (C).

(j) The gross moment of inertia is

$$I_g = \frac{bh^3}{12} = \frac{(12\text{ in})(28\text{ in})^3}{12}$$

$$= 21{,}952\text{ in}^4$$

From Eq. 50.45, the modulus of rupture is

$$f_r = 7.5\lambda\sqrt{f'_c} = (7.5)(1.0)\sqrt{3000\,\frac{\text{lbf}}{\text{in}^2}}$$

$$= 411\text{ lbf/in}^2\quad(0.411\text{ kip/in}^2)$$

The distance to the extreme tension fiber is

$$y_t = \frac{h}{2} = \frac{28\text{ in}}{2} = 14\text{ in}$$

Use Eq. 50.44.

$$M_{\text{cr}} = \frac{f_r I_g}{y_t} = \frac{\left(0.411\,\frac{\text{kip}}{\text{in}^2}\right)(21{,}952\text{ in}^4)}{(14\text{ in})\left(12\,\frac{\text{in}}{\text{ft}}\right)}$$

$$= \boxed{53.7\text{ ft-kips}\quad(54\text{ ft-kips})}$$

The answer is (D).

3. (a) In the positive moment region, $d = 27.5$ in, and $A_s = 3.0\text{ in}^2$. Therefore, from Eq. 50.26,

$$A_c = \frac{f_y A_s}{0.85 f'_c} = \frac{\left(60\,\frac{\text{kips}}{\text{in}^2}\right)(3\text{ in}^2)}{(0.85)\left(3\,\frac{\text{kips}}{\text{in}^2}\right)}$$

$$= 70.59\text{ in}^2$$

Assume the stress block is confined to the flange. Then, from Eq. 50.29,

$$a = \frac{A_c}{b} = \frac{70.59 \text{ in}}{30 \text{ in}} = 2.35 \text{ in}$$

(The assumption is correct.)

$$\lambda = \frac{a}{2} = \frac{2.35 \text{ in}}{2} = 1.18 \text{ in}$$

The nominal moment capacity is obtained from Eq. 50.27.

$$M_n = A_s f_y (d - \lambda)$$

$$= \frac{(3 \text{ in}^2)\left(60 \frac{\text{kips}}{\text{in}^2}\right)(27.5 \text{ in} - 1.18 \text{ in})}{12 \frac{\text{in}}{\text{ft}}}$$

$$= 394.8 \text{ ft-kips}$$

Determine the strength reduction factor.

$$c = \frac{a}{\beta_1} = \frac{2.35 \text{ in}}{0.85} = 2.76 \text{ in}$$

$$\epsilon_t = 0.003\left(\frac{d - c}{c}\right) = (0.003)\left(\frac{27.5 \text{ in} - 2.76 \text{ in}}{2.76 \text{ in}}\right)$$

$$= 0.027 > 0.005$$

Therefore, the section is tension controlled, and $\phi = 0.9$.

The design moment capacity (maximum positive moment) is

$$\phi M_n = (0.9)(394.8 \text{ ft-kips})$$

$$= \boxed{355.3 \text{ ft-kips} \quad (360 \text{ ft-kips})}$$

The answer is (B).

(b) In the negative moment region, $d = 27.5$ in, $A_s = 7.0 \text{ in}^2$, and $A_s' = 3.0 \text{ in}^2$.

Assuming the compression steel yields, from Eq. 50.66,

$$A_c = \frac{f_y A_s - f_s' A_s'}{0.85 f_c'}$$

$$= \frac{\left(60 \frac{\text{kips}}{\text{in}^2}\right)(7 \text{ in}^2) - \left(60 \frac{\text{kips}}{\text{in}^2}\right)(3 \text{ in}^2)}{(0.85)\left(3 \frac{\text{kips}}{\text{in}^2}\right)}$$

$$= 94.12 \text{ in}^2$$

Using Eq. 50.29,

$$a = \frac{A_c}{b_w} = \frac{94.12 \text{ in}^2}{12 \text{ in}} = 7.84 \text{ in}$$

$$\lambda = \frac{a}{2} = \frac{7.84 \text{ in}}{2} = 3.92 \text{ in}$$

$$c = \frac{a}{\beta_1} = \frac{7.84 \text{ in}}{0.85} = 9.22 \text{ in}$$

From Eq. 50.67,

$$\epsilon_s' = \left(\frac{0.003}{c}\right)(c - d') = \left(\frac{0.003}{9.22 \text{ in}}\right)(9.22 \text{ in} - 2.5 \text{ in})$$

$$= 0.0022$$

Use Eq. 50.68.

$$f_s' = E_s \epsilon_s' = \left(29{,}000 \frac{\text{kips}}{\text{in}^2}\right)(0.0022)$$

$$= 63.8 \text{ kips/in}^2$$

$$f_s' = f_y = \boxed{60 \text{ kips/in}^2} \quad \left[\begin{array}{l}\text{confirms assumption that} \\ \text{the compression steel yields}\end{array}\right]$$

The answer is (D).

(c) Divide the section into two elementary areas as shown. The computations are summarized in the following table.

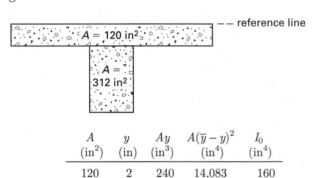

A (in^2)	y (in)	Ay (in^3)	$A(\bar{y} - y)^2$ (in^4)	I_0 (in^4)
120	2	240	14,083	160
312	17	5304	5417	17,576
totals 432		5544	19,500	17,736

$$\bar{y} = \frac{5544 \text{ in}^3}{432 \text{ in}^2} = 12.83 \text{ in}$$

$$I_g = 19{,}500 \text{ in}^4 + 17{,}736 \text{ in}^4$$

$$= \boxed{37{,}236 \text{ in}^4 \quad (37{,}000 \text{ in}^4)}$$

The answer is (B).

(d) From Eq. 50.38,

$$E_c = 57{,}000\sqrt{f_c'} = \frac{57{,}000\sqrt{3000 \frac{\text{lbf}}{\text{in}^2}}}{1000 \frac{\text{lbf}}{\text{kip}}} = 3122 \text{ kips/in}^2$$

From Eq. 50.37,

$$n = \frac{E_s}{E_c} = \frac{29{,}000 \ \frac{\text{kips}}{\text{in}^2}}{3122 \ \frac{\text{kips}}{\text{in}^2}}$$

$$= 9.29$$

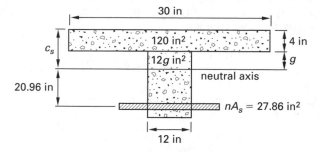

From the sketch of the transformed section,

$$(120 \ \text{in}^2)(g + 2 \ \text{in})$$

$$+ \frac{(12 \ \text{in})g^2}{2} = (27.86 \ \text{in}^2)(27.5 \ \text{in} - 4 \ \text{in} - g)$$

$$g = 2.54 \ \text{in}$$

The neutral axis is located at $c_s = 4 \ \text{in} + 2.54 \ \text{in} = 6.54 \ \text{in}$.

The transformed cracked moment of inertia is calculated using the parallel axis theorem.

$$I_{\text{cr}} = \left(\tfrac{1}{12}\right)(30 \ \text{in})(4 \ \text{in})^3 + \left(\tfrac{1}{3}\right)(12 \ \text{in})(2.54 \ \text{in})^3$$

$$+ \ (120 \ \text{in}^2)(4.54 \ \text{in})^2$$

$$+ \ (27.86 \ \text{in}^2)(20.96 \ \text{in})^2$$

$$= \boxed{14{,}938 \ \text{in}^4 \quad (15{,}000 \ \text{in}^4)}$$

The answer is (C).

(e) From Eq. 50.45,

$$f_r = 7.5\lambda\sqrt{f'_c} = \frac{(7.5)(1.0)\sqrt{3000 \ \frac{\text{lbf}}{\text{in}^2}}}{1000 \ \frac{\text{lbf}}{\text{kip}}}$$

$$= 0.41 \ \text{kip/in}^2$$

From Eq. 50.44,

$$M_{\text{cr}} = \frac{f_r I_g}{y_t} = \frac{\left(0.41 \ \frac{\text{kip}}{\text{in}^2}\right)(37{,}236 \ \text{in}^4)}{(30 \ \text{in} - 12.83 \ \text{in})\left(12 \ \frac{\text{in}}{\text{ft}}\right)}$$

$$= \boxed{74.10 \ \text{ft-kips} \quad (74 \ \text{ft-kips})}$$

The answer is (D).

(f) Use Eq. 50.44.

$$M_{\text{cr}} = \frac{f_r I_g}{y_t} = \frac{\left(0.41 \ \frac{\text{kip}}{\text{in}^2}\right)(37{,}236 \ \text{in}^4)}{(12.83 \ \text{in})\left(12 \ \frac{\text{in}}{\text{ft}}\right)}$$

$$= \boxed{99.16 \ \text{ft-kips} \quad (99 \ \text{ft-kips})}$$

The answer is (C).

(g) The maximum positive service moment is

$$M_s = \frac{wL^2}{24}$$

$$= \frac{\left(4 \ \frac{\text{kips}}{\text{ft}}\right)(30 \ \text{ft})^2}{24}$$

$$= 150 \ \text{ft-kips}$$

Therefore,

$$f_s = \frac{nM_s(d - c_s)}{I_{\text{cr}}}$$

$$= \frac{(9.29)(150 \ \text{ft-kips})\left(12 \ \frac{\text{in}}{\text{ft}}\right)(27.5 \ \text{in} - 6.54 \ \text{in})}{14{,}938 \ \text{in}^4}$$

$$= \boxed{23.46 \ \text{kips/in}^2 \quad (23 \ \text{kips/in}^2)}$$

The answer is (B).

(h)

$$c_c = d_c - \frac{d_b}{2} = 2.5 \ \text{in} - \frac{1.125 \ \text{in}}{2}$$

$$= 1.9375 \ \text{in}$$

$$f_{yt} = 60 \ \text{kips/in}^2$$

From Eq. 50.36,

$$s \leq 15 \left(\frac{40{,}000 \frac{\text{lbf}}{\text{in}^2}}{f_s} \right) - 2.5 c_c$$

$$\leq (15) \left(\frac{40{,}000 \frac{\text{lbf}}{\text{in}^2}}{23{,}460 \frac{\text{lbf}}{\text{in}^2}} \right) - (2.5)(1.9375 \text{ in})$$

$$\leq 20.7 \text{ in}$$

$$s \leq 12 \left(\frac{40{,}000 \frac{\text{lbf}}{\text{in}^2}}{f_s} \right) = (12) \left(\frac{40{,}000 \frac{\text{lbf}}{\text{in}^2}}{23{,}460 \frac{\text{lbf}}{\text{in}^2}} \right)$$

$$= \boxed{20.5 \text{ in} \quad (21 \text{ in}) \quad [\text{controls}]}$$

The answer is (D).

(i) The strength reduction factor for shear is 0.75. Therefore, from Eq. 50.51,

$$\phi V_n = \phi(V_c + V_s) = (0.75)(36.15 \text{ kips} + 60.50 \text{ kips})$$

$$= 72.49 \text{ kips}$$

The critical section is at a distance d from the face of the support. The shear at the critical section is

$$V_u = \frac{w_u}{2} \left(30 \text{ ft} - \frac{(2)(27.5 \text{ in})}{12 \frac{\text{in}}{\text{ft}}} \right)$$

$$= 12.71 w_u$$

Equating the maximum shear to the shear capacity,

$$w_u = \phi \frac{V_n}{V_u} = \frac{72.49 \text{ kips}}{12.71 \text{ ft}}$$

$$= \boxed{5.70 \text{ kips/ft} \quad (5.9 \text{ kips/ft})}$$

The answer is (B).

(j) Use Eq. 50.69 with top and bottom steel yielding. (λ was calculated in part (b).)

$$M_n = (A_s - A'_s)f_y(d - \lambda) + A'_s f'_s(d - d')$$

$$= \frac{(7 \text{ in}^2 - 3 \text{ in}^2)\left(60 \frac{\text{kips}}{\text{in}^2} \right)(27.5 \text{ in} - 3.92 \text{ in})}{12 \frac{\text{in}}{\text{ft}}}$$

$$+ \frac{(3 \text{ in}^2)\left(60 \frac{\text{kips}}{\text{in}^2} \right)(27.5 \text{ in} - 2.5 \text{ in})}{12 \frac{\text{in}}{\text{ft}}}$$

$$= 846.6 \text{ ft-kips} \quad [\text{global maximum}]$$

The maximum positive moment is

$$M_u = \frac{w_u L^2}{24}$$

$$= \frac{w_u (30 \text{ ft})^2}{24}$$

$$= 37.5 w_u$$

The maximum positive moment was calculated in part (a). Equating this value to the capacity ϕM_n for the positive moment region,

$$w_u = \frac{355.3 \text{ ft-kips}}{37.5 \text{ ft}^2} = \boxed{9.47 \text{ kips/ft} \quad (9.5 \text{ kips/ft})}$$

The maximum negative moment from this fixed-end beam is

$$M_u = \frac{w_u L^2}{12}$$

$$= \frac{w_u (30 \text{ ft})^2}{12}$$

$$= 75 w_u$$

Equate this value to the capacity ϕM_n for the negative moment region, using $\phi = 0.9$. (ϵ_t can be calculated as 0.0059.)

$$\phi M_n = M_u$$

$$(0.9)(846.6 \text{ ft-kips}) = 75 w_u$$

$$w_u = 10.2 \text{ kips/ft}$$

The positive region controls.

The answer is (C).

4. *step 1:* Values of $\rho f_y / f'_c$ near 0.18 typically lead to economical beams. Therefore, try a reinforcement ratio of

$$\rho = \frac{(0.18)\left(3000 \frac{\text{lbf}}{\text{in}^2} \right)}{50{,}000 \frac{\text{lbf}}{\text{in}^2}} = 0.0108$$

step 2: Solve for the required nominal moment strength.

$$M_u = (1.2)(50{,}000 \text{ ft-lbf}) + (1.6)(200{,}000 \text{ ft-lbf})$$

$$= 380{,}000 \text{ ft-lbf}$$

Assume the section is tension controlled.

$$M_n = \frac{M_u}{\phi} = \frac{380{,}000 \text{ ft-lbf}}{0.9} = 422{,}222 \text{ ft-lbf}$$

step 3: Substitute into Eq. 50.32 and solve for bd^2.

$$bd^2 = \frac{M_n}{\rho f_y \left(1 - \frac{\rho f_y}{1.7 f_c'}\right)}$$

$$= \frac{(422{,}222 \text{ ft-lbf})\left(12 \frac{\text{in}}{\text{ft}}\right)}{(0.0108)\left(50{,}000 \frac{\text{lbf}}{\text{in}^2}\right)}$$

$$\times \left(1 - \frac{(0.0108)\left(50{,}000 \frac{\text{lbf}}{\text{in}^2}\right)}{(1.7)\left(3000 \frac{\text{lbf}}{\text{in}^2}\right)}\right)$$

$$= 10{,}494 \text{ in}^3$$

step 4: Try $d = 1.75b$.

$$(1.75)^2 b^3 = 10{,}494 \text{ in}^3$$

$$b = \boxed{15.08 \text{ in} \quad (15.5 \text{ in})}$$

$$d = (1.75)(15.08 \text{ in})$$

$$= \boxed{26.39 \text{ in} \quad (27 \text{ in})}$$

step 5:

$$\text{actual } bd^2 = (15.5 \text{ in})(27 \text{ in})^2$$

$$= 11{,}300 \text{ in}^3 \quad [\text{close}]$$

No change in ρ is needed.

For this low reinforcement ratio, the section is going to be tension controlled. No formal check is necessary.

step 6: The steel area is

$$A_s = \rho bd = (0.0108)(15.5 \text{ in})(27 \text{ in})$$

$$= 4.52 \text{ in}^2$$

step 7: Use six no. 8 bars (4.74 in^2). These can all fit in a single layer.

step 8: Assume no. 4 stirrups.

$$d_c = 1.5 \text{ in} + \tfrac{1}{2} \text{ in} + 0.5 \text{ in} = \boxed{2.5 \text{ in}}$$

5. (a) *step 1:* Values of $\rho f_y/f_c'$ near 0.18 typically lead to economical beams. Therefore, try a reinforcement ratio of

$$\rho = \frac{(0.18)\left(4000 \frac{\text{lbf}}{\text{in}^2}\right)}{40{,}000 \frac{\text{lbf}}{\text{in}^2}}$$

$$= 0.018$$

step 2: Solve for the required nominal moment strength. Assume the section is tension controlled.

$$M_n = \frac{M_u}{\phi}$$

$$= \frac{400{,}000 \text{ ft-lbf}}{0.9}$$

$$= 444{,}444 \text{ ft-lbf}$$

step 3: Substitute into Eq. 50.32 and solve for bd^2.

$$bd^2 = \frac{M_n}{\rho f_y \left(1 - \frac{\rho f_y}{1.7 f_c'}\right)}$$

$$= \frac{(444{,}444 \text{ ft-lbf})\left(12 \frac{\text{in}}{\text{ft}}\right)}{(0.018)\left(40{,}000 \frac{\text{lbf}}{\text{in}^2}\right)}$$

$$\times \left(1 - \frac{(0.018)\left(40{,}000 \frac{\text{lbf}}{\text{in}^2}\right)}{(1.7)\left(4000 \frac{\text{lbf}}{\text{in}^2}\right)}\right)$$

$$= 8285 \text{ in}^3$$

step 4: Try $d = 1.75b$.

$$8285 \text{ in}^3 = b(1.75b)^2$$

$$b = \boxed{13.93 \text{ in} \quad (14 \text{ in})}$$

$$d = (1.75)(13.93 \text{ in}) = \boxed{24.38 \text{ in} \quad (24.5 \text{ in})}$$

(b) and (c) *step 5:* ρ is not substantially different.

step 6: The steel area is

$$A_s = \rho bd = (0.018)(14 \text{ in})(24.5 \text{ in})$$

$$= 6.17 \text{ in}^2$$

Check to see if the section is tension controlled.

$$c = \frac{A_s f_y}{0.85 f_c' b \beta_1} = \frac{(6.17 \text{ in}^2)\left(40{,}000 \frac{\text{lbf}}{\text{in}^2}\right)}{(0.85)\left(4000 \frac{\text{lbf}}{\text{in}^2}\right)(14 \text{ in})(0.85)}$$

$$= 6.13 \text{ in}$$

$$\epsilon_t = 0.003\left(\frac{d-c}{c}\right) = (0.003)\left(\frac{24.5 \text{ in} - 6.13 \text{ in}}{6.13 \text{ in}}\right)$$

$$= 0.00899 > 0.005$$

The section is tension controlled.

step 7: $\boxed{\text{Use four no. 11 bars (6.24 in}^2\text{) in one layer.}\\ \text{Use no. 3 stirrups.}}$

(d) Including room for reinforcing and stirrups, the overall beam depth will be

$$24.5 \text{ in} + \frac{1.410 \text{ in}}{2} + 0.375 \text{ in} + 1.5 \text{ in}$$
$$= \boxed{27.08 \text{ in} \quad (27.5 \text{ in})}$$

6. The effective beam width per ACI 318 Sec. 8.12.2 is

$$b_e = \text{minimum} \begin{cases} \frac{1}{4}l_{\text{beam}} \\ 8 \text{ in} + (16)(3 \text{ in}) = 56 \text{ in} \\ 8 \text{ in} + 22 \text{ in} = 30 \text{ in} \end{cases}$$

$$= 30 \text{ in}$$

Work with the following beam.

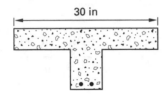

30 in

step 1: The modulus of elasticity for steel is

$$E_s = 2.9 \times 10^7 \text{ lbf/in}^2$$

From Eq. 48.2, the modulus of elasticity for the concrete is

$$E_c = 33w_c^{1.5}\sqrt{f_c'}$$
$$= (33)\left(145 \frac{\text{lbf}}{\text{in}^3}\right)^{1.5} \sqrt{3000 \frac{\text{lbf}}{\text{in}^2}}$$
$$= 3.16 \times 10^6 \text{ psi}$$

step 2: From Eq. 50.37,

$$n = \frac{E_s}{E_c} = \frac{2.9 \times 10^7 \frac{\text{lbf}}{\text{in}^2}}{3.16 \times 10^6 \frac{\text{lbf}}{\text{in}^2}} = 9.18$$

ACI 318 allows the use of the rounded value of n, but does not require it.

step 3: The expanded steel area is

$$nA_b = (9.18)(2)(1 \text{ in}^2) = 18.36 \text{ in}^2$$

step 4: Assume the neutral axis is below the flange.

step 5: (skip)

step 6: The area of concrete is

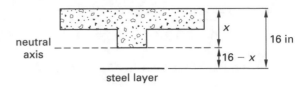

neutral axis
steel layer
x
16 − x
16 in

- stem to flange bottom:

$$A = (8 \text{ in})(x - 3 \text{ in}) = 8x - 24 \text{ in}^2$$

- all of flange:

$$A = (30 \text{ in})(3 \text{ in}) = 90 \text{ in}^2$$

Balance the moments of the transformed areas.

$$(8x - 24)\left(\frac{x-3}{2}\right)$$
$$+ (90)(x - 1.5) = (16 - x)(18.36)$$
$$= 4x^2 + 84.36x - 392.76$$
$$x = 3.93 \text{ in}$$
$$16 \text{ in} - x = 16 \text{ in} - 3.93 \text{ in} = 12.07 \text{ in}$$

step 7: The moment of inertia of the beam is

$$I_{\text{steel}} = (18.36 \text{ in}^2)(12.07 \text{ in})^2 = 2674.8 \text{ in}^4$$
$$I_{\text{concrete}} = \frac{(8 \text{ in})(0.92 \text{ in})^3}{3} + \frac{(30 \text{ in})(3 \text{ in})^3}{12}$$
$$+ (3 \text{ in})(30 \text{ in})(2.42 \text{ in})^2$$
$$= 596.7 \text{ in}^4$$
$$I_{\text{total}} = 2674.8 \text{ in}^4 + 596.7 \text{ in}^4$$
$$= 3271.4 \text{ in}^4$$

step 8: The moment capacity of the beam is

$$M = \left(20,000 \frac{\text{ft-lbf}}{\text{ft}}\right)\left(\frac{30 \text{ in}}{12 \frac{\text{in}}{\text{ft}}}\right)$$
$$= 50,000 \text{ ft-lbf}$$

step 9: (a) The stress in the concrete is

$$\sigma_{\text{concrete}} = \frac{(50,000 \text{ ft-lbf})\left(12 \frac{\text{in}}{\text{ft}}\right)(3.93 \text{ in})}{3271.4 \text{ in}^4}$$
$$= \boxed{720.8 \text{ lbf/in}^2}$$

(b) The stress in the steel is

$$\sigma_{\text{steel}} = \frac{(9.18)(50,000 \text{ ft-lbf}) \times \left(12 \frac{\text{in}}{\text{ft}}\right)(12.08 \text{ in})}{3271.4 \text{ in}^4}$$

$$= \boxed{20,339 \text{ lbf/in}^2}$$

7. (a) All loads are dead loads.

The total distributed load is $(30 \text{ ft})(2 \text{ kips/ft}) = 60$ kips.

Taking clockwise moments as positive,

$$\sum M_L = -(30 \text{ ft})R + (9 \text{ ft})(20 \text{ kips})$$
$$+ (24 \text{ ft})(60 \text{ kips}) + (60 \text{ kips})\left(\frac{30 \text{ ft}}{2}\right)$$
$$= 0$$
$$R = 84 \text{ kips}$$
$$\sum M_R = (30 \text{ ft})L - (60 \text{ kips})(6 \text{ ft})$$
$$- (20 \text{ kips})(21 \text{ ft}) - (60 \text{ kips})\left(\frac{30 \text{ ft}}{2}\right)$$
$$= 0$$
$$L = 56 \text{ kips}$$

Check: 84 kips + 56 kips = 20 kips + 60 kips + 60 kips = 140 kips [correct]

Taking clockwise moments as positive, the moment between the two center loads is

$$M_L = 56x - x^2 - (20)(x - 9)$$
$$\frac{dM}{dx} = 56 - 2x - 20 = 36 - 2x = 0$$
$$x = 18 \text{ ft} \quad [\text{at maximum moment}]$$
$$M_{18} = (56 \text{ kips})(18 \text{ ft}) - \frac{\left(2 \frac{\text{kips}}{\text{ft}}\right)(18 \text{ ft})^2}{2}$$
$$- (20 \text{ kips})(9 \text{ ft})$$
$$= 504 \text{ ft-kips} = M_w$$

Because only dead load is considered, the controlling load combination is $U = 1.4D$.

$$M_u = 1.4M_w = (1.4)(504 \text{ ft-kips}) = 705.6 \text{ ft-kips}$$

step 1: The maximum permitted reinforcement ratio corresponds to a net tensile strain of $\epsilon_t = 0.004$.

$$\rho_{\text{max}} = \left(\frac{0.85f'_c\beta_1}{f_y}\right)\left(\frac{0.003}{0.003 + 0.004}\right)$$
$$= \left(\frac{(0.85)\left(4000 \frac{\text{lbf}}{\text{in}^2}\right)(0.85)}{60,000 \frac{\text{lbf}}{\text{in}^2}}\right)\left(\frac{0.003}{0.003 + 0.004}\right)$$
$$= 0.0206$$

Use $\rho = 0.5\rho_{\text{max}}$.

step 2: Solve for the required nominal moment strength. Assume the section is tension controlled.

$$M_n = \frac{M_u}{\phi} = \frac{705,600 \text{ ft-lbf}}{0.9} = 784,000 \text{ ft-lbf}$$

step 3: Substitute into Eq. 50.32, and solve for bd^2.

$$bd^2 = \frac{M_n}{\rho f_y\left(1 - \frac{\rho f_y}{1.7f'_c}\right)}$$
$$= \frac{(784,000 \text{ ft-lbf})\left(12 \frac{\text{in}}{\text{ft}}\right)}{(0.0103)\left(60,000 \frac{\text{lbf}}{\text{in}^2}\right)}$$
$$\times \left(1 - \frac{(0.0103)\left(60,000 \frac{\text{lbf}}{\text{in}^2}\right)}{(1.7)\left(4000 \frac{\text{lbf}}{\text{in}^2}\right)}\right)$$
$$= 16,745 \text{ in}^3$$

step 4: Assume $d/b = 2.0$.

$$bd^2 = b(2b)^2 = 4b^3 = 16,745 \text{ in}^3$$
$$b = \boxed{16.12 \text{ in} \quad (16.0 \text{ in})}$$
$$d = \boxed{(2)(16 \text{ in}) = 32 \text{ in}}$$

(b) *step 5:* The reinforcement area is

$$A_s = \rho bd = (0.0103)(16 \text{ in})(32 \text{ in})$$
$$= 5.27 \text{ in}^2$$

(c) *step 6:* $\boxed{\text{Use four no. 11 bars } (6.24 \text{ in}^2) \text{ in one layer.}}$

Structural

Refer to Table 50.2. The required beam width with four no. 11 bars is 13.8 in.

beam width: 16 in > 13.8 in   [acceptable]

$$\text{actual } \rho = \frac{6.24 \text{ in}^2}{(16 \text{ in})(32 \text{ in})}$$
$$= 0.0122 < \rho_{\max} \quad \text{[acceptable]}$$

With this low reinforcement ratio, the section is tension controlled.

8. (a) *step 1:* Find ρ_{sb}. $\beta_1 = 0.85$ since $f'_c = 4000$ psi. From Eq. 50.33,

$$\rho_{sb} = \left(\frac{0.85\beta_1 f'_c}{f_y}\right)\left(\frac{87,000}{87,000 + f_y}\right)$$
$$= \left(\frac{(0.85)(0.85)\left(4000 \dfrac{\text{lbf}}{\text{in}^2}\right)}{60,000 \dfrac{\text{lbf}}{\text{in}^2}}\right)$$
$$\times \left(\frac{87,000}{87,000 + 60,000}\right)$$
$$= 0.0285$$

The minimum tensile strain at maximum steel is 0.004. The reinforcement ratio corresponding to $\epsilon_t = 0.004$ is

$$\rho = 0.714\rho_{sb} = (0.714)(0.0285) = 0.02035$$

The corresponding strength reduction factor is

$$\phi = 0.65 + \frac{250}{3}(\epsilon_t - 0.002)$$
$$= 0.65 + \left(\frac{250}{3}\right)(0.004 - 0.002)$$
$$= 0.817$$

step 2: Solve for the required nominal strength. The controlling load combination is

$$w_u = (1.2)\left(1 \frac{\text{kip}}{\text{ft}}\right) + (1.6)\left(2 \frac{\text{kips}}{\text{ft}}\right)$$
$$= 4.4 \text{ kips/ft} \quad (4400 \text{ lbf/ft})$$
$$M_u = \frac{w_u L^2}{8} = \frac{\left(4400 \dfrac{\text{lbf}}{\text{ft}}\right)(27 \text{ ft})^2}{8}$$
$$= 400,950 \text{ ft-lbf}$$
$$M_n = \frac{M_u}{\phi} = \frac{400,950 \text{ ft-lbf}}{0.817}$$
$$= 490,959 \text{ ft-lbf}$$

step 3: Substitute into Eq. 50.32 and solve for bd^2.

$$bd^2 = \frac{M_n}{\rho f_y \left(1 - \dfrac{\rho f_y}{1.7 f'_c}\right)}$$
$$= \frac{(490,959 \text{ ft-lbf})\left(12 \dfrac{\text{in}}{\text{ft}}\right)}{(0.02035)\left(60,000 \dfrac{\text{lbf}}{\text{in}^2}\right)}$$
$$\times \left(1 - \frac{(0.02035)\left(60,000 \dfrac{\text{lbf}}{\text{in}^2}\right)}{(1.7)\left(4000 \dfrac{\text{lbf}}{\text{in}^2}\right)}\right)$$
$$= 5881 \text{ in}^3$$

step 4: The beam width and depth are

$$b = \boxed{14 \text{ in} \quad \text{[given]}}$$
$$d = \sqrt{\frac{5881 \text{ in}^3}{14 \text{ in}}} = 20.5 \text{ in}$$

step 5: The maximum steel area is

$$A_{s,\max} = (0.02035)(14 \text{ in})(20.5 \text{ in})$$
$$= 5.84 \text{ in}^2 \quad \text{[maximum]}$$

This is a maximum value, and the steel chosen should be slightly less than this.

step 6: $\boxed{\text{Try four no. 11 bars.}}$

$$A_s = (4)(1.56 \text{ in}^2) = 6.24 \text{ in}^2$$

Since 6.24 in² > $A_{s,\max}$, increase beam depth to $\boxed{22 \text{ in.}}$

$$\rho = \frac{A_s}{bd} = \frac{6.24 \text{ in}^2}{(14 \text{ in})(22 \text{ in})}$$
$$= 0.02026 < 0.02035 \quad \text{[acceptable]}$$

step 7: Find the overall beam height.

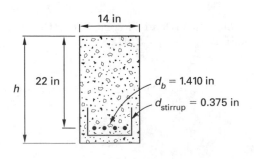

Structural

$$h = \text{beam depth} = 22 \text{ in} + \frac{1.410 \text{ in}}{2}$$
$$+ 0.375 \text{ in} + 1.5 \text{ in}$$
$$= \boxed{24.58 \text{ in} \quad (25 \text{ in})}$$

(b) *step 8:* The stress in the tension steel is assumed to be

$$f_s = \left(\tfrac{2}{3}\right)f_y = \left(\tfrac{2}{3}\right)\left(60{,}000 \ \frac{\text{lbf}}{\text{in}^2}\right)$$
$$= 40{,}000 \ \text{lbf/in}^2$$

There are 3 spaces between the four no. 11 bars. With a no. 3 stirrup and $^3/_4$ in side cover (interior service), the spacing of the four no. 11 bars is

$$s = \frac{14 \text{ in} - (2)(0.75 \text{ in}) - (2)(0.375 \text{ in}) - 1.41 \text{ in}}{3}$$
$$= 3.45 \text{ in}$$

From Eq. 50.36 with $f_s = 40{,}000 \ \text{lbf/in}^2$,

$$s_{\max} = 15\left(\frac{40{,}000}{f_s}\right) - 2.5c_c \le 12\left(\frac{40{,}000}{f_s}\right)$$
$$= (15)\left(\frac{40{,}000 \ \frac{\text{lbf}}{\text{in}^2}}{40{,}000 \ \frac{\text{lbf}}{\text{in}^2}}\right) - (2.5)(1.5 \text{ in} + 0.375 \text{ in})$$
$$< (12)\left(\frac{40{,}000 \ \frac{\text{lbf}}{\text{in}^2}}{40{,}000 \ \frac{\text{lbf}}{\text{in}^2}}\right)$$
$$= 10.3 \text{ in} < 12 \text{ in}$$

Since 3.45 in < 10.3 in, cracking will be acceptable for interior service.

(c) *step 9:* Use Eq. 50.44 to find the cracked moment of inertia.

$$I_g = \frac{bh^3}{12} = \frac{(14 \text{ in})(25 \text{ in})^3}{12} = 18{,}229 \text{ in}^4$$

$$M_{\max} = \frac{wL^2}{8} = \frac{\left(2 \ \frac{\text{kips}}{\text{ft}} + 1 \ \frac{\text{kip}}{\text{ft}}\right)(27 \text{ ft})^2}{8}$$
$$= 273.4 \text{ ft-kips} \quad [\text{unfactored}]$$
$$= (273.4 \text{ ft-kips})\left(12 \ \frac{\text{lbf}}{\text{kip}}\right)\left(1000 \ \frac{\text{lbf}}{\text{kip}}\right)$$
$$= 3.28 \times 10^6 \text{ in-lbf}$$

$$f_r = 7.5\lambda\sqrt{f_c'} = (7.5)(1.0)\sqrt{4000 \ \frac{\text{lbf}}{\text{in}^2}}$$
$$= 474.3 \text{ lbf/in}^2$$

$$y_t = \frac{h}{2} = \frac{25 \text{ in}}{2} = 12.5 \text{ in}$$

$$M_{\text{cr}} = \frac{f_r I_g}{y_t} = \frac{\left(474.3 \ \frac{\text{lbf}}{\text{in}^2}\right)(18{,}229 \text{ in}^4)}{12.5 \text{ in}}$$
$$= 6.92 \times 10^5 \text{ in-lbf}$$

$$\left(\frac{M_{\text{cr}}}{M_{\max}}\right)^3 = \left(\frac{6.92 \times 10^5 \text{ in-lbf}}{3.28 \times 10^6 \text{ in-lbf}}\right)^3$$
$$= 0.00939$$

To calculate I_{cr}, find the neutral axis. From Eq. 50.37,

$$n = \frac{E_s}{E_c} = \frac{2.9 \times 10^7 \ \frac{\text{lbf}}{\text{in}^2}}{3.6 \times 10^6 \ \frac{\text{lbf}}{\text{in}^2}} = 8.06$$

From Eq. 50.40,

$$c_s = \frac{nA_s}{b}\left(\sqrt{1 + \frac{2bd}{nA_s}} - 1\right)$$
$$= \left(\frac{(8.06)(6.24 \text{ in}^2)}{14 \text{ in}}\right)\left(\sqrt{1 + \frac{(2)(14 \text{ in})(22 \text{ in})}{(8.06)(6.24 \text{ in}^2)}} - 1\right)$$
$$= 9.48 \text{ in}$$

From Eq. 50.41,

$$I_{\text{cr}} = \frac{bc_s^3}{3} + nA_s(d - c_s)^2 = \frac{(14 \text{ in})(9.48 \text{ in})^3}{3}$$
$$+ (8.06)(6.24 \text{ in}^2)(22 \text{ in} - 9.48 \text{ in})^2$$
$$= 11{,}860 \text{ in}^3$$

From Eq. 50.43,

$$I_e = \left(\frac{M_{\text{cr}}}{M_a}\right)^3 I_g + \left(1 - \left(\frac{M_{\text{cr}}}{M_a}\right)^3\right)I_{\text{cr}} \le I_g$$
$$= (0.00939)(18{,}229 \text{ in}^4)$$
$$+ (1 - 0.00939)(11{,}860 \text{ in}^4)$$
$$= 11{,}920 \text{ in}^4$$

Structural

Find the instantaneous deflection for a uniform load.

$$\delta_{inst} = \frac{5wL^4}{384EI}$$

$$= \frac{(5)\left(3\ \dfrac{kips}{ft}\right)\left(1000\ \dfrac{lbf}{kip}\right)\left((27\ ft)\left(12\ \dfrac{in}{ft}\right)\right)^4}{(384)\left(3.6 \times 10^6\ \dfrac{lbf}{in^2}\right)(11{,}920\ in^4)\left(12\ \dfrac{in}{ft}\right)}$$

$$= \boxed{0.836\ in\quad \text{[with 100\% live load]}}$$

For a long-term loading, the percentage of the load to be sustained is

$$\frac{1+(0.30)(2)}{1+2} = 0.53\quad (53\%)$$

The factored long-term deflection is

$$\delta = (0.53)(0.836\ in) = 0.443\ in$$

$$\rho' = 0\quad \text{[no compression steel]}$$

$$\lambda = 2$$

$$\delta_{long\,term} = (2)(0.443\ in) = 0.886\ in$$

$$\delta_{total} = 0.836\ in + 0.886\ in$$

$$= \boxed{1.72\ in\quad \left[\begin{array}{c}\text{with 30\% of}\\\text{live load sustained}\end{array}\right]}$$

9. (a) The controlling factored load is

$$w_u = 1.2D + 1.6L = (1.2)\left(1300\ \frac{lbf}{ft}\right) + (1.6)\left(1900\ \frac{lbf}{ft}\right)$$

$$= 4600\ lbf/ft$$

The total factored load is $(28\ ft)(4600\ lbf/ft) = 128{,}800\ lbf$. Each reaction is

$$\frac{128{,}800\ lbf}{2} = 64{,}400\ lbf\quad \text{[factored]}$$

The column dimensions are unknown, so the reaction is placed at the face.

The centerline shear when the factored live load is positioned over half of the span is

$$V = 1.6\left(\frac{w_u\left(\dfrac{L}{2}\right)\left(\dfrac{L}{4}\right)}{L}\right) = \frac{1.6w_uL}{8}$$

$$= \frac{(1.6)\left(1.9\ \dfrac{kips}{ft}\right)(28\ ft)}{8}$$

$$= 10.64\ kips$$

Since the reaction induces compression, the critical section is located at a distance d from the face of the support. The shear envelope, with the shear at the critical section identified, is illustrated. (The effective depth has been taken as $d = 21\ in - 2.5\ in = 18.5\ in$. The critical shear, V_u, is found by interpolation.)

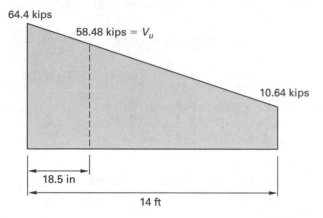

The critical section is located (from the face of the column) at

$$\frac{18.5\ in}{12\ \dfrac{in}{ft}} = 1.54\ ft$$

The nominal concrete shear strength is found by using Eq. 50.53.

$$V_c = 2\lambda\sqrt{f'_c}b_wd = \frac{(2)(1.0)\sqrt{3000\ \dfrac{lbf}{in^2}}(12\ in)(18.5\ in)}{1000\ \dfrac{lbf}{kip}}$$

$$= 24.32\ kips$$

Since $V_u > \phi V_c$, the beam needs extra shear reinforcement until V_u drops to ϕV_c. The steel shear contribution is given by Eq. 50.63.

$$V_s = \frac{V_u}{\phi} - V_c = \frac{58.48\ kips}{0.75} - 24.32\ kips$$

$$= \boxed{53.65\ kips}$$

Check to see if the beam should be redesigned. Use Eq. 50.57.

$$V_s \le 8\sqrt{f'_c}b_wd$$

$$V_{s,max} = 8\sqrt{3000\ \frac{lbf}{in^2}}(12\ in)(18.5\ in)$$

$$= 97{,}276\ lbf\quad \text{[acceptable]}$$

(b) No reinforcement is needed after V_u drops below $\frac{1}{2}\phi V_c$.

$$\tfrac{1}{2}\phi V_c = \left(\tfrac{1}{2}\right)(0.75)(24.32 \text{ kips}) = 9.12 \text{ kips}$$

Since 9.12 kips < 10.64 kips, stirrups are required over the entire span.

(c) Use Eq. 50.59 or Eq. 50.60 to determine the spacing.

$$4\sqrt{f'_c}b_w d = 4\sqrt{3000 \ \frac{\text{lbf}}{\text{in}^2}}(12 \text{ in})(18.5 \text{ in})$$
$$= 48{,}638 \text{ lbf}$$

Since $V_s > 48{,}638$ lbf, use Eq. 50.60.

$$s_{\max} = \frac{d}{4} = \frac{18.5 \text{ in}}{4} = 4.63 \text{ in}$$

However, even closer spacing may be needed. For no. 3 bars,

$$A_{\text{stirrup}} = (2)(0.11 \text{ in}^2) = 0.22 \text{ in}^2$$

From Eq. 50.64, the spacing is

$$s = \frac{A_{\text{stirrup}}f_{yt}d}{V_s}$$
$$= \frac{(0.22 \text{ in}^2)\left(60 \ \frac{\text{kips}}{\text{in}^2}\right)(18.5 \text{ in})}{53.66 \text{ kips}}$$
$$= \boxed{4.6 \text{ in}}$$

The minimum area of stirrups from Eq. 50.61 is also satisfied.

Use 4.5 in spacing starting $s/2 \approx 2.5$ in from the face of support.

10. (a) $\quad M_d = \dfrac{w_d L_{\text{span}}^2}{8} + Dx_D$

$$= \frac{\left(1.5 \ \frac{\text{kips}}{\text{ft}}\right)(18 \text{ ft})^2}{8} + (8 \text{ kips})(6 \text{ ft})$$
$$= 108.75 \text{ ft-kips}$$

$$M_l = \frac{w_l L_{\text{span}}^2}{8} + Lx_L$$
$$= \frac{\left(2 \ \frac{\text{kips}}{\text{ft}}\right)(18 \text{ ft})^2}{8} + (13 \text{ kips})(6 \text{ ft})$$
$$= 159 \text{ ft-kips}$$

The required flexural strength is

$$M_u = 1.2M_d + 1.6M_l$$
$$= (1.2)(108.75 \text{ ft-kips}) + (1.6)(159 \text{ ft-kips})$$
$$= \boxed{384.9 \text{ ft-kips}}$$

$$1.4D = (1.4)(108.75 \text{ ft-kips})$$
$$= 152.25 \text{ ft-kips} \quad [\text{does not control}]$$

(b) *step 1:* To achieve minimum beam depth as required by the problem, choose the maximum allowable reinforcement.

The maximum reinforcement ratio corresponding to $\epsilon_t = 0.004$ is

$$\rho_{\max} = \left(\frac{0.85f'_c \beta_1}{f_y}\right)\left(\frac{0.003}{0.003 + \epsilon_t}\right)$$
$$= \left(\frac{(0.85)\left(3000 \ \frac{\text{lbf}}{\text{in}^2}\right)(0.85)}{60{,}000 \ \frac{\text{lbf}}{\text{in}^2}}\right)\left(\frac{0.003}{0.003 + 0.004}\right)$$
$$= 0.0155$$

The corresponding strength reduction factor is

$$\phi = 0.65 + \frac{250}{3}(\epsilon_t - 0.002)$$
$$= 0.65 + \left(\frac{250}{3}\right)(0.004 - 0.002)$$
$$= 0.817$$

step 2: Solve for the required nominal moment strength.

$$M_n = \frac{M_u}{\phi} = \frac{384{,}900 \text{ ft-lbf}}{0.817}$$
$$= 471{,}306 \text{ ft-lbf}$$

Structural

step 3: Substitute into Eq. 50.32, and solve for bd^2.

$$bd^2 = \frac{M_n}{\rho f_y \left(1 - \dfrac{\rho f_y}{1.7 f_c'}\right)}$$

$$= \frac{(471{,}306 \text{ ft-lbf})\left(12 \dfrac{\text{in}}{\text{ft}}\right)}{(0.0155)\left(60{,}000 \dfrac{\text{lbf}}{\text{in}^2}\right)}$$

$$\times \left(1 - \frac{(0.0155)\left(60{,}000 \dfrac{\text{lbf}}{\text{in}^2}\right)}{(1.7)\left(3000 \dfrac{\text{lbf}}{\text{in}^2}\right)}\right)$$

$$= 7438 \text{ in}^3$$

step 4: b was given as 16 in.

$$d = \sqrt{\frac{7438 \text{ in}^3}{16 \text{ in}}}$$

$$= \boxed{21.56 \text{ in} \quad (21.5 \text{ in})}$$

(c) *step 5:* The steel area is

$$A_s = (0.0155)(16 \text{ in})(21.5 \text{ in}) = 5.33 \text{ in}^2$$

step 6: Select reinforcement.

$$\boxed{\text{Use two no. 10 bars and two no. 11 bars } (5.66 \text{ in}^2).}$$

The reinforcement ratio is

$$\rho = \frac{5.66 \text{ in}^2}{(16 \text{ in})(21.5 \text{ in})} = 0.0165 > \rho_{\max} = 0.0155$$

(At this point, it may be desirable to recalculate ρ and resize the beam.)

step 7: The depth of cover is

$$d_c = \frac{1.270 \text{ in}}{2} + 0.375 \text{ in} + 1.5 \text{ in}$$

$$= 2.51 \text{ in} \quad [\text{round to 2.5 in}]$$

(d) The controlling factored uniform load is

$$w_u = 1.2D + 1.6L$$

$$= (1.2)\left(1.5 \frac{\text{kips}}{\text{ft}}\right) + (1.6)\left(2 \frac{\text{kips}}{\text{ft}}\right)$$

$$= 5.0 \text{ kips/ft}$$

The controlling factored concentrated load is

$$W_u = 1.2D + 1.6L$$

$$= (1.2)(8 \text{ kips}) + (1.6)(13 \text{ kips})$$

$$= 30.4 \text{ kips}$$

ACI 318 Sec. 12.13.2.1 requires a hook.

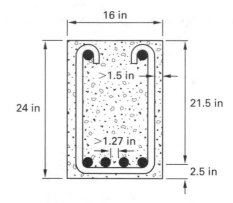

The factored reactions are

$$R_u = \frac{\left(5.0 \dfrac{\text{kips}}{\text{ft}}\right)(18 \text{ ft}) + (2)(30.4 \text{ kips})}{2} = 75.4 \text{ kips}$$

The shear diagram is

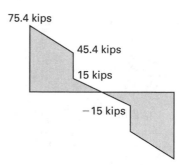

At a distance of 21.5 in from the support, the shear is

$$V_u = 75.4 \text{ kips} - \left(\frac{21.5 \text{ in}}{12 \dfrac{\text{in}}{\text{ft}}}\right)\left(5.0 \frac{\text{kips}}{\text{ft}}\right) = 66.4 \text{ kips}$$

Use Eq. 50.53.

$$V_c = 2\lambda\sqrt{f_c'}\, b_w d = (2)(1.0)\sqrt{3000 \frac{\text{lbf}}{\text{in}^2}}(16 \text{ in})(21.5 \text{ in})$$

$$= 37{,}683 \text{ lbf} \quad (37.7 \text{ kips})$$

The required steel strength is given by Eq. 50.63.

$$V_{s,\mathrm{req}} = \frac{V_u}{\phi} - V_c = \frac{66.4 \text{ kips}}{0.75} - 37.7 \text{ kips}$$

$$= 50.8 \text{ kips}$$

Check $V_{s,\mathrm{max}}$ using Eq. 50.57.

$$V_{s,\mathrm{max}} = 8\sqrt{f'_c}b_w d = 8\sqrt{3000 \frac{\mathrm{lbf}}{\mathrm{in}^2}}(16 \text{ in})(21.5 \text{ in})$$

$$= 150{,}733 \text{ lbf} \quad (144 \text{ kips}) \quad [\text{acceptable}]$$

Using $A_v = (2)(0.11 \text{ in}^2) = 0.22 \text{ in}^2$ for no. 3 bars, the spacing is

$$s = \frac{A_v f_{yt} d}{V_{s,\mathrm{req}}} = \frac{(2)(0.11 \text{ in}^2)\left(60{,}000 \frac{\mathrm{lbf}}{\mathrm{in}^2}\right)(21.5 \text{ in})}{50{,}800 \text{ lbf}}$$

$$= \boxed{5.59 \text{ in}}$$

Use 5.5 in spacing.

Start the first stirrup $s/2 = 2.75$ in from support face.

(It may be desirable to recalculate spacing required at $x = 3$ ft and continue that new spacing to $x = 6$ rather than using $s = 5$ in over the entire 6 ft.)

At $x = 6$ ft, from Eq. 50.63, the shear reinforcement required is

$$V_{s,\mathrm{req}} = \frac{V_u}{\phi} - V_c = \frac{15 \text{ kips}}{0.75} - 37.7 \text{ kips}$$

$$= -17.7 \text{ kips} \quad [\text{zero}]$$

Between $x = 6$ ft and $x = 12$ ft, only minimum reinforcement is required [ACI 318 Sec. 11.4.6.1].

11. (a) According to ACI 318 Sec. 10.7.1(a), a deep beam is one in which $l_n/h \leq 4$. In this case,

$$\frac{l_n}{h} = \frac{(3)(6.5 \text{ ft})}{6.5 \text{ ft}} = 3 < 4$$

Therefore, this is a deep beam.

(b) The entire deep beam is a D-region. It can be seen that the nodal zones at B and C are C-C-C nodal zones and the ones at A and D are C-C-T nodal zones. The nominal compression strength, F_{nn}, is given by ACI 318 Eq. A-7.

$$F_{nn} = f_{ce} A_{nz}$$

A_{nz} can be taken as the area of the bearing pads, which is $18 \text{ in} \times 20 \text{ in} = 360 \text{ in}^2$. The effective compressive strength of the concrete in the nodal zone, f_{ce}, can be determined from ACI 318 Eq. A-8.

$$f_{ce} = 0.85\beta_n f'_c$$

$\beta_n = 1.0$ in nodal zones bounded by struts or bearing areas or both [ACI 318 Sec. A.5.2.1]. The ϕ factor for a strut and tie model is 0.75 [ACI 318 Sec. A.2.6]. Therefore, for nodal zones B and C (C-C-C nodal zones), the design compression strength is

$$\phi F_{nn} = \phi f_{ce} A_{nz} = \phi(0.85\beta_n f'_c) A_{nz}$$

$$= (0.75)(0.85)(1.0)\left(4 \frac{\mathrm{kips}}{\mathrm{in}^2}\right)(360 \text{ in}^2)$$

$$= 918 \text{ kips} > 322 \text{ kips} \quad [\text{OK}]$$

$\beta_n = 0.80$ in nodal zones anchoring one tie [ACI 318 Sec. A.5.2.2]. For nodal zones A and D (C-C-T nodal zones),

$$\phi F_{nn} = \phi f_{ce} A_{nz} = \phi(0.85\beta_n f'_c) A_{nz}$$

$$= (0.75)(0.85)(0.8)\left(4 \frac{\mathrm{kips}}{\mathrm{in}^2}\right)(360 \text{ in}^2)$$

$$= 734 \text{ kips} > 322 \text{ kips} \quad [\text{OK}]$$

(c) From the free-body diagram, sum the moments about point A.

$$(322 \text{ kips})(78 \text{ in}) = F_{BC}\left(78 \text{ in} - \frac{w_c}{2} - \frac{w_t}{2}\right)$$

$$F_{BC} = (322 \text{ kips})\frac{78 \text{ in}}{78 \text{ in} - \frac{7.5 \text{ in}}{2} - \frac{9 \text{ in}}{2}}$$

$$= 360 \text{ kips}$$

The design strength of strut BC is

$$\phi F_{ns} = \phi f_{ce} A_{cs} \geq F_{BC} \quad [\text{ACI 318 Eq. A-2}]$$

$$= \phi(0.85\beta_s f'_c)w_c b \quad [\text{strut}] \quad [\text{ACI 318 Eq. A-3}]$$

$$= \phi(0.85\beta_n f'_c)w_c b \quad [\text{nodal zone}] \quad [\text{ACI 318 Eq. A-8}]$$

Since strut BC has a uniform cross section over its entire length, $\beta_s = 1.0$ per ACI 318 Sec. A.3.2.1. Also, $\beta_n = 1.0$, since the nodal zone at B is C-C-C [ACI 318 Sec. A.5.2.1]. Check the design strength of strut BC.

$$\phi F_{ns} = (0.75)(0.85)(1)\left(4 \frac{\mathrm{kips}}{\mathrm{in}^2}\right)(7.5 \text{ in})(20 \text{ in})$$

$$= 383 \text{ kips} > 360 \text{ kips} \quad [\text{OK}]$$

(d) From the free-body diagram of half of the beam, it can be seen that $F_{AD} = F_{BC}$.

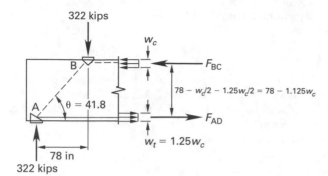

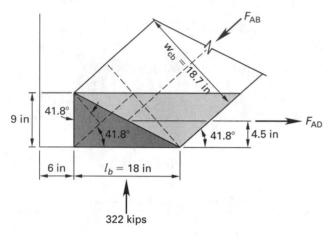

$$F_{AD} = 360 \text{ kips} \leq \phi F_{nt} = \phi A_{ts} f_y$$

$$A_{ts} = \frac{F_{AD}}{\phi f_y} = \frac{360 \text{ kips}}{(0.75)\left(60 \ \dfrac{\text{kips}}{\text{in}^2}\right)} = 8.0 \text{ in}^2$$

From ACI 318 Sec. 10.7.3 and Sec. 10.5.1, the minimum areas are

$$A_{ts,\min} = \frac{200bd}{f_y} = \frac{(200)(20 \text{ in})\left(\begin{array}{c}(6.5 \text{ ft})\left(12 \ \dfrac{\text{in}}{\text{ft}}\right) \\ - 4.5 \text{ in}\end{array}\right)}{60{,}000 \ \dfrac{\text{lbf}}{\text{in}^2}}$$

$$= 4.9 \text{ in}^2$$

$$A_{ts,\min} = \frac{3\sqrt{f_c'}}{f_y}bd$$

$$= \frac{3\sqrt{4000 \ \dfrac{\text{lbf}}{\text{in}^2}}}{60{,}000 \ \dfrac{\text{lbf}}{\text{in}^2}}(20 \text{ in})\left((6.5 \text{ ft})\left(12 \ \dfrac{\text{in}}{\text{ft}}\right) - 4.5 \text{ in}\right)$$

$$= 4.6 \text{ in}^2$$

Neither of the minimum values control. Therefore, use two layers of four no. 9 bars ($A_s = 8.0 \text{ in}^2$).

(e) Find the angle of strut AB with respect to tie AD.

$$\tan \theta = \frac{78 \text{ in} - \dfrac{7.5 \text{ in}}{2} - \dfrac{9 \text{ in}}{2}}{78 \text{ in}} = 0.8942$$

$$\theta = 41.8° > 25° \quad [\text{OK per ACI 318 Sec. A.2.5}]$$

$$F_{AB} = \frac{360 \text{ kips}}{\cos 41.8°} = 483 \text{ kips}$$

As shown in the problem illustrations, the width at the top of the strut, w_{ct}, is

$$w_{ct} = l_b \sin \theta + w_c \cos \theta$$
$$= (18 \text{ in})\sin 41.8° + (7.5 \text{ in})\cos 41.8°$$
$$= 17.6 \text{ in}$$

The width at the bottom of the strut, w_{cb}, is

$$w_{cb} = l_b \sin \theta + w_t \cos \theta$$
$$= (18 \text{ in})\sin 41.8° + (9 \text{ in})\cos 41.8°$$
$$= 18.7 \text{ in}$$

$\beta_s = 0.75$, assuming that reinforcement satisfying ACI 318 Sec. A.3.3.2 is provided. The design strength of the strut is based on the smaller of the two widths at the ends of the strut.

$$\phi F_{ns} = \phi 0.85 \beta_s f_c' w_{ct} b$$
$$= (0.75)(0.85)(0.75)\left(4 \ \dfrac{\text{kips}}{\text{in}^2}\right)(17.6 \text{ in})(20 \text{ in})$$
$$= 673 \text{ kips} > 483 \text{ kips} \quad [\text{OK}]$$

(f) The two layers of four no. 9 tie bars must be properly anchored at both ends of the beam.

The critical section for development of the tie reinforcement is determined in accordance with ACI 318 Sec. A.4.3.2. The distance, x, from the edge of the bearing plate (nodal zone) to the critical section is determined from geometry.

$$\tan 41.8° = \frac{4.5 \text{ in}}{x}$$
$$x = 5 \text{ in}$$

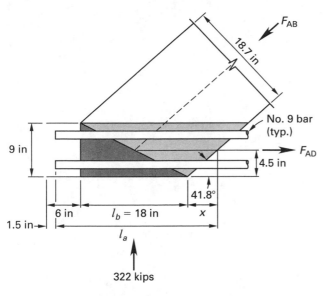

The available length for straight bar development is

$$24 \text{ in} - 1.5 \text{ in} + 5 \text{ in} = 27.5 \text{ in}$$

To develop a straight no. 9 bar, the required development length (see Chap. 55) is determined from

$$l_d = \left(\frac{3}{40}\right)\left(\frac{f_y}{\lambda\sqrt{f_c'}}\right)\left(\frac{\psi_t\psi_e\psi_s}{\dfrac{c + K_{tr}}{d_b}}\right)d_b \quad \text{[ACI 318 Eq. 12-1]}$$

$$= \left(\frac{3}{40}\right)\left(\frac{60{,}000 \ \dfrac{\text{lbf}}{\text{in}^2}}{1.0\sqrt{4000 \ \dfrac{\text{lbf}}{\text{in}^2}}}\right)\left(\frac{(1.0)(1.0)(1.0)}{\dfrac{2.0 \text{ in}}{1.128 \text{ in}}}\right)(1.128)$$

$$= 45.3 \text{ in} > 27.5 \text{ in}$$

Provide a standard hook at the ends of the no. 9 bars.

The development length of a no. 9 bar with a standard 90° hook is determined from ACI 318 Sec. 12.5.2.

$$l_{dh} = \frac{0.02\psi_e f_y d_b}{\lambda\sqrt{f_c'}}$$

$$= \frac{(0.02)(1.0)\left(60{,}000 \ \dfrac{\text{lbf}}{\text{in}^2}\right)(1.128 \text{ in})}{1.0\sqrt{4000 \ \dfrac{\text{lbf}}{\text{in}^2}}}$$

$$= 21.4 \text{ in} < 27.5 \text{ in} \quad \text{[OK]}$$

(g) Provide horizontal and vertical reinforcement in accordance with ACI 318 Sec. A.3.3.

For horizontal reinforcement, try two no. 5 bars (one bar on each face) on 12 in spacing.

$$\alpha_2 = \theta = 41.8°$$

$$\frac{A_{s2}\sin\alpha_2}{bs_2} = \frac{(2)(0.31 \text{ in}^2)\sin 41.8°}{(20 \text{ in})(12 \text{ in})}$$

$$= 0.0017$$

For vertical reinforcement, try two sets of two no. 4 overlapping ties on 12 in spacing.

$$\alpha_1 = 90° - \gamma_2 = 90° - 41.8°$$

$$= 48.2°$$

$$\frac{A_{s1}\sin\alpha_1}{bs_1} = \frac{(4)(0.2 \text{ in}^2)\sin 48.2°}{(20 \text{ in})(12 \text{ in})}$$

$$= 0.0025$$

Check ACI 318 Eq. A-4.

$$\sum\frac{A_{si}}{b_s s_i}\sin\alpha_i \geq 0.003 \quad \text{[ACI 318 Eq. A-4]}$$

$$0.0017 + 0.0025 = 0.0042 > 0.003 \quad \text{[OK]}$$

Structural

51 Reinforced Concrete: Slabs

PRACTICE PROBLEMS

Two-Way Slabs

1. A plan of a floor system supported on beams is shown. In addition to its own weight, the floor is subjected to uniformly distributed dead and live loads of 30 lbf/ft^2 and 50 lbf/ft^2, respectively. Material properties are $f'_c = 4000$ lbf/in^2 and $f_y = 60,000$ lbf/in^2. (a) Calculate the relative beam stiffness for beam A2-B2. (b) Determine if the thickness of the slab is adequate for deflection control. (c) Design the steel reinforcement for positive moment in beam A2-B2.

2. A flat plate with an edge beam is shown. In addition to its own weight, the floor slab is subjected to uniformly distributed dead and live loads of 20 lbf/ft^2 and 50 lbf/ft^2, respectively. Material properties are $f'_c = 4000$ lbf/in^2 and $f_y = 40,000$ lbf/in^2. (a) Determine the torsional constant, C, for the edge beam in span A2-A3. (b) Determine the adequacy of the slab thickness for deflection control. (c) Design the positive reinforcement needed in the column strip whose center lies on the axis A2-B2.

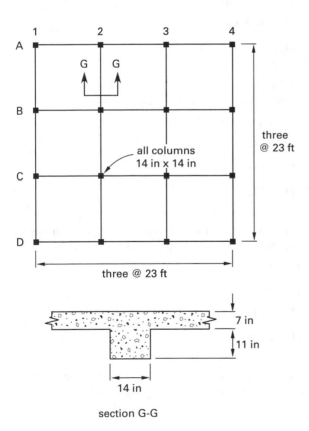

section G-G

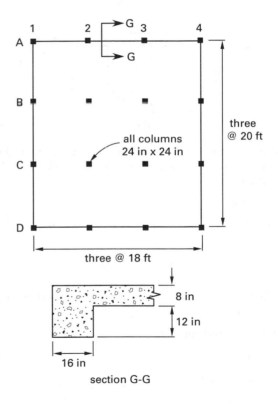

section G-G

3. (*Time limit: one hour*) The floor system shown is subjected, in addition to its own weight, to uniformly distributed dead and live loads of 20 lbf/ft^2 and 80 lbf/ft^2, respectively. Material properties are $f'_c = 3000$ lbf/in^2 and $f_y = 40{,}000$ lbf/in^2.

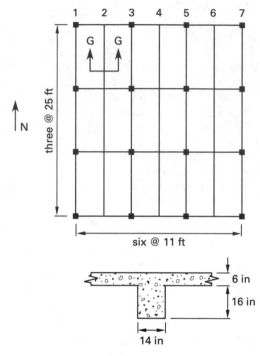

section G-G

(a) Assuming a single slab depth is used throughout, the minimum thickness required for deflection control is most nearly

(A) 5.0 in

(B) 5.5 in

(C) 6.0 in

(D) 7.5 in

For parts (b) through (j), assume the slab thickness is 6 in.

(b) The factored uniform load is most nearly

(A) 130 lbf/ft^2

(B) 160 lbf/ft^2

(C) 240 lbf/ft^2

(D) 350 lbf/ft^2

(c) The maximum live load that the slab can carry, as limited by shear, is most nearly

(A) 400 lbf/ft^2

(B) 470 lbf/ft^2

(C) 500 lbf/ft^2

(D) 600 lbf/ft^2

(d) The required spacing of no. 3 bars parallel to the N-S direction is most nearly

(A) 7 in

(B) 8 in

(C) 9 in

(D) 10 in

(e) The maximum positive moment in this floor system is most nearly

(A) 1.5 ft-kips/ft

(B) 1.6 ft-kips/ft

(C) 1.9 ft-kips/ft

(D) 2.1 ft-kips/ft

(f) The maximum negative moment in this floor system is most nearly

(A) 2.5 ft-kips/ft

(B) 2.6 ft-kips/ft

(C) 2.9 ft-kips/ft

(D) 3.1 ft-kips/ft

(g) The spacing of no. 4 bars for negative moment normal to axis 2 is most nearly

(A) 9 in

(B) 10 in

(C) 12 in

(D) 14 in

(h) The spacing of no. 6 bars for negative moment in the span between axes 2 and 3 is most nearly

(A) 12 in

(B) 16 in

(C) 18 in

(D) 27 in

(i) The spacing of no. 4 bars for positive moment in the span between axes 2 and 3 is most nearly

(A) 14 in

(B) 17 in

(C) 19 in

(D) 20 in

(j) The load per foot that should be used to design the beam located on axis 2 is most nearly

(A) 2.8 kips/ft

(B) 3.0 kips/ft

(C) 3.2 kips/ft

(D) 3.5 kips/ft

4. (*Time limit: one hour*) A flat plate is supported on 24 in × 24 in columns that are spaced 18 ft in the E-W direction and 22 ft in the N-S direction. In addition to its own weight, the floor is subjected to uniformly distributed dead and live loads of 20 lbf/ft^2 and 50 lbf/ft^2, respectively. Material properties are $f'_c = 4000$ lbf/in^2 and $f_y = 60,000$ lbf/in^2.

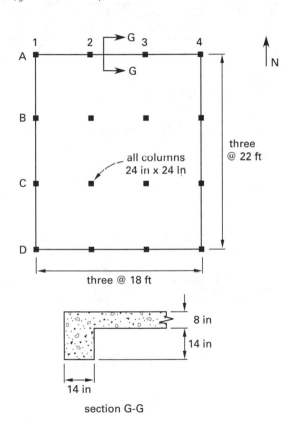

section G-G

(a) The torsional constant, C, for the edge beam is most nearly

(A) 9000 in^4

(B) 12,000 in^4

(C) 14,000 in^4

(D) 34,000 in^4

(b) The minimum thickness for deflection control is most nearly

(A) 6.0 in

(B) 7.3 in

(C) 8.0 in

(D) 8.5 in

(c) The minimum thickness if the edge beam is eliminated is most nearly

(A) 6.0 in

(B) 7.5 in

(C) 8.0 in

(D) 8.5 in

(d) The relative torsional stiffness, β_t, for the edge beam spanning in the E-W direction is most nearly

(A) 0.5

(B) 0.7

(C) 1.0

(D) 1.3

(e) The maximum positive moment in the column strip that runs from B2 to C2 is most nearly

(A) 40 ft-kips

(B) 48 ft-kips

(C) 52 ft-kips

(D) 62 ft-kips

(f) The maximum positive moment in the column strip that runs from B1 to C1 is most nearly

(A) 21 ft-kips

(B) 24 ft-kips

(C) 28 ft-kips

(D) 31 ft-kips

(g) The maximum positive moment in the N-S middle strips in panel B2-B3-C2-C3 is most nearly

(A) 28 ft-kips

(B) 32 ft-kips

(C) 38 ft-kips

(D) 40 ft-kips

(h) If drop panels that meet ACI 318 requirements are provided, the minimum slab thickness for deflection control (outside the drop panel) is most nearly

(A) 5.5 in

(B) 6.7 in

(C) 7.5 in

(D) 8.0 in

(i) The required total thickness for the drop panels is most nearly

(A) 7.0 in

(B) 7.5 in

(C) 8.5 in

(D) 9.0 in

(j) The minimum dimension in the E-W direction for interior drop panels is most nearly

(A) 3 ft

(B) 5 ft

(C) 6 ft

(D) 7 ft

SOLUTIONS

1. (a) The beam section is

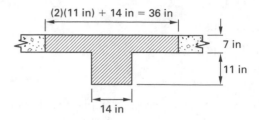

Use a table to calculate the moment of inertia of the beam.

A (in^2)	y (in)	Ay (in^3)	$A(y - \bar{y})^2$ (in^4)	I_o (in^4)
252	3.5	882	2936.8	1029.0
154	12.5	1925	4805.7	1552.8
totals 406		2807	7742.5	2581.8

$$\bar{y} = \frac{\sum Ay}{\sum A} = \frac{2807 \text{ in}^3}{406 \text{ in}^2} = 6.91 \text{ in}$$

$$I_b = \sum A(y - \bar{y})^2 + \sum I_o = 7742.5 \text{ in}^4 + 2581.8 \text{ in}^4$$
$$= 10{,}324 \text{ in}^4$$

The moment of inertia of the slab is

$$I_s = \frac{bh^3}{12} = \frac{(23 \text{ ft})\left(12 \frac{\text{in}}{\text{ft}}\right)(7 \text{ in})^3}{12} = 7889 \text{ in}^4$$

$$\alpha = \frac{I_b}{I_s} = \frac{10{,}324 \text{ in}^4}{7889 \text{ in}^4} = \boxed{1.31}$$

(b) $\alpha_m = 1.31$

$$l_n = (23 \text{ ft})\left(12 \frac{\text{in}}{\text{ft}}\right) - 14 \text{ in} = 262 \text{ in}$$

$$\beta = 1$$

Use Eq. 51.7.

$$t = \frac{l_n\left(0.8 + \dfrac{f_y}{200{,}000}\right)}{36 + 5\beta(\alpha_m - 0.2)} \geq 5 \text{ in}$$

$$= \frac{(262 \text{ in})\left(0.8 + \dfrac{60{,}000 \frac{\text{lbf}}{\text{in}^2}}{200{,}000 \frac{\text{lbf}}{\text{in}^2}}\right)}{36 + (5)(1)(1.31 - 0.2)}$$

$$= 6.94 \text{ in} < 7 \text{ in} \quad \boxed{\text{[adequate]}}$$

(c) The weight of the slab is

$$\left(\frac{7 \text{ in}}{12 \frac{\text{in}}{\text{ft}}}\right)\left(0.15 \frac{\text{kip}}{\text{ft}^3}\right) = 0.0875 \text{ kip/ft}^2$$

The factored uniform load is

$$w_u = 1.2D + 1.6L$$
$$= (1.2)\left(0.03 \frac{\text{kip}}{\text{ft}^2} + 0.0875 \frac{\text{kip}}{\text{ft}^2}\right)$$
$$+ (1.6)\left(0.05 \frac{\text{kip}}{\text{ft}^2}\right)$$
$$= 0.22 \text{ kip/ft}^2$$

The distances needed for the computation of M_o are

$$l_n = 23 \text{ ft} - \frac{14 \text{ in}}{12 \frac{\text{in}}{\text{ft}}} = 21.83 \text{ ft}$$
$$l_2 = 23 \text{ ft}$$

The additional weight of concrete below the slab must be included in the design load.

$$q_u = \left(0.22 \frac{\text{kip}}{\text{ft}^2}\right)(23 \text{ ft})$$
$$+ \frac{(1.2)(11 \text{ in})(14 \text{ in})\left(0.15 \frac{\text{kip}}{\text{ft}^3}\right)}{\left(12 \frac{\text{in}}{\text{ft}}\right)^2}$$
$$= 5.25 \text{ kips/ft}$$

From Eq. 51.3, the statical moment for a typical bay is

$$M_o = \frac{w_u l_2 l_n^2}{8} = \frac{q_u l_n^2}{8}$$
$$= \frac{\left(5.25 \frac{\text{kips}}{\text{ft}}\right)(21.83 \text{ ft})^2}{8}$$
$$= 312.7 \text{ ft-kips}$$

From Table 51.3, the positive design moment is

$$0.57 M_o = (0.57)(312.7 \text{ ft-kips}) = 178.2 \text{ ft-kips}$$

From Table 51.4(c), the portion of this moment assigned to the column strip is

$$M_{\text{column strip}} = 0.75(0.57 M_o) = (0.75)(178.2 \text{ ft-kips})$$
$$= 133.7 \text{ ft-kips}$$
$$\alpha\left(\frac{l_2}{l_1}\right) = (1.31)\left(\frac{23 \text{ ft}}{23 \text{ ft}}\right) = 1.31 > 1$$

Therefore, the beam takes 85% of the column strip moment. (See ACI 318 Sec. 13.6.5.1.)

$$M_u = (0.85)(133.7 \text{ ft-kips}) = 113.6 \text{ ft-kips}$$

Estimate $d = t - 2.5 \text{ in} = 15.5 \text{ in}$.

Assume $\lambda = (0.1)(15.5 \text{ in}) = 1.55 \text{ in}$ and the section is tension controlled. Use Eq. 50.28.

$$A_s = \frac{M_u}{\phi f_y(d - \lambda)}$$
$$= \frac{(113.6 \text{ ft-kips})\left(12 \frac{\text{in}}{\text{ft}}\right)}{(0.9)\left(60 \frac{\text{kips}}{\text{in}^2}\right)(15.5 \text{ in} - 1.55 \text{ in})}$$
$$= 1.81 \text{ in}^2$$

Use Eq. 50.26.

$$A_c = \frac{f_y A_s}{0.85 f_c'} = \frac{(1.81 \text{ in}^2)\left(60 \frac{\text{kips}}{\text{in}^2}\right)}{(0.85)\left(4 \frac{\text{kips}}{\text{in}^2}\right)}$$
$$= 31.9 \text{ in}^2$$
$$a = \frac{31.9 \text{ in}^2}{36 \text{ in}} = 0.89 \text{ in}$$
$$\lambda = \frac{a}{2} = \frac{0.89 \text{ in}}{2} = 0.445 \text{ in}$$

With a value of a this small, the section is going to be tension controlled.

Use Eq. 50.27 for the design moment capacity. $\phi = 0.9$.

$$\phi M_n = \phi A_s f_y(d - \lambda)$$
$$= \frac{(0.9)(1.81)\left(60 \frac{\text{kips}}{\text{in}^2}\right)(15.5 \text{ in} - 0.445 \text{ in})}{12 \frac{\text{in}}{\text{ft}}}$$
$$= 122.6 \text{ ft-kips}$$

Structural

Adjust the area of steel.

$$A_{s,\text{adjusted}} = \frac{M_u A_s}{\phi M_n} = \frac{(113.6 \text{ ft-kips})(1.81 \text{ in}^2)}{122.6 \text{ ft-kips}}$$

$$= 1.68 \text{ in}^2$$

Use three no. 7 bars ($A_s = 1.8 \text{ in}^2$).

2. (a) Refer to Fig. 51.4.

$$A = 0; B = 12 \text{ in}; 4t = (4)(8 \text{ in}) = 32 \text{ in}$$

$$12 \text{ in} < 32 \text{ in}$$

For the computation of C, the beam section is treated as the sum of two rectangles. The division that leads to the largest C is shown.

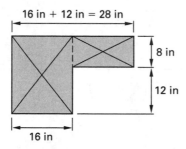

16 in + 12 in = 28 in

8 in

12 in

16 in

Use Eq. 51.6.

$$C = \sum \left(1 - 0.63\left(\frac{x}{y}\right)\right)\left(\frac{x^3 y}{3}\right)$$

$$= \left(1 - (0.63)\left(\frac{16 \text{ in}}{20 \text{ in}}\right)\right)\left(\frac{(16 \text{ in})^3(20 \text{ in})}{3}\right)$$

$$+ \left(1 - (0.63)\left(\frac{8 \text{ in}}{12 \text{ in}}\right)\right)\left(\frac{(8 \text{ in})^3(12 \text{ in})}{3}\right)$$

$$= \boxed{14{,}732 \text{ in}^4}$$

(b) From Table 51.5, without drop panels for exterior panels, for the edge beam A2-A3,

$$t \geq \frac{l_n}{36} = \frac{(20 \text{ ft})\left(12 \frac{\text{in}}{\text{ft}}\right) - (2 \text{ ft})\left(12 \frac{\text{in}}{\text{ft}}\right)}{36}$$

$$= 6.0 \text{ in} < 8 \text{ in} \quad \boxed{\text{[adequate]}}$$

(c) The weight of the slab is

$$\left(\frac{8 \text{ in}}{12 \frac{\text{in}}{\text{ft}}}\right)\left(0.15 \frac{\text{kip}}{\text{ft}^3}\right) = 0.1 \text{ kip/ft}^2$$

The controlling factored uniform load is

$$w_u = 1.2D + 1.6L$$

$$= (1.2)\left(0.1 \frac{\text{kip}}{\text{ft}^2} + 0.02 \frac{\text{kip}}{\text{ft}^2}\right) + (1.6)\left(0.05 \frac{\text{kip}}{\text{ft}^2}\right)$$

$$= 0.224 \text{ kip/ft}^2$$

The distances needed for the computation of M_o are

$$l_n = 20 \text{ ft} - 2 \text{ ft} = 18 \text{ ft} \quad \begin{bmatrix} \text{column face to column} \\ \text{face free distances} \end{bmatrix}$$

$$l_2 = 18 \text{ ft} \quad [\text{center-to-center}]$$

$$M_o = \frac{w_u l_2 l_n^2}{8} = \frac{\left(0.224 \frac{\text{kip}}{\text{ft}^2}\right)(18 \text{ ft})(18 \text{ ft})^2}{8}$$

$$= 163.3 \text{ ft-kips}$$

From Table 51.3, the positive moment is

$$0.50M_o = (0.50)(163.3 \text{ ft-kips}) = 81.65 \text{ ft-kips}$$

From Table 51.4(c),

$$M_{\text{column strip}} = 0.60(\text{positive moment})$$

$$= (0.60)(81.65 \text{ ft-kips})$$

$$= 48.99 \text{ ft-kips}$$

Estimate $d = 8 \text{ in} - 1 \text{ in} = 7 \text{ in}$.

Assume that the section is tension controlled and $\lambda = (0.1)(7 \text{ in}) = 0.7 \text{ in}$. Use Eq. 50.28.

$$A_s = \frac{M_u}{\phi f_y(d - \lambda)} = \frac{(48.99 \text{ ft-kips})\left(12 \frac{\text{in}}{\text{ft}}\right)}{(0.9)\left(40 \frac{\text{kips}}{\text{in}^2}\right)(7.0 \text{ in} - 0.7 \text{ in})}$$

$$= 2.59 \text{ in}^2$$

Use Eq. 50.26.

$$A_c = \frac{f_y A_s}{0.85 f'_c} = \frac{\left(40 \; \frac{\text{kips}}{\text{in}^2}\right)(2.59 \text{ in}^2)}{(0.85)\left(4 \; \frac{\text{kips}}{\text{in}^2}\right)}$$

$$= 30.47 \text{ in}^2$$

Since the column spacing is 18 ft in the short direction, the width of a column strip is

$$(0.25)(18 \text{ ft})(2) = 9 \text{ ft}$$

$$a = \frac{30.47 \text{ in}^2}{(9 \text{ ft})\left(12 \; \frac{\text{in}}{\text{ft}}\right)} = 0.28 \text{ in}$$

$$\lambda = \frac{a}{2} = \frac{0.28 \text{ in}}{2} = 0.14 \text{ in}$$

With the small value of a, the section is going to be tension controlled.

Use Eq. 50.27.

$$\phi M_n = \phi A_s f_y (d - \lambda)$$

$$= \frac{(0.9)(2.59)\left(40 \; \frac{\text{kips}}{\text{in}^2}\right)(7 \text{ in} - 0.14 \text{ in})}{12 \; \frac{\text{in}}{\text{ft}}}$$

$$= 53.3 \text{ ft-kips}$$

Adjust the area of steel.

$$A_{s,\text{adjusted}} = \frac{M_u A_s}{\phi M_n} = \frac{(48.99 \text{ ft-kips})(2.59 \text{ in}^2)}{53.3 \text{ ft-kips}}$$

$$= 2.38 \text{ in}^2$$

From ACI 318 Sec. 7.12.2, the minimum area of steel is

$$A_{s,\text{min}} = (0.002)(9 \text{ ft})\left(12 \; \frac{\text{in}}{\text{ft}}\right)(8 \text{ in})$$

$$= 1.73 \text{ in}^2 \quad [\text{does not control}]$$

Use twelve no. 4 bars ($A_s = 2.4 \text{ in}^2$).

3. (a) From Table 51.1, since this is a one-way slab system (long-to-short span ratio greater than 2),

$$l_n = (11 \text{ ft})\left(12 \; \frac{\text{in}}{\text{ft}}\right) - 14 \text{ in} = 118 \text{ in}$$

$$t \geq \frac{l_n}{24} = \frac{118 \text{ in}}{24} = \boxed{4.92 \text{ in} \quad (5.0 \text{ in})}$$

Since $f_y = 40$ ksi, the thickness obtained from Table 51.1 must be adjusted using Eq. 50.48. The ratio and adjusted thickness, respectively, are

$$0.4 + \frac{40{,}000 \; \frac{\text{lbf}}{\text{in}^2}}{100{,}000 \; \frac{\text{lbf}}{\text{in}^2}} = 0.8$$

$$t \geq (0.8)(4.92 \text{ in}) = 3.94 \text{ in}$$

The answer is (A).

(b) The weight of the slab is

$$\frac{(6 \text{ in})\left(150 \; \frac{\text{lbf}}{\text{ft}^3}\right)}{12 \; \frac{\text{in}}{\text{ft}}} = 75 \text{ lbf/ft}^2$$

The controlling factored uniform load is

$$w_u = 1.2D + 1.6L$$

$$= (1.2)\left(75 \; \frac{\text{lbf}}{\text{ft}^2} + 20 \; \frac{\text{lbf}}{\text{ft}^2}\right) + (1.6)\left(80 \; \frac{\text{lbf}}{\text{ft}^2}\right)$$

$$= \boxed{242 \text{ lbf/ft}^2 \quad (240 \text{ lbf/ft}^2)}$$

The answer is (C).

(c) $l_n = (11 \text{ ft})\left(12 \; \frac{\text{in}}{\text{ft}}\right) - 14 \text{ in} = 118 \text{ in}$

Calculate the shear capacity per foot ($b_w = 12$ in) of slab. Approximate the slab depth as $d = h - 1$ in $= 6$ in $- 1$ in $= 5$ in. From Eq. 50.53,

$$V_c = 2\lambda\sqrt{f'_c}\,b_w d = \frac{(2)(1.0)\sqrt{3000 \; \frac{\text{lbf}}{\text{in}^2}}(12 \text{ in})(5 \text{ in})}{1000 \; \frac{\text{lbf}}{\text{kip}}}$$

$$= 6.57 \text{ kips}$$

$$\phi V_c = (0.75)(6.57 \text{ kips}) = 4.93 \text{ kips}$$

Use Eq. 47.44 and equate the maximum shear from a load, w, to the capacity.

$$\frac{1.15wl_n}{2} = \frac{(1.15)w(118 \text{ in})}{(2)\left(12 \frac{\text{in}}{\text{ft}}\right)} = 4.93 \text{ kips}$$

$$w = 0.87 \text{ kip/ft}^2 \quad (870 \text{ lbf/ft}^2)$$

Subtract the total dead load weight and divide by the load factor of 1.6.

$$w_l = \frac{w - 1.2D}{1.6} = \frac{870 \frac{\text{lbf}}{\text{ft}^2} - (1.2)\left(75 \frac{\text{lbf}}{\text{ft}^2} + 20 \frac{\text{lbf}}{\text{ft}^2}\right)}{1.6}$$

$$= \boxed{473 \text{ lbf/ft}^2 \quad (470 \text{ lbf/ft}^2)}$$

The answer is (B).

(d) The temperature steel required for grade 40 reinforcement is given by ACI 318 Sec. 7.12.2 as

$$A_s = (0.002)(6 \text{ in})\left(12 \frac{\text{in}}{\text{ft}}\right) = 0.144 \text{ in}^2/\text{ft}$$

The number of bars per foot is

$$\frac{0.144 \frac{\text{in}^2}{\text{ft}}}{0.11 \frac{\text{in}^2}{\text{bar}}} = 1.309 \text{ bars/ft}$$

The spacing is

$$s = \frac{12 \frac{\text{in}}{\text{ft}}}{1.309 \frac{\text{bars}}{\text{ft}}} = \boxed{9.16 \text{ in} \quad (9 \text{ in})}$$

The answer is (C).

(e) $\quad l_n = \dfrac{118 \text{ in}}{12 \frac{\text{in}}{\text{ft}}} = 9.83 \text{ ft}$

From Eq. 47.43,

$$\frac{w_u l_n^2}{14} = \frac{\left(0.242 \frac{\text{kip}}{\text{ft}^2}\right)(9.83 \text{ ft})^2}{14}$$

$$= \boxed{1.67 \text{ ft-kips/ft} \quad (1.6 \text{ ft-kips/ft})}$$

The answer is (B).

(f) From Eq. 47.43,

$$\frac{w_u l_n^2}{10} = \frac{\left(0.242 \frac{\text{kip}}{\text{ft}^2}\right)(9.83 \text{ ft})^2}{10}$$

$$= \boxed{2.34 \text{ ft-kips/ft} \quad (2.5 \text{ ft-kips/ft})}$$

The answer is (A).

(g) $M_u = 2.34$ ft-kips/ft

Estimate $d = h - 1 \text{ in} = 6 \text{ in} - 1 \text{ in} = 5 \text{ in}$.

Assume $\lambda = (0.1)(5 \text{ in}) = 0.5 \text{ in}$ and that the section is tension controlled.

Use Eq. 50.28.

$$A_s = \frac{M_u}{\phi f_y(d - \lambda)} = \frac{(2.34 \text{ ft-kips})\left(12 \frac{\text{in}}{\text{ft}}\right)}{(0.9)\left(40 \frac{\text{kips}}{\text{in}^2}\right)(5 \text{ in} - 0.5 \text{ in})}$$

$$= 0.173 \text{ in}^2/\text{ft}$$

Compute M_n for this steel area. First, locate the centroid of the compression zone.

$$A_s f_y = 0.85 f_c' 2\lambda$$

$$\lambda = \frac{\left(0.173 \frac{\text{in}^2}{\text{ft}}\right)\left(40 \frac{\text{kips}}{\text{in}^2}\right)}{(0.85)\left(3 \frac{\text{kips}}{\text{in}^2}\right)\left(12 \frac{\text{in}}{\text{ft}}\right)(2)} = 0.113 \text{ in}$$

$$M_n = A_s f_y(d - \lambda)$$

$$= \frac{\left(0.173 \frac{\text{in}^2}{\text{ft}}\right)\left(40 \frac{\text{kips}}{\text{in}^2}\right)(5 \text{ in} - 0.114 \text{ in})}{12 \frac{\text{in}}{\text{ft}}}$$

$$= 2.82 \text{ ft-kips/ft}$$

$$\phi M_n = (0.9)\left(2.82 \frac{\text{ft-kips}}{\text{ft}}\right) = 2.54 \text{ ft-kips/ft}$$

The final adjusted steel area is

$$A_s = \left(0.173 \frac{\text{in}^2}{\text{ft}}\right)\left(\frac{2.34 \frac{\text{ft-kips}}{\text{ft}}}{2.54 \frac{\text{ft-kips}}{\text{ft}}}\right) = 0.16 \text{ in}^2/\text{ft}$$

The number of bars per foot is

$$\frac{0.16 \ \frac{\text{in}^2}{\text{ft}}}{0.2 \ \frac{\text{in}^2}{\text{bar}}} = 0.8 \ \text{bars/ft}$$

The spacing is

$$s = \frac{12 \ \frac{\text{in}}{\text{ft}}}{0.8 \ \frac{\text{bars}}{\text{ft}}}$$

$$= \boxed{15 \ \text{in} \quad (14 \ \text{in})}$$

The answer is (D).

(h) Scale the answer from the previous question.

$$s = \left(\frac{0.444 \ \text{in}^2}{0.2 \ \text{in}^2}\right)(13.79 \ \text{in})$$

$$= 30.61 \ \text{in}$$

The maximum spacing is

$$3t = (3)(6 \ \text{in})$$

$$= \boxed{18 \ \text{in} \quad [\text{controls}]}$$

The answer is (C).

(i) From Eq. 47.43,

$$\frac{w_u l_n^2}{16} = \frac{\left(0.242 \ \frac{\text{kip}}{\text{ft}^2}\right)(9.83 \ \text{ft})^2}{16} = 1.46 \ \text{ft-kips/ft}$$

$$A_{s,\text{min}} = (0.002)(6 \ \text{in})\left(12 \ \frac{\text{in}}{\text{ft}}\right)$$

$$= 0.144 \ \text{in}^2/\text{ft}$$

Use Eq. 50.28.

$$A_s = \frac{M_u}{\phi f_y(d - \lambda)} = \frac{(1.46 \ \text{ft-kips})\left(12 \ \frac{\text{in}}{\text{ft}}\right)}{(0.9)\left(40 \ \frac{\text{kips}}{\text{in}^2}\right)(5 \ \text{in} - 0.5 \ \text{in})}$$

$$= 0.108 \ \text{in}^2/\text{ft} \quad [\text{less than } A_{s,\text{min}}]$$

The number of bars per foot is

$$\frac{0.144 \ \frac{\text{in}^2}{\text{ft}}}{0.2 \ \frac{\text{in}^2}{\text{bar}}} = 0.72 \ \text{bars/ft}$$

$$s = \frac{12 \ \frac{\text{in}}{\text{ft}}}{0.72 \ \frac{\text{bars}}{\text{ft}}} = \boxed{16.67 \ \text{in} \quad (17 \ \text{in})}$$

[less than $3t$, so OK]

The answer is (B).

(j) From Eq. 47.44,

$$1.15\left(\frac{w_u l}{2}\right) + \frac{w_u l}{2} = \frac{2.15 w_u l}{2}$$

$$= \frac{(2.15)\left(0.242 \ \frac{\text{kip}}{\text{ft}^2}\right)(11 \ \text{ft})}{2}$$

$$= 2.86 \ \text{kips/ft}$$

The part of the weight of the beam not considered in the previous computation (including the 1.2 load factor) is

$$\frac{(1.2)(14 \ \text{in})(16 \ \text{in})\left(0.15 \ \frac{\text{kip}}{\text{ft}^3}\right)}{\left(12 \ \frac{\text{in}}{\text{ft}}\right)^2} = 0.28 \ \text{kip/ft}$$

The total factored load is

$$2.86 \ \frac{\text{kips}}{\text{ft}} + 0.28 \ \frac{\text{kip}}{\text{ft}} = \boxed{3.14 \ \text{kips/ft} \quad (3.2 \ \text{kips/ft})}$$

The answer is (C).

4. (a) For the computation of C, the beam section is treated as the sum of two rectangles. The division that leads to the largest C is shown.

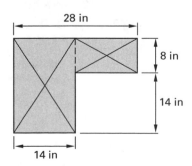

Use Eq. 51.6.

$$C = \sum \left(1 - 0.63 \left(\frac{x}{y}\right)\right)\left(\frac{x^3 y}{3}\right)$$
$$= \left(1 - (0.63)\left(\frac{8 \text{ in}}{14 \text{ in}}\right)\right)\left(\frac{(8 \text{ in})^3 (14 \text{ in})}{3}\right)$$
$$\quad + \left(1 - (0.63)\left(\frac{14 \text{ in}}{22 \text{ in}}\right)\right)\left(\frac{(14 \text{ in})^3 (22 \text{ in})}{3}\right)$$
$$= \boxed{13{,}584 \text{ in}^4 \quad (14{,}000 \text{ in}^4)}$$

The answer is (C).

(b) $l_n = 22 \text{ ft} - 2 \text{ ft} = 20 \text{ ft}$

From Table 51.5,

$$t \geq \frac{l_n}{33} = \frac{(20 \text{ ft})\left(12 \frac{\text{in}}{\text{ft}}\right)}{33} = \boxed{7.27 \text{ in} \quad (7.3 \text{ in})}$$

The answer is (B).

(c) From Table 51.5, the minimum thickness for the entire slab is

$$t \geq \frac{l_n}{30} = \frac{(20 \text{ ft})\left(12 \frac{\text{in}}{\text{ft}}\right)}{30} = \boxed{8.0 \text{ in}}$$

The answer is (C).

(d) $$I_s = \frac{bh^3}{12} = \frac{(18 \text{ ft})\left(12 \frac{\text{in}}{\text{ft}}\right)(8 \text{ in})^3}{12}$$
$$= 9216 \text{ in}^4$$

Use Eq. 51.5.

$$\beta_t = \frac{E_{cb} C}{2 E_{cs} I_s} = \frac{13{,}584 \text{ in}^4}{(2)(9216 \text{ in}^4)}$$
$$= \boxed{0.74 \quad (0.7)}$$

The answer is (B).

(e) The slab weight is

$$\frac{(8 \text{ in})\left(0.15 \frac{\text{kip}}{\text{ft}^3}\right)}{12 \frac{\text{in}}{\text{ft}}} = 0.1 \text{ kip/ft}^2$$

The factored uniform load for the unit area is

$$w_u = 1.2D + 1.6L$$
$$= (1.2)\left(0.1 \frac{\text{kip}}{\text{ft}^2} + 0.02 \frac{\text{kip}}{\text{ft}^2}\right) + (1.6)\left(0.05 \frac{\text{kip}}{\text{ft}^2}\right)$$
$$= 0.224 \text{ kip/ft}^2$$

The factors needed for the computation of M_o are

$$l_n = 22 \text{ ft} - 2 \text{ ft} = 20 \text{ ft}$$
$$l_2 = 18 \text{ ft}$$

The positive moment is

$$M_o = \frac{w_u l_2 l_n^2}{8} = \frac{\left(0.224 \frac{\text{kip}}{\text{ft}}\right)(18 \text{ ft})(20 \text{ ft})^2}{8}$$
$$= 201.6 \text{ ft-kips}$$

From Table 51.3,

$$0.35 M_o = (0.35)(201.6 \text{ ft-kips})$$
$$= 70.56 \text{ ft-kips}$$

From Table 51.4(c),

$$M_{\text{column strip}} = 0.60(\text{positive moment})$$
$$= (0.60)(70.56 \text{ ft-kips})$$
$$= \boxed{42.34 \text{ ft-kips} \quad (40 \text{ ft-kips})}$$

The answer is (A).

(f) This is an edge, so

$$l_2 = \frac{18 \text{ ft}}{2} = 9 \text{ ft}$$

From Table 51.4, since 60% of the moment is required, as in part (e), and since the loading is halved, the answer is 50% of the result in part (e).

$$(0.50)(42.34 \text{ ft-kips}) = \boxed{21.17 \text{ ft-kips} \quad (21 \text{ ft-kips})}$$

The answer is (A).

Structural

(g) From part (e), the total positive moment is 70.56 ft-kips. Since the column strip takes 60%, the middle strip takes 40%. The middle strip's positive moment is

$$(0.4)(70.56 \text{ ft-kips}) = \boxed{28.22 \text{ ft-kips} \quad (28 \text{ ft-kips})}$$

The answer is (A).

(h) From Table 51.5,

$$t_{\text{slab}} \geq \frac{l_n}{36} = \frac{(20 \text{ ft})\left(12 \dfrac{\text{in}}{\text{ft}}\right)}{36} = \boxed{6.67 \text{ in} \quad (6.7 \text{ in})}$$

The answer is (B).

(i) Per ACI 318 Sec. 13.2.5, at the drop panel location, the minimum thickness is

$$t_{\text{min}} \geq 1.25 t_{\text{slab}} = (1.25)(6.67 \text{ in}) = \boxed{8.34 \text{ in} \quad (8.5 \text{ in})}$$

The answer is (C).

(j) The minimum dimension is

$$(2)\left(\frac{18 \text{ ft}}{6}\right) = \boxed{6 \text{ ft}}$$

The answer is (C).

Structural

52 Reinforced Concrete: Short Columns

PRACTICE PROBLEMS

1. (*Time limit: one hour*) A short column in a building is subjected to a factored axial load of 600 kips. Moments are negligible. Material properties are $f'_c = 4000$ lbf/in², and $f_y = 60,000$ lbf/in². Design the column as a square tied column using a reinforcement ratio of 4%.

2. (*Time limit: one hour*) A short column is subjected to an axial load of $P_u = 650$ kips and a moment $M_u = 250$ ft-kips. Material properties are $f'_c = 4000$ lbf/in², and $f_y = 60,000$ lbf/in². Design a tied square column with approximately 4% steel. Use steel distributed in all four faces.

3. (*Time limit: one hour*) A short column is subjected to a factored axial load of $P_u = 850$ kips. Material properties are $f'_c = 4000$ lbf/in² and $f_y = 60,000$ lbf/in².

(a) Assume moments are negligible. If the column is designed as a tied column, the minimum gross cross-sectional area is most nearly

(A) 170 in²

(B) 190 in²

(C) 210 in²

(D) 220 in²

(b) Assume moments are negligible. If the column is designed as a spiral column, the minimum gross cross-sectional area is most nearly

(A) 150 in²

(B) 170 in²

(C) 180 in²

(D) 190 in²

(c) If the column is an 18 in × 18 in square tied column, the required area of steel is most nearly

(A) 6.0 in²

(B) 7.5 in²

(C) 8.5 in²

(D) 9.5 in²

(d) If the column is an 18 in × 18 in square tied column, the maximum moment that can act without affecting the design (for pure axial loading) is most nearly

(A) 50 ft-kips

(B) 130 ft-kips

(C) 200 ft-kips

(D) 250 ft-kips

(e) If the column is an 18 in × 18 in square tied column and no. 9 bars are used for the longitudinal reinforcement, the maximum spacing of no. 3 ties is most nearly

(A) 14 in

(B) 16 in

(C) 18 in

(D) 20 in

(f) If the column is a spiral circular column with an outside diameter of 20 in, the required area of steel is most nearly

(A) 4.4 in²

(B) 4.7 in²

(C) 5.0 in²

(D) 5.2 in²

(g) If the column is a circular spiral column with an outside diameter of 20 in, the maximum moment that can act without affecting the design (for pure axial loading) is most nearly

(A) 50 ft-kips

(B) 60 ft-kips

(C) 70 ft-kips

(D) 140 ft-kips

(h) If the column is a spiral circular column with an outside diameter of 20 in and the diameter of the spiral wire is $3/8$ in, the maximum spiral pitch is most nearly

(A) 1.8 in

(B) 2.3 in

(C) 2.5 in

(D) 3.0 in

(i) When minimum reinforcement is used, the gross cross-sectional area for a tied column is most nearly

(A) 330 in^2

(B) 350 in^2

(C) 380 in^2

(D) 410 in^2

(j) When minimum reinforcement is used, the gross cross-sectional area for a spiral column is most nearly

(A) 310 in^2

(B) 340 in^2

(C) 360 in^2

(D) 410 in^2

4. (*Time limit: one hour*) A 20 in × 20 in tied column is built with materials that have the properties of $f'_c = 4000 \text{ lbf/in}^2$ and $f_y = 60{,}000 \text{ lbf/in}^2$.

(a) The maximum axial load that can be designed for is most nearly

(A) 1600 kips

(B) 1800 kips

(C) 2000 kips

(D) 2100 kips

(b) Assume moments are negligible. For $P_u = 1000$ kips, the required steel area is most nearly

(A) 5.0 in^2

(B) 6.5 in^2

(C) 7.5 in^2

(D) 8.5 in^2

(c) For $P_u = 800$ kips, the largest moment that does not affect the design is most nearly

(A) 65 ft-kips

(B) 130 ft-kips

(C) 200 ft-kips

(D) 260 ft-kips

(d) For $P_u = 800$ kips and $M_u = 400$ ft-kips, the required steel area is most nearly

(A) 10 in^2

(B) 14 in^2

(C) 18 in^2

(D) 20 in^2

(e) When the steel ratio equals 4%, the maximum moment that can be resisted when $P_u = 800$ kips is most nearly

(A) 300 ft-kips

(B) 380 ft-kips

(C) 480 ft-kips

(D) 500 ft-kips

(f) When the steel ratio is 4%, the nonzero axial load at which the moment capacity is largest is most nearly

(A) 300 kips

(B) 400 kips

(C) 500 kips

(D) 600 kips

(g) Assuming the column is unbraced and the effective length factor is 1.5, the maximum clear height for which the column is short is most nearly

(A) 7.5 ft

(B) 8.0 ft

(C) 9.0 ft

(D) 10 ft

(h) If the column is reinforced with no. 11 bars, the appropriate transverse reinforcement is most nearly

    (A) no. 3 ties at 20 in

    (B) no. 4 ties at 20 in

    (C) no. 3 ties at 18 in

    (D) no. 4 ties at 24 in

(i) If the column is reinforced with eight no. 8 bars placed in four bundles of two bars at each corner, the appropriate transverse reinforcement is most nearly

    (A) no. 3 ties at 20 in

    (B) no. 4 ties at 16 in

    (C) no. 3 ties at 18 in

    (D) no. 4 ties at 24 in

(j) If the steel ratio is 3%, the maximum eccentricity for a load $P_u = 800$ kips is most nearly

    (A) 3.0 in

    (B) 4.6 in

    (C) 5.5 in

    (D) 6.5 in

5. Design a short spiral column to carry an axial dead load of 175 kips and an axial live load of 300 kips. Use $f'_c = 3000$ lbf/in^2 and $f_y = 40,000$ lbf/in^2.

6. Design a short square tied column to carry an axial dead load of 100 kips and an axial live load of 125 kips. Use $f'_c = 3500$ lbf/in^2 and $f_y = 40,000$ lbf/in^2. Use no. 5 bars or larger for the longitudinal reinforcement. Specify the ties.

SOLUTIONS

1. Use Eq. 52.11.

$$A_g = \frac{P_u}{\phi\beta\left(0.85f'_c(1 - \rho_g) + \rho_g f_y\right)}$$

$$= \frac{600 \text{ kips}}{(0.65)(0.80)\left((0.85)\left(4 \ \dfrac{\text{kips}}{\text{in}^2}\right)\right.}$$

$$\left. \times (1 - 0.04) + (0.04)\left(60 \ \dfrac{\text{kips}}{\text{in}^2}\right)\right)$$

$$= 203.7 \text{ in}^2$$

The cross-sectional dimension is

$$h = \sqrt{203.7 \text{ in}^2} = 14.27 \text{ in} \quad \boxed{(15 \text{ in})}$$

The steel area is

$$A_{\text{st}} = \rho_g A_g = (0.04)(203.8 \text{ in}^2)$$

$$= 8.15 \text{ in}^2 \quad \boxed{\begin{array}{l}\text{(Use eight no. 9 bars, which pro-}\\ \text{vide an area slightly smaller than}\\ \text{required.)}\end{array}}$$

The ties can be no. 3 since the bars are not larger than no. 10 and there are no bundles. The spacing of the ties is the smallest of (a) 16 diameters of the long bars, which is $(16)(^9/_8 \text{ in}) = 18$ in; (b) 48 diameters of the tie, which is $(48)(^3/_8 \text{ in}) = 18$ in; and (c) the minimum dimension of the cross section, $\boxed{15 \text{ in, which controls.}}$

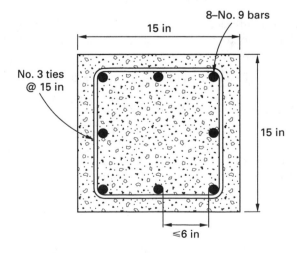

2. Start with a 16 in × 16 in section.

Use an interaction diagram. First, assuming $\phi = 0.65$, calculate

$$\frac{P_u}{\phi f'_c A_g} = \frac{650 \text{ kips}}{(0.65)\left(4 \dfrac{\text{kips}}{\text{in}^2}\right)(256 \text{ in}^2)}$$

$$= 0.98$$

$$\frac{M_u}{\phi f'_c A_g h} = \frac{(250 \text{ ft-kips})\left(12 \dfrac{\text{in}}{\text{ft}}\right)}{(0.65)\left(4 \dfrac{\text{kips}}{\text{in}^2}\right)(256 \text{ in}^2)(16 \text{ in})}$$

$$= 0.28$$

The value of γ can be estimated as

$$\gamma = \frac{16 \text{ in} - 5 \text{ in}}{16 \text{ in}} = 0.68 \quad [\text{use } 0.70]$$

Assume the steel will be distributed on all four faces. From the interaction diagram in App. 52.G (square, 4 ksi, 60 ksi, $\gamma = 0.70$), ρ_g is approximately 0.08. The column cross section must be increased. Try an 18 in × 18 in section.

$$\gamma = \frac{18 \text{ in} - 5 \text{ in}}{18 \text{ in}} = 0.72 \quad [\text{use } 0.70]$$

$$A_g = (18 \text{ in})(18 \text{ in}) = 324 \text{ in}^2$$

$$\frac{P_u}{\phi f'_c A_g} = \frac{650 \text{ kips}}{(0.65)\left(4 \dfrac{\text{kips}}{\text{in}^2}\right)(324 \text{ in}^2)}$$

$$= 0.771$$

$$\frac{M_u}{\phi f'_c A_g h} = \frac{(250 \text{ ft-kips})\left(12 \dfrac{\text{in}}{\text{ft}}\right)}{(0.65)\left(4 \dfrac{\text{kips}}{\text{in}^2}\right)(324 \text{ in}^2)(18 \text{ in})}$$

$$= 0.198$$

From the interaction diagram in App. 52.G, a steel ratio of approximately 4.3% is needed. This ratio is near the target of 4%.

The area of steel is calculated from Eq. 52.12.

$$A_{st} = \rho_g A_g = (0.043)(324 \text{ in}^2)$$

$$= 13.93 \text{ in}^2 \quad \boxed{[\text{Use twelve no. 10 bars.}]}$$

No. 10 bars have an area of 1.27 in² and a diameter of 1.27 in. With the chosen reinforcement,

$$\rho_g = \frac{n A_{\text{bar}}}{A_g} = \frac{(12)(1.27 \text{ in}^2)}{324 \text{ in}^2} = 0.047$$

From the interaction diagram in App. 52.G, the steel tensile strain, ϵ_t, corresponding to this column design is less than 0.002. Therefore, the section is compression controlled and $\phi = 0.65$, as assumed.

The ties can be no. 3 since the bars are not larger than no. 10 and there are no bundles. The spacing of the ties is the smallest of (a) 16 diameters of the long bars, which is $(16)(1.27 \text{ in}) = 20.32 \text{ in}$; (b) 48 diameters of the tie, which is $(48)\left(\frac{3}{8} \text{ in}\right) = 18 \text{ in}$; and (c) the minimum dimension of the cross section, which is 18 in. Tie spacing of $\boxed{18 \text{ in}}$ controls.

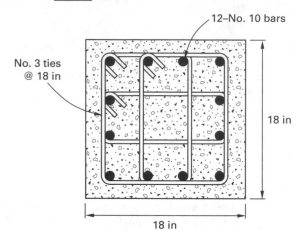

3. (a) $\phi = 0.65$ for tied column. Taking ρ_g as the maximum allowed, 0.08, and using Eq. 52.11,

$$A_g = \frac{P_u}{\phi\beta\left(0.85 f'_c(1 - \rho_g) + \rho_g f_y\right)}$$

$$= \frac{850 \text{ kips}}{(0.65)(0.80)\left((0.85)\left(4 \dfrac{\text{kips}}{\text{in}^2}\right) \right.}$$
$$\left. \times (1 - 0.08) + (0.08)\left(60 \dfrac{\text{kips}}{\text{in}^2}\right)\right)$$

$$= \boxed{206.2 \text{ in}^2 \quad (210 \text{ in}^2)}$$

The answer is (C).

(b) Taking ρ_g as the maximum allowed, 0.08, and using Eq. 52.11,

$$A_g = \frac{P_u}{\phi\beta\left(0.85 f'_c(1 - \rho_g) + \rho_g f_y\right)}$$

$$= \frac{850 \text{ kips}}{(0.75)(0.85)\left((0.85)\left(4 \dfrac{\text{kips}}{\text{in}^2}\right) \right.}$$
$$\left. \times (1 - 0.08) + (0.08)\left(60 \dfrac{\text{kips}}{\text{in}^2}\right)\right)$$

$$= \boxed{168.2 \text{ in}^2 \quad (170 \text{ in}^2)}$$

The answer is (B).

(c) $A_g = (18 \text{ in})^2 = 324 \text{ in}^2$

Using Eq. 52.11,

$$0.85f'_c(1 - \rho_g) + \rho_g f_y = \frac{P_u}{\phi \beta A_g}$$

$$(0.85)\left(4 \; \frac{\text{kips}}{\text{in}^2}\right)(1 - \rho_g)$$

$$+ \rho_g\left(60 \; \frac{\text{kips}}{\text{in}^2}\right) = \frac{850 \text{ kips}}{(0.65)(0.80)(324 \text{ in}^2)}$$

$$\rho_g = 0.0292$$

$$A_{\text{st}} = \rho_g A_g$$

$$= (0.0292)(324 \text{ in}^2)$$

$$= \boxed{9.46 \text{ in}^2 \quad (9.5 \text{ in}^2)}$$

The answer is (D).

(d) $$\left(\frac{e}{h}\right)_{\text{max}} = 0.1 \quad \text{[tied column]}$$

$$e = (0.1)(18 \text{ in}) = 1.8 \text{ in}$$

$$M = P_u e = \frac{(850 \text{ kips})(1.8 \text{ in})}{12 \; \dfrac{\text{in}}{\text{ft}}}$$

$$= \boxed{127.5 \text{ ft-kips} \quad (130 \text{ ft-kips})}$$

The answer is (B).

(e) Use the smallest of (a) 16 diameters of the long bars, which is $(16)(^9\!/_8 \text{ in}) = 18 \text{ in}$; (b) 48 diameters of the tie, which is $(48)(^3\!/_8 \text{ in}) = 18 \text{ in}$; and (c) the minimum dimension of the cross section, $\boxed{18 \text{ in, which controls.}}$

The answer is (C).

(f) From Eq. 52.11,

$$0.85f'_c(1 - \rho_g) + \rho_g f_y = \frac{P_u}{\phi \beta A_g}$$

$$(0.85)\left(4 \; \frac{\text{kips}}{\text{in}^2}\right)(1 - \rho_g)$$

$$+ \rho_g\left(60 \; \frac{\text{kips}}{\text{in}^2}\right) = \frac{850 \text{ kips}}{(0.75)(0.85)(314 \text{ in}^2)}$$

$$\rho_g = 0.0149$$

$$A_{\text{st}} = \rho_g A_g$$

$$= (0.0149)(314 \text{ in}^2)$$

$$= \boxed{4.68 \text{ in}^2 \quad (4.7 \text{ in}^2)}$$

The answer is (B).

(g) $$\left(\frac{e}{h}\right)_{\text{max}} = 0.05 \quad \text{[spiral column]}$$

$$e = (0.05)(20 \text{ in}) = 1.0 \text{ in}$$

$$M = P_u e = \frac{(850 \text{ kips})(1.0 \text{ in})}{12 \; \dfrac{\text{in}}{\text{ft}}}$$

$$= \boxed{70.8 \text{ ft-kips} \quad (70 \text{ ft-kips})}$$

The answer is (C).

(h) $$D_c = 20 \text{ in} - 3 \text{ in} = 17 \text{ in}$$

$$\frac{A_g}{A_c} = \left(\frac{D_g}{D_c}\right)^2 = \left(\frac{20 \text{ in}}{17 \text{ in}}\right)^2$$

$$= 1.384$$

Use Eq. 52.4.

$$\rho_s = 0.45\left(\frac{A_g}{A_c} - 1\right)\left(\frac{f'_c}{f_{yt}}\right)$$

$$= (0.45)(1.384 - 1)\left(\frac{4 \; \dfrac{\text{kips}}{\text{in}^2}}{60 \; \dfrac{\text{kips}}{\text{in}^2}}\right)$$

$$= 0.0115$$

Use Eq. 52.5.

$$s = \frac{4A_{\text{sp}}}{\rho_s D_c} = \frac{(4)(0.11 \text{ in}^2)}{(0.0115)(17 \text{ in})}$$

$$= \boxed{2.25 \text{ in} \quad (2.3 \text{ in})}$$

The answer is (B).

(i) Taking ρ_g as the minimum allowed, 0.01, and using Eq. 52.11,

$$A_g = \frac{P_u}{\phi \beta \left(0.85f'_c(1 - \rho_g) + \rho_g f_y\right)}$$

$$= \frac{850 \text{ kips}}{(0.65)(0.80)\left((0.85)\left(4 \; \dfrac{\text{kips}}{\text{in}^2}\right)\right.}$$

$$\left. \times (1 - 0.01) + (0.01)\left(60 \; \dfrac{\text{kips}}{\text{in}^2}\right)\right)$$

$$= \boxed{412.2 \text{ in}^2 \quad (410 \text{ in}^2)}$$

The answer is (D).

(j) Taking ρ_g as the minimum allowed, 0.01, and using Eq. 52.11,

$$A_g = \frac{P_u}{\phi\beta(0.85f'_c(1-\rho_g)+\rho_g f_y)}$$

$$= \frac{850 \text{ kips}}{(0.75)(0.85)\left((0.85)\left(4\ \dfrac{\text{kips}}{\text{in}^2}\right)\right.}$$
$$\left.\times (1-0.01)+(0.01)\left(60\ \dfrac{\text{kips}}{\text{in}^2}\right)\right)$$

$$= \boxed{336.2 \text{ in}^2 \quad (340 \text{ in}^2)}$$

The answer is (B).

4. (a) $P_u = \phi\beta A_g\left(0.85f'_c(1-\rho_g)+\rho_g f_y\right)$

$$= (0.65)(0.80)(400 \text{ in}^2)\left((0.85)\left(4\ \dfrac{\text{kips}}{\text{in}^2}\right)\right.$$
$$\left.\times (1-0.08)+(0.08)\left(60\ \dfrac{\text{kips}}{\text{in}^2}\right)\right)$$

$$= \boxed{1649.0 \text{ kips} \quad (1600 \text{ kips})}$$

The answer is (A).

(b) Use Eq. 52.11.

$$0.85f'_c(1-\rho_g)+\rho_g f_y = \frac{P_u}{\phi\beta A_g}$$

$$(0.85)\left(4\ \frac{\text{kips}}{\text{in}^2}\right)(1-\rho_g)$$
$$+ \rho_g\left(60\ \frac{\text{kips}}{\text{in}^2}\right) = \frac{1000 \text{ kips}}{(0.65)(0.80)(400 \text{ in}^2)}$$

$$\rho_g = 0.0248$$

$$A_{\text{st}} = (0.0248)(400 \text{ in}^2)$$

$$= \boxed{9.92 \text{ in}^2 \quad (8.5 \text{ in}^2)}$$

The answer is (D).

(c) $\quad \left(\dfrac{e}{h}\right)_{\text{max}} = 0.1 \quad$ [tied column]

$$e = (0.1)(20 \text{ in}) = 2.0 \text{ in}$$

$$M = P_u e = \frac{(800 \text{ kips})(2.0 \text{ in})}{12\ \dfrac{\text{in}}{\text{ft}}}$$

$$= \boxed{133.3 \text{ ft-kips} \quad (130 \text{ ft-kips})}$$

The answer is (B).

(d) Assume a compression-controlled section, $\phi = 0.65$. The value of γ can be estimated as

$$\gamma = \frac{20 \text{ in} - 5 \text{ in}}{20 \text{ in}} = 0.75$$

$$\frac{P_u}{A_g} = \frac{800 \text{ kips}}{400 \text{ in}^2} = 2.0 \text{ kips/in}^2$$

$$\frac{M_u}{A_g h} = \frac{(400 \text{ ft-kips})\left(12\ \dfrac{\text{in}}{\text{ft}}\right)}{(400 \text{ in}^2)(20 \text{ in})} = 0.6 \text{ kip/in}^2$$

From the interaction diagram in App. 52.H, the steel ratio is approximately 0.045. The area of steel is calculated from Eq. 52.12.

$$A_{\text{st}} = \rho_g A_g = (0.045)(400 \text{ in}^2) = \boxed{18 \text{ in}^2 \quad (20 \text{ in}^2)}$$

With this high amount of reinforcement, the section is compression controlled.

The answer is (D).

(e) From the interaction diagram in App. 52.H, when $\rho_g = 0.04$ and $P_u/A_g = 2 \text{ kips/in}^2$,

$$\frac{M_u}{A_g h} = 0.57 \text{ kips/in}^2$$

$$M_u = (0.192)\phi f'_c A_g h = \frac{\left(0.57\ \dfrac{\text{kips}}{\text{in}^2}\right)(400 \text{ in}^2)(20 \text{ in})}{12\ \dfrac{\text{in}}{\text{ft}}}$$

$$= \boxed{380 \text{ ft-kips}}$$

From the interaction diagram, the section is clearly compression controlled.

The answer is (B).

(f) Maximum moment occurs at the balanced point. At this point where $\rho_g = 0.04$, from App. 52.G,

$$\frac{P_u}{A_g} = 1.0 \text{ kip/in}^2$$

$$P_u = \left(1\ \frac{\text{kip}}{\text{in}^2}\right)(400 \text{ in}^2)$$

$$= \boxed{400 \text{ kips}}$$

The answer is (B).

(g) Use Eq. 52.2. For a short column,

$$\left(\frac{k_u l_u}{r}\right)_{max} = 22$$

$$r = 0.3h = (0.3)(20 \text{ in}) = 6 \text{ in} \quad \begin{bmatrix} \text{square tied} \\ \text{column} \end{bmatrix}$$

$$l_u = \frac{22r}{k_u} = \frac{(22)(6 \text{ in})}{(1.5)\left(12 \dfrac{\text{in}}{\text{ft}}\right)}$$

$$= \boxed{7.3 \text{ ft} \quad (7.5 \text{ ft})}$$

The answer is (A).

(h) Since longitudinal bars are larger than no. 10, no. 4 ties are required.

Use the smallest of (a) 16 diameters of the long bars, which is $(16)(1.41 \text{ in}) = 22.56 \text{ in}$; (b) 48 diameters of the tie, which is $(48)(0.5 \text{ in}) = 24 \text{ in}$; and (c) the minimum dimension of the cross section, 20 in, which controls.

The answer is (B).

(i) Since the longitudinal bars are bundled, no. 4 ties are required.

Use the smallest of (a) 16 diameters of the long bars, which is $(16)(1.0 \text{ in}) = 16 \text{ in}$; (b) 48 diameters of the tie, which is $(48)(0.5 \text{ in}) = 24 \text{ in}$; and (c) the minimum dimension of the cross section, 20 in. Therefore, the appropriate transverse reinforcement is no. 4 ties at 16 in.

The answer is (B).

(j) Assume the section is compression controlled.

$$\frac{P_u}{A_g} = 2.0 \text{ kips/in}^2$$

At $\rho_g = 0.03$ and $P_u/A_g = 2 \text{ kips/in}^2$, from App. 52.H,

$$\frac{M_u}{A_g h} = 0.46 \text{ kip/in}^2$$

$$M_u = \left(0.46 \frac{\text{kip}}{\text{in}^2}\right)(400 \text{ in}^2)(20 \text{ in})$$

$$= 3680 \text{ in-kips}$$

$$e = \frac{M_u}{P_u} = \frac{3680 \text{ in-kips}}{800 \text{ kips}} = \boxed{4.6 \text{ in}}$$

The answer is (B).

5. $\quad P_u = (1.2)(175{,}000 \text{ lbf}) + (1.6)(300{,}000 \text{ lbf})$

$$= 690{,}000 \text{ lbf}$$

Assume $\rho_g = 0.02$.

$$690{,}000 \text{ lbf} = A_g(0.85)(0.75)$$
$$\times \left((0.85)(1 - 0.02)\left(3000 \frac{\text{lbf}}{\text{in}^2}\right) \right.$$
$$\left. + \left(40{,}000 \frac{\text{lbf}}{\text{in}^2}\right)(0.02)\right)$$
$$A_g = 328.1 \text{ in}^2$$

$$A_g = \frac{\pi D_g^2}{4}$$

$$D_g = \sqrt{\frac{(4)(328.1 \text{ in}^2)}{\pi}} = 20.44 \text{ in}$$

Round D_g to the next half-inch, so that 20.44 in becomes 20.5 in.

Use $1\frac{1}{2}$ in cover. Assuming no. 8 bars and no. 3 spiral wire, the steel diameter is

$$20.5 \text{ in} - (2)(1.5 \text{ in}) - (2)(0.375 \text{ in}) - 1 \text{ in} = 15.75 \text{ in}$$

The core diameter is

$$20.5 \text{ in} - (2)(1.5 \text{ in}) = 17.5 \text{ in}$$

The required steel area is given by Eq. 52.12.

$$A_{st} = \rho_g A_g = (0.02)(328.1 \text{ in}^2) = 6.56 \text{ in}^2$$

Some possibilities are

> 15 no. 6 bars
>
> 11 no. 7 bars
>
> 9 no. 8 bars
>
> 7 no. 9 bars
>
> 6 no. 10 bars

Try 11 no. 7 bars. The steel circumference is

$$\pi(15.75 \text{ in}) = 49.48 \text{ in}$$

The clear spacing between bars is

$$\frac{49.48 \text{ in} - (11)(0.875 \text{ in})}{11 \text{ spaces}} = 3.62 \text{ in/space}$$

Since $3.62 \text{ in} > (1.5)(0.875 \text{ in}) = 1.31 \text{ in}$, the clear spacing between bars is acceptable.

Structural

Use Eq. 52.4.

$$\rho_s = (0.45)\left(\frac{A_g}{A_c} - 1\right)\left(\frac{f_c'}{f_{yt}}\right)$$

$$= (0.45)\left(\left(\frac{20.5 \text{ in}}{17.5 \text{ in}}\right)^2 - 1\right)\left(\frac{3000 \frac{\text{lbf}}{\text{in}^2}}{40,000 \frac{\text{lbf}}{\text{in}^2}}\right)$$

$$= 0.0126$$

With a spiral pitch of 2 in,

$$\rho_s = \frac{(4)(0.11 \text{ in}^2)}{(2 \text{ in})(17.5 \text{ in})} = 0.0126 \geq 0.0126 \quad \text{[acceptable]}$$

The clear spacing between spirals is

$$2 \text{ in} - 0.375 \text{ in} = 1.625 \text{ in}$$

Since $1.0 \text{ in} < 1.625 \text{ in} < 3 \text{ in}$, this is acceptable.

Use eleven no. 7 bars.

6. $P_u = (1.2)(100,000 \text{ lbf}) + (1.6)(125,000 \text{ lbf})$

$$= 320,000 \text{ lbf}$$

With $\phi = 0.65$ and $\rho_g = 0.02$,

$$320,000 \text{ lbf} = (0.80)(0.65)A_g$$
$$\times \left((0.85)\left(3500 \frac{\text{lbf}}{\text{in}^2}\right)(1 - 0.02)\right.$$
$$\left. + \left(40,000 \frac{\text{lbf}}{\text{in}^2}\right)(0.02)\right)$$
$$A_g = 165.6 \text{ in}^2$$

The column width is

$$\sqrt{165.6 \text{ in}^2} \approx \boxed{13 \text{ in square}}$$

The actual gross area is $A_g = (13 \text{ in})^2 = 169 \text{ in}^2$.

Assume no. 6 bars and no. 3 ties. With $1\frac{1}{2}$ in cover, the steel core size is

$$13 \text{ in} - (2)(1.5 \text{ in}) - (2)(0.375 \text{ in})$$
$$-(2)\left(\frac{0.75 \text{ in}}{2}\right) = 8.5 \text{ in}$$

The required steel area is

$$(0.02)(169 \text{ in}^2) = 3.38 \text{ in}^2$$

To distribute bars evenly around the column, the number of bars must be in multiples of four. Try 12 bars.

$$A_{\text{bar}} = \frac{3.38 \text{ in}^2}{12} = 0.28 \text{ in}^2$$

0.28 in^2 is too small since no. 5 bars or larger need to be used. Try eight bars.

$$A_{\text{bar}} = \frac{3.38 \text{ in}^2}{8} = 0.42 \text{ in}^2$$

Use eight no. 6 (0.44 in² each) bars.

The clear spacing between long bars is

$$\frac{(4)(8.5 \text{ in}) - (8)(0.75 \text{ in})}{8} = 3.5 \text{ in} \quad \text{[acceptable]}$$

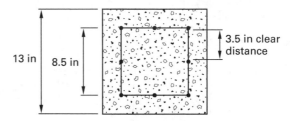

Use no. 3 ties. The tie spacing should be less than the minimum of

$$\left.\begin{cases} (48)\left(\frac{3}{8} \text{ in}\right) = 18 \text{ in} \\ (16)\left(\frac{6}{8} \text{ in}\right) = 12 \text{ in} \\ \text{gross width} = 13 \text{ in} \end{cases}\right\} = \boxed{12 \text{ in}}$$

With only eight bars, every alternate bar is supported. The spacing between bars is less than 6 in, so no additional cross ties are needed.

53 Reinforced Concrete: Long Columns

PRACTICE PROBLEMS

1. The frame shown is one of several parallel frames in an unbraced building. The frames are spaced on 20 ft centers and are constructed using 4000 lbf/in² concrete. All columns are 18 in × 18 in. Assume a pin connection between columns and footings. All slabs are two way. Compute the effective length factor, k_u, for the columns on the second story.

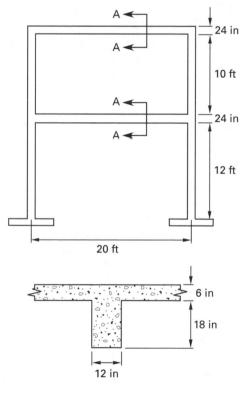

section A–A

(A) 1.1

(B) 1.2

(C) 1.5

(D) 1.8

2. The frame shown is part of a building that can be considered braced by the presence of stiff concrete walls surrounding the elevator shafts. The soil under the footings is soft, and a relative stiffness of $\Psi = 5$ is considered appropriate at the base. Other values of Ψ have been computed and are given in the illustration. Structural loadings for column B between points 0 and 1 are 400 kips dead load and 100 kips live load. The concrete modulus of elasticity, E_c, is 4000 kips/in². All columns are 16 in × 16 in square.

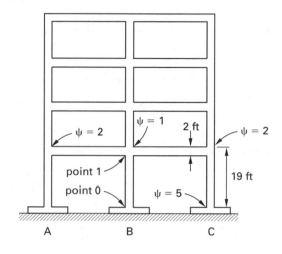

(a) The buckling load is most nearly

(A) 1100 kips

(B) 1600 kips

(C) 2200 kips

(D) 2800 kips

(b) The moment that should be used to design column B in the first story is most nearly

(A) 120 ft-kips

(B) 160 ft-kips

(C) 320 ft-kips

(D) 460 ft-kips

3. (*Time limit: one hour*) A six-floor (six-story) reinforced concrete building is supported on columns placed on a grid spaced 30 ft in the E-W direction and 20 ft in the N-S direction. The story height, measured from the top of one slab to the top of the slab above, is 14 ft. At the first level, the distance from the top of the footings to the top of the first story slab is 18 ft. The columns are 20 in × 20 in. Columns are connected in both directions by beams with an overall depth of 2 ft and a web width of 1 ft. The slab thickness is 6 in. For convenience, use $l_c = 18$ ft and $l_u = 16$ ft for the first-floor columns. All slabs are two way. 4000 lbf/in^2 concrete ($E = 4 \times 10^6$ lbf/in^2) is used throughout.

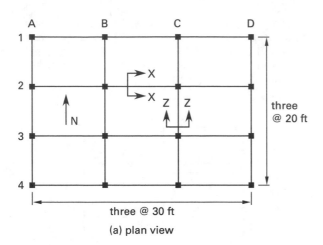

(a) plan view

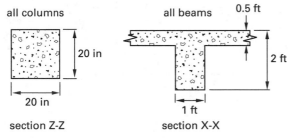

(b) column construction

(c) beam construction

(a) For the computation of the relative stiffness parameter, Ψ, the moment of inertia of the beams is most nearly

(A) 4800 in^4

(B) 6100 in^4

(C) 8600 in^4

(D) 14,000 in^4

In parts (b) through (h), assume the building is unbraced.

(b) The effective length factor for bending about the W-E axis in column B2 at the third level is most nearly

(A) 0.85

(B) 1.0

(C) 1.5

(D) 2.0

(c) The effective length factor for bending about the W-E axis in column B1 at the third level is most nearly

(A) 0.9

(B) 1.0

(C) 1.5

(D) 1.9

(d) Assuming the footing provides absolute ideal fixity at the base and using $\beta_{\text{dns}} = 0$, the effective length for bending about the W-E axis in column B2 at the first level is most nearly

(A) 14 ft

(B) 16 ft

(C) 19 ft

(D) 21 ft

(e) The buckling load about the W-E axis for column B2 in the first level is most nearly

(A) 2800 kips

(B) 4000 kips

(C) 4900 kips

(D) 5600 kips

(f) Assume the dead load is 0.1 kip/ft^2 and the live load is 0.06 kip/ft^2 on floors and the roof. Consider the load combination $U = 1.2D + 1.0L + 1.6W$. Neglecting reductions in the live load, the ΣP_u term used for the calculation of δ_s in the first level is most nearly

(A) 1100 kips

(B) 4500 kips

(C) 4900 kips

(D) 5800 kips

(g) Assume the unfactored wind load is 30 lbf/ft^2 and the load combination is $U = 1.2D + 1.0L + 1.6W$. The first-story column drift is 0.3 in. The foundation carries half of the first story's wind load. The value of Q at this level is most nearly

(A) 0.024

(B) 0.032

(C) 0.15

(D) 0.20

(h) The maximum axial load in the columns of the first level for which the maximum moment can be safely assumed to occur at one of the two ends is most nearly

(A) 990 kips

(B) 1600 kips

(C) 1900 kips

(D) 2400 kips

For parts (i) and (j), assume walls are added so that the building is braced.

(i) The effective length factor for bending about W-E in column B2 at the first level is most nearly

(A) 0.64

(B) 0.87

(C) 1.0

(D) 1.2

(j) Assume the unfactored axial forces from dead and live loads on a particular column are 200 kips and 50 kips, respectively. The load combination is $U = 1.2D + 1.6L$. The value of β_{dns} is most nearly

(A) 0.50

(B) 0.75

(C) 0.80

(D) 1.0

4. (*Time limit: one hour*) The single-story frame shown is built with concrete having a modulus of elasticity, E_c, of 3605 kips/in². The actions in the columns due to dead and live load are as shown. Column ends A, F, and E are fixed, but a realistic finite flexibility is to be considered.

Assume the frame is part of a building that is braced and the load combination is $U = 1.2D + 1.6L$.

(a) The moment, $M_{2,\text{ns}}$, in column AB is most nearly

(A) 120 ft-kips

(B) 140 ft-kips

(C) 160 ft-kips

(D) 220 ft-kips

(b) The effective length for column AB is most nearly

(A) 9.0 ft

(B) 11 ft

(C) 12 ft

(D) 13 ft

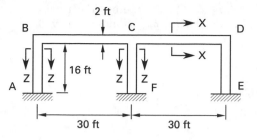

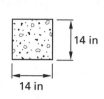

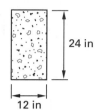

section Z-Z   section X-X

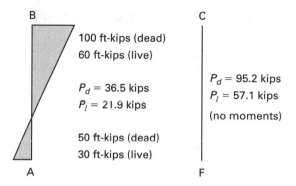

(c) The effective length of column CF is most nearly

(A) 10 ft

(B) 12 ft

(C) 13 ft

(D) 15 ft

(d) The factor β_{dns} that should be used to compute the buckling load for column AB is most nearly

(A) 0.00

(B) 0.45

(C) 0.56

(D) 0.65

(e) The buckling load for column AB is most nearly

(A) 1200 kips

(B) 1400 kips

(C) 1500 kips

(D) 1800 kips

Structural

(f) The parameter C_m for column AB (used to compute the amplification factor δ_{ns}) is most nearly

(A) 0.3

(B) 0.4

(C) 0.5

(D) 0.6

(g) The parameter C_m for column CF (used to compute the amplification factor δ_{ns}) is most nearly

(A) 0.4

(B) 0.5

(C) 0.6

(D) 1.0

(h) The buckling load for column CF is most nearly

(A) 1000 kips

(B) 1300 kips

(C) 1500 kips

(D) 2200 kips

(i) The amplification factor, δ_{ns}, in column CF is most nearly

(A) 0.9

(B) 1.0

(C) 1.2

(D) 1.5

(j) The amplification factor, δ_{ns}, in column AB is most nearly

(A) 0.44

(B) 1.0

(C) 1.1

(D) 1.3

SOLUTIONS

1. Since 18 in < (4)(6 in) [ACI 318 Sec. 8.12], the effective beam section is as follows.

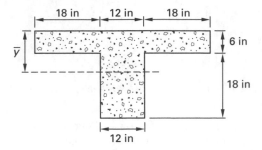

Compute I_b.

A (in^2)	y (in)	Ay (in^3)	$A(y-\overline{y})^2$ (in^4)	I_o (in^4)
288	3	864	7617.3	864
216	15	3240	10,156.4	5832
totals 504		4104	17,773.7	6696

$$\overline{y} = \frac{4104 \text{ in}^3}{504 \text{ in}^2} = 8.14 \text{ in}$$

$$I_b = 17{,}773.7 \text{ in}^4 + 6696 \text{ in}^4$$

$$= 24{,}470 \text{ in}^4$$

The column moment of inertia is

$$I_c = \frac{bh^3}{12} = \frac{(18 \text{ in})(18 \text{ in})^3}{12} = 8748 \text{ in}^4$$

Adjust for cracking and creep throughout the frame.

$$I_{be} = 0.35 I_b = (0.35)(24{,}470 \text{ in}^4) = 8564.5 \text{ in}^4$$

$$I_{ce} = 0.70 I_c = (0.70)(8748 \text{ in}^4) = 6123.6 \text{ in}^4$$

$$l_{c,1} = 12 \text{ ft} + \frac{24 \text{ in}}{(2)\left(12 \; \frac{\text{in}}{\text{ft}}\right)} = 13 \text{ ft}$$

$$l_{c,2} = 10 \text{ ft} + (2)\left(\frac{24 \text{ in}}{(2)\left(12 \; \frac{\text{in}}{\text{ft}}\right)}\right) = 12 \text{ ft}$$

Use Eq. 53.3.

$$\Psi = \frac{\displaystyle\sum_{\text{columns}} \frac{EI}{l_c}}{\displaystyle\sum_{\text{beams}} \frac{EI}{l}}$$

$$\Psi_{\text{top}} = \frac{\dfrac{6123.6 \text{ in}^4}{12 \text{ ft}}}{\dfrac{8564.5 \text{ in}^4}{20 \text{ ft}}} = 1.19$$

$$\Psi_{\text{bottom}} = \frac{\dfrac{6123.6 \text{ in}^4}{13 \text{ ft}} + \dfrac{6123.6 \text{ in}^4}{12 \text{ ft}}}{\dfrac{8564.5 \text{ in}^4}{20 \text{ ft}}} = 2.29$$

From Fig. 53.3(b),

$$k_u = \boxed{1.5}$$

The answer is (C).

2. (a) From Fig. 53.3, for $\Psi_A = 1$ and $\Psi_B = 5$, $k_b \approx 0.85$.

$$l_u = 19 \text{ ft} - 2 \text{ ft} = 17 \text{ ft} \quad \text{[clear distance]}$$

$$P_u = (1.2)(400 \text{ kips}) + (1.6)(100 \text{ kips}) = 640 \text{ kips}$$

$$P_u = 1.4D = (1.4)(400 \text{ kips}) = 560 \text{ kips} \quad \begin{bmatrix} \text{does not} \\ \text{control} \end{bmatrix}$$

There is no lateral loading. So,

$$M_1 = M_2 = 0$$

Using Eq. 53.11,

$$C_m = 0.6 + 0.4\left(\frac{M_1}{M_2}\right)$$

For $M_1 = M_2 = 0$, $C_m = 1$. Use Eq. 53.12.

$$M_{2,\text{min}} = P_u(0.6 + 0.03h)$$
$$= \frac{(640 \text{ kips})(0.6 \text{ in} + (0.03)(16 \text{ in}))}{12 \dfrac{\text{in}}{\text{ft}}}$$
$$= 57.6 \text{ ft-kips}$$

$$\beta_{\text{dns}} = \frac{(1.2)(400 \text{ kips})}{640 \text{ kips}} = 0.75$$

The moment of inertia of the column is

$$I_g = \frac{bh^3}{12} = \frac{(16 \text{ in})(16 \text{ in})^3}{12} = 5461.3 \text{ in}^4$$

Use Eq. 53.7.

$$EI = \frac{0.4E_cI_g}{1 + \beta_{\text{dns}}} = \frac{(0.4)\left(4000 \dfrac{\text{kips}}{\text{in}^2}\right)(5461.3 \text{ in}^4)}{1 + 0.75}$$
$$= 4.99 \times 10^6 \text{ kip-in}^2$$

The buckling load is given by Eq. 53.8.

$$P_c = \frac{\pi^2 EI}{(kl_u)^2} = \frac{\pi^2(4.99 \times 10^6 \text{ kip-in}^2)}{\left((0.85)(17 \text{ ft})\left(12 \dfrac{\text{in}}{\text{ft}}\right)\right)^2}$$
$$= \boxed{1638 \text{ kips} \quad (1600 \text{ kips})}$$

The answer is (B).

(b) The amplification factor is given by Eq. 53.10. Since $M_1 = M_2 = 0$, use $M_1/M_2 = 1$ and $C_m = 1$.

$$\delta_{\text{ns}} = \frac{C_m}{1 - \dfrac{P_u}{0.75P_c}} = \frac{1}{1 - \dfrac{640 \text{ kips}}{(0.75)(1638 \text{ kips})}} = 2.09$$

The design loads for column B are

$$P_u = 640 \text{ kips}$$
$$M_c = \delta_{\text{ns}}M_2 = (2.09)(57.6 \text{ ft-kips})$$
$$= \boxed{120.4 \text{ ft-kips} \quad (120 \text{ ft-kips})}$$

(The value of 2.09 for δ_{ns} is large and, as such, is indicative of a flexible column. An increase in the section size to reduce δ_{ns} would be necessary.)

The answer is (A).

3. (a) Per ACI 318 Sec. 8.12, the effective beam section is

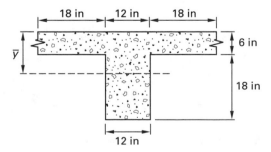

Find the location of the centroidal axis. Measure $\overline{y}$ from the upper surface of the slab.

Structural

$$A_1 = (6 \text{ in})(18 \text{ in} + 12 \text{ in} + 18 \text{ in}) = 288 \text{ in}^2$$

$$y_1 = \frac{6 \text{ in}}{2} = 3 \text{ in}$$

$$A_2 = (18 \text{ in})(12 \text{ in}) = 216 \text{ in}^2$$

$$y_2 = 6 \text{ in} + \frac{18 \text{ in}}{2} = 15 \text{ in}$$

$$\overline{y} = \frac{\sum A_i y_i}{\sum A_i} = \frac{(288 \text{ in}^2)(3 \text{ in}) + (216 \text{ in}^2)(15 \text{ in})}{288 \text{ in}^2 + 216 \text{ in}^2}$$
$$= 8.143 \text{ in}$$

$$I_{c,1} = \frac{bh^3}{12} = \frac{(18 \text{ in} + 12 \text{ in} + 18 \text{ in})(6 \text{ in})^3}{12} = 864 \text{ in}^4$$

$$I_{c,2} = \frac{bh^3}{12} = \frac{(12 \text{ in})(18 \text{ in})^3}{12} = 5832 \text{ in}^4$$

Use the parallel axis theorem to determine the beam's moment of inertia about the centroidal axis.

$$I_b = \sum \left(I_{c,i} + A_i (\overline{y} - y_i)^2 \right)$$
$$= 864 \text{ in}^4 + (288 \text{ in}^2)(8.143 \text{ in} - 3 \text{ in})^2$$
$$\quad + 5832 \text{ in}^4 + (216 \text{ in}^2)(8.143 \text{ in} - 15 \text{ in})^2$$
$$= 24{,}469.7 \text{ in}^4$$

For the computation of Ψ, the beam inertia is reduced by multiplying by 0.35.

$$I_{b,\text{effective}} = 0.35 I_b = (0.35)(24{,}469.7 \text{ in}^4)$$
$$= \boxed{8564.4 \text{ in}^4 \quad (8600 \text{ in}^4)}$$

The answer is (C).

(b)

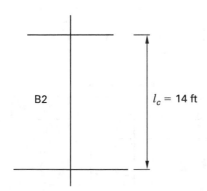

The effective moments of inertia are

$$\text{beam} = 8564.4 \text{ in}^4$$

$$I_{g,c} = \frac{bh^3}{12} = \frac{(20 \text{ in})(20 \text{ in})^3}{12} = 13{,}333.3 \text{ in}^4$$

$$\text{column} = 0.70 I_{g,c} = (0.70)(13{,}333.3 \text{ in}^4)$$
$$= 9333.3 \text{ in}^4$$

Since the conditions are identical at the top and the bottom, the values of Ψ_A and Ψ_B are the same. Use Eq. 53.3.

$$\Psi = \frac{\displaystyle\sum_{\text{columns}} \frac{EI}{l_c}}{\displaystyle\sum_{\text{beams}} \frac{EI}{l}} = \frac{(2)\left(\dfrac{9333.3 \text{ in}^4}{14 \text{ ft}}\right)}{(2)\left(\dfrac{8564.4 \text{ in}^4}{20 \text{ ft}}\right)}$$
$$= 1.56$$

$$k_u \approx \boxed{1.47 \quad (1.5)}$$

The answer is (C).

(c) In column B1, there is only one beam framing in the N-S direction (bending about W-E), so from Eq. 53.3,

$$\Psi = \frac{\displaystyle\sum_{\text{columns}} \frac{EI}{l_c}}{\displaystyle\sum_{\text{beams}} \frac{EI}{l}} = \frac{(2)\left(\dfrac{9333.3 \text{ in}^4}{14 \text{ ft}}\right)}{\dfrac{8564.4 \text{ in}^4}{20 \text{ ft}}}$$
$$= 3.11$$

$$k_u \approx \boxed{1.85 \quad (1.9)}$$

The answer is (D).

(d)

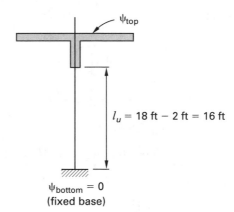

$$\Psi_{\text{top}} = \frac{\displaystyle\sum_{\text{columns}} \frac{EI}{l_c}}{\displaystyle\sum_{\text{beams}} \frac{EI}{l}} = \frac{\dfrac{9333.3 \text{ in}^4}{14 \text{ ft}} + \dfrac{9333.3 \text{ in}^4}{18 \text{ ft}}}{(2)\left(\dfrac{8564.4 \text{ in}^4}{20 \text{ ft}}\right)}$$
$$= 1.38$$

$$k_u = 1.2$$

The effective length is

$$k_u l_u = (1.2)(16 \text{ ft}) = \boxed{19.2 \text{ ft} \quad (19 \text{ ft})}$$

The answer is (C).

(e) $\beta_{dns} = 0$

Use Eq. 53.7.

$$EI = \frac{0.4 E_c I_g}{1 + \beta_{dns}} = \frac{(0.4)\left(4000 \ \dfrac{kips}{in^2}\right)(13{,}333.3 \ in^4)}{1 + 0}$$
$$= 2.13 \times 10^7 \ kip\text{-}in^2$$

Use Eq. 53.8.

$$P_c = \frac{\pi^2 EI}{(k_u l_u)^2} = \frac{\pi^2 (2.13 \times 10^7 \ kip\text{-}in^2)}{\left((1.2)(16 \ ft)\left(12 \ \dfrac{in}{ft}\right)\right)^2}$$
$$= \boxed{3960 \ kips \quad (4000 \ kips)}$$

The answer is (B).

(f) $\quad w_u = 1.2D + 1.0L + 1.6W$
$$= (1.2)\left(0.1 \ \dfrac{kip}{ft^2}\right) + (1.0)\left(0.06 \ \dfrac{kip}{ft^2}\right)$$
$$= 0.18 \ kip/ft^2$$

The area per floor is

$$(60 \ ft)(90 \ ft) = 5400 \ ft^2$$

Since the first level carries five stories and a roof,

$$\sum P_u = (6)\left(0.18 \ \dfrac{kip}{ft^2}\right)(5400 \ ft^2)$$
$$= \boxed{5832 \ kips \quad (5800 \ kips)}$$

The answer is (D).

(g) The ultimate shear force in the story is

$$V_{us} = \frac{(1.6)\left(30 \ \dfrac{lbf}{ft^2}\right)(90 \ ft)}{1000 \ \dfrac{lbf}{kip}}$$
$$\times \left((5)(14 \ ft) + \frac{18 \ ft}{2}\right)$$
$$= 341.3 \ kips$$

Use Eq. 53.1.

$$Q = \frac{\sum P_u \Delta_o}{V_{us} l_c} = \frac{(5832 \ kips)(0.3 \ in)}{(341.3 \ kips)(18 \ ft)\left(12 \ \dfrac{in}{ft}\right)}$$
$$= \boxed{0.024}$$

The answer is (A).

(h) The criterion to be satisfied is given by Eq. 53.15.

$$\frac{l_u}{r} \geq \frac{35}{\sqrt{\dfrac{P_u}{f'_c A_g}}}$$
$$r = (0.3)(20 \ in) = 6 \ in$$
$$l_u = 16 \ ft$$

$$\frac{l_u}{r} = \frac{(16 \ ft)\left(12 \ \dfrac{in}{ft}\right)}{6 \ in} = 32$$

$$\frac{P_u}{f'_c A_g} = \left(\frac{35}{32}\right)^2 = 1.196$$

$$P_u = \frac{\left(4000 \ \dfrac{lbf}{in^2}\right)(400 \ in^2)(1.196)}{1000 \ \dfrac{lbf}{kip}}$$
$$= \boxed{1914 \ kips \quad (1900 \ kips)}$$

The answer is (C).

(i) From part (d),

$$\Psi_{top} = 1.38$$
$$\Psi_{bottom} = 0$$

From the alignment chart,

$$k_b = \boxed{0.64}$$

The answer is (A).

(j) $\quad \beta_{dns} = \dfrac{\text{permanent factored axial load}}{\text{total factored axial load}}$
$$= \frac{(1.2)(200 \ kips)}{(1.2)(200 \ kips) + (1.6)(50 \ kips)}$$
$$= \boxed{0.75}$$

The answer is (B).

4. (a) $M_{2,ns} = (1.2)(100 \ ft\text{-}kips) + (1.6)(60 \ ft\text{-}kips)$
$$= \boxed{216 \ ft\text{-}kips \quad (220 \ ft\text{-}kips)}$$

The answer is (D).

(b) Compute Ψ at the top.

For the beams,

$$I_g = \frac{bh^3}{12} = \frac{(12 \ in)(24 \ in)^3}{12} = 13{,}824 \ in^4$$

Structural

To account for cracking,

$$I_b = 0.35 I_g = (0.35)(13{,}824 \text{ in}^4) = 4838.4 \text{ in}^4$$

For the columns,

$$I_g = \frac{bh^3}{12} = \frac{(14 \text{ in})(14 \text{ in})^3}{12} = 3201.3 \text{ in}^4$$

To account for cracking,

$$I_c = 0.70 I_g = (0.70)(3201.3 \text{ in}^4) = 2240.9 \text{ in}^4$$

Use Eq. 53.3.

$$\Psi_{\text{top}} = \frac{\displaystyle\sum_{\text{columns}} \frac{EI}{l_c}}{\displaystyle\sum_{\text{beams}} \frac{EI}{l}} = \frac{\dfrac{2240.9 \text{ in}^4}{17 \text{ ft}}}{\dfrac{4838.4 \text{ in}^4}{30 \text{ ft}}} = 0.82$$

At the base, the column is assumed fixed. In theory, this implies $\Psi_{\text{bottom}} = 0$. In practice, it is customary to substitute 1.0 for the theoretical value of zero to account for the finite flexibility at the "fixed end." Taking $\Psi_{\text{bottom}} = 1$, $k_b = 0.75$.

The effective length is

$$k_b l_u = (0.75)(16 \text{ ft}) = \boxed{12 \text{ ft}}$$

The answer is (C).

(c) Use Eq. 53.3.

$$\Psi_{\text{top}} = \frac{\displaystyle\sum_{\text{columns}} \frac{EI}{l_c}}{\displaystyle\sum_{\text{beams}} \frac{EI}{l}} = \frac{\dfrac{2240.9 \text{ in}^4}{17 \text{ ft}}}{(2)\left(\dfrac{4838.4 \text{ in}^4}{30 \text{ ft}}\right)} = 0.41$$

$$\Psi_{\text{bottom}} = 1$$
$$k_b = 0.72$$

$$k_b l_u = (0.72)(16 \text{ ft}) = \boxed{11.5 \text{ ft} \quad (12 \text{ ft})}$$

The answer is (B).

(d) $\quad \beta_{\text{dns}} = \dfrac{\text{permanent factored axial load}}{\text{total factored axial load}}$

$$= \frac{(1.2)(36.5 \text{ kips})}{(1.2)(36.5 \text{ kips}) + (1.6)(21.9 \text{ kips})}$$

$$= \boxed{0.56}$$

The answer is (C).

(e) Use Eq. 53.8.

$$P_c = \frac{\pi^2 EI}{(k_b l_u)^2}$$

From Eq. 53.7,

$$EI = \frac{0.4 E_c I_g}{1 + \beta_{\text{dns}}} = \frac{(0.4)\left(3605 \dfrac{\text{kips}}{\text{in}^2}\right)(3201.3 \text{ in}^4)}{1 + 0.56}$$

$$= 2.96 \times 10^6 \text{ kip-in}^2$$

$$P_c = \frac{\pi^2 EI}{(k_b l_u)^2} = \frac{\pi^2 (2.96 \times 10^6 \text{ kip-in}^2)}{\left((0.75)(16 \text{ ft})\left(12 \dfrac{\text{in}}{\text{ft}}\right)\right)^2}$$

$$= \boxed{1408.9 \text{ kips} \quad (1400 \text{ kips})}$$

The answer is (B).

(f) From Eq. 53.11,

$$C_m = 0.6 + 0.4\left(\frac{M_1}{M_2}\right)$$

$$\frac{M_1}{M_2} = -\frac{(1.2)(50 \text{ ft-kips}) + (1.6)(30 \text{ ft-kips})}{(1.2)(100 \text{ ft-kips}) + (1.6)(60 \text{ ft-kips})}$$

$$= -0.5$$

$$C_m = \boxed{0.4}$$

The answer is (B).

(g) In this column, $M_1 = M_2 = 0$. Therefore,

$$C_m = \boxed{1.0}$$

The answer is (D).

(h) $\quad \beta_{\text{dns}} = \dfrac{\text{permanent factored axial load}}{\text{total factored axial load}}$

$$= \frac{(1.2)(95.2 \text{ kips})}{(1.2)(95.2 \text{ kips}) + (1.6)(57.1 \text{ kips})}$$

$$= 0.56$$

From part (e), $EI = 2.96 \times 10^6 \text{ kip-in}^2$.

From part (c), $k_b = 0.72$.

Use Eq. 53.8.

$$P_c = \frac{\pi^2 EI}{(k_b l_u)^2} = \frac{\pi^2 (2.96 \times 10^6 \text{ kip-in}^2)}{\left((0.72)(16 \text{ ft})\left(12 \dfrac{\text{in}}{\text{ft}}\right)\right)^2}$$

$$= \boxed{1528.7 \text{ kips} \quad (1500 \text{ kips})}$$

The answer is (C).

(i) $P_u = (1.2)(95.2 \text{ kips}) + (1.6)(57.1 \text{ kips})$
        $= 205.6 \text{ kips}$

From Eq. 53.10,

$$\delta_{\text{ns}} = \frac{C_m}{1 - \dfrac{P_u}{0.75 P_c}} = \frac{1}{1 - \dfrac{205.6 \text{ kips}}{(0.75)(1528.7 \text{ kips})}}$$

$$= \boxed{1.219 \quad (1.2)}$$

The answer is (C).

(j) $P_u = (1.2)(36.5 \text{ kips}) + (1.6)(21.9 \text{ kips})$
        $= 78.8 \text{ kips}$

From part (e), $P_c = 1408.9$ kips.

From part (f), $C_m = 0.4$.

$$\delta_{\text{ns}} = \frac{C_m}{1 - \dfrac{P_u}{0.75 P_c}} = \frac{0.4}{1 - \dfrac{78.8 \text{ kips}}{(0.75)(1408.9 \text{ kips})}}$$

$$= 0.432 < 1.0$$

$$\delta_{\text{ns}} = \boxed{1.0}$$

The answer is (B).

Structural

54 Reinforced Concrete: Walls and Retaining Walls

PRACTICE PROBLEMS

1. A 14 ft high concrete bearing wall supports concrete roof beams spaced on 7 ft centers. Using the tributary area method, each end reaction is found to carry a dead load of 25 kips and a live load of 20 kips. The bearing width of the beam is 12 in. Both the top and bottom of the wall are restrained against translation and rotation. The concrete strength and the yield strength of steel are 4000 psi and 60,000 psi, respectively. The wall weight can be neglected. The ACI empirical formula is to be used. (a) What is the required wall thickness? (b) Is the bearing strength adequate if the wall is 7 in thick? (c) What is the vertical load capacity of a 7 in wall? (d) Design the vertical reinforcing steel. (e) Design the horizontal reinforcing steel.

2. (a) What is the vertical capacity of the wall described in Prob. 1 if the wall is not restrained against rotation at the top or bottom? Assume the wall thickness is 7 in. (b) Is the wall capacity still adequate?

3. (*Time limit: one hour*) The retaining wall shown supports a 400 lbf/ft^2 surcharge in addition to the active backfill loading. There is no batter on the visible face. 3000 psi concrete and 60,000 psi steel are used. The specific weights of the backfill and the concrete are 100 lbf/ft^3 and 150 lbf/ft^3, respectively. The active earth pressure coefficient is 0.5. The passive earth pressure coefficient is 2.0. (a) What is the factor of safety against overturning? (b) What is the minimum theoretical heel thickness? (c) Using a heel depth of 30 in and no. 8 reinforcing bars, what bar spacing is required in the heel?

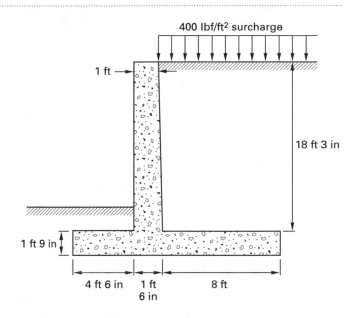

SOLUTIONS

1. (a) According to ACI 318 Sec. 14.5.3.1, the wall thickness, h, is the larger of 4 in or $^1/_{25}$ of the supported height or length.

$$\left(\frac{1}{25}\right)(14 \text{ ft})\left(12 \frac{\text{in}}{\text{ft}}\right) = \boxed{6.72 \text{ in}}$$

Use 7 in.

(b) The factored end reactions for each 7 ft section are

$$(1.2)(25 \text{ kips}) + (1.6)(20 \text{ kips}) = 62 \text{ kips}$$

The design bearing strength immediately under the beam ends is

$$\phi 0.85 f'_c A = (0.65)(0.85)\left(4 \frac{\text{kips}}{\text{in}^2}\right)(12 \text{ in})(7 \text{ in})$$

$$= 185.6 \text{ kips} \quad \boxed{[> 62 \text{ kips, so OK}]}$$

(c) $A_g = $ (horizontal wall length)(wall thickness)

According to ACI 318 Sec. 14.2.4, for the horizontal length of the wall to be considered effective for each concentrated load, the length cannot exceed the smaller of the center-to-center distance between loads and the width of the bearing plus four times the wall thickness.

The center-to-center spacing is

$$(7 \text{ ft})\left(12 \frac{\text{in}}{\text{ft}}\right) = 84 \text{ in}$$

The width of bearing plus four times the wall thickness, h, is

$$12 \text{ in} + (4)(7 \text{ in}) = 40 \text{ in}$$

The 40 in dimension controls. Use Eq. 54.1.

$$\phi P_{n,w} \le 0.55 \phi f'_c A_g \left(1 - \left(\frac{kl_c}{32h}\right)^2\right)$$

$$(0.55)(0.65)\left(4 \frac{\text{kips}}{\text{in}^2}\right)(40 \text{ in})(7 \text{ in})$$

$$\times \left(1 - \left(\frac{(0.8)(14 \text{ ft})\left(12 \frac{\text{in}}{\text{ft}}\right)}{(32)(7 \text{ in})}\right)^2\right)$$

$$= \boxed{256 \text{ kips}} \quad [> 62 \text{ kips, so OK}]$$

(d) This is a bearing wall. Per ACI 318 Sec. 14.3.2, Sec. 14.3.3, and Sec. 14.3.5, the steel will be as follows.

The maximum spacing of vertical and horizontal bars is

$$(3)(7 \text{ in}) = 21 \text{ in or } 18 \text{ in} \quad [18 \text{ in controls}]$$

The area of vertical steel is

$$(0.0012)\left(12 \frac{\text{in}}{\text{ft}}\right)(7 \text{ in}) = 0.101 \text{ in}^2/\text{ft}$$

$$\boxed{\text{Use no. 4 bars at 18 in (provides } 0.133 \text{ in}^2/\text{ft).}}$$

(e) The area of horizontal steel is

$$(0.002)\left(12 \frac{\text{in}}{\text{ft}}\right)(7 \text{ in}) = 0.168 \text{ in}^2/\text{ft}$$

$$\boxed{\text{Use no. 4 bars at 14 in (provides } 0.171 \text{ in}^2/\text{ft).}}$$

2. (a) The effective length factor, k, is 1.0 when the wall is not restrained against rotation at both ends. Use Eq. 54.1, the ACI empirical formula.

$$\phi P_{n,w} \le 0.55 \phi f'_c A_g \left(1 - \left(\frac{kl_c}{32h}\right)^2\right)$$

$$(0.55)(0.65)\left(4 \frac{\text{kips}}{\text{in}^2}\right)(40 \text{ in})(7 \text{ in})$$

$$\times \left(1 - \left(\frac{(1.0)(14 \text{ ft})\left(12 \frac{\text{in}}{\text{ft}}\right)}{(32)(7 \text{ in})}\right)^2\right)$$

$$= \boxed{175.2 \text{ kips}}$$

(b) $\boxed{175.2 \text{ kips} > 62 \text{ kips, so OK}}$

3. (a) $H = 18.25 \text{ ft} + 1.75 \text{ ft} = 20 \text{ ft}$

The soil pressure resultant is

$$R_a = \tfrac{1}{2}k_a \gamma H^2$$

$$= \left(\tfrac{1}{2}\right)(0.5)\left(100 \frac{\text{lbf}}{\text{ft}^3}\right)(20 \text{ ft})^2$$

$$= 10{,}000 \text{ lbf/ft}$$

R_a acts at 20 ft/3 = 6.67 ft up from the base.

The surcharge loading is

$$R_{q,h} = (0.5)\left(400 \frac{\text{lbf}}{\text{ft}^2}\right)(20 \text{ ft}) = 4000 \text{ lbf}$$

$R_{q,h}$ acts at 20 ft/2 = 10 ft up from the base.

Structural

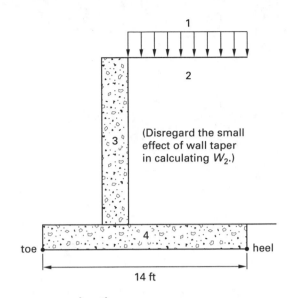

(Disregard the small effect of wall taper in calculating W_2.)

toe   heel

14 ft

element	length or area (ft/ft²)	q or γ (lbf/ft²)	F or W (lbf)	$r_{i,heel}$ (ft)
1	8	400	3200	
2	$8 \times 18.25 = 146$	100	14,600	
3	$1.25 \times 18.25 = 22.8$	150	3420	$\approx$
4	$14 \times 1.75 = 24.5$	150	3675	7
total			24,895	

element	F (lbf)	$14 - r_i = x_{i,toe}$ (ft)	moment about toe (ft-lbf)
1	3200	10	32,000
2	14,600	10	146,000
3	3420	5.5	18,810
4	3675	7	25,725
total			222,535

$$M_{OT} = \sum R_a y_a$$
$$= (10{,}000 \text{ lbf})(6.67 \text{ ft}) + (4000 \text{ lbf})(10 \text{ ft})$$
$$= 106{,}700 \text{ ft-lbf}$$

$$FS_{OT} = \frac{M_R}{M_{OT}} = \frac{222{,}535 \text{ ft-lbf}}{106{,}700 \text{ ft-lbf}}$$
$$= \boxed{2.09 \quad \text{[acceptable]}}$$

(Normally, a check of the factor of safety against sliding would also be made. $FS_{sliding}$ is inadequate in this problem, but a check was not requested.)

(b)   heel weight $= (8 \text{ ft})(1.75 \text{ ft})\left(150 \dfrac{\text{lbf}}{\text{ft}^3}\right)$
$$= 2100 \text{ lbf/ft} \quad \text{[per foot of width]}$$

$$V_u = (1.2)\left(2100 \frac{\text{lbf}}{\text{ft}} + 14{,}600 \frac{\text{lbf}}{\text{ft}}\right)$$
$$+ (1.6)\left(3200 \frac{\text{lbf}}{\text{ft}}\right)$$
$$= 25{,}160 \text{ lbf/ft}$$

The nominal shear strength of concrete is

$$\phi v_n = \phi v_c = \phi 2\lambda\sqrt{f'_c} = (0.75)(2)(1.0)\sqrt{3000 \frac{\text{lbf}}{\text{in}^2}}$$
$$= 82.1 \text{ lbf/in}^2$$

The required heel thickness, without requiring shear reinforcement and neglecting soil pressure beneath the footing, is

$$\frac{25{,}160 \dfrac{\text{lbf}}{\text{ft}}}{\left(82.1 \dfrac{\text{lbf}}{\text{in}^2}\right)\left(12 \dfrac{\text{in}}{\text{ft}}\right)} = 25.5 \text{ in} \quad \text{[round to 26 in]}$$

To include at least a 2 in cover, use a heel thickness of

$$t_b = \boxed{30 \text{ in}}$$

(c) The heel weight becomes

$$(8 \text{ ft})\left(\frac{30 \text{ in}}{12 \dfrac{\text{in}}{\text{ft}}}\right)(1)\left(150 \frac{\text{lbf}}{\text{ft}^3}\right) = 3000 \text{ lbf/ft}$$

Take moments about the stem face.

element	F (lbf)	distance (ft)	M (ft-lbf)
1	3200	4	12,800
2	14,600	4	58,400
heel	3000	4	12,000

The ultimate moment on the heel at the stem face is given by Eq. 54.10.

$$M_{u,base} = 1.2 M_{soil} + 1.2 M_{heel\,weight} + 1.6 M_{surcharge}$$
$$= (1.2)\left(58{,}400 \frac{\text{ft-lbf}}{\text{ft}} + 12{,}000 \frac{\text{ft-lbf}}{\text{ft}}\right)$$
$$+ (1.6)\left(12{,}800 \frac{\text{ft-lbf}}{\text{ft}}\right)$$
$$= 104{,}960 \text{ ft-lbf/ft}$$

Use Eq. 54.13, assuming the section is tension controlled.

$$M_n = \frac{M_u}{\phi} = \frac{104{,}960 \dfrac{\text{ft-lbf}}{\text{ft}}}{0.9}$$
$$= 116{,}622 \text{ ft-lbf/ft}$$

$$\left(116{,}622 \frac{\text{ft-lbf}}{\text{ft}}\right)$$
$$\times \left(12 \frac{\text{in}}{\text{ft}}\right) = \rho(12 \text{ in})(26 \text{ in})^2$$
$$\times \left(60{,}000 \frac{\text{lbf}}{\text{in}^2}\right)$$
$$\times \left(1 - \frac{\rho\left(60{,}000 \dfrac{\text{lbf}}{\text{in}^2}\right)}{(1.7)\left(3000 \dfrac{\text{lbf}}{\text{in}^2}\right)}\right)$$
$$\rho = 0.0030$$

Check.

$$\rho_{min} = \frac{200}{f_y} = \frac{200}{60,000 \ \frac{lbf}{in^2}} = 0.0033 \quad \text{[controls]}$$

Since ρ_{min} controls, the section is tension controlled.

Number 8 bars have an area of 0.79 in^2.

$$\text{spacing} = \frac{0.79 \ in^2}{(0.0033)(26 \ in)} = 9.2 \ in$$

Use one no. 8 bar every 9 in.

55 Reinforced Concrete: Footings

PRACTICE PROBLEMS

In the following problems, none of the loads are the result of wind or earthquake action. Unless specified in the problem, assume concrete has a specific weight of 150 lbf/ft³. All bars are uncoated.

1. A reinforced concrete column is supported at the center of a square footing. The footing is in contact with the bare ground 4 ft below grade. The gross allowable soil pressure is 2000 lbf/ft². The specific weight of the soil is 110 lbf/ft³. The footing thickness is 18 in. The axial loads, moments, and shears (directed along the top of the footing surface) acting on the footing through the column are

$$P_d = 100 \text{ kips}$$
$$P_l = 58 \text{ kips}$$
$$M_d = 20 \text{ ft-kips}$$
$$M_l = 40 \text{ ft-kips}$$
$$V_d = 2 \text{ kips} \quad [\text{to the right}]$$
$$V_l = 4 \text{ kips} \quad [\text{to the right}]$$

Determine the footing size. (Do not design the reinforcement.)

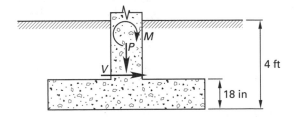

2. An 8 in thick concrete wall carries a service load of 22 kips/ft. As shown, the footing for the wall is located 5 ft below grade. 3000 psi concrete is used in the footing. At this depth, the allowable gross soil pressure is 5000 lbf/ft². The soil has a unit weight of 120 lbf/ft³. No. 4 bars are to be used for reinforcement. Take the ultimate load as 1.5 times the service load. Size the footing and determine the steel requirements. Do not check development length.

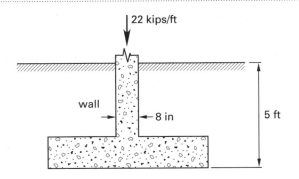

3. (*Time limit: one hour*) A 16 in thick, 10 ft square footing on grade supports a 12 in square column as shown. The net allowable soil pressure is 3000 lbf/ft².

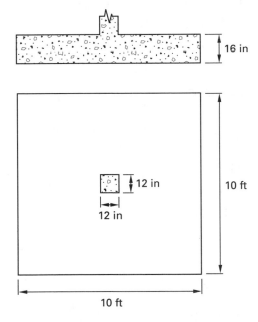

(a) If the column is not subjected to a moment, the maximum service load that the footing can transmit safely to the soil is most nearly

(A) 200 kips

(B) 250 kips

(C) 300 kips

(D) 350 kips

(b) If the column has a total axial service load (dead plus live) of 200 kips, the maximum moment about the centroid of the footing print that the footing can support is most nearly

(A) 140 ft-kips

(B) 170 ft-kips

(C) 190 ft-kips

(D) 220 ft-kips

(c) The area of the critical section resisting punching shear is most nearly

(A) 680 in^2

(B) 960 in^2

(C) 1200 in^2

(D) 1400 in^2

For parts (d) through (f), assume the column is subjected to a factored axial force of 340 kips and no moment.

(d) The punching shear stress at the critical section is most nearly

(A) 200 lbf/in^2

(B) 250 lbf/in^2

(C) 280 lbf/in^2

(D) 320 lbf/in^2

(e) The one-way shear stress at the critical section is most nearly

(A) 60 lbf/in^2

(B) 80 lbf/in^2

(C) 100 lbf/in^2

(D) 120 lbf/in^2

(f) The design moment at the critical section is most nearly

(A) 280 ft-kips

(B) 300 ft-kips

(C) 340 ft-kips

(D) 440 ft-kips

In parts (g) through (j), assume the loading on the footing consists of a factored load of 340 kips plus a moment of 150 ft-kips.

(g) The maximum punching shear stress at the critical section is most nearly

(A) 220 lbf/in^2

(B) 250 lbf/in^2

(C) 360 lbf/in^2

(D) 410 lbf/in^2

(h) The maximum one-way shear stress at the critical section is most nearly

(A) 86 lbf/in^2

(B) 97 lbf/in^2

(C) 100 lbf/in^2

(D) 120 lbf/in^2

(i) The design moment at the critical section is most nearly

(A) 350 ft-kips

(B) 380 ft-kips

(C) 410 ft-kips

(D) 510 ft-kips

(j) The number of no. 6 bars needed as reinforcement for the moment of part (i) is most nearly

(A) 12

(B) 14

(C) 19

(D) 24

4. (*Time limit: one hour*) A 10 ft × 14 ft rectangular footing supports a 12 in × 16 in reinforced concrete column placed at its center. The strength of the concrete for the column and the footing are 5000 lbf/in² and 3000 lbf/in², respectively. All steel is $f_y = 60{,}000$ lbf/in². The relative orientation of the column with respect to the footing is shown.

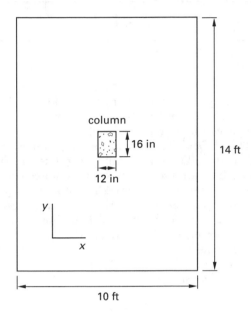

(a) Assume the column is subjected to a factored axial load of 650 kips. The minimum area of dowels needed to transfer the axial force from the column to the footing is most nearly

(A) 0 in²

(B) 1.0 in²

(C) 2.0 in²

(D) 4.0 in²

(b) Assume the column is subjected to axial compression only. If the dowels are no. 6 bars with 90° standard hooks, the minimum depth of the footing for which they will be fully anchored is most nearly

(A) 16 in

(B) 18 in

(C) 25 in

(D) 35 in

In parts (c) through (e), assume the footing is 24 in thick with an effective depth of 20 in.

(c) The minimum permitted amount of steel oriented parallel to the 10 ft side is most nearly

(A) 6.0 in²

(B) 7.3 in²

(C) 11 in²

(D) 12 in²

(d) As controlled by punching shear, the maximum factored axial load that can be resisted is most nearly

(A) 270 kips

(B) 480 kips

(C) 540 kips

(D) 610 kips

(e) As controlled by one-way shear, the maximum axial load that can be transferred by the footing is most nearly

(A) 300 kips

(B) 590 kips

(C) 670 kips

(D) 1200 kips

In parts (f) through (h), assume the column is subjected to a factored axial force of 400 kips and a factored moment of 600 ft-kips acting about the y-axis.

(f) The punching shear stress at the critical section is most nearly

(A) 140 lbf/in²

(B) 220 lbf/in²

(C) 300 lbf/in²

(D) 350 lbf/in²

(g) The one-way shear stress at the critical section is most nearly

(A) 21 lbf/in²

(B) 32 lbf/in²

(C) 49 lbf/in²

(D) 55 lbf/in²

(h) The design moment for bars parallel to the x-direction is most nearly

(A) 470 ft-kips

(B) 530 ft-kips

(C) 660 ft-kips

(D) 1100 ft-kips

(i) If the required steel area for bars parallel to the x-direction (see illustration) is 12 in² for some applied load, the spacing of no. 6 bars parallel to the x-direction in the region near the column is most nearly

(A) 4.0 in

(B) 4.5 in

(C) 5.2 in

(D) 6.0 in

(j) The available anchorage length for bars in the y-direction is most nearly

(A) 69 in

(B) 74 in

(C) 78 in

(D) 84 in

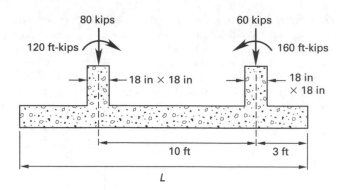

5. A square footing is to carry a live load of 240,000 lbf that is transmitted through a 16 in square column. (The column dead load can be disregarded, although the footing dead load cannot.) The allowable soil pressure is 4000 lbf/ft², and the top of the footing is level with the surrounding soil surface. Material properties are $f'_c = 3000$ lbf/in² and $f_y = 40,000$ lbf/in². The footing thickness is 20 in. Design the footing.

6. An 18 in square column supports a 200,000 lbf dead load and a 145,000 lbf live load. The allowable soil pressure is 4000 lbf/ft². Material properties are $f'_c = 3000$ lbf/in² and $f_y = 40,000$ lbf/in². The column steel consists of ten no. 9 bars. Design the footing.

7. A 10 ft wide combined footing carries the service loads and moments shown. Loads and moments are transmitted through 18 in square columns. The footing weight can be disregarded. (a) What dimension L will result in a uniform soil pressure? (b) Draw the shear and moment diagrams. (c) Does the footing need a layer of top steel?

8. A 14 in square column carries a live load of 200 kips and a live moment of 100 ft-kips, as shown. The column self-weight can be disregarded. The allowable soil pressure is 6000 lbf/ft². The concrete's compressive strength is 3000 lbf/in². The soil specific weight is 110 lbf/ft³. Number 4 bars are used for flexural reinforcement. The top of the footing is even with the soil surface. (a) Size the footing. (Do not design the steel reinforcement.) (b) What are the maximum and minimum pressures along the footing base? (c) What thickness of footing would be required if shear reinforcement was to be completely eliminated? (d) What is the maximum moment the footing must resist?

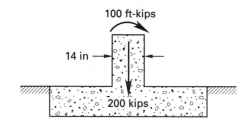

SOLUTIONS

1. The service loads are not factored when sizing the footing.

$$P_s = 100 \text{ kips} + 58 \text{ kips} = 158 \text{ kips}$$
$$V_s = 2 \text{ kips} + 4 \text{ kips} = 6 \text{ kips}$$
$$H = 4 \text{ ft}$$
$$h = 1.5 \text{ ft}$$

The service moment is

$$M_s = 20 \text{ ft-kips} + 40 \text{ ft-kips}$$
$$+ (2 \text{ kips} + 4 \text{ kips}) \left(\frac{18 \text{ in}}{12 \frac{\text{in}}{\text{ft}}} \right)$$
$$= 69 \text{ ft-kips}$$

From Eq. 55.1,

$$q_s = \frac{P_s}{A_f} + \gamma_c h + \gamma_s (H - h) + \frac{M_s \left(\frac{B}{2} \right)}{I_f}$$

$$2 \frac{\text{kips}}{\text{ft}^2} = \frac{158 \text{ kips}}{A_f} + \left(0.15 \frac{\text{kip}}{\text{ft}^3} \right) (1.5 \text{ ft})$$
$$+ \left(0.11 \frac{\text{kip}}{\text{ft}^3} \right) (4 \text{ ft} - 1.5 \text{ ft})$$
$$+ \frac{(69 \text{ ft-kips}) \left(\frac{B}{2} \right)}{I_f}$$

$$1.5 \frac{\text{kips}}{\text{ft}^2} = \frac{158 \text{ kips}}{A_f} + \frac{(34.5 \text{ ft-kips})B}{I_f}$$

Since the footing is square,

$$A_f = B^2$$
$$I_f = \frac{1}{12} B^4$$

Therefore,

$$1.5 \frac{\text{kips}}{\text{ft}^2} = \frac{158 \text{ kips}}{B^2} + \frac{414 \text{ ft-kips}}{B^3}$$

This is most easily solved by trial substitution.

$$\boxed{B = 11.4 \text{ ft} \quad [\text{use } 11.5 \text{ ft} \times 11.5 \text{ ft}]}$$

2. The ϕ factor for shear is 0.75. Therefore, from Eq. 55.8, the required depth of the footing for shear resistance is

$$d = \frac{q_u (B - t)}{2(q_u + \phi v_c)}$$

$$= \frac{\left(6.6 \frac{\text{kips}}{\text{ft}^2} \right) \left(5 \text{ ft} - \frac{8 \text{ in}}{12 \frac{\text{in}}{\text{ft}}} \right) \left(12 \frac{\text{in}}{\text{ft}} \right)}{(2) \left(6.6 \frac{\text{kips}}{\text{ft}^2} + (0.75) \left(15.77 \frac{\text{kips}}{\text{ft}^2} \right) \right)}$$

$$= \boxed{9.31 \text{ in}}$$

Using no. 4 bars in each direction,

$$h \geq 9.31 \text{ in} + \left(\frac{1}{2} \right)(0.5 \text{ in}) + 0.5 \text{ in} + 3 \text{ in}$$
$$= \boxed{13.06 \text{ in} \quad [\text{use } 13 \text{ in}]}$$

(The assumed value of h of 13 in is close to the h selected, so there is no need to revise the calculations for the width B.)

Determine the flexural steel.

$$d = 13 \text{ in} - 3 \text{ in} - 0.5 \text{ in} - 0.25 \text{ in} = 9.25 \text{ in}$$
$$t = \frac{8 \text{ in}}{12 \frac{\text{in}}{\text{ft}}} = 0.67 \text{ ft}$$

The moment at the critical section is

$$M_u = \frac{q_u \left(\frac{B}{2} - \frac{t}{2} \right)^2}{2} = \frac{q_u (B - t)^2}{8}$$
$$= \frac{\left(6.6 \frac{\text{kips}}{\text{ft}^2} \right) (5 \text{ ft} - 0.67 \text{ ft})^2}{8}$$
$$= 15.47 \text{ ft-kips/ft}$$

Assume $\lambda = 0.1d = (0.1)(9.25 \text{ in}) = 0.925 \text{ in}$.

Assume the section is tension controlled. From Eq. 55.24, the area of steel is

$$A_s = \frac{M_u}{\phi f_y (d - \lambda)}$$
$$= \frac{\left(15.47 \frac{\text{ft-kips}}{\text{ft}} \right) \left(12 \frac{\text{in}}{\text{ft}} \right)}{(0.9) \left(60 \frac{\text{kips}}{\text{in}^2} \right) (9.25 \text{ in} - 0.925 \text{ in})}$$
$$= 0.413 \text{ in}^2/\text{ft}$$

Structural

Check λ.

$$A_c = \frac{A_s f_y}{0.85 f'_c} = \frac{(0.413 \text{ in}^2)\left(60 \frac{\text{kips}}{\text{in}^2}\right)}{(0.85)\left(3 \frac{\text{kips}}{\text{in}^2}\right)} = 9.72 \text{ in}^2$$

$$\lambda = \frac{9.72 \text{ in}^2}{(12 \text{ in})(2)} = 0.405 \text{ in}$$

Check whether the section is tension controlled.

$$c = \frac{(0.405 \text{ in})(2)}{0.85} = 0.95 \text{ in}$$

$$\epsilon_t = 0.003\left(\frac{d-c}{c}\right) = (0.003)\left(\frac{9.25 \text{ in} - 0.95 \text{ in}}{0.95 \text{ in}}\right)$$
$$= 0.026 > 0.005$$

Therefore, the section is tension controlled.

The revised A_s is

$$A_s = \frac{\left(15.47 \frac{\text{ft-kips}}{\text{ft}}\right)\left(12 \frac{\text{in}}{\text{ft}}\right)}{(0.9)\left(60 \frac{\text{kips}}{\text{in}^2}\right)(9.25 \text{ in} - 0.405 \text{ in})}$$
$$= 0.39 \text{ in}^2/\text{ft}$$

Using ACI 318 Sec. 7.12.2.1, the minimum steel is

$$A_{s,\text{min}} = (0.0018)(12 \text{ in})(13 \text{ in}) = 0.281 \text{ in}^2/\text{ft}$$
$$\text{[does not control]}$$

The number of no. 4 bars per foot is

$$\frac{A_s}{A_b} = \frac{0.39 \frac{\text{in}^2}{\text{ft}}}{0.2 \text{ in}^2} = 1.95 \text{ bars/ft} \quad (2.0)$$

The spacing of no. 4 bars is

$$\frac{12 \text{ in}}{2.0} = 6 \text{ in}$$

The longitudinal (temperature) steel is

$$(0.0018)(5 \text{ ft})\left(12 \frac{\text{in}}{\text{ft}}\right)(13 \text{ in}) = 1.40 \text{ in}^2$$
$$\boxed{\text{Use eight no. 4 bars.}}$$

Space the outside bars to provide 3 in of cover.

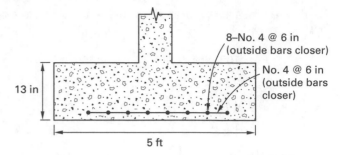

8–No. 4 @ 6 in (outside bars closer)

No. 4 @ 6 in (outside bars closer)

13 in

5 ft

(The next step would be to check development length and the need for hooked bars.)

3. (a) The maximum service load is

$$P_s = \left(3 \frac{\text{kips}}{\text{ft}^2}\right)(10 \text{ ft})(10 \text{ ft})$$
$$= \boxed{300 \text{ kips}}$$

The answer is (C).

(b) Since the footing is square,

$$I_f = \tfrac{1}{12}B^4 = \left(\tfrac{1}{12}\right)(10 \text{ ft})^4 = 833.3 \text{ ft}^4$$

Use Eq. 55.1. Disregard the correction for burial.

$$q_s = \frac{P_s}{A_f} + \gamma_c h + \gamma_s (H - h) + \frac{M_s\left(\frac{B}{2}\right)}{I_f}$$

$$3 \frac{\text{kips}}{\text{ft}^2} = \frac{200 \text{ kips}}{100 \text{ ft}^2} + \frac{M_s\left(\frac{10 \text{ ft}}{2}\right)}{833.3 \text{ ft}^4}$$

$$M_s = \boxed{166.7 \text{ ft-kips} \quad (170 \text{ ft-kips})}$$

The answer is (B).

(c) Leaving room for bars and cover, assume

$$d \approx h - 4 \text{ in} = 16 \text{ in} - 4 \text{ in} = 12 \text{ in}$$

$$b_1 = b_2 = \text{column size} + 2\left(\frac{d}{2}\right)$$
$$= 12 \text{ in} + (2)\left(\frac{12 \text{ in}}{2}\right)$$
$$= 24 \text{ in}$$

$$A_p = (4)(24 \text{ in})(12 \text{ in}) = \boxed{1152 \text{ in}^2 \quad (1200 \text{ in}^2)}$$

The answer is (C).

(d) Use Eq. 55.13 and Eq. 55.14.

$$R = \frac{P_u b_1 b_2}{A_f} = \frac{(340 \text{ kips})(24 \text{ in})^2}{(120 \text{ in})^2}$$

$$= 13.6 \text{ kips}$$

$$v_u = \frac{P_u - R}{A_p} = \frac{340 \text{ kips} - 13.6 \text{ kips}}{1152 \text{ in}^2}$$

$$= \boxed{0.283 \text{ kip/in}^2 \quad (280 \text{ lbf/in}^2)}$$

The answer is (C).

(e) The distance from the critical sections to the edge is

$$e = \frac{10 \text{ ft} - 1 \text{ ft}}{2} - 1 \text{ ft} = 3.5 \text{ ft}$$

$$q_u = \frac{340 \text{ kips}}{100 \text{ ft}^2} = 3.4 \text{ kips/ft}^2$$

$$v_u = \frac{\left(3.4 \dfrac{\text{kips}}{\text{ft}^2}\right)(3.5 \text{ ft})\left(1000 \dfrac{\text{lbf}}{\text{kip}}\right)}{(1 \text{ ft})\left(12 \dfrac{\text{in}}{\text{ft}}\right)^2}$$

$$= \boxed{82.64 \text{ lbf/in}^2 \quad (80 \text{ lbf/in}^2)}$$

The answer is (B).

(f) The critical section for flexure is at the face of the column, at 4.5 ft from the edge.

$$M_u = \left(\frac{340 \text{ kips}}{(10 \text{ ft})(10 \text{ ft})}\right)(10 \text{ ft})(4.5 \text{ ft})\left(\frac{4.5 \text{ ft}}{2}\right)$$

$$= \boxed{344.3 \text{ ft-kips} \quad (340 \text{ ft-kips})}$$

The answer is (C).

(g) From Eq. 55.16,

$$J = \left(\frac{db_1^3}{6}\right)\left(1 + \left(\frac{d}{b_1}\right)^2 + 3\left(\frac{b_2}{b_1}\right)\right)$$

$$= \left(\frac{(12 \text{ in})(24 \text{ in})^3}{6}\right)\left(1 + \left(\frac{12 \text{ in}}{24 \text{ in}}\right)^2 + (3)\left(\frac{24 \text{ in}}{24 \text{ in}}\right)\right)$$

$$= 117{,}504 \text{ in}^4$$

From Eq. 55.15,

$$\gamma_v = 1 - \frac{1}{1 + \frac{2}{3}\sqrt{\dfrac{b_1}{b_2}}} = 1 - \frac{1}{1 + \frac{2}{3}\sqrt{\dfrac{24 \text{ in}}{24 \text{ in}}}}$$

$$= 0.4$$

Use Eq. 55.13.

$$v_u = \frac{P_u - R}{A_p} + \frac{\gamma_v M_u(0.5b_1)}{J}$$

$$= \frac{340 \text{ kips} - 13.6 \text{ kips}}{1152 \text{ in}^2}$$

$$+ \frac{(0.4)(150 \text{ ft-kips})\left(12 \dfrac{\text{in}}{\text{ft}}\right)(0.5)(24 \text{ in})}{117{,}504 \text{ in}^4}$$

$$= \boxed{0.356 \text{ kip/in}^2 \quad (360 \text{ lbf/in}^2)}$$

The answer is (C).

(h) Determine the pressure distribution. Use Eq. 55.1. Disregard the correction for burial.

$$q_s = \frac{P_s}{A_f} + \gamma_c h + \gamma_s(H - h) \pm \frac{M_s\left(\dfrac{B}{2}\right)}{I_f}$$

$$= \frac{340 \text{ kips}}{100 \text{ ft}^2} \pm \frac{(150 \text{ ft-lbf})\left(\dfrac{10 \text{ ft}}{2}\right)}{833.3 \text{ ft}^4}$$

$$= 3.4 \frac{\text{kips}}{\text{ft}^2} \pm 0.9 \frac{\text{kip}}{\text{ft}^2}$$

$$= (4.3 \text{ kips/ft}^2, 2.5 \text{ kips/ft}^2)$$

At the critical section,

$$q = 4.3 \frac{\text{kips}}{\text{ft}^2} - \left(\frac{3.5 \text{ ft}}{10 \text{ ft}}\right)\left(4.3 \frac{\text{kips}}{\text{ft}^2} - 2.5 \frac{\text{kips}}{\text{ft}^2}\right)$$

$$= 3.67 \text{ kips/ft}^2$$

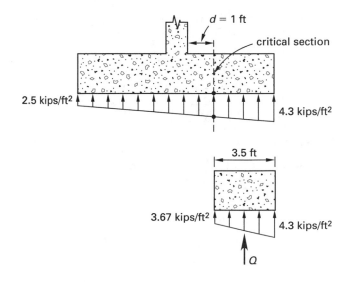

$$Q = \left(4.3 \; \frac{\text{kips}}{\text{ft}^2} + 3.67 \; \frac{\text{kips}}{\text{ft}^2}\right)\left(\frac{3.5 \text{ ft}}{2}\right)(10 \text{ ft})$$

$$= 139.5 \text{ kips}$$

$$v_u = \frac{(139.5 \text{ kips})\left(1000 \; \frac{\text{lbf}}{\text{kip}}\right)}{(10 \text{ ft})(1 \text{ ft})\left(12 \; \frac{\text{in}}{\text{ft}}\right)^2}$$

$$= \boxed{96.88 \text{ lbf/in}^2 \quad (97 \text{ lbf/in}^2)}$$

The answer is (B).

(i) From the free-body diagram,

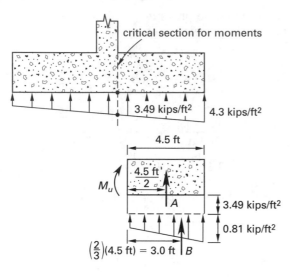

$$A = \left(3.49 \; \frac{\text{kips}}{\text{ft}^2}\right)(10 \text{ ft})(4.5 \text{ ft}) = 157.05 \text{ kips}$$

$$B = \left(0.81 \; \frac{\text{kip}}{\text{ft}^2}\right)(10 \text{ ft})(4.5 \text{ ft})\left(\tfrac{1}{2}\right) = 18.23 \text{ kips}$$

$$M_u = (157.05 \text{ kips})\left(\frac{4.5 \text{ ft}}{2}\right)$$

$$+ (18.23 \text{ kips})\left(\tfrac{2}{3}\right)(4.5 \text{ ft})$$

$$= \boxed{408.05 \text{ ft-kips} \quad (410 \text{ ft-kips})}$$

The answer is (C).

(j) Assume $\lambda = 0.1d$. Use Eq. 55.24.

$$A_s = \frac{M_u}{\phi f_y(d - \lambda)}$$

$$= \frac{(408.05 \text{ ft-kips})\left(12 \; \frac{\text{in}}{\text{ft}}\right)}{(0.9)\left(60 \; \frac{\text{kips}}{\text{in}^2}\right)(12 \text{ in} - 1.2 \text{ in})}$$

$$= 8.40 \text{ in}^2$$

The number of no. 6 bars is

$$\frac{A_s}{A_b} = \frac{8.4 \text{ in}^2}{0.44 \text{ in}^2} = \boxed{19.09 \quad (19)}$$

A second iteration from the beginning would refine this value.

The answer is (C).

4. (a) The area of concrete is

$$A_c = (16 \text{ in})(12 \text{ in}) = 192 \text{ in}^2$$

The factored load is given as $P_u = 650$ kips. Using Eq. 55.38, the bearing strength of the column is

$$0.85\phi f'_c A_c = (0.85)(0.65)\left(5 \; \frac{\text{kips}}{\text{in}^2}\right)(192 \text{ in}^2)$$

$$= 530.4 \text{ kips}$$

For the bearing strength of the footing, which uses a different concrete, the α factor may be used.

The aspect ratio of the column is 16 in/12 in = 1.333. The longest footing side length that can be used in the calculation of A_{ff} is

$$(1.333)(10 \text{ ft})\left(12 \; \frac{\text{in}}{\text{ft}}\right) = 160 \text{ in}$$

$$\alpha = \sqrt{\frac{A_{ff}}{A_c}} = \sqrt{\frac{(160 \text{ in})(120 \text{ in})}{192 \text{ in}^2}} = 10$$

Since α may not be greater than 2, use $\alpha = 2.0$. (The footing thickness is not known. However, it is likely that the $\alpha = 2.0$ limitation would control.) From Eq. 55.39, the bearing strength of the footing is

$$0.85\alpha f'_c A_c = (0.85)(2.0)(0.65)\left(3 \; \frac{\text{kips}}{\text{in}^2}\right)(192 \text{ in}^2)$$

$$= 636.5 \text{ kips}$$

The smaller value between the bearing strength of the column and the bearing strength of the footing is the controlling strength. Therefore, $\phi P_{n,b} = 530.4$ kips.

Since the $P_u > \phi P_{n,b}$, the excess that must be carried by the dowels is

$$650 \text{ kips} - 530.4 \text{ kips} = 119.6 \text{ kips}$$

Design the dowel steel. The dowel steel area reduces the concrete bearing surface area by the same amount. Subtract the concrete bearing strength from the steel yield strength.

$$119.6 \text{ kips} = \phi P_{n,\text{steel}} = \phi A_s (f_y - 0.85 f'_{c,\text{column}})$$

$$
\begin{aligned}
A_s &= \frac{\phi P_{n,\text{steel}}}{\phi (f_y - 0.85 f'_{c,\text{column}})} \\
&= \frac{119.6 \text{ kips}}{(0.65)\left(60 \dfrac{\text{kips}}{\text{in}^2} - (0.85)\left(5 \dfrac{\text{kips}}{\text{in}^2}\right)\right)} \\
&= 3.30 \text{ in}^2
\end{aligned}
$$

Check the minimum requirement using Eq. 55.35.

$$A_{\text{db,min}} = 0.005 A_c = (0.005)(192 \text{ in}^2) = 0.96 \text{ in}^2$$

Use 4 no. 9 bars (4.0 in^2).

$$\boxed{\text{Use } A_s = 4.0 \text{ in}^2.}$$

The answer is (D).

(b) Hooks are ineffective for compression. The development length of a bar in compression does not include the bend length. Use Eq. 55.36.

$$
\begin{aligned}
l_{\text{dc}} &= \left(\frac{0.02 d_b}{\lambda}\right)\left(\frac{f_y}{\sqrt{f'_c}}\right) \\
&= \left(\frac{(0.02)(0.75 \text{ in})}{1.0}\right)\left(\frac{60{,}000 \dfrac{\text{lbf}}{\text{in}^2}}{\sqrt{3000 \dfrac{\text{lbf}}{\text{in}^2}}}\right) \\
&= 16.43 \text{ in} \quad [\text{controls}] \\
&= 0.0003 d_b f_y = (0.0003)(0.75 \text{ in})\left(60{,}000 \dfrac{\text{kips}}{\text{in}^2}\right) \\
&= 13.5 \text{ in}
\end{aligned}
$$

Since the bar is hooked, it is necessary to add the radius of the bend to the cover and the steel layer depth.

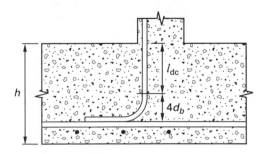

For no. 8 reinforcement,

$$
\begin{aligned}
h &= 16.43 \text{ in} + (4)(0.75 \text{ in}) + 2 \text{ in} + 3 \text{ in} \\
&= \boxed{24.43 \text{ in} \quad (25 \text{ in})}
\end{aligned}
$$

The answer is (C).

(c) The minimum steel permitted is

$$(0.0018)(14 \text{ ft})\left(12 \dfrac{\text{in}}{\text{ft}}\right)(24 \text{ in}) = \boxed{7.26 \text{ in}^2 \quad (7.3 \text{ in}^2)}$$

The answer is (B).

(d) Determine b_1 and b_2.

$$b_1 = 12 \text{ in} + 20 \text{ in} = 32 \text{ in}$$
$$b_2 = 16 \text{ in} + 20 \text{ in} = 36 \text{ in}$$

Use Eq. 55.12.

$$
\begin{aligned}
A_p &= 2(b_1 + b_2)d = (2)(32 \text{ in} + 36 \text{ in})(20 \text{ in}) \\
&= 2720 \text{ in}^2
\end{aligned}
$$

Use Eq. 55.14.

$$
\begin{aligned}
R &= \frac{P_u b_1 b_2}{A_f} = \frac{P_u(32 \text{ in})(36 \text{ in})}{(120 \text{ in})(168 \text{ in})} \\
&= 0.0571 P_u
\end{aligned}
$$

Use Eq. 55.13.

$$
\begin{aligned}
v_u &= \frac{P_u - R}{A_p} = \frac{P_u - 0.0571 P_u}{2720 \text{ in}^2} \\
&= (3.467 \times 10^{-4}) P_u
\end{aligned}
$$

The allowable punching shear stress is given by Eq. 55.17. $y = 2$.

$$
\begin{aligned}
\phi v_c &= \phi(2 + y)\lambda\sqrt{f'_c} = (0.75)(2 + 2)(1.0)\sqrt{3000 \dfrac{\text{lbf}}{\text{in}^2}} \\
&= 164.3 \text{ lbf/in}^2
\end{aligned}
$$

Equating v_u to ϕv_c,

$$P_u = \frac{164.3 \dfrac{\text{lbf}}{\text{in}^2}}{3.467 \times 10^{-4}} = \boxed{4.74 \times 10^5 \text{ lbf} \quad (480 \text{ kips})}$$

The answer is (B).

(e) The average soil pressure, q_u, is

$$q_u = \frac{P_u}{A_f} = \frac{P_u}{140 \text{ ft}^2} = 0.00714 P_u$$

$$e = \frac{\dfrac{(14 \text{ ft})\left(12 \dfrac{\text{in}}{\text{ft}}\right) - 16 \text{ in}}{2} - 20 \text{ in}}{12 \dfrac{\text{in}}{\text{ft}}} = 4.67 \text{ ft}$$

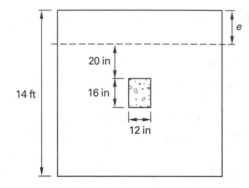

Use Eq. 55.11.

$$v_u = \frac{q_u e}{d} = \frac{(0.00714 P_u)(4.67 \text{ ft})\left(12 \dfrac{\text{in}}{\text{ft}}\right)}{20 \text{ in}} = 0.02 P_u$$

The allowable concrete stress is found using Eq. 55.5 and Eq. 55.6.

$$\phi v_c = \phi 2 \lambda \sqrt{f'_c} = \frac{(0.75)(2)(1.0)\sqrt{3000 \dfrac{\text{lbf}}{\text{in}^2}}\left(12 \dfrac{\text{in}}{\text{ft}}\right)^2}{1000 \dfrac{\text{lbf}}{\text{kip}}}$$

$$= 11.83 \text{ kips/ft}^2$$

Equating v_u to ϕv_c,

$$P_u = \frac{11.83 \dfrac{\text{kips}}{\text{ft}^2}}{0.02 \text{ ft}^2} = \boxed{591.5 \text{ kips} \quad (590 \text{ kips})}$$

The answer is (B).

(f) From Eq. 55.16,

$$J = \left(\frac{d b_1^3}{6}\right)\left(1 + \left(\frac{d}{b_1}\right)^2 + 3\left(\frac{b_2}{b_1}\right)\right)$$

$$= \left(\frac{(20 \text{ in})(32 \text{ in})^3}{6}\right)\left(1 + \left(\frac{20 \text{ in}}{32 \text{ in}}\right)^2 + (3)\left(\frac{36 \text{ in}}{32 \text{ in}}\right)\right)$$

$$= 520{,}530 \text{ in}^4$$

Use Eq. 55.14 and Eq. 55.15.

$$R = \frac{P_u b_1 b_2}{A_f} = \frac{(400 \text{ kips})(32 \text{ in})(36 \text{ in})}{(140 \text{ ft}^2)\left(12 \dfrac{\text{in}}{\text{ft}}\right)^2} = 22.86 \text{ kips}$$

$$\gamma_v = 1 - \frac{1}{1 + \frac{2}{3}\sqrt{\dfrac{b_1}{b_2}}} = 1 - \frac{1}{1 + \frac{2}{3}\sqrt{\dfrac{32 \text{ in}}{36 \text{ in}}}} = 0.386$$

Use Eq. 55.13.

$$v_u = \frac{P_u - R}{A_p} + \frac{\gamma_u M_u (0.5 b_1)}{J}$$

$$= \frac{400 \text{ kips} - 22.86 \text{ kips}}{2720 \text{ in}^2}$$

$$+ \frac{(0.386)(600 \text{ ft-kips})\left(12 \dfrac{\text{in}}{\text{ft}}\right)(0.5)(32 \text{ in})}{520{,}530 \text{ in}^4}$$

$$= \boxed{0.224 \text{ kip/in}^2 \quad (220 \text{ lbf/in}^2)}$$

The answer is (B).

(g) Determine the pressure distribution.

$$I_f = \tfrac{1}{12} L B^3 = \left(\tfrac{1}{12}\right)(14 \text{ ft})(10 \text{ ft})^3 = 1166.7 \text{ ft}^4$$

From Eq. 55.1,

$$q_u = \frac{P_u}{A_f} \pm \frac{M_s\left(\dfrac{B}{2}\right)}{I_f} = \frac{400 \text{ kips}}{140 \text{ ft}^2} \pm \frac{(600 \text{ ft-kips})(5 \text{ ft})}{1166.7 \text{ ft}^4}$$

$$= 2.86 \frac{\text{kips}}{\text{ft}^2} \pm 2.57 \frac{\text{kips}}{\text{ft}^2}$$

$$= (5.43 \text{ kips/ft}^2, 0.29 \text{ kip/ft}^2)$$

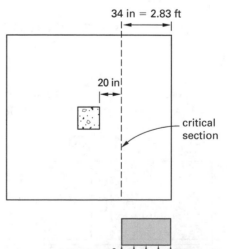

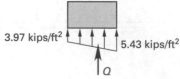

$$Q = \dfrac{\left(5.43 \ \dfrac{\text{kips}}{\text{ft}^2} + 3.97 \ \dfrac{\text{kips}}{\text{ft}^2}\right)(14 \ \text{ft})(2.83 \ \text{ft})}{2}$$

$$= 186.21 \ \text{kips}$$

$$v_u = \dfrac{186.21 \ \text{kips}}{(14 \ \text{ft})\left(12 \ \dfrac{\text{in}}{\text{ft}}\right)(20 \ \text{in})}$$

$$= \boxed{0.0554 \ \text{kip/in}^2 \quad (55 \ \text{lbf/in}^2)}$$

The answer is (D).

(h)

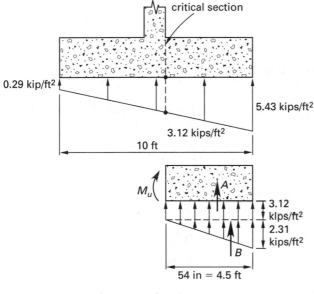

$$A = \left(3.12 \ \dfrac{\text{kips}}{\text{ft}^2}\right)(14 \ \text{ft})(4.5 \ \text{ft})$$

$$= 196.6 \ \text{kips}$$

$$B = \left(\tfrac{1}{2}\right)\left(2.31 \ \dfrac{\text{kips}}{\text{ft}^2}\right)(14 \ \text{ft})(4.5 \ \text{ft})$$

$$= 72.77 \ \text{kips}$$

$$M_u = (196.6 \ \text{kips})\left(\dfrac{4.5 \ \text{ft}}{2}\right)$$

$$+ \left(\tfrac{2}{3}\right)(72.77 \ \text{kips})(4.5 \ \text{ft})$$

$$= \boxed{660.7 \ \text{ft-kips} \quad (660 \ \text{ft-kips})}$$

The answer is (C).

(i) The ratio of the long side to the short side is

$$\beta = \dfrac{14 \ \text{ft}}{10 \ \text{ft}} = 1.4$$

Use Eq. 55.25.

$$A_1 = A_{\text{sd}}\left(\dfrac{2}{\beta + 1}\right) = (12 \ \text{in}^2)\left(\dfrac{2}{1.4 + 1}\right)$$

$$= 10 \ \text{in}^2$$

The number of no. 6 bars is

$$\dfrac{A_s}{A_b} = \dfrac{10 \ \text{in}^2}{0.44 \ \text{in}^2} = 22.73 \quad [23 \ \text{bars}]$$

The width of the band is 10 ft.

The spacing is

$$s = \dfrac{(10 \ \text{ft})\left(12 \ \dfrac{\text{in}}{\text{ft}}\right)}{23} = \boxed{5.2 \ \text{in}}$$

The answer is (C).

(j)

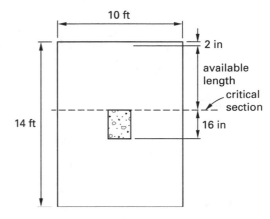

Use a 2 in tip cover.

$$\text{length} = \dfrac{(14 \ \text{ft})\left(12 \ \dfrac{\text{in}}{\text{ft}}\right) - 16 \ \text{in}}{2} - 2 \ \text{in} = \boxed{74 \ \text{in}}$$

The answer is (B).

5. *steps 1 and 2:* The unfactored (net) allowable soil pressure is

$$q_{a,\text{net}} = 4000 \ \dfrac{\text{lbf}}{\text{ft}^2} - \left(\dfrac{20 \ \text{in}}{12 \ \dfrac{\text{in}}{\text{ft}}}\right)\left(150 \ \dfrac{\text{lbf}}{\text{ft}^3}\right)$$

$$= 3750 \ \text{lbf/ft}^2$$

The required footing area is

$$A_f = \dfrac{240{,}000 \ \text{lbf}}{3750 \ \dfrac{\text{lbf}}{\text{ft}^2}} = 64.0 \ \text{ft}^2$$

$$\boxed{(8.0 \ \text{ft})(8.0 \ \text{ft})} = 64.0 \ \text{ft}^2 \quad [\text{governs}]$$

$$q'_{\text{actual}} = \dfrac{240{,}000 \ \text{lbf}}{64.0 \ \text{ft}^2} = 3750 \ \text{lbf/ft}^2$$

Assume a depth to reinforcement of $\boxed{d = 16 \ \text{in.}}$

The critical area is

$$(8.0 \text{ ft})^2 - \left(\frac{16 \text{ in} + 16 \text{ in}}{12 \frac{\text{in}}{\text{ft}}} \right)^2 = 56.89 \text{ ft}^2$$

The critical perimeter is

$$l_{\text{perim}} = \frac{(4)(16 \text{ in} + 16 \text{ in})}{12 \frac{\text{in}}{\text{ft}}} = 10.67 \text{ ft}$$

step 3: Since all of the load is live, the required ultimate shear is

$$V_u = (1.6) \left(3750 \frac{\text{lbf}}{\text{ft}^2} \right) (56.89 \text{ ft}^2)$$
$$= 341{,}340 \text{ lbf}$$

step 4: The allowable shear load is

$$\beta_c = \frac{16 \text{ in}}{16 \text{ in}} = 1$$

Use Eq. 55.17; $y = 2$.

$$V_c = (2 + y)\lambda \sqrt{f'_c} l_{\text{perim}} t$$
$$= \left((2 + 2)(1.0)\sqrt{3000 \frac{\text{lbf}}{\text{in}^2}} \right) \left(12 \frac{\text{in}}{\text{ft}} \right)^2$$
$$\times (10.67 \text{ ft}) \left(\frac{16 \text{ in}}{12 \frac{\text{in}}{\text{ft}}} \right)$$
$$= 448{,}834 \text{ lbf}$$

step 5: Use Eq. 55.5.

$$\phi V_c = (0.75)(448{,}834 \text{ lbf}) = 336{,}626 \text{ lbf}$$
$$V_u \approx \phi V_c \quad [\text{acceptable}]$$

(The footing is marginally undersized in capacity by 1.4%. If this is not acceptable, the footing should be increased in size.)

Square footings should also be checked for one-way shear.

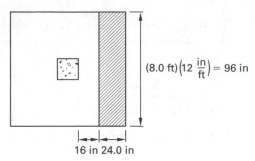

$$\text{critical area} = \frac{(24 \text{ in})(96 \text{ in})}{\left(12 \frac{\text{in}}{\text{ft}} \right)^2} = 16.0 \text{ ft}^2$$

$$V_u = (1.6) \left(3750 \frac{\text{lbf}}{\text{ft}^2} \right) (16.0 \text{ ft}^2)$$
$$= 96{,}000 \text{ lbf}$$

The allowable shear, using Eq. 55.6, is

$$V_c = 2\lambda \sqrt{f'_c} Bt = \left((2)(1.0)\sqrt{3000 \frac{\text{lbf}}{\text{in}^2}} \right) \left(12 \frac{\text{in}}{\text{ft}} \right)^2$$
$$\times (8.0 \text{ ft}) \left(\frac{16 \text{ in}}{12 \frac{\text{in}}{\text{ft}}} \right)$$
$$= 168{,}260 \text{ lbf}$$

$$\phi V_c = (0.75)(168{,}260 \text{ lbf}) = 126{,}195 \text{ lbf}$$
$$V_u < \phi V_c \quad [\text{acceptable}]$$

step 6:

$$l = \frac{8.0 \text{ ft} - \frac{16 \text{ in}}{12 \frac{\text{in}}{\text{ft}}}}{2} = 3.33 \text{ ft} \qquad [40.0 \text{ in}]$$

From Eq. 55.23,

$$M_u = \frac{q_u L l^2}{2} = \frac{(1.6) \left(3750 \frac{\text{lbf}}{\text{ft}^2} \right) (8.0 \text{ ft})(3.33 \text{ ft})^2}{2}$$
$$= 266{,}134 \text{ ft-lbf}$$

step 7: Solve for the required flexural steel using Eq. 50.32, assuming the section is tension controlled.

$$M_n = \frac{M_u}{\phi} = \frac{266{,}134 \text{ ft-lbf}}{0.9}$$
$$= 295{,}704 \text{ ft-lbf}$$
$$= \rho b d^2 f_y \left(1 - \frac{\rho f_y}{1.7 f'_c} \right)$$

$$\rho(8.0 \text{ ft})\left(12\,\tfrac{\text{in}}{\text{ft}}\right)(16\text{ in})^2\left(40{,}000\,\tfrac{\text{lbf}}{\text{in}^2}\right)$$

$$\times\left(1-\frac{\rho\left(40{,}000\,\tfrac{\text{lbf}}{\text{in}^2}\right)}{(1.7)\left(3000\,\tfrac{\text{lbf}}{\text{in}^2}\right)}\right)$$

$$=(295{,}704\text{ ft-lbf})\left(12\,\tfrac{\text{in}}{\text{ft}}\right)$$

$$\rho = 0.00371$$

The reinforcement ratio corresponding to $\epsilon_t = 0.005$ is

$$\rho = \frac{0.85\beta_1 f'_c}{f_y}\left(\frac{0.003}{0.003+0.005}\right)$$

$$=\left(\frac{(0.85)(0.85)\left(3000\,\tfrac{\text{lbf}}{\text{in}^2}\right)}{40{,}000\,\tfrac{\text{lbf}}{\text{in}^2}}\right)\left(\frac{0.003}{0.003+0.005}\right)$$

$$= 0.02$$

The section is tension controlled because $\rho = 0.00371 < 0.02$.

step 8:

$$A_s = (0.00371)(16\text{ in})(8.0\text{ ft})\left(12\,\tfrac{\text{in}}{\text{ft}}\right)$$

$$= 5.70\text{ in}^2$$

(If it is not necessary to know the reinforcement ratio, it is easier to use Eq. 55.24.)

$$\lambda = 0.1d = (0.1)(16\text{ in}) = 1.6\text{ in}$$

In this case, the section is tension controlled.

$$A_s = \frac{M_u}{\phi f_y(d-\lambda)}$$

$$=\frac{(266{,}134\text{ ft-lbf})\left(12\,\tfrac{\text{in}}{\text{ft}}\right)}{(0.9)\left(40{,}000\,\tfrac{\text{lbf}}{\text{in}^2}\right)(16\text{ in}-1.6\text{ in})}$$

$$= 6.16\text{ in}^2$$

Use 14 no. 6 bars (6.16 in²).

step 9: Minimum spacing is not a problem.

step 10: From Eq. 55.29, assuming straight bars with no hooks,

$$l_d = \frac{3d_b f_y \psi_t \psi_e}{50\lambda\sqrt{f'_c}} \ge 12\text{ in}$$

$$= \text{maximum}\begin{cases}\dfrac{(3)(0.75\text{ in})\left(40{,}000\,\tfrac{\text{lbf}}{\text{in}^2}\right)}{(50)(1.0)\sqrt{3000\,\tfrac{\text{lbf}}{\text{in}^2}}}\\[2pt] = 32.9\text{ in}\\ 12\text{ in}\end{cases}$$

$$32.9\text{ in} < 40.0\text{ in} - 2.0\text{ in} = 38.0\text{ in} \quad [\text{acceptable}]$$

6. This is similar to Prob. 5 except for the information about the column steel.

steps 1 and 2: Assume a 25 in thick footing. The approximate footing size is

$$A = \frac{200{,}000\text{ lbf}+145{,}000\text{ lbf}}{4000\,\tfrac{\text{lbf}}{\text{ft}^2}-\left(\tfrac{25\text{ in}}{12\,\tfrac{\text{in}}{\text{ft}}}\right)\left(150\,\tfrac{\text{lbf}}{\text{ft}^3}\right)}$$

$$= 93.6\text{ ft}^2$$

Try a 10 ft × 10 ft (100 ft²) footing.

(9.75 ft × 9.75 ft could also be used.)

$$q_{\text{actual}} = \frac{(1.2)(200{,}000\text{ lbf})+(1.6)(145{,}000\text{ lbf})}{100\text{ ft}^2}$$

$$= 4720\text{ lbf/ft}^2$$

Assume a depth to reinforcement of $d = 21$ in.

The critical area is

$$(10\text{ ft})^2 - \left(\frac{18\text{ in}+21\text{ in}}{12\,\tfrac{\text{in}}{\text{ft}}}\right)^2 = 89.44\text{ ft}^2$$

The critical perimeter is

$$(4)\left(\frac{18\text{ in}+21\text{ in}}{12\,\tfrac{\text{in}}{\text{ft}}}\right) = 13\text{ ft}$$

step 3: The ultimate shear is

$$V_u = (1.2)(200{,}000\text{ lbf})+(1.6)(145{,}000\text{ lbf})$$

$$= 472{,}000\text{ lbf}$$

step 4: Using Eq. 55.17, the allowable two-way shear is

$$y = 2 \quad [\text{minimum value}]$$

$$\phi V_c = \left((0.75)(2+2)(1.0)\sqrt{3000\ \frac{\text{lbf}}{\text{ft}^2}} \right) \left(12\ \frac{\text{in}}{\text{ft}} \right)^2$$

$$\times (13\ \text{ft}) \left(\frac{21\ \text{in}}{12\ \frac{\text{in}}{\text{ft}}} \right)$$

$$= 538{,}301\ \text{lbf}$$

step 5: Since $V_u < \phi V_c$, depth and size are acceptable. (The one-way shear should also be checked.)

step 6:

$$l = \frac{10\ \text{ft} - \dfrac{18\ \text{in}}{12\ \frac{\text{in}}{\text{ft}}}}{2} = 4.25\ \text{ft} \quad [51.0\ \text{in}]$$

Use Eq. 55.23.

$$M_u = \frac{q_u L l^2}{2} = \frac{\left(4720\ \frac{\text{lbf}}{\text{ft}^2} \right)(10\ \text{ft})(4.25\ \text{ft})^2}{2}$$

$$= 426{,}275\ \text{ft-lbf}$$

step 7: Assume the section is tension controlled.

$$M_n = \frac{M_u}{\phi} = \frac{(426{,}275\ \text{ft-lbf}) \left(12\ \frac{\text{in}}{\text{ft}} \right)}{0.90}$$

$$= 5.68 \times 10^6\ \text{in-lbf}$$

From Eq. 50.32,

$$M_n = \rho b d^2 f_y \left(1 - \frac{\rho f_y}{1.7 f_c'} \right)$$

$$5.68 \times 10^6\ \text{in-lbf} = \rho \left((10\ \text{ft}) \left(12\ \frac{\text{in}}{\text{ft}} \right) \right)(21\ \text{in})^2$$

$$\times \left(40{,}000\ \frac{\text{lbf}}{\text{in}^2} \right)$$

$$\times \left(1 - \frac{\rho \left(40{,}000\ \frac{\text{lbf}}{\text{in}^2} \right)}{(1.7) \left(3000\ \frac{\text{lbf}}{\text{in}^2} \right)} \right)$$

Solve for ρ.

$$\rho = 0.00274$$

With this low reinforcement ratio, the section is going to be tension controlled.

step 8:

$$A_s = (0.00274)(21\ \text{in})(10\ \text{ft}) \left(12\ \frac{\text{in}}{\text{ft}} \right)$$

$$= 6.90\ \text{in}^2$$

Use 12 no. 7 bars (7.2 in^2).

step 9: Minimum spacing is not a problem.

step 10: Assuming straight bars with no hooks, and using Eq. 55.30,

$$l_d = \frac{3 d_b f_y \psi_t \psi_e}{40 \lambda \sqrt{f_c'}} \geq 12\ \text{in}$$

$$= \text{maximum} \begin{cases} \dfrac{(3)(0.875\ \text{in}) \left(40{,}000\ \frac{\text{lbf}}{\text{in}^2} \right)}{(40)(1.0)\sqrt{3000\ \frac{\text{lbf}}{\text{in}^2}}} \\ = 47.9\ \text{in} \\ 12\ \text{in} \end{cases}$$

$$47.9\ \text{in} < 51.0\ \text{in} - 2.0\ \text{in} = 49.0\ \text{in} \quad [\text{acceptable}]$$

Since column steel was specified, check the need for dowel bars.

The gross column area is

$$A_g = (18\ \text{in})^2 = 324\ \text{in}^2$$

The bearing strength is

$$f_{\text{bearing}} = (0.85) \left(3000\ \frac{\text{lbf}}{\text{in}^2} \right) = 2550\ \text{lbf/in}^2$$

The bearing capacity of concrete is

$$\phi f_{\text{bearing}} A_g = (0.65) \left(2550\ \frac{\text{lbf}}{\text{in}^2} \right)(324\ \text{in}^2)$$

$$= 537{,}030\ \text{lbf}$$

Since 537,030 lbf > 472,000 lbf ($\phi P_n > P_u$), only the minimum dowel steel is required.

$$A_{s,\text{min}} = (0.005)(18\ \text{in})(18\ \text{in}) = 1.62\ \text{in}^2$$

Use four no. 6 bars (1.76 in^2).

Check the development length of the dowels in compression. From Eq. 55.36,

$$l_{dc} = \left(\frac{0.02d_b}{\lambda}\right)\left(\frac{f_y}{\sqrt{f_c'}}\right) \geq 0.0003d_b f_y \geq 8 \text{ in}$$

$$= \text{maximum} \begin{cases} \left(\dfrac{(0.02)(0.75 \text{ in})}{1.0}\right)\left(\dfrac{40{,}000 \frac{\text{lbf}}{\text{in}^2}}{\sqrt{3000 \frac{\text{lbf}}{\text{in}^2}}}\right) \\ \quad = 10.95 \text{ in} \\ (0.0003)(0.75 \text{ in})\left(40{,}000 \frac{\text{lbf}}{\text{in}^2}\right) \\ \quad = 9.0 \text{ in} \\ 8 \text{ in} \end{cases}$$

$$l_{dc} = 10.95 \text{ in} \quad \text{[controls]}$$

Since 10.95 in $< d = 21$ in, it is acceptable.

7. (a) Find the centroid of the load group.

$$\bar{x} = \frac{(3 \text{ ft})(60 \text{ kips}) + (13 \text{ ft})(80 \text{ kips})}{60 \text{ kips} + 80 \text{ kips}}$$

$$= 8.71 \text{ ft} \quad \text{[from right end]}$$

Calculate the eccentricities that would produce the same moments.

$$\epsilon_1 = \frac{M}{P} = \frac{120 \text{ ft-kips}}{80 \text{ kips}} = 1.5 \text{ ft}$$

$$\epsilon_2 = \frac{160 \text{ ft-kips}}{60 \text{ kips}} = 2.67 \text{ ft}$$

Move the loads.

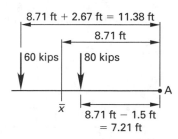

The total net vertical reaction must be 60 kips + 80 kips = 140 kips. This can be assumed to be at $\frac{1}{2}L$ if it is uniform.

$$\sum M_A = (60 \text{ kips})(11.38 \text{ ft}) + (80 \text{ kips})(7.21 \text{ ft})$$

$$- (140 \text{ kips})\left(\frac{L}{2}\right) = 0$$

$$L = \boxed{17.99 \text{ ft} \quad \text{[round to 18 ft]}}$$

Alternative Solution

Sum moments about the left free end.

$$(80 \text{ kips})(L - 13 \text{ ft}) + 120 \text{ ft-kips}$$

$$+ (60 \text{ kips})(L - 3 \text{ ft}) - 160 \text{ ft-kips}$$

$$- (140 \text{ kips})\left(\frac{L}{2}\right) = 0$$

$$L = \boxed{18 \text{ ft}}$$

(b) Calculate the soil pressure, q, then draw the shear and moment diagrams.

$$q = \frac{140 \text{ kips}}{18.0 \text{ ft}}$$

$$= 7.78 \text{ kips/ft} \quad [8 \text{ kips/ft}]$$

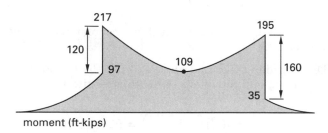

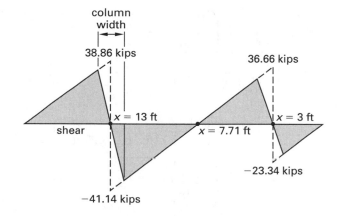

moment (ft-kips)

(c) | No top steel is needed because the entire footing is in single-curvature bending. |

8. (a) The eccentricity is

$$\epsilon = \frac{100 \text{ ft-kips}}{200 \text{ kips}} = 0.5 \text{ ft}$$

Assume a footing thickness of 24 in.

The net allowable soil pressure is

$$q_{a,\text{net}} = 6000 \frac{\text{lbf}}{\text{ft}^2} - \left(\frac{24 \text{ in}}{12 \frac{\text{in}}{\text{ft}}}\right)\left(150 \frac{\text{lbf}}{\text{ft}^3}\right) = 5700 \text{ lbf/ft}^2$$

For a footing with no moment,

$$A = \frac{200,000 \text{ lbf}}{5700 \frac{\text{lbf}}{\text{ft}^2}} = 35.09 \text{ ft}^2$$

$$B = \sqrt{35.09 \text{ ft}^2} = 5.92 \text{ ft} \quad [\text{round to 6 ft}]$$

$$q_{max} = 5700 \frac{\text{lbf}}{\text{ft}^2} = \left(\frac{200,000}{6L}\right)\left(1 + \frac{(6)(0.5)}{L}\right)$$

$$L = 8.03 \text{ ft}$$

$$\boxed{\text{Use a 6 ft} \times \text{8.25 ft footing.}}$$

(b) The pressures are

$$q_{max} = \left(\frac{200,000 \text{ lbf}}{(6 \text{ ft})(8.25 \text{ ft})}\right)\left(1 + \frac{(6 \text{ ft})(0.5)}{8.25 \text{ ft}}\right)$$
$$+ \left(\frac{24 \text{ in}}{12 \frac{\text{in}}{\text{ft}}}\right)\left(150 \frac{\text{lbf}}{\text{ft}^3}\right)$$
$$= \boxed{5810 \text{ lbf/ft}^2}$$

$$q_{min} = \left(\frac{200,000 \text{ lbf}}{(6 \text{ ft})(8.25 \text{ ft})}\right)\left(1 - \frac{(6 \text{ ft})(0.5)}{8.25 \text{ ft}}\right)$$
$$+ \left(\frac{24 \text{ in}}{12 \frac{\text{in}}{\text{ft}}}\right)\left(150 \frac{\text{lbf}}{\text{ft}^3}\right)$$
$$= \boxed{2871 \text{ lbf/ft}^2}$$

(c) Determine the minimum thickness required for one-way shear.

Disregarding dead load and factoring the live load, the maximum and minimum ultimate pressures are

$$q_{max} = (1.6)\left(5810 \frac{\text{lbf}}{\text{ft}^2}\right) = 9296 \text{ lbf/ft}^2$$

$$q_{min} = (1.6)\left(2871 \frac{\text{lbf}}{\text{ft}^2}\right) = 4594 \text{ lbf/ft}^2$$

Assume $d = 20$ in. The critical ultimate soil pressure is 20 in from the face of the column. The distance from the far (left) edge of the footing is

$$\frac{(8.25 \text{ ft})\left(12 \frac{\text{in}}{\text{ft}}\right)}{2} + \frac{14 \text{ in}}{2} + 20 \text{ in} = 76.5 \text{ in}$$

At the critical line, the soil pressure is

$$4594 \frac{\text{lbf}}{\text{ft}^2} + \left(9296 \frac{\text{lbf}}{\text{ft}^2} - 4594 \frac{\text{lbf}}{\text{ft}^2}\right)$$
$$\times \left(\frac{76.5 \text{ in}}{(8.25 \text{ ft})\left(12 \frac{\text{in}}{\text{ft}}\right)}\right)$$
$$= 8227 \text{ lbf/ft}^2$$

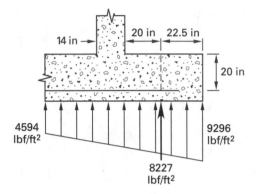

The required ultimate shear at the critical line is

$$V_u = \left(\frac{22.5 \text{ in}}{12 \frac{\text{in}}{\text{ft}}}\right)\left(8227 \frac{\text{lbf}}{\text{ft}^2}\right)$$
$$+ \left(\tfrac{1}{2}\right)\left(\frac{22.5 \text{ in}}{12 \frac{\text{in}}{\text{ft}}}\right)\left(9296 \frac{\text{lbf}}{\text{ft}^2} - 8227 \frac{\text{lbf}}{\text{ft}^2}\right)$$
$$= 16,428 \text{ lbf} \quad [\text{per foot of width}]$$

The ultimate shear strength is calculated from Eq. 55.6.

$$\phi v_c = \phi 2\lambda\sqrt{f_c'} = (0.75)(2)(1.0)\sqrt{3000 \frac{\text{lbf}}{\text{in}^2}}$$
$$= 82.2 \text{ lbf/in}^2$$

The depth required is

$$d = \frac{(16,428 \text{ lbf})\left(12 \frac{\text{in}}{\text{ft}}\right)}{\left(82.2 \frac{\text{lbf}}{\text{in}^2}\right)\left(12 \frac{\text{in}}{\text{ft}}\right)^2 (1 \text{ ft})}$$
$$= \boxed{16.7 \text{ in}}$$

Determine the minimum thickness required for two-way shear. The factored forces are

$$P_u = 200 \text{ kips}$$
$$M_u = 100 \text{ ft-kips}$$

Assume $d = 20$ in.

$$b_1 = b_2 = 14 \text{ in} + 20 \text{ in} = 34 \text{ in}$$

$$A_p = (b_1 + b_2)d = (34 \text{ in} + 34 \text{ in})(20 \text{ in}) = 1360 \text{ in}^2$$

$$R = \frac{P_u b_1 b_2}{A_f} = \frac{(200 \text{ kips})(34 \text{ in})^2}{(8.25 \text{ ft})(6 \text{ ft})\left(12 \frac{\text{in}}{\text{ft}}\right)^2} = 32.4 \text{ kips}$$

$$\gamma_v = 1 - \frac{1}{1 + \frac{2}{3}\sqrt{\frac{b_1}{b_2}}} = 0.4$$

$$J = \left(\frac{db_1^3}{6}\right)\left(1 + \left(\frac{d}{b_1}\right)^2 + 3\left(\frac{b_2}{b_1}\right)\right)$$

$$= \left(\frac{(20 \text{ in})(34 \text{ in})^3}{6 \text{ ft}}\right)\left(1 + \left(\frac{20 \text{ in}}{34 \text{ in}}\right)^2 + (3)\left(\frac{34 \text{ in}}{34 \text{ in}}\right)\right)$$

$$= 569{,}387 \text{ in}^4$$

$$v_u = \frac{P_u - R}{A_p} + \frac{\gamma_v M_u (0.5 b_1)}{J}$$

$$= \left(\begin{array}{c} \dfrac{200 \text{ kips} - 32.4 \text{ kips}}{1360 \text{ in}^2} \\[2mm] + \dfrac{(0.4)(100 \text{ ft-kips})\left(12 \frac{\text{in}}{\text{ft}}\right)(0.5)(34 \text{ in})}{569{,}387 \text{ in}^4} \end{array}\right)$$

$$\times \left(1000 \frac{\text{lbf}}{\text{kips}}\right)$$

$$= 137.6 \text{ lbf/in}^2$$

$$\phi v_c = \phi(2 + y)\lambda\sqrt{f'_c} = (0.75)(2 + 2)(1.0)\sqrt{3000 \frac{\text{lbf}}{\text{in}^2}}$$

$$= 164.3 \text{ lbf/in}^2$$

Since $164.3 \text{ lbf/in}^2 > 137.6 \text{ lbf/in}^2$, the thickness can be reduced. Through trial and error, the minimum required effective depth is

$$d = 18.2 \text{ in} \quad [\text{governs}]$$

The shear stress is

$$v_u = 163.5 \frac{\text{lbf}}{\text{in}^2} < 164.3 \frac{\text{lbf}}{\text{in}^2} \quad [\text{OK}]$$

The minimum footing depth is (using no. 4 bars for flexure)

$$h = 18.2 \text{ in} + (1.5)(0.5 \text{ in}) + 3 \text{ in} = \boxed{21.95 \text{ in} \quad (22 \text{ in})}$$

(d) The pressure at the critical flexure section is

$$4594 \frac{\text{lbf}}{\text{ft}^2} + \left(9296 \frac{\text{lbf}}{\text{ft}^2} - 4594 \frac{\text{lbf}}{\text{ft}^2}\right)$$

$$\times \left(\frac{(8.25 \text{ ft})\left(12 \frac{\text{in}}{\text{ft}}\right) - 20 \text{ in} - 22.5 \text{ in}}{(8.25 \text{ ft})\left(12 \frac{\text{in}}{\text{ft}}\right)}\right)$$

$$= 7277 \text{ lbf/ft}^2$$

The moment per unit length at the critical line is given by Eq. 55.23.

$$\frac{M_u}{L} = \frac{q_u l^2}{2}$$

$$= \left(\tfrac{1}{2}\right)\left(7277 \frac{\text{lbf}}{\text{ft}^2}\right)\left(\frac{42.5 \text{ in}}{12 \frac{\text{in}}{\text{ft}}}\right)^2$$

$$+ \left(\tfrac{1}{2}\right)\left(\frac{42.5 \text{ in}}{12 \frac{\text{in}}{\text{ft}}}\right)\left(9296 \frac{\text{lbf}}{\text{ft}^2} - 7277 \frac{\text{lbf}}{\text{ft}^2}\right)$$

$$\times \left(\tfrac{2}{3}\right)\left(\frac{42.5 \text{ in}}{12 \frac{\text{in}}{\text{ft}}}\right)$$

$$= \boxed{54{,}081 \text{ ft-lbf per foot of footing}}$$

Structural

56 Prestressed Concrete

PRACTICE PROBLEMS

1. Calculate the nominal moment strength of the cross section of the prestressed beam shown. $f'_c = 5000$ psi, $f_{pu} = 270,000$ psi (stress-relieved strands), and $f_{py} = 0.85f_{pu}$.

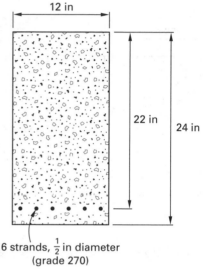

12 in

22 in 24 in

6 strands, $\frac{1}{2}$ in diameter
(grade 270)

2. The cross section of the simply supported pretensioned concrete beam is shown. The superimposed live and dead loads (in excess of beam weight) are as indicated. The beam has straight cables with initial prestress of 216 kips and final prestress after losses of 140 kips. The concrete is uncracked. The long-term deflection factor, λ, is 2.0, and the modulus of elasticity is 4×10^6 psi. (a) Calculate the centerline deflection of the beam immediately after the cables are cut. (b) Calculate the centerline deflection of the beam five years after the cables are cut.

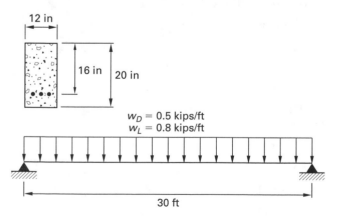

12 in

16 in 20 in

$w_D = 0.5$ kips/ft
$w_L = 0.8$ kips/ft

30 ft

3. (*Time limit: one hour*) The 12 in × 24 in prestressed beam shown carries a dead load of 3 kips/ft (which includes the beam's self-weight). The beam length is 20 ft. The tensile force in the prestressing steel is 250 kips.

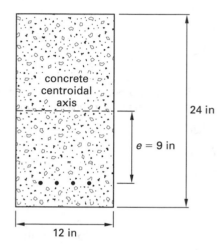

concrete centroidal axis

24 in

$e = 9$ in

12 in

(a) The moment of inertia of the cross section of the beam is most nearly

(A) 13,000 in⁴

(B) 14,000 in⁴

(C) 15,000 in⁴

(D) 16,000 in⁴

(b) The area of the cross section of the beam is most nearly

(A) 250 in²

(B) 290 in²

(C) 300 in²

(D) 320 in²

(c) The maximum bending moment due to dead load is most nearly

(A) 100 ft-kips

(B) 130 ft-kips

(C) 150 ft-kips

(D) 180 ft-kips

(d) The axial compression stress due to the prestress at the top fibers at the midspan of the beam is most nearly

(A) 0.47 ksi

(B) 0.57 ksi

(C) 0.77 ksi

(D) 0.87 ksi

(e) The axial compression stress due to the prestress at the bottom fibers at the midspan of the beam is most nearly

(A) 0.47 ksi

(B) 0.57 ksi

(C) 0.77 ksi

(D) 0.87 ksi

(f) The axial compression stress due to the prestress at the top fibers at the ends of the beam is most nearly

(A) 0.47 ksi

(B) 0.57 ksi

(C) 0.77 ksi

(D) 0.87 ksi

(g) The total stress at the top fibers at the midspan of the beam is most nearly

(A) 0.48 ksi

(B) 0.87 ksi

(C) 1.6 ksi

(D) 2.0 ksi

(h) The total stress at the bottom fibers at the midspan of the beam is most nearly

(A) 0.87 ksi

(B) 1.3 ksi

(C) 2.0 ksi

(D) 2.3 ksi

(i) The total stress at the top fibers at the ends of the beam is most nearly

(A) 0.48 ksi

(B) 0.87 ksi

(C) 1.1 ksi

(D) 1.6 ksi

(j) The total stress at the bottom fibers at the ends of the beam is most nearly

(A) 0.87 ksi

(B) 2.0 ksi

(C) 2.8 ksi

(D) 3.8 ksi

4. (*Time limit: one hour*) The cross section for a 12 m simply supported beam is shown. The initial tendon prestress is 1.10 GPa. $f_c' = 34.5$ MPa, and $f_{pu} = 1.725$ GPa. The total area of prestressing steel is 774 mm^2.

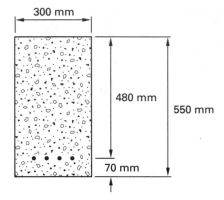

(a) The eccentricity for this section of the beam is most nearly

(A) 70 mm

(B) 210 mm

(C) 480 mm

(D) 550 mm

(b) The maximum bending moment due to the beam dead load is most nearly

(A) 60 kN·m

(B) 63 kN·m

(C) 70 kN·m

(D) 73 kN·m

(c) The initial force used for the initial prestressing is most nearly

(A) 800 kN

(B) 810 kN

(C) 840 kN

(D) 850 kN

(d) The concrete stress in the top of the beam at midspan after the tendons are cut is most nearly

(A) 1.8 MPa

(B) 4.6 MPa

(C) 5.2 MPa

(D) 12 MPa

(e) The concrete stress in the bottom of the beam at midspan after the tendons are cut is most nearly

(A) 4.6 MPa

(B) 5.2 MPa

(C) 10 MPa

(D) 12 MPa

(f) Assuming the losses in the tendons are 18%, the concrete stress at the top of the beam at midspan is most nearly

(A) 0.62 MPa

(B) 1.8 MPa

(C) 4.6 MPa

(D) 5.2 MPa

(g) Assuming the losses in the tendons are 18%, the concrete stress in the bottom of the beam at midspan is most nearly

(A) 5.2 MPa

(B) 4.6 MPa

(C) 9.1 MPa

(D) 9.8 MPa

(h) Assuming the allowable stress is $0.45f'_c$ in compression, the maximum allowable bending moment at midspan is most nearly

(A) 250 kN·m

(B) 290 kN·m

(C) 310 kN·m

(D) 340 kN·m

(i) Assuming the allowable stress is $0.5\sqrt{f'_c}$ in tension, the maximum allowable bending moment at midspan is most nearly

(A) 250 kN·m

(B) 270 kN·m

(C) 300 kN·m

(D) 330 kN·m

(j) Using the answers from parts (h) and (i), the maximum allowable uniform live load that the beam can support in addition to its own weight is most nearly

(A) 8.1 kN/m

(B) 8.6 kN/m

(C) 9.1 kN/m

(D) 10 kN/m

5. (*Time limit: one hour*) A prestressed hollow box cross section is shown. The prestressing is applied through twelve $1/2$ in diameter grade 270 seven-wire strand tendons. For these strands, $f_{py} = 0.95f_{pu}$, and the effective prestress after losses is $f_{se} = 148$ ksi. The specified strength of the concrete is 5000 psi. (a) Determine the flexural strength of the section. (b) Determine the cracking moment. (c) Does the relationship between the cracking moment and the flexural capacity satisfy the requirements of the ACI 318 code?

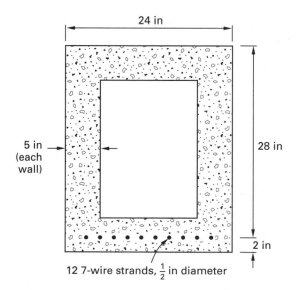

12 7-wire strands, $\frac{1}{2}$ in diameter

6. (*Time limit: one hour*) A pre-tensioned girder is constructed with 30 bonded strands. The girder spans 100 ft and is supported on simple supports. It carries a uniform live load of 540 lbf/ft as well as its own weight of 150 lbf/ft. The modular ratio is 7, $f'_c = 5000$ psi, $f'_{ci} = 3800$ psi, $f_{pu} = 250$ ksi, $f_y/f_{pu} = 0.85$, and $A_p = 0.144$ in²/strand. It is not necessary to check the tendon stress. (a) If the effective prestress before losses is 80% of f_{pu}, is the stress in the concrete acceptable immediately after transfer, before any live load is added? (b) If the effective prestress after losses is 60% of f_{pu}, is the concrete stress acceptable under service loading?

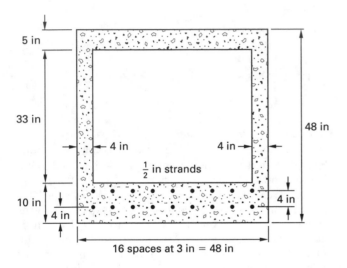

5 in

33 in

48 in

4 in   4 in

$\frac{1}{2}$ in strands

10 in

4 in   4 in

16 spaces at 3 in = 48 in

SOLUTIONS

1. For the nominal moment strength of the member's cross section,

$\gamma_p = 0.40$ for $f_{py}/f_{pu} = 0.85$ for stress-relieved strand

$\beta_1 = 0.80$ for $f'_c = 5000$ lbf/in²

$$\rho_p = \frac{A_{ps}}{bd_p} = \frac{(6)(0.153 \text{ in}^2)}{(12 \text{ in})(22 \text{ in})}$$

$$= 0.00348$$

For a fully prestressed member, Eq. 56.10 reduces to

$$f_{ps} = f_{pu}\left(1 - \left(\frac{\gamma_p}{\beta_1}\right)\rho_p\left(\frac{f_{pu}}{f'_c}\right)\right)$$

$$= \left(270 \ \frac{\text{kips}}{\text{in}^2}\right)$$

$$\times \left(1 - \left(\frac{0.40}{0.80}\right)(0.00348)\left(\frac{270 \ \frac{\text{kips}}{\text{in}^2}}{5 \ \frac{\text{kips}}{\text{in}^2}}\right)\right)$$

$$= 245 \text{ kips/in}^2$$

The depth of the compression block is

$$a = \frac{A_{ps}f_{ps}}{0.85bf'_c}$$

$$= \frac{(6)(0.153 \text{ in}^2)\left(245 \ \frac{\text{kips}}{\text{in}^2}\right)}{(0.85)(12 \text{ in})\left(5 \ \frac{\text{kips}}{\text{in}^2}\right)}$$

$$= 4.41 \text{ in}$$

The nominal moment strength is

$$M_n = A_{ps}f_{ps}\left(d_p - \frac{a}{2}\right)$$

$$= (0.918 \text{ in}^2)\left(245 \ \frac{\text{kips}}{\text{in}^2}\right)\left(\frac{22 \text{ in} - \frac{4.41 \text{ in}}{2}}{12 \ \frac{\text{in}}{\text{ft}}}\right)$$

$$= \boxed{371 \text{ ft-kips}}$$

2. (a) The gross moment of inertia is

$$I_g = \frac{bh^3}{12} = \frac{(12 \text{ in})(20 \text{ in})^3}{12} = 8000 \text{ in}^4$$

The eccentricity is

$$e = 16 \text{ in} - 10 \text{ in}$$
$$= 6 \text{ in}$$

The loads are not factored when calculating deflections. The beam's dead weight is

$$w_D = \frac{(12 \text{ in})(20 \text{ in})\left(150 \frac{\text{lbf}}{\text{ft}^3}\right)}{\left(12 \frac{\text{in}}{\text{ft}}\right)^2}$$
$$= 250 \text{ lbf/ft}$$

The deflection due to the pretensioning cables is

$$\delta_1 = \frac{-PeL^2}{8E_cI}$$
$$= \frac{-(216{,}000 \text{ lbf})(6 \text{ in})\left((30 \text{ ft})\left(12 \frac{\text{in}}{\text{ft}}\right)\right)^2}{(8)\left(4 \times 10^6 \frac{\text{lbf}}{\text{in}^2}\right)(8000 \text{ in}^4)}$$
$$= -0.656 \text{ in} \quad [\text{upward}]$$

The deflection due to the beam's own weight is

$$\delta_2 = \frac{5wL^4}{384E_cI}$$
$$= \frac{(5)\left(\dfrac{250 \frac{\text{lbf}}{\text{ft}}}{12 \frac{\text{in}}{\text{ft}}}\right)\left((30 \text{ ft})\left(12 \frac{\text{in}}{\text{ft}}\right)\right)^4}{(384)\left(4 \times 10^6 \frac{\text{lbf}}{\text{in}^2}\right)(8000 \text{ in}^4)}$$
$$= 0.142 \text{ in} \quad [\text{downward}]$$

The total deflection is

$$\delta = -0.656 \text{ in} + 0.142 \text{ in}$$
$$= \boxed{-0.514 \text{ in} \quad [\text{upward}]}$$

(b) The 5 yr deflection from the cable can be found proportionally.

$$\delta_1 = (2)\left(\frac{140 \text{ kips}}{216 \text{ kips}}\right)(-0.656 \text{ in})$$
$$= -0.85 \text{ in}$$

The deflection due to beam weight is

$$\delta_2 = \frac{\lambda 5wL^4}{384E_cI}$$
$$= \frac{(2)(5)\left(\dfrac{500 \frac{\text{lbf}}{\text{ft}} + 250 \frac{\text{lbf}}{\text{ft}}}{12 \frac{\text{in}}{\text{ft}}}\right)\left((30 \text{ ft})\left(12 \frac{\text{in}}{\text{ft}}\right)\right)^4}{(384)\left(4 \times 10^6 \frac{\text{lbf}}{\text{in}^2}\right)(8000 \text{ in}^4)}$$
$$= 0.854 \text{ in} \quad [\text{downward}]$$

The deflection due to the live load is found proportionally.

$$\left(\frac{800 \frac{\text{lbf}}{\text{ft}}}{250 \frac{\text{lbf}}{\text{ft}}}\right)(0.142 \text{ in}) = 0.454 \text{ in} \quad [\text{downward}]$$

The total deflection after 5 yr is

$$\delta = -0.85 \text{ in} + 0.854 \text{ in} + 0.454 \text{ in}$$
$$= \boxed{0.458 \text{ in} \quad [\text{downward}]}$$

This assumes a Class U section. This should be verified by computing extreme fiber stress, f_t, at service loads.

3. (a) $I = \dfrac{bh^3}{12} = \dfrac{(12 \text{ in})(24 \text{ in})^3}{12}$
$$= \boxed{13{,}824 \text{ in}^4 \quad (14{,}000 \text{ in}^4)}$$

The answer is (B).

(b) $A = (12 \text{ in})(24 \text{ in}) = \boxed{288 \text{ in}^2 \quad (290 \text{ in}^2)}$

The answer is (B).

(c) $M_{\max} = \dfrac{wL^2}{8} = \dfrac{\left(3 \frac{\text{kips}}{\text{ft}}\right)(20 \text{ ft})^2}{8} = \boxed{150 \text{ ft-kips}}$

The answer is (C).

(d) $f = \dfrac{-P}{A} = \dfrac{-250 \text{ kips}}{288 \text{ in}^2}$
$$= \boxed{-0.868 \text{ kip/in}^2 \quad (0.87 \text{ ksi})}$$

The answer is (D).

(e) $f = \dfrac{-P}{A} = \dfrac{-250 \text{ kips}}{288 \text{ in}^2}$

$\quad = \boxed{-0.868 \text{ kip/in}^2 \quad (0.87 \text{ ksi})}$

The answer is (D).

(f) $f = \dfrac{-P}{A} = \dfrac{-250 \text{ kips}}{288 \text{ in}^2}$

$\quad = \boxed{-0.868 \text{ kip/in}^2 \quad (0.87 \text{ ksi})}$

The answer is (D).

(g) Use Eq. 56.13. $y = c$ at the outer beam surface.

$$f_{\text{top}} = \frac{-P}{A} + \frac{Pec}{I} - \frac{M_s c}{I}$$

$$= \frac{-250 \text{ kips}}{288 \text{ in}^2}$$

$$+ \frac{(250 \text{ kips})(9 \text{ in})(12 \text{ in})}{13{,}824 \text{ in}^4}$$

$$- \frac{(150 \text{ ft-kips})\left(12 \dfrac{\text{in}}{\text{ft}}\right)(12 \text{ in})}{13{,}824 \text{ in}^4}$$

$$= \boxed{-0.477 \text{ kip/in}^2 \quad (0.48 \text{ ksi})}$$

The answer is (A).

(h) $f_{\text{bottom}} = \dfrac{-P}{A} - \dfrac{Pec}{I} + \dfrac{M_s c}{I}$

$$= \frac{-250 \text{ kips}}{288 \text{ in}^2} - \frac{(250 \text{ kips})(9 \text{ in})(12 \text{ in})}{13{,}824 \text{ in}^4}$$

$$+ \frac{(150 \text{ ft-kips})\left(12 \dfrac{\text{in}}{\text{ft}}\right)(12 \text{ in})}{13{,}824 \text{ in}^4}$$

$$= \boxed{-1.259 \text{ kips/in}^2 \quad (1.3 \text{ ksi})}$$

The answer is (B).

(i) $f_{\text{top}} = \dfrac{-P}{A} + \dfrac{Pec}{I}$

$$= \frac{-250 \text{ kips}}{288 \text{ in}^2} + \frac{(250 \text{ kips})(9 \text{ in})(12 \text{ in})}{13{,}824 \text{ in}^4}$$

$$= \boxed{1.085 \text{ kips/in}^2 \quad (1.1 \text{ ksi})}$$

The answer is (C).

(j) $f_{\text{bottom}} = \dfrac{-P}{A} - \dfrac{Pec}{I}$

$$= \frac{-250 \text{ kips}}{288 \text{ in}^2} - \frac{(250 \text{ kips})(9 \text{ in})(12 \text{ in})}{13{,}824 \text{ in}^4}$$

$$= \boxed{-2.821 \text{ kips/in}^2 \quad (2.8 \text{ ksi})}$$

The answer is (C).

Part (h) shows the critical midspan section to be Class U, which justifies the use of the gross moment of inertia in stress calculations.

4. (a) The eccentricity is

$$e = \frac{550 \text{ mm}}{2} - 70 \text{ mm}$$

$$= \boxed{205 \text{ mm} \quad (210 \text{ mm})}$$

The answer is (B).

(b) The beam's own loading is

$$w = \frac{(300 \text{ mm})(550 \text{ mm})\left(23.5 \dfrac{\text{kN}}{\text{m}^3}\right)}{\left(1000 \dfrac{\text{mm}}{\text{m}}\right)^2}$$

$$= 3.878 \text{ kN/m}$$

$$M_{\text{max}} = \frac{wL^2}{8}$$

$$= \frac{\left(3.878 \dfrac{\text{kN}}{\text{m}}\right)(12 \text{ m})^2}{8}$$

$$= \boxed{69.80 \text{ kN·m} \quad (70 \text{ kN·m})}$$

The answer is (C).

(c) $P_{\text{initial}} = \dfrac{(1.1 \text{ GPa})\left(10^9 \dfrac{\text{Pa}}{\text{GPa}}\right)(774 \text{ mm}^2)}{\left(1000 \dfrac{\text{mm}}{\text{m}}\right)^2 \left(1000 \dfrac{\text{N}}{\text{kN}}\right)}$

$$= \boxed{851.4 \text{ kN} \quad (850 \text{ kN})}$$

The answer is (D).

(d)
$$c = \frac{550 \text{ mm}}{2} = 275 \text{ mm}$$

$$A = (300 \text{ mm})(550 \text{ mm}) = 165\,000 \text{ mm}^2$$

$$I = \frac{bh^3}{12} = \frac{(300 \text{ mm})(550 \text{ mm})^3}{12}$$

$$= 4.159 \times 10^9 \text{ mm}^4$$

$$f_{\text{top}} = \frac{-P}{A} + \frac{Pec}{I} - \frac{M_s c}{I}$$

$$= \frac{-851\,400 \text{ N}}{165\,000 \text{ mm}^2}$$

$$+ \frac{(851\,400 \text{ N})(205 \text{ mm})(275 \text{ mm})}{4.159 \times 10^9 \text{ mm}^4}$$

$$- \frac{(69.80 \text{ kN·m})(275 \text{ mm}) \times \left(1000 \frac{\text{mm}}{\text{m}}\right)\left(1000 \frac{\text{N}}{\text{kN}}\right)}{4.159 \times 10^9 \text{ mm}^4}$$

$$= -5.16 \text{ MPa} + 11.54 \text{ MPa} - 4.61 \text{ MPa}$$

$$= \boxed{1.77 \text{ MPa} \quad (1.8 \text{ MPa})}$$

The answer is (A).

(e)
$$f_{\text{bottom}} = \frac{-P}{A} - \frac{Pec}{I} + \frac{M_s c}{I}$$

$$= \frac{-851\,400 \text{ N}}{165\,000 \text{ mm}^2}$$

$$- \frac{(851\,400 \text{ N})(205 \text{ mm})(275 \text{ mm})}{4.159 \times 10^9 \text{ mm}^4}$$

$$+ \frac{(69.80 \text{ kN·m})(275 \text{ mm}) \times \left(1000 \frac{\text{mm}}{\text{m}}\right)\left(1000 \frac{\text{N}}{\text{kN}}\right)}{4.159 \times 10^9 \text{ mm}^4}$$

$$= -5.16 \text{ MPa} - 11.54 \text{ MPa} + 4.61 \text{ MPa}$$

$$= \boxed{-12.09 \text{ MPa} \quad (12 \text{ MPa})}$$

The answer is (D).

(f) After an 18% loss, there is 82% remaining.

$$f_{\text{top}} = (-5.16 \text{ MPa})(0.82) + (11.54 \text{ MPa})(0.82)$$

$$- 4.61 \text{ MPa}$$

$$= \boxed{0.622 \text{ MPa} \quad (0.62 \text{ MPa})}$$

The answer is (A).

(g)
$$f_{\text{bottom}} = (-5.16 \text{ MPa})(0.82)$$

$$- (11.54 \text{ MPa})(0.82) + 4.61 \text{ MPa}$$

$$= \boxed{-9.08 \text{ MPa} \quad (9.1 \text{ MPa})}$$

The answer is (C).

(h)
$$f_{\text{top}} = \frac{-P}{A} + \frac{Pec}{I} - \frac{M_s c}{I}$$

$$(-0.45)(34.5 \text{ MPa}) = (0.82)\left(\frac{-851\,400 \text{ N}}{165\,000 \text{ mm}^2}\right)$$

$$+ \frac{(0.82)(851\,400 \text{ N}) \times (205 \text{ mm})(275 \text{ mm})}{4.159 \times 10^9 \text{ mm}^4}$$

$$- \frac{M(275 \text{ mm}) \times \left(1000 \frac{\text{mm}}{\text{m}}\right)\left(1000 \frac{\text{N}}{\text{kN}}\right)}{4.159 \times 10^9 \text{ mm}^4}$$

$$M = \boxed{313.9 \text{ kN·m} \quad (310 \text{ kN·m})}$$

The answer is (C).

(i)
$$f_{\text{bottom}} = \frac{-P}{A} - \frac{Pec}{I} + \frac{M_s c}{I}$$

$$0.5\sqrt{34.5 \text{ MPa}} = (0.82)\left(\frac{-851\,400 \text{ N}}{165\,000 \text{ mm}^2}\right)$$

$$- \frac{(0.82)(851\,400 \text{ N}) \times (205 \text{ mm})(275 \text{ mm})}{4.159 \times 10^9 \text{ mm}^4}$$

$$+ \frac{M(275 \text{ mm})\left(1000 \frac{\text{mm}}{\text{m}}\right) \times \left(1000 \frac{\text{N}}{\text{kN}}\right)}{4.159 \times 10^9 \text{ mm}^4}$$

$$M = \boxed{251.5 \text{ kN·m} \quad (250 \text{ kN·m})}$$

The answer is (A).

(j) From parts (h) and (i), the maximum bending moment is 251.5 kN·m.

$$M_{\text{max}} = \frac{wL^2}{8}$$

$$251.5 \text{ kN·m} = \frac{w(12 \text{ m})^2}{8}$$

$$w = 13.972 \text{ kN/m}$$

Structural

From part (b), the beam weight is 3.878 kN/m.

The allowable uniform load is

$$13.972 \ \frac{\text{kN}}{\text{m}} - 3.878 \ \frac{\text{kN}}{\text{m}} = \boxed{10.095 \ \text{kN/m} \quad (10 \ \text{kN/m})}$$

The answer is (D).

The tensile stress limitation in part (i) makes this a Class U section.

5. (a) The total area of the prestressing steel is

$$A_{\text{ps}} = (12)(0.153 \ \text{in}^2) = 1.836 \ \text{in}^2$$

From Eq. 56.11, the prestressing steel ratio is

$$\begin{aligned}
\rho_p &= \frac{A_{\text{ps}}}{bd_p} \\
&= \frac{1.836 \ \text{in}^2}{(24 \ \text{in})(28 \ \text{in})} \\
&= 0.0028 \quad \text{[round up]}
\end{aligned}$$

Since $f_{\text{py}}/f_{\text{pu}} > 0.9$, $\gamma_p = 0.28$ and $\beta_1 = 0.8$. The stress in the tendons at the attainment of the nominal moment strength is

$$\begin{aligned}
f_{\text{ps}} &= f_{\text{pu}}\left(1 - \left(\frac{\gamma_p}{\beta_1}\right)\left(\rho_p\left(\frac{f_{\text{pu}}}{f'_c}\right) + \left(\frac{d}{d_p}\right)(\omega - \omega')\right)\right) \\
&= \left(270 \ \frac{\text{kips}}{\text{in}^2}\right) \\
&\quad \times \left(1 - \left(\frac{0.28}{0.80}\right)\left((0.0028)\left(\frac{270 \ \frac{\text{kips}}{\text{in}^2}}{5 \ \frac{\text{kips}}{\text{in}^2}}\right)\right)\right) \\
&= 255.7 \ \text{kips/in}^2
\end{aligned}$$

The nominal moment strength is computed in the same manner as for a standard reinforced section. The total tension at maximum stress is

$$\begin{aligned}
T &= A_{\text{ps}}f_{\text{ps}} \\
&= (1.836 \ \text{in}^2)\left(255.7 \ \frac{\text{kips}}{\text{in}^2}\right) \\
&= 469.5 \ \text{kips}
\end{aligned}$$

The area of concrete required to balance this force is

$$\begin{aligned}
A_c &= \frac{T}{0.85f'_c} = \frac{469.5 \ \text{kips}}{(0.85)\left(5 \ \frac{\text{kips}}{\text{in}^2}\right)} \\
&= 110.5 \ \text{in}^2
\end{aligned}$$

Assume the depth of the equivalent rectangular stress block is within the 5 in thickness.

$$\begin{aligned}
a &= \frac{A_c}{b} \\
&= \frac{110.5 \ \text{in}^2}{24 \ \text{in}} \\
&= 4.60 \ \text{in}
\end{aligned}$$

The assumption is valid. The centroid of the compressed area is at $\lambda = a/2$.

$$\begin{aligned}
M_n &= T(d - \lambda) \\
&= \frac{(469.5 \ \text{kips})\left(28 \ \text{in} - \frac{4.60 \ \text{in}}{2}\right)}{12 \ \frac{\text{in}}{\text{ft}}} \\
&= 1006 \ \text{ft-kips}
\end{aligned}$$

Check to see if the section is tension controlled.

$$\begin{aligned}
\epsilon_t &= (0.003)\left(\frac{d - c}{c}\right) \\
&= (0.003)\left(\frac{28 \ \text{in} - 5.75 \ \text{in}}{5.75 \ \text{in}}\right) \\
&= 0.0116 > 0.005
\end{aligned}$$

Therefore, the section is tension controlled.

The design moment capacity is

$$\begin{aligned}
\phi M_n &= (0.90)(1006 \ \text{ft-kips}) \\
&= \boxed{905 \ \text{ft-kips}}
\end{aligned}$$

(b) The cracking moment, M_{cr}, is computed as the value that leads to a maximum tensile stress equal to the modulus of rupture. The gross moment of inertia may be used in the calculation.

$$A = (30 \text{ in})(24 \text{ in}) - (20 \text{ in})(14 \text{ in})$$
$$= 440 \text{ in}^2$$

$$I = \frac{b_1 h_1^3}{12} - \frac{b_2 h_2^3}{12}$$
$$= \left(\tfrac{1}{12}\right)\left((24 \text{ in})(30 \text{ in})^3 - (14 \text{ in})(20 \text{ in})^3\right)$$
$$= 44{,}667 \text{ in}^4$$

$y = $ distance from neutral axis to tension fiber
$$= 15 \text{ in}$$

$$P = f_{se} A_{ps} = \left(148 \ \frac{\text{kips}}{\text{in}^2}\right)(1.836 \text{ in}^2)$$
$$= 271.7 \text{ kips}$$

$e = $ eccentricity of tendons
$$= 13 \text{ in}$$

$$f_r = 7.5\lambda\sqrt{f_c'} = (7.5)(1.0)\sqrt{5000 \ \frac{\text{lbf}}{\text{in}^2}}$$
$$= 530 \text{ lbf/in}^2$$

$$f_c = \frac{-P}{A} + \frac{Pec}{I} - \frac{M_{cr}c}{I}$$

$$0.53 \ \frac{\text{kips}}{\text{in}^2} = \frac{-271.7 \text{ kips}}{440 \text{ in}^2}$$
$$+ \frac{(271.7 \text{ kips})(13 \text{ in})(-15 \text{ in})}{44{,}667 \text{ in}^4}$$
$$- \frac{M_{cr}(-15 \text{ in})\left(12 \ \frac{\text{in}}{\text{ft}}\right)}{44{,}667 \text{ in}^4}$$

$$M_{cr} = \boxed{579.1 \text{ ft-kips}}$$

(c) ACI 318 Sec. 18.8.2 requires that the design moment, ϕM_n, exceed the cracking moment by at least 20%.

$$\frac{\phi M_n}{M_{cr}} = \frac{905 \text{ ft-kips}}{579.1 \text{ ft-kips}}$$
$$= 1.56 \quad \boxed{[56\% \text{ increase, so OK}]}$$

6. (a) To determine if the stress in the concrete is acceptable immediately after transfer, follow these steps.

step 1: $w_D = 150 \text{ lbf/ft}$ [given]

step 2: $M_D = \dfrac{w_D L^2}{8} = \dfrac{\left(150 \ \frac{\text{lbf}}{\text{ft}}\right)(100 \text{ ft})^2}{8}$
$$= 187{,}500 \text{ ft-lbf}$$

step 3: Use the top as the reference. Neglect the tension holes.

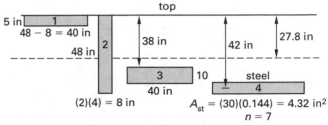

$$\bar{x} = \frac{\sum A_i \bar{x}_i}{\sum A_i}$$

$$(5 \text{ in})(40 \text{ in})(2.5 \text{ in}) + (48 \text{ in})(8 \text{ in})(24 \text{ in})$$
$$+ (10 \text{ in})(40 \text{ in})(38 \text{ in} + 5 \text{ in})$$
$$= \frac{+ (4.32 \text{ in}^2)(7)(42 \text{ in})}{(5 \text{ in})(40 \text{ in}) + (48 \text{ in})(8 \text{ in})}$$
$$+ (10 \text{ in})(40 \text{ in}) + (4.32 \text{ in}^2)(7)$$
$$= \frac{28{,}186 \text{ in}^3}{1014.2 \text{ in}^2}$$
$$= 27.8 \text{ in} \quad \text{[from top]}$$

step 4: Assume the entire beam is in compression. Set up a table to calculate the moment of inertia.

section	$I_c = \dfrac{bh^3}{12}$ (in^4)	A (in^2)	d (in)	Ad^2 (in^4)
1	416.7	200	25.3	128,018
2	73,728	384	3.8	5545
3	3333.3	400	15.2	92,416
4	0	30.24	14.2	6098
totals	77,478			232,077

$$I = 77{,}478 \text{ in}^4 + 232{,}077 \text{ in}^4$$
$$= 309{,}555 \text{ in}^4$$

step 5: $e = 42 \text{ in} - 27.8 \text{ in} = 14.2 \text{ in}$

step 6: Disregarding elastic shortening losses immediately after transfer (conservative assumption), and disregarding the moment caused by the beam weight at the end of the transfer zone, the initial prestress is

$$f_s = \text{initial prestress} - \text{losses}$$

$$= (0.8)\left(250,000\ \frac{\text{lbf}}{\text{in}^2}\right) - 0$$

$$= 200,000\ \text{lbf/in}^2$$

$$P = \left(200,000\ \frac{\text{lbf}}{\text{in}^2}\right)\left(0.144\ \frac{\text{in}^2}{\text{strand}}\right)$$

$$\times\ (30\ \text{strands})$$

$$= 864,000\ \text{lbf}$$

$$A_c = 200\ \text{in}^2 + 384\ \text{in}^2 + 400\ \text{in}^2$$

$$= 984\ \text{in}^2$$

$$c_{\text{top}} = 27.8\ \text{in}\quad[\text{for top fibers}]$$

$$c_{\text{bottom}} = 48\ \text{in} - 27.8\ \text{in}$$

$$= 20.2\ \text{in}\quad[\text{for bottom fibers}]$$

$$f_{\text{top}} = \frac{-P}{A} + \frac{Pec}{I}$$

$$= \frac{-864,000\ \text{lbf}}{984\ \text{in}^2}$$

$$+ \frac{(864,000\ \text{lbf})(14.2\ \text{in})(27.8\ \text{in})}{309,555\ \text{in}^4}$$

$$= 224\ \text{lbf/in}^2\quad[\text{tension in concrete}]$$

$$f_{\text{bottom}} = \frac{-P}{A} - \frac{Pec}{I}$$

$$= \frac{-864,000\ \text{lbf}}{984\ \text{in}^2} - \frac{(864,000\ \text{lbf})(14.2\ \text{in})(20.2\ \text{in})}{309,555\ \text{in}^4}$$

$$= -1678.6\ \text{lbf/in}^2\quad[\text{compression in concrete}]$$

step 7: The bending stress due to dead load is

$$f_{b,\text{top}} = \frac{Mc}{I}$$

$$= \frac{-(187,500\ \text{ft-lbf})(27.8\ \text{in})\left(12\ \frac{\text{in}}{\text{ft}}\right)}{309,555\ \text{in}^4}$$

$$= -202.1\ \text{lbf/in}^2\quad[\text{compression in concrete}]$$

$$f_{b,\text{bottom}} = \frac{(187,500\ \text{ft-lbf})(20.2\ \text{in})\left(12\ \frac{\text{in}}{\text{ft}}\right)}{309,555\ \text{in}^4}$$

$$= 146.8\ \text{lbf/in}^2\quad[\text{tension in concrete}]$$

step 8: $\quad f_{\text{top}} = 224\ \dfrac{\text{lbf}}{\text{in}^2} - 202.1\ \dfrac{\text{lbf}}{\text{in}^2}$

$$= 22\ \text{lbf/in}^2\quad[\text{tension}]$$

$$f_{\text{bottom}} = -1678.6\ \frac{\text{lbf}}{\text{in}^2} + 146.8\ \frac{\text{lbf}}{\text{in}^2}$$

$$= -1532\ \text{lbf/in}^2\quad[\text{compression}]$$

step 9: For a simply supported pre-tensioned concrete beam, the limiting stresses immediately after transfer are determined using ACI 318 Sec. 18.4.1.

For compression,

$$0.70f'_{\text{ci}} = (0.70)\left(3800\ \frac{\text{lbf}}{\text{in}^2}\right)$$

$$= 2660\ \text{lbf/in}^2 > 1506\ \text{lbf/in}^2\quad[\text{OK}]$$

For tension,

$$6\sqrt{f'_{\text{ci}}} = 6\sqrt{3800\ \frac{\text{lbf}}{\text{in}^2}}$$

$$= 370\ \text{lbf/in}^2 > 22\ \text{lbf/in}^2\quad[\text{OK}]$$

$$\boxed{\text{The stresses are acceptable.}}$$

(b) To check the concrete stress under service loading, two conditions must be evaluated: prestress plus the total (dead and live) load and prestress plus the sustained (dead) load. The prestress from step 6 is first reduced proportionally.

$$\frac{(0.6)\left(250,000\ \dfrac{\text{lbf}}{\text{in}^2}\right)}{(0.8)\left(250,000\ \dfrac{\text{lbf}}{\text{in}^2}\right)} = 0.75$$

Recalculate the stresses with the lower prestress.

$$f_{\text{top}} = (0.75)\left(224\ \frac{\text{lbf}}{\text{in}^2}\right) = 168\ \text{lbf/in}^2\quad[\text{tension}]$$

$$f_{\text{bottom}} = (0.75)\left(-1678.6\ \frac{\text{lbf}}{\text{in}^2}\right)$$

$$= -1259\ \text{lbf/in}^2\quad[\text{compression}]$$

$$M_{D+L} = \frac{wL^2}{8} = \frac{\left(150\ \dfrac{\text{lbf}}{\text{ft}} + 540\ \dfrac{\text{lbf}}{\text{ft}}\right)(100\ \text{ft})^2}{8}$$

$$= 862,500\ \text{ft-lbf}$$

$$f_{\text{top}} = \frac{-Pc}{I} = \frac{(-862,500\ \text{ft-lbf})(27.8\ \text{in})\left(12\ \dfrac{\text{in}}{\text{ft}}\right)}{309,555\ \text{in}^4}$$

$$= -929.5\ \text{lbf/in}^2\quad[\text{compression}]$$

$$f_{\text{bottom}} = \frac{Pc}{I} = \frac{(862,500\ \text{ft-lbf})(20.2\ \text{in})\left(12\ \dfrac{\text{in}}{\text{ft}}\right)}{309,555\ \text{in}^4}$$

$$= 675.4\ \text{lbf/in}^2\quad[\text{tension}]$$

$$f_{\text{top,total}} = 168\ \frac{\text{lbf}}{\text{in}^2} + \left(-929.5\ \frac{\text{lbf}}{\text{in}^2}\right)$$

$$= -761.5\ \text{lbf/in}^2\quad[\text{compression}]$$

$$f_{\text{bottom,total}} = -1259\ \frac{\text{lbf}}{\text{in}^2} + 675.4\ \frac{\text{lbf}}{\text{in}^2}$$

$$= -583.6\ \text{lbf/in}^2\quad[\text{compression}]$$

The maximum compressive stress in the beam is 761.5 lbf/in². The maximum permissible stress from ACI 318 Sec. 18.4.1 is $0.60f'_c$. Therefore, for the prestress plus the total (dead and live) load condition,

$$0.60f'_c = (0.60)\left(5000 \ \frac{\text{lbf}}{\text{in}^2}\right)$$
$$= 3000 \ \text{lbf/in}^2 > 761.5 \ \text{lbf/in}^2 \quad [\text{OK}]$$

The maximum compressive stress in the beam, 761.5 lbf/in², accounts for stress due to both dead and live loads. To determine the prestress plus the sustained (dead) load, the stress contribution from dead loads only must be known. The total stresses due to dead loads only are

$$f_{\text{top,total}} = 168 \ \frac{\text{lbf}}{\text{in}^2} - 202.1 \ \frac{\text{lbf}}{\text{in}^2}$$
$$= -34.1 \ \text{lbf/in}^2$$
$$f_{\text{bottom,total}} = -1259 \ \frac{\text{lbf}}{\text{in}^2} + 146.8 \ \frac{\text{lbf}}{\text{in}^2}$$
$$= -1112 \ \text{lbf/in}^2 \quad [\text{controls}]$$

From ACI 318 Sec. 18.4.1, the maximum permissible stress is $0.45f'_c$. Therefore, for the prestress plus the sustained (dead) load condition,

$$0.45f'_c = (0.45)\left(5000 \ \frac{\text{lbf}}{\text{in}^2}\right)$$
$$= 2250 \ \text{lbf/in}^2 > 1112 \ \text{lbf/in}^2 \quad [\text{OK}]$$

> The compressive stresses are acceptable. (There is no tensile stress.)

57 Composite Concrete and Steel Bridge Girders

PRACTICE PROBLEMS

1. (*Time limit: one hour*) A simple bridge span is shown. The deck of the bridge consists of a 7 in thick concrete slab supported by 88 ft W24 × 94 steel beams. Normal-weight concrete with a specific weight of 147 lbf/ft^3 is used. The concrete strength is 3000 psi. Use AASHTO specifications.

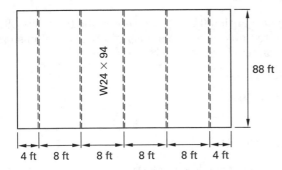

4 ft 8 ft 8 ft 8 ft 8 ft 4 ft

(a) The effective width for an interior girder is most nearly

(A) 42 in

(B) 84 in

(C) 96 in

(D) 260 in

(b) The effective width for an exterior girder is most nearly

(A) 42 in

(B) 48 in

(C) 88 in

(D) 96 in

(c) The modular ratio is most nearly

(A) 6

(B) 7

(C) 8

(D) 9

(d) Using a modular ratio of 9, the equivalent width (referred to a steel beam) of the effective width for an interior girder is most nearly

(A) 8.3 in

(B) 9.3 in

(C) 10 in

(D) 11 in

(e) Using a modular ratio of 9 and an effective width of 96 in, the centroidal distance of the transformed section measured from the bottom of the steel beam is most nearly

(A) 21 in

(B) 22 in

(C) 23 in

(D) 24 in

(f) The moment of inertia of the transformed section about the centroidal axis is most nearly

(A) 7700 in^4

(B) 8000 in^4

(C) 8300 in^4

(D) 8500 in^4

(g) The section modulus with respect to the top fiber is most nearly

(A) 910 in^3

(B) 930 in^3

(C) 950 in^3

(D) 1030 in^3

(h) The section modulus with respect to the bottom fiber is most nearly

(A) 290 in^3

(B) 340 in^3

(C) 380 in^3

(D) 420 in^3

(i) If a 1 in × 8 in plate is welded to the bottom flange of the steel beam, the centroid of the transformed section will

(A) move up

(B) move down

(C) not move

(D) move only after the concrete has failed (i.e., is 100% cracked)

(j) The addition of the 1 in × 8 in steel plate to the bottom flange will

(A) increase the moment of inertia about the x-axis

(B) decrease the moment of inertia about the x-axis

(C) have no effect on the moment of inertia about the y-axis

(D) decrease the moment of inertia about the y-axis

2. (*Time limit: one hour*) The steel support beams described in Prob. 1 are reinforced with a 1 in × 8 in steel plate welded to the bottom flange. After curing, the live load moment is 620 ft-kips. Normal weight concrete with a specific weight of 147 lbf/ft^3 (0.147 kip/ft^3) is used. Use AASHTO specifications to evaluate an interior girder with a span length of 39.3 ft.

(a) The moment of inertia of the transformed section of the composite section about the centroidal axis is most nearly

(A) 11,000 in^4

(B) 12,000 in^4

(C) 13,000 in^4

(D) 14,000 in^4

(b) The centroidal distance of the noncomposite welded steel section measured from the bottom of the steel plate is most nearly

(A) 8.3 in

(B) 10 in

(C) 11 in

(D) 12 in

(c) The moment of inertia of the noncomposite welded steel section about the noncomposite centroidal axis is most nearly

(A) 3700 in^4

(B) 3800 in^4

(C) 3900 in^4

(D) 4000 in^4

(d) The bending stress at the top steel fibers due to dead load in the case of unshored construction is most nearly

(A) 4.3 ksi

(B) 6.9 ksi

(C) 7.6 ksi

(D) 8.5 ksi

(e) The bending stress at the bottom fibers due to dead load in the case of unshored construction is most nearly

(A) 5.3 ksi

(B) 6.2 ksi

(C) 7.4 ksi

(D) 8.9 ksi

(f) The additional bending stress at the top fibers of the concrete slab due to live moment in the case of unshored construction is most nearly

(A) 0.5 ksi

(B) 0.6 ksi

(C) 0.7 ksi

(D) 0.8 ksi

(g) The additional bending stress at the top fibers of the steel beam due to live moment in the case of unshored construction is most nearly

(A) 1.5 ksi

(B) 1.9 ksi

(C) 2.4 ksi

(D) 3.8 ksi

(h) The additional bending stress at the bottom fibers of the steel plate due to live moment in the case of unshored construction is most nearly

(A) 14 ksi

(B) 15 ksi

(C) 20 ksi

(D) 22 ksi

(i) The bending stress at the top fibers of the concrete slab due to all loads in the case of shored construction is most nearly

(A) 0.2 ksi

(B) 0.8 ksi

(C) 1.3 ksi

(D) 2.3 ksi

(j) The bending stress at the bottom fibers of the steel plate due to all loads in the case of shored construction is most nearly

    (A) 11 ksi

    (B) 14 ksi

    (C) 17 ksi

    (D) 19 ksi

3. Calculate the ultimate moment capacity of the composite section shown. All pieces are A36 steel and the beam is W460 × 113 (metric). The concrete compressive strength is 20.7 MPa.

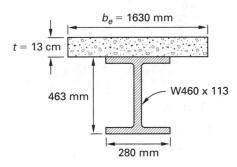

4. The beam described in Prob. 3 is changed such that the equivalent width of the concrete slab is 130 cm. The material properties are unchanged. Determine the ultimate moment capacity.

SOLUTIONS

1. (a) The effective flange width of a girder is defined by AASHTO as the "tributary width perpendicular to the axis of the member." For simple designs using slabs of uniform thickness, this is the girder spacing (8 ft).

$$b_{\text{eff}} = (8 \text{ ft})\left(12 \, \frac{\text{in}}{\text{ft}}\right) = \boxed{96 \text{ in}}$$

The answer is (C).

(b) Since the exterior girders support overhanging slabs of 4 ft, their tributary widths are also 8 ft.

$$b_{\text{eff}} = (8 \text{ ft})\left(12 \, \frac{\text{in}}{\text{ft}}\right) = \boxed{96 \text{ in}}$$

The answer is (D).

(c) The specific weight of the normal-weight concrete is 147 lbf/ft^3. Use Eq. 48.2 and Eq. 57.2.

$$n = \frac{E_s}{E_c} = \frac{E_s}{33 w_c^{1.5} \sqrt{f'_c}}$$

$$= \frac{29{,}000{,}000 \, \frac{\text{lbf}}{\text{in}^2}}{(33)\left(147 \, \frac{\text{lbf}}{\text{ft}^3}\right)^{1.5} \sqrt{3000 \, \frac{\text{lbf}}{\text{in}^2}}}$$

$$= \boxed{9}$$

(Same as in Table 57.1.)

The answer is (D).

(d) The equivalent width is

$$\frac{b_e}{n} = \frac{96 \text{ in}}{9} = \boxed{10.67 \text{ in} \quad (11 \text{ in})}$$

The answer is (D).

(e) Determine the centroidal location of an "all steel" composite beam.

$$y_b = \frac{\sum A_i y_i}{\sum A_i}$$

$$= \frac{(27.7 \text{ in}^2)\left(\frac{24.31 \text{ in}}{2}\right) + (7 \text{ in})(10.67 \text{ in})\left(24.31 \text{ in} + \frac{7 \text{ in}}{2}\right)}{27.7 \text{ in}^2 + (10.67 \text{ in})(7 \text{ in})}$$

$$= \boxed{23.57 \text{ in} \quad (24 \text{ in})}$$

The answer is (D).

(f)

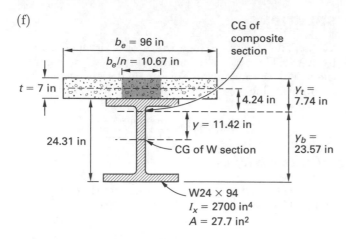

Use the parallel axis theorem.

$$I = I_x + Ad^2 = 2700 \text{ in}^4 + (27.7 \text{ in}^2)(11.42 \text{ in})^2$$
$$+ \frac{(10.67 \text{ in})(7 \text{ in})^3}{12}$$
$$+ (7 \text{ in})(10.67 \text{ in})(4.24 \text{ in})^2$$
$$= \boxed{7960.3 \text{ in}^4 \quad (8000 \text{ in}^4)}$$

The answer is (B).

(g) $S_t = \dfrac{I}{c} = \dfrac{7960.3 \text{ in}^4}{7.74 \text{ in}} = \boxed{1028.5 \text{ in}^3 \quad (1030 \text{ in}^3)}$

The answer is (D).

(h) $S_b = \dfrac{I}{c} = \dfrac{7960.3 \text{ in}^4}{23.57 \text{ in}} = \boxed{337.7 \text{ in}^3 \quad (340 \text{ in}^3)}$

The answer is (B).

(i) | The centroid of the transformed section will move down when a steel plate is added to the bottom flange. |

The answer is (B).

(j) | The moment of inertia about the x- and y-axes both will increase. |

The answer is (A).

---

2. (a)

$$y_b = \frac{\sum A_i y_i}{\sum A_i}$$

$$= \frac{\begin{array}{l}(27.7 \text{ in}^2)(13.16 \text{ in}) + (8 \text{ in}^2)(0.5 \text{ in}) \\ + (10.67 \text{ in})(7 \text{ in})(28.81 \text{ in})\end{array}}{27.7 \text{ in}^2 + 8 \text{ in}^2 + 74.69 \text{ in}^2}$$

$$= 22.83 \text{ in}$$

$$I = I_x + Ad^2$$
$$= 2700 \text{ in}^4 + (27.7 \text{ in}^2)(9.68 \text{ in})^2$$
$$+ \frac{(8 \text{ in})(1 \text{ in})^3}{12}$$
$$+ (8 \text{ in}^2)(22.33 \text{ in})^2$$
$$+ \frac{(10.67 \text{ in})(7 \text{ in})^3}{12}$$
$$+ (10.67 \text{ in})(7 \text{ in})(5.98 \text{ in})^2$$
$$= \boxed{12{,}261 \text{ in}^4 \quad (12{,}000 \text{ in}^4)}$$

The answer is (B).

(b) $\quad y_b = \dfrac{\sum A_i y_i}{\sum A_i}$

$$= \frac{(8 \text{ in}^2)(0.5 \text{ in}) + (27.7 \text{ in}^2)(13.16 \text{ in})}{8 \text{ in}^2 + 27.7 \text{ in}^2}$$

$$= \boxed{10.32 \text{ in} \quad (10 \text{ in})}$$

The answer is (B).

(c)

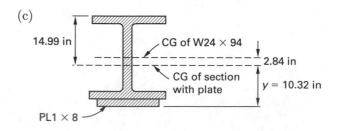

$$I = I_x + Ad^2$$

$$= 2700 \text{ in}^4 + (27.7 \text{ in}^2)(2.84 \text{ in})^2 + \frac{(8 \text{ in})(1 \text{ in})^3}{12}$$

$$+ (8 \text{ in}^2)(9.82 \text{ in})^2$$

$$= \boxed{3695.5 \text{ in}^4 \quad (3700 \text{ in}^4)}$$

The answer is (A).

(d) For construction without temporary shoring, the noncomposite dead load is carried by the steel section alone.

$$w_{\text{slab}} = \frac{(7 \text{ in})(8 \text{ ft})\left(0.147 \frac{\text{kip}}{\text{ft}^3}\right)}{12 \frac{\text{in}}{\text{ft}}}$$

$$= 0.69 \text{ kip/ft}$$

$$w_{\text{beam and plate}} = 0.121 \text{ kip/ft}$$

$$w_d = 0.69 \frac{\text{kip}}{\text{ft}} + 0.121 \frac{\text{kip}}{\text{ft}}$$

$$= 0.811 \text{ kip/ft}$$

$$M_d = \frac{w_d L^2}{8}$$

$$= \frac{\left(0.811 \frac{\text{kip}}{\text{ft}}\right)(39.3 \text{ ft})^2}{8}$$

$$= 156.6 \text{ ft-kips}$$

$$c = \frac{24.31 \text{ in}}{2} + 2.84 \text{ in} = 14.99 \text{ in}$$

$$f_{\text{top}} = \frac{M_d}{S_{t,\text{steel}}}$$

$$= \frac{(156.6 \text{ ft-kips})\left(12 \frac{\text{in}}{\text{ft}}\right)}{\dfrac{3695.5 \text{ in}^4}{14.99 \text{ in}}}$$

$$= \boxed{7.62 \text{ kips/in}^2 \quad (7.6 \text{ ksi})}$$

The answer is (C).

(e)
$$c = 10.32 \text{ in} \quad [\text{part (b)}]$$

$$f_{\text{bottom}} = \frac{M_d}{S_{b,\text{steel}}}$$

$$= \frac{(156.6 \text{ ft-kips})\left(12 \frac{\text{in}}{\text{ft}}\right)}{\dfrac{3695.5 \text{ in}^4}{10.32 \text{ in}}}$$

$$= \boxed{5.25 \text{ kips/in}^2 \quad (5.3 \text{ ksi})}$$

The answer is (A).

(f)
$$c = 1 \text{ in} + 24.31 \text{ in} + 7 \text{ in} - 22.83 \text{ in}$$

$$= 9.48 \text{ in}$$

The modular ratio, n, is 9.

$$f_{\text{top,concrete}} = \frac{M_l}{S_{t,\text{composite}}}$$

$$= \frac{(620 \text{ ft-kips})\left(12 \frac{\text{in}}{\text{ft}}\right)}{\dfrac{(9)(12{,}261 \text{ in}^4)}{9.48 \text{ in}}}$$

$$= \boxed{0.64 \text{ kip/in}^2 \quad (0.6 \text{ ksi})}$$

The answer is (B).

(g)
$$c = 1 \text{ in} + 24.31 \text{ in} - 22.83 \text{ in} = 2.48 \text{ in}$$

$$f_{\text{top,steel}} = \frac{M_l}{S_{t,\text{composite}}}$$

$$= \frac{(620 \text{ ft-kips})\left(12 \frac{\text{in}}{\text{ft}}\right)}{\dfrac{12{,}261 \text{ in}^4}{2.48 \text{ in}}}$$

$$= \boxed{1.50 \text{ kips/in}^2 \quad (1.5 \text{ ksi})}$$

The answer is (A).

(h)
$$c = 22.83 \text{ in} \quad [\text{part (a)}]$$

$$f_{\text{bottom,steel}} = \frac{M_l}{S_{b,\text{composite}}}$$

$$= \frac{(620 \text{ ft-kips})\left(12 \frac{\text{in}}{\text{ft}}\right)}{\dfrac{12{,}261 \text{ in}^4}{22.83 \text{ in}}}$$

$$= \boxed{13.85 \text{ kips/in}^2 \quad (14 \text{ ksi})}$$

The answer is (A).

(i) With temporary shoring, the steel beam and the wet concrete will be supported by temporary falsework until the concrete has cured.

$$f_{\text{top}} = \frac{M_l + M_d}{S_{t,\text{composite}}}$$

$$= \frac{(620 \text{ ft-kips} + 156.6 \text{ ft-kips})\left(12 \frac{\text{in}}{\text{ft}}\right)}{\dfrac{(9)(12{,}261 \text{ in}^4)}{9.48 \text{ in}}}$$

$$= \boxed{0.8 \text{ kip/in}^2 \quad (0.8 \text{ ksi})}$$

The answer is (B).

Structural

(j) $\quad f_{\text{bottom}} = \dfrac{M_l + M_d}{S_{b,\text{composite}}}$

$$= \dfrac{(620 \text{ ft-lbf} + 158.50 \text{ ft-lbf})\left(12 \dfrac{\text{in}}{\text{ft}}\right)}{\dfrac{12{,}261 \text{ in}^4}{22.83 \text{ in}}}$$

$$= \boxed{17.39 \text{ kips/in}^2 \quad (17 \text{ ksi})}$$

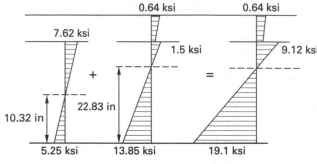

(a) without shores (unshored)

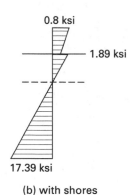

(b) with shores

The answer is (C).

3. Calculate the depth, a, of the stress block.

$$a = \dfrac{A_s F_y}{0.85 f'_c b_e}$$

$$= \dfrac{(14\,400 \text{ mm}^2)(248.3 \text{ MPa})}{(0.85)(20.7 \text{ MPa})(1630 \text{ mm})}$$

$$= 124.7 \text{ mm}$$

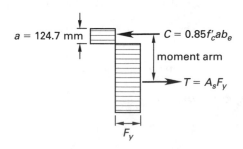

Since $a < t$, the slab is adequate.

$$C = 0.85 f'_c a b_e$$

$$= \dfrac{(0.85)(20.7 \text{ MPa})(124.7 \text{ mm})(1630 \text{ mm})}{\left(1000 \dfrac{\text{mm}}{\text{m}}\right)^2}$$

$$= 3.58 \text{ MN}$$

The moment arm is

$$\dfrac{d}{2} + t - \dfrac{a}{2} = \dfrac{463 \text{ mm}}{2} + 130 \text{ mm} - \dfrac{124.7 \text{ mm}}{2}$$

$$= 299.2 \text{ mm}$$

The ultimate composite moment capacity is

$$M_u = C(\text{moment arm})$$

$$= \dfrac{(3.58 \text{ MN})(299.3 \text{ mm})}{1000 \dfrac{\text{mm}}{\text{m}}}$$

$$= \boxed{1.07 \text{ MN·m}}$$

4.

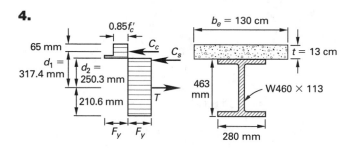

Calculate the depth, a, of the stress block.

$$a = \dfrac{A_s F_y}{0.85 f'_c b_e}$$

$$= \dfrac{(14\,400 \text{ mm}^2)(248.3 \text{ MPa})}{(0.85)(20.7 \text{ MPa})(1300 \text{ mm})}$$

$$= 156.3 \text{ mm}$$

Since $a > t$, the slab is inadequate. A portion of the compressive stress will be carried by the steel beam.

The compressive stress in the concrete is

$$C_c = 0.85 f'_c b_e t$$

$$= (0.85)(20.7 \text{ MPa})(1.3 \text{ m})(0.13 \text{ m})$$

$$= 2.97 \text{ MN}$$

However, the total tensile force is equal to the total compressive force.

$$T = C_c + C_s$$

The tensile force can also be calculated from the stress carried by the steel.

$$T = A_s F_y - C_s$$

Equate the two expressions for T and solve for C_s.

$$
\begin{aligned}
C_s &= \frac{A_s F_y - C_c}{2} \\
&= \frac{\dfrac{(14\,400 \text{ mm}^2)(248.3 \text{ MPa})}{\left(1000 \dfrac{\text{mm}}{\text{m}}\right)^2} - 2.97 \text{ MN}}{2} \\
&= 0.30 \text{ MN}
\end{aligned}
$$

Calculate the portion of the top flange that is needed to carry C_s.

$$
\begin{aligned}
d_f &= \frac{(0.30 \text{ MN})\left(1000 \dfrac{\text{mm}}{\text{m}}\right)}{(248.3 \text{ MPa})(0.280 \text{ m})} \\
&= 4.3 \text{ mm}
\end{aligned}
$$

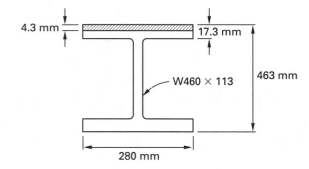

The height of the centroid of the tension portion of the steel beam, from the bottom, is

$$
\begin{aligned}
y &= \frac{(14\,400 \text{ mm}^2)\left(\dfrac{463 \text{ mm}}{2}\right) - (4.3 \text{ mm})(280 \text{ mm})\left(463 \text{ mm} - \dfrac{4.3 \text{ mm}}{2}\right)}{14\,400 \text{ mm}^2 - (280 \text{ mm})(4.3 \text{ mm})} \\
&= 210.57 \text{ mm}
\end{aligned}
$$

Therefore, the ultimate moment capacity of the composite section is

$$
\begin{aligned}
M_u &= C_c d_1 + C_s d_2 \\
&= \frac{(2.97 \text{ MN})(317.4 \text{ mm}) + (0.30 \text{ MN})(250.3 \text{ mm})}{1000 \dfrac{\text{mm}}{\text{m}}} \\
&= \boxed{1.02 \text{ MN·m}}
\end{aligned}
$$

58 Structural Steel: Introduction

PRACTICE PROBLEMS

There are no problems in this book corresponding to Chap. 58 of the *Civil Engineering Reference Manual*.

59 Structural Steel: Beams

PRACTICE PROBLEMS

Answer options for LRFD solutions are given in parentheses.

1. An A992 steel beam is loaded as shown. The beam carries a uniform load of 1400 lbf/ft over its entire span. (a) Determine the reactions. (b) Draw the shear and moment diagrams. (c) Choose the lightest W14 beam that can operate without developing the AISC critical buckling stress. (d) Specify the maximum unbraced length for the beam chosen. Use the ASD method.

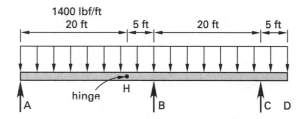

2. A 25 ft beam of A992 steel is simply supported at its left end and 7 ft from its right end. A load of 3000 lbf/ft is uniformly distributed over its entire length. Lateral support is provided only at the reactions. The beam's dead weight is initially estimated as 60 lbf/ft. Choose an economical W shape for this application. (Do not check shear stress.) Use either the ASD or LRFD method.

3. A beam spans 24 ft and has a deflection limited to $L/300$. The compression flange has complete lateral bracing. The beam carries a uniformly distributed load of 800 lbf/ft over its entire length, excluding the beam's self-weight. The beam's dead weight is initially estimated as 50 lbf/ft. Select an economical A992 W shape.

4. Select an A992 W shape with lateral support at $5^1/2$ ft intervals to span 20 ft and carry a uniformly distributed load of 1 kip/ft. The load includes a uniform dead load allowance of 40 lbf/ft. The maximum deflection is limited to $L/240$.

5. (*Time limit: one hour*) The steel beam shown carries a load of 1.4 kips/ft, excluding self-weight, over its entire span. Assume lateral bracing only at the reaction points.

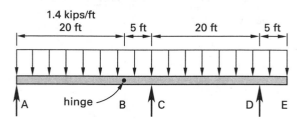

(Disregard the weight of the beam in the following questions.)

(a) The reaction (factored reaction, for LRFD) at A is most nearly

    (A) 12 kips (24 kips)

    (B) 14 kips (28 kips)

    (C) 16 kips (32 kips)

    (D) 18 kips (36 kips)

(b) The reaction at C is most nearly

    (A) 36 kips (58 kips)

    (B) 37 kips (60 kips)

    (C) 39 kips (62 kips)

    (D) 42 kips (67 kips)

(c) The reaction at D is most nearly

    (A) 13 kips (20 kips)

    (B) 18 kips (28 kips)

    (C) 19 kips (30 kips)

    (D) 22 kips (35 kips)

(d) The absolute maximum value of shear (factored shear, for LRFD) is most nearly

    (A) 12 kips (19 kips)

    (B) 14 kips (22 kips)

    (C) 18 kips (29 kips)

    (D) 21 kips (34 kips)

(e) The absolute maximum shear occurs closest to

    (A) support A

    (B) hinge B

    (C) support C

    (D) support D

(f) The absolute maximum moment is most nearly

(A) 70 ft-kips (140 ft-kips)

(B) 79 ft-kips (158 ft-kips)

(C) 88 ft-kips (176 ft-kips)

(D) 98 ft-kips (196 ft-kips)

(g) The absolute maximum moment occurs

(A) halfway between A and B

(B) at support C

(C) halfway between C and D

(D) at support D

(h) If the unbraced length is 25 ft, what is the lightest W18 section of A992 steel that can be used, based on moment-resisting capacity?

(A) $W18 \times 55$

(B) $W18 \times 60$

(C) $W18 \times 65$

(D) $W18 \times 76$

(i) For the lightest beam chosen in part (h), the actual maximum shear stress (ignoring the weight of the beam) is most nearly

(A) 2.0 ksi (3.2 ksi)

(B) 2.4 ksi (3.8 ksi)

(C) 2.7 ksi (4.3 ksi)

(D) 3.4 ksi (5.4 ksi)

(j) If the beam chosen in part (h) is constructed of 42 ksi steel, the maximum unbraced length that permits the development of a flexural strength of M_p is most nearly

(A) 60 in

(B) 96 in

(C) 121 in

(D) 240 in

6. (*Time limit: one hour*) A beam is constructed from a W shape and a C shape, both of A992 ($F_y = 50$ ksi) steel, as shown. Two equal loads remain 3 ft apart as they move across a 40 ft span. Sufficient lateral support is provided so that the critical bending stress is $0.7F_y$. The beam is not subjected to impact loading. Use the ASD method to solve the following problems.

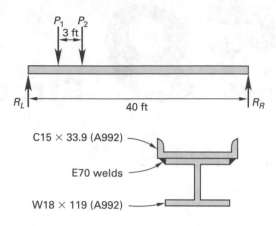

(a) The distance of the neutral axis of the beam cross section measured from the bottom of the section is most nearly

(A) 10.0 in

(B) 11.8 in

(C) 12.3 in

(D) 12.9 in

(b) The distance to the extreme compression fiber from the neutral axis is most nearly

(A) 10.6 in

(B) 11.3 in

(C) 12.0 in

(D) 12.4 in

(c) The distance to the extreme tension fiber from the neutral axis is most nearly

(A) 11.0 in

(B) 11.8 in

(C) 12.3 in

(D) 12.9 in

(d) The moment of inertia of the section (strong-axis bending) is most nearly

(A) 1950 in^4

(B) 2390 in^4

(C) 3020 in^4

(D) 3160 in^4

(e) For maximum moment to occur in the beam under load P_2, the position of the loads will be such that their resultant will be located at a distance (measured from the left support) equal to

  (A) 18.8 ft

  (B) 19.3 ft

  (C) 19.8 ft

  (D) 20.0 ft

(f) The maximum moment obtained as a function of each unknown moving load of P (in kips) is most nearly

  (A) $16.0P$ ft-kips

  (B) $17.1P$ ft-kips

  (C) $18.5P$ ft-kips

  (D) $19.1P$ ft-kips

(g) Considering moment only, the maximum allowable value of P is most nearly

  (A) 25 kips

  (B) 30 kips

  (C) 35 kips

  (D) 40 kips

(h) The absolute maximum value of the actual shear stress (anywhere in the beam) corresponding to the value of P obtained in part (g) is most nearly

  (A) 4.5 ksi

  (B) 5.0 ksi

  (C) 5.5 ksi

  (D) 6.0 ksi

(i) The minimum beam bearing length required at the reaction point from local web yielding criteria is most nearly

  (A) 0 in

  (B) 0.60 in

  (C) 1.1 in

  (D) 2.3 in

(j) The minimum beam bearing length required at the reaction point from web crippling criteria is most nearly

  (A) 0 in

  (B) 0.65 in

  (C) 1.0 in

  (D) 2.2 in

SOLUTIONS

1. (a) This is a determinate beam. Taking clockwise moments from the hinge to the left as positive,

$$\sum M_H = 20R_A - \left(\tfrac{1}{2}\right)\left(1400\ \frac{\text{lbf}}{\text{ft}}\right)(20\ \text{ft})^2 = 0$$

$$R_A = \boxed{14{,}000\ \text{lbf}}$$

$$V_H = 14{,}000\ \text{lbf} - (20\ \text{ft})\left(1400\ \frac{\text{lbf}}{\text{ft}}\right)$$

$$= -14{,}000\ \text{lbf}$$

$$R_B + R_C = (5\ \text{ft} + 20\ \text{ft} + 5\ \text{ft})\left(1400\ \frac{\text{lbf}}{\text{ft}}\right)$$

$$+\ 14{,}000\ \text{lbf}$$

$$= 56{,}000\ \text{lbf}$$

$$\sum M_C = 20R_B - (25\ \text{ft})(14{,}000\ \text{lbf})$$

$$-\left(\tfrac{1}{2}\right)\left(1400\ \frac{\text{lbf}}{\text{ft}}\right)(25\ \text{ft})^2$$

$$+\left(\tfrac{1}{2}\right)\left(1400\ \frac{\text{lbf}}{\text{ft}}\right)(5\ \text{ft})^2$$

$$= 0$$

$$R_B = \boxed{38{,}500\ \text{lbf}}$$

$$R_C = 56{,}000\ \text{lbf} - 38{,}500\ \text{lbf} = \boxed{17{,}500\ \text{lbf}}$$

(b) The shear and moment diagrams are as shown.

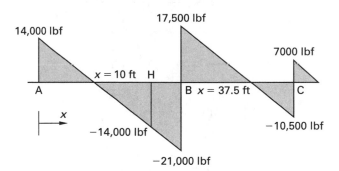

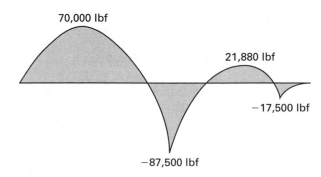

(c) The maximum moment is

$$M_{\max} = \frac{87{,}500\ \text{ft-lbf}}{1000\ \dfrac{\text{lbf}}{\text{kip}}} = 87.5\ \text{ft-kips}$$

A W14 must have an available strength, M_r/Ω, greater than the maximum required moment of 87.5 ft-kips to reach a critical flexural stress. From Table 3-2 in the *AISC Manual*, for a W14 × 38, $M_r/\Omega = 95.4$ ft-kips. Therefore, choose a W14 × 38.

(d) In order for the W14 × 38 to reach a critical flexural stress, the unbraced length, L_b, must be less than or equal to L_r. From Table 3-2 of the *AISC Manual*, for a W14 × 38, $L_r = 16.2$ ft. Therefore, the maximum unbraced length is 16.2 ft.

2. ASD Method Solution

Assume the weight of the beam is 60 lbf/ft.

The shear and moment diagrams are as shown.

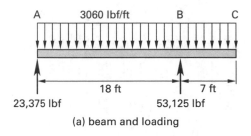

(a) beam and loading

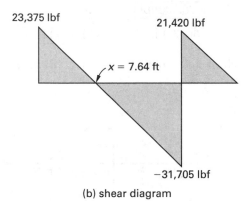

(b) shear diagram

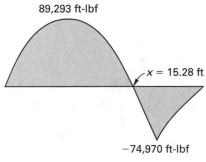

(c) moment diagram

Since $M_{\max}$ does not occur at the end of the cantilever span, C_b does not apply.

The unbraced length, L_b, is 18 ft, and this must be considered in addition to the flexural strength requirement of approximately 90 ft-kips. *AISC Manual* Table 3-2 cannot be used because $L_b > L_p$ for most of the beams with the

required flexural strength. For example, W14 × 26 (from *AISC Manual* Table 3-2) is an economical beam with a flexural strength of 100 ft-kips. However, its L_p value of 3.81 ft falls very short of 18 ft. *AISC Manual* Table 3-10 must be used when unbraced lengths are large. From the intersection of an unbraced length of 18 ft and an available moment of 90 ft-kips, move vertically upward to the solid line of $\boxed{\text{W12} \times 40.}$

LRFD Method Solution

$$w_u = 1.2w_D + 1.6w_L$$
$$= (1.2)\left(0.06 \ \frac{\text{kips}}{\text{ft}}\right) + (1.6)\left(3 \ \frac{\text{kips}}{\text{ft}}\right)$$
$$= 4.872 \ \text{kips/ft}$$
$$R_A = 37.2 \ \text{kips} \quad (37{,}200 \ \text{lbf})$$
$$R_B = 84.6 \ \text{kips} \quad (84{,}600 \ \text{lbf})$$
$$M_{\max} = 142.1 \ \text{ft-kips} \quad (142{,}100 \ \text{ft-lbf})$$

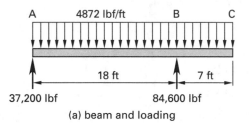

(a) beam and loading

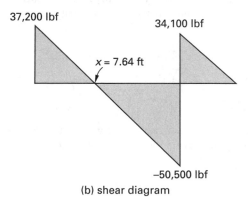

(b) shear diagram

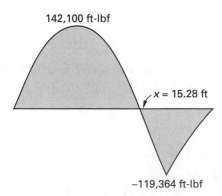

(c) moment diagram

From *AISC Manual* Table 3-10, choose a $\boxed{W12 \times 40}$ to meet the requirements of 142.1 ft-kips with an unbraced length of 18 ft.

3. *Common Solution*

step 1: Assume the beam weight is 50 lbf/ft.

$$w = w_{D+L} + w_{\text{beam}} = \frac{800 \ \frac{\text{lbf}}{\text{ft}} + 50 \ \frac{\text{lbf}}{\text{ft}}}{1000 \ \frac{\text{lbf}}{\text{kip}}}$$

$$= 0.85 \ \text{kip/ft}$$

step 2: The maximum allowable deflection is

$$y = \frac{L}{300} = \frac{(24 \ \text{ft})\left(12 \ \frac{\text{in}}{\text{ft}}\right)}{300}$$

$$= 0.96 \ \text{in}$$

step 3: Use App. 44.A, case 9. The required moment of inertia is

$$I_x = \frac{5wL^4}{384Ey} = \frac{(5)\left(0.85 \ \frac{\text{kip}}{\text{ft}}\right)(24 \ \text{ft})^4\left(12 \ \frac{\text{in}}{\text{ft}}\right)^3}{(384)\left(29{,}000 \ \frac{\text{kips}}{\text{in}^2}\right)(0.96 \ \text{in})}$$

$$= 227.9 \ \text{in}^4$$

step 4: From *AISC Manual* Table 3-3 (selection by I_x), try a W14 × 26, with $I_x = 245 \ \text{in}^4$.

ASD Method Solution

step 5: Check the available flexural strength.

$$w_a = w = 0.85 \ \text{kip/ft}$$

$$M_a = \frac{w_a L^2}{8} = \frac{\left(0.85 \ \frac{\text{kip}}{\text{ft}}\right)(24 \ \text{ft})^2}{8}$$

$$= 61.2 \ \text{ft-kips}$$

From *AISC Manual* Table 3-2, for a W14 × 26,

$$\frac{M_{px}}{\Omega} = 100 \ \text{ft-kips} > 61.2 \ \text{ft-kips} \quad \text{[OK]}$$

step 6: Check the required shear strength.

$$V_{u,\text{req}} = \frac{w_a L}{2} = \frac{\left(0.85 \ \frac{\text{kip}}{\text{ft}}\right)(24 \ \text{ft})}{2} = 10.2 \ \text{kips}$$

step 7: Check the available shear strength.

$$\frac{V_{nx}}{\Omega_v} = 70.9 \ \text{kips} > 10.2 \ \text{kips} \quad \text{[OK]}$$

The W14 × 26 is governed by deflection criteria, as the available moment and shear strengths are much greater than the required strengths.

$$\boxed{\text{Use a W14} \times 26.}$$

LRFD Method Solution

step 5: Check the required flexural strength.

$$w_u = 1.2w_D + 1.6w_L$$

$$= (1.2)\left(0.05 \ \frac{\text{kip}}{\text{ft}}\right) + (1.6)\left(0.8 \ \frac{\text{kip}}{\text{ft}}\right)$$

$$= 1.34 \ \text{kips/ft}$$

$$M_u = \frac{w_u L^2}{8} = \frac{\left(1.34 \ \frac{\text{kips}}{\text{ft}}\right)(24 \ \text{ft})^2}{8}$$

$$= 96.48 \ \text{ft-kips}$$

From *AISC Manual* Table 3-2, for a W14 × 26,

$$\phi_b M_{px} = 151 \ \text{ft-kips} > 96.48 \ \text{ft-kips} \quad \text{[OK]}$$

step 6: Check the required shear strength.

$$V_{u,\text{req}} = \frac{w_u L}{2} = \frac{\left(1.34 \ \frac{\text{kips}}{\text{ft}}\right)(24 \ \text{ft})}{2} = 16.08 \ \text{kips}$$

step 7: Check the available (or design) shear strength.

$$\phi_v V_{nx} = 106 \ \text{kips} > 16.08 \ \text{kips} \quad \text{[OK]}$$

The W14 × 26 is governed by deflection criteria, as the available moment and shear strengths are much greater than the required strengths.

$$\boxed{\text{Use a W14} \times 26.}$$

4. ASD Method Solution

step 1: The reactions are each

$$R = \frac{wL}{2} = \frac{\left(1\ \frac{\text{kip}}{\text{ft}}\right)(20\ \text{ft})}{2}$$
$$= 10\ \text{kips}$$

step 2: The required (i.e., maximum) moment is

$$M_a = M_{\max} = \frac{wL^2}{8} = \frac{\left(1\ \frac{\text{kip}}{\text{ft}}\right)(20\ \text{ft})^2}{8}$$
$$= 50\ \text{ft-kips}$$

step 3: From *AISC Manual* Table 3-2, try a W12 × 26 to meet 50 ft-kips. $L_p = 5.33$ ft, $L_r = 14.9$ ft, and $I_x = 204$ in^4.

$$\frac{M_{px}}{\Omega} = 92.8\ \text{ft-kips} > 50\ \text{ft-kips} \quad [\text{OK}]$$

step 4: Check for shear strength. The required shear strength is

$$V_a = R_a = 10\ \text{kips}$$

The available shear strength for a W12 × 26 is

$$\frac{V_{nx}}{\Omega} = 56.2\ \text{kips} > 10\ \text{kips} \quad [\text{OK}]$$

step 5: Check deflection. The allowable deflection is

$$y_{\text{allowable}} = \frac{L}{240} = \frac{(20\ \text{ft})\left(12\ \frac{\text{in}}{\text{ft}}\right)}{240} = 1\ \text{in}$$

The actual deflection is

$$y = \frac{5wL^4}{384EI} = \frac{(5)\left(1\ \frac{\text{kip}}{\text{ft}}\right)(20\ \text{ft})^4\left(12\ \frac{\text{in}}{\text{ft}}\right)^3}{(384)\left(29{,}000\ \frac{\text{kips}}{\text{in}^2}\right)(204\ \text{in}^4)}$$

$$= 0.609\ \text{in}$$

$$y < y_{\text{allowable}} \quad [\text{OK}]$$

$$\boxed{\text{Use a W12} \times 26.}$$

LRFD Method Solution

step 1: The reactions are each

$$R = \frac{wL}{2} = \frac{\left(1\ \frac{\text{kip}}{\text{ft}}\right)(20\ \text{ft})}{2}$$
$$= 10\ \text{kips}$$

step 2: The required flexural strength is

$$w_u = 1.2w_D + 1.6w_L$$
$$= (1.2)\left(0.04\ \frac{\text{kip}}{\text{ft}}\right) + (1.6)\left(0.96\ \frac{\text{kip}}{\text{ft}}\right)$$
$$= 1.584\ \text{kips/ft}$$

$$M_u = \frac{w_u L^2}{8} = \frac{\left(1.584\ \frac{\text{kips}}{\text{ft}}\right)(20\ \text{ft})^2}{8}$$
$$= 79.2\ \text{ft-kips}$$

step 3: From *AISC Manual* Table 3-2, try a W12 × 26 to meet 79.2 ft-kips. $L_p = 5.33$ ft, $L_r = 14.9$ ft, and $I_x = 204$ in^4.

$$\phi_b M_{px} = 140\ \text{ft-kips} > 79.2\ \text{ft-kips} \quad [\text{OK}]$$

step 4: Check for shear strength.

$$V_D = \frac{\left(0.04\ \frac{\text{kip}}{\text{ft}}\right)(20\ \text{ft})}{2} = 0.4\ \text{kips}$$

$$V_L = \frac{\left(0.96\ \frac{\text{kip}}{\text{ft}}\right)(20\ \text{ft})}{2} = 9.6\ \text{kips}$$

Therefore, the required shear strength is

$$V_u = (1.2)(0.4\ \text{kip}) + (1.6)(9.6\ \text{kips})$$
$$= 15.84\ \text{kips}$$

The design shear strength for a W12 × 26 is

$$\phi_v V_{nx} = 84.3\ \text{kips} > 15.84\ \text{kips} \quad [\text{OK}]$$

step 5: Check deflection. The allowable deflection is

$$y_{\text{allowable}} = \frac{L}{240} = \frac{(20\ \text{ft})\left(12\ \frac{\text{in}}{\text{ft}}\right)}{240} = 1\ \text{in}$$

The actual deflection is

$$y = \frac{5wL^4}{384EI} = \frac{(5)\left(1\,\dfrac{\text{kip}}{\text{ft}}\right)(20\text{ ft})^4\left(12\,\dfrac{\text{in}}{\text{ft}}\right)^3}{(384)\left(29{,}000\,\dfrac{\text{kips}}{\text{in}^2}\right)(204\text{ in}^4)}$$

$$= 0.609\text{ in}$$

$$y < y_{\text{allowable}} \quad [\text{OK}]$$

$$\boxed{\text{Use a W12} \times 26.}$$

5. ASD Method Solution

(a) Disregard the weight of the beam.

Taking clockwise moments as positive,

$$\sum M_{\text{hinge to left}} = R_A(20\text{ ft})$$
$$- \left(1.4\,\frac{\text{kips}}{\text{ft}}\right)(20\text{ ft})\left(\frac{20\text{ ft}}{2}\right) = 0$$
$$R_A = \boxed{14\text{ kips}}$$

The answer is (B).

(b) Disregard the weight of the beam.

$$\sum M_D = (14\text{ kips})(45\text{ ft})$$
$$- \left(1.4\,\frac{\text{kips}}{\text{ft}}\right)(45\text{ ft})\left(\frac{45\text{ ft}}{2}\right)$$
$$+ R_C(20\text{ ft}) + \left(1.4\,\frac{\text{kips}}{\text{ft}}\right)(5\text{ ft})\left(\frac{5\text{ ft}}{2}\right) = 0$$
$$R_C = \boxed{38.5\text{ kips} \quad (39\text{ kips})}$$

The answer is (C).

(c) The reaction at D is

$$R_D = \left(1.4\,\frac{\text{kips}}{\text{ft}}\right)(50\text{ ft}) - 14\text{ kips} - 38.5\text{ kips}$$
$$= \boxed{17.5\text{ kips} \quad (18\text{ kips})}$$

The answer is (B).

(d) Draw the shear and moment diagram.

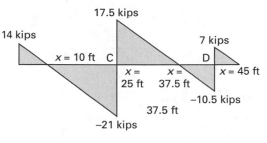

(a) shear diagram

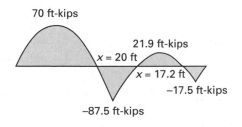

(b) moment diagram

From the shear diagram, the absolute maximum shear is $\boxed{21\text{ kips.}}$

The answer is (D).

(e) From the shear diagram, the absolute maximum shear occurs just to the left of $\boxed{\text{support C.}}$

The answer is (C).

(f) From the moment diagram, the absolute maximum moment is $\boxed{87.5\text{ ft-kips (88 ft-kips).}}$

The answer is (C).

(g) From the moment diagram, the absolute maximum moment occurs $\boxed{\text{at support C.}}$

The answer is (B).

(h) $L_b = 25$ ft; $M_{\text{max}} = 87.5$ ft-kips.

From *AISC Manual* Table 3-10, select a $\boxed{\text{W18} \times 76.}$

The allowable moment, M_n/Ω, of 272 ft-kips far exceeds the actual maximum moment of 87.5 ft-kips (excluding beam weight). There is ample reserve capacity for the additional moment due to the weight of the beam.

The answer is (D).

(i) For a W18 × 76,

$$d = 18.2\text{ in}$$
$$t_w = 0.425\text{ in}$$

The actual shear stress is

$$f_v = \frac{V}{A_w} = \frac{V}{dt_w} = \frac{21 \text{ kips}}{(18.2 \text{ in})(0.425 \text{ in})}$$
$$= \boxed{2.715 \text{ kips/in}^2 \quad (2.7 \text{ ksi})}$$

The answer is (C).

(j) In order to develop M_p, the unbraced length must be less than L_p. From Part 1 of the *AISC Manual*, for a W18 × 76, $r_y = 2.61$ in.

From Eq. 59.6,

$$L_p = 1.76 r_y \sqrt{\frac{E}{F_y}} = (1.76)(2.61 \text{ in}) \sqrt{\frac{29{,}000 \, \frac{\text{kips}}{\text{in}^2}}{42 \, \frac{\text{kips}}{\text{in}^2}}}$$
$$= \boxed{120.7 \text{ in} \quad (121 \text{ in})}$$

The answer is (C).

LRFD Method Solution

(a) Disregard the weight of the beam.

$$w_u = 1.6 w_L = (1.6)\left(1.4 \, \frac{\text{kips}}{\text{ft}}\right) = 2.24 \text{ kips/ft}$$

Taking clockwise moments as positive,

$$\sum M_{\text{hinge to left}} = R_A(20 \text{ ft})$$
$$- \left(2.24 \, \frac{\text{kips}}{\text{ft}}\right)(20 \text{ ft})\left(\frac{20 \text{ ft}}{2}\right) = 0$$
$$R_A = \boxed{22.4 \text{ kips} \quad (24 \text{ kips})}$$

The answer is (A).

(b) Disregard the weight of the beam.

$$\sum M_D = (22.4 \text{ kips})(45 \text{ ft})$$
$$- \left(2.24 \, \frac{\text{kips}}{\text{ft}}\right)(45 \text{ ft})\left(\frac{45 \text{ ft}}{2}\right)$$
$$+ R_C(20 \text{ ft}) + \left(2.24 \, \frac{\text{kips}}{\text{ft}}\right)(5 \text{ ft})\left(\frac{5 \text{ ft}}{2}\right) = 0$$
$$R_C = \boxed{61.6 \text{ kips} \quad (62 \text{ kips})}$$

The answer is (C).

(c) The reaction at D is

$$R_D = \left(2.24 \, \frac{\text{kips}}{\text{ft}}\right)(50 \text{ ft}) - 22.4 \text{ kips} - 61.6 \text{ kips}$$
$$= \boxed{28 \text{ kips}}$$

The answer is (B).

(d) Draw the shear and moment diagram.

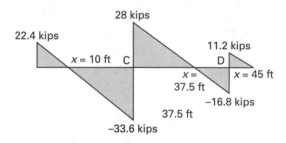

(a) shear diagram

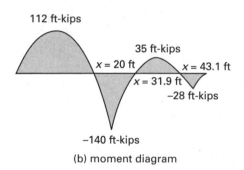

(b) moment diagram

From the shear diagram, the absolute maximum shear is $\boxed{33.6 \text{ kips} \ (34 \text{ kips}).}$

The answer is (D).

(e) From the shear diagram, the absolute maximum occurs just to the left of $\boxed{\text{support C.}}$

The answer is (C).

(f) From the moment diagram, the absolute maximum moment is $\boxed{140 \text{ ft-kips.}}$

The answer is (A).

(g) From the moment diagram, the absolute maximum moment occurs at $\boxed{\text{support C.}}$

The answer is (B).

(h) $L_r = 25$ ft; $M_{\max} = 140$ ft-kips.

From *AISC Manual* Table 3-10, select a $\boxed{\text{W18} \times 76.}$

The design moment, ϕM_n, of 410 ft-kips far exceeds the actual maximum (or required) moment of 140 ft-kips (excluding beam weight). There is ample reserve capacity for the additional moment due to the weight of the beam.

The answer is (D).

(i) For a W18 × 76,

$$d = 18.2 \text{ in}$$

$$t_w = 0.425 \text{ in}$$

The actual shear stress is

$$f_v = \frac{V_u}{dt_w} = \frac{33 \text{ kips}}{(18.2 \text{ in})(0.425 \text{ in})}$$

$$= \boxed{4.27 \text{ kips/in}^2 \quad (4.3 \text{ ksi})}$$

The answer is (C).

(j) In order to develop M_p, the unbraced length must be less than L_p. From Part 1 of the *AISC Manual* for a W18 × 76, $r_y = 2.61$ in.

From Eq. 59.6,

$$L_p = 1.76 r_y \sqrt{\frac{E}{F_y}} = (1.76)(2.61 \text{ in}) \sqrt{\frac{29{,}000 \dfrac{\text{kips}}{\text{in}^2}}{42 \dfrac{\text{kips}}{\text{in}^2}}}$$

$$= \boxed{120.7 \text{ in} \quad (121 \text{ in})}$$

The answer is (C).

6. (a) Locate the neutral axis (measured from the bottom of the beam). For a W18 × 119, $A = 35.1$ in^2, $d = 19.0$ in, and $I_x = 2190$ in^4. For a C15 × 33.9, $A = 10.0$ in^2, $b_f = 3.40$ in, $x = 0.788$ in, and $I_y = 8.07$ in^4.

$$y = \frac{\sum A_i y_i}{\sum A_i}$$

$$= \frac{(35.1 \text{ in}^2)\left(\dfrac{19.0 \text{ in}}{2}\right) + (10.00 \text{ in}^2)(19.0 \text{ in} + 0.788 \text{ in})}{35.1 \text{ in}^2 + 10.0 \text{ in}^2}$$

$$= \boxed{11.78 \text{ in} \quad (11.8 \text{ in})}$$

The answer is (B).

(b) The distance to the extreme compression fiber is

$$c = 19.0 \text{ in} + 3.40 \text{ in} - 11.78 \text{ in}$$

$$= \boxed{10.62 \text{ in} \quad (10.6 \text{ in})}$$

The answer is (A).

(c) The distance to the extreme tension fiber is

$$c = y = \boxed{11.78 \text{ in} \quad (11.8 \text{ in})}$$

The answer is (B).

(d) The channel properties are found in *AISC Manual* Table 1-5. The moment of inertia of the section is

$$I = I_{\text{W shape}} + I_{\text{channel}} = I_{o,W} + A_W d_1^2 + I_{o,c} + A_c d_2^2$$

$$= 2190 \text{ in}^4 + (35.1 \text{ in}^2)\left(\frac{19.0 \text{ in}}{2} - 11.78 \text{ in}\right)^2$$

$$+ 8.07 \text{ in}^4 + (10.00 \text{ in}^2)\left(\begin{array}{c} 19.0 \text{ in} + 0.788 \text{ in} \\ - 11.78 \text{ in} \end{array}\right)^2$$

$$= \boxed{3021.8 \text{ in}^4 \quad (3020 \text{ in}^4)}$$

The answer is (C).

(e) The moment is maximum when the loads are positioned as shown.

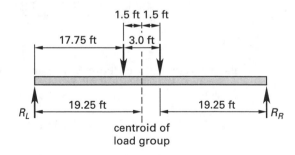

The maximum moment occurs when the resultant of the two loads is located $\boxed{19.25 \text{ ft} \ (19.3 \text{ ft})}$ from the left end of the beam.

The answer is (B).

(f) (This problem states that the beam is not subject to impact loading. But if it were, the load, P, would have been increased by 10–25%.)

Taking clockwise moments as positive,

$$\sum M_{\text{left support}} = P(17.75 \text{ ft}) + P(20.75 \text{ ft})$$
$$- R_R(40 \text{ ft}) = 0$$
$$R_R = 0.9625P$$
$$R_L = 2P - 0.9625P = 1.0375P$$
$$M_{\max} = (0.9625P)(19.25 \text{ ft})$$
$$= \boxed{18.53P \text{ ft-kips} \quad (18.5P \text{ ft-kips})}$$

The answer is (C).

(g) The critical bending stress is

$$F_b = 0.7F_y = (0.7)\left(50 \ \frac{\text{kips}}{\text{in}^2}\right)$$
$$= 35.0 \text{ kips/in}^2$$
$$f = \frac{Mc}{I}$$
$$35.0 \ \frac{\text{kips}}{\text{in}^2} = \frac{(18.53P \text{ ft-kips})\left(12 \ \frac{\text{in}}{\text{ft}}\right)(11.78 \text{ in})}{3021.8 \text{ in}^4}$$
$$P = \boxed{40.38 \text{ kips} \quad (40 \text{ kips})}$$

The answer is (D).

(h) Check the shear.

Placing the load group at one beam end,

$$V = R = 40.38 \text{ kips} + \frac{(40.38 \text{ kips})(40 \text{ ft} - 3 \text{ ft})}{40 \text{ ft}}$$
$$= 77.73 \text{ kips}$$

The area for shear (assuming the channel does not carry any shear) is

$$A_w = dt_w = (19.0 \text{ in})(0.655 \text{ in}) = 12.45 \text{ in}^2$$

The maximum shear stress is

$$f_v = \frac{V}{A_w} = \frac{77.73 \text{ kips}}{12.45 \text{ in}^2} = \boxed{6.24 \text{ kips/in}^2 \quad (6.0 \text{ ksi})}$$

The answer is (D).

(i) From *AISC Manual* Table 9-4, for a W18 × 119, the beam bearing constants are

$$R_1 = 79.8 \text{ kips}$$
$$R_2 = 21.8 \text{ kips/in}$$

Use Eq. 59.41. From a consideration of web yielding,

$$N_{\min} = \frac{R - R_1}{R_2} = \frac{77.73 \text{ kips} - 79.8 \text{ kips}}{21.8 \ \dfrac{\text{kips}}{\text{in}}}$$
$$= \boxed{-0.095 \text{ in} < 0}$$

The answer is (A).

(j) From *AISC Manual* Table 9-4, for a W18 × 119, the beam bearing constants are

$$R_3 = 131 \text{ kips}$$
$$R_4 = 10.1 \text{ kips/in}$$

Use Eq. 59.42. From a consideration of web crippling,

$$N_{\min} = \frac{R - R_3}{R_4} = \frac{77.73 \text{ kips} - 131 \text{ kips}}{10.1 \ \dfrac{\text{kips}}{\text{in}}}$$
$$= \boxed{-5.27 \text{ in} < 0}$$

The answer is (A).

Structural

60 Structural Steel: Tension Members

PRACTICE PROBLEMS

(Answer options for LRFD solutions are given in parentheses.)

1. A single-angle tension member $(L7 \times 4 \times {}^5/_8)$ of A36 steel has two gage lines in its long leg and one in its short leg for $^3/_4$ in diameter bolts arranged as shown. Determine the tensile strength of the member, neglecting block shear. Use either the ASD or LRFD method.

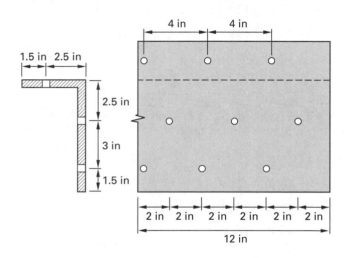

(A) 70 kips (140 kips)

(B) 110 kips (170 kips)

(C) 140 kips (210 kips)

(D) 180 kips (250 kips)

2. Design a plate eyebar of A36 steel to carry a live tensile load of 300 kips.

3. (*Time limit: one hour*) For the truss shown, member DE is made of $2L4 \times 3^1/_2 \times {}^1/_4$, long legs back-to-back (LLBB), A36 steel. All joints are assumed to be pin-connected with one gage line of $^3/_4$ in high-strength bolts through one leg of each angle. Each end of member DE is connected with four bolts on the gage line with an edge distance of $1^1/_4$ in and a $2^1/_4$ in pitch.

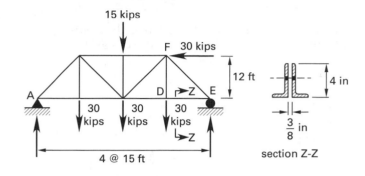

(a) The support reaction at E is most nearly

(A) 38 kips (42 kips)

(B) 47 kips (74 kips)

(C) 52 kips (62 kips)

(D) 58 kips (68 kips)

(b) The tensile force in member DE is most nearly

(A) 38 kips (42 kips)

(B) 46 kips (75 kips)

(C) 52 kips (62 kips)

(D) 58 kips (93 kips)

(c) The effective net area of member DE is most nearly

(A) 2.2 in^2

(B) 2.7 in^2

(C) 2.8 in^2

(D) 3.2 in^2

(d) The tensile strength of member DE, based on the fracture (rupture) criterion, is most nearly

(A) 58 kips (97 kips)

(B) 64 kips (103 kips)

(C) 69 kips (108 kips)

(D) 80 kips (120 kips)

(e) The tensile load strength of member DE, based on the yielding criterion, is most nearly

(A) 58 kips (97 kips)

(B) 68 kips (107 kips)

(C) 75 kips (114 kips)

(D) 78 kips (117 kips)

(f) In the computation of the block shear strength of member DE, the area in shear is most nearly

(A) 0.8 in^2

(B) 1.4 in^2

(C) 2.5 in^2

(D) 3.0 in^2

(g) In the computation of the block shear strength of member DE, the area in tension is most nearly

(A) 0.5 in^2

(B) 0.9 in^2

(C) 1.3 in^2

(D) 2.8 in^2

(h) The tensile strength of member DE, based on the block shear criterion, is most nearly

(A) 55 kips (82 kips)

(B) 58 kips (87 kips)

(C) 68 kips (102 kips)

(D) 73 kips (109 kips)

(i) The tensile strength of member DE, based on yielding, fracture, and block shear criteria, is most nearly

(A) 55 kips (82 kips)

(B) 58 kips (87 kips)

(C) 68 kips (102 kips)

(D) 78 kips (117 kips)

(j) The maximum slenderness ratio of member DE is most nearly

(A) 120

(B) 140

(C) 170

(D) 180

4. (*Time limit: one hour*) A 25 ft long W-shape member is to carry an axial live tensile load of 420 kips. It is assumed that there will be two lines of holes for $^7/_8$ in diameter bolts in each flange. There will be at least three bolts in each line.

(For parts (a) through (e), assume A36 steel is used.)

(a) The required gross area based on the yielding criterion is most nearly

(A) 10 in^2 (12 in^2)

(B) 15 in^2 (17 in^2)

(C) 19 in^2 (20 in^2)

(D) 25 in^2 (27 in^2)

(b) The required effective net area based on the fracture (rupture) criterion is most nearly

(A) 10 in^2 (12 in^2)

(B) 14 in^2 (15 in^2)

(C) 20 in^2 (28 in^2)

(D) 25 in^2 (33 in^2)

(c) The required net area based on the fracture (rupture) criterion is most nearly

(A) 10 in^2 (12 in^2)

(B) 16 in^2 (17 in^2)

(C) 20 in^2 (22 in^2)

(D) 25 in^2 (27 in^2)

(d) The minimum radius of gyration required to satisfy the preferred slenderness limit for the member is most nearly

(A) 0.55 in

(B) 0.75 in

(C) 0.85 in

(D) 1.0 in

(e) Using ASD only, among the four choices listed, the lightest W12 section suitable for this member is

(A) W12 × 50

(B) W12 × 65

(C) W12 × 79

(D) W12 × 106

(For parts (f) through (j), assume A572, grade 50 steel is used.)

(f) The required gross area based on the yielding criterion is most nearly

    (A) 10 in^2 (12 in^2)

    (B) 14 in^2 (15 in^2)

    (C) 20 in^2 (22 in^2)

    (D) 25 in^2 (28 in^2)

(g) The required net effective area based on the fracture criterion is most nearly

    (A) 10 in^2 (11 in^2)

    (B) 13 in^2 (14 in^2)

    (C) 20 in^2 (22 in^2)

    (D) 25 in^2 (28 in^2)

(h) The required net area based on the fracture criterion is most nearly

    (A) 10 in^2 (11 in^2)

    (B) 14 in^2 (15 in^2)

    (C) 20 in^2 (22 in^2)

    (D) 25 in^2 (28 in^2)

(i) The minimum radius of gyration required to satisfy the preferred slenderness limit for the member is most nearly

    (A) 0.55 in

    (B) 0.75 in

    (C) 0.85 in

    (D) 1.0 in

(j) Among the four choices listed, the lightest grade 50 W12 section suitable for this member is

    (A) W12 × 50

    (B) W12 × 65

    (C) W12 × 79

    (D) W12 × 106

SOLUTIONS

1. The gross area of the member is $A_g = 6.48$ in^2 [AISC Table 1-7].

The effective hole diameter includes a $^1/_8$ in allowance for clearance and manufacturing tolerances.

$$d_h = 0.75 \text{ in} + 0.125 \text{ in} = 0.875 \text{ in}$$

To calculate the net area, first find the net width using Eq. 60.5.

$$b_n = b - \sum d_h + \sum \frac{s^2}{4g}$$

The net width of the member must be evaluated by paths ABCD and ABECD.

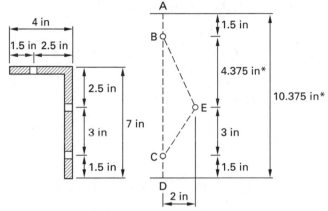

* distance through $\mathcal{C}$ of the angle.

Path ABCD does not have any diagonal runs.

$$b_{n,\text{ABCD}} = 10.375 \text{ in} - (2)(0.875 \text{ in}) = 8.625 \text{ in}$$

Path ABECD has two diagonal runs, BE and EC, for which the quantity $s^2/4g$ must be calculated. The pitch, s, is shown as 2 in, and the gage values are 4.375 in and 3 in, respectively.

For diagonal BE,

$$\frac{s^2}{4g} = \frac{(2 \text{ in})^2}{(4)(4.375 \text{ in})} = 0.229 \text{ in}$$

For diagonal EC,

$$\frac{s^2}{4g} = \frac{(2 \text{ in})^2}{(4)(3 \text{ in})} = 0.333 \text{ in}$$

$$b_{n,\text{ABECD}} = 10.375 \text{ in} - (3)(0.875 \text{ in}) + (1)(0.229 \text{ in})$$
$$+ (1)(0.333 \text{ in})$$
$$= 8.312 \text{ in} \quad [\text{controls}]$$

Therefore, the critical net area is given by Eq. 60.6.

$$A_n = b_{n,\text{ABECD}}t = (8.312 \text{ in})\left(\tfrac{5}{8} \text{ in}\right) = 5.195 \text{ in}^2$$
$$0.85A_g = (0.85)(6.48 \text{ in}^2) = 5.508 \text{ in}^2$$
$$5.195 \text{ in}^2 < 5.508 \text{ in}^2 \quad [\text{OK}]$$

The effective net area is given by Eq. 60.8. Since the tension load is transmitted to all elements, $U = 1.0$.

$$A_e = UA_n = (1)(5.195 \text{ in}^2) = 5.195 \text{ in}^2$$

To determine the tensile strength of the member in the net and gross sections, use either the ASD or the LRFD method.

ASD Method Solution

The allowable tensile strength, P_n/Ω_t, based on the net section, where $\Omega_t = 2.00$, is

$$P_n = F_uA_e = \left(58 \; \frac{\text{kips}}{\text{in}^2}\right)(5.195 \text{ in}^2) = 301.3 \text{ kips}$$

$$\frac{P_n}{\Omega_t} = \frac{301.3 \text{ kips}}{2.00} = 150.7 \text{ kips}$$

The allowable tensile strength, P_n/Ω_t, based on the gross section, where $\Omega_t = 1.67$, is

$$P_n = F_yA_g = \left(36 \; \frac{\text{kips}}{\text{in}^2}\right)(6.48 \text{ in}^2) = 233.3 \text{ kips}$$

$$\frac{P_n}{\Omega_t} = \frac{233.3 \text{ kips}}{1.67} = 139.7 \text{ kips}$$

The tensile capacity is the smaller value of
$\boxed{139.7 \text{ kips } (140 \text{ kips}).}$

LRFD Method Solution

The design tensile strength, $\phi_t P_n$, based on the net section, where $\phi = 0.75$, is

$$P_n = F_uA_e = \left(58 \; \frac{\text{kips}}{\text{in}^2}\right)(5.195 \text{ in}^2) = 301.3 \text{ kips}$$

$$\phi_t P_n = (0.75)(301.3 \text{ kips}) = 226.0 \text{ kips}$$

The design tensile strength, $\phi_t P_n$, based on the gross section, where $\phi = 0.90$, is

$$P_n = F_yA_g = \left(36 \; \frac{\text{kips}}{\text{in}^2}\right)(6.48 \text{ in}^2) = 233.3 \text{ kips}$$

$$\phi_t P_n = (0.90)(233.3 \text{ kips}) = 210.0 \text{ kips}$$

The tensile capacity is the smaller value of the two,
$\boxed{210 \text{ kips.}}$

The answer is (C).

2. Eyebar design is covered in the *AISC Specification* Sec. D6.

step 1: The required tensile strength is

$$P_a = 300 \text{ kips} \quad [\text{ASD}]$$
$$P_u = (1.6)(300 \text{ kips}) = 480 \text{ kips} \quad [\text{LRFD}]$$

step 2: For ASD, the allowable tensile yield strength, where $\Omega_t = 1.67$, is

$$\frac{F_yA_g}{\Omega_t} \geq 300 \text{ kips}$$

$$A_g = \frac{(300 \text{ kips})\Omega_t}{F_y} = \frac{(300 \text{ kips})(1.67)}{36 \; \frac{\text{kips}}{\text{in}^2}}$$

$$= 13.9 \text{ in}^2$$

For LRFD, the available tensile yield strength where $\phi_t = 0.90$, is

$$\phi_t F_yA_g \geq 480 \text{ kips}$$

$$A_g = \frac{480 \text{ kips}}{\phi_t F_y} = \frac{480 \text{ kips}}{(0.90)\left(36 \; \frac{\text{kips}}{\text{in}^2}\right)}$$

$$= 14.8 \text{ in}^2$$

step 3:

ASD Method Solution

Try a $1\frac{1}{2}$ in thick plate. The width is

$$b = \frac{A_g}{t} = \frac{13.9 \text{ in}^2}{1.5 \text{ in}} = 9.27 \text{ in}$$

Use PL $1\frac{1}{2} \times 9\frac{1}{2}$.

Check the tensile strength per AISC Sec. D2.

$$P_n = F_yA_g = \left(36 \; \frac{\text{kips}}{\text{in}^2}\right)(13.9 \text{ in}^2)$$
$$= 500.4 \text{ kips}$$

Using ASD,

$$\Omega_t = 1.67$$
$$\frac{P_n}{\Omega_t} = \frac{500.4 \text{ kips}}{1.67} = 299.6 \text{ kips}$$
$$\frac{P_n}{\Omega_t} \approx P_a \quad [\text{OK}]$$

Select a $1^3/4$ in thick plate.

$$A_g = \left(1\tfrac{3}{4} \text{ in}\right)\left(9\tfrac{1}{2} \text{ in}\right) = 16.625 \text{ in}^2$$

$$P_n = F_y A_g = \left(36 \ \frac{\text{kips}}{\text{in}^2}\right)(16.625 \text{ in}^2)$$
$$= 598.5 \text{ kips}$$

$$\frac{P_n}{\Omega_t} = \frac{598.5 \text{ kips}}{1.67} = 358.4 \text{ kips}$$

$$\frac{P_n}{\Omega_t} \geq P_a \quad [\text{OK}]$$

LRFD Method Solution

Select a $1^3/4$ in thick plate.

$$A_g = \left(1\tfrac{3}{4} \text{ in}\right)\left(9\tfrac{1}{2} \text{ in}\right) = 16.625 \text{ in}^2$$

$$P_n = F_y A_g = \left(36 \ \frac{\text{kips}}{\text{in}^2}\right)(16.625 \text{ in}^2)$$
$$= 598.5 \text{ kips}$$

$$\phi_t P_n = (0.9)(598.5 \text{ kips}) = 538.7 \text{ kips}$$

$$\phi_t P_n > P_u \quad [\text{OK}]$$

$$\boxed{\text{Use PL } 1^3/4 \times 9^1/2 \text{ PL.}}$$

step 4: Check the b/t ratio.

$$\frac{b}{t} = \frac{9.5 \text{ in}}{1.75 \text{ in}}$$
$$5.4 < 8 \quad [\text{OK}]$$

step 5: Calculate the minimum pin diameter.

$$d_{\min} = \tfrac{7}{8}b = \left(\tfrac{7}{8}\right)(9.5 \text{ in}) = 8.31 \text{ in}$$

step 6: Refer to *AISC Specification* Sec. D3.3. The maximum hole diameter is

$$d_{h,\max} \approx d_{\min} + \tfrac{1}{32} \text{ in}$$
$$= 8.31 \text{ in} + 0.03125 \text{ in}$$
$$= 8.34 \text{ in}$$

$$\boxed{\text{Make the hole 8.5 in in diameter.}}$$

step 7: The minimum value of w is

$$w_{\min} = \tfrac{2}{3}b = \left(\tfrac{2}{3}\right)(9.5 \text{ in}) = 6.33 \text{ in}$$

The maximum value of w is

$$w_{\max} = \tfrac{3}{4}b = \left(\tfrac{3}{4}\right)(9.5 \text{ in}) = 7.13 \text{ in}$$

Use $w = 6.5$ in.

The pin diameter is 8.5 in $-$ 0.03 in $=$ 8.47 in.

step 8: The minimum external diameter is

$$d_h + 2w = 8.5 \text{ in} + (2)(6.5 \text{ in}) = \boxed{21.5 \text{ in}}$$

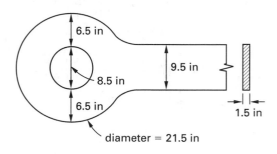

diameter = 21.5 in

3. (a) The length of member FF, is

$$\sqrt{(15 \text{ ft})^2 + (12 \text{ ft})^2} = 19.21 \text{ ft}$$

Find the support reaction E_y by summing moments about point A.

ASD Method Solution

For $\sum M_A = 0$ (clockwise moment positive),

$$(30 \text{ kips})(15 \text{ ft} + 30 \text{ ft} + 45 \text{ ft})$$
$$+ (15 \text{ kips})(30 \text{ ft}) - (30 \text{ kips})(12 \text{ ft}) - E_y(60 \text{ ft}) = 0$$
$$E_y = \boxed{46.5 \text{ kips} \quad (47 \text{ kips})}$$

LRFD Method Solution

For $\sum M_A = 0$ (clockwise moment positive),

$$(1.6)(30 \text{ kips})(15 \text{ ft} + 30 \text{ ft} + 45 \text{ ft})$$
$$+ (1.6)(15 \text{ kips})(30 \text{ ft}) - (1.6)(30 \text{ kips})(12 \text{ ft})$$
$$- E_y(60 \text{ ft}) = 0$$
$$E_y = \boxed{74.4 \text{ kips} \quad (74 \text{ kips})}$$

The answer is (B).

(b) Find the axial force in member DE by applying the method of joints at E.

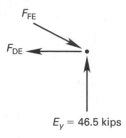

$E_y = 46.5$ kips

ASD Method Solution

$$\sum F_y = 46.5 \text{ kips} - F_{FE}\left(\frac{12 \text{ ft}}{19.21 \text{ ft}}\right) = 0$$
$$F_{FE} = 74.44 \text{ kips}$$

$$\sum F_x = F_{DE} - (74.44 \text{ kips})\left(\frac{15 \text{ ft}}{19.21 \text{ ft}}\right) = 0$$
$$F_{DE} = \boxed{58.13 \text{ kips} \quad (58 \text{ kips})}$$

LRFD Method Solution

$$\sum F_y = 74.44 \text{ kips} - F_{FE}\left(\frac{12 \text{ ft}}{19.21 \text{ ft}}\right) = 0$$
$$F_{FE} = 119.2 \text{ kips}$$

$$\sum F_x = F_{DE} - (119.2 \text{ kips})\left(\frac{15 \text{ ft}}{19.21 \text{ ft}}\right) = 0$$
$$F_{DE} = \boxed{93.08 \text{ kips} \quad (93 \text{ kips})}$$

The answer is (D).

(c) The gross area of 2L4 × 3¹/₂ × ¹/₄ is obtained from *AISC Manual* Table 1-7.

$$A_g = 3.62 \text{ in}^2$$

The angle thickness is $t = \frac{1}{4}$ in.

The diameter of the bolt hole is the bolt diameter plus ¹/₈ in.

$$d_h = \tfrac{3}{4} \text{ in} + \tfrac{1}{8} \text{ in} = \tfrac{7}{8} \text{ in}$$

The net area is given by Eq. 60.7.

$$A_n = A_g - d_h(2t) = 3.62 \text{ in}^2 - \left(\tfrac{7}{8} \text{ in}\right)(2)\left(\tfrac{1}{4} \text{ in}\right)$$
$$= 3.18 \text{ in}^2$$

Since the connection is made to only one leg of each of the angles, a reduction for shear lag is required. Referring to Table 60.1 or *AISC Specification* Table D3.1, two cases apply. Case 8, covering single angles with four or more fasteners, specifies a shear lag reduction coefficient, U, of 0.80. U can also be calculated from the connection geometry. The double-angle can be thought

of as two single angles in tension, each with its own eccentricity. Each angle's connection eccentricity, $\bar{x}$, is the distance from the plane of the connection to the shape's centroid. It can be read directly from the y-y axis properties of the L4 × 3¹/₂ × ¹/₄ angle (*AISC Manual* Table 1-7) as 0.897 in.

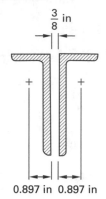

$\frac{3}{8}$ in

0.897 in   0.897 in

With four bolts and three pitch lengths, the length of the connection is

$$L = 3s = (3)(2.25 \text{ in}) = 6.75 \text{ in}$$

$$U = 1 - \frac{\bar{x}}{l} = 1 - \frac{0.897 \text{ in}}{6.75 \text{ in}} = 0.867$$

The larger of the two values may be used, so $U = 0.867$.

From Eq. 60.6, the effective net area is

$$A_e = UA_n = (0.867)(3.18 \text{ in}^2) = \boxed{2.76 \text{ in}^2 \quad (2.8 \text{ in}^2)}$$

The answer is (C).

(d) The tensile strength of member DE based on the fracture (rupture) criterion is given by Eq. 60.9 [*AISC Specification* Eq. D2-2].

$$P_n = F_u A_e = \left(58 \frac{\text{kips}}{\text{in}^2}\right)(2.76 \text{ in}^2) = 160.1 \text{ kips}$$

ASD Method Solution

The allowable tensile strength, where $\Omega_t = 2.00$, is

$$\frac{P_n}{\Omega_t} = \frac{160.1 \text{ kips}}{2.00} = \boxed{80.1 \text{ kips} \quad (80 \text{ kips})}$$

For member DE, $P_a < P_n/\Omega_t$, so 80 kips is OK.

LRFD Method Solution

The design tensile strength, where $\phi_t = 0.75$, is

$$\phi_t P_n = (0.75)(160.1 \text{ kips}) = \boxed{120.1 \text{ kips} \quad (120 \text{ kips})}$$

For member DE, $P_u < \phi_t P_n$, so 120 kips is OK.

The answer is (D).

(e) The tensile strength of member DE, based on the yielding criterion, is given by Eq. 60.8 [*AISC Specification* Eq. D2-1].

$$P_n = F_y A_g = \left(36 \ \frac{\text{kips}}{\text{in}^2}\right)(3.62 \ \text{in}^2) = 130.3 \ \text{kips}$$

ASD Method Solution

The allowable tensile strength, where $\Omega_t = 1.67$, is

$$\frac{P_n}{\Omega_t} = \frac{130.3 \ \text{kips}}{1.67} = \boxed{78.0 \ \text{kips}}$$

For member DE, $P_a < P_n/\Omega_t$, so 78.0 kips is OK.

LRFD Method Solution

The design tensile strength, where $\phi_t = 0.9$, is

$$\phi_t P_n = (0.9)(130.3 \ \text{kips}) = \boxed{117.3 \ \text{kips} \quad (117 \ \text{kips})}$$

For member DE, $P_u < \phi_t P_n$, so 117.3 kips is OK.

The answer is (D).

(f) For block shear strength computation of member DE, the area in shear is

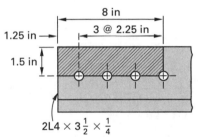

2L4 × 3½ × ¼

$$A_{nv} = 2(\text{length in shear} \times \text{thickness})$$
$$= (2)\left(8 \ \text{in} - (3.5 \ \text{in})\left(\tfrac{7}{8} \ \text{in}\right)\right)\left(\tfrac{1}{4} \ \text{in}\right)$$
$$= \boxed{2.47 \ \text{in}^2 \quad (2.5 \ \text{in}^2)}$$

The answer is (C).

(g) For block shear strength computation of member DE, the area in tension is

$$A_{nv} = 2(\text{length in tension} \times \text{thickness})$$
$$= (2)\left(1.5 \ \text{in} - (0.5 \ \text{in})\left(\tfrac{7}{8} \ \text{in}\right)\right)\left(\tfrac{1}{4} \ \text{in}\right)$$
$$= \boxed{0.53 \ \text{in}^2 \quad (0.5 \ \text{in}^2)}$$

The answer is (A).

(h) The tensile capacity of member DE is based on the block shear criterion. Use Eq. 60.10 [*AISC Specification* Eq. J4-5].

Stress is uniform, so $U_{bs} = 1.0$.

$$R_n = 0.6 F_u A_{nv} + U_{bs} F_u A_{nt} \leq 0.6 F_y A_{gv} + U_{bs} F_u A_{nt}$$
$$= (0.6)\left(58 \ \frac{\text{kips}}{\text{in}^2}\right)(2.47 \ \text{in}^2)$$
$$\quad + (1)\left(58 \ \frac{\text{kips}}{\text{in}^2}\right)(0.53 \ \text{in}^2)$$
$$\leq (0.6)\left(36 \ \frac{\text{kips}}{\text{in}^2}\right)(4 \ \text{in}^2)$$
$$\quad + (1)\left(58 \ \frac{\text{kips}}{\text{in}^2}\right)(0.53 \ \text{in}^2)$$
$$= 116.7 \ \text{kips} \leq 117.1 \ \text{kips}$$

For ASD, the allowable strength, R_n/Ω, where $\Omega = 2.00$, is 117 kips/2.00 = $\boxed{58.5 \ \text{kips} \ (58 \ \text{kips}).}$

For LRFD, the design strength ϕR_n, where $\phi = 0.75$, is $(0.75)(117 \ \text{kips}) = \boxed{87.8 \ \text{kips} \ (87 \ \text{kips}).}$

The capacity as limited by block shear is less than the capacity as limited by yielding (from part (e)), so block shear controls the member strength.

The answer is (B).

(i) The tensile capacity of member DE based on yielding, fracture, and block shear criteria, is the smallest of the three values. For ASD, the smallest P_n is $\boxed{58.5 \ \text{kips} \ (58 \ \text{kips}).}$ For LRFD, the smallest P_n is $\boxed{87.8 \ \text{kips} \ (87 \ \text{kips}).}$

The answer is (B).

(j) The length of member DE is

$$L = (15 \ \text{ft})\left(12 \ \frac{\text{in}}{\text{ft}}\right) = 180 \ \text{in}$$

From *AISC Manual* Table 1-15, for 2L4 × 3½ × ¼, LLBB with ⅜ in spacing, $r_y = 1.52$ in. This is greater than $r_x = 1.27$ in, also from *AISC Manual* Table 1-15, so the minimum radius of gyration is 1.27 in. (This value is given as 1.26 in in *AISC Manual* Table 1-7.) The maximum slenderness ratio is

$$\text{SR} = \frac{L}{r_{\min}} = \frac{L}{r_x} = \frac{180 \ \text{in}}{1.27 \ \text{in}} = \boxed{141.7 \quad (140)}$$

The answer is (B).

4. The axial tensile load to be carried by the W-shape member is $P_a = P_u = 420$ kips.

(a) The required gross area based on the yielding criterion can be found using Eq. 60.8 [*AISC Specification* Eq. D2-1].

ASD Method Solution

$$P_a \le \frac{P_n}{\Omega} = \frac{F_y A_g}{\Omega}$$

$$A_g \ge \frac{\Omega P_a}{F_y} = \frac{(1.67)(420 \text{ kips})}{36 \dfrac{\text{kips}}{\text{in}^2}} = 19.5 \text{ in}^2$$

LRFD Method Solution

$$P_u \le \phi P_n$$

$$1.6 P_a \le F_y A_g$$

$$A_g \ge \frac{1.6 P_a}{\phi F_y} = \frac{(1.6)(420 \text{ kips})}{(0.9)\left(36 \dfrac{\text{kips}}{\text{in}^2}\right)} = 20.7 \text{ in}^2$$

The answer is (C).

(b) The required effective net area based on the fracture criterion can be found using Eq. 60.9.

ASD Method Solution

$$\Omega_t = 2.00$$

$$P_a = \frac{P_n}{\Omega_t} = \frac{F_u A_e}{\Omega_t}$$

$$420 \text{ kips} = \frac{\left(58 \dfrac{\text{kips}}{\text{in}^2}\right) A_e}{2}$$

$$A_e = \boxed{14.48 \text{ in}^2 \quad (14 \text{ in}^2)}$$

LRFD Method Solution

$$\phi_t = 0.75$$

$$P_u = \phi_t P_n = \phi_t F_u A_e$$

$$(1.6)(420 \text{ kips}) = (0.75)\left(58 \dfrac{\text{kips}}{\text{in}^2}\right) A_e$$

$$A_e = \boxed{15.45 \text{ in}^2 \quad (16 \text{ in}^2)}$$

The answer is (B).

(c) From Table 60.1, the reduction coefficient is $U = 0.90$.

The required net area is based on the fracture criterion from Eq. 60.8.

ASD Method Solution

$$A_n = \frac{A_e}{U} = \frac{14.48 \text{ in}^2}{0.90} = \boxed{16.1 \text{ in}^2 \quad (16 \text{ in}^2)}$$

LRFD Method Solution

$$A_n = \frac{A_e}{U} = \frac{15.45 \text{ in}^2}{0.90} = \boxed{17.2 \text{ in}^2 \quad (17 \text{ in}^2)}$$

The answer is (B)

(d) Using Eq. 60.11, the minimum radius of gyration required to satisfy the preferred slenderness ratio of 300 is

$$r_{\min} = \frac{L}{\text{SR}} = \frac{(25 \text{ ft})\left(12 \dfrac{\text{in}}{\text{ft}}\right)}{300} = \boxed{1 \text{ in}}$$

The answer is (D).

(e) From Table 1-1 of the *AISC Manual*,

> W12 × 79 is the lightest suitable section among the four choices listed.

$$t_f = 0.735 \text{ in}$$

$$r_{\min} = 3.05 \text{ in} > \text{required 1 in}$$

$$A_g = 23.2 \text{ in}^2 > \text{required 20.7 in}^2 \quad \text{[yielding]}$$

$$> \text{required 16.1 in}^2 + (4)(0.735 \text{ in})(1 \text{ in})$$

$$= 19.04 \text{ in}^2 \quad \text{[fracture]}$$

The answer is (C).

For parts (f) through (j), A572 grade 50 steel is used.

$$F_y = 50 \text{ kips/in}^2$$

$$F_u = 65 \text{ kips/in}^2$$

(f) The required gross area, based on the yielding criterion, can be found using Eq. 60.8.

ASD Method Solution

$$\Omega_t = 1.67$$

$$P_a = \frac{P_n}{\Omega_t} = \frac{F_u A_e}{\Omega_t}$$

$$420 \text{ kips} = \frac{\left(50 \; \dfrac{\text{kips}}{\text{in}^2}\right) A_g}{1.67}$$

$$A_g = \boxed{14.03 \text{ in}^2 \quad (14 \text{ in}^2)}$$

LRFD Method Solution

$$\phi_t = 0.90$$

$$P_u = \phi_t P_n = \phi_t F_y A_g$$

$$(1.6)(420 \text{ kips}) = (0.90)\left(50 \; \frac{\text{kips}}{\text{in}^2}\right) A_g$$

$$A_g = \boxed{14.93 \text{ in}^2 \quad (15 \text{ in}^2)}$$

The answer is (B).

(g) The required effective net area, based on the fracture criterion, can be found using Eq. 60.9.

ASD Method Solution

$$\Omega_t = 2.00$$

$$P_a = \frac{P_n}{\Omega_t} = \frac{F_u A_e}{\Omega_t}$$

$$420 \text{ kips} = \frac{\left(65 \; \dfrac{\text{kips}}{\text{in}^2}\right) A_e}{2.0}$$

$$A_e = \boxed{12.92 \text{ in}^2 \quad (13 \text{ in}^2)}$$

LRFD Method Solution

$$\phi_t = 0.75$$

$$P_u = \phi_t P_n = \phi_t F_u A_e$$

$$(1.6)(420 \text{ kips}) = (0.75)\left(65 \; \frac{\text{kips}}{\text{in}^2}\right) A_e$$

$$A_e = \boxed{13.78 \text{ in}^2 \quad (14 \text{ in}^2)}$$

The answer is (B).

(h) The reduction coefficient is $U = 0.90$.

The required net area is based on the fracture criterion from Eq. 60.8.

ASD Method Solution

$$A_n = \frac{A_e}{U} = \frac{12.92 \text{ in}^2}{0.90} = \boxed{14.36 \text{ in}^2 \quad (14 \text{ in}^2)}$$

LRFD Method Soution

$$A_n = \frac{A_e}{U} = \frac{13.78 \text{ in}^2}{0.9} = \boxed{15.31 \text{ in}^2 \quad (15 \text{ in}^2)}$$

The answer is (B).

(i) Using Eq. 60.11, the minimum radius of gyration required to satisfy the preferred slenderness ratio of 300 is

$$r_{\min} = \frac{L}{\text{SR}} = \frac{(25 \text{ ft})\left(12 \; \dfrac{\text{in}}{\text{ft}}\right)}{300} = \boxed{1 \text{ in}}$$

The answer is (D).

(j) From Table 1-1 of the *AISC Manual*,

$$\boxed{\begin{array}{l} \text{W12} \times 65 \text{ is the lightest suitable section among the} \\ \text{four listed.} \end{array}}$$

$$t_f = 0.605 \text{ in}$$

$$r_{\min} = 3.02 \text{ in} > \text{required 1 in}$$

$$A_g = 19.1 \text{ in}^2 > \text{required } 14.0 \text{ in}^2 \quad \text{[yielding]}$$

$$> \text{required } 14.36 \text{ in}^2 + (4)(0.605 \text{ in})(1 \text{ in})$$

$$= 16.78 \text{ in}^2 \quad \text{[fracture]}$$

The answer is (B).

61 Structural Steel: Compression Members

PRACTICE PROBLEMS

(Answer options for LFRD solutions are given in parentheses.)

1. Select the lightest W section of A992 steel to serve as a column 30 ft long and to carry an axial live compressive load of 160 kips. The member is pinned at its top and bottom. The member's self-weight is estimated as 2000 lbf. It is supported at midheight in its weak direction.

(A) W8 × 40

(B) W10 × 39

(C) W12 × 45

(D) W14 × 48

2. (*Time limit: one hour*) The two-story, unbraced frame of A992 steel shown is one of several placed on 30 ft centers. All columns are pinned (rotation free and translation fixed) in the weak axis direction (i.e., perpendicular to the plane of the frame). The following loadings are applied.

roof dead load: 4 in thick reinforced concrete slab (unit weight = 150 lbf/ft³) and roofing material (weight = 6 lbf/ft²)
roof live load: 30 lbf/ft
floor dead load: 6 in thick reinforced concrete slab
floor live load: 90 lbf/ft²

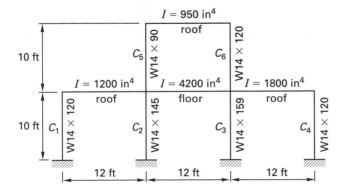

(a) Based on the framing member information provided, the effective length factor K_x of column C_2 for buckling about the strong axis is most nearly

(A) 0.70

(B) 0.85

(C) 1.0

(D) 1.2

(b) Based on the framing member information provided, the effective length factor K_x of column C_5 for buckling about the strong axis is most nearly

(A) 0.78

(B) 0.95

(C) 1.3

(D) 1.7

(c) The slenderness ratios of columns C_2 and C_5 with respect to the strong axis are most nearly

(A) 23 and 25, respectively

(B) 49 and 54, respectively

(C) 62 and 67, respectively

(D) 78 and 85, respectively

(d) The maximum slenderness ratio for column C_2 is most nearly

(A) 24

(B) 30

(C) 48

(D) 63

(e) The maximum slenderness ratio for column C_5 is most nearly

(A) 28

(B) 32

(C) 45

(D) 68

(f) The available (design) axial compressive load for column C_2 is most nearly

(A) 800 kips (1400 kips)

(B) 1000 kips (1600 kips)

(C) 1200 kips (1800 kips)

(D) 1400 kips (2000 kips)

(g) The available (design) axial compressive load for column C_5 is most nearly

(A) 550 kips (950 kips)

(B) 700 kips (1050 kips)

(C) 730 kips (1100 kips)

(D) 940 kips (1300 kips)

(h) The computed axial compressive load on column C_5 is most nearly

(A) 16 kips

(B) 78 kips

(C) 180 kips

(D) 220 kips

(i) The computed axial compressive load on column C_2 is most nearly

(A) 63 kips

(B) 78 kips

(C) 180 kips

(D) 220 kips

(j) Using ASD and considering axial loads only,

(A) Both columns C_2 and C_5 are adequate.

(B) Both columns C_2 and C_5 are inadequate.

(C) Column C_2 is adequate; column C_5 is inadequate.

(D) Column C_5 is adequate; column C_2 is inadequate.

3. (*Time limit: one hour*) A 20 ft high column, pinned at both ends, is constructed from an A992 channel and an A992 wide flange shape as shown. The column is constrained in the x-plane by bracing at midheight.

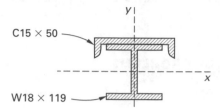

(a) The gross area of the column is most nearly

(A) 35 in^2

(B) 42 in^2

(C) 50 in^2

(D) 55 in^2

(b) The centroidal moment of inertia of the column for bending about the x-axis is most nearly

(A) 2700 in^4

(B) 2900 in^4

(C) 3100 in^4

(D) 3300 in^4

(c) The centroidal moment of inertia of the column for bending about the y-axis is most nearly

(A) 570 in^4

(B) 660 in^4

(C) 790 in^4

(D) 820 in^4

(d) The radius of gyration of the column with respect to the x-axis is most nearly

(A) 4.9 in

(B) 5.7 in

(C) 6.9 in

(D) 7.9 in

(e) The radius of gyration of the column with respect to the y-axis is most nearly

    (A) 2.9 in

    (B) 3.1 in

    (C) 3.6 in

    (D) 4.1 in

(f) The slenderness ratio of the column with respect to the x-axis is most nearly

    (A) 28

    (B) 30

    (C) 33

    (D) 41

(g) The slenderness ratio of the column with respect to the y-axis is most nearly

    (A) 28

    (B) 30

    (C) 33

    (D) 41

(h) The maximum slenderness ratio for the column is most nearly

    (A) 28

    (B) 30

    (C) 33

    (D) 41

(i) The available (design) critical stress for the column is most nearly

    (A) 16 ksi (30 ksi)

    (B) 20 ksi (34 ksi)

    (C) 24 ksi (36 ksi)

    (D) 28 ksi (42 ksi)

(j) The nominal axial compressive load on the column for both ASD and LRFD is most nearly

    (A) 1000 kips

    (B) 1500 kips

    (C) 2300 kips

    (D) 2600 kips

4. Select A992 interior columns of the moment frame shown. Assume columns are braced in the plane perpendicular to the frame and built in at the bottom. Sidesway is uninhibited in both directions. Girders are $W12 \times 96$.

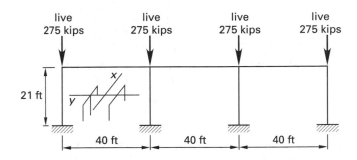

    (A) $W12 \times 53$

    (B) $W12 \times 65$

    (C) $W12 \times 72$

    (D) $W12 \times 87$

5. (*Time limit: one hour*) The truss shown is constructed of a variety of members, all of which are assumed to be pin-ended. Determine whether all members are adequate. If members are not adequate, specify replacements with the same nominal depth. Assume that the top chord is braced laterally at every joint.

top and bottom chords:	$W8 \times 31$ (A992)
diagonal bracing:	$L3^{1}/_{2} \times {}^{1}/_{2} \times {}^{5}/_{16}$ (pair)
vertical members:	$L4 \times 4 \times {}^{3}/_{8}$ (pair)

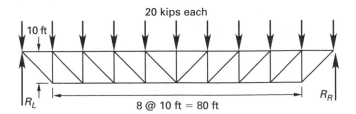

SOLUTIONS

1. Assume the column weight will be approximately 2 kips. The required axial strength is

$$160 \text{ kips} + 2 \text{ kips} = 162 \text{ kips} \quad [\text{ASD}]$$

$$(1.6)(160 \text{ kips}) + (1.2)(2 \text{ kips}) = 258 \text{ kips} \quad [\text{LRFD}]$$

The effective length factors are $K_x = K_y = 1.0$.

The unbraced lengths are $L_x = 30$ ft and $L_y = 15$ ft.

From *AISC Manual* Table 4-1, the following values were tabulated for $KL_y = 15$ ft.

	ASD, P_n/Ω_c (kips)	LRFD, $\phi_c P_n$ (kips)
W8 × 40	199	299
W10 × 39	188	282
W12 × 45	211	317
W14 × 48	221	332

Try the W10 × 39. The r_x/r_y ratio for the W10 × 39 is 2.16. The effective buckling length about the x-axis equals

$$\frac{KL_x}{\dfrac{r_x}{r_y}} = \frac{(1.0)(30 \text{ ft})}{2.16} = 13.9 \text{ ft}$$

This is less than KL_y, so the W10 × 39 is OK.

Select the $\boxed{\text{W10} \times 39.}$

The answer is (B).

2. (a) From Table 1-1 of the *AISC Manual*, for column C_2 (W14 × 145),

$$A = 42.7 \text{ in}^2$$
$$I_x = 1710 \text{ in}^4$$
$$r_x = 6.33 \text{ in}$$
$$r_y = 3.98 \text{ in}$$

For column C_5 (W14 × 90),

$$A = 26.5 \text{ in}^2$$
$$I_x = 999 \text{ in}^4$$
$$r_x = 6.14 \text{ in}$$
$$r_y = 3.70 \text{ in}$$

For both columns C_2 and C_5,

$$L_x = L_y = 10 \text{ ft}$$
$$K_y = 1.0 \quad [\text{pinned at both ends}]$$

For column C_2, $G_{\text{bottom}} = 1.0$ (rigid connection at footing). Use Eq. 61.4.

$$G_{\text{top}} = \frac{\sum\left(\dfrac{I}{L}\right)_c}{\sum\left(\dfrac{I}{L}\right)_b} = \frac{\dfrac{1710 \text{ in}^4}{10 \text{ ft}} + \dfrac{999 \text{ in}^4}{10 \text{ ft}}}{\dfrac{1200 \text{ in}^4}{12 \text{ ft}} + \dfrac{4200 \text{ in}^4}{12 \text{ ft}}}$$
$$= 0.602$$

From Fig. 61.2, the alignment chart for uninhibited sidesway, $K_x \approx \boxed{1.2.}$

The answer is (D).

(b) For column C_5, $G_{\text{bottom}} = 0.6$ (same as G_{top} for column C_2). Use Eq. 61.4.

$$G_{\text{top}} = \frac{\sum\left(\dfrac{I}{L}\right)_c}{\sum\left(\dfrac{I}{L}\right)_b} = \frac{\dfrac{999 \text{ in}^4}{10 \text{ ft}}}{\dfrac{950 \text{ in}^4}{12 \text{ ft}}}$$
$$= 1.262$$

From the alignment chart for uninhibited sidesway, $K_x \approx \boxed{1.3.}$

The answer is (C).

(c) Calculate the slenderness ratio with respect to the strong axis. For column C_2, using Eq. 61.5,

$$\text{SR} = \frac{K_x L_x}{r_x} = \frac{(1.2)(10 \text{ ft})\left(12 \, \dfrac{\text{in}}{\text{ft}}\right)}{6.33 \text{ in}}$$
$$= \boxed{22.75 \quad (23)}$$

For column C_5, using Eq. 61.5,

$$\text{SR} = \frac{K_x L_x}{r_x} = \frac{(1.3)(10 \text{ ft})\left(12 \, \dfrac{\text{in}}{\text{ft}}\right)}{6.14 \text{ in}}$$
$$= \boxed{25.41 \quad (25)}$$

The answer is (A).

(d) For column C_2, using Eq. 61.5,

$$SR = \frac{K_y L_y}{r_y} = \frac{(1)(10 \text{ ft})\left(12 \frac{\text{in}}{\text{ft}}\right)}{3.98 \text{ in}}$$

$$= 30.15 \quad (30)$$

From part (c),

$$\frac{K_x L_x}{r_x} = 22.75$$

Therefore, the maximum slenderness ratio is 30.15 $\boxed{(30)}$.

The answer is (B).

(e) For column C_5, using Eq. 61.5,

$$SR = \frac{K_y L_y}{r_y} = \frac{(1)(10 \text{ ft})\left(12 \frac{\text{in}}{\text{ft}}\right)}{3.70 \text{ in}}$$

$$= 32.43 \quad (32)$$

From part (c),

$$SR = \frac{K_x L_x}{r_x} = 25.41$$

Therefore, the maximum slenderness ratio will be 32.43 $\boxed{(32)}$.

The answer is (B).

(f) For column C_2,

$$KL = (1)(10 \text{ ft}) = 10 \text{ ft}$$

From *AISC Manual* Table 4-1, for a W14 × 145 with 50 ksi steel, the available axial compressive load for ASD is $P_n/\Omega_c = \boxed{1200 \text{ kips}}$, and $\phi_c P_n = \boxed{1800 \text{ kips}}$ for LRFD.

The answer is (C).

(g) For column C_5,

$$KL = (1.0)(10 \text{ ft}) = 10 \text{ ft}$$

From *AISC Manual* Table 4-1, for a W14 × 90 with 50 ksi steel, the available axial compressive load for ASD is $P_n/\Omega_c = \boxed{734 \text{ kips } (730 \text{ kips})}$, and for LRFD is $\phi_c P_n = \boxed{1100 \text{ kips}}$.

The answer is (C).

(h) With 15 ft on either side of the frame, the tributary area of the roof to one side (only) of column C_5 is

$$(30 \text{ ft})\left(\frac{12 \text{ ft}}{2}\right) = 180 \text{ ft}^2$$

The loadings are as follows.

roof DL		
roofing material		6 lbf/ft^2
concrete slab		
$\left(\dfrac{4 \text{ in}}{12 \frac{\text{in}}{\text{ft}}}\right)\left(150 \frac{\text{lbf}}{\text{ft}^3}\right) =$		50 lbf/ft^2
roof LL		30 lbf/ft^2
	total	86 lbf/ft^2

The total axial compressive load on column C_5 including the self-weight of 90 lbf/ft is

$$\text{(tributary area)}\text{(total loading)} + \text{column weight}$$

$$= (180 \text{ ft}^2)\left(86 \frac{\text{lbf}}{\text{ft}^2}\right) + \left(90 \frac{\text{lbf}}{\text{ft}}\right)(10 \text{ ft})$$

$$= \boxed{16{,}380 \text{ lbf} \quad (16 \text{ kips})}$$

The answer is (A).

(i) With 15 ft on either side of the frame, the tributary area of the second story floor on column C_2 is

$$(30 \text{ ft})\left(\frac{12 \text{ ft}}{2}\right) = 180 \text{ ft}^2$$

The loadings are as follows.

floor DL		
concrete slab		
$\left(\dfrac{6 \text{ in}}{12 \frac{\text{in}}{\text{ft}}}\right)\left(150 \frac{\text{lbf}}{\text{ft}^3}\right) =$		75 lbf/ft^2
floor LL		90 lbf/ft^2
	total	165 lbf/ft^2

The tributary area of the first story roof to the left of column C_2 is 180 ft^2. The total axial compressive load on column C_2 is

$$\text{(tributary area)}\text{(total load)} + \text{column weight}$$

$$+ \text{total axial compressive load on column } C_5$$

$$= (180 \text{ ft}^2)\left(165 \frac{\text{lbf}}{\text{ft}^2}\right) + (180 \text{ ft}^2)\left(86 \frac{\text{lbf}}{\text{ft}^2}\right)$$

$$+ \left(145 \frac{\text{lbf}}{\text{ft}}\right)(10 \text{ ft}) + 16{,}380 \text{ lbf}$$

$$= \boxed{63{,}010 \text{ lbf} \quad (63 \text{ kips})}$$

The answer is (A).

Structural

(j) Considering ASD and axial loads only,

column C_5: $P_{\text{req}} = 16.38$ kips $< P_a = 734$ kips [OK]

column C_2: $P_{\text{req}} = 63$ kips $< P_a = 1200$ kips [OK]

The answer is (C).

3. (a) From *AISC Manual* Table 1-1, for a W18 × 119,

$$A = 35.1 \text{ in}^2$$
$$d = 19.0 \text{ in}$$
$$I_x = 2190 \text{ in}^4$$
$$I_y = 253 \text{ in}^4$$

From *AISC Manual* Table 1-5, for a C15 × 50,

$$A = 14.7 \text{ in}^2$$
$$I_x = 404 \text{ in}^4$$
$$I_y = 11.0 \text{ in}^4$$
$$t_w = 0.716 \text{ in}$$
$$\overline{x} = 0.799 \text{ in}$$

The gross area of the column is the cross-sectional area of the column.

$$A_g = \sum A_i = 35.1 \text{ in}^2 + 14.7 \text{ in}^2$$
$$= \boxed{49.8 \text{ in}^2 \quad (50 \text{ in}^2)}$$

The answer is (C).

(b) To calculate the moment of inertia, the location of the centroid needs to be determined first, where $\overline{y}$ is measured from the bottom of the wide flange.

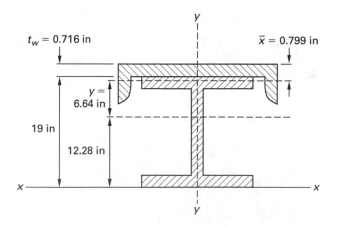

$$\overline{y} = \frac{\sum A_i \overline{y}_i}{\sum A_i}$$

$$= \frac{(35.1 \text{ in}^2)\left(\dfrac{19.0 \text{ in}}{2}\right)}{35.1 \text{ in}^2 + 14.7 \text{ in}^2}$$
$$\frac{+ (14.7 \text{ in}^2)(19.0 \text{ in} + 0.716 \text{ in} - 0.799 \text{ in})}{35.1 \text{ in}^2 + 14.7 \text{ in}^2}$$

$$= 12.28 \text{ in}$$

From the parallel axis theorem, the moment of inertia about the strong axis is

$$I_x = 2190 \text{ in}^4 + (35.1 \text{ in}^2)\left(12.28 \text{ in} - \frac{19.0 \text{ in}}{2}\right)^2$$
$$+ 11.0 \text{ in}^4 + (14.7 \text{ in}^2)\left(\begin{array}{c} \dfrac{19.0 \text{ in} + 0.716 \text{ in}}{2} \\ - 0.799 \text{ in} \\ - 12.28 \text{ in} \end{array}\right)^2$$

$$= \boxed{3120 \text{ in}^4 \quad (3100 \text{ in}^4)}$$

The answer is (C).

(c) The moment of inertia about the weak axis is

$$I_y = 253 \text{ in}^4 + 404 \text{ in}^4 = \boxed{657 \text{ in}^4 \quad (660 \text{ in}^4)}$$

The answer is (B).

(d) The radius of gyration with respect to the x-axis is

$$r_x = \sqrt{\frac{I_x}{A}} = \sqrt{\frac{3120 \text{ in}^4}{49.8 \text{ in}^2}}$$
$$= \boxed{7.92 \text{ in} \quad (7.9 \text{ in})}$$

The answer is (D).

(e) The radius of gyration with respect to the y-axis is

$$r_y = \sqrt{\frac{I_y}{A}} = \sqrt{\frac{657 \text{ in}^4}{49.8 \text{ in}^2}}$$
$$= \boxed{3.63 \text{ in} \quad (3.6 \text{ in})}$$

The answer is (C).

(f) For buckling about the x-axis, from Table 61.1, $K_x = 1$ (both ends pinned). The unbraced length, L_x, is 20 ft. The slenderness ratio of the column with respect to the x-axis is

$$\frac{K_x L_x}{r_x} = \frac{(1)(20 \text{ ft})\left(12 \, \dfrac{\text{in}}{\text{ft}}\right)}{7.91 \text{ in}}$$
$$= \boxed{30.34 \quad (30)}$$

The answer is (B).

(g) For buckling about the y-axis, from Table 61.1, $K_y = 1$ (both ends pinned). The unbraced length, L_y, is 10 ft. The slenderness ratio of the column with respect to the y-axis is

$$\frac{K_y L_y}{r_y} = \frac{(1)(10 \text{ ft})\left(12 \frac{\text{in}}{\text{ft}}\right)}{3.63 \text{ in}}$$

$$= \boxed{33.06 \quad (33)}$$

The answer is (C).

(h) The maximum slenderness ratio is $\boxed{33.06 \ (33).}$

The answer is (C).

(i) From *AISC Manual* Table 4-22, the available critical stress for $KL/r = 33.06$ and $F_y = 50$ ksi is

$$\frac{F_{cr}}{\Omega_c} = 27.7 \text{ ksi} \quad (28 \text{ ksi}) \quad \text{[ASD]}$$

$$\phi_c F_{cr} = 41.6 \text{ ksi} \quad (42 \text{ ksi}) \quad \text{[LRFD]}$$

The answer is (D).

(j) The nominal available axial compressive load is given by *AISC Specification* Eq. E3-1.

ASD Solution Method

From part (i),

$$\frac{F_{cr}}{\Omega_c} = 27.7 \text{ ksi}$$

$$\Omega_c - 1.67$$

$$F_{cr} = (1.67)\left(27.7 \frac{\text{kips}}{\text{in}^2}\right)$$

$$= 46.26 \text{ ksi}$$

From part (a), $A_g = 49.8$ in^2. Therefore, the nominal compressive load is

$$P_n = F_{cr} A_g = \left(46.26 \frac{\text{kips}}{\text{in}^2}\right)(49.8 \text{ in}^2)$$

$$= \boxed{2303.7 \text{ kips} \quad (2300 \text{ kips})}$$

LRFD Method Solution

From part (i),

$$\phi_c F_{cr} = 41.6 \text{ ksi}$$

$$\phi_c = 0.9$$

$$F_{cr} = \frac{41.6 \frac{\text{kips}}{\text{in}^2}}{\phi_c} = \frac{41.6 \frac{\text{kips}}{\text{in}^2}}{0.9} = 46.22 \text{ ksi}$$

From part (a), $A_g = 49.8$ in^2. Therefore, the nominal compressive load is

$$P_n = F_{cr} A_g = \left(46.22 \frac{\text{kips}}{\text{in}^2}\right)(49.8 \text{ in}^2)$$

$$= \boxed{2301.8 \text{ kips} \quad (2300 \text{ kips})}$$

The answer is (C).

4. This is a design problem with only a few options. Follow the design procedure given, modified as necessary to accommodate the given options.

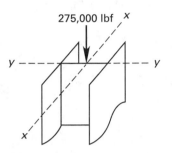

step 1: Assume the column will have a weight of approximately 87 lbf/ft (chosen as the heaviest shape in the options). The load to be carried by each column is

$$P_a = \frac{P_n}{\Omega_c} = \frac{275{,}000 \text{ lbf} + (21 \text{ ft})\left(87 \frac{\text{lbf}}{\text{ft}}\right)}{1000 \frac{\text{lbf}}{\text{kip}}}$$

$$= 276.8 \text{ kips} \quad \text{[ASD]}$$

$$P_u = \phi_c P_n = \frac{(1.6)(275{,}000 \text{ lbf}) + (1.2)(21 \text{ ft})\left(87 \frac{\text{lbf}}{\text{ft}}\right)}{1000 \frac{\text{lbf}}{\text{kip}}}$$

$$= 442.2 \text{ kips} \quad \text{[LRFD]}$$

step 2: The columns are built-in at the bottom and braced only at their tops, so $L_y = 21$ ft. The bracing method in the weak direction is not known, so Eq. 61.4 cannot be used. The tops of columns in moment frames are fixed against rotation. Sidesway (translation) is permitted (uninhibited). A generic value from Table 61.2 is used; $K_y = 1.2$. The effective length for minor axis bending is

$$K_y L_y = (1.2)(21 \text{ ft}) = 25.2 \text{ ft}$$

step 3: From *AISC Manual* Table 4-1, 50 ksi steel, $KL = 25$ ft, and $P_n/\Omega_c = 276$ kips or $\phi_c P_n = 442$ kips. Evaluating the four options, the W12 × 53 and W12 × 65 do not have the required capacity. Try the W12 × 72.

step 4: Check for buckling in the strong direction. For the W12 × 96 horizontal beams, from *AISC Manual* Tables 1-1, 3-3, or 4-1, $I_x = 833$ in⁴. For the W12 × 72, from *AISC Manual* Table 4-1, $I_x = 597$ in⁴, and $r_x/r_y = 1.76$.

Determine the effective length factor. Since the bottom end is built in, $G_{\text{bottom}} = 1.0$. For an interior column, from Eq. 61.4,

$$G_{\text{top}} = \frac{\sum\left(\dfrac{I}{L}\right)_c}{\sum\left(\dfrac{I}{L}\right)_b} = \frac{\dfrac{597 \text{ in}^4}{21 \text{ ft}}}{\dfrac{833 \text{ in}^4}{40 \text{ ft}} + \dfrac{833 \text{ in}^4}{40 \text{ ft}}} = 0.68$$

From the alignment chart, Fig. 61.2(b), with sidesway uninhibited, $K_x = 1.27$.

Use Eq. 61.12.

$$K_x L_x' = \frac{K_x L_x}{\dfrac{r_x}{r_y}} = \frac{(1.27)(21 \text{ ft})}{1.76} = 15.2 \text{ ft}$$

Since $K_x L_x' = 15.2$ ft $< K_y L_y = 25.2$ ft, the W12 × 72 column is adequate. The value of K_y cannot be checked because the framing in the x-direction is not known.

Select a ┃ W12 × 72. ┃

The answer is (C).

---

5. Neglect self-weight.

$$R_L = R_R = \frac{(9)(20 \text{ kips})}{2} = 90 \text{ kips}$$

• Top chords (in compression): W8 × 31 for $KL = 10$ ft.

The force in horizontal members is proportional to the moment across that panel. The moment on the truss is maximum at center.

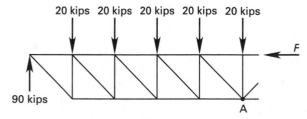

Taking clockwise moments as positive,

$$\sum M_A = (90 \text{ kips})(50 \text{ ft})$$
$$- (20 \text{ kips})(10 \text{ ft} + 20 \text{ ft} + 30 \text{ ft} + 40 \text{ ft})$$
$$- 10F$$
$$= 0$$
$$F = 250 \text{ kips}$$

For ASD, the required compressive strength is $P_a = 250$ kips. From *AISC Manual* Table 4-1, the available compressive strength is 211 kips, which is ┃ inadequate. ┃

Try a W8 × 40.

From *AISC Manual* Table 4-1, the available compressive strength is 273 kips, which is adequate.

For LRFD, $P_u = (1.6)(250 \text{ kips}) = 400$ kips. From *AISC Manual* Table 4-1, the available compressive strength is 317 kips, which is ┃ inadequate. ┃

Try a W8 × 40.

From *AISC Manual* Table 4-1, the available compressive strength is 410 kips, which is ┃ adequate. ┃

• Bottom chords (in tension): W8 × 31

Force is proportional to moment across the panel.

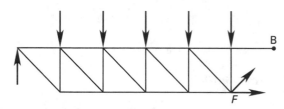

$$\sum M_B = (90 \text{ kips})(60 \text{ ft})$$
$$- (20 \text{ kips})(10 \text{ ft} + 20 \text{ ft} + 30 \text{ ft} + 40 \text{ ft} + 50 \text{ ft})$$
$$- 10F$$
$$= 0$$
$$F = 240 \text{ kips}$$

Try a W8 × 31.

For ASD, the required strength in tension is $P_a = 240$ kips. From *AISC Manual* Table 5-1, the available strength in tension is 273 kips, which is ┃ adequate. ┃

For LRFD, the required strength in tension is $P_u = (1.6)(240 \text{ kips}) = 384$ kips. From *AISC Manual* Table 5-1, the available strength in tension is 410 kips, which is ┃ adequate. ┃

• Diagonal bracing (in tension): L3¹⁄₂ × 3¹⁄₂ × ⁵⁄₁₆

The force in the diagonals is proportional to shear across the panel. At the end diagonal, $V = 90$ kips.

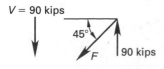

$$F = \frac{90 \text{ kips}}{\sin 45°} = 127.3 \text{ kips}$$

For ASD, the required axial strength in tension is $P_a = 127.3$ kips. From *AISC Manual* Table 5-8, the available axial strength in tension is 90.1 kips, which is inadequate.

Try $L3\frac{1}{2} \times 3\frac{1}{2} \times \frac{5}{16}$.

From *AISC Manual* Table 5-8, the available axial strength is 140 kips, which is adequate.

For LRFD, the required axial strength in tension is $P_u = (1.6)(127.3 \text{ kips}) = 203.8$ kips. From *AISC Manual* Table 5-8, the available axial strength in tension is 135 kips, which is inadequate.

Try $L3\frac{1}{2} \times 3\frac{1}{2} \times \frac{5}{16}$.

From *AISC Manual* Table 5-8, the available axial strength is 211 kips, which is adequate.

• Vertical members (in compression): $L4 \times 4 \times \frac{3}{8}$ for $KL = 10$ ft.

The force in the vertical member is proportional to the vertical force component in the adjacent diagonal.

$F = 90$ kips.

For ASD, the required compressive strength is $P_a = 90$ kips. From *AISC Manual* Table 4-8, the available compressive strength is 74.7 kips, which is inadequate.

Try a $L4 \times 4\frac{1}{2}$.

From *AISC Manual* Table 4-8, the available compressive strength is 96.6 kips, which is adequate.

For LRFD, the required compressive strength is $P_u = (1.6)(90 \text{ kips}) = 144$ kips. From *AISC Manual* Table 4-8, the available compressive strength is 112 kips, which is inadequate.

Try a $L4 \times 4\frac{1}{2}$.

From *AISC Manual* Table 4-8, the available compressive strength is 145 kips, which is adequate.

62 Structural Steel: Beam-Columns

PRACTICE PROBLEMS

(Answer options for LRFD solutions are given in parentheses.)

1. (*Time limit: one hour*) A W14 × 82 column carries an axial live load of 525 kips as shown. The 36 ft long W14 × 53 beams framing into it in the *x*-direction (shown) apply a live moment of 225 ft-kips. Identical 36 ft long beams (not shown) framing into the same joint in the *y*-direction prevent sidesway but do not apply any moment. The column is pinned at the bottom, permitting pivoting in both the *x*- and *y*-directions. All members are grade 50 steel. (a) Show that the column is inadequate. (b) Quickly determine if a grade 50 W14 × 132 column is adequate.

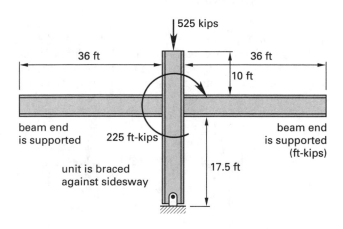

2. (*Time limit: one hour*) A tall column in an industrial storage building carries the vertical roof and lateral wind loads shown. The bottom end is built in, and the top end is fixed against translation, but free to rotate. The lateral wind loads are carried into a braced frame. The column is braced in its weak direction by a strut 24 ft from the bottom hinged end. There is no support of the column compression flange along its length. Make reasonable assumptions to simplify the structural analysis. (a) Select an economical A992 W24 section. Do not check shear strength. (b) Verify that the lateral column deflection is acceptable.

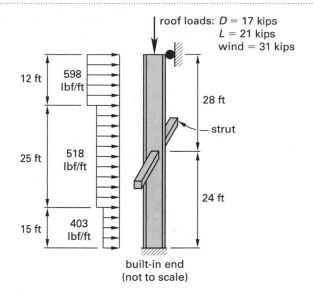

3. (*Time limit: one hour*) A 16 ft column is acted upon by an axial gravity load of 85 kips and a uniform wind load of 4 kips/ft as shown. The W12 × 58 column is made of A992 grade 50 steel. The lower end is built in. The upper end is not supported in the weak direction.

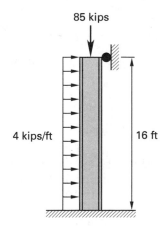

(a) The actual axial stress in the member is most nearly

    (A) 2.5 ksi

    (B) 4.0 ksi

    (C) 5.0 ksi

    (D) 6.5 ksi

(b) The critical slenderness ratio is most nearly

(A) 39

(B) 98

(C) 160

(D) 190

(c) The available (design) axial strength is most nearly

(A) 80 kips (130 kips)

(B) 90 kips (140 kips)

(C) 100 kips (150 kips)

(D) 110 kips (160 kips)

(d) The available (design) bending strength is most nearly

(A) 180 ft-kips (270 ft-kips)

(B) 190 ft-kips (280 ft-kips)

(C) 200 ft-kips (300 ft-kips)

(D) 220 ft-kips (340 ft-kips)

(e) The maximum moment with the 4 kips/ft wind load is most nearly

(A) 40 ft-kips

(B) 85 ft-kips

(C) 130 ft-kips

(D) 170 ft-kips

(f) The beam-column is subjected to

(A) small compression and is adequate

(B) large compression and is adequate

(C) small compression and is inadequate

(D) large compression and is inadequate

For parts (g) through (j), assume lateral bracing (pinned support) is provided in the weak direction at mid-height, as well as at the top of the member.

(g) The critical slenderness ratio is most nearly

(A) 40

(B) 100

(C) 160

(D) 190

(h) The available (design) axial strength is most nearly

(A) 290 kips (450 kips)

(B) 350 kips (515 kips)

(C) 460 kips (690 kips)

(D) 500 kips (725 kips)

(i) The available (design) bending stress with respect to the strong axis is most nearly

(A) 140 ft-kips (210 ft-kips)

(B) 180 ft-kips (260 ft-kips)

(C) 220 ft-kips (320 ft-kips)

(D) 250 ft-kips (380 ft-kips)

(j) The maximum lateral wind load the member can safely carry is most nearly

(A) 5.5 kips/ft (4.9 kips/ft)

(B) 6.1 kips/ft (5.7 kips/ft)

(C) 7.5 kips/ft (6.3 kips/ft)

(D) 9.5 kips/ft (8.7 kips/ft)

4. (*Time limit: one hour*) A W14 × 48 member of A992 steel carries the service loads shown. The member is part of a braced system and has support in the weak direction at mid-height, but it has support only at the top and bottom in the strong direction.

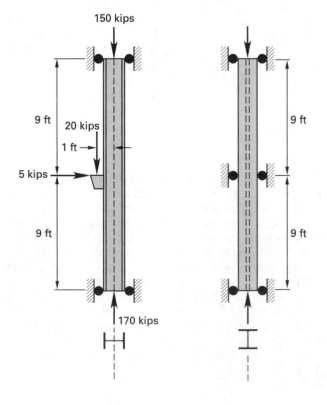

(a) The maximum primary bending moment about the strong axis occurs at

    (A) the top

    (B) one-quarter height from top

    (C) mid-height

    (D) three-quarter height from top

(b) The maximum primary bending moment about the strong axis is most nearly

    (A) 28 ft-kips

    (B) 33 ft-kips

    (C) 40 ft-kips

    (D) 43 ft-kips

(c) The slenderness ratio with respect to buckling about the weak axis is most nearly

    (A) 27

    (B) 37

    (C) 47

    (D) 57

(d) The slenderness ratio with respect to buckling about the strong axis is most nearly

    (A) 27

    (B) 37

    (C) 47

    (D) 57

(e) The critical slenderness ratio is most nearly

    (A) 27

    (B) 37

    (C) 47

    (D) 57

(f) The available (design) axial strength is most nearly

    (A) 250 kips (445 kips)

    (B) 290 kips (475 kips)

    (C) 335 kips (504 kips)

    (D) 420 kips (595 kips)

(g) The actual bending stress is most nearly

    (A) 5.5 ksi

    (B) 6.7 ksi

    (C) 10 ksi

    (D) 14 ksi

(h) Using ASD only, the ratio of actual axial strength to available axial strength is most nearly

    (A) 0.12

    (B) 0.37

    (C) 0.51

    (D) 0.68

(i) Using ASD only, the available bending strength is most nearly

    (A) 140 ft-kips

    (B) 160 ft-kips

    (C) 180 ft-kips

    (D) 200 ft-kips

(j) The beam-column is subjected to

    (A) small compression and is adequate

    (B) large compression and is adequate

    (C) small compression and is inadequate

    (D) large compression and is inadequate

Structural

SOLUTIONS

1. The properties of a W14 × 82 grade 50 column are

$$F_y = 50 \text{ kips/in}^2$$
$$I_x = 881 \text{ in}^4$$
$$I_y = 148 \text{ in}^4$$
$$r_x = 6.05 \text{ in}$$
$$r_y = 2.48 \text{ in}$$

The properties of W14 × 53 grade 50 beams are

$$F_y = 50 \text{ kips/in}^2$$
$$I_x = 541 \text{ in}^4$$

With no sidesway, $G_A = 10$ (pinned). Use Eq. 61.4 to obtain G_B.

$$G_B = \frac{\sum\left(\frac{I}{L}\right)_c}{\sum\left(\frac{I}{L}\right)_b} = \frac{\dfrac{881 \text{ in}^4}{10 \text{ ft}} + \dfrac{881 \text{ in}^4}{17.5 \text{ ft}}}{\dfrac{541 \text{ in}^4}{36 \text{ ft}} + \dfrac{541 \text{ in}^4}{36 \text{ ft}}} = 4.6$$

Use the alignment chart to determine $K_x = 0.94$.

Similarly, in the y-direction with no sidesway, $G_A = 10$ (pinned). Use Eq. 61.4 to obtain G_B.

$$G_B = \frac{\sum\left(\frac{I}{L}\right)_c}{\sum\left(\frac{I}{L}\right)_b} = \frac{\dfrac{148 \text{ in}^4}{10 \text{ ft}} + \dfrac{148 \text{ in}^4}{17.5 \text{ ft}}}{\dfrac{541 \text{ in}^4}{36 \text{ ft}} + \dfrac{541 \text{ in}^4}{36 \text{ ft}}} = 0.77$$

Use the alignment chart to determine $K_y = 0.85$.

The effective lengths are

$$K_y L_y = (0.85)(17.5 \text{ ft}) = 14.88 \text{ ft}$$
$$K_x L_x = (0.94)(17.5 \text{ ft}) = 16.45 \text{ ft}$$
$$\frac{K_x L_x}{\frac{r_y}{r_x}} = \frac{16.45 \text{ ft}}{\frac{6.05 \text{ in}}{2.48 \text{ in}}} = 6.74 \text{ ft} < 14.88 \text{ ft}$$

The y-direction controls.

ASD Method Solution

(a) The required axial compressive strength is

$$P_a = P_{\text{dead}} + P_{\text{live}} = \frac{(30 \text{ ft})\left(82 \ \dfrac{\text{lbf}}{\text{ft}}\right)}{1000 \ \dfrac{\text{lbf}}{\text{kip}}} + 525 \text{ kips}$$

$$= 527.5 \text{ kips}$$

The required moment strengths are

$$M_x = 225 \text{ ft-kips}$$
$$M_y = 0$$

Interpolate from *AISC Manual* Table 4-1. For a W14 × 82 with $K_y L = 14.88$ ft, the available strength is $P_n/\Omega_c = 493$ kips.

Check the limit from *AISC Specification* Eq. H1-1a. The ratio of the required and available strengths is

$$\frac{P_a}{\frac{P_n}{\Omega_c}} = \frac{527.5 \text{ kips}}{493 \text{ kips}} = 1.07$$

Since $1.07 > 1.0$, based on axial strength alone, the column is $\boxed{\text{inadequate.}}$

(b) Since this is a quick check, approximate the effective length conservatively as $K_y L_y \approx 18$ ft.

From *AISC Manual* Table 4-1, with an effective length of 18 ft, the available strength is $P_n \Omega_c = 912$ kips. From *AISC Specification* Eq. H1-1a,

$$\frac{P_a}{\frac{P_n}{\Omega_c}} = \frac{527.5 \text{ kips}}{912 \text{ kips}} = 0.578$$

Check for beam-column action. Since $0.578 > 0.2$, use Eq. 62.9. From *AISC Manual* Table 6-1, for a W14 × 132,

$$p = 1.10 \times 10^{-3} \text{ kips}^{-1}$$
$$b_x = 1.59 \times 10^{-3} \text{ (ft-kips)}^{-1}$$
$$b_y = 0$$

$$pP_a + b_x M_x + b_y M_y \leq 1.0$$

$$= \left(0.0011 \ \frac{1}{\text{kips}}\right)(527.5 \text{ kips})$$

$$+ \left(0.00159 \ \frac{1}{\text{ft-kips}}\right)(225 \text{ ft-kips}) + 0$$

$$= 0.938 \leq 1.0 \quad [\text{OK}]$$

A more precise analysis (using more precise end restraint coefficients) is not necessary to determine this, although a lighter structural member might be shown to be adequate.

$\boxed{\text{A W12} \times \text{132 column will support the axial load and the moment.}}$

LRFD Method Solution

(a) The required axial compressive strength is

$$P_u = 1.2 P_{\text{dead}} + 1.6 P_{\text{live}} = (1.2)\left(\dfrac{(30 \text{ ft})\left(82 \, \dfrac{\text{lbf}}{\text{ft}}\right)}{1000 \, \dfrac{\text{lbf}}{\text{kip}}}\right)$$

$$+ (1.6)(525 \text{ kips})$$

$$= 843.0 \text{ kips}$$

The required moment strengths are

$$M_x = (1.6)(225 \text{ ft-kips}) = 360 \text{ ft-kips}$$
$$M_y = 0$$

Interpolate from *AISC Manual* Table 4-1. For a W14 × 82 with $K_y L = 14.88$ ft, the available strength is $\phi_c P_n = 741$ kips.

Check the limit from *AISC Specification* Eq. H1-1a. The ratio of the required and design strengths is

$$\dfrac{P_a}{\phi_c P_n} = \dfrac{843 \text{ kips}}{741 \text{ kips}} = 1.14$$

Since $1.14 > 1.0$, based on axial strength alone, the column is $\boxed{\text{inadequate.}}$

(b) Since this is a quick check, approximate the effective length conservatively as $K_y L_y \approx 18$ ft.

From *AISC Manual* Table 4-1, with an effective length of 18 ft, the design strength is $\phi_c P_n - 1370$ kips. From *AISC Specification* Eq. H1-1a,

$$\dfrac{P_a}{\phi_c P_n} = \dfrac{843.0 \text{ kips}}{1370 \text{ kips}} = 0.615$$

Check for beam-column action. Since $0.615 > 0.2$, use Eq. 62.9. From *AISC Manual* Table 6-1, for a W14 × 132,

$$p = 0.730 \times 10^{-3} \text{ kips}^{-1}$$
$$b_x = 1.06 \times 10^{-3} \text{ (ft-kips)}^{-1}$$
$$b_y = 0$$

$$p P_a + b_x M_x + b_y M_y \leq 1.0$$

$$= \left(0.000730 \, \dfrac{1}{\text{kips}}\right)(843.0 \text{ kips})$$

$$+ \left(0.00106 \, \dfrac{1}{\text{ft-kips}}\right)(360 \text{ ft-kips}) + 0$$

$$= 0.997 < 1.0 \quad \text{[OK]}$$

A more precise analysis (using more precise end restraint coefficients) is not necessary to determine this, although a lighter structural member might be shown to be adequate.

$\boxed{\text{A W12 × 132 column will support the axial load and the moment.}}$

2. (a) Determine the effective length, L_x, of the column. The bottom is built in (fixed), and the top is free to rotate, but fixed against translation. The deflection curvature will be the same as a built-in/pinned combination. From Table 61.1, $K_x = 0.8$.

$$L_x = K_x L = (0.8)(28 \text{ ft} + 24 \text{ ft}) = 41.6 \text{ ft}$$

AISC Specification Chap. B, Sec. B2 defers to ASCE7 Chap. 2 for load factors. ASCE7 Sec. 2.4.1, combination 6, specifies the ASD load combination for roof live, dead, and wind loads. Particularly because of the large wind load component, other combinations result in smaller loadings.

$$P_a = D + 0.75W + 0.75 L_r \quad \text{[ASD]}$$

ASCE7 Sec. 2.3.2, combination 4, specifies the LRFD load combination for roof live, dead, and wind loads. Particularly because of the large wind load component, other combinations result in smaller loadings.

$$P_u = 1.2D + 1.6W + 0.5 L_r \quad \text{[LRFD]}$$

For ASD,

$$P_a = 17 \text{ kips} + (0.75)(31 \text{ kips}) + (0.75)(21 \text{ kips})$$
$$= 56 \text{ kips}$$

For LRFD,

$$P_u = (1.2)(17 \text{ kips}) + (1.6)(31 \text{ kips}) + (0.5)(21 \text{ kips})$$
$$= 80.5 \text{ kips}$$

Since the bottom end is built in and the top end is free to rotate, this column behaves like a propped cantilever in response to lateral loading. The difficulty in doing an exact analysis with the different wind intensities, say by superposition, is that the locations of the points of the maximum moments are different for the three different wind intensities. In order to incorporate the actual beam restraints without increasing the workload significantly, simplify the loading. (If the column analysis shows marginal adequacy, a more exact analysis can be performed.)

The various wind distribution magnitudes are not too dissimilar, so calculate an average wind loading.

$$w_{ave} = \frac{\begin{array}{c}(15 \text{ ft})\left(403 \dfrac{\text{lbf}}{\text{ft}}\right) + (25 \text{ ft})\left(518 \dfrac{\text{lbf}}{\text{ft}}\right) \\ + (12 \text{ ft})\left(598 \dfrac{\text{lbf}}{\text{ft}}\right)\end{array}}{15 \text{ ft} + 25 \text{ ft} + 12 \text{ ft}}$$

$$= 503.3 \text{ lbf/ft}$$

There are no dead or live lateral loads. The most extreme case (of load factors) for wind only comes from ASCE7, as before, but the load factors are not necessarily the same as for the vertical loading.

For ASD,

$$w_{a,lateral} = 1.00 w_{ave} = (1.00)\left(503.3 \frac{\text{lbf}}{\text{ft}}\right)$$

$$= 503.3 \text{ lbf/ft}$$

For LRFD,

$$w_{u,lateral} = 1.6 w_{ave} = (1.6)\left(503.3 \frac{\text{lbf}}{\text{ft}}\right)$$

$$= 805.3 \text{ lbf/ft}$$

Use App. 44.A, case 4.

For ASD,

$$M_{max} = \frac{9wL^2}{128} = \frac{(9)\left(503.3 \dfrac{\text{lbf}}{\text{ft}}\right)(52 \text{ ft})^2}{(128)\left(1000 \dfrac{\text{lbf}}{\text{kip}}\right)}$$

$$= 95.69 \text{ ft-kips}$$

For LRFD,

$$M_{max} = \frac{9wL^2}{128} = \frac{(9)\left(805.3 \dfrac{\text{lbf}}{\text{ft}}\right)(52 \text{ ft})^2}{(128)\left(1000 \dfrac{\text{lbf}}{\text{kip}}\right)}$$

$$= 153.11 \text{ ft-kips}$$

ASD Method Solution

$$P_a = 56 \text{ kips}$$
$$M_x = 95.69 \text{ ft-kips}$$
$$M_y = 0 \text{ ft-kips}$$

Try a W24 × 104 for $KL \approx 42$ ft. From *AISC Manual* Table 6-1,

$$p = 6.52 \times 10^{-3} \text{ kips}^{-1} = 0.00652 \frac{1}{\text{kips}}$$

$$b_x = 3.43 \times 10^{-3} \text{ (ft-kips)}^{-1} = 0.00343 \frac{1}{\text{ft-kip}}$$

$$b_y = 0$$

Assume $P_a/(P_n/\Omega_c) > 0.2$. Then, from Eq. 62.9,

$$pP_a + b_x M_x + b_y M_y \leq 1.0$$

$$= \left(0.00652 \frac{1}{\text{kips}}\right)(56 \text{ kips})$$

$$\quad + \left(0.00343 \frac{1}{\text{ft-kip}}\right)(95.69 \text{ ft-kips}) + 0$$

$$= 0.693 \leq 1.0 \text{ [adequate]}$$

LRFD Method Solution

$$P_u = 80.5 \text{ kips}$$
$$M_x = 153.11 \text{ ft-kips}$$
$$M_y = 0$$

Try a W24 × 104 for $KL \approx 42$ ft. From *AISC Manual* Table 6-1,

$$p = 4.34 \times 10^{-3} \text{ kips}^{-1} = 0.00434 \frac{1}{\text{kips}}$$

$$b_x = 2.28 \times 10^{-3} \text{ (ft-kips)}^{-1} = 0.00228 \frac{1}{\text{ft-kip}}$$

$$b_y = 0$$

Assume $P_u/\phi P_n > 0.2$. Then, from Eq. 62.9,

$$pP_a + b_x M_x + b_y M_y \leq 1.0$$

$$= \left(0.00434 \frac{1}{\text{kips}}\right)(80.5 \text{ kips})$$

$$\quad + \left(0.00228 \frac{1}{\text{ft-kips}}\right)(153.11 \text{ ft-kips}) + 0$$

$$= 0.698 \leq 1.0 \quad \text{[adequate]}$$

(b) Use App. 44.A, case 4 with the average wind force. The maximum deflection is

$$\Delta_{max} = \frac{w_{ave} L^4}{185 EI}$$

$$= \frac{\left(503.3 \dfrac{\text{lbf}}{\text{ft}}\right)(52 \text{ ft})^4 \left(12 \dfrac{\text{in}}{\text{ft}}\right)^2}{(185)\left(29 \times 10^6 \dfrac{\text{lbf}}{\text{in}^2}\right)(3100 \text{ in}^4)}$$

$$= 0.0319 \text{ ft}$$

$$\frac{\Delta_{max}}{L} = \frac{0.0319 \text{ ft}}{52 \text{ ft}} = 1/1630$$

$L/1630$ is less than the traditional guideline that deflection should not exceed $L/360$ for architectural and aesthetic reasons [*AISC Commentary*, Chap. L, Sec. L3]. The deflection is so small that it is not necessary to perform a more exact deflection analysis.

A $\boxed{\text{W24} \times 104}$ beam is adequate for deflection.

3. (a) For a W12 × 58, $A = 17.0 \text{ in}^2$. The actual axial stress is

$$f_a = \frac{P}{A} = \frac{85 \text{ kips}}{17 \text{ in}^2} = \boxed{5.0 \text{ ksi}}$$

The answer is (C).

(b) For a W12 × 58, $r_x = 5.28$ in, and $r_y = 2.51$ in. For buckling about the strong (*x-x*) axis, $K_x = 0.8$ (one end fixed, one end pinned). $L_x = 16$ ft. For buckling about the weak (*y-y*) axis, $K_y = 2.1$ (one end fixed and one end free). $L_y = 16$ ft.

$$\text{SR} = \text{larger of} \begin{cases} \dfrac{K_x L_x}{r_x} = \dfrac{(0.8)(16 \text{ ft})\left(12 \ \frac{\text{in}}{\text{ft}}\right)}{5.28 \text{ in}} = 29.1 \\[4mm] \dfrac{K_y L_y}{r_y} = \dfrac{(2.1)(16 \text{ ft})\left(12 \ \frac{\text{in}}{\text{ft}}\right)}{2.51 \text{ in}} = 160.6 \end{cases}$$

SR is $\boxed{160.6 \ (160).}$

The answer is (C).

(c) Use *AISC Manual* Table 4-1 for $KL = 34$ ft.

For ASD,

$$\frac{P_n}{\Omega_c} = \boxed{97 \text{ kips} \quad (100 \text{ kips})}$$

For LRFD,

$$\phi_c P_n = \boxed{146 \text{ kips} \quad (150 \text{ kips})}$$

The answer is (C).

(d) Use *AISC Manual* Table 3-10, for $L_b = 16$ ft.

For ASD,

$$\frac{M_n}{\Omega} = \boxed{187 \text{ ft-kips} \quad (190 \text{ ft-kips})}$$

For LRFD,

$$\phi M_n = \boxed{281 \text{ ft-kips} \quad (280 \text{ ft-kips})}$$

The answer is (B).

(e) The lateral wind load is 4.0 kips/ft. For a beam fixed at one end and supported at the other and that is carrying a uniformly distributed load, use *AISC Manual* Table 3-23, case 12.

$$M_{\max} = \frac{wL^2}{8} = \frac{\left(4 \ \frac{\text{kips}}{\text{ft}}\right)(16 \text{ ft})^2}{8}$$
$$= \boxed{128 \text{ ft-kips} \quad (130 \text{ ft-kips})}$$

The answer is (C).

(f) The axial strength ratio is

$$\frac{P_r}{P_c} = \frac{85 \text{ kips}}{97 \text{ kips}} = 0.876$$

$\boxed{\text{The beam-column is subjected to large compression.}}$

ASD Method Solution

$$P_r = P_a = 85 \text{ kips}$$

From part (c), $P_c = P_n/\Omega_c = 97$ kips.

From part (e), $M_c = M_a = 128$ ft-kips.

From part (d), $M_n/\Omega = 187$ ft-kips.

The axial strength ratio is

$$\frac{P_r}{P_c} = \frac{85 \text{ kips}}{97 \text{ kips}} = 0.876$$

Since $0.876 > 0.2$, use Eq. 62.3.

$$\frac{P_r}{P_c} + \left(\frac{8}{9}\right)\left(\frac{M_{rx}}{M_{cx}} + \frac{M_{ry}}{M_{cy}}\right) \le 1.0$$
$$\frac{85 \text{ kips}}{97 \text{ kips}} + \left(\frac{8}{9}\right)\left(\frac{128 \text{ ft-kips}}{187 \text{ ft-kips}}\right) = 1.48 > 1.0$$

The member is $\boxed{\text{inadequate.}}$

LRFD Method Solution

$$P_r = P_u = (1.6)(85 \text{ kips}) = 136 \text{ kips}$$

From part (c), $M_r = \phi_c P_n = 146$ kips.

From part (e),

$$M_c = M_u = (1.6)(128 \text{ ft-kips}) = 204.8 \text{ ft-kips}$$

From part (d), $\phi M_n = 281$ ft-kips.

The axial strength ratio is

$$\frac{P_r}{P_c} = \frac{136 \text{ kips}}{146 \text{ kips}} = 0.93$$

Since $0.93 > 0.2$, use Eq. 62.3.

$$\frac{P_r}{P_c} + \left(\frac{8}{9}\right)\left(\frac{M_{rx}}{M_{cx}} + \frac{M_{ry}}{M_{cy}}\right) \leq 1.0$$

$$\frac{136 \text{ kips}}{146 \text{ kips}} + \left(\frac{8}{9}\right)\left(\frac{204.8 \text{ ft-kips}}{281 \text{ ft-kips}}\right) = 1.58 > 1.0$$

The member is $\boxed{\text{inadequate.}}$

The answer is (D).

(g) For buckling about the strong (x-x) axis, $K_x = 0.8$ (one end fixed and the other end pinned). $L_x = 16$ ft. For buckling about the weak (y-y) axis, $K_y = 1.0$ (column with both ends pinned) and $L_y = 8$ ft.

From Eq. 61.5,

$$SR = \text{larger of} \begin{cases} \dfrac{K_x L_x}{r_x} = \dfrac{(0.8)(16 \text{ ft})\left(12 \frac{\text{in}}{\text{ft}}\right)}{5.28 \text{ in}} = 29.1 \\[3mm] \dfrac{K_y L_y}{r_y} = \dfrac{(1)(8 \text{ ft})\left(12 \frac{\text{in}}{\text{ft}}\right)}{2.51 \text{ in}} = 38.25 \end{cases}$$

SR is $\boxed{38.25 \ (40).}$

The answer is (A).

(h) Use *AISC Manual* Table 4-1 for $KL = 8$ ft.

For ASD,

$$\frac{P_n}{\Omega_c} = \boxed{459 \text{ kips} \quad (460 \text{ kips})}$$

For LRFD,

$$\phi_c P_n = \boxed{689 \text{ kips} \quad (690 \text{ kips})}$$

The answer is (C).

(i) From *AISC Manual* Table 4-1, for a 50 ksi W12 × 58 member, $L_p = 8.87$ ft, $L_r = 29.9$ ft, and $L_b = 8$ ft. Therefore, since $L_b < L_p$, AISC Table 3-2 can be used.

For ASD,

$$\frac{M_n}{\Omega} = \frac{M_p}{\Omega} = \boxed{216 \text{ ft-kips} \quad (220 \text{ ft-kips})}$$

For LRFD,

$$\phi M_n = \phi M_p = \boxed{324 \text{ ft-kips} \quad (320 \text{ ft-kips})}$$

The answer is (C).

(j) **ASD Method Solution**

$$P_r = P_a = 85 \text{ kips}$$

From part (h), $P_c = P_n/\Omega_c = 459$ kips.

From part (e),

$$M_{rx} = \frac{\left(w \frac{\text{kips}}{\text{ft}}\right)(16 \text{ ft})^2}{8} = 32w \text{ ft-kips}$$

From part (i),

$$\frac{P_r}{P_c} = \frac{85 \text{ kips}}{459 \text{ kips}} = 0.185$$

Since $0.185 < 0.2$, use Eq. 62.4.

$$\frac{P_r}{2P_c} + \frac{M_{rx}}{M_{cx}} + \frac{M_{ry}}{M_{cy}} \leq 1.0$$

$$\frac{85 \text{ kips}}{(2)(459 \text{ kips})} + \frac{32w \text{ ft-kips}}{216 \text{ ft-kips}} = 1.0$$

Solving for w gives $\boxed{6.125 \text{ kips/ft } (6.1 \text{ kips/ft}).}$

LRFD Method Solution

$$P_r = P_u = (1.6)(85 \text{ kips}) = 136 \text{ kips}$$

From part (h), $P_c = \phi_c P_n = 689$ kips.

From part (e),

$$M_{rx} = (1.6)(32w \text{ ft-kips}) = 51.2w \text{ ft-kips}$$

$$\frac{P_r}{P_c} = \frac{136 \text{ kips}}{689 \text{ kips}} = 0.197$$

From *AISC Manual* Table 3-10, for $L_b = 8$ ft, $\phi M_n = 324$ ft-kips. Since $0.197 < 0.2$, use Eq. 62.4.

$$\frac{P_r}{2P_c} + \frac{M_{rx}}{M_{cx}} + \frac{M_{ry}}{M_{cy}} \leq 1.0$$

$$\frac{136 \text{ kips}}{(2)(689 \text{ kips})} + \frac{51.2w \text{ ft-kips}}{324 \text{ ft-kips}} = 1.0$$

Solving for w gives $\boxed{5.70 \text{ kips/ft } (5.7 \text{ kips/ft}).}$

The answer is (B).

4. The notable features of this problem are that (a) the bracing is not at the same points for both directions, and (b) the lateral transverse loading causes the primary bending moment.

For a W14 × 48, $A = 14.1$ in^2, $r_x = 5.85$ in, $r_y = 1.91$ in, $S_x = 70.3$ in^3, $F'_y = 50$ ksi, $L_p = 6.75$ ft, and $L_r = 21.1$ ft.

(a) The maximum axial load, $P = 170$ kips, occurs in the lower half of the member. Determine the maximum primary bending moment. From the moment diagram, the maximum primary moment occurs at the mid-height of the member.

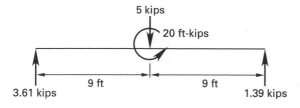

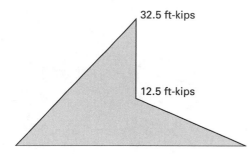

The answer is (C).

(b) From the moment diagram, the maximum primary bending moment is

$$M_x = \boxed{32.5 \text{ ft-kips} \quad (33 \text{ ft-kips})}$$

The moment about the weak axis is zero.

The answer is (B).

(c) For buckling about the weak (y-y) axis, $K_y = 1.0$ for both ends pinned. $L_y = 9$ ft.

$$\text{SR} = \frac{K_y L_y}{r_y} = \frac{(1)(9 \text{ ft})\left(12 \frac{\text{in}}{\text{ft}}\right)}{1.91 \text{ in}}$$
$$= \boxed{56.5 \quad (57)}$$

The answer is (D).

(d) For buckling about the strong (x-x) axis, from Table 61.1, $K_x = 1.0$ for both ends pinned. $L_x = 18$ ft.

$$\text{SR} = \frac{K_x L_x}{r_x} = \frac{(1)(18 \text{ ft})\left(12 \frac{\text{in}}{\text{ft}}\right)}{5.85 \text{ in}}$$
$$= \boxed{36.9 \quad (37)}$$

The answer is (B).

(e) The critical slenderness ratio is 57.

The answer is (D).

(f) From *AISC Manual* Table 4-1, $KL = 9$ ft. For ASD, $P_n/\Omega_c = \boxed{335 \text{ ksi.}}$ For LRFD, $\phi_c P_n = \boxed{504 \text{ ksi.}}$

The answer is (C).

(g) Calculate the bending stress.

$$f_{bx} = \frac{M_x}{S_x} = \frac{(32.5 \text{ ft-kips})\left(12 \frac{\text{in}}{\text{ft}}\right)}{70.3 \text{ in}^3}$$
$$= \boxed{5.55 \text{ kips/in}^2 \quad (5.5 \text{ ksi})}$$

The answer is (A).

(h) The actual axial stress is

$$\frac{P_r}{P_c} = \frac{170 \text{ kips}}{335 \text{ kips}} = \boxed{0.507 \quad (0.51)}$$

The answer is (C).

(i) The section is compact since $F_y' > F_y = 36$ ksi. The unsupported length of the compression flange is $L_b = 9$ ft. From *AISC Manual* Table 3-10,

$$M_a = \boxed{184 \text{ ft-kips} \quad (180 \text{ ft-kips})}$$

The answer is (C).

(j) Since $P_r/P_c = 0.507 > 0.2$, the member is subjected to large compression, and *AISC Specification* Eq. H1-1a applies. Use Eq. 62.3.

$$\frac{P_r}{P_c} + \left(\frac{8}{9}\right)\left(\frac{M_{rx}}{M_{cx}}\right) \leq 1.0$$

$$\frac{170 \text{ kips}}{335 \text{ kips}} + \left(\frac{8}{9}\right)\left(\frac{32.5 \text{ ft-kips}}{184 \text{ ft-kips}}\right) = 0.664 \leq 1.0$$

The beam-column is adequate.

The answer is (B).

63 Structural Steel: Plate Girders

PRACTICE PROBLEMS

1. A simply supported plate girder spans 40 ft and is constructed with $^3/_8$ in A36 steel plates. The flanges are 16 in wide, and the web is 48 in deep. A uniform, but unknown, load is carried along the girder's entire length. (a) Based on shear consideration only, determine the maximum uniformly distributed live load (in kips/ft) that the beam can carry assuming stiffeners are not used. (b) What is the nominal shear capacity of the beam? (c) A36 transverse stiffeners are used in pairs and placed every 4 ft. Design the stiffeners to develop the maximum shear capacity of the beam.

2. A welded A36 (specified minimum yield strength of 36 ksi) plate girder with a 60 ft span has to carry a uniformly distributed dead load of 2 kips/ft, plus its own weight, and a concentrated load of 135 kips at midspan. Lateral support is provided for the compression and tension flanges at beam ends and at midspan. No intermediate stiffeners are to be used. The web-to-flange weld size is $w = ^5/_{16}$ in. The length of bearing under the concentrated load is $N = 10$ in. (a) Design the girder cross section such that the overall girder depth does not exceed 60 in. (b) Determine if the bearing stiffeners are required, and if so, design them using 36 ksi steel.

3. (*Time limit: one hour*) A simply supported plate girder with a 60 ft span is composed of $1^1/_2$ in $\times$ 16 in flanges and a $^3/_8$ in $\times$ 64 in web. A36 steel is used. The girder supports a uniformly distributed service load of 2 kips/ft (including the girder weight) over the entire span and two concentrated service loads of 50 kips magnitude each, placed symmetrically one-third of the span from each end. Lateral supports for the compression flange are provided at the beam ends and the concentrated load points. Except for bearing stiffeners at the supports and concentrated loads, no transverse stiffeners are provided. Use ASD to analyze the plate girder.

(a) The actual value and the available limit of web depth-thickness ratio to prevent vertical buckling of the compression flange into the web are, respectively, most nearly

(A) 170 and 260

(B) 170 and 320

(C) 180 and 260

(D) 180 and 320

(b) The moment of inertia about the strong axis is most nearly

(A) 50,000 in^4

(B) 54,000 in^4

(C) 60,000 in^4

(D) 63,000 in^4

(c) The maximum service load shear is most nearly

(A) 85 kips

(B) 110 kips

(C) 130 kips

(D) 160 kips

(d) The maximum service load moment is most nearly

(A) 1600 ft-kips

(B) 1800 ft-kips

(C) 1900 ft-kips

(D) 2000 ft-kips

(e) The available bending strength (at the location of maximum computed bending stress) is most nearly

(A) 2200 ft-kips

(B) 2600 ft-kips

(C) 2900 ft-kips

(D) 4800 ft-kips

(f) The maximum computed shear stress due to service loads is most nearly

(A) 4.6 ksi

(B) 7.9 ksi

(C) 9.9 ksi

(D) 13 ksi

(g) The available shear stress is most nearly

(A) 1.6 ksi

(B) 2.9 ksi

(C) 4.5 ksi

(D) 8.2 ksi

(h) The plate girder is adequate for

(A) moment only

(B) shear only

(C) both moment and shear

(D) neither moment nor shear

(i) If the web plate thickness is changed to $^5/_8$ in, the plate girder will be adequate for

(A) moment only

(B) shear only

(C) both moment and shear

(D) neither moment nor shear

4. (*Time limit: one hour*) A simply supported welded plate girder having a 72 ft span carries a uniformly distributed load of 2 kips/ft and a concentrated load of 120 kips 24 ft from the left end. A572 grade 50 steel is used. The web is $^3/_8$ in thick and 48 in deep. Each flange is $1^1/_2$ in thick and 16 in wide. Bearing stiffeners are provided at the reaction points and at the concentrated load. Solve the problem using ASD.

(a) The maximum shear force the plate girder can carry without any intermediate stiffener is most nearly

(A) 84 kips

(B) 91 kips

(C) 96 kips

(D) 114 kips

(b) The maximum allowable spacing for the first intermediate stiffener from the left end is most nearly

(A) 51 in

(B) 56 in

(C) 60 in

(D) 64 in

(c) If the first intermediate stiffener is placed 48 in from the left end, the maximum allowable spacing for the next stiffener is most nearly

(A) 49 in

(B) 55 in

(C) 59 in

(D) 70 in

(d) Assuming a single plate is placed 48 in from the left end, the minimum size required for the first intermediate stiffener from the left end is most nearly

(A) 1.1 in^2

(B) 1.4 in^2

(C) 1.9 in^2

(D) 2.3 in^2

(e) The maximum bending stress at the concentrated load is most nearly

(A) 18 ksi

(B) 24 ksi

(C) 29 ksi

(D) 32 ksi

(f) Assuming that intermediate stiffeners are provided every 4 ft throughout the girder span, the actual shear stress and the available (i.e., ASD) tensile stress, respectively, at the concentrated load are most nearly

(A) 4.6 ksi and 32 ksi

(B) 5.8 ksi and 30 ksi

(C) 6.5 ksi and 33 ksi

(D) 7.3 ksi and 26 ksi

(g) The bearing stiffener size that will satisfy the width-thickness ratio requirement is most nearly

(A) $^1/_4$ in × 6 in

(B) $^3/_8$ in × 7 in

(C) $^3/_8$ in × $7^1/_2$ in

(D) $^1/_2$ in × $6^1/_2$ in

(h) Assuming two matched $^1/_2$ in × 7 in bars are used for end bearing stiffening, the total radius of gyration of the bearing stiffener is most nearly

(A) 2.0 in

(B) 2.6 in

(C) 3.8 in

(D) 4.1 in

Structural

(i) The computed stress in the end bearing stiffener is most nearly

- (A) 7.6 ksi
- (B) 8.4 ksi
- (C) 10 ksi
- (D) 18 ksi

(j) Assume that no intermediate stiffeners are to be used and only bearing stiffeners are to be provided at the beam ends and the concentrated load. The required minimum thickness of the web is most nearly

- (A) $^7/_{16}$ in
- (B) $^1/_2$ in
- (C) $^9/_{16}$ in
- (D) $^5/_8$ in

SOLUTIONS

1. (a) For ASTM A36 steel, the material and geometric properties are

$$F_y = 36 \text{ kips/in}^2$$
$$F_u = 58 \text{ kips/in}^2$$
$$t_w = \tfrac{3}{8} \text{ in}$$
$$t_f = \tfrac{3}{8} \text{ in}$$
$$d = 48 \text{ in}$$
$$b_f = 16 \text{ in}$$
$$h = 48 \text{ in}$$
$$\frac{h}{t_w} = \frac{48 \text{ in}}{0.375 \text{ in}} = 128$$

The web shear coefficient, C_v, is found from *AISC Specification* Sec. G2.1(b). Determine which of the following is applicable: Secs. G2.1(b)(i), G2.1(b)(ii), or G2.1(b)(iii). Since $h/t_w < 260$, per *AISC Specification* Sec. G2.1(i), the web plate buckling coefficient $k_v = 5$.

Only $h/t_w > 137\sqrt{k_v E/F_y}$ applies. Therefore, use Eq. 63.6 [*AISC Specification* Eq. G2-5] to determine C_v.

$$C_v = \frac{1.51 E k_v}{\left(\dfrac{h}{t_w}\right)^2 F_y} = \frac{(1.51)\left(29{,}000 \ \dfrac{\text{kips}}{\text{in}^2}\right)(5)}{(128)^2\left(36 \ \dfrac{\text{kips}}{\text{in}^2}\right)} = 0.371$$

Determine the nominal shear strength from Eq. 63.3 [*AISC Specification* Eq. G2-1].

$$\begin{aligned} V_n &= 0.6 F_y A_w C_v \\ &= (0.6)\left(36 \ \frac{\text{kips}}{\text{in}^2}\right)\left((48 \text{ in})(\tfrac{3}{8} \text{ in})\right)(0.371) \\ &= 144.2 \text{ kips} \end{aligned}$$

Find the available shear strength, V_c, using either the ASD or the LRFD method.

ASD Method Solution

$$V_c = \frac{V_n}{\Omega_v} = \frac{144.2 \text{ kips}}{1.67} = 86.3 \text{ kips}$$

The maximum shear occurs at the midpoint of the beam and is

$$V = \left(\tfrac{1}{2}\right)(40 \text{ ft})w = 20w$$

The maximum distributed load, w, is 86.3 kips/20 ft = 4.32 kips/ft.

Structural

Calculate the girder weight per foot.

$$\left(\frac{\frac{3}{8} \text{ in}}{12 \frac{\text{in}}{\text{ft}}}\right)\left(\frac{16 \text{ in} + 16 \text{ in} + 48 \text{ in}}{12 \frac{\text{in}}{\text{ft}}}\right)\left(\frac{490 \frac{\text{lbf}}{\text{ft}^3}}{1000 \frac{\text{lbf}}{\text{kip}}}\right)$$

$$= 0.1 \text{ kip/ft}$$

The maximum uniformly distributed live load the beam can carry is

$$w_{\text{load}} = 4.32 \frac{\text{kips}}{\text{ft}} - 0.1 \frac{\text{kip}}{\text{ft}} = \boxed{4.22 \text{ kips/ft}}$$

LRFD Method Solution

$$V_c = \phi_v V_n = (0.9)(144.2 \text{ kips}) = 129.8 \text{ kips}$$

The maximum shear occurs at the midpoint of the beam and is

$$V = \left(\tfrac{1}{2}\right)(40 \text{ ft})w = 20w$$

The maximum distributed load, w, will be 129.8 kips/ 20 ft = 6.49 kips/ft.

Calculate the girder weight per foot.

$$\left(\frac{\frac{3}{8} \text{ in}}{12 \frac{\text{in}}{\text{ft}}}\right)\left(\frac{16 \text{ in} + 16 \text{ in} + 48 \text{ in}}{12 \frac{\text{in}}{\text{ft}}}\right)\left(\frac{490 \frac{\text{lbf}}{\text{ft}^3}}{1000 \frac{\text{lbf}}{\text{kip}}}\right)$$

$$= 0.1 \text{ kip/ft}$$

Incorporating the load factors, the maximum uniformly distributed live load that the beam can carry is

$$w_{\text{load}} = \frac{6.49 \frac{\text{kips}}{\text{ft}} - (1.2)\left(0.1 \frac{\text{kip}}{\text{ft}}\right)}{1.6} = 3.98 \text{ kips/ft}$$

(b) ASD Method Solution

Determine if tension field action can be considered. The web and flange areas are

$$A_w = dt_w$$
$$= (48 \text{ in})\left(\tfrac{3}{8} \text{ in}\right)$$
$$= 18 \text{ in}^2$$

$$A_{fc} = A_{ft} = b_f t_f$$
$$= (16 \text{ in})\left(\tfrac{3}{8} \text{ in}\right)$$
$$= 6 \text{ in}^2$$

From *AISC Specification* Sec. G3.1, tension field action is not permitted when

$$\frac{2A_w}{A_{fc} + A_{ft}} > 2.5$$

$$\frac{(2)(18 \text{ in}^2)}{6 \text{ in}^2 + 6 \text{ in}^2} = 3 > 2.5$$

Tension field action is not permitted. The maximum shear capacity developed will be based on *AISC Specification* Sec. G2.1(b) only.

Given $a = 48$ in, $a/h = 1.0$. From Eq. 63.13 [*AISC Specification* Sec. G2], k_v is

$$k_v = 5 + \frac{5}{\left(\dfrac{a}{h}\right)^2}$$

$$= 5 + \frac{5}{(1)^2}$$

$$= 10$$

From *AISC Specification* Eq. G2-5,

$$1.37\sqrt{\frac{k_v E}{F_y}} = 1.37\sqrt{\frac{(10)\left(29{,}000 \frac{\text{kips}}{\text{in}^2}\right)}{36 \frac{\text{kips}}{\text{in}^2}}} = 123$$

This is a built-up shape, not a rolled shape. Since $h/t_w > 1.37\sqrt{k_v E/F_y}$, C_v is given by Eq. 63.6 [*AISC Specification* Eq. G2-5].

$$C_v = \frac{1.51 E k_v}{\left(\dfrac{h}{t_w}\right)^2 F_y} = \frac{(1.51)\left(29{,}000 \frac{\text{kips}}{\text{in}^2}\right)(10)}{(128)^2\left(36 \frac{\text{kips}}{\text{in}^2}\right)} = 0.742$$

V_n is given by Eq. 63.3 [*AISC Specification* Eq. G2-1].

$$V_n = 0.6 F_y A_w C_v$$

$$= (0.6)\left(36 \frac{\text{kips}}{\text{in}^2}\right)(48 \text{ in})(0.375 \text{ in})(0.742)$$

$$= 288.5 \text{ kips}$$

$$V_c = \frac{V_n}{\Omega}$$

$$= \frac{288.5 \text{ kips}}{1.67}$$

$$= \boxed{172.8 \text{ kips}}$$

LRFD Method Solution

Determine if tension field action can be considered. The web and flange areas are

$$A_w = dt_w = (48 \text{ in})(\tfrac{3}{8} \text{ in}) = 18 \text{ in}^2$$
$$A_{fc} = A_{ft} = b_f t_f = (16 \text{ in})(\tfrac{3}{8} \text{ in}) = 6 \text{ in}^2$$

From *AISC Specification* Sec. G3.1, tension field action is not permitted when

$$\frac{2A_w}{A_{fc} + A_{ft}} > 2.5$$

$$\frac{(2)(18 \text{ in}^2)}{6 \text{ in}^2 + 6 \text{ in}^2} = 3 > 2.5$$

Tension field action is not permitted. The maximum shear capacity developed will be based on *AISC Specification* Sec. G2.1(b) only.

Given $a = 48$ in, $a/h = 1.0$. From Eq. 63.13 [*AISC Specification* Sec. G2], k_v is

$$k_v = 5 + \frac{5}{\left(\frac{a}{h}\right)^2} = 5 + \frac{5}{(1)^2} = 10$$

From *AISC Specification* Eq. G2-5,

$$1.37\sqrt{\frac{k_v E}{F_y}} = 1.37\sqrt{\frac{(10)\left(29{,}000 \; \frac{\text{kips}}{\text{in}^2}\right)}{36 \; \frac{\text{kips}}{\text{in}^2}}} = 123$$

This is a built-up shape, not a rolled shape. Since $h/t_w > 1.37\sqrt{k_v E/F_y}$, C_v is given by Eq. 63.6 [*AISC Specification* Eq. G2-5].

$$C_v = \frac{1.51 E k_v}{\left(\frac{h}{t}\right)^2 F_y} = \frac{(1.51)\left(29{,}000 \; \frac{\text{kips}}{\text{in}^2}\right)(10)}{(128)^2\left(36 \; \frac{\text{kips}}{\text{in}^2}\right)} = 0.742$$

V_n is given by Eq. 63.3 [*AISC Specification* Eq. G2-1].

$$V_n = 0.6 F_y A_w C_v$$
$$= (0.6)\left(36 \; \frac{\text{kips}}{\text{in}^2}\right)(48 \text{ in})(0.375 \text{ in})(0.742)$$
$$= 288.5 \text{ kips}$$
$$V_c = \phi V_n = (0.9)(288.5 \text{ kips}) = 259.7 \text{ kips}$$

(c) The area of the stiffener is governed by *AISC Specification* Sec. G2.2, where the moment of inertia of the stiffener pair is greater than $at_w^3 j$, and j is given by

$$j = \frac{2.5}{\left(\frac{a}{h}\right)^2} - 2 = \frac{2.5}{(1)^2} - 2 = 0.5 \geq 0.5$$
$$a = 48 \text{ in}$$

Therefore, $I_{\text{pair}} \geq at_w^3 j$.

$$\frac{t_{\text{st}}(2b + t_w)^3}{12} \geq (48 \text{ in})(0.375 \text{ in})^3 (0.5) = 1.266 \text{ in}^4$$

Try $^3/_8$ in thick stiffeners. Solve for b with $t_{\text{st}} = t_w = {}^3/_8$ in.

$$b = 1.53 \text{ in}$$

Use 2 in $\times$ $^3/_8$ in stiffeners.

2. (a) *step 1:* Assume the weight of the plate girder will be approximately 350 lbf/ft (0.35 kip/ft). The distributed load, including the weight of the girder, is

$$w = 2 \; \frac{\text{kips}}{\text{ft}} + 0.35 \; \frac{\text{kip}}{\text{ft}} = 2.35 \text{ kips/ft}$$

For ASD,

The maximum shear is

$$V_a = V_{\max} = \tfrac{1}{2}(wL + P)$$
$$= (\tfrac{1}{2})\left(\left(2.35 \; \frac{\text{kips}}{\text{ft}}\right)(60 \text{ ft}) + 135 \text{ kips}\right)$$
$$= 138 \text{ kips}$$

The maximum moment is

$$M_a = M_{\max} = \left(\frac{L}{4}\right)\left(\frac{wL}{2} + P\right)$$
$$= \left(\frac{60 \text{ ft}}{4}\right)\left(\frac{\left(2.35 \; \frac{\text{kips}}{\text{ft}}\right)(60 \text{ ft})}{2} + 135 \text{ kips}\right)$$
$$= 3083 \text{ ft-kips}$$

For purposes of obtaining a first-iteration girder cross-section, assume an allowable bending stress of

$$F_b = 0.6 F_y = (0.6)\left(36 \; \frac{\text{kips}}{\text{in}^2}\right) = 21.6 \text{ ksi}$$

Structural

The required section modulus is

$$S_x = \frac{M_{max}}{F_b} = \frac{(3083 \text{ ft-kips})\left(12 \frac{\text{in}}{\text{ft}}\right)}{21.6 \frac{\text{kips}}{\text{in}^2}}$$

$$= 1713 \text{ in}^3$$

Try a girder with a $^3/_8$ in $\times$ 52 in web and $1^3/_4$ in $\times$ 18 in flange plates. The overall depth is

$$52 \text{ in} + (2)(1.75 \text{ in}) = 55.5 \text{ in} < 60 \text{ in} \quad \text{[OK]}$$
$$S_x = 1800 \text{ in}^3 > 1713 \text{ in}^3$$

For LRFD,

The maximum shear is

$$V_u = \frac{(1.2)\left(2.35 \frac{\text{kips}}{\text{ft}}\right)(60 \text{ ft}) + (1.6)(135 \text{ kips})}{2}$$

$$= 192.6 \text{ kips}$$

The maximum moment is

$$M_u = M_a = M_{max} = \left(\frac{L}{4}\right)\left(\frac{wL}{2} + P\right)$$

$$= \left(\frac{60 \text{ ft}}{4}\right)\left(\frac{(1.2)\left(2.35 \frac{\text{kips}}{\text{ft}}\right)(60 \text{ ft})}{2} + (1.6)(135 \text{ kips})\right)$$

$$= 4509 \text{ ft-kips}$$

For the purposes of obtaining a first-iteration girder cross-section, assume an allowable bending stress of

$$F_b = 0.6F_y = (0.6)\left(36 \frac{\text{kips}}{\text{in}^2}\right) = 21.6 \text{ ksi}$$

The required section modulus is

$$S_x = \frac{M_{max}}{F_b} = \frac{(4509 \text{ ft-kips})\left(12 \frac{\text{in}}{\text{ft}}\right)}{21.6 \frac{\text{kips}}{\text{in}^2}}$$

$$= 2505 \text{ in}^3$$

Try a girder with a $^3/_8$ in $\times$ 52 in web and $1^3/_4$ in $\times$ 18 in flange plates. The overall depth is

$$52 \text{ in} + (2)(1.75 \text{ in}) = 55.5 \text{ in} < 60 \text{ in} \quad \text{[OK]}$$
$$S_x = 1800 \text{ in}^3 > 2505 \text{ in}^3$$

step 2: Determine the web thickness required. Assumed web depth, h, is 52 in, and web thickness, t_w, is 0.375 in.

$$A_w = ht_w = (52 \text{ in})(0.375 \text{ in})$$
$$= 19.5 \text{ in}^2$$
$$\frac{h}{t_w} = \frac{52 \text{ in}}{0.375 \text{ in}} = 139$$

Since no intermediate stiffeners are to be used, from *AISC Specification* Sec. G2(b), $h/t_w = 260$, the minimum t_w is

$$\frac{52 \text{ in}}{260} = 0.2 \text{ in} < 0.375 \text{ in} \quad \text{[OK]}$$

Check the web shear stress. From *AISC Manual* Table 3-16a, interpolate for $h/t_w = 139$ and $a/h > 3$ (since no intermediate stiffeners are to be provided). $F_v < 4.5 \text{ ksi}(4.27 \text{ ksi actual})$ (ASD). The allowable vertical shear is

$$\frac{V_n}{\Omega} = F_v A_w = \left(4.5 \frac{\text{kips}}{\text{in}^2}\right)(19.5 \text{ in}^2)$$

$$= 87.8 \text{ kips} < 138 \text{ kips} \quad \text{[inadequate]}$$

Try a thicker web plate: $^7/_{16}$ in $\times$ 52 in.

$$A_w = ht_w = \left(\tfrac{7}{16} \text{ in}\right)(52 \text{ in}) = 22.75 \text{ in}^2$$
$$\frac{h}{t_w} = \frac{52 \text{ in}}{0.4375 \text{ in}} = 119$$

Using *AISC Manual* Table 3-16a with $h/t_w = 119$ and $a/h > 3$, $F_v < 6.0$ (5.9 ksi actual) (ASD). The allowable vertical shear is

$$\frac{V_n}{\Omega} = F_v A_w = \left(6.0 \frac{\text{kips}}{\text{in}^2}\right)(22.75 \text{ in}^2)$$

$$= 137 \text{ kips} < 138 \text{ kips} \quad \text{[inadequate]}$$

Try a thicker web plate: $^1/_2$ in $\times$ 52 in.

$$A_w = ht_w = \left(\tfrac{1}{2} \text{ in}\right)(52 \text{ in}) = 26 \text{ in}^2$$
$$\frac{h}{t_w} = \frac{52 \text{ in}}{0.5 \text{ in}} = 104$$

Using *AISC Manual* Table 3-16a with $h/t_w = 104$ and $a/h > 3$, $F_v = V_n/\Omega A_w \approx 8.0$ ksi. An initial estimate of the allowable vertical shear is

$$\frac{V_n}{\Omega} = F_v A_w = \left(8.0 \frac{\text{kips}}{\text{in}^2}\right)(26 \text{ in}^2)$$

$$= 208 \text{ kips} > 138 \text{ kips}$$

Since $a/h = 5$, not 3 as was used in *AISC Manual* Table 3-16a, determine the exact allowable (available) vertical shear strength from *AISC Specification* Sec. G.

$$k_v = 5 \quad [\textit{AISC Specification} \text{ Sec. G2.1b(ii)}]$$

$$1.37\sqrt{\frac{k_v E}{F_y}} = 1.37\sqrt{\frac{(5)\left(29{,}000 \ \dfrac{\text{kips}}{\text{in}^2}\right)}{36 \ \dfrac{\text{kips}}{\text{in}^2}}} = 86.9$$

$$139 > 86.9$$

Use Eq. 63.6.

$$C_v = \frac{1.51 E k_v}{\left(\dfrac{h}{t_w}\right)^2 F_y} = \frac{(1.51)\left(29{,}000 \ \dfrac{\text{kips}}{\text{in}^2}\right)(5)}{(104)^2 \left(36 \ \dfrac{\text{kips}}{\text{in}^2}\right)}$$

$$= 0.562$$

From Eq. 63.3,

$$V_n = 0.6 F_y A_w C_v \quad [\textit{AISC Specification} \text{ Eq. G2-1}]$$

$$= (0.6)\left(36 \ \frac{\text{kips}}{\text{in}^2}\right)(26 \ \text{in}^2)(0.562)$$

$$= 315.6 \ \text{kips} \quad (316 \ \text{kips})$$

For ASD, the available strength is

$$\frac{V_n}{\Omega} = \frac{316 \ \text{kips}}{1.67} = 189.2 \ \text{kips} \quad (189 \ \text{kips})$$

For LRFD, the design strength is

$$\phi_v V_n = (0.9)(316 \ \text{kips}) = 284.4 \ \text{kips} \quad (284 \ \text{kips})$$

The available capacity is greater than required, or 284 kips > 192.6 kips and 189 kips > 138 kips. Therefore, the assumed section is acceptable, and stiffeners are not required.

step 4: Check the bending stress.

step 4.1: Calculate the gross moment of inertia.

section	dimensions (in)	A (in^2)	y (in)	Ay^2 (in^4)	I_o (in^4)	I_{gr} (in^4)
web	$^1/_2 \times 52$	26	0	0	5859	5859
flange	$1^3/_4 \times 18$	31.5	26.875	22,751	8	22,759
flange	$1^3/_4 \times 18$	31.5	26.875	22,751	8	22,759
total		89		total		51,377

The distance from the neutral axis to the extreme fiber is

$$c = \frac{52 \ \text{in}}{2} + 1\tfrac{3}{4} \ \text{in} = 27.75 \ \text{in}$$

The section modulus of the plate girder cross section is

$$S_x = \frac{I}{c} = \frac{51{,}377 \ \text{in}^4}{27.75 \ \text{in}}$$

$$= 1851 \ \text{in}^3$$

step 4.2: Check for lateral-torsional buckling. From Eq. 59.8 [*AISC Specification* Sec. F2],

$$r_{ts} = \frac{b_f}{\sqrt{12\left(1 + \dfrac{h t_w}{6 b_f t_f}\right)}}$$

$$= \frac{18 \ \text{in}}{\sqrt{(12)\left(1 + \dfrac{(52 \ \text{in})\left(\frac{1}{2} \ \text{in}\right)}{(6)(18 \ \text{in})(1.75 \ \text{in})}\right)}}$$

$$= 4.87 \ \text{in}$$

Refer to the user note in *AISC Specification* Sec. F2.

$$L_r = \pi r_{ts} \sqrt{\frac{E}{0.7 F_y}}$$

$$= \pi (4.87 \ \text{in}) \sqrt{\frac{29{,}000 \ \dfrac{\text{kips}}{\text{in}^2}}{(0.7)\left(36 \ \dfrac{\text{kips}}{\text{in}^2}\right)}}$$

$$= 519 \ \text{in} \quad (43.25 \ \text{ft})$$

Calculate the gross moment of inertia about the y-axis (i.e., the weak axis). The centroidal moment of inertia is $bh^3/12$.

$$I_y = I_{y,w} + 2 I_{y,f} = \frac{(52 \ \text{in})(0.5 \ \text{in})^3}{12} + \frac{(2)(1.75 \ \text{in})(18)^3}{12}$$

$$= 1701.5 \ \text{in}^4$$

The radius of gyration about the y-axis is

$$r_y = \sqrt{\frac{I_y}{A}} = \sqrt{\frac{1701.5 \ \text{in}^4}{89 \ \text{in}^2}} = 4.37$$

From Eq. 59.6,

$$L_p = 1.76 r_y \sqrt{\frac{E}{F_y}}$$

$$= (1.76)(4.37 \ \text{in}) \sqrt{\frac{29{,}000 \ \dfrac{\text{kips}}{\text{in}^2}}{36 \ \dfrac{\text{kips}}{\text{in}^2}}}$$

$$= 218.3 \ \text{in}$$

Structural

$$L_p = \frac{218.3 \text{ in}}{12 \frac{\text{in}}{\text{ft}}} = 18.2 \text{ ft}$$

$L_b = 30$ ft $> L_p = 18.2$ ft. Therefore, lateral-torsional buckling occurs. Use *AISC Specification* Eq. F2-2 since $L_p < L_b \le L_r$, or 18.2 ft < 30 ft < 43.25 ft.

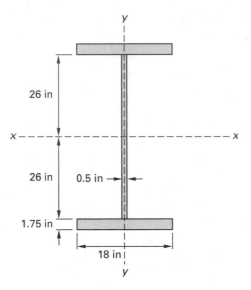

step 4.3: Check bending strength.

Z_x for the girder is not known. Therefore, determine if the section is compact in order to utilize $F_y = 36$ ksi, and $S_{xx} = 1851$ in³.

From *AISC Specification* Table B4.1, case 2, the section is compact if $b/t < 0.38\sqrt{E/F_y}$.

$$\frac{b}{t} = \frac{9.0 \text{ in}}{1.75 \text{ in}} = 5.14$$

$$< 0.38 \sqrt{\frac{29,000 \frac{\text{kips}}{\text{in}^2}}{36 \frac{\text{kips}}{\text{in}^2}}} = 10.8 \quad [\text{OK}]$$

$$Z_x = b_f t_f (h + t_f) + \frac{h^2 t_w}{4}$$
$$= (18 \text{ in})(1.75 \text{ in})(52 \text{ in} + 1.75 \text{ in})$$
$$+ \frac{(52 \text{ in})^2 (0.5 \text{ in})}{4}$$
$$= 2031 \text{ in}^3$$

The plastic bending moment, M_p, is

$$M_p = F_y Z_x$$
$$= \left(36 \frac{\text{kips}}{\text{in}^2}\right)(2031 \text{ in}^3)$$
$$= 73,116 \text{ in-kips}$$

Then, from Eq. 59.15 [*AISC Specification* Eq. F2-2], with $C_b = 1.0$ (conservatively, per *AISC Specification* Sec. F1),

$$M_n = C_b \left(M_p - (M_p - 0.7 F_y S_x)\left(\frac{L_b - L_p}{L_r - L_p}\right)\right) \le M_p$$

$$= (1.0) \left(\begin{array}{l} 73,116 \text{ in-kips} \\ \quad - \left(\begin{array}{l} 73,116 \text{ in-kips} \\ \quad - (0.7)\left(36 \frac{\text{kips}}{\text{in}^2}\right)(1851 \text{ in}^3) \end{array} \right) \\ \quad \times \left(\frac{360 \text{ in} - 218 \text{ in}}{519 \text{ in} - 218 \text{ in}}\right) \end{array} \right)$$

$$= 60,628 \text{ in-kips}$$

step 4.4: Check the required strength and the available strength.

ASD Method Solution

required strength $= M_a = 3083$ ft-kips   [from part (a)]

$$\text{available strength} = \frac{M_n}{\Omega_b} = \frac{60,628 \text{ in-kips}}{(1.67)\left(12 \frac{\text{in}}{\text{ft}}\right)}$$
$$= 3025 \text{ ft-kips}$$

Since $C_b = 1.0$ was previously conservatively selected, the available strength is essentially required strength.

The selection is sufficient.

LRFD Method Solution

required strength $= M_u = 4509$ ft-kips   [from part (a)]

$$\text{design strength} = \phi_b M_n = \frac{(0.90)(60,628 \text{ in-kips})}{12 \frac{\text{in}}{\text{ft}}}$$
$$= 4547 \text{ ft-kips}$$

The design strength exceeds the required strength.

The selection is sufficient.

(b) *step 5:* Design the bearing stiffeners.

step 5.1: Bearing stiffeners are required at unframed girder ends.

step 5.2: Check the bearing under concentrated load.

$$k = t_f + w = 1.75 \text{ in} + 0.3125 \text{ in} = 2.063 \text{ in}$$

Check web local yielding under the concentrated load (*AISC Specification* Eq. J10-2). From Eq. 59.33,

$$\frac{R}{t_w(N + 5k)} = \frac{135 \text{ kips}}{(0.5 \text{ in})(10 \text{ in} + (5)(2.063 \text{ in}))}$$
$$= 13.3 \text{ kips/in}^2$$

$$13.3 \; \frac{\text{kips}}{\text{in}^2} < 0.66 F_y = (0.66) \left(36 \; \frac{\text{kips}}{\text{in}^2} \right)$$
$$= 23.76 \; \text{kips/in}^2 \quad [\text{OK}]$$

So bearing stiffeners are not required for local web yielding.

Check web crippling under the concentrated load (*AISC Specification* Eq. J10-4).

$$\frac{N}{d} = \frac{10 \; \text{in}}{55.5 \; \text{in}} = 0.18 < 0.20$$

Bearing stiffeners are required if the concentrated load exceeds the value of R_n computed as

$$R_n = 0.80 t_w^2 \left(1 + 3 \left(\frac{N}{d} \right) \left(\frac{t_w}{t_f} \right)^{1.5} \right) \sqrt{\frac{EF_{yw} t_f}{t_w}}$$

$$= (0.80)(0.5 \; \text{in})^2$$
$$\times \left(1 + (3) \left(\frac{10 \; \text{in}}{55.5 \; \text{in}} \right) \left(\frac{0.5 \; \text{in}}{1.75 \; \text{in}} \right)^{1.5} \right)$$
$$\times \sqrt{\frac{\left(29{,}000 \; \frac{\text{kips}}{\text{in}^2} \right) \left(36 \; \frac{\text{kips}}{\text{in}^2} \right)(1.75 \; \text{in})}{0.5 \; \text{in}}}$$
$$= 413.9 \; \text{kips}$$

For ASD, the design strength is

$$\frac{R_n}{\Omega} = \frac{413.9 \; \text{kips}}{2.00} = 207.0 \; \text{kips}$$

For LRFD, the design strength is

$$\phi R_n = (0.75)(413.9 \; \text{kips}) = 310.4 \; \text{kips}$$

135 kips < 207.0 kips (ASD) and 310.4 kips (LRFD), so bearing stiffeners are not required.

Check sidesway web buckling (*AISC Specification* Eq. J10-6).

$$d_c = d - 2k = 55.5 \; \text{in} - (2)(2.063 \; \text{in})$$
$$= 51.37 \; \text{in}$$

Since the flanges are restrained against rotation at midpoint, the longest unbraced length is

$$L_b = \left(\frac{60 \; \text{ft}}{2} \right) \left(12 \; \frac{\text{in}}{\text{ft}} \right) = 360 \; \text{in}$$

$$\frac{\frac{d_c}{t_w}}{\frac{L_b}{b_f}} = \frac{\frac{51.37 \; \text{in}}{0.5 \; \text{in}}}{\frac{360 \; \text{in}}{18 \; \text{in}}}$$
$$= 5.14 > 2.3$$

Bearing stiffeners are not required under the concentrated load.

step 5.3: Determine the bearing stiffener size at the ends of the girder. Try two PL $7\frac{1}{2} \times \frac{1}{2}$ stiffeners.

Check the width-thickness ratio using Eq. 63.19.

$$\frac{b_{\text{st}}}{t_{\text{st}}} \leq \frac{95}{\sqrt{F_y}}$$

Therefore,

$$\frac{b}{t} = \frac{7.5 \; \text{in}}{0.5 \; \text{in}} = 15 < \frac{95}{\sqrt{F_y}} = \frac{95}{\sqrt{36 \; \frac{\text{kips}}{\text{in}^2}}}$$
$$= 15.8 \quad [\text{OK}]$$

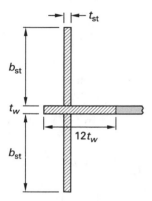

Check the compressive stress. For two stiffeners, one on either side of the web, the total stiffener length is

$$h = 2 b_{\text{st}} + t_w = (2)(7.5 \; \text{in}) + 0.5 \; \text{in} = 15.5 \; \text{in}$$
$$I = \frac{bh^3}{12} = \frac{(0.5 \; \text{in})(15.5 \; \text{in})^3}{12}$$
$$= 155.2 \; \text{in}^4$$
$$A_{\text{eff}} = 2 b_{\text{st}} t_{\text{st}} + 12 t_w^2$$
$$= (2)(7.5 \; \text{in})(0.5 \; \text{in}) + (12)(0.5 \; \text{in})^2$$
$$= 10.5 \; \text{in}^2$$
$$r = \sqrt{\frac{I}{A_{\text{eff}}}} = \sqrt{\frac{155.2 \; \text{in}^4}{10.5 \; \text{in}^2}}$$
$$= 3.84 \; \text{in}$$

The effective length of the stiffener is

$$Kl = 0.75 h = (0.75)(52 \; \text{in}) = 39 \; \text{in}$$
$$\frac{Kl}{r} = \frac{39 \; \text{in}}{3.84 \; \text{in}} = 10.2$$

From *AISC Manual* Table 4-22, $F_a = 21.4$ ksi.

The reaction at the beam end is 138 kips.

$$f_a = \frac{138 \text{ kips}}{10.5 \text{ in}^2} = 13.14 \text{ ksi} < 21.4 \text{ ksi} \quad [\text{OK}]$$

> Use (for bearing stiffeners) two $^1/_2$ in $\times$ $7^1/_2$ in $\times$ 4 ft $7^1/_4$ in bars, one on each side of the web, with close bearing on flanges receiving reactions.

3. (a) $\dfrac{h}{t_w} = \dfrac{64 \text{ in}}{0.375 \text{ in}} = \boxed{170.66 \quad [\text{use } 170]}$

The allowable limit on a web depth-thickness ratio is specified by *AISC Specification* Sec. G2 and Sec. F13.2, and is $\boxed{260}$ for unstiffened webs.

The answer is (A).

(b) The moment of inertia of the plate girder is

$$I = \sum \left(\frac{bh^3}{12} + Ad^2 \right)$$

$$= (2) \left(\begin{array}{l} \left(\frac{1}{12}\right)(16 \text{ in})(1.5 \text{ in})^3 \\[1mm] + \left(16 \text{ in}\right)(1.5 \text{ in})(32.75 \text{ in})^2 \end{array} \right)$$

$$+ \left(\frac{1}{12}\right)(0.375 \text{ in})(64 \text{ in})^3$$

$$= \boxed{59{,}684 \text{ in}^4 \quad (60{,}000 \text{ in}^4)}$$

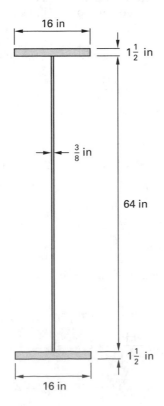

The answer is (C).

(c) Draw a shear and moment diagram.

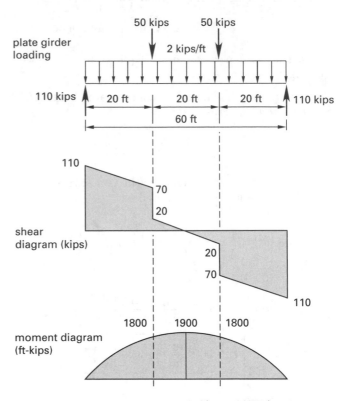

The maximum service load (i.e., ASD) shear is $\boxed{110 \text{ kips.}}$

The answer is (B).

(d) Using the moment diagram in the previous solution, the maximum service load moment (i.e., ASD) is $\boxed{1900 \text{ ft-kips.}}$

The answer is (C).

(e) For the web, from *AISC Specification* Table B4.1, case 9,

$$\lambda_r = 5.70\sqrt{\frac{E}{F_y}} = 5.70\sqrt{\frac{29{,}000 \ \frac{\text{kips}}{\text{in}^2}}{36 \ \frac{\text{kips}}{\text{in}^2}}}$$

$$= 162$$

Since $h/t_w = 64 \text{ in}/0.375 \text{ in} = 171 > \lambda_r$, the web is slender, and Sec. F5 of the *AISC Specification* applies. From Eq. 59.8 [*AISC Specification* Eq. F4-11],

$$a_w = \frac{h_c t_w}{b_{fc} t_{fc}} = \frac{(64 \text{ in})(0.375 \text{ in})}{(16 \text{ in})(1.5 \text{ in})} = 1.0$$

$$r_t \approx \frac{b_{fc}}{\sqrt{12\left(1 + \frac{1}{6}a_w\right)}} = \frac{16 \text{ in}}{\sqrt{(12)\left(1 + \left(\frac{1}{6}\right)(1.0)\right)}}$$

$$= 4.28 \text{ in}$$

From *AISC Specification* Eq. F5-6,

$$R_{pg} = 1 - \left(\frac{a_w}{1200 + 300 a_w}\right)\left(\frac{h_c}{t_w} - 5.7\sqrt{\frac{E}{F_y}}\right)$$

$$= 1 - \left(\frac{1.0}{1200 + (300)(1.0)}\right)$$

$$\times \left(\frac{64 \text{ in}}{0.375 \text{ in}} - 5.7\sqrt{\frac{29,000 \frac{\text{kips}}{\text{in}^2}}{36 \frac{\text{kips}}{\text{in}^2}}}\right)$$

$$= 0.994$$

$$S_{xc} = \frac{I_x}{c} = \frac{59,684 \text{ in}^4}{32 \text{ in} + 1.5 \text{ in}}$$

$$= 1782 \text{ in}^3$$

From *AISC Specification* Eq. F4-7,

$$L_p = 1.1 r_t \sqrt{\frac{E}{F_y}}$$

$$= \frac{(1.1)(4.28 \text{ in})\sqrt{\frac{29,000 \frac{\text{kips}}{\text{in}^2}}{36 \frac{\text{kips}}{\text{in}^2}}}}{12 \frac{\text{in}}{\text{ft}}}$$

$$= 11.14 \text{ ft}$$

From *AISC Specification* Eq. F5-5,

$$L_r = \pi r_t \sqrt{\frac{E}{0.7 F_y}}$$

$$= \frac{\pi(4.28 \text{ in})\sqrt{\frac{29,000 \frac{\text{kips}}{\text{in}^2}}{(0.7)\left(36 \frac{\text{kips}}{\text{in}^2}\right)}}}{12 \frac{\text{in}}{\text{ft}}}$$

$$= 38.01 \text{ ft}$$

From *AISC Specification* Eq. F5-1 for compression flange yielding,

$$M_n = R_{pg} F_y S_{xc} = \frac{(0.994)\left(36 \frac{\text{kips}}{\text{in}^2}\right)(1782 \text{ in})}{12 \frac{\text{in}}{\text{ft}}}$$

$$= 5314 \text{ ft-kips}$$

Since $L_p < L_b < L_r$, use *AISC Specification* Eq. F5-3 for lateral torsional buckling. The critical section is the middle portion of the beam, where $C_b \approx 1.0$.

$$F_{\text{cr}} = C_b\left(F_y - (0.3 F_y)\left(\frac{L_b - L_p}{L_r - L_p}\right)\right) \le F_y$$

$$= (1.0)\left(\begin{array}{c} 36 \frac{\text{kips}}{\text{in}^2} - (0.3)\left(36 \frac{\text{kips}}{\text{in}^2}\right) \\ \times \left(\frac{20 \text{ ft} - 11.14 \text{ ft}}{38.01 \text{ ft} - 11.14 \text{ ft}}\right)\end{array}\right)$$

$$= 32.44 \text{ ksi} \le 36 \text{ ksi}$$

$$M_n = R_{pg} F_y S_{xc} = \frac{(0.994)\left(32.44 \frac{\text{kips}}{\text{in}^2}\right)(1782 \text{ in})}{12 \frac{\text{in}}{\text{ft}}}$$

$$= 4788 \text{ ft-kips}$$

Since the compression flange is compact, compression flange buckling is not a consideration. Since the section modulus with respect to the compression flange is the same as that for the tension flange, tension flange yielding is not a consideration.

Therefore, the available bending strength is

$$\frac{M_n}{\Omega} = \frac{4788 \text{ ft-kips}}{1.67} = \boxed{2867 \text{ ft-kips} \quad (2900 \text{ ft-kips})}$$

The answer is (C).

(f) The maximum computed shear stress will be

$$f_v = \frac{V_{\max}}{h t_w} = \frac{110 \text{ kips}}{(64 \text{ in})(0.375 \text{ in})}$$

$$= \boxed{4.58 \text{ kips/in}^2 \quad (4.6 \text{ ksi})}$$

The answer is (A).

(g) For $h/t_w = 171$ and $a/h > 3$ (since no intermediate stiffeners are provided), and interpolating from *AISC Manual* Table 3-16a, the allowable (i.e., ASD) web shear stress is $F_v = \boxed{2.87 \text{ ksi } (2.9 \text{ ksi}).}$

The answer is (B).

(h) Since $M_r = 1900$ ft-kips $\le M_a = 2867$ ft-kips, $\boxed{\text{the plate girder is adequate for moments.}}$

From parts (g) and (h), $f_v > F_v$ (4.58 ksi > 2.87 ksi). Therefore, $\boxed{\text{the plate girder is inadequate for shear.}}$

The answer is (A).

(i) If $t_w = {}^5/_8$ in,

$$\frac{h}{t_w} = \frac{64 \text{ in}}{\frac{5}{8} \text{ in}} = 102.4$$

From *AISC Manual* Table 3-16a, the allowable web shear stress is $F_v = 7.96$ ksi (by interpolation). The maximum computed shear stress is

$$f_v = \frac{V_{\max}}{ht_w} = \frac{110 \text{ kips}}{(64 \text{ in})(0.625 \text{ in})}$$
$$= 2.75 \text{ kips/in}^2 < F_v \quad [\text{OK}]$$

> Since the thinner web was adequate for moment, the thicker web will also be adequate.

The answer is (C).

4. (a) The cross-sectional area of the web is

$$A_w = ht_w = (48 \text{ in})(\tfrac{3}{8} \text{ in}) = 18 \text{ in}^2$$
$$\frac{h}{t_w} = \frac{48 \text{ in}}{0.375 \text{ in}} = 128$$

If no intermediate stiffeners are used, $a/h > 3$. From *AISC Manual* Table 3-17a (by interpolation), $F_v = 5.36$ ksi $= f_v$. The maximum allowable shear without intermediate stiffeners is

$$V_a = F_v ht_w$$
$$= \left(5.36 \ \frac{\text{kips}}{\text{in}^2}\right)(48 \text{ in})(\tfrac{3}{8} \text{ in})$$
$$= \boxed{96.5 \text{ kips} \quad (96 \text{ kips})}$$

The answer is (C).

(b) $V_{\max} = 152$ kips   [at left support]

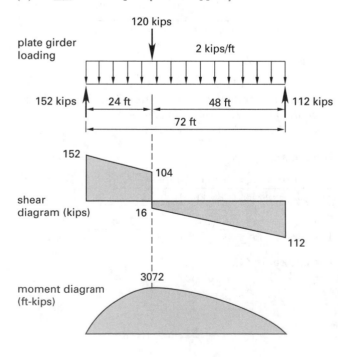

$$f_v = \frac{V_{\max}}{ht_w} = \frac{152 \text{ kips}}{(48 \text{ in})(0.375 \text{ in})}$$
$$= 8.44 \text{ kips/in}^2$$

Set $F_v = f_v = 8.44$ ksi.

Tension field action is not permitted in the end panel (*AISC Specification* Sec. G3.1). From *AISC Manual* Table 3-17a for $h/t_w = 128$ and $F_v = 8.44$ ksi (by interpolation), $a/h = 1.17$.

$$a = (1.17)(48 \text{ in})$$
$$= \boxed{56.2 \text{ in} \quad (56 \text{ in})}$$

The answer is (B).

(c) If the first intermediate stiffener is placed 48 in from the left end of the girder, the shear due to the service loads (i.e., ASD) at that location is

$$V = 152 \text{ kips} - \left(2 \ \frac{\text{kips}}{\text{ft}}\right)\left(\frac{48 \text{ in}}{12 \ \frac{\text{in}}{\text{ft}}}\right) = 144 \text{ kips}$$
$$f_v = \frac{V_{\max}}{ht_w} = \frac{144 \text{ kips}}{(48 \text{ in})(0.375 \text{ in})}$$
$$= 8.0 \text{ kips/in}^2$$

Set $F_v = f_v = 8.0$ ksi.

From *AISC Manual* Table 3-17a for $h/t_w = 128$ and $F_v = 8.0$ ksi (by interpolation), $a/h = 1.23$.

$$a = (1.23)(48 \text{ in})$$
$$= \boxed{59.0 \text{ in} \quad (59 \text{ in})}$$

The answer is (C).

(d) The first intermediate stiffener is placed 48 in from left end. Therefore, $a = 48$ in.

$$\frac{a}{h} = \frac{48 \text{ in}}{48 \text{ in}} = 1$$

As in part (c), use $a/h = 1$ and $h/t_w = 128$ to interpolate $F_v = 9.64$ ksi from *AISC Manual* Table 3-16a.

Use *AISC Specification* Sec. G2.1(b) to determine C_v, where $k_v = 5.0$.

Try *AISC Specification* Eq. G2-3. C_v will equal 1.0 if

$$\frac{h}{t_w} \le 1.10\sqrt{\frac{k_v E}{F_y}}$$

$$128 \le 1.10\sqrt{\frac{(5)\left(29{,}000\ \dfrac{\text{kips}}{\text{in}^2}\right)}{50\ \dfrac{\text{kips}}{\text{in}^2}}} = 59.2$$

$$128 \not\le 59.2 \quad \text{[inadequate]}$$

Try *AISC Specification* Eq. G2-4, which will apply for C_v if

$$1.10\sqrt{\frac{k_v E}{F_y}} < \frac{h}{t_w} \le 1.37\sqrt{\frac{k_v E}{F_y}}$$

$$59.2 < 128 \le 1.37\sqrt{\frac{(5)\left(29{,}000\ \dfrac{\text{kips}}{\text{in}^2}\right)}{50\ \dfrac{\text{kips}}{\text{in}^2}}} = 73.8$$

$$59.2 < 128 \not\le 73.8 \quad \text{[inadequate]}$$

Use *AISC Specification* Eq. G2-5.

$$C_v = \frac{1.51 E k_v}{\left(\dfrac{h}{t_w}\right)^2 F_y} = \frac{(1.51)\left(29{,}000\ \dfrac{\text{kips}}{\text{in}^2}\right)(5)}{(128)^2\left(50\ \dfrac{\text{kips}}{\text{in}^2}\right)}$$

$$= 0.267$$

The intermediate stiffener size is found using Eq. 63.17 [*AISC Specification* Eq. G3-3], where D_s is 2.4 for single plate stiffeners.

$$A_{\text{st}} = \frac{F_y}{F_{y,\text{st}}}\left(0.15 D_s h t_w (1 - C_v)\frac{V_r}{V_c} - 18 t_w^2\right)$$

$$= \left(\frac{50\ \dfrac{\text{kips}}{\text{in}^2}}{50\ \dfrac{\text{kips}}{\text{in}^2}}\right)$$

$$\times\left(\begin{array}{c} (0.15)(2.4)(48\ \text{in})(0.375\ \text{in}) \\[4pt] \times\ (1 - 0.267)\left(\dfrac{8.0\ \dfrac{\text{kips}}{\text{in}^2}}{9.64\ \dfrac{\text{kips}}{\text{in}^2}}\right) \\[4pt] -\ (18)(0.375\ \text{in})^2 \end{array}\right)$$

$$= \boxed{1.41\ \text{in}^2 \quad (1.4\ \text{in}^2)}$$

The answer is (B).

(e) From the moment diagram in part (b), the maximum service load moment at the concentrated load is $M_{\max} = 3072$ ft-kips.

Calculate the section modulus for a 48 in $\times$ $^3/_8$ in web and 16 in $\times$ $1^1/_2$ in flanges.

$$I_x = \frac{t_w h^3}{12} + 2\left(\frac{b_f t_f^3}{12}\right)$$

$$= 2 b_f t_f\left(\frac{h + t_w}{2}\right)^2$$

$$= \frac{\left(\frac{3}{8}\ \text{in}\right)(48\ \text{in})^3}{12} + (2)\left(\frac{(16\ \text{in})(1.5\ \text{in})^3}{12}\right)$$

$$\quad + (2)(16\ \text{in})(1.5\ \text{in})\left(\frac{48\ \text{in} + 1.5\ \text{in}}{2}\right)^2$$

$$= 32{,}868\ \text{in}^4$$

$$c = \frac{48\ \text{in}}{2} + 1.5\ \text{in} = 25.5\ \text{in}$$

$$S_x = \frac{I_x}{c} = \frac{32{,}868\ \text{in}^4}{25.5\ \text{in}} = 1288.9\ \text{in}^3$$

The maximum service load bending stress at the concentrated load is

$$f_b = \frac{M_{\max}}{S_x} = \frac{(3072\ \text{ft-kips})\left(12\ \dfrac{\text{in}}{\text{ft}}\right)}{1288.9\ \text{in}^3}$$

$$= \boxed{28.6\ \text{kips/in}^2 \quad (29\ \text{ksi})}$$

The answer is (C).

(f) From the shear diagram, the shear just to the left of the concentrated load is $V = 104$ kips. The computed service load shear stress is

$$f_v = \frac{V_{\max}}{h t_w} = \frac{104\ \text{kips}}{(48\ \text{in})(0.375\ \text{in})}$$

$$= \boxed{5.8\ \text{kips/in}^2}$$

Since the web is non-compact, Sec. F4 of the *AISC Specification* applies.

$$M_p = Z_x F_y = \left(b_f t_f (h + t_f) + \frac{h^2 t_w}{4}\right)F_y$$

$$= \left(\begin{array}{c}(16\ \text{in})(1.5\ \text{in})(48\ \text{in} + 1.5\ \text{in}) \\[4pt] + \dfrac{(48\ \text{in})^2(0.375\ \text{in})}{4}\end{array}\right)$$

$$\quad \times\left(50\ \frac{\text{kips}}{\text{in}^2}\right)\left(\frac{1\ \text{ft}}{12\ \text{in}}\right)$$

$$= 5850\ \text{ft-kips}$$

Structural

$$I_x = \frac{b_f(h + 2t_w)^3 - (b_f - t_f)h^3}{12}$$

$$= \frac{(16 \text{ in})\left(48 \text{ in} + (2)(1.5 \text{ in})\right)^3}{12}$$

$$= \frac{-(16 \text{ in} - 0.375 \text{ in})(48 \text{ in})^3}{12}$$

$$= 32{,}868 \text{ in}^4$$

$$S_{xt} = \frac{I}{c} = \frac{32{,}868 \text{ in}^4}{24 \text{ in} + 1.5 \text{ in}}$$

$$= 1289 \text{ in}^3$$

$$M_{yt} = S_{xt}F_y = (1289 \text{ in}^3)\left(50 \frac{\text{kips}}{\text{in}^2}\right)\left(\frac{1 \text{ ft}}{12 \text{ in}}\right)$$

$$= 5371 \text{ ft-kips}$$

$$\lambda_{pw} = 3.76\sqrt{\frac{E}{F_y}} = 3.76\sqrt{\frac{29{,}000 \frac{\text{kips}}{\text{in}^2}}{50 \frac{\text{kips}}{\text{in}^2}}}$$

$$= 90.6$$

$$\lambda_{rw} = 5.70\sqrt{\frac{E}{F_y}} = 5.70\sqrt{\frac{29{,}000 \frac{\text{kips}}{\text{in}^2}}{50 \frac{\text{kips}}{\text{in}^2}}}$$

$$= 137.3$$

$$\lambda = \frac{h}{t_w} = \frac{48 \text{ in}}{0.375 \text{ in}}$$

$$= 128$$

Since $\lambda > \lambda_{pw}$, the web plastification factor is given by *AISC Specification* Eq. F4-15b.

$$R_{pt} = \left(\frac{M_p}{M_{yt}} - \left(\frac{M_p}{M_{yt}} - 1\right)\left(\frac{\lambda - \lambda_{pw}}{\lambda_{rw} - \lambda_{pw}}\right)\right) \leq \frac{M_p}{M_{yt}}$$

$$\frac{M_p}{M_{yt}} = \frac{5850 \text{ ft-kips}}{5371 \text{ ft-kips}} = 1.09$$

$$R_{pt} = \left(1.09 - (1.09 - 1)\left(\frac{128 - 90.6}{137.3 - 90.6}\right)\right) \leq 1.09$$

$$= 1.018$$

The available moment is given by *AISC Specification* Eq. F4-14 and the Ω factor.

$$\frac{M_n}{\Omega} = \frac{R_{pt}M_{yt}}{\Omega} = \frac{(1.018)(5371 \text{ ft-kips})}{1.67} = 3274 \text{ ft-kips}$$

The available tensile stress is given by

$$\frac{M_n}{\Omega} = \frac{(3274 \text{ ft-kips})\left(12 \frac{\text{in}}{\text{ft}}\right)}{1289 \text{ in}^3}$$

$$= \boxed{30.48 \text{ kips/in}^2 \quad (30 \text{ ksi})}$$

The answer is (B).

(g) Using Eq. 63.19, the limiting width-thickness ratio for bearing stiffeners is

$$\frac{b_{st}}{t_{st}} = \frac{95}{\sqrt{F_y}} = \frac{95}{\sqrt{50 \frac{\text{kips}}{\text{in}^2}}} = 13.4$$

$\frac{1}{4}$ in × 6 in: $\quad \dfrac{b}{t} = \dfrac{6 \text{ in}}{0.25 \text{ in}} = 24 > 13.4$ [inadequate]

$\frac{3}{8}$ in × 7 in: $\quad \dfrac{b}{t} = \dfrac{7 \text{ in}}{0.375 \text{ in}} = 18.7 > 13.4$ [inadequate]

$\frac{3}{8}$ in × 7$\frac{1}{2}$ in: $\quad \dfrac{b}{t} = \dfrac{7.5 \text{ in}}{0.375 \text{ in}} = 20 > 13.4$ [inadequate]

$\boxed{\frac{1}{2} \text{ in} \times 7\frac{1}{2} \text{ in:}} \quad \dfrac{b}{t} = \dfrac{6.5 \text{ in}}{0.5 \text{ in}} = 13 < 13.4$ [OK]

The answer is (D).

(h) The clear distance between flanges is

$$h = 7 \text{ in} + 7 \text{ in} + \tfrac{3}{8} \text{ in} = 14.375 \text{ in}$$

$$I = \frac{bh^3}{12} = \frac{(0.5 \text{ in})(14.375 \text{ in})^3}{12} = 123.77 \text{ in}^4$$

$$A_{\text{eff}} = 2A_{st} + 12t_w^2$$

$$= (2)\left((7 \text{ in})(0.5 \text{ in})\right) + (12)(0.375 \text{ in})^2 = 8.69 \text{ in}^2$$

$$r = \sqrt{\frac{I}{A_{\text{eff}}}} = \sqrt{\frac{123.77 \text{ in}^4}{8.69 \text{ in}^2}}$$

$$= \boxed{3.77 \text{ in} \quad (3.8 \text{ in})}$$

The answer is (C).

(i) The effective length of the bearing stiffener is

$$KL = 0.75h = (0.75)(48 \text{ in}) = 36 \text{ in}$$

From part (h), the radius of gyration is $r = 3.77$ in.

$$\frac{KL}{r} = \frac{36 \text{ in}}{3.77 \text{ in}} = 9.55$$

From *AISC Specification* Eq. E3-4,

$$F_e = \frac{\pi^2 E}{\left(\frac{KL}{r}\right)^2} = \frac{\pi^2 \left(29{,}000 \frac{\text{kips}}{\text{in}^2}\right)}{(9.55)^2}$$

$$= 3138 \text{ kips/in}^2$$

$$\frac{KL}{r} < 4.71\sqrt{\frac{E}{F_y}} = 4.71\sqrt{\frac{29{,}000 \frac{\text{kips}}{\text{in}^2}}{50 \frac{\text{kips}}{\text{in}^2}}} = 113$$

$$9.55 < 113$$

Therefore, from Eq. 61.6, the flexural buckling stress is

$$F_{\text{cr}} = (0.658^{F_y/F_e})F_y \quad [\textit{AISC Specification} \text{ Eq. E3-2}]$$

$$= \left(0.658^{50 \text{ kips/in}^2/3138 \text{ kips/in}^2}\right)\left(50 \frac{\text{kips}}{\text{in}^2}\right)$$

$$= 49.7 \text{ kips/in}^2$$

The cross-sectional area of the "stiffener column" (shown in Fig. 63.3) for two end stiffeners is

$$A_c = (2 \text{ stiffeners})(7 \text{ in})(0.5 \text{ in})$$
$$\quad + (12)(0.375 \text{ in})(0.375 \text{ in})$$
$$= 8.6875 \text{ in}^2$$

The actual stress is

$$f_a = \frac{V_{\text{max}}}{A_c} = \frac{152 \text{ kips}}{8.6875 \text{ in}^2}$$
$$= \boxed{17.5 \text{ kips/in}^2 \quad (18 \text{ ksi})}$$

Since $f_a < F_{\text{cr}}$, this is adequate.

The answer is (D).

(j) For the left end panel,

$$a = (24 \text{ ft})\left(12 \frac{\text{in}}{\text{ft}}\right) = 288 \text{ in}$$
$$\frac{a}{h} = \frac{288 \text{ in}}{48 \text{ in}} = 6 > 3$$

For the right end panel,

$$a = (48 \text{ ft})\left(12 \frac{\text{in}}{\text{ft}}\right) = 576 \text{ in}$$
$$\frac{a}{h} = \frac{576 \text{ in}}{48 \text{ in}} = 12 > 3$$

If $t_w = {}^7\!/_{16}$ in, then

$$\frac{h}{t_w} = \frac{48 \text{ in}}{0.4375 \text{ in}} = 109.7$$
$$A_w = (48 \text{ in})(0.4375 \text{ in}) = 21 \text{ in}^2$$
$$f_v = \frac{V}{A_w} = \frac{152 \text{ kips}}{21 \text{ in}^2} = 7.24 \text{ kips/in}^2$$

From *AISC Manual* Table 3-17a, for $h/t_w = 110$ and $a/h > 3$, $F_v = 7.29$ ksi. Since $f_v < F_v$, this is adequate.

$$\boxed{t_w = {}^7\!/_{16} \text{ in}}$$

The answer is (A).

64 Structural Steel: Composite Beams

PRACTICE PROBLEMS

(Answer options for LRFD solutions are given in parentheses.)

1. A cross section of a composite beam is shown. The steel beam is an A992 W21 × 50, and the concrete strength is $f'_c = 4000$ psi. Assume shored construction and full composite action. Determine the resisting moment of the composite beam.

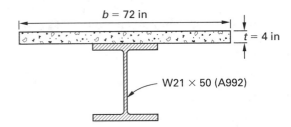

2. The floor framing system of an office building consists of A992 steel, simply supported beams of 40 ft span, spaced at 9 ft on center, and that act compositely with a 4 in thick concrete slab ($f'_c = 3$ ksi). The unit weight of concrete is 150 lbf/ft³. No temporary shoring is to be used during construction. The loads to be applied after the concrete hardens are: floor live (service) load = 100 psf, and ceiling live load = 10 psf. (a) Design an A992 interior beam such that the dead-load deflection, before the concrete cures, is limited to 2.5 in and the live load deflection is limited to $L/360$. (b) Design the shear connectors.

3. (*Time limit: one hour*) A building floor framing consists of A992 W21 × 68 beams of 30 ft span placed at 12 ft center-to-center spacing. The beams are attached to a 4 in thick concrete slab ($f'_c = 3$ ksi, unit weight = 150 lbf/ft³) through stud connectors. Assume shored construction, full composite action. All loads applied after the concrete has hardened are uniformly distributed live loads. The following questions relate to an interior beam.

(a) The effective concrete flange width is most nearly

- (A) 64 in
- (B) 74 in
- (C) 84 in
- (D) 90 in

(b) With full composite action, the ASD (LRFD) moment capacity of the composite beam is most nearly

- (A) 240 ft-kips (450 ft-kips)
- (B) 400 ft-kips (600 ft-kips)
- (C) 410 ft-kips (710 ft-kips)
- (D) 630 ft-kips (950 ft-kips)

(c) The unfactored dead load moment is most nearly

- (A) 68 ft-kips
- (B) 75 ft-kips
- (C) 88 ft-kips
- (D) 110 ft-kips

(d) The maximum unfactored (i.e., ASD) uniformly distributed live load that the composite beam can support is most nearly

- (A) 120 lbf/ft²
- (B) 180 lbf/ft²
- (C) 230 lbf/ft²
- (D) 410 lbf/ft²

(e) For full composite action, the total available horizontal shear, V_h, to be resisted between the point of maximum moment and the end of the beam is most nearly

- (A) 260 kips
- (B) 440 kips
- (C) 490 kips
- (D) 1000 kips

(f) In order to sustain full composite action, assuming ³/₄ in dia. × 3 in stud connectors are used, the total number of studs required for the entire beam is most nearly

- (A) 30
- (B) 50
- (C) 60
- (D) 120

4. (*Time limit: one hour*) A floor framing system of composite construction (unshored) consists of simply supported, W21 × 50 A992 steel beams of 40 ft span, equally spaced at 8 ft between beam centerlines, and a 3 in thick lightweight concrete slab ($f'_c = 3000$ psi, unit weight = 115 lbf/ft^3) on a 3 in deep formed steel deck with ribs running perpendicular to the beams. The average width of concrete rib, w_r, is 3 in. The steel beams are connected to the concrete slab by headed studs that have $^3/4$ in diameters and 5 in lengths. The beams support a service dead load of 65 lbf/ft^2 due to slab and deck, and a live load of 100 lbf/ft^2 (applied after the concrete has hardened). The following questions relate to an interior beam.

(a) The effective moment of inertia of the composite section is most nearly

  (A) 2600 in^4

  (B) 3200 in^4

  (C) 4100 in^4

  (D) 5100 in^4

(b) Assuming full composite action, the total horizontal shear, V_h, to be resisted at service loads between the point of maximum moment and the point of zero moment is most nearly

  (A) 370 kips

  (B) 410 kips

  (C) 440 kips

  (D) 740 kips

(c) Assuming 1 stud per rib, a conservative estimate of the stud reduction factor, $R_g R_p$, is most nearly

  (A) 0.51

  (B) 0.60

  (C) 0.64

  (D) 0.75

(d) Assuming full composite action under service loads, a conservative estimate of the number of studs required between the beam ends is most nearly

  (A) 70

  (B) 80

  (C) 86

  (D) 130

(e) Assuming partial composite action (75%) under service loads, the total number of studs required between the beam ends is most nearly

  (A) 54

  (B) 60

  (C) 66

  (D) 100

(f) Considering full composite action, the service live load deflection is most nearly

  (A) 0.34 in

  (B) 0.49 in

  (C) 0.61 in

  (D) 0.68 in

(g) Assuming partial composite action (75%) and studs provided accordingly, the service live load deflection is most nearly

  (A) 0.39 in

  (B) 0.54 in

  (C) 0.66 in

  (D) 0.88 in

5. (*Time limit: one hour*) A portion of a bridge is supported by a simply supported plate girder. Bridge loads are applied after construction and are resisted by composite action. The girder is 80 ft long. It will not be shored during construction. Flange support is provided at 20 ft intervals. The girder is loaded by a moving 40 kip load and a 1 kip/ft dead load (which includes an allowance for the concrete). The concrete strength is 3000 psi. The yield stress for all steel plates is 36 ksi. The modular ratio ($E_{steel}/E_{concrete}$) is 10. Impact loading is negligible. Assume a worst-case stiffener placement. Using ASD, determine if the plate girder is adequate.

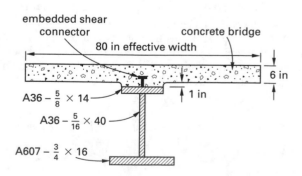

SOLUTIONS

1. (Refer to the Symbols section of the *AISC Manual* for notations used.)

Since full composite action is assumed, the PNA is at the top of the flange. From *AISC Manual* Table 3-19, $\sum Q_n$ equals 736 kips. The depth of the compressive stress block, a, is given by

$$a = \frac{\sum Q_n}{0.85 f_c' b} = \frac{736 \text{ kips}}{(0.85)\left(4 \ \frac{\text{kips}}{\text{in}^2}\right)(72 \text{ in})} = 3.0 \text{ in}$$

The distance to the compressive force, Y_2, is given by

$$Y_2 = t_s - \frac{a}{2} = 4 \text{ in} - \frac{3.0 \text{ in}}{2} = 2.5 \text{ in}$$

Therefore, from *AISC Manual* Table 3-19,

$$\frac{M_p}{\Omega_b} = 474 \text{ ft-kips} \quad \text{[ASD]}$$

$$\phi_b M_n = 712 \text{ ft-kips} \quad \text{[LRFD]}$$

2. (a) The weight of the 4 in slab is

$$\left(\frac{4 \text{ in}}{12 \ \frac{\text{in}}{\text{ft}}}\right)(9 \text{ ft})\left(150 \ \frac{\text{lbf}}{\text{ft}^3}\right) = 450 \text{ lbf/ft}$$

Assume a steel beam weight of 60 lbf/ft. The total dead load is

$$w_D = \frac{450 \ \frac{\text{lbf}}{\text{ft}} + 60 \ \frac{\text{lbf}}{\text{ft}}}{1000 \ \frac{\text{lbf}}{\text{kip}}} = 0.51 \text{ kip/ft}$$

$$M_D = \frac{w_D L^2}{8} = \frac{\left(0.51 \ \frac{\text{kip}}{\text{ft}}\right)(40 \text{ ft})^2}{8} = 102 \text{ ft-kips}$$

After the concrete has hardened,

$$\text{live load} + \text{ceiling load} = 100 \ \frac{\text{lbf}}{\text{ft}^2} + 10 \ \frac{\text{lbf}}{\text{ft}^2}$$
$$= 110 \text{ lbf/ft}^2$$

$$w_L = \frac{\left(110 \ \frac{\text{lbf}}{\text{ft}^2}\right)(9 \text{ ft})}{1000 \ \frac{\text{lbf}}{\text{kip}}} = 0.99 \text{ kip/ft}$$

$$M_L = \frac{w_L L^2}{8} = \frac{\left(0.99 \ \frac{\text{kip}}{\text{ft}}\right)(40 \text{ ft})^2}{8}$$
$$= 198 \text{ ft-kips}$$

Per *AISC Specification* Sec. I3.1a, the effective width of the concrete slab is the smaller of

$$b = \frac{L}{4} = \frac{(40 \text{ ft})\left(12 \ \frac{\text{in}}{\text{ft}}\right)}{4} = 120 \text{ in}$$

$$b = s = (9 \text{ ft})\left(12 \ \frac{\text{in}}{\text{ft}}\right) = 108 \text{ in} \quad \text{[governs]}$$

Since unshored construction is used, the dead load is resisted by the steel beam alone. Given a limiting dead load deformation of 2.5 in, the required moment of inertia for the beam is

$$I_{\text{req'd}} = \frac{(5)\left(0.51 \ \frac{\text{kip}}{\text{ft}}\right)(40 \text{ ft})^4\left(12 \ \frac{\text{in}}{\text{ft}}\right)^3}{(384)\left(29{,}000 \ \frac{\text{kips}}{\text{in}^2}\right)(2.5 \text{ in})} = 405 \text{ in}^4$$

The live load is resisted by the composite beam. Given a limiting live load deflection of 1 in, the required moment of inertia for the composite beam is

$$I_{\text{req'd}} = \frac{(5)\left(0.99 \ \frac{\text{kip}}{\text{ft}}\right)(40 \text{ ft})^4\left(12 \ \frac{\text{in}}{\text{ft}}\right)^3}{(384)\left(29{,}000 \ \frac{\text{kips}}{\text{in}^2}\right)(1.0 \text{ in})} = 1966 \text{ in}^4$$

From Table 3-20 in Part 3 of the *AISC Manual*, try a W21 × 48. The moment of inertia of the W21 × 48 equals 959 in^4, which is greater than the required moment of inertia of 405 in^4. Assuming Y_2 equals 2 in, and the PNA is at the top of the flange, the moment of inertia of the composite section equals 2030 in^4. This is greater than the required moment of inertia of 1966 in^4.

Check strength.

ASD Method Solution

The maximum moment is

$$M_{\max} = M_D + M_L = 102 \text{ ft-kips} + 198 \text{ ft-kips}$$
$$= 300 \text{ ft-kips}$$

Structural

The maximum shear is

$$V_{\max} = \frac{(w_D + w_L)L}{2}$$

$$= \frac{\left(0.51 \dfrac{\text{kip}}{\text{ft}} + 0.99 \dfrac{\text{kip}}{\text{ft}}\right)(40 \text{ ft})}{2}$$

$$= 30 \text{ kips}$$

From Table 3-19 of the *AISC Manual*, for the W21 × 48,

$$\frac{M_p}{\Omega_b} = 267 \text{ ft-kips} > M_D = 102 \text{ ft-kips}$$

Assuming the PNA is at the top of the flange, $\Sigma Q_n = 703$ kips. Therefore,

$$a = \frac{\Sigma Q_n}{0.85 f'_c b} = \frac{703 \text{ kips}}{(0.85)\left(3 \dfrac{\text{kips}}{\text{in}^2}\right)(108 \text{ in})} = 2.55 \text{ in}$$

$$Y_2 = t_s - \frac{a}{2} = 4 \text{ in} - \frac{2.55 \text{ in}}{2} = 2.73 \text{ in}$$

Interpolating gives

$$M_a = 460 \text{ ft-kips} > M_{\max} = 300 \text{ ft-kips}$$

Check shear strength.

From Table 3-2 of the *AISC Manual*,

$$\frac{V_{nx}}{\Omega_v} = 144 \text{ kips} > V_{\max} = 30 \text{ kips}$$

Therefore, use a W21 × 48.

LRFD Method Solution

The maximum moment is

$$1.4D = (1.4)\left(0.51 \frac{\text{kip}}{\text{ft}}\right) = 0.714 \text{ kip/ft}$$

$$1.2D + 1.6L = (1.2)\left(0.51 \frac{\text{kip}}{\text{ft}}\right)$$

$$+ (1.6)\left(0.99 \frac{\text{kip}}{\text{ft}}\right)$$

$$= 2.20 \text{ kips-ft}$$

$$M_u = \frac{\left(2.20 \dfrac{\text{kips}}{\text{ft}}\right)(40 \text{ ft})^2}{8} = 440 \text{ ft-kips}$$

$$M_D = \frac{\left(0.714 \dfrac{\text{kip}}{\text{ft}}\right)(40 \text{ ft})^2}{8} = 142.8 \text{ ft/kips}$$

$$M_{\max} = (1.2)(102 \text{ ft-kips}) + (1.6)(198 \text{ ft-kips})$$

$$= 439.2 \text{ ft-kips}$$

The maximum shear is

$$V_{\max} = \left((1.2)\left(0.51 \frac{\text{kip}}{\text{ft}}\right) + (1.6)\left(0.99 \frac{\text{kip}}{\text{ft}}\right)\right)\left(\frac{40 \text{ ft}}{2}\right)$$

$$= 43.9 \text{ kips}$$

From Table 3-19 of the *AISC Manual*, for the W21 × 48,

$$\phi M_p = 401 \text{ ft-kips} > M_D = 142.8 \text{ ft-kips} \quad \text{[OK]}$$

Assuming the PNA is at the top of the flange, $\Sigma Q_n = 707$ kips. Therefore,

$$a = \frac{\Sigma Q_n}{0.85 f'_c b} = \frac{707 \text{ kips}}{(0.85)\left(3 \dfrac{\text{kips}}{\text{in}^2}\right)(108 \text{ in})} = 2.57 \text{ in}$$

$$Y_2 = t_s - \frac{a}{2} = 4 \text{ in} - \frac{2.57 \text{ in}}{2} = 2.72 \text{ in}$$

Interpolating (between 679 ft-kips and 705 ft-kips from Table 3-19 of the *AISC Manual*) gives

$$\phi M_n = 690 \text{ ft-kips} > M_u = 440 \text{ ft-kips}$$

Check shear strength.

From Table 3-2 of the *AISC Manual*,

$$\phi V_n = 217 \text{ kips} > V_{\max} = 30 \text{ kips}$$

Therefore, use a W21 × 48.

(b) Determine the required number of shear studs.

As shown in part (a) of the solution, stiffness controls the design. From part (a), $I_{\text{req'd}} = 1966 \text{ in}^4$.

Given that the PNA is at the top of the flange, $Y_2 = 2.72 \text{ in}$.

From *AISC Manual* Table 3-20, the lower bound on the moment of inertia provided is interpolated (between 2120 in^4 and 2210 in^4) to be

$$I_{\text{prov'd}} = 2161 \text{ in}^4$$

Since this exceeds the required moment of inertia, the PNA can be lowered into the flange. Try the PNA at position 2.

$$\sum Q_n = 619 \text{ kips}$$

Therefore,

$$a = \frac{\sum Q_n}{0.85 f'_c b} = \frac{619 \text{ kips}}{(0.85)\left(3 \dfrac{\text{kips}}{\text{in}^2}\right)(108 \text{ in})} = 2.25 \text{ in}$$

$$Y_2 = t_s - \frac{a}{2} = 4 \text{ in} - \frac{2.25 \text{ in}}{2} = 2.88 \text{ in}$$

From *AISC Manual* Table 3-20, interpolating (between 2040 in^4 and 2130 in^4),

$$I_{\text{prov'd}} = 2108 \text{ in}^4 > I_{\text{req'd}} = 1966 \text{ in}^4 \quad \text{[OK]}$$

From *AISC Manual* Table 3-21, and assuming $^3/_4$ in studs, two weak studs per rib, $f'_c = 3$ ksi, and $q = 14.6$ kips/stud,

$$\frac{\sum W}{q} = \frac{532 \text{ kips}}{(2)\left(14.6 \dfrac{\text{kips}}{\text{stud}}\right)}$$

$$= 18.2 \quad \text{[say 20 studs on each side of the centerline]}$$

The total number of shear studs is 40 studs with 2 studs per rib.

Use a W21 × 48 beam with 20 $^3/_4$ in diameter × 3 in studs with two studs per rib on either side of the centerline.

3. (a) From *AISC Manual* Table 1-1 for a W21 × 68,

$$d = 21.1 \text{ in}$$
$$A_s = 20 \text{ in}^2$$
$$I_s = 1480 \text{ in}^4$$

The effective concrete flange width is the smaller of

$$b = \frac{L}{4} = \frac{(30 \text{ ft})\left(12 \dfrac{\text{in}}{\text{ft}}\right)}{4} = \boxed{90 \text{ in} \quad \text{[governs]}}$$
$$b = s = (12 \text{ ft})\left(12 \dfrac{\text{in}}{\text{ft}}\right) = 144 \text{ in}$$

The answer is (D).

(b) For full composite action, all of the concrete is in compression. Since the slab is 4 in thick, $Y2 = 4 \text{ in}/2 = 2$ in. From *AISC Manual* Table 3-19, for a W21 × 68 with $Y2 = 2$ in and PNA = TFL,

$$\frac{M_n}{\Omega_b} = \boxed{628 \text{ ft-kips} \quad (630 \text{ ft-kips})} \quad \text{[ASD]}$$

$$\phi_b M_n = \boxed{944 \text{ ft-kips} \quad (950 \text{ ft-kips})} \quad \text{[LRFD]}$$

The answer is (D).

(c) The dead load from the 4 in slab is

$$\left(\frac{4 \text{ in}}{12 \dfrac{\text{in}}{\text{ft}}}\right)\left(150 \dfrac{\text{lbf}}{\text{ft}^3}\right)(12 \text{ ft}) = 600 \text{ lbf/ft}$$

The W21 × 68 steel beam contributes 68 lbf/ft.

$$w_D = \frac{600 \dfrac{\text{lbf}}{\text{ft}} + 68 \dfrac{\text{lbf}}{\text{ft}}}{1000 \dfrac{\text{lbf}}{\text{kip}}}$$

$$= 0.668 \text{ kip/ft}$$

The unfactored maximum service dead load moment is

$$M_D = \frac{w_D L^2}{8} = \frac{\left(0.668 \dfrac{\text{kip}}{\text{ft}}\right)(30 \text{ ft})^2}{8}$$

$$= \boxed{75.2 \text{ ft-kips} \quad (75 \text{ ft-kips})}$$

The answer is (B).

(d) The moment capacity remaining for the live load is

$$M_L = M_R - M_D = 628 \text{ ft-kips} - 75.2 \text{ ft-kips}$$

$$= 552.8 \text{ ft-kips}$$

$$w_L = \frac{8 M_L}{L^2} = \frac{(8)(552.8 \text{ ft-kips})}{(30 \text{ ft})^2}$$

$$= 4.914 \text{ kips/ft}$$

The service live load capacity is

$$\frac{w_L}{s} = \frac{\left(4.914 \dfrac{\text{kips}}{\text{ft}}\right)\left(1000 \dfrac{\text{lbf}}{\text{kip}}\right)}{12 \text{ ft}}$$

$$= \boxed{409.5 \text{ lbf/ft}^2 \quad (410 \text{ lbf/ft}^2)}$$

The answer is (D).

(e) Find the total available horizontal shear. From *AISC Manual* Table 3-19, for PNA = TFL,

$$\sum Q_n = \boxed{1000 \text{ kips}}$$

The answer is (D).

(f) From *AISC Manual* Table 3-21, the available horizontal shear for one $3/4$ in diameter × 3 in stud is $Q_n = 17.2$ kips.

The number of shear connectors required on each side of the point of maximum moment is given by Eq. 64.6.

$$N = \frac{V_h}{Q_n} = \frac{1000 \text{ kips}}{17.2 \text{ kips}} = 58$$

The total number of shear connectors required for the entire beam is

$$2N = (2)(58) = \boxed{116 \quad (120)}$$

The answer is (D).

4. The nominal rib height of the steel deck is $h_r = 3$ in.

The thickness of the concrete slab above the steel deck is $t_o = 3$ in.

A conservative estimate of the thickness of the concrete in compression is $t = t_o = 3$ in.

From *AISC Manual* Table 1-1 for a W21 × 50,

$$d = 20.8 \text{ in}$$
$$A_s = 14.7 \text{ in}^2$$
$$I_s = 984 \text{ in}^4$$
$$S_s = 94.5 \text{ in}^3$$

The effective concrete flange width is the smaller of

$$b = \frac{L}{4} = \frac{(40 \text{ ft})\left(12 \frac{\text{in}}{\text{ft}}\right)}{4} = 120 \text{ in}$$

$$b = s = (8 \text{ ft})\left(12 \frac{\text{in}}{\text{ft}}\right) = 96 \text{ in} \quad \text{[governs]}$$

Assume $a = 3$ in.

$$Y2 = t_{\text{slab}} - \frac{a}{2} = 6 \text{ in} - \frac{3 \text{ in}}{2} = 4.5 \text{ in}$$

From *AISC Manual* Table 3-20, for a W21 × 50, $Y2 = 4.5$ in, PNA = location 1 (top of flange),

$$I_x = \boxed{2620 \text{ in}^4 \quad (2600 \text{ in}^4)}$$

The answer is (A).

(b) Evaluate the total horizontal shear for full composite action. Use *AISC Manual* Table 3-19.

For PNA at TFL,

$$\sum Q_n = \boxed{736 \text{ kips} \quad (740 \text{ kips})}$$

The answer is (D).

(c) From *AISC Specification* Sec. I3-2d, with ribs running perpendicular to the steel shape,

$$R_g = 1.0$$
$$R_p = 0.6 \quad \text{[across weak studs]}$$
$$R_g R_p = (1.0)(0.6) = \boxed{0.60}$$

The answer is (B).

(d) For a conservative estimate, all studs are assumed to be placed in weak locations. From *AISC Manual* Table 3-21, the available horizontal shear for each $3/4$ in diameter × 3 in stud is $Q_n = 17.2$ kips. The number of studs required (for full composite action) to resist the horizontal shear V_h on each side of the point of maximum moment is

$$N = \frac{V_h}{Q_n} = \frac{736 \text{ kips}}{17.2 \text{ kips}}$$
$$= 42.8 \quad (43)$$

The total number of studs required for the entire beam is

$$2N = (2)(43) = \boxed{86}$$

The answer is (C).

(e) 75% partial composite action means $V'_h / V_h = 0.75$.

$$V'_h = 0.75 V_h = (0.75)(736 \text{ kips}) = 552 \text{ kips}$$

The stud strength is $Q_n = 17.2$ kips.

The number of studs required to resist the horizontal shear V'_h on each side of the point of maximum moment is

$$N = \frac{V'_h}{Q_n} = \frac{552 \text{ kips}}{17.2 \text{ kips}}$$
$$= 32.1 \quad \text{[say 33]}$$

The total number of studs required for the entire beam is

$$2N = (2)(33) = \boxed{66}$$

The answer is (C).

(f) The service live load deflection for full composite action is

$$\Delta_L = \frac{5w_L L^4}{384 E I_{\mathrm{tr}}} = \frac{(5)\left(0.8\ \dfrac{\mathrm{kip}}{\mathrm{ft}}\right)(40\ \mathrm{ft})^4\left(12\ \dfrac{\mathrm{in}}{\mathrm{ft}}\right)^3}{(384)\left(29{,}000\ \dfrac{\mathrm{kips}}{\mathrm{in}^2}\right)(2620\ \mathrm{in}^4)}$$

$$= \boxed{0.606\ \mathrm{in} \quad (0.61\ \mathrm{in})}$$

The answer is (C).

(g) The service live load deflection for partial (75%) composite action (i.e., the composite percentage is $V'_h/V_h = 0.75$) from *AISC Commentary* Eq. C-I3-3 is

$$I_{\mathrm{equiv}} = I_s + \sqrt{\frac{V'_h}{V_h}}(I_{\mathrm{tr}} - I_s)$$

$$= 984\ \mathrm{in}^4 + \left(\sqrt{0.75}\right)(2620\ \mathrm{in}^4 - 984\ \mathrm{in}^4)$$

$$= 2401\ \mathrm{in}^4$$

$$I_{\mathrm{eff}} = 0.75 I_{\mathrm{equiv}} \quad [\textit{AISC Commentary Sec. I3.2}]$$

$$\Delta_{L,\mathrm{partial}} = \frac{5w_L L^4}{384 E I_{\mathrm{eff}}}$$

$$= \frac{(5)\left(0.8\ \dfrac{\mathrm{kip}}{\mathrm{ft}}\right)(40\ \mathrm{ft})^4\left(12\ \dfrac{\mathrm{in}}{\mathrm{ft}}\right)^3}{(384)\left(29{,}000\ \dfrac{\mathrm{kips}}{\mathrm{in}^2}\right)(0.75)(2401\ \mathrm{in}^4)}$$

$$= \boxed{0.882\ \mathrm{in} \quad (0.88\ \mathrm{in})}$$

The answer is (D).

5. Find the moment of inertia of the plate girder with and without the concrete flange.

shape	A (in^2)	$\overline{y}^a$ (in)	$\overline{y}A$ (in^3)	I_c (in^4)
$\frac{5}{8} \times 14$	8.75	41.0625^b	359.297	0.2848^c
$\frac{5}{16} \times 40$	12.5	20.75	259.375	1666.67
$\frac{3}{4} \times 16$	12	0.375	4.5	0.5625
subtotals	33.25		623.172	1667.52
transformed concrete	48.0	45.375	2178	144
totals	81.25 in^2		2801.2 in^3	1811.52 in^4

$^a \overline{y}$ measured from the bottom of the flange
$^b 0.75\ \mathrm{in} + 40.0\ \mathrm{in} + (^1/_2\ \mathrm{in})(^5/_8\ \mathrm{in})$
$^c bh^3/12 = (14\ \mathrm{in})(0.625\ \mathrm{in})^3/12$

Work with the steel acting alone before the concrete hardens.

$$\overline{y} = \frac{623.172\ \mathrm{in}^3}{33.25\ \mathrm{in}^2} = 18.74\ \mathrm{in}$$

$$I = 1667.52\ \mathrm{in}^4 + (12\ \mathrm{in}^2)(18.74\ \mathrm{in} - 0.375\ \mathrm{in})^2$$

$$+ (12.5\ \mathrm{in}^2)(20.75\ \mathrm{in} - 18.74\ \mathrm{in})^2$$

$$+ (8.75\ \mathrm{in}^2)(41.0625\ \mathrm{in} - 18.74\ \mathrm{in})^2$$

$$= 10{,}125.4\ \mathrm{in}^4$$

The distances to the extreme fibers are

$$c = \overline{y} = 18.74\ \mathrm{in} \quad [\text{tension}]$$

$$c = \tfrac{3}{4}\ \mathrm{in} + 40\ \mathrm{in} + \tfrac{5}{8}\ \mathrm{in} - 18.74\ \mathrm{in}$$

$$= 22.64\ \mathrm{in} \quad [\text{compression}]$$

The section modulus referred to the top of steel is

$$S_{xc} = \frac{I}{c} = \frac{10{,}125.4\ \mathrm{in}^4}{22.64\ \mathrm{in}} = 447.2\ \mathrm{in}^3$$

$$S_{xt} = \frac{10{,}125.4\ \mathrm{in}^4}{18.74\ \mathrm{in}} = 540.3\ \mathrm{in}^3$$

The maximum moment due to 1 kip/ft^2 service dead load (i.e., ASD) is

$$M_D = \frac{wL^2}{8} = \frac{\left(1\ \dfrac{\mathrm{kip}}{\mathrm{ft}}\right)(80\ \mathrm{ft})^2}{8} = 800\ \mathrm{ft\text{-}kips}$$

The shear due to service dead load (ASD) is

$$V_D = \frac{wL}{2} = \frac{\left(1\ \dfrac{\mathrm{kip}}{\mathrm{ft}}\right)(80\ \mathrm{ft})}{2} = 40\ \mathrm{kips}$$

Check the width-thickness ratios of the compression elements.

Check the width-thickness ratio of the web, per *AISC Specification* Sec. F13.2(b) since

$$\frac{a}{h} > 1.5$$

$$\left(\frac{h}{t_w}\right)_{max} = \frac{0.42E}{F_y} = \frac{(0.42)\left(29{,}000\ \frac{\text{kips}}{\text{in}^2}\right)}{36\ \frac{\text{kips}}{\text{in}^2}} = 338$$

$$\frac{h}{t} = \frac{40\ \text{in}}{\frac{5}{16}\ \text{in}} = 128$$

$$128 < 338 \quad \text{[acceptable]}$$

For the compression flange, considered stiffened along one edge, from *AISC Specification* Table B4.1, note (a), typical values of k_c are in the 0.35 to 0.76 range.

$$k_c = \frac{4}{\sqrt{\dfrac{h}{t}}} = \frac{4}{\sqrt{128}}$$

$$= 0.35$$

$$b = \left(\tfrac{1}{2}\right)(14\ \text{in}) = 7\ \text{in}$$

$$\frac{b}{t} = \frac{7\ \text{in}}{\frac{5}{8}\ \text{in}} = 11.2$$

From *AISC Specification* Table B4.1, case 2,

$$0.95\sqrt{\frac{k_c E}{F_L}} = 0.95\sqrt{\frac{(0.35)\left(29{,}000\ \frac{\text{kips}}{\text{in}^2}\right)}{(0.7)\left(36\ \frac{\text{kips}}{\text{in}^2}\right)}} = 19.1$$

$$11.2 < 19.1 \quad \text{[acceptable]}$$

Calculate h/t for the web.

$$\frac{h}{t} = \frac{40\ \text{in}}{\frac{5}{16}\ \text{in}} = 128$$

Compute λ_{pw} for the web. From *AISC Specification* Table B4.1, case 11,

$$\lambda_{pw} = \frac{\dfrac{h_c}{h_p}\sqrt{\dfrac{E}{F_y}}}{\left(0.54\left(\dfrac{M_p}{M_y}\right) - 0.09\right)^2} \le \lambda_{rw}$$

$$h_c = (2)\left(22.63\ \text{in} - \tfrac{5}{8}\ \text{in}\right) = 44.01\ \text{in}$$

Find the location of the PNA from the bottom of the top flange.

$$(14\ \text{in})\left(\tfrac{5}{8}\ \text{in}\right) + \text{PNA}\left(\tfrac{5}{16}\ \text{in}\right)$$

$$= (16\ \text{in})\left(\tfrac{3}{4}\ \text{in}\right) + (40 - \text{PNA})\left(\tfrac{5}{16}\ \text{in}\right)$$

$$\text{PNA} = 25.20\ \text{in from bottom of top flange}$$

$$h_p = (2)(25.20\ \text{in}) = 50.4\ \text{in}$$

$$\frac{h_c}{h_p} = \frac{44.01\ \text{in}}{50.40\ \text{in}} = 0.873$$

$$M_{yc} = S_{xc}F_y = (447.2\ \text{in}^3)\left(36\ \frac{\text{kips}}{\text{in}^2}\right) = 16{,}099\ \text{in-kips}$$

Compute M_p.

shape	A (in^2)	$\bar{y}$ (in)	$\bar{y}A$ (in^3)
top plate	$14 \times \frac{5}{8}$	8.75	25.51	223.2
flange	$25.2 \times \frac{5}{16}$	7.875	12.6	99.22
40 in	$14.8 \times \frac{5}{16}$	4.625	7.4	34.22
bottom plate	$16 \times \frac{3}{4}$	12	15.175	182.1
			$\sum = 539$	

Therefore, $Z = 539\ \text{in}^3$, and $M_p = (539\ \text{in}^3)(36\ \text{kips/in}^2) = 19{,}404\ \text{in-kips}$.

$$\frac{M_p}{M_{yc}} = \frac{19{,}404\ \text{in-kips}}{16{,}095\ \text{in-kips}} = 1.20$$

Compute λ_{pw} and λ_{rw}. From *AISC Specification* Table B4.1, case 11,

$$\lambda_{pw} = \frac{0.873\sqrt{\dfrac{29{,}000\ \frac{\text{kips}}{\text{in}^2}}{36\ \frac{\text{kips}}{\text{in}^2}}}}{\left((0.54)(1.20) - 0.09\right)^2} = 79.6$$

$$\lambda_{rw} = 5.70\sqrt{\frac{E}{F_y}} = 5.70\sqrt{\frac{29{,}000\ \frac{\text{kips}}{\text{in}^2}}{36\ \frac{\text{kips}}{\text{in}^2}}} = 162$$

Since $\lambda_{pw} < h/t_w < \lambda_{rw}$, the web is non-compact.

$$\lambda = \frac{h_c}{t_w} = \frac{44.01\ \text{in}}{\frac{5}{16}\ \text{in}} = 141$$

$$R_{pc} = \left(\frac{M_p}{M_{yc}} - \left(\frac{M_p}{M_{yc}} - 1\right)\left(\frac{\lambda - \lambda_{pw}}{\lambda_{rw} - \lambda_{pw}}\right)\right) \le \frac{M_p}{M_{yc}}$$

$$= 1.20 - (1.20 - 1)\left(\frac{141 - 79.6}{162 - 79.6}\right)$$

$$= 1.05 \le 1.20$$

$$r_t = \frac{b_{fc}}{12\left(\dfrac{h_o}{d} + \dfrac{1}{6}a_w\dfrac{h^2}{h_o d}\right)} \quad \text{[\textit{AISC Specification} Eq. F4-10]}$$

a_w is from *AISC Specification* Eq. F4-11.

$$a_w = \frac{h_c t_w}{b_{fc} t_{fc}} = \frac{(44.01 \text{ in})\left(\frac{5}{16} \text{ in}\right)}{(14 \text{ in})\left(\frac{5}{8} \text{ in}\right)}$$

$$= 1.572$$

b_{fc} is the compression flange width, t_{fc} is the compression flange thickness, and h_o is the distance between the flange centroids.

From *AISC Specification* Eq. F4-10, the value for r_t is

$$r_t = \frac{b_{fc}}{\sqrt{12\left(\dfrac{h_o}{d} + \dfrac{1}{6} a_w \dfrac{h^2}{h_o d}\right)}}$$

$$= \frac{14 \text{ in}}{\sqrt{(12)\left(\begin{array}{c} \dfrac{40.6875 \text{ in}}{41.375 \text{ in}} \\[2mm] + \left(\dfrac{1}{6}\right)(1.572)\left(\dfrac{(40 \text{ in})^2}{(40.6875 \text{ in})(41.375 \text{ in})}\right) \end{array}\right)}}$$

$$= 3.64 \text{ in}$$

$$L_p = 1.1 r_t \sqrt{\frac{E}{F_y}}$$

$$= (1.1)\left(\frac{3.64 \text{ in}}{12 \; \frac{\text{in}}{\text{ft}}}\right)\sqrt{\frac{29{,}000 \; \frac{\text{kips}}{\text{in}^2}}{36 \; \frac{\text{kips}}{\text{in}^2}}}$$

$$= 9.47 \text{ ft}$$

$$J \approx \sum \frac{bt^3}{3}$$

$$= \frac{\begin{array}{c} (14 \text{ in})\left(\frac{5}{8} \text{ in}\right)^3 + (40 \text{ in})\left(\frac{5}{16} \text{ in}\right)^3 \\[2mm] + (16 \text{ in})\left(\frac{3}{4} \text{ in}\right)^3 \end{array}}{3}$$

$$= 3.80 \text{ in}^4$$

$$\frac{S_{xt}}{S_{xc}} = \frac{540.3 \text{ in}^3}{447.2 \text{ in}^3} = 1.21$$

Since $S_{xt}/S_{xc} > 0.7$, then from *AISC Specification* Table B4.1, note b,

$$F_L = 0.7 F_y = (0.7)\left(36 \; \frac{\text{kips}}{\text{in}^2}\right) = 25.2 \text{ ksi}$$

The distance between the flange centroids is $h_o = 40.7$ in. Compute L_r.

$$L_r = 1.95 r_t \frac{E}{F_L} \sqrt{\frac{J}{S_{xc} h_o}}$$

$$\times \sqrt{1 + \sqrt{1 + 6.76\left(\frac{F_L S_{xc} h_o}{EJ}\right)^2}}$$

$$= (1.95)(3.64 \text{ in})\left(\frac{29{,}000 \; \frac{\text{kips}}{\text{in}^2}}{25.2 \; \frac{\text{kips}}{\text{in}^2}}\right)\sqrt{\frac{3.84 \text{ in}}{(447.2 \text{ in}^3)(40.7 \text{ in})}}$$

$$\times \sqrt{1 + \sqrt{1 + 6.76\left(\frac{\left(25.2 \; \frac{\text{kips}}{\text{in}^2}\right)(447.2 \text{ in}^3) \times (40.7 \text{ in})}{\left(29{,}000 \; \frac{\text{kips}}{\text{in}^2}\right)(3.80 \text{ in})}\right)^2}}$$

$$= 408.7 \text{ in} \quad (34.1 \text{ ft})$$

$$L_b = 20 \text{ ft}$$

Since $L_p < L_b < L_r$, the lateral-torsional buckling limit state is given by *AISC Specification* Eq. F4-2.

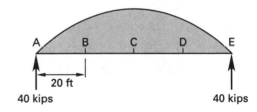

Compute C_b.

The moment equation is

$$M = \frac{-x^2}{2} + 40x \text{ ft-kips}$$

The critical unbraced segment is BC. The moments at the quarter points in segment BC are

$$x = 25 \text{ ft}, \quad M_{1/4} = 687.5 \text{ ft-kips}$$

$$x = 30 \text{ ft}, \quad M_{1/2} = 750 \text{ ft-kips}$$

$$x = 35 \text{ ft}, \quad M_{3/4} = 787.5 \text{ ft-kips}$$

$$x = 40 \text{ ft}, \quad M_{\max} = 800 \text{ ft-kips}$$

From *AISC Specification* Eq. F1-1,

$$C_b = \frac{12.5M_{\max}}{2.5M_{\max} + 3M_A + 4M_B + 3M_C} R_m \le 3.0$$

For singly symmetric cross sections subjected to single curvature, $R_m = 1.0$.

$$C_b = \left(\frac{(12.5)(800 \text{ ft-kips})}{(2.5)(800 \text{ ft-kips}) + (3)(687.5 \text{ ft-kips})} \right.$$
$$\left. + (4)(750 \text{ ft-kips}) + (3)(787.5 \text{ ft-kips}) \right)$$
$$\times (1.0)$$
$$= 1.06 \le 3.0$$

From *AISC Specification* Eq. F4-2,

$$M_n = C_b \left(\frac{R_{pc}M_{yc}}{} - (R_{pc}M_{yc} - F_L S_{xc})\left(\frac{L_b - L_p}{L_r - L_p}\right) \right)$$
$$\le R_{pc}M_{yc}$$

$$M_{yc} = F_y S_{xc} = \frac{\left(36 \frac{\text{kips}}{\text{in}^2}\right)(447.2 \text{ in}^3)}{12 \frac{\text{in}}{\text{ft}}}$$
$$= 1341.6 \text{ ft-kips}$$

$$R_{pc}M_{yc} = (1.05)(1341.6 \text{ ft-kips}) = 1408.7 \text{ ft-kips}$$

$$F_L S_{xc} = \frac{\left(25.2 \frac{\text{kips}}{\text{in}^2}\right)(447.2 \text{ in}^3)}{12 \frac{\text{in}}{\text{ft}}}$$
$$= 939.1 \text{ ft-kips}$$

$$\frac{L_b - L_p}{L_r - L_p} = \frac{20 \text{ ft} - 9.47 \text{ ft}}{34.1 \text{ ft} - 9.47 \text{ ft}} = 0.428$$

$$M_n = (1.06)\left(1408.7 \text{ ft-kips} - \left(\begin{array}{c} 1408.7 \text{ ft-kips} \\ - 939.1 \text{ ft-kips} \end{array} \right)(0.428) \right)$$
$$= 1280 \text{ ft-kips} \le 1408 \text{ ft-kips}$$

The limiting moment for the lateral-torsional buckling limit state equals 1280 ft-kips.

The limiting moment for compression flange local buckling is given by *AISC Specification* Eq. F4-12.

$$M_n = R_{pc}M_{yc} - (R_{pc}M_{yc} - F_L S_{xc})\left(\frac{\lambda - \lambda_{pf}}{\lambda_{rf} - \lambda_{pf}}\right)$$

From *AISC Specification* Table B4.1, case 7,

$$\lambda_{pf} = 0.38\sqrt{\frac{E}{F_y}} = 0.38\sqrt{\frac{29,000 \frac{\text{kips}}{\text{in}^2}}{36 \frac{\text{kips}}{\text{in}^2}}} = 10.8$$

$$M_n = 1408 \text{ ft-kips}$$
$$- (1408 \text{ ft-kips} - 939 \text{ ft-kips})\left(\frac{11.2 - 10.8}{19.1 - 10.8}\right)$$
$$= 1385 \text{ ft-kips}$$

The limiting moment for compression flange local buckling is 1385 ft-kips.

Since $S_{xt} > S_{xc}$, the tension flange yield need not be checked.

Lateral-torsional buckling is the limit state, and the nominal moment equals 1280 ft-kips.

$$\frac{M_n}{\Omega} = \frac{1280 \text{ ft-kips}}{1.67}$$
$$= 766 \text{ ft-kips} < M_a$$
$$= 800 \text{ ft-kips}$$

The steel cross section is not acceptable.

Check shear.

Since the stiffener spacing is assumed to be large, $k_v = 5$.

$$\frac{h}{t_w} = 128 > 1.37\sqrt{\frac{k_v E}{F_y}} = 1.37\sqrt{\frac{(5)\left(29,000 \frac{\text{kips}}{\text{in}^2}\right)}{36 \frac{\text{kips}}{\text{in}^2}}}$$
$$= 87$$

From *AISC Specification* Eq. G2-5,

$$C_v = \frac{1.51E k_v}{\left(\frac{h}{t_w}\right)^2 F_y} = \frac{(1.51)\left(29,000 \frac{\text{kips}}{\text{in}^2}\right)(5)}{(128)^2\left(36 \frac{\text{kips}}{\text{in}^2}\right)} = 0.371$$

From *AISC Specification* Eq. G2-1,

$$V_n = 0.6F_y A_w C_v$$
$$= (0.6)\left(36 \frac{\text{kips}}{\text{in}^2}\right)(40 \text{ in})\left(\tfrac{5}{16} \text{ in}\right)(0.371)$$
$$= 100 \text{ kips}$$

$$\frac{V_n}{\Omega} = \frac{100 \text{ kips}}{1.50} = 66.7 \text{ kips} > V_a = 40 \text{ kips} \quad \text{[acceptable]}$$

$$\frac{h}{t} = 128 > 3.76\sqrt{\frac{E}{F}} = 3.76\sqrt{\frac{29{,}000\,\frac{\text{kips}}{\text{in}^2}}{36\,\frac{\text{kips}}{\text{in}^2}}}$$

$$= 107$$

M_n is determined from the superposition of elastic stresses.

After the concrete hardens,

$$\overline{y} = \frac{2801.2 \text{ in}^3}{81.25 \text{ in}^2} = 34.48 \text{ in}$$

All of the concrete is in compression. The distances to the extreme fibers are

$$c = \begin{cases} \frac{3}{4}\text{ in} + 40\text{ in} + \frac{5}{8}\text{ in} - 34.48\text{ in} \\ \qquad = 6.895\text{ in} \quad \text{[steel in compression]} \\ \frac{3}{4}\text{ in} + 40\text{ in} + \frac{5}{8}\text{ in} + 1\text{ in} + 6\text{ in} - 34.48\text{ in} \\ \qquad = 13.895\text{ in} \quad \text{[concrete in compression]} \end{cases}$$

$$= 34.48 \text{ in} \quad \text{[steel in tension]}$$

$$\begin{aligned} I &= 1811.52 \text{ in}^4 + (8.75 \text{ in}^2)(41.0625 \text{ in} - 34.48 \text{ in})^2 \\ &\quad + (12.5 \text{ in}^2)(34.48 \text{ in} - 20.75 \text{ in})^2 \\ &\quad + (12 \text{ in}^2)(34.48 \text{ in} - 0.375 \text{ in})^2 \\ &\quad + (48 \text{ in}^2)(45.375 \text{ in} - 34.48 \text{ in})^2 \\ &= 24{,}202.5 \text{ in}^4 \end{aligned}$$

The transformed section modulus referred to the tension flange is

$$S = \frac{I}{c} = \frac{24{,}202.5 \text{ in}^4}{34.48 \text{ in}} = 701.9 \text{ in}^3$$

From the live load, the maximum shears and moments are

$$V_t = 40 \text{ kips} + \frac{\left(1\,\frac{\text{kip}}{\text{ft}}\right)(80 \text{ ft})}{2} = 80 \text{ kips}$$

$$M_L = (20 \text{ ft})(40 \text{ kips}) = 800 \text{ ft-kips}$$

$$\begin{aligned} M_t &= M_D + M_L = 800 \text{ ft-kips} + 800 \text{ ft-kips} \\ &= 1600 \text{ ft-kips} \end{aligned}$$

$$V = 80 \text{ kips} > 66.7 \text{ kips} \quad \text{[not acceptable]}$$

Check the compressive stress in steel.

Since there is no shoring, the dead and live loads are superimposed

$$\begin{aligned} f_b &= \frac{M_D c_{\text{steel}}}{I_{\text{steel}}} + \frac{M_L c_{\text{composite}}}{I_{\text{composite}}} \\ &= \frac{(800 \text{ ft-kips})\left(12\,\frac{\text{in}}{\text{ft}}\right)(22.64 \text{ in})}{10{,}125.4 \text{ in}^4} \\ &\quad + \frac{(800 \text{ ft-kips})\left(12\,\frac{\text{in}}{\text{ft}}\right)(6.895 \text{ in})}{24{,}202.5 \text{ in}^4} \\ &= 24.2 \text{ kips/in}^2 \end{aligned}$$

$$F_{b,\text{compression}} = \frac{F_y}{\Omega} = \frac{36\,\frac{\text{kips}}{\text{in}^2}}{1.67} = 21.6 \text{ kips/in}^2$$

$$24.2 \text{ kips/in}^2 > 21.6 \text{ kips/in}^2 \quad \text{[not acceptable]}$$

Check the tensile stress in steel.

$$F_{b,\text{tension}} = \frac{F_y}{\Omega} = \frac{36\,\frac{\text{kips}}{\text{in}^2}}{1.67} = 21.6 \text{ kips/in}^2$$

$$\begin{aligned} f_t &= \frac{(800 \text{ ft-kips})\left(12\,\frac{\text{in}}{\text{ft}}\right)(18.74 \text{ in})}{10{,}125.4 \text{ in}^4} \\ &\quad + \frac{(800 \text{ ft-kips})\left(12\,\frac{\text{in}}{\text{ft}}\right)(34.48 \text{ in})}{24{,}202.5 \text{ in}^4} \end{aligned}$$

$$= 31.4 \text{ kips/in}^2 > 21.6 \text{ kips/in}^2 \quad \begin{bmatrix} \text{not} \\ \text{acceptable} \end{bmatrix}$$

Check the concrete compressive stress.

From *AISC Specification* Sec. I2.2, $f_c < 0.45 f'_c$. With 3000 psi concrete,

$$F_c = \frac{(0.85)\left(3.0\,\frac{\text{kips}}{\text{in}^2}\right)}{1.67} = 1.53 \text{ kips/in}^2$$

Only the live load is applied after curing. Since the modular ratio ($E_{\text{steel}}/E_{\text{concrete}}$) is 10,

$$f_c = \frac{\dfrac{(800 \text{ ft-kips})\left(12\,\frac{\text{in}}{\text{ft}}\right)(13.895 \text{ in})}{24{,}202.5 \text{ in}^4}}{10}$$

$$= 0.55 \text{ kips/in}^2$$

$$0.55 \text{ kips/in}^2 < 1.35 \text{ kips/in}^2 \quad \text{[acceptable]}$$

The plate girder is $\boxed{\text{inadequate}}$ due to the excessive steel bending and shear stresses.

65 Structural Steel: Connectors

PRACTICE PROBLEMS

1. (*Time limit: one hour*) Two $^5/_8$ in plates are spliced as shown to form a member carrying an axial tensile force of unknown magnitude. Four $^7/_8$ in diameter bolts are used at center-to-center spacing of 3 in in both longitudinal and transverse directions. All holes are punched. The plate steel is A588, grade 50. Neglect block shear. Assume a C1.A faying surface.

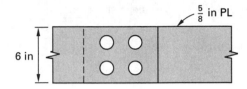

(a) The allowable (i.e., ASD) tensile load for the member, based on the yielding criterion alone, is most nearly

(A) 68 kips

(B) 80 kips

(C) 88 kips

(D) 110 kips

(b) The allowable (i.e., ASD) tensile load for the member, based on the fracture criterion alone, is most nearly

(A) 68 kips

(B) 80 kips

(C) 88 kips

(D) 110 kips

(c) Assuming that deformation around bolt holes is not a design consideration, the allowable (i.e., ASD) bearing stress in the member is most nearly

(A) 76 ksi

(B) 84 ksi

(C) 95 ksi

(D) 105 ksi

(d) Assuming an A325SC slip-critical connection, the allowable shear capacity (i.e., ASD) of the connection is most nearly

(A) 35 kips

(B) 41 kips

(C) 56 kips

(D) 68 kips

(e) Assuming an A325SC slip-critical connection, the allowable tensile load capacity of the member is most nearly

(A) 41 kips

(B) 68 kips

(C) 80 kips

(D) 87 kips

(f) Assuming an A325N bearing-type connection with threads in the shear plane, the shear capacity of the four bolts is most nearly

(A) 35 kips

(B) 58 kips

(C) 72 kips

(D) 90 kips

(g) Assuming an A325N bearing-type connection with threads in the shear plane, the allowable tensile load (i.e., ASD) capacity of the member is most nearly

(A) 41 kips

(B) 58 kips

(C) 80 kips

(D) 87 kips

(h) Assuming an A325X bearing-type connection with threads excluded from the shear plane, the shear capacity of the four bolts is most nearly

(A) 35 kips

(B) 41 kips

(C) 72 kips

(D) 90 kips

(i) Assuming an A3255X bearing-type connection with threads excluded from the shear plane, the maximum service tensile load the member can carry is most nearly

(A) 41 kips

(B) 72 kips

(C) 87 kips

(D) 90 kips

(j) The actual bearing stress in the member corresponding to its tensile capacity with a slip-critical connection is most nearly

(A) 17 kips/in^2

(B) 19 kips/in^2

(C) 20 kips/in^2

(D) 21 kips/in^2

2. (*Time limit: one hour*) In the bracket connection shown, all fasteners are $^3/_4$ in diameter bolts.

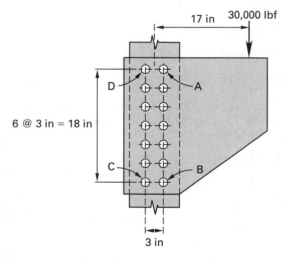

For parts (a) through (g), use a traditional elastic approach.

(a) The polar area moment of inertia of the bolt group is most nearly

(A) 190 in^4

(B) 240 in^4

(C) 290 in^4

(D) 300 in^4

(b) The torsional shear stress in bolts A, B, C, and D is most nearly

(A) 15,900 psi

(B) 18,700 psi

(C) 19,600 psi

(D) 20,300 psi

(c) The vertical component of the torsional shear stress in bolts A, B, C, and D is most nearly

(A) 2200 psi

(B) 2800 psi

(C) 3200 psi

(D) 3600 psi

(d) The horizontal component of the torsional shear stress in bolts A, B, C, and D is most nearly

(A) 12,500 psi

(B) 18,200 psi

(C) 18,900 psi

(D) 19,400 psi

(e) The direct vertical shear stress in the bolts is most nearly

(A) 4200 psi

(B) 4400 psi

(C) 4900 psi

(D) 5000 psi

(f) The total shear stress in bolts A and B is most nearly

(A) 16,200 psi

(B) 18,500 psi

(C) 21,000 psi

(D) 22,600 psi

(g) The total shear stress in bolts C and D is most nearly

(A) 15,200 psi

(B) 17,700 psi

(C) 18,900 psi

(D) 19,400 psi

For parts (h) through (j), use the tables in *AISC Manual* Part 7.

(h) Assuming A325SC bolts in standard holes, the allowable (i.e., ASD) load for the bolt group is most nearly

(A) 21,000 lbf

(B) 26,000 lbf

(C) 27,000 lbf

(D) 30,000 lbf

(i) Assuming A325N bolts in standard holes, the allowable (i.e., ASD) load for the bolt group is most nearly

(A) 22,900 lbf

(B) 27,300 lbf

(C) 38,400 lbf

(D) 42,900 lbf

(j) The connection is adequate for

(A) A325SC bolts only

(B) A325N bolts only

(C) both A325SC and A325N bolts

(D) neither A325SC nor A325N bolts

3. A bracing member made of a pair of L5 × 3 × ½ angles to carry a service tensile load of 140 kips is attached to a W12 column via a structural tee as shown. All connections are bearing-type and made with ⅞ in diameter A325 bolts with threads in the plane of shear. Neglecting block shear, determine the required number of bolts for (a) the connection between the bracing member and the structural tee and (b) the connection between the structural tee and the column.

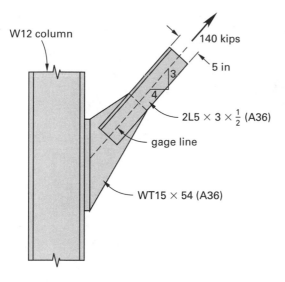

4. A structural tee tensile member is bolted to a W-shape column with eight A325X ¾ in diameter bolts as shown. The bolts pass through standard holes and are arranged in two vertical rows on a 3 in center-to-center pitch. Bolts are pretensioned using normal methods. All faying surfaces are clean and unpainted, exhibiting a tight mill scale. A static live load, P, is applied through the centroid of the eight-bolt group. The structural tee is oriented diagonally (3 vertical:4 horizontal). Disregard dead loading. Both slip-critical and bearing-type connections are being considered. Slip-critical connections are permitted to slip at the factored load levels, and any slippage will not result in geometry changes or significant load increases. All edge distances are adequate. The column, structural tee, and tee bolt base plate are rigid, their strengths are adequate, and they do not need to be checked.

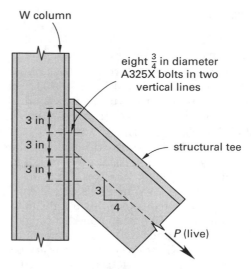

(a) According to an ASD analysis, what is most nearly the maximum live load, P, that this connection can support in a slip-critical connection?

(A) 25 kips

(B) 36 kips

(C) 40 kips

(D) 67 kips

(b) According to an LRFD analysis, what is most nearly the maximum unfactored live load, P, that this connection can support in a slip-critical connection?

(A) 34 kips

(B) 45 kips

(C) 63 kips

(D) 84 kips

(c) According to an ASD analysis, what is most nearly the maximum live load, *P*, that this connection can support in a bearing-type connection?

(A) 80 kips

(B) 120 kips

(C) 180 kips

(D) 220 kips

(d) According to an LRFD analysis, what is most nearly the maximum live load, *P*, that this connection can support in a bearing-type connection?

(A) 95 kips

(B) 140 kips

(C) 180 kips

(D) 230 kips

5. A tensile member is constructed by welding two A36 L4 × 3 × ¹/₂ angles long back-to-long back onto the web of an intermediate member cut from an A992 WF beam. E70 electrodes are used to form the maximum weld size permitted. The intermediate member is bolted to an A992 W-shape column with six A325X ⁷/₈ in diameter bolts as shown. The bolts pass through standard holes and are arranged in two vertical rows on a 3 in center-to-center pitch. All edge distances are 2 in. Bolts are pretensioned using normal methods. All faying surfaces are clean and unpainted, exhibiting a tight mill scale. A static live load of 140 kips is applied through the centroid of the six-bolt group. Disregard dead loading. Both slip-critical and bearing-type connections are being considered. Slip-critical connections are permitted to slip at the factored load levels, and any slippage will not result in geometry changes or significant load increases. The column, intermediate member, angles, and bolt base plate are rigid, their strengths are adequate, and they do not need to be checked.

(a) According to an ASD analysis, is the design adequate for a slip-critical connection?

(A) yes; adequate in bolt shear, bolt tension, and slip resistance

(B) no; inadequate in shear; adequate in tension and slip resistance

(C) no; inadequate in tension; adequate in shear and slip resistance

(D) no; inadequate in slip resistance; adequate in shear and tension

(b) According to an ASD analysis, is the design adequate for a bearing-type connection?

(A) yes; adequate in bolt shear and bolt tension

(B) no; adequate in shear; inadequate in tension

(C) no; inadequate in shear; adequate in tension

(D) no; inadequate in shear; inadequate in tension

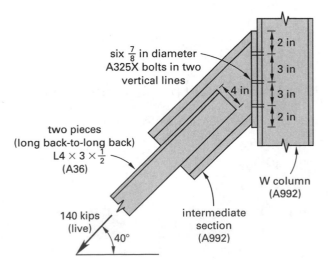

six ⁷/₈ in diameter
A325X bolts in two
vertical lines

2 in

3 in

4 in

3 in

2 in

two pieces
(long back-to-long back)
L4 × 3 × ¹/₂
(A36)

W column
(A992)

140 kips
(live)

40°

intermediate
section
(A992)

SOLUTIONS

1. (a) For A588 grade 50 steel, $F_y = 50$ ksi and $F_u = 70$ ksi.

The gross area of each plate is

$$A_g = bt = (6 \text{ in})\left(\tfrac{5}{8} \text{ in}\right) = 3.75 \text{ in}^2$$

The allowable (i.e., ASD) tensile load on the member based on the yielding criterion is

$$P_t = \frac{F_y A_g}{\Omega} = \frac{\left(50 \; \dfrac{\text{kips}}{\text{in}^2}\right)(3.75 \text{ in}^2)}{1.67}$$
$$= \boxed{112.3 \text{ kips} \quad (110 \text{ kips})}$$

The answer is (D).

(b) For $\tfrac{7}{8}$ in diameter bolts, the standard hole diameter for the purpose of analysis or design is

$$d_h = d_b + \tfrac{1}{8} \text{ in} = \tfrac{7}{8} \text{ in} + \tfrac{1}{8} \text{ in}$$
$$= 1 \text{ in}$$

The net area is

$$A_n = b_n t = \left(6 \text{ in} - (2)(1 \text{ in})\right)\left(\tfrac{5}{8} \text{ in}\right)$$
$$= 2.5 \text{ in}^2$$

Since $U = 1.0$, the effective net area is

$$A_e = U A_n = (1.0)(2.5 \text{ in}^2)$$
$$= 2.5 \text{ in}^2$$

The allowable (i.e., ASD) tensile load on the member based on the fracture criterion is

$$P_t = \frac{F_u A_e}{\Omega} = \frac{\left(70 \; \dfrac{\text{kips}}{\text{in}^2}\right)(2.5 \text{ in}^2)}{2.0}$$
$$= \boxed{87.5 \text{ kips} \quad (88 \text{ kips})}$$

The answer is (C).

(c) The allowable (i.e., ASD) bearing stress in the member is calculated from *AISC Specification* Eq. J3-6.

$$F_p = \frac{3.0 F_u}{\Omega} = \frac{(3.0)\left(70 \; \dfrac{\text{kips}}{\text{in}^2}\right)}{2}$$
$$= \boxed{105 \text{ ksi}}$$

The answer is (D).

(d) For each A325SC $\tfrac{7}{8}$ in diameter bolt in single shear, the allowable shear is 10.3 kips for ASD and 15.4 kips for LRFD [*AISC Manual* Table 7-3]. The allowable (i.e., ASD) shear capacity of the slip-critical connection is

$$(4)(10.3 \text{ kips}) = \boxed{41.2 \text{ kips} \quad (41 \text{ kips})}$$

The answer is (B).

(e) The tensile load capacity is the smallest of the results of parts (a), (b), and (d) (i.e., smallest of 112.5 kips, 87.5 kips, and 41.2 kips), which is $\boxed{41.2 \text{ kips} (41 \text{ kips}).}$

The answer is (A).

(f) For each A325N $\tfrac{7}{8}$ in diameter bolt in single shear, the allowable shear is 14.4 kips for ASD and 21.6 kips for LRFD [*AISC Manual* Table 7-1]. The allowable (i.e., ASD) shear capacity of the slip-critical connection is

$$(4)(14.4 \text{ kips}) = \boxed{57.6 \text{ kips} \quad (58 \text{ kips})}$$

The answer is (B).

(g) The allowable (i.e., ASD) tensile load capacity is the smallest of the results of parts (a), (b), and (f) (i.e., the smallest of 112.5 kips, 87.5 kips, and 57.6 kips), which is $\boxed{57.6 \text{ kips} (58 \text{ kips}).}$

The answer is (B).

(h) For each A325X $\tfrac{7}{8}$ in diameter bolt in single shear, the allowable shear is 18.0 kips [*AISC Manual* Table 7-1]. The shear capacity of the slip-critical connection is

$$(4)(18.0 \text{ kips}) = \boxed{72 \text{ kips}}$$

The answer is (C).

(i) The tensile load capacity is the smallest of the results of parts (a), (b), and (h) (i.e., the smallest of 112.5 kips, 87.5 kips, and 72 kips), which is $\boxed{72 \text{ kips.}}$

The answer is (B).

(j) From part (d) with a slip-critical connection, the tensile capacity of the member is 40.8 kips. The load per bolt is

$$P = \frac{41.2 \text{ kips}}{4} = 10.3 \text{ kips}$$

The actual bearing stress is

$$f_p = \frac{P}{dt} = \frac{10.3 \text{ kips}}{\left(\tfrac{7}{8} \text{ in}\right)\left(\tfrac{5}{8} \text{ in}\right)}$$
$$= \boxed{18.8 \text{ kips/in}^2 \quad (19 \text{ ksi})}$$

The answer is (B).

2. (a) Since the bolt group is symmetrical, the centroid is located between bolts 4 and 11.

For each bolt,

$$A_i = \left(\frac{\pi}{4}\right)(0.75 \text{ in})^2 = 0.4418 \text{ in}^2$$

$$J_i = \left(\frac{\pi}{32}\right)(0.75 \text{ in})^4 = 0.03106 \text{ in}^4$$

The solution is tabulated.

bolt	r (in)	r^2A (in^4)
3	3.354	4.97
2	6.185	16.90
1	9.124	36.78
	total	58.65
4	1.5	0.99

The total polar moment of inertia is

$$J = (14)(0.03106 \text{ in}^4) + (4)(58.65 \text{ in}^4) + (2)(0.99 \text{ in}^4)$$
$$= \boxed{237.0 \text{ in}^4 \quad (240 \text{ in}^4)}$$

The answer is (B).

(b) The torsional shear stress in bolts A, B, C, and D is

$$f_{v,t} = \frac{Tr}{J} = \frac{(30,000 \text{ lbf})(17 \text{ in})(9.124 \text{ in})}{237 \text{ in}^4}$$
$$= \boxed{19,634 \text{ lbf/in}^2 \quad (19,600 \text{ psi})}$$

The answer is (C).

(c) The vertical component of the torsional shear stress in bolts A, B, C, and D is

$$f_{v,t,y} = \left(\frac{1.5 \text{ in}}{9.124 \text{ in}}\right)\left(19,634 \frac{\text{lbf}}{\text{in}^2}\right)$$
$$= \boxed{3228 \text{ lbf/in}^2 \quad (3200 \text{ psi})}$$

The answer is (C).

(d) The horizontal component of the torsional shear stress in bolts A, B, C, and D is

$$f_{v,t,x} = \left(\frac{9 \text{ in}}{9.124 \text{ in}}\right)\left(19,634 \frac{\text{lbf}}{\text{in}^2}\right)$$
$$= \boxed{19,367 \text{ lbf/in}^2 \quad (19,400 \text{ psi})}$$

The answer is (D).

(e) The direct vertical shear stress in the bolts is

$$f_{v,d} = \frac{30,000 \text{ lbf}}{(14)(0.4418 \text{ in}^2)}$$
$$= \boxed{4850 \text{ lbf/in}^2 \quad (4900 \text{ psi})}$$

The answer is (C).

(f) The total shear stress in bolts A and B is

$$\sqrt{\left(19,367 \frac{\text{lbf}}{\text{in}^2}\right)^2 + \left(3228 \frac{\text{lbf}}{\text{in}^2} + 4850 \frac{\text{lbf}}{\text{in}^2}\right)^2}$$
$$= \boxed{20,984 \text{ lbf/in}^2 \quad (21,000 \text{ psi})}$$

The answer is (C).

(g) The total shear stress in bolts C and D is

$$\sqrt{\left(19,367 \frac{\text{lbf}}{\text{in}^2}\right)^2 + \left(3228 \frac{\text{lbf}}{\text{in}^2} - 4850 \frac{\text{lbf}}{\text{in}^2}\right)^2}$$
$$= \boxed{19,435 \text{ lbf/in}^2 \quad (19,400 \text{ psi})}$$

The answer is (D).

(h) Use *AISC Manual* Table 7-8, with a total number of fasteners in one vertical row, $n = 7$, and bolt pitch, $b = 3$ in. For a lever arm (eccentricity of load), $l = 16$ in, the coefficient, C, is 4.27. For $l = 18$ in, $C = 3.83$. By interpolation, for $l = 17$ in,

$$C = \frac{4.27 + 3.83}{2} = 4.05$$

From *AISC Manual* Table 7-3 for an A325SC $3/4$ in bolt in single shear (standard hole), the allowable (i.e., ASD) shear load, r_v, is 7.38 kips. The allowable eccentric load on the fastener group is

$$P = Cr_v = (4.05)(7.38 \text{ kips})\left(1000 \frac{\text{lbf}}{\text{kip}}\right)$$
$$= \boxed{29,889 \text{ lbf} \quad (30,000 \text{ lbf})}$$

The answer is (D).

Structural

(i) From part (h), $C = 4.05$. From *AISC Manual* Table 7-2 for an A325N $^3/_4$ in bolt in single shear (standard hole), the allowable (i.e., ASD) shear load, r_v, is 10.6 kips. The allowable eccentric load on the fastener group is

$$P = Cr_v = (4.05)(10.6 \text{ kips})\left(1000 \frac{\text{lbf}}{\text{kip}}\right)$$

$$= \boxed{42{,}930 \text{ lbf} \quad (42{,}900 \text{ lbf})}$$

The answer is (D).

(j) The actual eccentric load is 30,000 lbf. The allowable load for the A325SC bolt group is 29,889 lbf, which is less than 30,000 lbf. The allowable load for the A325N bolt group is 42,930 lbf, which is greater than 30,000 lbf. The connection is adequate for $\boxed{\text{A325N bolts only.}}$

The answer is (B).

3. (a) For the connection between the bracing member (double-angle) and the structural tee web, the bolts are in double shear. From *AISC Manual* Table 7-1, the capacity of one $^7/_8$ in diameter A325N bolt in double shear is 28.9 kips (ASD) and 43.3 kips (LRFD).

Bearing is on the web of the structural tee. For WT15 × 54, $t_w = 0.545$ in.

From *AISC Manual* Table 7-5, and assuming a 3 in bolt spacing in a standard type hole with $F_u = 58$ ksi, the capacity of one $^7/_8$ in diameter A325N bolt in bearing is 60.9 kips/in (ASD) and 91.4 kips/in (LRFD).

For ASD,

$$\left(60.9 \frac{\text{kips}}{\text{in}}\right)(0.545 \text{ in}) = 33.2 \text{ kips} > 28.9 \text{ kips}$$

For LRFD,

$$\left(91.4 \frac{\text{kips}}{\text{in}}\right)(0.545 \text{ in}) = 49.8 > 43.3 \text{ kips}$$

Bolt shear controls.

The required number of bolts is as follows.

For ASD,

$$N = \frac{140 \text{ kips}}{28.9 \frac{\text{kips}}{\text{bolt}}} = 4.84 \text{ bolts}$$

For LRFD,

$$N = \frac{(140 \text{ kips})(1.6)}{43.3 \frac{\text{kips}}{\text{bolt}}} = 5.17 \text{ bolts}$$

$\boxed{\text{Use six bolts with a 3 in pitch.}}$

(b) For the connection between the structural tee flange and the column flange, bolts are in combined shear and tension. The design of this connection is a trial-and-error procedure.

The total shear to be resisted by the bolt group is

$$V = \left(\tfrac{3}{5}\right)(140 \text{ kips}) = 84 \text{ kips}$$

The total tension to be resisted by the bolt group is

$$T = \left(\tfrac{4}{5}\right)(140 \text{ kips}) = 112 \text{ kips}$$

Assume eight $^7/_8$ in bolts. Area of $^7/_8$ in bolt is 0.601 in^2.

For ASD, the tension load per bolt is

$$\frac{\left(\tfrac{4}{5}\right)(140 \text{ kips})}{8} = 14.0 \text{ kips}$$

Shear load per bolt is

$$\frac{\left(\tfrac{3}{5}\right)(140 \text{ kips})}{8} = 10.5 \text{ kips}$$

For LRFD, the tension load per bolt is

$$\frac{\left(\tfrac{4}{5}\right)(140 \text{ kips})(1.6)}{8} = 22.4 \text{ kips}$$

Shear load per bolt is

$$\frac{\left(\tfrac{3}{5}\right)(140 \text{ kips})(1.6)}{8} = 16.8 \text{ kips}$$

Calculate $f_{v,\text{required}}$ shear stress per bolt.

For ASD,

$$\frac{10.5 \text{ kips}}{0.601 \text{ in}^2} = 17.5 \text{ kips/in}^2 \leq \frac{F_{nv}}{\Omega} \quad \text{[allowable]}$$

For LRFD,

$$\frac{16.8 \text{ kips}}{0.601 \text{ in}^2} = 28.0 \text{ kips/in}^2 \leq \phi F_{nv} \quad \text{[design]}$$

F_{nv} = nominal shear stress = 48 ksi from *AISC Specification* Table J3.2 assuming threads are not excluded from shear planes (ASD: $\Omega = 2.00$; LRFD: $\phi = 0.75$).

For combined tension and shear, for ASD,

$$\frac{F_{nv}}{\Omega} = \frac{48 \ \frac{\text{kips}}{\text{in}^2}}{2} = 24 \ \text{kips/in}^2$$

$$17.5 \ \text{kips/in}^2 < 24 \ \text{kips/in}^2 \quad [\text{OK}]$$

For LRFD,

$$\phi F_{nv} = (0.75)\left(48 \ \frac{\text{kips}}{\text{in}^2}\right) = 36 \ \text{kips/in}^2$$

$$28.0 \ \text{kips/in}^2 < 36 \ \text{kips/in}^2 \quad [\text{OK}]$$

Check combined tension and shear from *AISC Specification* Sec. J3. From AISC Eq. J3-3b, for ASD,

$$F'_{nt} = 1.3 F_{nt} - \left(\frac{\Omega F_{nt}}{F_{nv}}\right) F_v < F_{nt}$$

$$F_{nt} = 90 \ \text{kips/in}^2$$

From *AISC Specification* Table J3.2, the nominal tensile stress, F'_{nt}, is modified to include the effects of shearing stress.

For ASD,

$$F'_{nt} = (1.3)\left(90 \ \frac{\text{kips}}{\text{in}^2}\right) - \frac{(2)\left(90 \ \frac{\text{kips}}{\text{in}^2}\right)\left(17.5 \ \frac{\text{kips}}{\text{in}^2}\right)}{48 \ \frac{\text{kips}}{\text{in}^2}}$$

$$= 51.4 \ \text{kips/in}^2 < 90 \ \text{kips/in}^2 \quad [\text{OK}]$$

For LRFD, from AISC Eq. J3-3a,

$$F'_{nt} = 1.3 F_{nt} - \left(\frac{F_{nt}}{\phi F_{nv}}\right) F_v < F_{nt}$$

$$F_{nt} = 90 \ \text{kips/in}^2$$

Therefore,

$$F'_{nt} = (1.3)\left(90 \ \frac{\text{kips}}{\text{in}^2}\right) - \frac{\left(90 \ \frac{\text{kips}}{\text{in}^2}\right)\left(28 \ \frac{\text{kips}}{\text{in}^2}\right)}{(0.75)\left(48 \ \frac{\text{kips}}{\text{in}^2}\right)}$$

$$= 47 \ \text{kips/in}^2 < 90 \ \text{kips/in}^2 \quad [\text{OK}]$$

The allowable (i.e., ASD) tensile strength is

$$R_n = F'_{nt} A_b$$

$$= \left(51.4 \ \frac{\text{kips}}{\text{in}^2}\right)(0.601 \ \text{in}^2)$$

$$= 30.89 \ \text{kips}$$

$$\frac{R_n}{\Omega} = \frac{R_n}{2.0} = \frac{30.89 \ \text{kips}}{2.0} = 15.4 \ \text{kips}$$

$$15.4 \ \text{kips} > 14.0 \ \text{kips} \quad [\text{allowable}]$$

The design (i.e., LRFD) tensile stress is

$$R_n = F'_{nt} A_b$$

$$= \left(47 \ \frac{\text{kips}}{\text{in}^2}\right)(0.601 \ \text{in}^2)$$

$$= 28.25 \ \text{kips}$$

$$\phi R_n = 0.75 R_n$$

$$(0.75)(28.25 \ \text{kips}) = 21.2 \ \text{kips}$$

$$21.2 \ \text{kips} < 22.4 \ \text{kips} \quad [\text{design}]$$

Using the ASD methodology, 8 bolts would suffice. However, using the LRFD methodology 8 bolts would not be enough. Since the available strength of 21.2 kips is relatively close to the required strength of 22.4 kips, increasing 8 bolts to 10 bolts (i.e., 5 bolts on each side) is sufficient, using the LRFD methodology.

4. (a) Since the tee orientation is defined by a 3-4-5 triangle, the components of the applied live load can be determined from geometry. The tensile (horizontal, x-direction) component of the applied load is resisted by changes in bolt tension and is

$$R_t = \tfrac{4}{5} P = 0.8 P$$

The shear (vertical, y-direction) component of the applied load is resisted by friction between the faying surfaces and is

$$R_v = \tfrac{3}{5} P = 0.6 P$$

The required available connection slip resistance per bolt, $r_{a,v}$, is

$$r_{a,v} = \frac{R_v}{N_b} = \frac{0.6 P}{8} = 0.075 P$$

AISC Specification Eq. J3-4 and Eq. J3-5(b) are used to calculate the design slip resistance of a bolt that resists forces trying to simultaneously translate and separate two clamped surfaces. (The *AISC Manual* and *AISC Specification* are not consistent in the use of uppercase and lowercase r_n to designate the slip resistance per bolt.) For ASD, the actual tensile load on the entire bolt group, T_a, is $R_t = 0.8P$. The number of bolts, N_b, is 8. *AISC Specification* Sec. J3.8 prescribes the nominal values of some of the factors used in the equations. As described, the faying surfaces (clean, unpainted, normal mill scale) are class A. The mean slip coefficient, μ, for a class A surface is 0.35.

An A325X bolt is a standard A325 bolt with threads excluded (hence, the X) from the shear plane. The minimum pretension force, T_b, given in *AISC Specification* Table J3.1 for an A325 $^3/_4$ in bolt is 28 kips, although the bolts will be tensioned to higher values by normal installation methods. D_u is the ratio of the average installed bolt pretension and the minimum bolt pretension, T_b, equal to 1.13 for normal installation methods. The hole factor, h_{sc}, is 1.00 for standard sized holes. $N_s = 1$, as there is only one slip plane between the tee base and the WF flange.

Various values of the ASD factor of safety, Ω, are given in *AISC Specification* Sec. J3 for use with different loading modes. *AISC Specification* Sec. J3.7 specifies $\Omega = 2.00$ for bolts in combined tension and shear, while *AISC Specification* Sec. J3.8 specifies both $\Omega = 1.5$ and $\Omega = 1.76$ for slippage (i.e., faying) calculations. Since (by the problem statement) the connection allows slippage at factored service loadings, prevention of slip is a serviceability limit state, so $\Omega = 1.5$.

Combining *AISC Specification* Eq. J3-4 and Eq. J3-5(b), the allowable slip resistance per bolt is

$$\frac{r_n}{\Omega} = \left(\frac{\mu D_u h_{sc} T_b N_s}{\Omega}\right)\left(1 - \frac{1.5 T_u}{D_u T_b N_b}\right)$$
$$= \left(\frac{(0.35)(1.13)(1.00)(28 \text{ kips})(1)}{1.5}\right.$$
$$\left. \times \left(1 - \frac{(1.5)(0.8P)}{(1.13)(28 \text{ kips})(8)}\right)\right.$$
$$= 7.383 \text{ kips} - 0.035P \quad [P \text{ in kips}]$$

(The value of the first term, as defined by *AISC Specification* Eq. J3-4, can be read directly as 7.38 kips from *AISC Manual* Table 7.3.)

The required available slip resistance is $0.075P$.

$$r_{a,v} = \frac{r_n}{\Omega}$$
$$0.075P = 7.383 \text{ kips} - 0.035P$$
$$P = 67.12 \text{ kips} \quad (67 \text{ kips})$$

Check that the load is not limited by maximum available bolt tensile strength. From *AISC Manual* Table 7.2, the available tensile strength, $(r_n/\Omega)_{max}$, in a single A325 $^3/_4$ in bolt is 19.9 kips. The maximum connection loading as limited by bolt tension (the horizontal component of the load) is

$$(8 \text{ bolts})\left(\frac{5}{4}\right)(19.9 \text{ kips}) = 199 \text{ kips} > 67 \text{ kips}$$

Check that the load is not limited by maximum available bolt shear strength. From *AISC Manual* Table 7.1, using the "X" row for threads excluded from the shear plane, and the "S" row for single-shear, the available shear, r_n/Ω, in a single A325 $^3/_4$ in bolt is 13.3 kips. The maximum connection loading as limited by available bolt shear (the vertical component of the load) is

$$(8 \text{ bolts})\left(\frac{5}{3}\right)(13.3 \text{ kips}) = 177 \text{ kips} > 67 \text{ kips}$$

The maximum live load the connection can support is 67 kips.

The answer is (D).

(b) The components of the applied live load can be determined from geometry. Using a load factor of 1.6 for live loading, the factored tensile (horizontal, x-direction) component of the applied load is

$$R_{u,t} = (1.6)\tfrac{4}{5}P = 1.28P$$

The factored shear (vertical, y-direction) component of the applied load is

$$R_{u,v} = (1.6)\tfrac{3}{5}P = 0.96P$$

The required ultimate slip resistance per bolt, $r_{u,v}$, is

$$r_{u,v} = \frac{R_{u,v}}{N_b} = \frac{0.96P}{8} = 0.12P$$

Use *AISC Specification* Eq. J3-4 and Eq. J3-5(a). The factors are the same as for the ASD case except that the ultimate tensile load on the entire bolt group, T_u, which is $R_{u,t} = 1.28P$, is used. *AISC Specification* Sec. J3.8 specifies both $\phi = 1.00$ and $\phi = 0.85$ for slippage calculations. Since (by the problem statement) the connection allows slippage at factored service loadings, prevention of slip is a serviceability limit state, so $\phi = 1.00$.

Combining *AISC Specification* Eq. J3-4 and Eq. J3-5(a), the design slip resistance per bolt is

$$\phi r_n = \phi(\mu D_u h_{sc} T_b N_s)\left(1 - \frac{T_u}{D_u T_b N_b}\right)$$

$$= (1.0)\big((0.35)(1.13)(1.00)(28 \text{ kips})(1)\big)$$

$$\times \left(1 - \frac{1.28P}{(1.13)(28 \text{ kips})(8)}\right)$$

$$= 11.074 \text{ kips} - 0.056P \quad [P \text{ in kips}]$$

(The value of the first term, as defined by *AISC Specification* Eq. J3-4, can be read directly as 11.1 kips from *AISC Manual* Table 7.3.)

The required ultimate slip resistance, $R_{u,v}$, is $0.12P$.

$$R_{u,v} = \phi R_n$$

$$0.12P = 11.074 \text{ kips} - 0.056P$$

$$P = \boxed{62.92 \text{ kips} \quad (63 \text{ kips})}$$

The answer is (C).

(c) Tensile and shear bolt areas are based on the nominal shank diameter. The area of a single bolt is

$$A_b = \frac{\pi}{4}d^2 = \left(\frac{\pi}{4}\right)\left(\tfrac{3}{4} \text{ in}\right)^2 = 0.4418 \text{ in}^2$$

(This can also be read as 0.442 in² from *AISC Manual* Table 7.1.)

An A325X bolt is a standard A325 bolt with threads excluded (hence, the X) from the shear plane. From *AISC Specification* Table J3.2, for A325 bolts when threads are excluded from the shear plane, $F_{nt} = 90$ ksi, and $F_{nv} = 60$ ksi. *AISC Specification* Sec. 7 specifies $\Omega = 2.00$ for ASD bolts in combined shear and tension.

Shear basis

Using lowercase r to designate per-bolt values, the actual (required) shear stress, f_v, in a single bolt in a bearing-type connection is

$$f_v = \frac{r_v}{A_b} = \frac{R_v}{N_b A_b} = \frac{0.6P}{(8)(0.4418 \text{ in}^2)}$$

$$= 0.1698P$$

$$f_v < \frac{F_{nv}}{\Omega}$$

$$0.1698P < \frac{60 \ \dfrac{\text{kips}}{\text{in}^2}}{2}$$

$$P < 176.7 \text{ kips}$$

Tensile basis

From *AISC Specification* Sec. 7, combining *AISC Specification* Eq. J3-2 and Eq. J3-3(b), the nominal tensile strength of a shear- and tension-loaded bolt in bearing-type connection is

$$r_n = F'_{nt}A_b = \left(1.3F_{nt} - \frac{\Omega F_{nt}}{F_{nv}}f_v\right)A_b < F_{nt}A_b$$

$$\frac{r_n}{\Omega} = \left(1.3F_{nt} - \frac{\Omega F_{nt}}{F_{nv}}f_v\right)\frac{A_b}{\Omega} < \frac{F_{nt}A_b}{\Omega}$$

$$= \left(\begin{array}{c}(1.3)\left(90 \ \dfrac{\text{kips}}{\text{in}^2}\right) \\[2ex] -\left(\dfrac{(2.00)\left(90 \ \dfrac{\text{kips}}{\text{in}^2}\right)}{60 \ \dfrac{\text{kips}}{\text{in}^2}}\right)\left(\dfrac{0.075P}{0.4418 \text{ in}^2}\right)\end{array}\right)$$

$$\times \left(\frac{0.4418 \text{ in}^2}{2.00}\right) < \left(90 \ \frac{\text{kips}}{\text{in}^2}\right)\left(\frac{0.4418 \text{ in}^2}{2.00}\right)$$

$$= 25.85 \text{ kips} - 0.1125P < 19.88 \text{ kips}$$

The required available tensile strength is

$$r_t = \frac{R_t}{N_b} = \frac{0.8P}{8} = 0.1P$$

$$r_t = \frac{r_n}{\Omega}$$

$$0.1P = 25.85 \text{ kips} - 0.1125P$$

$$P = 121.6 \text{ kips} \quad (120 \text{ kips})$$

Since 121.6 kips < 176.7 kips, P is limited by tension.

Confirm:

$$r_n < F_{nt}A_b$$

$$\frac{r_n}{\Omega} < \frac{F_{nt}A_b}{\Omega}$$

$$\begin{array}{cc} 25.85 \text{ kips} & 12.17 \text{ kips} \\ - (0.1125)(121.6 \text{ kips}) & < 19.88 \text{ kips} \quad [\text{OK}] \end{array}$$

$$= $$

$$\boxed{\text{The maximum live load is 120 kips.}}$$

The answer is (B).

(d) From part (c), the area of a single bolt is 0.4418 in².

From *AISC Specification* Table J3.2, for A325 bolts when threads are excluded from the shear plane, $F_{nt} = 90$ ksi, and $F_{nv} = 60$ ksi. *AISC Specification* Sec. 7 specifies $\phi = 0.75$ for LRFD for combined shear and tension.

Shear basis

The shear stress in a single bolt is

$$f_v = \frac{r_{u,v}}{A_b} = \frac{0.12P}{A_b} = \frac{0.12P}{0.4418 \text{ in}^2}$$

$$= 0.2716P \text{ kips/in}^2$$

$$f_v < \phi F_{nv}$$

$$0.2716P \, \frac{\text{kips}}{\text{in}^2} < (0.75)\left(60 \, \frac{\text{kips}}{\text{in}^2}\right)$$

$$P < 165.7 \text{ kips}$$

Tension basis

From *AISC Specification* Sec. 7, combining *AISC Specification* Eq. J3-2 and Eq. J3-3(a), the tensile strength of a shear- and tension-loaded bolt in bearing-type connection is

$$r_n = F'_{nt}A_b = \left(1.3F_{nt} - \frac{F_{nt}}{\phi F_{nv}}f_v\right)A_b < F_{nt}A_b$$

$$\phi r_n = \phi\left(1.3F_{nt} - \frac{\Omega F_{nt}}{F_{nv}}f_v\right)A_b < \phi F_{nt}A_b$$

$$= (0.75)\left(\begin{array}{c}(1.3)\left(90 \, \dfrac{\text{kips}}{\text{in}^2}\right)\\[2mm] -\left(\dfrac{90 \, \frac{\text{kips}}{\text{in}^2}}{(0.75)\left(60 \, \frac{\text{kips}}{\text{in}^2}\right)}\right)\left(\dfrac{0.075P}{0.4418 \text{ in}^2}\right)\end{array}\right)$$

$$\times (0.4418 \text{ in}^2) < (0.75)\left(90 \, \frac{\text{kips}}{\text{in}^2}\right)(0.4418 \text{ in}^2)$$

$$= 38.77 \text{ kips} \quad 0.1125P < 29.82 \text{ kips}$$

The required ultimate (factored) tensile strength is

$$r_{u,t} = \frac{1.6R_t}{N_b} = \frac{(1.6)(0.8P)}{8} = 0.16P$$

$$r_{u,t} = \phi r_n$$

$$0.16P = 38.77 \text{ kips} - 0.1125P$$

$$P = 142.3 \text{ kips} \quad (140 \text{ kips})$$

Since 142.3 kips < 165.7 kips, P is limited by tension. Confirm:

$$r_n < F_{nt}A_b$$

$$\phi r_n < \phi F_{nt}A_b$$

$$= 38.77 \text{ kips} - (0.1125)(142.3 \text{ kips})$$

$$= 16.01 \text{ kips} < 29.82 \text{ kips} \quad [\text{OK}]$$

The maximum live load is 140 kips.

The answer is (B).

5. (a) The tensile (horizontal) component of the applied load is

$$R_t = (140 \text{ kips})\cos 40° = 107.2 \text{ kips}$$

The shear (vertical) component of the applied load is

$$R_v = (140 \text{ kips})\sin 40° = 90.0 \text{ kips}$$

The required available connection slip resistance per bolt, $r_{a,v}$, is

$$r_{a,v} = \frac{R_v}{N_b} = \frac{90 \text{ kips}}{6} = 15 \text{ kips}$$

AISC Specification Eq. J3-4 and Eq. J3-5(b) are used to calculate the design slip resistance of a bolt that resists forces trying to simultaneously translate and separate two clamped surfaces. For ASD, the actual tensile load on the entire bolt group, T_a, is $R_t = 107.2$ kips. The number of bolts, N_b, is 6. Clean, unpainted, normal mill scale faying surfaces are class A, and the mean slip coefficient, μ, is 0.35.

An A325X bolt is a standard A325 bolt with threads excluded (hence, the X) from the shear plane. The minimum pretension force, T_b, given in *AISC Specification* Table J3.1 for an A325 $^7/_8$ in bolt is 39 kips. D_u is the ratio of the average installed bolt pretension and the minimum bolt pretension, T_b, equal to 1.13 for normal installation methods. The hole factor, h_{sc}, is 1.00 for standard sized holes. $N_s = 1$, as there is only one slip plane between the intermediate member's bolt base plate and the WF flange.

Since the connection permits slippage at factored service loadings, prevention of slip is a serviceability limit state, so $\Omega = 1.5$.

Combining *AISC Specification* Eq. J3-4 and Eq. J3-5(b), the allowable slip resistance per bolt is

$$\frac{r_n}{\Omega} = \left(\frac{\mu D_u h_{sc} T_b N_s}{\Omega}\right)\left(1 - \frac{1.5 T_a}{D_u T_b N_b}\right)$$

$$= \left(\frac{(0.35)(1.13)(1.00)(39 \text{ kips})(1)}{1.5}\right)$$

$$\times \left(1 - \frac{(1.5)(107.2 \text{ kips})}{(1.13)(39 \text{ kips})(6)}\right)$$

$$= 4.03 \text{ kips}$$

Since the required available connection slip resistance per bolt, $r_{a,v}$, is 15 kips, the connection is inadequate in slip resistance. Other checks could be made, but option D is the only option with inadequate slip resistance.

The answer is (D).

Structural

(b) From *AISC Manual* Table 7.1, the cross-sectional area of a single $^7/_8$ in diameter bolt, A_b, is 0.601 in^2. From *AISC Specification* Table J3.2, for A325 bolts when threads are excluded from the shear plane, $F_{nt} = 90$ ksi, and $F_{nv} = 60$ ksi. *AISC Specification* Sec. 7 specifies $\Omega = 2.00$ for ASD for combined tension and shear.

Shear basis

From part (a), the vertical component of the applied force is 90 kips. The shear stress in a single bolt in a bearing-type connection is

$$f_v = \frac{r_v}{A_b} = \frac{R_v}{N_b A_b} = \frac{90 \text{ kips}}{(6)(0.601 \text{ in}^2)}$$

$$= 24.96 \text{ kips/in}^2$$

$$f_v < \frac{F_{nv}}{\Omega}$$

$$24.96 \ \frac{\text{kips}}{\text{in}^2} < \frac{60 \ \dfrac{\text{kips}}{\text{in}^2}}{2} = 30 \text{ kips/in}^2$$

> Since 24.96 ksi $<$ 30 ksi, the bolts are adequate in shear.

Tension basis

From *AISC Specification* Sec. 7, combining *AISC Specification* Eq. J3-2 and Eq. J3-3(b), the tensile strength of a shear- and tension-loaded bolt in bearing-type connection is

$$r_n = F'_{nt} A_b = \left(1.3 F_{nt} - \frac{\Omega F_{nt}}{F_{nv}} f_v \right) A_b < F_{nt} A_b$$

$$\frac{r_n}{\Omega} = \left(1.3 F_{nt} - \frac{\Omega F_{nt}}{F_{nv}} f_v \right) \frac{A_b}{\Omega} < \frac{F_{nt} A_b}{\Omega}$$

$$= \left(\begin{array}{c} (1.3)\left(90 \ \dfrac{\text{kips}}{\text{in}^2} \right) \\[2mm] - \left(\dfrac{(2.00)\left(90 \ \dfrac{\text{kips}}{\text{in}^2} \right)}{60 \ \dfrac{\text{kips}}{\text{in}^2}} \right)\left(24.96 \ \dfrac{\text{kips}}{\text{in}^2} \right) \end{array} \right)$$

$$\times \left(\frac{0.601 \text{ in}^2}{2.00} \right) < \left(90 \ \frac{\text{kips}}{\text{in}^2} \right)\left(\frac{0.601 \text{ in}^2}{2.00} \right)$$

$$= 12.66 \text{ kips} < 27.05 \text{ kips} \quad [\text{OK}]$$

From part (a), the required available connection slip resistance per bolt, $r_{a,v}$, is

$$r_v = \frac{R_v}{N_b} = \frac{90 \text{ kips}}{6} = 15 \text{ kips}$$

> Since 12.66 kips $<$ 15 kips, the bolts are inadequate in tension.

The answer is (B).

66 Structural Steel: Welding

PRACTICE PROBLEMS

1. Determine the size of the fillet weld required to connect the 1 in thick plate bracket to the column flange as shown. A572 steel is used for the bracket, and A992 steel is used for the column. Assume SMAW process with E70XX electrodes.

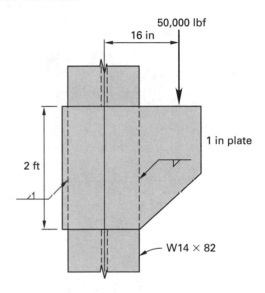

2. The connection between the column flange and plate is made with a $^3/_8$ in size fillet weld using E70 electrodes. Determine the capacity, P, of the connection if the process is (a) SMAW and (b) SAW.

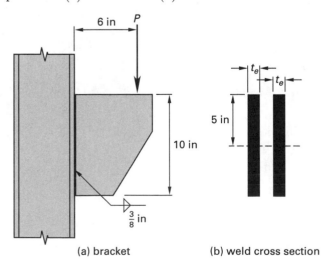

(a) bracket   (b) weld cross section

3. (*Time limit: one hour*) The fillet-welded lap joint shown is to be designed as a connection between the long leg of an L5 × 3 × $^1/_2$ tension member of A36 steel and a $^1/_2$ in thick gusset plate. The full strength of the angle is to be developed, minimizing the effect of eccentricity. E70 electrodes will be used. The gusset plate does not govern the design.

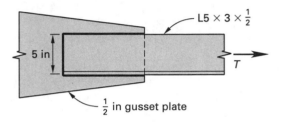

(a) The axial tensile capacity of the angle member is most nearly

    (A) 72 kips

    (B) 81 kips

    (C) 89 kips

    (D) 110 kips

(b) The minimum and maximum sizes of fillet weld that can be used are most nearly

    (A) $^1/_8$ in and $^3/_8$ in

    (B) $^3/_{16}$ in and $^7/_{16}$ in

    (C) $^1/_4$ in and $^1/_2$ in

    (D) $^5/_{16}$ in and $^5/_8$ in

For parts (c) through (g), assume an SMAW process with the minimum permissible weld size is used.

(c) The strength of the fillet weld per unit length is most nearly

    (A) 2.0 kips/in

    (B) 2.3 kips/in

    (C) 2.8 kips/in

    (D) 3.0 kips/in

(d) The load carried by the end (transverse) weld, P_2, is most nearly

(A) 11.4 kips

(B) 12.7 kips

(C) 13.9 kips

(D) 14.3 kips

(e) The load, P_3, carried by the bottom longitudinal weld is most nearly

(A) 33 kips

(B) 40 kips

(C) 46 kips

(D) 51 kips

(f) The load, P_1, carried by the top longitudinal weld is most nearly

(A) 17 kips

(B) 21 kips

(C) 26 kips

(D) 31 kips

(g) The required lengths of the top and bottom longitudinal welds are, respectively, most nearly

(A) 7 in and 17 in

(B) 8 in and 16 in

(C) 7 in and 18 in

(D) 8 in and 18 in

For parts (h) through (j), assume an SAW process with the minimum permissible weld size is used.

(h) The load, P_3, carried by the bottom longitudinal weld is most nearly

(A) 33 kips

(B) 38 kips

(C) 40 kips

(D) 43 kips

(i) The load, P_1, carried by the top longitudinal weld is most nearly

(A) 19 kips

(B) 21 kips

(C) 26 kips

(D) 31 kips

(j) The required lengths of the top and bottom longitudinal welds are, respectively, most nearly

(A) 5 in and 11 in

(B) 6 in and 11 in

(C) 5 in and 12 in

(D) 6 in and 12 in

4. (*Time limit: one hour*) The adequacy of a $^3/_8$ in fillet welded connection between a bracket and a column is to be checked. E70 electrodes are used. The column and the bracket do not control the design.

For parts (a) through (g), assume an effective throat thickness, t_e, of 1 in.

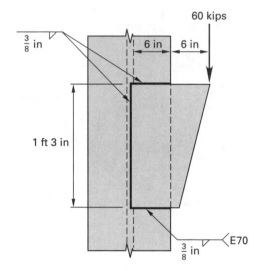

(a) The distance measured from the left end of the bracket plate to the centroid of the weld group is most nearly

(A) 1.1 in

(B) 1.3 in

(C) 1.9 in

(D) 2.1 in

(b) The polar moment of inertia of the weld group is most nearly

(A) 830 in^4

(B) 970 in^4

(C) 1100 in^4

(D) 1300 in^4

(c) The torsional moment acting on the weld group is most nearly

    (A) 410 in-kip

    (B) 470 in-kip

    (C) 520 in-kip

    (D) 640 in-kip

(d) The maximum stress in weld due to the torsional moment is most nearly

    (A) 3.5 ksi

    (B) 4.2 ksi

    (C) 5.4 ksi

    (D) 6.8 ksi

(e) The horizontal and vertical components of the maximum stress in weld due to the torsional moment are, respectively, most nearly

    (A) 3.9 ksi and 3.4 ksi

    (B) 4.6 ksi and 2.8 ksi

    (C) 5.0 ksi and 2.1 ksi

    (D) 5.7 ksi and 1.9 ksi

(f) The stress in weld due to direct shear is most nearly

    (A) 1.7 ksi

    (B) 2.2 ksi

    (C) 2.9 ksi

    (D) 3.2 ksi

(g) The resultant maximum stress (combination of torsional and direct shear stresses) in the weld group is most nearly

    (A) 3.5 ksi

    (B) 4.8 ksi

    (C) 5.8 ksi

    (D) 6.8 ksi

(h) Assuming that the SMAW process is used, the actual resistance per unit length of weld is most nearly

    (A) 3.8 kips/in

    (B) 4.7 kips/in

    (C) 5.6 kips/in

    (D) 6.8 kips/in

(i) Assuming that the SAW process is used, the shear resistance per unit length of weld is most nearly

    (A) 3.8 kips/in

    (B) 4.9 kips/in

    (C) 6.3 kips/in

    (D) 7.9 kips/in

(j) The connection is adequate for

    (A) both SMAW and SAW processes

    (B) neither SMAW nor SAW processes

    (C) a SMAW process only

    (D) a SAW process only

5. (*Time limit: one hour*) Two L4 × 3 × ¹⁄₂ angles are welded back to back to an intermediate wide-flange section that is bolted to a column as shown. All pieces are A36 steel except for the A992 wide flange. E70 electrodes are used to form the maximum weld size permitted. Specify the lengths and locations of the welds used to connect the angles to the intermediate section using a balanced configuration.

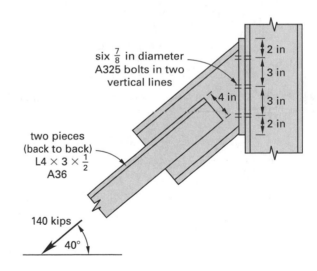

SOLUTIONS

1. *Elastic Analysis per AISC Manual Part 8:*

step 1: Assume the weld has thickness t. (All dimensions are in inches.)

step 2: Locate the centroid by inspection.

$$x_c = y_c = 0$$

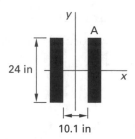

step 3: $I_x = (2)\left(\dfrac{t(24 \text{ in})^3}{12}\right) = 2304t$

step 4: From *AISC Manual* Table 1-1,

$$b_f = 10.1 \text{ in} \quad [\text{for W14} \times 82]$$

$$I_y = (2)\left(\dfrac{24t^3}{12} + (24t)\left(\dfrac{10.1 + t}{2}\right)^2\right)$$

$$\approx 1224t \quad [\text{assuming } t^2 = t^3 = 0]$$

step 5: $I_p = I_x + I_y = 2304t + 1224t = 3528t$

step 6: The maximum shear stress will occur at point A. The radial distance to point A is

$$c = \sqrt{\left(\dfrac{10.1 \text{ in}}{2}\right)^2 + (12 \text{ in})^2} = 13.02 \text{ in}$$

step 7: The applied or required moment (i.e., ASD) is

$$M = Pe = (50,000 \text{ lbf})(16 \text{ in})$$

$$= 800,000 \text{ in-lbf}$$

step 8: Use Eq. 66.13. The torsional shear stress is

$$r_m = \dfrac{Mc}{J} = \dfrac{(800,000 \text{ in-lbf})(13.02 \text{ in})}{3528t}$$

$$= 2952/t$$

$$r_{m,x} = \left(\dfrac{12 \text{ in}}{13.02 \text{ in}}\right)\left(\dfrac{2952}{t}\right) = 2721/t$$

$$[\text{to the right}]$$

$$r_{m,y} = \left(\dfrac{10.1 \text{ in}}{(2)(13.02 \text{ in})}\right)\left(\dfrac{2952}{t}\right)$$

$$= 1145/t \quad [\text{down}]$$

step 9: Use Eq. 66.14. The direct shear stress is

$$r_p = \dfrac{P}{A} = \dfrac{50,000 \text{ lbf}}{(2)(24t)}$$

$$= 1042/t \quad [\text{down}]$$

step 10: Use Eq. 66.15. The resultant shear stress at point A is

$$r_a = \left(\dfrac{1}{t}\right)\sqrt{(2721)^2 + (1145 + 1042)^2}$$

$$= 3491/t$$

step 11: Use Eq. 66.2. The allowable shear stress (i.e., ASD) in the weld is

$$\dfrac{F_w}{\Omega} = \dfrac{0.6 F_{u,\text{rod}}}{2} = \dfrac{(0.6)\left(70,000 \, \frac{\text{lbf}}{\text{in}^2}\right)}{2}$$

$$= 21,000 \text{ lbf/in}^2$$

Equating the required and allowable stresses,

$$\dfrac{3491}{t} = 21,000 \text{ lbf/in}^2$$

$$t = 0.166 \text{ in}$$

step 12: Use Eq. 66.1. The weld size is

$$w = \dfrac{0.166 \text{ in}}{0.707} = 0.23 \text{ in}$$

The flange thickness of a W14 $\times$ 82 is 0.855 in, and the thickness of the bracket plate is 1 in. Refer to Table 66.1. Since the thickness of the thicker part is over $^3/_4$ in, use a weld size of $\boxed{^5/_{16} \text{ in.}}$

AISC Method

Refer to *AISC Manual* Part 8, Table 8-4 (Eccentrically Loaded Weld Groups, Angle = 0°).

$$l = (2 \text{ ft})\left(12 \, \dfrac{\text{in}}{\text{ft}}\right) = 24 \text{ in}$$

$$e_x = al = 16 \text{ in}$$

$$a = \dfrac{16}{l} = \dfrac{16 \text{ in}}{24 \text{ in}} = 0.667$$

$$kl = b_f = 10.1 \text{ in}$$

$$k = \dfrac{10.1}{l} = \dfrac{10.1 \text{ in}}{24 \text{ in}} = 0.421$$

Use double interpolation with $a = 0.667$ and $k = 0.421$ to get $C = 1.9874$. From AISC Table 8-3, $C_1 = 1.0$ for E70 electrodes. Use Eq. 66.16.

From *AISC Specification* Table J2.5, $\Omega = 2.0$ for ASD, and $\phi = 0.75$ for LRFD.

For ASD,

$$D_m = \frac{\Omega P_a}{C C_1 l}$$
$$= \frac{(2.0)(50 \text{ kips})}{(1.9874)(1)(24 \text{ in})}$$
$$= 2.097 \quad \left[\text{say, } ^3/_{16} \text{ in}\right]$$

For LRFD,

$$D_m = \frac{P_u}{\phi C C_1 l}$$
$$= \frac{(1.6)(50 \text{ kips})}{(0.75)(1.9874)(1)(24)}$$
$$= 2.236 \quad \left[\text{say, } ^3/_{16} \text{ in}\right]$$

From Table 66.1, since the thickness of the thicker part is over $^3/_4$ in, the minimum weld size is $\boxed{^5/_{16} \text{ in.}}$

2. *Elastic Method:*

This is an eccentrically loaded welded connection where the weld group is subjected to combined shear and bending stresses.

$$w = \tfrac{3}{8} \text{ in}$$
$$L = 10 \text{ in}$$

For E70 electrodes, $F_u = 70 \text{ kips/in}^2$.

(a) *SMAW Process:* (All forces arc in kips. All dimensions are in inches.) Use Eq. 66.1.

$$t_e = 0.707w = (0.707)\left(\tfrac{3}{8} \text{ in}\right) = 0.265 \text{ in}$$

From Eq. 66.17, the shear stress is

$$r_p = \frac{P}{2Lt_e} = \frac{P}{(2)(10 \text{ in})(0.265 \text{ in})}$$
$$= 0.189P$$

The moment of inertia of the weld group is

$$I = \frac{2t_e L^3}{12} = \frac{(2)(0.265 \text{ in})(10 \text{ in})^3}{12}$$
$$= 44.167 \text{ in}^4$$

From Eq. 66.18, the maximum bending stress is

$$r_m = \frac{Mc}{I} = \frac{Pec}{I} = \frac{P(6 \text{ in})(5 \text{ in})}{44.167 \text{ in}^4}$$
$$= 0.679P$$

From Eq. 66.19, the maximum resultant stress is

$$r_a = \sqrt{r_p^2 + r_m^2}$$
$$= \sqrt{(0.189P)^2 + (0.679P)^2}$$
$$= 0.705P$$

The available shear stress for the weld is

$$F_v = \frac{0.6F_u}{\Omega} = \frac{(0.6)\left(70 \dfrac{\text{kips}}{\text{in}^2}\right)}{2.0} = 21 \text{ kips/in}^2$$

Equating the resultant stress to the allowable (i.e., ASD) stress,

$$0.705P = 21 \text{ kips/in}^2$$
$$P = \boxed{29.8 \text{ kips}}$$

(b) *SAW Process:* (All forces are in kips. All dimensions are in inches.) Use Eq. 66.1.

$$t_e = w = \tfrac{3}{8} \text{ in} = 0.375 \text{ in}$$

From Eq. 66.17, the shear stress is

$$r_p = \frac{P}{2Lt_e} = \frac{P}{(2)(10 \text{ in})(0.375 \text{ in})}$$
$$= 0.133P$$

The moment of inertia of the weld group is

$$I = \frac{2t_e L^3}{12} = \frac{(2)(0.375 \text{ in})(10 \text{ in})^3}{12}$$
$$= 62.5 \text{ in}^4$$

From Eq. 66.18, the maximum bending stress is

$$r_m = \frac{Mc}{I} = \frac{Pec}{I} = \frac{P(6 \text{ in})(5 \text{ in})}{62.5 \text{ in}^4}$$
$$= 0.48P$$

From Eq. 66.19, the maximum resultant stress is

$$r_a = \sqrt{r_p^2 + r_m^2}$$
$$= \sqrt{(0.133P)^2 + (0.48P)^2}$$
$$= 0.498P$$

Use Eq. 66.2. The allowable shear stress for the weld is

$$F_v = \frac{0.6F_u}{\Omega} = \frac{(0.6)\left(70\ \dfrac{\text{kips}}{\text{in}^2}\right)}{2.0} = 21\ \text{kips/in}^2$$

Equating the resultant stress to the allowable (i.e., ASD) stress,

$$0.498P = 21\ \text{kips/in}^2$$
$$P = \boxed{42.2\ \text{kips}}$$

AISC Method

Refer to *AISC Manual* Part 8, Table 8-4 (Special Case: Load Not in Plane of Weld Group).

$$l = (2)(5\ \text{in}) = 10\ \text{in}$$
$$e_x = al = 6\ \text{in}$$
$$a = \frac{6}{l} = \frac{6\ \text{in}}{10\ \text{in}} = 0.6$$
$$k = 0$$
$$C = 2.00 \quad \text{[table value]}$$

$C_1 = 1.0$ for E70 electrodes.

(a) *SMAW Process:*

$$t_e = 0.707w = (0.707)\left(\tfrac{3}{8}\ \text{in}\right) = 0.265\ \text{in}$$
$$D = \frac{t_e}{\frac{1}{16}\ \text{in}} = \frac{0.265\ \text{in}}{\frac{1}{16}\ \text{in}} = 4.24$$

For ASD,

$$P_a = \frac{DCC_1 l}{\Omega} = \frac{(4.24)(2.0)(1)(10\ \text{in})}{2}$$
$$= 42.4\ \text{kips}$$

For LRFD,

$$P_u = D\phi CC_1 l = (4.24)(0.75)(2.0)(1)(10\ \text{in})$$
$$= 63.6\ \text{kips}$$

(b) *SAW Process:*

$$t_e = w = \tfrac{3}{8}\ \text{in}$$
$$D = \frac{t_e}{\frac{1}{16}\ \text{in}} = \frac{\frac{3}{8}\ \text{in}}{\frac{1}{16}\ \text{in}} = 6$$

For ASD,

$$P_a = D_m CC_1 l = \frac{(6.0)(2.0)(1)(10\ \text{in})}{2}$$
$$= \boxed{60\ \text{kips}}$$

For LRFD,

$$P_u = D_m \phi CC_1 l = (6.0)(0.75)(2.0)(1)(10\ \text{in})$$
$$= \boxed{90\ \text{kips}}$$

3. (a) For an A36 steel angle member, $F_y = 36\ \text{kips/in}^2$ and $F_u = 58\ \text{kips/in}^2$.

From AISC Table 1-7, for L5 × 3 × $^1\!/_2$, $A_g = 3.75\ \text{in}^2$

From Table 60.2, since $l > 2w$, $U = 1.00$.

By Eq. 60.8,

$$A_e = UA_g = (1.00)(3.75\ \text{in}^2)$$
$$= 3.75\ \text{in}^2$$

The available strength to cause yielding is

$$P_t = \frac{F_y A_g}{\Omega} = \frac{\left(36\ \dfrac{\text{kips}}{\text{in}^2}\right)(3.75\ \text{in}^2)}{1.67}$$
$$= \boxed{80.8 \quad (81\ \text{kips}) \quad \text{[controls]}}$$

The force to cause fracture is

$$P_t = \frac{F_u A_e}{\Omega} = \frac{\left(58\ \dfrac{\text{kips}}{\text{in}^2}\right)(3.75\ \text{in}^2)}{2.0}$$
$$= 108.75\ \text{kips}$$

The answer is (B).

(b) From Table 66.1, the minimum size fillet weld is $\boxed{^3\!/_{16}\ \text{in}}$ for the thicker plate size of $^1\!/_2$ in. The maximum size of the fillet weld is

$$\tfrac{1}{2}\ \text{in} - \tfrac{1}{16}\ \text{in} = \boxed{\tfrac{7}{16}\ \text{in}}$$

The answer is (B).

(c) For a SMAW process with $F_{u,\text{rod}} = 70\ \text{kips/in}^2$,

$$w = \tfrac{3}{16}\ \text{in} \quad \text{[minimum size of weld]}$$

From Eq. 66.1 and Eq. 66.2,

$$
\begin{aligned}
R_w &= \frac{(0.60)(0.707w)F_{u,\text{rod}}}{\Omega} \\
&= \frac{(0.60)(0.707)\left(\frac{3}{16}\text{ in}\right)\left(70\ \dfrac{\text{kips}}{\text{in}^2}\right)}{2.0} \\
&= \boxed{2.78\text{ kips/in}\quad(2.8\text{ kips/in})}
\end{aligned}
$$

The answer is (C).

Refer to Fig. 66.5 for (d) through (j).

(d) $l_2 = d = 5$ in

From Eq. 66.8, the load carried by the end (transverse) weld is

$$
P_2 = R_w l_2 = \left(2.78\ \frac{\text{kips}}{\text{in}}\right)(5\text{ in}) = \boxed{13.9\text{ kips}}
$$

The answer is (C).

(e) From AISC Table 1-7, for L5 × 3 × ½, $y = 1.75$ in.

From part (a), $T = 81$ kips. From Eq. 66.7, the load carried by the bottom longitudinal weld is

$$
\begin{aligned}
P_3 &= T\left(1 - \frac{y}{d}\right) - \frac{P_2}{2} \\
&= (81\text{ kips})\left(1 - \frac{1.75\text{ in}}{5\text{ in}}\right) - \frac{13.9\text{ kips}}{2} \\
&= \boxed{45.7\text{ kips}\quad(46\text{ kips})}
\end{aligned}
$$

The answer is (C).

(f) From Eq. 66.9, the load carried by the top longitudinal weld is

$$
\begin{aligned}
P_1 &= T - P_2 - P_3 \\
&= 81\text{ kips} - 13.9\text{ kips} - 45.7\text{ kips} \\
&= \boxed{21.4\text{ kips}\quad(21\text{ kips})}
\end{aligned}
$$

The answer is (B).

(g) From Eq. 66.10, the required length (i.e., ASD) of the top longitudinal weld is

$$
\begin{aligned}
l_1 &= \frac{P_1}{R_w} = \frac{21.4\text{ kips}}{2.78\ \dfrac{\text{kips}}{\text{in}}} \\
&= \boxed{7.7\text{ in}\quad(8\text{ in})}
\end{aligned}
$$

From Eq. 66.11, the required length (i.e., ASD) of the bottom longitudinal weld is

$$
\begin{aligned}
l_3 &= \frac{P_3}{R_w} = \frac{45.7\text{ kips}}{2.78\ \dfrac{\text{kips}}{\text{in}}} \\
&= \boxed{16.44\text{ in}\quad(16\text{ in})}
\end{aligned}
$$

The answer is (B).

(h) For a SAW process with $F_{u,\text{rod}} = 70$ kips/in^2,

$$
w = \tfrac{3}{16}\text{ in}\quad[\text{minimum size of weld}]\ \le \tfrac{3}{8}\text{ in}
$$

From Eq. 66.1,

$$
t_e = w = \tfrac{3}{16}\text{ in}
$$

From Eq. 66.1 and Eq. 66.2,

$$
\begin{aligned}
R_w &= \frac{0.60 t_e F_{u,\text{rod}}}{\Omega} = \frac{(0.60)\left(\frac{3}{16}\text{ in}\right)\left(70\ \dfrac{\text{kips}}{\text{in}^2}\right)}{2.0} \\
&= 3.94\text{ kips/in}
\end{aligned}
$$

By Eq. 66.8, the load carried by the end (transverse) weld is

$$
\begin{aligned}
P_2 &= R_w l_2 = \left(3.94\ \frac{\text{kips}}{\text{in}}\right)(5\text{ in}) \\
&= 19.7\text{ kips}
\end{aligned}
$$

From Eq. 66.7 (from part (a), $T = 81$ kips), the load carried by the bottom longitudinal weld is

$$
\begin{aligned}
P_3 &= T\left(1 - \frac{y}{d}\right) - \frac{P_2}{2} \\
&= (81\text{ kips})\left(1 - \frac{1.75\text{ in}}{5\text{ in}}\right) - \frac{19.7\text{ kips}}{2} \\
&= \boxed{42.8\text{ kips}\quad(43\text{ kips})}
\end{aligned}
$$

The answer is (D).

(i) From Eq. 66.9, the load carried by the top longitudinal weld is

$$
\begin{aligned}
P_1 &= T - P_2 - P_3 \\
&= 81\text{ kips} - 19.7\text{ kips} - 42.8\text{ kips} \\
&= \boxed{18.5\text{ kips}\quad(19\text{ kips})}
\end{aligned}
$$

The answer is (A).

(j) By Eq. 66.10, the required length of the top longitudinal weld is

$$l_1 = \frac{P_1}{R_w} = \frac{18.5 \text{ kips}}{3.94 \frac{\text{kips}}{\text{in}}}$$

$$= \boxed{4.7 \text{ in} \quad (5 \text{ in})}$$

From Eq. 66.11, the required length of the bottom longitudinal weld is

$$l_3 = \frac{P_3}{R_w} = \frac{42.8 \text{ kips}}{3.94 \frac{\text{kips}}{\text{in}}}$$

$$= \boxed{10.86 \text{ in} \quad (11 \text{ in})}$$

The answer is (A).

4. The weld group is subjected to combined shear and torsion.

(a) The location of the centroid of the weld group is obtained by summing moments about the 15 in side.

$$\overline{x} = \frac{t_e\big((2)(6 \text{ in})(3 \text{ in}) + (15 \text{ in})(0)\big)}{t_e\big((2)(6 \text{ in}) + 15 \text{ in}\big)}$$

$$= \boxed{1.33 \text{ in} \quad (1.3 \text{ in})}$$

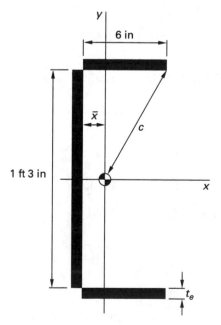

The answer is (B).

(b) The moment of inertia of the weld group about the x-axis (using the parallel axis equation, and neglecting the moment of inertia of the top and bottom welds about their own centroidal axes) is

$$I_x = \left(\tfrac{1}{12}\right)(1 \text{ in})(15 \text{ in})^3 + (2)\big((1 \text{ in})(6 \text{ in})(7.5 \text{ in})^2\big)$$

$$= 956.25 \text{ in}^4$$

The moment of inertia of the weld group about the y-axis is

$$I_y = (2)\left(\left(\tfrac{1}{12}\right)(1 \text{ in})(6 \text{ in})^3\right)$$

$$+ (2)(1 \text{ in})(6 \text{ in})(1.67 \text{ in})^2 + (1 \text{ in})(15 \text{ in})(1.33 \text{ in})^2$$

$$= 96 \text{ in}^4$$

The polar moment of inertia of the weld group about its centroid is

$$I_p = I_x + I_y = 956.25 \text{ in}^4 + 96 \text{ in}^4$$

$$= \boxed{1052.25 \text{ in}^4 \quad (1100 \text{ in}^4)}$$

The answer is (C).

(c) The torsional moment on the weld group is

$$M = Pe = (60 \text{ kips})(6 \text{ in} + 6 \text{ in} - 1.33 \text{ in})$$

$$= \boxed{640 \text{ in-kips}}$$

The answer is (D).

(d) The maximum radial distance is

$$x = 6 \text{ in} - 1.33 \text{ in} = 4.67 \text{ in}$$

$$y = \frac{15 \text{ in}}{2} = 7.5 \text{ in}$$

$$c = \sqrt{x^2 + y^2} = \sqrt{(4.67 \text{ in})^2 + (7.5 \text{ in})^2}$$

$$= 8.83 \text{ in}$$

The maximum stress in the weld due to the torsional moment is

$$r_m = \frac{Mc}{I_p} = \frac{(640 \text{ in-kips})(8.83 \text{ in})}{1052.25 \text{ in}^4}$$

$$= \boxed{5.37 \text{ kips/in}^2 \quad (5.4 \text{ ksi})}$$

The answer is (C).

(e) The horizontal and vertical components of the torsional stress in the weld are

$$r_{m,x} = \left(\frac{7.5 \text{ in}}{8.83 \text{ in}}\right)\left(5.37 \ \frac{\text{kips}}{\text{in}^2}\right)$$

$$= \boxed{4.56 \text{ kips/in}^2 \quad (4.6 \text{ ksi})}$$

$$r_{m,y} = \left(\frac{4.67 \text{ in}}{8.83 \text{ in}}\right)\left(5.37 \ \frac{\text{kips}}{\text{in}^2}\right)$$

$$= \boxed{2.84 \text{ kips/in}^2 \quad (2.8 \text{ ksi})}$$

The answer is (B).

(f) The direct shear stress in the weld is

$$r_p = \frac{P}{A} = \frac{60 \text{ kips}}{(27 \text{ in})(1 \text{ in})}$$

$$= \boxed{2.22 \text{ kips/in}^2 \quad (2.2 \text{ ksi})}$$

The answer is (B).

(g) Use Eq. 66.19. The resultant maximum stress is

$$r_a = \sqrt{r_m^2 + r_p^2}$$

$$= \sqrt{\left(4.56 \ \frac{\text{kips}}{\text{in}^2}\right)^2 + \left(2.22 \ \frac{\text{kips}}{\text{in}^2} + 2.84 \ \frac{\text{kips}}{\text{in}^2}\right)^2}$$

$$= \boxed{6.81 \text{ kips/in}^2 \quad (6.8 \text{ ksi})}$$

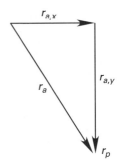

The answer is (D).

(h) For a SMAW process, with $w = \,^3/_8$ in, from Eq. 66.1 and Eq. 66.4, the shear strength of the weld per unit length is

$$R_w = (0.30)(0.707w)F_{u,\text{rod}}$$

$$= (0.30)(0.707)\left(\tfrac{3}{8} \text{ in}\right)\left(70 \ \frac{\text{kips}}{\text{in}^2}\right)$$

$$= \boxed{5.57 \text{ kips/in} \quad (5.6 \text{ kips/in})}$$

The answer is (C).

(i) For a SAW process, with $w = \,^3/_8$ in, from Eq. 66.1 and Eq. 66.4, the shear strength of the weld per unit length is

$$R_w = 0.30wF_{u,\text{rod}} = (0.30)\left(\tfrac{3}{8} \text{ in}\right)\left(70 \ \frac{\text{kips}}{\text{in}^2}\right)$$

$$= \boxed{7.87 \text{ kips/in} \quad (7.9 \text{ kips/in})}$$

The answer is (D).

(j) Use Eq. 66.2. The actual weld shear strength is

$$F_v = \frac{0.6F_{u,\text{rod}}}{\Omega} = \frac{(0.6)\left(70 \ \dfrac{\text{kips}}{\text{in}^2}\right)}{2.0} = 21 \text{ kips/in}^2$$

Assuming $t_e = 1$ in, from part (g), the computed shear stress (i.e., the required shear resistance, for ASD) is 6.81 kips/in^2.

Use Eq. 66.1. For a SMAW process with $w = \,^3/_8$ in,

$$t_e = (0.707)\left(\tfrac{3}{8} \text{ in}\right) = 0.265 \text{ in}$$

For this actual weld size, the required shear resistance (i.e., for ASD) is

$$f_v = \left(6.81 \ \frac{\text{kips}}{\text{in}^2}\right)\left(\frac{1 \text{ in}}{0.265 \text{ in}}\right) = 25.7 \text{ kips/in}^2$$

Since $f_v > F_v$, SMAW is not adequate.

For a SAW process with $w = \,^3/_8$ in,

$$t_e = \tfrac{3}{8} \text{ in} = 0.375 \text{ in}$$

For this actual weld size, the required shear resistance is

$$f_v = \left(6.81 \ \frac{\text{kips}}{\text{in}^2}\right)\left(\frac{1 \text{ in}}{0.375 \text{ in}}\right) = 18.2 \text{ kips/in}^2$$

Since $f_v < F_v$, SAW is adequate.

The connection is $\boxed{\text{adequate for a SAW process only.}}$

The answer is (D).

5. Each angle carries $(^1/_2)(140 \text{ kips}) = 70$ kips. Check the tensile strength using the ASD method.

$$A = 3.25 \text{ in}^2$$

$$F_a = \frac{36 \frac{\text{kips}}{\text{in}^2}}{1.67} = 21.6 \text{ kips/in}^2$$

$$f_a = \frac{70 \text{ kips}}{3.25 \text{ in}^2}$$

$$= 21.5 \text{ kips/in}^2 < 21.6 \text{ kips/in}^2 \quad [\text{acceptable}]$$

Since thickness $= ^1/_2$ in, the maximum weld thickness is

$$w = \tfrac{1}{2} \text{ in} - \tfrac{1}{16} \text{ in} = \tfrac{7}{16} \text{ in}$$
$$t_e = (0.707)(\tfrac{7}{16} \text{ in}) = 0.309 \text{ in}$$

For the angles, the thickness is 0.50 in. For E70 electrodes,

$$R_w = (0.309 \text{ in}) \left(\frac{(0.6)\left(70 \frac{\text{kips}}{\text{in}^2} \right)}{2.0} \right) = 6.49 \text{ kips/in}$$

$$d = 4 \text{ in}$$
$$y = 1.33 \text{ in}$$

From Eq. 66.8,

$$P_2 = R_w l_2 = \left(6.49 \frac{\text{kips}}{\text{in}} \right)(4 \text{ in}) = 26.0 \text{ kips}$$

From Eq. 66.7, Eq. 66.9, Eq. 66.10, and Eq. 66.11,

$$P_3 = T\left(1 - \frac{y}{d} \right) - \frac{P_2}{2}$$

$$= (70 \text{ kips})\left(1 - \frac{1.33 \text{ in}}{4 \text{ in}} \right) - \frac{26.0 \text{ kips}}{2}$$

$$= 33.7 \text{ kips}$$

$$l_3 = \frac{P_3}{R_w} = \frac{33.7 \text{ kips}}{6.49 \frac{\text{kips}}{\text{in}}} = \boxed{5.19 \text{ in}} \quad [\text{round to } 5\tfrac{1}{4} \text{ in}]$$

$$P_1 = T - P_2 - P_3 = 70 \text{ kips} - 26.0 \text{ kips} - 33.7 \text{ kips}$$

$$= 10.3 \text{ kips}$$

$$l_1 = \frac{P_1}{R_w} = \frac{10.3 \text{ kips}}{6.49 \frac{\text{kips}}{\text{in}}} = \boxed{1.59 \text{ in}} \quad [\text{round to } 1\tfrac{3}{4} \text{ in}]$$

Check the minimum weld length.

$$l_1 \geq \left\{ \begin{array}{l} 4 \times \tfrac{7}{16} \text{ in} = 1.75 \text{ in} \\ 1\tfrac{1}{2} \text{ in} \end{array} \right\} = 1.75 \text{ in} \quad [\text{acceptable}]$$

Structural

67 Properties of Masonry

PRACTICE PROBLEMS

There are no problems in this book corresponding to Chap. 67 of the *Civil Engineering Reference Manual*.

68 Masonry Walls

PRACTICE PROBLEMS

1. Determine the allowable vertical load of the masonry wall shown. Face shell mortar bedding is used.

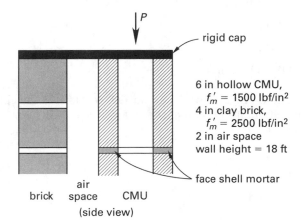

6 in hollow CMU,
$f'_m = 1500$ lbf/in^2
4 in clay brick,
$f'_m = 2500$ lbf/in^2
2 in air space
wall height = 18 ft

2. A 25 ft high, 10 in CMU wall is partially grouted and reinforced. The wall is nonloadbearing and is subjected to wind only. A one-third increase in allowable stresses is permitted for wind loading. The wall is simply supported at top and bottom, and reinforcement is placed in the center of the wall. Use $f'_m = 2000$ lbf/in^2 and $n = 16.1$.

(a) With reinforcement spaced at 24 in, determine the largest reinforcing bars that can be used such that the neutral axis falls within the face shell of the unit.

(A) no. 4 bars

(B) no. 5 bars

(C) no. 6 bars

(D) no. 7 bars

For parts (b) through (f), assume the wall has no. 6 vertical bars spaced 32 in apart.

(b) Determine the location of the neutral axis measured from the compression face.

(A) 1.0 in

(B) 1.3 in

(C) 1.7 in

(D) 2.0 in

(c) Determine the resisting moment assuming steel governs.

(A) 17,500 in-lbf/ft

(B) 20,000 in-lbf/ft

(C) 23,000 in-lbf/ft

(D) 24,500 in-lbf/ft

(d) Determine the allowable bending stress.

(A) 670 lbf/in^2

(B) 890 lbf/in^2

(C) 1000 lbf/in^2

(D) 1140 lbf/in^2

(e) Determine the resisting moment of the wall.

(A) 17,000 in-lbf/ft

(B) 22,000 in-lbf/ft

(C) 23,000 in-lbf/ft

(D) 29,000 in-lbf/ft

(f) Determine the shear capacity.

(A) 1500 lbf/ft

(B) 3400 lbf/ft

(C) 3900 lbf/ft

(D) 4300 lbf/ft

(g) Determine the area of steel required to produce a balanced design condition.

(A) 0.15 in^2/ft

(B) 0.18 in^2/ft

(C) 0.21 in^2/ft

(D) 0.25 in^2/ft

(h) Determine the maximum moment on the wall if it is subjected to a 25 lbf/ft^2 wind load.

(A) 23,400 in-lbf/ft

(B) 24,200 in-lbf/ft

(C) 24,800 in-lbf/ft

(D) 25,100 in-lbf/ft

(i) Using $\rho = 0.004$ and $d = 4.81$ in, determine the wind-induced service moment the wall can withstand.

(A) 24,000 in-lbf/ft

(B) 32,000 in-lbf/ft

(C) 38,000 in-lbf/ft

(D) 47,000 in-lbf/ft

For part (j), assume fully grouted masonry and that the wall must resist a moment of 20,000 in-lbf/ft.

(j) With no. 5 bars spaced at 16 in within a 10 in thick wall, determine the steel stress.

(A) 14,000 lbf/in^2

(B) 16,000 lbf/in^2

(C) 18,000 lbf/in^2

(D) 20,000 lbf/in^2

3. An 8 in hollow unreinforced, ungrouted concrete masonry wall, 12 ft high, is simply supported at top and bottom. Type M mortar-cement is used. For parts (a) through (e), $f'_m = 1800$ lbf/in^2. A one-third increase in allowable stress is permitted for wind loading.

(a) Determine the allowable stresses for permanent loads.

(A) $F_a = 390$ lbf/in^2, $F_b = 600$ lbf/in^2

(B) $F_a = 320$ lbf/in^2, $F_b = 800$ lbf/in^2

(C) $F_a = 520$ lbf/in^2, $F_b = 600$ lbf/in^2

(D) $F_a = 390$ lbf/in^2, $F_b = 800$ lbf/in^2

(b) Determine the maximum vertical load the wall can be designed for with a 2 in eccentricity.

(A) 6700 lbf/ft

(B) 7900 lbf/ft

(C) 10,000 lbf/ft

(D) 14,000 lbf/ft

(c) Determine the maximum wind load if the wall carries no vertical load. Neglect the self-weight of the wall.

(A) 9.4 lbf/ft^2

(B) 12 lbf/ft^2

(C) 16 lbf/ft^2

(D) 18 lbf/ft^2

For parts (d) and (e), the wall is subjected to an axial load of 5000 lbf/ft. There is no lateral load.

(d) Determine the allowable in-plane shear stress.

(A) 64 lbf/in^2

(B) 110 lbf/in^2

(C) 120 lbf/in^2

(D) 130 lbf/in^2

(e) Assuming the wall has no net tension, determine the maximum in-plane shear force the wall can be designed for.

(A) 900 lbf/ft

(B) 1000 lbf/ft

(C) 1100 lbf/ft

(D) 1300 lbf/ft

(f) Determine the minimum net area required to sustain an 18,000 lbf/ft axial load given that $f'_m = 1800$ lbf/in^2.

(A) 30 in^2/ft

(B) 39 in^2/ft

(C) 46 in^2/ft

(D) 52 in^2/ft

For parts (g) through (j), assume the wall is subjected to a vertical dead load, D, of 4500 lbf/ft ($e = 2.5$ in) and a lateral wind load, w, of 35 lbf/ft^2.

(g) Considering both wind and dead loads, determine the flexural compressive stress in the masonry.

(A) 120 lbf/in^2

(B) 140 lbf/in^2

(C) 160 lbf/in^2

(D) 180 lbf/in^2

(h) Considering both wind and dead loads, determine the masonry strength, f'_m, required.

(A) 890 lbf/in^2

(B) 950 lbf/in^2

(C) 1100 lbf/in^2

(D) 1500 lbf/in^2

(i) Determine the flexural tension capacity required.

    (A) 6 lbf/in^2

    (B) 10 lbf/in^2

    (C) 14 lbf/in^2

    (D) 15 lbf/in^2

(j) Determine the minimum required masonry strength, f'_m.

    (A) 890 lbf/in^2

    (B) 950 lbf/in^2

    (C) 1100 lbf/in^2

    (D) 1500 lbf/in^2

4. Design the joint reinforcement for a concrete masonry wall to carry a 20 lbf/ft^2 wind load. The wall spans 10 ft horizontally and is simply supported at both ends. Use allowable stress design. A one-third increase in allowable stress is permitted for wind loading. Assume the following.

- $f'_m = 1500$ lbf/in^2

- 8 in hollow units, laid with face shell mortar bedding, type S mortar

- The wall will be built using horizontal joint reinforcement every other course (spaced at 16 in).

- The wall carries no axial loads.

- The masonry modulus of elasticity is $E_m = 2.08 \times 10^6$ lbf/in^2.

- The steel modulus of elasticity is $E_s = 29,000,000$ lbf/in^2.

- Steel stress governs the design.

SOLUTIONS

1. Determine the section properties of the wall from App. 68.A.

$$A_n = 24 \text{ in}^2/\text{ft}$$

$$I = 139.3 \text{ in}^4/\text{ft}$$

$$r = 2.08 \text{ in}$$

$$\frac{h}{r} = \frac{(18 \text{ ft})\left(12 \frac{\text{in}}{\text{ft}}\right)}{2.08 \text{ in}} = 103.8$$

Since 103.8 is greater than 99, use Eq. 68.21.

$$F_a = \tfrac{1}{4}f'_m\left(\frac{70r}{h}\right)^2 = \left(\tfrac{1}{4}\right)\left(1500 \frac{\text{lbf}}{\text{in}^2}\right)\left(\frac{70}{103.8}\right)^2$$

$$= 170.5 \text{ lbf/in}^2$$

$$P_a = F_a A_n = \left(170.5 \frac{\text{lbf}}{\text{in}^2}\right)\left(24 \frac{\text{in}^2}{\text{ft}}\right) = 4092 \text{ lbf/ft}$$

Check buckling using Eq. 68.18 and Eq. 68.19.

$$E_m = 900f'_m = (900)\left(1500 \frac{\text{lbf}}{\text{in}^2}\right)$$

$$= 1,350,000 \text{ lbf/in}^2$$

$$P_e = \left(\frac{\pi^2 E_m I_n}{h^2}\right)\left(1 - 0.577\frac{e}{r}\right)^3$$

$$= \frac{\pi^2\left(1,350,000 \frac{\text{lbf}}{\text{in}^2}\right)\left(139.3 \frac{\text{in}^4}{\text{ft}}\right)}{\left((18 \text{ ft})\left(12 \frac{\text{in}}{\text{ft}}\right)\right)^2}$$

$$= 39,781 \text{ lbf/ft}$$

$$P \le \tfrac{1}{4}P_e$$

$$4092 \text{ lbf/ft} \le \left(\tfrac{1}{4}\right)\left(39,781 \frac{\text{lbf}}{\text{ft}}\right) = 9945 \text{ lbf/ft} \quad \text{[OK]}$$

$$P = \boxed{4092 \text{ lbf/ft}}$$

2. (a) 10 in is the nominal thickness. The actual thickness is 10 in − 0.375 in = 9.625 in.

$$kd = t_f = 1.375 \text{ in}$$

$$d = \frac{9.625 \text{ in}}{2} = 4.81 \text{ in}$$

$$k = \frac{t_f}{d} = \frac{1.375 \text{ in}}{4.81 \text{ in}} = 0.286$$

From Eq. 68.4,

$$k^2 + 2\rho nk - 2\rho n = 0$$
$$(0.286)^2 + 2\rho(16.1)(0.286) - 2\rho(16.1) = 0$$
$$\rho = 0.00356$$

Use $b = 24$ in, the vertical bar spacing.

$$A_s = \rho bd$$
$$= (0.00356)(24 \text{ in})(4.81 \text{ in})$$
$$= 0.411 \text{ in}^2 \quad [\text{for every 24 in}]$$

The area of a no. 6 bar is 0.44 in², which is too large.

The area of a no. 5 bar is 0.31 in², which is acceptable.

The answer is (B).

(b) Work on a per-foot-of-wall basis. From Table 68.3,

$$A_s = 0.166 \text{ in}^2/\text{ft}$$

On a per-foot basis, $b = 12$ in.

$$\rho = \frac{A_s}{bd} = \frac{0.166 \text{ in}^2}{(12 \text{ in})(4.81 \text{ in})} = 0.00288$$
$$\rho n = (0.00288)(16.1) = 0.0463$$

From Eq. 68.4,

$$k = \sqrt{2\rho n + (\rho n)^2} - \rho n$$
$$= \sqrt{(2)(0.0463) + (0.0463)^2} - 0.0463$$
$$= 0.262$$
$$kd = (0.262)(4.81 \text{ in}) = \boxed{1.26 \text{ in} \quad (1.3 \text{ in})}$$

The answer is (B).

(c) Use Eq. 68.5 and Eq. 68.7, and add a one-third increase in allowable stress for wind load.

$$F_s = \left(\tfrac{4}{3}\right)\left(24{,}000 \ \frac{\text{lbf}}{\text{in}^2}\right) = 32{,}000 \text{ lbf/in}^2$$
$$j = 1 - \frac{k}{3} = 1 - \frac{0.262}{3} = 0.913$$
$$M_s = A_s F_s jd$$
$$= \left(0.166 \ \frac{\text{in}^2}{\text{ft}}\right)\left(32{,}000 \ \frac{\text{lbf}}{\text{in}^2}\right)(0.913)(4.81 \text{ in})$$
$$= \boxed{23{,}328 \text{ in-lbf/ft} \quad (23{,}000 \text{ in-lbf/ft})}$$

The answer is (C).

(d) Use Eq. 68.22. Add a one-third increase in allowable stress for wind load.

$$F_b = \left(\tfrac{4}{3}\right)\tfrac{1}{3}f'_m = \left(\tfrac{4}{3}\right)\left(\tfrac{1}{3}\right)\left(2000 \ \frac{\text{lbf}}{\text{in}^2}\right)$$
$$= \boxed{889 \text{ lbf/in}^2 \quad (890 \text{ lbf/in}^2)}$$

The answer is (B).

(e) Use Eq. 68.6 and Eq. 68.8.

$$M_m = F_b bd^2\left(\frac{jk}{2}\right)$$
$$= \left(889 \ \frac{\text{lbf}}{\text{in}^2}\right)\left(12 \ \frac{\text{in}}{\text{ft}}\right)(4.81 \text{ in})^2\left(\frac{(0.913)(0.262)}{2}\right)$$
$$= 29{,}520 \text{ in-lbf/ft}$$
$$M_R = \text{lesser of } M_s \text{ and } M_m$$
$$= \boxed{23{,}328 \text{ in-lbf/ft} \quad (23{,}000 \text{ in-lbf/ft})}$$

Therefore, steel governs.

The answer is (C).

(f) Work on a per-foot-of-wall basis. Using Eq. 68.2,

$$F_v = \sqrt{f'_m} = \sqrt{2000 \ \frac{\text{lbf}}{\text{in}^2}}$$
$$= 44.7 \text{ lbf/in}^2 < 50 \text{ lbf/in}^2 \quad [\text{OK}]$$

Determine the core width for a 12 in stretcher block. Since there are two cells per core,

$$b = \frac{15.625 \text{ in} - 1.5 \text{ in}}{2} = 7.1 \text{ in}$$

For partially grouted walls, cells are grouted at vertical rebars. Bar spacing is 32 in on center. Therefore, the cell width is

$$b_w = \left(\frac{7.1 \text{ in}}{32 \text{ in}}\right)\left(12 \ \frac{\text{in}}{\text{ft}}\right) = 2.65 \text{ in/ft}$$

Use Eq. 68.16 with a one-third increase for wind loading.

$$V_R = \tfrac{4}{3}F_v\left(bt_{\text{fs}} + b_w(d - t_{\text{fs}})\right)$$
$$= \left(\tfrac{4}{3}\right)\left(44.7 \ \frac{\text{lbf}}{\text{in}^2}\right)\left(\left(12 \ \frac{\text{in}}{\text{ft}}\right)(1.375 \text{ in})\right.$$
$$\left. + \left(2.65 \ \frac{\text{in}}{\text{ft}}\right)(4.81 \text{ in} - 1.375 \text{ in})\right)$$
$$= \boxed{1526 \text{ lbf/ft} \quad (1500 \text{ lbf/ft})}$$

The answer is (A).

(g) From Eq. 68.10, considering the one-third increase for wind load,

$$\rho_{bal} = \frac{nF_b}{2F_s\left(n + \dfrac{F_s}{F_b}\right)}$$

$$= \frac{(16.1)\left(889 \, \dfrac{lbf}{in^2}\right)}{(2)\left(32{,}000 \, \dfrac{lbf}{in^2}\right)\left(16.1 + \dfrac{32{,}000 \, \dfrac{lbf}{in^2}}{889 \, \dfrac{lbf}{in^2}}\right)}$$

$$= 0.00429$$

$$A_{s,bal} = \rho_{bal} bd = (0.00429)\left(12 \, \frac{in}{ft}\right)(4.81 \, in)$$

$$= \boxed{0.248 \, in^2/ft \quad (0.25 \, in^2/ft)}$$

The answer is (D).

(h) The maximum moment of the wall is

$$M = \frac{wl^2}{8} = \frac{\left(25 \, \dfrac{lbf}{ft^2}\right)(25 \, ft)^2\left(12 \, \dfrac{in}{ft}\right)}{8}$$

$$= \boxed{23{,}438 \, in\text{-}lbf/ft \quad (23{,}400 \, in\text{-}lbf/ft)}$$

The answer is (A).

(i) The wind-induced service moment is calculated as follows.

$$\rho n = (0.004)(16.1) = 0.0644$$

$$k = \sqrt{2\rho n + (\rho n)^2} - \rho n$$

$$= \sqrt{(2)(0.0644) + (0.0644)^2} - 0.0644$$

$$= 0.300$$

$$j = 1 - \frac{k}{3} = 1 - \frac{0.300}{3} = 0.900$$

$$A_s = \rho bd = (0.004)\left(12 \, \frac{in}{ft}\right)(4.81 \, in)$$

$$= 0.231 \, in^2/ft$$

From Eq. 68.7, with the one-third increase for wind,

$$M_s = A_s F_s jd$$

$$= \left(0.231 \, \frac{in^2}{ft}\right)\left(32{,}000 \, \frac{lbf}{in^2}\right)(0.900)(4.81 \, in)$$

$$= \boxed{32{,}000 \, in\text{-}lbf/ft}$$

The answer is (B).

(j) From Table 68.3,

$$A_s = 0.230 \, in^2/ft$$

$$\rho = \frac{A_s}{bd} = \frac{0.230 \, \dfrac{in^2}{ft}}{\left(12 \, \dfrac{in}{ft}\right)(4.81 \, in)}$$

$$= 0.00398$$

$$\rho n = (0.00398)(16.1) = 0.0641$$

$$k = \sqrt{2\rho n + (\rho n)^2} - \rho n$$

$$= \sqrt{(2)(0.0641) + (0.0641)^2} - 0.0641$$

$$= 0.300$$

$$j = 1 - \frac{k}{3} = 1 - \frac{0.300}{3} = 0.900$$

$$f_s = \frac{M}{A_s jd} = \frac{20{,}000 \, \dfrac{in\text{-}lbf}{ft}}{\left(0.230 \, \dfrac{in^2}{ft}\right)(0.900)(4.81 \, in)}$$

$$= \boxed{20{,}087 \, lbf/in^2 \quad (20{,}000 \, lbf/in^2)}$$

The answer is (D).

3. (a) From App. 68.A, for an 8 in, single wythe, ungrouted wall,

$$r = 2.84 \, in$$

$$S_x = 81 \, in^3/ft$$

$$A_n = 30 \, in^2/ft$$

$$I_x = 334 \, in^4/ft$$

$$\frac{h}{r} = \frac{(12 \, ft)\left(12 \, \dfrac{in}{ft}\right)}{2.84 \, in} = 50.7$$

Since 50.7 is less than 99, use Eq. 68.20.

$$F_a = \tfrac{1}{4}f'_m\left(1 - \left(\frac{h}{140r}\right)^2\right)$$

$$= \left(\tfrac{1}{4}\right)\left(1800 \, \frac{lbf}{in^2}\right)\left(1 - \left(\frac{50.7}{140}\right)^2\right)$$

$$= \boxed{391 \, lbf/in^2 \quad (390 \, lbf/in^2)}$$

$$F_b = \tfrac{1}{3}f'_m = \left(\tfrac{1}{3}\right)\left(1800 \, \frac{lbf}{in^2}\right)$$

$$= \boxed{600 \, lbf/in^2}$$

The answer is (A).

Structural

(b) The maximum vertical load the wall can be designed for is calculated as follows.

$$f_a = \frac{P}{A_n} = \frac{P}{30 \ \dfrac{\text{in}^2}{\text{ft}}} = 0.033P$$

$$f_b = \frac{M}{S} = \frac{Pe}{S} = \frac{P(2 \ \text{in})}{81 \ \dfrac{\text{in}^3}{\text{ft}}} = 0.025P$$

From Eq. 68.17,

$$\frac{f_a}{F_a} + \frac{f_b}{F_b} \le 1$$

$$\frac{0.033P}{391 \ \dfrac{\text{in}^2}{\text{ft}}} + \frac{0.025P}{600 \ \dfrac{\text{in}^2}{\text{ft}}} = 1$$

$$P = 7932 \ \text{lbf/ft}$$

Check buckling criteria.

$$E_m = 900f'_m = (900)\left(1800 \ \frac{\text{lbf}}{\text{in}^2}\right) = 1{,}620{,}000 \ \text{lbf/in}^2$$

$$P \le \tfrac{1}{4}P_e = \tfrac{1}{4}\left(\frac{\pi^2 EI}{h^2}\right)\left(1 - 0.577\left(\frac{e}{r}\right)\right)^3$$

$$= (\tfrac{1}{4})\left(\frac{\pi^2\left(1{,}620{,}000 \ \dfrac{\text{lbf}}{\text{in}^2}\right)\left(334 \ \dfrac{\text{in}^4}{\text{ft}}\right)}{(12 \ \text{ft})^2\left(12 \ \dfrac{\text{in}}{\text{ft}}\right)^2}\right)$$

$$\times \left(1 - (0.577)\left(\frac{2 \ \text{in}}{2.84 \ \text{in}}\right)\right)^3$$

$$= 13{,}471 \ \text{lbf/ft}$$

Buckling does not control. Therefore,

$$P_{\max} = \boxed{7932 \ \text{lbf/ft} \quad (7900 \ \text{lbf/ft})}$$

The answer is (B).

(c) The maximum wind load, w, in lbf/ft^2 is determined as follows.

$$M = \frac{wh^2}{8} = \frac{w\left((12 \ \text{ft})\left(12 \ \dfrac{\text{in}}{\text{ft}}\right)\right)^2}{(8)\left(12 \ \dfrac{\text{in}}{\text{ft}}\right)}$$

$$= 216w \quad [\text{in in-lbf/ft}]$$

$$f_b = \frac{M}{S} = \frac{216w}{81 \ \dfrac{\text{in}^3}{\text{ft}}} = 2.67w \quad [\text{in lbf/in}^2]$$

From Table 68.1,

$$F_b = \left(\tfrac{4}{3}\right)\left(25 \ \frac{\text{lbf}}{\text{in}^2}\right) = 33.33 \ \text{lbf/in}^2$$

Set f_b equal to F_b. ($2.67w$ has units of lbf/in^2 when w has units of lbf/ft^2.)

$$2.67w = 33.33 \ \frac{\text{lbf}}{\text{in}^2}$$

$$w = \boxed{12.5 \ \text{lbf/ft}^2 \quad (12 \ \text{lbf/ft}^2)}$$

The answer is (B).

(d) F_v is the least of the following [*MSJC* Sec. 2.2.5.2].

- $1.5\sqrt{f'_m} = 1.5\sqrt{1800 \ \dfrac{\text{lbf}}{\text{in}^2}} = 63.6 \ \text{lbf/in}^2$

- $v + 0.45\left(\dfrac{N_v}{A_n}\right) = 37 \ \dfrac{\text{lbf}}{\text{in}^2} + \dfrac{(0.45)\left(5000 \ \dfrac{\text{lbf}}{\text{ft}}\right)}{30 \ \dfrac{\text{in}^2}{\text{ft}}}$

  $= 112 \ \text{lbf/in}^2$

- $120 \ \text{lbf/in}^2$

The first criterion controls.

$$F_v = \boxed{63.6 \ \text{lbf/in}^2 \quad (64 \ \text{lbf/in}^2)}$$

The answer is (A).

(e) From Eq. 68.37,

$$f_v = \frac{3V}{2A_n}$$

Set f_v equal to F_v.

$$V = \frac{2A_n F_v}{3}$$

$$= \frac{(2)\left(30 \ \dfrac{\text{in}^2}{\text{ft}}\right)\left(63.6 \ \dfrac{\text{lbf}}{\text{in}^2}\right)}{3}$$

$$= \boxed{1272 \ \text{lbf/ft} \quad (1300 \ \text{lbf/ft})}$$

The answer is (D).

(f) From part (a), since buckling does not control, $F_a = 391$ lbf/in^2.

$$F_a = \frac{P}{A_n}$$

$$391 \ \frac{\text{lbf}}{\text{in}^2} = \frac{18{,}000 \ \dfrac{\text{lbf}}{\text{ft}}}{A_n}$$

$$A_n = \boxed{46 \ \text{in}^2/\text{ft}}$$

The answer is (C).

(g) At the mid-height of the wall,

$$M = \frac{Pe}{2} + \frac{wh^2}{8}$$

$$= \frac{\left(4500 \ \dfrac{\text{lbf}}{\text{ft}}\right)(2.5 \ \text{in})}{2}$$

$$+ \frac{\left(35 \ \dfrac{\text{lbf}}{\text{ft}^2}\right)\left((12 \ \text{ft})\left(12 \ \dfrac{\text{in}}{\text{ft}}\right)\right)^2}{(8)\left(12 \ \dfrac{\text{in}}{\text{ft}}\right)}$$

$$= 13{,}185 \ \text{in-lbf/ft}$$

$$f_b = \frac{M}{S} = \frac{13{,}185 \ \dfrac{\text{in-lbf}}{\text{ft}}}{81 \ \dfrac{\text{in}^3}{\text{ft}}}$$

$$= \boxed{163 \ \text{lbf/in}^2 \quad (160 \ \text{lbf/in}^2)}$$

The answer is (C).

(h) The masonry strength is determined as follows.

$$F_a = \tfrac{4}{3}\left(\tfrac{1}{4}f'_m\left(1 - \left(\frac{h}{140r}\right)^2\right)\right)$$

$$= \left(\tfrac{4}{3}\right)\left(\tfrac{1}{4}f'_m\left(1 - \left(\frac{50.7}{140}\right)^2\right)\right)$$

$$= 0.290 f'_m$$

$$f_a = \frac{P}{A} = \frac{4500 \ \dfrac{\text{lbf}}{\text{ft}}}{30 \ \dfrac{\text{in}^2}{\text{ft}}} = 150 \ \text{lbf/in}^2$$

$$F_b = \tfrac{4}{3}\left(\tfrac{1}{3}f'_m\right) = 0.444 f'_m$$

$$\frac{f_a}{F_a} + \frac{f_b}{F_b} = 1$$

$$\frac{150 \ \dfrac{\text{lbf}}{\text{in}^2}}{0.290 f'_m} + \frac{163 \ \dfrac{\text{lbf}}{\text{in}^2}}{0.444 f'_m} = 1$$

$$f'_m = \boxed{884 \ \text{lbf/in}^2 \quad (890 \ \text{lbf/in}^2)}$$

The answer is (A).

(i) For dead and wind loads,

$$-f_a + f_b = \tfrac{4}{3}F_t$$

$$-150 \ \frac{\text{lbf}}{\text{in}^2} + 163 \ \frac{\text{lbf}}{\text{in}^2} = \tfrac{4}{3}F_t$$

$$F_t = 9.75 \ \text{lbf/in}^2$$

Considering dead load only, the moment at the top of the wall is

$$Pe = \left(4500 \ \frac{\text{lbf}}{\text{ft}}\right)(2.5 \ \text{in})$$

$$= 11{,}250 \ \text{in-lbf/ft}$$

$$f_b = \frac{M}{S} = \frac{11{,}250 \ \dfrac{\text{in-lbf}}{\text{ft}}}{81 \ \dfrac{\text{in}^3}{\text{ft}}}$$

$$= 139 \ \text{lbf/in}^2$$

$$-f_a + f_b = F_t$$

$$-150 \ \frac{\text{lbf}}{\text{in}^2} + 139 \ \frac{\text{lbf}}{\text{in}^2} = -11 \ \frac{\text{lbf}}{\text{in}^2} = F_t$$

There is no tension in the wall. The dead and wind load case controls.

$$F_{t,\text{req}} = \boxed{9.75 \ \text{lbf/in}^2 \quad (10 \ \text{lbf/in}^2)}$$

The answer is (B).

(j) For dead and wind loads, from part (h), $f'_m = 884$ lbf/in^2.

Check dead load only. From part (h),

$$f_a = 150 \ \text{lbf/in}^2$$

$$F_a = \frac{0.290 \ \dfrac{\text{lbf}}{\text{in}^2}}{\tfrac{4}{3}} = 0.218 f'_m$$

From part (i),

$$f_b = 139 \ \text{lbf/in}^2$$

$$F_b = \tfrac{1}{3}f'_m$$

$$\frac{f_a}{F_a} + \frac{f_b}{F_b} = 1$$

$$\frac{150 \ \dfrac{\text{lbf}}{\text{in}^2}}{0.218 f'_m} + \frac{139 \ \dfrac{\text{lbf}}{\text{in}^2}}{0.333 f'_m} = 1$$

$$f'_m = \boxed{1105 \ \text{lbf/in}^2 \quad (1100 \ \text{lbf/in}^2) \quad \text{[governs]}}$$

The answer is (C).

Structural

4. The allowable tension in steel wire joint reinforcement due to wind loading is

$$F_s = \left(\tfrac{4}{3}\right)\left(30{,}000 \ \frac{\text{lbf}}{\text{in}^2}\right) = 40{,}000 \ \text{lbf/in}^2$$

The allowable masonry compressive stress due to flexure from wind loading is

$$F_b = \tfrac{4}{3}\left(\tfrac{1}{3}f'_m\right) = \left(\tfrac{4}{3}\right)\left(\tfrac{1}{3}\right)\left(1500 \ \frac{\text{lbf}}{\text{in}^2}\right) = 667 \ \text{lbf/in}^2$$

$$n = \frac{E_s}{E_m} = \frac{29 \ \dfrac{\text{kips}}{\text{in}^2}}{2.08 \ \dfrac{\text{kips}}{\text{in}^2}} = 13.9$$

Determine the moment due to wind load. For simply supported conditions,

$$M = \frac{wl^2}{8}$$

$$= \left(\frac{\left(20 \ \dfrac{\text{lbf}}{\text{ft}^2}\right)(10 \ \text{ft})^2}{8}\right)\left(12 \ \frac{\text{in}}{\text{ft}}\right)$$

$$= 3000 \ \text{in-lbf/ft}$$

Determine the balanced reinforcement ratio, ρ_b.

$$\rho_b = \frac{nF_b}{2F_s\left(n + \dfrac{F_s}{F_b}\right)}$$

$$= \frac{(13.9)\left(667 \ \dfrac{\text{lbf}}{\text{in}^2}\right)}{(2)\left(40{,}000 \ \dfrac{\text{lbf}}{\text{in}^2}\right)\left(13.9 + \dfrac{40{,}000 \ \dfrac{\text{lbf}}{\text{in}^2}}{667 \ \dfrac{\text{lbf}}{\text{in}^2}}\right)}$$

$$= 0.0016$$

If the actual steel ratio, ρ, is less than ρ_b, steel stress governs the design.

Try W2.1 (8 gage) horizontal joint reinforcement every other course.

The effective depth of reinforcement is

$$d = \text{actual width of wall} - \tfrac{5}{8} \ \text{in cover}$$

$$= 7\tfrac{5}{8} \ \text{in} - \tfrac{5}{8} \ \text{in} = 7 \ \text{in}$$

$$A_s = 0.0206 \ \text{in}^2 \ \text{per wire}$$

(Since only one wire is in tension, this is the effective steel area.) With reinforcement every other course, the spacing is $(2)(8 \ \text{in}) = 16 \ \text{in}$.

$$\frac{\dfrac{0.0206 \ \text{in}^2}{16 \ \text{in}}}{12 \ \dfrac{\text{in}}{\text{ft}}} = 0.015 \ \text{in}^2/\text{ft}$$

$$\rho = \frac{A_s}{bd} = \frac{0.020 \ \text{in}^2}{(16 \ \text{in})(7 \ \text{in})} = 0.00018$$

Since $\rho < \rho_b$, steel governs, and the resisting moment of the wall is

$$M_r = A_s F_s jd$$

Determine j.

$$k = \sqrt{2n\rho + (n\rho)^2} - n\rho$$

$$= \sqrt{(2)(13.9)(0.00018) + \big((13.9)(0.00018)\big)^2}$$
$$\quad - (13.9)(0.00018)$$

$$= 0.068$$

$$j = 1 - \frac{k}{3}$$

$$= 1 - \frac{0.068}{3}$$

$$= 0.977$$

$$M_r = A_s F_s jd$$

$$= \left(0.015 \ \frac{\text{in}^2}{\text{ft}}\right)\left(40{,}000 \ \frac{\text{lbf}}{\text{in}^2}\right)(0.977)(7 \ \text{in})$$

$$= 4103 \ \text{in-lbf/ft}$$

Since $M_r > M$, the design is adequate with W2.1 joint reinforcement every other course.

69 Masonry Columns

PRACTICE PROBLEMS

1. A 16 in square concrete masonry column with four vertical no. 6 bars and an effective height of 18 ft supports an axial load of 75,000 lbf. Determine the required masonry strength.

$$r = 4.53 \text{ in}$$
$$S = 635.8 \text{ in}^3$$
$$A_n = 244.1 \text{ in}^2$$

2. A 32 in × 16 in clay brick column supports a compressive load of 350,000 lbf.

$$e = 4 \text{ in} \quad \text{[along the major axis]}$$
$$f'_m = 3000 \text{ lbf/in}^2$$
$$\text{effective height} = 10 \text{ ft}$$
$$r = 9.1 \text{ in}$$
$$g = 0.8$$

Determine the required steel area.

3. (*Time limit: one hour*) A 36 ft tall vertical concrete masonry column with the cross section shown is firmly connected to its foundation at the base but is unsupported at the top. A 30 lbf/ft^2 wind acts perpendicular to the column face uniformly over the entire length. A one-third increase in allowable stress is permitted for wind loading. $f'_m = 1500$ lbf/in^2, $f_y = 60,000$ lbf/in^2, and $n = 20$. Neglecting the axial stresses due to the dead load, determine if the stresses in the steel and masonry are acceptable.

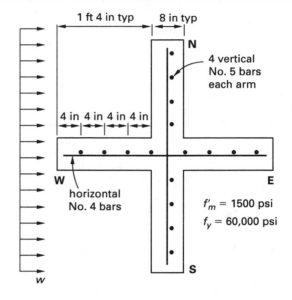

4. A brick column, 24 in × 24 in, is reinforced with four no. 5 bars. It has an effective height of 16 ft, $f'_m = 3000$ lbf/in^2, and a compressive load with 6 in eccentricity in the x-direction. Assume $g = 0.6$.

(a) Determine the radius of gyration, r.

(A) 5.9 in

(B) 6.2 in

(C) 6.8 in

(D) 7.0 in

(b) Determine the reinforcement ratio, ρ_t.

(A) 0.002

(B) 0.003

(C) 0.004

(D) 0.005

(c) Determine the axial capacity.

(A) 120,000 lbf

(B) 150,000 lbf

(C) 180,000 lbf

(D) 212,000 lbf

(d) Determine the moment capacity in the x-direction.

(A) 870,000 in-lbf

(B) 930,000 in-lbf

(C) 1,270,000 in-lbf

(D) 1,520,000 in-lbf

(e) Determine the axial stress.

(A) 320 lbf/in^2

(B) 380 lbf/in^2

(C) 420 lbf/in^2

(D) 500 lbf/in^2

(f) Determine the allowable compressive force.

(A) 300,000 lbf

(B) 400,000 lbf

(C) 500,000 lbf

(D) 600,000 lbf

For parts (g) through (j), assume an additional eccentricity of 4 in in the y-direction.

(g) Determine P_y.

(A) 160,000 lbf

(B) 190,000 lbf

(C) 230,000 lbf

(D) 280,000 lbf

(h) Determine P_a.

(A) 300,000 lbf

(B) 330,000 lbf

(C) 390,000 lbf

(D) 420,000 lbf

(i) Determine the maximum biaxial load.

(A) 140,000 lbf

(B) 170,000 lbf

(C) 200,000 lbf

(D) 230,000 lbf

(j) Determine the moment capacity in the y-direction.

(A) 980,000 in-lbf

(B) 1,030,000 in-lbf

(C) 1,120,000 in-lbf

(D) 1,200,000 in-lbf

5. A 32 in (nominal) square concrete masonry column is subjected to a dead load, D, of 180,000 lbf; a live load, L, of 140,000 lbf; and a design shear force of 30,000 lbf. The effects from an earthquake, E, perpendicular to a column face are as follows.

$$\text{bending moment} = 140,000 \text{ ft-lbf}$$

$$\text{axial compressive force} = 45,000 \text{ lbf}$$

Special inspection will be provided. Use $f'_m = 1500$ lbf/in^2, $h = 10$ ft, $r = 9.1$ in, $g = 0.4$, $n = 14$, and $d = 23$ in.

(a) Determine the design eccentricity for a load combination of $D + L$.

(A) 2.7 in

(B) 3.2 in

(C) 3.8 in

(D) 4.6 in

(b) Determine the design eccentricity for a load combination of $D + L + E$.

(A) 2.5 in

(B) 3.7 in

(C) 4.6 in

(D) 6.9 in

(c) Determine the design eccentricity for a load combination of $0.9D + E$.

(A) 4.3 in

(B) 5.2 in

(C) 6.9 in

(D) 8.1 in

(d) Determine the maximum compressive stress on the column for a load combination of $D + L + E$.

(A) 250 lbf/in^2

(B) 320 lbf/in^2

(C) 370 lbf/in^2

(D) 690 lbf/in^2

(e) Determine the maximum allowable axial load on the column. Neglect the steel's contribution.

(A) 370,000 lbf

(B) 420,000 lbf

(C) 460,000 lbf

(D) 510,000 lbf

(f) Considering the three load combinations $(D + L,$ $D + L + E,$ and $0.9D + E)$, determine the minimum steel area required. A one-third increase in allowable stress is permitted for earthquake loading.

(A) none

(B) 2 in^2

(C) 3 in^2

(D) 4 in^2

(g) Determine the design shear stress on the column.

(A) 35 lbf/in^2

(B) 41 lbf/in^2

(C) 50 lbf/in^2

(D) 57 lbf/in^2

(h) Determine the allowable shear stress.

(A) 39 lbf/in^2

(B) 48 lbf/in^2

(C) 50 lbf/in^2

(D) 57 lbf/in^2

(i) Determine the area of shear reinforcement required.

(A) none

(B) 0.11 in^2/ft

(C) 0.20 in^2/ft

(D) 0.27 in^2/ft

(j) If shear reinforcement is provided to resist 25% of the total design shear and two no. 5 bars are used per location, determine the minimum spacing of the shear reinforcement.

(A) 4 in

(B) 8 in

(C) 12 in

(D) 16 in

SOLUTIONS

1. The following calculations are used to determine the required masonry strength.

$$\frac{h}{r} = \frac{(18 \text{ ft})\left(12 \ \frac{\text{in}}{\text{ft}}\right)}{4.53 \text{ in}} = 47.68$$

Since $h/r < 99$,

$$F_a = \tfrac{1}{4} f'_m \left(1 - \left(\frac{h}{140r}\right)^2\right)$$

$$= \tfrac{1}{4} f'_m \left(1 - \left(\frac{47.68}{140}\right)^2\right)$$

$$= 0.221 f'_m$$

$$f_a = \frac{P}{A_n} = \frac{75{,}000 \text{ lbf}}{244.1 \text{ in}^2} = 307.3 \text{ lbf/in}^2$$

$$\text{minimum } e = 0.1t = (0.1)(15.625 \text{ in}) = 1.56 \text{ in}$$

$$e_k = \frac{t}{6} = \frac{15.625 \text{ in}}{6} = 2.6 \text{ in}$$

$$e < e_k$$

The section is in compression.

$$f_b = \frac{M}{S} = \frac{Pe}{S} = \frac{(75{,}000 \text{ lbf})(1.56 \text{ in})}{635.8 \text{ in}^3}$$

$$= 184.0 \text{ lbf/in}^2$$

$$F_b = \tfrac{1}{3} f'_m$$

$$\frac{f_a}{F_a} + \frac{f_b}{F_b} = 1$$

$$\frac{307.3 \ \frac{\text{lbf}}{\text{in}^2}}{0.221 f'_m} + \frac{184.0 \ \frac{\text{lbf}}{\text{in}^2}}{0.333 f'_m} = 1$$

$$\boxed{f'_m = 1943 \text{ lbf/in}^2}$$

2. The following calculations are used to determine the required steel area.

$$\frac{h}{r} = \frac{(10 \text{ ft})\left(12 \ \frac{\text{in}}{\text{ft}}\right)}{9.1 \text{ in}} = 13.19$$

Since $h/r < 99$,

$$F_a = \tfrac{1}{4} f'_m \left(1 - \left(\frac{h}{140r}\right)^2\right)$$

$$= \left(\tfrac{1}{4}\right)\left(3000 \ \frac{\text{lbf}}{\text{in}^2}\right)\left(1 - \left(\frac{13.19}{140}\right)^2\right)$$

$$= 743 \text{ lbf/in}^2$$

The maximum allowable load, P_a, is

$$P_a = F_a bt = \left(743 \; \frac{\text{lbf}}{\text{in}^2}\right)(32 \text{ in})(16 \text{ in}) = 380{,}416 \text{ lbf}$$

Since 350,000 lbf < 380,416 lbf, the column is not overloaded.

Per *MSJC Code*, the minimum eccentricity is

$$\rho = 0.1t = (0.1)(32 \text{ in}) = 3.2 \text{ in} \quad [\text{use 4 in}]$$

$$F_b = \tfrac{1}{3}f'_m = \left(\tfrac{1}{3}\right)\left(3000 \; \frac{\text{lbf}}{\text{in}^2}\right) = 1000 \text{ lbf/in}^2$$

$$\frac{P}{F_b bt} = \frac{350{,}000 \text{ lbf}}{\left(1000 \; \frac{\text{lbf}}{\text{in}^2}\right)(16 \text{ in})(32 \text{ in})} = 0.684$$

$$\frac{Pe}{F_b bt^2} = \frac{(350{,}000 \text{ lbf})(4 \text{ in})}{\left(1000 \; \frac{\text{lbf}}{\text{in}^2}\right)(16 \text{ in})(32 \text{ in})^2} = 0.0854$$

From App. 69.C, $n\rho_t = 0.15$.

$$E_m = 700f'_m = (700)\left(3000 \; \frac{\text{lbf}}{\text{in}^2}\right)$$

$$= 2{,}100{,}000 \text{ lbf/in}^2$$

$$n = \frac{E_s}{E_m} = \frac{29{,}000{,}000 \; \frac{\text{lbf}}{\text{in}^2}}{2{,}100{,}000 \; \frac{\text{lbf}}{\text{in}^2}} = 13.8$$

$$\rho_t = \frac{0.15}{n} = \frac{0.15}{13.8} = 0.0109$$

$$A_s = \rho_t bt = (0.0109)(16 \text{ in})(32 \text{ in}) = \boxed{5.58 \text{ in}^2}$$

($^1\!/_4$ in diameter lateral ties are also required per *MSJC* Sec. 2.1.6.5.)

3. The area projected to the wind per foot of column height is

$$\left(1 \; \frac{\text{ft}}{\text{ft}}\right)\left(\frac{16 \text{ in} + 8 \text{ in} + 16 \text{ in}}{12 \; \frac{\text{in}}{\text{ft}}}\right) = 3.333 \text{ ft}^2/\text{ft}$$

The distributed wind load per foot of height is

$$w = \left(30 \; \frac{\text{lbf}}{\text{ft}^2}\right)\left(3.333 \; \frac{\text{ft}^2}{\text{ft}}\right) = 100 \text{ lbf/ft}$$

The pier acts like a uniformly loaded horizontal cantilever beam. The moment at the base is

$$M = \tfrac{1}{2}wh^2 = \left(\tfrac{1}{2}\right)\left(100 \; \frac{\text{lbf}}{\text{ft}}\right)(36 \text{ ft})^2$$

$$= 64{,}800 \text{ ft-lbf}$$

The west half of the column will be in tension, and masonry cannot support tension. Assume the neutral axis will be in the east arm, placing the north-south steel in tension.

Disregard the effect of the horizontal no. 4 bars, which contribute to crack control. Use the most westerly end as the reference point to find the neutral axis.

The masonry in compression is A_1. (All dimensions are in inches.)

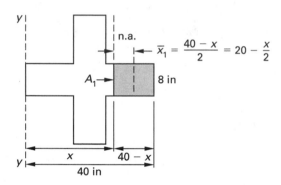

$$A_1 = (40 - x)(8) = 320 - 8x$$

$$\bar{x}_1 = \frac{40 - x}{2} = 20 - \frac{x}{2}$$

(x is measured from the y-y axis, but $\bar{x}$ is measured from the neutral axis.)

Each steel bar contributes a transformed area of

$$nA_b = (20)(0.31 \text{ in}^2) = 6.2 \text{ in}^2$$

Assume three bars in the east arm are in compression.

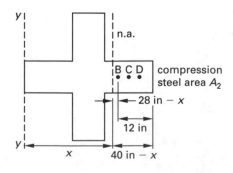

$A_{2B} = 6.2 \text{ in}^2$

$\bar{x}_{2B} = 40 - x - 12$

$\qquad = 28 - x \quad$ [from neutral axis]

$A_{2C} = 6.2 \text{ in}^2$

$\bar{x}_{2C} = 32 - x \quad$ [from neutral axis]

$A_{2D} = 6.2 \text{ in}^2$

$\bar{x}_{2D} = 36 - x \quad$ [from neutral axis]

The steel in tension is

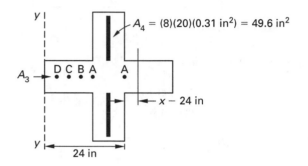

$A_{2A} = 6.2 \text{ in}^2$

$\bar{x}_{2A} = x - 24$

$A_{3A} = 6.2 \text{ in}^2$

$\bar{x}_{3A} = x - 16$

$A_{3B} = 6.2 \text{ in}^2$

$\bar{x}_{3B} = x - 12$

$A_{3C} = 6.2 \text{ in}^2$

$\bar{x}_{3C} = x - 8$

$A_{3D} = 6.2 \text{ in}^2$

$\bar{x}_{3D} = x - 4$

$A_4 = 49.6 \text{ in}^2$

$\bar{x}_4 = x - 20$

The neutral axis is located where

$$\sum_{\text{compression}} A_i \bar{x}_i = \sum_{\text{tension}} A_i \bar{x}_i$$

$$= (320 - 8x)\left(20 - \frac{x}{2}\right)$$
$$\qquad + (6.2)\big((28 - x) + (32 - x) + (36 - x)\big)$$

$$= (6.2)\big((x - 24) + (x - 16) + (x - 12)$$
$$\qquad + (x - 8) + (x - 4)\big) + (49.6)(x - 20)$$

$$= (6.2)(5x - 64) + (49.6)(x - 20)$$

$$4x^2 - 419.2x = -8384$$

$$x^2 - 104.8x = -2096$$

$$x = 26.90 \text{ in} \quad \left[\begin{array}{c} \text{round to } 27.0; \\ \text{location of neutral axis} \end{array}\right]$$

(The location of the neutral axis is not significantly affected by the initial assumption about the number of bars in compression.)

To find the centroidal moment of inertia, create a table.

element	A (in^2)	$\bar{x}$ (in)	$\bar{x}^2$ (in^2)	I (in^4)
A_1				$\dfrac{bh^3}{3} = 5859$
A_{2A}	6.2	3	9	55.8
A_{2B}	6.2	1	1	6.2
A_{2C}	6.2	5	25	155.0
A_{2D}	6.2	9	81	502.2
A_{3A}	6.2	11	121	750.2
A_{3B}	6.2	15	225	1395.0
A_{3C}	6.2	19	361	2238.2
A_{3D}	6.2	23	529	3279.8
A_4	49.6	7	49	$\underline{2430.2}$
			total	16,671.6 in^4

The extreme masonry stress is

$$f_m = \frac{Mc}{I} = \frac{(64{,}800 \text{ ft-lbf})\left(12 \dfrac{\text{in}}{\text{ft}}\right)(40 \text{ in} - 27 \text{ in})}{16{,}671.6 \text{ in}^4}$$

$$= 606.3 \text{ lbf/in}^2$$

The allowable bending stress due to wind loading is

$$\left(\tfrac{4}{3}\right)\left(\tfrac{1}{3}\right)\left(1500 \ \frac{\text{lbf}}{\text{in}^2}\right) = 667 \text{ lbf/in}^2 \quad \text{(psi)}$$

$$\boxed{\text{acceptable}}$$

Use bar A_{3D} to find the maximum steel stress.

$$f_{st} = \frac{(20)(64{,}800 \text{ ft-lbf})\left(12 \frac{\text{in}}{\text{ft}}\right)(23 \text{ in})}{16{,}671.6 \text{ in}^4}$$

$$= 21{,}455 \text{ lbf/in}^2$$

For grade 60 steel, f_{st} is limited to

$$(\tfrac{4}{3})(24{,}000 \text{ psi}) = 32{,}000 \text{ psi}$$

$$\boxed{\text{acceptable}}$$

4. (a) The radius of gyration is determined as follows.

$$I = \frac{bt^3}{12} = \frac{(23.625 \text{ in})^4}{12} = 25{,}960 \text{ in}^4$$

$$A = bt = (23.625 \text{ in})^2 = 558 \text{ in}^2$$

$$r = \sqrt{\frac{I}{A}} = \sqrt{\frac{25{,}960 \text{ in}^4}{558 \text{ in}^2}} = \boxed{6.8 \text{ in}}$$

The answer is (C).

(b) The reinforcement ratio is determined as follows.

$$A_s = 4\pi r^2 = 4\pi\left(\tfrac{5}{16} \text{ in}\right)^2 = 1.23 \text{ in}^2$$

$$\rho_t = \frac{A_s}{bt} = \frac{1.23 \text{ in}^2}{(23.625 \text{ in})(23.625 \text{ in})}$$

$$= \boxed{0.0022 \quad (0.002)}$$

The answer is (A).

(c) The axial capacity is determined as follows.

$$\frac{e}{t} = \frac{6 \text{ in}}{23.625 \text{ in}} = 0.25$$

$$E_m = 700 f'_m = (700)\left(3000 \frac{\text{lbf}}{\text{in}^2}\right)$$

$$= 2{,}100{,}000 \text{ lbf/in}^2$$

$$n = \frac{E_s}{E_m} = \frac{29{,}000{,}000 \frac{\text{lbf}}{\text{in}^2}}{2{,}100{,}000 \frac{\text{lbf}}{\text{in}^2}} = 13.8$$

$$n\rho_t = (13.8)(0.002) = 0.0276 \quad [\text{use } 0.03]$$

From App. 69.B,

$$\frac{P}{F_b bt} = 0.38$$

$$P = (0.38)(\tfrac{1}{3})\left(3000 \frac{\text{lbf}}{\text{in}^2}\right)(23.625 \text{ in})(23.625 \text{ in})$$

$$= \boxed{212{,}093 \text{ lbf} \quad (212{,}200 \text{ lbf})}$$

The answer is (D).

(d) The moment capacity in the x-direction is

$$M = Pe = (212{,}093 \text{ lbf})(6 \text{ in})$$

$$= \boxed{1{,}272{,}558 \text{ in-lbf} \quad (1{,}270{,}000 \text{ in-lbf})}$$

The answer is (C).

(e) The axial stress is

$$f_a = \frac{P}{A} = \frac{212{,}093 \text{ lbf}}{(23.625 \text{ in})^2} = \boxed{380 \text{ lbf/in}^2}$$

The answer is (B).

(f) Ignoring the steel, from Eq. 69.1,

$$P_a = 0.25 f'_m A_n \left(1 - \left(\frac{h}{140r}\right)^2\right)$$

$$= (0.25)\left(3000 \frac{\text{lbf}}{\text{in}^2}\right)(23.625 \text{ in})^2$$

$$\times \left(1 - \left(\frac{(16 \text{ ft})\left(12 \frac{\text{in}}{\text{ft}}\right)}{(140)(6.82 \text{ in})}\right)^2\right)$$

$$= \boxed{401{,}678 \text{ lbf} \quad (400{,}000 \text{ lbf})}$$

The answer is (B).

(g) P_y is determined as follows.

$$\frac{e}{t} = \frac{4 \text{ in}}{23.625 \text{ in}} = 0.17$$

$$n\rho_t = 0.03 \qquad [\text{from part (c)}]$$

From App. 69.B,

$$\frac{P}{F_b bt} = 0.5$$

$$P = 0.5 F_b bt = (0.5)\left(\tfrac{1}{3}\right)\left(3000 \ \frac{\text{lbf}}{\text{in}^2}\right)(23.625 \ \text{in})^2$$

$$= \boxed{279{,}070 \ \text{lbf} \quad (280{,}000 \ \text{lbf})}$$

The answer is (D).

(h) From Eq. 69.1,

$$P_a = \left(0.25 f'_m A_n + 0.65 A_{st} F_s\right)\left(1 - \left(\frac{h}{140r}\right)^2\right)$$

$$= \left((0.25)\left(3000 \ \frac{\text{lbf}}{\text{in}^2}\right)(558 \ \text{in}^2)\right.$$

$$\left. + (0.65)(1.23 \ \text{in}^2)\left(24{,}000 \ \frac{\text{lbf}}{\text{in}^2}\right)\right)$$

$$\times \left(1 - \left(\frac{(16 \ \text{ft})\left(12 \ \frac{\text{in}}{\text{ft}}\right)}{(140)(6.82 \ \text{in})}\right)^2\right)$$

$$= \boxed{419{,}989 \ \text{lbf} \quad (420{,}000 \ \text{lbf})}$$

The answer is (D).

(i) From Eq. 69.6,

$$\frac{1}{P_{\text{biaxial}}} = \frac{1}{P_x} + \frac{1}{P_y} - \frac{1}{P_o}$$

$$= \frac{1}{212{,}093 \ \text{lbf}} + \frac{1}{279{,}070 \ \text{lbf}} - \frac{1}{419{,}989 \ \text{lbf}}$$

$$P_{\text{biaxial}} = \boxed{168{,}998 \ \text{lbf} \quad (170{,}000 \ \text{lbf})}$$

The answer is (B).

(j) The moment capacity in the y-direction is

$$M_y = P_y e_y = (279{,}070 \ \text{lbf})(4 \ \text{in})$$

$$= \boxed{1{,}116{,}280 \ \text{in-lbf} \quad (1{,}120{,}000 \ \text{in-lbf})}$$

The answer is (C).

5. (a) The design eccentricity for dead and live loads is

$$e_{\text{minimum}} = 0.1t = (0.1)(32 \ \text{in})$$

$$= \boxed{3.2 \ \text{in}}$$

The answer is (B).

(b) The design eccentricity for dead and live loads and earthquake forces is determined as follows.

$$P = 180{,}000 \ \text{lbf} + 140{,}000 \ \text{lbf} + 45{,}000 \ \text{lbf}$$

$$= 365{,}000 \ \text{lbf}$$

$$e = \frac{M}{P} = \left(\frac{140{,}000 \ \text{ft-lbf}}{365{,}000 \ \text{lbf}}\right)\left(12 \ \frac{\text{in}}{\text{ft}}\right) = \boxed{4.6 \ \text{in}}$$

The answer is (C).

(c) The design eccentricity for $0.9D + E$ is determined as follows.

$$P = (0.9)(180{,}000 \ \text{lbf}) + 45{,}000 \ \text{lbf}$$

$$= 207{,}000 \ \text{lbf}$$

$$e = \frac{M}{P} = \left(\frac{140{,}000 \ \text{ft-lbf}}{207{,}000 \ \text{lbf}}\right)\left(12 \ \frac{\text{in}}{\text{ft}}\right)$$

$$= \boxed{8.1 \ \text{in}}$$

The answer is (D).

(d) The section modulus of the column is

$$S = \frac{I}{c} = \frac{r^2 A}{c} = \frac{r^2 b^2}{\frac{b}{2}} = 2r^2 b$$

$$= (2)(9.1 \ \text{in})^2(31.625 \ \text{in})$$

$$= 5238 \ \text{in}^3$$

The maximum compressive stress on the column is

$$\frac{P}{A} + \frac{M}{S} = \frac{180{,}000 \ \text{lbf} \ | \ 140{,}000 \ \text{lbf} + 45{,}000 \ \text{lbf}}{(31.625 \ \text{in})^2}$$

$$+ \frac{(140{,}000 \ \text{ft-lbf})\left(12 \ \frac{\text{in}}{\text{ft}}\right)}{5238 \ \text{in}^3}$$

$$= \boxed{685.7 \ \text{lbf/in}^2 \quad (690 \ \text{lbf/in}^2)}$$

The answer is (D).

(e) Use Eq. 69.1. Neglect steel.

$$P_a = 0.25 f'_m A_n\left(1 - \left(\frac{h}{140r}\right)^2\right)$$

$$= (0.25)\left(1500 \ \frac{\text{lbf}}{\text{in}^2}\right)(31.625 \ \text{in})^2$$

$$\times \left(1 - \left(\frac{(10 \ \text{ft})\left(12 \ \frac{\text{in}}{\text{ft}}\right)}{(140)(9.1 \ \text{in})}\right)^2\right)$$

$$= \boxed{371{,}725 \ \text{lbf} \quad (370{,}000 \ \text{lbf})}$$

The answer is (A).

(f) For $D + L$, with $b = t$,

$$\frac{P}{F_b bt} = \frac{320{,}000 \text{ lbf}}{\left(500 \ \frac{\text{lbf}}{\text{in}^2}\right)(31.625 \text{ in})^2} = 0.64$$

$$\frac{Pe}{F_b bt^2} = \frac{(320{,}000 \text{ lbf})(3.2 \text{ in})}{\left(500 \ \frac{\text{lbf}}{\text{in}^2}\right)(31.625 \text{ in})^3} = 0.065$$

From App. 69.A, $n\rho_t = 0.06$.

For $D + L + E$, increase F_b by one-third for load combinations including earthquake.

$$F_b = \left(\tfrac{4}{3}\right)\left(\tfrac{1}{3}\right)\left(1500 \ \frac{\text{lbf}}{\text{in}^2}\right) = 667 \text{ lbf/in}^2$$

$$\frac{P}{F_b bt} = \frac{365{,}000 \text{ lbf}}{\left(667 \ \frac{\text{lbf}}{\text{in}^2}\right)(31.625 \text{ in})^2} = 0.55$$

$$\frac{Pe}{F_b bt^2} = \frac{(365{,}000 \text{ lbf})(4.6 \text{ in})}{\left(667 \ \frac{\text{lbf}}{\text{in}^2}\right)(31.625 \text{ in})^3} = 0.080$$

From App. 69.A, $n\rho_t = 0.06$.

For $0.9D + E$,

$$\frac{P}{F_b bt} = \frac{207{,}000 \text{ lbf}}{\left(667 \ \frac{\text{lbf}}{\text{in}^2}\right)(31.625 \text{ in})^2} = 0.31$$

$$\frac{Pe}{F_b bt^2} = \frac{(207{,}000 \text{ lbf})(8.1 \text{ in})}{\left(667 \ \frac{\text{lbf}}{\text{in}^2}\right)(31.625 \text{ in})^3} = 0.079$$

From App. 69.A, $n\rho_t = 0$.

Use maximum steel of $n\rho_t = 0.06$.

$$\rho_t = \frac{n\rho_t}{n} = \frac{0.06}{14} = 0.0043$$

Since the code minimum of 0.0025 is less than 0.004, this is acceptable.

$$A_s = \rho_t bt = (0.0043)(31.625 \text{ in})^2$$
$$= \boxed{4.3 \text{ in}^2 \quad (4 \text{ in}^2)}$$

The answer is (D).

(g) The design shear stress on the column is

$$f_v = \frac{V}{bd} = \frac{30{,}000 \text{ lbf}}{(31.625 \text{ in})(23 \text{ in})}$$
$$= \boxed{41.2 \text{ lbf/in}^2 \quad (41 \text{ lbf/in}^2)}$$

The answer is (B).

(h) The allowable shear stress is given by Eq. 68.2.

$$F_v = \tfrac{4}{3}\sqrt{f'_m} = \tfrac{4}{3}\sqrt{1500 \ \frac{\text{lbf}}{\text{in}^2}} = 51.6 \text{ lbf/in}^2$$

The allowable shear stress is limited to 50 lbf/in² by *MSJC* Sec. 2.3.5.2.2.

$$50 \text{ lbf/in}^2 > 41 \text{ lbf/in}^2 \quad [\text{OK}]$$

The shear stress is $\boxed{50 \text{ lbf/in}^2.}$

The answer is (C).

(i) Since F_v is greater than f_v, $\boxed{\text{none is required.}}$

The answer is (A).

(j) Rearranging from Eq. 68.50, and increasing the allowable stress by one-third for earthquake loading,

$$s = \frac{A_v F_s d}{V} = \frac{(2)\left(\frac{\pi}{4}\right)\left(\frac{5}{16} \text{ in}\right)^2 \left(24{,}000 \ \frac{\text{lbf}}{\text{in}^2}\right)\left(\frac{4}{3}\right)(23 \text{ in})}{(0.25)(30{,}000 \text{ lbf})}$$
$$= 15.05 \text{ in}$$

Since the bars are typically spaced with each course, use no. 5 bars spaced at $\boxed{8 \text{ in.}}$

The answer is (B).

70 Properties of Solid Bodies

PRACTICE PROBLEMS

1. A spoked flywheel has an outside diameter of 60 in (1500 mm). The rim thickness is 6 in (150 mm) and the width is 12 in (300 mm). The cylindrical hub has an outside diameter of 12 in (300 mm), a thickness of 3 in (75 mm), and a width of 12 in (300 mm). The rim and hub are connected by six equally spaced cylindrical radial spokes, each having a diameter of 4.25 in (110 mm). All parts of the flywheel are ductile cast iron with a density of 0.256 lbm/in^3 (7080 kg/m^3). What is the rotational mass moment of inertia of the entire flywheel?

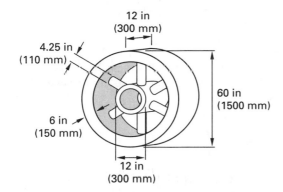

(A) 16,000 lbm-ft^2 (625 kg·m^2)

(B) 16,200 lbm-ft^2 (631 kg·m^2)

(C) 16,800 lbm-ft^2 (654 kg·m^2)

(D) 193,000 lbm-ft^2 (7500 kg·m^2)

SOLUTIONS

1. *Customary U.S. Solution*

From the hollow circular cylinder in App. 70.A, the mass moment of inertia of the rim is

$$I_{x,1} = \left(\frac{\pi\rho L}{2}\right)(r_o^4 - r_i^4)$$

$$= \frac{\left(\frac{\pi}{2}\right)\left(0.256\ \frac{\text{lbm}}{\text{in}^3}\right)(12\ \text{in})}{\left(12\ \frac{\text{in}}{\text{ft}}\right)^2}$$

$$\times \left(\left(\frac{60\ \text{in}}{2}\right)^4 - \left(\frac{60\ \text{in} - 12\ \text{in}}{2}\right)^4\right)$$

$$= 16{,}025\ \text{lbm-ft}^2$$

From the hollow circular cylinder in App. 70.A, the mass moment of inertia of the hub is

$$I_{x,2} = \left(\frac{\pi\rho L}{2}\right)(r_o^4 - r_i^4)$$

$$= \frac{\left(\frac{\pi}{2}\right)\left(0.256\ \frac{\text{lbm}}{\text{in}^3}\right)(12\ \text{in})}{\left(12\ \frac{\text{in}}{\text{ft}}\right)^2}$$

$$\times \left(\left(\frac{12\ \text{in}}{2}\right)^4 - \left(\frac{12\ \text{in} - 6\ \text{in}}{2}\right)^4\right)$$

$$= 41\ \text{lbm-ft}^2$$

The length of a cylindrical spoke is

$$L = 24\ \text{in} - 6\ \text{in} = 18\ \text{in}$$

The mass of a spoke is

$$m = \rho A L = \rho \pi r^2 L$$

$$= \left(0.256\ \frac{\text{lbm}}{\text{in}^3}\right)\pi\left(\frac{4.25\ \text{in}}{2}\right)^2(18\ \text{in})$$

$$= 65.37\ \text{lbm}$$

From the solid circular cylinder in App. 70.A, the mass moment of inertia of a spoke about its own centroidal axis is

$$I_z = \frac{m(3r^2 + L^2)}{12}$$

$$= \frac{(65.37 \text{ lbm})\left((3)\left(\frac{4.25 \text{ in}}{2}\right)^2 + (18 \text{ in})^2\right)}{(12)\left(12 \ \frac{\text{in}}{\text{ft}}\right)^2}$$

$$= 13 \text{ lbm-ft}^2$$

Use the parallel axis theorem, Eq. 70.12, to find the mass moment of inertia of a spoke about the axis of the flywheel.

$$d = \frac{12 \text{ in}}{2} + \frac{18 \text{ in}}{2} = 15 \text{ in}$$

$$I_{x,3 \text{ per spoke}} = I_c + md^2$$

$$= 13 \text{ lbm-ft}^2 + (65.37 \text{ lbm})\left(\frac{15 \text{ in}}{12 \ \frac{\text{in}}{\text{ft}}}\right)^2$$

$$= 115 \text{ lbm-ft}^2$$

The total for six spokes is

$$I_{x,3} = 6I_{x,3 \text{ per spoke}} = (6)(115 \text{ lbm-ft}^2)$$

$$= 690 \text{ lbm-ft}^2$$

Finally, the total rotational mass moment of inertia of the flywheel is

$$I = I_{x,1} + I_{x,2} + I_{x,3}$$

$$= 16{,}025 \text{ lbm-ft}^2 + 41 \text{ lbm-ft}^2 + 690 \text{ lbm-ft}^2$$

$$= \boxed{16{,}756 \text{ lbm-ft}^2 \quad (16{,}800 \text{ lbm-ft}^2)}$$

The answer is (C).

SI Solution

From the hollow circular cylinder in App. 70.A, the mass moment of inertia of the rim is

$$I_{x,1} = \left(\frac{\pi \rho L}{2}\right)(r_o^4 - r_i^4)$$

$$= \left(\frac{\pi \left(7080 \ \frac{\text{kg}}{\text{m}^3}\right)(0.3 \text{ m})}{2}\right)$$

$$\times \left(\left(\frac{1.5 \text{ m}}{2}\right)^4 - \left(\frac{1.5 \text{ m} - 0.30 \text{ m}}{2}\right)^4\right)$$

$$= 623.3 \text{ kg-m}^2$$

From the hollow circular cylinder in App. 70.A, the mass moment of inertia of the hub is

$$I_{x,2} = \left(\frac{\pi \rho L}{2}\right)(r_o^4 - r_i^4)$$

$$= \left(\frac{\pi \left(7080 \ \frac{\text{kg}}{\text{m}^3}\right)(0.3 \text{ m})}{2}\right)$$

$$\times \left(\left(\frac{0.3 \text{ m}}{2}\right)^4 - \left(\frac{0.3 \text{ m} - 0.15 \text{ m}}{2}\right)^4\right)$$

$$= 1.6 \text{ kg-m}^2$$

The length of a cylindrical spoke is

$$L = 0.6 \text{ m} - 0.15 \text{ m} = 0.45 \text{ m}$$

The mass of a spoke is

$$m = \rho AL = \rho \pi r^2 L$$

$$= \left(7080 \ \frac{\text{kg}}{\text{m}^3}\right) \pi \left(\frac{0.11 \text{ m}}{2}\right)^2 (0.45 \text{ m})$$

$$= 30.3 \text{ kg}$$

From the solid circular cylinder in App. 70.A, the mass moment of inertia of a spoke about its own centroidal axis is

$$I_z = \frac{m(3r^2 + L^2)}{12}$$

$$= \frac{(30.3 \text{ kg})\left((3)\left(\frac{0.11 \text{ m}}{2}\right)^2 + (0.45 \text{ m})^2\right)}{12}$$

$$= 0.53 \text{ kg-m}^2$$

Use the parallel axis theorem, Eq. 70.12, to find the mass moment of inertia of a spoke about the axis of the flywheel.

$$d = \frac{0.30 \text{ m}}{2} + \frac{0.45 \text{ m}}{2} = 0.375 \text{ m}$$

$$I_{x,3 \text{ per spoke}} = I_c + md^2$$

$$= 0.53 \text{ kg-m}^2 + (30.3 \text{ kg})(0.375 \text{ m})^2$$

$$= 4.8 \text{ kg-m}^2$$

The total for six spokes is

$$I_{x,3} = 6I_{x,3 \text{ per spoke}} = (6)(4.8 \text{ kg-m}^2)$$

$$= 28.8 \text{ kg-m}^2$$

The total rotational mass moment of inertia of the flywheel is

$$I = I_{x,1} + I_{x,2} + I_{x,3}$$
$$= 623.3 \text{ kg·m}^2 + 1.6 \text{ kg·m}^2 + 28.8 \text{ kg·m}^2$$
$$= \boxed{653.7 \text{ kg·m}^2 \quad (654 \text{ kg·m}^2)}$$

The answer is (C).

Transportation

71 Kinematics

PRACTICE PROBLEMS

Linear Particle Motion

1. A particle moves horizontally according to the formula $s = 2t^2 - 8t + 3$.

(a) When $t = 2$, what are the position, velocity, and acceleration?

(b) What are the linear displacement and total distance traveled between $t = 1$ and $t = 3$?

Uniform Acceleration

2. A jet plane acquires a speed of 180 mph (290 km/h) in 60 sec. What is its acceleration?

   (A) 4.4 ft/sec² (1.3 m/s²)

   (B) 6.3 ft/sec² (1.9 m/s²)

   (C) 8.9 ft/sec² (2.7 m/s²)

   (D) 12 ft/sec² (3.6 m/s²)

3. What is the acceleration of a train that increases its speed from 5 ft/sec to 20 ft/sec (1.5 m/s to 6 m/s) in 2 min?

   (A) 0.13 ft/sec² (0.038 m/s²)

   (B) 0.39 ft/sec² (0.12 m/s²)

   (C) 0.82 ft/sec² (0.25 m/s²)

   (D) 1.3 ft/sec² (0.39 m/s²)

4. A car traveling at 60 mph (100 km/h) applies its brakes and stops in 5 sec. What is its acceleration and distance traveled before stopping?

Projectile Motion

5. A projectile is fired at 45° from the horizontal with an initial velocity of 2700 ft/sec (820 m/s). Find the maximum altitude and range neglecting air friction.

6. A projectile is launched with an initial velocity of 900 ft/sec (270 m/s). The target is 12,000 ft (3600 m) away and 2000 ft (600 m) higher than the launch point. Air friction is to be neglected. At what angle should the projectile be launched?

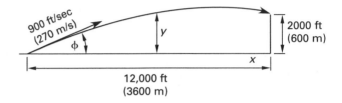

7. A baseball is hit at 60 ft/sec (20 m/s) and 36.87° from the horizontal. It strikes a fence 72 ft (22 m) away. Find the velocity components and the elevation above the origin at impact.

8. A bomb is dropped from a plane that is climbing at 30° and 600 ft/sec (180 m/s) while traveling at 12,000 ft (3600 m) altitude.

(a) What is the bomb's maximum altitude?

   (A) 12,500 ft (3700 m)

   (B) 13,000 ft (3900 m)

   (C) 13,400 ft (4000 m)

   (D) 14,200 ft (4300 m)

(b) How long will it take for the bomb to reach the ground from the release point?

   (A) 24 sec

   (B) 38 sec

   (C) 43 sec

   (D) 55 sec

Rotational Particle Motion

9. A point travels in a circle according to $\omega = 6t^2 - 10t$. At $t = 2$, the direction of motion is clockwise.

(a) What is the angular velocity at $t = 2$?

   (A) 4

   (B) 6

   (C) 8

   (D) 12

(b) What is the displacement between $t = 1$ and $t = 3$?

(A) 6 rad

(B) 10 rad

(C) 12 rad

(D) 15 rad

(c) What is the total angle turned through between $t = 1$ and $t = 3$? Assume $\theta(0) = 0$.

(A) 6 rad

(B) 8 rad

(C) 12 rad

(D) 15 rad

10. What is the linear speed of a point on the edge of a 14 in diameter (36 cm diameter) disk turning at 40 rpm?

(A) 2.4 ft/sec (0.75 m/s)

(B) 4.5 ft/sec (1.4 m/s)

(C) 7.8 ft/sec (2.3 m/s)

(D) 9.2 ft/sec (2.8 m/s)

11. What angular acceleration is required to increase an electric motor's speed from 1200 rpm to 3000 rpm in 10 sec?

(A) 10 rad/sec^2

(B) 14 rad/sec^2

(C) 18 rad/sec^2

(D) 25 rad/sec^2

12. An apparatus for determining the speed of a bullet consists of 2 paper disks mounted 5 ft (1.5 m) apart on a single horizontal shaft which is turning at 1750 rpm. A bullet pierces both disks at radius 6 in (15 cm) and an angle of 18° exists between each hole. What is the bullet velocity?

(A) 1700 ft/sec (510 m/s)

(B) 2100 ft/sec (630 m/s)

(C) 2400 ft/sec (720 m/s)

(D) 2900 ft/sec (880 m/s)

13. Disks B and C are in contact and rotate without slipping. A and B are splined together and rotate counterclockwise. Angular velocity and acceleration of disk C are 2 rad/sec (2 rad/s) and 6 rad/sec^2 (6 rad/s^2), respectively. What is the velocity and acceleration of point D?

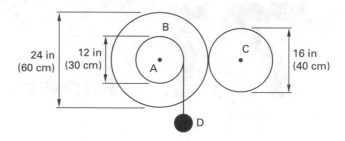

Relative Motion

14. The center of a wheel with an outer diameter of 24 in (610 mm) is moving at 28 mph (12.5 m/s). There is no slippage between the wheel and surface. A valve stem is mounted 6 in (150 mm) from the center. What are the velocity and direction of the valve stem when it is 45° from the horizontal?

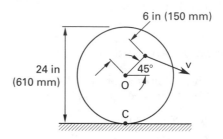

15. A balloon is 200 ft (60 m) above the ground and is rising at a constant 15 ft/sec (4.5 m/s). An automobile passes under it traveling along a straight and level road at 45 mph (72 km/h). How fast is the distance between them changing 1 sec later?

(A) 13 ft/sec (4.0 m/s)

(B) 34 ft/sec (10 m/s)

(C) 37 ft/sec (11 m/s)

(D) 51 ft/sec (15 m/s)

Rotation About a Fixed Axis

16. Find the velocities of points A and B with respect to point O if the wheel rolls without slipping. The axle that the wheel is attached to moves to the right at 10 ft/sec (3 m/s).

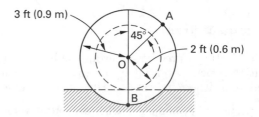

17. What is the velocity of point B with respect to point A in Prob. 16?

SOLUTIONS

1. (a) From the problem statement,

$$s(t) = 2t^2 - 8t + 3$$

Using Eq. 71.3 and Eq. 71.4,

$$v(t) = \frac{ds(t)}{dt} = 4t - 8$$

$$a(t) = \frac{dv(t)}{dt} = 4$$

At $t = 2$,

$$s = (2)(2)^2 - (8)(2) + 3 = \boxed{-5}$$
$$v = (4)(2) - 8 = \boxed{0}$$
$$a = \boxed{4}$$

(b) At $t = 1$,

$$s = (2)(1)^2 - (8)(1) + 3 = \boxed{-3}$$

At $t = 3$,

$$s = (2)(3)^2 - (8)(3) + 3 = \boxed{-3}$$

From Eq. 71.7,

$$\text{displacement} = s(t_2) - s(t_1) = -3 - (-3) = 0$$

The total distance traveled from $t = 1$ to $t = 3$ is

$$s = \int_1^3 |v(t)| \, dt = \int_1^3 |4t - 8| \, dt$$

$$= \int_1^2 (8 - 4t) \, dt + \int_2^3 (4t - 8) \, dt$$

$$= 2 + 2$$

$$= \boxed{4}$$

2. *Customary U.S. Solution*

$$a = \frac{\Delta v}{\Delta t} = \frac{\left(180 \; \frac{\text{mi}}{\text{hr}}\right)\left(5280 \; \frac{\text{ft}}{\text{mi}}\right)}{(60 \; \text{sec})\left(60 \; \frac{\text{sec}}{\text{min}}\right)\left(60 \; \frac{\text{min}}{\text{hr}}\right)}$$

$$= \boxed{4.4 \; \text{ft/sec}^2}$$

The answer is (A).

SI Solution

$$a = \frac{\Delta v}{\Delta t} = \frac{\left(290 \; \frac{\text{km}}{\text{h}}\right)\left(1000 \; \frac{\text{m}}{\text{km}}\right)}{(60 \; \text{s})\left(60 \; \frac{\text{s}}{\text{min}}\right)\left(60 \; \frac{\text{min}}{\text{h}}\right)}$$

$$= \boxed{1.3 \; \text{m/s}^2}$$

The answer is (A).

3. *Customary U.S. Solution*

Use Table 71.1.

$$a = \frac{v - v_0}{t} = \frac{20 \; \frac{\text{ft}}{\text{sec}} - 5 \; \frac{\text{ft}}{\text{sec}}}{(2 \; \text{min})\left(60 \; \frac{\text{sec}}{\text{min}}\right)}$$

$$= \boxed{0.125 \; \text{ft/sec}^2 \quad (0.13 \; \text{ft/sec}^2)}$$

The answer is (A).

SI Solution

Use Table 71.1.

$$a = \frac{v - v_0}{t} = \frac{6 \; \frac{\text{m}}{\text{s}} - 1.5 \; \frac{\text{m}}{\text{s}}}{(2 \; \text{min})\left(60 \; \frac{\text{s}}{\text{min}}\right)}$$

$$= \boxed{0.0375 \; \text{m/s}^2 \quad (0.038 \; \text{m/s}^2)}$$

The answer is (A).

4. *Customary U.S. Solution*

Assume uniform deceleration.

$$v_0 = \frac{\left(60 \; \frac{\text{mi}}{\text{hr}}\right)\left(5280 \; \frac{\text{ft}}{\text{mi}}\right)}{\left(60 \; \frac{\text{sec}}{\text{min}}\right)\left(60 \; \frac{\text{min}}{\text{hr}}\right)} = 88 \; \text{ft/sec}$$

$$a = \frac{v - v_0}{t} = \frac{0 \; \frac{\text{ft}}{\text{sec}} - 88 \; \frac{\text{ft}}{\text{sec}}}{5 \; \text{sec}} = \boxed{-17.6 \; \text{ft/sec}^2}$$

$$s = \tfrac{1}{2}t(v_0 + v) = \tfrac{1}{2}tv_0 = \left(\tfrac{1}{2}\right)(5 \; \text{sec})\left(88 \; \frac{\text{ft}}{\text{sec}}\right)$$

$$= \boxed{220 \; \text{ft}}$$

SI Solution

Assume uniform deceleration.

$$v_0 = \frac{\left(100 \ \frac{km}{h}\right)\left(1000 \ \frac{m}{km}\right)}{\left(60 \ \frac{s}{min}\right)\left(60 \ \frac{min}{h}\right)} = 27.78 \ m/s$$

$$a = \frac{v - v_0}{t} = \frac{0 \ \frac{m}{s} - 27.78 \ \frac{m}{s}}{5 \ s} = \boxed{-5.56 \ m/s^2}$$

$$s = \tfrac{1}{2}t(v_0 + v) = \tfrac{1}{2}tv_0 = \left(\tfrac{1}{2}\right)(5 \ s)\left(27.78 \ \frac{m}{s}\right)$$

$$= \boxed{69.5 \ m}$$

5. *Customary U.S. Solution*

Use Table 71.2.

$$H = \frac{v_0^2 \sin^2 \phi}{2g} = \frac{\left(2700 \ \frac{ft}{sec}\right)^2 (\sin 45°)^2}{(2)\left(32.2 \ \frac{ft}{sec^2}\right)}$$

$$= \boxed{56,599 \ ft}$$

$$R = \frac{v_0^2 \sin 2\phi}{g} = \frac{\left(2700 \ \frac{ft}{sec}\right)^2 (\sin(2)(45°))}{32.2 \ \frac{ft}{sec^2}}$$

$$= \boxed{226,398 \ ft}$$

SI Solution

Use Table 71.2.

$$H = \frac{v_0^2 \sin^2 \phi}{2g} = \frac{\left(820 \ \frac{m}{s}\right)^2 (\sin 45°)^2}{(2)\left(9.81 \ \frac{m}{s^2}\right)}$$

$$= \boxed{17\,136 \ m}$$

$$R = \frac{v_0^2 \sin 2\phi}{g} = \frac{\left(820 \ \frac{m}{s}\right)^2 (\sin(2)(45°))}{9.81 \ \frac{m}{s^2}}$$

$$= \boxed{68\,542 \ m}$$

6. *Customary U.S. Solution*

From Table 71.2, the x-distance is

$$x = (v_0 \cos \phi)t$$

Solve for t.

$$t = \frac{x}{v_0 \cos \phi} = \frac{12{,}000 \ ft}{\left(900 \ \frac{ft}{sec}\right)(\cos \phi)} = \frac{13.33}{\cos \phi}$$

From Table 71.2, the y-distance is

$$y = (v_0 \sin \phi)t - \tfrac{1}{2}gt^2$$

Substitute t and the given value of y.

$$2000 \ ft = \left(900 \ \frac{ft}{sec}\right)(\sin \phi)\left(\frac{13.33}{\cos \phi}\right)$$

$$- \left(\tfrac{1}{2}\right)\left(32.2 \ \frac{ft}{sec^2}\right)\left(\frac{13.33}{\cos \phi}\right)^2$$

Simplify.

$$1 = 6.0 \tan \phi - \frac{1.43}{\cos^2 \phi}$$

Use the following identity.

$$\frac{1}{\cos^2 \phi} = 1 + \tan^2 \phi$$

$$1 = 6.0 \tan \phi - 1.43 - 1.43 \tan^2 \phi$$

Simplify.

$$\tan^2 \phi - 4.20 \tan \phi + 1.70 = 0$$

Use the quadratic formula.

$$\tan \phi = \frac{4.20 \pm \sqrt{(4.20)^2 - (4)(1)(1.70)}}{(2)(1)}$$

$$= 3.75, \ 0.454 \quad [\text{radians}]$$

$$\phi = \tan^{-1} 3.75, \ \tan^{-1} 0.454$$

$$= \boxed{75.1°, \ 24.4°}$$

SI Solution

From Table 71.2, the x-distance is

$$x = (v_0 \cos \phi)t$$

Solve for t.

$$t = \frac{x}{v_0 \cos \phi} = \frac{3600 \ m}{\left(270 \ \frac{m}{s}\right)(\cos \phi)} = \frac{13.33}{\cos \phi}$$

From Table 71.2, the y-distance is

$$y = (v_0 \sin \phi)t - \tfrac{1}{2}gt^2$$

Substitute t and the given value of y.

$$600 \text{ m} = \left(270 \ \tfrac{\text{m}}{\text{s}}\right)(\sin \phi)\left(\frac{13.33}{\cos \phi}\right)$$
$$- \left(\tfrac{1}{2}\right)\left(9.81 \ \tfrac{\text{m}}{\text{s}^2}\right)\left(\frac{13.33}{\cos \phi}\right)^2$$

Simplify.

$$1 = 6.0 \tan \phi - \frac{1.45}{\cos^2 \phi}$$

Use the following identity.

$$\frac{1}{\cos^2 \phi} = 1 + \tan^2 \phi$$
$$1 = 6.0 \tan \phi - 1.45 - 1.45 \tan^2 \phi$$

Simplify.

$$\tan^2 \phi - 4.14 \tan \phi + 1.69 = 0$$

Use the quadratic formula.

$$\tan \phi = \frac{4.14 \pm \sqrt{(4.14)^2 - (4)(1)(1.69)}}{(2)(1)}$$
$$= 3.68, \ 0.459 \quad [\text{radians}]$$
$$\phi = \tan^{-1} 3.68, \ \tan^{-1} 0.459$$
$$= \boxed{74.8°, \ 24.7°}$$

7. *Customary U.S. Solution*

Neglect air friction. From Table 71.2,

$$v_x(t) = v_0 \cos \phi = \left(60 \ \tfrac{\text{ft}}{\text{sec}}\right)(\cos 36.87°)$$
$$= \boxed{48 \text{ ft/sec} \quad [\text{constant}]}$$

$$t = \frac{s}{v_x} = \frac{72 \text{ ft}}{48 \ \tfrac{\text{ft}}{\text{sec}}} = 1.5 \text{ sec}$$

$$v_y(t) = v_0 \sin \phi - gt$$
$$= \left(60 \ \tfrac{\text{ft}}{\text{sec}}\right)(\sin 36.87°) - \left(32.2 \ \tfrac{\text{ft}}{\text{sec}^2}\right)(1.5 \text{ sec})$$
$$= \boxed{-12.3 \text{ ft/sec}}$$

$$y = (v_0 \sin \phi)t - \tfrac{1}{2}gt^2$$
$$= \left(60 \ \tfrac{\text{ft}}{\text{sec}}\right)(\sin 36.87°)(1.5 \text{ sec})$$
$$- \left(\tfrac{1}{2}\right)\left(32.2 \ \tfrac{\text{ft}}{\text{sec}^2}\right)(1.5 \text{ sec})^2$$
$$= \boxed{17.78 \text{ ft}}$$

SI Solution

Neglect air friction. From Table 71.2,

$$v_x(t) = v_0 \cos \phi = \left(20 \ \tfrac{\text{m}}{\text{s}}\right)(\cos 36.87°)$$
$$= \boxed{16 \text{ m/s} \quad [\text{constant}]}$$

$$t = \frac{s}{v_x} = \frac{22 \text{ m}}{16 \ \tfrac{\text{m}}{\text{s}}} = 1.375 \text{ s}$$

$$v_y(t) = v_0 \sin \phi - gt$$
$$= \left(20 \ \tfrac{\text{m}}{\text{s}}\right)(\sin 36.87°) - \left(9.81 \ \tfrac{\text{m}}{\text{s}^2}\right)(1.375 \text{ s})$$
$$= \boxed{-1.49 \text{ m/s}}$$

$$y = (v_0 \sin \phi)t - \tfrac{1}{2}gt^2$$
$$= \left(20 \ \tfrac{\text{m}}{\text{s}}\right)(\sin 36.87°)(1.375 \text{ s})$$
$$- \left(\tfrac{1}{2}\right)\left(9.81 \ \tfrac{\text{m}}{\text{s}^2}\right)(1.375 \text{ s})^2$$
$$= \boxed{7.227 \text{ m}}$$

8. *Customary U.S. Solution*

(a) Using Table 71.2, the bomb's maximum altitude is

$$H = z + \frac{v_0^2 \sin^2 \phi}{2g}$$
$$= 12{,}000 \text{ ft} + \frac{\left(600 \ \tfrac{\text{ft}}{\text{sec}}\right)^2 (\sin 30°)^2}{(2)\left(32.2 \ \tfrac{\text{ft}}{\text{sec}^2}\right)}$$
$$= \boxed{13{,}398 \text{ ft} \quad (13{,}400 \text{ ft})}$$

The answer is (C).

(b) Let t_1 be the time the bomb takes to reach the maximum altitude.

$$t_1 = \tfrac{1}{2}\left(\frac{2v_0 \sin \phi}{g}\right) = \frac{\left(600 \ \tfrac{\text{ft}}{\text{sec}}\right)(\sin 30°)}{32.2 \ \tfrac{\text{ft}}{\text{sec}^2}} = 9.32 \text{ sec}$$

Transportation

Let t_2 be the time the bomb takes to fall from H.

$$t_2 = \sqrt{\frac{2H}{g}} = \sqrt{\frac{(2)(13{,}398 \text{ ft})}{32.2 \ \dfrac{\text{ft}}{\text{sec}^2}}} = 28.85 \text{ sec}$$

$$t = t_1 + t_2 = 9.32 \text{ sec} + 28.85 \text{ sec}$$

$$= \boxed{38.17 \text{ sec} \quad (38 \text{ sec})}$$

The answer is (B).

SI Solution

(a) Using Table 71.2, the bomb's maximum altitude is

$$H = z + \frac{v_0^2 \sin^2 \phi}{2g}$$

$$= 3600 \text{ m} + \frac{\left(180 \ \dfrac{\text{m}}{\text{s}}\right)^2 (\sin 30^\circ)^2}{(2)\left(9.81 \ \dfrac{\text{m}}{\text{s}^2}\right)}$$

$$= \boxed{4012.8 \text{ m} \quad (4000 \text{ m})}$$

The answer is (C).

(b) Let t_1 be the time the bomb takes to reach the maximum altitude.

$$t_1 = \frac{1}{2}\left(\frac{2v_0 \sin \phi}{g}\right) = \frac{\left(180 \ \dfrac{\text{m}}{\text{s}}\right)(\sin 30^\circ)}{9.81 \ \dfrac{\text{m}}{\text{s}^2}} = 9.17 \text{ s}$$

Let t_2 be the time the bomb takes to fall from H.

$$t_2 = \sqrt{\frac{2H}{g}} = \sqrt{\frac{(2)(4012.8 \text{ m})}{9.81 \ \dfrac{\text{m}}{\text{s}^2}}} = 28.60 \text{ s}$$

$$t = t_1 + t_2 = 9.17 \text{ s} + 28.60 \text{ s}$$

$$= \boxed{37.77 \text{ s} \quad (38 \text{ s})}$$

The answer is (B).

9. (a) The angular velocity is

$$\omega(t) = 6t^2 - 10t$$

$$\omega(2) = (6)(2)^2 - (10)(2) = \boxed{4}$$

The answer is (A).

(b) The displacement is

$$\theta = \theta(3) - \theta(1) = \int_1^3 \omega(t)\, dt$$

$$= \int_1^3 (6t^2 - 10t)\, dt = 2t^3 - 5t^2 + C \Big|_1^3$$

$$[C = 0 \text{ since } \theta(0) = 0]$$

$$= (2)(3)^3 - (5)(3)^2 - (2)(1)^3 + (5)(1)^2$$

$$= \boxed{12 \text{ rad}}$$

The answer is (C).

(c) To find the total distance traveled, check for sign reversals in $\omega(t)$ over the interval $t = 1$ to $t = 3$.

$$6t^2 - 10t = 0$$

$$\text{sign reversal at } t = \tfrac{5}{3}$$

The total angle turned is

$$\int_1^3 |\omega(t)|\, dt = \int_1^{5/3} (10t - 6t^2)\, dt + \int_{5/3}^3 (6t^2 - 10t)\, dt$$

$$= \boxed{15.26 \text{ rad} \quad (15 \text{ rad})}$$

The answer is (D).

10. *Customary U.S. Solution*

Use Eq. 71.19.

$$\text{v}(t) = \omega r = (2\pi f)r = d\pi f$$

$$= \frac{(14 \text{ in})\pi\left(40 \ \dfrac{\text{rev}}{\text{min}}\right)}{\left(12 \ \dfrac{\text{in}}{\text{ft}}\right)\left(60 \ \dfrac{\text{sec}}{\text{min}}\right)}$$

$$= \boxed{2.44 \text{ ft/sec} \quad (2.4 \text{ ft/sec})}$$

The answer is (A).

SI Solution

Use Eq. 71.19.

$$\text{v}(t) = \omega r = (2\pi f)r = d\pi f$$

$$= \frac{(0.36 \text{ m})\pi\left(40 \ \dfrac{\text{rev}}{\text{min}}\right)}{60 \ \dfrac{\text{s}}{\text{min}}}$$

$$= \boxed{0.754 \text{ m/s} \quad (0.75 \text{ m/s})}$$

The answer is (A).

Transportation

11. Convert 1200 rpm and 3000 rpm into radians per second. There are 2π radians per complete revolution.

$$\omega_1 = 2\pi f_1 = \frac{\left(2\pi \frac{\text{rad}}{\text{rev}}\right)\left(1200 \frac{\text{rev}}{\text{min}}\right)}{60 \frac{\text{sec}}{\text{min}}}$$

$$= 125.66 \text{ rad/sec}$$

$$\omega_2 = 2\pi f_2 = \frac{\left(2\pi \frac{\text{rad}}{\text{rev}}\right)\left(3000 \frac{\text{rev}}{\text{min}}\right)}{60 \frac{\text{sec}}{\text{min}}}$$

$$= 314.16 \text{ rad/sec}$$

$$\alpha = \frac{\omega_2 - \omega_1}{\Delta t} = \frac{314.16 \frac{\text{rad}}{\text{sec}} - 125.66 \frac{\text{rad}}{\text{sec}}}{10 \text{ sec}}$$

$$= \boxed{18.85 \text{ rad/sec}^2 \quad (18 \text{ rad/sec}^2)}$$

The answer is (C).

12. *Customary U.S. Solution*

$$f = \frac{1750 \frac{\text{rev}}{\text{min}}}{60 \frac{\text{sec}}{\text{min}}} = 29.167 \text{ rev/sec}$$

$$\theta = (18°)\left(\frac{1 \text{ rev}}{360°}\right) = 0.05 \text{ rev}$$

$$t = \frac{\theta}{f} = \frac{0.05 \text{ rev}}{29.167 \frac{\text{rev}}{\text{sec}}} = 0.001714 \text{ sec}$$

$$\text{v} = \frac{s}{t} = \frac{5 \text{ ft}}{0.001714 \text{ sec}} = \boxed{2916.7 \text{ ft/sec} \quad (2900 \text{ ft/sec})}$$

The answer is (D).

SI Solution

$$\omega = \frac{\left(1750 \frac{\text{rev}}{\text{min}}\right)\left(2\pi \frac{\text{rad}}{\text{rev}}\right)}{60 \frac{\text{s}}{\text{min}}} = 183.26 \text{ rad/s}$$

$$\theta = (18°)\left(\frac{2\pi \text{ rad}}{360°}\right) = 0.314 \text{ rad}$$

$$t = \frac{\theta}{\omega} = \frac{0.314 \text{ rad}}{183.26 \frac{\text{rad}}{\text{s}}} = 0.001713 \text{ s}$$

$$\text{v} = \frac{s}{t} = \frac{1.5 \text{ m}}{0.001713 \text{ s}} = \boxed{875.66 \text{ m/s} \quad (880 \text{ m/s})}$$

The answer is (D).

13. *Customary U.S. Solution*

$$\omega_\text{B} = \left(\frac{16 \text{ in}}{24 \text{ in}}\right)\omega_\text{C} = \left(\frac{16 \text{ in}}{24 \text{ in}}\right)\left(2 \frac{\text{rad}}{\text{sec}}\right) = 1.333 \text{ rad/sec}$$

$$\alpha_\text{B} = \left(\frac{16 \text{ in}}{24 \text{ in}}\right)\alpha_\text{C} = \left(\frac{16 \text{ in}}{24 \text{ in}}\right)\left(6 \frac{\text{rad}}{\text{sec}^2}\right) = 4 \text{ rad/sec}^2$$

Since A and B are splined together,

$$\omega_\text{A} = \omega_\text{B}$$

$$\alpha_\text{A} = \alpha_\text{B}$$

$$\text{v}_\text{D} = r_\text{A}\omega_\text{A} = (0.5 \text{ ft})\left(1.333 \frac{\text{rad}}{\text{sec}}\right)$$

$$= \boxed{0.6665 \text{ ft/sec}}$$

$$\alpha_\text{D} = r_\text{A}\alpha_\text{A} = (0.5 \text{ ft})\left(4 \frac{\text{rad}}{\text{sec}^2}\right)$$

$$= \boxed{2 \text{ ft/sec}^2}$$

SI Solution

$$\omega_\text{B} = \left(\frac{40 \text{ cm}}{60 \text{ cm}}\right)\omega_\text{C} = \left(\frac{40 \text{ cm}}{60 \text{ cm}}\right)\left(2 \frac{\text{rad}}{\text{s}}\right) = 1.333 \text{ rad/s}$$

$$\alpha_\text{B} = \left(\frac{40 \text{ cm}}{60 \text{ cm}}\right)\alpha_\text{C} = \left(\frac{40 \text{ cm}}{60 \text{ cm}}\right)\left(6 \frac{\text{rad}}{\text{s}^2}\right) = 4 \text{ rad/s}^2$$

Since A and B are splined together,

$$\omega_\text{A} = \omega_\text{B}$$

$$\alpha_\text{A} = \alpha_\text{B}$$

$$\text{v}_\text{D} = r_\text{A}\omega_\text{A} = (0.15 \text{ m})\left(1.333 \frac{\text{rad}}{\text{s}}\right)$$

$$= \boxed{0.2 \text{ m/s}}$$

$$\alpha_\text{D} = r_\text{A}\alpha_\text{A} = (0.15 \text{ m})\left(4 \frac{\text{rad}}{\text{s}^2}\right)$$

$$= \boxed{0.6 \text{ m/s}^2}$$

14. *Customary U.S. Solution*

The angular velocity of the wheel is

$$\omega = \frac{\text{v}_\text{C}}{r}$$

$$= \frac{\left(28 \frac{\text{mi}}{\text{hr}}\right)\left(5280 \frac{\text{ft}}{\text{mi}}\right)\left(12 \frac{\text{in}}{\text{ft}}\right)}{(12 \text{ in})\left(60 \frac{\text{sec}}{\text{min}}\right)\left(60 \frac{\text{min}}{\text{hr}}\right)}$$

$$= 41.07 \text{ rad/sec}$$

Transportation

The distance from the valve stem to the instant center of the point of contact of the wheel and the surface is determined from the law of cosines for the triangle defined by the valve stem, instant center, and center of wheel.

$$l^2 = r_{wheel}^2 + r_{stem}^2 - 2r_{wheel}r_{stem}\cos\phi$$
$$= (12 \text{ in})^2 + (6 \text{ in})^2 - (2)(12 \text{ in})(6 \text{ in})(\cos 135°)$$
$$l = 16.79 \text{ in}$$

Use Eq. 71.52.

$$v = l\omega$$
$$= \frac{(16.79 \text{ in})\left(41.07 \dfrac{\text{rad}}{\text{sec}}\right)}{12 \dfrac{\text{in}}{\text{ft}}}$$
$$= \boxed{57.46 \text{ ft/sec}}$$

The direction is perpendicular to the direction of l.

SI Solution

The angular velocity of the wheel is

$$\omega = \frac{v_C}{r} = \frac{12.5 \dfrac{\text{m}}{\text{s}}}{0.305 \text{ m}} = 40.98 \text{ rad/s}$$

The distance from the valve stem to the instant center at the point of contact of the wheel and the surface is determined from the law of cosines for the triangle defined by the valve stem, instant center, and center of wheel.

$$l^2 = r_{wheel}^2 + r_{stem}^2 - 2r_{wheel}r_{stem}\cos\phi$$
$$= (0.305 \text{ m})^2 + (0.150 \text{ m})^2$$
$$\qquad - (2)(0.305 \text{ m})(0.150 \text{ m})(\cos 135°)$$
$$l = 0.425 \text{ m}$$

Use Eq. 71.52.

$$v = l\omega$$
$$= (0.425 \text{ m})\left(40.98 \dfrac{\text{rad}}{\text{s}}\right)$$
$$= \boxed{17.4 \text{ m/s}}$$

The direction is perpendicular to the direction of l.

15. *Customary U.S. Solution*

$$v_{car} = \frac{\left(45 \dfrac{\text{mi}}{\text{hr}}\right)\left(5280 \dfrac{\text{ft}}{\text{mi}}\right)}{\left(60 \dfrac{\text{sec}}{\text{min}}\right)\left(60 \dfrac{\text{min}}{\text{hr}}\right)}$$
$$= 66 \text{ ft/sec}$$

The separation distance after 1 sec is

$$s(1) = \sqrt{\begin{array}{l}\left(200 \text{ ft} + \left(15 \dfrac{\text{ft}}{\text{sec}}\right)(1 \text{ sec})\right)^2 \\ + \left(\left(66 \dfrac{\text{ft}}{\text{sec}}\right)(1 \text{ sec})\right)^2\end{array}}$$
$$= 224.9 \text{ ft}$$

The separation velocity is the difference in components of the car's and balloon's velocities along a mutually parallel line. Use the separation vector as this line.

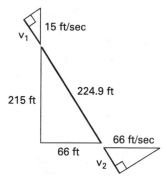

(not to scale)

$$v_1 = \left(15 \dfrac{\text{ft}}{\text{sec}}\right)\left(\dfrac{215 \text{ ft}}{224.9 \text{ ft}}\right) = 14.34 \text{ ft/sec}$$
$$v_2 = \left(66 \dfrac{\text{ft}}{\text{sec}}\right)\left(\dfrac{66 \text{ ft}}{224.9 \text{ ft}}\right) = 19.37 \text{ ft/sec}$$
$$\Delta v = v_1 + v_2 = 14.34 \dfrac{\text{ft}}{\text{sec}} + 19.37 \dfrac{\text{ft}}{\text{sec}}$$
$$= \boxed{33.71 \text{ ft/sec} \quad (34 \text{ ft/sec})}$$

The answer is (B).

SI Solution

$$v_{car} = \frac{\left(72 \dfrac{\text{km}}{\text{h}}\right)\left(1000 \dfrac{\text{m}}{\text{km}}\right)}{\left(60 \dfrac{\text{s}}{\text{min}}\right)\left(60 \dfrac{\text{min}}{\text{h}}\right)}$$
$$= 20 \text{ m/s}$$

The separation distance after 1 second is

$$s(1) = \sqrt{\left(60 \text{ m} + \left(4.5 \text{ } \frac{\text{m}}{\text{s}}\right)(1 \text{ s})\right)^2 + \left(\left(20 \text{ } \frac{\text{m}}{\text{s}}\right)(1 \text{ s})\right)^2}$$
$$= 67.53 \text{ m}$$

The separation velocity is the difference in components of the car's and balloon's velocities along a mutually parallel line. Use the separation vector as this line.

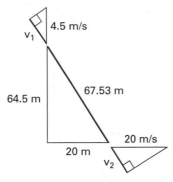

(not to scale)

$$v_1 = \left(4.5 \text{ } \frac{\text{m}}{\text{s}}\right)\left(\frac{64.5 \text{ m}}{67.53 \text{ m}}\right) = 4.298 \text{ m/s}$$

$$v_2 = \left(20 \text{ } \frac{\text{m}}{\text{s}}\right)\left(\frac{20 \text{ m}}{67.53 \text{ m}}\right) = 5.923 \text{ m/s}$$

$$\Delta v = v_1 + v_2 = 4.298 \text{ } \frac{\text{m}}{\text{s}} + 5.923 \text{ } \frac{\text{m}}{\text{s}}$$
$$= \boxed{10.22 \text{ m/s} \quad (10 \text{ m/s})}$$

The answer is (B).

16. *Customary U.S. Solution*

From Eq. 71.52,

$$\omega = \frac{v_0}{r_{\text{inner}}} = \frac{10 \text{ } \frac{\text{ft}}{\text{sec}}}{2 \text{ } \frac{\text{ft}}{\text{rad}}} = 5 \text{ rad/sec}$$

$$v_{\text{A/O}} = r_{\text{outer}}\omega = \left(3 \text{ } \frac{\text{ft}}{\text{rad}}\right)\left(5 \text{ } \frac{\text{rad}}{\text{sec}}\right)$$
$$= \boxed{\begin{array}{l} 15 \text{ ft/sec, } 45° \text{ below the horizontal,} \\ \text{to the right} \end{array}}$$

$$v_{\text{B/O}} = r_{\text{outer}} \omega = \boxed{15 \text{ ft/sec, horizontal, to the left}}$$

SI Solution

From Eq. 71.52,

$$\omega = \frac{v_0}{r_{\text{inner}}} = \frac{3 \text{ } \frac{\text{m}}{\text{s}}}{0.6 \text{ } \frac{\text{m}}{\text{rad}}} = 5 \text{ rad/s}$$

$$v_{\text{A/O}} = r_{\text{outer}}\omega = \left(0.9 \text{ } \frac{\text{m}}{\text{rad}}\right)\left(5 \text{ } \frac{\text{rad}}{\text{s}}\right)$$
$$= \boxed{\begin{array}{l} 4.5 \text{ m/s, } 45° \text{ below the horizontal,} \\ \text{to the right} \end{array}}$$

$$v_{\text{B/O}} = r_{\text{outer}}\omega = \boxed{4.5 \text{ m/s, horizontal, to the left}}$$

17. *Customary U.S. Solution*

From the law of cosines,

$$|\text{AB}|^2 = r_{\text{A}}^2 + r_{\text{B}}^2 - 2r_{\text{A}}r_{\text{B}} \cos \phi$$
$$= (3 \text{ ft})^2 + (3 \text{ ft})^2 - (2)(3 \text{ ft})^2 (\cos 135°)$$
$$= 30.728 \text{ ft}^2$$
$$|\text{AB}| = 5.543 \text{ ft}$$

$$v_{\text{B/A}} = |\text{AB}|\omega = (5.543 \text{ ft})\left(5 \text{ } \frac{\text{rad}}{\text{sec}}\right)$$
$$= \boxed{\begin{array}{l} 27.72 \text{ ft/sec, } 22.5° \text{ above the horizontal,} \\ \text{to the left} \end{array}}$$

SI Solution

From the law of cosines,

$$|\text{AB}|^2 = r_{\text{A}}^2 + r_{\text{B}}^2 - 2r_{\text{A}}r_{\text{B}} \cos \phi$$
$$= (0.9 \text{ m})^2 + (0.9 \text{ m})^2 - (2)(0.9 \text{ m})^2 (\cos 135°)$$
$$= 2.7655 \text{ m}^2$$
$$|\text{AB}| = 1.663 \text{ m}$$

$$v_{\text{B/A}} = |\text{AB}|\omega = (1.663 \text{ m})\left(5 \text{ } \frac{\text{rad}}{\text{s}}\right)$$
$$= \boxed{\begin{array}{l} 8.315 \text{ m/s, } 22.5° \text{ above the horizontal,} \\ \text{to the left} \end{array}}$$

72 Kinetics

PRACTICE PROBLEMS

Centripetal and Centrifugal Forces

1. Calculate the superelevation in percent necessary on a curve with a 6000 ft (1800 m) radius so that at 60 mph (100 km/h) cars will not have to rely on friction to stay on the roadway.

(A) 4%

(B) 7%

(C) 11%

(D) 14%

2. The 8.05 lbm (3.65 kg) object is rotating at 20 ft/sec (6 m/s). What is the angle between the pole and the wire if the radius of the path is 4 ft (1.2 m)?

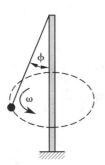

(A) 46°

(B) 58°

(C) 72°

(D) 79°

3. A 10 lbm (5 kg) mass is tied to a 2 ft (50 cm) string and whirled at 5 rev/sec horizontally to the ground.

(a) Find the centripetal acceleration.

(A) 1400 ft/sec² (350 m/s²)

(B) 1600 ft/sec² (400 m/s²)

(C) 1800 ft/sec² (450 m/s²)

(D) 2000 ft/sec² (490 m/s²)

(b) Find the centrifugal force.

(A) 490 lbf (1900 N)

(B) 610 lbf (2500 N)

(C) 750 lbf (2900 N)

(D) 820 lbf (3100 N)

(c) Find the centripetal force.

(A) 490 lbf (1900 N)

(B) 610 lbf (2500 N)

(C) 750 lbf (2900 N)

(D) 820 lbf (3100 N)

(d) Find the angular momentum.

(A) 39 ft-lbf-sec (39 J·s)

(B) 44 ft-lbf-sec (44 J·s)

(C) 53 ft-lbf-sec (53 J·s)

(D) 71 ft-lbf-sec (71 J·s)

Friction

4. A 100 lbm (50 kg) body has an initial velocity of 12.88 ft/sec (3.93 m/s) while moving on a plane with a coefficient of friction of 0.2. What distance will it travel before coming to rest?

(A) 13 ft (3.9 m)

(B) 15 ft (4.5 m)

(C) 18 ft (5.4 m)

(D) 24 ft (7.2 m)

5. A box is dropped onto a conveyor belt moving at 10 ft/sec (3 m/s). The coefficient of friction between the box and the belt is 0.333. How long will it take before the box stops slipping on the belt?

(A) 0.48 sec

(B) 0.67 sec

(C) 0.93 sec

(D) 1.2 sec

6. A motorcycle and rider weigh 400 lbm (200 kg). They travel horizontally around the inside of a hollow, right-angle cylinder of 100 ft (30 m) inside diameter. What is the coefficient of friction that will allow a speed of 40 mph (60 km/h)?

(A) 0.3

(B) 0.4

(C) 0.5

(D) 0.6

7. When a force acts on a 10 lbm (5 kg) body initially at rest, a speed of 12 ft/sec (3.6 m/s) is attained in 36 ft (11 m). What is the force if the coefficient of friction is 0.25 and the acceleration is constant?

(A) 1.8 lbf (8.6 N)

(B) 2.4 lbf (13 N)

(C) 3.1 lbf (15 N)

(D) 4.6 lbf (22 N)

8. A 130 lbm (60 kg) block slides up a 22.62° incline with a coefficient of friction of 0.1 and an initial velocity of 30 ft/sec (9 m/s). How far will the block slide up the incline before coming to rest?

(A) 18 ft (5.4 m)

(B) 29 ft (8.7 m)

(C) 42 ft (13 m)

(D) 57 ft (17 m)

9. A 100 lbm (50 kg) block is acted upon by a 100 lbf (400 N) force while resting on a horizontal surface with a coefficient of friction of 0.2. A 50 lbm (25 kg) block sits on top of the 100 lbm (50 kg) block. What is the minimum coefficient of friction between the 50 lbm and 100 lbm (25 kg and 50 kg) blocks for there to be no slipping?

Rigid Body Motion

10. A constant force of 20 lbf (90 N) is applied to a 100 lbm (50 kg) door supported on rollers at A and B.

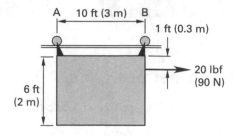

(a) What is the acceleration of the door?

(b) What are the reactions at A and B?

(c) Where would the 20 lbf (90 N) force have to be applied if the reactions at A and B were to be equal?

Constrained Motion

11. A solid sphere rolls without slipping down a 30° incline, starting from rest. What is its speed after 2 sec?

(A) 17 ft/sec (5.1 m/s)

(B) 23 ft/sec (7.0 m/s)

(C) 34 ft/sec (10 m/s)

(D) 86 ft/sec (26 m/s)

12. Object A weighs 10 lbm (5 kg) and rests on a frictionless plane with a 36.87° slope (from the horizontal). Object B weighs 20 lbm (10 kg). What is the velocity of object B 3 sec after release?

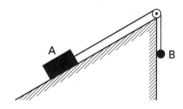

(A) 14 ft/sec (4.2 m/s)

(B) 22 ft/sec (6.6 m/s)

(C) 38 ft/sec (11 m/s)

(D) 45 ft/sec (14 m/s)

Impulse and Momentum

13. Sand is dropping at the rate of 560 lbm/min (250 kg/min) onto a conveyor belt moving with a velocity of 3.2 ft/sec (0.98 m/s). What force is required to keep the belt moving?

(A) 0.93 lbf (4.1 N)

(B) 1.2 lbf (5.3 N)

(C) 2.3 lbf (10 N)

(D) 4.8 lbf (14 N)

14. What is the impulse imparted to a 0.4 lbm (0.2 kg) baseball that approaches the batter at 90 ft/sec (30 m/s) and leaves at 130 ft/sec (40 m/s)?

(A) 1.8 lbf-sec (9.4 N·s)

(B) 2.7 lbf-sec (14 N·s)

(C) 4.1 lbf-sec (21 N·s)

(D) 8.9 lbf-sec (46 N·s)

15. At what velocity will a 1000 lbm (500 kg) gun mounted on wheels recoil if a 2.6 lbm (1.2 kg) projectile is propelled to 2100 ft/sec (650 m/s)?

(A) 3.7 ft/sec (1.1 m/s)

(B) 4.8 ft/sec (1.4 m/s)

(C) 5.5 ft/sec (1.6 m/s)

(D) 15 ft/sec (4.5 m/s)

16. A 0.15 lbm (60 g) bullet traveling 2300 ft/sec (700 m/s) embeds itself in a 9 lbm (4.5 kg) wooden block. What will be the block's initial velocity?

(A) 24 ft/sec (7.2 m/s)

(B) 38 ft/sec (9.2 m/s)

(C) 66 ft/sec (20 m/s)

(D) 110 ft/sec (33 m/s)

17. A nozzle discharges 40 gal/min (0.15 m³/min) of water at 60 ft/sec (20 m/s). Find the total force required to hold a flat vertical plate in front of the nozzle. Assume that there is no splashback.

(A) 10 lbf (50 N)

(B) 20 lbf (100 N)

(C) 40 lbf (200 N)

(D) 80 lbf (400 N)

18. In a water turbine, 100 gal/sec (0.4 m³/s) impinge on a stationary blade. The water is turned through an angle of 160° and exits at 57 ft/sec (17 m/s). The water impinges at 60 ft/sec (20 m/s).

(a) Find the force exerted by the blade on the stream.

(b) What is the angle from the horizontal of the force?

Impacts

19. An electron (0.0005486 atomic mass units) collides with a hydrogen atom (1.007277 atomic mass units) initially at rest. The electron's initial velocity is 1500 ft/sec (500 m/s). Its final velocity is 65 ft/sec (20.2 m/s) with a path 30° from its original path. Find the velocity of the hydrogen atom in the x-direction if it recoils 1.2° from the original path of the electron.

(A) 0.043 ft/sec (0.013 m/s)

(B) 0.79 ft/sec (0.26 m/s)

(C) 4.6 ft/sec (1.4 m/s)

(D) 53 ft/sec (16 m/s)

20. Two freight cars weighing 5 tons (5000 kg) each roll toward each other and couple. The left car has a velocity of 5 ft/sec (1.5 m/s) and the right car has a velocity of 4 ft/sec (1.2 m/s) prior to the impact. What is the velocity of the two cars coupled together after the impact?

(A) 0 ft/sec (0 m/s)

(B) 0.5 ft/sec (0.15 m/s)

(C) 1.0 ft/sec (0.30 m/s)

(D) 2.0 ft/sec (0.60 m/s)

21. Two 2 lbm billiard balls collide as shown. The coefficient of restitution is 0.8. What are the velocities after impact?

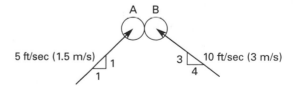

22. A 10 lbm (5 kg) pendulum is released from rest and strikes a 50 lbm (25 kg) block. The coefficient of restitution is 0.7. The block slides on a frictionless surface.

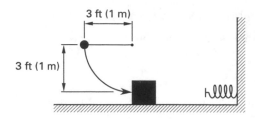

(a) Find the velocity of the pendulum at impact.

    (A) 14 ft/sec (4.4 m/s)

    (B) 26 ft/sec (7.8 m/s)

    (C) 38 ft/sec (11 m/s)

    (D) 51 ft/sec (15 m/s)

(b) Find the tension in the cord at impact.

    (A) 10 lbf (50 N)

    (B) 20 lbf (100 N)

    (C) 30 lbf (150 N)

    (D) 40 lbf (200 N)

(c) Find the block's velocity immediately after impact.

    (A) 2.2 ft/sec (0.66 m/s)

    (B) 3.9 ft/sec (1.3 m/s)

    (C) 4.6 ft/sec (1.4 m/s)

    (D) 5.8 ft/sec (1.7 m/s)

(d) Find the required spring constant to stop the block with less than 6 in (15 cm) deflection.

    (A) 54 lbf/ft (1200 N/m)

    (B) 62 lbf/ft (1400 N/m)

    (C) 75 lbf/ft (1600 N/m)

    (D) 96 lbf/ft (1800 N/m)

23. A contractor needs to drag a heavy object horizontally across the floor and closer to the east wall. To do so, the contractor connects the object through snatch blocks and a dead lead to a wire rope "come-along." The object requires a net force of 2000 lbf to maintain motion once the object starts sliding. Most nearly, what initial tension in the wire rope is needed to start the object moving?

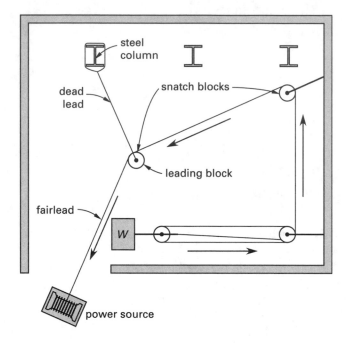

    (A) 660 lbf

    (B) 820 lbf

    (C) 1800 lbf

    (D) 2200 lbf

SOLUTIONS

1. *Customary U.S. Solution*

The tangential velocity is

$$v_t = \frac{\left(60 \ \frac{mi}{hr}\right)\left(5280 \ \frac{ft}{mi}\right)}{\left(60 \ \frac{min}{hr}\right)\left(60 \ \frac{sec}{min}\right)} = 88 \ ft/sec$$

From Eq. 72.29, the superelevation is

$$\tan \phi = \frac{v_t^2}{gr} = \frac{\left(88 \ \frac{ft}{sec}\right)^2}{\left(32.2 \ \frac{ft}{sec^2}\right)(6000 \ ft)} = \boxed{0.04 \quad (4\%)}$$

The answer is (A).

SI Solution

The tangential velocity is

$$v_t = \frac{\left(100 \ \frac{km}{h}\right)\left(1000 \ \frac{m}{km}\right)}{\left(60 \ \frac{min}{h}\right)\left(60 \ \frac{s}{min}\right)} = 27.78 \ m/s$$

From Eq. 72.29, the superelevation is

$$\tan \phi = \frac{v_t^2}{gr} = \frac{\left(27.78 \ \frac{m}{s}\right)^2}{\left(9.81 \ \frac{m}{s^2}\right)(1800 \ m)} = \boxed{0.0437 \quad (4\%)}$$

The answer is (A).

2. *Customary U.S. Solution*

The net acceleration vector is directed along the string. Using Eq. 72.29,

$$\phi = \arctan \frac{a_c}{a_n} = \arctan \frac{v_t^2}{gr}$$

$$= \arctan \frac{\left(20 \ \frac{ft}{sec}\right)^2}{\left(32.2 \ \frac{ft}{sec^2}\right)(4 \ ft)}$$

$$= \boxed{72.15° \quad (72°) \quad \begin{bmatrix} \text{independent of} \\ \text{object's mass} \end{bmatrix}}$$

The answer is (C).

SI Solution

The net acceleration vector is directed radially along the string. Using Eq. 72.29,

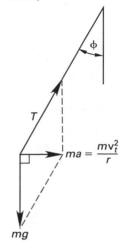

$$\frac{mv_t^2}{r} \cos \phi = mg \sin \phi$$

$$\tan \phi = \frac{v_t^2}{rg}$$

$$\phi = \arctan \frac{v_t^2}{gr}$$

$$= \arctan \frac{\left(6 \ \frac{m}{s}\right)^2}{\left(9.81 \ \frac{m}{s^2}\right)(1.2 \ m)}$$

$$= \boxed{71.89° \quad (72°) \quad \begin{bmatrix} \text{independent of} \\ \text{object's mass} \end{bmatrix}}$$

The answer is (C).

3. *Customary U.S. Solution*

(a) The tangential velocity is

$$v_t = \omega r = 2\pi f r$$

$$= \left(2\pi \ \frac{rad}{rev}\right)\left(5 \ \frac{rev}{sec}\right)(2 \ ft)$$

$$= 62.83 \ ft/sec$$

The centripetal acceleration (i.e., the acceleration normal to the path of motion) is

$$a_n = \frac{v_t^2}{r} = \frac{\left(62.83 \ \frac{ft}{sec}\right)^2}{2 \ ft}$$

$$= \boxed{1974 \ ft/sec^2 \quad (2000 \ ft/sec^2)}$$

The answer is (D).

(b) From Eq. 72.12, the centrifugal force is

$$F_{\text{centrifugal}} = \frac{ma_n}{g_c} = \frac{(10 \text{ lbm})\left(1974 \dfrac{\text{ft}}{\text{sec}^2}\right)}{32.2 \dfrac{\text{lbm-ft}}{\text{lbf-sec}^2}}$$

$$= \boxed{613.0 \text{ lbf} \quad (610 \text{ lbf}) \quad [\text{directed outward}]}$$

The answer is (B).

(c) The centripetal force is

$$|\mathbf{F}_{\text{centripetal}}| = |\mathbf{F}_{\text{centrifugal}}|$$

$$F_{\text{centripetal}} = \boxed{613.0 \text{ lbf} \quad (610 \text{ lbf}) \quad [\text{directed inward}]}$$

The answer is (B).

(d) From Eq. 72.8, the angular momentum is

$$\mathbf{h} = \frac{\mathbf{r} \times m\mathbf{v}_t}{g_c}$$

$$h = \frac{rm\mathbf{v}_t}{g_c} \quad [\text{since } \mathbf{r} \perp \mathbf{v}_t]$$

$$= \frac{(2 \text{ ft})(10 \text{ lbm})\left(62.83 \dfrac{\text{ft}}{\text{sec}}\right)}{32.2 \dfrac{\text{lbm-ft}}{\text{lbf-sec}^2}}$$

$$= \boxed{39.02 \text{ ft-lbf-sec} \quad (39 \text{ ft-lbf-sec})}$$

The answer is (A).

SI Solution

(a) The tangential velocity is

$$\text{v}_t = \omega r = 2\pi f r$$

$$= \left(2\pi \dfrac{\text{rad}}{\text{rev}}\right)\left(5 \dfrac{\text{rev}}{\text{s}}\right)(0.5 \text{ m})$$

$$= 15.708 \text{ m/s}$$

The centripetal acceleration (i.e., the acceleration normal to the path of motion) is

$$a_n = \frac{\text{v}_t^2}{r} = \frac{\left(15.708 \dfrac{\text{m}}{\text{s}}\right)^2}{0.5 \text{ m}} = \boxed{493.5 \text{ m/s}^2 \quad (490 \text{ m/s}^2)}$$

The answer is (D).

(b) From Eq. 72.12, the centrifugal force is

$$F_{\text{centrifugal}} = ma_n = (5 \text{ kg})\left(493.5 \dfrac{\text{m}}{\text{s}^2}\right)$$

$$= \boxed{2467.5 \text{ N} \quad (2500 \text{ N}) \quad [\text{directed outward}]}$$

The answer is (B).

(c) The centripetal force is

$$|\mathbf{F}_{\text{centripetal}}| = |\mathbf{F}_{\text{centrifugal}}|$$

$$F_{\text{centripetal}} = \boxed{2467.5 \text{ N} \quad (2500 \text{ N}) \quad [\text{directed inward}]}$$

The answer is (B).

(d) From Eq. 72.8, the angular momentum is

$$\mathbf{h} = \mathbf{r} \times m\mathbf{v}_t$$

$$h = rm\mathbf{v}_t \quad [\text{since } \mathbf{r} \perp \mathbf{v}_t]$$

$$= (0.5 \text{ m})(5 \text{ kg})\left(15.708 \dfrac{\text{m}}{\text{s}}\right)$$

$$= \boxed{39.27 \text{ J·s} \quad (39 \text{ J·s})}$$

The answer is (A).

4. *Customary U.S. Solution*

The deceleration is

$$a = \frac{F_f}{m} = \frac{\dfrac{fN}{g_c}}{\dfrac{m}{g_c}} = \frac{\dfrac{fmg}{g_c}}{\dfrac{m}{g_c}}$$

$$= fg = (0.2)\left(32.2 \dfrac{\text{ft}}{\text{sec}^2}\right)$$

$$= 6.44 \text{ ft/sec}^2$$

From Table 71.1, the skidding distance is

$$s_{\text{skidding}} = \frac{\text{v}^2}{2a} = \frac{\left(12.88 \dfrac{\text{ft}}{\text{sec}}\right)^2}{(2)\left(6.44 \dfrac{\text{ft}}{\text{sec}^2}\right)} = \boxed{12.88 \text{ ft} \quad (13 \text{ ft})}$$

The answer is (A).

SI Solution

The deceleration is

$$a = \frac{F_f}{m} = \frac{fN}{m} = \frac{fmg}{m}$$

$$= fg = (0.2)\left(9.81 \dfrac{\text{m}}{\text{s}^2}\right)$$

$$= 1.962 \text{ m/s}^2$$

From Table 71.1, the skidding distance is

$$s_{\text{skidding}} = \frac{\text{v}^2}{2a} = \frac{\left(3.93 \dfrac{\text{m}}{\text{s}}\right)^2}{(2)\left(1.962 \dfrac{\text{m}}{\text{s}^2}\right)} = \boxed{3.936 \text{ m} \quad (3.9 \text{ m})}$$

The answer is (A).

Transportation

5. *Customary U.S. Solution*

The deceleration is

$$a = \frac{F_f}{\frac{m}{g_c}} = \frac{fN}{\frac{m}{g_c}} = \frac{\frac{fmg}{g_c}}{\frac{m}{g_c}}$$

$$= fg = (0.333)\left(32.2 \ \frac{ft}{sec^2}\right)$$

$$= 10.72 \ ft/sec^2$$

From Table 71.1,

$$\Delta t = \frac{v}{a} = \frac{10 \ \frac{ft}{sec}}{10.72 \ \frac{ft}{sec^2}} = \boxed{0.93 \ sec}$$

The answer is (C).

SI Solution

The deceleration is

$$a = \frac{F_f}{m} = \frac{fN}{m} = \frac{fmg}{m}$$

$$= fg = (0.333)\left(9.81 \ \frac{m}{s^2}\right)$$

$$= 3.267 \ m/s^2$$

From Table 71.1,

$$\Delta t = \frac{v}{a} = \frac{3 \ \frac{m}{s}}{3.267 \ \frac{m}{s^2}} = \boxed{0.918 \ s \quad (0.93 \ s)}$$

The answer is (C).

6. *Customary U.S. Solution*

The tangential velocity is

$$v_t = \frac{\left(40 \ \frac{mi}{hr}\right)\left(5280 \ \frac{ft}{mi}\right)}{\left(60 \ \frac{min}{hr}\right)\left(60 \ \frac{sec}{min}\right)}$$

$$= 58.67 \ ft/sec$$

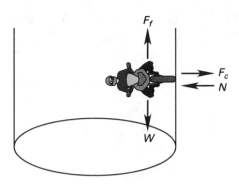

The normal force is calculated from Eq. 72.15.

$$N = F_c = ma_n = \frac{mv_t^2}{g_c r}$$

The work, W, is equal to the frictional force, given by Eq. 72.21.

$$W = F_f = fN$$

So, from Eq. 72.21, the coefficient of friction is

$$f = \frac{W}{N} = \frac{\frac{mg}{g_c}}{\frac{mv_t^2}{g_c r}} = \frac{gr}{v_t^2}$$

$$= \frac{\left(32.2 \ \frac{ft}{sec^2}\right)(50 \ ft)}{\left(58.67 \ \frac{ft}{sec}\right)^2}$$

$$= \boxed{0.468 \quad (0.5)}$$

The answer is (C).

SI Solution

The tangential force is

$$v_t = \frac{\left(60 \ \frac{km}{h}\right)\left(1000 \ \frac{m}{km}\right)}{\left(60 \ \frac{min}{h}\right)\left(60 \ \frac{s}{min}\right)}$$

$$= 16.67 \ m/s$$

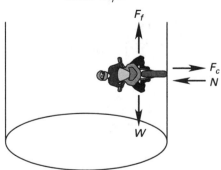

The normal force is calculated from Eq. 72.15.

$$N = F_c = ma_n = \frac{mv_t^2}{r}$$

The work, W, is equal to the frictional force, given by Eq. 72.21.

$$W = F_f = fN$$

So, from Eq. 72.21, the coefficient of friction is

$$f = \frac{W}{N} = \frac{mg}{\dfrac{mv_t^2}{r}} = \frac{gr}{v_t^2}$$

$$= \frac{\left(9.81 \ \frac{\text{m}}{\text{s}^2}\right)(15 \ \text{m})}{\left(16.67 \ \frac{\text{m}}{\text{s}}\right)^2}$$

$$= \boxed{0.53 \quad (0.5)}$$

The answer is (C).

---

7. *Customary U.S. Solution*

From Table 71.1, the uniform acceleration is

$$a = \frac{\text{v}^2}{2s} = \frac{\left(12 \ \frac{\text{ft}}{\text{sec}}\right)^2}{(2)(36 \ \text{ft})} = 2 \ \text{ft/sec}^2$$

Using Eq. 72.21, the frictional force is

$$F_f = fN = fW = (0.25)(10 \ \text{lbf})$$
$$= 2.5 \ \text{lbf}$$

Therefore, the force is

$$F = F_f + \frac{ma}{g_c} = 2.5 \ \text{lbf} + \frac{(10 \ \text{lbm})\left(2 \ \frac{\text{ft}}{\text{sec}^2}\right)}{32.2 \ \frac{\text{lbm-ft}}{\text{lbf-sec}^2}}$$

$$= \boxed{3.12 \ \text{lbf} \quad (3.1 \ \text{lbf})}$$

The answer is (C).

SI Solution

From Table 71.1, the uniform acceleration is

$$a = \frac{\text{v}^2}{2s} = \frac{\left(3.6 \ \frac{\text{m}}{\text{s}}\right)^2}{(2)(11 \ \text{m})} = 0.589 \ \text{m/s}^2$$

Using Eq. 72.21, the frictional force is

$$F_f = fN = fW = (0.25)(5 \ \text{kg})\left(9.81 \ \frac{\text{m}}{\text{s}^2}\right)$$
$$= 12.26 \ \text{N}$$

Therefore, the force is

$$F = F_f + ma = 12.26 \ \text{N} + (5 \ \text{kg})\left(0.589 \ \frac{\text{m}}{\text{s}^2}\right)$$

$$= \boxed{15.21 \ \text{N} \quad (15 \ \text{N})}$$

The answer is (C).

---

8. *Customary U.S. Solution*

The acceleration is

$$a = g \sin\theta + fg \cos\theta$$
$$= \left(32.2 \ \frac{\text{ft}}{\text{sec}^2}\right)(\sin 22.62°)$$
$$\quad + (0.1)\left(32.2 \ \frac{\text{ft}}{\text{sec}^2}\right)(\cos 22.62°)$$
$$= 15.36 \ \text{ft/sec}^2$$

From Table 71.1, the sliding distance is

$$s = \frac{\text{v}_0^2}{2a} = \frac{\left(30 \ \frac{\text{ft}}{\text{sec}}\right)^2}{(2)\left(15.36 \ \frac{\text{ft}}{\text{sec}^2}\right)} = \boxed{29.30 \ \text{ft} \quad (29 \ \text{ft})}$$

The answer is (B).

SI Solution

The acceleration is

$$a = g \sin\theta + fg \cos\theta$$
$$= \left(9.81 \ \frac{\text{m}}{\text{s}^2}\right)(\sin 22.62°) + (0.1)\left(9.81 \ \frac{\text{m}}{\text{s}^2}\right)(\cos 22.62°)$$
$$= 4.679 \ \text{m/s}^2$$

From Table 71.1, the sliding distance is

$$s = \frac{\text{v}_0^2}{2a} = \frac{\left(9 \ \frac{\text{m}}{\text{s}}\right)^2}{(2)\left(4.679 \ \frac{\text{m}}{\text{s}^2}\right)} = \boxed{8.656 \ \text{m} \quad (8.7 \ \text{m})}$$

The answer is (B).

---

9. *Customary U.S. Solution*

The friction force is given by Eq. 72.21.

$$F_{f(1,G)} = (W_1 + W_2)f_{1,G} = (100 \ \text{lbf} + 50 \ \text{lbf})(0.2)$$

$$= 30 \ \text{lbf} \quad \begin{bmatrix} < F; \ \text{blocks move together at accel-} \\ \text{eration } a, \ \text{assuming no slipping} \end{bmatrix}$$

The acceleration is

$$a = \frac{(F - F_{f(1,G)})g_c}{m_1 + m_2}$$

$$= \frac{(100 \text{ lbf} - 30 \text{ lbf})\left(32.2 \frac{\text{lbm-ft}}{\text{lbf-sec}^2}\right)}{100 \text{ lbm} + 50 \text{ lbm}}$$

$$= 15.03 \text{ ft/sec}^2$$

If block 2 is not slipping,

$$F_{f(1,2)} = f_{1,2} W_2 = \frac{m_2 a}{g_c}$$

Therefore, the coefficient of friction between blocks 1 and 2 will be

$$f_{1,2} = \frac{a}{g} = \frac{15.03 \frac{\text{ft}}{\text{sec}^2}}{32.2 \frac{\text{ft}}{\text{sec}^2}} = \boxed{0.467}$$

SI Solution

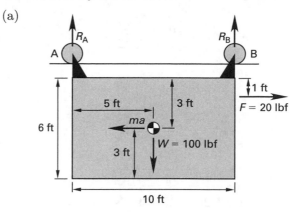

The frictional force is given by Eq. 72.21.

$$F_{f(1,G)} = (W_1 + W_2)f_{1,G}$$

$$= (50 \text{ kg} + 25 \text{ kg})\left(9.81 \frac{\text{m}}{\text{s}^2}\right)(0.2)$$

$$= 147.2 \text{ N} \quad \begin{bmatrix} < F; \text{ blocks move together at accel-} \\ \text{eration } a, \text{ assuming no slipping} \end{bmatrix}$$

The acceleration is

$$a = \frac{F - F_{f(1,G)}}{m_1 + m_2} = \frac{400 \text{ N} - 147.2 \text{ N}}{50 \text{ kg} + 25 \text{ kg}}$$

$$= 3.371 \text{ m/s}^2$$

If block 2 is not slipping,

$$F_{f(1,2)} = f_{1,2} W_2 = m_2 a$$

Therefore, the coefficient of friction between blocks 1 and 2 will be

$$f_{1,2} = \frac{a}{g} = \frac{3.371 \frac{\text{m}}{\text{s}^2}}{9.81 \frac{\text{m}}{\text{s}^2}} = \boxed{0.344}$$

10. *Customary U.S. Solution*

(a)

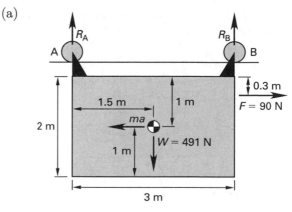

From Eq. 72.12, the acceleration is

$$a_x = \frac{F_x g_c}{m} = \frac{(20 \text{ lbf})\left(32.2 \frac{\text{lbm-ft}}{\text{lbf-sec}^2}\right)}{100 \text{ lbm}}$$

$$= \boxed{6.44 \text{ ft/sec}^2}$$

(b) The inertial force, ma, is also 20 lbf.

$$\sum M_A = -(5 \text{ ft})(100 \text{ lbf}) + (10 \text{ ft})R_B$$
$$+ (1 \text{ ft})(20 \text{ lbf}) - (3 \text{ ft})(20 \text{ lbf}) = 0$$

$$R_B = \boxed{54 \text{ lbf} \quad [\text{upward}]}$$

$$\sum F_y = R_A + 54 \text{ lbf} - 100 \text{ lbf} = 0$$

$$R_A = \boxed{46 \text{ lbf} \quad [\text{upward}]}$$

(c) Let y be the distance (positive upwards) from the center of gravity to F's line of action for $R_A = R_B = R$.

$$\sum M_{\text{CG}} = W(0) - (5 \text{ ft})R + (5 \text{ ft})R + y(20 \text{ lbf}) = 0$$

$$y = 0$$

Apply the force in line with the center of gravity.

SI Solution

(a)

From Eq. 72.12, the acceleration is

$$a_x = \frac{F_x}{m} = \frac{90 \text{ N}}{50 \text{ kg}} = \boxed{1.8 \text{ m/s}^2}$$

(b) The inertial force is also 90 N.

$$\sum M_A = -(1.5 \text{ m})(491 \text{ N}) + (3 \text{ m})R_B$$
$$+ (0.3 \text{ m})(90 \text{ N}) - (1 \text{ m})(90 \text{ N}) = 0$$
$$R_B = \boxed{266.5 \text{ N} \quad [\text{upward}]}$$
$$\sum F_y = R_A + 266.5 \text{ N} - 491 \text{ N} = 0$$
$$R_A = \boxed{224.5 \text{ N} \quad [\text{upward}]}$$

(c) Let y be the distance (positive upward) from the center of gravity to F's line of action for $R_A = R_B = R$.

$$\sum M_{CG} = W(0) - (1.5 \text{ m})R + (1.5 \text{ m})R + y(90 \text{ N}) = 0$$
$$y = 0$$

$$\boxed{\text{Apply the force in line with the center of gravity.}}$$

11. *Customary U.S. Solution*

$$F_f = \frac{mg}{g_c}\sin\phi - \frac{ma}{g_c} \qquad [\text{Eq. I}]$$

For constrained motion,

$$F_f r = I_0 \left(\frac{\alpha}{g_c}\right) = \tfrac{2}{5}\left(\frac{mr^2}{g_c}\right)\left(\frac{a}{r}\right)$$
$$= \tfrac{2}{5}\left(\frac{mar}{g_c}\right)$$
$$F_f = \tfrac{2}{5}\left(\frac{ma}{g_c}\right) \qquad [\text{Eq. II}]$$

From Eq. I and Eq. II, $mg\sin\phi - ma = (2/5)ma$. The sphere's acceleration will be

$$a = \tfrac{5}{7}g\sin\phi = \left(\tfrac{5}{7}\right)\left(32.2 \frac{\text{ft}}{\text{sec}^2}\right)(\sin 30°)$$
$$= 11.5 \text{ ft/sec}^2$$

Therefore, the sphere speed is

$$v = at = \left(11.5 \frac{\text{ft}}{\text{sec}^2}\right)(2 \text{ sec}) = \boxed{23 \text{ ft/sec}}$$

The answer is (B).

SI Solution

$$F_f = mg\sin\phi - ma \qquad [\text{Eq. I}]$$

For constrained motion,

$$F_f r = I_0\alpha = \left(\tfrac{2}{5}mr^2\right)\left(\frac{a}{r}\right) = \tfrac{2}{5}mar$$
$$F_f = \tfrac{2}{5}ma \qquad [\text{Eq. II}]$$

From Eq. I and Eq. II, $mg\sin\phi - ma = (2/5)ma$. The sphere's acceleration will be

$$a = \tfrac{5}{7}g\sin\phi = \left(\tfrac{5}{7}\right)\left(9.81 \frac{\text{m}}{\text{s}^2}\right)(\sin 30°)$$
$$= 3.504 \text{ m/s}^2$$

Therefore, the sphere speed is

$$v = at = \left(3.504 \frac{\text{m}}{\text{s}^2}\right)(2 \text{ s}) = \boxed{7.008 \text{ m/s} \quad (7.0 \text{ m/s})}$$

The answer is (B).

12. *Customary U.S. Solution*

Let B's acceleration and velocity be a and v, positive upwards. Then A's acceleration and velocity are $-a$ and $-v$, respectively.

For B: $\quad T - \dfrac{m_B g}{g_c} = \dfrac{m_B a}{g_c}$

$$T = \dfrac{m_B a + m_B g}{g_c}$$

For A: $\quad T - \dfrac{m_A g\sin\phi}{g_c} = \dfrac{m_A(-a)}{g_c}$

$$T = \dfrac{m_A g\sin\phi - m_A a}{g_c}$$

Find the acceleration.

$$a = \frac{m_A g\sin\phi - m_B g}{m_B + m_A}$$
$$= \frac{\begin{aligned}&(10 \text{ lbm})\left(32.2 \frac{\text{ft}}{\text{sec}^2}\right)(\sin 36.87°)\\ &\quad - (20 \text{ lbm})\left(32.2 \frac{\text{ft}}{\text{sec}^2}\right)\end{aligned}}{20 \text{ lbm} + 10 \text{ lbm}}$$
$$= -15.03 \text{ ft/sec}^2$$

Therefore, the velocity of object B is

$$v = at = \left(-15.03 \ \frac{\text{ft}}{\text{sec}^2}\right)(3 \text{ sec})$$

$$= \boxed{-45.09 \ \text{ft/sec} \quad (45 \ \text{ft/sec}) \quad [\text{downward}]}$$

The answer is (D).

SI Solution

Let B's acceleration and velocity be a and v, positive upwards. Then A's acceleration and velocity are $-a$ and $-v$, respectively.

For B: $\quad T - m_B g = m_B a$

$$T = m_B a + m_B g$$

For A: $\quad T - m_A g \sin \phi = m_A(-a)$

$$T = m_A g \sin \phi - m_A a$$

Find the acceleration.

$$a = \frac{m_A g \sin \phi - m_B g}{m_A + m_B}$$

$$= \frac{(5 \text{ kg})\left(9.81 \ \frac{\text{m}}{\text{s}^2}\right)(\sin 36.87°) - (10 \text{ kg})\left(9.81 \ \frac{\text{m}}{\text{s}^2}\right)}{5 \text{ kg} + 10 \text{ kg}}$$

$$= -4.578 \ \text{m/s}^2$$

Therefore, the velocity of object B is

$$v = at = \left(-4.578 \ \frac{\text{m}}{\text{s}^2}\right)(3 \text{ s})$$

$$= \boxed{-13.7 \ \text{m/s} \quad (14 \ \text{m/s}) \quad [\text{downward}]}$$

The answer is (D).

13. *Customary U.S. Solution*

From Eq. 72.57(b), the force is

$$F = \frac{\dot{m}\Delta v}{g_c} = \frac{\left(560 \ \frac{\text{lbm}}{\text{min}}\right)\left(3.2 \ \frac{\text{ft}}{\text{sec}}\right)}{\left(32.2 \ \frac{\text{lbm-ft}}{\text{lbf-sec}^2}\right)\left(60 \ \frac{\text{sec}}{\text{min}}\right)}$$

$$= \boxed{0.9275 \ \text{lbf} \quad (0.93 \ \text{lbf})}$$

The answer is (A).

SI Solution

From Eq. 72.57(a), the force is

$$F = \dot{m}\Delta v = \frac{\left(250 \ \frac{\text{kg}}{\text{min}}\right)\left(0.98 \ \frac{\text{m}}{\text{s}}\right)}{60 \ \frac{\text{s}}{\text{min}}}$$

$$= \boxed{4.083 \ \text{N} \quad (4.1 \ \text{N})}$$

The answer is (A).

14. *Customary U.S. Solution*

From Eq. 72.54, the impulse is

$$|\mathbf{Imp}| = \frac{|\Delta \mathbf{p}|}{g_c} = \frac{|\mathbf{p}_2 - \mathbf{p}_1|}{g_c} = \frac{m v_2 - (-m v_1)}{g_c}$$

$$= \frac{(0.4 \text{ lbm})\left(90 \ \frac{\text{ft}}{\text{sec}} + 130 \ \frac{\text{ft}}{\text{sec}}\right)}{32.2 \ \frac{\text{lbm-ft}}{\text{lbf-sec}^2}}$$

$$= \boxed{2.73 \ \text{lbf-sec} \quad (2.7 \ \text{lbf-sec})}$$

The answer is (B).

SI Solution

From Eq. 72.54, the impulse is

$$|\mathbf{Imp}| = |\Delta \mathbf{p}| = |\mathbf{p}_2 - \mathbf{p}_1| = m v_2 - (-m v_1)$$

$$= (0.2 \text{ kg})\left(40 \ \frac{\text{m}}{\text{s}} + 30 \ \frac{\text{m}}{\text{s}}\right)$$

$$= \boxed{14 \ \text{N·s}}$$

The answer is (B).

15. *Customary U.S. Solution*

Momentum is always conserved.

$$\Delta p_{\text{gun}} = -\Delta p_{\text{proj}}$$

$$\frac{m_{\text{gun}} v_{\text{gun}}}{g_c} = \frac{-m_{\text{proj}} v_{\text{proj}}}{g_c}$$

$$v_{\text{gun}} = -\frac{m_{\text{proj}} v_{\text{proj}}}{m_{\text{gun}}} = \frac{-(2.6 \text{ lbm})\left(2100 \ \frac{\text{ft}}{\text{sec}}\right)}{1000 \ \text{lbm}}$$

$$= \boxed{-5.46 \ \text{ft/sec} \quad (5.5 \ \text{ft/sec}) \quad \begin{bmatrix} \text{opposite projectile} \\ \text{direction} \end{bmatrix}}$$

The answer is (C).

SI Solution

Momentum is always conserved.

$$\Delta p_{gun} = -\Delta p_{proj}$$

$$m_{gun} v_{gun} = -m_{proj} v_{proj}$$

$$v_{gun} = -\frac{m_{proj} v_{proj}}{m_{gun}} = \frac{-(1.2 \text{ kg})\left(650 \text{ }\frac{m}{s}\right)}{500 \text{ kg}}$$

$$= \boxed{-1.56 \text{ m/s} \quad (1.6 \text{ m/s})} \quad \left[\begin{array}{c} \text{opposite projectile} \\ \text{direction} \end{array}\right]$$

The answer is (C).

16. *Customary U.S. Solution*

In absence of external forces, $\Delta \mathbf{p} = 0$.

$$\frac{m_{bullet} v_{bullet}}{g_c} = \frac{m_{(bullet+block)} v_{(bullet+block)}}{g_c}$$

$$v_{(bullet+block)} = \frac{m_{bullet} v_{bullet}}{m_{(bullet+block)}}$$

$$= \frac{(0.15 \text{ lbm})\left(2300 \text{ }\frac{ft}{sec}\right)}{0.15 \text{ lbm} + 9 \text{ lbm}}$$

$$= \boxed{37.7 \text{ ft/sec} \quad (38 \text{ ft/sec})}$$

The answer is (B).

SI Solution

In absence of external forces, $\Delta \mathbf{p} = 0$.

$$m_{bullet} v_{bullet} = m_{(bullet+block)} v_{(bullet+block)}$$

$$v_{(bullet+block)} = \frac{m_{bullet} v_{bullet}}{m_{(bullet+block)}} = \frac{(0.06 \text{ kg})\left(700 \text{ }\frac{m}{s}\right)}{4.5 \text{ kg} + 0.06 \text{ kg}}$$

$$= \boxed{9.21 \text{ m/s} \quad (9.2 \text{ m/s})}$$

The answer is (B).

17. *Customary U.S. Solution*

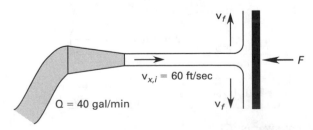

The mass flow rate is

$$\dot{m} = \rho Q$$

$$= \frac{\left(62.4 \text{ }\frac{lbm}{ft^3}\right)\left(40 \text{ }\frac{gal}{min}\right)}{\left(60 \text{ }\frac{sec}{min}\right)\left(7.48 \text{ }\frac{gal}{ft^3}\right)}$$

$$= 5.56 \text{ lbm/sec}$$

Neglecting gravity, water is turned through an angle of $90°$ in equal portions in all directions.

For every direction other than along the x-axis, equal amounts of water are directed in opposite senses; no net force is applied in these directions. From Eq. 72.57,

$$F_x = \frac{\dot{m}\Delta v_x}{g_c} = \frac{-\dot{m}v_{x,i}}{g_c} = \frac{-\left(5.56 \text{ }\frac{lbm}{sec}\right)\left(60 \text{ }\frac{ft}{sec}\right)}{32.2 \text{ }\frac{lbm\text{-}ft}{lbf\text{-}sec^2}}$$

$$= \boxed{-10.36 \text{ lbf} \quad (10 \text{ lbf})} \quad \left[\begin{array}{c} \text{opposite water} \\ \text{direction} \end{array}\right]$$

The answer is (A).

SI Solution

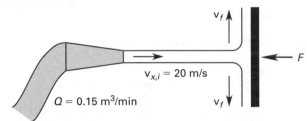

The mass flow rate is

$$\dot{m} = \rho Q$$

$$= \frac{\left(1000 \text{ }\frac{kg}{m^3}\right)\left(0.15 \text{ }\frac{m^3}{min}\right)}{60 \text{ }\frac{s}{min}}$$

$$= 2.5 \text{ kg/s}$$

Neglecting gravity, water is turned through an angle of $90°$ in equal portions in all directions. For every direction other than along the x-axis, equal amounts of water are directed in opposite senses; no net force is applied in these directions. From Eq. 72.57,

$$F_x = \dot{m}\Delta v_x = -\dot{m}v_{x,i} = -\left(2.5 \text{ }\frac{kg}{s}\right)\left(20 \text{ }\frac{m}{s}\right)$$

$$= \boxed{-50 \text{ N} \quad (50 \text{ N})} \quad [\text{opposite water direction}]$$

The answer is (A).

Transportation

18. *Customary U.S. Solution*

(a) The mass flow rate is

$$\dot{m} = \rho Q = \frac{\left(62.4 \ \frac{\text{lbm}}{\text{ft}^3}\right)\left(100 \ \frac{\text{gal}}{\text{sec}}\right)}{7.48 \ \frac{\text{gal}}{\text{ft}^3}}$$

$$= 834.2 \ \text{lbm/sec}$$

Let inward directions be positive.

The outward momentum flow rates are

$$\dot{p}_{\text{out},x} = \dot{m}\text{v}_{\text{out}} \cos \phi$$

$$= \left(834.2 \ \frac{\text{lbm}}{\text{sec}}\right)\left(57 \ \frac{\text{ft}}{\text{sec}}\right)(\cos 160°)$$

$$= -44{,}682 \ \text{lbm-ft/sec}^2$$

$$\dot{p}_{\text{out},y} = \dot{m}\text{v}_{\text{out}} \sin \phi$$

$$= \left(834.2 \ \frac{\text{lbm}}{\text{sec}}\right)\left(57 \ \frac{\text{ft}}{\text{sec}}\right)(\sin 160°)$$

$$= 16{,}263 \ \text{lbm-ft/sec}^2$$

The inward momentum flow rate is

$$\dot{p}_{\text{in},x} = \dot{m}\text{v}_{\text{in}} = \left(834.2 \ \frac{\text{lbm}}{\text{sec}}\right)\left(60 \ \frac{\text{ft}}{\text{sec}}\right)$$

$$= 50{,}052 \ \text{lbm-ft/sec}^2$$

The forces are

$$F_x = \frac{\Delta \dot{p}_x}{g_c} = \frac{\dot{p}_{\text{out},x} - \dot{p}_{\text{in},x}}{g_c}$$

$$= \frac{-44{,}682 \ \frac{\text{lbm-ft}}{\text{sec}^2} - 50{,}052 \ \frac{\text{lbm-ft}}{\text{sec}^2}}{32.2 \ \frac{\text{lbm-ft}}{\text{lbf-sec}^2}}$$

$$= -2942 \ \text{lbf}$$

$$F_y = \frac{\Delta \dot{p}_y}{g_c} = \frac{16{,}263 \ \frac{\text{lbm-ft}}{\text{sec}^2} - 0 \ \frac{\text{lbm-ft}}{\text{sec}^2}}{32.2 \ \frac{\text{lbm-ft}}{\text{lbf-sec}^2}} = 505.0 \ \text{lbf}$$

$$F_{\text{blade}} = \sqrt{F_x^2 + F_y^2} = \sqrt{(-2942 \ \text{lbf})^2 + (505 \ \text{lbf})^2}$$

$$= \boxed{2985 \ \text{lbf}}$$

(b) From the horizontal component of the force, the angle is

$$\phi = \arctan \frac{F_y}{F_x} = \arctan \frac{505 \ \text{lbf}}{-2942 \ \text{lbf}}$$

$$= -9.74° \quad \text{[second quadrant]}$$

$$= \boxed{-170.26°, \text{ counterclockwise from horizontal}}$$

SI Solution

(a) The mass flow rate is

$$\dot{m} = \rho Q = \left(1000 \ \frac{\text{kg}}{\text{m}^3}\right)\left(0.4 \ \frac{\text{m}^3}{\text{s}}\right)$$

$$= 400 \ \text{kg/s}$$

Let inward directions be positive.

The outward momentum flow rates are

$$\dot{p}_{\text{out},x} = \dot{m}\text{v}_{\text{out}} \cos \phi$$

$$= \left(400 \ \frac{\text{kg}}{\text{s}}\right)\left(17 \ \frac{\text{m}}{\text{s}}\right)(\cos 160°)$$

$$= -6390 \ \text{N}$$

$$\dot{p}_{\text{out},y} = \dot{m}\text{v}_{\text{out}} \sin \phi$$

$$= \left(400 \ \frac{\text{kg}}{\text{s}}\right)\left(17 \ \frac{\text{m}}{\text{s}}\right)(\sin 160°)$$

$$= 2326 \ \text{N}$$

The inward momentum flow rate is

$$\dot{p}_{\text{in},x} = \dot{m}\text{v}_{\text{in}} = \left(400 \ \frac{\text{kg}}{\text{s}}\right)\left(20 \ \frac{\text{m}}{\text{s}}\right) = 8000 \ \text{N}$$

The forces are

$$F_x = \Delta \dot{p}_x = \dot{p}_{\text{out},x} - \dot{p}_{\text{in},x} = -6390 \ \text{N} - 8000 \ \text{N}$$

$$= -14\,390 \ \text{N}$$

$$F_y = \Delta \dot{p}_y = 2326 \ \text{N} - 0 \ \text{N} = 2326 \ \text{N}$$

$$F_{\text{blade}} = \sqrt{F_x^2 + F_y^2} = \sqrt{(-14\,390 \ \text{N})^2 + (2326 \ \text{N})^2}$$

$$= \boxed{14\,577 \ \text{N}}$$

Transportation

(b) From the horizontal component of the force, the angle is

$$\phi = \arctan \frac{F_y}{F_x} = \arctan \frac{2326 \text{ N}}{-14\,390 \text{ N}}$$

$$= -9.18° \quad \text{[second quadrant]}$$

$$= \boxed{-170.82°, \text{ counterclockwise from horizontal}}$$

19. *Customary U.S. Solution*

Since the electron is deflected from its original path, the collision is oblique. The ratio of the masses is

$$\frac{m_H}{m_e} = \frac{1.007277u}{0.0005486u} = 1836$$

Kinetic energy may or may not be conserved in this collision; there is insufficient information to make the determination. Momentum is always conserved, regardless of the axis along which it is evaluated. Consider the original path of the electron to be parallel to the x-axis. Then, by conservation of momentum in the x-direction,

$$m_e v_{e,x} + m_H v_{H,x} = m_e v'_{e,x} + m_H v'_{H,x}$$

Recognizing that the x-component of v'_H was requested and $v_{H,x} = 0$, substitute the ratio of masses.

$$v_{e,x} = v'_e \cos \theta_e + 1836 v'_{H,x}$$

$$1500 \frac{\text{ft}}{\text{sec}} = \left(65 \frac{\text{ft}}{\text{sec}}\right)(\cos 30°) + 1836 v'_{H,x}$$

$$v'_{H,x} = \boxed{0.786 \text{ ft/sec} \quad (0.79 \text{ ft/sec})}$$

The answer is (B).

SI Solution

Since the electron is deflected from its original path, the collision is oblique. The ratio of the masses is

$$\frac{m_H}{m_e} = \frac{1.007277u}{0.0005486u} = 1836$$

Kinetic energy may or may not be conserved in this collision; there is insufficient information to make the determination. Momentum is always conserved, regardless of the axis along which it is evaluated. Consider the original path of the electron to be parallel to the x-axis. Then, by conservation of momentum in the x-direction,

$$m_e v_{e,x} + m_H v_{H,x} = m_e v'_{e,x} + m_H v'_{H,x}$$

Recognizing that the x-component of v'_H was requested and $v_{H,x} = 0$, substitute the ratio of masses.

$$v_{e,x} = v'_e \cos \theta_e + 1836 v'_{H,x}$$

$$500 \frac{\text{m}}{\text{s}} = \left(20.2 \frac{\text{m}}{\text{s}}\right)(\cos 30°) + 1836 v'_{H,x}$$

$$v'_{H,x} = \boxed{0.26 \text{ m/s}}$$

The answer is (B).

20. *Customary U.S. Solution*

$$m_{\text{left}} v_{\text{left}} + m_{\text{right}} v_{\text{right}} = m_{\text{couple}} v_{\text{couple}}$$

$$v_{\text{couple}} = \frac{(5 \text{ tons})\left(5 \frac{\text{ft}}{\text{sec}}\right) + (5 \text{ tons})\left(-4 \frac{\text{ft}}{\text{sec}}\right)}{10 \text{ tons}}$$

$$= \boxed{0.5 \text{ ft/sec} \quad \text{[to the right]}}$$

The answer is (B).

SI Solution

$$m_{\text{left}} v_{\text{left}} + m_{\text{right}} v_{\text{right}} = m_{\text{couple}} v_{\text{couple}}$$

$$v_{\text{couple}} = \frac{(5000 \text{ kg})\left(1.5 \frac{\text{m}}{\text{s}}\right) + (5000 \text{ kg})\left(-1.2 \frac{\text{m}}{\text{s}}\right)}{10\,000 \text{ kg}}$$

$$= \boxed{0.15 \text{ m/s} \quad \text{[to the right]}}$$

The answer is (B).

21. *Customary U.S. Solution*

$$v_{A,y} = \left(5 \frac{\text{ft}}{\text{sec}}\right)(\sin 45°) = 3.536 \text{ ft/sec}$$

$$v_{A,x} = \left(5 \frac{\text{ft}}{\text{sec}}\right)(\cos 45°) = 3.536 \text{ ft/sec}$$

$$v_{B,y} = \left(10 \frac{\text{ft}}{\text{sec}}\right)\left(\frac{3}{5}\right) = 6 \text{ ft/sec}$$

$$v_{B,x} = \left(10 \frac{\text{ft}}{\text{sec}}\right)\left(-\frac{4}{5}\right) = -8 \text{ ft/sec}$$

The force of impact is in the x-direction only.

$$v'_{A,y} = v_{A,y}$$

$$v'_{B,y} = v_{B,y}$$

In the x-direction,

$$e = \frac{v'_{A,x} - v'_{B,x}}{v_{B,x} - v_{A,x}} = \frac{v'_{A,x} - v'_{B,x}}{-8 \ \frac{\text{ft}}{\text{sec}} - 3.536 \ \frac{\text{ft}}{\text{sec}}}$$

$$v'_{A,x} - v'_{B,x} = (0.8)\left(-8 \ \frac{\text{ft}}{\text{sec}} - 3.536 \ \frac{\text{ft}}{\text{sec}}\right)$$

$$= -9.229 \ \text{ft/sec} \qquad \text{[Eq. I]}$$

$$m_A v_{A,x} + m_B v_{B,x} = m_A v'_{A,x} + m_B v'_{B,x}$$

Since $m_A = m_B$,

$$v'_{A,x} + v'_{B,x} = 3.536 \ \frac{\text{ft}}{\text{sec}} + \left(-8 \ \frac{\text{ft}}{\text{sec}}\right) = -4.464 \ \text{ft/sec}$$

$$\text{[Eq. II]}$$

Solving Eq. I and Eq. II simultaneously,

$$v'_{A,x} = -6.846 \ \text{ft/sec}$$
$$v'_{B,x} = 2.382 \ \text{ft/sec}$$

$$v'_A = \sqrt{(v'_{A,x})^2 + (v'_{A,y})^2}$$

$$= \sqrt{\left(-6.846 \ \frac{\text{ft}}{\text{sec}}\right)^2 + \left(3.536 \ \frac{\text{ft}}{\text{sec}}\right)^2}$$

$$= \boxed{7.705 \ \text{ft/sec}}$$

$$v'_B = \sqrt{(v'_{B,x})^2 + (v'_{B,y})^2}$$

$$= \sqrt{\left(2.382 \ \frac{\text{ft}}{\text{sec}}\right)^2 + \left(6 \ \frac{\text{ft}}{\text{sec}}\right)^2}$$

$$= \boxed{6.456 \ \text{ft/sec}}$$

$$\phi_A = \arctan \frac{3.536 \ \frac{\text{ft}}{\text{sec}}}{-6.846 \ \frac{\text{ft}}{\text{sec}}} = \boxed{152.7°}$$

$$\phi_B = \arctan \frac{6 \ \frac{\text{ft}}{\text{sec}}}{2.382 \ \frac{\text{ft}}{\text{sec}}} = \boxed{68.3°}$$

SI Solution

$$v_{A,y} = \left(1.5 \ \frac{\text{m}}{\text{s}}\right)(\sin 45°) = 1.061 \ \text{m/s}$$

$$v_{A,x} = \left(1.5 \ \frac{\text{m}}{\text{s}}\right)(\cos 45°) = 1.061 \ \text{m/s}$$

$$v_{B,y} = \left(3 \ \frac{\text{m}}{\text{s}}\right)\left(\frac{3}{5}\right) = 1.8 \ \text{m/s}$$

$$v_{B,x} = \left(3 \ \frac{\text{m}}{\text{s}}\right)\left(-\frac{4}{5}\right) = -2.4 \ \text{m/s}$$

The force of impact is in the x-direction only.

$$v'_{A,y} = v_{A,y}$$
$$v'_{B,y} = v_{B,y}$$

In the x-direction,

$$e = \frac{v'_{A,x} - v'_{B,x}}{v_{B,x} - v_{A,x}} = \frac{v'_{A,x} - v'_{B,x}}{-2.4 \ \frac{\text{m}}{\text{s}} - 1.061 \ \frac{\text{m}}{\text{s}}}$$

$$v'_{A,x} - v'_{B,x} = (0.8)\left(-2.4 \ \frac{\text{m}}{\text{s}} - 1.061 \ \frac{\text{m}}{\text{s}}\right)$$

$$= -2.769 \ \text{m/s} \qquad \text{[Eq. I]}$$

$$m_A v_{A,x} + m_B v_{B,x} = m_A v'_{A,x} + m_B v'_{B,x}$$

Since $m_A = m_B$,

$$v'_{A,x} + v'_{B,x} = 1.061 \ \frac{\text{m}}{\text{s}} + \left(-2.4 \ \frac{\text{m}}{\text{s}}\right) = -1.339 \ \text{m/s}$$

$$\text{[Eq. II]}$$

Solving Eq. I and Eq. II simultaneously,

$$v'_{A,x} = -2.054 \ \text{m/s}$$
$$v'_{B,x} = 0.715 \ \text{m/s}$$

$$v'_A = \sqrt{(v'_{A,x})^2 + (v'_{A,y})^2}$$

$$= \sqrt{\left(-2.054 \ \frac{\text{m}}{\text{s}}\right)^2 + \left(1.061 \ \frac{\text{m}}{\text{s}}\right)^2}$$

$$= \boxed{2.312 \ \text{m/s}}$$

$$v'_B = \sqrt{(v'_{B,x})^2 + (v'_{B,y})^2}$$

$$= \sqrt{\left(0.715 \ \frac{\text{m}}{\text{s}}\right)^2 + \left(1.8 \ \frac{\text{m}}{\text{s}}\right)^2}$$

$$= \boxed{1.937 \ \text{m/s}}$$

$$\phi_A = \arctan \frac{1.061 \ \frac{\text{m}}{\text{s}}}{-2.054 \ \frac{\text{m}}{\text{s}}} = \boxed{152.7°}$$

$$\phi_B = \arctan \frac{1.8 \ \frac{\text{m}}{\text{s}}}{0.715 \ \frac{\text{m}}{\text{s}}} = \boxed{68.34°}$$

Transportation

22. *Customary U.S. Solution*

(a) Energy is conserved.

$$mgh = \tfrac{1}{2}mv^2$$

$$v = \sqrt{2gh} = \sqrt{(2)\left(32.2\ \frac{\text{ft}}{\text{sec}^2}\right)(3\ \text{ft})}$$

$$= \boxed{13.9\ \text{ft/sec}\quad(14\ \text{ft/sec})}$$

The answer is (A).

(b) The tension in the cord resists the pendulum's weight and the centrifugal force.

$$F_c = \frac{mv^2}{g_c r} = \frac{(10\ \text{lbm})\left(13.9\ \dfrac{\text{ft}}{\text{sec}}\right)^2}{\left(32.2\ \dfrac{\text{lbm-ft}}{\text{lbf-sec}^2}\right)(3\ \text{ft})} = 20\ \text{lbf}$$

$$T = W + F_c = 10\ \text{lbf} + 20\ \text{lbf} = \boxed{30\ \text{lbf}}$$

The answer is (C).

(c) The block's velocity immediately after impact is found as follows.

$$e = \frac{v_1' - v_2'}{v_2 - v_1} = \frac{v_1' - v_2'}{0\ \dfrac{\text{ft}}{\text{sec}} - 13.9\ \dfrac{\text{ft}}{\text{sec}}}$$

$$v_1' - v_2' = (0.7)\left(-13.9\ \frac{\text{ft}}{\text{sec}}\right) = -9.73\ \text{ft/sec} \quad \text{[Eq. I]}$$

$$m_1 v_1 + m_2 v_2 = m_1 v_1' + m_2 v_2'$$

$$(10\ \text{lbm})v_1' + (50\ \text{lbm})v_2' = 139\ \text{ft/sec} \quad \text{[Eq. II]}$$

Solving Eq. I and Eq. II simultaneously,

$$v_1' = -5.79\ \text{ft/sec}$$

$$v_2' = \boxed{3.94\ \text{ft/sec}\quad(3.9\ \text{ft/sec})}$$

The answer is (B).

(d) The required spring constant, k, is

$$\frac{mv^2}{2g_c} = \frac{kx^2}{2}$$

$$k = \frac{mv^2}{g_c x^2} = \frac{(50\ \text{lbm})\left(3.94\ \dfrac{\text{ft}}{\text{sec}}\right)^2}{\left(32.2\ \dfrac{\text{lbm-ft}}{\text{lbf-sec}^2}\right)(0.5\ \text{ft})^2}$$

$$= \boxed{96.4\ \text{lbf/ft}\quad(96\ \text{lbf/ft})}$$

The answer is (D).

SI Solution

(a) Energy is conserved.

$$mgh = \frac{mv^2}{2}$$

$$v = \sqrt{2gh} = \sqrt{(2)\left(9.81\ \frac{\text{m}}{\text{s}^2}\right)(1\ \text{m})}$$

$$= \boxed{4.43\ \text{m/s}\quad(4.4\ \text{m/s})}$$

The answer is (A).

(b) The tension in the cord resists the pendulum's weight and the centrifugal force.

$$F_c = \frac{mv^2}{r} = \frac{(5\ \text{kg})\left(4.43\ \dfrac{\text{m}}{\text{s}}\right)^2}{1\ \text{m}} = 98.1\ \text{N}$$

$$T = W + F_c = (5\ \text{kg})\left(9.81\ \frac{\text{m}}{\text{s}^2}\right) + 98.1\ \text{N}$$

$$= \boxed{147\ \text{N}\quad(150\ \text{N})}$$

The answer is (C).

(c) The block's velocity immediately after impact is found as follows.

$$e = \frac{v_1' - v_2'}{v_2 - v_1} = \frac{v_1' - v_2'}{0\ \dfrac{\text{m}}{\text{s}} - 4.43\ \dfrac{\text{m}}{\text{s}}}$$

$$v_1' - v_2' = (0.7)\left(-4.43\ \frac{\text{m}}{\text{s}}\right) = -3.1\ \text{m/s} \quad \text{[Eq. I]}$$

$$m_1 v_1 + m_2 v_2 = m_1 v_1' + m_2 v_2'$$

$$(5\ \text{kg})v_1' + (25\ \text{kg})v_2' = 22.15\ \text{m/s} \quad \text{[Eq. II]}$$

Solving Eq. I and Eq. II simultaneously,

$$v_1' = -1.845\ \text{m/s}$$

$$v_2' = \boxed{1.255\ \text{m/s}\quad(1.3\ \text{m/s})}$$

The answer is (B).

(d) The required spring constant, k, is

$$\frac{mv^2}{2} = \frac{kx^2}{2}$$

$$k = \frac{mv^2}{x^2} = \frac{(25\ \text{kg})\left(1.255\ \dfrac{\text{m}}{\text{s}}\right)^2}{(0.15\ \text{m})^2}$$

$$= \boxed{1750\ \text{N/m}\quad(1800\ \text{N/m})}$$

The answer is (D).

23. The running block (the block attached to the object) has 3 falls (cables) running from it. Therefore, the mechanical advantage of this arrangement is 3:1. Once the block is moving, the tension in the cable is $(^1/_3)(2000 \text{ lbf}) = 667 \text{ lbf}$. However, since the coefficient of static friction is higher than the coefficient of dynamic friction, the initial cable tension will be slightly higher.

The answer is (B).

Transportation

73 Roads and Highways: Capacity Analysis

PRACTICE PROBLEMS

Design Speeds/Volume Parameters

1. (*Time limit: one hour*) Commuter trains travel between stations spaced 1.0 mi apart. A train arrives every 5 min, and the trains can attain a maximum speed of 80 mph. Only one train is permitted between stations at a time. Each train has 5 cars, with a maximum capacity of 220 people per car. When the train stops, it stays in the station for as long as is required to maintain the schedule. The uniform acceleration of the train is 5.5 ft/sec^2. Deceleration is 4.4 ft/sec^2. (a) What is the top speed of the train? (b) What is the maximum capacity of the line in people per hour? (c) What is the average train speed? (d) What is the average running speed? (e) In regard to the answer in part (d), explain whether the basis is time or spacing. (f) How much time would be saved between stations if the maximum speed was increased to 100 mph?

2. A rural Class I two-lane highway has a design speed of 60 mph in both directions. One particular segment of the highway is a 3% uphill grade lasting for 2 mi with 40% of that length unavailable for passing. There are no access points along the segment. The lane width is 11 ft. The shoulders are 2 ft wide on the uphill side and 6 ft wide on the downhill side. In the peak 60 min of travel, vehicular volume is near the maximum for level of service B for level terrain. The traffic distribution is 5% trucks, 15% buses, and 80% cars. 70% of the traffic goes uphill during the peak hour, while the remaining 30% travels in the opposite direction.

(a) What is the highest possible two-way flow rate for extended lengths of a level, two-lane highway without flow becoming oversaturated?

(A) 1500 pcph each lane

(B) 1700 pcph each lane

(C) 3200 pcph both lanes combined

(D) 6400 pcph both lanes combined

(b) Using average travel speed (ATS) criteria, what is the uphill grade adjustment factor for a specific upgrade?

(A) 0.91

(B) 0.93

(C) 0.97

(D) 0.99

(c) Using average travel speed criteria, what is the heavy vehicle adjustment factor for a specific upgrade going uphill?

(A) 0.50

(B) 0.71

(C) 0.85

(D) 0.91

(d) Using average travel speed criteria, what is the passenger car equivalent volume for the uphill lane during the peak 15 min period?

(A) 750 pcph

(B) 1150 pcph

(C) 1270 pcph

(D) 1800 pcph

(e) Using the design speed as the base free-flow speed, what is the uphill free flow speed?

(A) 51 mph

(B) 53 mph

(C) 55 mph

(D) 57 mph

(f) Assuming a free flow speed of 57 mph, what is the average travel speed on the uphill lane?

(A) 35 mph

(B) 37 mph

(C) 42 mph

(D) 46 mph

(g) What is the percent time spent following for this segment?

(A) 45%

(B) 55%

(C) 65%

(D) 75%

(h) What treatments might be made to improve the level of service?

(A) adding a median and providing curbs along the lane edge

(B) adding a sidewalk, adding a traffic signal, and improving the pavement markings

(C) adding roadway lighting and reducing the speed limit

(D) realigning the roadway to improve sight distance and adding a passing lane

3. One lane of a two-lane highway was observed for an hour during the day. The following data were gathered.

average distance between front bumpers of successive cars	80 ft
spot mean speed	30 mph
space mean speed	29 mph

(a) What is the average headway?

(A) 1.9 sec/veh

(B) 2.7 sec/veh

(C) 2.8 sec/veh

(D) 42 sec/veh

(b) What is the density in vehicles per mile?

(A) 30 vpm

(B) 66 vpm

(C) 180 vpm

(D) 2000 vpm

(c) What is the traffic volume in vehicles per hour?

(A) 1300 vph

(B) 1350 vph

(C) 1900 vph

(D) 2900 vph

(d) What is the maximum (ideal) capacity in both directions sustainable for short distances?

(A) 2800 pcph

(B) 3200 pcph

(C) 4400 pcph

(D) 4800 pcph

(e) Which is a more accurate parameter of traffic capacity: volume or density? Why?

(A) Density is more accurate: Only density is a function of cars in a given length of roadway.

(B) Density is more accurate: Only density has units of time.

(C) Volume is more accurate: Only volume is a function of cars in a given length of roadway.

(D) Volume is more accurate: Only volume has units of time.

(f) What is the maximum (ideal) capacity of one lane of a four-lane freeway?

(A) 1900 pcph

(B) 2400 pcph

(C) 4400 pcph

(D) 8800 pcph

Intersection Design

4. The level intersection of First Street and Main Street, located in a suburb with negligible pedestrian and bicycle activity, is being investigated. The peak-hour factor on all approaches is 0.85. The signal cycle is 60 sec, two-phase. On Main Street, there is one 20 ft lane in each direction, and parking is prohibited. First Street is one-way westbound with two 11 ft lanes and parking lanes on both sides. There are 40 parking maneuvers per hour. Traffic travels equally in the two lanes of First Street. Turning on red is prohibited. All of the amber time is considered lost. There is no residual delay.

parameter	First (westbound)	Main (northbound)	Main (southbound)
green time	27 sec	27 sec	27 sec
amber time	3 sec	3 sec	3 sec
trucks	4%	5%	5%
buses	0%	0%	0%
left turns	10%	10%	0%
right turns	10%	0%	10%
volume	1240 vph	500 vph	350 vph

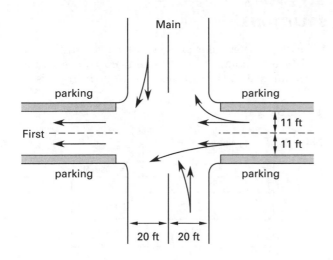

(a) What is the saturation flow rate for level of service E on First Street?

(A) 1900 vph

(B) 2700 vph

(C) 2900 vph

(D) 3800 vph

(b) What is the lane group capacity on First Street?

(A) 800 vph

(B) 1200 vph

(C) 1300 vph

(D) 2900 vph

(c) What is the maximum delay for level of service E?

(A) 35 sec

(B) 55 sec

(C) 60 sec

(D) 80 sec

(d) Determine the First Street lane group volume-capacity ratio. Will the intersection operation be satisfactory?

(A) 1.0; satisfactory

(B) 1.0; unsatisfactory

(C) 1.1; satisfactory

(D) 1.1; unsatisfactory

(e) Determine the maximum approach volume for First Street for satisfactory operation.

(A) 1130 vph

(B) 1260 vph

(C) 1330 vph

(D) 1480 vph

(f) What is the level of service for the given volume on First Street?

(A) C

(B) D

(C) E

(D) F

5. Two streets in a town of 250,000 people intersect at a stop sign: a 36 ft side street and a 44 ft arterial. There have been many complaints from the side street users that traffic signals are needed at the intersection. Use information from the *Manual on Uniform Traffic Control Devices*.

(a) What factors should be taken into consideration when deciding whether to signalize the intersection?

(A) warrants 1, 2, and 6

(B) warrants 1, 2, and 8

(C) warrants 1 through 7

(D) all warrants as applicable

(b) Give a logical step-by-step description of how the question should be investigated.

(A) Start with warrant 6, then work only the warrants needed.

(B) Work through the warrants for which adverse data is known.

(C) Work through the warrants in forward order.

(D) Follow warrant 8 for peak hour analysis, then review warrants for a stop sign.

(c) Explain the criteria for arriving at a recommendation.

(A) Compare actual conditions to the warrant criteria.

(B) Use the *Highway Capacity Manual* to design the signal.

(C) Install a stop beacon and see if the situation improves.

(D) Recommend further study by a traffic engineer.

Transportation

(d) If a signal is required, what conditions would justify an actuated signal?

   (A) Only pedestrian criteria are satisfied, and there are no turning movements.

   (B) There are low, fluctuating, or unbalanced traffic volumes.

   (C) The intersection has poor visibility.

   (D) The traffic volumes are too low for a pretimed signal.

Economic Justification of Highway Safety Features

6. A benefit/cost analysis is to be performed to justify the installation of safety-related road improvements such as flexible barriers and breakaway poles. What records could be used to obtain an economic value for property damage prevented and human lives saved by the safety improvements?

   (A) police records, telephone directories, and voting records

   (B) school bus incident reports, municipal building permits, and zoning records

   (C) hospital emergency room reports and paramedic files

   (D) insurance records, state disability fund records, accepted federal standards (e.g., OSHA), and property damage reports on file with police

SOLUTIONS

1. (a) For the standard case in which the station spacing is great enough to allow the train to accelerate to maximum speed, the distance and time diagram is

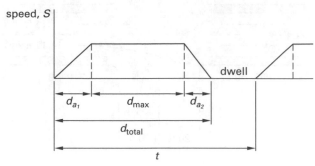

distance, d, and time, t

The first step is to check the acceleration and deceleration distance against the distance between stations to determine if the train can reach maximum speed. The acceleration distance is

$$
d_{a_1} = \frac{S_{max}^2 - S_0^2}{2a_1} = \frac{\left(\left(80\ \frac{mi}{hr}\right)^2 - \left(0\ \frac{mi}{hr}\right)^2\right) \times \left(5280\ \frac{ft}{mi}\right)^2}{(2)\left(5.5\ \frac{ft}{sec^2}\right)\left(3600\ \frac{sec}{hr}\right)^2}
$$

$$
= 1251.6\ ft
$$

The deceleration distance is

$$
d_{a_2} = \frac{S_0^2 - S_{max}^2}{2a_2} = \frac{\left(\left(0\ \frac{mi}{hr}\right)^2 - \left(80\ \frac{mi}{hr}\right)^2\right) \times \left(5280\ \frac{ft}{mi}\right)^2}{(2)\left(-4.4\ \frac{ft}{sec^2}\right)\left(3600\ \frac{sec}{hr}\right)^2}
$$

$$
= 1564.5\ ft
$$

The sum of the acceleration and deceleration distances is

$$
d_{a_1} + d_{a_2} = 1251.6\ ft + 1564.5\ ft = 2816.1\ ft
$$

This distance is less than the distance between stations (5280 ft). Therefore, the train can reach maximum speed between stations.

$$
\boxed{80\ \text{mph}}
$$

(b) The line's maximum capacity is

$$\left(\frac{60 \frac{\text{min}}{\text{hr}}}{5 \frac{\text{min}}{\text{train}}}\right)\left(5 \frac{\text{cars}}{\text{train}}\right)\left(220 \frac{\text{passengers}}{\text{car}}\right)$$
$$= \boxed{13{,}200 \text{ passengers/hr}}$$

(c) The average travel speed between stations includes station dwell. It is a parameter used to set schedules.

$$S_{\text{travel,avg}} = \frac{d_{\text{total}}}{t_{\text{total}}}$$
$$= \frac{(1 \text{ mi})\left(60 \frac{\text{min}}{\text{hr}}\right)}{5 \text{ min}}$$
$$= \boxed{12 \text{ mph}}$$

(d) The average running speed is not affected by the station dwell time. Running time includes only time for which the train is in motion.

Subtracting the acceleration and deceleration distances from the total distance between two stations yields the distance that is traveled at constant maximum speed.

$$d_{\text{max}} = d_{\text{total}} - d_{a_1} - d_{a_2}$$
$$= 5280 \text{ ft} - 1251.6 \text{ ft} - 1564.5 \text{ ft}$$
$$= 2463.9 \text{ ft}$$

The travel time between stations is the sum of times needed for acceleration, travel at constant maximum speed, and deceleration.

$$t_{a_1} = \frac{S_{\text{max}} - S_0}{a_1} = \frac{\left(80 \frac{\text{mi}}{\text{hr}} - 0 \frac{\text{mi}}{\text{hr}}\right)\left(5280 \frac{\text{ft}}{\text{mi}}\right)}{\left(5.5 \frac{\text{ft}}{\text{sec}^2}\right)\left(3600 \frac{\text{sec}}{\text{hr}}\right)}$$
$$= 21.3 \text{ sec}$$

$$t_{\text{max}} = \frac{d_{\text{max}}}{S} = \frac{(2463.9 \text{ ft})\left(3600 \frac{\text{sec}}{\text{hr}}\right)}{\left(80 \frac{\text{mi}}{\text{hr}}\right)\left(5280 \frac{\text{ft}}{\text{mi}}\right)}$$
$$= 21.0 \text{ sec}$$

$$t_{a_2} = \frac{S_0 - S_{\text{max}}}{a_2} = \frac{\left(0 \frac{\text{mi}}{\text{hr}} - 80 \frac{\text{mi}}{\text{hr}}\right)\left(5280 \frac{\text{ft}}{\text{mi}}\right)}{\left(-4.4 \frac{\text{ft}}{\text{sec}^2}\right)\left(3600 \frac{\text{sec}}{\text{hr}}\right)}$$
$$= 26.7 \text{ sec}$$

The average running speed is

$$S_{\text{running,avg}} = \frac{d_{\text{total}}}{t_{a_1} + t_{\text{max}} + t_{a_2}}$$
$$= \frac{(1 \text{ mi})\left(3600 \frac{\text{sec}}{\text{hr}}\right)}{21.3 \text{ sec} + 21 \text{ sec} + 26.7 \text{ sec}}$$
$$= \boxed{52.2 \text{ mph}}$$

(e) $\boxed{\text{The average travel speed over a specified distance during a time period is termed } \textit{space mean speed.}}$

(f) The acceleration distance for the increased maximum speed is

$$d_{a_1} = \frac{S_{\text{max}}^2 - S_0^2}{2a_1}$$
$$= \frac{\left(\left(100 \frac{\text{mi}}{\text{hr}}\right)^2 - \left(0 \frac{\text{mi}}{\text{hr}}\right)^2\right) \times \left(5280 \frac{\text{ft}}{\text{mi}}\right)^2}{(2)\left(5.5 \frac{\text{ft}}{\text{sec}^2}\right)\left(3600 \frac{\text{sec}}{\text{hr}}\right)^2}$$
$$= 1955.6 \text{ ft}$$

The deceleration distance is

$$d_{a_2} = \frac{\left(\left(0 \frac{\text{mi}}{\text{hr}}\right)^2 - \left(100 \frac{\text{mi}}{\text{hr}}\right)^2\right) \times \left(5280 \frac{\text{ft}}{\text{mi}}\right)^2}{(2)\left(-4.4 \frac{\text{ft}}{\text{sec}^2}\right)\left(3600 \frac{\text{sec}}{\text{hr}}\right)^2}$$
$$= 2444.4 \text{ ft}$$

The distance traveled at constant maximum speed is the distance between stations minus the acceleration and deceleration distances.

$$d_{\text{max}} = d_{\text{total}} - d_{a_1} - d_{a_2}$$
$$= 5280 \text{ ft} - 1955.6 \text{ ft} - 2444.4 \text{ ft}$$
$$= 880 \text{ ft}$$

The travel time between stations is the sum of times for acceleration, travel at maximum speed, and deceleration.

Transportation

From part (d), the total travel time between stations at 80 mph is

$$t_{a1} + t_{max} + t_{a2} = 21.3 \text{ sec} + 21.0 \text{ sec} + 26.7 \text{ sec}$$
$$= 69 \text{ sec}$$

At 100 mph,

$$t_{a_1} = \frac{S_{max} - S_0}{a_1} = \frac{\left(100 \frac{\text{mi}}{\text{hr}} - 0 \frac{\text{mi}}{\text{hr}}\right)\left(5280 \frac{\text{ft}}{\text{mi}}\right)}{\left(5.5 \frac{\text{ft}}{\text{sec}^2}\right)\left(3600 \frac{\text{sec}}{\text{hr}}\right)}$$
$$= 26.7 \text{ sec}$$

$$t_{max} = \frac{d_{max}}{S_{max}} = \frac{(880 \text{ ft})\left(3600 \frac{\text{sec}}{\text{hr}}\right)}{\left(100 \frac{\text{mi}}{\text{hr}}\right)\left(5280 \frac{\text{ft}}{\text{mi}}\right)}$$
$$= 6 \text{ sec}$$

$$t_{a_2} = \frac{S_0 - S_{max}}{a_2} = \frac{\left(0 \frac{\text{mi}}{\text{hr}} - 100 \frac{\text{mi}}{\text{hr}}\right)\left(5280 \frac{\text{ft}}{\text{mi}}\right)}{\left(-4.4 \frac{\text{ft}}{\text{sec}^2}\right)\left(3600 \frac{\text{sec}}{\text{hr}}\right)}$$
$$= 33.3 \text{ sec}$$

$$t_{a_1} + t_{max} + t_{a_2} = 26.7 \text{ sec} + 6 \text{ sec} + 33.3 \text{ sec}$$
$$= 66 \text{ sec}$$

The time that is saved at 100 mph maximum speed versus 80 mph maximum speed is 69 sec − 66 sec = $\boxed{3 \text{ sec.}}$

2. (a) Refer to *HCM* p. 20-3 or 20-10. The capacity for a two-lane highway is 1700 pcph. The combined (i.e., both ways) capacity is $\boxed{3200 \text{ pcph.}}$ This would occur at LOS E.

The answer is (C).

(b) From *HCM* p. 20-14, "Any upgrade of three percent or more and a length of 0.6 mi or more must be analyzed as a specific upgrade." Relevant parameters for specific upgrades are different from those for general rolling terrain. From *HCM* p. 12-16, the total maximum volume at LOS B is 780 pcph in both directions. From *HCM* Exh. 20-13, the grade adjustment factor for a 3% grade 2 mi long and a directional flow of (0.70)(780 pcph) = 546 pcph is $\boxed{0.97.}$

The answer is (C).

(c) This is a specific upgrade. From *HCM* Exh. 20-15, for a directional flow of 546 pcph, 3% grade, and 2 mi length, the passenger car equivalent of trucks is 5.9. Trucks and buses are grouped together. There are no

RVs. From *HCM* Eq. 20-4, the heavy vehicle adjustment factor is

$$f_{HV} = \frac{1}{1 + P_T(E_T - 1) + P_R(E_R - 1)}$$
$$= \frac{1}{1 + (0.15 + 0.05)(5.9 - 1)}$$
$$= \boxed{0.505 \quad (0.50)}$$

The answer is (A).

(d) From *HCM* p. 12-16, the total maximum volume at LOS B is 780 pcph in both directions. The uphill volume is $V = (0.7)(780 \text{ pcph}) = 546 \text{ pcph}$. The default value of the peak hour factor (PHF) in rural areas is 0.88 (*HCM* p. 12-10).

From *HCM* Eq. 20-12,

$$v_d = \frac{V}{(\text{PHF})f_G f_{HV}} = \frac{546 \frac{\text{pc}}{\text{hr}}}{(0.88)(0.97)(0.505)}$$
$$= \boxed{1267 \text{ pc/hr} \quad (1270 \text{ pcph})}$$

The answer is (C).

(e) The base free flow speed (BFFS) is the design speed of 60 mph. From *HCM* Exh. 20-5, for a lane width of 11 ft and a shoulder of 2 ft, the lane and shoulder width adjustment factor is 3.0 mph. From *HCM* Exh. 20-6, the adjustment for access point density is 0.0.

From *HCM* Eq. 20-2,

$$\text{FFS} = \text{BFFS} - f_{LS} - f_A$$
$$= 60 \text{ mph} - 3.0 \text{ mph} - 0.0 \text{ mph}$$
$$= \boxed{57 \text{ mph}}$$

The answer is (D).

(f) For the opposing direction,

$$V_o = 780 \text{ pcph} - 546 \text{ pcph} = 234 \text{ pcph}$$
$$\text{PHF} = 0.88 \text{ (rural areas)}$$
$$f_G = 1.00 \text{ (downhill, } HCM \text{ p. 20-14)}$$
$$E_T = 1.7 \text{ (} HCM \text{ Exh. 20-9, trucks,}$$
$$V_o < 300 \text{ pcph, level terrain}$$
$$\text{used for specific downgrade)}$$

$$f_{HV} = \frac{1}{1 + P_T(E_T - 1)} = \frac{1}{1 + (0.15 + 0.05)(1.7 - 1)}$$

$$= 0.877$$

$$v_o = \frac{V_o}{(PHF)f_G f_{HV}} = \frac{234 \text{ pcph}}{(0.88)(1.00)(0.877)}$$

$$= 303 \text{ pcph}$$

$$v_p = v_d + v_o = 1267 \text{ pcph} + 303 \text{ pcph} = 1570 \text{ pcph}$$

From *HCM* Exh. 20-19 for opposing demand flow rate = 303 pc/h and 40% no passing zones, the adjustment for no passing zones is $f_{np} \approx 2.27$ mi/hr.

From *HCM* Eq. 20-15, the average travel speed, ATS, along a specific upgrade is

$$ATS_d = FFS_d - 0.0125(v_d + v_o) - f_{np}$$

$$= 57 \frac{\text{mi}}{\text{hr}} - (0.0125)(1570 \text{ pcph}) - 2.27 \frac{\text{mi}}{\text{hr}}$$

$$= \boxed{35.11 \text{ mi/hr} \quad (35 \text{ mph})}$$

The answer is (A).

(g) Repeat parts (b) through (e) for time spent following.

From *HCM* Exh. 20-14 for 3% grade, 2 mi length, and directional flow rate of 546 pc/h, the grade adjustment factor for time spent following is $f_{Gf} = 0.95$.

Trucks and buses are grouped together. $P_T = 0.15 + 0.5 = 0.20$. There are no RVs. From *HCM* Exh. 20-16, $E_{Tf} = 1$.

$$f_{HVf} = \frac{1}{1 + P_T(E_{Tf} - 1)} = \frac{1}{1 + (0.20)(1 - 1)}$$

$$= 1.0$$

$$v_{pf} = \frac{v}{(PHF)f_{Gf} f_{HVf}} = \frac{780 \frac{\text{pc}}{\text{h}}}{(0.88)(0.95)(1.0)}$$

$$= 933 \text{ pc/h}$$

The basic percent time spent following is

$$BPTSF = (1 - e^{-0.000879 v_{pf}}) \times 100\%$$

$$= (1 - e^{-(0.000879)(933)}) \times 100\%$$

$$= 55.96\%$$

From *HCM* Exh. 20-12 for two-way flow of $v_{pf} = 933$ pc/h, 40% no passing zones, and 70%/30% directional split, the adjustment factor for directionality and percentage of no-passing zones is interpolated (between 800 and 1400) as $f_{d/np} = 9.41\%$.

The percent time spent following is

$$PTSF = BPTSF + f_{d/np} = 55.96\% + 9.41\%$$

$$= \boxed{65.37\% \quad (65\%)}$$

The answer is (C).

(h) Curbs, traffic signals, and reductions in speed will reduce the capacity. Improving the sight distance and reducing the percent of no-passing distance will allow the trucks to pull over and will improve the level of service.

The answer is (D).

3. (a) Spot speed is an instantaneous value at a particular point and is not a function of the number of vehicles on the highway. Use space mean speed.

$$29 \text{ mph} = \left(29 \frac{\text{mi}}{\text{hr}}\right) \left(\frac{5280 \frac{\text{ft}}{\text{mi}}}{3600 \frac{\text{sec}}{\text{hr}}}\right) = 42.53 \text{ ft/sec}$$

From Eq. 73.12,

$$\text{headway}_{\text{sec/veh}} = \frac{\text{spacing}_{\text{ft/veh}}}{\text{space mean speed}_{\text{ft/sec}}}$$

$$= \frac{80 \frac{\text{ft}}{\text{veh}}}{42.53 \frac{\text{ft}}{\text{sec}}}$$

$$= \boxed{1.88 \text{ sec/veh} \quad (1.9 \text{ sec/veh})}$$

The answer is (A).

(b) $$\text{density} = \frac{5280 \frac{\text{ft}}{\text{mi}}}{80 \frac{\text{ft}}{\text{veh}}} = \boxed{66 \text{ veh/mi} \quad (66 \text{ vpm})}$$

The answer is (B).

(c) From Eq. 73.13,

$$v_{\text{vph}} = \frac{3600 \frac{\text{sec}}{\text{hr}}}{\text{headway}_{\text{sec/veh}}} = \frac{3600 \frac{\text{sec}}{\text{hr}}}{1.88 \frac{\text{sec}}{\text{veh}}}$$

$$= \boxed{1914 \text{ veh/hr} \quad (1900 \text{ vph})}$$

The answer is (C).

Transportation

(d) From the *HCM* p. 20-3, the maximum stable capacity is $\boxed{3200-3400 \text{ pcph (two directions)}}$.

The answer is (B).

(e) Only volume has units of time. Density is a function of the number of cars in a given distance, but has nothing to do with speed. The cars could be stopped and still have a high density.

$$\boxed{\text{Capacity based on volume is more accurate.}}$$

The answer is (D).

(f) From *HCM* Exh. 13-3 or p. 13-4,

$$\boxed{2400 \text{ passenger cars/hr} \quad (2400 \text{ pcph})}$$

The free-flow speed has little effect on the maximum capacity.

The answer is (B).

4. (a) Since the lost time includes the amber time, the effective green time is the actual green time.

Treat First Street as one-lane group.

Lane width [HCM Exh. 16-7]:

$$f_W = 1 + \frac{W-12}{30} = 1 + \frac{11-12}{30} = 0.967$$

Heavy vehicles [HCM Exh. 16-7]:

$$\%\text{HV} = \%\text{trucks} = 4\%$$
$$f_{\text{HV}} = \frac{100}{100 + \%\text{HV}(E_T - 1)} = \frac{100}{100 + (4\%)(2-1)}$$
$$= 0.962$$

Grade [HCM Exh. 16-7]:

$$\%G = 0$$
$$f_g = 1.0$$

Parking [HCM Exh. 16-7]:

$$f_P = \frac{N - 0.1 - \dfrac{18N_m}{3600}}{N} = \frac{2 - 0.1 - \dfrac{(18)(40)}{3600}}{2} = 0.85$$

Bus blockage [HCM Exh. 16-7]:

$$N_b = 0$$
$$f_{\text{bb}} = 1.0$$

Type of area [HCM Exh. 16-7]:

The area is a suburb, so $f_a = 1.0$.

Lane utilization [HCM Exh. 16-7]:

No information given, so uniform flow assumed.

$$f_{\text{LU}} = 1.0$$

Left turns [HCM Exh. 16-7]:

For shared lanes,

$$f_{\text{LT}} = \frac{1}{1.0 + 0.05 P_{\text{LT}}} = \frac{1}{1.0 + (0.05)(0.10)} = 0.995$$

Right turns [HCM Exh. 16-7]:

For shared lanes,

$$f_{\text{RT}} = 1.0 - (0.15)P_{\text{RT}} = 1.0 - (0.15)(0.10) = 0.985$$

Pedestrian bicycle blockage:

There is negligible bicycle and pedestrian traffic, so

$$f_{\text{Lpb}} = f_{\text{Rpb}} = 1.0$$

Therefore, the saturation flow rate is given by Eq. 73.33.

$$s = s_o N f_w f_{\text{HV}} f_g f_p f_{\text{bb}} f_a f_{\text{LU}} f_{\text{RT}} f_{\text{LT}} f_{\text{Lpb}} f_{\text{Rpb}}$$
$$= \left(1900 \; \frac{\text{vph}}{\text{lane}}\right)(2 \text{ lanes})(0.967)(0.962)(1.000)(0.85)$$
$$\quad \times (1.000)(1.00)(0.985)(0.995)(1.00)(1.00)$$
$$= \boxed{2944 \text{ vph} \quad (2900 \text{ vph})}$$

The answer is (C).

(b) The lane group capacity [*HCM* Eq. 16-6] is

$$c = \frac{sg}{C} = \left(2944 \; \frac{\text{veh}}{\text{hr}}\right)\left(\frac{27 \text{ sec}}{60 \text{ sec}}\right)$$
$$= \boxed{1325 \text{ vph} \quad (1300 \text{ vph})}$$

The answer is (C).

(c) Maximum delay for level of service E is $\boxed{80 \text{ sec}}$ [*HCM* Exh. 16-2].

The answer is (D).

Transportation

(d) Use the demand flow rate, v_p, to find the volume-capacity ratio for the lane group.

$$v_p = \frac{v}{\text{PHF}} = \frac{1240 \ \frac{\text{veh}}{\text{hr}}}{0.85} = 1459 \ \text{vph}$$

$$\frac{v}{c} = \frac{1459 \ \frac{\text{veh}}{\text{hr}}}{1325 \ \frac{\text{veh}}{\text{hr}}}$$

$$= \boxed{1.1}$$

The First Street approach is oversaturated, and the operation will be $\boxed{\text{unsatisfactory.}}$

The answer is (D).

(e) With v_p at capacity, $v_p = c$.

$$v = c(\text{PHF}) = \left(1325 \ \frac{\text{veh}}{\text{hr}}\right)(0.85)$$

$$= \boxed{1126 \ \text{vph} \quad (1130 \ \text{vph})}$$

The answer is (A).

(f) The level of service is found by the intersection approach delay, found by *HCM* Eq. 16-9. d_1 is given by *HCM* Eq. 16-11, where X is limited to 1.0, per the *HCM*. d_2 is given by *HCM* Eq. 16-12 for all values of X. $d_3 = 0$ since there is no initial queue (i.e., no residual delay).

$$d = d_1(\text{PF}) + d_2 + d_3$$

$$= 0.50 \, C \left(\frac{\left(1 - \frac{g}{C}\right)^2}{1 - X\left(\frac{g}{C}\right)}\right) \left(\frac{(1 - P)f_{\text{PA}}}{1 - \frac{g}{C}}\right)$$

$$+ \, 900 \, T\left((X - 1) + \sqrt{(X - 1)^2 + \frac{8kIX}{cT}}\right) + 0$$

$$= (0.50)(60 \ \text{sec}) \left(\frac{\left(1 - \frac{27 \ \text{sec}}{60 \ \text{sec}}\right)^2}{1 - (1.0)\left(\frac{27 \ \text{sec}}{60 \ \text{sec}}\right)}\right)$$

$$\times \left(\frac{\left(1 - \frac{27 \ \text{sec}}{60 \ \text{sec}}\right)(1.0)}{1 - \frac{27 \ \text{sec}}{60 \ \text{sec}}}\right)$$

$$+ \, (900)(0.25) \left(\begin{array}{c}(1.1 - 1) \\ + \sqrt{\begin{array}{c}(1.1 - 1)^2 \\ + \frac{(8)(0.50)(1.0)(1.1)}{(1325)(0.25)}\end{array}}\end{array}\right)$$

$$+ \, 0$$

$$= 73.3 \ \text{sec}$$

From *HCM* Exh. 16-2, the level of service is $\boxed{\text{E.}}$

The answer is (C).

5. (a) Although the geometric design of the intersection is specified, no information about any aspect of intersection performance is given. All warrants are theoretically applicable in this instance.

The answer is (D).

(b) All aspects of intersection performance should be evaluated. Work through the warrants in forward order.

The answer is (C).

(c) Signalization is an art, not a science. The warrants are necessary but not sufficient conditions for signalization. Therefore, a strict comparison of warrants to actual performance is only part of the procedure; a traffic engineer should perform further study.

The answer is (D).

(d) Traffic-actuated controllers should be considered when

- there are low, fluctuating, or unbalanced traffic volumes
- there are high side street flows and delays during peak hours only
- only the accident or crosswalk warrant is satisfied
- only one direction of the two-way traffic is to be controlled

The answer is (B).

6. Accident data can be obtained from the following.

- police records (property damage only)
- municipal records
- commercial insurance records
- state disability funds
- expected remaining earnings estimates
- accepted federal standards (e.g., OSHA)

The answer is (D).

Transportation

74 Bridges: Condition and Rating

PRACTICE PROBLEMS

There are no problems in this book corresponding to Chap. 74 of the *Civil Engineering Reference Manual*.

75 Vehicle Dynamics and Accident Analysis

PRACTICE PROBLEMS

Collisions

1. Two cars are moving at 60 mph in the same direction and in the same lane. The cars are separated by one car length (20 ft) for each 10 mph. The coefficient of friction (skidding) between the tires and the roadway is 0.6. The reaction time for each driver is 0.5 sec.

(a) If the lead car hits a parked truck, what is the speed of the second car when it hits the first (stationary) car?

(A) 13 mph

(B) 24 mph

(C) 38 mph

(D) 47 mph

(b) If the lead car hits a parked truck and comes to an abrupt halt, at what speed does the rule of one car length of spacing for every 10 mph of speed become unsafe?

(A) 23 mph

(B) 33 mph

(C) 55 mph

(D) 85 mph

(c) At 60 mph, what should the rule for every 10 mph actually be?

(A) 30 ft

(B) 40 ft

(C) 50 ft

(D) 60 ft

2. You have been hired by the owner of a damaged house to investigate a car crash that caused the damage. From the police report, you learn that the car was traveling down a 3% grade at an unknown speed. The skid marks are 185 ft (56 m) long, and the pavement was dry at the time of the accident. The police report estimates that the initial speed of the car was 25 mph (40 km/h). From an inspection, you know that the car's tires were new. The house owner does not believe the

estimate of initial speed. You have the following data from tests performed on dry, level roadways.

initial speed in mph (km/h)	30 (50)	40 (60)	50 (800)	60 (100)
coefficient of friction	0.59	0.51	0.45	0.35

(a) Find the minimum initial speed of the car.

(A) 49 mph (78 kph)

(B) 50 mph (81 kph)

(C) 56 mph (89 kph)

(D) 74 mph (120 kph)

(b) If the police report was mistaken in assuming the road surface was dry, what was the minimum initial speed of the car?

(A) 39 mph (63 kph)

(B) 41 mph (66 kph)

(C) 43 mph (69 kph)

(D) 45 mph (72 kph)

3. You are hired as a consultant to give expert evidence in a court action arising from a vehicle collision. The two vehicles collided head-on while traveling on a curve tangent with a 4% grade. Vehicle A skidded 195 ft (59.4 m) downhill before colliding with vehicle B. Vehicle B skidded 142 ft (43.3 m). The police report estimates that both vehicles were traveling at 25 mph (40 kph) prior to the collision, based on vehicle deformation. Assume a coefficient of friction of 0.48.

(a) What was the speed of vehicle A prior to the application of brakes?

(A) 56 mph (90 kph)

(B) 57 mph (91 kph)

(C) 61 mph (97 kph)

(D) 62 mph (99 kph)

(b) What was the speed of vehicle B prior to the application of the brakes?

(A) 53 mph (86 kph)

(B) 57 mph (91 kph)

(C) 61 mph (97 kph)

(D) 62 mph (99 kph)

(c) Which of your assumptions (impact speed or coefficient of friction) produces the greatest possible error in your initial speed calculations?

(A) The impact speed is more sensitive to error.

(B) The impact speed and coefficient of friction are equally sensitive to error.

(C) The impact speed is more sensitive to error over 50 mph (80 kph).

(D) The coefficient of friction is more sensitive to error.

Accident Data Analysis

4. (*Time limit: one hour*) Four intersections and five segments of highway have been evaluated using prior years' accident data. (a) Calculate the number of accidents per million vehicles for each intersection. Rank the intersections in order of highest to lowest need for improvement. (b) For the highway segments, calculate the number of accidents per year per million vehicle miles. Rank the segments in order of highest to lowest need for improvement. (c) Explain why calculations in (a) and (b) should be done. (d) Assume that the intersections varied in terms of the fractions of accidents that were fatal, injury, and property damage only. How would you compare the safety needs at those intersections?

intersection	ADT	accidents/year
A	820	4
B	1200	5
C	1070	7
D	1400	6

highway segment	ADT	accidents/year	length mi (km)
1	1900	1	1.50 (2.41)
2	2000	14	1.35 (2.17)
3	5500	18	4.50 (7.24)
4	3000	11	0.53 (0.85)
5	4000	30	2.48 (4.00)

SOLUTIONS

1. (a) Convert 60 mph to ft/sec.

$$\left(60 \ \frac{\text{mi}}{\text{hr}}\right) \left(\frac{5280 \ \frac{\text{ft}}{\text{mi}}}{3600 \ \frac{\text{sec}}{\text{hr}}}\right) = 88 \ \text{ft/sec}$$

The initial separation is

$$(20 \ \text{ft})(6) = 120 \ \text{ft}$$

The separation after reaction time is

$$120 \ \text{ft} - (0.5 \ \text{sec})\left(88 \ \frac{\text{ft}}{\text{sec}}\right) = 76 \ \text{ft}$$

From Newton's second law,

$$F_f = \text{frictional retarding force} = ma$$
$$= w\mu = mg\mu$$
$$ma = mg\mu$$
$$a = g\mu = \left(32.2 \ \frac{\text{ft}}{\text{sec}^2}\right)(0.6) = 19.32 \ \text{ft/sec}^2$$
$$v_o = 88 \ \text{ft/sec}$$
$$s = 76 \ \text{ft}$$
$$a = -19.32 \ \text{ft/sec}^2$$
$$v = \sqrt{v_o^2 + 2as}$$
$$= \sqrt{\left(88 \ \frac{\text{ft}}{\text{sec}}\right)^2 + (2)\left(-19.32 \ \frac{\text{ft}}{\text{sec}^2}\right)(76 \ \text{ft})}$$
$$= \boxed{69.3 \ \text{ft/sec} \quad (47 \ \text{mph})}$$

The answer is (D).

(b) The rule is one car length (20 ft) per 10 mph. Convert 10 mph to ft/sec.

$$\left(10 \ \frac{\text{mi}}{\text{hr}}\right) \left(\frac{5280 \ \frac{\text{ft}}{\text{mi}}}{3600 \ \frac{\text{sec}}{\text{hr}}}\right) = 14.67 \ \text{ft/sec}$$

Working backward, knowing that terminal velocity must be zero,

$$0 = \sqrt{\begin{array}{c} v_o^2 + (2)\left(-19.32 \ \frac{\text{ft}}{\text{sec}^2}\right) \\ \times \left((20 \ \text{ft})\left(\frac{v_o}{14.67 \ \frac{\text{ft}}{\text{sec}}}\right) - (0.5 \ \text{sec})v_o\right) \end{array}}$$

$$v_o^2 - 33.36v_o = 0$$
$$v_o = \boxed{33.36 \ \text{ft/sec} \quad (23 \ \text{mph})}$$

The answer is (A).

(c) If a similar simple car-length rule is to apply at 60 mph,

$$0 = \left(88 \; \frac{ft}{sec}\right)^2 - (2)\left(19.32 \; \frac{ft}{sec^2}\right)$$

$$\times \left(d \left(\frac{88 \; \frac{ft}{sec}}{14.67 \; \frac{ft}{sec}}\right) - (0.5 \; sec)\left(88 \; \frac{ft}{sec}\right)\right)$$

$$d = \boxed{40 \; ft} \quad \begin{bmatrix} \text{approximately two car lengths} \\ \text{for every 10 mph} \end{bmatrix}$$

The answer is (B).

2. (a) Use Eq. 75.25.

$$s_b = \frac{v_{mph}^2}{30(f + G)}$$

$$185 \; ft = \frac{v_{mph}^2}{30(f - 0.03)}$$

By trial and error, using the test values of f,

$$v \geq \boxed{49 \; mph \quad (78 \; kph)}$$

The initial speed is clearly higher than this, as damage to the house means the speed was not zero at impact. While the damage might be used to estimate the impact speed, the skid marks alone are sufficient to disprove the police report estimate of initial speed.

The answer is (A).

(b) Use Table 75.1 for a wet road surface. Interpolate values of f.

v	interpolated f	s_b
39 mph	0.333	167.3
41 mph	0.331	186.2
43 mph	0.324	209.6
45 mph	0.320	232.8

$$v \geq \boxed{41 \; mph \quad (66 \; kph)}$$

The answer is (B).

3. (a) 25 mph is the terminal speed at time of accident (not the original speed).

Vehicle A

If there had been no collision, and using Eq. 75.25, the skidding distance would have been

$$s_{v_1} = \frac{v_1^2}{30(f + G)} = \frac{v_1^2}{(30)(0.48 - 0.04)}$$

$$= v_1^2/13.2$$

At 25 mph, the skidding distance would be

$$s_{25,1} = \frac{\left(25 \; \frac{mi}{hr}\right)^2}{13.2} = 47.3 \; ft$$

The actual skid was 195 ft, and the car was still moving at 25 mph when the collision occurred. Determine the skid distance without a collision.

$$s_{v_1} = 195 \; ft + 47.35 \; ft$$

$$= 242.35 \; ft$$

Then,

$$242.35 \; ft = \frac{v_1^2}{13.2}$$

$$v_1 = \boxed{56.6 \; mi/hr \quad (57 \; mph) \quad (91 \; kph)}$$

The answer is (B).

Vehicle B

(b) Follow the same analysis as used for Vehicle A.

$$s_{v_2} = \frac{v_2^2}{(30)(0.48 + 0.04)} = v_2^2/15.6$$

$$s_{25,2} = \frac{\left(25 \; \frac{mi}{hr}\right)^2}{15.6} = 40.06 \; ft$$

$$s_{v_2} = 142 \; ft + 40.06 \; ft = 182.06 \; ft$$

$$182.06 \; ft = \frac{v_2^2}{15.6}$$

$$v_2 = \boxed{53 \; mph \quad (86 \; kph)}$$

The answer is (A).

(c) The speed of impact and coefficient of friction are the only two variables. Vary each by 10% and see how velocities change.

If the speed at impact is $(1 + 0.10)(25 \; mph) = 27.5 \; mph$,

$$v_1 = 57.7 \; mph \quad [1.9\% \; change]$$

If the coefficient of friction is $(1 + 0.10)(0.48) = 0.528$,

$$v_1 = 59.0 \text{ mph} \quad [4.2\% \text{ change}]$$

The coefficient of friction is more sensitive to error.

The answer is (D).

4. (a) Use Eq. 75.26.

$$R_{\text{RMEV}} = \frac{(\text{no. accidents})(10^6)}{(\text{ADT})(\text{no. of years})\left(365 \frac{\text{days}}{\text{yr}}\right)}$$

A: $\dfrac{\left(4 \frac{1}{\text{yr}}\right)(1{,}000{,}000)}{\left(820 \frac{1}{\text{day}}\right)\left(365 \frac{\text{days}}{\text{yr}}\right)}$

$= 13.4 \text{ accidents/MEV}$

B: $\dfrac{\left(5 \frac{1}{\text{yr}}\right)(1{,}000{,}000)}{\left(1200 \frac{1}{\text{day}}\right)\left(365 \frac{\text{days}}{\text{yr}}\right)}$

$= 11.4 \text{ accidents/MEV}$

C: $\dfrac{\left(7 \frac{1}{\text{yr}}\right)(1{,}000{,}000)}{\left(1070 \frac{1}{\text{day}}\right)\left(365 \frac{\text{days}}{\text{yr}}\right)}$

$= 17.9 \text{ accidents/MEV}$

D: $\dfrac{\left(6 \frac{1}{\text{yr}}\right)(1{,}000{,}000)}{\left(1400 \frac{1}{\text{day}}\right)\left(365 \frac{\text{days}}{\text{yr}}\right)}$

$= 11.7 \text{ accidents/MEV}$

Ranking from highest to lowest,

C, A, D, B

(b) Use Eq. 75.27. (Use 10^6 instead of 10^8 because the ranking is per million vehicle miles, not 100 million.)

$$R_{\text{HMVM}} = \frac{(\text{no. accidents})(10^6)}{(\text{ADT})(\text{no. of years})\left(365 \frac{\text{days}}{\text{yr}}\right)L_{\text{mi}}}$$

1: $\dfrac{\left(1 \frac{1}{\text{yr}}\right)(1{,}000{,}000)}{\left(1900 \frac{1}{\text{day}}\right)\left(365 \frac{\text{days}}{\text{yr}}\right)(1.5 \text{ mi})}$

$= 0.96 \text{ accidents/mi-yr}$

2: $\dfrac{\left(14 \frac{1}{\text{yr}}\right)(1{,}000{,}000)}{\left(2000 \frac{1}{\text{day}}\right)\left(365 \frac{\text{days}}{\text{yr}}\right)(1.35 \text{ mi})}$

$= 14.2 \text{ accidents/mi-yr}$

3: $\dfrac{\left(18 \frac{1}{\text{yr}}\right)(1{,}000{,}000)}{\left(5500 \frac{1}{\text{day}}\right)\left(365 \frac{\text{days}}{\text{yr}}\right)(4.5 \text{ mi})}$

$= 1.99 \text{ accidents/mi-yr}$

4: $\dfrac{\left(11 \frac{1}{\text{yr}}\right)(1{,}000{,}000)}{\left(3000 \frac{1}{\text{day}}\right)\left(365 \frac{\text{days}}{\text{hr}}\right)(0.53 \text{ mi})}$

$= 18.95 \text{ accidents/mi-yr}$

5: $\dfrac{\left(30 \frac{1}{\text{yr}}\right)(1{,}000{,}000)}{\left(4000 \frac{1}{\text{day}}\right)\left(365 \frac{\text{days}}{\text{hr}}\right)(2.48 \text{ mi})}$

$= 8.29 \text{ accidents/mi-yr}$

Ranking from highest to lowest,

4, 2, 5, 3, 1

(c) Calculations are performed in order to rank highway improvement priorities. However, this assumes that all locations produce accidents of equal costs. Actually, some locations may produce a higher fraction of fatalities than others.

(d) The need for safety improvements at various intersections could be ranked by cumulative cost (i.e., value) of the accidents occurring over a fixed period of time.

Transportation

76 Flexible Pavement Design

PRACTICE PROBLEMS

Marshall Mixture Testing

1. Results of Marshall mix testing are shown. An asphalt pavement is to be designed for medium traffic using aggregate with a $^3/_4$ in nominal size. What is the appropriate percentage of asphalt for the mix?

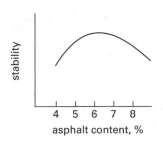

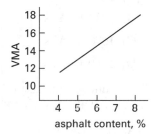

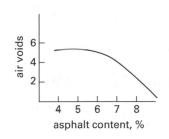

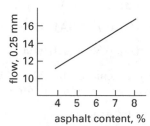

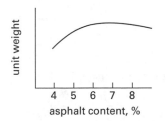

Mixture Properties

2. The mass of a core of asphalt concrete in air is 1150 g. The mass of the core in water is 498.3 g. What is the core's density in lbm/ft^3?

3. The specific gravity of a voidless mixture (i.e., the maximum theoretical specific gravity) of an asphalt concrete is 2.550. The components are specified as follows.

material	specific gravity	apparent specific gravity	by weight (%)
asphalt cement	1.020	–	6.3
limestone dust	2.820	2.650	13.7
sand	2.650	2.905	30.4
gravel	2.650	2.873	49.6

(a) What is the air void content if the bulk specific gravity of the mixture is 2.340?

(A) 8.0%

(B) 8.2%

(C) 8.5%

(D) 8.6%

(b) What is the bulk specific gravity of the aggregate?

(A) 2.67

(B) 2.69

(C) 2.73

(D) 2.75

(c) What is the VMA of the mixture?

(A) 16%

(B) 17%

(C) 18%

(D) 19%

(d) What is the effective specific gravity of the aggregate?

(A) 2.80

(B) 2.82

(C) 2.84

(D) 2.86

(e) What is the asphalt absorption?

(A) 1.90%

(B) 2.05%

(C) 2.13%

(D) 2.18%

(f) What is the effective asphalt content of the mixture?

(A) 4.3%

(B) 4.5%

(C) 4.7%

(D) 5.0%

(g) What is the VFA of the mixture?

(A) 52.4%

(B) 53.8%

(C) 54.3%

(D) 54.6%

(h) What is the apparent bulk specific gravity of the aggregate?

(A) 2.4

(B) 2.6

(C) 2.7

(D) 2.8

(i) How much would the percent air voids of the mixture change if the weight of the asphalt is increased by 2%?

(A) 0.05% decrease

(B) 0.14% decrease

(C) 0.14% increase

(D) no change

(j) How much would the VMA of the mixture change if the weight of the asphalt is increased by 2%?

(A) 0.1% decrease

(B) 0.2% decrease

(C) 0.1% increase

(D) no change

AASHTO Pavement Design

4. The following data apply to a particular road.

initial ADT	5000 vpd
growth rate	4%/yr
design period	20 yr
fraction of truck traffic	0.25

loadometer data

axle load (lbf)	number of axles out of 1250 vehicles
single-axle trucks	
2000	800
6000	922
14,000	1851
20,000	37
double-axle trucks	
16,000	417
20,000	608
34,000	84

(a) Assuming a structural number of 3, what is the 18 kip equivalent single axle loading (ESALs)?

(A) 1000 ESALs

(B) 1030 ESALs

(C) 1070 ESALs

(D) 1240 ESALs

(b) Assuming a structural number of 3, calculate the 18 kip equivalent single axle load per truck.

(A) 0.61 ESAL/truck

(B) 0.63 ESAL/truck

(C) 0.78 ESAL/truck

(D) 0.82 ESAL/truck

(c) Calculate the 20 yr AADT (i.e., the ADT in an average year) in one direction.

(A) 2700 vpd

(B) 3000 vpd

(C) 3700 vpd

(D) 4000 vpd

(d) Calculate the total number of trucks in one direction for the 20 yr design period.

(A) 5.2×10^6 trucks

(B) 6.0×10^6 trucks

(C) 6.8×10^6 trucks

(D) 7.3×10^6 trucks

(e) Calculate the total 18 kip single-axle load applications over the design period.

(A) 5.6×10^6

(B) 6.1×10^6

(C) 7.9×10^6

(D) 8.3×10^6

(f) Calculate the effective roadbed modulus for use in design, given the following data.

season	roadbed soil modulus (psi)
winter (Dec–Feb)	16,000
spring (Mar–Apr)	2000
summer (May–Sep)	6000
fall (Oct–Nov)	4000

(A) 2500 psi

(B) 2800 psi

(C) 3250 psi

(D) 3730 psi

(g) Using the AASHTO design charts, a standard deviation of 0.35, a reliability of 0.9, and a design serviceability loss of 2.0, calculate the required structural number for a rural divided highway.

(A) 4

(B) 5

(C) 6

(D) 7

(h) If the minimum plant-mix asphalt layer thickness is 4 in, and the required structural number is 6, calculate the appropriate crushed-stone base layer thickness if it is placed directly on the subgrade. Use $m = 1$ for all drainage coefficients.

(A) 15 in

(B) 20 in

(C) 25 in

(D) 30 in

(i) For a 6 in plant-mix asphalt surface placed over an 8 in crushed stone base, calculate the required subbase thickness for a subbase with a layer coefficient of 0.08 and a required structural number of 6. Use $m = 1$ for all drainage coefficients.

(A) 20 in

(B) 24 in

(C) 28 in

(D) 32 in

(j) Given the following minimum thicknesses and unit costs and a required structural number of 6, design the most economical pavement section.

material	layer coefficient	minimum thickness	unit cost
asphalt plant mix	0.44 1/in	4 in	$0.17/in
crushed stone base	0.14 1/in	6 in	$0.04/in
granular subbase	0.08 1/in	–	$0.01/in

(A) asphalt concrete, 4 in; base, 6 in; subbase, 28 in

(B) asphalt concrete, 4 in; base, 6 in; subbase, 43 in

(C) asphalt concrete, 6 in; base, 6 in; subbase, 28 in

(D) asphalt concrete, 4 in; base, 8 in; subbase, 30 in

SOLUTIONS

1.
$$P_{\text{air voids}} = 7.0\%$$
$$P_{\text{stability}} = 6.5\%$$
$$P_{\text{unit weight}} = 7.0\%$$
$$P_{\text{ave}} = \frac{7.0\% + 6.5\% + 7.0\%}{3} = 6.83\%$$

Check that VMA ≥ 13.5.   [OK]

Check that $8 < \text{flow} < 16$.   [OK]

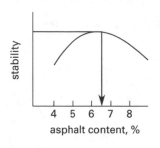
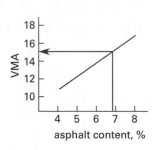

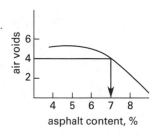
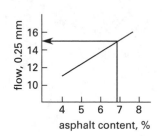

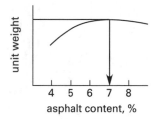

$$\boxed{\text{Select } 6.8\% \text{ AC.}}$$

2. Use Eq. 14.7 and Eq. 76.7.
$$\rho = G\rho_{\text{H}_2\text{O}}$$
$$= \left(\frac{m_{\text{air}}}{m_{\text{air}} - m_{\text{H}_2\text{O}}}\right)\rho_{\text{H}_2\text{O}}$$
$$= \left(\frac{1150 \text{ g}}{1150 \text{ g} - 498.3 \text{ g}}\right)\left(62.4 \text{ } \frac{\text{lbm}}{\text{ft}^3}\right)$$
$$= \boxed{110.11 \text{ lbm/ft}^3}$$

3. (a) Use Eq. 76.19. For the air voids,

$$\text{VTM} = \frac{G_{\text{mm}} - G_{\text{mb}}}{G_{\text{mm}}} \times 100\%$$
$$= \frac{2.550 - 2.340}{2.550} \times 100\%$$
$$= \boxed{8.235\% \quad (8.2\%)}$$

The answer is (B).

(b) Use Eq. 76.6. The aggregate bulk specific gravity is

$$G_{\text{sb}} = \frac{P_{\text{LD}} + P_S + P_G}{\dfrac{P_{\text{LD}}}{G_{\text{LD}}} + \dfrac{P_S}{G_S} + \dfrac{P_G}{G_G}}$$
$$= \frac{13.7\% + 30.4\% + 49.6\%}{\dfrac{13.7\%}{2.820} + \dfrac{30.4\%}{2.650} + \dfrac{49.6\%}{2.650}}$$
$$= \boxed{2.674 \quad (2.67)}$$

The answer is (A).

(c) Use Eq. 76.18. The VMA is

$$\text{VMA} = 100\% - \frac{G_{\text{mb}}P_s}{G_{\text{sb}}} \times 100\%$$
$$= 100\% - \left(\frac{(2.340)(1 - 0.063)}{2.674}\right) \times 100\%$$
$$= \boxed{18.00\% \quad (18\%)}$$

The answer is (C).

(d) Use Eq. 76.13. The effective specific gravity of aggregate is

$$G_{\text{se}} = \frac{100\% - P_b}{\dfrac{100\%}{G_{\text{mm}}} - \dfrac{P_b}{G_b}}$$
$$= \frac{100\% - 6.3\%}{\dfrac{100\%}{2.550} - \dfrac{6.3\%}{1.020}}$$
$$= \boxed{2.836 \quad (2.84)}$$

The answer is (C).

Transportation

(e) Use Eq. 76.16. The asphalt absorption is

$$P_{ba} = \frac{G_b(G_{se} - G_{sb})}{G_{sb}G_{se}} \times 100\%$$

$$= (1.020)\left(\frac{2.836 - 2.674}{(2.674)(2.836)}\right) \times 100\%$$

$$= \boxed{2.179\% \quad (2.18\%)}$$

The answer is (D).

(f) Use Eq. 76.17. The effective asphalt content is

$$P_{be} = P_b - \frac{P_{ba}P_s}{100\%}$$

$$= 6.3\% - \left(\frac{2.179\%}{100\%}\right)(93.7\%)$$

$$= \boxed{4.258\% \quad (4.3\%)}$$

The answer is (A).

(g) Use Eq. 76.20. The VFA is

$$VFA = \frac{VMA - P_a}{VMA} \times 100\%$$

$$= \frac{18.00\% - 8.235\%}{18.00\%} \times 100\%$$

$$= \boxed{54.25\% \quad (54.3\%)}$$

The answer is (C).

(h) Use Eq. 76.6. The apparent bulk specific gravity of aggregate is

$$G_{sa} = \frac{P_{LD} + P_S + P_G}{\dfrac{P_{LD}}{G_{LD,a}} + \dfrac{P_S}{G_{S,a}} + \dfrac{P_G}{G_{G,a}}}$$

$$= \frac{13.7\% + 30.4\% + 49.6\%}{\dfrac{13.7\%}{2.650} + \dfrac{30.4\%}{2.905} + \dfrac{49.6\%}{2.873}}$$

$$= \boxed{2.848 \quad (2.8)}$$

The answer is (D).

(i) Assume a 100 lbm mix. The weight of AC is

$$(6.3 \text{ lbm})(1.02) = 6.426 \text{ lbm}$$

The new weight of the mix is 100.126 lbm.

$$P_b = \frac{6.426 \text{ lbm}}{100.126 \text{ lbm}} \times 100\%$$

$$= 6.4\%$$

$$P_s = \frac{93.7 \text{ lbm}}{100.126 \text{ lbm}} \times 100\%$$

$$= 93.6\%$$

Use Eq. 76.14.

$$G_{mm} = \frac{100\%}{\dfrac{P_s}{G_{se}} + \dfrac{P_b}{G_b}}$$

$$= \frac{100\%}{\dfrac{93.6\%}{2.836} + \dfrac{6.4\%}{1.02}}$$

$$= 2.546$$

Use Eq. 76.19.

$$VTM = \frac{G_{mm} - G_{mb}}{G_{mm}} \times 100\%$$

$$= \frac{2.546 - 2.340}{2.546} \times 100\%$$

$$= 8.09\%$$

The change is

$$8.23\% - 8.09\% = \boxed{0.14\% \text{ decrease}}$$

The answer is (B).

(j) Use Eq. 76.18.

$$VMA = 100\% - \frac{G_{mb}P_s}{G_{sb}}$$

$$= 100\% - \frac{(2.340)(93.6\%)}{2.674}$$

$$= 18.091\%$$

The change is

$$18.091\% - 18.000\% = \boxed{0.091\% \quad (0.1\% \text{ increase})}$$

The answer is (C).

4. (a) Use App. 76.A and App. 76.B. Use nonlinear interpolation for greatest accuracy.

axle load
(kips)	equiv. factor	×	no. axles	= ESALs
	single axle			
2	0.0003		800	0.24
6	0.017		922	15.67
14	0.399		1851	738.55
20	1.49		37	55.13
	double axle			
16	0.070		417	29.19
20	0.162		608	98.50
34	1.11		84	93.24
			total	1030.52 (1030)

The answer is (B).

(b) $\dfrac{1030.52 \text{ ESALs}}{1250 \text{ trucks}} = \boxed{\begin{array}{c}0.824 \text{ ESAL/truck} \\ (0.82 \text{ ESAL/truck})\end{array}}$

The answer is (D).

(c) The initial ADT in one direction is 5000 vpd/2 = 2500 vpd, which grows at 4% per year for 20 years.
From App. 87.B,

$$(F/A, 4\%, 20) = 29.7781$$

The AADT in one direction is

$$\frac{(29.7781)(2500 \text{ vpd})}{20 \text{ years}} = \boxed{3722 \text{ vpd} \quad (3700 \text{ vpd})}$$

The answer is (C).

(d) $(3722 \text{ vpd})(0.25) = 931 \text{ trucks/day}$

The total number of trucks in one direction for a 20 yr period is

$$\text{AADT} = \left(931 \; \frac{\text{trucks}}{\text{day}}\right)(20 \text{ yr})\left(365 \; \frac{\text{days}}{\text{yr}}\right)$$
$$= \boxed{6{,}796{,}300 \text{ trucks} \quad (6.8 \times 10^6 \text{ trucks})}$$

The answer is (C).

(e) $w_{18} = (6{,}796{,}300 \text{ trucks})\left(0.824 \; \frac{\text{ESAL}}{\text{truck}}\right)$
$$= \boxed{5{,}600{,}151 \quad (5.6 \times 10^6)}$$

The answer is (A).

(f) The effective M_R is calculated as shown.

month	roadbed soil modulus, M_R (psi)	relative damage u_f
January	16,000	0.021
February	16,000	0.021
March	2000	2.6
April	2000	2.6
May	6000	0.2
June	6000	0.2
July	6000	0.2
August	6000	0.2
September	6000	0.2
October	4000	0.52
November	4000	0.52
December	16,000	0.021
summation: $\Sigma u_f =$		7.30

average: $\bar{u}_f = \dfrac{\Sigma u_f}{n} = \dfrac{7.30}{12} = 0.61$

effective roadbed soil resilient modulus,
M_R (psi) = __3730__ (corresponds to $\bar{u}_f$)

$$\boxed{M_R = 3730 \text{ psi}}$$

The answer is (D).

Transportation

(g) Using $w_{18} = 5.6 \times 10^6$, $M_R = 3730$ psi, $R = 0.9$, $S_o = 0.35$, and $\Delta\text{PSI} = 2$,

$$\boxed{\text{SN} \approx 6}$$

The answer is (C).

(h) Use Eq. 76.33. Use $a_1 = 0.44 \; {}^1/_{\text{in}}$ and $a_2 = 0.14 \; {}^1/_{\text{in}}$.

$$\text{SN} = D_1 a_1 + D_2 a_2 m_2$$
$$6 = (4 \text{ in})\left(0.44 \; \frac{1}{\text{in}}\right) + D_2\left(0.14 \; \frac{1}{\text{in}}\right)(1)$$
$$D_2 = \boxed{30.3 \text{ in} \quad (30 \text{ in})}$$

The answer is (D).

(i) Use Eq. 76.33.

$$\text{SN} = D_1 a_1 + D_2 a_2 m_2 + D_3 a_3 m_3$$
$$6 = (6 \text{ in})\left(0.44 \; \frac{1}{\text{in}}\right) + (8 \text{ in})\left(0.14 \; \frac{1}{\text{in}}\right)(1)$$
$$\qquad + D_3\left(0.08 \; \frac{1}{\text{in}}\right)(1)$$
$$D_3 = \boxed{28 \text{ in}}$$

The answer is (C).

(j) The ratios of the layer coefficients can be used to show that it is least expensive to use subbase in place of aggregate and asphalt in order to satisfy the structural number requirements. The total SN is 6.

For asphalt, select 4 in to minimize layer cost.

$$(4 \text{ in})\left(0.44 \; \frac{1}{\text{in}}\right) = 1.76 = \text{SN}_1^* \quad \text{[not used]}$$

For aggregate, select 6 in to minimize layer cost.

$$(4 \text{ in})\left(0.44 \; \frac{1}{\text{in}}\right) + (6 \text{ in})\left(0.14 \; \frac{1}{\text{in}}\right) = 2.60 = \text{SN}_2^*$$

The required D_3 is

$$\frac{6 \text{ in} - 2.6 \text{ in}}{0.08} = 42.5 \text{ in} \quad (43 \text{ in subbase})$$

$$\boxed{\begin{array}{l} 4 \text{ in AC} \\ 6 \text{ in base} \\ 43 \text{ in subbase} \end{array}}$$

The answer is (B).

Transportation

77 Rigid Pavement Design

PRACTICE PROBLEMS

Concrete Paving Mixtures

1. (*Time limit: one hour*) A concrete mixture is made up of materials with the following properties.

material	absolute volume fraction	bulk specific gravity (oven dry)	absorption (%)
sand	0.25	2.60	1.0
gravel	0.50	2.70	1.5
cement	0.10	3.15	–
water	0.15	–	–

(a) What is the volumetric water-cement ratio?

   (A) 1.2:1

   (D) 1.4.1

   (C) 1.5:1

   (D) 1.7:1

(b) What is the gravimetric water-cement ratio?

   (A) 0.3:1

   (B) 0.5:1

   (C) 0.7:1

   (D) 0.9:1

(c) What is the water-cement ratio?

   (A) 4.8 gal/sack

   (B) 5.4 gal/sack

   (C) 5.7 gal/sack

   (D) 6.0 gal/sack

(d) If the moisture content of the sand is 3.0%, what is the weight of free water in the sand used for a 1 ft^3 batch?

   (A) 0.7 lbf excess

   (B) 0.8 lbf excess

   (C) 0.9 lbf excess

   (D) 0.8 lbf deficit

(e) If the moisture content is 0.5%, what is the weight of free water in the gravel for a 1 ft^3 batch?

   (A) 0.7 lbf excess

   (B) 0.8 lbf excess

   (C) 0.9 lbf excess

   (D) 0.8 lbf deficit

(f) Using the results in parts (d) and (e), calculate the weight of the "added water."

   (A) 8.7 lbf

   (B) 8.9 lbf

   (C) 9.1 lbf

   (D) 9.4 lbf

(g) What is the weight of the materials in 1 ft^3 of the mix?

   (A) 144 lbf

   (B) 148 lbf

   (C) 155 lbf

   (D) 161 lbf

(h) If 0.25 ft^3 of the mixture weighs 37 lbf, what percent air is in the mixture?

   (A) 3.2%

   (B) 4.5%

   (C) 4.9%

   (D) 5.3%

(i) Another concrete mixture has the proportions of cement to sand and gravel of 1:2.5:3.5 by weight, and the water-cement ratio is 6 gal/sack. The specific gravities of the components are unchanged. What is the ratio by weight of water to cement?

   (A) 0.5

   (B) 0.7

   (C) 0.9

   (D) 1.1

(j) What is the volume of all ingredients in a one-sack batch of the mixture described in part (i)?

(A) 3.5 ft^3

(B) 3.8 ft^3

(C) 4.2 ft^3

(D) 4.7 ft^3

2. How much water must be added to the following materials to obtain 6 gal net of mixing water?

material	weight (lbf)	moisture content (%)	absorption capacity (%)
damp gravel	75.0	3.4	1.50
damp sand	46.0	4.5	0.90

3. The batch weights for a concrete mixture include 1344 lbf damp sand with a moisture content of 4.24%, and 2260 lbf of damp gravel with a moisture content of 2.10%. The absorption capacity of the sand is 0.90%, and the absorption capacity of the gravel is 1.50%. 220 lbf of water are added to the mixture. What is the net mixing water?

Rigid Pavement Design

4. (*Time limit: one hour*) A jointed, reinforced concrete pavement using doweled joints and asphalt shoulders without dowels is being designed with the AASHTO rigid pavement procedure. A reliability of 90% is needed. The change in pavement serviceability index is specified as 2.0. The slab length is 50 ft. Two 12 ft lanes are tied together. The pavement is to be constructed in an environment with average drainage. The subgrade is a sandy clay with a CBR of 7. The friction factor, F, is 1.8. At 28 days, the concrete has a compressive strength of 4000 psi with a modulus of rupture of 700 psi. The traffic in the design lane is projected to be 30×10^6 ESALs. An 8 in cement-treated subbase will be used. Grade 40 steel and $1/2$ in tie bars will be used.

(a) What subgrade k-value would you use in the design?

(A) 50 lbf/in^3

(B) 100 lbf/in^3

(C) 150 lbf/in^3

(D) 200 lbf/in^3

(b) What is the effective k-value you would use in the design?

(A) 400 lbf/in^3

(B) 500 lbf/in^3

(C) 600 lbf/in^3

(D) 700 lbf/in^3

(c) Estimate the concrete modulus of elasticity, E_c, for use in the design.

(A) 3.0×10^6 lbf/in^2

(B) 3.2×10^6 lbf/in^2

(C) 3.6×10^6 lbf/in^2

(D) 3.8×10^6 lbf/in^2

(d) What is the required slab thickness?

(A) 10.0 in

(B) 11.0 in

(C) 12.0 in

(D) 13.0 in

(e) What size dowel bars are to be used?

(A) no. 8

(B) no. 10

(C) no. 12

(D) no. 14

(f) How much longitudinal steel is required, expressed in terms of percentage of area?

(A) 0.15%

(B) 0.18%

(C) 0.20%

(D) 0.24%

(g) How much transverse steel is required, expressed in terms of percentage of area?

(A) 0.05%

(B) 0.07%

(C) 0.10%

(D) 0.14%

(h) At what distance from the free longitudinal edge should the first tie bar be placed?

(A) 8 in

(B) 12 in

(C) 18 in

(D) 26 in

(i) What is the required decrease in slab thickness if traffic is decreased by 40% and all other original factors remain unchanged?

 (A) 0.5 in

 (B) 1.0 in

 (C) 1.5 in

 (D) 2.0 in

(j) By how many inches must the pavement thickness be increased if the subbase is untreated and all other original factors remain unchanged?

 (A) 0.5 in

 (B) 1.0 in

 (C) 1.5 in

 (D) 2.0 in

SOLUTIONS

1. (a) The water-cement ratio by volume is

$$\frac{w}{c} = \frac{0.15}{0.10} = \boxed{1.5 \quad (1.5{:}1)}$$

The answer is (C).

(b) Work with 1 cm³ of concrete. The water-cement ratio by weight is found as follows.

Convert to weight using specific gravity.

The weight of the cement is

$$(3.15)(0.10 \text{ cm}^3)\left(1 \; \frac{\text{g}}{\text{cm}^3}\right) = 0.315 \text{ g}$$

The weight of the water is

$$(1.0)(0.15 \text{ cm}^3)\left(1 \; \frac{\text{g}}{\text{cm}^3}\right) = 0.15 \text{ g}$$

Therefore,

$$\frac{w}{c} = \frac{0.15 \text{ g}}{0.315 \text{ g}} - \boxed{0.476 \quad (0.5{:}1)}$$

The answer is (B).

(c) The standard cement sack weighs 94 lbf. The water-cement ratio in gal/sack is

$$\left(\frac{94 \; \dfrac{\text{lbf}}{\text{sack}}}{8.34 \; \dfrac{\text{lbf}}{\text{gal}}}\right)(0.476) = \boxed{5.36 \text{ gal/sack}}$$

The answer is (B).

(d) With 3% moisture content in sand, the free water in 1 ft³ of the concrete mixture is

$$\text{weight of sand} = (0.25 \text{ ft}^3)(2.60)\left(62.4 \; \frac{\text{lbf}}{\text{ft}^3}\right)$$

$$= 40.56 \text{ lbf}$$

$$\text{free water} = (40.56 \text{ lbf})(0.03 - 0.01)$$

$$= \boxed{0.81 \text{ lbf excess}}$$

The answer is (B).

Transportation

(e) With a moisture content of 0.5%, the free water in the gravel is

$$\text{weight gravel} = (0.5 \text{ ft}^3)(2.70)\left(62.4 \ \frac{\text{lbf}}{\text{ft}^3}\right) = 84.24 \text{ lbf}$$

$$\text{free water} = (84.24 \text{ lbf})(0.005 - 0.015)$$
$$= \boxed{-0.84 \text{ lbf deficit}}$$

The answer is (D).

(f) Calculate the added water.

The required mixing water is

$$(0.15 \text{ ft}^3)\left(62.4 \ \frac{\text{lbf}}{\text{ft}^3}\right) = 9.36 \text{ lbf}$$

The net water is

$$9.36 \text{ lbf} - 0.81 \text{ lbf} + 0.84 \text{ lbf} = \boxed{9.39 \text{ lbf}}$$

The answer is (D).

(g) The weight of 1 ft^3 is

ingredient	calculation	weight (lbf)
sand (dry)		40.56
gravel (dry)		84.24
water (net)		9.39
cement	$(0.1 \text{ ft}^3)(3.15)\left(62.4 \ \frac{\text{lbf}}{\text{ft}^3}\right) =$	19.66
absorbed water	$(40.56 \text{ lbf})(0.01) + (84.24 \text{ lbf})(0.015) =$	1.67
	total	$\boxed{155.52 \text{ lbf}}$

The answer is (C).

(h) The air content is

$$\frac{0.25 \text{ ft}^3}{37 \text{ lbf}} = \frac{x}{155.52 \text{ lbf}}$$
$$x = 1.051 \text{ ft}^3$$
$$\% \text{ air} = \frac{1.051 \text{ ft}^3 - 1.000 \text{ ft}^3}{1.051 \text{ ft}^3}$$
$$= \boxed{4.85\% \quad (4.9\%)}$$

The answer is (C).

(i) The proportion of water is

$$\frac{w}{c} = \left(6 \ \frac{\text{gal}}{\text{sack}}\right)\left(8.34 \ \frac{\text{lbf}}{\text{gal}}\right)\left(\frac{1 \text{ sack}}{94 \text{ lbf}}\right) = \boxed{0.532 \quad (0.5)}$$

The answer is (A).

(j) For a one-sack batch,

ingredient	weight (lbf)	volume (ft^3) $= \dfrac{\text{weight}}{(\text{SG})\gamma_w}$
cement	94 lbf	$\dfrac{94}{(3.15)(62.4)} = 0.478$
water	$(94 \text{ lbf})(0.532) = 50 \text{ lbf}$	$\dfrac{50}{(1.00)(62.4)} = 0.801$
sand	$(94 \text{ lbf})(2.5) = 235 \text{ lbf}$	$\dfrac{235}{(2.60)(62.4)} = 1.448$
gravel	$(94 \text{ lbf})(3.5) = 329 \text{ lbf}$	$\dfrac{329}{(2.70)(62.4)} = 1.953$
		total $\boxed{4.680 \text{ ft}^3}$

The answer is (D).

2. The net mixing water required is

$$(6 \text{ gal})\left(8.34 \ \frac{\text{lbf}}{\text{gal}}\right) = 50.04 \text{ lbf}$$

The free water in gravel is

$$\left(\frac{75 \text{ lbf}}{1.034}\right)(0.034 - 0.015) = 1.38 \text{ lbf}$$

The free water in sand is

$$\left(\frac{46 \text{ lbf}}{1.045}\right)(0.045 - 0.009) = 1.58 \text{ lbf}$$

The total free water is 1.38 lbf + 1.58 lbf = 2.96 lbf.
The added water is

$$50.04 \text{ lbf} - 2.96 \text{ lbf} = \boxed{47.08 \text{ lbf}}$$

3. The free water is

$$\left(\frac{1344 \text{ lbf}}{1.0424}\right)(0.0424 - 0.009)$$
$$+ \left(\frac{2260 \text{ lbf}}{1.0210}\right)(0.021 - 0.015) = 56.34 \text{ lbf}$$

The net mixing water is

$$220 \text{ lbf} + 56.34 \text{ lbf} = \boxed{276.34 \text{ lbf}}$$

4. (a) From Fig. 77.1, with a CBR of 7, the subgrade k-value is $\boxed{150 \text{ lbf/in}^3}$.

The answer is (C).

Transportation

(b) From Table 77.8, interpolating for $k_{\text{subgrade}} = 150\ \text{lbf/in}^3$, the effective k is

$$\left(\tfrac{1}{2}\right)\left(520\ \frac{\text{lbf}}{\text{in}^3} + 830\ \frac{\text{lbf}}{\text{in}^3}\right) = \boxed{675\ \text{lbf/in}^3 \quad (700\ \text{lbf/in}^3)}$$

The answer is (D).

(c) Use Eq. 77.4.

$$E_c = 57{,}000\sqrt{f'_c}$$
$$= 57{,}000\sqrt{4000\ \frac{\text{lbf}}{\text{in}^2}}$$
$$= \boxed{3.6 \times 10^6\ \text{lbf/in}^2}$$

The answer is (C).

(d) Find the slab thickness from Fig. 77.2.

$$\text{CBR} = 7$$
$$k_{\text{subgrade}} = 150\ \text{lbf/in}^3$$
$$k_{\text{effective}} = 675\ \text{lbf/in}^3$$
$$E_c = 3.6 \times 10^6\ \text{lbf/in}^2$$
$$J = 4 \quad [\text{shoulders without tie bar dowels}]$$
$$C_d = 1.0 \quad [\text{average drainage}]$$
$$R = 90\% \quad [\text{reliability}]$$
$$S_o = 0.35 \quad [\text{typical rigid pavement}]$$
$$\Delta\text{PSI} = 2.0$$

The slab thickness, D, is $\boxed{12.0\ \text{in.}}$

The answer is (C).

(e) The dowel bar diameter is

$$\frac{D}{8} \approx \frac{12.0\ \text{in}}{8} = 1.50\ \text{in}$$
$$= \boxed{1\,\tfrac{5}{8}\ \text{in} \quad [\text{or no. 14}]}$$

(There is no no. 12 bar.)

The answer is (D).

(f) With grade 40 steel, the working stress is

$$f_s = 0.75f_y = (0.75)\left(40\ \frac{\text{kips}}{\text{in}^2}\right)$$
$$= 30\ \text{kips/in}^2$$

The friction factor is given as $F = 1.8$.

The percentage of longitudinal steel is given by Eq. 77.5.

$$P_t = \frac{LF}{2f_s} \times 100\% = \frac{(50\ \text{ft})(1.8)}{(2)\left(30{,}000\ \frac{\text{lbf}}{\text{in}^2}\right)} \times 100\%$$
$$= \boxed{0.15\%}$$

The answer is (A).

(g) Since the two lanes are tied, both must be considered for horizontal steel. The width between untied edges is 24 ft.

From Eq. 77.5, the area of transverse steel is

$$P_t = \frac{LF}{2f_s} \times 100\% = \frac{(24\ \text{ft})(1.8)}{(2)\left(30{,}000\ \frac{\text{lbf}}{\text{in}^2}\right)} \times 100\%$$
$$= \boxed{0.072\% \quad (0.07\%)}$$

The answer is (B).

(h) For a distance to the closest free edge from the tied longitudinal joint of 12 ft and a slab thickness of 12 in, with $\tfrac{1}{2}$ in tie bars, from Fig. 77.3, the distance is $\boxed{\text{approximately 26 in.}}$

The answer is (D).

(i) The new traffic volume is

$$(30 \times 10^6)(1 - 0.40) = 18 \times 10^6$$

The new thickness is

$$D = 11\ \text{in}$$

The decrease is

$$12\ \text{in} - 11\ \text{in} = \boxed{1.0\ \text{in}}$$

The answer is (B).

(j)
$$\text{CBR} = 7$$
$$k_{\text{subgrade}} = 150\ \text{lbf/in}^3$$

Using Table 77.7 and double interpolation for $k = 150$ lbf/in^3 and $t = 8$ in,

$$k_{\text{effective}} = 205 \text{ lbf/in}^3$$

$$E_c = 3.6 \times 10^6 \text{ lbf/in}^2$$

$$S'_c = 700 \text{ lbf/in}^2$$

$$J = 4$$

$$C_d = 1.0$$

$$(\text{match line} \approx 75)$$

$$\Delta\text{PSI} = 2$$

$$R = 90\%$$

$$S_o = 0.35$$

$$\text{ESALs} = 30 \times 10^6$$

From Fig. 77.2,

$$D \approx 12.5 \text{ in}$$

The thickness increase is

$$12.5 \text{ in} - 12.0 \text{ in} = \boxed{0.5 \text{ in}}$$

The answer is (A).

78 Plane Surveying

PRACTICE PROBLEMS

Surveying Methods

1. Which of the following are common methods of conducting topographic surveys?

I.   transit-tape and transit-stadia

II.   total station

III. global positioning system (GPS)

IV.   metes and bounds

V.   photogrammetry

(A) II, III, and V

(B) I, II, III, and V

(C) II, III, IV, and V

(D) I, II, III, IV, and V

Errors

2. Four level circuits were run over four different routes to determine the elevation of a benchmark. The observed elevations and probable errors for each circuit are as given. What is the most probable value of the benchmark's elevation?

route 1:	745.08 ± 0.03
route 2:	745.22 ± 0.01
route 3:	745.45 ± 0.09
route 4:	745.17 ± 0.05

Leveling

3. A level is placed where the height of the instrument above the datum is 294.43 ft. The rod is held on a point 5000 ft away, and a rod reading of 4.63 ft is recorded. (a) What is the effect of the earth's curvature on the rod reading? (b) What is the effect of atmospheric refraction on the rod reading?

4. Complete the survey notes from a leveling session.

station	+ BS	HI	− FS	elevation
BMA	5.54			100.60
1+00			2.88	
2+00			5.56	
3+00			4.37	
4+00			2.22	
TP1			7.89	
	6.34			
5+00			10.89	
6+00			11.12	
7+00			10.45	
8+00			12.61	
TP2			9.32	
	1.45			
9+00			5.82	
10+00			6.99	
11+00			5.34	
12+00			4.98	
TP3			5.67	
	2.38			
13+00			6.24	
14+00			6.76	
BMB			4.90	

5. A transit with an interval factor of 100 and an instrument factor of 1.0 is used to take the stadia sights recorded. The height of the instrument above the ground was 4.8 ft. The ground elevation where the instrument was set up was 297.8 ft. (a) What is the horizontal distance between points A and B? (b) What is the elevation of point B?

point	azimuth	rod interval $(R_2 - R_1)$	middle hair reading	vertical angle
A	42.17°	3.22	5.7	−6.3°
B	222.17°	2.60	10.9	+4.17°

6. Recorded data from a leveling circuit is shown. (a) What is the elevation of BM11? (b) What is the elevation of BM12?

station	backsight	foresight	elevation
BM10	4.64		179.65
TP1	5.80	5.06	
TP2	2.25	5.02	
BM11	6.02	5.85	
TP3	8.96	4.34	
TP4	8.06	3.22	
TP5	9.45	3.71	
TP6	12.32	2.02	
BM12		1.98	

Traverses

7. The balanced latitudes and departure of the legs of a closed traverse were determined as shown. What is the traverse area?

leg	latitude	departure
AB	N 350	E 0
BC	N 550	E 600
CD	S 250	E 1200
DE	S 750	E 200
EF	S 550	W 1100
FA	N 650	W 900

8. Use the compass rule to balance and adjust (to the nearest 0.1) the traverse.

leg	bearing	length
AB	N	500
BC	N 45.00° E	848.6
CD	S 69.45° E	854.4
DE	S 11.32° E	1019.8
EF	S 79.70° W	1118.0
FA	N 54.10° W	656.8

9. A five-leg closed traverse is taped and scoped in the field, but obstructions make it impossible to collect all readings. It is known that the general direction of leg EA is east-west. Complete the field notes.

leg	north azimuth	distance
AB	106.22°	1081.3
BC	195.23°	1589.5
CD	247.12°	1293.7
DE	332.37°	———
EA	———	1737.9

Photogrammetry

10. (*Time limit: one hour*) A camera with a focal length of 6.0 in and a picture size of 12 in × 12 in is used to photograph an area from an altitude of 4000 ft above sea level.

(a) If the average ground elevation is 650 ft above sea level, what is the average scale of the photo?

(A) 1 in = 560 ft

(B) 1 in = 600 ft

(C) 1:6000

(D) 1:6500

(b) What is the ground area covered by a single photograph?

(A) 1030 ac

(B) 1500 ac

(C) 2000 ac

(D) 2070 ac

(c) If overlap and sidelap are 60% and 20%, respectively, what is the ground spacing between exposures?

(A) 2670 ft

(B) 2672 ft

(C) 2678 ft

(D) 2778 ft

(d) With the overlap as given in part (c), what is the ground spacing between flight lines?

(A) 5236 ft

(B) 5336 ft

(C) 5346 ft

(D) 5357 ft

(e) If the aircraft flies at a ground speed of 300 mph along the flight line outlined, what is the time interval (to the nearest 0.1 sec) between exposures along the flight line?

(A) 6.0 sec

(B) 6.1 sec

(C) 6.2 sec

(D) 8.1 sec

(f) Two points at an elevation of 665 ft appear on a photograph. The distance between these points is 89 mm on the photograph. What is the ground distance, in feet, between the two points?

(A) 1557 ft

(B) 1647 ft

(C) 1737 ft

(D) 1947 ft

(g) At what elevation above sea level should the plane be flown to obtain photography at an average scale of 1:10,000 if the terrain to be photographed ranges from 1000 ft to 1200 ft above sea level?

(A) 6000 ft

(B) 6050 ft

(C) 6100 ft

(D) 6150 ft

(h) Two points lie 5.68 in apart on a photo taken with a lens whose focal length is 12 in. A map with a scale of 1 in = 1680 ft shows the same points 0.54 in apart. What is the scale of the photo?

(A) 1 in = 160 ft

(B) 1 in = 200 ft

(C) 1:1500

(D) 1:2000

(i) For the same photo and map in part (h), at what elevation above sea level was the plane flying if the elevation of the points is 889 ft above sea level?

(A) 2000 ft

(B) 2500 ft

(C) 2800 ft

(D) 3000 ft

(j) For the same photo and map in part (h), at what elevation above sea level was the plane flying if the same two points are 0.98 in apart on the map and the ground elevation is 889 ft above sea level?

(A) 4400 ft

(B) 4500 ft

(C) 4600 ft

(D) 4700 ft

11. An aerial survey of a construction site produces nadir imagery taken from a height of 900 m above a flat landscape. The displacement of a tall feature is measured as 9.5 mm. The distance from the nadir (principal point) of the photo is measured as 122 mm. Most nearly, how tall is the feature?

(A) 20 m

(B) 40 m

(C) 60 m

(D) 70 m

Topographic Maps

12. Two maps of different scales show the same points. On one map, which has a scale of 1 in:$^1/_4$ mi, the points are 5 in apart. On the other map, the points are 8 in apart. What is the scale of the second map?

(A) 1 in:0.156 mi

(B) 1 in:0.2 mi

(C) 1 in:0.4 mi

(D) 1 in:6.4 mi

13. A line on a topographic map connecting points of equal elevation is known as

(A) a meridian line

(B) a stadia line

(C) an isogonic line

(D) a contour line

SOLUTIONS

1. Metes and bounds is an archaic method of describing the boundaries of a plot or parcel based on topographic features, directions, and distances. It was widely used in prior centuries as part of the recordation process. All of the other methods are still in use.

The answer is (B).

2. $(0.03)^2 + (0.01)^2 + (0.09)^2 + (0.05)^2 = 0.0116$

The weights to be applied are

$$\frac{0.0116}{(0.03)^2} = 12.89$$

$$\frac{0.0116}{(0.01)^2} = 116.0$$

$$\frac{0.0116}{(0.09)^2} = 1.43$$

$$\frac{0.0116}{(0.05)^2} = 4.64$$

The most probable elevation is

$$745 + \frac{\begin{array}{c}(12.89)(0.08) + (116)(0.22) \\ + (1.43)(0.45) + (4.64)(0.17)\end{array}}{12.89 + 116 + 1.43 + 4.64} = \boxed{745.21}$$

3. (a) Use Eq. 78.14.

$$h_c = \left(2.39 \times 10^{-8} \; \frac{1}{\text{ft}}\right) x_\text{ft}^2$$

$$= \left(2.39 \times 10^{-8} \; \frac{1}{\text{ft}}\right)(5000 \text{ ft})^2$$

$$= \boxed{0.5975 \text{ ft}}$$

(b) Use Eq. 78.15.

$$h_r = \left(3.3 \times 10^{-9} \; \frac{1}{\text{ft}}\right) x_\text{ft}^2$$

$$= \left(3.3 \times 10^{-9} \; \frac{1}{\text{ft}}\right)(5000 \text{ ft})^2$$

$$= \boxed{0.0825 \text{ ft}}$$

4. The completed survey notes are as follows.

station	+ BS	HI	− FS	elevation
BMA	5.54	106.14		100.60
1+00			2.88	103.26
2+00			5.56	100.58
3+00			4.37	101.77
4+00			2.22	103.92
TP1			7.89	98.25
	6.34			
5+00		104.59	10.89	93.70
6+00			11.12	93.47
7+00			10.45	94.14
8+00			12.61	91.98
TP2			9.32	95.27
	1.45			
9+00		96.72	5.82	90.90
10+00			6.99	89.73
11+00			5.34	91.38
12+00			4.98	91.74
TP3			5.67	91.05
	2.38			
13+00		93.43	6.24	87.19
14+00			6.76	86.67
BMB			4.90	88.53
total BS	15.71	total TP	27.78	

Check.

$$100.6 + 15.71 - 27.78 = 88.53 \quad [\text{OK}]$$

5. (a) Use Eq. 78.12.

$$x = K(R_2 - R_1)\cos^2 \theta + C \cos \theta$$

$$H_\text{scope-A} = (100)(3.22 \text{ ft})\left(\cos^2(-6.3°)\right)$$
$$+ (1)\left(\cos(-6.3°)\right)$$
$$= 319.1 \text{ ft}$$

$$H_\text{scope-B} = (100)(2.6 \text{ ft})(\cos^2 4.17°) + (1 \text{ ft})(\cos 4.17°)$$
$$= 259.6 \text{ ft}$$

Since the azimuths differ by 180°, the telescope is in line with points A and B.

$$AB = 319.1 \text{ ft} + 259.6 \text{ ft} = \boxed{578.7 \text{ ft}}$$

(b) Use Eq. 78.13.

$$y = \tfrac{1}{2}K(R_2 - R_1)\sin 2\theta + C \sin \theta$$

$$V_\text{scope-B} = \left(\tfrac{1}{2}\right)(100)(2.60 \text{ ft})\left(\sin(2)(4.17°)\right)$$
$$+ (1 \text{ ft})(\sin 4.17°)$$
$$= 18.9 \text{ ft}$$

$$\text{elevation B} = 297.8 \text{ ft} + 18.9 \text{ ft} + 4.8 \text{ ft} - 10.9 \text{ ft}$$
$$= \boxed{310.6 \text{ ft}}$$

Transportation

6. (a) elevation BM11 $= 179.65 + \sum BS - \sum FS$

$$= 179.65 + 4.64 + 5.80 + 2.25$$
$$- (5.06 + 5.02 + 5.85)$$
$$= \boxed{176.41}$$

(b) elevation BM12 $= 179.65 + 57.5 - 31.2$
$$= \boxed{205.95}$$

7. While it is not necessary, convert the latitudes and departures into (x, y) coordinates. Assume point A is at $(0, 0)$. Then,

$$B_x = 0 + 0 = 0$$
$$B_y = 0 + 350 = 350$$

point	(x, y)
A	$(0, 0)$
B	$(0, 350)$
C	$(600, 900)$
D	$(1800, 650)$
E	$(2000, -100)$
F	$(900, -650)$

Use Eq. 78.36.

$$A = \frac{1}{2}\left|\sum_{i=1}^{n} y_i(x_{i-1} - x_{i+1})\right|$$

$$\frac{0}{0} \times \frac{0}{350} \times \frac{600}{900} \times \frac{1800}{650} \times \frac{2000}{-100} \times \frac{900}{-650} \times \frac{0}{0}$$

$$\sum \text{full line products} = -1,090,000$$
$$\sum \text{dotted line products} = 3,040,000$$
$$\text{area} = \frac{1}{2}\left| -1,090,000 - 3,040,000\right|$$
$$= \boxed{2,065,000}$$

8. Calculate the interior angles.

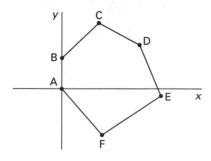

step 1: The interior angles are

A: $90° - 54.10° + 90° = 125.90°$

B: $180° - 45° = 135°$

C: $45° + 69.45° = 114.45°$

D: $90° - 69.45° + 90° + 11.32° = 121.87°$

E: $90° - 11.32° + 90° - 79.70° = 88.98°$

F: $79.70° + 54.10° = 133.8°$

step 2: The sum of the interior angles is 720°. For a six-sided polygon, the sum of the interior angles should be

$$(6 - 2)(180°) = 720°$$

Since the sum of the angles is exactly 720°, no angle balancing is necessary. However, there may be a closure error.

For leg AB,

$$\text{latitude} = 500$$
$$\text{departure} = 0$$

For leg BC,

$$\text{latitude} = (848.6)(\cos 45°) = 600$$
$$\text{departure} = (848.6)(\sin 45°) = 600$$

Similarly, the following table is prepared (rounding to the nearest 0.1).

leg	latitude	departure
AB	+500.0	0
BC	+600.0	+600.0
CD	−299.9	+800.0
DE	−1000.0	+200.2
EF	−199.9	−1100.0
FA	+385.1	−532.0
totals	−14.7	−31.8

The total perimeter length is 4997.6 (round to 5000).

The latitude correction to line AB is

$$\left(\frac{500}{5000}\right)(14.7) = 1.47$$

[round to nearest 0.1 to get 1.5]

The departure correction to line AB is

$$\left(\frac{500}{5000}\right)(31.8) = 3.18 \quad \text{[round to 3.2]}$$

The corrected latitude and departure of leg AB is

$$\text{latitude:} \quad 500 + 1.5 = 501.5$$
$$\text{departure:} \quad 0 + 3.2 = 3.2$$

Transportation

Similarly, the following table is prepared.

leg	latitude	departure
AB	501.5	3.2
BC	602.5	605.4
CD	-297.4	805.4
DE	-997.0	206.7
EF	-196.6	-1092.9
FA	387.0	-527.8
totals	0.0	0.0

The angles and lengths calculated from these latitudes and departures will not be the same as the given angles and lengths. Therefore, corrected angles and lengths must be calculated.

For leg AB, the corrected angle is

$$\arctan \frac{3.2}{501.5} = 0.37°$$

The corrected bearing is N 0.37° E.

The corrected length is

$$\sqrt{(3.2)^2 + (501.5)^2} = 501.5$$

The following table is similarly prepared.

	corrected	
leg	bearing	length
AB	N 0.37° E	501.5
BC	N 45.14° E	854.1
CD	S 69.73° E	858.6
DE	S 11.71° E	1018.2
EF	S 79.80° W	1110.4
FA	N 53.75° W	654.5

9. The plot of the available data shows two possible locations for point E. However, only one is essentially west of point A. Point E' is ruled out since E'A is not easterly.

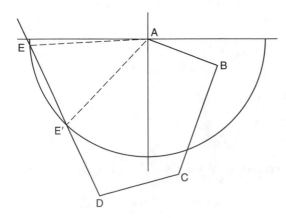

The latitudes and departures of the first three legs are

leg	latitude	departure
AB	-302.0	1038.3
BC	-1533.7	-417.6
CD	-503.0	-1191.9
totals	-2338.7	-571.2

The slope of line DE is

$$-\tan(332.37° - 270°) = -1.91$$

The coordinates of point D are $(-571.2, -2338.7)$. The equation of line DE is

$$y = mx + b$$
$$-2338.7 = (-1.91)(-571.2) + b$$
$$b = -3429.7$$
$$y = -1.91x - 3429.7$$

The equation of a circle centered at point A with a radius of 1737.9 is

$$x^2 + y^2 = (1737.9)^2$$

Since point E is on both the line and the circle,

$$x^2 + (-1.91x - 3429.7)^2 = (1737.9)^2$$
$$x^2 + 3.65x^2 + 13{,}101.5x + (3429.7)^2 = (1737.9)^2$$
$$x^2 + 2817.5x = -1.88 \times 10^6$$

This has solutions $x = -1085.2$ and -1732.4. The larger (negative) is -1732.4.

From the line equation,

$$y = -(1.91)(-1732.4) - 3429.7 = -120.8$$

The angle of line EA is

$$\arctan \frac{-120.8}{-1732.4} \approx 4.0°$$

The north azimuth of line EA is

$$90° - 4.0° = \boxed{86°}$$

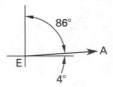

The length of line DE is

$$\sqrt{(-1732.4 + 571.2)^2 + (-120.8 + 2338.7)^2} = \boxed{2503.5}$$

10. (a) Use Eq. 78.42.

$$S = \frac{f}{H_{MSL} - E_{MSL}} = \frac{6 \text{ in}}{4000 \text{ ft} - 650 \text{ ft}}$$
$$= 1 \text{ in}/558 \text{ ft} \quad (1 \text{ in}/560 \text{ ft})$$

The average scale of the photo is $\boxed{1 \text{ in} = 560 \text{ ft.}}$

The answer is (A).

(b) The ground area is

$$\frac{\left((12 \text{ in})\left(558 \frac{\text{ft}}{\text{in}}\right)\right)^2}{43{,}560 \frac{\text{ft}^2}{\text{ac}}} = \boxed{1029.3 \text{ ac} \quad (1030 \text{ ac})}$$

The answer is (A).

(c) The end spacing is

$$\text{end spacing} = (1 - 0.60)(12 \text{ in})\left(558 \frac{\text{ft}}{\text{in}}\right)$$
$$= \boxed{2678.4 \text{ ft} \quad (2678 \text{ ft})}$$

The answer is (C).

(d) The side spacing is

$$\text{side spacing} = (1 - 0.20)(12 \text{ in})\left(558 \frac{\text{ft}}{\text{in}}\right)$$
$$= \boxed{5356.8 \text{ ft} \quad (5357 \text{ ft})}$$

The answer is (D).

(e) The time interval is

$$t = \frac{d}{v} = \frac{(2678.4 \text{ ft})\left(3600 \frac{\text{sec}}{\text{hr}}\right)}{\left(300 \frac{\text{mi}}{\text{hr}}\right)\left(5280 \frac{\text{ft}}{\text{mi}}\right)}$$
$$= \boxed{6.1 \text{ sec}}$$

The answer is (B).

(f) Use Eq. 78.42.

$$S = \frac{f}{H_{MSL} - E_{MSL}} = \frac{6 \text{ in}}{4000 \text{ ft} - 665 \text{ ft}}$$
$$= 1 \text{ in}/555.8 \text{ ft}$$
$$S = \frac{\text{photo distance}}{\text{true distance}}$$
$$\text{true distance} = \frac{\text{photo distance}}{S}$$
$$= \frac{89 \text{ mm}}{\left(25.4 \frac{\text{mm}}{\text{in}}\right)\left(\frac{1 \text{ in}}{555.8 \text{ ft}}\right)}$$
$$= \boxed{1947.5 \text{ ft} \quad (1947 \text{ ft})}$$

The answer is (D).

(g) The focal length is

$$f = \frac{6 \text{ in}}{12 \frac{\text{in}}{\text{ft}}} = 0.5 \text{ ft}$$

The average terrain altitude is 1100 ft MSL. Use Eq. 78.42, and let H be the altitude of the plane above mean sea level (MSL).

$$S = \frac{f}{H_{MSL} - E_{MSL}}$$
$$\frac{1}{10{,}000} = \frac{0.5 \text{ ft}}{H_{MSL} - 1100 \text{ ft}}$$
$$H_{MSL} = \boxed{6100 \text{ ft MSL}}$$

The answer is (C).

(h) Calculate the real distance, then find the photo's scale.

$$\text{true distance} = \frac{\text{map distance}}{\text{scale}} = \frac{0.54 \text{ in}}{\frac{1 \text{ in}}{1680 \text{ ft}}}$$
$$= 907.2 \text{ ft}$$
$$\text{photo scale} = \frac{\text{photo distance}}{\text{true distance}} = \frac{5.68 \text{ in}}{907.2 \text{ ft}}$$
$$= 1 \text{ in}/159.7 \text{ ft} \quad (1 \text{ in}/160 \text{ ft})$$

The photo's scale is $\boxed{1 \text{ in} = 160 \text{ ft.}}$

The answer is (A).

Transportation

(i) Use Eq. 78.42. The plane's elevation above mean sea level (MSL) is

$$S = \frac{f}{H_{MSL} - E_{MSL}}$$

$$\frac{1 \text{ in}}{159.7 \text{ ft}} = \frac{12 \text{ in}}{H_{MSL} - 889 \text{ ft}}$$

$$H_{MSL} = \boxed{2805.4 \text{ ft} \quad (2800 \text{ ft MSL})}$$

The answer is (C).

(j) Calculate the real distance, then find the photo's scale to determine the plane's elevation.

$$\text{true distance} = \frac{\text{map distance}}{\text{map scale}} = \frac{0.98 \text{ in}}{\dfrac{1 \text{ in}}{1680 \text{ ft}}}$$

$$= 1646.4 \text{ ft}$$

$$\text{photo scale} = \frac{\text{photo distance}}{\text{true distance}} = \frac{5.68 \text{ in}}{1646.4 \text{ ft}} = \frac{1 \text{ in}}{289.9 \text{ ft}}$$

Use Eq. 78.42. The plane's elevation above mean sea level (MSL) is

$$S = \frac{f}{H_{MSL} - E_{MSL}}$$

$$\frac{1 \text{ in}}{289.9 \text{ ft}} = \frac{12 \text{ in}}{H_{MSL} - 889 \text{ ft}}$$

$$H_{MSL} = \boxed{4367.8 \text{ ft} \quad (4400 \text{ ft})}$$

The answer is (A).

11. r is the radial distance from the nadir to the top of the object, H is the altitude above the base of the object, and d is the length of the displaced object from base to top. The height of the object photographed in a nadir image is

$$h = \frac{dH}{r} = \frac{(9.5 \text{ mm})(900 \text{ m})}{122 \text{ mm}} = \boxed{70.08 \text{ m} \quad (70 \text{ m})}$$

The answer is (D).

12. The new scale is

$$\left(\frac{1}{4} \frac{\text{mi}}{\text{in}}\right)\left(\frac{5 \text{ in}}{8 \text{ in}}\right) = \boxed{0.156 \text{ mi/in} \quad (1 \text{ in}{:}0.156 \text{ mi})}$$

The answer is (A).

13. Contour lines indicate points of equal elevation on a topographic map.

The answer is (D).

79 Horizontal, Compound, Vertical, and Spiral Curves

PRACTICE PROBLEMS

Vertical Sag Curves

1. (*Time limit: one hour*) A sag vertical curve is shown.

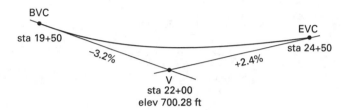

(a) At what station is the low point located?

(A) sta 22+15

(B) sta 22+36

(C) sta 22+68

(D) sta 22+79

(b) What is the elevation of the low point?

(A) 701 ft

(B) 702 ft

(C) 703 ft

(D) 704 ft

(c) What is the elevation of sta 21+25.04?

(A) 701 ft

(B) 704 ft

(C) 706 ft

(D) 708 ft

(d) Based on comfort, what is the highest design speed for this curve?

(A) 60 mph

(B) 64 mph

(C) 68 mph

(D) 72 mph

(e) What is the rate of grade change per station?

(A) 1.0 %/sta

(B) 1.1 %/sta

(C) 1.4 %/sta

(D) 1.6 %/sta

(f) If 20 ft of clearance are required, the lowest elevation that the bottom of an overhead bridge deck may have at sta 23+16 is most nearly

(A) 724 ft

(B) 726 ft

(C) 728 ft

(D) 730 ft

(g) What would the curve length have to be if the bridge deck in part (f) were kept at the same elevation, but 22 ft (6.7 m) of clearance were required? Assume that the stationing and elevation of the BVC are kept the same.

(A) 660 ft

(B) 680 ft

(C) 700 ft

(D) 710 ft

(h) For the curve solved in part (g), how much additional excavation is required at sta 22+00?

(A) 1.0 ft

(B) 1.9 ft

(C) 2.5 ft

(D) 2.9 ft

(i) What is the AASHTO K-value for the longer curve?

(A) 120.8 ft/%

(B) 121.8 ft/%

(C) 122.8 ft/%

(D) 123.8 ft/%

(j) At 65 mph, stopping sight distance based on braking distance for the longer curve is most nearly

(A) 270 ft

(B) 520 ft

(C) 650 ft

(D) 880 ft

2. A falling grade of 4% meets a rising grade of 5% in a sag curve. At the start of the curve, the elevation is 123.06 ft at sta 4034+20. At sta 4040+20, there is an overpass with an underside elevation of 134.06 ft. If the curve is designed to afford a clearance of 15 ft under the overpass, calculate the required length.

3. (*Time limit: one hour*) Two highways are planned, each running perpendicular to the other. The geometric details of the underpass have been determined. The design process is continuing for the perpendicular overpass, which must maintain a 25 ft clearance (measured between highway surface centerlines). (a) What is the minimum station where the new highway can be constructed? (b) Locate the maximum station where the new highway can be constructed. (c) What is the station of the lowest point on the existing curve? (d) What is the elevation of the lowest point on the existing curve?

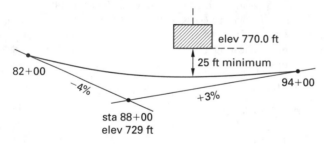

4. (*Time limit: one hour*) A new highway can go either over or under (perpendicular to) an existing highway at sta 75+00. The elevation of the existing highway at sta 75+00 is 510.00 ft. 22 ft of clearance (between the centerlines of the two road surfaces) are required if the new highway goes over the existing highway. 20 ft of clearance are required if the new highway goes under the existing highway. (a) What is the longest vertical curve that will pass under the existing highway? (b) What is the shortest vertical curve that will pass over the existing highway?

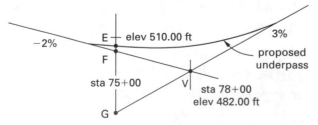

5. (*Time limit: one hour*) The grades on a vertical curve intersect at sta 105+00 and elevation 350 ft. A drain grating exists at sta 105+00 and elevation 358.30 ft. (a) If a curve length of 1900 ft is used, what is the elevation of the BVC? (b) If a length of 1900 ft is used, what is the elevation at sta 106+40? (c) If a curve length of 1900 ft is used, what is the elevation of the lowest point on the curve? (d) If a curve length of 1900 ft is used, what is the sta of the BVC? (e) If a curve length of 1900 ft is used, what is the station of the EVC? (f) What is the minimum continuous curve length such that no point on the curve will be lower than the drain grating?

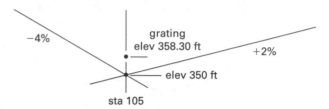

Vertical Crest Curves

6. An existing length of road consists of a rising gradient of 1 in 20 followed by a vertical parabolic crest curve 300 ft long, and then a falling gradient of 1 in 40. The curve joins both gradients tangentially, and the elevation of the highest point on the curve is 173.07 ft. Visibility is to be improved over this stretch of road by replacing the curve with another parabolic curve 600 ft long. (a) Find the depth of excavation required at the midpoint of the curve. (b) Tabulate the elevations of points at 100 ft intervals along the new curve.

7. (*Time limit: one hour*) An equal-tangent vertical crest curve has the following geometric properties.

BVC sta	110+00
PVI sta	116+00
PVI elev	1262
EVC	122+00
ascending grade	+3%
descending grade	−2%

(a) Find the centerline roadway elevation of sta 110+50.
(b) Find the centerline roadway elevation of sta 115+50.
(c) The road continues on to a horizontal curve to the right. Superelevation is 8%. Assume the curve is in full superelevation throughout. The road is 60 ft wide at all points. The horizontal curve is supported by three pile bents symmetrical about the curve's PVI. The pile bents are located at stations 115+50, 116+00, and 116+50. Each pile bent consists of seven piles symmetrical about the road centerline, placed 10 ft apart. The tops of the piles are 3.5 ft below the roadway. What are the elevations of the pile heads at stations 115+50, 116+00, and 116+50?

Horizontal Curves

8. Three points (A, B, and C) were selected on the centerline of an existing horizontal circular curve. The instrument was set horizontally at point B. Stadia readings were taken on a vertical staff at points A and C. The instrument had constants of 100 and 0. The instrument axis was 4.7 ft above the road at point B. (a) What is the radius of the curve? (b) What is the line-of-sight gradient AB? (c) What is the line-of-sight gradient BC?

staff at	horizontal bearing	stadia readings (ft)		
A	0.00°	10.038	6.221	2.403
C	211.14°	7.236	5.778	4.320

9. (*Time limit: one hour*) Two straight tangents of a two-lane highway intersect at point B. The tangents have azimuths of 270° and 110°, respectively. They are to be joined by a circular curve that must pass through a point D, 350 ft from point B. The azimuth of line BD is 260°.

(a) The required radius is most nearly

 (A) 7280 ft

 (B) 7330 ft

 (C) 7650 ft

 (D) 7900 ft

(b) The tangent lengths are most nearly

 (A) 1280 ft

 (B) 1300 ft

 (C) 1340 ft

 (D) 1380 ft

(c) The length of curve is most nearly

 (A) 2000 ft

 (B) 2500 ft

 (C) 3000 ft

 (D) 3500 ft

(d) The central angle for a 500 ft arc is most nearly

 (A) 1°

 (B) 2°

 (C) 3°

 (D) 4°

(e) If the curve is laid out by the tangent offset method, the tangent distance required to place the stake located an arc distance of 100 ft from the PC is most nearly

 (A) 80 ft

 (B) 90 ft

 (C) 100 ft

 (D) 200 ft

(f) Assuming that snow and ice are not normally encountered on the highway, the maximum design speed for this curve without relying on friction is most nearly

 (A) 55 mph

 (B) 65 mph

 (C) 75 mph

 (D) 100 mph

(g) Assuming a design speed of 70 mph, the passing sight distance for this curve is most nearly

 (A) 2000 ft

 (B) 2500 ft

 (C) 3000 ft

 (D) 3500 ft

(h) What would the maximum superelevation rate be if snow and ice were commonly encountered on the freeway?

 (A) 0.02

 (B) 0.04

 (C) 0.06

 (D) 0.10

(i) Assume lane widths are 12 ft and the crown cross-slope is most nearly 0.02 ft/ft. The transition rate is 1:200. A vehicle is traveling out of the curve. Over what distance (from the PC or PT) should all superelevation be removed from the alignment and the crown be fully developed?

 (A) 130 ft

 (B) 140 ft

 (C) 170 ft

 (D) 190 ft

Transportation

(j) What is the approximate minimum stopping sight distance for this curve for a vehicle traveling at 70 mph?

(A) 730 ft

(B) 800 ft

(C) 850 ft

(D) 900 ft

10. (*Time limit: one hour*) An existing 6° horizontal circular curve connects a PC and PT$_1$ as shown. It is desired to avoid having vehicles pass too close to a historical monument, so a proposal has been made to relocate the curve 120 ft back. The PC will remain the same. (a) What is the PC station? (b) What is the radius of curvature of the new curve? (c) What is the stationing of the new curve's PT?

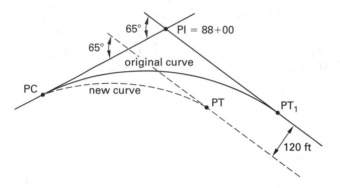

11. The tangent of the horizontal curve shown is relocated 10 ft west of its original position. The PC remains the same. (a) What is the original curve radius? (b) What is the new curve radius?

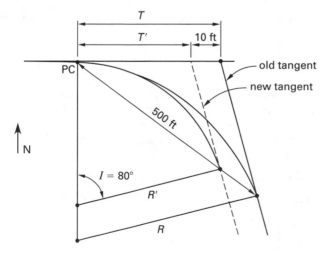

12. (*Time limit: one hour*) The tangent of a horizontal curve is parallel to a railroad line 150 ft away. The two curve tangents intersect at sta 182+27.52. Other information is provided in the illustration. Determine the (a) station of the PC, (b) the station of the PT, (c) the horizontal sightline offset (HSO), (d) the external ordinate (E), and (e) the station of the railroad line's intersection with the highway.

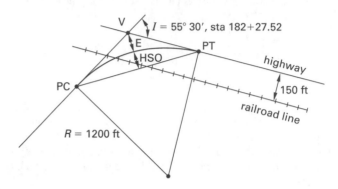

13. (*Time limit: one hour*) A freeway is being constructed to pass close to an existing first-order USGS survey monument. The freeway can be on either side of the monument. The freeway has a 75 ft right-of-way on each side of the curve centerline. A 1 ft clearance from the monument (which is considered to be a dimensionless point) to the edge of the right-of-way is required. The clearance is radial, measured along a line from the curve center to the monument. Find the range of radii of circular arcs that will provide the required clearance.

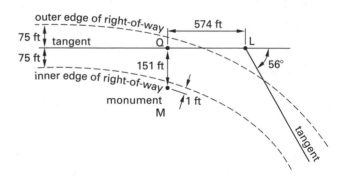

Reverse Curves

14. (*Time limit: one hour*) A historical monument must be avoided near a proposed highway. A reverse curve is being considered. Both curves will have the same curvature. What is the station of the (a) PC, (b) PRC, and (c) PT? (d) What is the deflection angle from the PC to the midpoint (MPC) of the first curve? (e) What are the disadvantages associated with reverse curves?

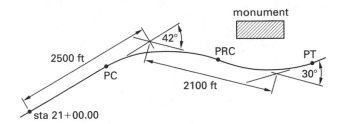

Sight Distance

15. (*Time limit: one hour*) An unsignalized blind intersection is shown. It is desired to prohibit parking near the intersection in order to provide adequate sight distance into the intersection. Vehicles travel on the major street at 35 mph or slower 85% of the time. PIEV time is 1.0 sec for the average driver. Vehicles are 20 ft long. Drivers' eyes are 5 ft behind the front bumpers and from the far side of the vehicle. Assume wet pavement and worn tires. Vehicles accelerate according to the data given. (a) Determine the unknown no parking length, L, needed to allow adequate sight distance. (b) Make reasonable assumptions about the intersection geometry and parking parameters. Approximately, how many 24 ft parallel parking places will be lost?

distance traveled (ft)	time (sec)
10	2.3
30	3.7
50	6.2
80	8.5
100	11.8

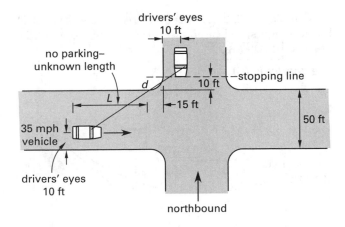

SOLUTIONS

1. (a) Use Eq. 79.46.

$$R = \frac{G_2 - G_1}{L} = \frac{2.4\% - (-3.2\%)}{5 \text{ sta}}$$

$$= 1.12 \text{ \%/sta}$$

The low point is given by Eq. 79.48.

$$x = \frac{-G_1}{R} = \frac{3.2\%}{1.12 \frac{\%}{\text{sta}}} = 2.857 \text{ sta}$$

The low point's station is

$$(\text{sta } 19+50) + 2.857 \text{ sta} = \boxed{\text{sta } 22+35.71 \quad (\text{sta } 22+36)}$$

The answer is (B).

(b) The elevation of the BVC is

$$700.28 \text{ ft} + (3.2\%)(2.5 \text{ sta}) = 708.28 \text{ ft}$$

The elevation of the low point is

$$y = \frac{Rx^2}{2} + G_1 x + \text{elev}_{\text{BVC}}$$

$$= \frac{\left(1.12 \frac{\%}{\text{sta}}\right)(2.857 \text{ sta})^2}{2}$$

$$+ (-3.2\%)(2.857 \text{ sta}) + 708.28 \text{ ft}$$

$$= \boxed{703.71 \text{ ft} \quad (704 \text{ ft})}$$

The answer is (D).

(c) The distance from the BVC to sta 21+25.04 is

$$x = (\text{sta } 21+25.04) - (\text{sta } 19+50)$$

$$= 1.7504 \text{ sta} \quad (1.75 \text{ sta})$$

The elevation of sta 21+25.04 is

$$y = \frac{Rx^2}{2} + G_1 x + \text{elev}_{\text{BVC}}$$

$$= \frac{\left(1.12 \frac{\%}{\text{sta}}\right)(1.75 \text{ sta})^2}{2}$$

$$+ (-3.2\%)(1.75 \text{ sta}) + 708.28 \text{ ft}$$

$$= \boxed{704.40 \text{ ft} \quad (704 \text{ ft})}$$

The answer is (B).

(d) The highest comfort speed is found from Eq. 79.58.

$$L = \frac{|(G_2 - G_1)|v^2}{46.5}$$

$$v = \sqrt{\frac{(500 \text{ ft})(46.5)}{|(2.4\% - (-3.2\%)|}}$$

$$= \boxed{64.43 \text{ mph} \quad (64 \text{ mph})}$$

The answer is (B).

(e) The rate of grade change is

$$R = \boxed{1.12 \text{ \%/sta} \quad (1.1 \text{ \%/sta})}$$

The answer is (B).

(f) At sta 23+16,

$$x = (\text{sta } 23{+}16) - (\text{sta } 19{+}50)$$

$$= 3.66 \text{ sta}$$

$$y = \frac{Rx^2}{2} + G_1 x + \text{elev}_{\text{BVC}}$$

$$= \frac{\left(1.12 \dfrac{\%}{\text{sta}}\right)(3.66 \text{ sta})^2}{2}$$

$$+ (-3.2\%)(3.66 \text{ sta}) + 708.28 \text{ ft}$$

$$= 704.07 \text{ ft}$$

20 ft of clearance are needed.

$$\text{elevation} = y + 20 \text{ ft} = 704.07 \text{ ft} + 20 \text{ ft}$$

$$= \boxed{724.07 \text{ ft} \quad (724 \text{ ft})}$$

The answer is (A).

(g) At sta 23+16,

$$x = (\text{sta } 23{+}16) - (\text{sta } 19{+}50)$$

$$= 3.66 \text{ sta}$$

$$R = \frac{2.4\% - (-3.2\%)}{L} = \frac{5.6\%}{L}$$

$$y = \frac{Rx^2}{2} + G_1 x + \text{elev}_{\text{BVC}}$$

$$724.07 \text{ ft} - 22 \text{ ft} = \frac{\left(\dfrac{5.6}{L} \dfrac{\%}{\text{sta}}\right)(3.66 \text{ sta})^2}{2}$$

$$+ (-3.2\%)(3.66 \text{ sta}) + 708.28 \text{ ft}$$

$$L = \boxed{6.82 \text{ sta} \quad (680 \text{ ft})}$$

The answer is (B).

(h) For curve 1,

$$y = \frac{Rx^2}{2} + G_1 x + \text{elev}_{\text{BVC}}$$

$$= \frac{\left(1.12 \dfrac{\%}{\text{sta}}\right)(2.5 \text{ sta})^2}{2}$$

$$+ (-3.2\%)(2.5 \text{ sta}) + 708.28 \text{ ft}$$

$$= 703.78 \text{ ft}$$

For curve 2,

$$y = \frac{Rx^2}{2} + G_1 x + \text{elev}_{\text{BVC}}$$

$$= \frac{\left(\dfrac{5.6}{6.82} \dfrac{\%}{\text{sta}}\right)(2.5 \text{ sta})^2}{2}$$

$$+ (-3.2\%)(2.5 \text{ sta}) + 708.28 \text{ ft}$$

$$= 702.85 \text{ ft}$$

The additional excavation needed is

$$703.78 \text{ ft} - 702.85 \text{ ft} = \boxed{0.93 \text{ ft} \quad (1.0 \text{ ft})}$$

The answer is (A).

(i) From Eq. 79.57,

$$K = \frac{L}{A} = \frac{682 \text{ ft}}{|-3.2\% - 2.4\%|} = \frac{682 \text{ ft}}{5.6\%}$$

$$= \boxed{121.79 \text{ ft/\%} \quad (121.8 \text{ ft/\%})}$$

The answer is (B).

(j) Use Table 79.2 (*AASHTO Green Book* Exh. 3-1). For a design speed of 65 mph, the design stopping distance is $\boxed{645 \text{ ft} (650 \text{ ft})}$.

The answer is (C).

2.

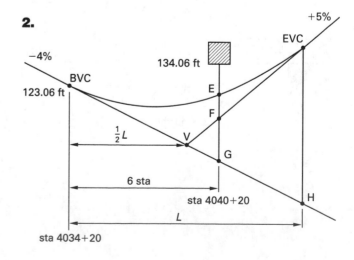

step 1: The elevation at point E is

$$134.06 \text{ ft} - 15 \text{ ft} = 119.06 \text{ ft}$$

step 2: The elevation at point G is

$$123.06 \text{ ft} - \left(4 \frac{\text{ft}}{\text{sta}}\right)(4040.2 \text{ sta} - 4034.2 \text{ sta})$$
$$= 99.06 \text{ ft}$$

$$\text{distance EG} = 119.06 \text{ ft} - 99.06 \text{ ft} = 20 \text{ ft}$$

step 3: $\angle$FVG is a $5\% - (-4\%) = 9\%$ divergence. Therefore,

$$\text{distance EVC-H} = (9)\left(\tfrac{1}{2}L\right)$$
$$= 4.5L$$

step 4: Measuring all x distances from BVC, and since parabolic offsets from a line (BVC-H in this case) are proportional to x^2,

$$\frac{\text{EG}}{(6 \text{ sta})^2} = \frac{\text{EVC-H}}{L^2}$$
$$\frac{20 \text{ ft}}{(6 \text{ sta})^2} = \frac{4.5L}{L^2}$$
$$L = \boxed{8.1 \text{ sta}}$$

3. Define the curve mathematically.

$$L = 94 \text{ sta} - 82 \text{ sta} = 12 \text{ sta}$$
$$\text{elev}_{\text{BVC}} = 729 \text{ ft} + (88 \text{ sta} - 82 \text{ sta})\left(4 \frac{\text{ft}}{\text{sta}}\right)$$
$$= 753 \text{ ft}$$

The rate of grade per station is

$$R = \frac{G_2 - G_1}{L} = \frac{3\% - (-4\%)}{12 \text{ sta}}$$
$$= \tfrac{7}{12} \, \%/\text{sta}$$
$$y = \tfrac{7}{24}x^2 - 4x + 753 \text{ ft}$$

The maximum elevation that provides clearance is $770 \text{ ft} - 25 \text{ ft} = 745 \text{ ft}$. Solve for x.

$$745 \text{ ft} = \tfrac{7}{24}x^2 - 4x + 753 \text{ ft}$$
$$x = 11.29 \text{ sta}, \ 2.43 \text{ sta}$$

(a) The minimum station is

$$82 + 2.43 = \boxed{\text{sta } 84+43}$$

(b) The maximum station is

$$82 + 11.29 = \boxed{\text{sta } 93+29}$$

(c) The turning point is at

$$x = \frac{-(-4\%)}{\dfrac{7}{12} \dfrac{\%}{\text{sta}}} = 6.86 \text{ ft}$$
$$\text{location} = (\text{sta } 82+00) + (\text{sta } 6+86)$$
$$= \boxed{\text{sta } 88+86}$$

(d) The elevation of the turning point is

$$y = \left(\frac{7}{24} \frac{\%}{\text{sta}}\right)(6.86 \text{ sta})^2 - (4\%)(6.86 \text{ sta}) + 753 \text{ ft}$$
$$= \boxed{739.29 \text{ ft}}$$

4. (a) Underpass—20 ft clearance:

Refer to Fig. 79.11.

step 1: The elevation at E is

$$\text{elev}_{\text{E}} = 510 \text{ ft} - 20 \text{ ft} = 490 \text{ ft}$$

step 2: The elevation at G is

$$\text{elev}_{\text{G}} = \text{elev}_{\text{V}} - (\text{slope})x$$
$$= 482 \text{ ft} - (3\%)(3 \text{ sta})$$
$$= 473 \text{ ft}$$
$$\text{EG} = \text{elev}_{\text{E}} - \text{elev}_{\text{G}} = 490 \text{ ft} - 473 \text{ ft}$$
$$= 17 \text{ ft}$$

step 3: The elevation at EF is

$$\text{EF} = \text{elev}_{\text{E}} - \text{elev}_{\text{V}} + \Delta\text{elev}_{\text{V-F}}$$
$$= 490 \text{ ft} - 482 \text{ ft} + (-2\%)(3 \text{ sta})$$
$$= 2 \text{ ft}$$

step 4: The offset from the grade line is proportional to the square of the distance from the BVC (or EVC) for a parabolic curve, since for a parabola,

$$y = ax^2$$
$$\frac{y_1}{x_1^2} = \frac{y_2}{x_2^2}$$
$$\frac{17 \text{ ft}}{\left(\dfrac{L}{2} + 3 \text{ sta}\right)^2} = \frac{2 \text{ ft}}{\left(\dfrac{L}{2} - 3 \text{ sta}\right)^2}$$
$$L = \boxed{12.2 \text{ sta}}$$

(b) Overpass—22 ft clearance:

step 1: The elevation at E is

$$\text{elev}_{\text{E}} = 510 \text{ ft} + 22 \text{ ft} = 532 \text{ ft}$$

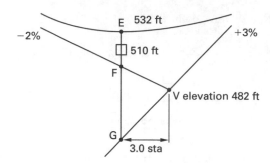

step 2: The length at EG is

$$EG = 532 \text{ ft} - 482 \text{ ft} + (3 \text{ sta})(3\%)$$
$$= 59 \text{ ft}$$

step 3: The length at EF is

$$EF = 532 \text{ ft} - 482 \text{ ft} - (3 \text{ sta})(2\%)$$
$$= 44 \text{ ft}$$

step 4: The length of the shortest vertical curve is

$$\frac{59 \text{ ft}}{\left(\dfrac{L}{2} + 3 \text{ sta}\right)^2} = \frac{44 \text{ ft}}{\left(\dfrac{L}{2} - 3 \text{ sta}\right)^2}$$
$$L = \boxed{82.0 \text{ sta}}$$

5. (a) The rate of change per station is

$$R = \frac{G_2 - G_1}{L} = \frac{2\% - (-4\%)}{19 \text{ sta}}$$
$$= 0.316 \text{ \%/sta}$$

$$\text{distance from BVC to V} = \frac{19 \text{ sta}}{2}$$
$$= 9.5 \text{ sta}$$

$$\text{elev}_{BVC} = 350 \text{ ft} + (4\%)(9.5 \text{ sta})$$
$$= \boxed{388 \text{ ft}}$$

(b) The elevation at sta 106+40 is

$$y = \left(\frac{0.316}{2}\right)x^2 - 4x + 388 \text{ ft}$$

The number of stations from BVC to station 106.40 is $106.40 \text{ sta} - (105 \text{ sta} - 9.5 \text{ sta}) = 10.9 \text{ sta}$.

$$y = \left(\frac{0.316}{2} \frac{\%}{\text{sta}}\right)(10.9 \text{ sta})^2 - (4\%)(10.9 \text{ sta}) + 388 \text{ ft}$$
$$= \boxed{363.17 \text{ ft}}$$

(c) Use Eq. 79.48. The minimum point occurs at

$$x = \frac{-G_1}{R} = \frac{-(-4\%)}{0.316 \dfrac{\%}{\text{sta}}}$$
$$= 12.66 \text{ sta}$$
$$y = \left(\frac{0.316}{2} \frac{\%}{\text{sta}}\right)(12.66 \text{ sta})^2 - (4\%)(12.66 \text{ sta})$$
$$+ 388 \text{ ft}$$
$$= \boxed{362.68 \text{ ft}}$$

(d) and (e) The stations of BVC and EVC are

$$105 \text{ sta} \pm \frac{19 \text{ sta}}{2} = \boxed{\text{sta } 95+50, \text{ sta } 114+50}$$

The road must go through the grating, so the low point is at the vertex.

(f) With a symmetrical (equal tangent) parabolic curve, the lowest point will not be at the vertex unless $G_1 = G_2$. Therefore, the curve cannot be equal tangent.

Each tangent of an unequal curve can be thought of as an equal-tangent curve with one of the grades equal to zero (i.e., $g = 0\%$).

BVC to V,

$$G_1 = -4\%$$
$$G_2 = 0$$
$$L_{BVC-V} = \text{sta}_V - \text{sta}_{BVC} = 105 - \text{sta}_{BVC}$$
$$R = \frac{0 - (-4\%)}{L_{BVC-V}} = \frac{4\%}{L_{BVC-V}}$$

The minimum elevation is 358.30 ft at station 105. Use Eq. 79.48.

$$x = \frac{-G_1}{R} = \frac{-(-4\%)}{\dfrac{4\%}{L_{BVC-V}}}$$
$$= L_{BVC-V}$$

The elevation of BVC is

$$\text{elev}_V + |G_1|x = 350 \text{ ft} + 4L_{\text{BVC-V}}$$

$$358.30 \text{ ft} = \frac{4L^2}{2L} - 4L + 350 \text{ ft} + 4L$$

$$8.30 \text{ ft} = 2L$$

$$L_{\text{BVC-V}} = 4.15 \text{ sta}$$

$$\text{location}_{\text{BVC}} = 105 \text{ sta} - 4.15 \text{ sta}$$

$$= 100.85 \text{ sta} \quad (\text{sta } 100+85)$$

$$\text{elev}_{\text{BVC}} = 350 \text{ ft} + (4\%)(4.15 \text{ sta})$$

$$= 366.60 \text{ ft}$$

For V to EVC, work with a mirror image to avoid $x = 0$ problems.

$$G_1 = -2\%$$

$$G_2 = 0$$

$$L_{\text{V-EVC}} = \text{sta}_V - \text{sta}_{\text{EVC}}$$

$$= 105 \text{ sta} - \text{sta}_{\text{EVC}}$$

$$R = \frac{0 - (-2\%)}{L_{\text{V-EVC}}} = \frac{2\%}{L_{\text{V-EVC}}}$$

$$x = \frac{-(-2\%)}{\dfrac{2\%}{L_{\text{V-EVC}}}} = L_{\text{V-EVC}}$$

$$\text{elev}_{\text{EVC}} = 350 \text{ ft} + 2L_{\text{V-EVC}}$$

$$358.30 \text{ ft} = \left(\frac{2}{2L}\right)L^2 - 2L + 350 \text{ ft} + 2L$$

$$L = 8.30 \text{ sta}$$

Working with the actual curve,

$$\text{location}_{\text{EVC}} = 105 \text{ sta} + 8.30 \text{ sta}$$

$$= 113.30 \text{ sta} \quad (\text{sta } 113+30)$$

$$\text{elev}_{\text{EVC}} = 350 \text{ ft} + \left(2 \frac{\text{ft}}{\text{sta}}\right)(8.30 \text{ sta}) = 366.6 \text{ ft}$$

$$L_{\text{BVC-EVC}} = 4.15 \text{ sta} + 8.30 \text{ sta} = \boxed{12.45 \text{ sta}}$$

This is the same answer that you would get without recognizing that the curve was not symmetrical. However, the premise would be wrong, and the curve equation $y(x)$ would be incorrect.

6. (a) Use Eq. 79.46.

$$R_1 = \frac{G_2 - G_1}{L} = \frac{-2.5\% - 5\%}{3 \text{ sta}}$$

$$= -2.5 \ \%/\text{sta}$$

Use Eq. 79.48.

$$x_{\text{turning point}} = \frac{-G_1}{R} = \frac{-5\%}{-2.5 \ \dfrac{\%}{\text{sta}}}$$

$$= 2 \text{ sta} \quad [\text{distance to turning point}]$$

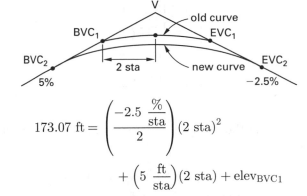

$$173.07 \text{ ft} = \left(\frac{-2.5 \ \dfrac{\%}{\text{sta}}}{2}\right)(2 \text{ sta})^2$$

$$+ \left(5 \ \frac{\text{ft}}{\text{sta}}\right)(2 \text{ sta}) + \text{elev}_{\text{BVC1}}$$

$$\text{elev}_{\text{BVC1}} = 168.07 \text{ ft}$$

The equation of curve 1 is

$$y_1 = -1.25x^2 + 5x + 168.07 \text{ ft}$$

The elevation at the midpoint of curve 1 is

$$y = \left(-1.25 \ \frac{\%}{\text{sta}}\right)(1.5 \text{ sta})^2 + (5\%)(1.5 \text{ sta})$$

$$+ 168.07 \text{ ft}$$

$$= 172.76 \text{ ft}$$

BVC$_1$ and BVC$_2$ are separated by 150 ft (1.5 sta). The elevation at BVC$_2$ is

$$168.07 \text{ ft} - \left(5 \ \frac{\text{ft}}{\text{sta}}\right)(1.5 \text{ sta}) = 160.57 \text{ ft}$$

$$R_2 = \frac{-2.5\% - 5\%}{6 \text{ sta}} = -1.25 \ \%/\text{sta}$$

$$y_2 = -0.625x^2 + 5x + 160.57 \text{ ft}$$

The elevation at the midpoint of curve 2 is

$$y_2 = \left(-0.625 \ \frac{\%}{\text{sta}}\right)(3 \text{ sta})^2 + (5\%)(3 \text{ sta}) + 160.57 \text{ ft}$$

$$= 169.95 \text{ ft}$$

The midpoint cut is

$$172.76 \text{ ft} - 169.95 \text{ ft} = \boxed{2.81 \text{ ft}}$$

(Equation 79.49 can be used to solve for M directly for each curve. The difference in M values is 2.8 ft.)

Transportation

(b) Measuring x from BVC_2,

$$
\begin{aligned}
y_0 &= 160.57 \text{ ft} \\
y_1 &= 164.95 \text{ ft} \\
y_2 &= 168.07 \text{ ft} \\
y_3 &= 169.95 \text{ ft} \\
y_4 &= 170.57 \text{ ft} \\
y_5 &= 169.95 \text{ ft} \\
y_6 &= 168.07 \text{ ft}
\end{aligned}
$$

7. (a) Find the elevation of BVC and the rate of grade change, R.

$$
\text{elev}_{BVC} = 1262 \text{ ft} - (3\%)(116 \text{ sta} - 110 \text{ sta})
$$
$$
= 1244 \text{ ft}
$$
$$
L = 122 \text{ sta} - 110 \text{ sta} = 12 \text{ sta}
$$
$$
R = \frac{G_2 - G_1}{L} = \frac{-2\% - 3\%}{12 \text{ sta}} = -0.417 \text{ \%/sta}
$$

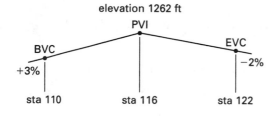

For the first half-station, $x = 0.5$ sta.

$$
y = \left(\tfrac{1}{2}\right)\left(-0.417 \, \frac{\%}{\text{sta}}\right)(0.5 \text{ sta})^2 + (3\%)(0.5 \text{ sta})
$$
$$
+ 1244 \text{ ft}
$$
$$
= \boxed{1245.45 \text{ ft}}
$$

(b) At $x = 5.5$ sta,

$$
y_{115.5} = \left(\frac{-0.417\%}{2 \text{ sta}}\right)(5.5 \text{ sta})^2 + (3\%)(5.5 \text{ sta})
$$
$$
+ 1244 \text{ ft}
$$
$$
= \boxed{1254.2 \text{ ft}}
$$

(c) Label the piles A through G. D is the centerline pile.

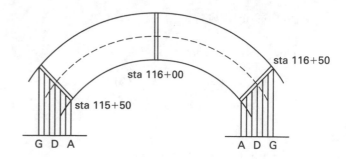

The elevation of the centerline pile (pile D) is found from part (b), less 3.5 ft.

This is an 8% superelevation, so for each 10 ft run, the rise is 0.8 ft.

For the first bent at sta 115+50, the top of pile D is at

$$
\begin{aligned}
D:&\quad 1254.2 \text{ ft} - 3.5 \text{ ft} = 1250.7 \text{ ft} \\
E:&\quad 1250.7 \text{ ft} + 0.8 \text{ ft} = 1251.5 \text{ ft} \\
F:&\quad 1250.7 \text{ ft} + (2)(0.8 \text{ ft}) = 1252.3 \text{ ft} \\
G:&\quad 1250.7 \text{ ft} + (3)(0.8 \text{ ft}) = 1253.1 \text{ ft} \\
C:&\quad 1250.7 \text{ ft} - 0.8 \text{ ft} = 1249.9 \text{ ft} \\
B:&\quad 1250.7 \text{ ft} - (2)(0.8 \text{ ft}) = 1249.1 \text{ ft} \\
A:&\quad 1250.7 \text{ ft} - (3)(0.8 \text{ ft}) = 1248.3 \text{ ft}
\end{aligned}
$$

bent at station	A	B	C	D	E	F	G
115+50	1248.3	1249.1	1249.9	1250.7	1251.5	1252.3	1253.1
116	1248.6	1249.4	1250.2	1251.0	1251.8	1252.6	1253.4
116+50	1248.8	1249.6	1250.4	1251.2	1252.0	1252.8	1253.6

8. (a) This is essentially a horizontal curve.

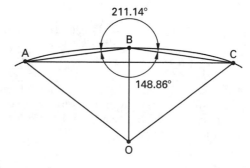

The 211.14° angle is shown. The interior angle is 360° − 211.14° = 148.86°.

The distances AB and BC can be found from the stadia readings.

$$AB = (100)(10.038 \text{ ft} - 2.403 \text{ ft}) + 0 = 763.5 \text{ ft}$$
$$BC = (100)(7.236 \text{ ft} - 4.320 \text{ ft}) = 291.6 \text{ ft}$$

Use the law of cosines to calculate distance AC.

$$\begin{aligned}(AC)^2 &= (AB)^2 + (BC)^2 - 2(AB)(BC)\cos\phi \\ &= (763.5 \text{ ft})^2 + (291.6 \text{ ft})^2 \\ &\quad - (2)(763.5 \text{ ft})(291.6 \text{ ft})(\cos 148.86°)\end{aligned}$$
$$AC = 1024.2 \text{ ft}$$

Angles BAC and BCA can be found from the law of sines.

$$\frac{\sin \angle \text{BAC}}{\text{BC}} = \frac{\sin \angle \text{BCA}}{\text{AB}} = \frac{\sin \angle \text{ABC}}{\text{AC}}$$
$$\frac{\sin \angle \text{BAC}}{291.6 \text{ ft}} = \frac{\sin \angle \text{BCA}}{763.5 \text{ ft}} = \frac{\sin 148.86°}{1024.2 \text{ ft}}$$
$$\angle \text{BAC} = 8.47°$$
$$\angle \text{BCA} = 22.67°$$

($\angle$BCA is not needed for the solution but can be used for checking.)

Check.

$$\begin{aligned}\angle \text{BAC} + \angle \text{BCA} + \angle \text{ABC} &= 8.47° + 22.67° + 148.86° \\ &= 180° \quad [\text{OK}]\end{aligned}$$
$$\text{DC} = \left(\tfrac{1}{2}\right)(291.6 \text{ ft}) = 145.8 \text{ ft}$$
$$\angle \text{BOC} = 2\angle \text{BAC}$$

From the deflection angle principles,

$$\angle \text{DOC} = \tfrac{1}{2}\angle \text{BOC} = \angle \text{BAC}$$

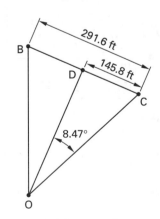

$$\angle \text{DOC} = \angle \text{BAC} = 8.47°$$
$$R = \text{OC} = \frac{145.8 \text{ ft}}{\sin 8.47°} = \boxed{989.9 \text{ ft}}$$

(b) Gradient AB is

$$\text{grade AB} = \frac{6.221 \text{ ft} - 4.7 \text{ ft}}{763.5 \text{ ft}}$$
$$= \boxed{+0.00199 \text{ ft/ft}}$$

(c) Gradient BC is

$$\text{grade BC} = \frac{4.7 \text{ ft} - 5.778 \text{ ft}}{291.6 \text{ ft}}$$
$$= \boxed{-0.00370 \text{ ft/ft}}$$

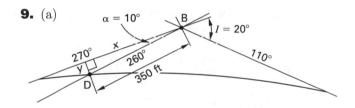

9. (a)

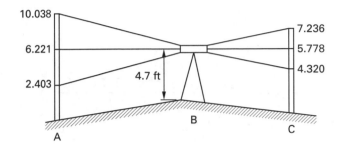

Refer to Fig. 79.6.

$$m = 350 \text{ ft}$$
$$\alpha = 10°$$
$$\cos \alpha = \frac{x}{350 \text{ ft}}$$
$$x = 344.68 \text{ ft}$$
$$\sin \alpha = \frac{y}{350 \text{ ft}}$$
$$y = 60.78 \text{ ft}$$
$$\gamma = 90° - \frac{I}{2} - \alpha = 90° - \frac{20°}{2} - 10°$$
$$= 70°$$
$$\phi = 180° - \arcsin \frac{\sin 70°}{\cos 10°}$$
$$= 107.41°$$
$$\theta = 180° - \gamma - \phi$$
$$= 180° - 70° - 107.41°$$
$$= 2.59°$$
$$\frac{\sin \theta}{m} = \frac{\sin \phi \cos \dfrac{I}{2}}{R}$$

$$\frac{\sin 2.59°}{350 \text{ ft}} = \frac{(\sin 107.41°)\left(\cos \dfrac{20°}{2}\right)}{R}$$

$$R = \boxed{7278.21 \text{ ft} \quad (7280 \text{ ft})}$$

The answer is (A).

(b) Use Eq. 79.4.

$$T = R \tan \frac{I}{2} = (7278.21 \text{ ft})\left(\tan \frac{20°}{2}\right)$$

$$= \boxed{1283.34 \text{ ft} \quad (1280 \text{ ft})}$$

The answer is (A).

(c) Use Eq. 79.8 and Eq. 79.10.

$$L = \frac{100IR}{5729.6} = \frac{(100 \text{ ft})(20°)(7278.21 \text{ ft})}{5729.6 \text{ ft}}$$

$$= \boxed{2540.56 \text{ ft} \quad (2500 \text{ ft})}$$

The answer is (B).

(d) Refer to Fig. 79.3. Use Eq. 79.14. To set out a 500 ft arc, the central angle would be

$$\beta = \frac{(360°)(\text{arc length})}{2\pi R} = \frac{(360°)(500 \text{ ft})}{2\pi(7278.21 \text{ ft})}$$

$$= \boxed{3.936° \quad (4°)}$$

The answer is (D).

(e) Use Eq. 79.8. To lay out the first stake at a 100 ft arc distance,

$$\beta = D = \frac{5729.578 \text{ ft}}{R} = \frac{5729.578 \text{ ft}}{7278.21 \text{ ft}}$$

$$= 0.7872°$$

From Eq. 79.13,

$$\alpha = \frac{\beta}{2} = \frac{0.7872°}{2} = 0.3936°$$

From Eq. 79.21 and Eq. 79.23,

$$\text{tangent distance} = NQ \cos \alpha = 2R \sin \alpha \cos \alpha$$

$$= (2)(7278.21 \text{ ft})(\sin 0.3936°)$$

$$\times (\cos 0.3936°)$$

$$= \boxed{99.99 \text{ ft} \quad (100 \text{ ft})}$$

The answer is (C).

(f) When snow and ice are not present, the superelevation is generally limited to 0.08–0.12. Use Eq. 79.35 with 0.10.

$$\tan \phi_{max} = \frac{v^2}{gR}$$

$$v = \sqrt{\tan \phi gR}$$

$$= \frac{\sqrt{(0.10)\left(32.2 \dfrac{\text{ft}}{\text{sec}^2}\right)(7278.21 \text{ ft})}\left(3600 \dfrac{\text{sec}}{\text{hr}}\right)}{5280 \dfrac{\text{ft}}{\text{mi}}}$$

$$= \boxed{104 \text{ mph} \quad (100 \text{ mph})}$$

The answer is (D).

(g) Use *AASHTO Green Book* Exh. 3-7. Given a design speed of 70 mph, the passing sight distance is $\boxed{2480 \text{ ft} \ (2500 \text{ ft}).}$

The answer is (B).

(h) The maximum superelevation rate in areas encountering snow and ice is typically between $\boxed{0.06}$ and 0.08.

The answer is (C).

(i) Remove all of the superelevation by $^2/_3 L$, and develop the crown on the tangent runout. Use Eq. 79.40.

$$SRR = \frac{1}{200}$$

$$T_R = \frac{wp}{SRR} = \frac{(12 \text{ ft})\left(0.02 \dfrac{\text{ft}}{\text{ft}}\right)}{\dfrac{1}{200}}$$

$$= 48 \text{ ft}$$

Use Eq. 79.41.

$$L = \frac{we}{SRR}$$

$$\tfrac{2}{3}L = \tfrac{2}{3}\left(\frac{we}{SRR}\right) = \left(\tfrac{2}{3}\right)(12 \text{ ft})\left(\frac{0.06 \dfrac{\text{ft}}{\text{ft}}}{\dfrac{1}{200}}\right)$$

$$= 96 \text{ ft}$$

$$48 \text{ ft} + 96 \text{ ft} = \boxed{144 \text{ ft} \quad (140 \text{ ft})}$$

The answer is (B).

(j) From Table 79.2, the minimum stopping sight distance is $\boxed{727.6 \text{ ft} \ (730 \text{ ft}).}$

The answer is (A).

Transportation

10. (a) Calculate the radius.

$$R = \left(\frac{360°}{6°}\right)\left(\frac{100 \text{ ft}}{2\pi}\right) = 954.93 \text{ ft}$$

Calculate the original back tangent length.

$$T = (954.93 \text{ ft})\left(\tan \frac{65°}{2}\right) = 608.36 \text{ ft}$$
$$\text{location of PC} = \text{sta PI} - T$$
$$= (\text{sta } 88{+}00) - (\text{sta } 6{+}8.36)$$
$$= \boxed{\text{sta } 81{+}91.64}$$

(b) The back tangent length decreases.

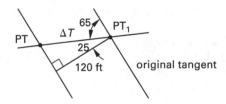

$$\Delta T = \frac{120 \text{ ft}}{\cos 25°} = 132.41 \text{ ft}$$

The new back tangent is $608.36 \text{ ft} - 132.41 \text{ ft} = 475.95 \text{ ft}$.

$$R_{\text{new}} = \frac{475.95 \text{ ft}}{\tan \dfrac{65°}{2}} = \boxed{747.09 \text{ ft}}$$

(c) The length of the curve is

$$L = (747.09 \text{ ft})(65°)\left(\frac{2\pi}{360°}\right) = 847.55 \text{ ft}$$

PT is located at

$$\text{sta PC} + L = (\text{sta } 81{+}91.64) + (\text{sta } 8{+}47.55)$$
$$= \boxed{\text{sta } 90{+}39.19}$$

11. (a) First, find the angle, θ.

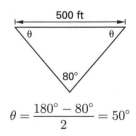

$$\theta = \frac{180° - 80°}{2} = 50°$$

Then, from the sine rule,

$$\frac{500 \text{ ft}}{\sin 80°} = \frac{R}{\sin 50°}$$
$$R = \boxed{388.93 \text{ ft}}$$

(b) Find the tangent length.

$$T = (388.93 \text{ ft})\left(\tan \frac{80°}{2}\right) = 326.35 \text{ ft}$$
$$T' = T - 10 \text{ ft} = 326.35 \text{ ft} - 10 \text{ ft} = 316.35 \text{ ft}$$

The new curve radius is

$$R' = \frac{316.3 \text{ ft}}{\tan 40°} = \boxed{377.0 \text{ ft}}$$

12. (a) The tangent length is

$$T = (1200 \text{ ft})\left(\tan \frac{55.5°}{2}\right) = 631.35 \text{ ft}$$
$$\begin{array}{l}\text{location} \\ \text{of PC}\end{array} = (\text{sta } 182{+}27.52) - (\text{sta } 6{+}31.35)$$
$$= \boxed{\text{sta } 175{+}96.17}$$

(b) The length of the curve is

$$L = (1200 \text{ ft})(55.5°)\left(\frac{2\pi}{360°}\right) = 1162.39 \text{ ft}$$

PT is located at

$$\text{sta PC} + L = (\text{sta } 175{+}96.17) + (\text{sta } 11{+}62.39)$$
$$= \boxed{\text{sta } 187{+}58.56}$$

(c) The horizontal signature offset is

$$\text{HSO} = (1200 \text{ ft})\left(1 - \cos \frac{55.5°}{2}\right)$$
$$= \boxed{138.01 \text{ ft}}$$

(d) From Eq. 79.5,

$$E = R \tan \frac{I}{2} \tan \frac{I}{4}$$
$$= (1200 \text{ ft})\left(\tan \frac{55.5°}{2}\right)\left(\tan \frac{55.5°}{4}\right)$$
$$= \boxed{155.95 \text{ ft}}$$

(e) Refer to the following diagram. Find the angle, β.

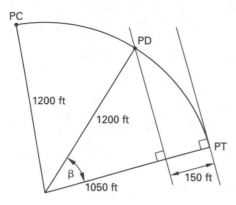

$$\beta = \arccos \frac{1050 \text{ ft}}{1200 \text{ ft}} = 28.96°$$

The curve length from PD to PT is

$$L = R\beta\left(\frac{2\pi}{360°}\right) = (1200 \text{ ft})(28.96°)\left(\frac{2\pi}{360°}\right)$$

$$= 606.54 \text{ ft}$$

$$\text{sta}_{PD} = \text{sta}_{PT} - \text{arc length}$$

$$= 18{,}758.56 \text{ ft} - 606.54 \text{ ft}$$

$$= 18{,}152.02 \text{ ft}$$

$$= \boxed{\text{sta } 181{+}52.02}$$

13. Refer to the following diagram. Notice that the 1 ft clearance is not in line with segment QM.

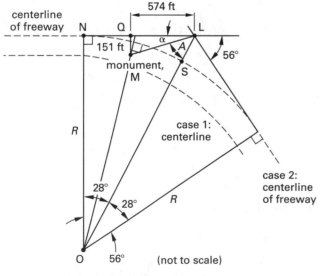

(not to scale)

First, determine the unknown angles and distances.

$$\alpha = \arctan \frac{QM}{QL} = \arctan \frac{151 \text{ ft}}{574 \text{ ft}} = 14.739°$$

For triangle NLO, the angle at vertex L is

$$\angle NLO = 180° - 90° - 28° = 62°$$

$$\angle A = 62° - \alpha = 62° - 14.739° = 47.261°$$

$$LM = \sqrt{(QL)^2 + (QM)^2} = \sqrt{(574 \text{ ft})^2 + (151 \text{ ft})^2}$$

$$= 593.53 \text{ ft}$$

From Eq. 79.5,

$$LS = R\tan\frac{I}{2}\tan\frac{I}{4}$$

$$= R\tan\frac{56°}{2}\tan\frac{56°}{4}$$

$$= 0.13257R$$

$$c = \text{distance } OL = R + LS = 1.13257R$$

Case 1: The monument is between the freeway center-line and point O. Work with triangle OML. From the law of cosines,

$$(OM)^2 = (LM)^2 + (OL)^2 - 2(LM)(OL)\cos A$$

$$LM = 593.53 \text{ ft}$$

$$OL = 1.13257R$$

$$A = 47.261°$$

$$OM = R - \frac{\text{road width}}{2} - \text{clearance}$$

$$= R - \frac{150 \text{ ft}}{2} - 1 \text{ ft}$$

$$= R - 76 \text{ ft}$$

$$(R - 76 \text{ ft})^2 = (593.53 \text{ ft})^2 + (1.13257R)^2$$

$$- (2)(593.53 \text{ ft})(1.13257R)(\cos 47.261°)$$

By trial and error, equation solver, or other appropriate means,

$$R_1 = \boxed{2108.35 \text{ ft}}$$

Case 2: The freeway centerline is between the monument and point O.

$$OM = R_2 + 76 \text{ ft}$$

$$(R + 76 \text{ ft})^2 = (593.53 \text{ ft})^2 + (1.13257R)^2$$

$$- (2)(593.53 \text{ ft})(1.13257R)(\cos 47.261°)$$

$$R_2 = \boxed{3405.01 \text{ ft}}$$

14. (a) Refer to the following diagram.

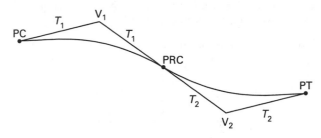

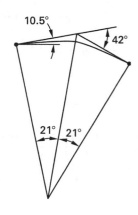

Since the two intersection angles are not the same, the tangents are not the same.

$$T_1 + T_2 = 2100 \text{ ft}$$

$$T_1 = R \tan \frac{42°}{2} = 0.384R$$

$$T_2 = R \tan \frac{30°}{2} = 0.268R$$

$$(0.384 + 0.268)R = 2100 \text{ ft}$$

$$R = 3220.9 \text{ ft}$$

$$T_1 = (0.384)(3220.9 \text{ ft}) = 1236.8 \text{ ft}$$

$$T_2 = (0.268)(3220.9 \text{ ft}) = 863.2 \text{ ft}$$

$$L_1 = (3220.9 \text{ ft})(42°)\left(\frac{2\pi}{360°}\right)$$

$$= 2361.04 \text{ ft}$$

$$L_2 = (3220.9 \text{ ft})(30°)\left(\frac{2\pi}{360°}\right)$$

$$= 1686.46 \text{ ft}$$

Station of V_1 is 21+25 = 46+00. Therefore, the station of the PC is

$$\text{sta PC} = (\text{sta } 46{+}00) - (\text{sta } 12{+}36.8)$$

$$= \boxed{\text{sta } 33{+}63.20}$$

(b) The station of the PRC is

$$\text{sta PRC} = \text{sta PC} + L_1$$

$$= (\text{sta } 33{+}63.20) + (\text{sta } 23{+}61.04)$$

$$= \boxed{\text{sta } 57{+}24.24}$$

(c) The station of the PT is

$$\text{sta PT} = \text{sta PRC} + L_2$$

$$= (\text{sta } 57{+}24.24) + (\text{sta } 16{+}86.46)$$

$$= \boxed{\text{sta } 74{+}10.70}$$

(d) The deflection angle from PC to MPC is

$$\text{deflection angle} = \left(\tfrac{1}{2}\right)\left(\left(\tfrac{1}{2}\right)(42°)\right) = \boxed{10.5°}$$

(e) The hazards are

> • driving off the road due to distraction or inattention
> • tipping over (whipping) at high speeds
> • difficulty in getting proper superelevation

15. (a) Since the angles (with respect to an E-W line) made by the diagonal between drivers' eyes are relatively acute (i.e., less than 15°), the diagonal and E-W distances are essentially the same. The offset and other distances are not used in this first analysis. At 35 mph, the stopping sight distance is given by Eq. 79.43. Assume the worst case (wet pavement, worn tires). From Table 75.1,

$$f = 0.34$$

Case 1: An eastbound car sees a southbound car appear suddenly in front and skids to a stop.

If sight is into the intersection only (that is, drivers cannot see each other until the other is into the intersection), add 2.3 sec (the time to accelerate 10 ft from a stop into the intersection).

$$S = (1.47)(1 \text{ sec} + 2.3 \text{ sec})(35 \text{ mph}) + \frac{(35 \text{ mph})^2}{(30)(0.34)}$$

$$= 289.9 \text{ ft}$$

Case 2: After entering the intersection, the southbound car sees the eastbound car (whose driver is not watching) and accelerates across the intersection.

Starting from a stop, to clear the intersection completely, the southbound car will have to travel 10 ft + 50 ft + 20 ft = 80 ft, where 20 ft is the car length. From the acceleration data, this will take 8.5 sec. By the time the vehicle has traveled the first 10 ft and has become vulnerable, 2.3 sec will have elapsed. The time that the southbound vehicle is vulnerable to being hit is

$$t = 8.5 \text{ sec} - 2.3 \text{ sec} = 6.2 \text{ sec}$$

Transportation

The distance traveled by the eastbound car at 35 mph is

$$(1.47)(6.2 \text{ sec})(35 \text{ mph}) \approx 319 \text{ ft}$$

Since 319 ft > 289.9 ft,

$$d \approx \boxed{319 \text{ ft}}$$

(b) The calculation of the number of parking spaces lost must take into consideration the length of interior parking places (given as 24 ft), the length of the end parking place (say 22 ft), a no-parking clear zone at the corner (say 15 ft), the southbound lane width (say 12 ft) and position of the car, and perhaps the distance between the drivers' eyes and the front/side of the vehicle (say 5 ft).

Simplistically, the number of 24 ft spaces lost is approximately

$$1 + \frac{319 \text{ ft} - 22 \text{ ft} - 15 \text{ ft} - 12 \text{ ft} + 5 \text{ ft} + 5 \text{ ft}}{24 \text{ ft}}$$

$$= \boxed{12.7 \text{ spaces} \quad (13 \text{ spaces})}$$

80 Construction Earthwork

PRACTICE PROBLEMS

1. Which of the following is a true statement regarding a neat line on an excavation drawing?

(A) The contractor will not normally be paid for excavating beyond the neat line.

(B) The contractor may not travel over or disturb soil beyond the neat line.

(C) The contractor must return the surface beyond the neat line to its original vegetated condition.

(D) The contractor may not store construction materials, vehicles, or spoils beyond the neat line.

2. A planimeter is used to determine the areas of subgrade cut (excavation) cross sections along a construction roadway. Most nearly, what is the total volume excavated?

station (100s of ft)	area from planimeter (ft^2)
16+50	0
16+84	410
16+98	780
17+08	820
17+36	250
17+45	0

(A) 20,000 ft^3

(B) 30,000 ft^3

(C) 40,000 ft^3

(D) 50,000 ft^3

3. 80 yd^3 of bank run soil is excavated and stockpiled before being transported and subsequently compacted. The soil has a swell factor of 0.35 and a shrinkage factor of 0.10. The final volume of the compacted soil is most nearly

(A) 72 yd^3

(B) 97 yd^3

(C) 100 yd^3

(D) 109 yd^3

4. Which of the following formulas for bank volume (BCY), loose volume (LCY), compacted volume (CCY), swell factor (S), shrinkage factor (D), and load factor (LF) is NOT correct?

(A) $LCY = (1 + S) \times BCY$

(B) $BCY = LF \times LCY$

(C) $CCY = LCY \div (1 + S)$

(D) $CCY = BCY \times (1 - D)$

5. 15 tons of a material with a specific weight of 120 lbf/ft^3, an internal friction angle of 30°, and a natural moisture content of 6% are dropped by an overhead conveyor onto a level surface. The resulting pile will NOT be higher than

(A) 4.3 ft

(B) 6.3 ft

(C) 8.4 ft

(D) 9.1 ft

6. The optimum moisture content of a fill is 10% with a compacted specific weight of 118 lbf/ft^3. The fill material is delivered onsite with a moisture content of 6%. What weight of water must be added per cubic foot of compacted soil to achieve the optimum moisture content?

(A) 1.9 lbf/ft^3

(B) 2.7 lbf/ft^3

(C) 4.3 lbf/ft^3

(D) 6.4 lbf/ft^3

7. Green's theorem (Green's law) is the basis for determining the

(A) geodetic location of a benchmark

(B) distance between points along a sewer trunk line

(C) surface area of a construction site to be scarified

(D) volume of a lake to be drained

8. Haul road crowns should ideally have a slope of

(A) 0%

(B) 1%

(C) 3%

(D) 6%

9. A 50 ton crane with four extendable outriggers weighs 90,000 lbf. The pad of each outrigger has a diameter of 24 in. The crane is set up on type A soil (OSHA designation, unconfined compressive strength of 1.5 tons/ft^2). Approximately what area of cribbing is needed under each outrigger?

(A) 10 ft^2

(B) 15 ft^2

(C) 30 ft^2

(D) 60 ft^2

10. If a mechanically spliced $^1/_2$ in wire rope has a vertical lifting capacity of 1.5 tons, what is its approximate maximum capacity when used as a vertical basket?

(A) 0.8 tons

(B) 1.5 tons

(C) 3.0 tons

(D) 4.5 tons

11. A planimeter consists of two 14 in articulated arms (the pole arm and a tracer arm) with a 3.0 in diameter wheel centered along the length of the tracer arm. While tracing out a closed area, the planimeter wheel turns 7.65 revolutions. Most nearly, what is the area of the area traced?

(A) 210 in^2

(B) 330 in^2

(C) 670 in^2

(D) 1000 in^2

12. Measurements on a planimeter are read from a dial meter and the tracer wheel's vernier. A result is shown. What number is represented?

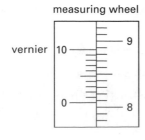

(A) 2800

(B) 2804

(C) 2916

(D) 3809

13. A reservoir with vertical sides and a rectangular plan view is 520 ft wide and 1740 ft long. The depth slopes uniformly in the long direction, starting at 35 ft and ending at 80 ft. Most nearly, what is the volume of the reservoir?

(A) 52 MG

(B) 180 MG

(C) 320 MG

(D) 390 MG

14. In order to minimize disturbance causing possible subsequent erosion, motorized equipment driven over stream banks should have tires that exert a ground pressure of less than

(A) 6 psi

(B) 25 psi

(C) 50 psi

(D) 75 psi

15. Which of the following installation instructions for erosion control features is INCORRECTLY stated?

(A) Install a silt fence, straw bale barriers, and other erosion control methods along the toe of the slope below the high water level of the receiving stream channel.

(B) Straw bale barriers should only be installed where sediment-laden water can pond, allowing sediment to settle out.

(C) Install a gravel bag berm where it can intercept and slow the flow of sediment-laden sheet flow.

(D) Install a silt fence on the down-slope side of a straw bale barrier closest to the receiving stream channel.

16. A balanced dirt job is one in which

(A) earth moving is performed and paid for in the same billing cycle

(B) cut and fill volumes are equal

(C) as much fill is brought onto the construction site as is required to fill excavations

(D) top soil is retained for subsequent reuse on the site

17. A planimeter is used to trace the shoreline contour lines corresponding to difference depths of a reservoir. The map being traced has a scale of 1 in = 100 ft. Most nearly, what is the volume of the reservoir?

depth contour (ft)	planimeter area (in^2)
0	220
5	160
10	110
15	85
20	20
23 (maximum depth)	0

(A) 12×10^6 ft^3

(B) 17×10^6 ft^3

(C) 24×10^6 ft^3

(D) 32×10^6 ft^3

18. By doubling the length of the tracer arm on an adjustable planimeter, the wheel roll vernier dial will

(A) decrease by a factor of 4

(B) decrease by a factor of 2

(C) increase by a factor of 2

(D) increase by a factor of 4

SOLUTIONS

1. *Neat lines* are used to show limits of excavation. Excavation beyond the neat line may occur, but the contractor is generally not paid unless prior arrangements have been made.

The answer is (A).

2. Use the average end area method. Determine the distance between stations and the average area between them. For example, the distance between the first two stations is 84 ft − 50 ft = 34 ft. The average area between them is $(^1/_2)(0 + 410 \text{ ft}^2) = 205 \text{ ft}^2$. The volume excavated is approximately

$$V = dA = (34 \text{ ft})(205 \text{ ft}^2) = 6970 \text{ ft}^3$$

The following table is prepared similarly.

station (100s of ft)	area from planimeter (ft^2)	distance (ft)	average area (ft^2)	volume (ft^3)
16+50	0			
		34	205	6970
16+84	410			
		14	595	8330
16+98	780			
		10	800	8000
17+08	820			
		28	535	14,980
17+36	250			
		9	125	1125
17+45	0			

The total volume of subgrade excavation is

$$V = 6970 \text{ ft}^3 + 8330 \text{ ft}^3 + 8000 \text{ ft}^3$$
$$+ 14{,}980 \text{ ft}^3 + 1125 \text{ ft}^3$$
$$= \boxed{39{,}405 \text{ ft}^3 \quad (40{,}000 \text{ ft}^3)}$$

The answer is (C).

3. Swell is measured with respect to the banked condition. The stockpiled and transported volume will be

$$V_l = (1 + 0.35)(80 \text{ yd}^3) = 108 \text{ yd}^3$$

Shrinkage is also measured with respect to the bank condition. The compacted volume will be

$$V_c = (1 - 0.1)(80 \text{ yd}^3) = \boxed{72 \text{ yd}^3}$$

The answer is (A).

4. None of the factors (swell, shrinkage, or load) associates loose volumes to compacted volumes.

The answer is (C).

5. The material will ideally form a cone-shaped pile. The volume of a cone is

$$V = \frac{\text{weight}}{\text{density}} = \frac{(15 \text{ tons})\left(2000 \; \frac{\text{lbf}}{\text{ton}}\right)}{120 \; \frac{\text{lbf}}{\text{ft}^3}}$$

$$= 250 \text{ ft}^3$$

The sides of the cone will be inclined at the natural angle of repose, equal to the angle of internal friction.

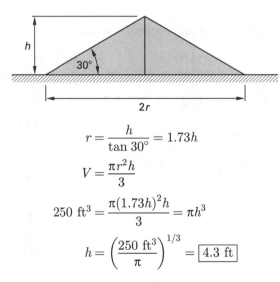

$$r = \frac{h}{\tan 30°} = 1.73h$$

$$V = \frac{\pi r^2 h}{3}$$

$$250 \text{ ft}^3 = \frac{\pi (1.73h)^2 h}{3} = \pi h^3$$

$$h = \left(\frac{250 \text{ ft}^3}{\pi}\right)^{1/3} = \boxed{4.3 \text{ ft}}$$

The answer is (A).

6. Moisture content is calculated on a per unit weight basis. Consider 1 ft^3 of the optimally compacted soil. The soil solids weight is

$$W_s = \frac{W_t}{1+w} = \frac{118 \text{ lbf}}{1+0.10} = 107.3 \text{ lbf}$$

The soil is delivered with a moisture content deficit of $10\% - 6\% = 4\%$. The moisture weight deficit is

$$W_w = w_{\text{deficit}} W_s = (0.04)(107.3 \text{ lbf})$$
$$= \boxed{4.29 \text{ lbf} \quad (4.3 \text{ lbf})}$$

The answer is (C).

7. *Green's theorem* describes mathematically how a planimeter can be used to calculate the area of an irregular two-dimensional shape.

The answer is (C).

8. A haul road crown slope of $\boxed{3\%}$ is considered ideal. Lower crowns allow water to pool on the road and flow to low spots where tires can sustain damage. High crowns cause uneven tire wear.

The answer is (C).

9. Although there are four outriggers, when a crane begins to tip, all of the load can be transferred to a single outrigger. Maximum loading on the outrigger pads will occur when the outriggers are not extended (i.e., do not have a substantial lever arm).

The area of each outrigger pad is

$$A_{\text{pad}} = \frac{\pi}{4} d^2 = \frac{\left(\frac{\pi}{4}\right)(24 \text{ in})^2}{\left(12 \; \frac{\text{in}}{\text{ft}}\right)^2} = 3.1 \text{ ft}^2$$

The total weight of the crane and maximum load is

$$W = (50 \text{ tons})\left(2000 \; \frac{\text{lbf}}{\text{ton}}\right) + 90{,}000 \text{ lbf} = 190{,}000 \text{ lbf}$$

The compressive strength of type A soil is 1.5 tsf.

The minimum support area is

$$A = \frac{190{,}000 \text{ lbf}}{\left(1.5 \; \frac{\text{tons}}{\text{ft}^2}\right)\left(2000 \; \frac{\text{lbf}}{\text{ton}}\right)} = \boxed{63.3 \text{ ft}^2 \quad (60 \text{ ft}^2)}$$

The answer is (D).

10.

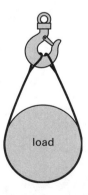

The capacity of the sling when used as a vertical basket will depend on the sling angle. When the wire rope legs are nearly vertical, the basket will be at maximum capacity, doubling the single-wire capacity of 1.5 tons. Therefore, the maximum capacity is $\boxed{3.0 \text{ tons.}}$

The answer is (C).

11. The area traced out by a planimeter is the length of the tracer arm times the circumference of the wheel times the number of revolutions that the wheel makes as the area is traced completely once.

$$A = 2\pi r N L = \pi d N L = \pi(3 \text{ in})(7.65)(14 \text{ in})$$
$$= \boxed{1009 \text{ in}^2 \quad (1000 \text{ in}^2)}$$

The answer is (D).

12.

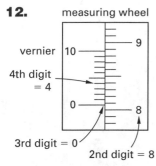 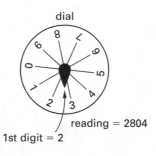

reading = 2804

The dial has passed 2, but has not yet reached 3. Therefore, the most significant digit is 2. The "0" mark on the measuring wheel vernier has just passed 8, which is the next second most significant digit. The "0" mark has not reached the first graduation on the measuring wheel past "8," so the next most significant digit is 0. The two closest aligning lines on the vernier are 4, which is the least significant digit. The number is $\boxed{2804.}$

The answer is (B).

13. Use the average end area formula. The area of the deep end of the reservoir is

$$A_{\text{deep}} = wd = (520 \text{ ft})(80 \text{ ft}) = 41,600 \text{ ft}^2$$

The area of the shallow end is

$$A_{\text{shallow}} = wd = (520 \text{ ft})(35 \text{ ft}) = 18,200 \text{ ft}^2$$

The volume of the reservoir is

$$V = \tfrac{1}{2}(A_1 + A_2)l$$
$$= \left(\tfrac{1}{2}\right)(41,600 \text{ ft}^2 + 18,200 \text{ ft}^2)(1740 \text{ ft})$$
$$\times \left(\frac{7.48 \; \frac{\text{gal}}{\text{ft}^3}}{1,000,000 \; \frac{\text{gal}}{\text{MG}}} \right)$$
$$= \boxed{389.2 \text{ MG} \quad (390 \text{ MG})}$$

The answer is (D).

14. Wide or high flotation tires, dual tires, bogie axle systems, tracked machines, and lightweight equipment

can be used to reduce contact pressure to $\boxed{5\text{--}6 \text{ psi}}$ in order to minimize disturbance.

The answer is (A).

15. Erosion control features should not be submerged, so they must be installed *above* the high water level of any receiving water channels.

The answer is (A).

16. Earthwork is balanced by equal volumes of cuts and fills.

The answer is (B).

17. Use the average end-area method with the ends working vertically downward. Between a depth of 0 ft and 5 ft, the average lake area is

$$A_{0\text{-}5} = \left(\tfrac{1}{2}\right)(220 \text{ in}^2 + 160 \text{ in}^2)\left(100 \; \frac{\text{ft}}{\text{in}}\right)^2$$
$$= 1,900,000 \text{ ft}^2$$

The volume of the water between depths 0 ft and 5 ft is

$$V_{0\text{-}5} = Ah = (1,900,000 \text{ ft}^2)(5 \text{ ft}) = 9,500,000 \text{ ft}^3$$

The following table is prepared similarly.

depth contour	planimeter area (in^2)	area (ft^2)	volume (ft^3)
0	220		
		1,900,000	9,500,000
5	160		
		1,350,000	6,750,000
10	110		
		975,000	4,875,000
15	85		
		525,000	2,625,000
20	20		
		100,000	300,000
23 $\left(\begin{array}{c}\text{maximum}\\\text{depth}\end{array}\right)$	0		

The total volume is the sum of the layer volumes.

$$V_t = 9,500,000 \text{ ft}^3 + 6,750,000 \text{ ft}^3 + 4,875,000 \text{ ft}^3$$
$$+ 2,625,000 \text{ ft}^3 + 300,000 \text{ ft}^3$$
$$= \boxed{24,050,000 \text{ ft}^3 \quad (24 \times 10^6 \text{ ft}^3)}$$

The answer is (C).

Construction

18. The area traced out by a planimeter is

$$A = 2\pi r N L = \pi d N L$$

The area, A, of a measured section does not change, so if the tracer's arm length, L, is doubled, the number of wheel rolls, N, will $\boxed{\text{decrease by a factor of 2.}}$

The answer is (B).

81 Construction Staking and Layout

PRACTICE PROBLEMS

1. What is the meaning of the lath stake markings?

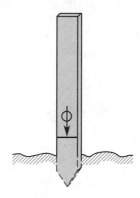

(A) Soil should be graded up to the level of the horizontal line.

(B) There is a buried pipe below.

(C) The stake is on the centerline of the roadway.

(D) A sewer invert should be installed at this level.

2. What is the meaning of the lath stake markings?

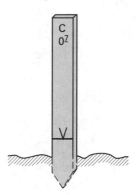

(A) The finished elevation is 0.7 ft below the existing surface.

(B) The finished elevation is 0.7 ft below the horizontal line.

(C) The finished elevation is at the horizontal line.

(D) The finished elevation 0.7 ft toward the centerline is at the horizontal line.

3. What is the meaning of the lath stake markings?

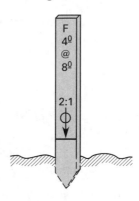

(A) 4 ft of fill is needed 8 ft away.

(B) 8 ft of fill is needed 4 ft away.

(C) 4 ft of fill is needed here; follow a 2:1 slope for 8 ft.

(D) 8 ft of fill is needed here; follow a 2:1 slope down 4 ft.

4. What is the meaning of the lath stake markings?

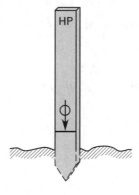

(A) The horizontal line corresponds to the highest point of the grade.

(B) A stake is on the centerline of a high pressure gas line.

(C) The stake is located where the grade becomes horizontal.

(D) Excess soil may be heaped at this point.

5. What is the meaning of the sideways "K" mark on this lath stake?

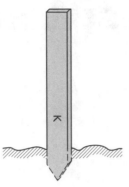

(A) Cut is required.

(B) Fill is required.

(C) Neither cut nor fill is required because this is an undisturbed grade.

(D) Cut is required on one side of the stake; fill is required on the other side.

6. A stake is encountered with the following markings.

DS 34-B
FL DIKE
ELEV 1073.15
$10^{\underline{00}}$ CL GRT

Which of the following is NOT a correct interpretation of this stake?

(A) This is a drainage stake corresponding to plan feature 34-B.

(B) The stake is along the flow line of a dike.

(C) The elevation of the flow line invert is 1073.15 ft.

(D) The centerline of the grate is 10 ft from the stake.

7. A stake is encountered with the following markings.

CS B-3
A150+40
$22^{\underline{00}}$ LT
C $0^{\underline{25}}$
$4^{\underline{00}}$ FL

Which of the following is NOT a correct interpretation of this stake?

(A) This is a curb stake corresponding to curb type B-3.

(B) The reference point is 22 ft to the left of the curb flow line.

(C) A cut of 0.25 ft is required to locate the curb flow line.

(D) The flow line is located 4 ft away from the reference point.

8. Which of the following marks would you most likely find on the BACK of a lath slope stake?

(A) 2:1

(B) 117+50

(C) SE 0.03

(D) F $0^{\underline{9}}$

9. The location of a concrete box culvert outlet is to be identified by a lath grade stake. The elevation outlet invert is 12 ft higher than the existing ground level, and the invert is located 8 ft from the reference point. What would be a reasonable marking on a lath grade stake?

(A) F $12^{\underline{00}}$
  $8^{\underline{00}}$

(B) C $12^{\underline{00}}$
  $8^{\underline{00}}$

(C) F $8^{\underline{00}}$
  $12^{\underline{00}}$

(D) F $12^{\underline{03}}$
  $8^{\underline{00}}$

10. A lath stake with the following markings is observed in the vicinity of a pipe feature.

$15^{\underline{00}}$
21+50
18×40
CMP w/2
FL END

Which interpretation is not consistent with these markings?

(A) The reference point is 15 ft from the pipe.

(B) The pipe is 40 ft long and 18 in in diameter.

(C) The pipe is steel-reinforced concrete.

(D) The pipe has two flared ends.

11. Prior to setting elevation stakes on a small construction site, a surveyor is most likely to

(A) obtain topographic maps from the National Geodetic Survey

(B) establish a local control

(C) correct azimuths for declination

(D) research recorded deeds and locate parcel boundaries

Construction

12. What is the preferred method to avoid negative stations in a construction document?

(A) Station 0+00 is selected some significant distance from the project boundary.

(B) The starting station within the project boundary is arbitrarily selected as 0+00.

(C) Stationing is reset to the secondary 0+00 point when a negative station occurs.

(D) The negative stations are reported as positive numbers with the designation "R" added to indicate reverse stationing.

13. Which feature is indicated by the line ABCC′B′A′ on the following roadway construction drawing?

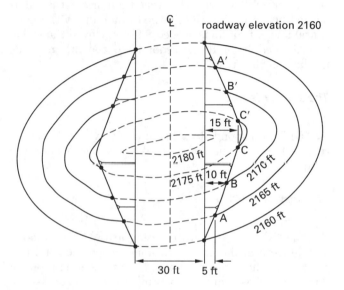

(A) safety fence

(B) loose rock barrier

(C) detention embankment dam

(D) brow interception ditch

14. Elm Street and 23rd Street intersect at sta 5+30 and at an angle of 62° (118°) as shown. Stationing is along the centerline of Elm Street. The pavement width for each street is 36 ft. The radii to the edges of the pavements are both 30 ft. Where should the PC curb return (point A on the illustration) be staked?

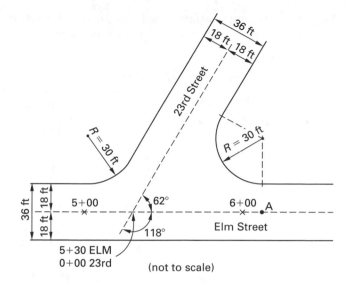

(A) sta 5+79

(B) sta 5+97

(C) sta 6+10

(D) sta 6+30

15. A grade stake is to be set to show a cut to the flow line of a pipe. The height of the instrument is 472.36 ft, the elevation of the flow line is 462.91 ft, and the ground rod at the point of the stake is 5.1 ft. The grade stake is to be set to provide a reference once the ground is disturbed through excavation. Determine the grade stake marking that will indicate a half-foot cut to the flow line.

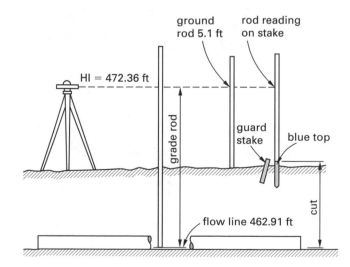

(A) C 3′11″

(B) C 4′2″

(C) C 4′6″

(D) C 4′11″

16. In laying out features on a jobsite for the day's construction, a robotic total station should be set up

(A) at a free station

(B) directly over a control point

(C) at a resection point

(D) any of the above

SOLUTIONS

1. A crow's foot or arrow drawn through a circle means that the horizontal line represents the final grade. An operator of excavation equipment will add fill up to the grade line.

The answer is (A).

2. Since the crow's foot does not have a circle, the horizontal line represents a reference point, not the finished elevation. The finished elevation is 0.7 ft below the horizontal line. Cutting is required.

The answer is (B).

3. The crow's foot with the circled tail means that fill is needed here to bring the finished grade up to the horizontal line. 4 ft of fill is needed 8 ft away from the stake. 8 ft divided by 4 ft corresponds to the 2:1 gradient marked on the stake.

The answer is (A).

4. "HP" corresponds to a hinge point, where the grade (e.g., 2:1) on one side of the stake becomes level on the other side.

The answer is (C).

5. The "K" is an upward-pointing arrow (crow's foot). The horizontal line indicates some unspecified finished grade, and the ∧ can be thought of as meaning a pile of dirt (i.e., fill is required to bring the grade up to the desired level.)

The answer is (B).

6. These markings would be used on a drainage stake that identifies feature 34-B on the drainage plan. The feature is the intersection of the dike flow line and the grate centerline, which is located 10 ft from the reference point. The elevation of the reference point is 1073.15.

The answer is (C).

7. These markings would be used on a curb stake that identifies feature B-3, probably a curb type. The reference point is 22 ft to the left of a marked line, probably the roadway stationing line. (The curb flow line is being located with this stake, so the flow line itself cannot be used as a reference.) The flow line is 4 ft away, requiring a cut of 0.25 ft.

The answer is (B).

8. Slope (2:1) and fill (F 0$\underline{0}$) need to be seen by the grading operator, so those markings would appear on the front, facing the traveled way. Neither the super-elevation nor the station are useful to the grading operator. It is common for superelevation to be written on the side, while the station is written on the back.

The answer is (B).

9. 12 ft of fill can be expected to settle. The amount of settlement should be added to the fill specification.

The answer is (D).

10. The stake markings are interpreted to mean the reference point is 15 ft from the pipe, at station 21+50. The pipe diameter is 18 in, and the length is 40 ft. Corrugated metal pipe with two flared ends is to be used.

The answer is (C).

11. All elevation stakes must be referenced to a single point. This point can be a federally recorded control point, but most construction sites are distant from such points. Therefore, a surveyor will establish some local feature on or close to the jobsite that can be used as a reference. The other alternatives, while having utility depending on the situation, are not specifically relevant to setting elevation stakes.

The answer is (B).

12. In order to avoid negative stations, the starting station should be outside of the project boundary. Alternatively, the starting station within the project boundary can be arbitrarily selected as some positive number, such as 10+00 or 100+00.

The answer is (A).

13. From the contours, the roadway travels through a cut. The line appears to be a brow interception ditch with overside (slope) drains. Roadside gutters are not shown.

The answer is (D).

14. Work with the points identified on the illustration.

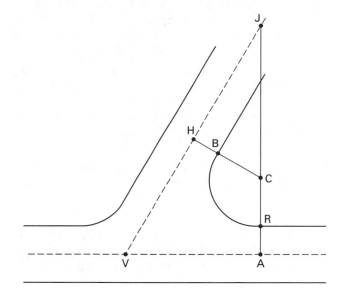

Next, work with triangle CHJ.

$$CH = CB + BH = 30 \text{ ft} + 18 \text{ ft} = 48 \text{ ft}$$
$$\angle C = \angle V = 62°$$
$$CJ = \frac{CH}{\cos \angle C} = \frac{48 \text{ ft}}{\cos 62°} = 102.24 \text{ ft}$$

Then, work with triangle VJA.

$$CA = CR + RA = 30 \text{ ft} + 18 \text{ ft} = 48 \text{ ft}$$
$$JA = CJ + CA = 102.24 \text{ ft} + 48 \text{ ft} = 150.24 \text{ ft}$$
$$VA = \frac{JA}{\tan \angle V} = \frac{150.24 \text{ ft}}{\tan 62°} = 79.88 \text{ ft}$$

The PC curb return should be staked at

$$\text{sta A} = \text{sta V} + VA = (5+30) + 79.88$$
$$= \boxed{6+9.88 \quad (\text{sta } 6+10)}$$

The answer is (C).

15. Flow line and invert are synonymous. With the grade rod set at the flow line, the grade rod reading would be

$$\text{grade rod} = HI - \text{flow elevation}$$
$$= 472.36 \text{ ft} - 462.91 \text{ ft}$$
$$= 9.45 \text{ ft}$$

Since the ground rod is 5.1 ft, the depth of excavation is

$$\text{cut} = 9.45 \text{ ft} - 5.1 \text{ ft} = 4.35 \text{ ft}$$

Construction

However, during rough grading, cuts (and fills) are generally measured just to the nearest half-foot. This means that the distance from the top of the grade stake to the flow line should be in whole multiples of 0.5 ft. So the cut distance could be 4.5 ft, 5.0 ft, 5.5 ft, and so on.

The grade stake must protrude above the ground an inch or two, so the distance from the top of the grade stake to the flow line elevation will be 4.35 ft plus the protrusion distance. A cut distance (from the grade stake) of 4.5 ft would place the top of the stake about 0.1 ft above the ground, which is adequate.

The corresponding ground stake rod reading will be less than 5.1 ft. If the grade stake is set for a cut of 4.5 ft, the grade stake rod reading will decrease and will be 5.1 ft − (4.5 ft − 4.35 ft) = 4.95 ft.

The stake is driven so that the rod reading is 4.95. The top of the stake is marked with blue keel (to indicate a grade stake), and the guard stake is marked $\boxed{\text{C } 4'6''}$.

The answer is (C).

16. The total station is normally set up directly over a control point. If the instrument is set up in an out-of-harm's-way location (sometimes referred to as *free station* or *resection*) its positioning should be validated by shooting at least two other offsite control points and at least one control point on the jobsite.

The answer is (D).

82 Building Codes and Materials Testing

PRACTICE PROBLEMS

1. Which of the following item(s) usually does NOT need to be remediated or abated prior to most building demolitions?

I.   asbestos

II.   bird droppings

III. rodent infestations

IV.   radon

V.   lead paint

VI.   mold

VII. fluorescent light bulbs

   (A) IV

   (B) I and VI

   (C) II and III

   (D) II, III, and VII

2. Which entity (organization, jurisdiction, agency, etc.) must issue a permit when construction will take place in or near regulated water features such as wetlands?

   (A) National Park Service (NPS)

   (B) Environmental Protection Agency (EPA)

   (C) Army Corp of Engineers (USACE)

   (D) National Oceanic and Atmospheric Administration (NOAA)

3. Which of the following are normal specifications for concrete?

I.   Classification: ASTM C150, type II, grey

II.   Aggregate type: fine and coarse per ASTM C33

III. Water: clean, potable, and not detrimental to concrete

IV.   28-day compressive strength: 1500 psi

V.   Maximum water to cement ratio: 0.45

VI.   Slump: 2–4 in

VII. Air entrainment: 5–15%

   (A) I, III, IV, and VI

   (B) II, III, VI, and VII

   (C) I, II, III, IV, and VII

   (D) I, II, III, V, and VI

4. Which division(s) of the Construction Specifications Institute's MasterFormat™ deal(s) with construction concrete?

   (A) division 2

   (B) division 3

   (C) divisions 4–7

   (D) divisions 13–15

5. A contractor at a remote job site notices the concrete batch just received contains lumps. What should the contractor do?

   (A) reject the load

   (B) add retardant

   (C) ask for the concrete to be rotated at high speed in the mixer truck drum

   (D) increase the duration of vibratory devices following placement

6. An activity is coded as 0793-31 62 13.16 using the Construction Specifications Institute's post-2004 MasterFormat. This coding represents

   (A) clearing and grubbing land

   (B) demolition of an existing structure

   (C) shoring an excavation

   (D) driving a deep foundation pile

Construction

7. Which of the following does NOT generally correspond to the term "substantially complete" as used with a newly-constructed building?

(A) The building is suitable for occupancy and the intended purposes.

(B) All inspections have been completed.

(C) All punch-list (punch-out) work has been completed.

(D) All systems are ready to use, including walkways, parking areas, grading, and landscaping.

8. When using the *International Building Code* (IBC), when must provisions contained in various appendices be adhered to?

(A) when the IBC refers to them

(B) when they are relevant to a geographical area

(C) when they contain "equal to or equivalent" provisions

(D) when they are specifically adopted by an agency

9. When does a copyrighted model building code (e.g., the IBC) enter the public domain so that copyright protection is no longer applicable?

(A) seven years after publication

(B) when it is adopted by a public agency (e.g., a state)

(C) when a new edition is published

(D) never

10. What word refers to the means by which disabled people use and enjoy the same features as able-bodied people?

(A) accessibility

(B) equivalent access

(C) fair use

(D) reasonable accommodation

11. "Means of egress" refers to the

(A) features used by people to enter and exit a structure

(B) design of windows, doors, elevators, and fire escapes

(C) visual, tactile, and audible signals used to guide people in an emergency

(D) path of travel to an exit, the exit itself, and the exit discharge

12. How might a building code differ from a fire code?

(A) A building code limits the geometry of the building envelope, while a fire code limits the number of people that can occupy the building.

(B) A building code specifies the number, size, and location of exits in the building, while the fire code specifies when the exits must be left unblocked.

(C) A building code determines what the building may be utilized for, while the fire code specifies the duration of such use.

(D) A building code specifies when fire sprinklers must be installed, while a fire code specifies what types of sprinklers may be used.

13. When may a consulting professional engineer (PE) sign an inspection report normally signed by an agency inspector?

(A) when the engineer uses the same codes and criteria that would be used by an inspector

(B) when the engineer is working for the agency

(C) when the engineer is being paid by the agency

(D) when the engineer is an agent of the agency

14. Hotels accommodating transient occupants are categorized by the *International Building Code* (IBC) as occupancy group

(A) R-1

(B) R-2

(C) R-3

(D) R-4

15. A child-care facility continuously (around the clock) accommodating more than five children is categorized by the *International Building Code* (IBC) as occupancy group

(A) I-2

(B) I-4

(C) R-3

(D) R-4

16. Most likely, the provisions of which code would be used to determine the diameter of conduit required to bring electrical service wires to an HVAC unit in a bathroom?

    (A) *National Electric Code* (NEC)

    (B) *International Plumbing Code* (IPC)

    (C) *International Mechanical Code* (IMC)

    (D) *International Building Code* (IBC)

17. For the purpose of electric service design, if a room contains only a shower, how would the room normally be classified?

    (A) washroom

    (B) bathroom

    (C) toilet room

    (D) shower room

18. Devices that reduce the possibility of electrocution in wet and damp areas are known as

    (A) circuit breakers

    (B) automatic trip sensors

    (C) alternate path fusible links

    (D) ground fault circuit interrupters

19. According to the *International Building Code* (IBC), what is the minimum compressive strength of concrete used in footings?

    (A) 1200 psi

    (B) 1500 psi

    (C) 2000 psi

    (D) 2500 psi

20. According to the *International Building Code* (IBC), what is the minimum thickness of a concrete slab that is cast directly on the ground?

    (A) $3\frac{1}{2}$ in

    (B) 4 in

    (C) $4\frac{1}{2}$ in

    (D) 6 in

SOLUTIONS

1. Asbestos, lead paint, and mold are well-known environmental hazards that require special handling. Less well-known hazards that must also be abated include significant accumulations of bird droppings (to remove histoplasmosis related bacteria), rodent infestations (to prevent the rodents from moving elsewhere), and fluorescent light bulbs which contain mercury. Radon is a gas that is emitted from some geological features (i.e., from the ground), and though hazardous, radon is not abatable.

The answer is (A).

2. Title 404 of the Clean Water Act specifies that the Army Corp of Engineers (USACE) issues permits and sets limits on construction water (runoff) turbidity, extent of wetland disturbance, and destruction of riparian vegetation and fisheries.

The answer is (C).

3. Concrete with cured strengths of less than 2000 psi might be acceptable for setting fence posts, but 2500–3500 psi concrete is typical for general construction. Air entrainment percentages will depend on the nature of the project, but a range of 5–15% is too large to ensure consistent performance. All of the other items are normal specifications for concrete.

The answer is (D).

4. In both pre- and post-2004 versions of MasterFormat, only division 3 of the Construction Specifications Institute's MasterFormat deals with concrete. Division 1 deals with general requirements, division 2 deals with site construction, and divisions 4–16 deal with architectural (not engineering) features.

The answer is (B).

5. Concrete that has begun to set up is beyond remediation. The contractor should reject the load.

The answer is (A).

6. The activity designation is translated as follows.

    project 793
    level 1 activity category 31
    level 2 activity category 62
    level 3 activity category 13

In the MasterFormat specifications, level 1 category 31 is earthwork, level 2 category 62 is driven piles, and level 3 category 13 is concrete piles.

The answer is (D).

Construction

7. A substantially completed building does not need to be "complete and perfect," such that all punch-list work has been completed.

The answer is (C).

8. Model codes become regulatory when they are adopted by a city, county, state, or other agency. Appendices must specifically be adopted to be enforceable.

The answer is (D).

9. When public agencies (e.g., states) adopt model building codes and designate them as actual laws, the laws enter public domain. The model codes themselves, prior to their adoption as laws, are not in the public domain, and the authoring entity retains copyright to the model code itself.

The answer is (B).

10. "Accessibility" refers to the accommodation of disabled people in structures, building entrances, parking areas, elevators, and restrooms.

The answer is (A).

11. "Egress" refers only to exiting. "Means of egress" refers to the path of travel to an exit, the exit itself, and the exit discharge (i.e., the path to a safe area outside).

The answer is (D).

12. A building code establishes design standards based on intended occupancy and use. Building codes implicitly limit use (activities and numbers of people). Local ordinances limit hours of use. Fire codes may contain design details, but design features are generally elected by architects and engineers. Fire codes specify how design features are to be used, maintained, and tested.

The answer is (B).

13. An engineer must be acting as an agent of the agency (city, county, etc.) in order to approve construction on the agency's behalf. An engineer may be working for an agency in various capacities (e.g., plan checking) but not specifically allowed to sign off on inspections. If an engineer is acting under the direction of the agency, with or without payment, the engineer may sign inspection documents.

The answer is (D).

14. Per IBC Sec. 310.1, the occupancy group encompassing hotels, motels, and boarding houses is R-1.

The answer is (A).

15. Per IBC Sec. 308.3, the occupancy group is I-2. The category would be R-3 for five or fewer children.

The answer is (A).

16. NFPA's *National Electric Code* (NEC) specifies the numbers and sizes of conductors, as well as the types of conduit that are used to bring them to equipment. The *International Building Code* (IBC) refers to the NEC, but does not contain actual provisions.

The answer is (A).

17. The room would normally be considered a bathroom. The *National Electric Code* (NEC) definition of bathroom includes the words, "a urinal, a shower, a bidet, or similar plumbing fixtures."

The answer is (B).

18. Ground fault interrupters (GFIs) or ground fault circuit interrupters (GFCIs) protect against electrical shock in kitchens, bathrooms, and other damp areas and are required by code.

The answer is (D).

19. According to IBC Sec. 1807.1.6.2, the minimum strength of concrete used in footings is 2500 psi.

The answer is (D).

20. According to IBC Sec. 1910.1, the minimum thickness of slabs on grade is $3^1/_2$ in.

The answer is (A).

Construction

83 Construction and Jobsite Safety

PRACTICE PROBLEMS

1. During a particular year being audited, a company experienced 15 injuries and illnesses. 400,000 hours were worked by all employees during the year. What is the injury incidence rate for safety reporting?

(A) 3.75

(B) 7.50

(C) 12.3

(D) 26.7

2. After performing a site survey, a soils engineer reports that the site soil seems to "fall apart just by looking at it." This soil is probably

(A) type A

(B) type B

(C) type C

(D) type D

3. What does the crane hand signal mean?

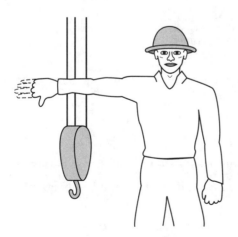

(A) lower the boom

(B) lower the load

(C) raise the boom and lower the load

(D) lower the boom and raise the load

4. Which of the following written certifications is NOT required before steel erection can begin on a construction site?

(A) the foundation concrete has attained 75% of its design strength

(B) pedestrian traffic on the construction site will be controlled

(C) access roads within the site are suitable for derricks and cranes

(D) traffic on roads approaching the site has been rerouted or diverted for crane and derrick approach.

5. At a construction site adjacent to a highway, who is required to wear safety vests or other highly visible clothing?

(A) only flaggers exposed to public vehicular traffic

(B) all flaggers exposed to all traffic

(C) all construction workers exposed to all traffic

(D) all personnel, including supervisory staff and visitors

6. In addition to protective goggles, what personal protective equipment is required when welding pipe in a deep trench?

(A) fall protection harness

(B) hard hat

(C) respirator

(D) hearing protection

7. When may plywood be substituted for upright planks as structural members in shoring systems?

(A) when the duration of use is less than eight hours

(B) when the thickness is greater than $3/4$ in and the plywood is an exterior-glue variety

(C) when the plywood carries a "structural" designation

(D) never

8. Excavations in a type C soil must be sloped no steeper than

    (A) $^1/_2$H:1V

    (B) 1H:1V

    (C) $1^1/_2$H:1V

    (D) 2H:1V

9. What is the minimum distance from the edge of a trench that spoils (excavation material) can be stored?

    (A) 1 ft

    (B) 2 ft

    (C) 3 ft

    (D) 5 ft

10. Factors to consider when designing a protective support system for deep trenching include

    (A) soil classification

    (B) OSHA's analysis of site conditions

    (C) water content of soil

    (D) adjacency of swamps, lakes, and reservoirs

11. Which of the following would NOT normally be considered by OSHA to be a confined space?

    (A) trench

    (B) sewer trunk line

    (C) manhole

    (D) pit

12. For trenching in any type of soil when there is sloping toward the cut, how far must support or shield systems extend above the top of the vertical cut side?

    (A) 6 in

    (B) 12 in

    (C) 18 in

    (D) 24 in

13. Remotely located, unattended excavations may be left open (unfilled)

    (A) when no one is working in the area

    (B) when the excavation has been barricaded or covered

    (C) when a watcher is present

    (D) never

14. Water is observed freely seeping from the exposed sides of a cut. The soil type is most likely to be

    (A) type A

    (B) type B

    (C) type C

    (D) type D

15. Excavation site emergency rescue equipment (respirators, harnesses and lifelines, and basket stretchers) is required by OSHA when

    (A) hazardous atmospheres exist

    (B) hazardous atmospheres have existed previously

    (C) the atmosphere has not been tested

    (D) there is no competent person

16. The first person to enter an excavation should normally be

    (A) a competent person

    (B) any trained person

    (C) a newly hired person

    (D) a supervisor

17. Construction workers who enter a trench deeper than 8 ft must always wear

    (A) a respirator

    (B) an extraction harness

    (C) a hard hat

    (D) eye protection

18. Acceptable means of entering and exiting trenches do NOT include

    (A) ramps

    (B) ladders

    (C) derricks used with fall arrest harnesses

    (D) jumping down into and climbing out

19. In order to enter an excavation site without atmosphere-supplying respirators, the minimum oxygen content (by volume) must be

    (A) 16.9%

    (B) 17.0%

    (C) 19.5%

    (D) 20.5%

Construction

20. Which of the following does NOT constitute a protective system for excavation?

(A) sloping or benching sides of the trench

(B) trench shields

(C) shoring

(D) ladders

21. Which of the following is NOT normally necessary prior to excavating a trench with a backhoe?

(A) contacting the utility company or property owner

(B) ensuring underground utilities are protected and supported

(C) determining the soil type by a competent person

(D) notifying the local OSHA office

22. How often should an excavation be inspected by a competent person?

(A) hourly

(B) daily

(C) weekly

(D) continuously

23. The maximum allowable slope for an excavation in type B soil less than 20 ft deep is most nearly

(A) 30°

(B) 45°

(C) 60°

(D) 90°

24. When a competent person performs an inspection of an excavation site, the inspection does NOT need to include checking for

(A) likelihood of collapse and cave-in

(B) hazardous atmosphere

(C) hazardous condition

(D) first-responder response time

25. A sliding trench shield is moved progressively through a trench as work follows a new sewer trunk line. When are employees allowed inside a trench shield being moved?

(A) when the shield is being moved by a crane

(B) when the shield is moving horizontally through the trench without use of motorized or mechanically assisted devices

(C) when each employee is attached to the shield

(D) never

26. A construction worker may enter an excavation without a protective system when the excavation is

(A) entirely in soil type A

(B) less than 5 ft deep with no indication of a potential cave-in

(C) free of standing water

(D) none of the above

27. A slot trench 4 ft deep must have an exit every

(A) 20 ft

(B) 25 ft

(C) 50 ft

(D) none required

28. Testing for flammable gas is NOT normally required prior to entering an excavation if the excavation is adjacent to a

(A) landfill site

(B) water reservoir

(C) gas station

(D) swampland

29. How far should ladders extend above the edge of an excavation greater than 4 ft deep?

(A) 1 ft

(B) 2 ft

(C) 3 ft

(D) 4 ft

30. What is the approximate minimum top (surface) width of a trench in type A soil that is 10 ft deep and 4 ft wide at the bottom?

(A) 12 ft

(B) 14 ft

(C) 19 ft

(D) 24 ft

31. The *four-foot rule*

(A) refers to a means of escape from an excavation

(B) says that a means of escape must be provided if an excavation is more than 4 ft deep

(C) dictates an exit must be within 25 ft of every worker

(D) all of the above

Construction

32. Approximately how much weight can a heavy-duty scaffold plank support?

(A) 25 psf

(B) 50 psf

(C) 75 psf

(D) 100 psf

33. An uncleated 10 ft long scaffold plank must extend over its support at least

(A) 6 in

(B) 8 in

(C) 10 in

(D) 12 in

34. What is the maximum scaffold height above which employees must be protected from falling by guardrails or fall arrest systems?

(A) 4 ft

(B) 6 ft

(C) 10 ft

(D) 12 ft

35. How close to an overhead 110 V power line may a scaffold be erected?

(A) 3 ft

(B) 5 ft

(C) 8 ft

(D) 10 ft

36. Which of the following may be placed on the ground to support scaffold frame legs?

(A) 2 in × 10 in pieces of lumber

(B) 8 in concrete masonry units

(C) 12 in × 12 in pieces of plywood

(D) none of the above

37. When a floor opening is protected by perimeter guardrail, wire rope cables, the cables must be flagged every

(A) 5 ft

(B) 6 ft

(C) 8 ft

(D) 10 ft

38. What distance may the upper surface of a guardrail be above the walking/working surface?

(A) 36–39 in

(B) 36–45 in

(C) 38–42 in

(D) 39–45 in

39. Hole covers must be installed on a floor opening without guardrails

(A) when the hole is not monitored by another employee on the same floor

(B) when work on that level is completed

(C) when the level with the hole is occupied

(D) by the end of the shift

40. What is the maximum height that a fall can be arrested by a net?

(A) 6 ft

(B) 10 ft

(C) 15 ft

(D) 30 ft

41. How many construction workers can tie onto a vertical lifeline NOT inside an elevator shaft?

(A) 1

(B) 2

(C) 3

(D) as many as there are attachment points

42. The ratio of loading on a floor hole cover to its capacity must not exceed

(A) 0.20

(B) 0.25

(C) 0.33

(D) 0.50

43. Wall openings must have guardrails if they open onto a drop of 6 ft or more, and if the lower edge is less than what distance above the floor level?

(A) 30 in

(B) 36 in

(C) 39 in

(D) 42 in

Construction

44. Employees working on concrete formwork must be protected from falling if they are higher than

(A) 6 ft

(B) 10 ft

(C) 16 ft

(D) 24 ft

45. What load (per person) must a fall anchorage be able to support?

(A) 2500 lbf

(B) 5000 lbf

(C) 7500 lbf

(D) 10,000 lbf

46. Anchorages for fall protection should be set higher than

(A) the floor

(B) waist height

(C) head height

(D) the D-ring location

47. Ends of reinforcing steel projecting horizontally must be protected by caps or covers when the

(A) rebar is no. 4 or smaller

(B) rebar extends 24 in or more

(C) location is high enough that a falling or tripping construction worker could become impaled upon it

(D) ends will be exposed longer during periods of darkness

48. Identification tags on alloy steel chain slings and synthetic slings are NOT required to show

(A) web material

(B) rated capacity

(C) grade

(D) date of manufacture

49. When two flaggers direct traffic approaching in two opposing directions, all of the following are required EXCEPT

(A) a red flag or sign

(B) a high-visibility vest or shirt

(C) advance traffic warning signs

(D) communication devices

50. The swing radius of a crane's cab and counterweight must be

(A) barricaded

(B) within the tracks of the machine

(C) wider than the tracks of the machine

(D) patrolled outside by the oiler

51. The OSHA threshold limit value (TLV) by volume for eight hours of carbon monoxide exposure is

(A) 10 ppm

(B) 25 ppm

(C) 35 ppm

(D) 50 ppm

52. While working, equipment such as cranes, forklifts, backhoes, and scaffolding must maintain a minimum clearance from overhead power lines of at least

(A) 6 ft

(B) 10 ft

(C) 15 ft

(D) 20 ft

53. Hand signals between crane operators, oilers, and ground personnel

(A) must be agreed upon in advance

(B) must be agreed to by the project administrator

(C) are specified by the crane industry

(D) are specified by OSHA and/or ANSI

54. Which of the following will NOT form a flammable or explosive mixture with air?

(A) fly ash or portland cement dust

(B) coal dust

(C) wheat dust

(D) sawdust

55. Portland cement

(A) may cause chemical burns

(B) is toxic in cement-rich mixtures

(C) remains chemically active for years

(D) can be recovered by pulverizing concrete

Construction

56. Which of the following is NOT true of flashing arrow signs used to channel traffic around construction zones?

(A) Lights should be dimmed during the day and brightened during the night.

(B) Lights should not be glaring to drivers of oncoming truck traffic.

(C) Lights should be visible from a minimum or greater distance.

(D) Lights powered by solar panels require more careful aiming than generator-powered lights.

57. A brand of expandable earplugs claims an NRR of 35. If a worker is exposed to sound with 100 dBA of noise, what will be the OSHA equivalent exposure?

(A) 14 dB

(B) 52 dB

(C) 65 dB

(D) 86 dB

58. A worker is exposed to 92 dBA sound for 8 hours. What is the worker's time-weighted average exposure level?

(A) 92 dB

(B) 94 dB

(C) 96 dB

(D) 98 dB

59. What OSHA regulation governs cranes in construction?

(A) 1910.180

(B) 1910.184

(C) 1926.550

(D) 1926.556

60. What is the boom angle of a truck-mounted mobile hydraulic crane?

(A) the vertical angle between the horizontal and the centerline of the boom

(B) the vertical angle between the vertical and the centerline of the boom

(C) the horizontal angle between the vertical plane of the boom and the vertical plane of the truck tail

(D) the horizontal angle between the vertical plane of the boom and the vertical plane of containing the cab

61. How many randomly-located individual strands of a hoisting rope may be broken before the rope must be replaced?

(A) 2

(B) 4

(C) 5

(D) 6

62. In regards to cranes and derricks, what is generally NOT included in the definition of a critical lift?

(A) a lift that exceeds 70% of the crane's capacity

(B) a lift that is on the critical path of the lift schedule

(C) a lift that is close to power lines

(D) a lift that moves high-value materials

63. A crane's capacity generally does NOT consider

(A) use of outriggers

(B) tire pressure

(C) counterweighting

(D) wind speed and air temperature

64. What is NOT generally a reason for lubricating wire hoisting rope?

(A) to reduce internal friction between the individual strands

(B) to reduce internal friction between the strand bundles and the core

(C) to reduce external corrosion

(D) to reduce external friction between the rope and sheaves

65. For a mobile hydraulic crane, the load radius is the distance from the vertical centerline of the load to the vertical centerline of the

(A) boom pin

(B) jib pin

(C) center pin

(D) centroid of outriggers, tires, or supports

Construction

66. Complete the following sentence: A spreader bar is used with a crane...

(A) under the outriggers to distribute support load over a larger area

(B) to distribute the load between two cranes in a multi-crane lift

(C) to distribute the load between multiple rope segments when lifting wide objects

(D) by the crane rigger to straighten tangled hoisting rope

67. What is the danger of eccentric boom reeving?

(A) boom twisting

(B) decreased jib capacity

(C) increased wire rope wear

(D) increased sheave wear

68. What does "duty cycle operations" ("duty cycle work") mean for crane lifting jobs?

(A) shifts of no more than two hours between rest breaks

(B) lifts that are known to be 50% or less than maximum crane capacity

(C) repetitive lifts that are identical or nearly so

(D) lifts that are limited by hoist motor temperature rise

SOLUTIONS

1. The rate is

$$\frac{\text{(no. of injuries)}(200{,}000\text{ hr})}{\text{number of hours worked}} = \frac{(15)(200{,}000\text{ hr})}{400{,}000\text{ hr}}$$
$$= \boxed{7.50}$$

The answer is (B).

2. Type C soils have the lowest unconfined compressive strength of OSHA's A-B-C rating system [OSHA 29 CFR 1926 Subpart P, App. A], and tend to easily fall apart.

The answer is (C).

3. Hand signals in the crane industry are standardized. This signal means to lower the boom and raise the load. If the fingers were not shown (a fist was being made), the signal would mean to lower the boom only [OSHA 29 CFR 1926.550(a)(4)].

The answer is (D).

4. OSHA describes the written certifications that the controlling contractor must provide to the steel erector before erection may begin. However, the controlling contractor's responsibility ends at the boundaries of the jobsite. Therefore, rerouting or diverting traffic is not required [OSHA 29 CFR 1926.752].

The answer is (D).

5. Only flaggers who work outside of the construction zone, as delineated by cones and barricades, are required by OSHA to wear highly visible clothing. Within the construction zone, high-visibility apparel is not considered to be personal protective equipment (PPE) [OSHA 29 CFR 1926.201].

The answer is (A).

6. OSHA does not normally consider an open trench to be a confined space. Although there are some combinations of base metals and rods that produce toxic fumes and would require respirators, these are rare in normal construction work, so welding in a trench would not normally result in the accumulation of toxic gases or the need for a respirator. A hard hat is always required [OSHA 29 CFR 1926.100].

The answer is (B).

Construction

7. Plywood may never be used as a structural member in shoring. It may be included between the soil and the uprights in order to distribute the load [OSHA 29 CFR 1926.652(c)(1) and CFR 1926 Subpart P, App. C].

The answer is (D).

8. A type C soil must be sloped no steeper than $1^1/_2$H:1V [OSHA 29 CFR 1926 Subpart P, App. B].

The answer is (C).

9. Spoils can be stored a minimum distance of 2 ft from the edge of a trench [OSHA 29 CFR 1926.651(j)(2)].

The answer is (B).

10. OSHA does not provide site analysis services. The contractor's competent person would do so, taking many factors implicitly into consideration. However, the OSHA regulations are primarily organized by soil classification types A, B, and C and should be used when designing a protective support system [OSHA 29 CFR 1926.651(k)(1)].

The answer is (A).

11. OSHA does not automatically consider trenches to be confined spaces. However, frequent testing for oxygen content and toxic gases is required, and respirators may still be required [OSHA 29 CFR 1926.651(g) and 1926.146(b)].

The answer is (A).

12. Support or shield systems must extend 18 in above the top of the vertical cut side when there is sloping toward the cut [OSHA 29 CFR 1926 Subpart P, App. B, B-1.1, B-1.2, and B-1.3].

The answer is (C).

13. Remotely located, unattended excavations may be left open (unfilled) when the excavation has been barricaded or covered [OSHA 29 CFR 1926.501(b)(1)].

The answer is (B).

14. Type C soil is most likely to allow water to freely seep from the exposed sides of a cut [OSHA 29 CFR 1926 Subpart P, App. A(b)].

The answer is (C).

15. Per OSHA, emergency rescue equipment is required when a hazardous atmosphere exists or can be reasonably expected to exist. Options (C) and (D) are OSHA code violations [OSHA 29 CFR 1926.651(g)(2)(i)].

The answer is (A).

16. The first person to enter an excavation must be a *trained person*. The responsibilities of a *competent person* (as defined by OSHA) include soil analysis, assigning employee duties, and designing protective systems. However, the first person to enter an excavation does not necessarily have to be a competent person.

The answer is (B).

17. Construction workers must always wear a hard hat. An extraction harness is required for employees entering bell-bottomed holes and pier shafts [OSHA 29 CFR 1926.100 and 1926.651(g)(2)(ii)].

The answer is (C).

18. Hoisting by derricks with fall arrest harnesses is not an acceptable method for entering and exiting trenches. It is acceptable to jump down into and climb out of trenches less than or equal to 4 ft deep, as well as to use ramps and ladders [OSHA 29 CFR 1926.651(c)(2)].

The answer is (C).

19. The oxygen content of an excavation site must be 19.5% in order to enter it without atmosphere-supplying respirators [OSHA 29 CFR 1926.651(g)(1)(i)].

The answer is (C).

20. A protective system for excavation protects against collapse and excavation. Ladders do not offer such protection [OSHA 29 CFR 1926.652].

The answer is (D).

21. OSHA is not notified prior to excavating [OSHA 29 CFR 1926.651(b)].

The answer is (D).

22. An excavation should be inspected by a competent person daily [OSHA 29 CFR 1926.651(k)(1)].

The answer is (B).

23. For a type B soil, the maximum allowable slope is 45° [OSHA 29 CFR 1926 Subpart P, App. B, Table B-1].

The answer is (B).

24. The daily inspection by a competent person does not have to determine how fast emergency personnel can arrive at the site [OSHA 29 CFR 1926.32(f) and 1926.651(k)(1)].

The answer is (D).

25. Employees are never allowed inside a trench shield when it is being moved [OSHA 29 CFR 1926.652(g)(1)(iv)].

The answer is (D).

26. A construction worker may enter an excavation without a protective system when the excavation is less than 5 ft deep with no indication of a potential cave-in [OSHA 29 CFR 1926.652(a)(1)(ii)].

The answer is (B).

27. A slot excavation 4 ft deep must not require workers to travel more than 25 ft to an exit. Therefore, the excavation must have an exit every 50 ft [OSHA 29 CFR 1926.651(c)(2)].

The answer is (C).

28. Landfills and swamplands both generate methane gas, and gas stations may have volatile hydrocarbon vapor build up. Although an adjacent reservoir may be a concern for seepage, it would not trigger a test for flammable gas content [OSHA 29 CFR 1926.651(g)(1)(ii)].

The answer is (B).

29. Ladders should extend 3 ft above the edge of an excavation greater than 4 ft deep [OSHA 29 CFR 1926.1053(b)(1)].

The answer is (C).

30. According to OSHA 29 CFR 1926 Subpart P, App. B, the maximum slope for type A soil is $3/4$ ft horizontally for every 1 ft vertically. For a depth of 10 ft, the sloped side would extend $(10 \text{ ft})(3/4 \text{ ft/ft}) = 7.5$ ft. The total surface width is

$$w = 7.5 \text{ ft} + 4 \text{ ft} + 7.5 \text{ ft} = \boxed{19 \text{ ft}}$$

The answer is (C).

31. The four-foot rule says that a means of escape must be provided if an excavation is more than 4 ft deep and that an exit must be within 25 ft of every worker [OSHA 29 CFR 1926.651(c)(2)].

The answer is (D).

32. Per OSHA, a scaffold plank with a heavy-duty rating can support 75 psf uniformly distributed over its entire surface [OSHA 29 CFR 1926 Subpart L, App. A (1)(c)].

The answer is (C).

33. Scaffold planks 10 ft long must extend at least 6 in and not more than 12 in over their supports [OSHA 29 CFR 1926.451(b)(4) and 1926(b)(5)(i)].

The answer is (A).

34. The maximum scaffold height above which employees must be protected from falling by guardrails or fall arrest systems is 10 ft [OSHA 29 CFR 1926.451(g)(1)].

The answer is (C).

35. A scaffold may be erected as close as 3 ft to a 110 V overhead power line [OSHA 29 CFR 1926.451(f)(6)].

The answer is (A).

36. Scaffold legs must be on base plates that sit on wooden planks ("mud sills") running between them. Portland cement, asphalt concrete, and other solid surfaces that prevent settling or displacement may be used without mudsills [OSHA 29 CFR 1926.451(c)(2)].

The answer is (D).

37. The cables must be flagged every 6 ft [OSHA 29 CFR 1926.502(f)(2)(i)].

The answer is (B).

38. The upper surface of a guardrail must be at least 42 in, plus or minus 3 in [OSHA 29 CFR 1926.502(b)(1) and 502(g)(3)(i)].

The answer is (D).

39. A floor opening does not need guardrails when it is monitored by an employee. When a cover is not in place, a floor hole must be constantly attended to by someone, or it must be protected by a removable standard railing [OSHA 29 CFR 1926.502(h)].

The answer is (A).

Construction

40. 30 ft is the maximum height a fall can be arrested by a net [OSHA 29 CFR 1926.502(c)(1)].

The answer is (D).

41. Multiple employees may hook up to a horizontal lifeline. However, each employee must have a separate vertical lifeline unless they are working in an elevator shaft and other conditions have been met [OSHA 29 CFR 1926.502(d)(10)(i) and (ii)].

The answer is (A).

42. Hole covers must be capable of supporting twice the expected weight of workers and equipment [OSHA 29 CFR 1926.501(i)(1)].

The answer is (D).

43. If a wall opening opens onto a drop of 6 ft or more and its lower edge is less than 39 in above the floor level, it must have a guardrail [OSHA 29 CFR 1926.501(b)(14)].

The answer is (C).

44. Employees working on concrete formwork must be protected from falling if they are higher than 6 ft [OSHA 29 CFR 1926.501(b)(1)].

The answer is (A).

45. A fall anchorage must be able to support a 5000 lbf load per person [OSHA 29 CFR 1926.502(d)(15)].

The answer is (B).

46. Anchorages should be as high as possible to limit the falling distance. Anchorages should be placed center of the back, shoulder height, or above head level. Waist belts are no longer permitted, so waist-level attachment points are also not allowed [OSHA 29 CFR 1910.66 App. C and 1926.502(d)(17)].

The answer is (C).

47. Ends of reinforcing steel projecting horizontally must be protected by caps or covers when the location is high enough that a falling or tripping construction worker could become impaled upon it [OSHA 29 CFR 1926.701(b)].

The answer is (C).

48. Identification tags on alloy steel chain slings and synthetic slings are not required to show the date of manufacture [OSHA 29 CFR 1910.184(e)(1) or 1926.251(b)(1)].

The answer is (D).

49. Communication devices are not required when two flaggers are directing traffic in two directions [OSHA 29 CFR 1926.201(a) and MUTCD 6C-5(a)].

The answer is (D).

50. The swing radius must be barricaded to prevent entrance into a dangerous area during crane operation. Only the oiler is permitted to enter the protected area in order to perform his/her duties [OSHA 29 CFR 1926.550(a)(9)]. However, the oiler does not have to patrol the area.

The answer is (A).

51. The OSHA threshold limit value (TLV) as well as the PEL by volume for carbon monoxide is 50 ppm [OSHA 1910.1000 Table Z-1] The NIOSH limit is 35 ppm and the ACGIH limit is 25 ppm.

The answer is (D).

52. Cranes, forklifts, backhoes, and scaffolding must maintain a minimum clearance of at least 10 ft from overhead power lines [OSHA 29 CFR 1926.550(a)(15)(i)].

The answer is (B).

53. Hand signals between crane operators, oilers, and ground personnel are specified by OSHA and/or ANSI [OSHA 29 CFR 1926.550(a)(4)].

The answer is (D).

54. Dust explosions require a combustible material. Fly ash and cement dust will not form a flammable or explosive mixture when exposed to air.

The answer is (A).

55. Portland cement is caustic, abrasive, and hygroscopic (i.e., absorbs moisture) even without the addition of lime in mortar mixes. Employees exposed to wet concrete should be protected in order to avoid cement burns. Once hardened, cement becomes somewhat chemically inert as well as inaccessible (i.e., unlikely to be in contact with skin for prolonged periods).

The answer is (A).

56. Lights should be brightened during the day and dimmed at night (so as not to blind approaching drivers) [MUTCD 6F-28].

The answer is (A).

Construction

57. Use Eq. 83.7, the OSHA formula for an A-weighted measurement.

$$\Delta L = \frac{NRR - 7}{2} = \frac{35 - 7}{2} = 14 \text{ dB}$$

The exposure is

$$100 \text{ dB} - 14 \text{ dB} = 86 \text{ dB}$$

The answer is (D).

58. The reference duration can be calculated or determined from Table 83.4. From Table 83.4, the reference duration for 92 dBA is 6 hr. The dose is

$$D = \frac{8 \text{ hr}}{6 \text{ hr}} \times 100\% = 133\%$$

Using Eq. 83.6, the TWA is

$$\begin{aligned} TWA &= 90 \text{ dB} + 16.61 \log_{10} \frac{D}{100\%} \\ &= 90 \text{ dB} + 16.61 \log_{10} \frac{133\%}{100\%} \\ &= 92.06 \text{ dB} \quad (92 \text{ dB}) \end{aligned}$$

The answer is (A).

59. OSHA 29 CFR 1926.550 governs cranes in construction.

The answer is (C).

60. The boom angle is the vertical angle between the horizontal and the centerline of the boom, defined as the line between the sheaves and heel pins.

The answer is (A).

61. Wire rope is to be taken out of service when six randomly distributed wires are broken [OSHA 29 CFR 1926.550(a)(7)(i)].

The answer is (D).

62. There are several definitions of a critical lift used in the construction industry. NIOSH defines a critical lift as one with the hoisted load approaching the crane's maximum capacity (70% to 90%); lifts involving two or more cranes; personnel being hoisted; and special hazards such as lifts within an industrial plant, cranes on floating barges, loads lifted close to power-lines, and lifts in high winds or with other adverse environmental conditions present.

The U.S. Army Corps of Engineers defines a critical lift to include lifts made out of the view of the operator (blind picks), and lifts involving nonroutine or technically difficult rigging arrangements. The Department of Energy further defines a critical lift to include lifting high value, unique, irreplaceable, hazardous, explosive, or radioactive loads.

The Construction Safety Association of Ontario includes lifts in congested areas, lifts involving turning or flipping the load where "shock loading" and/or "side loading" may occur, lifts where the load weight is not known, lifts in poor soil or unknown ground condition, and lifts involving unstable pieces.

The critical path of a project is not part of the definition of critical lift.

The answer is (B).

63. When supported by its outriggers, a crane is lifted off of its tires. However, for pick-and-carry operations, the tire's should be inflated to the manufacturer's specifications. Wind speed is a factor in cancelling lifts. However, wind speed and air temperature are not generally noted on load charts.

The answer is (D).

64. Wire rope is lubricated to reduce corrosion and internal friction. Since wire rope does not slide over a fixed sheave, the forces on the rope and sheave are compressive.

The answer is (D).

65. The radius is always the distance from the center of gravity of the load to the center of rotation, which is almost always the center pin.

The answer is (C).

66. A spreader bar is used to distribute the lifting force to ends or edges of long or wide loads.

The answer is (C).

67. Eccentric reeving occurs when the hoist line is not centered over the boom tip. This causes torque (twisting) in the boom.

The answer is (A).

68. In duty cycle operations, the loads are known and/or can be maintained safely below the rated capacity for the crane. The phrase "safely below the rated capacity of the crane" usually means the total load being handled does not exceed 75% of the rated capacity of the crane. The lifting tasks needed to support duty cycle work should involve loads of less than 50% of the rated capacity of the crane.

The answer is (B).

Construction

84 Electrical Systems and Equipment

PRACTICE PROBLEMS

1. The full-load phase current drawn by a 440 V (rms) 60 hp (total) three-phase induction motor having a full-load efficiency of 86% and a full-load power factor of 76% is most nearly

(A) 52 A

(B) 78 A

(C) 160 A

(D) 700 A

2. A 200 hp, three-phase, four-pole, 60 Hz, 440 V (rms) squirrel-cage induction motor operates at full load with an efficiency of 85%, power factor of 91%, and 3% slip.

(a) The speed is most nearly

(A) 1500 rpm

(B) 1700 rpm

(C) 2300 rpm

(D) 5800 rpm

(b) The torque developed is most nearly

(A) 180 ft-lbf

(B) 450 ft-lbf

(C) 600 ft-lbf

(D) 690 ft-lbf

(c) The line current is most nearly

(A) 34 A

(B) 150 A

(C) 250 A

(D) 280 A

3. A factory's induction motor load draws 550 kW at 82% power factor. What is most nearly the size of an additional synchronous motor required to produce 250 hp and raise the power factor to 95%? The line voltage is 220 V (rms).

(A) 140 kVA

(B) 230 kVA

(C) 240 kVA

(D) 330 kVA

4. The nameplate of an induction motor lists 960 rpm as the full-load speed. The frequency the motor was designed for is most nearly

(A) 24 Hz

(B) 34 Hz

(C) 48 Hz

(D) 50 Hz

SOLUTIONS

1. From Eq. 84.67, the full-load phase current is

$$I_p = \frac{P_p}{\eta V_p \cos\phi} = \frac{\left(\frac{60\text{ hp}}{3}\right)\left(0.7457 \times 10^3\ \frac{\text{W}}{\text{hp}}\right)}{(0.86)(440\text{ V})(0.76)}$$

$$= \boxed{51.86\text{ A}\quad(52\text{ A})}$$

The answer is (A).

2. (a) From Eq. 84.76 and Eq. 84.78,

$$n_r = n_{\text{synchronous}}(1 - s) = \left(\frac{120f}{p}\right)(1 - s)$$

$$= \left(\frac{(2)\left(60\ \frac{\text{sec}}{\text{min}}\right)(60\text{ Hz})}{4}\right)(1 - 0.03)$$

$$= \boxed{1746\text{ rpm}\quad(1700\text{ rpm})}$$

The answer is (B).

(b) From Eq. 84.71, the torque developed is

$$T = \frac{5252P}{n_r} = \frac{\left(550\ \frac{\text{ft-lbf}}{\text{hp-sec}}\right)\left(60\ \frac{\text{sec}}{\text{min}}\right)(200\text{ hp})}{(2\pi)\left(1746\ \frac{\text{rev}}{\text{min}}\right)}$$

$$= \boxed{602\text{ ft-lbf}\quad(600\text{ ft-lbf})}$$

The answer is (C).

(c) From Eq. 84.67, the line current is

$$I_l = \frac{P_t}{\sqrt{3}\eta V_l \cos\phi}$$

$$= \frac{(200\text{ hp})\left(0.7457 \times 10^3\ \frac{\text{W}}{\text{hp}}\right)}{(\sqrt{3})(0.85)(440\text{ V})(0.91)}$$

$$= \boxed{253\text{ A}\quad(250\text{ A})}$$

The answer is (C).

3. Draw the power triangle.

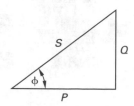

The original power angle is

$$\phi_i = \arccos 0.82 = 34.92°$$

$$P_1 = 550\text{ kW}$$

From Eq. 84.56,

$$Q_i = P_1 \tan\phi_i = (550\text{ kW})(\tan 34.92°) = 384.0\text{ kVAR}$$

The new conditions are

$$P_2 = (250\text{ hp})\left(0.7457\ \frac{\text{kW}}{\text{hp}}\right) = 186.4\text{ kW}$$

$$\phi_f = \arccos 0.95 = 18.19°$$

Since both motors perform real work,

$$P_f = P_1 + P_2 = 550\text{ kW} + 186.4\text{ kW} = 736.4\text{ kW}$$

The new reactive power is

$$Q_f = P_f \tan\phi_f = (736.4\text{ kW})(\tan 18.19°)$$

$$= 242.0\text{ kVAR}$$

The change in reactive power is

$$\Delta Q = 384.0\text{ kVAR} - 242.0\text{ kVAR} = 142\text{ kVAR}$$

Synchronous motors used for power factor correction are rated by apparent power.

$$S = \sqrt{(\Delta P)^2 + (\Delta Q)^2}$$

$$= \sqrt{(186.4\text{ kW})^2 + (142\text{ kVAR})^2}$$

$$= \boxed{234.3\text{ kVA}\quad(230\text{ kVA})}$$

The answer is (B).

4. From Eq. 84.76 and Eq. 84.78, the frequency is

$$f = \frac{pn_{\text{synchronous}}}{120} = \frac{pn_{\text{actual}}}{(2)\left(60 \; \frac{\text{sec}}{\text{min}}\right)(1-s)}$$

The slip and number of poles are unknown. Assume $s = 0$ and $p = 4$.

$$f = \frac{(4)\left(960 \; \frac{\text{rev}}{\text{min}}\right)}{(2)\left(60 \; \frac{\text{sec}}{\text{min}}\right)(1-0)} = 32 \text{ Hz}$$

32 Hz is not close to any frequency in commercial use. Try $p = 6$.

$$f = \frac{(6)\left(960 \; \frac{\text{rev}}{\text{min}}\right)}{(2)\left(60 \; \frac{\text{sec}}{\text{min}}\right)(1-0)} = 48 \text{ Hz}$$

With a 4% slip, $f = 50$ Hz.

$$\boxed{50 \text{ Hz (European)}}$$

The answer is (D).

85 Instrumentation and Measurements

PRACTICE PROBLEMS

1. The temperature of a steel girder during a fire test is measured with a type 404 platinum RTD and simple two-wire bridge, as shown. The characteristics of the RTD are: resistance, R, 100 Ω at 0°C; temperature coefficient, α, 0.00385 1/°C. The values of the bridge resistances are $R_1 = 1000 \ \Omega$, and $R_3 = 1000 \ \Omega$. When the meter is nulled out during a test, $R_2 = 376 \ \Omega$. The lead resistances are each 100 Ω. Most nearly, what is the temperature of the beam during the test?

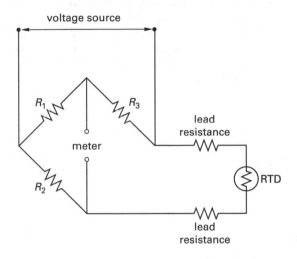

(A) 130°C

(B) 175°C

(C) 200°C

(D) 450°C

2. A copper-constantan thermocouple is used with an ice bath to measure the temperature of waste sludge during digestion. The voltage is proportional to the difference in temperature between the junctions. When the thermocouple is calibrated, it generates a voltage of 5.07×10^{-3} V when the temperature of the hot junction is 108.9°C and the reference temperature is at 0°C. Most nearly, what is the temperature of waste sludge when the voltage is 1.83×10^{-3} V?

(A) 40°C

(B) 45°C

(C) 50°C

(D) 55°C

3. A strain gauge with a gage factor of 2.0 and resistance of 120 Ω experiences a microstrain of 500. Most nearly, what is the percentage change in resistance?

(A) 0.00001%

(B) 0.001%

(C) 0.1%

(D) 1000%

4. A strain gauge has a gage factor of 3.0 and a nominal (strain-free) resistance of 350 Ω. The gauge is bonded to the top of a rectangular aluminum bar 12 cm long, 2 cm high, and 4 cm wide. The aluminum has a modulus of elasticity of 73 GPa. The bar is loaded as a cantilever beam. The distance from the applied load to the center of the strain gauge is 10 cm. The instrumentation consists of a simple Wheatstone bridge consisting of two 1000 Ω resistors, a variable resistor, and the strain gauge. Lead resistance is negligible. Most nearly, what is the change in resistance of the strain gauge when the free end of the aluminum bar is loaded vertically by a 1 kg mass?

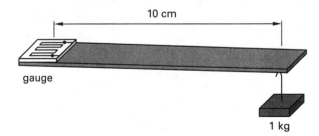

(A) 0.00054 Ω

(B) 0.0053 Ω

(C) 0.048 Ω

(D) 0.50 Ω

5. A cylindrical 2 in diameter solid shaft used to power an aerator blade is mounted as a stationary horizontal cantilever and instrumented with a strain gauge, R_1, bonded to its upper surface. The strain gauge is connected to an uncompensated bridge circuit where R_2 is 160 Ω and R_3 is 2500 Ω. R_4 is the adjustable resistance with a resistance of 2505 Ω when the shaft is unloaded, and 2511.7 Ω when the shaft tip 24 in from the midpoint of the strain gauge is loaded with vertical force of 1100 lbf. The gage factor of the strain gauge is 2.10. Disregard slip ring and lead resistances. Most nearly, what is the modulus of elasticity of the shaft?

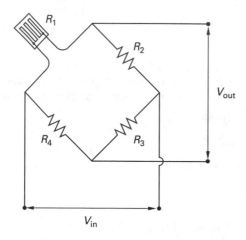

(A) 12×10^6 lbf/in^2

(B) 18×10^6 lbf/in^2

(C) 23×10^6 lbf/in^2

(D) 26×10^6 lbf/in^2

6. A 45° (rectangular) strain gauge rosette is bonded to the web of a large A36 steel built-up girder beam. The web's modulus of elasticity is 200 GPa, and Poisson's ratio is 0.3. The microstrains are $\epsilon_A = 650$, $\epsilon_B = -300$, and $\epsilon_C = 480$.

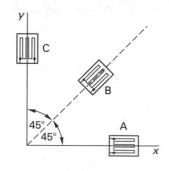

(a) Most nearly, what are the principal strains in the web?

(A) 840, 290

(B) 1400, −300

(C) 2300, 1200

(D) 2900, −610

(b) Most nearly, what are the principal stresses in the web?

(A) 2.5 MPa, 0.17 MPa

(B) 2.9 MPa, 0.57 MPa

(C) 3.0 MPa, 0.28 MPa

(D) 3.3 MPa, 0.57 MPa

7. Diesel engines in a waste-to-energy plant are powered by a gaseous mixture of methane and other digestion gases. The combustion conditions in each cylinder are continuously monitored by pressure and temperature transducers. A particular cylinder's indicator diagram is shown. The scale selected in the monitoring software is set to 111 kPa/mm. Most nearly, what is the mean effective pressure in the cylinder?

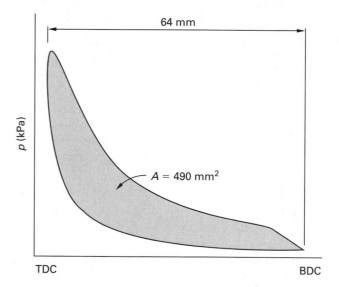

(A) 7.7 kPa

(B) 15 kPa

(C) 360 kPa

(D) 850 kPa

SOLUTIONS

1. When the meter is nulled out, the resistance of the leg containing the RTD is

$$R_4 = R_{\text{RTD}} + 2R_{\text{lead}} = \frac{R_2 R_3}{R_1}$$

$$= \frac{(376 \ \Omega)(1000 \ \Omega)}{1000 \ \Omega}$$

$$= 376 \ \Omega$$

The resistance of the RTD is

$$R_{\text{RTD}} = R_4 - 2R_{\text{lead}} = 376 \ \Omega - (2)(100 \ \Omega)$$

$$= 176 \ \Omega$$

The relationship between temperature and resistance for an RTD is

$$R_T = R_0(1 + AT + BT^2)$$

For a platinum RTD with $\alpha = 0.00385 \ 1/°\text{C}$, the resistance is

$$R_T = R_0 \left(\begin{array}{c} 1 + (3.9083 \times 10^{-3})\,T \\ + (-5.775 \times 10^{-7})\,T^2 \end{array} \right)$$

$$\approx R_0 \left(1 + (3.9083 \times 10^{-3})\,T \right)$$

$$176 \ \Omega \approx (100 \ \Omega) \left(1 + (3.9083 \times 10^{-3})\,T \right)$$

$$T \approx \boxed{194.5°\text{C} \quad (200°\text{C})}$$

(If the squared term is kept, the temperature is 200.3°C.)

The answer is (C).

2. During calibration, the temperature difference is $108.9°\text{C} - 0°\text{C} = 108.9°\text{C}$. The thermoelectric constant is

$$k_T = \frac{V}{\Delta T} = \frac{5.07 \times 10^{-3} \ \text{V}}{108.9°\text{C}} = 4.656 \times 10^{-5} \ \text{V}/°\text{C}$$

Use Eq. 85.9. The temperature of the waste sludge is

$$T = T_{\text{ref}} + \frac{V}{k_T}$$

$$= 0°\text{C} + \frac{1.83 \times 10^{-3} \ \text{V}}{4.656 \times 10^{-5} \ \dfrac{\text{V}}{°\text{C}}}$$

$$= \boxed{39.3°\text{C} \quad (40°\text{C})}$$

The answer is (A).

3. Use Eq. 85.11.

$$\frac{\Delta R_g}{R_g} = (\text{GF})\epsilon = (2.0)(500 \times 10^{-6})$$

$$= \boxed{0.001 \quad (0.1\%)}$$

The answer is (C).

4. The moment of inertia of the bar in bending is

$$I = \frac{bh^3}{12} = \frac{(4 \ \text{cm})(2 \ \text{cm})^3}{12} = 2.667 \ \text{cm}^4$$

The stress in the bar is

$$\sigma = \frac{Mc}{I} = \frac{mglc}{I}$$

$$= \frac{(1 \ \text{kg}) \left(9.81 \ \dfrac{\text{m}}{\text{s}^2} \right)(10 \ \text{cm}) \left(\dfrac{2 \ \text{cm}}{2} \right) \left(100 \ \dfrac{\text{cm}}{\text{m}} \right)^2}{2.667 \ \text{cm}^4}$$

$$= 367{,}829 \ \text{Pa}$$

The strain is

$$\epsilon = \frac{\sigma}{E} = \frac{367{,}829 \ \text{Pa}}{(73 \ \text{GPa}) \left(10^9 \ \dfrac{\text{Pa}}{\text{GPa}} \right)} = 5.04 \times 10^{-6}$$

Use Eq. 85.11.

$$\Delta R_g = \epsilon(\text{GF})R_g = (5.04 \times 10^{-6})(3.0)(350 \ \Omega)$$

$$= \boxed{0.00529 \ \Omega \quad (0.0053 \ \Omega)}$$

The answer is (B).

5. First, use the information from the unloaded case to determine the unstressed resistance of the strain gauge. Since the bridge numbering does not correspond to Fig. 85.6, rewrite Eq. 85.16 using the ratio of the vertical legs, top to bottom.

$$\frac{R_1}{R_4} = \frac{R_2}{R_3}$$

The unstressed resistance of the strain gauge is

$$R_1 = \frac{R_4 R_2}{R_3} = \frac{(2505 \ \Omega)(160 \ \Omega)}{2500 \ \Omega} = 160.32 \ \Omega$$

The resistance of the strain gauge when the shaft is loaded is

$$R_1 = \frac{R_4 R_2}{R_3} = \frac{(2511.7 \ \Omega)(160 \ \Omega)}{2500 \ \Omega} = 160.7488 \ \Omega$$

Now, use Eq. 85.11 to determine the strain when the shaft is loaded.

$$\epsilon = \frac{\Delta R_g}{(GF)R_g} = \frac{160.7488 \ \Omega - 160.32 \ \Omega}{(2.10)(160.32 \ \Omega)}$$

$$= 0.00127364$$

The shaft's moment of inertia as a beam is

$$I = \frac{\pi r^4}{4} = \frac{\pi \left(\frac{2 \ \text{in}}{2}\right)^4}{4} = 0.7854 \ \text{in}^4$$

The bending stress in the shaft is

$$\sigma = \frac{Mc}{I} = \frac{Flc}{I} = \frac{(1100 \ \text{lbf})(24 \ \text{in})\left(\frac{2 \ \text{in}}{2}\right)}{0.7854 \ \text{in}^4}$$

$$= 33{,}613 \ \text{lbf/in}^2$$

The modulus of elasticity is

$$E = \frac{\sigma}{\epsilon} = \frac{33{,}613 \ \dfrac{\text{lbf}}{\text{in}^2}}{0.00127364}$$

$$= \boxed{26.39 \times 10^6 \ \text{lbf/in}^2 \quad (26 \times 10^6 \ \text{lbf/in}^2)}$$

The answer is (D).

6. (a) Use Table 85.6. (The strain gauge designations are changed.) Work in microstrains (μm/m). The principal strains are

$$\epsilon_p, \epsilon_q = \frac{1}{2}\left(\epsilon_A + \epsilon_C \pm \sqrt{2(\epsilon_A - \epsilon_B)^2 + 2(\epsilon_B - \epsilon_C)^2}\right)$$

$$= \left(\tfrac{1}{2}\right)\left(650 + 480 \pm \sqrt{\begin{array}{c}(2)\left(650 - (-300)\right)^2 \\ + (2)(-300 - 480)^2\end{array}}\right)$$

$$= \left(\tfrac{1}{2}\right)(1130 \pm 1738)$$

$$= \boxed{1434, -304 \quad (1430, -300)}$$

The answer is (B).

(b) Use Table 85.6. Work in microstrains (μm/m). The principal stresses are

$$\sigma_p, \sigma_q = \frac{E}{2}\left(\begin{array}{c}\dfrac{\epsilon_A + \epsilon_C}{1 - \nu} \pm \dfrac{1}{1 + \nu} \\ \times \sqrt{2(\epsilon_A - \epsilon_B)^2 + 2(\epsilon_B - \epsilon_C)^2}\end{array}\right)$$

$$= \left(\frac{200 \ \text{GPa}}{2}\right)\left(10^9 \ \frac{\text{Pa}}{\text{GPa}}\right)$$

$$\times \left(\begin{array}{c}\dfrac{650 + 480}{1 - 0.3} \pm \left(\dfrac{1}{1 + 0.3}\right) \\ \times \sqrt{\begin{array}{c}(2)\left(650 - (-300)\right)^2 \\ + (2)(-300 - 480)^2\end{array}}\end{array}\right)$$

$$\times \left(10^{-6} \ \frac{\text{m}}{\mu\text{m}}\right)$$

$$= (1000)(1614 \pm 1337) \ \text{Pa}$$

$$= 2.95 \times 10^6 \ \text{Pa}, 0.277 \times 10^6 \ \text{Pa}$$

$$\boxed{\begin{array}{l}\text{The principal stresses are 2.95 MPa and} \\ \text{0.277 MPa} \quad (3.0 \ \text{MPa and 0.28 MPa}).\end{array}}$$

The answer is (C).

7. The average height of the pressure plot is

$$\bar{h} = \frac{A}{w} = \frac{490 \ \text{mm}^2}{64 \ \text{mm}} = 7.656 \ \text{mm}$$

The average pressure is

$$\bar{p} = k\bar{h} = \left(111 \ \frac{\text{kPa}}{\text{mm}}\right)(7.656 \ \text{mm})$$

$$= \boxed{849.8 \ \text{kPa} \quad (850 \ \text{kPa})}$$

The answer is (D).

86 Project Management, Budgeting, and Scheduling

PRACTICE PROBLEMS

1. The activities that constitute a project are listed. The project starts at $t = 0$. (a) Draw an activity-on-node critical path network. (b) Indicate the critical path. (c) What is the earliest finish? (d) What is the latest finish? (e) What is the slack along the critical path? (f) What is the float along the critical path?

activity	predecessors	successors	duration
start	–	A	0
A	start	B, C, D	7
B	A	G	6
C	A	E, F	5
D	A	G	2
E	C	H	13
F	C	H, I	4
G	D, B	I	18
H	E, F	finish	7
I	F, G	finish	5
finish	H, I	–	0

2. PERT activities constituting a short project are listed with their characteristic completion times. If the project starts on day 15, what is the probability that the project will be completed on or before day 42?

activity	predecessors	successors	t_{min}	$t_{most\ likely}$	t_{max}
start	–	A	0	0	0
A	start	B, D	1	2	5
B	A	C	7	9	20
C	B	D	5	12	18
D	A, C	finish	2	4	7
finish	D	–	0	0	0

3. Listed is a set of activities and sequence requirements to start a warehouse construction project. Prepare an activity-on-arc project diagram.

activity	letter code	code of immediate predecessor
move-in	A	
job layout	B	A
excavations	C	B
make-up forms	D	A
shop drawing, order rebar	E	A
erect forms	F	C, D
rough in plumbing	G	F
install rebar	H	E, F
pour, finish concrete	I	G, H

4. Listed is a set of activities, sequence requirements, and estimated activity times required for the renewal of a pipeline.

(a) Prepare an activity-on-arc PERT project diagram.

activity	letter code	code of immediate predecessor	activity time requirement (days)
assemble crew for job	A		10
use old line to build inventory	B	D	28
measure and sketch old line	C	A	2
develop materials list	D	C	1
erect scaffold	E	D	2
procure pipe	F	D	30
procure valves	G	D	45
deactivate old line	H	B	1
remove old line	I	E	6
prefabricate new pipe	J	F	5
place valves	K	H, I	1
place new pipe	L	J	6
weld pipe	M	I, L	2
connect valves	N	G, K	1
insulate	O	M, N	4
pressure test	P	G, K	1
remove scaffold	Q	O	1
clean up and turn over to operating crew	R	P, Q	1

(b) There is additional information in the form of optimistic, most likely, and pessimistic time estimates for the project. Compute the expected mean time and the variance, σ^2, for the activities. Which activities have the greatest uncertainty in their completion schedules?

activity code	optimistic time, t_{min}	most likely time, $t_{most\ likely}$	pessimistic time, t_{max}
A	8	10	12
B	26	26.5	36
C	1	2	3
D	0.5	1	1.5
E	1.5	1.63	4
F	28	28	40
G	40	42.5	60
H	1	1	1
I	4	6	8
J	4	4.5	8
K	0.5	0.9	2
L	5	5.25	10
M	1	2	3
N	0.5	1	1.5
O	3	3.75	6
P	1	1	1
Q	1	1	1
R	1	1	1

(c) Suppose that, due to penalties in the contract, each day the pipeline renewal project can be shortened is worth $100. Which of the following possibilities would you follow and why?

- Shorten $t_{most\ likely}$ of activity B by 4 days at a cost of $100.

- Shorten $t_{most\ likely}$ of activity G by 5 days at a cost of $50.

- Shorten $t_{most\ likely}$ of activity O by 2 days at a cost of $150.

- Shorten $t_{most\ likely}$ of activity O by 2 days by drawing resources from activity N, thereby lengthening its $t_{most\ likely}$ by 2 days.

5. Activities constituting a bridge construction project are listed. (a) Draw an activity-on-node network showing the critical path. (b) Compute ES, EF, LS, and LF for each of the activities. Assume a target time for completing the project that, for the bridge, is 3 days after the EF time.

activity		immediate predecessors	duration (days)
A	start		0
B		A	4
C		B	2
D		C	4
E		D	6
F		C	1
G		F	2
H		F	3
I		D	2
J		D, G	4
K		I, J, H	10
L		K	3
M		L	1
N		L	2
O		L	3
P		E	2
Q		P	1
R		C	1
S		O, T	2
T		M, N	3
U		T	1
V		Q, R	2
W		V	5
X	finish	S, U, W	0

6. Prepare an activity-on-arc diagram for the activities and sequence requirements given in Prob. 5.

7. Which of the following describes a preparation difference between construction drawings (i.e., prepared by architects and used by bidders, contractors, and inspectors) and design drawings (prepared by engineers and designers)?

(A) Construction drawings are prepared on vellum, while design drawings are prepared on paper.

(B) Construction drawings are copies of originals, while design drawings are originals.

(C) Construction drawings are typically larger in size than design drawings.

(D) Construction drawings typically have blue lines (i.e., are prepared by the diazo blueprint process) while design drawings have black lines (i.e., are prepared by CAD and xerographic processes).

8. Record drawings (also known as "as-builts") of a buried sewer line installation that will be submitted to the client should be certified by the

(A) contractor

(B) building official

(C) architect

(D) engineer

9. Which of the following details would NOT normally be found in the specifications for a large roadway contract requiring subgrade preparation?

(A) proof testing

(B) limits and boundaries

(C) scarification

(D) moisture conditioning

10. In a bid document, the engineer allowed bidders to install either cast-in-place or precast concrete culverts and headers, with the requirement that submitted bids include a description and separate price (bid) for the method used. This is an example of bid

(A) alternatives

(B) enhancement

(C) diversification

(D) programming

11. Which of the following items identified in a bid take-off would normally NOT be bid on the basis of weight?

I.   shotcrete or gunite

II.   water pipe

III.   riprap

IV.   reinforcing steel

V.   clearing and grubbing

VI.   grout

(A) II, IV, and VI

(B) I, II, III, and VI

(C) I, II, V, and VI

(D) I, II, III, IV, and V

12. An inexperienced estimator wishes to determine the approximate cost of compacting soil by means of a sheepsfoot roller making 4 passes with 12 in lifts. How should the estimator proceed?

(A) contact the state contractor's license board

(B) survey equipment rental costs, labor rates, and cost-of-living indices

(C) use published resources and cost data

(D) contact the local chapter of the Associated General Contractors of America

13. The *Building Cost Index* (BCI) and *Construction Cost Index* (CCI) published by the Engineering News-Record are based on which cost elements?

I.   carpentry labor

II.   brick laying labor

III.   structural steel shapes

IV. 2×4 lumber

V.   portland cement

VI.   heavy machinery

(A) II, V, and VI

(B) II, IV, V, and VI

(C) I, II, III, IV, and V

(D) I, II, III, IV, V, and VI

14. The table shown contains year-by-year cost indices. A job was bid in 1998. It was built over a period of three years, 2007, 2008, and 2009. What is the average escalation factor most nearly?

year	cost index
1997	1539
1998	1601
1999	1665
2000	1731
2001	1800
2002	1873
2003	1948
2004	2026
2005	2107
2006	2191
2007	2279
2008	2370
2009	2465

(A) 1.4

(B) 1.5

(C) 1.6

(D) 1.7

15. A contractor uses unbalanced bidding to bid a four-phase highway repaving project. The unbalanced bid results in phase billings of 38%, 12%, 42%, and 8%, respectively. The billings are roughly evenly spaced along the construction period. The increase in profit seen by the contractor from this practice is most nearly

(A) 0

(B) 0.7%

(C) 1.1%

(D) 1.6%

16. A construction firm has an annual fixed overhead (burden) cost of $500,000. During the year, the actual overhead was $550,000. The firm's gross billings during the year were $6,000,000. Each of the jobs billed had been estimated with a 10% allowance for overhead and profit. Assume the jobs were accurately bid and accurately managed. The firm's actual profit was most nearly

(A) $45,000

(B) $50,000

(C) $100,000

(D) $105,000

17. A project manager finds a loose sheet of binder paper at a construction job site. The paper contains various numbers, some with abbreviations of BF, CF, SF, WLD, and WHD. The manager should probably make an effort to give the sheet of paper to the

(A) engineer/architect

(B) contractor

(C) utility installer

(D) building inspector

18. A dump-hauler has a purchase price of $80,000. Freight for delivery is $1500. Tires are an additional $15,000. The hauler is expected to operate 1100 hours per year, and its useful life is estimated at 10 years. After 10 years, the salvage value is expected to be minimal. Interest, insurance, storage, fuel, oil, lubricants, filters, repairs, and sales taxes are estimated at $21,000 per year. Tires have an estimated life of 3500 hr. What is the before-tax estimated hourly cost of ownership and operation, excluding operator labor cost?

(A) $25/hr

(B) $31/hr

(C) $35/hr

(D) $40/hr

19. Which of the following financial ratios is stated INCORRECTLY?

(A) gross profit margin ratio = gross profit ÷ sales

(B) operating profit margin ratio = earnings after interest and taxes ÷ sales

(C) net profit margin ratio = net income ÷ sales

(D) return on assets = net income ÷ total assets

20. Given the following table of precedence relationships, how many nodes would be needed to form an activity-on-branch critical path network?

activity	description	predecessors
A	site clearing	–
B	removal of trees	–
C	general excavation	A
D	grading general area	A
E	excavation for utility trenches	B, C
F	placing formwork and reinforcement for concrete	B, C
G	installing sewer lines	D, E
H	installing other utilities	D, E
I	pouring concrete	F, G

(A) 6

(B) 7

(C) 9

(D) 11

21. Given the following table of precedence relationships, how many nodes would be needed to form an activity-on-node critical path network?

activity	description	predecessors
A	site clearing	–
B	removal of trees	–
C	general excavation	A
D	grading general area	A
E	excavation for utility trenches	B, C
F	placing formwork and reinforcement for concrete	B, C
G	installing sewer lines	D, E
H	installing other utilities	D, E
I	pouring concrete	F, G

(A) 6

(B) 7

(C) 9

(D) 11

22. A project is described by the following precedence table. The project manager wants to decrease the normal project time by 3 days. Most nearly, how much extra will it cost to reduce the project completion time by 3 days?

task	prede-cessors	normal time (days)	crash time	normal cost (per day)	crash cost (per day)
A	–	4	2	100	200
B	A	5	4	80	90
C	A	2	1	110	130
D	B	3	1	50	90
E	C	5	3	70	150
F	E	2	1	90	120
G	D, F	3	2	60	120

(A) 90

(B) 120

(C) 130

(D) 160

23. An engineering design firm has estimated the following billing scenarios for the following year.

billed amount	probability
$0–$500,000	0.60
$500,001–$1,000,000	0.25
$1,000,001–$1,500,000	0.10
$1,500,001–$2,000,000	0.05

What is the expected billing amount for the upcoming year?

(A) $550,000

(B) $625,000

(C) $690,000

(D) $740,000

24. An engineering design firm has a year-long lease on a bucket crane used to inspect construction details at second- and third-floor levels. Into which general ledger account should the cost of the lease be accumulated?

(A) rents

(B) amortization for leasehold

(C) depreciation

(D) overhead

25. A design firm contracts with a client using a standard 6% cost-plus arrangement to prepare a design for a footbridge and to manage the construction project. The owner has budgeted $750,000 for the design and construction. The design firm estimates that the bridge will cost the client $600,000. The bid-winning contractor estimates that the bridge materials and labor will cost $450,000 and bids the job at $550,000. The actual cost to the contractor to build the bridge as originally designed is $470,000. Along the way, the owner agrees to change orders, increasing the project billing by $40,000. Most nearly, what is the design firm's total billing to the owner?

(A) $570,000

(B) $610,000

(C) $630,000

(D) $650,000

26. What type of Gantt chart is illustrated?

	Jan	Feb	Mar	Apr	May	Jun	Jul	Aug	Sep
task 1	▬▬	▬							
task 2			▬▬						
task 3			▬						
task 4					▬▬	▬			
task 5					▬				
task 6								▬▬	

(A) milestones Gantt

(B) Gantt with dependencies

(C) baseline Gantt

(D) timeline Gantt

27. A subcontractor's contract requires the general contractor to provide an area for lay-down. The lay-down area will be best used for

(A) parking heavy equipment and subcontractor employee vehicles

(B) receiving delivered materials to be used by the subcontractor

(C) space for subcontractor personnel during rest periods

(D) pre-installation assembly of overhead fire sprinkler pipes, electrical conduit, and standpipes

28. The main parties to a construction project contract include the

I.   client

II. engineer-architect

III. OSHA

IV. prime contractor

V.   subcontractors

VI. local building official

(A) I, II, and IV

(B) I, II, IV, and V

(C) I, II, IV, V, and VI

(D) I, II, III, IV, V, and VI

29. The term "design-build" means the

(A) design firm designs the project and the client builds it

(B) design firm both designs and builds the project

(C) client designs the project and the contractor builds it

(D) contractor designs the project and the subcontractors build it

30. The term "fast-track" refers to

(A) constructing different areas of the project site in parallel

(B) a project delivery method where the sequencing of construction activities allows some portions of the project to begin before the design is completed on other portions of the project

(C) working multiple shifts, overtime, weekends, and holidays to decrease time to completion

(D) providing financial incentives to the contractor and subcontractors to beat the published schedule

31. The term "turnkey" refers to

(A) a contractor providing full service to the owner, including such things as obtaining financing, planning and costing, cost management, shell construction, and interior finishing

(B) an OSHA inspection in which all areas of the jobsite are inspected

(C) a responsibility-sharing partnership between an engineer, contractor, and architect

(D) a financial lender's right to resell the construction loan to third party investors

32. Which of the following duties would normally NOT be a responsibility of the estimating department within a general contractor's organization?

(A) obtaining bid documents

(B) securing subcontractor/material quotations

(C) project cost accounting

(D) delivering competitive or negotiated proposals

33. Under normal circumstances, an architect-engineer will have all of the following authorities EXCEPT to

(A) interpret contract questions from the prime contractor

(B) judge a subcontractor's performance and condemn defective work

(C) stop field operations

(D) issue certificates of occupancy

34. What is the term used to describe a complete listing of all the materials and items of work that will be required for a project?

(A) materials resource plan

(B) specifications detail list

(C) quantity survey

(D) resource foundation accounting

35. Which two categories of projects have different bidding procedures, rules, and regulations?

(A) commercial and residential

(B) public and private

(C) transportation (bridge and highway) and nontransportation

(D) critical (school, hospital, emergency services, etc.) and noncritical

36. A contractor's decision to bid is affected by the bidding climate, a term that does NOT apply to

(A) the contractor's bonding capacity

(B) the owner's reputation and financial soundness

(C) the completion date

(D) construction loan interest rates

37. Which of the following is the most reliable source of productivity data for use in bidding by a contractor?

(A) U.S. Department of Labor

(B) professional construction/contractor associations

(C) compilations from RSMeans and Engineering News-Record

(D) internal historical data

38. Following the estimating process, a contractor normally adds a markup or margin. This is an allowance for

(A) profit

(B) general overhead

(C) contingency

(D) all of the above

39. All of the following are examples of indirect costs EXCEPT

(A) payroll taxes for construction labor

(B) equipment rentals for the project

(C) corporate income taxes

(D) worker's compensation insurance

40. Which statement about large construction equipment depreciation is true?

(A) Tires are normally depreciated separately from the chassis.

(B) The asset life selected must be 3, 5, or 15 years depending on the asset class.

(C) The salvage value must be included in the depreciation calculation.

(D) Repairs to the equipment increases the depreciation basis (i.e., is added to the depreciated cost).

41. What is the primary implication of the phrase, "50 minute hour"?

(A) Employees must be permitted to take their breaks and lunch.

(B) Heavy equipment is not 100% productive.

(C) Vacation and illness decrease employee productivity.

(D) Contractors must train their employees in safety procedures.

42. Contingency is something that should be added to a(n)

I.   owner's construction loan amount

II. design engineering firm's fixed-fee design contract

III. prime's bid to an owner

IV. subcontractor's bid to the general contractor

V.   inspector's schedule of inspection time

VI. lessee's move-in schedule

(A) I, III, and IV

(B) I, II, III, and IV

(C) II, III, IV, and V

(D) I, II, III, IV, V, and VI

43. When the bids are all opened, the owner will normally award the contract to the lowest

(A) available bidder

(B) qualified bidder

(C) responsible bidder

(D) bonded bidder

44. When a contractor is awarded a job on a cost-plus-fee basis, with the fee being 6% of the reimbursable expenses, what percentage of the contractor's total expenses will typically be reimbursed by the owner?

(A) 100%

(B) less than 100%

(C) more than 100%

(D) any of the above

45. The amount of a phase payment to the contractor withheld by the owner is known as

(A) retainage

(B) recovery

(C) incentive deposit

(D) overhead withholding

46. When a project is completed late, the amount of money specified by the contract to be paid to the owner by the contractor (or, withheld by the owner from the contractor's final payment) is known as

(A) responsibility penalties

(B) essence time payments

(C) completion recovery charges

(D) liquidated damages

47. A surety is

(A) an agreement that assigns liability for another party's debt, default, or failure in duty

(B) an agreement that outlines the conditions when liability for the debt, default, or failure in duty of another is assumed by the guarantor

(C) a party that assumes liability for the debt, default, or failure in duty of another party

(D) a deposit of money that is used to cover liability for the debt, default, or failure in duty of another party

48. Which of the following is NOT synonymous with surety bond?

(A) contract surety bond

(B) contract bond

(C) construction contract bond

(D) construction insurance

49. Once default has been established triggering a surety bond, the surety has an obligation to

(A) pay the face amount of the bond

(B) reimburse the owner's expenses in financing the remainder of the project

(C) pay the owner the balance of the construction contract amount not yet paid out

(D) do all steps necessary to complete the project

50. Which of the following is NOT a common type of bond on a construction project?

(A) bid bond

(B) performance bond

(C) payment bond

(D) schedule bond

51. A mechanic's lien is a

(A) claim on the recorded title of property

(B) public filing notice of a supplier's right to receive payment for material

(C) public filing that notifies the owner of the supplier's participation in the job

(D) public filing that notifies the owner of a default of the general contractor

52. Which of the following statements regarding the Miller Act is FALSE?

(A) The Miller Act was enacted in 1890.

(B) The Miller Act requires a performance and payment bond on all federal projects greater than $25,000.

(C) The Miller Act requires the bond to be 100% of the contract amount.

(D) The Miller Act protects first and second tier subcontractors only.

53. When the contract documents require a "Best Rating" of the surety, this means the

(A) contract is rated A-1

(B) surety must be rated A++

(C) surety must be rated by a company named "Best"

(D) surety must have the ability to bond 125% of the contract amount

54. Which of the following surety ratings means that the surety is in liquidation?

(A) C−

(B) E

(C) F

(D) S

55. Which of the following is NOT a characteristic of accrual accounting?

(A) income is counted when earned, not when received

(B) expenses are counted when paid, not when incurred

(C) income and expenses can be accumulated per project

(D) requires more detailed records and complex processes

56. Which one of the following financial equations is correct?

(A) assets = liabilities + net worth

(B) assets = cash + net worth

(C) net worth = net worth + liabilities

(D) assets = cash + accumulated depreciation

57. A contractor's liquidity ratio in three consecutive years is 1.01, 1.05, and 1.09, respectively. It is probably true that the contractor is

(A) becoming more profitable

(B) increasingly drawing on his/her line of credit

(C) accumulating cash

(D) receiving payments from clients increasingly earlier

58. Which of the following descriptions regarding materials purchasing and delivery to a job site is NOT correct?

(A) FOB (free on board) indicates that the seller puts the materials onboard the freight carrier free of expense to the buyer, with freight paid to the FOB point designated. Title to the material passes to the buyer when the seller consigns the material to the freight company.

(B) CIF (cost, insurance, freight) indicates that the purchase price includes the cost of goods, customary insurance, and freight to the buyer's destination. Title passes when the seller delivers the merchandise to the carrier and forwards to the buyer the bill of lading, insurance policy, and receipt showing payment of freight.

(C) COD (collect on delivery) indicates that the title passes to the buyer, if the buyer is to pay the transportation, at the time the goods are received by the carrier. However, the seller reserves the right to receive payment before surrender of possessions to the buyer.

(D) None of the above.

SOLUTIONS

1. (a) The activity-on-node critical path network is as follows.

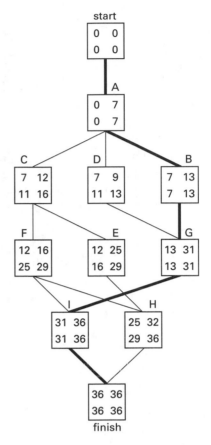

ES (earliest start) Rule: The earliest start time for an activity leaving a particular node is equal to the largest of the earliest finish times for all activities entering the node.

LF (latest finish) Rule: The latest finish time for an activity entering a particular node is equal to the smallest of the latest start times for all activities leaving the node.

The activity is critical if the earliest start equals the latest start.

(b) The critical path is $\boxed{\text{A-B-G-I.}}$

(c) The earliest finish is $\boxed{36.}$

(d) The latest finish is $\boxed{36.}$

(e) The slack along the critical path is $\boxed{0.}$

(f) The float along the critical path is $\boxed{0.}$

2. From Eq. 86.2,

$$\mu = \tfrac{1}{6}\left(t_{\min} + 4t_{\text{most likely}} + t_{\max}\right)$$

Use Eq. 86.3 to find the variance, which is the square of the standard deviation, σ.

$$\sigma^2 = \left(\tfrac{1}{6}(t_{max} - t_{min})\right)^2$$

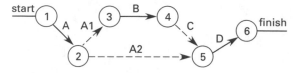

The critical path is A-A1-B-C-D.

The following probability calculations assume that all activities are independent. Use the following theorems for the sum of independent random variables and use the normal distribution for T (project time).

$$\mu_{total} = t_A + t_B + t_C + t_D$$
$$\sigma^2_{total} = \sigma^2_A + \sigma^2_B + \sigma^2_C + \sigma^2_D$$

The variance is 10.52778, and the standard deviation is 3.244654. (See the following table.)

$$\mu_{total} = 43.83333$$
$$\sigma^2_{total} = 10.52778$$
$$\sigma_{total} = 3.244654$$

$$z = \frac{t - \mu_{total}}{\sigma}$$
$$= \left|\frac{42 - 43.83333}{3.244654}\right|$$
$$= 0.565$$

From the normal table, the probability of finishing for $T \leq 42$ is 0.286037 $\boxed{(28.6\%)}$.

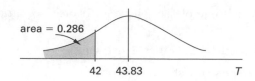

3. Draw an activity-on-arc diagram for the project.

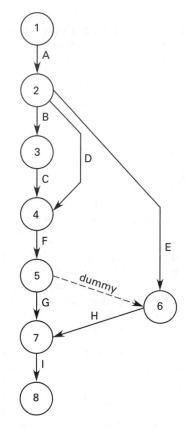

(A dummy activity cannot be replaced with any other activity letter without duplicating that activity.)

Table for Sol. 2

no.	name	activity exp. time	variance	earliest start	latest start	earliest finish	latest finish	slack LS − ES
1	A	2.33333	0.44444	15	15	17.3333	17.3333	0
2	A1	0	0	17.3333	17.3333	17.3333	17.3333	0
3	A2	0	0	17.3333	39.6667	17.3333	39.6667	22.3333
4	B	10.5000	4.69444	17.3333	17.3333	27.8333	27.8333	0
5	C	11.8333	4.69444	27.8333	27.8333	39.6667	39.6667	0
6	D	4.16667	0.69444	39.6667	39.6667	43.8333	43.8333	0

expected completion time = 43.83333 (43.83)

4. (a) Draw an activity-on-arc PERT diagram for the project. (See *Illustration for Sol. 4(a)*.)

(b) A table is the easiest way to determine which activities have the greatest uncertainty in their completion schedules.

activity	variance	mean time
A	0.44	10
B	2.77	28
C	0.11	2
D	0.027	1
E	0.17	2
F	4.0	30
G	11.11	45
H	0	1
I	0.44	6
J	0.44	5
K	0.0625	1
L	0.69	6
M	0.11	2
N	0.027	1
O	0.25	4
P	0	1
Q	0	1
R	0	1

Activities B, F, and G have the three largest variances, so these activities have the greatest uncertainties.

(c)

- Activity B is not on the critical path, so shortening it will not shorten the project.

- Activity G is on the critical path, so it should be shortened. Shortening it 5 days may change the critical path, however.

  path D-G-K: $1 + 45 = 6$ days

  path D-E-I-K: $1 + 2 + 6 = 9$ days

  path D-B-H-I-K: $1 + 28 + 1 + 6 = 36$ days

Since the shortened path D-G-K has a length of 46 days − 5 days = 41 days and is still the longest path from D to K, it should be shortened.

- Activity O is on the critical path. The cost of the crash schedule is $150. Savings would be (2)($100) = $200. This activity should be shortened.

- The current path length is

  N-O-Q-R: $1 + 4 + 1 = 6$ days

The new paths would be

  N-P-Q-R: $3 + 1 + 1 = 5$ days

  N-O-Q-R: $3 + 2 + 1 = 6$ days

  N-O-R: $(3 + 2) + (4 - 2) = 7$ days

N-O-R would become part of the critical path and would be longer than the original critical path. Therefore, it is not acceptable.

Illustration for Sol. 4(a)

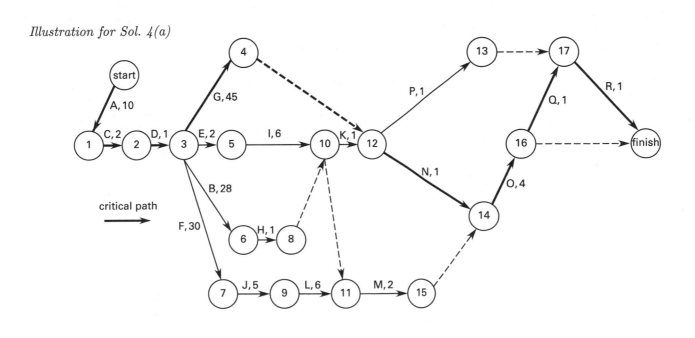

5. Draw the activity-on-node CPM network showing the critical path.

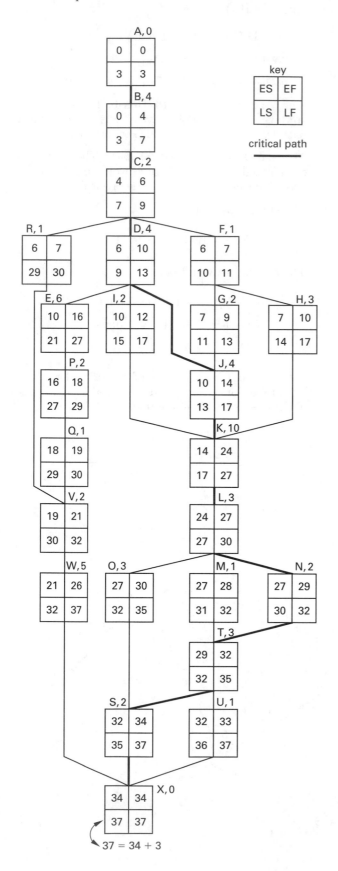

6. Draw a PERT diagram. (See *Illustration for Sol. 6*.)

7. The industry standard size for drawings prepared by architects is 36 in × 48 in (914 mm × 1219 mm), but this size sheet is rarely used in engineering design. The engineering design industry primarily uses 24 in × 36 in (610 mm × 914 mm), although 22 in × 34 in (559 × 864 mm) paper is also used. Half-size drawings are frequently used during reviews and submittals because they are less bulky (i.e., fit inside briefcases and file cabinets) and can be easily copied.

The answer is (C).

8. The engineer of record certifies engineering design. This is commonly referred to as "stamping" or "sealing" and includes dating and signing the design documents. Architects do not certify engineering work.

The answer is (D).

9. Construction documents, not the specifications, show the limits (in plan view) and boundaries of the subgrade preparation area. Other details of the preparation are found in the specifications.

The answer is (B).

10. *Bid alternatives* are used to give bidders the option of tailoring the methodology and materials to their expertise, experience, and capabilities.

The answer is (A).

11. Shotcrete (gunite) and clearing and grubbing are measured per square foot or square yard. Water pipe is measured by unit length or pieces. Grout is measured by volume. All of the other items are bid by weight.

The answer is (C).

12. Construction costs data are published regularly by several sources, including Engineering News-Record and RS Means. Other sources and some state transportation departments make similar data available.

The answer is (C).

13. The Engineering News-Record's (ENR's) *Building Cost Index* is developed from the costs of 68.38 hours of skilled labor at the 20-city average of bricklayers, carpenters, and structural ironworkers rates, plus 25 cwt of standard structural steel shapes at the mill price prior to 1996; as well as the fabricated 20-city price from 1996, plus 1.128 tons of portland cement at the 20-city price, plus 1088 board-ft of 2 × 4 lumber at the 20-city price.

Illustration for Sol. 6

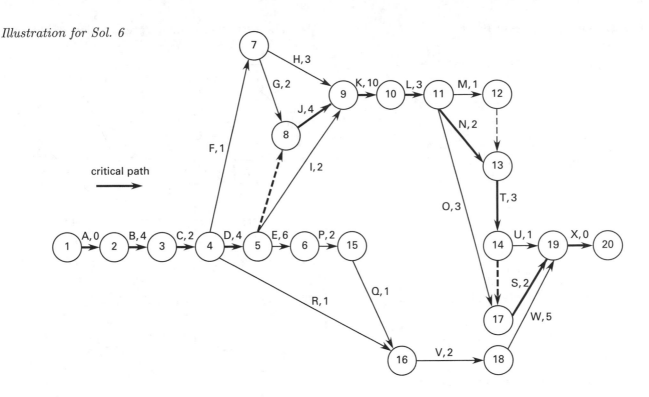

ENR's *Construction Cost Index* is developed from the costs of 200 hours of common labor at the 20-city average of common labor rates, plus 25 cwt of standard structural steel shapes at the mill price prior to 1996; as well as the fabricated 20-city price from 1996, plus 1.128 tons of portland cement at the 20-city price, plus 1088 board-ft of 2 × 4 lumber at the 20-city price. Neither index includes heavy machinery.

The answer is (C).

14. The average cost index over the construction period is

$$\frac{2279 + 2370 + 2465}{3} = 2371.3$$

Divide by the cost index from 1998 to determine the average escalation factor.

$$\frac{2371.3}{1601} = \boxed{1.48 \quad (1.5)}$$

The answer is (B).

15. Unbalanced bidding is used by contractors to improve cash flow by distorting cost in certain periods. While the overall bid remains competitive by virtue of the total cost, the payments are structured to suit the contractor's needs. Unbalanced bidding is not illegal, and it does not increase the contractor's gross profit, but it may be unethical as the contractor must make untruthful cost statements in certain periods. The opportunity value of having the money earlier rather than later, as well as the time value of money, cannot be determined without knowing the contractor's marginal rate of return and the actual timing of the payments. In this problem, such information is clearly not available. Therefore, the question is answered from a general viewpoint—no increase in profit.

The answer is (A).

16. The jobs had been billed to create an allowance for profit and overhead of

$$(0.10)(\$6,000,000) = \$600,000$$

The actual profit is the allowance less the actual overhead cost.

$$\$600,000 - \$550,000 = \boxed{\$50,000}$$

The answer is (B).

17. The abbreviations most likely mean board feet (BF), cubic feet (CF), square feet (SF), width × length × depth (WLD), and width × height × depth (WHD). These would appear together on a materials or cost estimation sheet, probably for ordering concrete and formwork lumber. Such a sheet would have been prepared by the contractor.

The answer is (B).

18. The delivered price (less tires) is

$$C = \$80{,}000 + \$1500 = \$81{,}500$$

The hourly cost of ownership is

$$H_1 = \frac{\$81{,}500}{(10 \text{ yr})\left(1100 \ \dfrac{\text{hr}}{\text{yr}}\right)} = 7.41 \ \$/\text{hr}$$

The hour cost of tire usage is

$$H_2 = \frac{\$15{,}000}{3500 \text{ hr}} = 4.29 \ \$/\text{hr}$$

The hourly cost of operation is

$$H_3 = \frac{\$21{,}000}{1100 \text{ hr}} = 19.09 \ \$/\text{hr}$$

The total hourly cost of ownership and operation is

$$\begin{aligned} H &= H_1 + H_2 + H_3 \\ &= 7.41 \ \frac{\$}{\text{hr}} + 4.29 \ \frac{\$}{\text{hr}} + 19.09 \ \frac{\$}{\text{hr}} \\ &= \boxed{30.79 \ \$/\text{hr} \quad (\$31/\text{hr})} \end{aligned}$$

The answer is (B).

19. The operating profit margin is calculated before interest and taxes, not after.

The answer is (B).

20. The activity-on-branch representation of this project is

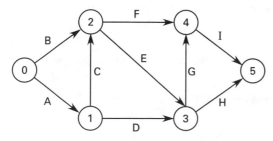

There are $\boxed{6}$ nodes.

The answer is (A).

21. The activity-on-node representation of this project is

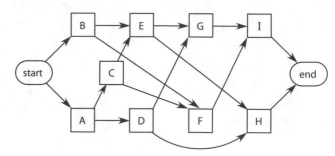

Counting the starting and ending dummy activities, there are $\boxed{11}$ nodes.

The answer is (D).

22. Draw the activity-on-node network.

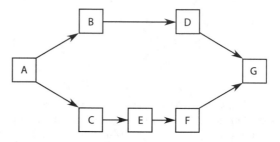

Path ABDG has a normal length of $4 + 5 + 3 + 3 = 15$.

Path ACEFG has a normal length of $4 + 2 + 5 + 2 + 3 = 16$.

Path ACEFG is the longest, and it is the critical path.

Determine the incremental costs of crashing.

activity	incremental cost (per day)
A	$200 - 100 = 100$
B	$90 - 80 = 10$
C	$130 - 110 = 20$
D	$90 - 50 = 40$
E	$150 - 70 = 80$
F	$120 - 90 = 30$
G	$120 - 60 = 60$

In order to shorten the project, an activity along the critical path must be crashed. Activity C has the lowest incremental cost of activities A, C, E, F, and G. For an incremental cost of 20, the duration can be reduced by 1.

At this point, both paths have the same length, and reducing one without reducing the other will not decrease the project duration. So, find the lowest incremental cost activity in each path. In path ABDG, activity B can reduce the path length by 1 at a cost of 10. In path ACEFG, with activity C already having been crashed, activity F can reduce the path length by 1 at a cost of 30.

Now, both paths have been reduced by 1 and have the same length. Each path must be reduced by one additional day. Crashing activity G at a cost of 60 will shorten both paths by 1.

The total cost of reducing the project duration by 3 days is

$$20 + 10 + 30 + 60 = \boxed{120}$$

The answer is (B).

23. Take the midpoint of each range.

billed amount	midpoint	probability
$0–$500,000	$250,000	0.60
$500,001–$1,000,000	$750,000	0.25
$1,000,001–$1,500,000	$1,250,000	0.10
$1,500,001–$2,000,000	$1,750,000	0.05

The expected billing amount is

$$E = (\$250{,}000)(0.60) + (\$750{,}000)(0.25)$$
$$+ (\$1{,}250{,}000)(0.10) + (\$1{,}750{,}000)(0.05)$$
$$= \boxed{\$550{,}000}$$

The answer is (A).

24. In a general ledger account, "rents" almost always refers to the expense of rental (leased) office space. Leasehold improvements are semi-permanent improvements made to rented (leased) office space, and their costs are amortized over time. Only an owner, not a lessee, can depreciate an asset. The crane is used for all projects, and like other generic costs, its expense should be accumulated into overhead. The design firm may also decide to allocate portions of the crane lease expense to the overhead accounts of each of the projects.

The answer is (D).

25. The total amount billed by the contractor through the design firm to the client is $550,000 + $40,000 = $590,000. This amount is the "cost" to the client, to which the design firm adds 6% for its efforts.

$$(1.06)(\$590{,}000) = \boxed{\$625{,}400 \quad (\$630{,}000)}$$

The answer is (C).

26. This is a basic timeline Gantt. No milestones, dependencies, or baselines are included.

The answer is (D).

27. Although a lay-down area could be used for almost anything, it is usually a storage area. Alternative B is the most general and, therefore, the best answer.

The answer is (B).

28. The client, engineer-architect, and the primary contractor are the main parties to a construction project contract. OSHA and the local building official have implicit authority over a project, but they are not parties to the contract. Subcontractors have contracts with the general contractor, but they are not usually parties to the project contract.

The answer is (A).

29. *Design-build* is a process where the client interacts only with a single entity, the "design-builder." While the design-builder is usually the general contractor, it can also be the design firm or a partnership consisting of the design firm and contractor.

The answer is (B).

30. *Fast-track* means that different phases of the design and construction overlap.

The answer is (B).

31. A *turnkey* is a contractor who provides full-service to the owner.

The answer is (A).

32. The estimation process does not include tracking costs or recording actual expenses.

The answer is (C).

33. The architect-engineer normally has the contractual authority to interpret the construction documents, judge quality, and stop work under unsafe and other conditions. Only the local building official can issue a certificate of occupancy.

The answer is (D).

34. The term *quantity survey* is synonymous with the terms *take-off survey* and *take-off report*.

The answer is (C).

Systems, Mgmt, and Professional

35. Private projects have bidding procedures normally established by the owner and AE. Public projects follow various procurement statutes developed by federal, state, county, and municipal governments.

The answer is (B).

36. The *bidding climate* is affected by the following factors: bonding capacity considerations, location of project, severity of contractual terms (contractor responsibilities and liabilities), identity of the owner and owner's financial status, identity of the architect/engineer, nature and size of project as it relates to company experience and equipment, labor conditions and availability, and completion date. Construction loan interest rates will affect an owner's decision to build the project, but the contractor does not generally care how the owner finances the project as long as it is not through nonpayment.

The answer is (D).

37. Productivity information is available from many sources; however, public and commercially available information is less reliable to bidding than internal historical data.

The answer is (D).

38. A contractor usually adds a markup or margin to the profit, general overhead, and contingency following the estimating process.

The answer is (D).

39. Indirect costs include such items as payroll taxes, insurance, employee fringe benefits (e.g., health, vacation, employee insurance, and 401K), employer contributions to social security, unemployment insurance, workers compensation insurance, and public liability and property damage insurance. Income taxes are incurred on profit, after all income and expenses (direct and indirect) have been accounted for.

The answer is (C).

40. Tires are a substantial part of the original cost of the equipment, but they do not last as long as the equipment. In fact, tires for some very large items of equipment (e.g., quarry trucks) have short life cycles. Therefore, tires are normally depreciated separately from the chassis.

The answer is (A).

41. On an average, heavy equipment is productive only approximately 50 minutes per hour. The remaining time is consumed by nonproductive activities such as greasing, fueling, minor repairs, maintenance, shifting position, operator delays, and waiting in line.

The answer is (B).

42. All of the items are subject to unforeseen delays; therefore, contingency should be applied to each.

The answer is (D).

43. In open competitive bidding by public and private owners, the bid will be awarded to the lowest responsible bidder.

The *lowest responsible bidder* is the lowest bidder whose offer best responds in quality, fitness, and capacity to fulfill the particular requirements of the proposed project, and who can fulfill these requirements with the qualifications needed to complete the job in accordance with the terms of the contract.

The answer is (C).

44. In a cost-plus-fee contract, 100% of a contractor's reimbursable expenses are paid to the contractor. However, it is very unlikely that all expenses incurred by the contractor will be reimbursable under the contract. For example, the contractor may incur an expense by flying first class to a distant meeting while the contract only provides for coach or economy seating. Therefore, only a portion of the expenses will be reimbursable.

The answer is (B).

45. *Retainage* is a percentage, usually 10%, of the payment to the contractor that is withheld from each payment. The owner may reduce retainage below 10% after substantial completion. The retainage is paid by owner to contractor after final completion. Retainage serves as an incentive for a contractor to complete all work per contract.

The answer is (A).

46. *Liquidated damages* are specified in the contract as a penalty for late completion. The amount is specified (e.g., per day of lateness) and is agreed to by the contractor at the time the agreement is entered into. The amount is established in advance in lieu of a determination of actual damages suffered by the owner. Liquidated damages must be a reasonable penalty based on a forecast of actual damages the owner would suffer if the contractor is in breach of contract by finishing late.

The answer is (D).

47. A *surety* is a party that assumes liability for the debt, default, or failure in duty of another.

The answer is (C).

48. A *surety bond* is not an insurance policy. A surety bond is a contract that outlines the conditions of the assumed liability by a surety.

The answer is (D).

49. If the contractor fails to fulfill his/her contractual obligations, the surety must assume the obligations of the contractor and see that the contract is completed, including paying all costs up to the face amount of the bond. This requires the surety to do more than merely provide money to get the project completed—the surety is responsible for finishing the contract. The contract can be completed in one of three ways: (1) Complete the contract itself through a completion contractor. (2) Select a new contractor to contract directly with the owner. (3) Allow the owner to complete the work with the surety paying the costs.

The answer is (D).

50. The contractor's obligation to maintain a reasonable schedule is contained in the contract agreement, and a separate bond is seldom, if ever, issued for that purpose. When the schedule is covered by a bond, it will be part of the performance bond. Other types of legitimate, but uncommon, bonds include bonds to release retainage, to discharge liens or claims (commonly referred to as bonds for "bonding over a lien"), and to indemnify owners against liens, license bonds, self-insurers' workers' compensation bonds, and union wage bonds.

The answer is (D).

51. A *mechanic's lien* is a right created by law to secure payment for work performed and material furnished in the improvement of real property. It is secured by the real property itself and will be recorded (e.g., with the county recorder) against the title or deed for a property.

The answer is (A).

52. The Miller Act was enacted in 1935. One of its provisions is that the subcontractor cannot claim on the payment bond until 90 days after the last day labor was performed on job.

The answer is (A).

53. When the contract documents require a "Best Rating" of the surety, the surety must be rated by the A. M. Best company. *A. M. Best Insurance Reports* contains financial ratings for insurance and surety companies.

The answer is (C).

54. The standard A. M. Best ratings for secure sureties are

- A++ and A+ (superior)
- A and A− (excellent)
- B++ and B+ (very good)

The standard A. M. Best ratings for vulnerable sureties are

- B and B− (fair)
- C++ and C+ (marginal)
- C and C− (weak)
- D (poor)
- E (under regulatory supervision)
- F (in liquidation)
- S (rating suspended)

The answer is (C).

55. Accrual accounting counts income and expenses when incurred, not when paid.

The answer is (B).

56. The net worth of a company is the difference between assets and liabilities.

$$\text{net worth} = \text{assets} - \text{liabilities}$$

This is algebraically the same as

$$\text{assets} = \text{liabilities} + \text{net worth}$$

The answer is (A).

57. Liquidity is a contractor's ability to meet short-term debt. One of the components of liquidity is cash, and another is assets that can quickly be converted into cash. Given the increasing liquidity ratio, it is probably true that the contractor is accumulating cash.

The answer is (C).

58. FOB indicates that the seller puts the goods on board the common carrier free of expense to the buyer, with freight paid to the FOB point designated (e.g., "FOB jobsite" or "FOB storage yard"). Title goes to the buyer when the carrier delivers the goods at the place indicated, not when the seller consigns the material to the freight company.

The answer is (A).

87 Engineering Economic Analysis

PRACTICE PROBLEMS

1. At 6% effective annual interest, how much will be accumulated if $1000 is invested for ten years?

2. At 6% effective annual interest, what is the present worth of $2000 that becomes available in four years?

3. At 6% effective annual interest, how much should be invested to accumulate $2000 in 20 years?

4. At 6% effective annual interest, what year-end annual amount deposited over seven years is equivalent to $500 invested now?

5. At 6% effective annual interest, what will be the accumulated amount at the end of ten years if $50 is invested at the end of each year for ten years?

6. At 6% effective annual interest, how much should be deposited at the start of each year for ten years (a total of 10 deposits) in order to empty the fund by drawing out $200 at the end of each year for ten years (a total of 10 withdrawals)?

7. At 6% effective annual interest, how much should be deposited at the start of each year for five years to accumulate $2000 on the date of the last deposit?

8. At 6% effective annual interest, how much will be accumulated in ten years if three payments of $100 are deposited every other year for four years, with the first payment occurring at $t = 0$?

9. $500 is compounded monthly at a 6% nominal annual interest rate. How much will have accumulated in five years?

10. What is the effective annual rate of return on an $80 investment that pays back $120 in seven years?

11. A new machine will cost $17,000 and will have a resale value of $14,000 after five years. Special tooling will cost $5000. The tooling will have a resale value of $2500 after five years. Maintenance will be $2000 per year. The effective annual interest rate is 6%. What will be the average annual cost of ownership during the next five years?

12. An old covered wooden bridge can be strengthened at a cost of $9000, or it can be replaced for $40,000. The present salvage value of the old bridge is $13,000. It is estimated that the reinforced bridge will last for 20 years, will have an annual cost of $500, and will have a salvage value of $10,000 at the end of 20 years. The estimated salvage value of the new bridge after 25 years is $15,000. Maintenance for the new bridge would cost $100 annually. The effective annual interest rate is 8%. Which is the best alternative?

13. A firm expects to receive $32,000 each year for 15 years from sales of a product. An initial investment of $150,000 will be required to manufacture the product. Expenses will run $7530 per year. Salvage value is zero, and straight-line depreciation is used. The income tax rate is 48%. What is the after-tax rate of return?

14. A public works project has initial costs of $1,000,000, benefits of $1,500,000, and disbenefits of $300,000. (a) What is the benefit/cost ratio? (b) What is the excess of benefits over costs?

15. A speculator in land pays $14,000 for property that he expects to hold for ten years. $1000 is spent in renovation, and a monthly rent of $75 is collected from the tenants. (Use the year-end convention.) Taxes are $150 per year, and maintenance costs are $250 per year. What must be the sale price in ten years to realize a 10% rate of return?

16. What is the effective annual interest rate for a payment plan of 30 equal payments of $89.30 per month when a lump sum payment of $2000 would have been an outright purchase?

17. A depreciable item is purchased for $500,000. The salvage value at the end of 25 years is estimated at $100,000. What is the depreciation in each of the first three years using the (a) straight line, (b) sum-of-the-years' digits, and (c) double-declining balance methods?

18. Equipment that is purchased for $12,000 now is expected to be sold after ten years for $2000. The estimated maintenance is $1000 for the first year, but it is expected to increase $200 each year thereafter. The effective annual interest rate is 10%. What are the (a) present worth and (b) annual cost?

19. A new grain combine with a 20-year life can remove seven pounds of rocks from its harvest per hour. Any rocks left in its output hopper will cause $25,000 damage in subsequent processes. Several investments are available to increase the rock-removal capacity, as listed in the table. The effective annual interest rate is 10%. What should be done?

rock removal rate	probability of exceeding rock removal rate	required investment to achieve removal rate
7	0.15	0
8	0.10	$15,000
9	0.07	$20,000
10	0.03	$30,000

20. (*Time limit: one hour*) A mechanism that costs $10,000 has operating costs and salvage values as given. An effective annual interest rate of 20% is to be used.

year	operating cost	salvage value
1	$2000	$8000
2	$3000	$7000
3	$4000	$6000
4	$5000	$5000
5	$6000	$4000

(a) What is the economic life of the mechanism? (b) Assuming that the mechanism has been owned and operated for four years already, what is the cost of owning and operating the mechanism for one more year?

21. (*Time limit: one hour*) A salesperson intends to purchase a car for $50,000 for personal use, driving 15,000 miles per year. Insurance for personal use costs $2000 per year, and maintenance costs $1500 per year. The car gets 15 miles per gallon, and gasoline costs $1.50 per gallon. The resale value after five years will be $10,000. The salesperson's employer has asked that the car be used for business driving of 50,000 miles per year and has offered a reimbursement of $0.30 per mile. Using the car for business would increase the insurance cost to $3000 per year and maintenance to $2000 per year. The salvage value after five years would be reduced to $5000. If the employer purchased a car for the salesperson to use, the initial cost would be the same, but insurance, maintenance, and salvage would be $2500, $2000, and $8000, respectively. The salesperson's effective annual interest rate is 10%. (a) Is the reimbursement offer adequate? (b) With a reimbursement of $0.30 per mile, how many miles must the car be driven per year to justify the employer buying the car for the salesperson to use?

22. (*Time limit: one hour*) Alternatives A and B are being evaluated. The effective annual interest rate is 10%. What alternative is economically superior?

	alternative A	alternative B
first cost	$80,000	$35,000
life	20 years	10 years
salvage value	$7000	0
annual costs		
years 1–5	$1000	$3000
years 6–10	$1500	$4000
years 11–20	$2000	0
additional cost		
year 10	$5000	0

23. (*Time limit: one hour*) A car is needed for three years. Plans A and B for acquiring the car are being evaluated. An effective annual interest rate of 10% is to be used. Which plan is economically superior?

Plan A: lease the car for $0.25/mile (all inclusive)

Plan B: purchase the car for $30,000; keep the car for three years; sell the car after three years for $7200; pay $0.14 per mile for oil and gas; pay other costs of $500 per year

24. (*Time limit: one hour*) Two methods are being considered to meet strict air pollution control requirements over the next ten years. Method A uses equipment with a life of ten years. Method B uses equipment with a life of five years that will be replaced with new equipment with an additional life of five years. Capacities of the two methods are different, but operating costs do not depend on the throughput. Operation is 24 hours per day, 365 days per year. The effective annual interest rate for this evaluation is 7%.

	method A	method B	
	years 1–10	years 1–5	years 6–10
installation cost	$13,000	$6000	$7000
equipment cost	$10,000	$2000	$2200
operating cost per hour	$10.50	$8.00	$8.00
salvage value	$5000	$2000	$2000
capacity (tons/yr)	50	20	20
life	10 years	5 years	5 years

(a) What is the uniform annual cost per ton for each method? (b) Over what range of throughput (in units of tons/yr) does each method have the minimum cost?

25. (*Time limit: one hour*) A transit district has asked for your assistance in determining the proper fare for its bus system. An effective annual interest rate of 7% is to be used. The following additional information was compiled for your study.

cost per bus	$60,000
bus life	20 years
salvage value	$10,000
miles driven per year	37,440
number of passengers per year	80,000
operating cost	$1.00 per mile in the first year, increasing $0.10 per mile each year thereafter

(a) If the fare is to remain constant for the next 20 years, what is the break-even fare per passenger? (b) If the transit district decides to set the per-passenger fare at $0.35 for the first year, by what amount should the per-passenger fare go up each year thereafter such that the district can break even in 20 years? (c) If the transit district decides to set the per-passenger fare at $0.35 for the first year and the per-passenger fare goes up $0.05 each year thereafter, what additional governmental subsidy (per passenger) is needed for the district to break even in 20 years?

26. Make a recommendation to your client to accept one of the following alternatives. Use the present worth comparison method. (Initial costs are the same.)

Alternative A: a 25 year annuity paying $4800 at the end of each year, where the interest rate is a nominal 12% per annum

Alternative B: a 25 year annuity paying $1200 every quarter at 12% nominal annual interest

27. A firm has two alternatives for improvement of its existing production line. The data are as follows.

	alternative A	alternative B
initial installment cost	$1500	$2500
annual operating cost	$800	$650
service life	5 years	8 years
salvage value	0	0

Determine the best alternative using an interest rate of 15%.

28. Two mutually exclusive alternatives requiring different investments are being considered. The life of both alternatives is estimated at 20 years with no salvage values. The minimum rate of return that is considered acceptable is 4%. Which alternative is best?

	alternative A	alternative B
investment required	$70,000	$40,000
net income per year	$5620	$4075
rate of return on total investment	5%	8%

29. Compare the costs of two plant renovation schemes, A and B. Assume equal lives of 25 years, no salvage values, and interest at 25%. Make the comparison on the basis of (a) present worth, (b) capitalized cost, and (c) annual cost.

	alternative A	alternative B
first cost	$20,000	$25,000
annual expenditure	$3000	$2500

30. With interest at 8%, obtain the solutions to the following to the nearest dollar. (a) A machine costs $18,000 and has a salvage value of $2000. It has a useful life of 8 years. What is its book value at the end of 5 years using straight line depreciation? (b) Using data from part (a), find the depreciation in the first three years using the sinking fund method. (c) Repeat part (a) using double declining balance depreciation to find the first five years' depreciation.

31. A chemical pump motor unit is purchased for $14,000. The estimated life is 8 years, after which it will be sold for $1800. Find the depreciation in the first two years by the sum-of-the-years' digits method. Calculate the after-tax depreciation recovery using 15% interest with 52% income tax.

32. A soda ash plant has the water effluent from processing equipment treated in a large settling basin. The settling basin eventually discharges into a river that runs alongside the basin. Recently enacted environmental regulations require all rainfall on the plant to be diverted and treated in the settling basin. A heavy rainfall will cause the entire basin to overflow. An uncontrolled overflow will cause environmental damage and heavy fines. The construction of additional height on the existing basic walls is under consideration.

Data on the costs of construction and expected costs for environmental cleanup and fines are shown. Data on 50 typical winters have been collected. The soda ash plant management considers 12% to be their minimum rate of return, and it is felt that after 15 years the plant will be closed. The company wants to select the alternative that minimizes its total expected costs.

additional basin height (ft)	number of winters with basin overflow	expense for environmental clean up per year	construction cost
0	24	$550,000	0
5	14	$600,000	$600,000
10	8	$650,000	$710,000
15	3	$700,000	$900,000
20	1	$800,000	$1,000,000
	50		

33. A wood processing plant installed a waste gas scrubber at a cost of $30,000 to remove pollutants from the exhaust discharged into the atmosphere. The scrubber has no salvage value and will cost $18,700 to operate next year, with operating costs expected to increase at the rate of $1200 per year thereafter. When should the company consider replacing the scrubber? Money can be borrowed at 12%.

34. Two alternative piping schemes are being considered by a water treatment facility. Head and horsepower are reflected in the hourly cost of operation. On the basis of a 10-year life and an interest rate of 12%, determine the number of hours of operation for which the two installations will be equivalent.

	alternative A	alternative B
pipe diameter	4 in	6 in
head loss for required flow	48 ft	26 ft
size motor required	20 hp	7 hp
energy cost per hour of operation	$0.30	$0.10
cost of motor installed	$3600	$2800
cost of pipes and fittings	$3050	$5010
salvage value at end of 10 years	$200	$280

35. An 88% learning curve is used with an item whose first production time was 6 weeks. How long will it take to produce the fourth item? How long will it take to produce the sixth through fourteenth items?

36. (*Time limit: one hour*) A company is considering two alternatives, only one of which can be selected.

alternative	initial investment	salvage value	annual net profit	life
A	$120,000	$15,000	$57,000	5 yr
B	$170,000	$20,000	$67,000	5 yr

The net profit is after operating and maintenance costs, but before taxes. The company pays 45% of its year-end profit as income taxes. Use straight line depreciation. Do not use investment tax credit. Find the best alternative if the company's minimum attractive rate of return is 15%.

37. (*Time limit: one hour*) A company is considering the purchase of equipment to expand its capacity. The equipment cost is $300,000. The equipment is needed for 5 years, after which it will be sold for $50,000. The company's before-tax cash flow will be improved $90,000 annually by the purchase of the asset. The corporate tax rate is 48%, and straight line depreciation will be used. The company will take an investment tax credit of 6.67%. What is the after-tax rate of return associated with this equipment purchase?

38. (*Time limit: one hour*) A 120-room hotel is purchased for $2,500,000. A 25-year loan is available for 12%. The year-end convention applies to loan payments. A study was conducted to determine the various occupancy rates.

occupancy	probability
65% full	0.40
70%	0.30
75%	0.20
80%	0.10

The operating costs of the hotel are as follows.

taxes and insurance	$20,000 annually
maintenance	$50,000 annually
operating	$200,000 annually

The life of the hotel is figured to be 25 years when operating 365 days per year. The salvage value after 25 years is $500,000.

Neglect tax credit and income taxes. Determine the average rate that should be charged per room per night to return 15% of the initial cost each year.

39. (*Time limit: one hour*) A company is insured for $3,500,000 against fire and the insurance rate is $0.69/$1000. The insurance company will decrease the rate to $0.47/$1000 if fire sprinklers are installed. The initial cost of the sprinklers is $7500. Annual costs are $200; additional taxes are $100 annually. The system life is 25 years. What is the rate of return on this investment?

40. (*Time limit: one hour*) Heat losses through the walls in an existing building cost a company $1,300,000 per year. This amount is considered excessive, and two alternatives are being evaluated. Neither of the alternatives will increase the life of the existing building beyond the current expected life of 6 years, and neither of the alternatives will produce a salvage value. Improvements can be depreciated.

Alternative A: Do nothing, and continue with current losses.

Alternative B: Spend $2,000,000 immediately to upgrade the building and reduce the loss by 80%. Annual maintenance will cost $150,000.

Alternative C: Spend $1,200,000 immediately. Then, repeat the $1,200,000 expenditure 3 years from now. Heat loss the first year will be reduced 80%. Due to deterioration, the reduction will be 55% and 20% in the second and third years. (The pattern is repeated starting after the second expenditure.) There are no maintenance costs.

All energy and maintenance costs are considered expenses for tax purposes. The company's tax rate is 48%, and straight line depreciation is used. 15% is regarded as the effective annual interest rate. Evaluate each alternative on an after-tax basis, and recommend the best alternative.

41. (*Time limit: one hour*) You have been asked to determine if a 7-year-old machine should be replaced. Give a full explanation for your recommendation. Base your decision on a before-tax interest rate of 15%.

The existing machine is presumed to have a 10-year life. It has been depreciated on a straight line basis from its original value of $1,250,000 to a current book value of $620,000. Its ultimate salvage value was assumed to be $350,000 for purposes of depreciation. Its present salvage value is estimated at $400,000, and this is not expected to change over the next 3 years. The current operating costs are not expected to change from $200,000 per year.

A new machine costs $800,000, with operating costs of $40,000 the first year, and increasing by $30,000 each year thereafter. The new machine has an expected life of 10 years. The salvage value depends on the year the new machine is retired.

year retired	salvage
1	$600,000
2	$500,000
3	$450,000
4	$400,000
5	$350,000
6	$300,000
7	$250,000
8	$200,000
9	$150,000
10	$100,000

SOLUTIONS

1.

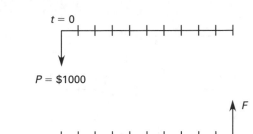

$i = 6\%$ a year

Using the formula from Table 87.1,

$$F = P(1+i)^n = (\$1000)(1+0.06)^{10}$$
$$= \boxed{\$1790.85}$$

Using the factor from App. 87.B, $(F/P, i, n) = 1.7908$ for $i = 6\%$ a year and $n = 10$ years.

$$F = P(F/P, 6\%, 10) = (\$1000)(1.7908)$$
$$= \boxed{\$1790.80}$$

2.

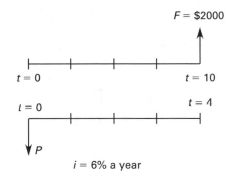

$i = 6\%$ a year

Using the formula from Table 87.1,

$$P = \frac{F}{(1+i)^n} = \frac{\$2000}{(1+0.06)^4}$$
$$= \boxed{\$1584.19}$$

Using the factor from App. 87.B, $(P/F, i, n) = 0.7921$ for $i = 6\%$ a year and $n = 4$ years.

$$P = F(P/F, 6\%, 4) = (\$2000)(0.7921)$$
$$= \boxed{\$1584.20}$$

3.

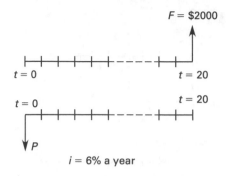

$$i = 6\% \text{ a year}$$

Using the formula from Table 87.1,

$$P = \frac{F}{(1+i)^n} = \frac{\$2000}{(1+0.06)^{20}}$$

$$= \boxed{\$623.61}$$

Using the factor from App. 87.B, $(P/F, i, n) = 0.3118$ for $i = 6\%$ a year and $n = 20$ years.

$$P = F(P/F, 6\%, 20) = (\$2000)(0.3118)$$
$$= \boxed{\$623.60}$$

4.

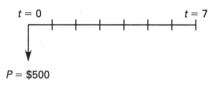

$$i = 6\% \text{ a year}$$

Using the formula from Table 87.1,

$$A = P\left(\frac{i(1+i)^n}{(1+i)^n - 1}\right) = (\$500)\left(\frac{(0.06)(1+0.06)^7}{(1+0.06)^7 - 1}\right)$$

$$= \boxed{\$89.57}$$

Using the factor from App. 87.B, $(A/P, i, n) = 0.17914$ for $i = 6\%$ a year and $n = 7$ years.

$$A = P(A/P, 6\%, 7) = (\$500)(0.17914)$$
$$= \boxed{\$89.57}$$

5.

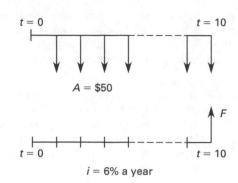

$$i = 6\% \text{ a year}$$

Using the formula from Table 87.1,

$$F = A\left(\frac{(1+i)^n - 1}{i}\right)$$

$$= (\$50)\left(\frac{(1+0.06)^{10} - 1}{0.06}\right)$$

$$= \boxed{\$659.04}$$

Using the factor from App. 87.B, $(F/A, i, n) = 13.181$ for $i = 6\%$ a year and $n = 10$ years.

$$F = A(F/A, 6\%, 10) = (\$50)(13.181)$$
$$= \boxed{\$659.05}$$

6.

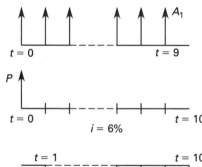

$$i = 6\%$$

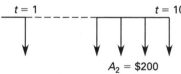

$$A_2 = \$200$$

From Table 87.1, for each cash flow diagram,

$$P = A_1 + A_1\left(\frac{(1+0.06)^9 - 1}{(0.06)(1+0.06)^9}\right)$$

$$= A_2\left(\frac{(1+0.06)^{10} - 1}{(0.06)(1+0.06)^{10}}\right)$$

Therefore, for $A_2 = \$200$,

$$A_1 + A_1\left(\frac{(1+0.06)^9 - 1}{(0.06)(1+0.06)^9}\right)$$

$$= (\$200)\left(\frac{(1+0.06)^{10} - 1}{(0.06)(1+0.06)^{10}}\right)$$

$$7.80A_1 = \$1472.02$$

$$A_1 = \boxed{\$188.72}$$

Using factors from App. 87.B,

$$(P/A, 6\%, 9) = 6.8017$$

$$(P/A, 6\%, 10) = 7.3601$$

$$A_1 + A_1(6.8017) = (\$200)(7.3601)$$

$$7.8017A_1 = \$1472.02$$

$$A_1 = \frac{\$1472.02}{7.8017}$$

$$= \boxed{\$188.68}$$

7.

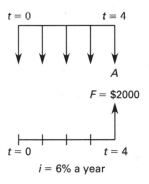

From Table 87.1,

$$F = A\left(\frac{(1+i)^n - 1}{i}\right)$$

Since the deposits start at the beginning of each year, five deposits are made that contribute to the final amount. This is equivalent to a cash flow that starts at $t = -1$ without a deposit and has a duration (starting at $t = -1$) of 5 years.

$$F = A\left(\frac{(1+i)^n - 1}{i}\right) = \$2000$$

$$= A\left(\frac{(1+0.06)^5 - 1}{0.06}\right)$$

$$\$2000 = 5.6371A$$

$$A = \frac{\$2000}{5.6371}$$

$$= \boxed{\$354.79}$$

Using factors from App. 87.B,

$$F = A\big((F/P, 6\%, 4) + (F/A, 6\%, 4)\big)$$

$$\$2000 = A(1.2625 + 4.3746)$$

$$A = \boxed{\$354.79}$$

8.

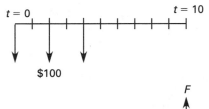

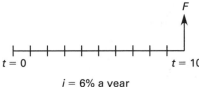

From Table 87.1, $F = P(1 + i)^n$. If each deposit is considered as P, each will accumulate interest for periods of 10, 8, and 6 years.

Therefore,

$$F = (\$100)(1+0.06)^{10} + (\$100)(1+0.06)^8$$
$$+ (\$100)(1+0.06)^6$$
$$= (\$100)(1.7908 + 1.5938 + 1.4185)$$
$$= \boxed{\$480.31}$$

Using App. 87.B,

$$(F/P, i, n) = 1.7908 \text{ for } i = 6\% \text{ and } n = 10$$
$$= 1.5938 \text{ for } i = 6\% \text{ and } n = 8$$
$$= 1.4185 \text{ for } i = 6\% \text{ and } n = 6$$

By summation,

$$F = (\$100)(1.7908 + 1.5938 + 1.4185)$$
$$= \boxed{\$480.31}$$

9.

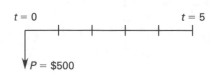

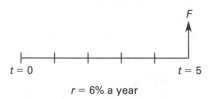

$r = 6\%$ a year

Since the deposit is compounded monthly, the effective interest rate should be calculated from Eq. 87.51.

$$i = \left(1 + \frac{r}{k}\right)^k - 1 = \left(1 + \frac{0.06}{12}\right)^{12} - 1$$
$$= 0.061678 \quad (6.1678\%)$$

From Table 87.1,

$$F = P(1 + i)^n = (\$500)(1 + 0.061678)^5 = \boxed{\$674.43}$$

To use App. 87.B, interpolation is required.

$i\%$	factor F/P
6	1.3382
6.1678	desired
7	1.4026

$$\Delta(F/P) = \left(\frac{6.1678\% - 6\%}{7\% - 6\%}\right)(1.4026 - 1.3382)$$
$$= 0.0108$$

Therefore,

$$F/P = 1.3382 + 0.0108 = 1.3490$$
$$F = P(F/P, 6.1677\%, 5) = (\$500)(1.3490)$$
$$= \boxed{\$674.50}$$

10.

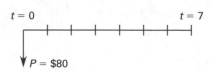

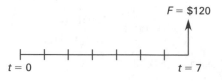

From Table 87.1,

$$F = P(1 + i)^n$$

Therefore,

$$(1 + i)^n = F/P$$
$$i = (F/P)^{1/n} - 1 = \left(\frac{\$120}{\$80}\right)^{1/7} - 1$$
$$= 0.0596 \approx \boxed{6\%}$$

From App. 87.B,

$$F = P(F/P, i\%, 7)$$
$$(F/P, i\%, 7) = F/P = \frac{\$120}{\$80}$$
$$= 1.5$$

Searching App. 87.B,

$$(F/P, i\%, 7) = 1.4071 \text{ for } i = 5\%$$
$$= 1.5036 \text{ for } i = 6\%$$
$$= 1.6058 \text{ for } i = 7\%$$

Therefore, $i \approx \boxed{6\%}.$

11.

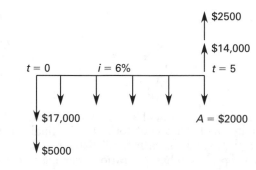

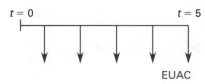

Annual cost of ownership, EUAC, can be obtained by the factors converting P to A and F to A.

$$P = \$17,000 + \$5000$$
$$= \$22,000$$
$$F = \$14,000 + \$2500$$
$$= \$16,500$$
$$\text{EUAC} = A + P(A/P, 6\%, 5) - F(A/F, 6\%, 5)$$
$$(A/P, 6\%, 5) = 0.23740$$
$$(A/F, 6\%, 5) = 0.17740$$
$$\text{EUAC} = \$2000 + (\$22,000)(0.23740)$$
$$- (\$16,500)(0.17740)$$
$$= \boxed{\$4295.70}$$

12.

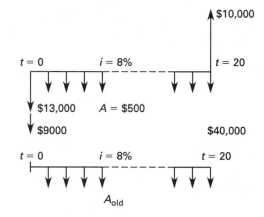

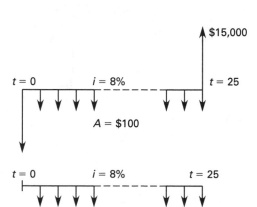

Consider the salvage value as a benefit lost (cost).

$$\text{EUAC}_{\text{old}} = \$500 + (\$9000 + \$13,000)(A/P, 8\%, 20)$$
$$- (\$10,000)(A/F, 8\%, 20)$$
$$(A/P, 8\%, 20) = 0.1019$$
$$(A/F, 8\%, 20) = 0.0219$$
$$\text{EUAC}_{\text{old}} = \$500 + (\$22,000)(0.1019)$$
$$- (\$10,000)(0.0219)$$
$$= \$2522.80$$

Similarly,

$$\text{EUAC}_{\text{new}} = \$100 + (\$40,000)(A/P, 8\%, 25)$$
$$- (\$15,000)(A/F, 8\%, 25)$$
$$(A/P, 8\%, 25) = 0.0937$$
$$(A/F, 8\%, 25) = 0.0137$$
$$\text{EUAC}_{\text{new}} = \$100 + (\$40,000)(0.0937)$$
$$- (\$15,000)(0.0137)$$
$$= \$3642.50$$

Therefore, the new bridge is more costly.

$$\boxed{\text{The best alternative is to strengthen the old bridge.}}$$

13.

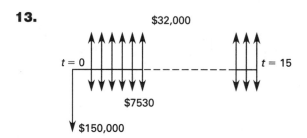

The annual depreciation is

$$D = \frac{C - S_n}{n} = \frac{\$150,000}{15}$$
$$= \$10,000/\text{year}$$

The taxable income is

$$\$32,000 - \$7530 - \$10,000 = \$14,470/\text{year}$$

Taxes paid are

$$(\$14,470)(0.48) = \$6945.60/\text{year}$$

The after-tax cash flow is

$$\$32{,}000 - \$7530 - \$6945.60 = \$17{,}524.40$$

The present worth of the alternate is zero when evaluated at its ROR.

$$0 = -\$150{,}000 + (\$17{,}524.40)(P/A, i\%, 15)$$

Therefore,

$$(P/A, i\%, 15) = \frac{\$150{,}000}{\$17{,}524.40} = 8.55949$$

Searching App. 87.B, this factor matches $i = 8\%$.

$$\boxed{\text{ROR} = 8\%}$$

14. The conventional benefit/cost ratio is

$$B/C = \frac{B - D}{D}$$

(a) The benefit/cost ratio will be

$$B/C = \frac{\$1{,}500{,}000 - \$300{,}000}{\$1{,}000{,}000} = \boxed{1.2}$$

(b) The excess of benefits over cost is $\boxed{\$200{,}000.}$

15. The annual rent is

$$(\$75)\left(12 \ \frac{\text{months}}{\text{year}}\right) = \$900$$

$$P = P_1 + P_2 = \$15{,}000$$

$$A_1 = -\$900$$

$$A_2 = \$250 + \$150 = \$400$$

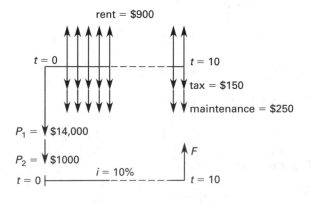

Use App. 87.B.

$$F = (\$15{,}000)(F/P, 10\%, 10)$$
$$+ (\$400)(F/A, 10\%, 10)$$
$$- (\$900)(F/A, 10\%, 10)$$

$$(F/P, 10\%, 10) = 2.5937$$
$$(F/A, 10\%, 10) = 15.937$$

$$F = (\$15{,}000)(2.5937) + (\$400)(15.937)$$
$$- (\$900)(15.937)$$
$$= \boxed{\$30{,}937}$$

16.

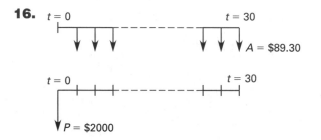

From Table 87.1,

$$P = A\left(\frac{(1 + i)^n - 1}{i(1 + i)^n}\right)$$

$$\frac{(1 + i)^{30} - 1}{i(1 + i)^{30}} = \frac{\$2000}{\$89.30} = 22.40$$

By trial and error,

$i\%$	$(1 + i)^{30}$	$\dfrac{(1 + i)^{30} - 1}{i(1 + i)^{30}}$
10	17.45	9.42
6	5.74	13.76
4	3.24	17.28
2	1.81	22.37

2% per month is close.

$$i = (1 + 0.02)^{12} - 1 = \boxed{0.2682 \quad (26.82\%)}$$

17. (a) Use the straight line method, Eq. 87.25.

$$D = \frac{C - S_n}{n}$$

Each year depreciation will be the same.

$$D = \frac{\$500{,}000 - \$100{,}000}{25} = \boxed{\$16{,}000}$$

(b) Use Eq. 87.27.

$$T = \tfrac{1}{2}n(n+1) = \left(\tfrac{1}{2}\right)(25)(25+1) = 325$$

Sum-of-the-years' digits (SOYD) depreciation can be calculated from Eq. 87.28.

$$D_j = \frac{(C - S_n)(n - j + 1)}{T}$$

$$D_1 = \frac{(\$500{,}000 - \$100{,}000)(25 - 1 + 1)}{325}$$

$$= \boxed{\$30{,}769}$$

$$D_2 = \frac{(\$500{,}000 - \$100{,}000)(25 - 2 + 1)}{325}$$

$$= \boxed{\$29{,}538}$$

$$D_3 = \frac{(\$500{,}000 - \$100{,}000)(25 - 3 + 1)}{325}$$

$$= \boxed{\$28{,}308}$$

(c) The double declining balance (DDB) method can be used. By Eq. 87.32,

$$D_j = dC(1 - d)^{j-1}$$

Use Eq. 87.31.

$$d = \frac{2}{n} = \frac{2}{25}$$

$$D_1 = \left(\tfrac{2}{25}\right)(\$500{,}000)\left(1 - \tfrac{2}{25}\right)^0 = \boxed{\$40{,}000}$$

$$D_2 = \left(\tfrac{2}{25}\right)(\$500{,}000)\left(1 - \tfrac{2}{25}\right)^1 = \boxed{\$36{,}800}$$

$$D_3 = \left(\tfrac{2}{25}\right)(\$500{,}000)\left(1 - \tfrac{2}{25}\right)^2 = \boxed{\$33{,}856}$$

18.

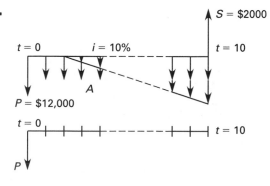

(a) A is \$1000, and G is \$200 for $t = n - 1 = 9$ years. With $F = S = \$2000$, the present worth is

$$P = \$12{,}000 + A(P/A, 10\%, 10) + G(P/G, 10\%, 10)$$
$$\quad - F(P/F, 10\%, 10)$$

$$= \$12{,}000 + (\$1000)(6.1446) + (\$200)(22.8913)$$
$$\quad - (\$2000)(0.3855)$$

$$= \boxed{\$21{,}952}$$

```
t = 0                                    t = 10
 |————|———|———|——— — — — ——|———|———
      |   |   |            |   |
      ↓   ↓   ↓            ↓   ↓
  A
```

(b) The annual cost is

$$A = (\$12{,}000)(A/P, 10\%, 10) + \$1000$$
$$\quad + (\$200)(A/G, 10\%, 10)$$
$$\quad - (\$2000)(A/F, 10\%, 10)$$

$$= (\$12{,}000)(0.1627) + \$1000 + (\$200)(3.7255)$$
$$\quad - (\$2000)(0.0627)$$

$$= \boxed{\$3572.10}$$

19. An increase in rock removal capacity can be achieved by a 20-year loan (investment). Different cases available can be compared by equivalent uniform annual cost (EUAC).

$$\begin{aligned}
\text{EUAC} &= \text{annual loan cost} \\
&\quad + \text{expected annual damage} \\
&= \text{cost}\,(A/P, 10\%, 20) \\
&\quad + (\$25{,}000)(\text{probability})
\end{aligned}$$

$$(A/P, 10\%, 20) = 0.1175$$

A table can be prepared for different cases.

rock removal rate	cost ($)	annual loan cost ($)	expected annual damage ($)	EUAC ($)
7	0	0	3750	3750.00
8	15,000	1761.90	2500	4261.90
9	20,000	2349.20	1750	4099.20
10	30,000	3523.80	750	4273.80

$$\boxed{\text{It is cheapest to do nothing.}}$$

20. Calculate the cost of owning and operating for each year.

$$A_1 = (\$10{,}000)(A/P, 20\%, 1) + \$2000$$
$$- (\$8000)(A/F, 20\%, 1)$$

$$(A/P, 20\%, 1) = 1.2$$
$$(A/F, 20\%, 1) = 1.0$$

$$A_1 = (\$10{,}000)(1.2) + \$2000 - (\$8000)(1.0)$$
$$= \$6000$$

$$A_2 = (\$10{,}000)(A/P, 20\%, 2) + \$2000$$
$$+ (\$1000)(A/G, 20\%, 2)$$
$$- (\$7000)(A/F, 20\%, 2)$$

$$(A/P, 20\%, 2) = 0.6545$$
$$(A/G, 20\%, 2) = 0.4545$$
$$(A/F, 20\%, 2) = 0.4545$$

$$A_2 = (\$10{,}000)(0.6545) + \$2000$$
$$+ (\$1000)(0.4545) - (\$7000)(0.4545)$$
$$= \$5818$$

$$A_3 = (\$10{,}000)(A/P, 20\%, 3) + \$2000$$
$$+ (\$1000)(A/G, 20\%, 3)$$
$$- (\$6000)(A/F, 20\%, 3)$$

$$(A/P, 20\%, 3) = 0.4747$$
$$(A/G, 20\%, 3) = 0.8791$$
$$(A/F, 20\%, 3) = 0.2747$$

$$A_3 = (\$10{,}000)(0.4747) + \$2000$$
$$+ (\$1000)(0.8791)$$
$$- (\$6000)(0.2747)$$
$$= \$5977.90$$

$$A_4 = (\$10{,}000)(A/P, 20\%, 4)$$
$$+ \$2000 + (\$1000)(A/G, 20\%, 4)$$
$$- (\$5000)(A/F, 20\%, 4)$$

$$(A/P, 20\%, 4) = 0.3863$$
$$(A/G, 20\%, 4) = 1.2762$$
$$(A/F, 20\%, 4) = 0.1863$$

$$A_4 = (\$10{,}000)(0.3863) + \$2000$$
$$+ (\$1000)(1.2762) - (\$5000)(0.1863)$$
$$= \$6207.70$$

$$A_5 = (\$10{,}000)(A/P, 20\%, 5) + \$2000$$
$$+ (\$1000)(A/G, 20\%, 5)$$
$$- (\$4000)(A/F, 20\%, 5)$$

$$(A/P, 20\%, 5) = 0.3344$$
$$(A/G, 20\%, 5) = 1.6405$$
$$(A/F, 20\%, 5) = 0.1344$$

$$A_5 = (\$10{,}000)(0.3344) + \$2000$$
$$+ (\$1000)(1.6405) - (\$4000)(0.1344)$$
$$= \$6446.90$$

(a) Since the annual owning and operating cost is smallest after two years of operation, it is advantageous to sell the mechanism after the second year.

> The economic life is two years.

(b) After four years of operation, the owning and operating cost of the mechanism for one more year will be

$$A = \$6000 + (\$5000)(1 + i) - \$4000$$
$$i = 0.2 \quad (20\%)$$
$$A = \$6000 + (\$5000)(1.2) - \$4000$$
$$= \boxed{\$8000}$$

21. To find out if the reimbursement is adequate, calculate the business-related expense.

Charge the company for business travel.

$$\begin{aligned}
\text{insurance:} \quad & \$3000 - \$2000 = \$1000 \\
\text{maintenance:} \quad & \$2000 - \$1500 = \$500 \\
\text{drop in salvage value:} \quad & \$10{,}000 - \$5000 = \$5000
\end{aligned}$$

The annual portion of the drop in salvage value is

$$A = (\$5000)(A/F, 10\%, 5)$$
$$(A/F, 10\%, 5) = 0.1638$$
$$A = (\$5000)(0.1638) = \$819/\text{year}$$

(a) The annual cost of gas is

$$\left(\frac{50{,}000 \text{ mi}}{15 \frac{\text{mi}}{\text{gal}}}\right)\left(\frac{\$1.50}{\text{gal}}\right) = \$5000$$

$$\text{EUAC per mile} = \frac{\$1000 + \$500 + \$819 + \$5000}{50{,}000 \text{ mi}}$$
$$= \boxed{\$0.14638/\text{mi}}$$

Since the reimbursement per mile was \$0.30 and since \$0.30 > \$0.14638, the reimbursement is adequate.

Systems, Mgmt, and Professional

(b) Next, determine (with reimbursement) how many miles the car must be driven to break even.

If the car is driven M miles per year,

$$\left(\frac{\$0.30}{1\ \text{mi}}\right) M = (\$50,000)(A/P, 10\%, 5) + \$2500$$
$$+ \$2000 - (\$8000)(A/F, 10\%, 5)$$
$$+ \left(\frac{M}{15\ \frac{\text{mi}}{\text{gal}}}\right)(\$1.50)$$

$$(A/P, 10\%, 5) = 0.2638$$
$$(A/F, 10\%, 5) = 0.1638$$
$$0.3M = (\$50,000)(0.2638) + \$2500 + \$2000$$
$$- (\$8000)(0.1638) + 0.1M$$
$$0.2M = \$16,379.60$$
$$M = \frac{\$16,379.60}{\frac{\$0.20}{1\ \text{mi}}} = \boxed{81,898\ \text{mi}}$$

22.

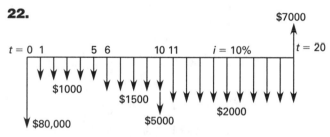

$$P_{\text{A}} = \$80,000 + (\$1000)(P/A, 10\%, 5)$$
$$+ (\$1500)(P/A, 10\%, 5)(P/F, 10\%, 5)$$
$$+ (\$2000)(P/A, 10\%, 10)(P/F, 10\%, 10)$$
$$+ (\$5000)(P/F, 10\%, 10)$$
$$- (\$7000)(P/F, 10\%, 20)$$

$$(P/A, 10\%, 5) = 3.7908$$
$$(P/F, 10\%, 5) = 0.6209$$
$$(P/A, 10\%, 10) = 6.1446$$
$$(P/F, 10\%, 10) = 0.3855$$
$$(P/F, 10\%, 20) = 0.1486$$
$$P_{\text{A}} = \$80,000 + (\$1000)(3.7908)$$
$$+ (\$1500)(3.7908)(0.6209)$$
$$+ (\$2000)(6.1446)(0.3855)$$
$$+ (\$5000)(0.3855)$$
$$- (\$7000)(0.1486)$$
$$= \$92,946.15$$

Since the lives are different, compare by EUAC.

$$\text{EUAC(A)} = (\$92,946.14)(A/P, 10\%, 20)$$
$$= (\$92,946.14)(0.1175)$$
$$= \$10,921$$

Similarly, evaluate alternative B.

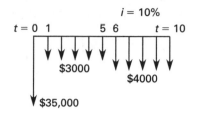

$$P_{\text{B}} = \$35,000 + (\$3000)(P/A, 10\%, 5)$$
$$+ (\$4000)(P/A, 10\%, 5)(P/F, 10\%, 5)$$
$$(P/A, 10\%, 5) = 3.7908$$
$$(P/F, 10\%, 5) = 0.6209$$
$$P_{\text{B}} = \$35,000 + (\$3000)(3.7908)$$
$$+ (\$4000)(3.7908)(0.6209)$$
$$= \$55,787.23$$
$$\text{EUAC(B)} = (\$55,787.23)(A/P, 10\%, 10)$$
$$= (\$55,787.23)(0.1627)$$
$$= \$9077$$

Since EUAC(B) < EUAC(A),

$$\boxed{\text{Alternative B is economically superior.}}$$

23. For both cases, if the annual cost is compared with a total annual mileage of M,

$$A_{\text{A}} = \$0.25M$$
$$A_{\text{B}} = (\$30,000)(A/P, 10\%, 3) + \$0.14M$$
$$+ \$500 - (\$7200)(A/F, 10\%, 3)$$
$$(A/P, 10\%, 3) = 0.4021$$
$$(A/F, 10\%, 3) = 0.3021$$
$$A_{\text{B}} = (\$30,000)(0.4021) + \$0.14M + \$500$$
$$- (\$7200)(0.3021)$$
$$= \$10,387.88 + \$0.14M$$

For an equal annual cost $A_A = A_B$,

$$\$0.25M = \$10{,}387.88 + \$0.14M$$
$$\$0.11M = \$10{,}387.88$$
$$M = 94{,}435$$

An annual mileage would be $M = 94{,}435$ mi.

For an annual mileage less than that, $A_A < A_B$.

Plan A is economically superior until that mileage is exceeded.

---

24. (a) Method A:

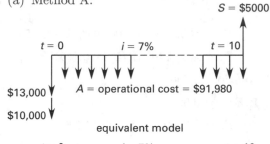

24 hr/day

365 days/yr

total of $(24)(365) = 8760$ hr/yr

$10.50 operational cost/hr

total of $(8760)(\$10.50) = \$91{,}980$ operational cost/yr

$$A = \$91{,}980 + (\$23{,}000)(A/P, 7\%, 10)$$
$$\quad - (\$5000)(A/F, 7\%, 10)$$
$$(A/P, 7\%, 10) = 0.1424$$
$$(A/F, 7\%, 10) = 0.0724$$
$$A = \$91{,}980 + (\$23{,}000)(0.1424)$$
$$\quad - (\$5000)(0.0724)$$
$$= \$94{,}893.20/\text{yr}$$

Therefore, the uniform annual cost per ton each year will be

$$\frac{\$94{,}893.20}{50 \text{ ton}} = \boxed{\$1897.86}$$

Method B:

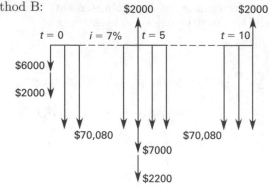

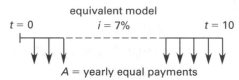

8760 hr/yr

$8 operational cost/hr

total of $70,080 operational cost/yr

$$A = \$70{,}080 + (\$6000 + \$2000)$$
$$\quad \times (A/P, 7\%, 10)$$
$$\quad + (\$7000 + \$2200 - \$2000)$$
$$\quad \times (P/F, 7\%, 5)(A/P, 7\%, 10)$$
$$\quad - (\$2000)(A/F, 7\%, 10)$$
$$(A/P, 7\%, 10) = 0.1424$$
$$(A/F, 7\%, 10) = 0.0724$$
$$(P/F, 7\%, 5) = 0.7130$$
$$A = \$70{,}080 + (\$8000)(0.1424)$$
$$\quad + (\$7200)(0.7130)(0.1424)$$
$$\quad - (\$2000)(0.0724)$$
$$= \$71{,}805.42/\text{yr}$$

Therefore, the uniform annual cost per ton each year will be

$$\frac{\$71{,}805.42}{20 \text{ ton}} = \boxed{\$3590.27}$$

(b) The following table can be used to determine which alternative is least expensive for each throughput range.

tons/yr	cost of using A		cost of using B		cheapest
0–20	$94,893	(1×)	$71,805	(1×)	B
20–40	$94,893	(1×)	$143,610	(2×)	A
40–50	$94,893	(1×)	$215,415	(3×)	A
50–60	$189,786	(2×)	$215,415	(3×)	A
60–80	$189,786	(2×)	$287,220	(4×)	A

ENGINEERING ECONOMIC ANALYSIS **87-15**

25.
$$A_e = (\$60{,}000)(A/P, 7\%, 20) + A$$
$$+ G(P/G, 7\%, 20)(A/P, 7\%, 20)$$
$$- (\$10{,}000)(A/F, 7\%, 20)$$

$$(A/P, 7\%, 20) = 0.0944$$

$$A = (37{,}440 \text{ mi})\left(\frac{\$1.0}{1 \text{ mi}}\right) = \$37{,}440$$

$$G = 0.1A = (0.1)(\$37{,}440) = \$3744$$

$$(P/G, 7\%, 20) = 77.5091$$

$$(A/F, 7\%, 20) = 0.0244$$

$$A_e = (\$60{,}000)(0.0944) + \$37{,}440$$
$$+ (\$3744)(77.5091)(0.0944)$$
$$- (\$10{,}000)(0.0244)$$

$$= \$70{,}254.32$$

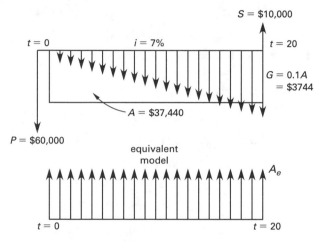

(a) With 80,000 passengers a year, the break-even fare per passenger would be

$$\text{fare} = \frac{A_e}{80{,}000} = \frac{\$70{,}254.32}{80{,}000} = \boxed{\$0.878/\text{passenger}}$$

(b) The passenger fare should go up each year by

$$\$0.878 = \$0.35 + G(A/G, 7\%, 20)$$

$$G = \frac{\$0.878 - \$0.35}{7.3163}$$

$$= \boxed{\$0.072 \text{ increase per year}}$$

(c) As in part (b), the subsidy should be

$$\text{subsidy} = \text{cost} - \text{revenue}$$

$$P = \$0.878 - \big(\$0.35 + (\$0.05)(A/G, 7\%, 20)\big)$$

$$= \$0.878 - \big(\$0.35 + (\$0.05)(7.3163)\big)$$

$$= \boxed{\$0.162}$$

26. Use the present worth comparison method.

$$P(A) = (\$4800)(P/A, 12\%, 25)$$
$$= (\$4800)(7.8431)$$
$$= \$37{,}646.88$$

$$(4 \text{ quarters})(25 \text{ years}) = 100 \text{ compounding periods}$$

$$P(B) = (\$1200)(P/A, 3\%, 100)$$
$$= (\$1200)(31.5989)$$
$$= \$37{,}918.68$$

$$\boxed{\text{Alternative B is economically superior.}}$$

27. Use the equivalent uniform annual cost method.

$$\text{EUAC(A)} = (\$1500)(A/P, 15\%, 5) + \$800$$
$$= (\$1500)(0.2983) + \$800$$
$$= \$1247.45$$

$$\text{EUAC(B)} = (\$2500)(A/P, 15\%, 8) + \$650$$
$$= (\$2500)(0.2229) + \$650$$
$$= \$1207.25$$

$$\boxed{\text{Alternative B is economically superior.}}$$

28. The data given imply that both investments return 4% or more. However, the increased investment of $30,000 may not be cost effective. Do an incremental analysis.

$$\text{incremental cost} = \$70{,}000 - \$40{,}000 = \$30{,}000$$

$$\text{incremental income} = \$5620 - \$4075 = \$1545$$

$$0 = -\$30{,}000 + (\$1545)(P/A, i\%, 20)$$

$$(P/A, i\%, 20) = 19.417$$

$$i \approx 0.25\% < 4\%$$

$$\boxed{\text{Alternative B is economically superior.}}$$

(The same conclusion could be reached by taking the present worths of both alternatives at 4%.)

Systems, Mgmt, and Professional

PPI • www.ppi2pass.com

29. (a)   The present worth comparison is

$$P(A) = (-\$3000)(P/A, 25\%, 25) - \$20,000$$
$$= (-\$3000)(3.9849) - \$20,000$$
$$= -\$31,954.70$$
$$P(B) = (-\$2500)(3.9849) - \$25,000$$
$$= -\$34,962.25$$

$$\boxed{\text{Alternative A is better.}}$$

(b)   The capitalized cost comparison is

$$\text{CC(A)} = \$20,000 + \frac{\$3000}{0.25} = \$32,000$$
$$\text{CC(B)} = \$25,000 + \frac{\$2500}{0.25} = \$35,000$$

$$\boxed{\text{Alternative A is better.}}$$

(c)   The annual cost comparison is

$$\text{EUAC(A)} = (\$20,000)(A/P, 25\%, 25) + \$3000$$
$$= (\$20,000)(0.2509) + \$3000$$
$$= \$8018.00$$
$$\text{EUAC(B)} = (\$25,000)(0.2509) + \$2500$$
$$= \$8772.50$$

$$\boxed{\text{Alternative A is better.}}$$

30. (a)   The depreciation in the first 5 years is

$$\text{BV} = \$18,000 - (5)\left(\frac{\$18,000 - \$2000}{8}\right)$$
$$= \boxed{\$8000}$$

(b)   With the sinking fund method, the basis is

$$(\$18,000 - \$2000)(A/F, 8\%, 8)$$
$$= (\$18,000 - \$2000)(0.0940)$$
$$= \$1504$$
$$D_1 = (\$1504)(1.000) = \boxed{\$1504}$$
$$D_2 = (\$1504)(1.0800) = \boxed{\$1624}$$
$$D_3 = (\$1504)(1.0800)^2 = \boxed{\$1754}$$

(c)   Using double declining balance depreciation, the first five years' depreciation is

$$D_1 = \left(\tfrac{2}{8}\right)(\$18,000) = \boxed{\$4500}$$
$$D_2 = \left(\tfrac{2}{8}\right)(\$18,000 - \$4500) = \boxed{\$3375}$$
$$D_3 = \left(\tfrac{2}{8}\right)(\$18,000 - \$4500 - \$3375) = \boxed{\$2531}$$
$$D_4 = \left(\tfrac{2}{8}\right)(\$18,000 - \$4500 - \$3375 - \$2531)$$
$$= \boxed{\$1898}$$
$$D_5 = \left(\tfrac{2}{8}\right)(\$18,000 - \$4500 - \$3375$$
$$\qquad - \$2531 - \$1898)$$
$$= \boxed{\$1424}$$
$$\text{BV} = \$18,000 - \$4500 - \$3375 - \$2531$$
$$\qquad - \$1898 - \$1424$$
$$= \boxed{\$4272}$$

31. The after-tax depreciation recovery is

$$T = \left(\tfrac{1}{2}\right)(8)(9) = 36$$
$$D_1 = \left(\tfrac{8}{36}\right)(\$14,000 - \$1800) = \$2711$$
$$\Delta D = \left(\tfrac{1}{36}\right)(\$14,000 - \$1800) = \$339$$
$$D_2 = \$2711 - \$339 = \$2372$$
$$\text{DR} = (0.52)(\$2711)(P/A, 15\%, 8)$$
$$\qquad - (0.52)(\$339)(P/G, 15\%, 8)$$
$$= (0.52)(\$2711)(4.4873)$$
$$\qquad - (0.52)(\$339)(12.4807)$$
$$= \boxed{\$4125.74}$$

32. Use the equivalent uniform annual cost method to find the best alternative.

$$(A/P, 12\%, 15) = 0.1468$$

$$\text{EUAC}_{5\,\text{ft}} = (\$600,000)(0.1468)$$
$$\qquad + \left(\tfrac{14}{50}\right)(\$600,000) + \left(\tfrac{8}{50}\right)(\$650,000)$$
$$\qquad + \left(\tfrac{3}{50}\right)(\$700,000) + \left(\tfrac{1}{50}\right)(\$800,000)$$
$$= \$418,080$$
$$\text{EUAC}_{10\,\text{ft}} = (\$710,000)(0.1468)$$
$$\qquad + \left(\tfrac{8}{50}\right)(\$650,000) + \left(\tfrac{3}{50}\right)(\$700,000)$$
$$\qquad + \left(\tfrac{1}{50}\right)(\$800,000)$$
$$= \$266,228$$

$$\text{EUAC}_{15\,\text{ft}} = (\$900{,}000)(0.1468)$$
$$+ \left(\tfrac{3}{50}\right)(\$700{,}000) + \left(\tfrac{1}{50}\right)(\$800{,}000)$$
$$= \$190{,}120$$

$$\text{EUAC}_{20\,\text{ft}} = (\$1{,}000{,}000)(0.1468)$$
$$+ \left(\tfrac{1}{50}\right)(\$800{,}000)$$
$$= \$162{,}800$$

$$\boxed{\text{Build to 20 ft.}}$$

33. Assume replacement after 1 year.

$$\text{EUAC}(1) = (\$30{,}000)(A/P, 12\%, 1) + \$18{,}700$$
$$= (\$30{,}000)(1.12) + \$18{,}700$$
$$= \$52{,}300$$

Assume replacement after 2 years.

$$\text{EUAC}(2) = (\$30{,}000)(A/P, 12\%, 2)$$
$$+ \$18{,}700 + (\$1200)(A/G, 12\%, 2)$$
$$= (\$30{,}000)(0.5917) + \$18{,}700$$
$$+ (\$1200)(0.4717)$$
$$= \$37{,}017$$

Assume replacement after 3 years.

$$\text{EUAC}(3) = (\$30{,}000)(A/P, 12\%, 3)$$
$$+ \$18{,}700 + (\$1200)(A/G, 12\%, 3)$$
$$= (\$30{,}000)(0.4163) + \$18{,}700$$
$$+ (\$1200)(0.9246)$$
$$= \$32{,}299$$

Similarly, calculate to obtain the numbers in the following table.

years in service	EUAC
1	$52,300
2	$37,017
3	$32,299
4	$30,207
5	$29,152
6	$28,602
7	$28,335
8	$28,234
9	$28,240
10	$28,312

$$\boxed{\text{Replace after 8 yr.}}$$

34. Since the head and horsepower data are already reflected in the hourly operating costs, there is no need to work with head and horsepower.

Let N = no. of hours operated each year.

$$\text{EUAC(A)} = (\$3600 + \$3050)(A/P, 12\%, 10)$$
$$- (\$200)(A/F, 12\%, 10) + 0.30N$$
$$= (\$6650)(0.1770) - (\$200)(0.0570) + 0.30N$$
$$= 1165.65 + 0.30N$$

$$\text{EUAC(B)} = (\$2800 + \$5010)(A/P, 12\%, 10)$$
$$+ (\$280)(A/F, 12\%, 10) + 0.10N$$
$$= (\$7810)(0.1770) - (\$280)(0.0570) + 0.10N$$
$$= 1366.41 + 0.10N$$

$$\text{EUAC(A)} = \text{EUAC(B)}$$
$$1165.65 + 0.30N = 1366.41 + 0.10N$$
$$N = \boxed{1003.8 \text{ hr}}$$

35. (a) From Eq. 87.88,

$$\frac{T_2}{T_1} = 0.88 = 2^{-b}$$
$$\log 0.88 = -b \log 2$$
$$-0.0555 = -(0.3010)b$$
$$b = 0.1843$$

$$T_4 = (6)(4)^{-0.1843} = \boxed{4.65 \text{ wk}}$$

(b) From Eq. 87.89,

$$T_{6-14} = \left(\frac{T_1}{1-b}\right)\left(\left(n_2 + \tfrac{1}{2}\right)^{1-b} - \left(n_1 - \tfrac{1}{2}\right)^{1-b}\right)$$
$$= \left(\frac{6}{1 - 0.1843}\right)$$
$$\times \left(\left(14 + \tfrac{1}{2}\right)^{1-0.1843} - \left(6 - \tfrac{1}{2}\right)^{1-0.1843}\right)$$
$$= \left(\frac{6}{0.8157}\right)(8.857 - 4.017)$$
$$= \boxed{35.6 \text{ wk}}$$

36. First check that both alternatives have an ROR greater than the MARR. Work in thousands of dollars. Evaluate alternative A.

$$P(A) = -\$120 + (\$15)(P/F, i\%, 5)$$
$$+ (\$57)(P/A, i\%, 5)(1 - 0.45)$$
$$+ \left(\frac{\$120 - \$15}{5}\right)(P/A, i\%, 5)(0.45)$$
$$= -\$120 + (\$15)(P/F, i\%, 5)$$
$$+ (\$40.8)(P/A, i\%, 5)$$

Try 15%.

$$P(A) = -\$120 + (\$15)(0.4972) + (\$40.8)(3.3522)$$
$$= \$24.23$$

Try 25%.

$$P(A) = -\$120 + (\$15)(0.3277) + (\$40.8)(2.6893)$$
$$= -\$5.36$$

Since $P(A)$ goes through 0,

$$(ROR)_A > MARR = 15\%$$

Next, evaluate alternative B.

$$P(B) = -\$170 + (\$20)(P/F, i\%, 5)$$
$$+ (\$67)(P/A, i\%, 5)(1 - 0.45)$$
$$+ \left(\frac{\$170 - \$20}{5}\right)(P/A, i\%, 5)(0.45)$$
$$= -\$170 + (\$20)(P/F, i\%, 5)$$
$$+ (\$50.35)(P/A, i\%, 5)$$

Try 15%.

$$P(B) = -\$170 + (\$20)(0.4972) + (\$50.35)(3.352)$$
$$= \$8.72$$

Since $P(B) > 0$ and will decrease as i increases,

$$(ROR)_B > 15\%$$

ROR > MARR for both alternatives.

Do an incremental analysis to see if it is worthwhile to invest the extra $\$170 - \$120 = \$50$.

$$P(B - A) = -\$50 + (\$20 - \$15)(P/F, i\%, 5)$$
$$+ (\$50.35 - \$40.8)(P/A, i\%, 5)$$

Try 15%.

$$P(B - A) = -\$50 + (\$5)(0.4972)$$
$$+ (\$9.55)(3.3522)$$
$$= -\$15.50$$

Since $P(B - A) < 0$ and would become more negative as i increases, the ROR of the added investment is greater than 15%.

$$\boxed{\text{Alternative A is superior.}}$$

37. Use the year-end convention with the tax credit. The purchase is made at $t = 0$. However, the credit is received at $t = 1$ and must be multiplied by $(P/F, i\%, 1)$.

$$P = -\$300,000 + (0.0667)(\$300,000)(P/F, i\%, 1)$$
$$+ (\$90,000)(P/A, i\%, 5)(1 - 0.48)$$
$$+ \left(\frac{\$300,000 - \$50,000}{5}\right)(P/A, i\%, 5)(0.48)$$
$$+ (\$50,000)(P/F, i\%, 5)$$
$$= -\$300,000 + (\$20,000)(P/F, i\%, 1)$$
$$+ (\$46,800)(P/A, i\%, 5)$$
$$+ (\$24,000)(P/A, i\%, 5)$$
$$+ (\$50,000)(P/F, i\%, 5)$$

By trial and error,

i	P
10%	$17,616
15%	−$20,412
12%	$1448
13%	−$6142
12 ¼%	−$479

$$\boxed{i \text{ is between } 12\% \text{ and } 12^{1}/_{4}\%.}$$

38. Find the annual loan payment.

$$\text{payment} = (\$2,500,000)(A/P, 12\%, 25)$$
$$= (\$2,500,000)(0.1275)$$
$$= \$318,750$$

$$\text{distributed profit} = (0.15)(\$2,500,000)$$
$$= \$375,000$$

After paying all expenses and distributing the 15% profit, the remainder should be 0.

$$0 = \text{EUAC} = \$20{,}000 + \$50{,}000 + \$200{,}000$$
$$+ \$375{,}000 + \$318{,}750 - \text{annual receipts}$$
$$- (\$500{,}000)(A/F, 15\%, 25)$$
$$= \$963{,}750 - \text{annual receipts}$$
$$- (\$500{,}000)(0.0047)$$

This calculation assumes $i = 15\%$, which equals the desired return. However, this assumption only affects the salvage calculation, and since the number is so small, the analysis is not sensitive to the assumption.

$$\text{annual receipts} = \$961{,}400$$

The average daily receipts are

$$\frac{\$961{,}400}{365} = \$2634$$

Use the expected value approach. The average occupancy is

$$(0.40)(0.65) + (0.30)(0.70) + (0.20)(0.75)$$
$$+ (0.10)(0.80) = 0.70$$

The average number of rooms occupied each night is

$$(0.70)(120 \text{ rooms}) = 84 \text{ rooms}$$

The minimum required average daily rate per room is

$$\frac{\$2634}{84} = \boxed{\$31.36}$$

39. To find the rate of return, first find the annual savings.

$$\frac{\text{annual}}{\text{savings}} = \left(\frac{0.69 - 0.47}{1000}\right)(\$3{,}500{,}000) = \$770$$
$$P = -\$7500 + (\$770 - \$200 - \$100)$$
$$\times (P/A, i\%, 25) = 0$$
$$(P/A, i\%, 25) = 15.957$$

Searching the tables and interpolating,

$$i \approx \boxed{3.75\%}$$

40. Evaluate alternative A, working in millions of dollars.

$$P(A) = -(\$1.3)(1 - 0.48)(P/A, 15\%, 6)$$
$$= (\$1.3)(0.52)(3.7845)$$
$$= -\$2.56 \quad \text{[millions]}$$

Use straight line depreciation.

Evaluate alternative B.

$$D_j = \frac{2}{6} = 0.333$$
$$P(B) = -\$2 - (\$0.20)(\$1.3)(1 - 0.48)(P/A, 15\%, 6)$$
$$- (\$0.15)(1 - 0.48)(P/A, 15\%, 6)$$
$$+ (\$0.333)(0.48)(P/A, 15\%, 6)$$
$$= -\$2 - (\$0.20)(\$1.3)(0.52)(3.7845)$$
$$- (\$0.15)(0.52)(3.7845)$$
$$+ (\$0.333)(0.48)(3.7845)$$
$$= -\$2.202 \quad \text{[millions]}$$

Next, evaluate alternative C.

$$D_j = \frac{1.2}{3} = 0.4$$
$$P(C) = -(\$1.2)\big(1 + (P/F, 15\%, 3)\big)$$
$$- (\$0.20)(\$1.3)(1 - 0.48)$$
$$\times \big((P/F, 15\%, 1) + (P/F, 15\%, 4)\big)$$
$$- (\$0.45)(\$1.3)(1 - 0.48)$$
$$\times \big((P/F, 15\%, 2) + (P/F, 15\%, 5)\big)$$
$$- (\$0.80)(\$1.3)(1 - 0.48)$$
$$\times \big((P/F, 15\%, 3) + (P/F, 15\%, 6)\big)$$
$$+ (\$0.4)(\$0.48)(P/A, 15\%, 6)$$
$$= -(\$1.2)(1.6575)$$
$$- (\$0.20)(\$1.3)(0.52)(0.8696 + 0.5718)$$
$$- (\$0.45)(\$1.3)(0.52)(0.7561 + 0.4972)$$
$$- (\$0.80)(\$1.3)(0.52)(0.6575 + 0.4323)$$
$$+ (\$0.4)(0.48)(3.7845)$$
$$= -\$2.428 \quad \text{[millions]}$$

$$\boxed{\text{Alternative B is superior.}}$$

41. This is a replacement study. Since production capacity and efficiency are not a problem with the defender, the only question is when to bring in the challenger.

Since this is a before-tax problem, depreciation is not a factor, nor is book value.

The cost of keeping the defender one more year is

$$\text{EUAC(defender)} = \$200{,}000 + (0.15)(\$400{,}000)$$
$$= \$260{,}000$$

For the challenger,

$$\text{EUAC(challenger)}$$
$$= (\$800{,}000)(A/P, 15\%, 10) + \$40{,}000$$
$$+ (\$30{,}000)(A/G, 15\%, 10)$$
$$- (\$100{,}000)(A/F, 15\%, 10)$$
$$= (\$800{,}000)(0.1993) + \$40{,}000$$
$$+ (\$30{,}000)(3.3832)$$
$$- (\$100{,}000)(0.0493)$$
$$= \$296{,}006$$

Since the defender is cheaper, keep it. The same analysis next year will give identical answers. Therefore, keep the defender for the next 3 years, at which time the decision to buy the challenger will be automatic.

Having determined that it is less expensive to keep the defender than to maintain the challenger for 10 years, determine whether the challenger is less expensive if retired before 10 years.

If retired in 9 years,

$$\text{EUAC(challenger)} = (\$800{,}000)(A/P, 15\%, 9) + \$40{,}000$$
$$+ (\$30{,}000)(A/G, 15\%, 9)$$
$$- (\$150{,}000)(A/F, 15\%, 9)$$
$$= (\$800{,}000)(0.2096)$$
$$+ \$40{,}000 + (\$30{,}000)(3.0922)$$
$$- (\$150{,}000)(0.0596)$$
$$= \$291{,}506$$

Similar calculations yield the following results for all the retirement dates.

n	EUAC
10	$296,000
9	$291,506
8	$287,179
7	$283,214
6	$280,016
5	$278,419
4	$279,909
3	$288,013
2	$313,483
1	$360,000

Since none of these equivalent uniform annual costs are less than that of the defender, it is not economical to buy and keep the challenger for any length of time.

$$\boxed{\text{Keep the defender.}}$$

88 Professional Services, Contracts, and Engineering Law

PRACTICE PROBLEMS

1. List the different forms of company ownership. What are the advantages and disadvantages of each?

2. Define the requirements for a contract to be enforceable.

3. What standard features should a written contract include?

4. Describe the ways a consulting fee can be structured.

5. What is a retainer fee?

6. Which of the following organizations is NOT a contributor to the standard design and construction contract documents developed by the Engineers Joint Contract Documents Committee (EJCDC)?

    (A) National Society of Professional Engineers

    (B) Construction Specifications Institute

    (C) Associated General Contractors of America

    (D) American Institute of Architects

7. To be affected by the Fair Labor Standards Act (FLSA) and be required to pay minimum wage, construction firms working on bridges and highways must generally have

    (A) 1 or more employees

    (B) 2 or more employees and annual gross billings of $500,000

    (C) 10 or more employees and be working on federally funded projects

    (D) 50 or more employees and have been in business for longer than 6 months

8. A "double-breasted" design firm

    (A) has errors and omissions as well as general liability insurance coverage

    (B) is licensed to practice in both engineering and architecture

    (C) serves both union and nonunion clients

    (D) performs post-construction certification for projects it did not design

9. The phrase "without expressed authority" means which of the following when used in regards to partnerships of design professionals?

    (A) Each full member of a partnership is a general agent of the partnership and has complete authority to make binding commitments, enter into contracts, and otherwise act for the partners within the scope of the business.

    (B) The partnership may act in a manner that it considers best for the client, even though the client has not been consulted.

    (C) Only plans, specifications, and documents that have been signed and stamped (sealed) by the authority of the licensed engineer may be relied upon.

    (D) Only officers to the partnership may obligate the partnership.

10. A limited partnership has 1 managing general partner, 2 general partners, 1 silent partner, and 3 limited partners. If all partners cast a single vote when deciding on an issue, how many votes will be cast?

    (A) 1

    (B) 3

    (C) 4

    (D) 7

11. Which of the following are characteristics of a limited liability corporation (LLC)?

I.   limited liability for all members

II.   no taxation as an entity (no double taxation)

III.   more than one class of stock

IV.   limited to fewer than 25 members

V.   fairly easy to establish

VI.   no "continuity of life" like regular corporation

    (A) I, II, and IV

    (B) I, II, III, and VI

    (C) I, II, III, IV, and V

    (D) I, III, IV, V, and VI

12. Which of the following statements is FALSE in regard to joint ventures?

(A) Members of a joint venture may be any combination of sole proprietorships, partnerships, and corporations.

(B) A joint venture is a business entity separate from its members.

(C) A joint venture spreads risk and rewards, and it pools expertise, experience, and resources. However, bonding capacity is not aggregated.

(D) A joint venture usually dissolves after the completion of a specific project.

13. Which of the following construction business types can have unlimited shareholders?

I.   S corporation

II.   LLC

III. corporation

IV. sole proprietorship

(A) II and III

(B) I, II, and III

(C) I, III, and IV

(D) I, II, III, and IV

14. The phrase "or approved equal" allows a contractor to

(A) substitute one connection design for another

(B) substitute a more expensive feature for another

(C) replace an open-shop subcontractor with a union subcontractor

(D) install a product whose brand name and model number are not listed in the specifications

15. Cities, other municipalities, and departments of transportation often have standard specifications, in addition to the specifications issued as part of the construction document set, that cover such items as

(A) safety requirements

(B) environmental requirements

(C) concrete, fire hydrant, manhole structures, and curb requirements

(D) procurement and accounting requirements

16. What is intended to prevent a contractor from bidding on a project and subsequently backing out after being selected for the project?

(A) publically recorded bid

(B) property lien

(C) surety bond

(D) proposal bond

17. Which of the following is illegal, in addition to being unethical?

(A) bid shopping

(B) bid peddling

(C) bid rigging

(D) bid unbalancing

18. Which of the following is NOT normally part of a construction contract?

I.   invitation to bid

II.   instructions to bidders

III.   general conditions

IV.   supplementary conditions

V.   liability insurance policy

VI.   technical specifications

VII. drawings

VIII. addenda

IX.   proposals

X.   bid bond

XI.   agreement

XII. performance bond

XIII. labor and material payment bond

XIV. nondisclosure agreement

(A) I

(B) II

(C) V

(D) XIV

19. Once a contract has been signed by the owner and contractor, changes to the contract

(A) cannot be made

(B) can be made by the owner, but not by the contractor

(C) can be made by the contractor, but not by the owner

(D) can be made by both the owner and the contractor

20. A constructive change is a change to the contract that can legally be construed to have been made, even though the owner did not issue a specific, written change order. Which of the following situations is normally a constructive change?

(A) request by the engineer-architect to install OSHA-compliant safety features

(B) delay caused by the owner's failure to provide access

(C) rework mandated by the building official

(D) expense and delay due to adverse weather

21. A bid for foundation construction is based on owner supplied soil borings showing sandy clay to a depth of 12 ft. However, after the contract has been assigned and during construction, the backhoe encounters large pieces of concrete buried throughout the construction site. This situation would normally be referred to as

(A) concealed conditions

(B) unexplained features

(C) unexpected characteristics

(D) hidden detriment

22. If a contract has a value engineering clause and a contractor suggests to the owner that a feature or method be used to reduce the annual maintenance cost of the finished project, what will be the most likely outcome?

(A) The contractor will be able to share one time in the owner's expected cost savings.

(B) The contractor will be paid a fixed amount (specified by the contract) for making suggestion, but only if the suggestion is accepted.

(C) The contract amount will be increased by some amount specified in the contract.

(D) The contractor will receive an annuity payment over some time period specified in the contract.

23. A contract has a value engineering clause that allows the parties to share in improvements that reduce cost. The contractor had originally planned to transport concrete on site for a small pour with motorized wheelbarrows. On the day of the pour, however, a concrete pump is available and is used, substantially reducing the contractor's labor cost for the day. This is an example of

(A) value engineering whose benefit will be shared by both contractor and owner

(B) efficient methodology whose benefit is to the contractor only

(C) value engineering whose benefit is to the owner only

(D) cost reduction whose benefit will be shared by both contractor and laborers

24. A material breach of contract occurs when

(A) the contractor uses material not approved by the contract to use

(B) the contractor's material order arrives late

(C) the owner becomes insolvent

(D) the contractor installs a feature incorrectly

25. When an engineer stops work on a job site after noticing unsafe conditions, the engineer is acting as a(n)

I. agent

II. local official

III. OSHA safety inspector

IV. competent person

(A) I

(B) I and IV

(C) II and III

(D) III and IV

26. While performing duties pursuant to a contract for professional services with an owner/developer, a professional engineer gives erroneous instruction to a subcontractor hired by the prime contractor. To whom would the subcontractor look for financial relief?

I. the professional engineer

II. the owner/developer

III. the subcontractor's bonding company

IV. the prime contractor

(A) III

(B) IV

(C) I and II

(D) I, II, III, and IV

27. A professional engineer employed by a large, multinational corporation is the lead engineer in designing and producing a consumer product that proves injurious to some purchasers. What can the engineer expect in the future?

(A) legal action against him/her from consumers

(B) termination and legal action against him/her from his employer

(C) legal action against him/her from the U.S. Consumer Protection Agency

(D) thorough review of his/her work, legal support from the employer, and possible termination

28. A professional engineer in private consulting practice makes a calculation error that causes the collapse of a structure during the construction process. What can the engineer expect in the future?

(A) cancellation of his/her Errors & Omissions insurance

(B) criminal prosecution

(C) loss of his/her professional engineering license

(D) incarceration

29. A professional engineer is hired by a homeowner to design a septic tank and leach field. The septic tank fails after 18 years of operation. What sentence best describes what comes next?

(A) The homeowner could pursue a tort action claiming a septic tank should retain functionality longer than 18 years.

(B) The engineer will be protected by a statute of limitations law.

(C) The homeowner may file a claim with the engineer's original bonding company.

(D) The engineer is ethically bound to provide remediation services to the homeowner.

SOLUTIONS

1. The three different forms of company ownership are the (1) sole proprietorship, (2) partnership, and (3) corporation.

A *sole proprietor* is his or her own boss. This satisfies the proprietor's ego and facilitates quick decisions, but unless the proprietor is trained in business, the company will usually operate without the benefit of expert or mitigating advice. The sole proprietor also personally assumes all the debts and liabilities of the company. A sole proprietorship is terminated upon the death of the proprietor.

A *partnership* increases the capitalization and the knowledge base beyond that of a proprietorship, but offers little else in the way of improvement. In fact, the partnership creates an additional disadvantage of one partner's possible irresponsible actions creating debts and liabilities for the remaining partners.

A *corporation* has sizable capitalization (provided by the stockholders) and a vast knowledge base (provided by the board of directors). It keeps the company and owner liability separate. It also survives the death of any employee, officer, or director. Its major disadvantage is the administrative work required to establish and maintain the corporate structure.

2. To be legal, a contract must contain an *offer*, some form of *consideration* (which does not have to be equitable), and an *acceptance* by both parties. To be enforceable, the contract must be voluntarily entered into, both parties must be competent and of legal age, and the contract cannot be for illegal activities.

3. A written contract will identify both parties, state the purpose of the contract and the obligations of the parties, give specific details of the obligations (including relevant dates and deadlines), specify the consideration, state the boilerplate clauses to clarify the contract terms, and leave places for signatures.

4. A consultant will either charge a fixed fee, a variable fee, or some combination of the two. A one-time fixed fee is known as a *lump-sum fee*. In a *cost plus fixed fee* contract, the consultant will also pass on certain costs to the client. Some charges to the client may depend on other factors, such as the salary of the consultant's staff, the number of days the consultant works, or the eventual cost or value of an item being designed by the consultant.

5. A *retainer* is a (usually) nonreturnable advance paid by the client to the consultant. While the retainer may be intended to cover the consultant's initial expenses until the first big billing is sent out, there does not need to be any rational basis for the retainer. Often, a small retainer is used by the consultant to qualify the client (i.e., to make sure the client is not just shopping around and getting free initial consultations) and as a security deposit (to make sure the client does not change consultants after work begins).

6. The Engineers Joint Contract Documents Committee (EJCDC) consists of the National Society of Professional Engineers, the American Council of Engineering Companies (formerly the American Consulting Engineers Council), the American Society of Civil Engineers, Construction Specifications Institute, and the Associated General Contractors of America. The American Institute of Architects is not a member, and it has its own standardized contract documents.

The answer is (D).

7. A business in the construction industry must have two or more employees and a minimum annual gross sales volume of $500,000 to be subject to the Fair Labor Standards Act (FLSA). Individual coverage also applies to employees whose work regularly involves them in commerce between states (i.e., interstate commerce). Any person who works on, or otherwise handles, goods moving in interstate commerce, or who works on the expansion of existing facilities of commerce, is individually subject to the protection of the FLSA and the current minimum wage and overtime pay requirements, regardless of the sales volume of the employer.

The answer is (B).

8. A *double-breasted* design firm serves both union and nonunion clients. When union-affiliated companies find themselves uncompetitive in bidding on nonunion projects, the company owners may decide to form and operate a second company that is *open shop* (i.e., employees are not required to join a union as a condition of employment). Although there are some restrictions requiring independence of operation, the common ownership of two related firms is legal.

The answer is (C).

9. *Without expressed authority* means each member of a partnership has full authority to obligate the partnership (and the other partners).

The answer is (A).

10. Only general partners can vote in a partnership. Both silent and limited partners share in the profit and benefits of the partnership, but they only contribute financing and do not participate in the management. The identities of silent partners are often known only to a few, whereas limited partners are known to all.

The answer is (B).

11. Limited liability corporations have limited liability for all members, no double taxation, more than one type of stock, and no continuity of life. They are not limited in members, and they are comparatively fairly difficult to establish.

The answer is (B).

12. One of the reasons for forming joint ventures is that the bonding capacity is aggregated. Even if each contractor cannot individually meet the minimum bond requirements, the total of the bonding capacities may be sufficient.

The answer is (C).

13. Both normal corporations and limited liability corporations (LLCs) can have unlimited shareholders. S corporations and sole proprietorships are for individuals.

The answer is (A).

14. When the specifications include a nonstructural, brand-named article and the accompanying phrase "or approved equal," the contractor can substitute something with the same functionality, even though it is not the brand-named article. "Or approved equal" would not be used with a structural detail such as a connection design.

The answer is (D).

15. Municipalities that experience frequent construction projects within their boundaries have standard specifications that are included by reference in every project's construction document set. This document set would cover items such as concrete, fire hydrants, manhole structures, and curb requirements.

The answer is (C).

16. *Proposal bonds*, also known as *bid bonds*, are insurance policies payable to the owner in the event that the contractor backs out after submitting a qualified bid.

The answer is (D).

17. *Bid rigging*, also known as *price fixing*, is an illegal arrangement between contractors to control the bid prices of a construction project or to divide up customers or market areas. *Bid shopping* before or after the bid letting is where the general contractor tries to secure better subcontract proposals by negotiating with the subcontractors. *Bid peddling* is done by the subcontractor to try to lower its proposal below the lowest proposal. *Bid unbalancing* is where a contractor pushes the payment for some expense items to prior construction phases in order to improve cash flow.

The answer is (C).

18. The construction contract includes many items, some explicit only by reference. The contractor may be required to carry liability insurance, but the policy itself is between the contractor and its insurance company, and is not normally part of the construction contract.

The answer is (C).

19. Changes to the contract can be made by either the owner or the contractor. The method for making such changes is indicated in the contract. Almost always, the change must be agreed to by both parties in writing.

The answer is (D).

20. A *constructive change* to the contract is the result of an action or lack of action of the owner or its agent. If the project is delayed by the owner's failure to provide access, the owner has effectively changed the contract.

The answer is (B).

21. *Concealed conditions* are also known as *changed conditions* and *differing site conditions*. Most, but not all, contracts have provisions dealing with changed conditions. Some place the responsibility to confirm the site conditions before bidding on the contractor. Others detail the extent of changed conditions that will trigger a review of reimbursable expenses.

The answer is (A).

22. Changes to a structure's performance, safety, appearance, or maintenance that benefit the owner in the long run will be covered by the value engineering clause of a contract. Normally, the contractor is able to share in cost savings in some manner by receiving a payment or credit to the contract.

The answer is (A).

23. The problem gives an example of efficient methodology, where the benefit is to the contractor only. It is not an example of value engineering, as the change affects the contractor, not the owner. Performance, safety, appearance, and maintenance are unaffected.

The answer is (B).

24. A *material breach of the contract* is a significant event that is grounds for cancelling the contract entirely. Typical triggering events include failure of the owner to make payments, the owner causing delays, the owner declaring bankruptcy, the contractor abandoning the job, or the contractor getting substantially off schedule.

The answer is (C).

25. An engineer with the authority to stop work gets his/her authority from the agency clause of the contract for professional services with the owner/developer. It is unlikely that the local building department or OSHA would have authorized the engineer to act on their behalves. It is also possible that the engineer may have been designated as the jobsite's "competent person" (as required by OSHA) for one or more aspects of the job, with authority to implement corrective action.

The answer is (B).

26. In the absence of other information, the subcontractor only has a contractual relationship with the prime contractor, so the request for financial relief would initially go to the prime contractor. The prime contractor would ask for relief from the owner/developer, who in turn may want relief from the professional engineer. The subcontractor's bonding company's role is to guarantee the subcontractor's work product against subcontractor errors, not engineering errors.

The answer is (B).

27. Manufacturers of consumer products are generally held responsible for the safety of their products. Individual team members are seldom held personally accountable, and when they are, there is evidence of intentional wrongdoing and/or gross incompetence. The manufacturer will organize and absorb the costs of the legal defense. If the engineer is fired, it is likely this will occur after the case is closed.

The answer is (D).

28. A professional engineer is not held to the standard of perfection, so making a calculation error is neither a criminal act nor sufficient evidence of incompetence to result in loss of license. Without criminal prosecution for fraud or other wrongdoing, it is unlikely that the engineer would ever see the inside of a jail. However, it is likely that the engineer's bonding company will cancel his/her policy (after defending the case).

The answer is (A).

29. Satisfactory operation for 18 years is proof that the design was partially, if not completely, adequate. There is no ethical obligation to remediate normal wear-and-tear, particularly when the engineer was not involved in how the septic system was used or maintained. While the homeowner can threaten legal action and even file a complaint, in the absence of a warranty to the contrary, the engineer is probably well-protected by a statute of limitations.

The answer is (B).

89 Engineering Ethics

PRACTICE PROBLEMS

(Each problem has two parts. Determine whether the situation is (or can be) permitted legally. Then, determine whether the situation is permitted ethically.)

1. (a) Was it legal and/or ethical for an engineer to sign and seal plans that were not prepared by him or prepared under his responsible direction, supervision, or control?

(b) Was it legal and/or ethical for an engineer to sign and seal plans that were not prepared by him but were prepared under his responsible direction, supervision, and control?

2. Under what conditions would it be legal and/or ethical for an engineer to rely on the information (e.g., elevations and amounts of cuts and fills) furnished by a grading contractor?

3. Was it legal and/or ethical for an engineer to alter the soils report prepared by another engineer for his client?

4. Under what conditions would it be legal and/or ethical for an engineer to assign work called for in his contract to another engineer?

5. A licensed professional engineer was convicted of a felony totally unrelated to his consulting engineering practice.

(a) What actions would you recommend be taken by the state registration board?

(b) What actions would you recommend be taken by the professional or technical society (e.g., ASCE, ASME, IEEE, NSPE, and so on)?

6. An engineer came across some work of a predecessor. After verifying the validity and correctness of all assumptions and calculations, the engineer used the work. Under what conditions would such use be legal and/or ethical?

7. A building contractor made it a policy to provide cell phones to the engineers of projects he was working on. Under what conditions could the engineers accept the phones?

8. An engineer designed a tilt-up slab building for a client. The design engineer sent the design out to another engineer for checking. The checking engineer sent the plans to a concrete contractor for review. The concrete contractor made suggestions that were incorporated into the recommendations of the checking contractor. These recommendations were subsequently incorporated into the plans by the original design engineer. What steps must be taken to keep the design process legal and/or ethical?

9. A consulting engineer registered his corporation as "John Williams, P.E. and Associates, Inc." even though he had no associates. Under what conditions would this name be legal and/or ethical?

10. When it became known that a chemical plant was planning on producing a toxic product, an engineer employed by the plant wrote to the local newspaper condemning the chemical plant's action. Under what conditions would the engineer's action be legal and/or ethical?

11. An engineer signed a contract with a client. The fee the client agreed to pay was based on the engineer's estimate of time required. The engineer was able to complete the contract satisfactorily in half the time he expected. Under what conditions would it be legal and/or ethical for the engineer to keep the full fee?

12. After working on a project for a client, the engineer was asked by a competitor of the client to perform design services. Under what conditions would it be legal and/or ethical for the engineer to work for the competitor?

13. Two engineers submitted bids to a prospective client for a design project. The client told engineer A how much engineer B had bid and invited engineer A to beat the amount. Under what conditions could engineer A legally/ethically submit a lower bid?

14. A registered civil engineer specializing in well-drilling, irrigation pipelines, and farmhouse sanitary systems took a booth at a county fair located in a farming town. By a random drawing, the engineer's booth was located next to a hog-breeder's booth, complete with live (prize) hogs. The engineer gave away helium balloons with his name and phone number to all visitors to the booth. Did the engineer violate any laws/ethical guidelines?

15. While in a developing country supervising construction of a project an engineer designed, the engineer discovered the client's project manager was treating local workers in an unsafe and inhumane (but, for that country, legal) manner. When the engineer objected, the client told the engineer to mind his own business. Later, the local workers asked the engineer to participate in a walkout and strike with them.

(a) What legal/ethical positions should the engineer take?

(b) Should it have made any difference if the engineer had or had not yet accepted any money from the client?

16. While working for a client, an engineer learns confidential knowledge of a proprietary production process being used by the client's chemical plant. The process is clearly destructive to the environment, but the client will not listen to the objections of the engineer. To inform the proper authorities will require the engineer to release information that was gained in confidence. Is it legal and/or ethical for the engineer to expose the client?

17. While working for an engineering design firm, an engineer was moonlighting as a soils engineer. At night, the engineer used the employer's facilities to analyze and plot the results of soils tests. He then used his employer's computers to write his reports. The equipment and computers would otherwise be unused. Under what conditions could the engineer's actions be considered legal and/or ethical?

18. Ethical codes and state legislation forbidding competitive bidding by design engineers are

(A) enforceable in some states

(B) not enforceable on public (nonfederal) projects

(C) enforceable for projects costing less than $5 million dollars

(D) not enforceable

SOLUTIONS

Introduction to the Solutions

Case studies in law and ethics can be interpreted in many ways. The problems presented are simple thumbnail outlines. In most real cases, there will be more facts to influence a determination than are presented in the case scenarios. In some cases, a state may have specific laws affecting the determination; in other cases, prior case law will have been established.

The determination of whether an action is legal can be made in two ways. The obvious interpretation of an illegal action is one that violates a specific law or statute. An action can also be *found to be illegal* if it is judged in court to be a breach of a written, verbal, or implied contract. Both of these approaches are used in the following solutions.

These answers have been developed to teach legal and ethical principles. While being realistic, they are not necessarily based on actual incidents or prior case law.

1. (a) Stamping plans for someone else is illegal. The registration laws of all states permit a registered engineer to stamp/sign/seal only plans that were prepared by him personally or were prepared under his direction, supervision, or control. This is sometimes called being in *responsible charge*. The stamping/signing/sealing, for a fee or gratis, of plans produced by another person, whether that person is registered or not and whether that person is an engineer or not, is illegal.

(b) The act is unethical. An illegal act, being a concealed act, is intrinsically unethical. In addition, stamping/signing/sealing plans that have not been checked violates the rule contained in all ethical codes that requires an engineer to protect the public.

2. Unless the engineer and contractor worked together such that the engineer had personal knowledge that the information was correct, accepting the contractor's information is illegal. Not only would using unverified data violate the state's registration law (for the same reason that stamping/signing/sealing unverified plans in Prob. 1 was illegal), but the engineer's contract clause dealing with assignment of work to others would probably be violated.

The act is unethical. An illegal act, being a concealed act, is intrinsically unethical. In addition, using unverified data violates the rule contained in all ethical codes that requires an engineer to protect the client.

3. It is illegal to alter a report to bring it "more into line" with what the client wants unless the alterations represent actual, verified changed conditions. Even when the alterations are warranted, however, use of the unverified remainder of the report is a violation of the state registration law requiring an engineer only to stamp/sign/seal plans developed by or under him. Furthermore, this would be a case of fraudulent misrepresentation unless the originating engineer's name was removed from the report.

Unless the engineer who wrote the original report has given permission for the modification, altering the report would be unethical.

4. Assignment of engineering work is legal (1) if the engineer's contract permitted assignment, (2) all prerequisites (i.e., notifying the client) were met, and (3) the work was performed under the direction of another licensed engineer.

Assignment of work is ethical (1) if it is not illegal, (2) if it is done with the awareness of the client, and (3) if the assignor has determined that the assignee is competent in the area of the assignment.

5. (a) The registration laws of many states require a hearing to be held when a licensee is found guilty of unrelated, but nevertheless unforgivable, felonies (e.g., moral turpitude). The specific action (e.g., suspension, revocation of license, public censure, and so on) taken depends on the customs of the state's registration board.

(b) By convention, it is not the responsibility of technical and professional organizations to monitor or judge the personal actions of their members. Such organizations do not have the authority to discipline members (other than to revoke membership), nor are they immune from possible retaliatory libel/slander lawsuits if they publicly censure a member.

6. The action is legal because, by verifying all the assumptions and checking all the calculations, the engineer effectively does the work. Very few engineering procedures are truly original; the fact that someone else's effort guided the analysis does not make the action illegal.

The action is probably ethical, particularly if the client and the predecessor are aware of what has happened (although it is not necessary for the predecessor to be told). It is unclear to what extent (if at all) the predecessor should be credited. There could be other extenuating circumstances that would make referring to the original work unethical.

7. Gifts, per se, are not illegal. Unless accepting the phones violates some public policy or other law, or is in some way an illegal bribe to induce the engineer to favor the contractor, it is probably legal to accept the phones.

Ethical acceptance of the phones requires (among other considerations) that (1) the phones be required for the job, (2) the phones be used for business only, (3) the phones are returned to the contractor at the end of the

job, and (4) the contractor's and engineer's clients know and approve of the transaction.

8. There are two issues: (1) the assignment and (2) the incorporation of work done by another. To avoid a breach, the contracts of both the design and checking engineers must permit the assignments. To avoid a violation of the state registration law requiring engineers to be in responsible charge of the work they stamp/sign/seal, both the design and checking engineers must verify the validity of the changes.

To be ethical, the actions must be legal and all parties (including the design engineer's client) must be aware that the assignments have occurred and that the changes have been made.

9. The name is probably legal. If the name was accepted by the state's corporation registrar, it is a legally formatted name. However, some states have engineering registration laws that restrict what an engineering corporation may be named. For example, all individuals listed in the name (e.g., "Cooper, Williams, and Somerset—Consulting Engineers") may need to be registered. Whether having "Associates" in the name is legal depends on the state.

Using the name is unethical. It misleads the public and represents unfair competition with other engineers running one-person offices.

10. Unless the engineer's accusation is known to be false or exaggerated, or the engineer has signed an agreement (confidentiality, nondisclosure, and so on) with his employer forbidding the disclosure, the letter to the newspaper is probably not illegal.

The action is probably unethical. (If the letter to the newspaper is unsigned it is a concealed action and is definitely unethical.) While whistle-blowing to protect the public is implicitly an ethical procedure, unless the engineer is reasonably certain that manufacture of the toxic product represents a hazard to the public, he has a responsibility to the employer. Even then, the engineer should exhaust all possible remedies to render the manufacture nonhazardous before blowing the whistle. Of course, the engineer may quit working for the chemical plant and be as critical as the law allows without violating engineer-employer ethical considerations.

11. Unless the engineer's payment was explicitly linked in the contract to the amount of time spent on the job, taking the full fee would not be illegal or a breach of the contract.

An engineer has an obligation to be fair in estimates of cost, particularly when the engineer knows no one else is providing a competitive bid. Taking the full fee would be ethical if the original estimate was arrived at logically and was not meant to deceive or take advantage of the

Systems, Mgmt, and Professional

client. An engineer is permitted to take advantage of economies of scale, state-of-the-art techniques, and break-through methods. (Similarly, when a job costs more than the estimate, the engineer may be ethically bound to stick with the original estimate.)

12. In the absence of a nondisclosure or noncompetition agreement or similar contract clause, working for the competitor is probably legal.

Working for both clients is unethical. Even if both clients know and approve, it is difficult for the engineer not to "cross-pollinate" his work and improve one client's position with knowledge and insights gained at the expense of the other client. Furthermore, the mere appearance of a conflict of interest of this type is a violation of most ethical codes.

13. In the absence of a sealed-bid provision mandated by a public agency and requiring all bids to be opened simultaneously (and the award going to the lowest bidder), the action is probably legal.

It is unethical for an engineer to undercut the price of another engineer. Not only does this violate a standard of behavior expected of professionals, it unfairly benefits one engineer because a similar chance is not given to the other engineer. Even if both engineers are bidding openly against each other (in an auction format), the client must understand that a lower price means reduced service. Each reduction in price is an incentive to the engineer to reduce the quality or quantity of service.

14. It is generally legal for an engineer to advertise his services. Unless the state has relevant laws, the engineer probably did not engage in illegal actions.

Most ethical codes prohibit unprofessional advertising. The unfortunate location due to a random drawing might be excusable, but the engineer should probably refuse to participate. In any case, the balloons are a form of unprofessional advertising, and as such, are unethical.

15. (a) As stated in the scenario statement, the client's actions are legal for that country. The fact that the actions might be illegal in another country is irrelevant. Whether or not the strike is legal depends on the industry and the laws of the land. Some or all occupations (e.g., police and medical personnel) may be forbidden to strike. Assuming the engineer's contract does not prohibit participation, the engineer should determine the legality of the strike before making a decision to participate.

If the client's actions are inhumane, the engineer has an ethical obligation to withdraw from the project. Not doing so associates the profession of engineering with human misery.

(b) The engineer has a contract to complete the project for the client. (It is assumed that the contract between the engineer and client was negotiated in good faith, that the engineer had no knowledge of the work conditions prior to signing, and that the client did not falsely induce the engineer to sign.) Regardless of the reason for withdrawing, the engineer is breaching his contract. In the absence of proof of illegal actions by the client, withdrawal by the engineer requires a return of all fees received. Even if no fees have been received, withdrawal exposes the engineer to other delay-related claims by the client.

16. A contract for an illegal action cannot be enforced. Therefore, any confidentiality or nondisclosure agreement that the engineer has signed is unenforceable if the production process is illegal, uses illegal chemicals, or violates laws protecting the environment. If the production process is not illegal, it is not legal for the engineer to expose the client.

Society and the public are at the top of the hierarchy of an engineer's responsibilities. Obligations to the public take precedence over the client. If the production process is illegal, it would be ethical to expose the client.

17. It is probably legal for the engineer to use the facilities, particularly if the employer is aware of the use. (The question of whether the engineer is trespassing or violating a company policy cannot be answered without additional information.)

Moonlighting, in general, is not ethical. Most ethical codes prohibit running an engineering consulting business while receiving a salary from another employer. The rationale is that the moonlighting engineer is able to offer services at a much lower price, placing other consulting engineers at a competitive disadvantage. The use of someone else's equipment compounds the problem since the engineer does not have to pay for using the equipment, and so does not have to charge any clients for it. This places the engineer at an unfair competitive advantage compared to other consultants who have invested heavily in equipment.

18. Ethical bans on competitive bidding are not enforceable. The National Society of Professional Engineers' (NSPE) ethical ban on competitive bidding was struck down by the U.S. Supreme Court in 1978 as a violation of the Sherman Antitrust Act of 1890.

The answer is (D).

90 Engineering Licensing in the United States

PRACTICE PROBLEMS

There are no problems in this book corresponding to Chap. 90 of the *Civil Engineering Reference Manual*.

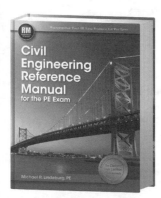
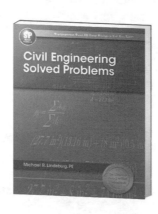
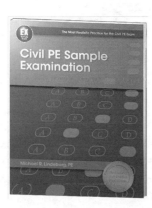

Coverage of the Civil PE Exam Depth Sections
For more information, visit www.ppi2pass.com.

Construction

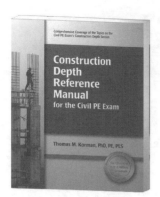

Construction Depth Reference Manual for the Civil PE Exam
Thomas M. Korman, PhD, PE, PLS

- Clear, easy-to-understand explanations of construction engineering concepts
- 35 example problems
- 178 equations, 29 tables, 74 figures, and 5 appendices
- An easy-to-use index

Six-Minute Solutions for Civil PE Exam Construction Problems
Elaine Huang, PE

- 100 challenging multiple-choice problems
- 20 morning and 80 afternoon session problems
- Coverage of exam-adopted design standards
- Step-by-step solutions

Geotechnical

Soil Mechanics and Foundation Design: 201 Solved Problems
Liiban A. Affi, PE

- A comprehensive review for the Civil PE exam's geotechnical depth section and the California GE exam
- More than 200 solved problems
- Step-by-step solutions

Six-Minute Solutions for Civil PE Exam Geotechnical Problems
Bruce A. Wolle, MSE, PE

- 100 challenging multiple-choice problems
- 20 morning and 80 afternoon session problems
- Coverage of exam-adopted codes
- Step-by-step solutions

Structural

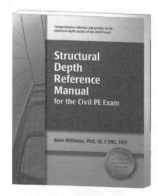

Structural Depth Reference Manual for the Civil PE Exam
Alan Williams, PhD, SE, C Eng, FICE

- Clear, easy-to-understand explanations of structural engineering concepts
- More than 120 example and practice problems
- Dozens of detailed graphics

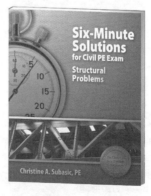

Six-Minute Solutions for Civil PE Exam Structural Problems
Christine A. Subasic, PE

- 100 challenging multiple-choice problems
- 20 morning and 80 afternoon session problems
- Coverage of exam-adopted codes
- Step-by-step solutions

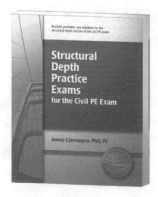

Structural Depth Practice Exams for the Civil PE Exam
James Giancaspro, PhD, PE

- Two, 40-problem multiple-choice practice exams
- Problems that closely match the topics, level of difficulty, and reference standards on the exam
- Detailed solutions

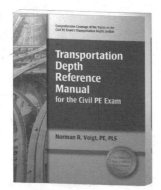
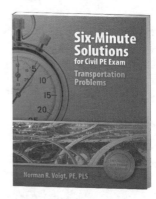
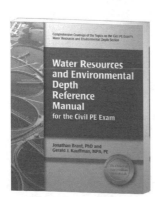
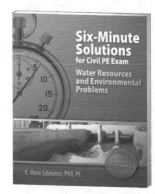

Supplement Your Review With Design Principles

For more information, visit www.ppi2pass.com.

Comprehensive Reference and Practice Materials

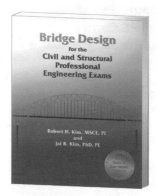

Bridge Design for the Civil and Structural PE Exams

Robert H. Kim, MSCE, PE; Jai B. Kim, PhD, PE

- An overview of key bridge design principles
- 5 design examples
- 2 practice problems
- Step-by-step solutions

Concrete Design for the Civil and Structural PE Exams

C. Dale Buckner, PhD, PE, SECB

- Comprehensive concrete design review
- A complete overview of the relevant codes and standards
- 37 practice problems
- Easy-to-use tables, figures, and concrete design nomenclature

Steel Design for the Civil PE and Structural SE Exams

Frederick S. Roland, PE, SECB, RA, CFEI, CFII

- Comprehensive overview of the key elements of structural steel design and analysis
- Side-by-side LRFD and ASD solutions
- More than 50 examples and 35 practice problems
- Solutions cite specific *Steel Manual* sections, equations, or table numbers

Timber Design for the Civil and Structural PE Exams

Robert H. Kim, MSCE, PE; Jai B. Kim, PhD, PE

- A complete overview of the relevant codes and standards
- 40 design examples
- 6 scenario-based practice problems
- Easy-to-use tables, figures, and timber design nomenclature